Advances in Experimental Medicine and Biology

Volume 1492

Advances in Experimental Medicine and Biology provides a platform for scientific contributions in the main disciplines of the biomedicine and the life sciences. This series publishes thematic volumes on contemporary research in the areas of microbiology, immunology, neurosciences, biochemistry, biomedical engineering, genetics, physiology, and cancer research. Covering emerging topics and techniques in basic and clinical science, it brings together clinicians and researchers from various fields.

Advances in Experimental Medicine and Biology has been publishing exceptional works in the field for over 40 years, and is indexed in SCOPUS, Medline (PubMed), EMBASE, Reaxys, EMBiology, the Chemical Abstracts Service (CAS), and Pathway Studio.

2023 CiteScore: 5.9

Nima Rezaei • Heliya Ziaei
Editors

Oral Immunology

Springer

Editors
Nima Rezaei
Universal Scientific Education
and Research Network (USERN)
Göteborg, Sweden

Heliya Ziaei
Universal Scientific Education
and Research Network (USERN)
Los Angeles, CA, USA

ISSN 0065-2598 ISSN 2214-8019 (electronic)
Advances in Experimental Medicine and Biology
ISBN 978-3-032-03175-4 ISBN 978-3-032-03176-1 (eBook)
https://doi.org/10.1007/978-3-032-03176-1

This Springer imprint is published by the registered company Springer Nature Switzerland AG
The registered company address is: Gewerbestrasse 11, 6330 Cham, Switzerland

This book would not have been possible without the continuous encouragement and support of our families.
We dedicate it to our parents, Mina and Abolghasem, Sima and Mohsen, hoping that advancing the science of oral immunity will improve human health care and help build a brighter future for all.

Preface

"Diseases of the teeth may cause pain in the head, ears, and even the heart." — Avicenna (980–1037 CE)

"Mouth and tongue ulcers may signify disease of liver or stomach." — Al-Razi (*864–935 CE)*—Interpreted from Kitab al-Hawi

For centuries, physicians and scientists have realized that oral cavity is not simply a site of eating or speech, but it is deeply connected to other organs of the body. Long before the advent of advanced molecular diagnostics and performing large-scale epidemiological studies, Avicenna and Al-Razi realized that oral conditions could reveal or even influence systemic diseases. In modern times, physicians like Osler, Mayo, and Satcher have echoed the same message, nowadays backed by decades of immunological and clinical evidence.

"The mouth is the mirror of the body." — Dr. William Osler (1849–1919)

"People who keep their teeth live longer." — Dr. Charles Mayo (1865–1939)

"Oral health is integral to overall health." — Dr. David Satcher (b. 1941)

The oral cavity is more than an anatomical entrance to the digestive tract; it is a dynamic and complex immunological interface with complex microbial interactions and systemic signaling. The oral cavity reflects the health of the cardiovascular system, gut, brain, and endocrine system, and in many cases, it contributes to their dysfunction. Oral inflammation, microbiome dysbiosis, and mucosal immune disruptions are not isolated events. They are part of a broader understanding of the human body functioning as an integrated whole.

Despite centuries of insight regarding the body's interconnected systems, as science has advanced, its different fields and specialties have also become increasingly isolated. With every discipline diving deeper into its own complexities, the distance between fields has grown, and scientists and clinicians have become increasingly focused on their specialties and subspecialties, sometimes at the expense of broader integration. Dentistry gradually drifted apart from medicine, immunology became increasingly molecular, and clinical and basic sciences progressed along parallel yet disconnected paths. Specialization has brought depth but at the cost of connection. As the distance between disciplines has widened, the full picture of oral-systemic health has been gradually overlooked. This book is an attempt to restore not only the connection between oral and systemic health, but also to integrate many dimensions of oral immunology, from foundational immune biology and disease mechanisms to craniofacial development, immune therapies, and diagnostic innovations. *Oral Immunity* gathers the insights of multiple disciplines to reframe the oral cavity as a critical component of overall health.

In the 38 chapters that follow, researchers and clinicians from different disciplines come together to explore how oral immunity influences and is influenced by systemic disorders. This volume invites the reader to see the oral cavity not in isolation, but as a central player in human health and disease. This book is divided into five interconnected sections, each exploring a different dimension of oral immunity and its systemic implications. Part I introduces the reader to the architecture of oral immune defenses, including the roles of epithelial cells, microbial communities, and bacterial metabolites such as short-chain fatty acids (Chapters 1 to 4).

Part II focuses on the bidirectional relationship between oral and systemic health. The chapters explore how immune dysfunction in the oral cavity can affect systemic conditions such as diabetes, metabolic syndrome, neuroinflammation, autoimmune diseases, and adverse pregnancy outcomes. Conversely, this section also explores how systemic diseases may alter oral health-related immune responses. This section offers mechanistic insights and clinical context to better understand the interconnected nature of these conditions (Chapters 5 to 15).

Part III shifts the focus to the immunopathological mechanisms underlying various oral diseases and cancer. Chapters of this section cover in-depth discussions on inflammasomes and their role in chronic periodontal disease and oral malignancies. They further examine the significance of proinflammatory biomarkers and the immunological features of the head and neck tumor microenvironment, such as their implications for prognosis and immunotherapy responsiveness. The section also includes a chapter on in silico analysis of squamous cell carcinoma, covering the integration of computational tools in identifying molecular therapeutic targets. In addition, it addresses the immune-mediated pathology of medication-related osteonecrosis of the jaw (MRONJ) and peri-implantitis. Moreover, the role of innate immune gene variants is explored in relation to periodontitis susceptibility. These chapters discuss how chronic and dysregulated immune responses not only lead to localized oral pathologies but also contribute to malignant transformations, tissue breakdown, and systemic complications. They highlight critical opportunities for early diagnosis and targeted therapeutic interventions (Chapters 16 to 23).

Part IV highlights the connection of immunology with oral and craniofacial development. The section starts with exploring how immune mechanisms, from embryonic stages of development and maternal-fetal interactions, to postnatal immune dysregulations. The following chapters address the role of the immune system in oral and craniofacial bone metabolism, orthodontic treatment, and enamel formation, illustrating how immune signaling coordinates tissue growth, remodeling, and structural integrity in the craniofacial region (Chapters 24 to 29).

Finally, Part V explores modern diagnostic and therapeutic strategies, including immunomodulation, stem cell therapy, salivary diagnostics, and the immunological impact of medicinal plants. These chapters reflect the translational potential of oral immunology, from modulating immune tolerance and enhancing bone regeneration to identifying salivary biomarkers with potential diagnostic values for systemic diseases. Together, *Oral Immunity* provides a comprehensive view of the field and highlights a central argument: oral health is not merely dental, but deeply immunological, systemic, and crucial to overall human health.

This book is the result of collaboration among many disciplines and perspectives. Its chapters are written not only by dentists or immunologists but also by researchers in the fields of microbiology, neuroscience, developmental biology, and regenerative medicine. Each field brings a different perspective and what unites them is a shared understanding that oral diseases are not limited to a narrow specialty, but arise from, and conversely contribute to, a complex immune network with implications that extend far beyond the oral cavity.

In a time when scientific fields are becoming increasingly specialized, *Oral Immunity* aims to reconnect them. It does not claim to offer a complete picture, like the "blind men in Rumi's parable of the elephant, each feeling only a part," but it does invite the reader to step back from isolated parts and begin to understand how they fit into a more connected whole. This volume encourages a shift in how we think about oral health, not as a local concern, but as a vital, immunological conversation between the oral cavity and the rest of the body.

We sincerely appreciate all the contributors who devoted their time, expertise, and insight into this project. Their work reflects not only the current understanding of oral immunity, but also its potential to reshape how we approach disease, diagnostics, and therapy in the years ahead.

Göteborg, Sweden — Nima Rezaei
Los Angeles, CA, USA — Heliya Ziaei

Contents

About the Editors

Nima Rezaei MD, PHD gained his medical degree (MD) from Tehran University of Medical Sciences and subsequently obtained an MSc in Molecular and Genetic Medicine and a PhD in Clinical Immunology and Human Genetics from the University of Sheffield, UK. He also spent a short-term fellowship of Pediatric Clinical Immunology and Bone Marrow Transplantation in the Newcastle General Hospital. Professor Rezaei is now the Full Professor of Immunology and Vice Dean of Research and Technologies, School of Medicine, Tehran University of Medical Sciences, and the co-founder and Head of the Research Center for Immunodeficiencies. He is also the Founder of the Universal Scientific Education and Research Network (USERN). Prof. Rezaei supervised more than 200 research projects and designed and participated in several international collaborative projects. Prof. Rezaei is the editor, editorial assistant, or editorial board member of more than 50 international journals. He has edited more than 100 international books, presented more than 1000 lectures in congresses/meetings, and published more than 1500 scientific papers in international journals.

Heliya Ziaei DDS earned her Doctor of Dental Surgery (DDS) degree from Tehran University of Medical Sciences in 2020 and is currently completing her PhD in Craniofacial Molecular Biology at the Herman Ostrow School of Dentistry, University of Southern California. She has contributed to several interdisciplinary research projects in the areas of oral biology, craniofacial developmental science, and regenerative medicine. Dr. Ziaei served as Executive Director and Deputy of International Affairs at the Universal Scientific Education and Research Network (USERN). She has co-authored several peer-reviewed scientific articles and book chapters and has presented more than 20 lectures at academic conferences and meetings.

Part I

Introduction to Oral Immunity and Host-Microbe Interactions

Introduction to Oral Immunity

1

Heliya Ziaei and Nima Rezaei

Abstract

The oral immune system functions as a primary line of defense, composed of oral epithelial barriers, salivary antimicrobial factors, and various innate and adaptive immune components to prevent pathogen entry. Resident immune cells in oral tissues help maintain tolerance to commensal microorganisms while simultaneously responding to harmful external stimuli and contributing to systemic immune regulation. This chapter provides a comprehensive overview of the immunological components and their functions in the oral cavity, emphasizing the dual role of maintaining tolerance to commensal microbes and dietary antigens while initiating protective responses against pathogens. Any disruptions in this balance, such as oral dysbiosis or immune dysregulation, can lead to the development of local inflammatory conditions; it may also contribute to systemic immune disturbances and related pathologies. Immune mechanisms also regulate craniofacial development and postnatal bone remodeling and regeneration, mainly through cytokine-mediated signaling pathways and interactions between stem cells and immune cells. Several local and systemic immunological pathways are often dysregulated in oral inflammatory conditions, which makes them important therapeutic targets. Therapeutic strategies targeting these pathways include immune checkpoint inhibitors, microbiome-directed interventions, stem cell-based therapies, and salivary diagnostics for real-time and noninvasive immune profiling. These offer promising approaches for restoring oral and systemic immune balance. Finally, this chapter has reviewed recent technological advances, such as single-cell RNA sequencing (scRNA-seq), spatial transcriptomics, metagenomics, and multi-omics integration, in the context of oral immunity. These novel techniques are transforming oral immunology, since they enable high-resolution characterization of cellular, microbial, and molecular interactions, and support the transition toward establishing more precise diagnosis and treatment plans. These findings suggest that oral immunity plays a critical role in linking local mucosal

H. Ziaei
Universal Scientific Education and Research Network (USERN), Los Angeles, CA, USA

N. Rezaei (✉)
Universal Scientific Education and Research Network (USERN), Göteborg, Sweden

Network of Immunity in Infection, Malignancy and Autoimmunity (NIIMA), Universal Scientific Education and Research Network (USERN), Tehran, Iran

Research Center for Immunodeficiencies, Children's Medical Center, Tehran University of Medical Sciences, Tehran, Iran

Department of Immunology, School of Medicine, Tehran University of Medical Sciences, Tehran, Iran
e-mail: Rezaei_nima@tums.ac.ir

N. Rezaei, H. Ziaei (eds.), *Oral Immunology*, Advances in Experimental Medicine and Biology 1492,
https://doi.org/10.1007/978-3-032-03176-1_1

defense and systemic immune responses. Therefore, understanding oral immune mechanisms in health and inflammatory conditions is important for revealing disease pathogenesis and guiding targeted interventions.

Keywords

Oral immunity · Oral immunology · Oral inflammation

1.1 Introduction

The oral cavity represents one of the most immunologically complex sites of the body with a unique anatomical structure. Since it serves as an important primary point of contact with the external environment, it is constantly exposed to various stimuli such as food particles, pathogenic and commensal microorganisms, potentially harmful substances, and environmental antigens. This continuous exposure requires a highly responsive and specialized immune defense system, serving as the primary immunological defense barrier, with diverse innate and adaptive immunological mechanisms (Deo and Deshmukh 2019; Feller et al. 2013).

The oral cavity, unlike many other immune sites, must simultaneously balance two apparently contradictory roles: facilitating interactions with the external environment with tolerance to benign antigens and providing a protective barrier against pathogens (Wambre and Jeong 2018). The dual challenge of preserving homeostasis with commensal microbes, dietary antigens, or any benign stimulus while protecting against harmful pathogens highlights the immune system's remarkable ability to distinguish between these stimuli to maintain the balance between protection and tolerance (Jang et al. 2023). This homeostatic immune regulation is reached through a dynamic and rich network of innate and adaptive immune components containing various immune cells, including macrophages, Langerhans cells, salivary antimicrobial peptides, antigen-presenting cells (APCs), neutrophils, lymphocytes, etc., in the oral mucosa, gingival crevicular fluid (GCF), and tonsillar tissue (Cocksedge et al. 2023; Li et al. 2022). The failure of this balance can lead to chronic inflammation, which affects the oral cavity and might also contribute to systemic inflammatory conditions.

Anatomically, the oral cavity has a complex organization that includes keratinized and nonkeratinized mucosal surfaces, soft and hard tissues, teeth, and major and minor salivary glands. These tissues are continuously subjected to dynamic changes in the oral cavity, such as constant exposure to mechanical trauma (from chewing and food intake), pH fluctuations, temperature alterations, and changes in microbial colonization, necessitating rapid and coordinated immune responses (Cocksedge et al. 2023; Li et al. 2022).

In addition to its structural complexity, the oral cavity has one of the most diverse and densely populated microbial ecosystems in the body. The oral microbiome plays an important role both in the maintenance of immune tolerance in healthy conditions and in triggering immune responses when dysregulated (Li et al. 2022). However, disruptions in this equilibrium, referred to as oral dysbiosis, can initiate or exacerbate a variety of local and systemic inflammatory diseases, most commonly periodontitis, peri-implantitis, oral mucosal diseases, and even oral malignancies (Thomas et al. 2021).

The oral mucosa, particularly the oral junctional epithelium (JE) around the teeth, is an important site for immune interactions, playing roles as both a physical barrier and an immunological gateway (Belkaid and Hand 2014). This tissue uses pattern recognition receptors (PRRs) to recognize microbe-associated molecular patterns (MAMPs), which, if stimulated, initiate signaling cascades that lead to immune activation or tolerance depending on the condition (Belkaid and Hand 2014). The saliva also plays a crucial role in mucosal immune defense of the oral cavity by transporting a range of antimicrobial peptides, immunoglobulins (Ig) (e.g., secretory IgA (sIgA)), lactoferrin, lysozyme, and other enzymes that also contribute to maintaining microbial homeostasis (Fábián et al. 2012).

Understanding these basic principles is essential for exploring how oral immune function interacts with microbial communities, contributes to the pathogenesis of systemic disease processes, and serves as a target for therapeutic interventions. This introductory chapter will provide a brief overview of the broader topics relevant to oral immunity, such as immunobiology of oral tissues, host–microbe interactions, the bidirectional relationship between oral and systemic immunity, immune regulation during oral and craniofacial development and bone remodeling, and technological advances in the field. Additionally, novel therapeutic approaches, such as immunomodulation, salivary diagnostics, and stem cell-based therapeutic approaches, are highlighted as key innovative methods in the field of oral immunology.

1.2 Oral Mucosal Immunity

As briefly discussed in the previous section, diverse anatomical structures, immune cell types, and signaling pathways in the oral cavity are continuously interacting together to maintain immunological homeostasis in this highly dynamic environment.

First of all, diverse mucosal surfaces lining the oral cavity provide a structural barrier against pathogens and harmful stimuli. The stratified squamous epithelium acts as the first line of defense against pathogens by providing a mechanical barrier (Şenel 2021). A junctional epithelium (JE) is a nonkeratinized epithelial layer, located within the gingival sulcus, which provides an interface between the gingival connective tissue and teeth surfaces. This tissue exhibits high but selective permeability, allowing the passage of GCF containing immune cells and molecules that support local immune defense (Nakamura 2017; Theodoro et al. 2023). However, due to this permeability, the JE may be a potential point of entry for pathogens as well, especially when epithelial health and integrity are compromised due to dysbiosis, inflammation, or early stages of gingivitis and periodontal diseases (Bosshardt and Lang 2005).

In addition to epithelial layers themselves, specialized intercellular junctions, including tight junctions and desmosomes, also play a crucial role in preserving tissue integrity. Tight junctions, which are located at the most apical region of epithelial cells, restrict paracellular transportation and form a barrier that enhances cell–cell adhesiveness, further preventing the translocation of harmful antigens (Wan et al. 2018; Wang et al. 2019). Desmosomes are also present in the oral mucosa as anchoring intermediate filaments at deeper intercellular junctions between adjacent epithelial cells, in contrast to more superficial tight junctions. Desmosomes provide mechanical strength and cohesion—especially important given the constant mechanical stress in the oral cavity due to daily oral functions such as mastication and speech (Raza and Ghafoor 2020; Squier and Kremer 2001).

On top of structural defense, in order to adapt and respond to constant mechanical, chemical, and microbial stressors in the oral cavity, oral epithelial cells have various specialized receptors for detecting pathogens. These receptors are pattern recognition receptors (PRRs), including toll-like receptors (TLRs) and NOD-like receptors (NLRs), which detect microbe-associated molecular patterns (MAMPs) (Groeger and Meyle 2019). Stimulation of these receptors activates downstream inflammatory signaling pathways, leading to the secretion of various immunological molecules such as antimicrobial peptides, cytokines, and chemokines. In general, disruptions of any of these structures or constant stimulation of these receptors might lead to a shift in the immunological balance toward inflammatory conditions such as gingivitis and periodontitis (Zhu et al. 2022).

Beyond the structural barriers, the oral mucosa also contains a dense network of innate immune cells, acting as rapid responders to harmful stimuli. These cells, such as macrophages, neutrophils, and Langerhans and dendritic cells, reside in different regions of the oral tissue. Macrophages residing in the lamina propria secrete both pro- and anti-inflammatory cytokines. Neutrophiles are commonly found in GCF, passing through JE, acting as a first-line defense against microbes. Dendritic cells and their subset, Langerhans cells, acting as antigen-presenting cells (APCs), are present in the oral epithelium; these cells have a key role in initiating adaptive immune responses (Hovav 2014; Lombardi et al. 1993; Metcalfe et al. 2022).

While these cells and related molecules in the innate immune defense offer rapid protection, the oral immune system also relies on adaptive mechanisms to provide long-lasting and highly specific immunity. The adaptive immune system of the oral cavity facilitates antigen-specific defense and immunological memory. Various types of T helper (Th) cells, such as Th1 and Th17, mostly linked to inflammation, and T regulatory cells (Tregs), linked to immune regulation, are abundantly present in the gingival tissue and tonsils (Wang et al. 2023; Xue et al. 2022; Zhang et al. 2021). B cells and their differentiated form, plasma cells, are found in the mucosa-associated lymphoid tissue (MALT), such as tonsils, and in the gingival connective tissue. These cells of the adaptive immune system mostly secrete secretory IgA (sIgA) in the oral mucosa, which helps in neutralizing pathogens and inhibiting pathogen adherence (Cerutti et al. 2010; Johansen and Brandtzaeg 2004).

Another structural and immunologically active component of the oral cavity is Waldeyer's ring, which is an oral-associated lymphoid structure including the palatine tonsils, adenoids, and lingual tonsils. This ring plays a central role in immune surveillance as the first line of defense against inhaled and ingested antigens. This structure activates innate and adaptive immune responses and specifically plays a central role in detecting inhaled or ingested antigens and producing antibodies, contributing to immune memory (Talks et al. 2024). Together, all these innate and adaptive immune responses in the oral cavity maintain an immunological balance, poised to defend against pathogens while preserving tolerance.

1.3 The Oral Microbiome and Host–Microbiome Interactions

As briefly mentioned earlier, the oral microbiome is one of the most heterogeneous microbial communities in the body, harboring over 700 types of bacterial species (e.g., *Streptococcus*, *Actinomyces*, *Veillonella*, etc.), fungi, viruses, and archaea. Commensal microbes reside in the body without causing any harmful effects and play essential roles in functions such as immune system priming, nutrient processing, and resistance against the colonization of pathogenic microbes (Caselli et al. 2020; Rajasekaran et al. 2024). Under specific environmental or immunological disturbances, commensal microbes may shift toward a pathobiont phenotype, contributing to oral inflammatory diseases, such as gingivitis and periodontitis (Moreno et al. 2023).

The host immune system continuously produces specific immunological factors such as sIgA and defensins, thus regulating the microbial community in order to maintain a diverse yet balanced microbiome. On the other hand, commensal microorganisms contribute to immune system development by interacting with PRRs without triggering inflammation, thereby establishing tolerance and preventing excessive immune responses to harmless environmental stimuli. The mechanisms of immune tolerance and activation have been discussed in different chapters of the book *Oral Immunity*. Various environmental shifts, such as poor oral hygiene, systemic diseases, and diet, can push the microbiota toward dysbiosis (Chu and Mazmanian 2013; Liu et al. 2022a; Mantis et al. 2011).

Recent research has also highlighted the role of short-chain fatty acids (SCFAs) in oral microbial balance and immune activation. While SCFAs normally modulate immune responses and help maintain immune homeostasis, their excessive accumulation or imbalance can promote inflammatory processes (Du et al. 2024).

Oral microbial dysbiosis is strongly associated with the pathology of various oral and inflammatory diseases, such as gingivitis, periodontitis, peri-implantitis, and dental caries. In periodontitis, a microbial shift toward *Porphyromonas gingivalis*, *Tannerella forsythia*, and *Treponema denticola* leads to immunological responses and tissue destruction. Dental caries also mainly occurs due to a microbial shift toward acidogenic species such as *Streptococcus mutans* and the loss of protective bacterial species (Abdulkareem et al. 2023; Zhang et al. 2022). Therefore, understanding the immunological dynamics of the oral microbiome provides insights into the mechanisms of disease initiation and progression and helps identify potential therapeutic targets for restoring microbial homeostasis.

1.4 The Oral–Systemic Immune Axis

Emerging evidence strongly suggests a bidirectional relationship between oral and systemic health, where oral health not only serves as a critical indicator of underlying immunological alterations in the body but also contributes to the development and progression of systemic diseases through mechanisms such as microbial translocation, chronic inflammation, and immune dysregulation (Kapila 2021).

Several systemic diseases have established manifestations in the oral cavity, such as autoimmune diseases, diabetes mellitus, cardiovascular disorders, neurodegenerative conditions, atherosclerosis, and rheumatoid arthritis. Certain manifestations, including gingivitis, periodontitis, oral mucosal lesions, or salivary dysfunction, can arise in various types of systemic disorders primarily due to underlying immunological alterations (Hajishengallis and Chavakis 2021; Natarajan et al. 2025; Rautemaa et al. 2007). On the other hand, oral dysbiosis and its chronic inflammatory conditions can lead to systemic disorders through the release and circulation of pro-inflammatory cytokines, microbial translocation, and immune dysregulation (Georges et al. 2022). Even low-grade chronic oral inflammation can contribute to systemic inflammatory burden. This bidirectional interaction represents an interesting yet complex field of study, as the oral cavity not only reflects systemic immune status but also actively modulates immune responses beyond the local environment. The complexity results from the multifactorial nature of both oral and systemic disorders, diverse immunological profiles of patients with these conditions, individual variations in host genetics, microbial communities, and the effect of environmental factors (Kapila 2021; Natarajan et al. 2025). Although this interaction is multifaceted and influenced by numerous variables, certain oral–systemic health connections have robust evidence linking oral inflammation to systemic disorders such as diabetes, atherosclerosis, and adverse pregnancy outcomes.

As an example of these bidirectional relationships, diabetes mellitus, in cases of poor management, can cause impaired neutrophil function and elevated secreted cytokine levels in the gingiva; on the other hand, periodontitis can exacerbate insulin resistance through immunological pathways involving tumor necrosis factor (TNF)-α and interleukin (IL)-1β (Hatanaka et al. 2007; Shi et al. 2023). Neurodegenerative disorders also exemplify the systemic effect of chronic oral inflammation, since oral dysbiosis contributes to chronic systemic inflammation. In chronic oral inflammation, secretion of pro-inflammatory cytokines such as IL-6, their release into the bloodstream, and their ability to cross the blood–brain barrier (BBB) can lead to neuroinflammation in the brain and potentially accelerate neurodegenerative processes (Ashique et al. 2024; Nicholson and Landry 2022; Yang et al. 2022). One example of the oral manifestation of systemic disease with an inflammatory origin is Sjögren's syndrome, in which autoimmune destruction of the salivary glands causes xerostomia. Another example is the oral manifestations of Crohn's disease, including oral ulcers, mainly due to the chronic inflammation characteristic of the disease (Ashique et al. 2024; Nicholson and Landry 2022; Yang et al. 2022).

Oral manifestations sometimes serve as early indicators of systemic diseases, while effective management of oral inflammatory conditions may help prevent or mitigate the burden of systemic inflammatory and immune-mediated disorders. Therefore, studying and recognizing the interconnection between oral and systemic health and immunity provides critical insights and establishes a foundation for integrative diagnostic and therapeutic approaches.

1.5 Immune Regulation in Craniofacial Development and Bone Remodeling

Craniofacial development and bone formation are mainly driven by neural crest cell migration and differentiation under the control of genetic factors and morphogenetic cues during embryonic and postnatal development (Roth et al. 2021). During these processes, various immune-related signaling cues and molecules such as cytokines, chemokines, and growth factors are proposed to be actively involved in the regulation of cranial neural crest cell migration and differentiation and in determining the skeletal architecture (Olesnicky Killian et al. 2009; Roth et al. 2021). Impaired immune function due to maternal infections or elevated pro-inflammatory cytokines and chemokines has been shown to impair proper craniofacial morphogenesis and contribute to neurodevelopmental disorders (Han et al. 2021). Cytokines such as TNF-α, interferon-γ (IFN-γ), and IL-1β, which are released from activated T lymphocytes and macrophages, can affect the regenerative capacity of mesenchymal stem cells (MSCs) (Liu et al. 2011). These secreted pro-inflammatory molecules may induce apoptosis, inhibit proliferation, and suppress the differentiation of mesenchymal stem cells (Liu et al. 2011; Mallis et al. 2021).

Additionally, toll-like receptor (TLR) activation alters MSC immunoregulation and function (Li et al. 2023). The activation of TLR4 on dental follicle stem cells (DFSCs) has been shown to impair their immunosuppressive capacity (Li et al. 2023). For instance, treatment of DFSCs with a TLR4 agonist such as lipopolysaccharide (LPS) significantly reduces their ability to suppress peripheral blood mononuclear cell (PBMC) proliferation in coculture assays, which indicates a functional shift away from immune regulation (Li et al. 2023). These findings suggest that the activation of innate immune receptors under inflammatory conditions can alter local immune tolerance mechanisms and compromise the therapeutic potential of MSCs during development or tissue repair. In general, under normal conditions, craniofacial MSCs support immune tolerance and tissue repair; however, TLR4 activation, especially during bacterial infection or chronic inflammation, can shift them toward a pro-inflammatory or immunoregulatory-deficient phenotype (Yang et al. 2017). This has implications for oral and craniofacial development, orthodontic remodeling, and the success rate of regenerative therapies.

An example of chronic inflammatory conditions that can alter the immunoregulatory function of craniofacial MSCs is periodontitis. Based on a recent study, MSCs derived from inflamed periodontal tissues, such as inflamed periodontal ligament stem cells (iPDLSCs), have shown a reduced ability to inhibit T cell proliferation and Th17 differentiation (Liu et al. 2012). In a coculture system, iPDLSCs showed significantly altered immunomodulatory properties compared to normal PDLSCs, including inhibited suppression of T cell proliferation and reduced induction of regulatory T cells. This decreased immunosuppressive function might be associated with the altered expression of key immunoregulatory molecules, including prostaglandin E2 (PGE2), indoleamine 2,3-dioxygenase (IDO), and human leukocyte antigen-G (HLA-G) (Wang et al. 2020). Additionally, inflammatory cytokines such as TNF-α and IL-1β, present in inflamed periodontal tissues, may further compromise the inhibition of effector T cell subsets mediated by MSCs. Th17 and its signature cytokine IL-17A have been implicated in the pathogenesis of periodontitis and peri-implantitis. Since Th17 cells play an important role in regulating both inflammatory responses and tissue remodeling at oral mucosal surfaces, their dysregulation may lead to the progression of periodontal disease and impair MSC-based regenerative processes (Huang et al. 2021; Mazurek-Mochol et al. 2023; Wang et al. 2023).

Another example of inflammatory conditions affecting the regenerative potential of dental MSCs is a previous study showing that inflamed dental tissues exhibit reduced osteogenic, dentinogenic, adipogenic, and neurogenic differentiation capacity. The regenerative potential of cells derived from inflamed dental pulp was successfully restored through an immunomodulatory mechanism (Hong et al. 2020).

Several studies have suggested that maternal immune status plays a critical role in the morphogenesis of craniofacial structures during embryonic development; maternal infections and associated inflammatory responses can trigger epigenetic modifications in the fetal genome with variable defects at different stages of development (Bhagirath et al. 2021)

Maternal infections can elevate systemic cytokines, such as TNF-α, IL-1β, and IFN-γ, which may interfere with critical developmental processes in the fetus (Bhagirath et al. 2021; Elgueta et al. 2022). Various inflammatory mediators can cross the placenta and cause detrimental effects in the developing embryo. For instance, there is a link between maternal infection and inflammation and nonsyndromic orofacial clefts (Bhagirath et al. 2021; Kwon et al. 2022). Development of cleft lip and palate occurs during the first trimester of gestation, a period during which maternal infectious diseases, including respiratory infections, influenza, and urinary tract infections, have been epidemiologically associated with an increased prevalence of nonsyndromic orofacial clefts (Bhagirath et al. 2021; Inchingolo et al. 2022). The underlying mechanisms leading to orofacial clefts may involve inflammatory cascades and epigenetic modifications of *IRF6* and *PVRL1*, genes critical for epithelial differentiation and intercellular adhesion that have demonstrated strong associations with orofacial clefts (Martinelli et al. 2020). Moreover, evidence suggests that hyperthermia induced by maternal infections during pregnancy may be another etiology of orofacial clefts, highlighting the potential protective role of maternal vaccination strategies during gestation (Hashmi et al. 2010; Peterka et al. 1994). Altogether, these inflammatory signals can alter gene expression patterns essential for craniofacial symmetry and organogenesis, increasing the risk of congenital anomalies such as cleft lip and palate.

An example of a maternal infection affecting embryonic development is maternal *Rubella* infection, which can result in craniofacial malformations, such as triangular face, micrognathia, cleft palate, microcephaly, and midfacial hypoplasia (Chattannavar et al. 2023). Another example is maternal cytomegalovirus (CMV) infection, which can be transmitted to the fetus, potentially causing congenital CMV infection. These infections may lead to microcephaly, craniofacial abnormalities, and tooth developmental anomalies (Boppana et al. 2013; Jaskoll et al. 2008). The pathophysiology of these conditions might involve disturbed neural crest cell migration and osteogenic differentiation mediated by pro-inflammatory cytokine production, but further research is needed to elucidate these mechanisms.

Thus, disruptions to immunological factors during embryogenesis, particularly those associated with maternal viral infections, may result in craniofacial malformations; however, in adulthood, the immune system continues to play an important role in regulating bone formation, remodeling, homeostasis, and repair. Bone remodeling is immune-regulated, as, under normal homeostatic conditions, immune cells such as regulatory T cells, macrophages, and neutrophils favor bone healing and maintenance, while, during chronic inflammation, the release of pro-inflammatory cytokines promotes bone resorption (Xiong et al. 2022; Yang and Liu 2021).

Several immunological factors affect osteoblast and osteoclast differentiation and activation. For example, IL-3 is a cytokine that has been shown to indirectly inhibit osteoblast differentiation from mesenchymal stem cells while also influencing osteoclastogenesis by enhancing the expression of receptor activator of nuclear factor kappa B (NF-κB) ligand (RANKL), thereby promoting bone resorption (Ehrlich et al. 2005; Singh et al. 2018). RANKL binds to RANK on pre-osteoclasts, driving osteoclastogenesis.

Other immunological regulators of bone homeostasis are regulatory T cells (Tregs) and B cells. Tregs inhibit osteoclast differentiation and activity by secreting anti-inflammatory cytokines such as IL-10 and by direct cell–cell contact with osteoclasts, thereby protecting against excessive bone resorption (Lei et al. 2015; Wang et al. 2024). On the other hand, B cells produce osteoprotegerin (OPG), a decoy receptor for RANKL, thereby preventing RANKL from binding to its receptor RANK on osteoclast precursors, thus inhibiting osteoclastogenesis.

Regarding bone repair after surgery or bone injury, during the healing process, inflammatory cytokines such as IL-6 and TNF-α are upregulated, initiating the recruitment of MSCs to the site of injury and their differentiation into osteoprogenitor cells (Lei et al. 2015). IL-6, in particular, has been shown to stimulate osteoclastogenesis and is involved in the acute-phase response of bone injury, playing an important role in the early stages of bone healing (Feng et al. 2017; Prystaz et al. 2016).

During orthodontic tooth movement (OTM), constant mechanical stress induces an inflammatory response, leading to the formation of bone on the tension side and resorption of bone on the compression side. *During OTM, compression and tension* sides exhibit differential and distinct expression of immunological modulators, leading to enhanced osteoblast activity and bone formation on the tension side while triggering osteoclast activity and bone resorption on the compression side. These include higher expression of TNF-α and RANKL on the compression side but higher expression of IL-10, OPG, and osteocalcin (OCN) on the tension side. During OTM, various cells in the region, including periodontal ligament (PDL) cells, macrophages, and T cells, are activated to release inflammatory cytokines such as IL-1β and TNF-α, leading to localized bone remodeling (Garlet et al. 2007; Jeon et al. 2021)

After orthodontic force withdrawal, pro-inflammatory cytokines such as RANKL may persist in supporting tissues of the teeth, potentially leading to relapse through sustained osteoclast activation and remodeling (Alhasyimi et al. 2024)

Altogether, these insights emphasize that immunological regulation plays a critical role in the development of various craniofacial structures during the embryonic and postnatal stages of development. From the potential effect of maternal inflammatory alterations on neural crest cell migration and differentiation during embryogenesis to the postnatal remodeling of the alveolar bone in response to local inflammatory microenvironments and mechanical loading, immunological pathways function in coordination with genetic cues, endocrine regulation, and mechanotransduction pathways to orchestrate craniofacial development, skeletal formation and patterning, and long-term tissue homeostasis. Understanding these immunological interactions is essential for addressing developmental disorders, optimizing orthodontic therapies, and managing immune-related bone pathologies in the oral and craniofacial region.

1.6 Innovations in Immunotherapeutics and Regenerative Approaches in Oral Immunology

1.6.1 Probiotics, Prebiotics, and Microbiome Modulation Strategies

Growing evidence about the oral microbiome's impact on health and disease has led to the development of microbial modulation approaches as potential therapeutic strategies. Specifically, interventions using probiotics (beneficial bacteria), prebiotics (nondigestible fibers that feed these bacteria), and targeted microbiome manipulations characterize novel adjunctive modalities to standard treatment protocols in the management of oral immunological conditions (Shimabukuro et al. 2021).

Recent research has indicated that in periodontitis patients, probiotic supplementation, particularly with *Lactobacillus reuteri* and *Bifidobacterium*, alongside standard periodontal therapy, significantly improves clinical outcomes. Adjunct probiotic administration has been shown to reduce pathogenic bacteria, lower pro-inflammatory factors such as TNF-α and IL-1β, and improve gingival health status, including decreased pocket depth (Sachelarie et al. 2025; Shimabukuro et al. 2021).

Prebiotics are selectively fermentable substrates that specifically promote the growth and metabolic activity of commensal microorganisms linked to host health benefits, while suppressing pathogenic microorganisms associated with periodontal inflammation. Recent studies have also suggested that prebiotics, such as nitrate, beta-methyl-D-galactoside, and *N*-acetyl-D-mannosamine, have the capacity to selectively modulate oral microbial communities by promoting beneficial commensal species. This microbiome modulation is characterized by increased abundance of commensal genera such as *Streptococcus*, *Neisseria*, and *Rothia*, and by reduced levels of periodontopathic microorganisms, including *Fusobacterium nucleatum* and *Porphyromonas gingivalis*. The resulting microbial community restructuring appears to favor immunoregulatory phenotypes that may contribute to oral immune homeostasis (Yu et al. 2024; Zhou et al. 2024).

These findings suggest that prebiotics and probiotics can help restore oral microbial balance and modulate immune responses, offering a promising adjunctive approach to oral immune homeostasis and periodontal treatment. However, further research is necessary to reveal the specific mechanisms underlying these modulations and to optimize prebiotic formulations and probiotic administrations for therapeutic application.

Recent studies have demonstrated that probiotics and prebiotics modulate oral immune responses by several mechanisms, including upregulation of the Treg population, inhibition of pro-inflammatory signaling pathways, and disruption of pathogen biofilm development, contributing to the maintenance of mucosal immune homeostasis (Liu et al. 2022b; Smits et al. 2005). Certain probiotics enhance Treg populations, which are important for the maintenance of immune tolerance and homeostasis and for inhibiting inflammation in the oral mucosa. Probiotics also play a role in stimulating dendritic cells, which secrete IL-10 and transforming growth factor-β (TGF-β), promoting the differentiation of T cells into Tregs. Probiotics can also inhibit pathogen adhesion and prevent the formation of pathogen biofilms, thereby indirectly regulating immune homeostasis (Jeon et al. 2012; Mazziotta et al. 2023). All these immune-modulatory effects of probiotics are dose- and strain-specific and are impacted by host factors such as oral microbiome composition, age, and health status, thus highlighting the need for further research and the development of personalized and precision probiotic therapies.

Oral microbiota transplantation (OMT) represents a novel therapeutic approach, inspired by the established concept of fecal microbiota transplantation (FMT), aimed at the restoration of oral microbial homeostasis in dysbiotic conditions such as periodontitis and dental caries. This approach starts with an antimicrobial pretreatment of recipient oral mucosal surfaces, followed by the application of donor-derived healthy plaque samples to facilitate colonization by beneficial microbial communities in the oral cavity. A successful OMT treatment depends on multiple factors, including immunomodulatory effects on host responses and sustained stability of the transferred microbiota in the long term, especially when targeting chronic inflammatory diseases in the oral cavity (Khor et al. 2021; Nascimento 2017). Despite the theoretical potential of OMT as a targeted therapy for oral diseases, its clinical application is yet to be standardized, established, and evaluated for long-term clinical outcomes through controlled trials. In general, this approach holds promise as a potential strategy to reestablish beneficial oral microbiota, thus preventing the recurrence or progression of inflammatory conditions in the oral cavity.

The main goal of all methods discussed in this section is to restore microbial eubiosis, suppress pathological inflammation, and support immune tolerance in the oral microenvironment.

1.6.2 Immunotherapy and Targeted Approaches in Oral Diseases

In recent years, immunotherapy has emerged as a novel strategy in the management of diverse oral diseases, notably oral malignancies and hypersensitivity reactions to food antigens. Recent advances in understanding tumor immunology have facilitated the application of immunotherapeutic interventions, primarily as an adjunct therapy in oral malignancies, with the main aim of enhancing both antitumor immune responses and treatment outcomes.

In oral squamous cell carcinoma (OSCC), programmed cell death protein 1 (PD-1), which is a receptor on T cells, and programmed cell death ligand 1 (PD-L1), a protein expressed on some cancer cells, are upregulated. Excessive expression of these proteins leads to the inhibition of T cell activation and, thereby, tumor immune evasion and cancer progression (Lin et al. 2024). Inhibitors of these checkpoints, such as nivolumab and pembrolizumab, have been shown to enhance T cell immunity and tumor regression in clinical trials, resulting in promising clinical outcomes in terms of survival rate and prognosis, especially for recurrent or metastatic head and neck SCCs (Botticelli et al. 2021; Lin et al. 2024). Cytotoxic T-lymphocyte-associated protein 4 (CTLA-4) is expressed on T cells, and its excessive expression prevents the immune system from eliminating malignant cells. Combination immunotherapy against CTLA-4 and PD-1 has shown promising therapeutic outcomes in preclinical models of OSCC (Dorta-Estremera et al. 2019; Mei et al. 2020).

Cancer vaccines and monoclonal antibodies are other therapies for oral cancers that target tumor-associated antigens, such as epidermal growth factor receptor (EGFR), which is usually overexpressed in head and neck cancers and is responsible for tumor cell growth, division, and differentiation (Boeckx et al. 2013; Pidone et al. 2014). The application of monoclonal antibodies is also well established in cancer treatment; in OSCCs, one of the primary targets is EGFR. Monoclonal antibodies, such as cetuximab, panitumumab, and nimotuzumab, function by inhibiting ligands from binding to EGFR and therefore blocking downstream signaling pathways, thereby preventing to tumor progression and invasion (Pidone et al. 2014; Tamimi et al. 2023). Another immunotherapeutic strategy for targeting specific antigens associated with oral cancer progression and development is the use of cancer vaccines. However, the application and design of oral cancer vaccines are still in the early stages of research and are not yet widely approved or available.

In oral cancer, the tumor microenvironment (TME) plays a critical role in tumor progression, immune evasion, and therapy resistance. TME components are known to suppress antitumor immunity and promote both angiogenesis and cancer progression. Remodeling of the TME is one of the main future strategies for oral cancer immunotherapy, especially targeting immunosuppressive components such as myeloid-derived suppressor cells (MDSCs) and tumor-associated macrophages (TAMs). The main aim of TME remodeling approaches is to enhance the effectiveness of existing and well-established therapies by reducing the tumor-promoting inflammatory microenvironment and restoring immune surveillance in the oral cavity (Babu et al. 2024).

Beyond targeting oral malignancies, immunotherapeutic strategies have also been applied in managing food allergies, which involve oral immune responses to food antigens. In oral immunotherapy (OIT), controlled allergen exposure protocols stimulate immunological tolerance through continuous desensitization to food antigens (Pouessel and Lezmi 2023). The immunological mechanism behind OIT includes modification of T cells, characterized by decreased allergen-specific Th2 cell activity and expansion of Treg populations, resulting in enhanced suppressive capacity (Kulis et al. 2018).

Maintaining a balance between immune stimulation and tolerance is crucial in both cancer and food allergy immunotherapies to optimize the efficacy of treatment, while minimizing adverse inflammatory or immunosuppressive effects.

1.6.3 Stem Cell and Regenerative Immunotherapies

MSCs derived from different oral tissues exhibit significant anti-inflammatory and immunomodulatory properties. These cells, including MSCs derived from dental pulp stem cells (DPSCs), periodontal ligament stem cells (PDLSCs), and gingival tissue, have the ability to regulate immune responses through mechanisms such as suppressing T cell proliferation and stimulating a Treg profile (Li et al. 2014). These immunomodulatory mechanisms make oral-derived MSCs potentially promising for periodontal regeneration and the management of immune-mediated disorders affecting the oral cavity. The therapeutic efficacy of oral-derived MSCs is strongly influenced by the microenvironment surrounding the region. For instance, stimulation of TLRs can have context-dependent effects on MSC function; for instance, activation of TLR4 has been shown to improve the immunosuppressive characteristics of dental follicle stem cells (DFSCs) while reducing those of DPSCs (DelaRosa et al. 2012). Moreover, exposure to pro-inflammatory cytokines can impair both the differentiation potential and immunomodulatory properties of MSCs, thereby limiting their regenerative efficacy in inflamed tissues (Han et al. 2022).

DPSCs also have clinical implications beyond their applications in oral regeneration. Recent research has suggested their therapeutic potential in mitigating systemic inflammatory conditions such as colitis, arthritis, and neuroinflammatory disorders through immune suppression mechanisms. The underlying molecular mechanisms include the secretion of IL-10, modulation of macrophage polarization toward anti-inflammatory phenotypes, and indoleamine 2,3-dioxygenase (IDO) activity (Maccaferri et al. 2024). These data highlight the important regenerative and immunomodulatory characteristics of oral-derived MSCs, emphasizing their therapeutic value beyond local tissue repair applications. These cells show significant potential for periodontal tissue regeneration while simultaneously emerging as promising candidates for systemic immunomodulation in the management of various chronic inflammatory conditions.

1.6.4 Salivary Diagnostics and Biomarkers

Saliva is a potential noninvasive diagnostic bioliquid with potential advantages over conventional blood sampling due to its ease of access and rich immunological composition (Kumar et al. 2023). This biofluid contains various immune markers, including pro-inflammatory cytokines, such as IL-1β, IL-6, TNF-α, chemokines, and sIgA, which reflect both localized oral immune activity and systemic inflammatory responses (Al-Majid et al. 2018).

Omics approaches in salivary diagnostics include proteomics, genomics, and metabolomics, which are used to analyze biomarkers in saliva. Genomic and transcriptomic methods are used to detect genetic polymorphisms and gene expression profiles from host and microbial sources (Bonne and Wong 2012). Proteomic analyses are used to identify proteins relevant to oral or systemic diseases present in the saliva, including enzymes, antibodies, and cytokines. This approach is feasible due to enhanced protein stabilization methods (Bonne and Wong 2012; Umapathy et al. 2023). Metabolomic analysis of saliva has identified specific metabolite profiles that reflect metabolic dysregulation associated with various oral or systemic diseases. These metabolites can potentially serve as indicators of disease presence and progression, offering a complementary approach to genomic and proteomic analyses in salivary diagnostics (Hyvärinen et al. 2021, 2023).

Together, these multi-omics techniques enable noninvasive detection of oral and systemic conditions through saliva sampling. Multiple immune-related protein markers with diagnostic significance have been identified by proteomic analysis of saliva. In periodontitis, elevated levels of matrix metalloproteinase-8 (MMP-8) and calprotectin are associated with the severity and progression of disease (Al-Majid et al. 2018). Regarding oral cancer, salivary levels of IL-8 and CD44 are being investigated as potential biomarkers for early detection of cancer (Sahibzada et al. 2017).

The development of chairside salivary diagnostic devices is enabled by advancements in microfluidics, biosensors, and molecular detection technologies. These devices facilitate real-time monitoring of inflammatory biomarkers in saliva to simplify immediate clinical decision-making. In the management of periodontitis, for instance, these technologies help in assessing treatment efficacy, timing of intervention, and disease activity (Chai et al. 2025; Haririan et al. 2021; Punyadeera and Slowey 2019). Generally, salivary immune profiles are increasingly recognized as valuable tools for guiding personalized therapeutic strategies in various types of diseases. Distinct patterns of inflammatory biomarkers in patients can predict therapeutic responses and enable the selection of targeted anti-inflammatory agents based on individual immune phenotypes (Bonne and Wong 2012; Dongiovanni et al. 2023).

Clinical application of salivary diagnostic technologies is a key advancement toward precision-based oral medicine, which facilitates immune-guided, patient-specific therapeutic strategies. Using this approach, the timing and type of therapeutic intervention and treatment monitoring can be personalized in various diseases, such as periodontal disease, oral cancers, and autoimmune disorders such as Sjögren's syndrome (Kaczor-Urbanowicz et al. 2016). In general, these technologies represent a paradigm shift in dental practice in terms of transition from symptom-driven management to targeted immune modulation as well as initiation of early, individualized preventive and therapeutic treatments.

1.7 Technological Advances Transforming Oral Immunology

1.7.1 Single-Cell RNA Sequencing (scRNA-seq) and Spatial Transcriptomics

Single-cell RNA sequencing (scRNA-seq) and spatial transcriptomics are recent techniques that have dramatically revolutionized our understanding of tissue-specific immune landscapes at a the single-cell level, and oral immunology is no exception. These cutting-edge technologies allow researchers to dissect the cellular heterogeneity of oral tissues, such as the gingiva, dental pulp, oral mucosa, and periodontal tissues, at a high resolution. Using these techniques, complex immune cell subsets and their spatial organization within oral tissues in healthy and inflammatory conditions can be revealed (Cantalupo et al. 2024; Krivanek et al. 2020). This comprehensive cellular mapping provides critical insights into oral immune dynamics, which may facilitate the discovery of disease-specific immunological biomarkers and potential therapeutic targets. By mapping inflammatory niches, immune-tolerant zones, and cellular crosstalk in oral diseases like periodontitis and oral cancers, these tools are opening new avenues for biomarker discovery and targeted immunotherapy. In recent studies, the integration of single-cell transcriptomic data with spatial localization has enabled the precise identification of inflammatory niches and specific cell–cell interactions between oral mucosal cells and immune cells. It also facilitates observing changes in the local immune architecture in inflammatory conditions such as periodontitis, oral lichen planus, and oral cancers (Easter et al. 2024; Lee et al. 2022). As the field advances, these emerging technologies are expected to advance our understanding of chronic oral inflammation and immune dysregulation and to support the development of precision immunotherapies tailored to the distinct immune landscape of the oral cavity.

1.7.2 Metagenomics, In Silico Modeling, and Multi-omics Integration

Recent developments in metagenomic approaches and integrated multi-omics analysis have transformed our understanding of host–microbe–immune interactions in the oral cavity. Advanced metagenomic technologies such as 16S rRNA sequencing have enabled comprehensive taxonomic characterization of oral microbial communities associated with pathological conditions like periodontitis, dental caries, and peri-implantitis (Wen et al. 2025). These techniques have resulted in establishing a significant correlation between alterations in oral microbial composition and alterations in inflammatory biomarkers and immune dysfunction in chronic oral diseases (Baker 2023; Xu and Gunsolley 2014).

Moreover, in silico systems biology approaches are now used to analyze complex immune gene networks and how they respond to various factors, such as therapeutic responses, thereby predicting therapeutic results. For example, in OSCC, analysis of transcriptomic datasets from The Cancer Genome Atlas (TCGA) and Gene Expression Omnibus (GEO) has helped identify immune-related biomarkers of oral cancer, such as PD-L1 and CTLA-4, and has supported simulation of immune microenvironment modification strategies (Keerti Kumar et al. 2022; Taherkhani et al. 2024). Machine learning algorithms may further help in distinguishing immune phenotypes and predicting clinical outcomes based on molecular signatures.

As discussed in previous sections, integrated multi-omics approaches now combine single-cell transcriptomics, metagenomics, and proteomics to offer comprehensive insights into oral immune environments by linking cellular phenotypes, microbial composition, and immune signaling pathways. For example, the identification of biomarkers associated with periodontitis severity and immune dysregulation in OSCC is facilitated by combining salivary proteomic profiling with metagenomic data. These integrated approaches enhance our ability to characterize disease states, holding promise for the prediction of immunotherapeutic responses and providing insights for timely drug repurposing in oral malignancies (Belstrøm et al. 2017; Dumitrescu et al. 2025; Wu et al. 2024).

Collectively, these emerging technologies are gradually redefining oral immunology as a data-driven, precision-guided field, which bridges basic science with clinical applications, thus paving the way for personalized patient care.

1.8 Conclusion

Oral immunity forms a dynamic interface between the external environment and the host immune system, combining physical barriers, immune surveillance of the microbiota, and mechanisms that maintain immunological tolerance. The oral cavity functions not only as a site of local immune defense but also as a key contributor to systemic immunological balance. Its mucosal architecture, epithelial barriers, and intricate networks of immune cells all work together to simultaneously preserve tolerance to dietary antigens and commensals while initiating protective responses against pathogens. Disruptions in this finely regulated balance, such as oral dysbiosis, can lead to chronic oral inflammation, including periodontitis, and may even contribute to systemic inflammatory responses and related disorders.

Importantly, the oral microbiome serves as both a mediator and a target of the immune system. Commensal microbial communities are involved in regulating and training the immune system and its homeostasis, whereas pathogenic shifts stimulate inflammatory responses with both local and systemic effects. These interactions extend beyond the oral cavity, influencing immune responses in organs such as the brain, heart, and pancreas, thereby emphasizing the clinical relevance of the oral–systemic immune axis.

Additionally, the immune system plays an important role in craniofacial development, bone remodeling, and tissue regeneration. The interactions between immune cells, stem cells, inflammatory cytokines, and epigenetic factors trigger several developmental and regenerative processes.

Moreover, advancements in immunotherapeutics, from microbiome-directed strategies to stem cell therapies and immune checkpoint modulation, offer promising approaches for managing oral and systemic diseases. New technologies such as single-cell transcriptomics, spatial mapping, and multi-omics integration are also improving our understanding of the oral immune system and its cellular and molecular interactions at a high resolution.

This review has outlined the fundamentals of oral mucosal immunity and host–microbe interactions, highlighted clinical applications in systemic diseases, cancers, and regenerative medicine, and reviewed technological innovations that are driving the field forward.

References

Abdulkareem A, Al-Taweel F, Al-Sharqi A, Gul S, Sha A, Chapple I (2023) Current concepts in the pathogenesis of periodontitis: from symbiosis to dysbiosis. J Oral Microbiol 15:2197779. https://doi.org/10.1080/20002297.2023.2197779

Alhasyimi AA, Nugraha AP, Ayub A, Utami TW, Gerardo TS, Abshori NF, Khumaidi MA, Tallei TE, Taslim NA, Kim B, Tjandrawinata RR, Tsopmo A, Nurkolis F (2024) Mechanistic Elucidation of green seaweed compounds in orthodontic relapse management via RANKL/TNF-α-mediated ROS/Keap1/Nrf2 signaling: In silico and Ex Vivo studies. J Agric Food Res 18: 101396. https://doi.org/10.1016/j.jafr.2024.101396

Al-Majid A, Alassiri S, Rathnayake N, Tervahartiala T, Gieselmann D-R, Sorsa T (2018) Matrix Metalloproteinase-8 as an inflammatory and prevention biomarker in periodontal and Peri-implant diseases. Int J Dent 2018(1):7891323. https://doi.org/10.1155/2018/7891323

Ashique S, Mohanto S, Ahmed MG, Mishra N, Garg A, Chellappan DK, Omara T, Iqbal S, Kahwa I (2024) Gut-brain axis: a cutting-edge approach to target neurological disorders and potential synbiotic application. Heliyon 10(13):e34092. https://doi.org/10.1016/j.heliyon.2024.e34092

Babu S, Manavalan MJ, Jasmine SH, Krishnan M (2024) Tumor microenvironment in oral squamous cell carcinoma. Implications for novel therapies, Oral Oncol Rep 12:100666. https://doi.org/10.1016/j.oor.2024.100666. ISSN 2772–9060

Baker J (2023) Illuminating the oral microbiome and its host interactions: recent advancements in omics and bioinformatics technologies in the context of oral microbiome research. FEMS Microbiol Rev 47. https://doi.org/10.1093/femsre/fuad051

Belkaid Y, Hand TW (2014) Role of the microbiota in immunity and inflammation. Cell 157(1):121–141. https://doi.org/10.1016/j.cell.2014.03.011

Belstrøm D, Constancias F, Liu Y, Yang L, Moses D, Schuster S, Kohli G, Jakobsen T, Holmstrup P, Givskov M (2017) Metagenomic and metatranscriptomic analysis of saliva reveals disease-associated microbiota in patients with periodontitis and dental caries. Npj Biofilms Microb 3. https://doi.org/10.1038/s41522-017-0031-4

Bhagirath AY, Medapati MR, de Jesus VC, Yadav S, Hinton M, Dakshinamurti S, Atukorallaya D (2021) Role of maternal infections and inflammatory responses on craniofacial development. Front Oral Health 2. https://www.frontiersin.org/journals/oral-health/articles/10.3389/froh.2021.735634

Boeckx C, Baay M, Wouters A, Specenier P, Vermorken JB, Peeters M, Lardon F (2013) Anti-epidermal growth factor receptor therapy in head and neck squamous cell carcinoma: focus on potential molecular mechanisms of drug resistance. Oncologist 18(7):850–864. https://doi.org/10.1634/theoncologist.2013-0013

Bonne N, Wong D (2012) Salivary biomarker development using genomic, proteomic and Metabolomic approaches. Genome Med 4:82. https://doi.org/10.1186/gm383

Boppana SB, Ross SA, Fowler KB (2013) Congenital cytomegalovirus infection: clinical outcome. Clin Infect Dis 57(suppl_4):S178–S181. https://doi.org/10.1093/cid/cit629

Bosshardt DD, Lang NP (2005) The junctional epithelium: from health to disease. J Dent Res 84(1):9–20. https://doi.org/10.1177/154405910508400102

Botticelli A, Cirillo A, Strigari L, Valentini F, Cerbelli B, Scagnoli S, Cerbelli E, Zizzari IG, Rocca CD, D'Amati G, Polimeni A, Nuti M, Merlano MC, Mezi S, Marchetti P (2021) Anti–PD-1 and anti–PD-L1 in head and neck cancer: a network meta-analysis. Front Immunol 12:705096. https://www.frontiersin.org/journals/immunology/articles/10.3389/fimmu.2021.705096

Cantalupo P, Diacou A, Park S, Soman V, Chen J, Glenn D, Chandran U, Clark D (2024) Single-cell RNA-seq reveals a resolving immune phenotype in the oral mucosa. IScience 27(9):110735. https://doi.org/10.1016/j.isci.2024.110735

Caselli E, Fabbri C, D'Accolti M, Soffritti I, Bassi C, Mazzacane S, Franchi M (2020) Defining the oral microbiome by whole-genome sequencing and resistome analysis: the complexity of the healthy picture. BMC Microbiol 20(1):120. https://doi.org/10.1186/s12866-020-01801-y

Cerutti A, Chen K, Chorny A (2010) Immunoglobulin responses at the mucosal Interface. Annu Rev Immunol 29:273–293. https://doi.org/10.1146/annurev-immunol-031210-101317

Chai Z, Yu Y, Cheng P, Zhao L, Petrovic B, Li A, Xu F, Li Y, You M (2025) Recent advances of oral fluids-based point-of-care testing platforms for oral disease diagnosis. Transl Dent Res 1(1):100003. https://doi.org/10.1016/j.tdr.2024.100003

Chattannavar G, Bansal A, Kekunnaya R (2023) Facial dysmorphism in congenital rubella syndrome. J Am Assoc Pediatr Ophthalmol Strabismus 28. https://doi.org/10.1016/j.jaapos.2023.08.020

Chu H, Mazmanian S (2013) Innate immune recognition of the microbiota promotes host-microbial symbiosis. Nat Immunol 14:668–675. https://doi.org/10.1038/ni.2635

Cocksedge SP, Causer AJ, Winyard PG, Jones AM, Bailey SJ (2023) Oral temperature and pH influence dietary nitrate metabolism in healthy adults. Nutrients 15(3). https://doi.org/10.3390/nu15030784

DelaRosa O, Dalemans W, Lombardo E (2012) Toll-like receptors as modulators of mesenchymal stem cells. Front Immunol 3. https://doi.org/10.3389/fimmu.2012.00182

Deo P, Deshmukh R (2019) Oral microbiome: unveiling the fundamentals. J Oral Maxillof Pathol 23:122–128. https://doi.org/10.4103/jomfp.JOMFP_304_18

Dongiovanni P, Meroni M, Casati S, Goldoni R, Thomaz DV, Kehr NS, Galimberti D, Del Fabbro M, Tartaglia GM (2023) Salivary biomarkers: novel noninvasive tools to diagnose chronic inflammation. Int J Oral Sci 15(1):27. https://doi.org/10.1038/s41368-023-00231-6

Dorta-Estremera S, Hegde V, Slay R, Sun R, Yanamandra A, Nicholas C, Nookala S, Galvan Sierra G, Curran M, Sastry K (2019) Targeting interferon signaling and CTLA-4 enhance the therapeutic efficacy of anti-PD-1 immunotherapy in preclinical model of HPV+ oral cancer. J Imm Ther Can 7. https://doi.org/10.1186/s40425-019-0728-4

Du Y, He C, An Y, Huang Y, Zhang H, Fu W, Wang M, Shan Z, Xie J, Yang Y, Zhao B (2024) The role of short chain fatty acids in inflammation and body health. Int J Mol Sci 25(13). https://doi.org/10.3390/ijms25137379

Dumitrescu R, Bolchis V, Fratila A, Jumanca D, Buzatu B, Sava-Rosianu R, Alexa V, Atena G, Balean O (2025) The global trends and advances in oral microbiome research on oral squamous cell carcinoma: a systematic review. Microorganisms 13:373. https://doi.org/10.3390/microorganisms13020373

Easter QT, Fernandes Matuck B, Beldorati Stark G, Worth CL, Predeus AV, Fremin B, Huynh K, Ranganathan V, Ren Z, Pereira D, Rupp BT, Weaver T, Miller K, Perez P, Hasuike A, Chen Z, Bush M, Qu X, Lee J et al (2024) Single-cell and spatially resolved interactomics of tooth-associated keratinocytes in periodontitis. Nat Commun 15(1):5016. https://doi.org/10.1038/s41467-024-49037-y

Ehrlich LA, Chung HY, Ghobrial I, Choi SJ, Morandi F, Colla S, Rizzoli V, Roodman GD, Giuliani N (2005) IL-3 is a potential inhibitor of osteoblast differentiation in multiple myeloma. Blood 106(4):1407–1414. https://doi.org/10.1182/blood-2005-03-1080

Elgueta D, Murgas P, Riquelme E, Yang G, Cancino G (2022) Consequences of viral infection and cytokine production during pregnancy on brain development in offspring. Front Immunol 13:816619. https://doi.org/10.3389/fimmu.2022.816619

Fábián T, Hermann P, Beck A, Fejérdy P, Fábián G (2012) Salivary defense proteins: their network and role in innate and acquired oral immunity. Int J Mol Sci 13:4295–4320. https://doi.org/10.3390/ijms13044295

Feller L, Altini M, Khammissa RAG, Chandran R, Bouckaert M, Lemmer J (2013) Oral mucosal immunity. Oral Surg Oral Med Oral Pathol Oral Radiol 116(5):576–583. https://doi.org/10.1016/j.oooo.2013.07.013

Feng W, Liu H, Luo T, Liu D, Du J, Sun J, Wang W, Han X, Yang K, Guo J, Amizuka N, Li M (2017) Combination of IL-6 and sIL-6R differentially regulate varying levels of RANKL-induced osteoclastogenesis through NF-κB, ERK and JNK signaling pathways. Sci Rep 7(1):41411. https://doi.org/10.1038/srep41411

Garlet TP, Coelho U, Silva JS, Garlet GP (2007) Cytokine expression pattern in compression and tension sides of the periodontal ligament during orthodontic tooth movement in humans. Eur J Oral Sci 115(5):355–362. https://doi.org/10.1111/j.1600-0722.2007.00469.x

Georges FM, Do NT, Seleem D (2022) Oral dysbiosis and systemic diseases. Front Dent Med 3. https://www.frontiersin.org/journals/dental-medicine/articles/10.3389/fdmed.2022.995423

Groeger S, Meyle J (2019) Oral mucosal epithelial cells. Front Immunol 10. https://www.frontiersin.org/journals/immunology/articles/10.3389/fimmu.2019.00208

Hajishengallis G, Chavakis T (2021) Local and systemic mechanisms linking periodontal disease and inflammatory comorbidities. Nat Rev Immunol 21(7):426–440. https://doi.org/10.1038/s41577-020-00488-6

Han V, Patel S, Jones H, Dale R (2021) Maternal immune activation and neuroinflammation in human neurodevelopmental disorders. Nat Rev Neurol 17:1–16. https://doi.org/10.1038/s41582-021-00530-8

Han Y, Yang J, Fang J, Zhou Y, Candi E, Wang J, Hua D, Shao C, Shi Y (2022) The secretion profile of mesenchymal stem cells and potential applications in treating human diseases. Signal Transduct Target Ther 7(1):92. https://doi.org/10.1038/s41392-022-00932-0

Haririan H, Andrukhov O, Laky M, Rausch-Fan X (2021) Saliva as a source of biomarkers for periodontitis and Periimplantitis. Front Dent Med 2. https://www.frontiersin.org/journals/dental-medicine/articles/10.3389/fdmed.2021.687638

Hashmi S, Gallaway M, Waller D, Langlois P, Hecht J (2010) Maternal fever during early pregnancy and the risk of oral clefts. Birth Defect Res Part A Clin Mol Teratol 88:186–194. https://doi.org/10.1002/bdra.20646

Hatanaka E, Monteagudo P, Marrocos M, Campa A (2007) Neutrophils and monocytes as potentially important sources of proinflammatory cytokines in diabetes. Clin Exp Immunol 146:443–447. https://doi.org/10.1111/j.1365-2249.2006.03229.x

Hong H, Chen X, Li K, Wang N, Li M, Yang B, Yu X, Wei X (2020) Dental follicle stem cells rescue the regenerative capacity of inflamed rat dental pulp through a paracrine pathway. Stem Cell Res Ther 11(1):333. https://doi.org/10.1186/s13287-020-01841-1

Hovav A-H (2014) Dendritic cells of the oral mucosa. Mucosal Immunol 7(1):27–37. https://doi.org/10.1038/mi.2013.42

Huang N, Dong H, Luo Y, Shao B (2021) Th17 cells in periodontitis and its regulation by A20. Front Immunol 12. https://www.frontiersin.org/journals/immunology/articles/10.3389/fimmu.2021.742925

Hyvärinen E, Savolainen M, Mikkonen JJW, Kullaa AM (2021) Salivary metabolomics for diagnosis and monitoring diseases: challenges and possibilities. Meta 11(9):587. https://doi.org/10.3390/metabo11090587

Hyvärinen E, Solje E, Vepsäläinen J, Kullaa A, Tynkkynen T (2023) Salivary metabolomics in the diagnosis and monitoring of neurodegenerative dementia. Meta 13:233. https://doi.org/10.3390/metabo13020233

Inchingolo AM, Fatone M, Malcangi G, Avantario P, Piras F, Patano A, Pede C, Netti A, Ciocia A, Ruvo E, Viapiano F, Palmieri G, Campanelli M, Mancini A, Settanni V, Carpentiere V, Marinelli G, Latini G, Rapone B, Dipalma G (2022) Modifiable risk factors of non-syndromic orofacial clefts: a systematic review. Children 9:1846. https://doi.org/10.3390/children9121846

Jang H, Matsuoka M, Freire M (2023) Oral mucosa immunity: ultimate strategy to stop spreading of pandemic viruses. Front Immunol 14. https://www.frontiersin.org/journals/immunology/articles/10.3389/fimmu.2023.1220610

Jaskoll T, Abichaker G, Jangaard N, Bringas P, Melnick M (2008) Cytomegalovirus inhibition of embryonic mouse tooth development: a model of the human amelogenesis imperfecta phenocopy. Arch Oral Biol 53(5):405–415. https://doi.org/10.1016/j.archoralbio.2007.11.014

Jeon S, Kayama H, Ueda Y, Takahashi T, Asahara T, Tsuji H, Tsuji N, Kiyono H, Ma J, Kusu T, Okumura R, Hara H, Yoshida H, Yamamoto M, Nomoto K, Takeda K (2012) Probiotic bifidobacterium breve induces IL-10-producing Tr1 cells in the colon. PLoS Pathog 8:e1002714. https://doi.org/10.1371/journal.ppat.1002714

Jeon H, Teixeira H, Tsai A (2021) Mechanistic insight into orthodontic tooth movement based on animal studies: a critical review. J Clin Med 10:1733. https://doi.org/10.3390/jcm10081733

Johansen F-E, Brandtzaeg P (2004) Transcriptional regulation of the mucosal IgA system. Trends Immunol 25(3):150–157. https://doi.org/10.1016/j.it.2004.01.001

Kaczor-Urbanowicz K, Carreras-Presas C, Aro K, Tu M, Garcia-Godoy F, Wong D (2016) Saliva diagnostics – current views and directions. Exp Biol Med 242. https://doi.org/10.1177/1535370216681550

Kapila Y (2021) Oral health's inextricable connection to systemic health: special populations bring to bear multimodal relationships and factors connecting periodontal disease to systemic diseases and conditions. Periodontology 2000(87):11–16. https://doi.org/10.1111/prd.12398

Keerti Kumar Y, Kumar A, Ajay Kumar S (2022) A systems biology approach towards oral cancer using computational tools and techniques. Chemom Intell Lab Syst 231:104709. https://doi.org/10.1016/j.chemolab.2022.104709

Khor B, Snow M, Herrman E, Ray N, Mansukhani K, Patel K, Said-Al-Naief N, Maier T, Machida C (2021) Interconnections between the oral and gut microbiomes: reversal of microbial dysbiosis and the balance between systemic health and disease. Microorganisms 9:496. https://doi.org/10.3390/microorganisms9030496

Krivanek J, Soldatov RA, Kastriti ME, Chontorotzea T, Herdina AN, Petersen J, Szarowska B, Landova M, Matejova VK, Holla LI, Kuchler U, Zdrilic IV, Vijaykumar A, Balic A, Marangoni P, Klein OD, Neves VCM, Yianni V, Sharpe PT et al (2020) Dental cell type atlas reveals stem and differentiated cell types in mouse and human teeth. Nat Commun 11(1):4816. https://doi.org/10.1038/s41467-020-18512-7

Kulis MD, Patil SU, Wambre E, Vickery BP (2018) Immune mechanisms of oral immunotherapy. J Allergy Clin Immunol 141(2):491–498. https://doi.org/10.1016/j.jaci.2017.12.979

Kumar P, Gupta S, Das B (2023) Saliva as a potential non-invasive liquid biopsy for early and easy diagnosis/prognosis of head and neck cancer. Transl Oncol 40:101827. https://doi.org/10.1016/j.tranon.2023.101827

Kwon H-K, Choi GB, Huh JR (2022) Maternal inflammation and its ramifications on fetal neurodevelopment. Trends Immunol 43(3):230–244. https://doi.org/10.1016/j.it.2022.01.007

Lee H, Joo J-Y, Sohn D, Kang J, Yu Y, Park H, Kim YH (2022) Single-cell RNA sequencing reveals rebalancing of immunological response in patients with periodontitis after non-surgical periodontal therapy. J Transl Med 20. https://doi.org/10.1186/s12967-022-03702-2

Lei H, Schmidt-Bleek K, Dienelt A, Reinke P, Volk H-D (2015) Regulatory T cell-mediated anti-inflammatory effects promote successful tissue repair in both indirect and direct manners. Front Pharmacol 6. https://www.frontiersin.org/journals/pharmacology/articles/10.3389/fphar.2015.00184

Li Z, Jiang C-M, An S, Cheng Q, Huang Y-F, Wang Y-T, Gou Y-C, Xiao L, Yu W-J, Wang J (2014) Immunomodulatory properties of dental tissue-derived mesenchymal stem cells. Oral Dis 20(1):25–34. https://doi.org/10.1111/odi.12086

Li X, Liu Y, Yang X, Li C, Song Z (2022) The oral microbiota: community composition, influencing factors, pathogenesis, and interventions. Front Microbiol 13. https://www.frontiersin.org/journals/microbiology/articles/10.3389/fmicb.2022.895537

Li P, Ou Q, Shi S, Shao C (2023) Immunomodulatory properties of mesenchymal stem cells/dental stem cells and their therapeutic applications. Cell Mol Immunol 20(6):558–569. https://doi.org/10.1038/s41423-023-00998-y

Lin X, Kang K, Chen P, Zeng Z, Li G, Xiong W, Yi M, Xiang B (2024) Regulatory mechanisms of PD-1/PD-L1 in cancers. Mol Cancer 23(1):108. https://doi.org/10.1186/s12943-024-02023-w

Liu Y, Wang L, Kikuiri T, Akiyama K, Chen C, Xu X, Yang R, Chen W, Wang S, Shi S (2011) Mesenchymal stem cell–based tissue regeneration is governed by recipient T lymphocytes via IFN-γ and TNF-α. Nat Med 17(12):1594–1601. https://doi.org/10.1038/nm.2542

Liu D, Xu J, Liu O, Fan Z, Liu Y, Wang F, Ding G, Wei F, Zhang C, Wang S (2012) Mesenchymal stem cells derived from inflamed periodontal ligaments exhibit impaired immunomodulation. J Clin Periodontol 39(12):1174–1182. https://doi.org/10.1111/jcpe.12009

Liu Y, Qv W, Ma Y, Zhang Y, Ding C, Chu M, Chen F (2022a) The interplay between oral microbes and immune responses. Front Microbiol:13. https://www.frontiersin.org/journals/microbiology/articles/10.3389/fmicb.2022.1009018

Liu Y, Wang J, Wu C (2022b) Modulation of gut microbiota and immune system by probiotics, pre-biotics, and post-biotics. Front Nutr 8. https://www.frontiersin.org/journals/nutrition/articles/10.3389/fnut.2021.634897

Lombardi T, Hauser C, Budtz-Jörgensen E (1993) Langerhans cells: structure, function and role in oral pathological conditions. J Oral Pathol Med 22(5): 193–202. https://doi.org/10.1111/j.1600-0714.1993.tb01056.x

Maccaferri M, Pisciotta A, Carnevale G, Salvarani C, Pignatti E (2024) Human dental pulp stem cells modulate pro-inflammatory macrophages both through cell-to-cell contact and paracrine signaling. Front Immunol 15. https://www.frontiersin.org/journals/immunology/articles/10.3389/fimmu.2024.1440974

Mallis P, Michalopoulos E, Chatzistamatiou T, Giokas CS (2021) Interplay between mesenchymal stromal cells and immune system: clinical applications in immune-related diseases. Explor Immunol 1:112–139. https://doi.org/10.37349/ei.2021.00010

Mantis NJ, Rol N, Corthésy B (2011) Secretory IgA's complex roles in immunity and mucosal homeostasis in the gut. Mucosal Immunol 4(6):603–611. https://doi.org/10.1038/mi.2011.41

Martinelli M, Palmieri A, Carinci F, Scapoli L (2020) Non-syndromic cleft palate: an overview on human genetic and environmental risk factors. Front Cell Dev Biol 8. https://doi.org/10.3389/fcell.2020.592271

Mazurek-Mochol M, Serwin K, Bonsmann T, Kozak M, Piotrowska K, Czerewaty M, Safranow K, Pawlik A (2023) Expression of interleukin 17A and 17B in gingival tissue in patients with periodontitis. J Clin Med 12:4614. https://doi.org/10.3390/jcm12144614

Mazziotta C, Tognon M, Martini F, Torreggiani E, Rotondo JC (2023) Probiotics mechanism of action on immune cells and beneficial effects on human health. Cells 12(1). https://doi.org/10.3390/cells12010184

Mei Z, Huang J, Qiao B, Lam AK (2020) Immune checkpoint pathways in immunotherapy for head and neck squamous cell carcinoma. Int J Oral Sci 12(1):16. https://doi.org/10.1038/s41368-020-0084-8

Metcalfe S, Anselmi N, Escobar A, Visser MB, Kay JG (2022) Innate phagocyte polarization in the oral cavity. Front Immunol 12. https://www.frontiersin.org/journals/immunology/articles/10.3389/fimmu.2021.768479

Moreno CM, Boeree E, Freitas CMT, Weber KS (2023) Immunomodulatory role of oral microbiota in inflammatory diseases and allergic conditions. Front Allergy 4. https://www.frontiersin.org/journals/allergy/articles/10.3389/falgy.2023.1067483

Nakamura M (2017) Histological and immunological characteristics of the junctional epithelium. Jpn Den Sci Rev 54. https://doi.org/10.1016/j.jdsr.2017.11.004

Nascimento M (2017) Oral microbiota transplant: a potential new therapy for oral diseases. J Calif Dent Assoc 45:565–568. https://doi.org/10.1080/19424396.2017.12222506

Natarajan P, Madanian S, Marshall S (2025) Investigating the link between oral health conditions and systemic diseases: a cross-sectional analysis. Sci Rep 15(1): 10476. https://doi.org/10.1038/s41598-025-92523-6

Nicholson J, Landry K (2022) Oral dysbiosis and neurodegenerative diseases: correlations and potential causations. Microorganisms 10. https://doi.org/10.3390/microorganisms10071326

Olesnicky Killian EC, Birkholz DA, Artinger KB (2009) A role for chemokine signaling in neural crest cell migration and craniofacial development. Dev Biol 333(1):161–172. https://doi.org/10.1016/j.ydbio.2009.06.031

Peterka M, Tvrdek M, Likovský Z, Peterkova R, Farahani M (1994) Maternal hyperthermia and infection as one of possible causes of orofacial clefts. Acta Chir Plast 36:114–118

Pidone F, Noguti J, Oshima C, Ribeiro D (2014) Effective targeting of the epidermal growth factor receptor (EGFR) for treating Oral cancer: a promising approach. Anticancer Res 34:1547–1552

Pouessel G, Lezmi G (2023) Oral immunotherapy for food allergy: translation from studies to clinical practice? World Allergy Organ J 16(2):100747. https://doi.org/10.1016/j.waojou.2023.100747

Prystaz K, Vikman A, Kaiser K, Heidler V, Kroner J, Haffner-Luntzer M, Ignatius A (2016) Role of interleukin-6 in the early and late fracture healing phase. Bone Abstract. https://doi.org/10.1530/boneabs.5.P54

Punyadeera, C., & Slowey, P. D. (2019). Chapter 22 – Saliva as an emerging biofluid for clinical diagnosis and applications of MEMS/NEMS in salivary diagnostics. In K. Subramani & W. Ahmed (Eds.), Nanobiomaterials in clinical dentistry. 2nd (pp. 543–565) Elsevier. doi:https://doi.org/10.1016/B978-0-12-815886-9.00022-X

Rajasekaran JJ, Krishnamurthy HK, Bosco J, Jayaraman V, Krishna K, Wang T, Bei K (2024) Oral microbiome: a review of its impact on oral and systemic health. Microorganisms 12(9). https://doi.org/10.3390/microorganisms12091797

Rautemaa R, Lauhio A, Cullinan MP, Seymour GJ (2007) Oral infections and systemic disease—an emerging problem in medicine. Clin Microbiol Infect 13(11): 1041–1047. https://doi.org/10.1111/j.1469-0691.2007.01802.x

Raza NUH, Ghafoor S (2020) Desmosomal protein regulation and clinical implications in oral mucosal tissues. J Pak Med Assoc 70:1425–1431. https://doi.org/10.5455/JPMA.15798

Roth D, Bayona F, Baddam P, Graf D (2021) Craniofacial development: neural crest in molecular embryology. Head Neck Pathol 15. https://doi.org/10.1007/s12105-021-01301-z

Sachelarie L, Ioana S, Ignat-Romanul I, Iurcov R, Potra Cicalau GI, Todor L (2025) Probiotic therapy as an adjuvant in the treatment of periodontal disease: an innovative approach. Medicina 61:126. https://doi.org/10.3390/medicina61010126

Sahibzada H, Sultan Z, Khan R, Naseem M, Siddiqi K, Mali M, Zafar M (2017) Salivary IL-8, IL-6 and TNF-α as potential diagnostic biomarkers for oral cancer. Diagnostics 7. https://doi.org/10.3390/diagnostics7020021

Şenel S (2021) An overview of physical, microbiological and immune barriers of oral mucosa. Int J Mol Sci 22: 7821. https://doi.org/10.3390/ijms22157821

Shi N, Kong C, Yuan L, Liu L, Zhao K, Lü J, Wang X (2023) The bidirectional relationship between periodontitis and diabetes: new prospects for stem cell-derived exosomes. Biomed Pharmacother 165: 115219. https://doi.org/10.1016/j.biopha.2023.115219

Shimabukuro N, de S Cataruci AC, Ishikawa KH, de Oliveira BE, Kawamoto D, Ando-Suguimoto ES, Albuquerque-Souza E, Nicoli JR, Ferreira CM, de Lima J, Bueno MR, da Silva LBR, Silva PHF, Messora MR, Camara NOS, Simionato MRL, Mayer MPA (2021) *Bifidobacterium* strains present distinct effects on the control of alveolar bone loss in a periodontitis experimental model. Front Pharmacol 12. https://www.frontiersin.org/journals/pharmacology/articles/10.3389/fphar.2021.713595

Singh K, Piprode V, Mhaske ST, Barhanpurkar-Naik A, Wani MR (2018) IL-3 differentially regulates membrane and soluble RANKL in osteoblasts through metalloproteases and the JAK2/STAT5 pathway and improves the RANKL/OPG ratio in adult mice. J Immunol 200(2):595–606. https://doi.org/10.4049/jimmunol.1601528

Smits HH, Engering A, van der Kleij D, de Jong EC, Schipper K, van Capel TMM, Zaat BAJ, Yazdanbakhsh M, Wierenga EA, van Kooyk Y, Kapsenberg ML (2005) Selective probiotic bacteria induce IL-10–producing regulatory T cells in vitro by modulating dendritic cell function through dendritic cell–specific intercellular adhesion molecule 3–grabbing nonintegrin. J Allergy Clin Immunol 115(6):1260–1267. https://doi.org/10.1016/j.jaci.2005.03.036

Squier CA, Kremer MJ (2001) Biology of Oral mucosa and esophagus. JNCI Monogr 2001(29):7–15. https://doi.org/10.1093/oxfordjournals.jncimonographs.a003443

Taherkhani A, Moradifar G, Bayat Z, Samadi P, Jamshidi S (2024) A systems biology approach to uncover key genes and pathways mediating progression of leukoplakia to primary oral squamous cell carcinoma. J Oral Maxillofacial Surg Med Pathol 36(1):120–132. https://doi.org/10.1016/j.ajoms.2023.04.005

Talks B, Mather M, Chahal M, Coates M, Clatworthy M, Haniffa M (2024) Mapping human immunity and the education of Waldeyer's ring. Annu Rev Genomics Hum Genet 25. https://doi.org/10.1146/annurev-genom-120522-012938

Tamimi A, Tamimi A, Sorkheh F, Asl SM, Ghafari A, Karimi AG, Erabi G, Pourmontaseri H, Deravi N (2023) Monoclonal antibodies for the treatment of squamous cell carcinoma: a literature review. Cancer Rep 6(5):e1802. https://doi.org/10.1002/cnr2.1802

Theodoro LH, Garcia VG, Ervolino E, Holcroft J, McCulloch CA, Ganss B (2023) Role of junctional epithelium in maintaining dento-gingival adhesion and periodontal health. Front Dental Med 4. https://www.frontiersin.org/journals/dental-medicine/articles/10.3389/fdmed.2023.1144537

Thomas C, Minty M, Vinel A, Canceill T, Loubières P, Burcelin R, Kaddech M, Blasco-Baque V, Laurencin-Dalicieux S (2021) Oral microbiota: a major player in the diagnosis of systemic diseases. Diagnostics 11(8). https://doi.org/10.3390/diagnostics11081376

Umapathy D, Manickam Natarajan P, Bhuminathan S (2023) Review insights on salivary proteomics biomarkers in oral cancer detection and diagnosis. Molecules 28:5283. https://doi.org/10.3390/molecules28135283

Wambre E, Jeong D (2018) Oral tolerance development and maintenance. Immunol Allerg Clin 38(1):27–37. https://doi.org/10.1016/j.iac.2017.09.003

Wan H, Gadmor H, Brown L (2018) Cell-cell interactions in the oral mucosa: tight junctions and gap junctions. In: Bergmeier LA (ed) Oral mucosa in health and disease: a concise handbook. Springer, pp 19–30. https://doi.org/10.1007/978-3-319-56065-6_2

Wang S-S, Tang Y-L, Pang X, Zheng M, Tang Y-J, Liang X-H (2019) The maintenance of an oral epithelial barrier. Life Sci 227:129–136. https://doi.org/10.1016/j.lfs.2019.04.029

Wang M, Xie J, Wang C, Zhong D, Xie L, Fang H (2020) Immunomodulatory properties of stem cells in periodontitis: current status and future prospective. Stem Cells Int 2020(1):9836518. https://doi.org/10.1155/2020/9836518

Wang Y, Xue N, Wang Z, Zeng X, Ji N, Chen Q (2023) Targeting Th17 cells: a promising strategy to treat oral mucosal inflammatory diseases. Front Immunol 14. https://doi.org/10.3389/fimmu.2023.1236856

Wang X, Sun B, Wang Y, Gao P, Song J, Chang W, Xiao Z, Xi Y, Li Z, An F, Yan C (2024) Research progress of targeted therapy regulating Th17/Treg balance in bone immune diseases. Front Immunol 15. https://www.frontiersin.org/journals/immunology/articles/10.3389/fimmu.2024.1333993

Wen C-C, Lin JCY, Salamanca E, Teng N-C, Lin Y-J, Wu Y-F, Chang H-M, Chang W-J (2025) Full-length 16S rRNA sequencing reveals microbial characteristics in supragingival plaque of periodontitis patients. J Dent Sci 20. https://doi.org/10.1016/j.jds.2025.03.040

Wu X, Yang X, Dai Y, Zhao Z, Zhu J, Guo H, Yang R (2024) Single-cell sequencing to multi-omics: technologies and applications. Biomark Res 12(1): 110. https://doi.org/10.1186/s40364-024-00643-4

Xiong Y, Mi B-B, Lin Z, Hu Y-Q, Yu L, Zha K-K, Panayi AC, Yu T, Chen L, Liu Z-P, Patel A, Feng Q, Zhou S-H, Liu G-H (2022) The role of the immune microenvironment in bone, cartilage, and soft tissue regeneration: from mechanism to therapeutic opportunity. Mil Med Res 9(1):65. https://doi.org/10.1186/s40779-022-00426-8

Xu P, Gunsolley J (2014) Application of metagenomics in understanding oral health and disease. Virulence 5. https://doi.org/10.4161/viru.28532

Xue N, Wang Y, Cheng H, Liang H, Fan X, Zuo F, Zeng X, Ji N, Chen Q (2022) Regulatory T cell therapy suppresses inflammation of oral mucosa. Front Immunol 13. https://www.frontiersin.org/journals/immunology/articles/10.3389/fimmu.2022.1009742

Yang N, Liu Y (2021) The role of the immune microenvironment in bone regeneration. Int J Med Sci 18(16): 3697–3707. https://doi.org/10.7150/ijms.61080

Yang R, Yu T, Zhou Y (2017) Interplay between craniofacial stem cells and immune stimulus. Stem Cell Res Ther 8(1):147. https://doi.org/10.1186/s13287-017-0607-1

Yang J, Ran M, Li H, Lin Y, Ma K, Yang Y, Fu X, Yang S (2022) New insight into neurological degeneration: inflammatory cytokines and blood–brain barrier. Front Mol Neurosci 15. https://www.frontiersin.org/journals/molecular-neuroscience/articles/10.3389/fnmol.2022.1013933

Yu X, Devine D, Vernon J (2024) Manipulating the diseased oral microbiome: the power of probiotics and prebiotics. J Oral Microbiol 16. https://doi.org/10.1080/20002297.2024.2307416

Zhang Y, Guo J, Jia R (2021) Treg: a promising immunotherapeutic target in Oral diseases. Front Immunol 12. https://www.frontiersin.org/journals/immunology/articles/10.3389/fimmu.2021.667862

Zhang JS, Chu C-H, Yu OY (2022) Oral microbiome and dental caries development. Dent J 10(10). https://doi.org/10.3390/dj10100184

Zhou P, Chen C, Patil S, Dong S (2024) Unveiling the therapeutic symphony of probiotics, prebiotics, and postbiotics in gut-immune harmony. Front Nutr 11. https://www.frontiersin.org/journals/nutrition/articles/10.3389/fnut.2024.1355542

Zhu X, Huang H, Zhao L (2022) PAMPs and DAMPs as the bridge between periodontitis and atherosclerosis: the potential therapeutic targets. Front Cell Dev Biol 10. https://doi.org/10.3389/fcell.2022.856118

Oral Innate Immunity and the Role of Oral Epithelial Cells

2

Wipawee Nittayananta

Abstract

The oral mucosal epithelium is a complex structure that protects the underlying tissues from environmental exposure. Epithelial cells lining the mucosal surfaces of the oral cavity play an important role in innate immunity. They consist of stratified squamous epithelia that not only serve as a physical barrier but also contribute to innate immunity through the production of antimicrobial peptides and cytokines to prevent the invasion of oral pathogens. The oral cavity is constantly exposed to a great number of microorganisms. To fight against infection and maintain oral tissue homeostasis, the expression and function of cell surface receptors such as toll-like receptors (TLRs) are essential. TLRs belong to the family of pattern recognition receptors (PRRs) that recognize molecular structures categorized as "pathogen-associated molecular patterns" (PAMPs) present in microorganisms. Binding to TLRs triggers the activation of downstream signaling pathways and stimulates both innate and adaptive immune responses to suppress infection. Epithelial cells also engage distinct innate immune pathways that are implicated in the pathogenesis of oral cancer. The anticancer strategies of innate immunity include the development of an epithelial barrier, direct cytotoxicity, and presentation to various immune cells. Immune cells may have a dual impact by either promoting or suppressing cancer cell growth. Future research should investigate the factors that determine whether epithelial and immune cells promote or inhibit tumor growth. Although the mechanisms underlying this balance are not fully understood, enhancing innate immunity to combat oral infections and cancer holds potential as a strategy for early disease detection and improved treatment outcomes. Additionally, individual variations in host innate immune responses, contributing to oral cancer development, are another aspect of the complexity that requires further research.

Keywords

Oral mucosal immunity · Toll-like receptors (TLRs) · Innate immune response · Oral epithelial barrier · Oral cancer immunology

2.1 Introduction

Innate immunity is a nonspecific defense mechanism that serves as the first line of defense in the body against pathogens. It responds rapidly, typically within hours of pathogen exposure, without involving immunological memory. This defense

W. Nittayananta (✉)
Faculty of Dentistry, Bangkokthonburi University, Bangkok, Thailand
e-mail: wipawee.nit@bkkthon.ac.th

N. Rezaei, H. Ziaei (eds.), *Oral Immunology*, Advances in Experimental Medicine and Biology 1492,
https://doi.org/10.1007/978-3-032-03176-1_2

mechanism is categorized into two phases: immediate and induced. The immediate phase, occurring within 0–4 h of infection, relies on soluble antimicrobial peptides and complement system proteins. In contrast, the induced phase, occurring between 4 and 96 h after infection, is characterized by the activation and recruitment of immune cells, such as neutrophils, monocytes, macrophages, and natural killer (NK) cells, which function to eliminate pathogens and infected host cells (Taguchi and Mukai 2019).

Innate immunity plays a critical role in defending the host against infections and maintaining homeostasis to support overall health. In the oral cavity, it contributes to immune surveillance and preservation of oral health by maintaining homeostasis (Challacombe and Naglik 2006). Impairment in oral innate immune function can increase susceptibility to infections and may contribute to the development of oral malignancies.

2.2 Role of Oral Epithelial Cells in Innate Immunity

2.2.1 The Oral Epithelium as a Physical Barrier

The oral epithelium is a structurally complex tissue that functions as the primary barrier in the innate immune defense system, protecting the body from physical, chemical, and microbial insults. It separates the underlying tissues from the external environment, thus playing an important role in innate immunity. It is composed of two main layers: a superficial stratified squamous epithelium and a deeper lamina propria. The composition of the epithelial barrier varies between the keratinized and nonkeratinized oral mucosa. In the keratinized oral mucosa, such as the gingiva and hard palate, the epithelium comprises four layers: the stratum basale, stratum spinosum, stratum granulosum, and stratum corneum. On the other hand, in the nonkeratinized epithelium, such as in the alveolar and buccal mucosa, the stratified layers are composed of the stratum basale, followed by the stratum filamentosum and stratum distendum. The structural integrity and thickness of the squamous epithelium prevent pathogen infiltration into the connective tissues, reinforced by closely adherent cells connected by tight junctions and maintained through continuous cellular differentiation.

Oral epithelial cells undergo a tightly regulated differentiation process, leading to the formation of structural proteins that create a protective barrier essential for maintaining tissue integrity. These cells are interconnected by transmembrane adhesion proteins, such as keratin filaments, anchored to the plasma membrane by desmosomes, forming a robust three-dimensional matrix. Beyond their role as a physical barrier, oral epithelial cells also create a chemical barrier by secreting antimicrobial peptides to prevent microbial adhesion and colonization.

The squamous epithelium contains resident immune cells, such as Langerhans cells, which are a specialized subset of dendritic cells characterized by the expression of langerin, a cell surface receptor (Stoitzner and Romani 2011). Additional mucosal innate immune cells, such as dendritic cells and macrophages, play pivotal roles in modulating immune responses. Dendritic cells present antigens to T lymphocytes, connecting innate and adaptive immunity, while macrophages focus on pathogen recognition and phagocytosis (Iwasaki 2007; Kelsall 2008; Modlin 2012). Both cell types detect pathogens through pattern recognition receptors (PRRs) that identify conserved pathogen-associated molecular patterns (PAMPs). Their functions vary across tissues due to differences in PRR expression and antigen exposure (Kelsall 2008), modulating local and adaptive immune responses crucial for effective protection against mucosal infections.

In summary, the oral epithelium serves as a complex barrier interface between the external environment—including saliva, antimicrobial peptides, and resident microbiota—and immune cells within the underlying tissues, safeguarding oral health (Fig. 2.1).

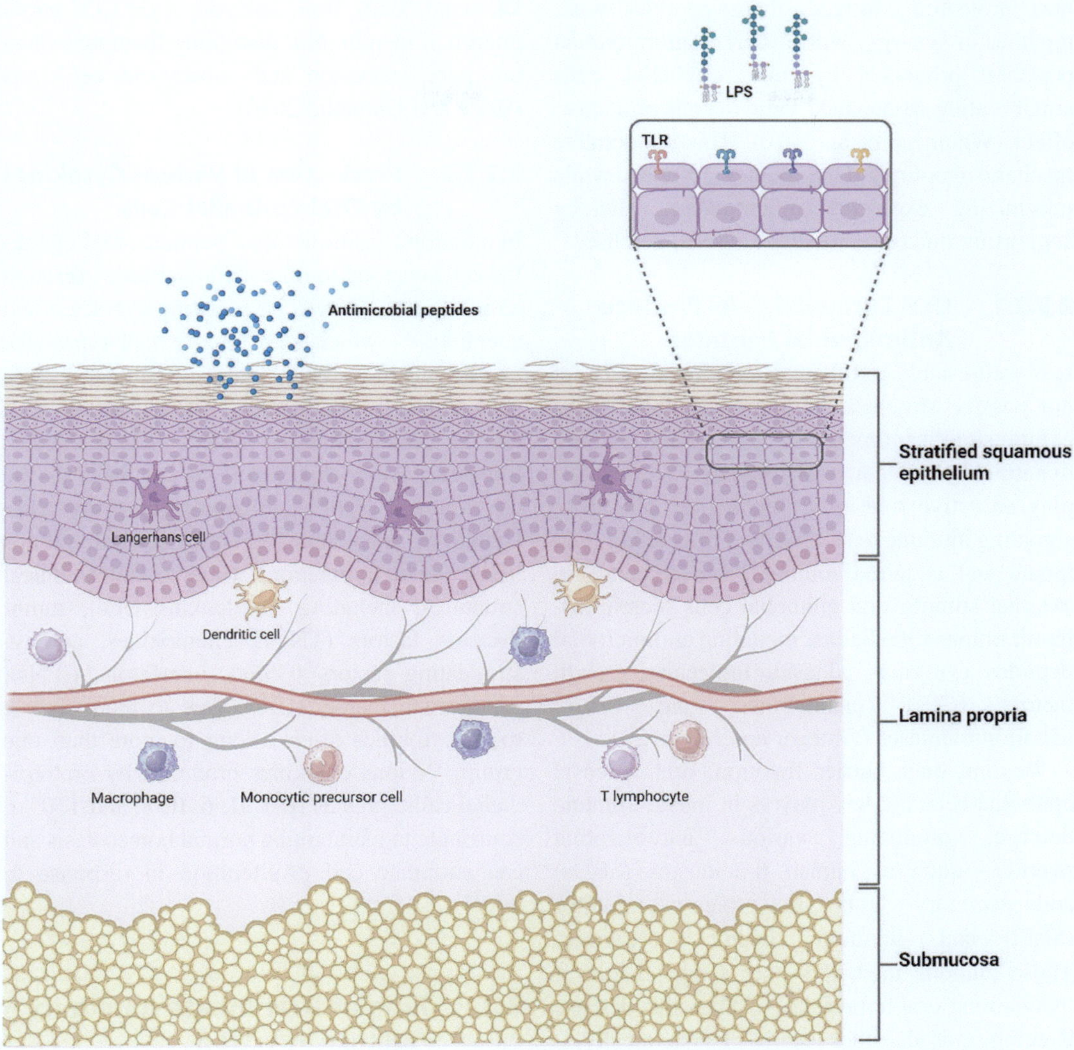

Fig. 2.1 Epithelial cells lining the mucosal surfaces of the oral cavity are essential components of the innate immune system. These cells form a complex structure of the stratified squamous epithelium that serves as a barrier, protecting underlying tissues from the external environment and resident microbiota. In addition to acting as a physical barrier, epithelial cells also provide chemical defense by producing antimicrobial peptides that inhibit the adhesion and invasion of microbes. This protective function is facilitated by the expression of cell surface receptors, such as toll-like receptors (TLRs), which recognize pathogen-associated molecular patterns (PAMPs), including lipopolysaccharides (LPSs) present on microorganisms. (Graphical figure drawn by Asst. Prof. Aornrutai Promsong, Ph.D. using BioRender (https://BioRender.com/g66f862))

2.2.2 Oral Epithelial Cells Provide a Chemical Barrier

Epithelial cells in the oral cavity are continuously exposed to various microorganisms and their by-products. Owing to their role as a chemical barrier, these cells constantly produce a variety of antimicrobial peptides and proteins in response to microbial presence. These antimicrobial peptides are key components of the innate immune system, helping to control microbial invaders and maintain a healthy balance within the mucosal microbiota (Weinberg et al. 2020). Importantly, no antimicrobial peptide is solely responsible for host protection. Instead, these peptides work together in synergy, with different antimicrobial peptides produced by oral epithelial cells collaborating to enhance their overall protective effect (Weinberg et al. 2020). This cooperative mechanism optimizes the immune response while minimizing excessive inflammation, thereby supporting mucosal integrity and host defense.

2.2.2.1 Oral Epithelial Cells Produce Antimicrobial Peptides

It is well-established that oral epithelial cells are not passive structural components; rather, they exhibit significant metabolic activity and respond dynamically to their environment. These cells play an active role in responding to infections, signaling immune responses, and integrating both innate and acquired immunity. In response to external stimuli, oral epithelial cells secrete different immune mediators, including antimicrobial peptides, cytokines, adhesion molecules, growth factors (GFs), chemokines, and matrix metalloproteinases (Groeger and Meyle 2015).

Beyond their barrier function, oral mucosal epithelial cells are key players in innate immune defense, producing various antimicrobial peptides—such as human β-defensins (hBDs) and secretory leukocyte protease inhibitor (SLPI)—and immunomodulatory cytokines. These immune mediators play a crucial role in maintaining oral homeostasis. When this defense is compromised, either locally or systemically, it can lead to oral infections or malignancies.

The antimicrobial peptides produced by oral epithelial cells exhibit broad-spectrum antimicrobial activity and additional immunomodulatory functions. In humans, there are two main classes of antimicrobial peptides: hBDs and cathelicidins, such as LL-37. The defensin family includes several members, including hBD-1 to hBD-5 (Li et al. 2006). Even in the healthy gingiva, hBD-2 and hBD-3 are expressed, suggesting that their production is influenced by chronic exposure to oral commensal bacteria (Dale et al. 2001). LL-37 plays a key role in protecting against infectious diseases by modulating immune cells and regulating the secretion of inflammatory molecules (Brice and Diamond 2020). Both defensins and LL-37 target microbial membranes, disrupting them and causing pore formation and subsequent cell lysis (Brice and Diamond 2020).

2.2.2.2 Production of Various Cytokines by Oral Epithelial Cells

In addition to antimicrobial peptides, oral epithelial cells respond to microbial challenges through an interactive process by secreting cytokines and chemokines, which alert various cell types like neutrophils. Mucosal cytokines can have both pro- and anti-inflammatory effects, acting as a crucial communication system between cells. Imbalances in cytokine levels can lead to the development of inflammatory diseases. Cytokines are classified into several groups based on their functions, origins, and chemical structures, including interleukins (ILs), tumor necrosis factors (TNFs), chemokines, colony-stimulating factors (CSFs), interferons (IFNs), and growth factors (GFs). Due to their diverse roles, cytokines often belong to more than one group. Various cytokines produced by oral epithelial cells, such as IL-1, IL-6, IL-8, and TNF-α, contribute to maintaining normal homeostasis and can stimulate cell proliferation in response to injury.

2.3 Role of Oral Innate Immunity Against Infection

The epithelial cells lining the mucosal surfaces of the oral cavity are composed of a stratified squamous epithelium, which plays a key role in innate immunity (Squier and Kremer 2001). This structure functions as both a physical and chemical barrier, preventing microbial invasion. Structurally, the physical barrier is maintained by intercellular junctions, closely adherent epithelial cells, and an ongoing process of differentiation. Functionally, the chemical barrier is characterized by the secretion of innate immune factors by epithelial cells, such as hBD-2 and SLPI (Joly et al. 2009; Wen et al. 2011), which protect mucosal membranes from pathogen colonization and invasion (Holmgren and Czerkinsky 2005;

Kimball et al. 2006). Additionally, previous studies have shown that oral epithelial cells secrete high levels of cytokines and chemokines, including TNF-α, IL-2, and IL-4, which act as pro-inflammatory and immunoregulatory factors (Fakhry et al. 2013). These mediators contribute to the activation and regulation of immune and inflammatory processes and facilitate communication between innate and adaptive immune mechanisms (Abbas et al. 2007). When mucosal innate immunity is impaired, as seen in patients with diabetes, cancer, and HIV infection, individuals become more susceptible to opportunistic infections, such as oral candidiasis (Meyer et al. 2004; Fidel Jr 2011).

2.3.1 Role of Toll-Like Receptors (TLRs) in Innate Immunity

Oral epithelial cells express cell surface receptors called TLRs. TLRs are cellular effectors and transmembrane proteins, extensively expressed in eukaryotic cells, and are essential for detecting fungal, bacterial, and viral pathogens. Not only are these receptors essential for immune defense, but they also act as the first line of recognition in the host's response to pathogens (Sasai and Yamamoto 2013). In humans, the TLR family consists of ten members, including TLR1–10. Some TLRs are located on the cell surface, where they primarily recognize extracellular pathogens, while others are located within intracellular compartments such as endosomes or endolysosomes, where they detect components of intracellular pathogens (Lester and Li 2014; Shen et al. 2023; Kawasaki and Kawai 2014). All TLRs exhibit specific ligands, except TLR10, whose specific ligand has not yet been discovered.

To date, mRNA expression for all ten TLRs has been identified in oral epithelial cells (Groeger and Meyle 2019). Beklen et al. (2008) analyzed tissue samples from patients and validated the expression of TLR1–9, along with TLR10, in the oral mucosal epithelium using immunohistochemical techniques. Their findings revealed that, with the exception of TLR7 and TLR8, all TLRs showed statistically significant differences in expression between inflamed periodontal tissues and healthy controls.

Similarly, Sugawara et al. (2006) demonstrated that TLRs are expressed in the normal oral mucosal epithelium, with TLR2 and TLR4 being particularly prominent, as shown by immunohistochemical analysis. Further supporting this, Lukova et al. (2020) showed that primary cultured oral epithelial cells express TLR2 and TLR4, confirmed through PCR, flow cytometry, and immunostaining. Importantly, inflamed oral epithelial tissues demonstrate higher expression of TLRs on the cell membrane than those observed in healthy tissues.

The oral cavity is continuously exposed to a diverse array of microorganisms, making TLRs' expression and function crucial for maintaining oral tissue homeostasis. TLRs recognize molecular structures classified as PAMPs, which are commonly found across a wide range of microorganisms (Akira and Sato 2003), thereby classifying them as PRRs. Upon ligand binding, TLRs activate downstream signaling pathways that play a critical role in both innate and adaptive immune responses.

Several TLRs, including TLR1, TLR2, TLR4, TLR5, TLR6, and TLR10, are expressed on the cell surface, where they recognize microorganisms and their associated ligands. In contrast, TLR3, TLR7, TLR8, and TLR9 are located within cytosolic endosomal compartments, where they detect microbial components that have penetrated the host cell membrane (Cario et al. 2002). TLR2 forms heterodimers with either TLR1 or TLR6 to recognize a range of microbial components, including peptidoglycans, lipopeptides, and lipoproteins. In contrast, TLR4 specifically recognizes lipopolysaccharide (LPS) from Gram-negative bacteria (Poltorak et al. 1998; Takeuchi et al. 2002). TLR3 is responsible for detecting double-stranded RNA (dsRNA), a molecular pattern commonly associated with viral infections. TLR5 is specialized in recognizing bacterial flagellin, while TLR7 and TLR8 respond to imidazoquinolines and single-stranded RNA (ssRNA). TLR9 detects unmethylated CpG motifs present in bacterial and viral DNA (Alexopoulou et al. 2001; Bauer et al. 2001; Hayashi et al. 2001; Hemmi et al. 2002; Hemmi et al. 2000; Jurk et al. 2002).

2.3.2 Role of Other Cell Surface Receptors

In addition to TLRs, cytoplasmic nucleotide-binding oligomerization domain (NOD)-like receptors (NLRs) and protease-activated receptors (PARs) are also part of the PRR family, recognizing microbial components classified as PAMPs. PARs belong to the G protein-coupled receptor (GPCR) family and include four identified subtypes: from PAR-1 through PAR-4. Activation of these receptors involves proteolytic cleavage of their N-terminal extracellular domain by specific proteinases. This cleavage exposes a new N-terminal sequence that acts as a tethered ligand, initiating several signaling cascades upon binding to the receptor (Bunnett 2006; Knecht et al. 2007; Lourbakos et al. 1998). While all PARs function via the same mechanism, they are activated by different proteinases and exhibit distinct distributions and biological activities (Coughlin and Camerer 2003). PARs play essential roles in wound healing, inflammation, hemostasis, thrombosis, cancer progression, and embryonic development (Gieseler et al. 2013). TLRs, NODs, and PARs interact in a complex network, coordinating various immune responses.

2.4 Oral Innate Immunity in Infection

The innate immune system identifies pathogens by recognizing PAMPs, which are distinct molecules specifically expressed by microbial organisms. This recognition occurs through PRRs located on the cell surface (Diamond and Kanneganti 2022; Kanneganti 2020). When PAMPs bind to PRRs, they trigger the expression of cytokines, chemokines, and adhesion molecules by innate immune cells. These responses are aimed at controlling the infection, facilitating pathogen clearance, and, if the infection persists, activating the adaptive immune response (Sonnenberg and Hepworth 2019).

2.4.1 Oral Innate Immunity in Response to Bacteria

The epithelial compartment functions not only as a physical barrier against pathogens but also actively participates in innate immune defense mechanisms (Dale 2002; Darveau et al. 1997; Tonetti 1997). Epithelial cells are constantly exposed to bacterial products from biofilms on the tooth surface and bacteria attached to mucosal surfaces. In response, these cells interact with bacteria by secreting antimicrobial peptides, cytokines, and chemokines, thus alerting various immune cell types, especially neutrophils. As key components of innate immunity, these peptides play an important role in early host defense, helping to control microbial invaders and maintain a healthy microbial balance within the mucosa (Weinberg et al. 1998). Along with human β-defensins (hBDs), which exhibit antibacterial properties, LL-37 is the only identified human cathelicidin with a broad range of immune functions, including antibacterial activity and chemokine induction. LL-37, a cationic antimicrobial peptide, is produced by various cell types, including epithelial cells (Hua et al. 2014). Therefore, the epithelium not only acts as a protective barrier but also actively responds to environmental stimuli, participates in infection defense, signals further host responses, and integrates innate and adaptive immune responses.

A previous study has highlighted the critical role of the chemokine IL-8 in oral health. IL-8 facilitates the migration of activated immune cells, promotes immune cell adhesion and tissue remodeling, and facilitates angiogenesis (Campbell et al. 2013). Elevated levels of IL-8 have been detected in the crevicular fluid of patients with severe periodontitis, even at healthy sites (Goutoudi et al. 2012). Schueller et al. (2017) reported a correlation between basal IL-8 levels and the composition of oral bacteria, indicating a connection between the oral microbiota and the host's inflammatory state. Additionally, their study found connections between IL-8 levels, nutrition, personal oral hygiene, and the oral microbiota (Schueller et al. 2017). The identification of specific bacteria in healthy individuals with high IL-8 levels may serve as an important biomarker, potentially indicating the onset of oral diseases (Schueller et al. 2017).

2.4.2 Oral Innate Immunity in Response to Fungal Infections

2.4.2.1 *Candida* Infection

Candida albicans, a common commensal microorganism found in about half of the world's population, typically resides within the mucus layer of mucosal tissues. However, under certain conditions, *C. albicans* may come into direct contact with host epithelial cells, leading to adhesion of the fungus to these cells. The outcome of this interaction depends on the specific *C. albicans* strain and the host's physiological status and immune condition. In some cases, this attachment can result in superficial colonization, epithelial invasion, and subsequent onset of disease and activation of the immune system.

Oral candidiasis is a common oral infection primarily caused by the *Candida* species, with *Candida albicans* being the most prevalent. It interacts with oral epithelial cells by adherence, invasion, and induction of cell damage (Weindl et al. 2010). Non-*albicans Candida* species, including *C. glabrata*, *C. tropicalis*, *C. parapsilosis*, *C. guilliermondii*, and *C. krusei*, are increasingly recognized as opportunistic pathogens affecting the oral mucosal environment (Paoletti et al. 2013; Samaranayake et al. 2009). The role of the host's oral innate immunity in the progression of oral candidiasis remains incompletely understood. However, previous studies have indicated that defensins, particularly β-defensins, help protect against *Candida* infections (Feng et al. 2005).

2.4.2.2 Epithelial Responses to *C. albicans*

Epithelial cells play a crucial role in host defense against *Candida* infection. Epithelial cells are capable of recognizing both the yeast and hyphal forms of *Candida albicans*; however, only hyphae can invade the cells and trigger endocytosis. Recent studies have shown that epithelial cells are primarily activated by Candidalysin, a newly discovered peptide toxin (Naglik et al. 2017). Hypha formation is crucial for inducing host cell damage and triggering immune responses, primarily through the secretion of the cytolytic peptide toxin Candidalysin. Although *C. albicans* adhesion and invasion initiate fungal recognition and activate signaling pathways, these events alone are insufficient to cause epithelial damage or fully activate the innate immune response (Moyes et al. 2010; Wachtler et al. 2011). It is the secretion of Candidalysin by *C. albicans* hyphae that induces epithelial damage and activates innate immunity, leading to the expression of pro-inflammatory cytokines, chemokines, and antimicrobial peptides necessary for immune cell recruitment (Moyes et al. 2010; Moyes et al. 2016; Conti et al. 2009).

The endocytosis of *C. albicans* by epithelial cells is mediated through multiple receptors. The innate host defense system recognizes *C. albicans* through PRRs belonging to the TLR, C-type lectin receptor (CLR), and NOD-like receptor (NLR) families (Netea et al. 2008; Bryant and Fitzgerald 2009). Primarily, the recognition of *C. albicans* by epithelial cells is facilitated by TLR1–4 and TLR6 (Netea et al. 2006; Weindl et al. 2007; Weindl et al. 2010; Pinheiro et al. 2018). TLR4 has been shown to be upregulated during oral candidiasis (Weindl et al. 2007; Rosentul et al. 2014). These PRRs mediate host immune defense by activating antimicrobial responses to control the infection, which include the elevated production of specific antimicrobial peptides such as defensins that contribute to pathogen clearance. Another key role of the innate immune system is to trigger an appropriate adaptive immune response against the invading pathogen (Weindl et al. 2010). Therefore, PRRs modulate both innate immune responses and adaptive immunity during *Candida* infections (Weindl et al. 2010; Mora-Montes et al. 2011).

Mucosal surfaces contribute to the innate immune defense against *C. albicans* by secreting a variety of antimicrobial peptides. Antimicrobial peptides are classified into two major categories: α-defensins, predominantly released by neutrophils, and β-defensins, mainly produced by epithelial cells (Diamond and Ryan 2011). Recently, α-defensin 6, produced by intestinal epithelial cells, has been observed to play an inhibitory role in the pathogenicity of *C. albicans* by preventing both epithelial invasion and biofilm formation (Chairatana and Nolan 2017). Human epithelial cells express four β-defensins including hBD-1, hBD-2, hBD-3, and hBD-4 (Diamond and Ryan 2011; Hua et al. 2014). hBD-1 and hBD-4 are constitutively expressed by epithelial cells, in contrast to hBD-2 and hBD-3, which are normally expressed at lower levels but are strongly upregulated in response to microbial infection or cellular stress (Diamond and Ryan 2011). hBD-2 and hBD-3 are present in human buccal epithelial cells (Sawaki et al. 2002) and are more strongly upregulated in response to *C. albicans* hyphae compared to yeast forms (Feng et al. 2005). Defensins also act as chemoattractants, recruiting T cells, dendritic cells, and neutrophils to the site of infection (Vylkova et al. 2006). In addition to defensins, LL-37, which is found in the oral cavity, can prevent *C. albicans* adhesion to epithelial cells by binding to fungal cell wall components such as mannans, chitin, and glucans (Tsai et al. 2011).

2.4.3 Oral Innate Immunity in Viral Infections

During viral infections, innate immune signaling is stimulated upon pathogen recognition through PRRs, including TLRs and retinoic acid-inducible gene I-like receptors (RLRs). These PRRs are crucial for sensing RNA virus infections, establishing an antiviral state that inhibits viral replication in infected cells and decreases the susceptibility of neighboring cells to infection. The signaling pathways activated by these receptors result in the production and secretion of critical immune mediators, including interferons (IFNs), cytokines, chemokines, and surfactant proteins. These mediators are vital for initiating immune responses, drawing immune cells—including neutrophils, macrophages, and natural killer cells—to the site of infection, where they work together to eliminate the pathogen.

Furthermore, the innate immune system plays a central role in shaping the adaptive immune response. Dendritic cells, which present antigens from pathogens to naïve T cells, are essential for inducing the activation, proliferation, and differentiation of antigen-specific T cells into functional effector cells. By delivering co-stimulatory signals and cytokines, dendritic cells also influence the polarization of T cell effector functions and support the development of long-lived memory T cells, thereby enhancing immunity against future exposures to the same pathogen (Weber 2021).

However, some viruses, due to their high sequence variability, can evade immune detection by altering the antigens expressed by infected cells and disrupting immune responses (Dos Santos 2021; Balamurugan et al. 2017). Viruses can manipulate host genetic material to impair antiviral signaling, hinder PRR expression, and suppress cytokine production (Chan and Gack 2016). Additionally, viruses can reduce immune effectiveness by downregulating major histocompatibility complex I (MHC-I) molecules (Pymm et al. 2017; Schmiedel and Mandelboim 2017) and inhibiting the cytotoxic activity of T cells (Wieland et al. 2017).

2.4.3.1 Oral Innate Immunity in HIV Infection

HIV transmission most commonly occurs through contact with the genital and rectal mucosa, which are key viral entry points, leading to extensive research on mucosal immune responses to the virus. In contrast, HIV transmission via the oral mucosa has been less commonly reported. Oral innate immunity serves as the first line of defense against HIV infection and plays an important role in preventing the virus from entering mucosal surfaces. However, HIV infection may affect the function of oral epithelial cells in innate immunity (Steele and Fidel Jr 2002; Egusa et al. 2005; Egusa et al. 2006), as evidenced by alterations in salivary secretory leukocyte protease inhibitor (SLPI) expression in HIV-infected individuals (Nittayananta et al. 2013a). It has been suggested that HIV may influence SLPI expression through infection-independent interactions on the cell surface (Jana et al. 2005). Studies have shown that SLPI concentration increases in the presence of HIV infection (Jana et al. 2005; Lin et al. 2004; Baqui et al. 1999). Specifically, Nittayananta et al. (2013a) found that HIV infection affects both transcriptional and translational expression levels of SLPI in the oral cavity. Their research revealed that while oral SLPI mRNA levels were elevated in HIV-infected individuals, SLPI protein levels in stimulated saliva were significantly reduced. This suggests that HIV may alter posttranscriptional regulation, impacting mRNA translation or protein synthesis. Additionally, the study identified CD4 T-cell count and HIV viral load as factors influencing oral SLPI expression (Nittayananta et al. 2013a). Therefore, HIV-induced impairment of oral epithelial innate immunity leads to changes in SLPI expression, which can contribute to the occurrence of opportunistic infections and malignancies in HIV-positive individuals (Clifford et al. 2005; Gillison 2009; Nittayananta et al. 2010).

HIV infection has also been associated with various indirect alterations in the innate immune responses of oral epithelial cells (Challacombe and Naglik 2006). In oral epithelial cell lines, HIV exposure has been associated with increased expression levels of hBD-2 and hBD-3 (Quiñones-Mateu et al. 2003). Moreover, Nittayananta et al. (2013b) reported that HIV-infected patients exhibit significantly higher concentrations of hBD-2 protein in saliva, regardless of whether it is unstimulated or stimulated, compared to noninfected controls. This increase in hBD-2 levels may have functional significance, as hBD-2 and hBD-3 have been shown to inhibit HIV replication in vitro (Quiñones-Mateu et al. 2003).

2.4.3.2 Oral Innate Immunity in HPV Infection

Human papillomaviruses (HPVs) are responsible for various cancers, including cervical and oropharyngeal cancers, contributing to around 5% of all cancers globally. During infection, HPV replication triggers the DNA damage response (DDR), which facilitates the progression of the viral life cycle. Despite the presence of DNA damage, HPV-infected cells can continue to proliferate, which contributes to the persistence of the virus. Simultaneously, HPV employs various mechanisms to inhibit the innate immune response, creating a conducive environment for viral replication and cancer progression. HPV requires both DDR activation and inhibition of host innate immunity to effectively replicate and maintain its life cycle (Spriggs and Laimins 2017; Galloway and Laimins 2015; Hong and Laimins 2013; Moody and Laimins 2010; Beglin et al. 2009; Doorbar et al. 2012).

DNA damage typically triggers an innate immune response, including interferon production, which can attenuate cellular proliferation and induce antiviral defense mechanisms. HPV's ability to suppress the DDR helps it evade the immune system and continuing its life cycle. Disruption of DDR in HPV-infected cells has been shown to block the HPV life cycle, making it a potential therapeutic target for controlling and treating HPV-associated cancers. Additionally, reactivating the innate immune response could serve as a novel approach in immunotherapy for HPV-related cancers.

The modulation of DDR by HPV proteins, specifically E6, E7, and E1, is well documented. For example, HPV16 genomes can activate DDR independently of E6 and E7 expression. However, upregulation of E6, E7, E2, and E1 can suppress the innate immune response. Notably, disrupting the expression of E6 or E7 does not significantly affect DDR in HPV16-infected cells but does restore the innate immune response. Reactivating the innate immune response in cells with persistent DDR could enhance immune responses, offering a potential strategy to strengthen immunotherapy for HPV-associated cancers.

2.4.3.3 Oral Innate Immunity in SARS-CoV-2 Infection

Coronavirus disease 2019 (COVID-19) is caused by severe acute respiratory syndrome coronavirus 2 (SARS-CoV-2), which is characterized by its positive-sense, single-stranded RNA (ssRNA) genome of approximately 30 kb. This genome encodes four structural proteins, including spike (S), envelope (E), membrane (M), and nucleocapsid (N) proteins (Troyano-Hernáez et al. 2021). The S protein is crucial for the virus's ability to enter host cells, as it regulates fusion with the host cell membrane after binding to the host cell receptor, angiotensin-converting enzyme 2 (ACE2). The S protein is cleaved into two subunits, S1 and S2, which are essential for viral entry (Hoffmann et al. 2020; Jackson et al. 2022). S1 contains the receptor-binding domain (RBD), which mediates attachment to ACE2, while the S2 subunit anchors the spike proteins to the viral membrane (Nejat et al. 2023).

As previously mentioned, the innate immune system directly fights against infection and triggers the adaptive immune response, as the body's first line of defense. Upon infection, SARS-CoV-2 is recognized by a variety of PRRs expressed by epithelial cells and innate immune cells, such as alveolar macrophages (Fung and Liu 2019). These PRRs detect PAMPs, including the ssRNA genome of SARS-CoV-2, which is recognized by TLRs, specifically TLR2, TLR3, TLR4, TLR7, TLR8, and TLR9, localized to the cell membrane and endosomal membranes (Cervantes-Barragan et al. 2007; Mantovani et al. 2023). TLR2 and TLR4 are involved in recognizing viral proteins, such as the SARS-CoV-2 envelope (E) protein, and in triggering inflammatory cytokine production (Tyrkalska et al. 2023; Zheng et al. 2021). Additionally, intracellular sensors like melanoma differentiation-associated protein 5 (MDA-5) and retinoic acid-inducible gene I (RIG-I) detect double-stranded RNA (dsRNA), a by-product of viral replication. Other cytosolic receptors, including NLRs, activate inflammasome complexes in response to intracellular danger signals, further triggering an inflammatory response (Jung and Lee 2021; Yamada and Takaoka 2023).

Activation of these PRRs is crucial for initiating a robust and efficient immune response against SARS-CoV-2 infection. However, excessive or prolonged activation can result in harmful inflammation. Overactivation of TLRs and NLRs, due to viral PAMPs and damage-associated molecular patterns (DAMPs) released from infected or lysed cells, can lead to an exaggerated production of pro-inflammatory cytokines and chemokines, contributing to tissue damage (Mantovani et al. 2023). In particular, TLR4 has been implicated in driving hyperinflammation in COVID-19. Studies have shown that severe COVID-19 is associated with impaired type I IFN responses and increased levels of pro-inflammatory cytokines, including TNF-α and IL-6, which are regulated by nuclear factor kappa B (NF-κB) signaling. A defective antiviral type-I IFN response allows viral replication to continue unchecked, resulting in the accumulation of viral particles and damaged cells. This unchecked inflammation can progress to a cytokine storm, exacerbating the severity of COVID-19 symptoms, particularly in severely ill patients (Hadjadj et al. 2020; Aboudounya and Heads 2021).

2.5 Oral Innate Immunity in Cancer

Oral cancer, despite advances in treatment, remains a significant oral health issue worldwide, with oral squamous cell carcinoma (OSCC) representing over 90% of all oral cancers (Feller 2012). Over the last two decades, the overall 5-year survival rate for oral cancer has remained around 50% (Bell et al. 2007; Kim and Ahn 2024). OSCC primarily affects individuals over the age of 65, but its incidence is increasingly being observed in younger populations, including adolescents, which has been rising each year. Several factors contribute to the development of OSCC, including tobacco use, alcohol consumption, betel nut chewing, and infection with HPV (Kwak et al. 2016).

2.5.1 Innate Immune Biomarkers, Inflammation, and Cancer Susceptibility

Previous studies have demonstrated that the expression of human β-defensins (hBDs) and secretory leukocyte protease inhibitor (SLPI) is altered in OSCC compared to normal oral epithelium (Joly et al. 2009; Kesting et al. 2009; Wen 2011). Pro-inflammatory cytokines such as TNF-α, IL-6, and IL-8 also play critical roles in regulating inflammatory responses and cancer development (Brailo et al. 2012), maintaining a central role in host protective immunity. Thus, changes in their expression can significantly influence host defense processes.

Epithelial cells produce various pro-inflammatory cytokines and chemokines, including IL-1, IL-6, IL-8, and TNF-α, with IL-8 expression levels often increased in the saliva of patients with oral carcinomas. IL-8 has been suggested as a potential biomarker for detecting oral and oropharyngeal SCCs (St John et al. 2004). Furthermore, infections caused by pathogenic bacteria including *Porphyromonas gingivalis* and *Aggregatibacter actinomycetemcomitans* can modulate cytokine expression. There is a well-established link between bacterial infections and cancer, with bacterial components capable of upregulating immunomodulatory receptors on malignant cells, potentially fostering malignant transformation in a microenvironment with impaired immune responses.

Oral cancer involves various immune cells, including macrophages, mast cells, and innate lymphoid cells (ILCs), which infiltrate the tumor and alter cytokine signaling (Bouaoud et al. 2021). Chronic inflammation, often regulated by innate immune cells, plays a significant role in promoting oncogenesis by releasing growth factors, aiding in metastasis, and inhibiting cell death pathways. However, the innate immune system also exerts anticancer effects through functions like maintaining the epithelial barrier, directly inducing cytotoxicity, and presenting antigens to adaptive immune cells, including T and B cells (Manjili et al. 2022).

The proportion of immune cells is higher in oral cancer tissues than in healthy oral tissues, and these cells can have dual roles in either promoting or suppressing cancer cells. The tumor-promoting activities of immune cells include reducing cancer immunity, enhancing blood vessel formation, and facilitating metastasis, while their antitumor functions include recruiting immune cells to eliminate cancer cells (Peña-Romero and Orenes-Piñero 2022). The opposing effects of inflammation can simultaneously support and inhibit oral cancer progression (Ni et al. 2022). This dual role involves complex signaling mechanisms, including cytokines like IL-1, IL-6, and TGF-β and chemokines like interferons and CCL2 (Kwaśniak et al. 2019), which trigger intracellular pathways like NF-kB, signal transducer and activator of transcription 3 (STAT3), and mitogen-activated protein kinase (MAPK). These pathways influence gene expression and cell fate decisions. While these signaling cascades are well-documented, the initial triggers determining whether these responses are pro- or anti-tumorigenic remain unclear, adding to the complexity of oral cancer development. Further research is needed to elucidate these mechanisms, especially given the varying responses observed in individual patients.

2.5.2 Role of Toll-Like Receptors in the Pathogenesis of Oral Cancer

By maintaining tissue homeostasis in the oral cavity, innate immunity plays an important role in managing both infectious and neoplastic processes (Joly et al. 2009; Kesting et al. 2009; Wen et al. 2011; Brailo et al. 2012). TLRs are key molecules in innate immunity that recognize PAMPs and initiate inflammatory responses to eliminate infected cells, thereby preventing the spread of infection. These receptors are integral to the body's defense mechanisms, as they can trigger the release of pro-inflammatory cytokines and other mediators critical for immune responses.

Moreover, TLRs are being studied for their potential involvement in cancer development. Certain polymorphisms in TLRs can increase susceptibility to oral cancer. Identifying these TLRs and the genotypes associated with cancer development offers a promising avenue for using them as biomarkers for diagnosing or predicting the progression of oral cancer. Research has shown that TLR7 and TLR5 are implicated in the suppression of oral cancer, while TLR4 and TLR2 are associated with its progression. Thus, TLRs represent an important target for future therapeutic strategies aimed at preventing or treating oral cancer, with the potential to design treatments that modulate TLR signaling to enhance immune responses or mitigate tumor progression (Bhardwaj et al. 2024).

2.5.3 Innate Immune Biomarkers in Oral Cancer Therapy

The ability to enhance the body's natural ability to fight cancer, particularly OSCC, can potentially improve treatment outcomes without triggering systemic inflammation. Recent advancements in cancer immunotherapy have suggested that immune checkpoint inhibitors might be effective in managing OSCC (Kujan et al. 2020). These inhibitors work by reactivating the immune system to target and destroy cancer cells more effectively. Additionally, activating PRRs or inhibiting immunosuppressive cytokines could improve treatment efficacy by boosting the immune response while minimizing unwanted inflammation.

Researchers are also exploring innovative approaches, such as interleukin-loaded nanoparticles, to deliver targeted therapy directly to tumor sites. This strategy aims to enhance the immune response specifically at the site of cancer, reducing the impact on other tissues and limiting side effects (Singh et al. 2019).

Moreover, there is growing interest in developing innate immune biomarkers that can detect early signs of cancer development caused by immune dysregulation in the tumor microenvironment. These biomarkers could help differentiate between patients who are likely to respond positively to immunotherapy and those who may not as well as predict the likelihood of cancer recurrence following surgery. Monitoring innate immune biomarkers in blood or saliva could pave the way for more personalized cancer treatments, allowing for tailored therapies based on an individual's immune system profile. This approach could significantly advance the field of cancer immunotherapy, enabling more precise and effective treatments for OSCC and other cancers.

2.6 Conclusion

The epithelial cells lining the oral cavity play a vital role in innate immunity. These cells not only act as a physical barrier but also produce antimicrobial peptides and cytokines that help prevent the invasion of oral pathogens. To resolve infections and maintain tissue homeostasis, the expression and function of cell surface receptors such as TLRs are crucial for detecting PAMPs found in microbes and triggering downstream signaling pathways that activate both innate and adaptive immune responses. Epithelial cells also participate in specific innate immune pathways that can influence the development of oral cancer. Thus, enhancing innate immunity presents a promising strategy for the early detection and treatment of oral infections and cancers. Additionally, individual differences in the innate immune response may contribute to the initiation and progression of oral cancer, highlighting the need for continued research into these complex interactions.

References

Abbas A, Lichtman A, Pillai S (2007) Cellular and molecular immunology, 6th edn. Saunders, Elsevier, pp 267–301

Aboudounya MM, Heads RJ (2021) COVID-19 and toll-like receptor 4 (TLR4): SARS-CoV-2 may bind and activate TLR4 to increase ACE2 expression. Facilitat Entry Caus Hyperinflam Mediat Inflamm 1:8874339

Akira S, Sato S (2003) Toll-like receptors and their signaling mechanisms. Scand J Infect Dis 35(9):555–562

Alexopoulou L, Holt AC, Medzhitov R, Flavell RA (2001) Recognition of double-stranded RNA and activation of NF-kappaB by toll-like receptor 3. Nature 413(6857): 732–738

Balamurugan A, Claiborne D, Ng HL, Yang OO (2017) HIV-1 epitope variability is associated with T cell receptor repertoire instability and breadth. J Virol 91: e00771–e00717

Baqui AA, Meiller TF, Falkler WA Jr (1999) Enhanced secretory leukocyte protease inhibitor in human immunodeficiency virus type 1-infected patients. Clin Diagn Lab Immunol 6:808–811

Bauer S, Kirschning CJ, Häcker H, Redecke V, Hausmann S, Akira S, Wagner H, Lipford GB (2001) Human TLR9 confers responsiveness to bacterial DNA via species-specific CpG motif recognition. Proc Natl Acad Sci USA 98(16):9237–9242

Beglin M, Melar-New M, Laimins LA (2009) Human Papillomaviruses and the interferon response. J Interf Cytokine Res 29:629–635. https://doi.org/10.1089/jir.2009.0075

Beklen A, Hukkanen M, Richardson R, Konttinen YT (2008) Immunohistochemical localization of toll-like receptors 1-10 in periodontitis. Oral Microbiol Immunol 23(5):425–431. https://doi.org/10.1111/j.1399-302X.2008.00448.x

Bell RB, Kademani D, Homer L, Dierks EJ, Potter BE (2007) Tongue cancer: is there a difference in survival compared with other subsites in the oral cavity? J Oral Maxillofac Surg 65(2):229–236

Bhardwaj A, Prasad D, Mukherjee S (2024) Role of toll-like receptor in the pathogenesis of oral cancer. Cell Biochem Biophys 82(1):91–105

Bouaoud J, Foy JP, Tortereau A, Michon L, Lavergne V, Gadot N, Boyault S, Valantin J, De Souza G, Zrounba P, Bertolus C, Bendriss-Vermare N, Saintigny P (2021) Early changes in the immune microenvironment of oral potentially malignant disorders reveal an unexpected association of M2 macrophages with oral cancer free survival. Onco Targets Ther 10(1):1944554

Brailo V, Vucicevic-Boras V, Lukac J, Biocina-Lukenda-D, Zilic-Alajbeg I, Milenovic A, Ml B (2012) Salivary and serum interleukin 1 beta, interleukin 6 and tumor necrosis factor alpha in patients with leukoplakia and oral cancer. Med Oral Patol Oral Cir Buca 17(1):e10–e15

Brice DC, Diamond G (2020) Activities of human host defense peptides. Curr Med Chem 27(9):1420–1443

Bryant C, Fitzgerald KA (2009) Molecular mechanisms involved in inflammasome activation. Trends Cell Biol 19:455–464

Bunnett NW (2006) Protease-activated receptors: how proteases signal to cells to cause inflammation and pain. Semin Thromb Hemost (Suppl. 1):39–48

Campbell LM, Maxwell PJ, Waugh DJ (2013) Rationale and means to target pro-inflammatory interleukin-8 (CXCL8) signaling in cancer. Pharmaceuticals 6: 929–959

Cario E, Brown D, McKee M, Lynch-Devaney K, Gerken G, Podolsky DK (2002) Commensal-associated molecular patterns induce selective toll-like receptor-trafficking from apical membrane to cytoplasmic compartments in polarized intestinal epithelium. Am J Pathol 160(1):165–173

Cervantes-Barragan L, Züst R, Weber F, Spiegel M, Lang KS, Akira S, Thiel V, Ludewig B (2007) Control of coronavirus infection through plasmacytoid dendritic-cell-derived type I interferon. Blood 109(3): 1131–1137

Chairatana P, Nolan EM (2017) Human α-Defensin 6: a small peptide that self-assembles and protects the host by entangling microbes. Acc Chem Res 50(4):960–967

Challacombe SJ, Naglik JR (2006) The effects of HIV infection on oral mucosal immunity. Adv Dent Res 19:29–35

Chan YK, Gack MU (2016) Viral evasion of intracellular DNA and RNA sensing. Nat Rev Microbiol 14:360–373

Clifford GM, Polesel J, Rickenbach M, Dal Maso L, Keiser O, Kofler A, Rapiti E, Levi F, Jundt G, Fisch T, Bordoni A, De Weck D, Franceschi S (2005) Swiss HIV cohort. Cancer risk in the Swiss HIV cohort study: associations with immunodeficiency, smoking, and highly active antiretroviral therapy. J Natl Cancer Inst 97(6):425–432

Conti HR, Shen F, Nayyar N, Stocum E, Sun JN, Lindemann MJ, Ho AW, Hai JH, Yu JJ, Jung JW, Filler SG, Masso-Welch P, Edgerton M, Gaffen SL (2009) Th17 cells and IL-17 receptor signaling are essential for mucosal host defense against oral candidiasis. J Exp Med 206(2):299–311

Coughlin SR, Camerer E (2003) PARticipation in inflammation. J Clin Invest 111:25–27

Dale BA (2002) Periodontal epithelium: a newly recognized role in health and disease. Periodontology 30:70–78

Dale BA, Kimball JR, Krisanaprakornkit S, Roberts F, Robinovitch M, O'Neal R, Valore EV, Ganz T, Anderson GM, Weinberg A (2001) Localized antimicrobial peptide expression in human gingiva. J Periodontal Res 36(5):285–294

Darveau RP, Tanner A, Page RC (1997) The microbial challenge in periodontitis. Periodontology 2000(14): 12–32

Diamond MS, Kanneganti TD (2022) Innate immunity: the first line of defense against SARS-CoV-2. Nat Immunol 23:165–176

Diamond G, Ryan L (2011) Beta-defensins: what are they really doing in the oral cavity? Oral Dis 17:628–635. https://doi.org/10.1111/j.1601-0825.2011.01799.x

Doorbar J, Quint W, Banks L, Bravo IG, Stoler M, Broker TR, Stanley MA (2012) The biology and life-cycle of human papillomaviruses. Vaccine 30(Suppl 5):F55–F70

Dos Santos WG (2021) Impact of virus genetic variability and host immunity for the success of COVID-19 vaccines. BioMed Pharmacother 136:111272

Egusa H, Nikawa H, Makihira S, Jewett A, Yatani H, Hamada T (2005) Intercellular adhesion molecule 1-dependent activation of interleukin 8 expression in Candida albicans-infected human gingival epithelial cells. Infect Immun 73(1):622–626

Egusa H, Nikawa H, Makihira S, Yatani H, Hamada T (2006) *In vitro* mechanisms of interleukin-8-mediated responses of human gingival epithelial cells to *Candida albicans* infection. Int J Med Microbiol 296:301–311

Fakhry C, Marks MA, Gilman RH, Cabrerra L, Yori P, Kosek M, Gravitt PE (2013) Comparison of the immune microenvironment of the oral cavity and cervix in healthy women. Cytokine 64(2):597–604

Feller L (2012) Oral squamous cell carcinoma: epidemiology, clinical presentation and treatment. J Cancer Ther 3:263–268

Feng Z, Jiang B, Chandra J, Ghannoum M, Nelson S, Weinberg A (2005) Human beta-defensins: differential activity against candidal species and regulation by *Candida albicans*. J Dent Res 84:445–450

Fidel PL Jr (2011) Candida-host interactions in HIV disease: implications for oropharyngeal candidiasis. Adv Dent Res 23:45–49

Fung TS, Liu DX (2019) Human coronavirus: host-pathogen interaction. Annu Rev Microbiol 73:529–557. https://doi.org/10.1146/annurev-micro-020518-115759. Epub 2019 Jun 21. PMID: 31226023

Galloway DA, Laimins LA (2015) Human papillomaviruses: shared and distinct pathways for pathogenesis. Curr Opin Virol 14:87–92

Gieseler F, Ungefroren H, Settmacher U, Hollenberg MD, Kaufmann R (2013) Proteinase-activated receptors (PARs) – focus on receptor-receptor-interactions and their physiological and pathophysiological impact. Cell Commun Signal 11:86. https://doi.org/10.1186/1478-811X-11-86

Gillison ML (2009) Oropharyngeal cancer: a potential consequence of concomitant HPV and HIV infection. Curr Opin Oncol 21:439–444

Groeger S, Meyle J (2019) Oral Mucosal Epithelial Cells. Front Immunol 10:208. https://doi.org/10.3389/fimmu.2019.00208

Goutoudi P, Diza E, Arvanitidou M (2012) Effect of periodontal therapy on crevicular fluid interleukin-6 and interleukin-8 levels in chronic periodontitis. Int J Dent 2012:362905. https://doi.org/10.1155/2012/362905

Groeger SE, Meyle J (2015) Epithelial barrier and oral bacterial infection. Periodontol 69(1):46–67

Hadjadj J, Yatim N, Barnabei L, Corneau A, Boussier J, Smith N, Péré H, Charbit B, Bondet V, Chenevier-Gobeaux C, Breillat P, Carlier N, Gauzit R, Morbieu C, Pène F, Marin N, Roche N, Szwebel TA, Merkling SH, Treluyer JM, Veyer D, Mouthon L, Blanc C, Tharaux PL, Rozenberg F, Fischer A, Duffy D, Rieux-Laucat F, Kernéis S, Terrier B (2020) Impaired type I interferon activity and inflammatory responses in severe COVID-19 patients. Science 369(6504):718–724

Hayashi F, Smith KD, Ozinsky A, Hawn TR, Yi EC, Goodlett DR, Eng JK, Akira S, Underhill DM, Aderem A (2001) The innate immune response to bacterial flagellin is mediated by toll-like receptor 5. Nature 410(6832):1099-1103

Hemmi H, Takeuchi O, Kawai T, Kaisho T, Sato S, Sanjo H, Matsumoto M, Hoshino K, Wagner H, Takeda K, Akira S (2000) A toll-like receptor recognizes bacterial DNA. Nature 408(6813):740–745

Hemmi H, Kaisho T, Takeuchi O, Sato S, Sanjo H, Hoshino K, Horiuchi T, Tomizawa H, Takeda K, Akira S (2002) Small anti-viral compounds activate immune cells via the TLR7 MyD88-dependent signaling pathway. Nat Immunol 3(2):196–200

Hoffmann M, Kleine-Weber H, Pöhlmann S (2020) A multibasic cleavage site in the spike protein of SARS-CoV-2 is essential for infection of human lung cells. Mol Cell 78(4):779–784.e5

Holmgren J, Czerkinsky C (2005) Mucosal immunity and vaccines. Nat Med 11:S45–S53

Hong S, Laimins L (2013) Regulation of the life cycle of HPVs by differentiation and the DNA damage response. Future Microbiol 8:1547–1557

Hua X, Yuan X, Tang X, Li Z, Pflugfelder SC, Li DQ (2014) Human corneal epithelial cells produce antimicrobial peptides LL-37 and β-defensins in response to heat-killed Candida albicans. Ophthalmic Res 51(4):179–186

Iwasaki A (2007) Mucosal dendritic cells. Annu Rev Immunol 25:381–418

Jackson CB, Farzan M, Chen B, Choe H (2022) Mechanisms of SARS-CoV-2 entry into cells. Nat Rev Mol Cell Biol 23:3–20

Jana NK, Gray LR, Shugars DC (2005) Human immunodeficiency virus type 1 stimulates the expression and production of secretory leukocyte protease inhibitor (SLPI) in oral epithelial cells: a role for SLPI in innate mucosal immunity. J Virol 79:6432–6440

Joly S, Compton LM, Pujol C, Kurago ZB, Guthmiller JM (2009) Loss of human beta-defensin 1, 2, and 3 expression in oral squamous cell carcinoma. Oral Microbiol Immunol 24:353–360

Jung HE, Lee HK (2021) Current understanding of the innate control of toll-like receptors in response to SARS-CoV-2 infection. Viruses 13. https://doi.org/10.3390/v13112132

Jurk M, Heil F, Vollmer J, Schetter C, Krieg AM, Wagner H, Lipford G, Bauer S (2002) Human TLR7 or TLR8 independently confer responsiveness to the antiviral compound R-848. Nat Immunol 3(6):499

Kanneganti T (2020) Intracellular innate immune receptors: life inside the cell. Immunol Rev 297:5–12

Kawasaki T, Kawai T (2014) Toll-like receptor signaling pathways. Front Immunol 5:461

Kelsall B (2008) Recent progress in understanding the phenotype and function of intestinal dendritic cells and macrophages. Mucosal Immunol 1:460–469

Kesting MR, Loeffelbein DJ, Hasler RJ, Wolff KD, Rittig A, Schulte M, Hirsch T, Wagenpfeil S, Jacobsen F, Steinstraesser L (2009) Expression profile of human beta-defensin 3 in oral squamous cell carcinoma. Cancer Investig 27(5):575–581

Kim MJ, Ahn KM (2024) Prognostic factors of oral squamous cell carcinoma: the importance of recurrence and pTNM stage. Maxillof Plast Reconstruct Surg 46(1):8

Kimball JR, Nittayananta W, Klausner M, Chung WO, Dale BA (2006) Antimicrobial barrier of an in vitro oral epithelial model. Arch Oral Biol 51:775–783

Knecht W, Cottrell GS, Amadesi S, Mohlin J, Skåregärde A, Gedda K, Peterson A, Chapman K, Hollenberg MD, Vergnolle N, Bunnett NW (2007) Trypsin IV or mesotrypsin and p23 cleave protease-activated receptors 1 and 2 to induce inflammation and hyperalgesia. J Biol Chem 282(36):26089–26100

Kujan O, van Schaijik B, Farah CS (2020) Immune checkpoint inhibitors in oral cavity squamous cell carcinoma and oral potentially malignant disorders: a systematic review. Cancers (Basel) 12(7):1937

Kwak HH, Kim IR, Kim HJ, Park BS, Yu SB (2016) a-Mangostin induces apoptosis and cell cycle arrest in oral squamous cell carcinoma cell. Evi Based Comple Alter Med 2016. https://doi.org/10.1155/2016/5352412

Kwaśniak K, Czarnik-Kwaśniak J, Maziarz A, Aebisher D, Zielińska K, Karczmarek-Borowska B, Tabarkiewicz J (2019) Scientific reports concerning the impact of interleukin 4, interleukin 10 and transforming growth factor β on cancer cells. Cent Eur J Immunol 44(2):190–200

Lester SN, Li K (2014) Toll-like receptors in antiviral innate immunity. J Mol Biol 426:1246–1264

Li T, Xie J, He Y, Fan H, Baril L, Qiu Z, Zuo Y (2006) Long-term persistence of robust antibody and cytotoxic T cell responses in recovered patients infected with SARS coronavirus. PLoS One 1(1):e24

Lin AL, Johnson DA, Stephan KT, Yeh CK (2004) Salivary secretory leukocyte protease inhibitor increases in HIV infection. J Oral Pathol Med 33:410–416

Lourbakos A, Chinni C, Thompson P, Potempa J, Travis J, Mackie EJ, Pike RN (1998) Cleavage and activation of proteinase-activated receptor-2 on human neutrophils by gingipain-R from porphyromonas gingivalis. FEBS Lett 435(1):45–48

Lukova OA, Zaslavskaya MI, Makhrova TV, Kropotov VS, Kitaeva EV (2020) Expression of toll-like and adhesive receptors on epithelial cells of the oral mucosa in periodontitis. Klin Lab Diagn 65(10): 645–648. https://doi.org/10.18821/0869-2084-2020-65-10-645-648

Manjili SH, Isbell M, Ghochaghi N, Perkinson T, Manjili MH (2022) Multifaceted functions of chronic inflammation in regulating tumor dormancy and relapse. Semin Cancer Biol 78:17–22

Mantovani S, Oliviero B, Varchetta S, Renieri A, Mondelli MU (2023) TLRs: innate immune sentries against SARS-CoV-2 infection. Int J Mol Sci 24(9):8065

Meyer JE, Harder J, Görögh T, Weise JB, Schubert S, Janssen D, Maune S (2004) Human beta-defensin-2 in oral cancer with opportunistic Candida infection. Anticancer Res 24(2B):1025–1030

Modlin RL (2012) Innate immunity: ignored for decades, but not forgotten. J Invest Dermatol 132:882–886

Moody CA, Laimins L (2010) Human papillomavirus oncoproteins: pathways to transformation. Nat Rev Cancer 10:550–560

Mora-Montes HM, Netea MG, Ferwerda G, Lenardon MD, Brown GD, Mistry AR, Kullberg BJ, O'Callaghan CA, Sheth CC, Odds FC, Brown AJ, Munro CA, Gow NA (2011) Recognition and blocking of innate immunity cells by Candida albicans chitin. Infect Immun 79(5):1961–1970

Moyes DL, Runglall M, Murciano C, Shen C, Nayar D, Thavaraj S, Kohli A, Islam A, Mora-Montes H, Challacombe SJ, Naglik JR (2010) A biphasic innate immune MAPK response discriminates between the yeast and hyphal forms of Candida albicans in epithelial cells. Cell Host Microbe 8(3):225-235

Moyes DL, Wilson D, Richardson JP, Mogavero S, Tang SX, Wernecke J, Höfs S, Gratacap RL, Robbins J, Runglall M, Murciano C, Blagojevic M, Thavaraj S, Förster TM, Hebecker B, Kasper L, Vizcay G, Iancu SI, Kichik N, Häder A, Kurzai O, Luo T, Krüger T, Kniemeyer O, Cota E, Bader O, Wheeler RT, Gutsmann T, Hube B, Naglik JR (2016) Candidalysin is a fungal peptide toxin critical for mucosal infection. Nature 532(7597):64–68

Naglik JR, König A, Hube B, Gaffen SL (2017) Candida albicans-epithelial interactions and induction of mucosal innate immunity. Curr Opin Microbiol 40:104–112

Nejat R, Torshizi MF, Najafi DJ (2023) S protein, ACE2 and host cell proteases in SARS-CoV-2 cell entry and infectivity; is soluble ACE2 a two blade sword? Narrat Rev Vacc (Basel) 11(2):204

Netea MG, Gow NA, Munro CA, Bates S, Collins C, Ferwerda G, Hobson RP, Bertram G, Hughes HB, Jansen T, Jacobs L, Buurman ET, Gijzen K, Williams DL, Torensma R, McKinnon A, MacCallum DM, Odds FC, Van der Meer JW, Brown AJ, Kullberg BJ (2006) Immune sensing of Candida albicans requires cooperative recognition of mannans and glucans by lectin and toll-like receptors. J Clin Invest 116(6): 1642–1650

Netea MG, Brown GD, Kullberg BJ, Gow NA (2008) An integrated model of the recognition of Candida albicans by the innate immune system. Nat Rev Microbiol 6:67–78

Ni Y, Low JT, Silke J, O'Reilly LA (2022) Digesting the role of JAK-STAT and cytokine signaling in oral and gastric cancers. Front Immunol 29(13):835997

Nittayananta W, Talungchit S, Jaruratanasirikul S, Silpapojakul K, Chayakul P, Nilmanat A, Pruphetkaew N (2010) Effects of long-term use of HAART on oral health status of HIV-infected subjects. J Oral Pathol Med 39(5):397–406

Nittayananta W, Kemapunmanus M, Yangngam S, Talungchit S, Sriplung H (2013a) Expression of oral secretory leukocyte protease inhibitor in HIV-infected subjects with long-term use of antiretroviral therapy. J Oral Pathol Med 42:208–215

Nittayananta W, Kemapunmanus M, Amornthatree K, Talungchit S, Sriplung H (2013b) Oral human beta-defensin 2 in HIV-infected subjects with long-term use of antiretroviral therapy. J Oral Pathol Med 42:53–60

Paoletti I, Fusco A, Grimaldi E, Perillo L, Coretti L, Di Domenico M, Cozza V, Contaldo M, Serpico R, Guida A, Donnarumma G (2013) Assessment of host defence mechanisms induced by Candida species. Int J Immunopathol Pharmacol 26(3):663–672

Peña-Romero AC, Orenes-Piñero E (2022) Dual effect of immune cells within tumour microenvironment: pro- and anti-tumour effects and their triggers. Cancers (Basel) 14(7):1681

Pinheiro CR, Coelho AL, de Oliveira CE, Gasparoto TH, Garlet GP, Silva JS, Santos C.F, Cavassani KA, Hogaboam CM, Campanelli AP (2018) Recognition of *Candida albicans* by gingival fibroblasts: the role of TLR2, TLR4/CD14, and MyD88. Cytokine 106:67–75

Poltorak A, He X, Smirnova I, Liu MY, Van Huffel C, Du X, Birdwell D, Alejos E, Silva M, Galanos C, Freudenberg M, Ricciardi-Castagnoli P, Layton B, Beutler B (1998) Defective LPS signaling in C3H/HeJ and C57BL/10ScCr mice: mutations in Tlr4 gene. Science 282(5396):2085–2088

Pymm P, Illing PT, Ramarathinam SH, O'Connor GM, Hughes VA, Hitchen C, Price DA, Ho BK, McVicar DW, Brooks AG, Purcell AW, Rossjohn J, Vivian JP (2017) MHC-I peptides get out of the groove and enable a novel mechanism of HIV-1 escape. Nat Struct Mol Biol 24(4):387–394

Quiñones-Mateu ME, Lederman MM, Feng Z, Chakraborty B, Weber J, Rangel HR, Marotta ML, Mirza M, Jiang B, Kiser P, Medvik K, Sieg SF, Weinberg A (2003) Human epithelial beta-defensins 2 and 3 inhibit HIV-1 replication. AIDS 17(16):F39–F48

Rosentul DC, Delsing CE, Jaeger M, Plantinga TS, Oosting M, Costantini I, Venselaar H, Joosten LA, van der Meer JW, Dupont B, Kullberg BJ, Sobel JD, Netea MG (2014) Gene polymorphisms in pattern recognition receptors and susceptibility to idiopathic recurrent vulvovaginal candidiasis. Front Microbiol 23(5):483

Samaranayake LP, Keung Leung W, Jin L (2009) Oral mucosal fungal infections. Periodontol 2000(49): 39–59

Sasai M, Yamamoto M (2013) Pathogen recognition receptors: ligands and signaling pathways by toll-like receptors. Int Rev Immunol 32(2):116–133

Sawaki K, Mizukawa N, Yamaai T, Fukunaga J, Sugahara T (2002) Immunohistochemical study on expression of alpha-defensin and beta-defensin-2 in human buccal epithelia with candidiasis. Oral Dis 8(1):37–41

Schmiedel D, Mandelboim O (2017) Disarming cellular alarm systems manipulation of stress-induced NKG2D ligands by human herpesviruses. Front Immunol 8:390

Schueller K, Riva A, Pfeiffer S, Berry D, Somoza V (2017) Members of the oral microbiota are associated with IL-8 release by gingival epithelial cells in healthy individuals. Front Microbiol 8:416. https://doi.org/10.3389/fmicb.2017.00416

Shen J, Fan J, Zhao Y, Jiang D, Niu Z, Zhang Z, Cao G (2023) Innate and adaptive immunity to SARS-CoV-2 and predisposing factors. Front Immunol 14:115932

Singh AP, Biswas A, Shukla A, Maiti P (2019) Targeted therapy in chronic diseases using nanomaterial-based drug delivery vehicles. Signal Transduct Target Ther 4: 33. https://doi.org/10.1038/s41392-019-0068-3

Sonnenberg GF, Hepworth MR (2019) Functional interactions between innate lymphoid cells and adaptive immunity. Nat Rev Immunol 19:599–613

Spriggs CC, Laimins LA (2017) Human papillomavirus and the DNA damage response: exploiting host repair pathways for viral replication. Viruses 9(8):232

Squier CA, Kremer MJ (2001) Biology of oral mucosa and esophagus. J Natl Cancer Inst Monogr 29:7–15

St John MA, Li Y, Zhou X, Denny P, Ho CM, Montemagno C, Shi W, Qi F, Wu B, Sinha U, Jordan R, Wolinsky L, Park NH, Liu H, Abemayor E, Wong DT (2004) Interleukin 6 and interleukin 8 as potential biomarkers for oral cavity and oropharyngeal squamous cell carcinoma. Arch Otolaryngol Head Neck Surg 130(8):929–935

Steele C, Fidel PL Jr (2002) Cytokine and chemokine production by human oral and vaginal epithelial cells in response to Candida albicans. Infect Immun 70:577–583

Stoitzner P, Romani N (2011) Langerin, the "Catcher in the Rye": an important receptor for pathogens on Langerhans cells. Eur J Immunol 41:2526–2529

Sugawara Y, Uehara A, Fujimoto Y, Kusumoto S, Fukase K, Shibata K et al (2006) Toll-like receptors, NOD1, and NOD2 in Oral epithelial cells. J Dent Res 85(6):524–529. https://doi.org/10.1177/154405910608500609

Taguchi T, Mukai K (2019) Innate immunity signalling and membrane trafficking. Curr Opin Cell Biol 59:1–7

Takeuchi O, Sato S, Horiuchi T, Hoshino K, Takeda K, Dong Z, Modlin RL, Akira S (2002) Cutting edge: role of toll-like receptor 1 in mediating immune response to microbial lipoproteins. J Immunol 169(1):10–14

Tonetti MS (1997) Molecular factors associated with compartmentalization of gingival immune responses and transepithelial neutrophil migration. J Periodontal Res 32:104–109

Troyano-Hernáez P, Reinosa R, Holguín Á (2021) Evolution of SARS-CoV-2 envelope, membrane, nucleocapsid, and spike structural proteins from the beginning of the pandemic to September 2020: a global and regional approach by epidemiological week. Viruses 13(2):243

Tsai PW, Yang CY, Chang HT, Lan CY (2011) Human antimicrobial peptide LL-37 inhibits adhesion of Candida albicans by interacting with yeast cell-wall carbohydrates. PLoS One 6:e17755

Tyrkalska SD, Martínez-López A, Pedoto A, Candel S, Cayuela ML, Mulero V (2023) The spike protein of SARS-CoV-2 signals via Tlr2 in zebrafish. Dev Comp Immunol 140:104626

Vylkova S, Li XS, Berner JC, Edgerton M (2006) Distinct antifungal mechanisms: beta-defensins require Candida albicans Ssa1 protein, while Trk1p mediates activity of cysteine-free cationic peptides. Antimicrob Agents Chemother 50:324–331

Wachtler B, Wilson D, Haedicke K, Dalle F, Hube B (2011) From attachment to damage: defined genes of Candida albicans mediate adhesion, invasion and damage during interaction with oral epithelial cells. PLoS One 6:e17046

Weber F (2021) Encyclopedia of virology. Antiviral innate immunity: introduction. Elsevier, Amsterdam, pp 577–583

Weinberg A, Krisanaprakornkit S, Dale BA (1998) Epithelial antimicrobial peptides: review and significance for oral applications. Crit Rev Oral Biol Med 9:399–414

Wen J, Nikitakis NG, Chaisuparat R, et al. (2011) Secretory leukocyte protease inhibitor (SLPI) expression and tumor invasion in oral squamous cell carcinoma. Am J Pathol 178: 2866–78

Weinberg A, Tugizov S, Pandiyan P, Jin G, Rakshit S, Vyakarnam A, Naglik JR (2020) Innate immune mechanisms to oral pathogens in oral mucosa of HIV-infected individuals. Oral Dis 26(Suppl 1):69–79

Weindl G, Naglik JR, Kaesler S, Biedermann T, Hube B, Korting HC, Schaller M (2007) Human epithelial cells establish direct antifungal defense through TLR4-mediated signaling. J Clin Invest 117(12):3664–3672

Weindl G, Wagener J, Schaller M (2010) Epithelial cells and innate antifungal defense. J Dent Res 89:666–675

Wen J, Nikitakis NG, Chaisuparat R, Greenwell-Wild T, Gliozzi M, Jin W, Adli A, Moutsopoulos N, Wu T, Warburton G, Wahl SM (2011) Secretory leukocyte protease inhibitor (SLPI) expression and tumor invasion in oral squamous cell carcinoma. Am J Pathol 178(6):2866–2878

Wieland D, Hofmann M, Thimme R (2017) Overcoming CD8+ T cell exhaustion in viral hepatitis: lessons from the mouse model and clinical perspectives. Dig Dis 35: 334–338

Yamada T, Takaoka A (2023) Innate immune recognition against SARS-CoV-2. Inflamm Regen 43:7

Zheng M, Karki R, Williams EP, Yang D, Fitzpatrick E, Vogel P, Jonsson CB, Kanneganti TD (2021) TLR2 senses the SARS-CoV-2 envelope protein to produce inflammatory cytokines. Nat Immunol 22(7):829–838

Examining Oral Microbial Communities at the Gingival Interface During Health and Inflammatory Diseases

3

Loreto Abusleme, Marion Arce, and Nicolas Dutzan

Abstract

The subgingival environment is a unique ecological niche where host–microbe interactions continuously occur, ultimately determining the balance that characterizes periodontal health or the disequilibrium featured in periodontal diseases. Importantly, knowledge of the subgingival microbiome has expanded dramatically due to the extensive use of next-generation sequencing techniques in both periodontal health and diseases, making it necessary to summarize and contextualize evidence from these types of studies. Consequently, this chapter first explores the anatomical and histological characteristics of the subgingival area and provides an overview of the microbial communities and immune surveillance mechanisms unraveling in this space and during periodontal health. Then, focus is placed on better defining periodontal inflammation as a main driver of microbial dysbiosis during disease. The subgingival microbiome in gingivitis and periodontitis is further analyzed, focusing on the environmental changes triggered by periodontal inflammation, as inflammation and tissue destruction lead to dramatic environmental changes that are associated with distinct microbial shifts. The microbial transition from health to gingivitis is explored in detail, discussing evidence stemming from studies dedicated to natural and experimental gingivitis. Finally, the dynamics of subgingival microbial communities in periodontitis and their connection with inflammation are explored, acknowledging the bidirectional nature of this relationship and highlighting its importance in determining health and disease in this setting.

Keywords

Oral microbiome · Periodontal health · Gingivitis · Periodontitis · Subgingival microbiome

L. Abusleme (✉)
Laboratory of Oral Microbiology and Immunology, Faculty of Dentistry, University of Chile, Santiago, Chile

Department of Pathology and Oral Medicine, Faculty of Dentistry, University of Chile, Santiago, Chile

Clinical Research Unit in Periodontology and Implants, Faculty of Dentistry, University of Chile, Santiago, Chile
e-mail: loreto.abusleme@odontologia.uchile.cl

M. Arce · N. Dutzan
Laboratory of Oral Microbiology and Immunology, Faculty of Dentistry, University of Chile, Santiago, Chile

Clinical Research Unit in Periodontology and Implants, Faculty of Dentistry, University of Chile, Santiago, Chile

Department of Conservative Dentistry, Faculty of Dentistry, University of Chile, Santiago, Chile

N. Rezaei, H. Ziaei (eds.), *Oral Immunology*, Advances in Experimental Medicine and Biology 1492,
https://doi.org/10.1007/978-3-032-03176-1_3

3.1 The Subgingival Environment and Host–Microbe Interactions During Periodontal Health

A large body of work has been dedicated to understanding the immune and microbiome shifts in the subgingival compartment that occur during periodontal diseases. However, less attention has been given to the host immune response and microbial communities that sustain periodontal health, which are fundamental to the underlying regulatory mechanisms that constantly maintain homeostasis at this critical barrier site. To start unraveling this interaction, it is necessary to recognize the anatomical and histological features of the subgingival region, as these are key to understanding the intimate relationship between the subgingival microbiota and host immune mechanisms. Moreover, these mechanisms might explain microbial composition and assembly in health.

3.1.1 Anatomical and Histological Characteristics of the Subgingival Environment

The gingival tissues are structured to act as the first line of defense, displaying distinct structural and physiological features that promote protection (Schroeder and Listgarten 1997). In periodontal health, the close anatomical relationship between the gingiva and the tooth creates a defined subgingival environment known as the gingival sulcus, which consists of a rigid wall (dental enamel), a soft wall (sulcular epithelium), and the junctional epithelium (JE). Connective tissue lies below this epithelium, which is highly vascularized and populated by resident stromal and immune cells, primarily neutrophils, lymphocytes, and macrophages (Konkel and Moutsopoulos 2018) (Fig. 3.1). JE comprises nonkeratinized, stratified squamous epithelial cells with high turnover rates. This layer is characterized by continuous coronal cell migration and rapid cell exfoliation into the gingival sulcus, maintaining the protective integrity of the gingiva. In humans, JE is about 2 mm in height and is composed of flat cells oriented parallel to the tooth and originates from a layer of cuboidal basal cells resting on a basement membrane (Listgarten 1966; Nanci and Bosshardt 2006; Schroeder and Listgarten 1997). Unlike the sulcular and oral epithelium, JE comprises only a few layers of epithelial cells. It is typically wider in its coronal region and gradually narrows toward its apical region, creating an environment

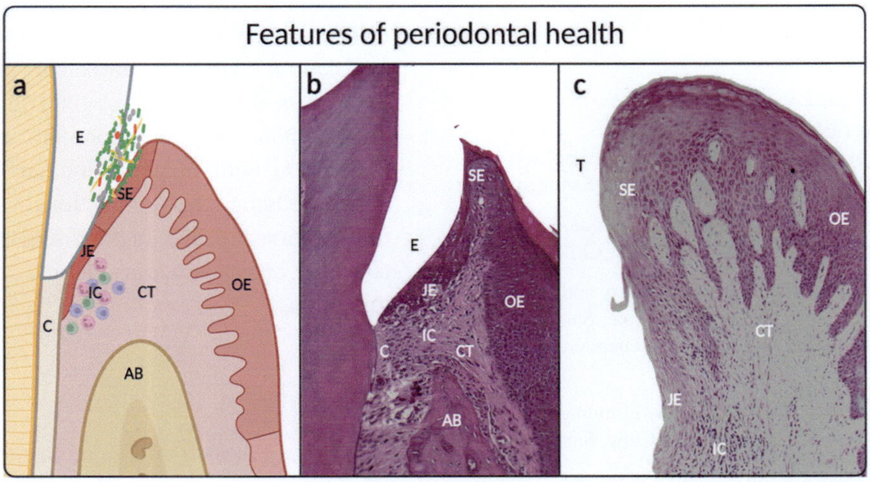

Fig. 3.1 Main histological features of periodontal tissues in health. (**a**) Schematic representation of the healthy subgingival environment consisting of an enamel (E), a sulcular epithelium (SE), and a junctional epithelium (JE), along with the anatomy of associated structures, including a cementum (C), infiltrating cells (ICs), connective tissue (CT), an alveolar bone (AB), and the oral epithelium (OE). (**b**) Histological section of the mouse periodontium with all structures that are shown in panel (**a**). (**c**) Histological section of human soft tissue structures shown in panel (**a**), including the space where the tooth (T) should be located. (Created with BioRender)

of close contact between commensal microbes and immune cells (Fig. 3.1). This structure is essential for maintaining a healthy periodontium, as it supports effective immune surveillance and response in an area highly susceptible to microbial exposure (Nanci and Bosshardt 2006; Schroeder and Listgarten 1997).

Beneath the JE, the connective tissue primarily consists of a type I collagen fiber network. Even in the absence of clinical inflammation, this tissue contains a variety of immune cells that form the first line of defense against microorganisms (Kinane and Lappin 2002). An elevated number of neutrophils has been described within immune cells, and small foci of lymphocytes, macrophages, and some plasma cells can be observed in the connective tissue beneath the junctional epithelium. This region also contains the subepithelial plexus of venules (Kinane and Lappin 2002; Schroeder and Listgarten 1997; Schroeder and Munzel-Pedrazzoli 1973). During periodontal health, the gingival compartment is composed, on average, of 4% junctional epithelium, 27% oral gingival epithelium, and 69% connective tissue that includes a cellular infiltrate occupying about 3%–6% of the gingival volume (Schroeder and Munzel-Pedrazzoli 1973).

JE plays a vital role in the subgingival environment, facilitating close interactions between microbial communities and host tissues. Structural or compositional changes in JE are detected by resident structural and immune cells, which can initiate an inflammatory response. If sustained, this response may lead to tissue destruction and histological alterations within these structures.

3.1.2 Subgingival Microbial Communities During Periodontal Health

In recent years, our knowledge about oral microbial communities has significantly expanded. The advent of high-throughput sequencing techniques has revealed unprecedented microbial richness and diversity in the oral cavity, only second to that of the distal gastrointestinal tract (Segata et al. 2011). Through the seminal observations generated by the Human Microbiome Project, it has been possible to appreciate how the ecological determinants at each body site influence microbial structure and composition (Costello et al. 2009) and also how the oral microbiome is distinct from the microbial communities inhabiting other locations. Moreover, within the oral cavity, there is notable variation in the microbial assemblies recovered from different intraoral locations (Human Microbiome Project 2012), as these microenvironments exhibit unique anatomical and environmental features influencing the biogeography at this site (Baker et al. 2024; Mark Welch et al. 2016; Proctor et al. 2020). Hence, a major distinction can be made between mucosal- and tooth-associated communities, not only in terms of taxonomy but also spatial organization and biomass (Human Microbiome Project 2012; Mark Welch et al. 2020). These differences are probably due to the nature of the surface (hard versus soft tissue), the desquamation of the epithelium, exposure to salivary flow, and mechanical forces, among other factors (Baker et al. 2024; Proctor et al. 2018; Simon-Soro et al. 2013). One of the key properties of tooth-associated microbial communities is their ability to assemble and mature over extended periods of time, as they are less subject to turnover compared to communities that assemble on oral mucosal surfaces (Borisy and Valm 2021).

3.1.2.1 General Features of Microbial Communities on Tooth Surfaces

Tooth-associated microbial communities are defined based on their location relative to the gingival margin, commonly referred to as supragingival and subgingival plaque, and have a well-recognized role in maintaining oral homeostasis (Chang et al. 2021; Valm 2019). Supra- and subgingival plaque are organized as complex biofilms, and their architectural structure started to unravel several years ago through microscopy (Listgarten 1976, 1994). However, the resolution of these analyses has dramatically improved with multispectral imaging techniques such as combinatorial labeling and spectral imaging fluorescence in situ hybridization (CLASI-FISH) (Valm et al. 2012; Valm et al. 2016). The development of such technology has made it possible to distinguish individual taxa, often down to the genus, and sometimes even the species level of several microbial constituents at once (Borisy and Valm 2021; Valm et al. 2016). For instance, architectural analyses of supragingival plaque collected from healthy individuals have provided vital insights into the spatial arrangement of these communities. These observations have revealed how individual species organize into specific types of assemblies, such as hedgehog structures with corncob-like tips, characterized by long filaments of *Corynebacterium* spp. decorated by *Streptococcus* spp. and *Porphyromonas* spp., among other bacterial taxa (Mark Welch et al. 2016). These types of organizations seem to be conserved in other hominids, also reflecting the long history of the evolution of these microbial arrangements (Fellows Yates et al. 2021). In contrast, most of our knowledge of subgingival microbial communities associated with periodontal health comes from landmark studies by Listgarten (1976, 1994). These studies visualized the biofilm immediately coronal to the junctional epithelium, suggesting that more filamentous forms were present in this region compared to supragingival areas, indicating a notable difference in microbial composition and architecture. These studies were based on microscopical techniques that did not allow for the taxonomical identification of bacteria, and, thus, the distribution of specific genera and/or species within these biofilms associated with the healthy gingiva remains largely unexplored.

3.1.2.2 Main Constituents and Ecology of Subgingival Microbial Communities During Periodontal Health

Subgingival microbial communities are intimately linked to the etiopathogenesis of periodontal diseases (Papapanou et al. 2018a; Valm 2019). However, to fully appreciate the dysbiotic shifts that subgingival microbial communities undergo during periodontal diseases, it is first necessary to identify the traits that characterize the subgingival microbiota in periodontal health. The subgingival microbiome is mainly constituted by bacteria but also harbors smaller proportions of archaea, eukaryotes (mainly fungi and protozoa), and viruses (Diaz 2021). It is increasingly recognized that these less abundant microbes might play a role in the ecology of the oral microbiome (Baker et al. 2024). However, their importance in the overall balance from health to disease in the periodontal environment remains unclear.

Bacteria are highly diverse and abundant in the subgingival environment; early molecular studies were fundamental in expanding our appreciation of bacterial richness at this site (Kroes et al. 1999; Kumar et al. 2005), indicating that around 500 different species could be found. During periodontal health, we and others have shown that bacterial richness (e.g., number of observed taxa) in subgingival communities ranges from approximately 50 up to 250 bacterial species at a given site (Abusleme et al. 2013; Abusleme et al. 2021; Bik et al. 2010; Griffen et al. 2012), which is significantly lower than what is observed in periodontal diseases (Abusleme et al. 2021). Estimators of bacterial diversity, which is a more complex ecological concept as it encompasses not only the number of observed taxa but also its relative abundance (Lozupone and Knight 2008), have been used to characterize the subgingival environment. These estimators reveal that subgingival communities associated with periodontal health exhibit lower diversity than those during gingivitis (e.g., gingival inflammation without destruction of tooth-supporting tissues (Caton et al. 2018)) but comparable diversity to periodontitis (e.g., inflammation with destruction of tooth-supporting structures (Caton et al. 2018)) (as reviewed in Abusleme et al. (2021)). On the other hand, delving into the taxonomical identity of the most abundant bacteria inhabiting the subgingival environment during health, *Streptococcus* spp., which include *S. sanguinis*, *S. gordonii*, and *S. mitis*, as well as *Actinomyces* spp., including *A. oris* and *A. naeslundii*, are traditionally found in these communities. These bacteria are known for their role as early colonizers of tooth-associated

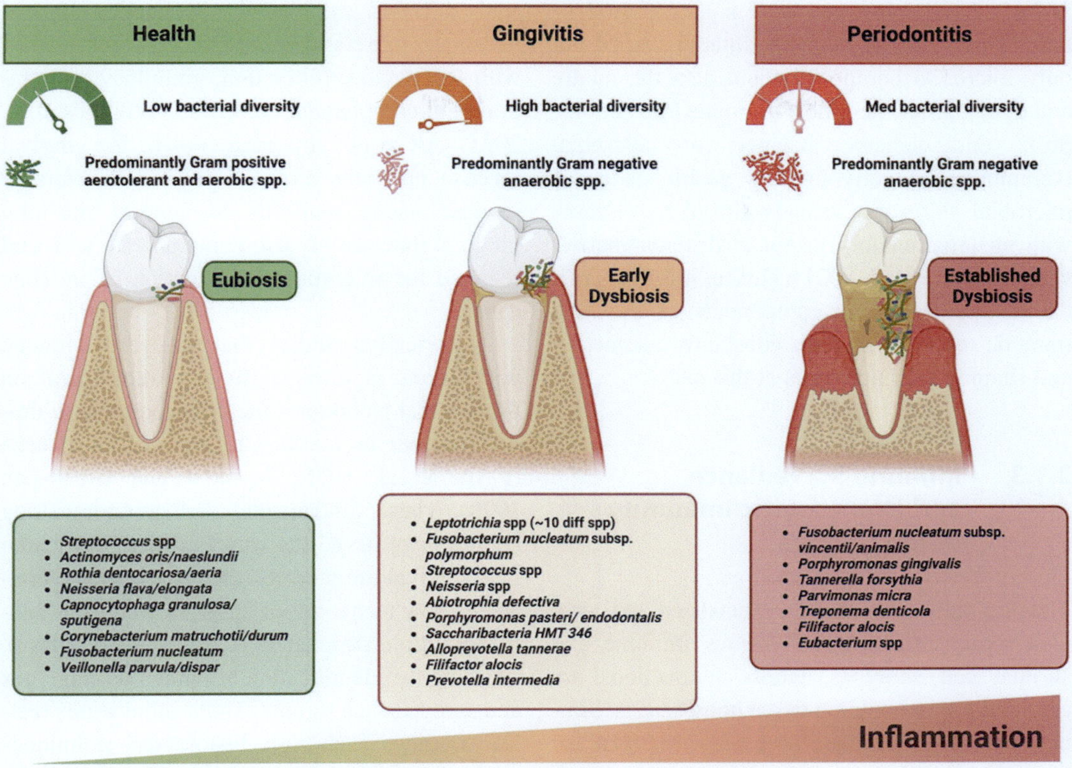

Fig. 3.2 Key microbiological attributes in periodontal health, gingivitis, and periodontitis. The dials indicate the differences in bacterial diversity between the three periodontal conditions. The teeth representations illustrate the conditions' main features and the corresponding changes in microbial communities. Signature bacterial species for each condition are highlighted within text shapes. Inflammation increases progressively from health to periodontitis. Color code: *green*=health, *orange*=gingivitis, and *red*=periodontitis. (Created with BioRender by Abusleme (2025). Link: https://BioRender.com/r84s479)

surfaces (Abranches et al. 2018; Baker et al. 2024; Socransky et al. 1998). Other health-associated genera are *Rothia* (*R. dentocariosa*, *R. aeria*), *Neisseria* (*N. flava*, *N. elongata*), and *Capnocytophaga* (*C. granulosa*, *C. sputigena*) (Eren et al. 2014). The genus *Corynebacterium* is also an important constituent of these communities and has a structural and metabolic role in tooth-associated biofilms, fundamentally represented by the species *C. matruchotii* and *C. durum* (Abusleme et al. 2021; Mark Welch et al. 2016). Strict anaerobes also have relevant representation at the gingival crevice during health, such as *Fusobacterium nucleatum* (subsp. *vicentii*, *nucleatum*, and *animalis*) and *Veillonella* (*V. parvula* and *V. dispar*), among others (Abusleme et al. 2013; Griffen et al. 2012). These microbial profiles and other features of health-associated communities are summarized in Fig. 3.2.

Importantly, classical periodontal pathobionts such as *Porphyromonas gingivalis*, *Treponema denticola*, *Tannerella forsythia*, and *Filifactor alocis* can be routinely detected in subgingival communities associated with periodontal health, albeit at a much lower detection frequency and relative abundance (Abusleme et al. 2021). This is relevant as it highlights the fact that these pathobionts are normal inhabitants of the subgingival environment, increasing in proportion as inflammation and tissue destruction progress.

Subgingival microbial communities and the host immune response are in constant dialogue as both have coevolved over a long period of time. This interplay has undoubtedly shaped not only microbial composition but also the nature and dynamics of these host–immune interactions. Still, this delicate balance of microbial communities directly impacts health status, as microbial shifts are strongly linked to diseases with an inflammatory origin at different barrier sites (Ansaldo et al. 2021). Hence, it is also critical to appreciate the components of the host immune response that are constantly interacting and shaping the microbiota at this site.

3.1.3 Immune Surveillance and Homeostatic Immunity in the Gingiva

The oral mucosa is a highly specialized and complex tissue composed of various immune and stromal cell subtypes, whose interactions are essential for maintaining tissue homeostasis. Similar to other barrier sites, these cells are organized into four primary compartments: endothelial, epithelial, fibroblastic, and immune cells (Williams et al. 2021). These compartments are continuously tailored by commensal microorganisms and local environmental cues, including food, airborne particles, and mechanical stimuli (Dutzan et al. 2017; Moutsopoulos and Konkel 2018). This dynamic environment requires oral mucosal cells to perform finely tuned, localized immune surveillance, safeguarding tissue integrity and function.

In the healthy gingiva, immune function is not limited to leukocytes. Different studies have characterized the responses of gingival epithelial cells and fibroblasts when challenged by various microorganisms (Aggor et al. 2020; Jang et al. 2017; Kim et al. 2024). Epithelial cells can secrete antimicrobial peptides, inflammatory cytokines, and chemokines under stimulation, and recent studies using single-cell RNA sequencing (scRNA-seq) have identified an exclusive gingival epithelial population that, in health, expresses an immune-related gene profile (Easter et al. 2024; Kim et al. 2024; Williams et al. 2021). Like epithelial cells, healthy gingival fibroblast subsets also express genes associated with leukocyte proliferation, granulocyte migration, and complement activation (Williams et al. 2021). All in all, structural cells of the gingival mucosa not only serve as physical barriers or secrete matrix scaffolds to support the oral gingival mucosa but also communicate with and regulate the migration and function of immune cells.

Historically, studies that analyzed immune populations in gingival tissues were based on histological techniques that allowed using a limited number of markers to identify those cells (Payne et al. 1975; Seymour and Greenspan 1979). These fundamental and groundbreaking studies have paved the way for modern research into gingival immune cell characterization. Nowadays, new technologies have allowed us to analyze multiple parameters at the single-cell level, allowing the identification of numerous cell types and subtypes in a sample. These innovations reaffirm findings from early histological techniques while expanding the knowledge of immune cells residing in the gingival tissues and their interactions with stromal and epithelial cells in both health and disease (Easter et al. 2024; Monasterio et al. 2024; Williams et al. 2021). Using these new technologies, studies have found that in the healthy gingiva, immune cells could be divided into clusters of natural killer T (T/NK), B/plasma, and granulocyte/myeloid cells. The majority of immune cells are T cells, subdivided into T helper, T cytotoxic, mucosal-associated invariant T (MAIT), γδ T, and NK/T cells. The second largest population is myeloid cells, which include myeloid dendritic cells, macrophages, and neutrophils. Mast cells, plasmacytoid dendritic cells (pDCs), B cells, and plasma cells are present in small proportions in these healthy tissues (Williams et al. 2021). Compared to other oral mucosal areas, the gingiva has a more enriched immune cell compartment, with a higher proportion of $CD45^+$ cells, attributed to the increased presence of granulocytes and primarily neutrophils (Dutzan et al. 2016; Williams et al. 2021).

3.1.3.1 Role of Neutrophils in Maintaining Gingival Tissue Homeostasis

Neutrophils continuously migrate from blood vessels within the lamina propria to the junctional epithelium, guided by a chemokine gradient secreted by oral keratinocytes and partially induced by the subgingival microbiota (Tonetti et al. 1998; Zenobia et al. 2013). Studies have shown that germ-free (GF) mice have reduced but continuous transmigration of neutrophils than wild-type mice (Zenobia et al. 2013). Another factor regulating neutrophil recruitment to the gingiva is the continuous tissue damage by mastication, which activates T helper 17 (Th17) cell immunity and subsequently enhances neutrophil recruitment (Dutzan et al. 2017). This constant trigger and ubiquitous presence of neutrophils in the gingiva suggests that these cells play a pivotal role in maintaining oral mucosal health by providing a rapid and efficient immune response and by participating in the resolution of inflammation and tissue repair (Silva et al. 2023; Uriarte and Hajishengallis 2023). The critical role of neutrophils in gingival homeostasis becomes evident from observations that different congenital neutrophil deficiencies are associated with severe periodontal destruction in early life (Silva et al. 2019). An example of these inborn deficiencies is leukocyte adhesion deficiency I (LAD-I), an autosomal recessive primary immunodeficiency due to mutations in the *ITGB2* gene, which encodes the subunit CD18 of the β2 integrins. Integrins facilitate the firm adhesion of neutrophils to the endothelium by binding to intercellular adhesion molecule (ICAM)-1 and ICAM-2, thereby mediating the crawling and transmigration of neutrophils (Rosenzweig and Holland 2004). Patients with LAD-I have severe neutropenia in gingival and other tissues, producing a spectrum of manifestations ranging from recurrent infections and immunopathology of the skin and mucosal surfaces to life-threatening complications. One of the most frequent manifestations is a severe periodontal disease with early onset, severe inflammation, and rapid loss of teeth-supporting bone, which commonly leads to tooth loss (Moutsopoulos et al. 2014; Silva et al. 2019). LAD-I periodontitis is a microbe-triggered inflammatory disease with enhanced interleukin (IL)-23/IL-17-mediated immune response (Moutsopoulos et al. 2014; Moutsopoulos et al. 2017). Physiologically, the lack of neutrophils is detected in tissues by a mechanism known as neutrostat. In this mechanism, IL-23/IL-17 triggers neutrophil recruitment by induction of granulocyte colony-stimulating factor (G-CSF) and chemokines that attract neutrophils. After neutrophils enter the tissue and execute their functions, they undergo apoptosis, and tissue macrophages engulf the apoptotic neutrophils, downregulating the IL-23 responses (a mechanism known as efferocytosis) (Uriarte and Hajishengallis 2023). In LAD-I, the absence of neutrophils in gingival tissues leads to the constant activation of the IL-23 response that induces IL-17 overproduction and immunopathology (Moutsopoulos et al. 2014; Moutsopoulos et al. 2017).

Similar to the detrimental effects of the absence of neutrophils, excessive neutrophil infiltration and activation are also linked to periodontal inflammation. In fact, during periodontitis, studies have demonstrated an increased presence of these cells, which are activated by fibrin deposition in the periodontium to mediate tissue destruction via the release of reactive oxygen species (ROS) and neutrophil extracellular traps (NETs) (Dutzan et al. 2016; Silva et al. 2021). This evidence highlights the importance of a tightly regulated neutrophil response for maintaining tissue homeostasis.

3.1.3.2 Th17 Cells: Critical Mediators of Gingival Homeostasis and Pathogenesis

IL-17, specifically IL-17A, has a well-recognized role in immune surveillance in the oral mucosa and has been implicated as a driver of periodontitis immunopathology (Abusleme and Moutsopoulos 2017; Gaffen and Moutsopoulos 2020). In human gingival tissues, the primary source of IL-17 during health and disease is Th17 cells (Dutzan et al. 2018; Dutzan et al. 2016). In the context of homeostasis, a small subset of T helper cells, representing approximately 1%, produce IL-17, and the signals that shape their network in gingival tissues differ from other barrier sites, such as the skin and intestinal mucosa (Dutzan et al. 2017; Dutzan et al. 2016). The accumulation of Th17 cells in the gingiva is not influenced by the presence of commensal bacteria. Notably, the quantity of Th17 cells remains comparable between germ-free (GF) and specific pathogen-free (SPF) mice, suggesting that factors beyond the microbiome influence their accumulation in the gingiva (Dutzan et al. 2017). In fact, Th17 accumulation in the gingiva is driven by low but constant mechanical stimulation from mastication. This mechanical stimulation induces IL-6 production from the epithelial cells, a cytokine cue critical for physiological Th17 cell accumulation in gingival mucosal tissues (Dutzan et al. 2017).

The importance of Th17 cells in maintaining oral mucosal homeostasis becomes evident in patients with Th17 cell deficiencies. Hyperimmunoglobulin E (hyper IgE) syndrome (HIES) is caused by a dominant-negative mutation in the gene that codes for signal transducer and activator of transcription 3 (STAT3), a key transcription factor for Th17 cell development (Holland et al. 2007; Milner et al. 2008). Phosphorylation of STAT3 induces its translocation to the nucleus, regulating the expression of genes such as *RORC*, *IL17A*, and *IL23R* (Kane et al. 2014; Yang et al. 2007). Patients with loss-of-function mutations of STAT3 and reduced frequencies of Th17 cells have a spectrum of symptoms, including recurrent oropharyngeal candidiasis (Chandesris et al. 2012). This evidence, combined with the observation that mutations in other key genes involved in Th17 cell development also lead to candidiasis, underscores the critical role of Th17 cells in maintaining antifungal immunity within the oral mucosa in humans (Gaffen and Moutsopoulos 2020).

In contrast to oral candidiasis, impaired Th17 cell development does not cause an evident periodontal phenotype. On the contrary, multiple studies have consistently reported elevated levels of IL-17 and accumulation of Th17 cells in periodontitis lesions (Cardoso et al. 2009; Dutzan et al. 2018; Dutzan et al. 2016; Moutsopoulos et al. 2012). During periodontitis, Th17 cells adopt a pathological phenotype, secreting interferon (IFN)-γ and G-CSF together with IL-17 (Dutzan et al. 2018). Th17 expansion is also a characteristic of ligature-induced periodontitis (LIP), a widely used mouse model of the disease (Abe and Hajishengallis 2013; Dutzan et al. 2018). Studies in this model suggest a pathogenic role for Th17 cells, as genetic or pharmacological interventions targeting Th17 cells or IL-17 have been shown to prevent immunopathology, highlighting their potential as therapeutic targets (Dutzan et al. 2018; Eskan et al. 2012; Tsukasaki et al. 2018). Consistent with these results in animal models, HIES patients have reduced periodontal inflammation (bleeding on probing) and tissue destruction (clinical attachment loss) compared to the general population (Dutzan et al. 2018). Collectively, this evidence demonstrates that the IL-17/Th17 pathway plays a pivotal role in regulating infection and inflammation within the oral mucosa.

3.2 Periodontal Inflammation as a Driver of Microbial Dysbiosis During Disease

3.2.1 Impact of Inflammation and Tissue Destruction During Gingivitis and Periodontitis on the Subgingival Microbiome

One of the main ecological forces that shapes microbial composition in the subgingival environment is periodontal inflammation (Curtis et al. 2020). A dysregulated inflammatory response is central to periodontal disease (Papapanou et al. 2018a), triggering an ecological change that fuels microbial dysbiosis. In fact, the subgingival microbiome has been described as "inflammophIlic" (Hajishengallis 2014), as these microbial communities thrive during inflammation, perpetuating a positive feedback loop that further dysregulates the host immune response and leads to more tissue breakdown. Importantly, each periodontal and peri-implant condition exhibits different profiles of inflammatory and destructive events, which may also shape a distinct microbiome (Abusleme et al. 2021; Diaz et al. 2016).

3.2.1.1 An Overview of Environmental Changes Triggered by Inflammation During Periodontal Diseases

Gingivitis and periodontitis are characterized by an exaggerated inflammatory response, in which bacterial dysbiosis triggers an augmented host immune reaction (Murakami et al. 2018). Changes in microbial communities trigger tissue destruction by exacerbating the host's immune response. This initiates a cascade of events, leading to the degradation of the underlying connective tissue, beginning around blood vessels and then spreading into adjacent regions, resulting in the structural and functional disintegration of the gingiva (Nanci and Bosshardt 2006). For instance, the host produces enzymes, such as matrix metalloproteinases, leading to connective tissue breakdown. This is a necessary process initiated and controlled by the host to facilitate tissue recession away from the microbial communities undergoing dysbiotic changes (Tsukasaki et al. 2018). Once the immune and inflammatory processes are initiated, various host-derived mediators such as proteases, cytokines, prostaglandins, and host enzymes are released from leukocytes or structural cells of the tissue (Kinane 2001). Molecules derived from the adjacent periodontal tissues are important constituents of gingival crevicular fluid (GCF), further enriching its composition. Besides, the GCF flow rate gradually increases from health to gingivitis and periodontitis (Barros et al. 2016; Ozkavaf et al. 2000). Changes in GCF composition and quantity during inflammation, together with the presence of blood cells as a part of tissue inflammation and disruption, can modulate the subgingival bacterial community (Curtis et al. 2020; ter Steeg et al. 1987). In response to these shifts in available nutrients, there is noticeable enrichment of proteolytic, asaccharolytic, and anaerobic bacterial species that can thrive and take advantage of the inflammatory environment (Hajishengallis 2014; Marsh 2003). The role of inflammation in regulating the subgingival microbiota becomes evident in studies of periodontitis progression in experimental models (Hasturk et al. 2006; Lee et al. 2016). Active control of inflammation has been shown to result in a reversal of the dysbiotic shifts observed in these experimental models. These microbial changes were accompanied by regeneration following previous tissue destruction (Hasturk et al. 2006; Lee et al. 2016). These findings highlight the key principles regarding the role of inflammation in host–microbiome interactions within the subgingival microenvironment.

3.2.1.2 Subgingival Microbial Shifts During Gingivitis

Microbially induced gingivitis is one of the most common forms of periodontal disease, affecting more than 90% of the global population (Chapple et al. 2018). The role of subgingival microbial accumulation as the main driver of gingival inflammation has been widely accepted since the 1960s, when landmark studies demonstrated a causality link through the experimental gingivitis model in humans (Loe et al. 1965; Loe et al. 1967; Theilade et al. 1966). In this model, oral hygiene measures were suspended for up to 21 days, leading to dramatic changes in microbial communities and subsequent gingival inflammation. In this setting, experimental approaches were directed to demonstrate the causative role of bacteria in triggering gingival inflammation at early stages by employing different antibiotic rinses (Loe et al. 1967). These seminal results revealed that using an antibiotic rinse without proper hygiene measures (a 5-day period) reduced the amount of gingival exudate and the number of leukocytes recovered from patients (Loe et al. 1967), providing early evidence of the overall importance of bacteria in the initiation of gingival inflammation in humans.

Additionally, these studies also revealed the microbial dynamics underlying plaque accumulation through microscopic examination of bacterial smears. They revealed dramatic microbial shifts that had already been established, which were even much more profound by the time clinical gingivitis was diagnosed (Loe et al. 1965). Before inducing experimental gingivitis, initial observations described a predominance of Gram-positive cocci and small rods. Upon induction, there was significant enrichment of filamentous bacteria followed by a later increase in the proportion of vibrios and spirochaetes (Loe et al. 1965). The authors then made a relevant observation, indicating that the health-associated microbiota (Gram-positive cocci and small rods) was consistently reduced in proportion but was never replaced by other taxa. This notion was later reexamined and further expanded in light of more comprehensive studies performed with high-throughput sequencing, confirming that inflammatory changes associated with periodontal disease lead to a microbial succession without replacing health-associated species (Abusleme et al. 2013; Abusleme et al. 2021; Diaz et al. 2016). Early studies also recognized that the microbiota associated with gingivitis is distinct from that of periodontal health and periodontitis. They found that some periodontitis-associated species are already present in the microbial communities during gingivitis, supporting the concept of a microbial transition underlying the progression from gingivitis to periodontitis (Moore et al. 1987). In fact, a reanalysis of 16S rRNA-based sequencing datasets from studies that included microbial samples obtained from periodontal health, gingivitis, and periodontitis reveals that some bacterial species associated with periodontitis are also frequently detected during gingivitis, such as *Parvimonas micra*, *Fusobacterium nucleatum* subsp. *polymorphum*, *Porphyromonas endodontalis*, *Saccharibacteria* HMT 346, *Alloprevotella tannerae*, *Filifactor alocis*, and *Prevotella intermedia*, among others (Abusleme et al. 2021).

Other relevant insights into the ecology and microbial dynamics of the microbiome during gingivitis stem from sequencing studies. For instance, bacterial richness and diversity in gingivitis are significantly higher than in health and periodontitis (Abusleme et al. 2021; Iniesta et al. 2023; Park et al. 2015). This finding might reinforce the idea that gingivitis should be framed as an intermediate but distinct dysbiotic state, in which health-associated species are still able to colonize albeit at lower proportions than before and periodontitis-associated species are starting to bloom as inflammation provides an additional nutrient source, contributing to increased bacterial diversity. Moreover, a number of species thrive during gingivitis, such as several representatives of the genus *Leptotrichia*, an anaerobic Gram-negative bacilli, including *L. hongkonghensis*, *L. buccalis*, *L. wadei*, and several other unnamed members of this genus (Abusleme et al. 2021; Huang et al. 2014; Nowicki et al. 2018; Schincaglia et al. 2017). Other species include *Fusobacterium nucleatum* subsp. *polymorphum*, *Streptococcus oralis*, *Streptococcus cristatus*, *Abiotrophia defectiva*, *Neisseria sicca*, *Neisseria flava/subflava*, and *Porphyromonas pasteri*, among other taxa (Abusleme et al. 2021; Huang et al. 2014; Park et al. 2015). These microbial shifts and other features of gingivitis-associated communities are summarized in Fig. 3.2.

Recent studies have delved into the complexity of the host–microbe interactions underlying gingivitis development, taking advantage of the experimental gingivitis model in humans. An important aspect that has been further explored is the heterogeneity of the host immune response during gingivitis induction. Studies have shown that individuals can exhibit variation in the type, magnitude, and temporal dynamics of select inflammatory mediators, leading to their classification as high, low, and slow responders (Bamashmous et al. 2021). This variation is also associated with particular shifts in the microbial communities, such as an expansion of *Streptococcus* spp. in the slow responders, particularly *S. sanguinis* and *S. oralis*, both usually linked to periodontal health (Bamashmous et al. 2021). This finding may indicate that there is a "tipping point" for commensals, beyond which they may contribute to immune dysregulation at the periodontal interface. Also, related studies have illustrated that microbial dysbiosis during experimental gingivitis can influence physically distant control of tooth-associated niches within the mouth, leading to an early microbial shift (Kerns et al. 2023). Similarly, microbiome shifts associated with experimental gingivitis are not restricted to supra- and subgingival plaque but also extend to other intraoral sites and fluids such as salivary and tongue microbiomes (Hall et al. 2023). These findings also suggest that environmental stressors in the oral cavity may disrupt biogeography and specialization patterns, not only during established disease, as previously known (Socransky and Haffajee 2005), but also during the early stages of gingival inflammation.

3.2.1.3 Subgingival Microbial Communities in Periodontitis and Their Relationship with Inflammation

Periodontal inflammation during gingivitis can persist over time and may progress to periodontitis, in which dysregulated host inflammatory responses become more destructive, affecting the supporting structures of the tooth and potentially leading to tooth loss (Papapanou et al. 2018b). Central to the current definition of periodontitis is its association with the presence of dysbiotic subgingival microbial communities (Papapanou et al. 2018b). For decades, various studies have been dedicated to characterizing the bacterial composition associated with periodontitis, which has been considerably improved using open-ended molecular methods such as sequencing-based techniques, as reviewed extensively by Diaz et al. (2016). The microbial shifts that characterize periodontitis are typically marked by an expansion of well-defined Gram-negative obligate anaerobes, such as *Fusobacterium nucleatum* (subsp. *vicentii* and *animalis*), *Parvimonas micra*, *Porphyromonas gingivalis*, *Tannerella forsythia*, *Treponema denticola*, and some still uncultured representatives, such as *Treponema* sp. 237 and *Fusobacterium* sp. HMT 203. Additionally, even Gram-positive anaerobes, such as *Filifactor alocis*, and various members of the genus *Eubacterium* (Abusleme et al. 2021) are part of these microbial shifts. These microbial shifts and other features of periodontitis-associated communities are summarized in Fig. 3.2.

Beyond taxonomical changes, functional properties are altered in these microbial communities, particularly in terms of the expression of genes encoding virulence factors. Notably, these alterations are more significant in putative commensals than in traditional periodontal pathobionts (Yost et al. 2015), suggesting that the entire microbial community may contribute to periodontitis pathogenesis. Another aspect of microbial dysbiosis refers to the changes in the structural organization of these communities, which, during periodontitis, seem to display distinct features with periodontal pathobionts forming unique spatial arrangements. Still, to our knowledge, this aspect has been (Zijnge et al. 2010), highlighting the necessity for further research.

The bidirectional relationship between periodontal inflammation and the microbiota has been the subject of ongoing investigations. For instance, bleeding on probing (BoP), which is the main clinical measure of gingival inflammation, has been positively associated with the presence of specific bacterial taxa, such as spirochaetes and *Prevotella intermedia* (Armitage et al. 1982; Socransky et al. 1991). Moreover, BoP correlates with a significant increase in the total bacterial load during periodontitis (Abusleme et al. 2013), periodontitis progression (Tanner et al. 2007), and various pro-inflammatory mediators (Teles et al. 2024; Teles et al. 2010).

Host-mediated tissue destruction during periodontitis plays a critical role in shaping the subgingival microbiome, as several host-derived molecules can act as known drivers of microbial dysbiosis. As previously mentioned, there is a marked increase in the amount of GCF during periodontitis, leading to increased available nutrients for microbial communities at this site (Barros et al. 2016; Ozkavaf et al. 2000). Several studies have been dedicated to dissecting the individual effects of specific nutrients present in GCF and periodontal pockets on the microbiota. Serum is one of the most studied factors; its addition in the context of various in vitro subgingival microbiome models leads to enrichment of classical periodontitis-associated pathobionts (Baraniya et al. 2020; Lamont et al.

2021; Naginyte et al. 2019). Albumin has also been shown to serve as a nutrient that the subgingival microbiota can readily utilize, as several pathobionts exhibit proteolytic activity that supports their growth in this pro-inflammatory environment (Wei et al. 1999). Additionally, increased availability of micronutrients during inflammation, such as potassium, can enhance the virulence and pathogenic potential of the periodontitis-associated microbial community (Yost et al. 2017).

Besides examining specific nutrients and molecules enriched during inflammation, studies have also explored the contribution of key host immune pathways to shaping the subgingival microbiome. In experimental periodontitis models, various treatments to control inflammation are capable of decreasing the total bacterial load while simultaneously controlling and even reverting tissue destruction (Hasturk et al. 2006; Lee et al. 2016; Mooney et al. 2021; Moutsopoulos et al. 2014; Shehabeldin et al. 2024). Moreover, the reduction of inflammation directly affects the microbiota by reducing its total bacterial load and shifting its composition back to resemble health-associated communities (Lee et al. 2016; Mooney et al. 2021; Shehabeldin et al. 2024). This evidence further confirms the direct influence of inflammation on the subgingival microbiome during periodontitis.

Importantly, the microbiota also exerts an effect in dysregulating select host immune pathways and further fostering inflammation. An example of this immune dysregulation is the strategies employed by *P. gingivalis*, which evades phagocytosis by manipulating complement receptor C5aR and TLR2 signaling, protecting *P. gingivalis* and other bacterial taxa and further promoting subgingival dysbiosis (Maekawa et al. 2014). On the other hand, lipopolysaccharide (LPS) is able to translocate from the subgingival microbiome to the connective tissue in leukocyte adhesion deficiency (LAD) periodontitis, and these LAD microbial communities mirror the upregulation of IL-23 observed in this disease in both in vitro and in vivo settings (Moutsopoulos et al. 2015). Moreover, recent findings have demonstrated that bacterial flagellin can trigger a pro-inflammatory response dominated by IL-23 from epithelial cells during periodontitis (Kim et al. 2024). This illustrates another contribution of microbial-mediated immune dysregulation to this disease.

3.3 Concluding Remarks

The subgingival environment is an ecosystem of unique characteristics, as it is a site of constant exposure to microbial communities. It consists of both hard and soft tissues, lined by a thin epithelium that delimits this anatomical space, separating it from external stimuli. Subgingival microbial communities and host immune responses have coevolved together for several years. Multiple immune surveillance mechanisms are dedicated to sustaining a balanced relationship with the microbiota. However, during disease, this equilibrium is lost, featuring microbial dysbiosis accompanied by dysregulation of innate and adaptive immune responses that lead to periodontal inflammation during gingivitis and destruction of supporting tissues during periodontitis. Then, inflammation and tissue destruction lead to dramatic environmental changes that are associated with distinct microbial shifts during gingivitis and periodontitis. Our knowledge about these subgingival communities has expanded tremendously due to the advent and widespread use of next-generation sequencing techniques, allowing us to confirm previous observations while adding novel, important findings. These types of analyses have reinforced the concept that the whole microbial community, rather than just a few known pathobionts, acts as the primary driver of immune dysregulation. Advances in research now allow us to appreciate more clearly the microbial transition from health to gingivitis to periodontitis. Notably, during gingivitis, microbial dysbiosis features increased bacterial diversity, even higher than that in periodontitis, and enrichment of predominantly Gram-negative anaerobic bacterial taxa that are different from those that dominate during periodontitis. In periodontitis, which is a later disease state, as sustained inflammation and tissue destruction have changed this landscape even further, a selection of a different array of bacterial species occurs with a subsequent reduction in bacterial diversity, allowing the dysbiotic changes within these communities to become established. Lastly, to understand the microbial dysbiosis underlying periodontal disease, it is necessary to consider and acknowledge the bidirectional relationship between periodontal inflammation and the subgingival microbiota, which will ultimately determine periodontal health and disease.

Acknowledgments This work was partly funded by FONDECYT regular grants 1231728 (to LA) and 1231350 (to ND) from ANID, courtesy of the Chilean government. Also, this work was partially supported by National Institutes of Health (NIH)/NIDCR (to LA), and MA is a recipient of the scholarship ANID 2022-21221003 courtesy of the Chilean government.

References

Abe T, Hajishengallis G (2013) Optimization of the ligature-induced periodontitis model in mice. J Immunol Methods 394(1-2):49–54. https://doi.org/10.1016/j.jim.2013.05.002

Abranches J, Zeng L, Kajfasz JK, Palmer SR, Chakraborty B, Wen ZT, Richards VP, Brady LJ, Lemos JA (2018) Biology of oral streptococci. Microbiol Spectr 6(5). https://doi.org/10.1128/microbiolspec.GPP3-0042-2018

Abusleme L, Moutsopoulos NM (2017) IL-17: overview and role in oral immunity and microbiome. Oral Dis 23(7):854–865. https://doi.org/10.1111/odi.12598

Abusleme L, Dupuy AK, Dutzan N, Silva N, Burleson JA, Strausbaugh LD, Gamonal J, Diaz PI (2013) The subgingival microbiome in health and periodontitis and its relationship with community biomass and inflammation. ISME J 7(5):1016–1025. https://doi.org/10.1038/ismej.2012.174

Abusleme L, Hoare A, Hong BY, Diaz PI (2021) Microbial signatures of health, gingivitis, and periodontitis. Periodontol 86(1):57–78. https://doi.org/10.1111/prd.12362

Aggor FEY, Break TJ, Trevejo-Nuñez G, Whibley N, Coleman BM, Bailey RD, Kaplan DH, Naglik JR, Shan W, Shetty AC, McCracken C, Durum SK, Biswas PS, Bruno VM, Kolls JK, Lionakis MS, Gaffen SL (2020) Oral epithelial IL-22/STAT3 signaling licenses IL-17-mediated immunity to oral mucosal candidiasis. Sci Immunol 5(48). https://doi.org/10.1126/sciimmunol.aba0570

Ansaldo E, Farley TK, Belkaid Y (2021) Control of immunity by the microbiota. Annu Rev Immunol 39:449–479. https://doi.org/10.1146/annurev-immunol-093019-112348

Armitage GC, Dickinson WR, Jenderseck RS, Levine SM, Chambers DW (1982) Relationship between the percentage of subgingival spirochetes and the severity of periodontal disease. J Periodontol 53(9):550–556. https://doi.org/10.1902/jop.1982.53.9.550

Baker JL, Mark Welch JL, Kauffman KM, McLean JS, He X (2024) The oral microbiome: diversity, biogeography and human health. Nat Rev Microbiol 22(2): 89–104. https://doi.org/10.1038/s41579-023-00963-6

Bamashmous S, Kotsakis GA, Kerns KA, Leroux BG, Zenobia C, Chen D, Trivedi HM, McLean JS, Darveau RP (2021) Human variation in gingival inflammation. Proc Natl Acad Sci USA 118(27). https://doi.org/10.1073/pnas.2012578118

Baraniya D, Naginyte M, Chen T, Albandar JM, Chialastri SM, Devine DA, Marsh PD, Al-Hebshi NN (2020) Modeling Normal and Dysbiotic subgingival microbiomes: effect of nutrients. J Dent Res 99(6): 695–702. https://doi.org/10.1177/0022034520902452

Barros SP, Williams R, Offenbacher S, Morelli T (2016) Gingival crevicular fluid as a source of biomarkers for periodontitis. Periodontol 70(1):53–64. https://doi.org/10.1111/prd.12107

Bik EM, Long CD, Armitage GC, Loomer P, Emerson J, Mongodin EF, Nelson KE, Gill SR, Fraser-Liggett CM, Relman DA (2010) Bacterial diversity in the oral cavity of 10 healthy individuals. ISME J 4(8): 962–974. https://doi.org/10.1038/ismej.2010.30

Borisy GG, Valm AM (2021) Spatial scale in analysis of the dental plaque microbiome. Periodontol 86(1): 97–112. https://doi.org/10.1111/prd.12364

Cardoso CR, Garlet GP, Crippa GE, Rosa AL, Junior WM, Rossi MA, Silva JS (2009) Evidence of the presence of T helper type 17 cells in chronic lesions of human periodontal disease. Oral Microbiol Immunol 24(1): 1–6. https://doi.org/10.1111/j.1399-302X.2008.00463.x

Caton JG, Armitage G, Berglundh T, Chapple ILC, Jepsen S, Kornman KS, Mealey BL, Papapanou PN, Sanz M, Tonetti MS (2018) A new classification scheme for periodontal and peri-implant diseases and conditions – introduction and key changes from the 1999 classification. J Periodontol 89(Suppl 1):S1–S8. https://doi.org/10.1002/JPER.18-0157

Chandesris MO, Melki I, Natividad A, Puel A, Fieschi C, Yun L, Thumerelle C, Oksenhendler E, Boutboul D, Thomas C, Hoarau C, Lebranchu Y, Stephan JL, Cazorla C, Aladjidi N, Micheau M, Tron F, Baruchel A, Barlogis V, Picard C (2012) Autosomal dominant STAT3 deficiency and hyper-IgE syndrome: molecular, cellular, and clinical features from a French national survey. Medicine (Baltimore) 91(4):e1–e19. https://doi.org/10.1097/MD.0b013e31825f95b9

Chang AM, Kantrong N, Darveau RP (2021) Maintaining homeostatic control of periodontal epithelial tissue.

Periodontol 86(1):188–200. https://doi.org/10.1111/prd.12369
Chapple ILC, Mealey BL, Van Dyke TE, Bartold PM, Dommisch H, Eickholz P, Geisinger ML, Genco RJ, Glogauer M, Goldstein M, Griffin TJ, Holmstrup P, Johnson GK, Kapila Y, Lang NP, Meyle J, Murakami S, Plemons J, Romito GA, Yoshie H (2018) Periodontal health and gingival diseases and conditions on an intact and a reduced periodontium: consensus report of workgroup 1 of the 2017 world workshop on the classification of periodontal and Peri-implant diseases and conditions. J Clin Periodontol 45 (Suppl 20):S68–S77. https://doi.org/10.1111/jcpe.12940
Costello EK, Lauber CL, Hamady M, Fierer N, Gordon JI, Knight R (2009) Bacterial community variation in human body habitats across space and time. Science 326(5960):1694–1697. https://doi.org/10.1126/science.1177486
Curtis MA, Diaz PI, Van Dyke TE (2020) The role of the microbiota in periodontal disease. Periodontol 83(1): 14–25. https://doi.org/10.1111/prd.12296
Diaz PI (2021) Subgingival fungi, archaea, and viruses under the omics loupe. Periodontol 85(1):82–89. https://doi.org/10.1111/prd.12352
Diaz PI, Hoare A, Hong BY (2016) Subgingival microbiome shifts and community dynamics in periodontal diseases. J Calif Dent Assoc 44(7):421–435. https://www.ncbi.nlm.nih.gov/pubmed/27514154
Dutzan N, Konkel JE, Greenwell-Wild T, Moutsopoulos NM (2016) Characterization of the human immune cell network at the gingival barrier. Mucosal Immunol 9(5): 1163–1172. https://doi.org/10.1038/mi.2015.136
Dutzan N, Abusleme L, Bridgeman H, Greenwell-Wild T, Zangerle-Murray T, Fife ME, Bouladoux N, Linley H, Brenchley L, Wemyss K, Calderon G, Hong BY, Break TJ, Bowdish DME, Lionakis MS, Jones SA, Trinchieri G, Diaz PI, Belkaid Y, Moutsopoulos NM (2017) On-going mechanical damage from mastication drives homeostatic Th17 cell responses at the oral barrier. Immunity 46(1):133–147. https://doi.org/10.1016/j.immuni.2016.12.010
Dutzan N, Kajikawa T, Abusleme L, Greenwell-Wild T, Zuazo CE, Ikeuchi T, Brenchley L, Abe T, Hurabielle C, Martin D, Morell RJ, Freeman AF, Lazarevic V, Trinchieri G, Diaz PI, Holland SM, Belkaid Y, Hajishengallis G, Moutsopoulos NM (2018) A dysbiotic microbiome triggers TH17 cells to mediate oral mucosal immunopathology in mice and humans. Sci Transl Med 10(463). https://doi.org/10.1126/scitranslmed.aat0797
Easter QT, Fernandes Matuck B, Beldorati Stark G, Worth CL, Predeus AV, Fremin B, Huynh K, Ranganathan V, Ren Z, Pereira D, Rupp BT, Weaver T, Miller K, Perez P, Hasuike A, Chen Z, Bush M, Qu X, Lee J, Byrd KM (2024) Single-cell and spatially resolved interactomics of tooth-associated keratinocytes in periodontitis. Nat Commun 15(1):5016. https://doi.org/10.1038/s41467-024-49037-y
Eren AM, Borisy GG, Huse SM, Mark Welch JL (2014) Oligotyping analysis of the human oral microbiome. Proc Natl Acad Sci USA 111(28):E2875–E2884. https://doi.org/10.1073/pnas.1409644111
Eskan MA, Jotwani R, Abe T, Chmelar J, Lim J-H, Liang S, Ciero PA, Krauss JL, Li F, Rauner M, Hofbauer LC, Choi EY, Chung K-J, Hashim A, Curtis MA, Chavakis T, Hajishengallis G (2012) The leukocyte integrin antagonist Del-1 inhibits IL-17-mediated inflammatory bone loss. Nat Immunol 13(5):465–473. https://doi.org/10.1038/ni.2260
Fellows Yates JA, Velsko IM, Aron F, Posth C, Hofman CA, Austin RM, Parker CE, Mann AE, Nagele K, Arthur KW, Arthur JW, Bauer CC, Crevecoeur I, Cupillard C, Curtis MC, Dalen L, Diaz-Zorita Bonilla M, Diez Fernandez-Lomana JC, Drucker DG, Warinner C (2021) The evolution and changing ecology of the African hominid oral microbiome. Proc Natl Acad Sci USA 118(20). https://doi.org/10.1073/pnas.2021655118
Gaffen SL, Moutsopoulos NM (2020) Regulation of host-microbe interactions at oral mucosal barriers by type 17 immunity. Sci Immunol 5(43). https://doi.org/10.1126/sciimmunol.aau4594
Griffen AL, Beall CJ, Campbell JH, Firestone ND, Kumar PS, Yang ZK, Podar M, Leys EJ (2012) Distinct and complex bacterial profiles in human periodontitis and health revealed by 16S pyrosequencing. ISME J 6(6): 1176–1185. https://doi.org/10.1038/ismej.2011.191
Hajishengallis G (2014) The inflammophilic character of the periodontitis-associated microbiota. Mol Oral Microbiol 29(6):248–257. https://doi.org/10.1111/omi.12065
Hall MW, Wellappuli NC, Huang RC, Wu K, Lam DK, Glogauer M, Beiko RG, Senadheera DB (2023) Suspension of oral hygiene practices highlights key bacterial shifts in saliva, tongue, and tooth plaque during gingival inflammation and resolution. ISME Commun 3(1):23. https://doi.org/10.1038/s43705-023-00229-5
Hasturk H, Kantarci A, Ohira T, Arita M, Ebrahimi N, Chiang N, Petasis NA, Levy BD, Serhan CN, Van Dyke TE (2006) RvE1 protects from local inflammation and osteoclast- mediated bone destruction in periodontitis. FASEB J 20(2):401–403. https://doi.org/10.1096/fj.05-4724fje
Holland SM, DeLeo FR, Elloumi HZ, Hsu AP, Uzel G, Brodsky N, Freeman AF, Demidowich A, Davis J, Turner ML, Anderson VL, Darnell DN, Welch PA, Kuhns DB, Frucht DM, Malech HL, Gallin JI, Kobayashi SD, Whitney AR, Grimbacher B (2007) STAT3 mutations in the hyper-IgE syndrome. N Engl J Med 357(16):1608–1619. https://doi.org/10.1056/NEJMoa073687
Huang S, Li R, Zeng X, He T, Zhao H, Chang A, Bo C, Chen J, Yang F, Knight R, Liu J, Davis C, Xu J (2014) Predictive modeling of gingivitis severity and susceptibility via oral microbiota. ISME J 8(9):1768–1780. https://doi.org/10.1038/ismej.2014.32

Human Microbiome Project, C. (2012). Structure, function and diversity of the healthy human microbiome. Nature, 486(7402), 207-214. doi: https://doi.org/10.1038/nature11234

Iniesta M, Chamorro C, Ambrosio N, Marin MJ, Sanz M, Herrera D (2023) Subgingival microbiome in periodontal health, gingivitis and different stages of periodontitis. J Clin Periodontol 50(7):905–920. https://doi.org/10.1111/jcpe.13793

Jang JY, Song I-S, Baek KJ, Choi Y, Ji S (2017) Immunologic characteristics of human gingival fibroblasts in response to oral bacteria. J Periodontal Res 52(3): 447–457. https://doi.org/10.1111/jre.12410

Kane A, Deenick EK, Ma CS, Cook MC, Uzel G, Tangye SG (2014) STAT3 is a central regulator of lymphocyte differentiation and function. Curr Opin Immunol 28: 49–57. https://doi.org/10.1016/j.coi.2014.01.015

Kerns KA, Bamashmous S, Hendrickson EL, Kotsakis GA, Leroux BG, Daubert DD, Roberts FA, Chen D, Trivedi HM, Darveau RP, McLean JS (2023) Localized microbially induced inflammation influences distant healthy tissues in the human oral cavity. Proc Natl Acad Sci USA 120(41): e2306020120. https://doi.org/10.1073/pnas.2306020120

Kim TS, Ikeuchi T, Theofilou VI, Williams DW, Greenwell-Wild T, June A, Adade EE, Li L, Abusleme L, Dutzan N, Yuan Y, Brenchley L, Bouladoux N, Sakamachi Y, Genomics NN, Computational Biology C, Palmer RJ Jr, Iglesias-Bartolome R, Trinchieri G, Moutsopoulos NM (2024) Epithelial-derived interleukin-23 promotes oral mucosal immunopathology. Immunity 57(4):859–875 e811. https://doi.org/10.1016/j.immuni.2024.02.020

Kinane DF (2001) Causation and pathogenesis of periodontal disease. Periodontol 2000(25):8–20. https://doi.org/10.1034/j.1600-0757.2001.22250102.x

Kinane DF, Lappin DF (2002) Immune processes in periodontal disease: a review. Ann Periodontol 7(1): 62–71. https://doi.org/10.1902/annals.2002.7.1.62

Konkel JE, Moutsopoulos NM (2018) Unique tailoring of Th17 at the gingival oral mucosal barrier. J Dent Res 97(2):128–131. https://doi.org/10.1177/0022034517736030

Kroes I, Lepp PW, Relman DA (1999) Bacterial diversity within the human subgingival crevice. Proc Natl Acad Sci USA 96(25):14547–14552. https://doi.org/10.1073/pnas.96.25.14547

Kumar PS, Griffen AL, Moeschberger ML, Leys EJ (2005) Identification of candidate periodontal pathogens and beneficial species by quantitative 16S clonal analysis. J Clin Microbiol 43(8):3944–3955. https://doi.org/10.1128/JCM.43.8.3944-3955.2005

Lamont EI, Gadkari A, Kerns KA, To TT, Daubert D, Kotsakis G, Bor B, He X, McLean JS (2021) Modified SHI medium supports growth of a disease-state subgingival polymicrobial community in vitro. Mol Oral Microbiol 36(1):37–49. https://doi.org/10.1111/omi.12323

Lee CT, Teles R, Kantarci A, Chen T, McCafferty J, Starr JR, Brito LC, Paster BJ, Van Dyke TE (2016) Resolvin E1 reverses experimental periodontitis and Dysbiosis. J Immunol 197(7):2796–2806. https://doi.org/10.4049/jimmunol.1600859

Listgarten MA (1966) Electron microscopic study of the gingivo-dental junction of man. Am J Anat 119(1): 147–177. https://doi.org/10.1002/aja.1001190109

Listgarten MA (1976) Structure of the microbial flora associated with periodontal health and disease in man. A light and electron microscopic study. J Periodontol 47(1):1–18. https://doi.org/10.1902/jop.1976.47.1.1

Listgarten MA (1994) The structure of dental plaque. Periodontol 2000(5):52–65. https://doi.org/10.1111/j.1600-0757.1994.tb00018.x

Loe H, Theilade E, Jensen SB (1965) Experimental gingivitis in man. J Periodontol (1930) 36:177–187. https://doi.org/10.1902/jop.1965.36.3.177

Loe H, Theilade E, Jensen SB, Schiott CR (1967) Experimental gingivitis in man. 3. Influence of antibiotics on gingival plaque development. J Periodontal Res 2(4): 282–289. https://doi.org/10.1111/j.1600-0765.1967.tb01901.x

Lozupone CA, Knight R (2008) Species divergence and the measurement of microbial diversity. FEMS Microbiol Rev 32(4):557–578. https://doi.org/10.1111/j.1574-6976.2008.00111.x

Maekawa T, Krauss JL, Abe T, Jotwani R, Triantafilou M, Triantafilou K, Hashim A, Hoch S, Curtis MA, Nussbaum G, Lambris JD, Hajishengallis G (2014) Porphyromonas gingivalis manipulates complement and TLR signaling to uncouple bacterial clearance from inflammation and promote dysbiosis. Cell Host Microbe 15(6):768–778. https://doi.org/10.1016/j.chom.2014.05.012

Mark Welch JL, Rossetti BJ, Rieken CW, Dewhirst FE, Borisy GG (2016) Biogeography of a human oral microbiome at the micron scale. Proc Natl Acad Sci USA 113(6):E791–E800. https://doi.org/10.1073/pnas.1522149113

Mark Welch JL, Ramirez-Puebla ST, Borisy GG (2020) Oral microbiome geography: micron-scale habitat and niche. Cell Host Microbe 28(2):160–168. https://doi.org/10.1016/j.chom.2020.07.009

Marsh PD (2003) Are dental diseases examples of ecological catastrophes? Microbiol (Reading) 149(Pt 2): 279–294. https://doi.org/10.1099/mic.0.26082-0

Milner JD, Brenchley JM, Laurence A, Freeman AF, Hill BJ, Elias KM, Kanno Y, Spalding C, Elloumi HZ, Paulson ML, Davis J, Hsu A, Asher AI, O'Shea J, Holland SM, Paul WE, Douek DC (2008) Impaired T(H)17 cell differentiation in subjects with autosomal dominant hyper-IgE syndrome. Nature 452(7188): 773–776. https://doi.org/10.1038/nature06764

Monasterio G, Morales RA, Bejarano DA, Abalo XM, Fransson J, Larsson L, Schlitzer A, Lundeberg J, Das S, Villablanca EJ (2024) A versatile tissue-rolling technique for spatial-omics analyses of the entire murine gastrointestinal tract. Nat Protoc 19(10): 3085–3137. https://doi.org/10.1038/s41596-024-01001-2

Mooney EC, Holden SE, Xia XJ, Li Y, Jiang M, Banson CN, Zhu B, Sahingur SE (2021) Quercetin preserves oral cavity health by mitigating inflammation and microbial dysbiosis. Front Immunol 12:774273. https://doi.org/10.3389/fimmu.2021.774273

Moore LV, Moore WE, Cato EP, Smibert RM, Burmeister JA, Best AM, Ranney RR (1987) Bacteriology of human gingivitis. J Dent Res 66(5):989–995. https://doi.org/10.1177/00220345870660052401

Moutsopoulos NM, Konkel JE (2018) Tissue-specific immunity at the oral mucosal barrier. Trends Immunol 39(4):276–287. https://doi.org/10.1016/j.it.2017.08.005

Moutsopoulos NM, Kling HM, Angelov N, Jin W, Palmer RJ, Nares S, Osorio M, Wahl SM (2012) Porphyromonas gingivalis promotes Th17 inducing pathways in chronic periodontitis. J Autoimmun 39(4):294–303. https://doi.org/10.1016/j.jaut.2012.03.003

Moutsopoulos NM, Konkel J, Sarmadi M, Eskan MA, Wild T, Dutzan N, Abusleme L, Zenobia C, Hosur KB, Abe T, Uzel G, Chen W, Chavakis T, Holland SM, Hajishengallis G (2014) Defective neutrophil recruitment in leukocyte adhesion deficiency type I disease causes local IL-17-driven inflammatory bone loss. Sci Transl Med 6(229):229ra240. https://doi.org/10.1126/scitranslmed.3007696

Moutsopoulos NM, Chalmers NI, Barb JJ, Abusleme L, Greenwell-Wild T, Dutzan N, Paster BJ, Munson PJ, Fine DH, Uzel G, Holland SM (2015) Subgingival microbial communities in leukocyte adhesion deficiency and their relationship with local immunopathology. PLoS Pathog 11(3):e1004698. https://doi.org/10.1371/journal.ppat.1004698

Moutsopoulos NM, Zerbe CS, Wild T, Dutzan N, Brenchley L, DiPasquale G, Uzel G, Axelrod KC, Lisco A, Notarangelo LD, Hajishengallis G, Notarangelo LD, Holland SM (2017) Interleukin-12 and Interleukin-23 blockade in leukocyte adhesion deficiency type 1. N Engl J Med 376(12):1141–1146. https://doi.org/10.1056/NEJMoa1612197

Murakami S, Mealey BL, Mariotti A, Chapple ILC (2018) Dental plaque–induced gingival conditions. J Periodontol 89(S1):S17–S27. https://doi.org/10.1002/JPER.17-0095

Naginyte M, Do T, Meade J, Devine DA, Marsh PD (2019) Enrichment of periodontal pathogens from the biofilms of healthy adults. Sci Rep 9(1):5491. https://doi.org/10.1038/s41598-019-41882-y

Nanci A, Bosshardt DD (2006) Structure of periodontal tissues in health and disease. Periodontol 2000(40): 11–28. https://doi.org/10.1111/j.1600-0757.2005.00141.x

Nowicki EM, Shroff R, Singleton JA, Renaud DE, Wallace D, Drury J, Zirnheld J, Colleti B, Ellington AD, Lamont RJ, Scott DA, Whiteley M (2018) Microbiota and metatranscriptome changes accompanying the onset of gingivitis. mBio 9(2). https://doi.org/10.1128/mBio.00575-18

Ozkavaf A, Aras H, Huri CB, Mottaghian-Dini F, Tozum TF, Etikan I, Yamalik N, Caglayan F (2000) Relationship between the quantity of gingival crevicular fluid and clinical periodontal status. J Oral Sci 42(4): 231–238. https://doi.org/10.2334/josnusd.42.231

Papapanou PN, Sanz M, Buduneli N, Dietrich T, Feres M, Fine DH, Flemmig TF, Garcia R, Giannobile WV, Graziani F, Greenwell H, Herrera D, Kao RT, Kebschull M, Kinane DF, Kirkwood KL, Kocher T, Kornman KS, Kumar PS, Tonetti MS (2018a) Periodontitis: consensus report of workgroup 2 of the 2017 world workshop on the classification of periodontal and peri-implant diseases and conditions. J Clin Periodontol 45(Suppl 20):S162–S170. https://doi.org/10.1111/jcpe.12946

Papapanou PN, Sanz M, Buduneli N, Dietrich T, Feres M, Fine DH, Flemmig TF, Garcia R, Giannobile WV, Graziani F, Greenwell H, Herrera D, Kao RT, Kebschull M, Kinane DF, Kirkwood KL, Kocher T, Kornman KS, Kumar PS, Tonetti MS (2018b) Periodontitis: consensus report of workgroup 2 of the 2017 world workshop on the classification of periodontal and Peri-implant diseases and conditions. J Periodontol 89(Suppl 1):S173–S182. https://doi.org/10.1002/JPER.17-0721

Park OJ, Yi H, Jeon JH, Kang SS, Koo KT, Kum KY, Chun J, Yun CH, Han SH (2015) Pyrosequencing analysis of subgingival microbiota in distinct periodontal conditions. J Dent Res 94(7):921–927. https://doi.org/10.1177/0022034515583531

Payne WA, Page RC, Ogilvie AL, Hall WB (1975) Histopathologic features of the initial and early stages of experimental gingivitis in man. J Periodontal Res 10(2):51–64. https://doi.org/10.1111/j.1600-0765.1975.tb00008.x

Proctor DM, Fukuyama JA, Loomer PM, Armitage GC, Lee SA, Davis NM, Ryder MI, Holmes SP, Relman DA (2018) A spatial gradient of bacterial diversity in the human oral cavity shaped by salivary flow. Nat Commun 9(1):681. https://doi.org/10.1038/s41467-018-02900-1

Proctor DM, Shelef KM, Gonzalez A, Davis CL, Dethlefsen L, Burns AR, Loomer PM, Armitage GC, Ryder MI, Millman ME, Knight R, Holmes SP, Relman DA (2020) Microbial biogeography and ecology of the mouth and implications for periodontal diseases. Periodontol 82(1):26–41. https://doi.org/10.1111/prd.12268

Rosenzweig SD, Holland SM (2004) Phagocyte immunodeficiencies and their infections. J Allergy Clin Immunol 113(4):620–626. https://doi.org/10.1016/j.jaci.2004.02.001

Schincaglia GP, Hong BY, Rosania A, Barasz J, Thompson A, Sobue T, Panagakos F, Burleson JA, Dongari-Bagtzoglou A, Diaz PI (2017) Clinical, immune, and microbiome traits of gingivitis and peri-

implant mucositis. J Dent Res 96(1):47–55. https://doi.org/10.1177/0022034516668847
Schroeder HE, Listgarten MA (1997) The gingival tissues: the architecture of periodontal protection. Periodontol 2000(13):91–120. https://doi.org/10.1111/j.1600-0757.1997.tb00097.x
Schroeder HE, Munzel-Pedrazzoli S (1973) Correlated morphometric and biochemical analysis of gingival tissue. Morphometric model, tissue sampling and test of stereologic procedures. J Microsc 99(3):301–329. https://doi.org/10.1111/j.1365-2818.1973.tb04629.x
Segata N, Izard J, Waldron L, Gevers D, Miropolsky L, Garrett WS, Huttenhower C (2011) Metagenomic biomarker discovery and explanation. Genome Biol 12(6): R60. https://doi.org/10.1186/gb-2011-12-6-r60
Seymour GJ, Greenspan JS (1979) The phenotypic characterization of lymphocyte subpopulations in established human periodontal disease. J Periodontal Res 14(1):39–46. https://doi.org/10.1111/j.1600-0765.1979.tb00216.x
Shehabeldin M, Gao J, Cho Y, Chong R, Tabib T, Li L, Smardz M, Gaffen SL, Diaz PI, Lafyatis R, Little SR, Sfeir C (2024) Therapeutic delivery of CCL2 modulates immune response and restores host-microbe homeostasis. Proc Natl Acad Sci USA 121(36): e2400528121. https://doi.org/10.1073/pnas.2400528121
Silva LM, Brenchley L, Moutsopoulos NM (2019) Primary immunodeficiencies reveal the essential role of tissue neutrophils in periodontitis. Immunol Rev 287(1):226–235. https://doi.org/10.1111/imr.12724
Silva LM, Doyle AD, Greenwell-Wild T, Dutzan N, Tran CL, Abusleme L, Juang LJ, Leung J, Chun EM, Lum AG, Agler CS, Zuazo CE, Sibree M, Jani P, Kram V, Martin D, Moss K, Lionakis MS, Castellino FJ, Moutsopoulos NM (2021) Fibrin is a critical regulator of neutrophil effector function at the oral mucosal barrier. Science 374(6575):eabl5450. https://doi.org/10.1126/science.abl5450
Silva LM, Kim TS, Moutsopoulos NM (2023) Neutrophils are gatekeepers of mucosal immunity. Immunol Rev 314(1):125–141. https://doi.org/10.1111/imr.13171
Simon-Soro A, Tomas I, Cabrera-Rubio R, Catalan MD, Nyvad B, Mira A (2013) Microbial geography of the oral cavity. J Dent Res 92(7):616–621. https://doi.org/10.1177/0022034513488119
Socransky SS, Haffajee AD (2005) Periodontal microbial ecology. Periodontol 2000(38):135–187. https://doi.org/10.1111/j.1600-0757.2005.00107.x
Socransky SS, Haffajee AD, Smith C, Dibart S (1991) Relation of counts of microbial species to clinical status at the sampled site. J Clin Periodontol 18(10): 766–775. https://doi.org/10.1111/j.1600-051x.1991.tb00070.x
Socransky SS, Haffajee AD, Cugini MA, Smith C, Kent RL Jr (1998) Microbial complexes in subgingival plaque. J Clin Periodontol 25(2):134–144. https://doi.org/10.1111/j.1600-051x.1998.tb02419.x
Tanner AC, Kent R Jr, Kanasi E, Lu SC, Paster BJ, Sonis ST, Murray LA, Van Dyke TE (2007) Clinical characteristics and microbiota of progressing slight chronic periodontitis in adults. J Clin Periodontol 34(11):917–930. https://doi.org/10.1111/j.1600-051X.2007.01126.x
Teles R, Sakellari D, Teles F, Konstantinidis A, Kent R, Socransky S, Haffajee A (2010) Relationships among gingival crevicular fluid biomarkers, clinical parameters of periodontal disease, and the subgingival microbiota. J Periodontol 81(1):89–98. https://doi.org/10.1902/jop.2009.090397
Teles F, Martin L, Patel M, Hu W, Bittinger K, Kallan MJ, Chandrasekaran G, Cucchiara AJ, Giannobile WV, Stephens D, Kantarci A (2024) Gingival crevicular fluid biomarkers during periodontitis progression and after periodontal treatment. J Clin Periodontol. https://doi.org/10.1111/jcpe.14061
ter Steeg PF, Van der Hoeven JS, de Jong MH, van Munster PJ, Jansen MJ (1987) Enrichment of subgingival microflora on human serum leading to accumulation of Bacteroides species Peptostreptococci and Fusobacteria. Antonie Van Leeuwenhoek 53(4): 261–272. https://doi.org/10.1007/BF00393933
Theilade E, Wright WH, Jensen SB, Loe H (1966) Experimental gingivitis in man. II. A longitudinal clinical and bacteriological investigation. J Periodontal Res 1:1–13. https://doi.org/10.1111/j.1600-0765.1966.tb01842.x
Tonetti MS, Imboden MA, Lang NP (1998) Neutrophil migration into the gingival sulcus is associated with Transepithelial gradients of Interleukin-8 and ICAM-1. J Periodontol 69(10):1139–1147. https://doi.org/10.1902/jop.1998.69.10.1139
Tsukasaki M, Komatsu N, Nagashima K, Nitta T, Pluemsakunthai W, Shukunami C, Iwakura Y, Nakashima T, Okamoto K, Takayanagi H (2018) Host defense against oral microbiota by bone-damaging T cells. Nat Commun 9(1):701. https://doi.org/10.1038/s41467-018-03147-6
Uriarte SM, Hajishengallis G (2023) Neutrophils in the periodontium: interactions with pathogens and roles in tissue homeostasis and inflammation. Immunol Rev 314(1):93–110. https://doi.org/10.1111/imr.13152
Valm AM (2019) The structure of dental plaque microbial communities in the transition from health to dental caries and periodontal disease. J Mol Biol 431(16): 2957–2969. https://doi.org/10.1016/j.jmb.2019.05.016
Valm AM, Mark Welch JL, Borisy GG (2012) CLASI-FISH: principles of combinatorial labeling and spectral imaging. Syst Appl Microbiol 35(8):496–502. https://doi.org/10.1016/j.syapm.2012.03.004
Valm AM, Oldenbourg R, Borisy GG (2016) Multiplexed spectral imaging of 120 different fluorescent labels. PLoS One 11(7):e0158495. https://doi.org/10.1371/journal.pone.0158495
Wei GX, van der Hoeven JS, Smalley JW, Mikx FH, Fan MW (1999) Proteolysis and utilization of albumin by enrichment cultures of subgingival microbiota. Oral

Microbiol Immunol 14(6):348–351. https://doi.org/10.1034/j.1399-302x.1999.140603.x

Williams DW, Greenwell-Wild T, Brenchley L, Dutzan N, Overmiller A, Sawaya AP, Webb S, Martin D, Genomics NN, Computational Biology C, Hajishengallis G, Divaris K, Morasso M, Haniffa M, Moutsopoulos NM (2021) Human oral mucosa cell atlas reveals a stromal-neutrophil axis regulating tissue immunity. Cell 184(15):4090–4104 e4015. https://doi.org/10.1016/j.cell.2021.05.013

Yang XO, Panopoulos AD, Nurieva R, Chang SH, Wang D, Watowich SS, Dong C (2007) STAT3 regulates cytokine-mediated generation of inflammatory helper T cells. J Biol Chem 282(13):9358–9363. https://doi.org/10.1074/jbc.C600321200

Yost S, Duran-Pinedo AE, Teles R, Krishnan K, Frias-Lopez J (2015) Erratum to: functional signatures of oral dysbiosis during periodontitis progression revealed by microbial metatranscriptome analysis. Genome Med 7(1):111. https://doi.org/10.1186/s13073-015-0231-6

Yost S, Duran-Pinedo AE, Krishnan K, Frias-Lopez J (2017) Potassium is a key signal in host-microbiome dysbiosis in periodontitis. PLoS Pathog 13(6): e1006457. https://doi.org/10.1371/journal.ppat.1006457

Zenobia C, Luo XL, Hashim A, Abe T, Jin L, Chang Y, Jin ZC, Sun JX, Hajishengallis G, Curtis MA, Darveau RP (2013) Commensal bacteria-dependent select expression of CXCL2 contributes to periodontal tissue homeostasis. Cell Microbiol 15(8):1419–1426. https://doi.org/10.1111/cmi.12127

Zijnge V, van Leeuwen MB, Degener JE, Abbas F, Thurnheer T, Gmur R, Harmsen HJ (2010) Oral biofilm architecture on natural teeth. PLoS One 5(2): e9321. https://doi.org/10.1371/journal.pone.0009321

The Role of Short-Chain Fatty Acids Produced by Oral Bacteria in Modulating Oral Immunity

4

Hiromasa Tsuda and Yoshikazu Mikami

Abstract

Dental plaque is a complex biofilm composed of various bacterial species. These bacteria metabolize nutrients, producing high concentrations of short-chain fatty acids (SCFAs) as metabolic by-products, including acetate, propionate, butyrate, isobutyrate, isovalerate, and others. SCFAs are also produced by colonic bacteria. Given that the concentration of SCFAs is significantly higher in the colon than in dental plaque, most of the existing literature focuses on the intestinal context. Following the discovery of SCFA receptors, transporters, and their histone deacetylase (HDAC) inhibitory activities, research on the effects of SCFAs on the immune system has gained increasing attention. Additionally, certain SCFAs induce cell death in periodontal tissues, leading to the release of intracellular molecules that act as damage-associated molecular patterns (DAMPs), thereby influencing the immune system. In this chapter, we first describe oral bacteria that produce SCFAs and compare the concentrations of these SCFAs in the oral cavity with those in the intestinal tract. Then, we provide an overview of the SCFA receptors, the transporters required for their transmembrane passage, and the histone deacetylase inhibitory effects of certain SCFAs. We then discuss how these mechanisms contribute to, or are hypothesized to influence, the effects of SCFAs on various immune cells. After reviewing various forms of cell death capable of triggering DAMP release, we will discuss the potential role of SCFA-induced gingival epithelial cell death in promoting the release of DAMPs and its contribution to gingival inflammation.

Keywords

Short-chain fatty acids (SCFAs) · Cell death · Damage-associated molecular patterns (DAMPs) · Innate immunity · Epithelial cells

4.1 Introduction

Periodontal diseases are chronic inflammatory conditions of gingival tissues, periodontal ligament, alveolar bone, and cementum. The accumulation of diverse species of bacteria on the margins of the gingival sulcus forms dental plaque (also referred to as dental biofilm). Until recently, it was believed that molecules comprising the bacterial surface layer, such as

H. Tsuda (✉)
Department of Biochemistry, Nihon University School of Dentistry, Tokyo, Japan
e-mail: tsuda.hiromasa@nihon-u.ac.jp

Y. Mikami
Division of Microscopic Anatomy, Niigata University Graduate School of Medical and Dental Sciences, Niigata, Japan

N. Rezaei, H. Ziaei (eds.), *Oral Immunology*, Advances in Experimental Medicine and Biology 1492,
https://doi.org/10.1007/978-3-032-03176-1_4

lipopolysaccharides (LPSs), were primarily responsible for periodontal disease. However, emerging evidence suggests that, in addition to LPSs and other surface molecules, bacterial metabolites produced within dental plaque play a major role in the pathogenesis of periodontitis. Dental plaque consists of a vast number of oral bacterial cells embedded in a matrix of supportive substrates, including glucans and fructans. Short-chain fatty acids (SCFAs) are the main metabolites produced by Gram-negative anaerobic periodontopathic bacteria such as *Porphyromonas gingivalis*, *Fusobacterium nucleatum*, and *Treponema denticola*. These bacteria produce high concentrations of SCFAs. SCFAs, also called volatile fatty acids, are aliphatic carboxylic acids with fewer than six carbon atoms, such as formic acid, acetic acid, propionic acid, butyric acid, and valeric acid (Cummings et al. 2004). In addition, isobutyrate and isovalerate, which have hydrocarbon branches, are also included in this category. These metabolites are the main end products of nondigestive carbohydrate and protein fermentation by anaerobic bacteria (Macfarlane and Macfarlane 2003). Since SCFAs have a pKa value of 4.6–4.9, they predominantly exist in their anionic forms (Cummings et al. 1987, 2004; Bergman 1990) under physiological conditions. Therefore, hereafter, short-chain fatty acids will be referred to in their salt forms.

Anaerobic bacteria in the intestinal tracts also produce high concentrations of SCFAs (Cummings et al. 1987; Topping and Clifton 2001; Macfarlane and Macfarlane 2003; Flint et al. 2015). Recent studies have revealed not only the role of SCFAs in maintaining intestinal tract immunity but also the presence of SCFA receptors and transporters (Cassatella 1995; Koh 2016; Sivaprakasam et al. 2017). In addition, certain short-chain fatty acids (SCFAs) contribute to the regulation of gene expression and protein acetylation, a post-translational modification that modulates protein function (Sealy and Chalkley 1978; Eberharter and Becker 2002; Felsenfeld and Groudine 2003; Lin et al. 2015; Thomas and Denu 2021; van der Hee and Wells 2021). This suggests that SCFAs may exert a broad spectrum of biological effects on host tissues.

Research on the effects of SCFAs on immunity, with particular focus on the oral cavity, has been primarily conducted by Kurita-Ochiai et al. (1997, 1998, 2001, 2002, 2003) but is still in its early stages. In this chapter, we will first review SCFAs produced by oral bacteria, discussing SCFA receptors, the inhibitory effects of SCFAs on histone deacetylases (HDACs), and their impact on immunity, as evidenced by studies on intestinal and other immune responses. SCFAs, such as butyrate, have been shown to induce cell death in gingival epithelial cells and neutrophils, leading to the release of intracellular components known as damage-associated molecular patterns (DAMPs) into the extracellular environment. These SCFAs have the potential to influence oral immunity by interacting with the surrounding immune and nonimmune cells within the oral environment. Considering these facts, the mechanism of DAMP release via SCFA-induced cell death and subsequent immunological effects of DAMPs will also be outlined. Finally, we propose the potential immunomodulatory effects of oral bacteria-derived SCFAs on oral immunity.

4.2 Production, Utilization, and Regional Concentration of SCFAs

4.2.1 Production and Fate of SCFAs by the Gut Microbiota

It is estimated that the human intestinal tract accommodates on the order of 10^14 bacteria (Sender et al. 2016). *Firmicutes*, *Bacteroidetes*, *Actinobacteria*, *Proteobacteria*, *Fusobacteria*, and *Verrucomicrobia* are the main phyla in the large intestine (Hou et al. 2022), among which *Bacteroides*, *Firmicutes*, and *Actinobacteria* are the main producers of SCFAs (Duncan et al. 2002; Rey et al. 2010; Louis 2014; Koh 2016). SCFA production largely depends on dietary composition. While dietary carbohydrates serve as the primary substrates for major SCFAs such as acetate, propionate, and butyrate, branched-chain SCFAs, including isobutyrate and isovalerate, are produced through the fermentation of branched-chain amino acids (Smith and Macfarlane 1997). SCFA concentrations in the large intestine are relatively high, although they vary depending on diet, gut microbiota composition, and transit time through the large intestine. SCFA concentrations at the proximal and distal colon have been reported to be 70–140 mM and 20–70 mM, respectively (typically within the millimolar (mM) range of tens to a hundred) (Topping and Clifton 2001; Macfarlane and Macfarlane 2003; Wong et al. 2006). The reduction in SCFA concentration from the proximal to the distal colon is primarily due to increased SCFA absorption by intestinal cells. This process is mediated by solute carrier family 5 member 8 (SLC5A8), a sodium-coupled monocarboxylate transporter (SMCT), and SLC16A1 (Coady 2004; Miyauchi et al. 2004; Thangaraju et al. 2006), a proton-coupled monocarboxylate transporter. These transporters facilitate SCFA uptake into intestinal epithelial cells (Galić et al. 2003). Once absorbed in the intestine, SCFAs are transferred into the bloodstream, while less than 5% are excreted in the feces (Lange et al. 2023; Rekha et al. 2024). Among SCFAs, butyrate is primarily used as an energy source for colon epithelial cells (Rekha et al. 2024). SCFAs that are not metabolized in the colon epithelium are transported into the portal vein, where they serve as energy substrates for the liver (Schönfeld and Wojtczak 2016; Lange et al. 2023; Rekha et al. 2024). However, acetate is not oxidized in the liver, making circulating acetate the most abundant SCFA in the systemic circulation (Lange et al. 2023; Rekha et al. 2024). Reported intravenous concentrations of acetate, propionate, and butyrate in patients are 98–143, 3.8–5.4, and 0.5–3.3 μmol/L, respectively (Lange et al. 2023).

4.2.2 SCFA Production by Bacteria in Dental Plaque

Firmicutes, *Bacteroidetes*, *Actinobacteria*, *Proteobacteria*, *Fusobacteria*, and *Spirochaetes* are the main bacteria in the dental plaque microbiota (Chen and Jiang 2014; Moon and Lee 2016; Rosier et al. 2018; Mark Welch et al. 2019; Hou et al. 2022), among which the first four produce acetate, propionate, and butyrate through saccharide and amino acid fermentation (Koh 2016; Wu et al. 2022). The concentrations of butyrate and propionate in the dental plaque of periodontal disease patients are reported to be 2.6–14 and 9.5–44 mM, respectively (Singer and Buckner 1981; Niederman et al. 1997; Pöllänen et al. 2003). Additionally, SCFA concentrations in the culture supernatants of dental plaque bacteria are quantified and shown in Table 4.1 (Asayama et al. 2024). These reports support that plaque bacteria can produce high concentrations of SCFAs typically within the 1–10 mM range. However, the absorption and consumption of SCFAs within periodontal tissues in direct contact with dental plaque remain to be explored, necessitating further studies.

4.3 SCFA Receptors

Accumulating knowledge about the effects of SCFAs has led to further research into their mechanisms of action, resulting in the identification of several short-chain fatty acid receptors. These receptors belong to the G protein-coupled receptor (GPCR) family, which has seven transmembrane domains. Each SCFA receptor and its known signaling pathways are explained in this subsection, and additional cellular functions are summarized in Table 4.2.

4.3.1 Free Fatty Acid Receptor 2 (FFAR2, GPR43)

FFAR2, also known as GPR43, is a GPCR that couples with multiple G protein subunits including $G_{\alpha q}$ and $G_{\alpha i/o}$ and G_{β}, and G_{γ}. $G_{\alpha q}$ signaling activates phospholipase C and produces inositol 1,4,5-trisphosphates (IP_3) and diacylglycerol (DG). IP_3 binds to ion channel-coupled IP_3 receptors and induces calcium ion release from the endoplasmic reticulum, resulting in upregulation of intracellular calcium concentrations. Signaling through $G_{\alpha i/o}$ inhibits protein kinase A (PKA) activities and reduces cyclic adenosine monophosphate (cAMP) production.

Table 4.1 SCFA concentrations in the supernatant of bacterial cultures of periodontopathic bacteria

	SCFA (mM)							
Periodontopathic bacteria	Butyrate	Propionate	Isobutyrate	Isovalerate	Acetate	Formate	Lactate	Succinate
Porphyromonas gingivalis	24.386	9.065	15.841	13.159	11.802	0.641	0.000	0.963
Fusobacterium nucleatum	32.414	7.365	8.219	0.000	11.452	8.824	0.000	0.202
Prevotella intermedia	0.000	3.729	0.000	1.755	24.065	15.680	0.000	6.996
Prevotella nigrescens	0.000	3.464	0.000	2.737	17.183	8.337	0.000	8.675
Aggregatibacter actinomycetemcomitans	0.000	0.000	0.000	0.000	12.723	2.676	9.257	2.661

Each periodontopathic bacterium was cultured in brain heart infusion broth supplemented with 5 ppm hemin and 0.5 ppm menadione at 37 °C. SCFA concentrations in the bacterial culture supernatants were determined using gas chromatography. SCFA concentrations in uncultured broth were subtracted

Table 4.2 List of each SCFA receptor and its known signaling pathways with additional cellular functions

Receptors						
Names	Aliases	Ligand SCFAs	Gα	2nd messengers	Target organs	Reported functions
FFAR2	GPR43	Formate, acetate, propionate, and butyrate	$G_{\alpha q}$, $G_{\alpha i/o}$	Ca^{2+}↑ and cAMP↓	Intestine	PYY and GLP-1 secretion from L-cells
						Facillitating the developpment
					Immune system	Regulation of leukocyte migration, cytokine secretion, and the colonic T_{reg} cell number during inflammation
					Adipose tissue	Regulation of adipogenesis, adipose insulin signaling, and LPL activity
					Pancreas	Insulin secretion
						Facillitating the developpment
					Liver	Prevention of NAFLD and NASH progression
					Brain	Regulation of microglial morphology and function
FFAR3	GPR41	Formate, acetate, propionate, and butyrate	$G_{\alpha i/o}$	cAMP↓	Small intestine	PYY and GLP-1 secretion from L-cells
					Adipose tissue	Leptin production in adipocytes
					Peripheral nerve	Enhancement NA release, heart rate, and energy expenditure
						Modulation of tympathetic neuronal differentiation
					Pancreas	Inhibition of glucose-stimulated insulin secretion of β-cells
					Bone marrow	DC generation
					Thymus	Promotion of T_{reg} differentiation
					Brain	IGN activation
HCAR2	GPR109A	Butyrate	$G_{\alpha i/o}$	cAMP↓	Adipose tissue	Fatty acid release↓
					Large intestine	NLRP3 inflammasome↑
						Diffrentiation of Treg cells and IL-10 producing T cells↑
					Brain	Microglial response to amyloid deposition↑
OR51E2	Olfr78 *mouse ortholog	Acetate and propionate	$G_{\alpha olf}$, $G_{\alpha s}$	cAMP↑	Kidney	Renin secretion↑
					Smooth muscle	Hypotensice response
						Airway smooth muccle remodeling and proliferation
					Small intestine	PPY secretion↑
OR51E1	Olfr558 *mouse ortholog	Butyrate	$G_{\alpha olf}$, $G_{\alpha s}$	cAMP↑	Prostate	Cell growth of lymph note carcinoma↓
						Cell death marker (p21, p27, p53, and annexin V)↑
					Small intestine	GLP-1 secretion↑

SCFA short-chain fatty acid

FFAR2 is reported to be expressed in the small and large intestines, immune cells (e.g., leukocytes, mast cells, and colonic T-regulatory (T_{reg}) cells), pancreas, liver, and brain microglial cells. The binding affinities between human FFAR2 and its ligands follow the order: acetate = propionate > butyrate formate (Le Poul et al. 2003; Koh 2016; van der Hee and Wells 2021).

4.3.2 Free Fatty Acid Receptor 3 (FFAR3, GPR41)

FFAR3, also referred to as GPR41, couples with only $G_{\alpha i/o}$ as an α-subunit and reduces intracellular cAMP. It has been reported that FFAR3 is expressed in the small and large intestine epithelium, adipose tissue, peripheral nerve, pancreas, bone marrow, thymus, and brain. The binding affinity of human FFAR3 for its ligands is in the order of propionate > butyrate > > acetate (Le Poul et al. 2003; Koh 2016; van der Hee and Wells 2021).

4.3.3 Hydroxycarboxylic Acid Receptor 2 (HCAR2, GPR109A)

HCAR2, also known as GPR109A, exclusively couples with the $G_{\alpha i/o}$ α-subunit and decreases the intracellular cAMP level (Tunaru et al. 2003; Wise et al. 2003; Taggart et al. 2005; Ahmed et al. 2009; Thangaraju et al. 2009; Singh 2014; Macia et al. 2015). This receptor, originally identified as the niacin receptor, has since been found to bind to butyrate and β-hydroxybutyric acid, a ketone body. It was also found that this receptor is not activated by acetate and propionate (Thangaraju et al. 2009; Singh 2014).

4.3.4 Olfactory Receptor Family 51 Subfamily E Member 2 (OR51E2, Olfr78)

OR51E2, a human ortholog of a mouse olfactory chemosensory receptor named Olfr78, is expressed not only in the olfactory epithelium but also in many other tissues (Pluznick et al. 2013; Fleischer 2015; Aisenberg et al. 2016). OR51E2 couples with $G_{\alpha olf}$ in the olfactory epithelium and with $G_{\alpha s}$ in other tissues. It is reported that OR51E2 is expressed in the kidney, smooth muscles, and small intestine. It plays a role in SCFA-mediated signaling beyond the olfactory system, especially in renal and vascular physiology. Acetate and propionate but not butyrate act as ligands for OR51E2, and the binding of these ligands to receptors activates adenylate cyclase and results in an increase in intracellular cAMP concentrations.

4.3.5 Olfactory Receptor Family 51 Subfamily E Member 1 (OR51E1, Olfr558)

Human OR51E1 is a mouse functional ortholog of a mouse olfactory receptor Olfr558 (Saito et al. 2009; Halperin et al. 2019; Pronin and Slepak 2021). Ligands of this G protein-coupled receptor, such as butyrate (Halperin et al. 2019) and medium-chain fatty acids (Pronin and Slepak 2021), activate adenylate cyclase through the $G_{\alpha s}$ protein, resulting in an increase in intracellular cAMP. This receptor is reported to be expressed in prostate cancer cells (Pronin and Slepak 2021) and the small intestine.

4.4 HDAC Inhibitory Effects of SCFAs

In the last subsection, a variety of SCFA receptors were introduced. SCFAs bind to their receptors on the cell surface and initiate signaling pathways that alter cellular phenotypes. In addition to receptor-mediated signaling, SCFAs also exert epigenetic effects by which they alter cellular function and gene expression, particularly through inhibiting histone deacetylases (HDACs). Butyrate was first identified as a compound that can inhibit HDACs, leading to histone hyperacetylation (Boffa et al. 1978; Reeves and Candido 1978; Sealy and Chalkley 1978). Propionate, valerate, and other types of SCFAs, are also identified as HDAC inhibitors, although they express HDAC activities to a lesser degree than that of butyrate (Hinnebusch et al. 2002; Kiefer et al. 2006). SCFAs specifically inhibit activities of class I HDACs (HDAC1, HDAC2, HDAC3, and HDAC8) and class IIa HDACs (HDAC4, HDAC5, HDAC7, and HDAC9) at millimolar concentrations (Davie 2003). It is

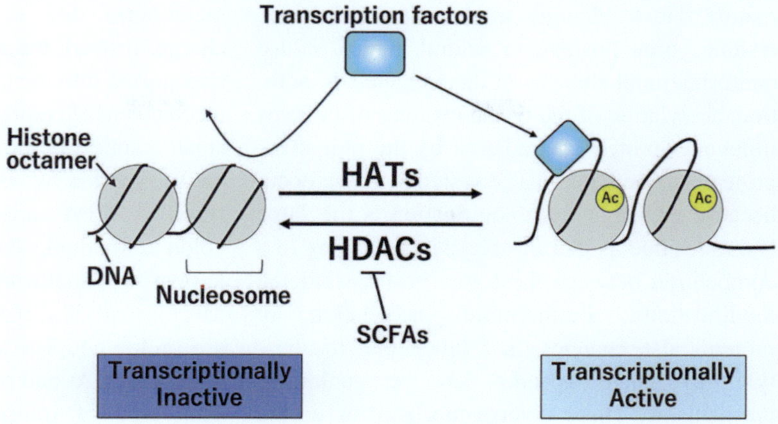

Fig. 4.1 Histone acetylation status controls gene expression

hypothesized that SCFAs inhibit HDACs by directly binding to the zinc ion at their active site (Davie 2003; Weinmann and Ottow 2004). Histone acetylation and deacetylation play a crucial role in controlling gene expressions, which will be discussed in the following subsections. This mechanism involves controlled chemical modifications of histones, independent of DNA sequences, and is referred to as epigenetics (Eberharter and Becker 2002; Felsenfeld and Groudine 2003). These HDAC inhibitory activities of SCFAs also affect the posttranslational modifications of many intracellular proteins. These modifications are believed to play a crucial role in mediating the cellular effects of SCFAs.

4.4.1 HDAC Inhibitory Effects of SCFAs Activate Gene Expression

A nucleosome consists of a histone octamer, which is composed of a tetramer of histones H3 and H4, and two dimers of histones H2A and H2B. DNA wraps around an octameric histone approximately twice to form the nucleosome structure, nucleosomes (Felsenfeld and Groudine 2003). In this status, negatively charged DNAs are tightly wrapped around histones, which have positively charged lysine residues on the N-terminal tails of histones. Acetylation of the lysine residues on the histone tails by histone acetyltransferases (HATs) reduces the interactions between DNA and histones by neutralizing the positive charges of lysine residues, resulting in more relaxed, open chromatin structures (Fig. 4.1). These relaxed chromatin structures allow transcription factors and other regulatory proteins easier access to the enhancer and promoter regions of DNA. Acetylated histones are therefore associated with active transcription (Eberharter and Becker 2002; Felsenfeld and Groudine 2003). As described above, some SCFAs such as butyrate and propionate exert HDAC inhibitory effects. Despite the HDAC suppression by SCFAs, HAT activity remains intact, leading to increased histone acetylation (Fig. 4.1). Thus, SCFAs can activate gene expression by inducing hyperacetylation, subsequently leading to a more open chromatin state.

4.4.2 HDAC Inhibitory Effects of SCFAs Also Increase Acetylation of Intracellular Proteins

HATs acetylate the lysine residues of not only histones but also those of nonhistone proteins. In addition, HDACs also remove acetyl groups from these acetylated lysine residues (Narita et al. 2019; Shvedunova and Akhtar 2022). Therefore, HATs and HDACs are also referred to as lysine acetyltransferases (KATs) and lysine deacetylases (KDACs), respectively. Acetylation of lysine residues on intracellular proteins alters the binding affinity of proteins to their binding partners and also alters the catalytic activities and substrate specificities of the enzymes. This is mainly due to changes in the charge of lysine residues from positive to neutral, which induce conformational changes in the enzymes. In addition, acetylation of the lysine residues of proteins influences protein degradation by the ubiquitin–proteasome and autophagy systems. These occur because the target of ubiquitination is the same lysine residue as that of acetylation, leading to a competition between these two posttranslational modifications. Furthermore, acetylation of proteins also changes their subcellular localization by their specific or yet unknown mechanisms. These diverse mechanisms during acetylation of nonhistone proteins reportedly result in a huge variety of physiological and pathological alterations, including changes in metabolism, autophagy, cell proliferation, repair from DNA damage, and signal transduction.

4.4.3 SCFA Transporters

As HDACs are intracellular enzymes, SCFAs should cross the cell membrane to exert their HDAC inhibitory effect. However, almost all SCFAs exist in their anionic forms (Cummings et al. 1987, 2004; Bergman 1990), which cannot passively diffuse through the cytoplasmic membranes due to their hydrophilicity and charge. Instead, the anionic forms of SCFAs are transported into the cell through carrier-mediated mechanisms (Sivaprakasam et al. 2017), whereas small amounts of the protonated forms of SCFAs are transported by simple diffusion. So far, three types of SCFA carriers have been found in the colon epithelium. The first type, SLC16A1 (also known as monocarboxylate transporter 1 or MCT1) and SLC16A3 (also known as MCT4) are proton-coupled transporters that cotransport anionic SCFAs and protons in the same direction. This type of transporter requires binding of CD147, a ubiquitously expressed chaperone protein, to facilitate their role in SCFA transportation (Kirk et al. 2000). The second type, SLC5A8 (also known as sodium-coupled monocarboxylate transporter 1 or SMCT1) is a sodium-coupled transporter of SCFAs that cotransports anionic SCFAs and sodium in the same direction. The third type, adenosine triphosphate (ATP)-binding cassette superfamily G member 2 (ABCG2), which is also known as breast cancer resistance protein (BCRP), is an SCFA efflux pump that requires ATP for its transportation (Enokizono 2007; Drozdzik 2014). The localization and roles of SCFA transporters are mainly examined in the intestine and cancer cells (Table 4.3), and

Table 4.3 Localization of the SCFA transporter to the colon epithelium and distribution of the SCFA transporter mRNA expressed in the immune system

Transporters		colon epithelium			
Names	Aliases	localization	import/ export	Immune cell types that express mRNA of SCFA transpoters[a]	Substrates
Slc16a1	Mct1	Apical/ basolateral	Import (api)/ export (baso)	Pro-B-cells, mast cells, double-negative thymocytes, naïve CD8$^+$ T-cell, effector CD8$^+$ T cell, NK T-cell, ILC2s	Acetate, propionate, butyrate, lacate, pyruvate, β-hydroxybutyrate
Slc16a3	Mct4	Basolateral	Export	Neutrophols, Basophills, Eosinophills, Ly6 monocytes, Mφ	Lactate, pyruvate
Slc5a8	Smct1	Apical	Import	Thymic medullary epithelial cells, NK T-cells, neutrophils	Acetate, propionate, butyrate, lactate,pyruvate, valerate, nicotinate, β-hydroxybutyrate
Abcg2	Bcrp	Apical	Export	Lymph node endothelial cells, bone marrow cells, NK T-cells, Mφ, ILC3s, double-positive thymocytes	Butyrate

SCFA short-chain fatty acid

[a]Expressions of SCFA transporter mRNA were examined using the Gene Skyline database, which presents the expression profiles of a selected gene, of the Immunological Genome Project (https://www.immgen.org/Databrowser19/DatabrowserPage.html)

the expression of these SCFA transporters in immune cells has not been well-reported at the protein level. Therefore, the expression of each SCFA transporter in immune cells is summarized in Table 4.3. These were examined using the Gene Skyline database from the Immunological Genome Project.

4.5 Effects of SCFAs on the Function of Peripheral Immune Cells

Evidence supporting the effects of SCFAs on peripheral immune cells has been reported. While most of the current research focuses on SCFAs derived from the gut microbiota, it is important to note that high concentrations of SCFAs are also present in dental plaque, which consists of a large number of oral bacteria. This suggests that similar effects may occur in the oral cavity. In this section, we will first introduce the effects of SCFAs on monocytes/macrophages, lymphocytes, and neutrophils.

4.5.1 Monocytes/Macrophages

Monocytes originate from hematopoietic stem cells in the bone marrow. Upon exiting the bloodstream, they differentiate into macrophages in response to macrophage colony-stimulating factor (M-CSF) and into dendritic cells when exposed to granulocyte–macrophage colony-stimulating factor (GM-CSF). These cells are involved in the phagocytosis of pathogenic microorganisms, antigen presentation, and the modulation of inflammatory responses. Therefore, these cells play important roles in innate immunity.

Millard et al. (2002) and Wang et al. (2008) demonstrated in vitro that butyrate treatment reduces functional differentiation of dendritic cells and macrophages. Activated macrophages are often divided into two types: M1 and M2. M1-type macrophages are activated by inflammatory stimuli, such as LPS, and secrete pro-inflammatory cytokines. On the other hand, M2-type macrophages are activated by interleukin (IL)-4 and IL-13 and play a critical role in tissue repair after inflammation. LPS stimulation induces production of nitric oxide, IL-6, and IL-12p40 in macrophages, hallmark indicators of M1 macrophage polarization. Treatment with butyrate or propionate resulted in a dose-dependent reduction of LPS-induced M1 polarization, whereas acetate treatment did not produce this effect (Chang et al. 2014). Since the diminished effects of butyrate and propionate treatments were also observed in cells treated with trichostatin A, an HDAC inhibitor, it can be inferred that the M1-suppressive effects of these SCFAs are attributable to the HDAC inhibitory properties of butyrate and propionate (Chang et al. 2014).

4.5.2 Neutrophils

Neutrophils are the most abundant immune cells in circulation, and serve as the first line of defense against bacterial invasion. These cells phagocytose microorganisms and eliminate them by two mechanisms: bacterial digestion using lysosomal enzymes and reactive oxygen species (ROS) production. In 2004, Brinkmann et al. discovered web-like structures composed of filamentous DNA and intracellular molecules, known as neutrophil extracellular traps (NETs). They demonstrated that these structures can degrade virulence factors and inactivate bacteria (Brinkmann et al. 2004). In addition, activated neutrophils release cytokines such as tumor necrosis factor-α (TNF-α), IL-1β, IL-8, nitric oxide, and eicosanoids (e.g., prostaglandin E_2) to recruit immune cells and activate inflammation (Cassatella 1995). Therefore, neutrophils play a crucial role in the first line of defense against the invasion of microorganisms.

It has been demonstrated that SCFAs, such as acetate, propionate, and butyrate, increase neutrophil disposition to the infection site by upregulation of their recruitment and activation. These SCFAs upregulate L-selectin expression on the surface of neutrophils, facilitating their rolling, adherence to endothelial cells, and

Fig. 4.2 NETosis

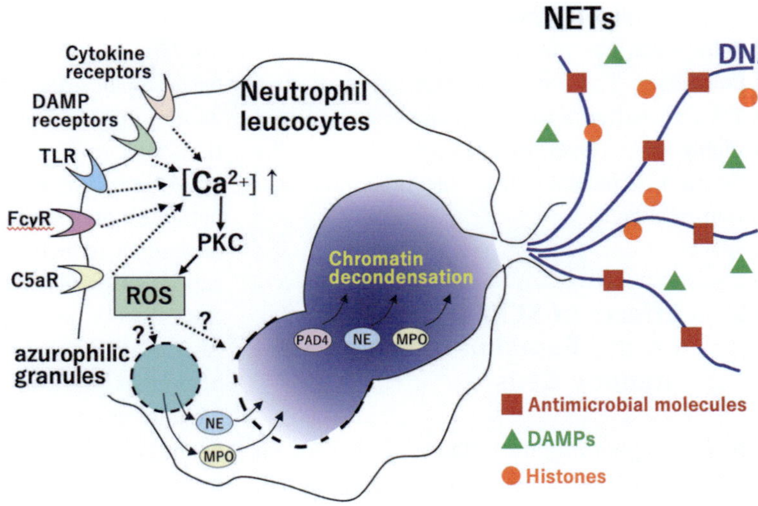

transmigration (Le Poul et al. 2003; Sina et al. 2009; Vinolo et al. 2009). This activation is dependent on FFAR2, a G protein-coupled SCFA receptor (Le Poul et al. 2003; Sina et al. 2009). Dahlstrand Rudin et al. (2021)found that the bacterial culture supernatant of *F. nucleatum* activates human neutrophil chemotaxis, an effect mediated through the FFAR2 receptor. Therefore, SCFAs produced by periodontopathic bacteria may cause neutrophil accumulation in the gingival sulcus. SCFAs also control cytokine production by neutrophils. Butyrate and propionate reduce LPS-induced TNF-α production in human neutrophils (Tedelind 2007). Vinolo et al. (2011) also demonstrated that the HDAC inhibitory activities of propionate and butyrate attenuate LPS-induced upregulation of TNF-α, Cinc-2αβ, and nitric oxide in rat neutrophils.

Neutrophils use other crucial mechanism to combat bacterial invasion, which are referred to as NETs (Yang et al. 2016; Wigerblad and Kaplan 2023). NETs are web-like structures composed of chromatin, including DNA and histones, as well as various antimicrobial molecules (Wigerblad and Kaplan 2023; Yang et al. 2016) (Fig. 4.2). NETs trap invading pathogens, while the release of antimicrobial proteins facilitates bacterial killing. This process is often accompanied by neutrophil cell death, with trapped bacteria remaining trapped within NETs. Recently, Íñiguez-Gutiérrez et al. (2020) have reported that treatment with acetate (> 100 μM), propionate (> 80 μM), and butyrate (> 20 μM) induces human NET formation. Among these, acetate-induced NET formation was attenuated in the presence of ((S)-3-(2-(3-chlorophenyl)phenyl)butanoic acid (CATPB)), a FFAR2 antagonist, genistein, a blocker of $G_{\alpha q}$ protein that acts downstream of FFAR2, and diphenyleneiodonium chloride (DPI), an NADPH oxidase inhibitor. These results suggest that NET formation induced by acetic acid requires the involvement of FFAR2 and the production of reactive oxygen species (ROS) mediated by NADPH oxidase. However, the authors did not investigate the effect of these inhibitors on NET formation induced by propionate or butyrate. Although NET formation has been observed in the periodontal pockets and gingival crevicular fluid of periodontitis patients (Vitkov et al. 2009), a portion of these NETs may be induced by SCFAs, produced by dental plaque bacteria.

4.5.3 Lymphocytes

Lymphocytes primarily play a critical role in adaptive immunity (Kim 2021). Among them, natural killer (NK) cells directly attack virus-infected cells and cancer cells. B lymphocytes differentiate into plasma cells that produce antigen-specific antibodies. Helper T lymphocytes ($CD4^+$ T cells) activate other immune cells including B lymphocytes. Cytotoxic T lymphocytes ($CD8^+$ T cells) recognize and eliminate virus-infected, cancerous, or abnormal cells by distinguishing between self and non-self-antigens. Regulatory T cells (T_{regs}) play a crucial role in suppressing immune responses and preventing the onset of autoimmune diseases, thereby maintaining the balance of the immune system.

Butyrate has been reported to induce p53-independent apoptosis of lymphocytes (Kurita-Ochiai et al. 1997, 1998, 2000). Kurita-Ochiai et al. examined the proapoptotic activities of butyrate using Jurkat T cells and demonstrated that mitochondrial ROS generation, ceramide production, and c-Jun N-terminal kinase activation are required for caspase-3, −8, and − 9 activation, leading to apoptosis execution (Kurita-Ochiai et al. 2001, 2003). These reports suggest that butyrate may suppress the immune response by inducing T-cell apoptosis. Interestingly, the apoptosis of human T cells induced by butyrate was partially inhibited when cocultured with gingival fibroblasts. This protective effect was mediated by the secretion of IL-6 and IL-11, which conferred resistance to butyrate-induced cytotoxicity (Kurita-Ochiai et al. 2002). Therefore, it is likely that fibroblasts in the lamina propria play an important role in protecting the host immune system against the immunosuppressive effects of butyrate-producing bacteria.

4.6 SCFAs' Effects on Innate Immunity Through Cell Death and Subsequent DAMP Release

The release of intracellular molecules into the extracellular space after cell death that involves the destruction of the cell membrane is a well-documented phenomenon. Some of these molecules are known to influence immune defense mechanisms and are referred to as damage-associated molecular patterns (DAMPs). Given reports that short-chain fatty acids (SCFAs) can induce cell death accompanied by membrane destruction, this section will introduce the mechanisms that lead to such cell death and the subsequent release of DAMPs. The role of SCFAs in the induction of gingival epithelial cell death and the release of DAMPs will be discussed in the final subsection.

4.6.1 Damage-Associated Molecular Patterns

In the innate immune system, pathogen-derived molecules are detected through pattern recognition receptors (PRRs) and their corresponding ligands are classified as pathogen-associated molecular patterns (PAMPs) such as LPS, a bacterial endotoxin. Recent research has demonstrated that molecules released from damaged or necrotic cells are similarly detected by innate immune receptors, many of which are also involved in the recognition of PAMPs. These molecules, released from damaged cells, are referred to as damage-associated molecular patterns (DAMPs). The binding of DAMPs to their receptors is involved in physiological inflammatory, immune, and tissue repair responses and is associated with pathological conditions such as chronic inflammation, autoimmune diseases, and sterile inflammation. Table 4.4 shows a list of typical DAMPs and their corresponding receptors and functions. High mobility group box 1 (HMGB1), which is one of the most extensively studied DAMPs, is discussed in detail.

4.6.1.1 High Mobility Group Box 1 (HMGB1)

HMGB1, a 25-kDa protein composed of 215 amino acids in humans, has been identified as a ubiquitously expressed nonhistone nuclear protein that plays a crucial role in the packaging of chromosomal DNA (Andersson and Tracey 2011; Tang et al. 2023). However, under specific conditions, such as during the production of ROS associated with inflammation and tissue damage, this protein can translocate into the cytosol and be secreted into the extracellular environment (Tang et al. 2010; Andersson and Tracey 2011).

Table 4.4 DAMPs and DAMP-sensing receptors in inflammation

DAMPs	DAMP-sensing receptors	Pro-inflammatory functions
HMGB1	RAGE, TLR2, TLR4, TLR9	DCs, Mφ activation
		Pro-inflammatory cytokine production
HMGN1	TLR4	Leucocyte recruitment
		DC maturation
HSPs	CD91, FEEL1, SREC1, TLR2, TLR4, TREM1, TREM2	Monocyte and neutrophil recruitment
		DC maturation
SAP130	MINCLE	Mφ activation,
		Neutrophil recruitment
		Cytokine induction
S100s	RAGE, TLR4	Leukocyte recruitment
		Cytokine induction
Histones	TLR2, TLR4	Pro-inflammatory cytokine production
		Chemokine production
ATP	P2X1R, P2X3R, P2X7R, P2Y2R, P2Y6R, P2Y12R	DCs, Mφ activation
		Inflammasome activation
dsDNA/dsRNA	TLR3, TLR7, TLR9, AIM2	DCs activation
		Inflammasome activation
		Cytokine induction
Histone-DNA	CLEC2D, TLR9, cGAS, AIM2, IFI16	Pro-inflammatory cytokine production

SCFA short-chain fatty acid, *DAMP* damage-associated molecular pattern, *HMGB1* high mobility group box 1, *RAGE* receptor for advanced glycation end products, *TLRs* toll-like receptors, *DC* dendritic cell, *Mφ* macrophages, *HMGN1* high mobility group nucleosome-binding domain 1, *HSPs* heat shock proteins, *CD91* cluster of differentiation 91, *FEEL1* fasciclin, EGF-like, laminin-type EGF-like, and link domain-containing scavenger receptor-1, *SREC1* scavenger receptor expressed by endothelial cells, *TREM1* tagging receptor expressed on myeloid cells 1, *SAP130* sin3A-associated protein 130, *MINCLE* macrophage-inducible C-type lectin, *P2X1R* P2X1 receptor, *P2Y2R* P2Y2 receptor, *dsDNA* double-stranded DNA, *dsRNA* double-stranded RNA, *AIM2* absent in melanoma 2, *CLEC2D* c-type lectin domain family 2 member D, *cGAS* cyclic GMP-AMP synthase, *IFI16* interferon gamma inducible protein 16

HMGB1 is released both actively and passively to extracellular spaces.

Active release of HMGB1 during LPS stimulation of macrophages is reported by Bonaldi (2003; Kwak et al. 2020; Chen et al. 2022). LPS-treated macrophages induce HMGB1 acetylation by HATs at two nuclear localization sequences (NLSs) rich in lysine residues. This acetylation causes HMGB1 dissociation from chromatin, resulting in its translocation into the cytoplasm through exportin-1. In the cytoplasm, acetylated HMGB1 accumulates in lysosomes and is subsequently released into the extracellular environment through lysosomal exocytosis. HMGB1 is also passively released during cell death processes that involve membrane damage, such as necrosis, pyroptosis, necroptosis, and NETosis in addition to late apoptotic cells (Ma 2024).

The secretion and function of HMGB1 are dependent upon its redox state. Human HMGB1 has three cysteine residues at C23, C45, and C106, and the redox status of these residues determines its receptor interactions and activity (Fig. 4.3). There are three types of HMGB1 redox forms: all-thiol type HMGB1 (AT-HMGB1), disulfide-type HMGB1 (DS-HMGB1), and fully oxidized HMGB1. AT-HMGB1 retains reduced thiol groups at all three cysteine residues. It binds to the receptor for advanced glycation end products (RAGE) and induces pro-inflammatory cytokine and chemokine production via p38 mitogen-activated protein kinase (MAPK) and nuclear factor kappa B (NF-κB) signaling to enhance inflammation and immune responses (Bianchi and Manfredi 2007; Harris et al. 2012). AT-HMGB1 also binds to C–X–C motif chemokine receptor type 4 (CXCR4) as its receptor.

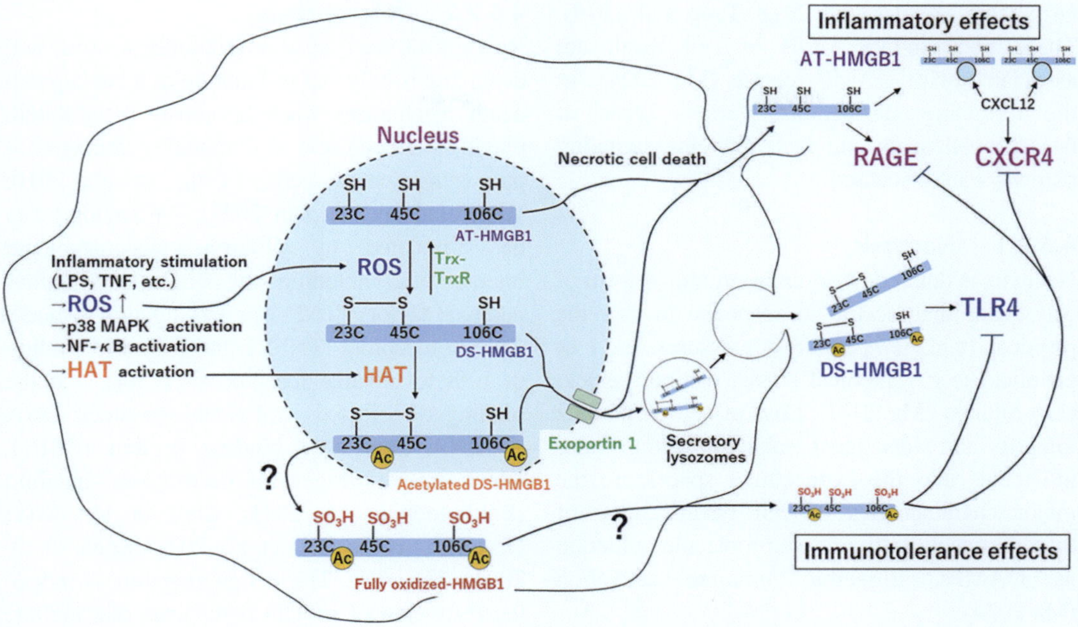

Fig. 4.3 HMGB1 oxidization and different effects. *HMGB1* high mobility group box 1

This binding requires C–X–C motif chemokine ligand 12 (CXCL12) and augments the chemotactic effects of CXCL12 (Schiraldi et al. 2012). DS-HMGB1 has an intramolecular disulfide bond between C23 and C45 and is more prone to translocate from the nucleus to the cytoplasm compared to AT-HMGB1, thereby facilitating extracellular secretion of the molecule (Kwak et al. 2019, 2020). DS-HMGB1 binds to toll-like receptor 4 and induces the production of pro-inflammatory cytokines and chemokines by monocytic cells (Yang et al. 2016). In contrast to the previously characterized pro-inflammatory role of HMGB1, fully oxidized HMGB1, in which all three cysteine residues are fully oxidized, inhibits the cytokine- and chemokine-inducing activities of other HMGB1 redox forms, thereby promoting immune tolerance (Kazama et al. 2008; Kwak et al. 2020). AT-HMGB1 and DS-HMGB1 are released during necrotic cell death and inflammatory stimulation, respectively (Kwak et al. 2020).

4.6.2 Cell Death Mechanisms That Have a Potential Role in Inducing DAMP Release

Kerr et al. (1972) introduced apoptosis as a form of programmed cell death in 1972 and distinguished it from necrosis by its unique morphological features (Tang et al. 2019). Apoptosis, characterized by the retention of the cell membrane and subsequent phagocytosis by phagocytic cells, has a minimal effect on surrounding cells (Amarante-Mendes et al. 2018; Tang et al. 2019). In contrast, necrosis, characterized by accidental cell death and cell membrane rupture, results in the release of cellular contents, thereby activating the surrounding immune cells and triggering inflammation (Amarante-Mendes et al. 2018; Tang et al. 2019). Recent research has reported the existence of controlled forms of necrotic cell death, including necroptosis, pyroptosis, and NETosis, which exhibit controlled molecular mechanisms despite their necrosis-like features

(Amarante-Mendes et al. 2018; Tang et al. 2019). These necrosis-like forms of cell death are associated with DAMP release (Ma 2024). In the following subsections, various types of necrotic cell death and their inducing signaling pathways are discussed.

4.6.2.1 Necrosis

Necrosis is defined as an unprogrammed form of cell death that occurs in response to extreme physical (e.g., temperatures, pressures, etc.) or chemical (e.g., chemical stress, osmotic shock, etc.) injuries (Ma 2024). Loss of cell membrane integrity and subsequent leakage of intracellular molecules into the extracellular spaces are the major characteristics of cell death. Some of these extracellularly released molecules function as DAMPs, triggering immune activation (Ma 2024).

4.6.2.2 Necroptosis

Necroptosis is a type of programmed necrotic cell death and is believed to function as a backup cell death mechanism when apoptosis is inhibited, playing a crucial role in eliminating damaged or pathogen-infected cells (Yang et al. 2016; Wigerblad and Kaplan 2023). Necroptosis can be stimulated by different receptor–ligand interactions, including the binding of tumor necrosis factors (TNFs) or Fas ligands to death domain receptors (TNFR1 and Fas), the binding of LPS to toll-like receptor 4 (TLR4), and the binding of infected viral double-stranded DNA or RNA to Z-DNA-binding protein (ZBP1), which are all triggers of necroptosis signaling (Bertheloot et al. 2021; Choi et al. 2019; Grootjans 2017; Seo et al. 2021; Speir et al. 2021) (Fig. 4.4). These signaling inputs induce the formation of protein complexes that include

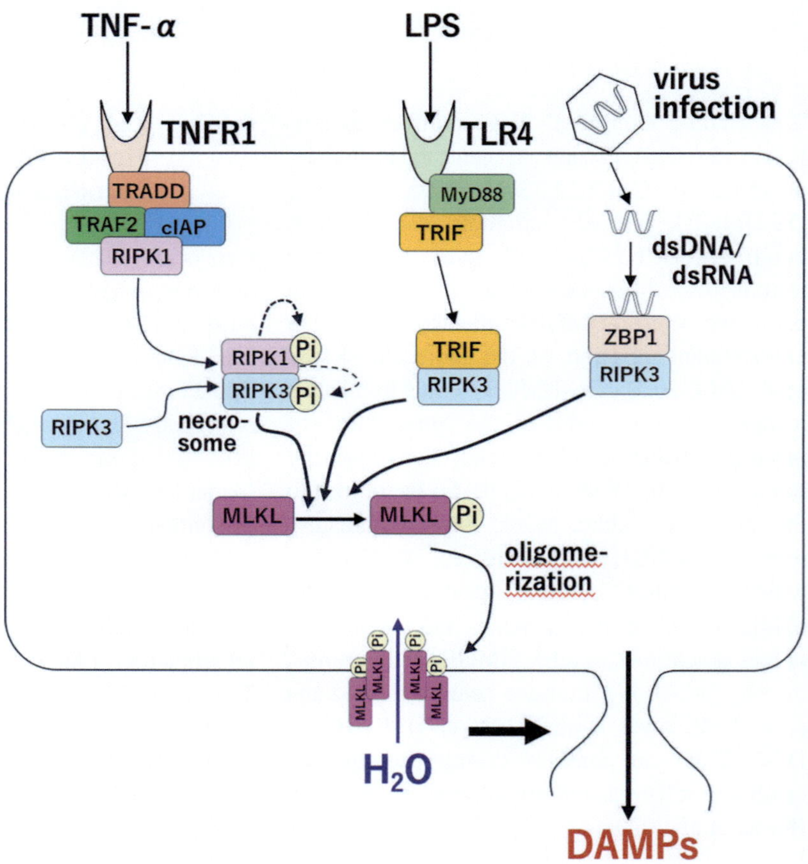

Fig. 4.4 Necroptosis

activated receptor-interacting protein kinase 3 (RIPK3). Activated RIPK3 phosphorylates mixed lineage kinase domain-like protein (MLKL), and this phosphorylation induces conformational changes in MLKL, resulting in its oligomerization and formation of pores on cellular membranes. Pore formation on the cell membrane facilitates the influx of extracellular water, resulting in cell membrane disruption and subsequent release of DAMPs (Huang et al. 2024).

4.6.2.3 Pyroptosis

Pyroptosis is a type of programmed necrotic cell death associated with the processing and secretion of IL-1β and IL-18, and it is typically triggered by intracellular bacterial, viral, or fungal infections (Broz and Dixit 2016; Amarante-Mendes et al. 2018; Tang et al. 2019) (Fig. 4.5). Therefore, pyroptosis plays an essential role in defending against microbial infections. The cell death in this process often occurs in two sequential steps (Broz and Dixit 2016). In the first step, known as the primary phase, NF-κB activation and interferon regulatory factor (IRF) induction through pattern recognition receptor signaling upregulate the production of proteins, which are essential for inflammasome formation. These essential proteins include gasdermin D (GSDMD), pro-IL-1β, and pro-IL-18 (Broz and Dixit 2016; Kovacs and Miao 2017; Amarante-Mendes et al. 2018). In the second step, pattern recognition receptor activation forms inflammasomes by recruiting apoptosis-associated speck-like protein containing a CARD (ASC) and pro-caspase-1 (Broz and Dixit 2016; Kovacs and Miao 2017; Amarante-Mendes et al. 2018). These proteins are essential for subsequent inflammasome activation and pyroptotic cell death. Inflammasome formation results in the cleavage of pro-caspase-1 to activate caspase-1, which, in turn, cleaves GSDMD. The N-terminal fragments of GSDMD undergo oligomerization and are inserted into cellular membranes to form pores. Activated caspase-1 also cleaves pro-IL-1β and pro-IL-18 to produce mature IL-1β and IL-18. These mature cytokines are then extracellularly released through GDSM-formed cell membrane pores; this process is called the canonical pyroptosis pathway (Yang

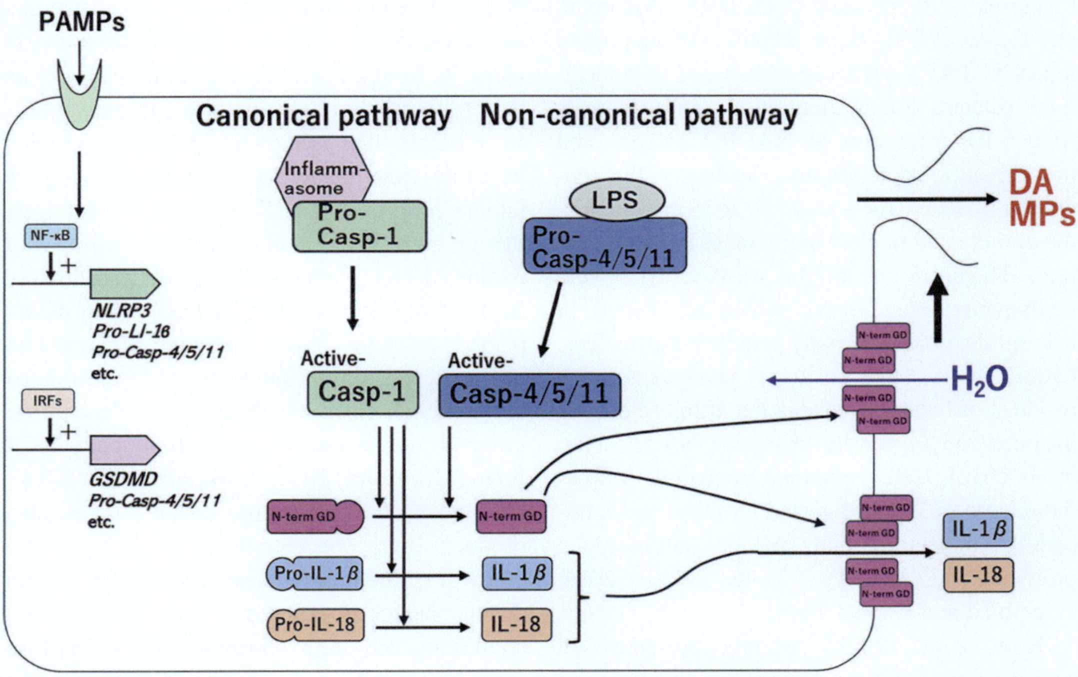

Fig. 4.5 Pyroptosis

et al. 2016). Pore formation in cell membranes allows extracellular water influx, leading to membrane disruption and subsequent DAMP release (Broz and Dixit 2016; Kovacs and Miao 2017; Amarante-Mendes et al. 2018). Alternatively, cytosolic LPS triggers activation of caspase-4 (human), caspase-5 (human), and caspase-11 (mouse), and this activation cleaves GSDMD to form pores on cell membranes, similar to the process described in the canonical pathway. This mechanism is known as the noncanonical pyroptosis pathway, occurring independently of inflammasome activation (Broz and Dixit 2016; Kovacs and Miao 2017; Amarante-Mendes et al. 2018; Huang et al. 2024).

4.6.2.4 NETosis (ETosis)

NETosis is a type of programmed cell death that accompanies the formation of NETs (Yang et al. 2016; Wigerblad and Kaplan 2023). Since this type of cell death has also been observed in other cell types, such as eosinophils, it is sometimes referred to as ETosis (Guimarães-Costa et al. 2012). NETosis can be triggered by the activation of cytokine receptors, TLRs, Fcγ receptors, complement receptors, and DAMPs (Yang et al. 2016; Tang et al. 2019; Wigerblad and Kaplan 2023). These stimulI activate protein kinase C (PKC) and cause an increase in intracellular calcium ion concentration. These signals trigger the formation of NADPH oxidase and the activation of neutrophils, leading to the generation of ROS. As a result of ROS production, the granule and nuclear membranes undergo rupture. Therefore, neutrophil elastase (NE) and myeloperoxidase (MPO), which are stored in azurophilic granules, easily penetrate the nucleus. Citrullination of the positively charged arginine residues of histone weakens the affinity between histones and negatively charged DNA (Leshner et al. 2012). Cell membrane breakdown allows the release of DNA, digested proteins, and intracellular molecules that include antimicrobial proteins and DAMPs (Yang et al. 2016; Wigerblad and Kaplan 2023).

Kim et al. (2023) recently presented an extremely interesting paper on the role of NETosis in the pathology of periodontitis. They demonstrated that neutrophils infiltrate the gingival mucosa in the early stages of disease induction, when they release NETs, leading to periodontal tissue inflammation and alveolar bone destruction in an experimental periodontitis mouse model. They also demonstrated that the extracellularly released histones accompanying NETs upregulate the IL-17/Th17 response. Moreover, they established a significant correlation between clinical attachment loss and NET components detected in the gingival crevicular fluid of periodontitis patients. This underscores the pivotal role of NETosis as the key cell death mechanism in the onset and progression of periodontal disease.

4.6.3 Exploring the Impact of SCFAs on Gingival Tissue Cell Mortality and the Ensuing Release of DAMPs

Bacteria in dental plaques obtain nutrients from different sources, based on the stage and location of plaque formation. In the early stages of dental plaque formation (supragingival plaque), they obtain nutrients from saliva. As plaque matures and forms in the subgingival regions, bacteria use nutrients from gingival crevicular fluid (GCF) as their primary source (Jakubovics 2015) (Fig. 4.6). In contrast, the gingival epithelium obtains nutrients from the tissue fluid in the gingival lamina propria and GCF (Fig. 4.6). Therefore, the gingival epithelium in direct contact with mature dental plaque often faces nutritionally depleted conditions, since bacteria within dental plaque consume large amounts of available nutrients. In addition, smoking exacerbates nutrient depletion since it reduces blood flow to the gingival lamina propria and subsequently GCF flow (Morozumi 2004; Mokeem et al. 2014). Because bacteria in mature dental plaques also produce high concentrations of SCFAs, the gingival epithelium in close contact with mature dental plaques is exposed to both nutrient-poor conditions and high concentrations of SCFAs (Fig. 4.6). Tsuda et al. (2010) demonstrated that treating human gingival epithelial Ca9–22 cells

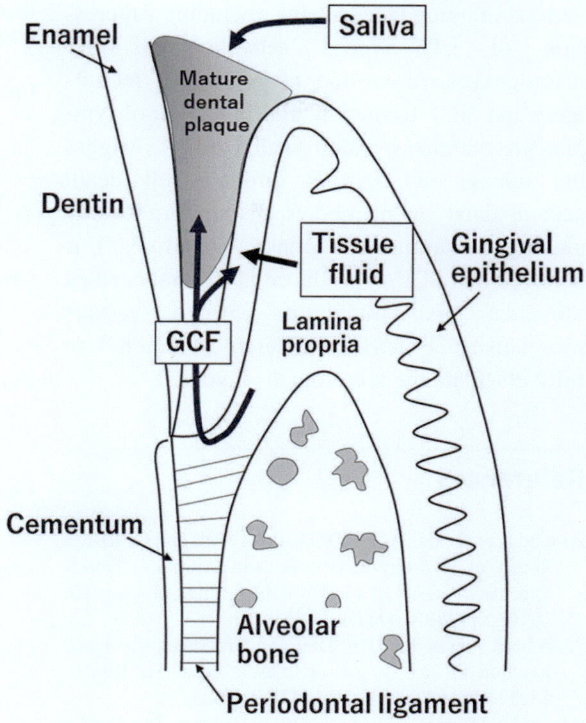

Fig. 4.6 Nutrient supply routes of dental plaque bacteria and the gingival epithelium

with sodium butyrate under low nutrient (1% fetal bovine serum (FBS)) conditions activates AMP-dependent protein kinase (AMPK) and induces autophagy-mediated cell death (Evans et al. 2017). Although partial amount of dead cells are characterized by apoptotic featurs, large amount of dead cells exhibited necrosis-like features with cellular membrane disruption, followed by the release of DAMPs, such as HMGB1 and SAP130 (Fujiwara et al. 2021; Kurosawa et al. 2023; Miyake et al. 2024). Ebe et al. (2011) also demonstrated that butyrate treatment of Ca9–22 cells induced necrosis-like cell death and subsequent HMGB1 release. These results were obtained from in vitro experiments using cell lines. The authors also examined HMGB1 localization using human clinical samples and observed that HMGB1 was localized to the nucleus of healthy gingival epithelium (Ebe et al. 2011). In contrast, in the pocket epithelium, HMGB1 was localized to the cytoplasm and extracellular spaces of gingival epithelial cells (Ebe et al. 2011). Given the presence of butyrate at high concentrations (10 mM) within the periodontal pockets of patients with periodontitis, it is hypothesized that the spatial redistribution of HMGB1 may be attributed to the necrotic death of gingival epithelial cells induced by butyrate. Taken together, the impact of SCFAs produced by oral microorganisms on gingival epithelial tissues may affect immune responses through cell membrane-disrupted epithelial cell death and subsequent release of DAMPs, such as HMGB1. This effect of SCFAs on the gingiva may potentially induce gingivitis.

4.7 Conclusions

In this chapter, we focus on the high concentrations of SCFAs produced by oral bacteria within dental plaques and their impact on oral immunity. SCFAs exert their effects not only through receptor-mediated pathways but also by inhibiting the deacetylation of lysine residues on proteins. This inhibition enhances histone acetylation, thereby contributing to the epigenetic regulation of gene transcription. SCFAs modulate various immune mechanisms, including suppression of LPS-induced activation of M1 macrophages, promotion of neutrophil recruitment and NET formation, and induction of lymphocyte apoptosis. Additionally, SCFAs trigger the release of DAMPs through cell death accompanied by membrane disruption, thereby altering the immune response. Therefore, it is believed that SCFAs produced in the oral cavities influence oral immunity through various mechanisms. Extensive research is required to fully elucidate the mechanisms involved.

References

Ahmed K et al (2009) GPR109A, GPR109B and GPR81, a family of Hydroxy-carboxylic acid receptors. Trends Pharmacol Sci 30(11):557–562. https://doi.org/10.1016/j.tips.2009.09.001

Aisenberg WH et al (2016) Defining an olfactory receptor function in airway smooth muscle cells. Sci Rep 6:38231. https://doi.org/10.1038/srep38231

Amarante-Mendes GP et al (2018) Pattern recognition receptors and the host cell death molecular machinery. Front Immunol 9:2379. https://doi.org/10.3389/fimmu.2018.02379

Andersson U, Tracey KJ (2011) HMGB1 is a therapeutic target for sterile inflammation and infection. Annu Rev Immunol 29:139–162. https://doi.org/10.1146/annurev-immunol-030409-101323

Asayama T et al (2024) Possible roles of short-chain fatty acids produced by oral bacteria in the development of alveolar Osteitis. J Oral Sci 66(2):102–106. https://doi.org/10.2334/josnusd.23-0410

Bergman EN (1990) Energy contributions of volatile fatty acids from the gastrointestinal tract in various species. Physiol Rev 70(2):567–590. https://doi.org/10.1152/physrev.1990.70.2.567

Bertheloot D et al (2021) Necroptosis, pyroptosis and apoptosis: an intricate game of cell death. Cell Mol Immunol 18(5):1106–1121. https://doi.org/10.1038/s41423-020-00630-3

Bianchi ME, Manfredi AA (2007) High-mobility group box 1 (HMGB1) protein at the crossroads between innate and adaptive immunity. Immunol Rev 220:35–46. https://doi.org/10.1111/j.1600-065X.2007.00574.x

Boffa LC et al (1978) Suppression of histone deacetylation in vivo and in vitro by sodium butyrate. J Biol Chem 253(10):3364–3366

Bonaldi T et al (2003) Monocytic cells hyperacetylate chromatin protein HMGB1 to redirect it towards secretion. EMBO J 22(20):5551–5560. https://doi.org/10.1093/emboj/cdg516

Brinkmann V et al (2004) Neutrophil extracellular traps kill bacteria. Science 303(5663):1532–1535. https://doi.org/10.1126/science.1092385

Broz P, Dixit VM (2016) Inflammasomes: mechanism of assembly, regulation and signalling. Nat Rev Immunol 16(7):407–420. https://doi.org/10.1038/nri.2016.58

Cassatella MA (1995) The production of cytokines by polymorphonuclear neutrophils. Immunol Today 16(1):21–26. https://doi.org/10.1016/0167-5699(95)80066-2

Chang PV et al (2014) The microbial metabolite butyrate regulates intestinal macrophage function via histone deacetylase inhibition. Proc Natl Acad Sci 111(6):2247–2252. https://doi.org/10.1073/pnas.1322269111

Chen H, Jiang W (2014) Application of high-throughput sequencing in understanding human oral microbiome related with health and disease. Front Microbiol 5:508. https://doi.org/10.3389/fmicb.2014.00508

Chen R et al (2022) The mechanism of HMGB1 secretion and release. Exp Mol Med 54(2):91–102. https://doi.org/10.1038/s12276-022-00736-w

Choi ME et al (2019) Necroptosis: a crucial pathogenic mediator of human disease. JCI Insight 4(15):128834. https://doi.org/10.1172/jci.insight.128834

Coady MJ et al (2004) The human tumour suppressor gene SLC5A8 expresses a Na+-Monocarboxylate cotransporter. J Physiol 557(Pt 3):719–731. https://doi.org/10.1113/jphysiol.2004.063859

Cummings JH et al (1987) Short chain fatty acids in human large intestine, portal, hepatic and venous blood. Gut 28(10):1221–1227. https://doi.org/10.1136/gut.28.10.1221

Cummings JH et al (eds) (2004) Physiological and clinical aspects of short-chain fatty acids, 1st edn. Cambridge University Press

Dahlstrand Rudin A et al (2021) Short chain fatty acids released by Fusobacterium Nucleatum are neutrophil Chemoattractants acting via free fatty acid receptor 2 (FFAR2). Cell Microbiol 23(8):e13348. https://doi.org/10.1111/cmi.13348

Davie J (2003) Inhibition of histone deacetylase activity by butyrate. J Nutr 133(7):2485S–2493S. https://doi.org/10.1093/jn/133.7.2485S

Drozdzik M (2014) Et al. 'protein abundance of clinically relevant multidrug transporters along the entire length of the human intestine'. Mol Pharm 11(10):3547–3555. https://doi.org/10.1021/mp500330y

Duncan SH et al (2002) Acetate utilization and Butyryl coenzyme a (CoA):acetate-CoA transferase in butyrate-producing bacteria from the human large intestine. Appl Environ Microbiol 68(10):5186–5190. https://doi.org/10.1128/AEM.68.10.5186-5190.2002

Ebe N et al (2011) Pocket epithelium in the pathological setting for HMGB1 release. J Dent Res 90(2):235–240. https://doi.org/10.1177/0022034510385688

Eberharter A, Becker PB (2002) Histone acetylation: a switch between repressive and permissive chromatin. Second in review series on chromatin dynamics.

EMBO Rep 3(3):224–229. https://doi.org/10.1093/embo-reports/kvf053

Enokizono J et al (2007) Regional expression and activity of breast cancer resistance protein (Bcrp/Abcg2) in mouse intestine: overlapping distribution with sulfotransferases. Drug Metab Dispos 35(6):922–928. https://doi.org/10.1124/dmd.106.011239

Evans M et al (2017) Combined effects of starvation and butyrate on autophagy-dependent gingival epithelial cell death. J Periodontal Res 52(3):522–531. https://doi.org/10.1111/jre.12418

Felsenfeld G, Groudine M (2003) Controlling the double helix. Nature 421(6921):448–453. https://doi.org/10.1038/nature01411

Fleischer J et al (2015) Expression of odorant receptor Olfr78 in Enteroendocrine cells of the colon. Cell Tissue Res 361(3):697–710. https://doi.org/10.1007/s00441-015-2165-0

Flint HJ, Duncan SH, Scott KP, & Louis P (2015) Links between diet, gut microbiota composition and gut metabolism Proc Nutr Soc 74:13–22. https://doi.org/10.1017/S0029665114001463

Fujiwara Y et al (2021) Reactive oxygen species-dependent release of damage-associated molecular patterns from human gingival epithelial Ca9-22 cells during butyrate or propionate exposure. J Oral Sci 63(2):195–197. https://doi.org/10.2334/josnusd.20-0411

Galić S et al (2003) The loop between helix 4 and helix 5 in the Monocarboxylate transporter MCT1 is important for substrate selection and protein stability. Biochem J 376(Pt 2):413–422. https://doi.org/10.1042/BJ20030799

Grootjans S et al (2017) Initiation and execution mechanisms of necroptosis: an overview. Cell Death Differ 24(7):1184–1195. https://doi.org/10.1038/cdd.2017.65

Guimarães-Costa AB et al (2012) ETosis: a microbicidal mechanism beyond cell death. J Parasitol Res 2012:929743. https://doi.org/10.1155/2012/929743

Halperin KVL et al (2019) Characterizing novel olfactory receptors expressed in the murine renal cortex. Am J Physiol Renal Physiol 317(1):F172–F186. https://doi.org/10.1152/ajprenal.00624.2018

Harris HE et al (2012) HMGB1: a multifunctional Alarmin driving autoimmune and inflammatory disease. Nat Rev Rheumatol 8(4):195–202. https://doi.org/10.1038/nrrheum.2011.222

Hinnebusch BF et al (2002) The effects of short-chain fatty acids on human colon cancer cell phenotype are associated with histone hyperacetylation. J Nutr 132(5):1012–1017. https://doi.org/10.1093/jn/132.5.1012

Hou K et al (2022) Microbiota in health and diseases. Signal Transduct Target Ther 7(1):135. https://doi.org/10.1038/s41392-022-00974-4

Huang Y et al (2024) DAMP sensing and sterile inflammation: intracellular, intercellular and inter-organ pathways. Nat Rev Immunol 24. https://doi.org/10.1038/s41577-024-01027-3

Íñiguez-Gutiérrez L et al (2020) Physiological concentrations of short-chain fatty acids induce the formation of neutrophil extracellular traps in vitro. Int J Immunopathol Pharmacol 34:2058738420958949. https://doi.org/10.1177/2058738420958949

Jakubovics NS (2015) Saliva as the sole nutritional source in the development of multispecies communities in dental plaque. Microbiol Spect 3(3)

Kazama H et al (2008) Induction of immunological tolerance by apoptotic cells requires caspase-dependent oxidation of high-mobility group Box-1 protein. Immunity 29(1):21–32. https://doi.org/10.1016/j.immuni.2008.05.013

Kerr JFR et al (1972) Apoptosis: a basic biological phenomenon with wide-ranging implications in tissue kinetics. Br J Cancer 26(4):239–257

Kiefer J et al (2006) Mixtures of SCFA, composed according to physiologically available concentrations in the gut lumen, modulate histone acetylation in human HT29 colon cancer cells. Br J Nutr 96(5):803–810. https://doi.org/10.1017/bjn20061948

Kim CH (2021) Control of lymphocyte functions by gut microbiota-derived short-chain fatty acids. Cell Mol Immunol 18(5):1161–1171. https://doi.org/10.1038/s41423-020-00625-0

Kim TS et al (2023) Neutrophil extracellular traps and extracellular histones potentiate IL-17 inflammation in periodontitis. J Exp Med 220(9):e20221751. https://doi.org/10.1084/jem.20221751

Kirk P et al (2000) CD147 is tightly associated with lactate transporters MCT1 and MCT4 and facilitates their cell surface expression. EMBO J 19(15):3896–3904. https://doi.org/10.1093/emboj/19.15.3896

Koh A et al (2016) From dietary fiber to host physiology: short-chain fatty acids as key bacterial metabolites. Cell 165(6):1332–1345. https://doi.org/10.1016/j.cell.2016.05.041

Kovacs SB, Miao EA (2017) Gasdermins: effectors of Pyroptosis. Trends Cell Biol 27(9):673–684. https://doi.org/10.1016/j.tcb.2017.05.005

Kurita-Ochiai T, Fukushima K et al (1997) Butyric acid-induced apoptosis of murine Thymocytes, splenic T cells, and human Jurkat T cells. Infect Immun 65(1):35–41. https://doi.org/10.1128/iai.65.1.35-41.1997

Kurita-Ochiai T, Ochiai K et al (1998) Volatile fatty acid, metabolic by-product of Periodontopathic bacteria, induces apoptosis in WEHI 231 and RAJI B lymphoma cells and splenic B cells. Infect Immun 66(6):2587–2594. https://doi.org/10.1128/IAI.66.6.2587-2594.1998

Kurita-Ochiai T, Ochiai K et al (2000) Butyric-acid-induced apoptosis in murine Thymocytes and splenic T- and B-cells occurs in the absence of P53. J Dent Res 79(12):1948–1954. https://doi.org/10.1177/00220345000790120501

Kurita-Ochiai T, Ochiai K et al (2001) Butyric acid-induced T-cell apoptosis is mediated by Caspase-

8 and -9 activation in a Fas-independent manner. Clin Diagn Lab Immunol 8(2):325–332. https://doi.org/10.1128/CDLI.8.2.325-332.2001

Kurita-Ochiai T et al (2002) Human gingival fibroblasts rescue butyric acid-induced T-cell apoptosis. Infect Immun 70(5):2361–2367. https://doi.org/10.1128/IAI.70.5.2361-2367.2002

Kurita-Ochiai T, Amano S et al (2003) Cellular events involved in butyric acid-induced T cell apoptosis. J Immunol 171(7):3576–3584. https://doi.org/10.4049/jimmunol.171.7.3576

Kurosawa Y et al (2023) Butyrate-treatment induces gingival epithelial cell death in a three-dimensional gingival-connective tissue hybrid co-culture system. J Dental Sci 18(2):893–897. https://doi.org/10.1016/j.jds.2022.08.034

Kwak MS et al (2019) Peroxiredoxin-mediated disulfide bond formation is required for nucleocytoplasmic translocation and secretion of HMGB1 in response to inflammatory stimuli. Redox Biol 24:101203. https://doi.org/10.1016/j.redox.2019.101203

Kwak MS et al (2020) Immunological significance of HMGB1 post-translational modification and redox biology. Front Immunol 11:1189. https://doi.org/10.3389/fimmu.2020.01189

Lange O et al (2023) Short-chain fatty acids-a product of the microbiome and its participation in two-way communication on the microbiome-host mammal line. Curr Obes Rep 12(2):108–126. https://doi.org/10.1007/s13679-023-00503-6

Le Poul E et al (2003) Functional characterization of human receptors for short chain fatty acids and their role in Polymorphonuclear cell activation. J Biol Chem 278(28):25481–25489. https://doi.org/10.1074/jbc.M301403200

Leshner M et al (2012) PAD4 mediated histone Hypercitrullination induces heterochromatin Decondensation and chromatin unfolding to form neutrophil extracellular trap-like structures. Front Immunol 3:307. https://doi.org/10.3389/fimmu.2012.00307

Lin MY, De Zoete MR, Van Putten JP, & Strijbis K (2015) Redirection of epithelial immune responses by short-chain fatty acids through inhibition of histone deacetylases Front Immunol 554. https://doi.org/10.3389/fimmu.2015.00554

Louis P et al (2014) The gut microbiota, bacterial metabolites and colorectal cancer. Nat Rev Microbiol 12(10):661–672. https://doi.org/10.1038/nrmicro3344

Ma M et al (2024) DAMPs and DAMP-sensing receptors in inflammation and diseases. Immunity 57(4):752–771. https://doi.org/10.1016/j.immuni.2024.03.002

Macfarlane S, Macfarlane GT (2003) Regulation of short-chain fatty acid production. Proc Nutr Soc 62(1):67–72. https://doi.org/10.1079/PNS2002207

Macia L et al (2015) Metabolite-sensing receptors GPR43 and GPR109A facilitate dietary fibre-induced gut homeostasis through regulation of the Inflammasome. Nat Commun 6:6734. https://doi.org/10.1038/ncomms7734

Mark Welch JL, Dewhirst FE et al (2019) Biogeography of the oral microbiome: the site-specialist hypothesis. Ann Rev Microbiol 73:335–358. https://doi.org/10.1146/annurev-micro-090817-062503

Millard AL et al (2002) Butyrate affects differentiation, maturation and function of human monocyte-derived dendritic cells and macrophages. Clin Exp Immunol 130(2):245–255. https://doi.org/10.1046/j.0009-9104.2002.01977.x

Miyake K et al (2024) Reactive oxygen species generation required for autophagy induction during butyrate- or propionate-induced release of damage-associated molecular patterns from dying gingival epithelial Ca9-22 cells. J Oral Sci 66(2):125–129. https://doi.org/10.2334/josnusd.23-0421

Miyauchi S et al (2004) Functional identification of SLC5A8, a tumor suppressor down-regulated in colon cancer, as a Na(+)-coupled transporter for short-chain fatty acids. J Biol Chem 279(14):13293–13296. https://doi.org/10.1074/jbc.C400059200

Mokeem SA et al (2014) Influence of smoking on clinical parameters and gingival crevicular fluid volume in patients with chronic periodontitis. Oral Health Dental Manag 13(2):469–473

Moon J-H, Lee J-H (2016) Probing the diversity of healthy oral microbiome with bioinformatics approaches. BMB Rep 49(12):662–670. https://doi.org/10.5483/bmbrep.2016.49.12.164

Morozumi T et al (2004) Smoking cessation increases gingival blood flow and gingival Crevicular fluid. J Clin Periodontol 31(4):267–272. https://doi.org/10.1111/j.1600-051X.2004.00476.x

Narita T et al (2019) Functions and mechanisms of non-histone protein acetylation. Nat Rev Mol Cell Biol 20(3):156–174. https://doi.org/10.1038/s41580-018-0081-3

Niederman R et al (1997) Short-chain carboxylic acid concentration in human gingival Crevicular fluid. J Dent Res 76(1):575–579. https://doi.org/10.1177/00220345970760010801

Pluznick JL et al (2013) Olfactory receptor responding to gut microbiota-derived signals plays a role in renin secretion and blood pressure regulation. Proc Natl Acad Sci USA 110(11):4410–4415. https://doi.org/10.1073/pnas.1215927110

Pöllänen MT et al (2003) Structure and function of the tooth-epithelial Interface in health and disease. Periodontol 31:12–31. https://doi.org/10.1034/j.1600-0757.2003.03102.x

Pronin A, Slepak V (2021) Ectopically expressed olfactory receptors OR51E1 and OR51E2 suppress proliferation and promote cell death in a prostate cancer cell line. J Biol Chem 296:100475. https://doi.org/10.1016/j.jbc.2021.100475

Reeves R, Candido EP (1978) Turnover of histone acetyl groups in cultured cells is inhibited by sodium butyrate. FEBS Lett 91(1):117–120. https://doi.org/10.1016/0014-5793(78)80030-1

Rekha K et al (2024) Short-chain fatty acid: an updated review on signaling, metabolism, and therapeutic effects. Crit Rev Food Sci Nutr 64(9):2461–2489. https://doi.org/10.1080/10408398.2022.2124231

Rey FE et al (2010) Dissecting the in vivo metabolic potential of two human gut Acetogens. J Biol Chem 285(29):22082–22090. https://doi.org/10.1074/jbc.M110.117713

Rosier BT et al (2018) Resilience of the oral microbiota in health: mechanisms that prevent Dysbiosis. J Dent Res 97(4):371–380. https://doi.org/10.1177/0022034517742139

Saito H et al (2009) Odor coding by a mammalian receptor repertoire. Sci Signal 2(60):ra9. https://doi.org/10.1126/scisignal.2000016

Schiraldi M et al (2012) HMGB1 promotes recruitment of inflammatory cells to damaged tissues by forming a complex with CXCL12 and signaling via CXCR4. J Exp Med 209(3):551–563. https://doi.org/10.1084/jem.20111739

Schönfeld P, Wojtczak L (2016) Short- and medium-chain fatty acids in energy metabolism: the cellular perspective. J Lipid Res 57(6):943–954. https://doi.org/10.1194/jlr.R067629

Sealy L, Chalkley R (1978) The effect of sodium butyrate on histone modification. Cell 14(1):115–121. https://doi.org/10.1016/0092-8674(78)90306-9

Sender R et al (2016) Revised estimates for the number of human and bacteria cells in the body. PLoS Biol 14(8): e1002533. https://doi.org/10.1371/journal.pbio.1002533

Seo J et al (2021) Necroptosis molecular mechanisms: recent findings regarding novel necroptosis regulators. Exp Mol Med 53(6):1007–1017. https://doi.org/10.1038/s12276-021-00634-7

Shvedunova M, Akhtar A (2022) Modulation of cellular processes by histone and non-histone protein acetylation. Nat Rev Mol Cell Biol 23(5):329–349. https://doi.org/10.1038/s41580-021-00441-y

Sina C et al (2009) G protein-coupled receptor 43 is essential for neutrophil recruitment during intestinal inflammation. J Immunol 183(11):7514–7522. https://doi.org/10.4049/jimmunol.0900063

Singer RE, Buckner BA (1981) Butyrate and propionate: important components of toxic dental plaque extracts. Infect Immun 32(2):458–463. https://doi.org/10.1128/iai.32.2.458-463.1981

Singh N et al (2014) Activation of Gpr109a, receptor for niacin and the commensal metabolite butyrate, suppresses colonic inflammation and carcinogenesis. Immunity 40(1):128–139. https://doi.org/10.1016/j.immuni.2013.12.007

Sivaprakasam S et al (2017) Short-chain fatty acid transporters: role in colonic homeostasis. In: Comprehensive physiology. Wiley, pp 299–314. https://doi.org/10.1002/cphy.c170014

Smith EA, Macfarlane GT (1997) Dissimilatory amino acid metabolism in human colonic bacteria. Anaerobe 3(5):327–337. https://doi.org/10.1006/anae.1997.0121

Speir M et al (2021) Targeting RIP kinases in chronic inflammatory disease. Biomolecules 11(5):646. https://doi.org/10.3390/biom11050646

Taggart AKP et al (2005) (D)-Beta-Hydroxybutyrate inhibits adipocyte lipolysis via the nicotinic acid receptor PUMA-G. J Biol Chem 280(29):26649–26652. https://doi.org/10.1074/jbc.C500213200

Tang D et al (2010) Endogenous HMGB1 regulates autophagy. J Cell Biol 190(5):881–892. https://doi.org/10.1083/jcb.200911078

Tang D, Kang R, Berghe TV, Vandenabeele P, Kroemer G (2019) The molecular machinery of regulated cell death. Cell Res 29(5):347–364. https://doi.org/10.1038/s41422-019-0164-5

Tang D, Kang R, Zeh HJ, Lotze MT (2023) The multifunctional protein HMGB1: 50 years of discovery. Nat Rev Immunol 23(12):824–841. https://doi.org/10.1038/s41577-023-00894-6

Tedelind S et al (2007) Anti-inflammatory properties of the short-chain fatty acids acetate and propionate: a study with relevance to inflammatory bowel disease. World J Gastroenterol 13(20):2826–2832. https://doi.org/10.3748/wjg.v13.i20.2826

Thangaraju M et al (2006) SLC5A8 triggers tumor cell apoptosis through pyruvate-dependent inhibition of histone deacetylases. Cancer Res 66(24): 11560–11564. https://doi.org/10.1158/0008-5472.CAN-06-1950

Thangaraju M, Cresci GA et al (2009) GPR109A is a G-protein-coupled receptor for the bacterial fermentation product butyrate and functions as a tumor suppressor in colon. Cancer Res 69(7):2826–2832. https://doi.org/10.1158/0008-5472.CAN-08-4466

Thomas SP, & Denu JM (2021) Short-chain fatty acids activate acetyltransferase p300 eLife. https://doi.org/10.7554/eLife.72171

Topping DL, Clifton PM (2001) Short-chain fatty acids and human colonic function: roles of resistant starch and nonstarch polysaccharides. Physiol Rev 81(3): 1031–1064. https://doi.org/10.1152/physrev.2001.81.3.1031

Tsuda H et al (2010) Butyrate, a bacterial metabolite, induces apoptosis and Autophagic cell death in gingival epithelial cells. J Periodontal Res 45(5): 626–634. https://doi.org/10.1111/j.1600-0765.2010.01277.x

Tunaru S et al (2003) PUMA-G and HM74 are receptors for nicotinic acid and mediate its anti-Lipolytic effect. Nat Med 9(3):352–355. https://doi.org/10.1038/nm824

van der Hee B, Wells JM (2021) Microbial regulation of host physiology by short-chain fatty acids. Trends Microbiol 29(8):700–712. https://doi.org/10.1016/j.tim.2021.02.001

Vinolo MA, Rodrigues HG, Hatanaka E, Hebeda CB, Farsky SH, Curi R (2009) Short-chain fatty acids stimulate the migration of neutrophils to inflammatory sites. Clin Sci 117(9):331–338

Vinolo MA, Rodrigues HG, Hatanaka E, Sato FT, Sampaio SC, Curi R (2011) Suppressive effect of

short-chain fatty acids on production of proinflammatory mediators by neutrophils. J Nutr Biochem 22(9):849–855
Vitkov L et al (2009) Extracellular neutrophil traps in periodontitis. J Periodontal Res 44(5):664–672. https://doi.org/10.1111/j.1600-0765.2008.01175.x
Wang B et al (2008) Butyrate inhibits functional differentiation of human monocyte-derived dendritic cells. Cell Immunol 253(1–2):54–58. https://doi.org/10.1016/j.cellimm.2008.04.016
Weinmann H, Ottow E (2004) Recent advances in medicinal chemistry of histone deacetylase inhibitors. Annu Rep Med Chem 39:185–196. https://doi.org/10.1016/S0065-7743(04)39015-9
Wigerblad G, Kaplan MJ (2023) Neutrophil extracellular traps in systemic autoimmune and autoinflammatory diseases. Nat Rev Immunol 23(5):274–288. https://doi.org/10.1038/s41577-022-00787-0
Wise A et al (2003) Molecular identification of high and low affinity receptors for nicotinic acid. J Biol Chem 278(11):9869–9874. https://doi.org/10.1074/jbc.M210695200
Wong JMW et al (2006) Colonic health: fermentation and short chain fatty acids. J Clin Gastroenterol 40(3):235–243. https://doi.org/10.1097/00004836-200603000-00015
Wu L et al (2022) Porphyromonas Gingivalis induces increases in branched-chain amino acid levels and exacerbates liver injury through Livh/Livk. Front Cell Infect Microbiol 12:776996. https://doi.org/10.3389/fcimb.2022.776996
Yang H et al (2016) New insights into neutrophil extracellular traps: mechanisms of formation and role in inflammation. Front Immunol 7:302. https://doi.org/10.3389/fimmu.2016.00302

Part II

Oral-Systemic Health Connections

Periodontitis and Systemic Health: An Interconnected Relationship

5

Daniel Diehl

Abstract

The oral–systemic health connection refers to the bidirectional relationship between oral health and overall systemic health. Emerging research has shown that periodontal diseases, such as gingivitis and periodontitis, are linked to a range of systemic conditions, including cardiovascular disease (CVD), diabetes mellitus, adverse pregnancy outcomes (APOs), and cancer. The pathogenesis of periodontitis involves complex host–microbe interactions, which can lead to periodontal tissue destruction and systemic inflammation. Bacterial translocation and the systemic spread of pro-inflammatory mediators, such as lipopolysaccharides and cytokines, are explored as potential pathways linking oral to systemic health.

In this chapter, the evidence from epidemiological and clinical studies supporting these associations is reviewed, addressing the limitations and controversies in existing research and emphasizing the mechanistic evidence for a possible causal relationship. Additionally, it discusses the impact of improving periodontal health on managing systemic conditions, presenting periodontal treatment as a potential adjunctive therapy to reduce systemic disease risk and severity.

Keywords

Periodontitis · Systemic inflammation · Diabetes · Cancer · Arthritis

Abbreviations

AGEs	Advanced Glycation End Products
APCAs	Anti-Citrullinated Protein Antibodies
APOs	Adverse Pregnancy Outcomes
CVD	Cardiovascular Disease
CRP	C-Reactive Protein
GCF	Gingival Crevicular Fluid
GSK-3β/BH4/ eNOS/Nrf2	Glycogen Synthase Kinase-3 Beta /Tetrahydrobiopterin / Endothelial Nitric Oxide Synthase /Nuclear Factor Erythroid 2-Related Factor 2
HbA1C	Hemoglobin A1c (Glycated Hemoglobin)
HNC	Head and Neck Cancer
ICAM-1	Intercellular Adhesion Molecule 1
IL	Interleukin
JAK3	Janus Kinase 3
LBW	Low Birth Weight

D. Diehl (✉)
Department of Periodontology, School of Dentistry, Faculty of Health, Herdecke University, Witten, Germany
e-mail: Daniel.Diehl@uni-wh.de

N. Rezaei, H. Ziaei (eds.), *Oral Immunology*, Advances in Experimental Medicine and Biology 1492,
https://doi.org/10.1007/978-3-032-03176-1_5

LDL	Low-Density Lipoprotein
LPS	Lipopolysaccharide
MAPK	Mitogen-Activated Protein Kinase
MCP-1	Monocyte Chemoattractant Protein-1
NF-κB	Nuclear Factor Kappa B
NO	Nitric Oxide
OSCC	Oral Squamous Cell Carcinoma
PPAD	Peptidylarginine Deiminase
PAI-1	Plasminogen Activator Inhibitor-1
PBMCs	Peripheral Blood Mononuclear Cells
PD	Periodontal Disease
PD-1	Programmed Death Protein 1
PD-L1	Programmed Death-Ligand 1
PE	Preeclampsia
PGE2	Prostaglandin E2
PI3K	Phosphoinositide 3-Kinase
PTEN	Phosphatase and Tensin Homolog
RA	Rheumatoid Arthritis
RCT	Randomized Controlled Trial
RAGE	Receptor for Advanced Glycation End Products
ROS	Reactive Oxygen Species
STAT3	Signal Transducer and Activator of Transcription 3
TLR	Toll-Like Receptor
TNF-α	Tumor Necrosis Factor Alpha
VCAM-1	Vascular Cell Adhesion Molecule 1

5.1 Introduction

Periodontitis is a chronic inflammatory disease that affects the supporting structures of the teeth, leading to progressive destruction of the periodontal ligament, alveolar bone, and, eventually, tooth loss. Over the past several decades, researchers have increasingly recognized periodontitis as more than just an oral health issue—it is now widely regarded as a condition that can have systemic health implications (Linden and Herzberg 2013). Numerous studies have demonstrated associations between periodontitis and a wide array of systemic diseases, including cardiovascular disease (CVD), diabetes mellitus, rheumatoid arthritis (RA), adverse pregnancy outcomes (APOs), and even cancer (Martínez-García and Hernández-Lemus 2021).

The connection between periodontitis and systemic diseases is still a topic of debate because it is difficult to prove a direct cause-and-effect relationship. The main issue is distinguishing between mere association and actual causation. While numerous studies have shown that people with periodontitis are more likely to develop systemic diseases, it does not necessarily indicate that periodontitis is the direct cause of these diseases. This distinction is important because proving causation has significant impacts on both medical and dental practices, influencing how practitioners can prevent, manage, and treat these conditions. An association refers to a statistical relationship between two conditions or variables where they occur together more often than would be expected by chance. On the other hand, causation implies a direct cause-and-effect relationship, where changes in one condition lead to changes in the other.

In epidemiological studies, researchers often identify associations by comparing the prevalence or incidence of systemic diseases in individuals with and without periodontitis. For example, numerous observational studies have found that individuals with periodontitis are more likely to have cardiovascular disease or diabetes (Jepsen et al. 2018; Sanz et al. 2018). However, these findings do not necessarily indicate that periodontitis is the cause of these diseases. Several other factors could contribute to the observed association, including shared risk factors such as smoking, diet, socioeconomic status, and genetics. Additionally, systemic inflammation or immune dysregulation, common in both conditions, could explain the link between periodontitis and systemic diseases (Hajishengallis 2022).

On the other hand, establishing causality is far more complex. Research must demonstrate that periodontitis directly contributes to the development or progression of a systemic disease rather than merely being associated with it. This typically requires evidence from randomized controlled trials (RCTs) or longitudinal studies that show a temporal relationship—i.e., that periodontitis precedes the development of a systemic disease. It also necessitates a precise biological mechanism that explains how periodontal inflammation could influence systemic health (Winning and Linden 2017). In 1965, Austin Bradford Hill proposed a set of criteria to help distinguish between mere associations and authentic causal relationships in epidemiology. These criteria, known as the Bradford Hill criteria, include strength of association, consistency across studies, specificity, temporality, biological gradient (dose–response relationship), plausibility, coherence, experimental evidence, and analogy. Applying these criteria to the relationship between periodontitis and systemic diseases can help clarify whether the observed associations are likely causal.

However, not all of the Bradford Hill criteria are easily met in the context of periodontitis and systemic diseases. For example, specificity refers to the idea that a particular exposure should lead to a specific outcome. In the case of periodontitis, the condition is associated with a wide range of systemic diseases, making it difficult to pinpoint a single causal relationship. Similarly, while experimental evidence from RCTs supports the idea that periodontal treatment can reduce systemic markers of inflammation, it is unclear whether this leads to long-term improvements in systemic disease outcomes (Martínez-García and Hernández-Lemus 2021). This chapter will give an overview of the current experimental and clinical evidence centered around the association between periodontitis and systemic diseases.

5.2 Immunopathogenesis of Periodontitis: Local and Systemic Implications

5.2.1 Periodontal Inflammation

The pathogenesis of periodontitis is multifactorial, driven by a combination of host susceptibility and a dysbiotic polymicrobial community. Host vulnerability is influenced by genetic, epigenetic, behavioral, and environmental factors that affect immune responses and bacterial colonization in the pathogenic biofilm. The microbial communities involved trigger inflammation and exploit it to extract nutrients for survival. Key virulence factors, such as lipopolysaccharides (LPSs) from Gram-negative bacteria, suppress antimicrobial functions and enhance the pro-inflammatory and tissue-destructive immune response, allowing the bacteria to persist and aggravate the disease (Hajishengallis 2014). Thus, host–microbe interactions in periodontal disease are both complex and dynamic. Toll-like receptors (TLRs), primarily TLR2 and TLR4 present on periodontal cells, recognize endotoxins from periodontal pathogens. Binding to these receptors triggers signaling cascades that promote the production of pro-inflammatory cytokines like interleukin (IL)-1β, IL-6, and prostaglandin E2 (PGE2). In susceptible individuals, a cycle of increasing inflammation, tissue breakdown, and improved ecological conditions for pathogenic bacteria occurs, leading to gradual attachment loss and, ultimately, tooth loss. Based on the pathogenesis of periodontitis, various theories have been developed on how periodontitis may be linked to other diseases (Hajishengallis and Chavakis 2021).

5.2.2 Bacterial Translocation During Periodontitis

Periodontal tissues are highly vascularized. During the inflammatory breakdown of the periodontal barrier, periodontal pathogens or their by-products can enter the bloodstream, a process known as bacteremia. Once in the systemic circulation, these pathogens can potentially disseminate to distant tissues, where they may contribute to various disease processes. For instance, *Porphyromonas gingivalis* has been detected in atherosclerotic plaques, indicating a possible connection between bacterial translocation and the development of cardiovascular diseases. Similarly, periodontal pathogens have been found in the amniotic fluid of women experiencing preterm labor, suggesting that bacterial translocation could be involved in adverse pregnancy outcomes (Jepsen et al. 2018). However, to date, there is limited experimental evidence of their presence in a viable state (Rafferty et al. 2011). Even though periodontal bacteria have developed immune subversive mechanisms to thrive within periodontal pockets, once they are separated from the biofilm and enter the bloodstream, they may become more accessible to the immune system. Furthermore, most periodontal pathogens require anaerobic conditions to survive, which makes it doubtful that they could effectively move to aerobic, high-oxygen tissues. This is supported by the fact that most existing research has not grown periodontal pathogens from the tissue in question; instead, they have only been able to amplify parts of their genome using quantitative polymerase chain reaction (PCR) (León et al. 2007; Mikuls et al. 2014; Siddiqui et al. 2019).

Nevertheless, bacterial components such as outer membrane proteins and endotoxins like lipopolysaccharides (LPSs) may also spread systemically (Geerts et al. 2002). In addition to hematogenous spread, viable periodontal bacteria may travel via nonblood pathways. For instance, swallowed bacteria from the oral cavity may translocate through the oro-digestive tract and colonize other sites, such as the gut. Indeed, the presence of periodontal pathogens has been confirmed in the gut microbiota of patients with inflammatory bowel disease (IBD) (Sun et al. 2024).

5.2.3 Systemic Inflammation and Periodontitis

On an observational level, there is compelling evidence that periodontitis can trigger low-grade systemic inflammation, which may contribute to various comorbidities. Patients with severe periodontitis exhibit higher levels of pro-inflammatory markers, such as IL-1β, IL-6, C-reactive protein (CRP), fibrinogen, and increased neutrophil counts, compared to healthy individuals (D'Aiuto et al. 2004; Nibali et al. 2007; Pink et al. 2015). Conversely, successful periodontal treatment can reduce systemic inflammatory markers (D'Aiuto et al. 2004; Preshaw et al. 2020). The clinical literature states that white blood cell counts in periodontitis can be 25–30% higher than in healthy controls, while systemic acute-phase protein levels may be up to 100% higher (Kweider et al. 1993; Leivadaros et al. 2005; Loos et al. 2000). This periodontitis-driven systemic inflammation is believed to arise from the spread of periodontal bacteria and inflammatory mediators into the bloodstream. As mentioned above, the ulcerated epithelium in periodontal pockets may allow bacteria and their toxins, such as lipopolysaccharides (LPSs) or proteases, to enter the circulation, leading to bacteremia. This can lead to the release of pro-inflammatory immune mediators by red and white blood cells and tissue cells at distant sites, like endothelial cells and hepatocytes. Local pro-inflammatory molecules, such as interleukin-1 (IL-1), tumor necrosis factor-alpha (TNF-α), interleukin-6 (IL-6), and prostaglandin E2 (PGE2), may also enter the bloodstream and contribute to systemic inflammatory effects.

5.2.4 Maladaptive Immunity in Periodontitis

Trained immunity is a process in which innate immune cells, such as monocytes and neutrophils, undergo epigenetic and metabolic reprogramming after an initial exposure to pathogens or inflammatory signals. This results in an enhanced and often maladaptive response to subsequent challenges (Li et al. 2022). Unlike adaptive immunity, which relies on antigen-specific memory, trained immunity is mediated by persistent functional changes in innate immune cells, leading to a heightened inflammatory state. While beneficial in acute infections, such training can become maladaptive in chronic inflammatory conditions like periodontitis, contributing to exaggerated immune responses that cause collateral tissue damage (Hajishengallis et al. 2022).

In periodontitis, maladaptive trained immunity is pivotal in sustaining chronic inflammation and exacerbating tissue destruction. Myeloid cells, mainly monocytes, are central to this process. During periodontitis, these cells can acquire an aberrant trained state due to repeated exposure to periodontal pathogens like *Porphyromonas gingivalis* and other microbial components, leading to a long-lasting pro-inflammatory phenotype. This maladaptive training results in an exaggerated production of pro-inflammatory cytokines, such as IL-1β, TNF-α, and IL-6, which not only drive local tissue destruction but also promote systemic inflammation, even after the initial infection has been defeated (Li et al. 2023).

Neutrophils, the most abundant immune cells in the oral cavity, also exhibit signs of maladaptive training in periodontitis. Traditionally considered short-lived and terminally differentiated, these cells can acquire a prolonged lifespan and altered functional capacity under chronic inflammatory conditions (Kalafati et al. 2023). Trained neutrophils display enhanced degranulation and reactive oxygen species (ROS) production, contributing to the breakdown of periodontal tissues and the amplification of local inflammation. Furthermore, these dysfunctional neutrophils can paradoxically act as protective and pathogenic elements in the periodontal environment, reflecting their dual role in health and disease (Vitkov et al. 2023).

The impact of maladaptive immunity in periodontitis is not confined to the oral cavity. Recent studies have suggested that the epigenetic and metabolic reprogramming of myeloid progenitors in the bone marrow, a process called maladaptive myelopoiesis, can contribute to systemic inflammatory comorbidities such as cardiovascular diseases and diabetes (Li et al. 2022). Periodontitis has been shown to induce systemic changes in hematopoiesis, producing myeloid cells with a heightened pro-inflammatory phenotype. This process can link periodontitis to distant organ inflammation, creating a mechanistic bridge between oral health and systemic diseases (Hajishengallis et al. 2022).

Clonal hematopoiesis, an age-associated condition characterized by the expansion of hematopoietic stem cells carrying specific mutations, has been proposed as another contributor to the systemic effects of maladaptive immunity in periodontitis. In this context, mutated stem cells that give rise to trained myeloid cells may exacerbate both local and systemic inflammation, thereby potentially connecting periodontitis with other age-related inflammatory conditions (Hajishengallis et al. 2022).

5.3 Impact of Periodontal Inflammation on Systemic Diseases

Periodontal inflammation has been associated with several systemic conditions, including cardiovascular disease, diabetes mellitus, rheumatoid arthritis, adverse pregnancy outcomes, and cancer. The evidence linking periodontitis to these diseases is both epidemiological and mechanistic.

5.3.1 Cardiovascular Diseases

Epidemiological studies have consistently shown an association between periodontitis and an increased risk of cardiovascular events. A systematic review and meta-analysis of observational studies including over 80,000 patients found a more than twofold increased risk of coronal heart disease in periodontitis patients (Bahekar et al. 2007). Moreover, a systematic review by Dietrich et al. (2013) identified six cohort studies, which demonstrated that individuals with clinically diagnosed or severe periodontitis face a higher risk of experiencing a first coronary event compared to those with no or less severe periodontitis. However, the risk estimates differ across the studies due to population characteristics and heterogeneous definitions of periodontitis (Mendez et al. 1998; Sim et al. 2008; Xu and Lu 2011).

Atherosclerosis, the buildup of plaques in the arteries, is a critical factor in cardiovascular disease. Inflammation plays a central role in the development of atherosclerosis, and periodontitis contributes to this process by releasing pro-inflammatory cytokines and translocating periodontal pathogens. These factors can promote endothelial dysfunction, a condition in which the inner lining of blood vessels becomes impaired, leading to an increased risk of plaque formation (Endemann and Schiffrin 2004). Moreover, the ATHEROREMO-IVUS (Association To Help Examine Remodeling Effects of Medication and Other Factors in Intravascular Ultrasound) study found an association between high antibody levels against periodontal pathogens and a lower extent of positive atheromatous plaque remodeling (de Boer et al. 2014). Lipopolysaccharides from periodontal pathogens like *Porphyromonas gingivalis* can induce the release of these inflammatory factors via a TLR–nuclear factor kappa B (NF-κB)-dependent axis (Xie et al. 2020). Endotoxins from periodontal pathogens activate a network of inflammatory cytokines, including TNF-α, IL-1, IL-6, and IL-8, and trigger their release. These cytokines infiltrate the endothelial layer, promoting the expression of adhesion molecules such as intercellular adhesion molecule 1 (ICAM-1), vascular cell adhesion molecule 1 (VCAM-1), and P-selectin, further exacerbating endothelial inflammation and dysfunction by promoting leukocyte and monocyte adhesion to the blood vessel wall. The increased endothelial permeability leads to the accumulation of low-density lipoprotein (LDL) in the intima of the vessels. Once attached to the endothelial surface, monocytes and T lymphocytes are recruited to the arterial intima via monocyte chemoattractant protein-1 (MCP-1) and interferon-γ (IFN-γ)-responsive chemokines (Gu et al. 1998). In the arterial intima, these monocytes differentiate into scavenger receptor-expressing macrophages, able to take up substantial quantities of LDL—a process responsible for foam cell formation. Periodontitis is also linked to increased reactive oxygen species (ROS) production and reduced nitric oxide (NO) bioavailability. ROS disrupts the NO signaling pathway, leading to decreased NO levels, which impairs endothelium-dependent relaxation. Studies have shown that *P. gingivalis* infection leads to increased mitochondrial ROS production and decreased salivary NO levels (Bullon et al. 2011; Xu et al. 2021). Furthermore, endothelial cells exposed to periodontitis exhibit impaired vasorelaxation responses due to disruptions in the glycogen synthase kinase-3 beta/tetrahydrobiopterin/endothelial nitric oxide synthase/nuclear factor erythroid 2-related factor 2 (GSK-3β/BH4/eNOS/Nrf2) pathways, suggesting potential therapeutic targets for mitigating periodontitis-induced endothelial dysfunction (Sampath et al. 2021).

As atherosclerotic plaque formation progresses, fibrinogen and fibrin are incorporated, serving as a framework for muscle cell and macrophage proliferation—the major cellular components of atheromatous plaques. Periodontitis patients exhibit increased serum levels of fibrinogen, indicating another mechanistic link to how periodontitis may contribute to atherosclerosis (Ana et al. 2013; Chandy et al. 2017).

Under physiological conditions, plasmin, formed by plasminogen activators (PAs), tissue-type PAs, and urokinase-type PAs, mediates fibrin clearance from the blood vessels. Plasminogen activator inhibitor-1 (PAI-1) constitutes the primary inhibitor of PA. Consequently, an excess of PAI-1 has been identified as an evident risk factor for ischemic cardiovascular disease and myocardial infarction (Hamsten et al. 1985; Thögersen et al. 1998). Moreover, a transgenic mouse model showed that PAI-1-deficient mice exhibit reduced atherosclerotic plaque progression (Eitzman et al. 2000). In the context of periodontitis, it is worth noting that periodontitis patients exhibit elevated levels of serum PAI-1 activity (Bizzarro et al. 2007).

Overall, periodontitis appears to contribute to endothelial dysfunction through inflammatory cytokine release, increased endothelial permeability, and oxidative stress. This leads to a complex interplay of vascular changes that heighten the risk of cardiovascular diseases.

5.3.2 Diabetes Mellitus

Diabetes mellitus is well-recognized as a significant systemic risk factor for developing periodontitis. Diabetic patients not only develop periodontitis more frequently but also experience more severe forms of the disease compared to nondiabetics, particularly as they age. The severity and progression of periodontitis are closely linked to the patient's glycosylated hemoglobin (HbA1C) level. Patients with an HbA1C >9% have up to three times the risk of severe periodontal disease compared to those with lower levels (Preshaw and Bissett 2019). In contrast, individuals with reasonable glycemic control do not have a significantly higher risk of periodontitis than nondiabetics. Chronic hyperglycemia, characteristic of diabetes, leads to an accumulation of advanced glycation end products (AGEs) (Goldin et al. 2006). AGEs bind to the receptor for advanced glycation end products (RAGE), activating pro-inflammatory transcription factors like NF-κB and NADPH oxidases. This activation leads to the dysregulation of inflammatory mediators and increased production of reactive oxygen species, which exacerbate the inflammatory processes and disease progression in periodontitis as well as other diabetes-related complications (Chuah et al. 2013).

Conversely, periodontal inflammation also affects glycemic control in both diabetics and nondiabetics. Severe periodontitis is associated with higher HbA1C and fasting blood glucose levels. Additionally, prediabetics with periodontitis have up to a 33% increased risk of developing type 2 diabetes compared to those with healthy gingival tissue. Conversely, meta-analyses of clinical studies have found a significant reduction of HbA1C levels, ranging from 0.37 to 0.50%, following comprehensive periodontal treatment (Liew et al. 2013; Wang et al. 2014). However, mechanistic evidence for this relationship remains limited. Periodontitis appears to increase systemic inflammation, as evidenced by elevated levels of peripheral inflammatory mediators such as IL-1β, TNF-α, IL-6, and CRP. At a cellular level, these mediators significantly inhibit insulin receptor signaling. TNF and IL-1, for example, have been shown to reduce the autophosphorylation of the insulin receptor, leading to a dramatic decrease in the phosphorylation of insulin receptor substrate-1 (IRS-1), which leads to decreased insulin-mediated glucose uptake in fat and muscle cells, complicating glycemic control in affected diabetics (He et al. 2006; Hotamisligil et al. 1994; Jager et al. 2007). Unsurprisingly, diabetes-related complications, such as nephropathy, proteinuria, diabetic foot syndrome, and retinopathy, are also significantly associated with the presence and severity of periodontitis (Preshaw and Bissett 2019). Furthermore, a recent cross-sectional cohort study in the United States has revealed a 27% increased risk of all-cause mortality in diabetics with periodontitis (Li et al. 2024).

5.3.3 Rheumatoid Arthritis

Recent research has established significant associations between periodontitis and various rheumatic diseases, with rheumatoid arthritis (RA) being the most extensively studied. RA is a systemic autoimmune disease affecting roughly 0.5% of the global population, characterized by chronic joint inflammation, pain, disability, and increased mortality (Almutairi et al. 2021). The disease's exact cause is unknown, but it is believed to result from genetic and environmental factors that disrupt the function of the immune system at mucosal surfaces, such as the gut and periodontium (Holers et al. 2018; Möller et al. 2020). This disruption triggers an autoimmune response by producing rheumatoid factor and anti-citrullinated protein antibodies (ACPAs). ACPAs bind to citrullinated proteins in the joints, forming immune complexes with rheumatoid factor that activate synovial macrophages and dendritic cells, releasing pro-inflammatory cytokines and enzymes (Uysal et al. 2010).

Peptidylarginine deiminase (PPAD) enzymes, especially PAD2 and PAD4, catalyze the conversion of arginine residues in proteins to citrulline, a process known as citrullination. This posttranslational modification generates neo-epitopes recognized by ACPAs, thereby promoting immune responses that drive RA pathology (Klareskog et al. 2013). The presence of PAD enzymes in inflamed synovial tissue and their elevated activity in RA patients suggests their critical role in initiating and perpetuating the autoimmune response in RA. The autoimmune response in RA begins with the presentation of citrullinated peptides by antigen-presenting cells to $CD4^+$ T cells, which then activate B cells to produce ACPAs. These ACPAs form immune complexes with citrullinated proteins in the synovial joints, triggering complement activation, pro-inflammatory cytokine release (e.g., TNF-α, IL-6), and the recruitment of inflammatory cells. This inflammatory milieu results in synovial hyperplasia, cartilage destruction, and bone erosion, ultimately maintaining a feedforward loop of inflammation, joint damage, and further generation of citrullinated autoantigens (Curran et al. 2020).

Utilizing data from the Third National Health and Nutrition Examination Survey (NHANES III), De Pablo et al. (2008) demonstrated a significant association between RA and PD in the US population. Moreover, epidemiological evidence suggests that patients with long-standing, active RA have a significantly higher incidence of periodontitis than healthy individuals (Dissick et al. 2010; Pischon et al. 2008).

The most substantial mechanistic evidence linking periodontitis to RA is the discovery of (PPADs) from *P. gingivalis*. Like their human counterpart, PPADs generate citrullinated peptides and modulate inflammation by targeting molecules like bradykinin, C5a, and epidermal growth factor, thereby exacerbating local inflammation in the periodontium (Gawron et al. 2017; Quirke et al. 2014). Unlike human PAD enzymes, PPADs can citrullinate C-terminal arginine residues, generating novel epitopes that trigger immune responses. Essential RA-related proteins such as fibrinogen, α-enolase, and vimentin are C-terminally citrullinated by PPADs and gingipains (Laugisch et al. 2016). Both enzymes are detectable in gingival crevicular fluid (GCF), suggesting that they have access to substrates essential for developing autoantibodies in RA patients. Further evidence for this connection is supported by the observation that PPAD activity is increased in both RA patients and individuals with periodontitis, suggesting a shared pathogenic mechanism. This dual role of PPADs—promoting citrullination and enhancing inflammation—further supports their involvement in linking periodontitis to RA.

Another periodontal pathogen, *Aggregatibacter actinomycetemcomitans*, also has mechanistic ties to RA. This bacterium secretes a leukotoxin, which induces hypercitrullination in neutrophils. Moreover, the leukotoxin induces extracellular trap formation, releasing the hypercitrullinated cargo at periodontitis sites. Antibodies against *A. actinomycetemcomitans* and its leukotoxin are strongly associated with ACPAs and rheumatoid factor in RA patients, further supporting a connection between periodontitis and RA (Konig et al. 2016; Mukherjee et al. 2018).

5.3.4 Adverse Pregnancy Outcomes

Adverse pregnancy outcomes (APOs) can manifest in several ways, including preterm birth, preeclampsia (PE), and low birth weight (LBW). Preterm birth, defined as delivery before 37 weeks of gestation, is the leading cause of perinatal death and morbidity. With around 1 in 10 pregnancies, the preterm delivery rate represents not only an emotional burden on families but also a large economic burden (Goldenberg et al. 2008). The costs of prematurity are significant, particularly in terms of neonatal intensive care and long-term health complications. In the United States, the annual cost of preterm birth is estimated to be $36 billion (Figuero et al. 2020).

Preeclampsia (PE) is another major APO, characterized by maternal hypertension and proteinuria after the 20th week of gestation. Affecting 5–8% of women, preeclampsia is the third cause of maternal death and is responsible for 30% of preterm births, according to WHO

(Khan et al. 2006; MacKay et al. 2005). With multiple subtypes, rather than a singular disease, it constitutes a complex syndrome and its exact causes remain unclear. Its pathogenesis is centered around abnormal placentation, combined with a set of genetic and environmental factors. PE is often viewed as a two-step process, beginning with abnormal placental development early in pregnancy, which leads to maternal vascular dysfunction. This dysfunction triggers an imbalance in angiogenic factors, increased inflammation, and heightened vascular resistance, ultimately resulting in elevated maternal blood pressure and the hallmark symptoms of PE (Ma'ayeh et al. 2020).

Low birth weight (LBW) is defined as birth weight below 2500 grams. This condition is a significant public health concern and a key indicator of infant mortality and morbidity. More than 20 million newborns are affected by LBW every year, with incidence approximating 15% (Okwaraji et al. 2024). Maternal factors such as age, nutritional status, and stress have been identified as contributing to LBW, though a substantial number of cases remain unexplained (Blencowe et al. 2019).

Since an early study by Offenbacher et al. (1996) revealed periodontitis as a risk factor for LBW, the link between periodontitis and adverse pregnancy outcomes has been extensively researched. As of today, numerous studies, including at least six systematic reviews with meta-analyses, have identified a strong correlation between periodontitis and outcomes like preterm birth, low birth weight, and preeclampsia (Sanz et al. 2013; Xiong et al. 2006). Although these studies often have methodological limitations—including inconsistent definitions of periodontitis and APOs, failure to control for confounding factors, and inadequate sample size or data analysis—the findings consistently suggest that periodontitis may increase the risk of adverse pregnancy outcomes. This increased risk is likely mediated by systemic inflammation, bacterial dissemination, and resulting hormonal disruptions.

Fusobacterium nucleatum, a Gram-negative anaerobic bacterium commonly recognized as a periodontal pathogen, has the ability to colonize the placenta. This is facilitated by outer membrane proteins like *Fusobacterium* adhesion A (FadA), fibroblast activation protein 2 (Fap2), and radiation genes (RadD) (Saadaoui et al. 2021). It has therefore been identified in various placental and fetal tissues linked to adverse pregnancy outcomes, including amniotic fluid, fetal membranes, cord blood, and neonatal gastric aspirates (Figuero et al. 2020; Gonzales-Marin et al. 2013; Han et al. 2009). The putative mechanism linking *F. nucleatum* dissemination to preterm birth is mediated by TLRs. In the mouse placenta, the pathogen is detected by TLR4, leading to neutrophil infiltration and preterm fetal death days after infection (Vander Haar et al. 2018).

A systemic hallmark feature of pregnancy is the elevation of systemic inflammatory mediators like TNF-α, IL-1β, and IL-6 in maternal serum, myometrium, and amniotic fluid. Together, these mediators contribute to the rise in prostaglandin E2 (PGE2), which ultimately helps trigger labor. However, when these inflammatory mediators increase excessively within the fetal placental unit, they can lead to premature membrane rupture and uterine contractions, causing preterm birth or miscarriage (Latorre Uriza et al. 2018). Interestingly, gingival crevicular fluids of women with preterm birth exhibit higher amounts of these cytokines than those of women with full-term birth (Perunovic et al. 2016). Since these mediators may also accumulate systemically in periodontitis patients, it is reasonable to suggest that these mediators may act as a link between periodontal disease and adverse pregnancy outcomes.

5.3.5 Neurodegenerative Diseases

The potential association between periodontitis and neurodegenerative diseases has garnered increasing attention over the past decade, driven by emerging research that connects systemic inflammation and microbial dysbiosis to cognitive decline and other neurodegenerative conditions. Various studies have sought to understand the pathways through which periodontal disease might contribute to neurodegenerative disorders such as Alzheimer's disease (AD), Parkinson's disease, and other forms of cognitive impairment, proposing the oral–gut–brain (GBA) axis as a central mechanism in this interaction (Doifode et al. 2021; Lu et al. 2022).

Several epidemiological studies have provided evidence supporting the association between periodontitis and neurodegenerative diseases. For instance, a large-scale population-based study from South Korea found that individuals with a history of periodontitis had a significantly higher risk of developing neurodegenerative diseases such as dementia and Alzheimer's disease (Kim et al. 2022). Another systematic review concluded that while there is substantial evidence of an association, the causative link between periodontitis and neurodegeneration remains to be definitively established due to potential confounding factors, such as age, lifestyle, and genetic predispositions (Alvarenga et al. 2021). However, a case–control study by Leira et al. (2020) demonstrated elevated levels of neurodegeneration markers, such as tau protein and amyloid-β, in individuals with periodontitis, further reinforcing the hypothesis that periodontitis may act as a distal trigger of neurodegeneration. The study found that these markers were significantly higher in periodontitis patients, indicating that periodontal inflammation might contribute to the early pathological changes observed in neurodegenerative conditions (Leira et al. 2020).

The systemic inflammatory response to periodontitis is characterized by cytokines and pro-inflammatory mediators like IL-6, TNF-α, and CRP, which are also implicated in neuroinflammation, a common feature of neurodegenerative diseases (Ng et al. 2018). As animal studies suggest, systemic inflammation triggered by bacterial LPS induces the expression of the same cytokines in rodent brains. More importantly, similar neuroinflammation can be triggered by ligature-induced periodontitis in mice, indicating that the periodontitis-associated inflammatory response may lead to measurable neuroinflammation (Furutama et al. 2020; Godbout et al. 2005).

On the other hand, keystone periodontal pathogens have been repeatedly found in post-mortem histopathological examinations of deceased Alzheimer's patients (Poole et al. 2013; Riviere et al. 2002). In fact, infection with *Fusobacterium nucleatum* or *Porphyromonas gingivalis* leads to increased pro-inflammatory cytokine expression associated with increased amyloid-β deposition in rodent brains, mediated via the TLR4/myeloid differentiation primary response 88/c-Jun N-terminal kinase (TLR4/MYD88/JNK) pathway (Dominy et al. 2019; Ilievski et al. 2018; Olsen 2021; Wu et al. 2022).

Another proposed mechanism linking periodontitis to neurodegenerative diseases is the disruption of the oral–gut–brain axis (GBA). The GBA is a complex, bidirectional communication network involving the central nervous system, the autonomic nervous system, the enteric nervous system, and the hypothalamic–pituitary–adrenal axis. It links the brain to the gastrointestinal system, integrating neural, hormonal, and immune pathways. The gut microbiota plays a key role in this communication, influencing brain function through hormonal signaling and immune responses. Dysbiosis, or an imbalanced gut microbiota, is linked to various neuropsychiatric conditions like anxiety, depression, and autism (Carabotti et al. 2015).

Oral dysbiosis caused by periodontal pathogens can alter the composition of the gut microbiome, leading to gut dysbiosis, which may, in turn, impact brain health. The altered gut microbiota composition and the disrupted gut barrier function may contribute to systemic inflammation and enhanced neuroinflammation in the brain. This indicates that the periodontitis-related oral microbiota can modulate AD pathology via the gut–brain axis (Lu et al. 2022). This is consistent with findings from human studies that link gut dysbiosis to conditions such as Alzheimer's disease, suggesting that the oral–gut–brain axis may be a crucial pathway for understanding how periodontitis contributes to neurodegenerative diseases (Bercik et al. 2012).

The growing body of research suggests a complex interplay between periodontal disease and neurodegenerative conditions, mediated through systemic inflammation, microbial dysbiosis, and the oral–gut–brain axis. Although the exact causal mechanisms are still under investigation, current evidence supports a bidirectional relationship in which periodontal health may influence neurodegenerative processes, and vice versa. Addressing periodontal health could, therefore, be a potential strategy for reducing the risk or slowing the progression of neurodegenerative diseases, warranting further research and clinical consideration.

5.3.6 Cancer

Accounting for nearly 25% of global deaths from noncommunicable diseases worldwide, cancer represents a significant public health and economic burden on society (Bray et al. 2021). Traditionally, cancer has been regarded as a genetic disease, with many cases arising from unmodifiable risks, such as the random accumulation of genetic mutations. However, a significant proportion of malignancies can be attributed to modifiable risk factors, including tobacco use and alcohol. Interestingly, increasing evidence also suggests that cancer can be driven by chronic infection and inflammation (Plummer et al. 2016). In vitro studies, along with research in animal models and human populations, have demonstrated a link between prolonged microbial exposure, subsequent inflammation, and the development of certain cancers. A well-known example is chronic *Helicobacter pylori* infection, which is associated with an increased risk of gastric cancer (Yu et al. 2023).

Considering the mechanistic and epidemiological evidence for the perio-systemic link and the overwhelming evidence for inflammation-driven malignancies, it is reasonable to hypothesize the negative influence of periodontitis on cancer development and progression. Indeed, multiple large-scale epidemiological studies and a recent meta-analysis have reported a statistically significant connection between periodontitis and overall cancer risk. In a population-based retrospective cohort study over the course of 10 years, periodontitis patients exhibited a 2.2 times higher cumulative cancer incidence compared to the control group after adjusting for various demographic, lifestyle, and medical factors. Moreover, a meta-analysis found a significant hazard ratio of 1.14 for total cancer risk in periodontitis patients (Corbella et al. 2018; Michaud et al. 2016).

Given the periodontium's proximity to the oral cavity and the frequent exposure of oral mucosal tissues to periodontal pathogens, epidemiological studies have consistently focused on and demonstrated an elevated risk of head and neck cancers (HNCs) in individuals with periodontitis. A large prospective cohort study involving nearly 20,000 nonsmoking US men reported a 2.3-fold increase in the risk of oropharyngeal cancer among those with self-reported periodontitis (Michaud et al. 2016). Similar findings were reported from a South Korean study involving 424 patients, including 146 oral squamous carcinoma (OSCC) patients and 278 controls matched by age, sex, and smoking. After adjusting for confounders like alcohol use, physical activity, and metabolic factors, the results showed that periodontitis increased the risk of OSCC by 3.7 times (Shin et al. 2019). The association was stronger in males, older individuals, and those with more significant tooth loss, indicating a quasi-causal relationship. Moreover, the risk did not vary significantly across tumor stages or locations, suggesting that periodontitis is an independent risk factor for OSCC.

Histological analyses of cancer tissues have identified periodontitis-associated bacteria, including *Peptostreptococcus*, *Filifactor*, *Catonella*, and *Parvimonas*. However, the most frequent and pathogenic bacteria are *Porphyromonas gingivalis* and *Fusobacterium nucleatum* (Estemalik et al. 2017; Mesa et al. 2022; Nejman et al. 2020; Zhao et al. 2017). The role of these bacteria in carcinogenesis may involve several molecular mechanisms.

The most obvious mechanism is related to chronic inflammation driven by pathogen-associated molecular patterns (PAMPs) from these bacteria. These PAMPs activate TLRs, mainly TLR4, which, in turn, activate NF-κB and mitogen-activated protein kinase (MAPK) pathways. This triggers a pro-inflammatory environment that enhances cell proliferation, migration, and resistance to apoptosis, facilitating tumorigenesis. Indeed, the phosphoinositide 3-kinase /AKT/NF-κB (PI3K/AKT/NF-κB) pathway is responsible for the inactivation of PTEN (phosphatase and tensin homolog), a substantial tumor suppressor gene, promoting invasiveness in lung cancer cells (Akca et al. 2011). In colorectal cancer cells, pro-inflammatory cytokines have also been shown to promote cell growth by synergistically activating signal transducer and activator of transcription 3 (STAT3) and NF-κB (De Simone et al. 2015).

Second, periodontitis-associated bacteria can modulate epigenetic alterations. *F. nucleatum* and *P. gingivalis* can downregulate the p53 tumor suppressor and Ku70 repair pathways through histone H2A phosphorylation, which promotes increased OSCC cell proliferation (Geng et al. 2020). Epigenetic changes have also been observed in laryngeal squamous carcinoma specimens exposed to *F. nucleatum*. The researchers found extensive upregulation of Janus kinase 3 (JAK3) mediated by gene methylation in response to *F. nucleatum*, thereby upregulating the JAK/STAT and PI3K/AKT pathways responsible for cell proliferation (Yuan et al. 2024).

Additionally, periodontal bacteria can upregulate immune checkpoint molecules. Immune checkpoints, including PD-1, CTLA-4, LAG3, TIM3, TIGIT, and CD40, are inhibitory or stimulatory receptors that regulate immune responses and prevent overactivation (Muñoz-Medel et al. 2024). A major hallmark of cancer is that tumors exploit these checkpoints by upregulating inhibitory signals to evade detection by the immune system. In the last decade, immune checkpoint inhibitors, such as anti-programmed death protein 1 (PD-1)/programmed death-ligand 1 (PD-L1) therapies, have successfully treated cancers like melanoma and lung cancer and have become a significant focus in biomedical research.

Fractions of *P. gingivalis* cells have been shown to induce the expression of B7-H1 (PD-L1) and B7-DC (PD-L2), two major inhibitory checkpoint proteins. This process inhibits T-cell activation and survival, thereby promoting immune evasion and tumor progression (Groeger et al. 2017, 2022; Katz et al. 2011). On the other hand, *P. gingivalis* has been shown to facilitate immune evasion of tumor cells from colon and oral carcinomas by upregulating PD-L1, thereby preventing $CD8^+$ T-lymphocyte activity (Groeger et al. 2017, 2020; Ren et al. 2023).

Lastly, the maladaptive immune cell priming witnessed in periodontitis patients may contribute to the development of nonsolid malignancies. Even though the evidence is scarce, it is reasonable to suggest that the sustained and excessive recruitment of polymorphonuclear neutrophils (PMNs) and peripheral blood mononuclear cells (PBMCs) at the site of inflammation may trigger a state of chronic oxidative stress with elevated levels of ROS in the bone marrow. This chronic inflammation may induce mutagenesis due to persistent inflammation-induced oxidative damage to DNA in the hematopoietic stem cell niche (Hasselbalch 2013; Kryston et al. 2011). Moreover, studies have shown that activated neutrophils may contribute to cancer metastasis by binding to the endothelium, forming complexes with cancer cells to facilitate their adhesion (McDonald et al. 2009). On the other hand, these findings are contrasted by the fact that maladaptive immunity and autoinflammation have been substantially associated with reduced cancer incidence, due to improved immunosurveillance (Pol et al. 2022).

5.4 Future Directions and Research

While substantial progress has been made in understanding the associations between periodontitis and systemic diseases, many questions remain. Future research should establish causal relationships, identify the most effective interventions for reducing systemic disease risk through periodontal treatment, and further

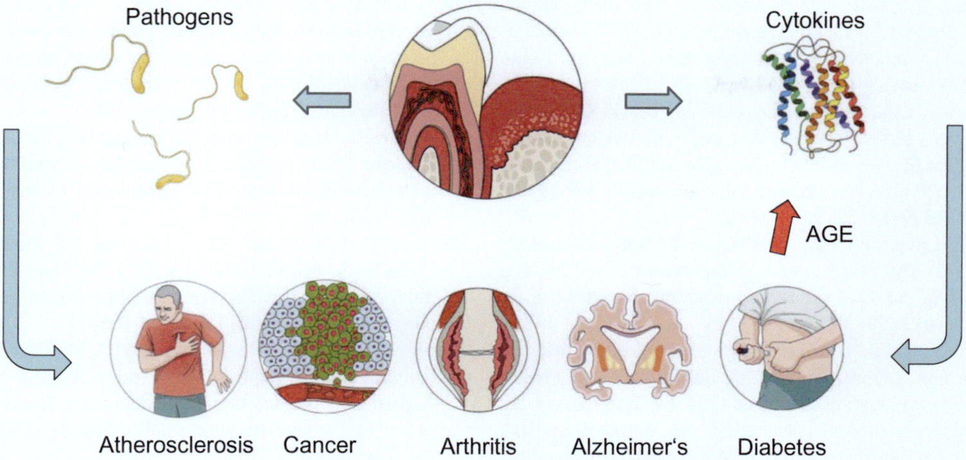

Fig. 5.1 Periodontal inflammation and bacterial dysbiosis may result in systemic dissemination of pathogens, their by-products, and inflammatory mediators. This may contribute to the pathogenesis of systemic diseases like cardiovascular disease, diabetes, adverse pregnancy outcomes, and cancer. Periodontal treatment may help reduce systemic disease burden

elucidate the underlying mechanisms linking periodontal disease to systemic health.

Long-term clinical trials and mechanistic studies will be essential for advancing our understanding of the interplay between oral and systemic health. By addressing these knowledge gaps, researchers and clinicians can work together to develop strategies for improving patient care and preventing systemic diseases through better management of periodontal health.

5.5 Clinical Implications

The potential link between periodontal inflammation and systemic diseases (summarized in Fig. 5.1) highlights the importance of oral health in preventive medicine. Regular dental checkups and periodontal treatments may play a crucial role in reducing the risk of systemic diseases, particularly in individuals with preexisting conditions such as diabetes or cardiovascular disease. A collaborative approach between dental and medical professionals is essential. Dentists should be aware of their patients' systemic health conditions, and physicians should recognize the potential impact of oral health on systemic diseases. This interdisciplinary approach and enhancing patient education can help ensure comprehensive care and better outcomes for patients.

References

Akca H, Demiray A, Tokgun O, Yokota J (2011) Invasiveness and anchorage independent growth ability augmented by PTEN inactivation through the PI3K/AKT/NFkB pathway in lung cancer cells. Lung Cancer 73(3):302–309

Almutairi K, Nossent J, Preen D, Keen H, Inderjeeth C (2021) The global prevalence of rheumatoid arthritis: a meta-analysis based on a systematic review. Rheumatol Int 41(5):863–877

Alvarenga MOP, Frazão DR, Matos IGD, Bittencourt LO, Fagundes NCF, Rösing CK et al (2021) Is there any association between neurodegenerative diseases and periodontitis? A systematic review. Front Aging Neurosci 13:651437

Ana P, Draginja K, Dimitrije M, Ivan M, Mariola S (2013) The markers of systemic inflammation in patients with chronic periodontitis: leukocytes, C-reactive protein and fibrinogen. World J Prev Med 1(3):43–49

Bahekar AA, Singh S, Saha S, Molnar J, Arora R (2007) The prevalence and incidence of coronary heart disease is significantly increased in periodontitis: a meta-analysis. Am Heart J 154(5):830–837

Bercik P, Collins S, Verdu E (2012) Microbes and the gut-brain axis. Neurogastroenterol Motil 24(5):405–413

Bizzarro S, Van Der Velden U, Ten Heggeler J, Leivadaros E, Hoek F, Gerdes V et al (2007) Periodontitis is characterized by elevated PAI-1 activity. J Clin Periodontol 34(7):574–580
Blencowe H, Krasevec J, De Onis M, Black RE, An X, Stevens GA et al (2019) National, regional, and worldwide estimates of low birthweight in 2015, with trends from 2000: a systematic analysis. Lancet Glob Health 7(7):e849–e860
Bray F, Laversanne M, Weiderpass E, Soerjomataram I (2021) The ever-increasing importance of cancer as a leading cause of premature death worldwide. Cancer 127(16):3029–3030
Bullon P, Cordero MD, Quiles JL, Morillo JM, del Carmen Ramirez-Tortosa M, Battino M (2011) Mitochondrial dysfunction promoted by Porphyromonas gingivalis lipopolysaccharide as a possible link between cardiovascular disease and periodontitis. Free Radic Biol Med 50(10):1336–1343
Carabotti M, Scirocco A, Maselli MA, Severi C (2015) The gut-brain axis: interactions between enteric microbiota, central and enteric nervous systems. Ann Gastroenterol 28(2):203–209
Chandy S, Joseph K, Sankaranarayanan A, Issac A, Babu G, Wilson B, Joseph J (2017) Evaluation of C-reactive protein and fibrinogen in patients with chronic and aggressive periodontitis: a clinicobiochemical study. J Clin Diagn Res 11(3):ZC41
Chuah YK, Basir R, Talib H, Tie TH, Nordin N (2013) Receptor for advanced glycation end products and its involvement in inflammatory diseases. Int J Inflamm 2013(1):403460
Corbella S, Veronesi P, Galimberti V, Weinstein R, Del Fabbro M, Francetti L (2018) Is periodontitis a risk indicator for cancer? A meta-analysis. PLoS One 13(4):e0195683
Curran AM, Naik P, Giles JT, Darrah E (2020) PAD enzymes in rheumatoid arthritis: pathogenic effectors and autoimmune targets. Nat Rev Rheumatol 16(6): 301–315
D'Aiuto F, Parkar M, Andreou G, Suvan J, Brett PM, Ready D, Tonetti MS (2004) Periodontitis and systemic inflammation: control of the local infection is associated with a reduction in serum inflammatory markers. J Dent Res 83(2):156–160
de Boer SP, Cheng JM, Rangé H, Garcia-Garcia HM, Heo JH, Akkerhuis KM et al (2014) Antibodies to periodontal pathogens are associated with coronary plaque remodeling but not with vulnerability or burden. Atherosclerosis 237(1):84–91
De Pablo P, Dietrich T, McAlindon TE (2008) Association of periodontal disease and tooth loss with rheumatoid arthritis in the US population. J Rheumatol 35(1): 70–76
De Simone V, Franzè E, Ronchetti G, Colantoni A, Fantini M, Di Fusco D et al (2015) Th17-type cytokines, IL-6 and TNF-α synergistically activate STAT3 and NF-kB to promote colorectal cancer cell growth. Oncogene 34(27):3493–3503
Dietrich T, Sharma P, Walter C, Weston P, Beck J (2013) The epidemiological evidence behind the association between periodontitis and incident atherosclerotic cardiovascular disease. J Clin Periodontol 40(s14):S70–S84. https://doi.org/10.1111/jcpe.12062
Dissick A, Redman RS, Jones M, Rangan BV, Reimold A, Griffiths GR et al (2010) Association of periodontitis with rheumatoid arthritis: a pilot study. J Periodontol 81(2):223–230
Doifode T, Giridharan VV, Generoso JS, Bhatti G, Collodel A, Schulz PE et al (2021) The impact of the microbiota-gut-brain axis on Alzheimer's disease pathophysiology. Pharmacol Res 164:105314
Dominy SS, Lynch C, Ermini F, Benedyk M, Marczyk A, Konradi A et al (2019) Porphyromonas gingivalis in Alzheimer's disease brains: evidence for disease causation and treatment with small-molecule inhibitors. Sci Adv 5(1):eaau3333
Eitzman DT, Westrick RJ, Xu Z, Tyson J, Ginsburg D (2000) Plasminogen activator inhibitor-1 deficiency protects against atherosclerosis progression in the mouse carotid artery. Blood J Am Soc Hematol 96(13):4212–4215
Endemann DH, Schiffrin EL (2004) Endothelial dysfunction. J Am Soc Nephrol 15(8):1983–1992
Estemalik J, Demko C, Bissada NF, Joshi N, Bodner D, Shankar E, Gupta S (2017) Simultaneous detection of oral pathogens in subgingival plaque and prostatic fluid of men with periodontal and prostatic diseases. J Periodontol 88(9):823–829
Figuero E, Han YW, Furuichi Y (2020) Periodontal diseases and adverse pregnancy outcomes: mechanisms. Periodontol 83(1):175–188. https://doi.org/10.1111/prd.12295
Furutama D, Matsuda S, Yamawaki Y, Hatano S, Okanobu A, Memida T et al (2020) IL-6 induced by periodontal inflammation causes neuroinflammation and disrupts the blood–brain barrier. Brain Sci 10(10):679
Gawron K, Montgomery A, Łazarz-Bartyzel K, Bereta G, Chomyszyn-Gajewska M, Venables P, Potempa J (2017) Porphyromonas gingivalis peptidyl arginine deiminase: a unique bacterial PAD with implications for periodontal disease and rheumatoid arthritis. In: Protein deimination in human health and disease, pp 99–135
Geerts SO, Nys M, De Mol P, Charpentier J, Albert A, Legrand V, Rompen EH (2002) Systemic release of endotoxins induced by gentle mastication: association with periodontitis severity. J Periodontol 73(1):73–78
Geng F, Zhang Y, Lu Z, Zhang S, Pan Y (2020) Fusobacterium nucleatum caused DNA damage and promoted cell proliferation by the Ku70/p53 pathway in oral cancer cells. DNA Cell Biol 39(1):144–151
Godbout J, Chen J, Abraham J, Richwine A, Berg B, Kelley K, Johnson R (2005) Exaggerated neuroinflammation and sickness behavior in aged mice after activation of the peripheral innate immune system. FASEB J 19(10):1329–1331

Goldenberg RL, Culhane JF, Iams JD, Romero R (2008) Epidemiology and causes of preterm birth. Lancet 371(9606):75–84

Goldin A, Beckman JA, Schmidt AM, Creager MA (2006) Advanced glycation end products. Circulation 114(6): 597–605

Gonzales-Marin C, Spratt DA, Allaker RP (2013) Maternal oral origin of Fusobacterium nucleatum in adverse pregnancy outcomes as determined using the 16S–23S rRNA gene intergenic transcribed spacer region. J Med Microbiol 62(1):133–144

Groeger S, Jarzina F, Mamat U, Meyle J (2017) Induction of B7-H1 receptor by bacterial cells fractions of Porphyromonas gingivalis on human oral epithelial cells: B7-H1 induction by Porphyromonas gingivalis fractions. Immunobiology 222(2):137–147

Groeger S, Denter F, Lochnit G, Schmitz M, Meyle J (2020) Porphyromonas gingivalis Cell Wall components induce programmed death ligand 1 (PD-L1) expression on human Oral carcinoma cells by a receptor-interacting protein kinase 2 (RIP2)-dependent mechanism. Infect Immun 88(5):e00051–e00020

Groeger S, Wu F, Wagenlehner F, Dansranjav T, Ruf S, Denter F, Meyle J (2022) PD-L1 up-regulation in prostate cancer cells by Porphyromonas gingivalis. Front Cell Infect Microbiol 12:935806

Gu L, Okada Y, Clinton SK, Gerard C, Sukhova GK, Libby P, Rollins BJ (1998) Absence of monocyte chemoattractant protein-1 reduces atherosclerosis in low density lipoprotein receptor–deficient mice. Mol Cell 2(2):275–281

Hajishengallis G (2014) Periodontitis: from microbial immune subversion to systemic inflammation. Nat Rev Immunol 15:30–44. https://doi.org/10.1038/nri3785

Hajishengallis G (2022) Interconnection of periodontal disease and comorbidities: evidence, mechanisms, and implications. Periodontol 89(1):9–18

Hajishengallis G, Chavakis T (2021) Local and systemic mechanisms linking periodontal disease and inflammatory comorbidities. Nat Rev Immunol 21(7):426–440

Hajishengallis G, Li X, Divaris K, Chavakis T (2022) Maladaptive trained immunity and clonal hematopoiesis as potential mechanistic links between periodontitis and inflammatory comorbidities. Periodontol 89(1): 215–230

Hamsten A, Wiman B, De Faire U, Blombäck M (1985) Increased plasma levels of a rapid inhibitor of tissue plasminogen activator in young survivors of myocardial infarction. N Engl J Med 313(25): 1557–1563

Han YW, Shen T, Chung P, Buhimschi IA, Buhimschi CS (2009) Uncultivated bacteria as etiologic agents of intra-amniotic inflammation leading to preterm birth. J Clin Microbiol 47(1):38–47

Hasselbalch HC (2013) Chronic inflammation as a promotor of mutagenesis in essential thrombocythemia, polycythemia vera and myelofibrosis. A human inflammation model for cancer development? Leuk Res 37(2):214–220

He J, Usui I, Ishizuka K, Kanatani Y, Hiratani K, Iwata M et al (2006) Interleukin-1α inhibits insulin signaling with phosphorylating insulin receptor substrate-1 on serine residues in 3T3-L1 adipocytes. Mol Endocrinol 20(1):114–124

Holers VM, Demoruelle MK, Kuhn KA, Buckner JH, Robinson WH, Okamoto Y et al (2018) Rheumatoid arthritis and the mucosal origins hypothesis: protection turns to destruction. Nat Rev Rheumatol 14(9): 542–557

Hotamisligil GS, Murray DL, Choy LN, Spiegelman BM (1994) Tumor necrosis factor alpha inhibits signaling from the insulin receptor. Proc Natl Acad Sci 91(11): 4854–4858

Ilievski V, Zuchowska PK, Green SJ, Toth PT, Ragozzino ME, Le K et al (2018) Chronic oral application of a periodontal pathogen results in brain inflammation, neurodegeneration and amyloid beta production in wild type mice. PLoS One 13(10):e0204941

Jager J, Grémeaux T, Cormont M, Le Marchand-Brustel Y, Tanti J-F (2007) Interleukin-1β-induced insulin resistance in adipocytes through down-regulation of insulin receptor substrate-1 expression. Endocrinology 148(1): 241–251

Jepsen S, Caton JG, Albandar JM, Bissada NF, Bouchard P, Cortellini P et al (2018) Periodontal manifestations of systemic diseases and developmental and acquired conditions: consensus report of workgroup 3 of the 2017 world Workshop on the classification of periodontal and Peri-implant diseases and conditions. J Clin Periodontol 45:S219–S229

Kalafati L, Hatzioannou A, Hajishengallis G, Chavakis T (2023) The role of neutrophils in trained immunity. Immunol Rev 314(1):142–157

Katz J, Onate MD, Pauley KM, Bhattacharyya I, Cha S (2011) Presence of Porphyromonas gingivalis in gingival squamous cell carcinoma. Int J Oral Sci 3(4): 209–215. https://doi.org/10.4248/IJOS11075

Khan KS, Wojdyla D, Say L, Gülmezoglu AM, Van Look PF (2006) WHO analysis of causes of maternal death: a systematic review. Lancet 367(9516):1066–1074

Kim S-R, Son M, Kim Y-R, Kang H-K (2022) Risk of dementia according to the severity of chronic periodontitis in Korea: a nationwide retrospective cohort study. Epidemiol Health 44

Klareskog L, Lundberg K, Malmström V (2013) Autoimmunity in rheumatoid arthritis: citrulline immunity and beyond. Adv Immunol 118:129–158

Konig MF, Abusleme L, Reinholdt J, Palmer RJ, Teles RP, Sampson K et al (2016) Aggregatibacter actinomycetemcomitans–induced hypercitrullination links periodontal infection to autoimmunity in rheumatoid arthritis. Sci Transl Med 8(369):369ra176-369ra176

Kryston TB, Georgiev AB, Pissis P, Georgakilas AG (2011) Role of oxidative stress and DNA damage in

human carcinogenesis. Mutat Res 711(1):193–201. https://doi.org/10.1016/j.mrfmmm.2010.12.016
Kweider M, Lowe G, Murray G, Kinane D, McGowan D (1993) Dental disease, fibrinogen and white cell count; links with myocardial infarction? Scott Med J 38(3): 73–74
Latorre Uriza C, Velosa-Porras J, Roa NS, Quiñones Lara SM, Silva J, Ruiz AJ, Escobar Arregoces FM (2018) Periodontal disease, inflammatory cytokines, and PGE2 in pregnant patients at risk of preterm delivery: a pilot study. Infect Dis Obstet Gynecol 2018(1): 7027683
Laugisch O, Wong A, Sroka A, Kantyka T, Koziel J, Neuhaus K et al (2016) Citrullination in the periodontium—a possible link between periodontitis and rheumatoid arthritis. Clin Oral Investig 20:675–683
Leira Y, Carballo Á, Orlandi M, Aldrey JM, Pías-Peleteiro JM, Moreno F et al (2020) Periodontitis and systemic markers of neurodegeneration: a case–control study. J Clin Periodontol 47(5):561–571
Leivadaros E, Van der Velden U, Bizzarro S, Heggeler JM, Gerdes VE, Hoek FJ et al (2005) A pilot study into measurements of markers of atherosclerosis in periodontitis. J Periodontol 76(1):121–128
León R, Silva N, Ovalle A, Chaparro A, Ahumada A, Gajardo M et al (2007) Detection of Porphyromonas gingivalis in the amniotic fluid in pregnant women with a diagnosis of threatened premature labor. J Periodontol 78(7):1249–1255
Li X, Wang H, Yu X, Saha G, Kalafati L, Ioannidis C et al (2022) Maladaptive innate immune training of myelopoiesis links inflammatory comorbidities. Cell 185(10):1709–1727. e1718
Li Y, Chen Y, Cai G, Ni Q, Geng Y, Wang T et al (2023) Roles of trained immunity in the pathogenesis of periodontitis. J Periodontal Res 58(5):864–873
Li W, Peng J, Shang Q, Yang D, Zhao H, Xu H (2024) Periodontitis and the risk of all-cause and cause-specific mortality among US adults with diabetes: a population-based cohort study. J Clin Periodontol 51(3):288–298
Liew AKC, Punnanithinont N, Lee YC, Yang J (2013) Effect of non-surgical periodontal treatment on HbA1c: a meta-analysis of randomized controlled trials. Aust Dent J 58(3):350–357
Linden GJ, Herzberg MC (2013) Periodontitis and systemic diseases: a record of discussions of working group 4 of the joint EFP/AAP Workshop on periodontitis and systemic diseases. J Periodontol 84:S20–S23
Loos BG, Craandijk J, Hoek FJ, Dillen PMW, Van Der Velden U (2000) Elevation of systemic markers related to cardiovascular diseases in the peripheral blood of periodontitis patients. J Periodontol 71(10):1528–1534
Lu J, Zhang S, Huang Y, Qian J, Tan B, Qian X et al (2022) Periodontitis-related salivary microbiota aggravates Alzheimer's disease via gut-brain axis crosstalk. Gut Microbes 14(1):2126272
Ma'ayeh M, Rood KM, Kniss D, Costantine MM (2020) Novel interventions for the prevention of preeclampsia. Curr Hypertens Rep 22(2):17. https://doi.org/10.1007/s11906-020-1026-8
MacKay AP, Berg CJ, Duran C, Chang J, Rosenberg H (2005) An assessment of pregnancy-related mortality in the United States. Paediatr Perinat Epidemiol 19(3): 206–214
Martínez-García M, Hernández-Lemus E (2021) Periodontal inflammation and systemic diseases: an overview. Front Physiol 12:709438
McDonald B, Spicer J, Giannais B, Fallavollita L, Brodt P, Ferri LE (2009) Systemic inflammation increases cancer cell adhesion to hepatic sinusoids by neutrophil mediated mechanisms. Int J Cancer 125(6):1298–1305
Mendez MV, Scott T, LaMorte W, Vokonas P, Menzoian JO, Garcia R (1998) An association between periodontal disease and peripheral vascular disease. Am J Surg 176(2):153–157
Mesa F, Mesa-López MJ, Egea-Valenzuela J, Benavides-Reyes C, Nibali L, Ide M et al (2022) A new comorbidity in periodontitis: Fusobacterium nucleatum and colorectal cancer. Medicina 58(4):546
Michaud D, Kelsey K, Papathanasiou E, Genco C, Giovannucci E (2016) Periodontal disease and risk of all cancers among male never smokers: an updated analysis of the health professionals follow-up study. Ann Oncol 27(5):941–947
Mikuls TR, Payne JB, Yu F, Thiele GM, Reynolds RJ, Cannon GW et al (2014) Periodontitis and Porphyromonas gingivalis in patients with rheumatoid arthritis. Arthritis Rheumatol 66(5):1090–1100
Möller B, Kollert F, Sculean A, Villiger PM (2020) Infectious triggers in periodontitis and the gut in rheumatoid arthritis (RA): a complex story about association and causality. Front Immunol 11:1108
Mukherjee A, Jantsch V, Khan R, Hartung W, Fischer R, Jantsch J et al (2018) Rheumatoid arthritis-associated autoimmunity due to aggregatibacter actinomycetemcomitans and its resolution with antibiotic therapy. Front Immunol 9:2352
Muñoz-Medel M, Pinto MP, Goralsky L, Caceres M, Villarroel-Espíndola F, Manque P et al (2024) Porphyromonas gingivalis, a bridge between oral health and immune evasion in gastric cancer. Front Oncol 14:1403089
Nejman D, Livyatan I, Fuks G, Gavert N, Zwang Y, Geller LT et al (2020) The human tumor microbiome is composed of tumor type–specific intracellular bacteria. Science 368(6494):973–980
Ng A, Tam WW, Zhang MW, Ho CS, Husain SF, McIntyre RS, Ho RC (2018) IL-1β, IL-6, TNF-α and CRP in elderly patients with depression or Alzheimer's disease: systematic review and meta-analysis. Sci Rep 8(1):12050
Nibali L, D'Aiuto F, Griffiths G, Patel K, Suvan J, Tonetti MS (2007) Severe periodontitis is associated with systemic inflammation and a dysmetabolic status: a case–

control study. J Clin Periodontol 34(11):931–937. https://doi.org/10.1111/j.1600-051X.2007.01133.x

Offenbacher S, Katz V, Fertik G, Collins J, Boyd D, Maynor G et al (1996) Periodontal infection as a possible risk factor for preterm low birth weight. J Periodontol 67:1103–1113

Okwaraji YB, Krasevec J, Bradley E, Conkle J, Stevens GA, Gatica-Domínguez G et al (2024) National, regional, and global estimates of low birthweight in 2020, with trends from 2000: a systematic analysis. Lancet 403(10431):1071–1080

Olsen I (2021) Porphyromonas gingivalis-induced neuroinflammation in alzheimer's disease. Front Neurosci 15:691016

Perunovic ND, Rakic MM, Nikolic LI, Jankovic SM, Aleksic ZM, Plecas DV et al (2016) The association between periodontal inflammation and labor triggers (elevated cytokine levels) in preterm birth: a cross-sectional study. J Periodontol 87(3):248–256

Pink C, Kocher T, Meisel P, Dörr M, Markus MR, Jablonowski L et al (2015) Longitudinal effects of systemic inflammation markers on periodontitis. J Clin Periodontol 42(11):988–997

Pischon N, Pischon T, Kröger J, Gülmez E, Kleber BM, Bernimoulin JP et al (2008) Association among rheumatoid arthritis, oral hygiene, and periodontitis. J Periodontol 79(6):979–986

Plummer M, de Martel C, Vignat J, Ferlay J, Bray F, Franceschi S (2016) Global burden of cancers attributable to infections in 2012: a synthetic analysis. Lancet Glob Health 4(9):e609–e616

Pol J, Paillet J, Plantureux C, Kroemer G (2022) Beneficial autoimmunity and maladaptive inflammation shape epidemiological links between cancer and immune-inflammatory diseases. Oncoimmunology 11(1): 2029299

Poole S, Singhrao SK, Kesavalu L, Curtis MA, Crean S (2013) Determining the presence of periodontopathic virulence factors in short-term postmortem Alzheimer's disease brain tissue. J Alzheimer's Dis 36(4):665–677

Preshaw PM, Bissett SM (2019) Periodontitis and diabetes. Br Dent J 227(7):577–584

Preshaw PM, Taylor JJ, Jaedicke KM, De Jager M, Bikker JW, Selten W et al (2020) Treatment of periodontitis reduces systemic inflammation in type 2 diabetes. J Clin Periodontol 47(6):737–746

Quirke A-M, Lugli EB, Wegner N, Hamilton BC, Charles P, Chowdhury M et al (2014) Heightened immune response to autocitrullinated Porphyromonas gingivalis peptidylarginine deiminase: a potential mechanism for breaching immunologic tolerance in rheumatoid arthritis. Ann Rheum Dis 73(1):263–269

Rafferty B, Jönsson D, Kalachikov S, Demmer RT, Nowygrod R, Elkind MS et al (2011) Impact of monocytic cells on recovery of uncultivable bacteria from atherosclerotic lesions. J Intern Med 270(3):273–280

Ren J, Han X, Lohner H, Hoyle RG, Li J, Liang S, Wang H (2023) P. Gingivalis infection upregulates PD-L1 expression on dendritic cells, suppresses CD8+ T-cell responses, and aggravates oral cancer. Cancer Immunol Res 11(3):290–305

Riviere GR, Riviere K, Smith K (2002) Molecular and immunological evidence of oral Treponema in the human brain and their association with Alzheimer's disease. Oral Microbiol Immunol 17(2):113–118

Saadaoui M, Singh P, Al Khodor S (2021) Oral microbiome and pregnancy: a bidirectional relationship. J Reprod Immunol 145:103293

Sampath C, Okoro EU, Gipson MJ, Chukkapalli SS, Farmer-Dixon CM, Gangula PR (2021) Porphyromonas gingivalis infection alters Nrf2-phase II enzymes and nitric oxide in primary human aortic endothelial cells. J Periodontol 92(7):e54–e65

Sanz M, Kornman K, Workshop WG et al (2013) Periodontitis and adverse pregnancy outcomes: consensus report of the joint EFP/AAP Workshop on periodontitis and systemic diseases. J Periodontol 84:S164–S169

Sanz M, Ceriello A, Buysschaert M, Chapple I, Demmer RT, Graziani F et al (2018) Scientific evidence on the links between periodontal diseases and diabetes: consensus report and guidelines of the joint workshop on periodontal diseases and diabetes by the international diabetes federation and the European Federation of Periodontology. Diabetes Res Clin Pract 137:231–241

Shin Y, Choung H, Lee J, Rhyu I, Kim H (2019) Association of periodontitis with oral cancer: a case-control study. J Dent Res 98(5):526–533

Siddiqui H, Eribe Ribs EK, Singhrao SK, Olsen I (2019) High throughput sequencing detects gingivitis and periodontal oral bacteria in Alzheimer's disease autopsy brains. J Neurosci Res 1(1):3

Sim SJ, Kim HD, Moon JY, Zavras AI, Zdanowicz J, Jang SJ et al (2008) Periodontitis and the risk for non-fatal stroke in Korean adults. J Periodontol 79(9): 1652–1658

Sun B, Wang Y, Wu M, Feng G, Guo T (2024) Key periodontal pathogens may mediate potential pathogenic relationships between periodontitis and crohn's disease. BMC Oral Health 24:668

Thögersen AM, Jansson J-H et al (1998) High plasminogen activator inhibitor and tissue plasminogen activator levels in plasma precede a first acute myocardial infarction in both men and women: evidence for the fibrinolytic system as an independent primary risk factor. Circulation 98(21):2241–2247

Uysal H, Nandakumar KS, Kessel C, Haag S, Carlsen S, Burkhardt H, Holmdahl R (2010) Antibodies to citrullinated proteins: molecular interactions and arthritogenicity. Immunol Rev 233(1):9–33

Vander Haar EL, So J, Gyamfi-Bannerman C, Han YW (2018) Fusobacterium nucleatum and adverse pregnancy outcomes: epidemiological and mechanistic evidence. Anaerobe 50:55–59

Vitkov L, Herrmann M, Knopf J (2023) Oral neutrophils-the good, the bad, and the ugly. Front Immunol 14: 1225210

Wang X, Han X, Guo X, Luo X, Wang D (2014) The effect of periodontal treatment on hemoglobin a1c levels of diabetic patients: a systematic review and meta-analysis. PLoS One 9(9):e108412

Winning L, Linden GJ (2017) Periodontitis and systemic disease: association or causality? Curr Oral Health Rep 4:1–7

Wu H, Qiu W, Zhu X, Li X, Xie Z, Carreras I et al (2022) The periodontal pathogen Fusobacterium nucleatum exacerbates Alzheimer's pathogenesis via specific pathways. Front Aging Neurosci 14:912709

Xie M, Tang Q, Yu S, Sun J, Mei F, Zhao J, Chen L (2020) Porphyromonas gingivalis disrupts vascular endothelial homeostasis in a TLR-NF-κB axis dependent manner. Int J Oral Sci 12(1):28

Xiong X, Buekens P, Fraser W, Beck J, Offenbacher S (2006) Periodontal disease and adverse pregnancy outcomes: a systematic review. BJOG Int J Obstet Gynaecol 113(2):135–143

Xu F, Lu B (2011) Prospective association of periodontal disease with cardiovascular and all-cause mortality: NHANES III follow-up study. Atherosclerosis 218(2):536–542

Xu T, Dong Q, Luo Y, Liu Y, Gao L, Pan Y, Zhang D (2021) Porphyromonas gingivalis infection promotes mitochondrial dysfunction through Drp1-dependent mitochondrial fission in endothelial cells. Int J Oral Sci 13(1):28

Yu B, Xiang L, Peppelenbosch MP, Fuhler GM (2023) Overlapping cytokines in H. Pylori infection and gastric cancer: a tandem meta-analysis. Front Immunol 14: 1125658

Yuan X, Lau H-C, Huang H, Hsueh C-Y, Gong H, Zhou L (2024) Integrative methylome and transcriptome analysis reveals epigenetic regulation of Fusobacterium nucleatum in laryngeal cancer. Microb Genom 10(3): 001221

Zhao H, Chu M, Huang Z, Yang X, Ran S, Hu B et al (2017) Variations in oral microbiota associated with oral cancer. Sci Rep 7(1):11773

6 The Oral Microbiome, Oral Hyperpermeability, and Periodontitis in Metabolic Syndrome

Amine Nehaoua, Amin Gasmi, Sadaf Noor, and Asma Gasmi Benahmed

Abstract

The links between the oral microbiome, oral hyperpermeability, periodontitis, and metabolic syndrome (MetS) are good examples of the oral–systemic health connection. The oral microbiota is involved in host defense and metabolism. Oral dysbiosis disrupts periodontal homeostasis, increases the permeability of the oral mucosa, and triggers systemic inflammation, thus leading to MetS. On the other hand, metabolic disorders can change the environmental conditions in the oral cavity and exacerbate dysbiosis and periodontitis. Oral hyperpermeability is evidently a key player in this oral–systemic connection, allowing passage of oral bacteria, their by-products, and inflammatory mediators into the systemic circulation. It triggers the low-grade, chronic inflammation commonly seen in MetS and may exacerbate insulin resistance and other metabolic changes. Thus, periodontitis—a chronic inflammatory disease—is both a result and a cause of MetS. The classification of periodontitis as both a diagnostic marker and a treatment target in metabolic risk enhances the opportunity for the assessment and control of MetS. New-generation therapies targeting oral permeability and dysbiosis, such as barrier-strengthening agents, precision probiotics, and advanced antimicrobial agents, are potential strategies to address the oral–systemic health link. The concept of oral–systemic healthcare, incorporating both dental and medical care, is an emerging model of care delivery systems. Such strategies, including salivary risk management (SRM), targeted interventions, and patient education, may contribute to the enhancement of general health, particularly among individuals with or at risk of MetS.

Keywords

Oral microbiome · Periodontitis · Metabolic syndrome · Oral hyperpermeability · Oral–systemic health

A. Nehaoua
Research Department, Precision Health International, Ain Oulmene, Algeria

Laboratory of Physical Activity Sciences and Public Health, Mohamed Lamine Debaghine Sétif 2 University, Setif, Algeria

A. Gasmi (✉)
Société Francophone de Nutrithérapie et de Nutrigénétique Appliquée, Villeurbanne, France

International Institute of Nutrition and Micronutrition Sciences, Saint-Étienne, France

S. Noor
Institute of Molecular Biology and Biotechnology, Bahauddin Zakariya University, Multan, Pakistan

A. G. Benahmed
Académie Internationale de Médecine Dentaire Intégrative, Paris, France

N. Rezaei, H. Ziaei (eds.), *Oral Immunology*, Advances in Experimental Medicine and Biology 1492,
https://doi.org/10.1007/978-3-032-03176-1_6

6.1 Introduction

Metabolic syndrome (MetS) is a complex of closely related physiological, biochemical, clinical, and metabolic factors that have an established causal relationship with cardiovascular disease, type 2 diabetes mellitus (T2DM), and all-cause mortality (Hayden 2023; Saklayen 2018). This complex condition is characterized by the presence of at least three of the following five risk factors: abdominal obesity, raised blood pressure, raised fasting plasma glucose, raised serum triglycerides, and reduced high-density lipoprotein (HDL) cholesterol (Ahima 2024; Alberti et al. 2009). The prevalence of MetS is increasing in the global population, with data pointing to the fact that about one in four adults in the world suffer from this condition, which is a major public health concern (O'Neill and O'Driscoll 2015). Even though the causes of MetS are genetic predisposition, physical inactivity, and an unhealthy diet, recent studies have indicated that oral health can be considered an important, yet often neglected, risk factor in the development and progression of this pathology (Mohamed et al. 2023; Nibali et al. 2013). (Kilian et al. 2016). The complex interaction between oral health and other systemic metabolic disorders, including MetS, is highlighted by the oral–systemic health connection (Genco and Sanz 2020).

The oral microbiota, a diverse population of microbes living in the oral cavity, is central to both oral and overall health (Dewhirst et al. 2010; Santacroce et al. 2023). In the healthy state, the oral microbiota is in symbiotic equilibrium with the host, contributing to the host's defense and tissue homeostasis (Lamont et al. 2018). Nonetheless, any disturbance in this equilibrium, referred to as dysbiosis, can trigger a chain of biofilm-related events that not only impact oral health but also cause systemic conditions (Hajishengallis 2015). Oral hyperpermeability, or what has been described as "leaky mouth," is another significant factor in the oral–systemic health connection (Hajishengallis 2014a). This condition is marked by increased permeability of the oral mucosa through which oral bacteria, their by-products, and inflammatory cytokines get transported to the systemic circulation (Olsen and Yamazaki 2019). Oral hyperpermeability has been compared to another more researched topic known as intestinal hyperpermeability or "leaky gut," where the idea of several entry points for inflammatory agents into the bloodstream is proposed (Aleman et al. 2023; Michaud et al. 2017).

Among all oral diseases, periodontitis, a chronic inflammatory disease, of the supporting structures of the teeth, is identified as an important moderator between oral health and MetS (Nibali et al. 2013; Ray 2023). Characterized by destruction of the periodontal ligament and alveolar bone, periodontitis not only affects oral function but also contributes to low-grade, chronic systemic inflammation (Hajishengallis 2015). The association between periodontitis and MetS has been a focus of growing interest in the last few years, as research evidence has indicated that these two conditions may be reciprocally associated (Genco and Sanz 2020). The purpose of this chapter is to present a complex analysis of the interconnection between the oral microbiome, oral hyperpermeability, periodontitis, and MetS (Fig. 6.1).

6.2 The Oral Microbiome in Metabolic Syndrome

6.2.1 The Connection Between Oral Health and Metabolic Syndrome

There is growing evidence that links oral health to MetS since both diseases share common inflammatory pathways and risk factors. MetS is characterized by obesity, hypertension, dyslipidemia, and insulin resistance, all of which are associated with a high risk of cardiovascular disease and type 2 diabetes mellitus (T2DM). Gingivitis and periodontal diseases are inflammatory conditions affecting the gingival tissue and tooth-supporting structures and are associated with systemic inflammation and MetS. Some of

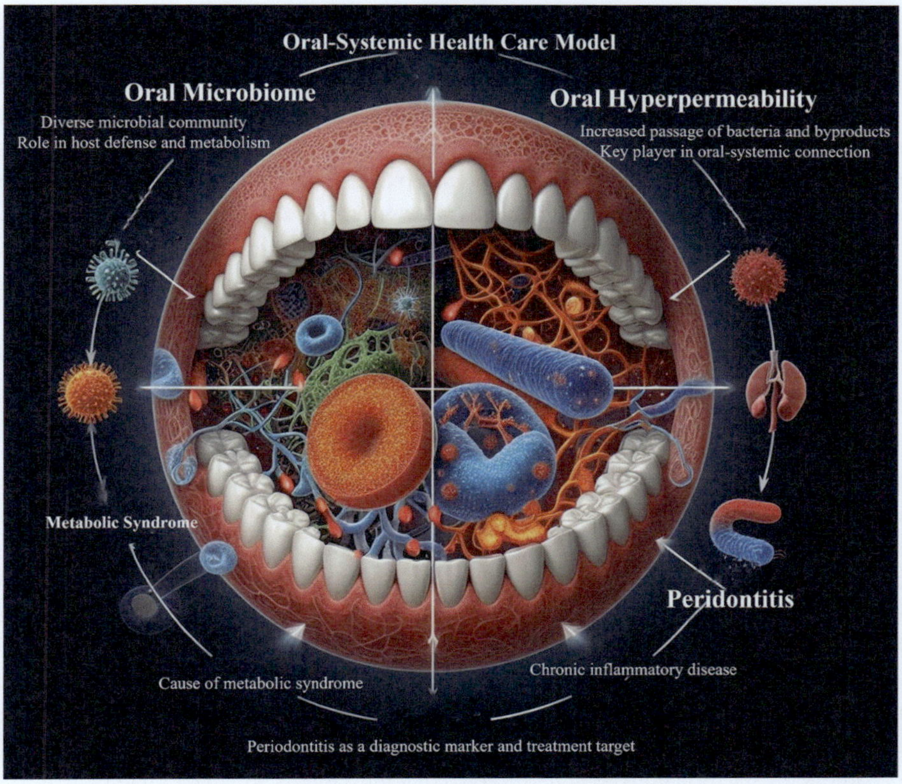

Fig. 6.1 Schematic depicting the oral–systemic healthcare model, illustrating the interconnected roles of the oral microbiome, oral hyperpermeability, periodontitis, and metabolic syndrome

the inflammatory cytokines released during periodontal diseases can be transported into the bloodstream and hence cause the general inflammation associated with MetS. This is backed by research that has indicated high levels of inflammatory indicators such as C-reactive protein (CRP) in individuals with periodontitis and MetS. Furthermore, it is also evident that periodontal pathogens worsen insulin resistance, which is one of the criteria of MetS through inflammation that affects glucose homeostasis (Ghanem and Nagy 2024).

In addition, the bidirectional interaction between diabetes and periodontal disease is well-established. Diabetes compromises the body's immune system, making it more vulnerable to infections. On the other hand, periodontal disease has a negative impact on the glycemic control of diabetic patients, resulting in the creation of a cycle of disease progression. Periodontal disease has been reported to have a positive relationship with MetS, and its proper management is associated with improved glycemic control (Chweidan et al. 2024). Some current investigations also indicate that metabolic changes associated with MetS, including dyslipidemia and hypertension, may affect oral health. For instance, hypertension may influence blood circulation to the gingiva and worsen periodontal disease. Likewise, changes in lipid profiles in MetS may be held responsible for plaque formation and gingivitis, thus supporting the relationship between oral health and MetS (Yamazaki et al. 2024).

Recent research has highlighted the involvement of adipokines, proteins secreted by the adipose tissue, in mediating the interaction between obesity, a key feature of MetS, and periodontal disease. Pro-inflammatory cytokines including leptin and adiponectin have been reported to play a role in inflammation and are altered in obesity and periodontal disease. Such dysregulation may foster a chronic pro-inflammatory state, exacerbating both disorders (Checa-Ros et al. 2024). The link between oral health and MetS is an intricate one and includes inflammation, reciprocal effects, and shared risk factors. Oral health is an important component of MetS; therefore, more studies are required to establish the mechanisms involved and to enhance the treatment approaches.

6.2.2 The Oral Microbiome: Composition and Function

The human oral microbiome is a complex and ever-changing microbial community, which is believed to contain between 700 and 1000 bacterial species (Dewhirst et al. 2010; Souak et al. 2023). This diverse biofilm is essential for oral homeostasis, protection of the host, and modulation of host responses and systemic disease (Lamont et al. 2018). In a state of health, the oral microbiome is a biofilm in which commensal as well as pathogenic bacteria coexist in harmony with the host (Kilian et al. 2016). The healthy oral microbiome consists of several phyla such as Firmicutes, Bacteroidetes, Proteobacteria, Actinobacteria, Spirochaetes, and Fusobacteria (Wade 2013). In these phyla, numerous genera and species are involved in the overall microbial diversity of the oral cavity, among which the most frequent ones include *Streptococcus*, *Veillonella*, *Granulicatella*, *Neisseria*, and *Haemophilus* (Zaura et al. 2009). These microorganisms adhere to form biofilms on various tissues of the oral cavity: teeth, gingival sulcus, tongue, and oral mucosa, each with its specific microflora (Abdulhadi et al. 2024; Kolenbrander et al. 2010; Santacroce et al. 2023).

The oral microbiome plays various functional roles in maintaining oral health. Commensal bacteria are involved in oral homeostasis through their ability to outcompete the pathogenic species for available nutrients and colonization sites on the surface of the teeth and gingival tissue, thereby excluding pathogenic microorganisms from the oral cavity (Marsh 2018). Moreover, some oral bacteria are involved in the synthesis of other related products like nitric oxide precursors that play a role in cardiovascular health regulation (Chopra et al. 2024; Hezel and Weitzberg 2015). Nevertheless, changes in the composition and role of the oral microbiota, known as dysbiosis, can have dramatic effects on oral and general health (Hajishengallis 2015; Talapko et al. 2024). Dysbiosis is defined by the reduction of microbial richness and the dominance of pathogenic microorganisms accompanied by the prevalence of pathogenic species and the decrease of commensal species (Lamont et al. 2018). This new microbial status is increasingly associated with different systemic disorders such as MetS (Chowdhury et al. 2023; Nibali et al. 2013).

The connection between oral dysbiosis and MetS is quite well-substantiated by the literature. It has been established that people with MetS have a different oral microbial profile than those with no MetS (Demmer et al. 2017; Li et al. 2023). For instance, Zeigler et al. (2012) reported that individuals with MetS had increased levels of periodontal pathogens such as *Porphyromonas gingivalis* and *Treponema denticola* compared to healthy individuals. There are various ways through which oral dysbiosis is implicated in the development of MetS. One of them is oral bacteria and their by-products, including lipopolysaccharides (LPSs), entering the bloodstream (Olsen and Yamazaki 2019). Some of these bacterial components can activate toll-like receptors (TLRs) and other pattern recognition receptors (PRRs), initiating the synthesis of pro-inflammatory cytokines (Hajishengallis 2015).

Some specific oral bacteria have been selectively associated with MetS. The periodontal pathogen *Porphyromonas gingivalis* has been demonstrated to induce insulin resistance in animal models by promoting systemic inflammation and altering the composition of the gut microbiota (Arimatsu et al. 2014; Jia et al. 2023; Niu et al. 2024). Likewise, another oral pathogen, *Fusobacterium nucleatum*, has been associated with adipose tissue inflammation and insulin resistance in obese individuals (Anderson et al. 2018; Chang et al. 2023; Rahman et al. 2023). The oral microbiome may affect MetS through changes in nutrient and energy metabolism. For instance, some oral bacteria can change the nitrate metabolism affecting blood pressure regulation, which is an essential aspect of MetS (Bryan et al. 2017; Chai et al. 2024). Furthermore, the synthesis of short-chain fatty acids (SCFAs) by oral bacteria could affect systemic metabolism and inflammation, but the underlying pathways are not fully understood (Koren et al. 2021).

6.2.3 Microbial Translocation and Systemic Effects

One of the most important pathogenic pathways that connects oral dysbiosis with MetS and other systemic diseases is microbial translocation from the oral cavity to the systemic circulation (Olsen and Yamazaki 2019). This process involves the movement of oral bacteria, their metabolites, and inflammatory agents across the oral mucosal barrier into the systemic circulation, possibly triggering systemic inflammation and metabolism (Hajishengallis 2015; Tan et al. 2023). The processes that allow microbial translocation are complex and depend on both host and microbial factors. There are two major pathways for bacterial translocation, one of which involves disruption of the integrity of the oral mucosal barrier, which is usually related to periodontal inflammation (Tomás et al. 2012; Vitkov et al. 2023). Periodontal pockets are deep invaginations of gingival tissues, creating an ideal environment for bacterial growth. The mucosal lining of these pockets is often ulcerated, providing direct access to bacteria and their by-products into the systemic circulation (Hujoel et al. 2001; Purno and Namrata 2023).

LPS, a major outer membrane component of Gram-negative bacteria, is a potent pro-inflammatory molecule, which has a major role in the systemic effects of oral microbial translocation (Olsen 2016). Research has shown that LPS levels in the bloodstream are increased in patients with periodontitis and are related to inflammatory markers (Pussinen et al. 2007; Zeng et al. 2024). The link between MetS and circulating LPS levels is well-established, with LPS linked to insulin resistance, dyslipidemia, and inflammation in the adipose tissue (Barchetta et al. 2023; Cani et al. 2007). Another way that oral dysbiosis affects the body is through a phenomenon called bacteremia, in which viable oral bacteria are transported to the bloodstream. Chewing and tooth brushing can cause transient bacteremia in healthy individuals, with a higher frequency and intensity in those with poor oral hygiene or periodontal inflammation (Forner et al. 2006; Titiz Yurdakal et al. 2024). Although these bacteria are usually rapidly cleared by the immune system, constant or prolonged circulation of oral pathogens in the bloodstream may lead to chronic low-grade inflammation and metabolic imbalance (Hajishengallis 2015).

The vascular endothelium is another key target of translocated oral microbial products. Endothelial activation and dysfunction provoked by LPSs play a critical role in the pathogenesis of atherosclerosis, which is considered one of the most significant comorbidities of MetS (Stoll et al. 2004b; Xue et al. 2023). Exposure of endothelial cells to LPSs induces expression of adhesion molecules and release of pro-inflammatory cytokines that recruit leukocytes and provoke vascular inflammation (Hajishengallis 2015). This endothelial dysfunction not only participates in the advancement of atherosclerosis but also interferes with insulin-mediated vasodilation, which, in turn, increases insulin resistance (Lopes-Virella and Virella 2023; Muniyappa and Sowers 2013). Additionally, research has been focused on the potential of oral microbial translocation to alter the gut microbiota, as a way of showing how oral dysbiosis can affect metabolic health. Oral bacteria that are swallowed can colonize the gut and change their microbial profile or function (Atarashi et al. 2017; Elzayat et al. 2023).

6.3 Oral Hyperpermeability and Systemic Inflammation

6.3.1 The Concept of Oral Hyperpermeability

Oral hyperpermeability, also known as the "leaky mouth," is a relatively new concept that is central to the oral–systemic health connection, especially to MetS (Olsen and Yamazaki 2019). This condition refers to increased permeability of the oral mucosal barrier, which leads to increased translocation of oral bacteria, bacterial by-products, and pro-inflammatory molecules into the bloodstream (Hajishengallis 2014b). The oral mucosa is an important barrier that separates the outer environment from the deeper tissues of the oral cavity, and it has great importance in the regulation of oral mucosal homeostasis and protection against pathogenic agents and toxic substances (Di Sabatino et al. 2023; Squier and Kremer 2001). This barrier function is maintained through the interactions of structural and functional layers such as the stratified squamous epithelium, the basement membrane, and the underlying connective tissue (Hegde et al. 2023; Presland and Dale 2000).

The epithelial layer of the oral mucosa is composed of closely packed cells that are bound by intercellular junctions such as tight junctions, adherent junctions, and desmosomes (Groeger and Meyle 2015). These junctional complexes control paracellular permeability and are important for the proper function of the mucosal barrier. For instance, tight junctions are especially involved in the regulation of the permeability of molecules and ions across cells by providing a selective barrier to the cells that are involved in the maintenance of tissue homeostasis (Anderson and Van Itallie 2009). Apart from the epithelial layer, the oral mucosa employs additional protective mechanisms. Saliva secretion is continuous and ensures that the oral cavity is cleansed, maintaining a favorable pH and delivering antimicrobial proteins (Dawes et al. 2015). In addition, the oral mucosa is equipped with an elaborate immune system comprising dendritic cells, macrophages, and lymphocytes that are involved in mucosal immunity and defense (Abdullah and Mohtarrudin 2023; Novak et al. 2008).

However, several factors can disrupt the protective mechanisms that negatively affect the stability of the oral mucosal barrier and increase its permeability. It has been established that the main cause of hyperpermeability in the mouth is longstanding inflammation, especially periodontal disease (Hajishengallis 2014b). Inflammation may also affect the intercellular junctions, which may cause alterations in paracellular permeability (Groeger and Meyle 2015). Tumor necrosis factor-alpha (TNF-α) and interleukin-1β (IL-1β) have been shown to reduce the expression of tight junction proteins, thereby compromising the barrier's integrity (Al-Sadi et al. 2009). Microbial dysbiosis in the oral cavity can also contribute to oral hyperpermeability through various factors such as degradation of the oral epithelial barrier. Some oral pathogens like *Porphyromonas gingivalis* secrete proteases, which have been shown to cleave intercellular junction proteins, enhancing epithelial permeability (Katz et al. 2000; Vitkov et al. 2023). In addition, bacterial products such as LPSs can alter the morphology and function of epithelial cells and therefore exacerbate barrier dysfunction (Guo et al. 2013; Rogers et al. 2023).

Several environmental factors are also known to influence the permeability of the oral mucosa. For example, tobacco use involves enhanced oral permeability, which could be due to oxidative stress that is capable of damaging the epithelial barrier (Gajendra et al. 2023; Ryder et al. 1998). Similarly, alcohol consumption increases mucosal permeability due to the alteration of tight junction structure and function (Kuo et al. 2024; Waszkiewicz et al. 2013). Diets also have an impact on oral mucosal permeability, and this is one of the reasons that people are advised to take healthy diets.

6.3.2 Oral Hyperpermeability in Metabolic Syndrome

Given the strong correlation between oral hyperpermeability and metabolic syndrome, the oral cavity serves as a pivotal interface influencing both the pathophysiology and therapeutic strategies for metabolic disorders (Olsen and Yamazaki 2019). Oral hyperpermeability is one of the ways through which MetS develops, since it helps in the translocation of bacteria and inflammatory mediators into the systemic circulation (Hajishengallis 2014b). This low-grade, sustained, systemic inflammation is generally described as "metaflammation" and is the cornerstone of MetS, since it is responsible for the emergence of insulin resistance, dyslipidemia, and cardiovascular pathology (Hotamisligil 2017).

It has been shown that levels of circulating bacterial endotoxins are increased in patients with MetS and are linked to systemic inflammation and insulin resistance (Bergheim and Moreno-Navarrete 2024; Lassenius et al. 2011). For example, Pussinen et al. (2011) observed that higher serum LPS activity was an independent predictor of the risk of further MetS development. This indicates that the translocation of bacterial by-products from the oral cavity may be a significant cause of metabolic disorder development and progression. The consequences of oral hyperpermeability are not limited to bacterial translocation at the systemic level. The increased permeability also permits the passage of greater molecules such as the pro-inflammatory cytokines resulting from oral dysbiosis and periodontal inflammation, as those mentioned by Hajishengallis (2015). Some of these cytokines, such as TNF-α, IL-6, and IL-1β, can independently cause insulin resistance and metabolic disorder in tissues (Shoelson et al. 2006; Tylutka et al. 2024).

In the adipose tissue, oral-derived inflammatory mediators can further aggravate inflammation and thus alter insulin signaling and secretion of adipokines (Ouchi et al. 2011). This not only leads to systemic insulin resistance but also enhances a pro-inflammatory environment that can negatively affect the oral mucosal barrier and develop a cycle of inflammation and metabolic disturbance (Genco and Sanz 2020). Being the primary site of endotoxin clearance, the liver is especially sensitive to alterations in oral hyperpermeability. Increased exposure to bacterial by-products and inflammatory mediators can directly contribute to the development of hepatic inflammation, which is a characteristic feature of nonalcoholic fatty liver disease (NAFLD), a condition closely linked to MetS (Buzzetti et al. 2016). The reduced hepatic uptake of insulin also leads to an impaired lipid profile, which, in addition to other metabolic disturbances of the syndrome, worsens the symptoms of MetS (Tilg and Moschen 2010).

Pancreatic β-cells are also vulnerable to increased concentrations of inflammatory mediators and bacterial by-products, resulting from oral hyperpermeability. Prolonged exposure to these factors may compromise the function of β-cells and stimulate apoptosis, thereby accelerating the transition from insulin resistance to T2DM (Donath and Shoelson 2011; Khin et al. 2023). This reveals the significance of maintaining the integrity of the oral mucosal barrier to prevent pancreatic β-cell dysfunction. Another important target of the systemic effects of oral hyperpermeability is the vascular endothelium. Increased exposure to bacterial by-products and inflammatory mediators can contribute to endothelial activation and dysfunction, a major step in the formation of atherosclerosis and hypertension (Cimmino et al. 2023; Stoll et al. 2004a). This endothelial dysfunction not only leads to cardiovascular disease associated with MetS but also impairs insulin-mediated vasodilation, which, in turn, aggravates insulin resistance (Muniyappa and Sowers 2013).

Surprisingly, the connection between oral hyperpermeability and MetS seems to be bidirectional. Although increased oral permeability augments the development and progression of MetS, the metabolic disturbances associated with this syndrome impair outer medullary blood flow (OMBF) (Genco and Sanz 2020).

For example, elevated blood glucose levels due to insulin resistance can result in advanced glycation end product (AGE) formation; AGEs have been reported to adversely affect the integrity of the epithelial barrier and stimulate inflammation (Vlassara and Uribarri 2014).

6.4 Periodontitis as a Mediator of Metabolic Syndrome

6.4.1 Pathophysiology of Periodontitis

Periodontitis is a chronic inflammatory disease of the supporting tissues of the teeth with irreversible attachment loss and destruction of the periodontal ligament and alveolar bone (Pihlstrom et al. 2005). This condition is one of the most important oral health concerns worldwide, with severe periodontitis impacting 10–15% of the adult population (Kassebaum et al. 2014). Periodontitis has a multifactorial etiology, affected by various factors such as the oral microbiota, the host immune system, and the environment (Hajishengallis 2015). Periodontitis is mainly characterized by changes in the subgingival microbiota, and its initiation and progression are mainly attributed to microbial dysbiosis (Hajishengallis et al. 2015). In healthy periodontal tissues, the periodontal microbiome involves a complex population of microorganisms that peacefully coexist with the host, exerting several beneficial effects on the nearby tissues (Ray 2023; Socransky and Haffajee 2005). However, several conditions can interfere with this normal homeostasis, and, as a consequence, the pathogenic species start to dominate and dysbiosis of the microbiota occurs (Chu et al. 2023; Hajishengallis and Lamont 2012). Periodontal pathogens, which are frequently addressed as red complex bacteria, are *Porphyromonas gingivalis*, *Tannerella forsythia*, and *Treponema denticola* (Socransky et al. 1998). These organisms have multiple factors that allow them to survive in the subgingival area, avoid host immune response, and cause tissue damage (Holt and Ebersole 2005; Shahoumi et al. 2023). For example, *P. gingivalis* synthesizes gingipains, which are highly active proteolytic enzymes that can degrade host proteins, suppress the immune response, and promote tissue destruction (Chen et al. 2023; Imamura 2003). The host immune response is an extremely significant contributor to the development of periodontitis. At first, innate immunity detects the microbial threat using pattern recognition receptors (PRRs), such as TLRs, to identify pathogen-associated molecular patterns (PAMPs) (Mahanonda and Pichyangkul 2007). This recognition leads to the synthesis of pro-inflammatory cytokines and chemokines, which attract neutrophils and other pro-inflammatory immune cells to the site of infection (Garlet 2010).

Although acute inflammation plays a protective role by targeting and destroying microbial invaders, chronic inflammation leads to tissue breakdown. Periodontitis is characterized by the transition from neutrophil-driven acute inflammation to chronic inflammation that is dominated by lymphocytes and macrophages (Cekici et al. 2014). This chronic inflammatory state leads to the synthesis of matrix metalloproteinases (MMPs) and pro-inflammatory cytokines, such as IL-1β, TNF-α, IL-6, and prostaglandin E2 (PGE2), which are involved in the degradation of periodontal tissues (Graves and Cochran 2003) through both direct and indirect mechanisms. Some bacterial products like LPSs and proteases have a direct toxic effect on periodontal tissues (Basic and Dahlén 2023; Isola et al. 2023). However, the majority of tissue destruction is mediated by the host's immune response to invading pathogens. Neutrophils also produce reactive oxygen species (ROS) and proteases that, while part of the immune system, can also promote tissue injury and inflammation (Scott and Krauss 2012).

Periodontitis is further advanced through periodontal pockets that are deepened gingival crevices resulting from the loss of periodontal attachment and alveolar bone (Isola et al. 2023; Pihlstrom et al. 2005). These pockets create the most ideal conditions for anaerobic bacterial growth, further fueling dysbiosis and perpetuating the cycle of inflammation and tissue degradation (Hajishengallis et al. 2015).

6.4.2 Periodontitis and Metabolic Syndrome

The association between periodontitis and MetS has recently become a focus of interest, and the existing literature points to a bidirectional relationship between these two conditions. Cross-sectional investigations have also shown a significantly higher prevalence of periodontitis in individuals with MetS, and vice versa, indicating that patients with periodontal diseases are more prone to developing MetS (Nibali et al. 2013). A meta-analysis by Nibali et al. (2013) found that individuals with MetS had a significantly higher risk of periodontitis compared to those without MetS. This association persisted even after adjusting for potential confounders including age, sex, and smoking history. Similarly, a longitudinal study by Morita et al. (2010) demonstrated that individuals with periodontitis had a higher risk of developing MetS over a 4-year follow-up period.

There is also considerable evidence regarding the association between both conditions. Chaffee and Weston (2010) conducted a meta-analysis examining the relationship between periodontitis and obesity, finding a significant association between the two. Moreover, in their systematic review, Martin-Cabezas et al. (2016) reported a significant relationship between periodontitis and hypertension. Overall, a significant increase in the incidence of stroke in patients with diabetes was reported.

As mentioned previously, periodontitis and MetS are bidirectionally interconnected. MetS may put an individual at a higher risk of periodontitis; at the same time, periodontal inflammation may worsen metabolic dysfunction (Genco and Sanz 2020). This interaction is best explained by several pathological processes that relate periodontal infection to metabolic dysfunctions in the body. The translocation of oral bacteria, particularly lipopolysaccharides (LPSs), into the bloodstream triggers inflammatory pathways in adipocytes and hepatocytes, leading to insulin resistance and metabolic dysfunction (Olsen and Yamazaki 2019). In their study, Amar et al. (2009) showed that oral infection with *Porphyromonas gingivalis*, a periodontal pathogen, caused systemic inflammation and impaired glucose tolerance in mice. The authors observed higher serum concentrations of TNF-α, IL-6, and CRP as well as insulin resistance in skeletal muscle glucose uptake. These results offer the molecular basis for the association between periodontal infection and metabolic dysfunction. Untreated periodontal inflammation is a low-grade inflammation, causing chronic systematic inflammation with increased levels of cytokines like TNF-α, IL-6, and IL-1β (Loos 2005). These pro-inflammatory mediators are involved in the pathophysiology of MetS. For example, TNF-α contributes to insulin resistance through alteration of the insulin signaling pathways in the adipose tissue and skeletal muscles (Hotamisligil et al. 1993; Szukiewicz 2023).

A meta-analysis by Borgnakke et al. (2013) found that individuals with periodontitis had a higher risk of developing T2DM compared to those without periodontitis, and this study supports the hypothesis that increased age is associated with a higher risk of developing heart failure. The factors by which this relationship is driven are complex. Several mechanisms have been proposed to explain the link between periodontitis and insulin resistance, including chronic inflammation, oxidative stress, and dysregulation of adipokine secretion (Preshaw et al. 2012). In addition, periodontal pathogens and their by-products can also reduce insulin sensitivity and glucose metabolism in target organs (Păunică et al. 2023; Su et al. 2023).

6.4.3 Periodontitis as a Biomarker and Therapeutic Target

There has been increasing interest in studying the role of periodontitis in identifying and treating MetS. This dual perspective provides promising opportunities for increasing the chances of early detection, appropriate classification, and management of individuals who are either predisposed to or already affected by MetS.

Periodontitis has some advantages when it comes to being considered as a biomarker. Clinical oral examinations are generally minimally invasive and can be carried out frequently, and they may yield important information concerning the general health of an individual (Genco and Sanz 2020). Some previous investigations have examined the potential use of periodontal parameters as a MetS biomarker. In the study conducted by Benguigui et al. (2010), clinical attachment loss (CAL) was significantly linked to the number of MetS components. The authors suggested that periodontal examination might be used as a low-cost screening method to determine individuals with potential MetS. In the same way, Kushiyama et al. (2009) observed a significant relationship between the community periodontal index (CPI) and MetS prevalence, which could serve as a biomarker for metabolic risk assessment.

Another potential approach for linking oral health to metabolic risk is the employment of salivary biomarkers. Saliva is a complex fluid that contains molecules relevant to oral and systemic health. Its collection method is noninvasive and simple for repeated testing (Yoshizawa et al. 2013). In another study, Salminen et al. (2015) reported that salivary matrix metalloproteinase-8 (MMP-8), a marker of periodontal inflammation, was found to be significantly associated with MetS, regardless of conventional cardiovascular risk factors. This indicates that there might be a possibility of using other salivary biomarkers to complement the existing metabolic risk assessment, in addition to clinical measures.

Other microbial markers that have been examined for periodontitis have also been considered as biomarkers of metabolic risk. In a cross-sectional study, Demmer et al. (2015) collected subgingival biofilm samples and discovered that periodontal pathogens are related to insulin resistance and impaired fasting glucose (IFG). The authors suggested that the subgingival microbiota might be used as a new diagnostic marker for diabetes. Periodontal inflamed surface area (PISA), which reflects the extent of inflammation of periodontal tissues, has been suggested as a more comprehensive measure of periodontal inflammation (Nesse et al. 2008; Tang et al. 2023). Iwasaki et al. (2018) observed that PISA was positively and significantly correlated with the number of MetS components, indicating its possibility as a quantitative biomarker of metabolic risk.

Several clinical trials have assessed the effects of periodontal treatment on at least one aspect of MetS. A systematic review and meta-analysis by Teeuw et al. (2014) found that periodontal treatment resulted in modest but significant improvements in glycemic control in T2DM patients, as evidenced by reductions in HbA1c levels. The observed improvement in glycemic control was greater than the effect of periodontal interventions on pain reduction during the study. While the benefits of periodontal therapy on other aspects of MetS have been less conclusive, they are quite encouraging. D'Aiuto et al. (2018) conducted a randomized controlled trial showing that 6 months of intensive periodontal treatment (IPT) resulted in significant enhancement of endothelial function, decreased systemic inflammation, and reduced blood pressure in patients with severe periodontitis. From these results, it can be concluded that periodontal treatments might have additional effects on other parts of the body.

There are several possible ways through which periodontal therapy can lead to improved metabolic control, potentially due to a decrease in systemic inflammation reflected by a reduction in the levels of CRP and IL-6 after periodontal treatment (D'Aiuto et al. 2004; Matsuda et al. 2024). This lowering of systemic inflammation may, in turn, enhance insulin sensitivity as well as vascular health. Moreover, periodontal therapy decreases bacterial load in the oral cavity, thereby minimizing the translocation of periodontal pathogens and their by-products into the systemic circulation. Such a decrease in bacteremia and endotoxin exposure might be useful for metabolic control in different tissues, including the liver and adipose tissue (Olsen and Yamazaki 2019).

The term "periodontal medicine" has been developed to describe the branch of science that studies the relationship between periodontal diseases and systemic diseases such as MetS (Kuraji et al. 2023; Lee 2023; Williams and Offenbacher 2000). It focuses on the concept of linking oral disease prevention and treatment to overall healthcare delivery. For patients with MetS, it might include routine periodontal examination, effective treatment of periodontal inflammation, and integration of dental and medical treatment. New therapeutic approaches increasingly emphasize the oral–systemic health relationship. For example, supplementation with oral probiotics was found to influence both the oral and gut microbiota and may have positive effects on the periodontal and metabolic status (Kobayashi et al. 2017). Similarly, locally applied anti-inflammatory substances to periodontal tissues can be used to control both oral and systemic inflammation with fewer side effects on the body.

6.5 Therapeutic Approaches and Clinical Applications

6.5.1 Integrated Oral–Systemic Health Strategies

There is a significant need for a combined approach that can target both conditions and seek to enhance the general well-being of patients by tackling the common causes of diseases, lowering inflammation levels, and enhancing the overall management of patients. At the center of this model is the cooperation between dentists and other doctors, such as primary care doctors, endocrinologists, and cardiologists, among others. This model of care delivery makes it possible for patients to receive both oral and other general health evaluation and intervention services (Genco and Sanz 2020).

In a study by Lamster et al. (2008), the authors showed that incorporating oral health assessments into primary care practice is possible and may have positive outcomes. The authors identified that implementation of simple oral health screening and risk assessment during primary care visits increased detection of patients at risk of both oral and systemic diseases, such as diabetes and cardiovascular diseases. Similarly, integration of metabolic risk assessment in dental care would improve identification of MetS at an early stage. Supporting this approach, Friman et al. (2015) demonstrated that dental hygienists could identify diabetes risk by examining the periodontal status and administering the diabetes risk questionnaire, suggesting the important role of dental professionals in metabolic risk assessment.

Ramseier et al. (2020) conducted a systematic review on lifestyle interventions and their effects on periodontal and cardiometabolic health, concluding that dietary modification and increased physical activity were particularly associated with positive changes in both aspects. Of these, nutritional care is one of the most promising approaches for improving both oral and systemic health. There is a clear link between diet, oral health, and MetS, and they cyclically influence one another. A diet high in sugar and refined carbohydrates has been linked to poor oral health and metabolic disorders (Hujoel 2009; Kaur 2023). On the other hand, periodontal inflammation and tooth loss may influence dietary patterns and nutrient intake as well, which may further intensify metabolic imbalance.

Woelber et al. (2017) proved that a low-carbohydrate, anti-inflammatory diet resulted in a significant decrease in gingival bleeding and probing depths in the context of periodontal management. Beyond oral health, these dietary modifications may also have other positive effects on metabolic health, thus supporting the role of nutritional interventions in treating oral and systemic inflammation at the same time. Oral probiotics are now considered one of the novel methods of managing oral and systemic health. Probiotics are defined as living microorganisms that have significant health benefits when consumed in sufficient quantities (Hill et al. 2014). In the context of oral health, probiotics have been found to have the potential to decrease the inflammation of gingival tissues and also reform the composition of oral microbes (Teughels et al. 2011).

Tekce et al. (2015) conducted a randomized controlled clinical trial to compare the effects of using a probiotic lozenge containing *Lactobacillus reuteri* in combination with scaling and root planing (SRP) compared to SRP alone. They reported that periodontal parameters were improved and pro-inflammatory cytokine levels were decreased in the group that received the probiotic lozenge. These observations indicate that oral probiotics might help mitigate both local and systemic inflammation. Oral probiotics may have broader systemic benefits, which are not limited to only periodontal health. In their study, Sabatini et al. (2017) also showed that multi-strain probiotic supplements had positive effects on glycemic control and decreased the markers of oxidative stress in patients with T2DM and chronic periodontitis. These results underscore the possibility of using oral probiotics as a potential treatment strategy to target both oral and metabolic diseases at the same time.

Another area that may be targeted for manipulation of oral and systemic health is prebiotics—nondigested food components that selectively stimulate the growth and/or activity of beneficial microorganisms (Gibson et al. 2017). Regarding oral prebiotics, the current studies are quite limited; however, the results indicate that arginine and nitrate may have positive effects on oral and cardiovascular health (Rosier et al. 2018). Keller (2021) demonstrated that increased dietary nitrate intake enhances oral nitrate-reducing capacity and lowers blood pressure in healthy individuals. Given the crucial role of nitric oxide in vascular function and the emerging link between oral nitrate-reducing bacteria and cardiovascular health, this area holds significant promise for further investigation, particularly in the context of MetS. Understanding how dietary nitrate influences oral bacteria and, subsequently, cardiovascular health in MetS patients could open avenues for novel preventative and therapeutic strategies.

6.5.2 Innovative Therapies Targeting Oral Permeability and Dysbiosis

In light of the oral–systemic health connection, new therapies aimed at modulating oral permeability and dysbiosis have been developed as approaches to treat both oral and metabolic diseases at the same time. These approaches are designed to attempt reinstating the integrity of the oral mucosal barrier, recognizing the oral microbiome, and decreasing the systemic inflammation that may provide novel therapeutic strategies for the prevention and treatment of MetS.

6.5.2.1 Barrier-Enhancing Agents

The idea of enhancing the oral mucosal barrier to lessen the systemic effects of oral inflammation and microbial infiltration has recently attracted researchers' interest. Many agents and approaches have been used to promote the restoration and functionality of the oral epithelial layer and its associated structures. One potential approach is the application of natural compounds with barrier-protective effects. For instance, some nonessential polyphenols present in green tea, such as epigallocatechin-3-gallate (EGCG), have been reported to have positive impacts on the barrier function of the oral mucosa. In their study, Liao et al. (2020) showed that EGCG improved the expression of tight junction proteins and decreased inflammation in gingival epithelial cells infected by periodontal pathogens. These results indicate that natural compounds offer a relatively safe method for enhancing oral barrier function.

The use of advanced tissue engineering models to study oral mucosal biology and regeneration is a novel concept for investigating mucosal barrier function in periodontal diseases. Imai et al. (2023) created a three-dimensional 3D, tissue-engineered oral mucosal model using human oral keratinocytes and fibroblasts. Compared to the traditional two-dimensional cell culture system, this model demonstrated enhanced barrier properties and more physiologically relevant inflammatory responses. Therefore, it can be used to understand oral mucosal biology and design and test new therapeutic interventions.

Another strategy that can be employed by nanoparticle delivery systems is improving the integrity of the oral mucosal barrier. A study by Moulari et al. (2013) examined the application of chitosan-coated nanoparticles to deliver anti-inflammatory drugs to the oral mucosa. The authors reported that this approach increased drug retention and its anti-inflammatory properties and proposed the use of this method in the treatment of periodontal diseases, thereby potentially reducing epithelial damage and helping preserve mucosal integrity. All in all, another approach for the modulation of oral mucosal immunity is the improvement of barrier function and the decrease of inflammation.

6.5.2.2 Microbiome Modulation

Optimization of the oral microbiome as a way of minimizing dysbiosis is one of the main objectives of new approaches to oral–systemic health. These approaches are intended to suppress pathogenic species, while encouraging the growth of beneficial bacteria that can enhance oral and metabolic health. Precision probiotics are a more developed approach to target the microbiome. In contrast to conventional probiotics that contain nonspecific bacterial strains, precision probiotics are developed to target specific microbial imbalances. Mattos-Graner et al. (2014) conducted a study on the clinical trial of a genetically modulated *Streptococcus mutans* strain that secretes an antibacterial peptide that is lethal to cariogenic bacteria. The authors stated that this approach resulted in a highly significant decrease in the cariogenic bacteria in the oral cavity, while still maintaining the balance of the microbiota.

Inspired by the success of fecal microbiota transplantation for gut dysbiosis, researchers have been exploring the potential use of oral microbiome transplantation (OMT) for treating oral diseases. Pozhitkov et al. (2015) examined the possibility of using oral microbiota transplantation to treat periodontitis. Although it was a small pilot study, it pointed to the possibility of changing the microbial community in a direction that would be healthier for the body, which is a new avenue in managing oral dysbiosis.

Another novel strategy for targeting the oral microbiome is the use of next-generation antimicrobial peptides. These peptides, which are frequently based on natural host defense proteins, are capable of selectively eliminating pathogenic bacterial strains with little impact on the rest of the microbial population. Xu et al. (2020) described another study that created a new antimicrobial peptide with an affinity for *Porphyromonas gingivalis*, one of the major periodontal pathogens. The peptide proved to be effective against *P. gingivalis*, while its effect on the normal oral flora was negligible, thus indicating the possibility of selectively using the antimicrobial peptide in treating periodontal disease.

Photodynamic therapy (PDT) has been identified as one of the most effective non-antibiotic approaches to control oral dysbiosis. This technique involves the use of light-sensitive photosensitizing agents to selectively destroy pathogenic organisms. Akram et al. (2016) conducted a systematic review and meta-analysis to assess the clinical effectiveness of adjunctive PDT in nonsurgical periodontal therapy and reported a significant improvement in the clinical parameters in favor of combination therapy. The selective action of PDT may have advantages over other broad-spectrum antimicrobial approaches, since it eliminates certain pathogens, while minimizing the influence on the general microbial community.

Disruption of quorum sensing, the "language" that bacteria use to coordinate their activities such as the regulation of biofilm formation, is another novel strategy to address oral dysbiosis. For instance, quorum-sensing inhibitors can reduce the pathogenic activity of bacteria, without broadly disrupting the commensal microbiome.

Exploiting the symbiotic relationship between the oral and gut microbiota provides novel approaches to oral and systemic diseases. A recent study by Koren et al. (2021) has demonstrated that certain oral commensal bacteria could be delivered orally to alter the oral microbiota and decrease the inflammation in the periodontal tissues of mice. The communication between oral and gut microbiomes underlines the possibility of synergistic strategies in the modulation of microbiomes for the treatment of MetS.

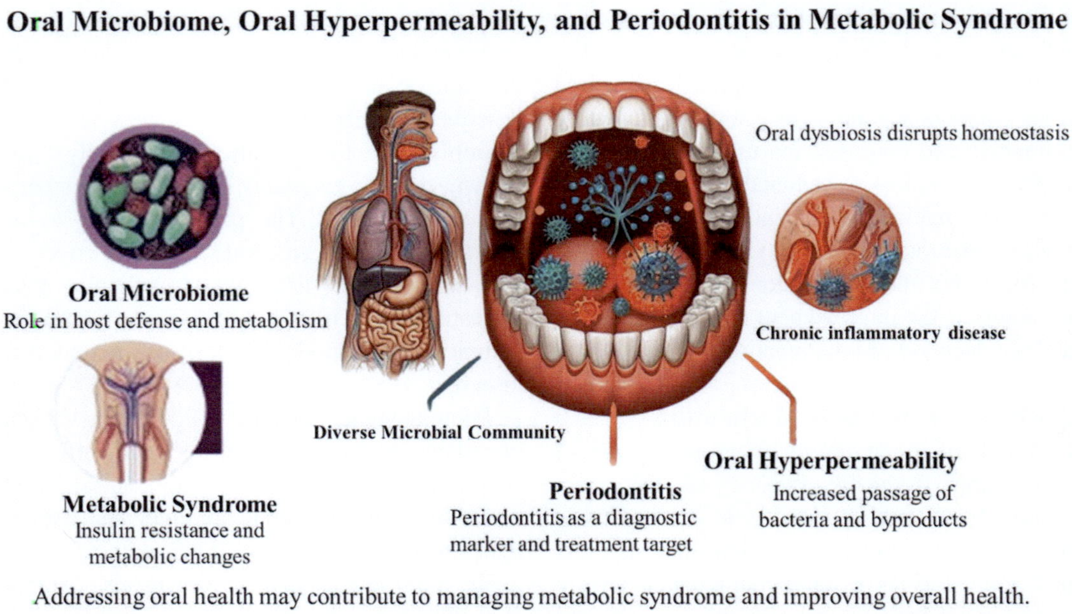

Fig. 6.2 Visual representation of the oral-systemic connection between the oral microbiome, oral hyperpermeability, periodontitis, and MetS. The image depicts how oral dysbiosis disrupts homeostasis, leading to oral hyperpermeability and systemic inflammation, ultimately contributing to MetS. It emphasizes the role of periodontitis as a diagnostic marker and therapeutic target. *MetS* metabolic syndrome

6.6 Conclusions

There are links between the oral microbiome, oral hyperpermeability, periodontitis, and MetS, illustrating the connection between oral and general health (Fig. 6.2). Dysbiosis of the oral microbiota may lead to periodontal inflammation, increased permeability of the oral mucosa, and systemic and chronic low-grade inflammation, which contribute to the pathogenesis of MetS. On the other hand, MetS can also change oral conditions, which intensify dysbiosis and inflammation. Oral hyperpermeability is a process that allows the movement of bacteria, metabolites, and inflammatory molecules through the oral mucosa into the bloodstream, which is associated with chronic inflammation observed in MetS. Periodontitis acts as both a risk factor and a consequence of MetS, which requires the management of both diseases at the same time. New ideas such as oral permeability and dysbiosis may hold promise for understanding and addressing oral–systemic health links. Barrier-enhancing agents, precision probiotics, and new antimicrobial strategies may offer the potential to reestablish oral microbial homeostasis and decrease systemic inflammation. Further studies should be designed with the main aim of identifying the interaction between oral dysbiosis, inflammation, and metabolic disturbances and evaluating the effectiveness of multimodal interventions in the long term.

References

Abdulhadi AMH, Amer I, Mohammed MK (2024) Gram positive bacteria and their distributions according anatomical site in oral cavity and effects on oral health. Int J Health Sci *8*(2):180–193

Abdullah M, Mohtarrudin N (2023) Oral immune system. In: Immunology for dentistry. Wiley, Hoboken, pp 13–25

Ahima RS (2024) Overview of metabolic syndrome. In: Metabolic syndrome: a comprehensive textbook. Springer, Cham, pp 3–14

Akram Z, Abduljabbar T, Abu Hassan MI, Javed F, Vohra F (2016) Cytokine profile in chronic periodontitis patients with and without obesity: a systematic review and meta-analysis. Dis Markers 2016(1):4801418

Al-Sadi R, Boivin M, Ma T (2009) Mechanism of cytokine modulation of epithelial tight junction barrier. Front Biosci 14:2765

Alberti KG, Eckel RH, Grundy SM, Zimmet PZ, Cleeman JI, Donato KA, Fruchart J-C, James WPT, Loria CM, Smith SC Jr (2009) Harmonizing the metabolic syndrome: a joint interim statement of the international diabetes federation task force on epidemiology and prevention; national heart, lung, and blood institute; American heart association; world heart federation; international atherosclerosis society; and international association for the study of obesity. Circulation 120(16):1640–1645

Aleman RS, Moncada M, Aryana KJ (2023) Leaky gut and the ingredients that help treat it: a review. Molecules 28(2):619

Amar S, Wu S-C, Madan M (2009) Is Porphyromonas gingivalis cell invasion required for atherogenesis? Pharmacotherapeutic implications. J Immunol 182(3): 1584–1592

Anderson JM, Van Itallie CM (2009) Physiology and function of the tight junction. Cold Spring Harb Perspect Biol 1(2):a002584

Anderson AC, Rothballer M, Altenburger MJ, Woelber JP, Karygianni L, Lagkouvardos I, Hellwig E, Al-Ahmad A (2018) In-vivo shift of the microbiota in oral biofilm in response to frequent sucrose consumption. Sci Rep 8(1):14202

Arimatsu K, Yamada H, Miyazawa H, Minagawa T, Nakajima M, Ryder MI, Gotoh K, Motooka D, Nakamura S, Iida T (2014) Oral pathobiont induces systemic inflammation and metabolic changes associated with alteration of gut microbiota. Sci Rep 4(1):4828

Atarashi K, Suda W, Luo C, Kawaguchi T, Motoo I, Narushima S, Kiguchi Y, Yasuma K, Watanabe E, Tanoue T (2017) Ectopic colonization of oral bacteria in the intestine drives TH1 cell induction and inflammation. Science 358(6361):359–365

Barchetta I, Cimini FA, Sentinelli F, Chiappetta C, Di Cristofano C, Silecchia G, Leonetti F, Baroni MG, Cavallo MG (2023) Reduced lipopolysaccharide-binding protein (LBP) levels are associated with non-alcoholic fatty liver disease (NAFLD) and adipose inflammation in human obesity. Int J Mol Sci 24(24): 17174

Basic A, Dahlén G (2023) Microbial metabolites in the pathogenesis of periodontal diseases: a narrative review. Front Oral Health 4:1210200

Benguigui C, Bongard V, Ruidavets JB, Chamontin B, Sixou M, Ferrières J, Amar J (2010) Metabolic syndrome, insulin resistance, and periodontitis: a cross-sectional study in a middle-aged French population. J Clin Periodontol 37(7):601–608

Bergheim I, Moreno-Navarrete JM (2024) The relevance of intestinal barrier dysfunction, antimicrobial proteins and bacterial endotoxin in metabolic dysfunction-associated steatotic liver disease. Eur J Clin Investig 54(7):e14224

Borgnakke WS, Ylöstalo PV, Taylor GW, Genco RJ (2013) Effect of periodontal disease on diabetes: systematic review of epidemiologic observational evidence. J Periodontol 84:S135–S152

Bryan NS, Tribble G, Angelov N (2017) Oral microbiome and nitric oxide: the missing link in the management of blood pressure. Curr Hypertens Rep 19:1–8

Buzzetti E, Pinzani M, Tsochatzis EA (2016) The multiple-hit pathogenesis of non-alcoholic fatty liver disease (NAFLD). Metabolism 65(8):1038–1048

Cani PD, Amar J, Iglesias MA, Poggi M, Knauf C, Bastelica D, Neyrinck AM, Fava F, Tuohy KM, Chabo C (2007) Metabolic endotoxemia initiates obesity and insulin resistance. Diabetes 56(7):1761–1772

Cekici A, Kantarci A, Hasturk H, Van Dyke TE (2014) Inflammatory and immune pathways in the pathogenesis of periodontal disease. Periodontol 64(1):57–80

Chaffee BW, Weston SJ (2010) Association between chronic periodontal disease and obesity: a systematic review and meta-analysis. J Periodontol 81(12): 1708–1724

Chai X, Liu L, Chen F (2024) Oral nitrate-reducing bacteria as potential probiotics for blood pressure homeostasis. Front Cardiovasc Med 11:1337281

Chang Y-R, Cheng W-C, Hsiao Y-C, Su G-W, Lin S-J, Wei Y-S, Chou H-C, Lin H-P, Lin G-Y, Chan H-L (2023) Links between oral microbiome and insulin resistance: involvement of MAP kinase signaling pathway. Biochimie 214:134–144

Checa-Ros A, Hsueh W-C, Merck B, González-Torres H, Bermúdez V, D'Marco L (2024) Obesity and oral health: the link between adipokines and periodontitis. touchREV Endocrinol 20(1):25

Chen WA, Dou Y, Fletcher HM, Boskovic DS (2023) Local and systemic effects of Porphyromonas gingivalis infection. Microorganisms 11(2):470

Chopra A, Franco-Duarte R, Rajagopal A, Choowong P, Soares P, Rito T, Eberhard J, Jayasinghe TN (2024) Exploring the presence of oral bacteria in non-oral sites of patients with cardiovascular diseases using whole metagenomic data. Sci Rep 14(1):1476

Chowdhury A, Edem D, Raj R, Nain P, Joglekar M, Verma V, Kant R (2023) Gut microbiome supplementation as therapy for metabolic syndrome. World J Diabetes 14(10):1502

Chu J, Feng S, Guo C, Xue B, He K, Li L (2023) Immunological mechanisms of inflammatory diseases caused by gut microbiota dysbiosis: a review. Biomed Pharmacother 164:114985

Chweidan H, Rudyuk N, Tzur D, Goldstein C, Almoznino G (2024) Statistical methods and machine learning algorithms for investigating metabolic syndrome in temporomandibular disorders: a nationwide study. Bioengineering 11(2):134

Cimmino G, Muscoli S, De Rosa S, Cesaro A, Perrone MA, Selvaggio S, Selvaggio G, Aimo A, Pedrinelli R, Mercuro G (2023) Evolving concepts in the pathophysiology of atherosclerosis: from endothelial dysfunction to thrombus formation through multiple shades of inflammation. J Cardiovasc Med 24(Supplement 2):e156–e167

D'Aiuto F, Parkar M, Andreou G, Suvan J, Brett PM, Ready D, Tonetti MS (2004) Periodontitis and systemic inflammation: control of the local infection is associated with a reduction in serum inflammatory markers. J Dent Res 83(2):156–160

D'Aiuto F, Gkranias N, Bhowruth D, Khan T, Orlandi M, Suvan J, Masi S, Tsakos G, Hurel S, Hingorani AD (2018) Systemic effects of periodontitis treatment in patients with type 2 diabetes: a 12 month, single-centre, investigator-masked, randomised trial. Lancet Diabetes Endocrinol 6(12):954–965

Dawes C, Pedersen AL, Villa A, Ekström J, Proctor GB, Vissink A, Aframian D, McGowan R, Aliko A, Narayana N (2015) The functions of human saliva: a review sponsored by the world workshop on oral medicine VI. Arch Oral Biol 60(6):863–874

Demmer R, Jacobs D Jr, Singh R, Zuk A, Rosenbaum M, Papapanou P, Desvarieux M (2015) Periodontal bacteria and prediabetes prevalence in ORIGINS: the oral infections, glucose intolerance, and insulin resistance study. J Dent Res 94(9_suppl):201S–211S

Demmer RT, Breskin A, Rosenbaum M, Zuk A, LeDuc C, Leibel R, Paster B, Desvarieux M, Jacobs DR Jr, Papapanou PN (2017) The subgingival microbiome, systemic inflammation and insulin resistance: the oral infections, glucose intolerance and insulin resistance study. J Clin Periodontol 44(3):255–265

Dewhirst FE, Chen T, Izard J, Paster BJ, Tanner AC, Yu W-H, Lakshmanan A, Wade WG (2010) The human oral microbiome. J Bacteriol 192(19):5002–5017

Di Sabatino A, Santacroce G, Rossi CM, Broglio G, Lenti MV (2023) Role of mucosal immunity and epithelial–vascular barrier in modulating gut homeostasis. Intern Emerg Med 18(6):1635–1646

Donath MY, Shoelson SE (2011) Type 2 diabetes as an inflammatory disease. Nat Rev Immunol 11(2):98–107

Elzayat H, Mesto G, Al-Marzooq F (2023) Unraveling the impact of gut and oral microbiome on gut health in inflammatory bowel diseases. Nutrients 15(15):3377

Forner L, Larsen T, Kilian M, Holmstrup P (2006) Incidence of bacteremia after chewing, tooth brushing and scaling in individuals with periodontal inflammation. J Clin Periodontol 33(6):401

Friman G, Hultin M, Nilsson GH, Wårdh I (2015) Medical screening in dental settings: a qualitative study of the views of authorities and organizations. BMC Res Notes 8:1–9

Gajendra S, McIntosh S, Ghosh S (2023) Effects of tobacco product use on oral health and the role of oral healthcare providers in cessation: a narrative review. Tob Induc Dis 21:1

Garlet GP (2010) Destructive and protective roles of cytokines in periodontitis: a re-appraisal from host defense and tissue destruction viewpoints. J Dent Res 89(12):1349–1363

Genco RJ, Sanz M (2020) Clinical and public health implications of periodontal and systemic diseases: an overview. Periodontol 83(1):7–13

Ghanem AS, Nagy AC (2024) Oral health's role in diabetes risk: a cross-sectional study with sociodemographic and lifestyle insights. Front Endocrinol 15:1342783

Gibson GR, Hutkins R, Sanders ME, Prescott SL, Reimer RA, Salminen SJ, Scott K, Stanton C, Swanson KS, Cani PD (2017) Expert consensus document: The International Scientific Association for Probiotics and Prebiotics (ISAPP) consensus statement on the definition and scope of prebiotics. Nat Rev Gastroenterol Hepatol 14(8):491–502

Graves DT, Cochran D (2003) The contribution of interleukin-1 and tumor necrosis factor to periodontal tissue destruction. J Periodontol 74(3):391–401

Groeger SE, Meyle J (2015) Epithelial barrier and oral bacterial infection. Periodontol 69(1):46–67

Guo S, Al-Sadi R, Said HM, Ma TY (2013) Lipopolysaccharide causes an increase in intestinal tight junction permeability in vitro and in vivo by inducing enterocyte membrane expression and localization of TLR-4 and CD14. Am J Pathol 182(2): 375–387

Hajishengallis G (2014a) Immunomicrobial pathogenesis of periodontitis: keystones, pathobionts, and host response. Trends Immunol 35(1):3–11

Hajishengallis G (2014b) The inflammophilic character of the periodontitis-associated microbiota. Mol Oral Microbiol 29(6):248–257

Hajishengallis G (2015) Periodontitis: from microbial immune subversion to systemic inflammation. Nat Rev Immunol 15(1):30–44

Hajishengallis G, Lamont RJ (2012) Beyond the red complex and into more complexity: the polymicrobial synergy and dysbiosis (PSD) model of periodontal disease etiology. Mol Oral Microbiol 27(6):409–419

Hajishengallis G, Lamont RJ, Graves DT (2015) The enduring importance of animal models in understanding periodontal disease. Virulence 6(3):229–235

Hayden MR (2023) Overview and new insights into the metabolic syndrome: risk factors and emerging variables in the development of type 2 diabetes and cerebrocardiovascular disease. Medicina 59(3):561

Hegde P, Rao R, Prabhu S, Reddy S (2023) Defence mechanisms of Oral and genital mucosae. In: Sexually transmissible oral diseases. Wiley, Hoboken, pp 73–81

Hezel M, Weitzberg E (2015) The oral microbiome and nitric oxide homoeostasis. Oral Dis 21(1):7–16

Hill C, Guarner F, Reid G, Gibson GR, Merenstein DJ, Pot B, Morelli L, Canani RB, Flint HJ, Salminen S (2014) Expert consensus document: The International Scientific Association for Probiotics and Prebiotics consensus statement on the scope and appropriate use of the term probiotic. Nat Rev Gastroenterol Hepatol 11:506–514

Holt S, Ebersole J (2005) Porphyromonas gingivalis, Treponema denticola, and Tannerella forsythia: the 'red complex', a prototype polybacterial pathogenic consortium in periodontitis. Periodontology 2000 38(1): 72–122

Hotamisligil GS (2017) Inflammation, metaflammation and immunometabolic disorders. Nature 542(7640): 177–185

Hotamisligil GS, Shargill NS, Spiegelman BM (1993) Adipose expression of tumor necrosis factor-α: direct role in obesity-linked insulin resistance. Science 259(5091):87–91

Hujoel P (2009) Dietary carbohydrates and dental-systemic diseases. J Dent Res 88(6):490–502

Hujoel P, White B, Garcia R, Listgarten M (2001) The dentogingival epithelial surface area revisited. J Periodontal Res 36(1):48–55

Imai H, Yamashita Y, Takasu H, Fujita K, Ono T, Hirota M, Mitsudo K (2023) Accuracy and influencing factors of maxillary and mandibular repositioning using pre-bent locking plates: a prospective study. Br J Oral Maxillofac Surg 61(10):659–665

Imamura T (2003) The role of gingipains in the pathogenesis of periodontal disease. J Periodontol 74(1): 111–118

Isola G, Santonocito S, Lupi SM, Polizzi A, Sclafani R, Patini R, Marchetti E (2023) Periodontal health and disease in the context of systemic diseases. Mediat Inflamm 2023(1):9720947

Iwasaki M, Taylor GW, Awano S, Yoshida A, Kataoka S, Ansai T, Nakamura H (2018) Periodontal disease and pneumonia mortality in haemodialysis patients: a 7-year cohort study. J Clin Periodontol 45(1):38–45

Jia S, Li X, Du Q (2023) Host insulin resistance caused by Porphyromonas gingivalis-review of recent progresses. Front Cell Infect Microbiol 13:1209381

Kassebaum N, Bernabé E, Dahiya M, Bhandari B, Murray C, Marcenes W (2014) Global burden of severe periodontitis in 1990–2010: a systematic review and meta-regression. J Dent Res 93(11):1045–1053

Katz J, Sambandam V, Wu JH, Michalek SM, Balkovetz DF (2000) Characterization of Porphyromonas gingivalis-induced degradation of epithelial cell junctional complexes. Infect Immun 68(3):1441–1449

Kaur H (2023) Diet, nutrition and oral health. Int Dent J Stud Res 11(1):1

Keller RM (2021) Investigating the mechanisms for the effects of nitrate treatment on exercise performance and energy metabolism. https://ir.library.oregonstate.edu/concern/graduate_thesis_or_dissertations/tm70n322q?locale=en

Khin PP, Lee JH, Jun H-S (2023) Pancreatic beta-cell dysfunction in type 2 diabetes. Eur J Inflamm 21: 1721727X231154152

Kilian M, Chapple I, Hannig M, Marsh P, Meuric V, Pedersen A, Tonetti M, Wade W, Zaura E (2016) The oral microbiome – an update for oral healthcare professionals. Br Dent J 221(10):657–666

Kobayashi R, Kobayashi T, Sakai F, Hosoya T, Yamamoto M, Kurita-Ochiai T (2017) Oral administration of Lactobacillus gasseri SBT2055 is effective in preventing Porphyromonas gingivalis-accelerated periodontal disease. Sci Rep 7(1):545

Kolenbrander PE, Palmer RJ Jr, Periasamy S, Jakubovics NS (2010) Oral multispecies biofilm development and the key role of cell–cell distance. Nat Rev Microbiol 8(7):471–480

Koren N, Zubeidat K, Saba Y, Horev Y, Barel O, Wilharm A, Heyman O, Wald S, Eli-Berchoer L, Shapiro H (2021) Maturation of the neonatal oral mucosa involves unique epithelium-microbiota interactions. Cell Host Microbe 29(2):197–209.e195

Kuo CH, Wu LL, Chen HP, Yu J, Wu CY (2024) Direct effects of alcohol on gut-epithelial barrier: unraveling the disruption of physical and chemical barrier of the gut-epithelial barrier that compromises the host–microbiota interface upon alcohol exposure. J Gastroenterol Hepatol 39:1247

Kuraji R, Shiba T, Dong TS, Numabe Y, Kapila YL (2023) Periodontal treatment and microbiome-targeted therapy in management of periodontitis-related nonalcoholic fatty liver disease with oral and gut dysbiosis. World J Gastroenterol 29(6):967

Kushiyama M, Shimazaki Y, Yamashita Y (2009) Relationship between metabolic syndrome and periodontal disease in Japanese adults. J Periodontol 80(10): 1610–1615

Lamont RJ, Koo H, Hajishengallis G (2018) The oral microbiota: dynamic communities and host interactions. Nat Rev Microbiol 16(12):745–759

Lamster IB, Lalla E, Borgnakke WS, Taylor GW (2008) The relationship between oral health and diabetes mellitus. J Am Dent Assoc 139:19S–24S

Lassenius MI, Pietiläinen KH, Kaartinen K, Pussinen PJ, Syrjänen J, Forsblom C, Pörsti I, Rissanen A, Kaprio J, Mustonen J (2011) Bacterial endotoxin activity in human serum is associated with dyslipidemia, insulin resistance, obesity, and chronic inflammation. Diabetes Care 34(8):1809–1815

Lee S-W (2023) Microbiological links between periodontitis and systemic diseases: a brief review. Oral Biol Res 47(3):81–94

Li Y, Qian F, Cheng X, Wang D, Wang Y, Pan Y, Chen L, Wang W, Tian Y (2023) Dysbiosis of oral microbiota and metabolite profiles associated with type 2 diabetes mellitus. Microbiol Spectr 11(1):e03796-03722

Liao S, Tang Y, Chu C, Lu W, Baligen B, Man Y, Qu Y (2020) Application of green tea extracts epigallocatechin-3-gallate in dental materials: recent progress and perspectives. J Biomed Mater Res A 108(12):2395–2408

Loos BG (2005) Systemic markers of inflammation in periodontitis. J Periodontol 76:2106–2115

Lopes-Virella MF, Virella G (2023) Diabetes and atherosclerosis. In: Diabetes and cardiovascular disease. Springer, Cham, pp 257–306

Mahanonda R, Pichyangkul S (2007) Toll-like receptors and their role in periodontal health and disease. Periodontology 2000 43(1):41–55

Marsh P (2018) In sickness and in health—what does the oral microbiome mean to us? An ecological perspective. Adv Dent Res 29(1):60–65

Martin-Cabezas R, Seelam N, Petit C, Agossa K, Gaertner S, Tenenbaum H, Davideau J-L, Huck O (2016) Association between periodontitis and arterial hypertension: a systematic review and meta-analysis. Am Heart J 180:98–112

Matsuda S, Shintani T, Miyagawa T, Yumoto H, Komatsu Y, Dewake N, Iwata T, Nagano T, Morozumi T, Goto R (2024) Effect of periodontal treatment on reducing chronic inflammation in systemically healthy patients with periodontal disease. Am J Med 137(3):273–279.e272

Mattos-Graner RO, Klein MI, Smith DJ (2014) Lessons learned from clinical studies: roles of mutans streptococci in the pathogenesis of dental caries. Curr Oral Health Rep 1:70–78

Michaud DS, Fu Z, Shi J, Chung M (2017) Periodontal disease, tooth loss, and cancer risk. Epidemiol Rev 39(1):49–58

Mohamed SM, Shalaby MA, El-Shiekh RA, El-Banna HA, Emam SR, Bakr AF (2023) Metabolic syndrome: risk factors, diagnosis, pathogenesis, and management with natural approaches. Food Chem Adv 3:100335

Morita T, Yamazaki Y, Mita A, Takada K, Seto M, Nishinoue N, Sasaki Y, Motohashi M, Maeno M (2010) A cohort study on the association between periodontal disease and the development of metabolic syndrome. J Periodontol 81(4):512–519

Moulari B, Beduneau A, Pellequer Y, Lamprecht A (2013) Nanoparticle targeting to inflamed tissues of the gastrointestinal tract. Curr Drug Deliv 10(1):9–17

Muniyappa R, Sowers JR (2013) Role of insulin resistance in endothelial dysfunction. Rev Endocr Metab Disord 14:5–12

Nesse W, Abbas F, Van Der Ploeg I, Spijkervet FKL, Dijkstra PU, Vissink A (2008) Periodontal inflamed surface area: quantifying inflammatory burden. J Clin Periodontol 35(8):668–673

Nibali L, Tatarakis N, Needleman I, Tu Y-K, D'Aiuto F, Rizzo M, Donos N (2013) Association between metabolic syndrome and periodontitis: a systematic review and meta-analysis. J Clin Endocrinol Metabol 98(3): 913–920

Niu C, Lv W, Zhu X, Dong Z, Yuan K, Jin Q, Zhang P, Li P, Mao M, Dong T (2024) Intestinal translocation of live Porphyromonas gingivalis drives insulin resistance. J Dent Res 103(2):197–207

Novak N, Haberstok J, Bieber T, Allam J-P (2008) The immune privilege of the oral mucosa. Trends Mol Med 14(5):191–198

O'Neill S, O'Driscoll L (2015) Metabolic syndrome: a closer look at the growing epidemic and its associated pathologies. Obes Rev 16(1):1–12

Olsen I (2016) The oral microbiome in health and disease. In: Oral infections and general health: from molecule to chairside. Springer, Cham, pp 97–114

Olsen I, Yamazaki K (2019) Can oral bacteria affect the microbiome of the gut? J Oral Microbiol 11(1): 1586422

Ouchi N, Parker JL, Lugus JJ, Walsh K (2011) Adipokines in inflammation and metabolic disease. Nat Rev Immunol 11(2):85–97

Păunică I, Giurgiu M, Dumitriu AS, Păunică S, Pantea Stoian AM, Martu M-A, Serafinceanu C (2023) The bidirectional relationship between periodontal disease and diabetes mellitus—a review. Diagnostics 13(4): 681

Pihlstrom BL, Michalowicz BS, Johnson NW (2005) Periodontal diseases. Lancet 366(9499):1809–1820

Pozhitkov AE, Leroux BG, Randolph TW, Beikler T, Flemmig TF, Noble PA (2015) Towards microbiome transplant as a therapy for periodontitis: an exploratory study of periodontitis microbial signature contrasted by oral health, caries and edentulism. BMC Oral Health 15:1–11

Preshaw PM, Alba AL, Herrera D, Jepsen S, Konstantinidis A, Makrilakis K, Taylor R (2012) Periodontitis and diabetes: a two-way relationship. Diabetologia 55:21–31

Presland RB, Dale BA (2000) Epithelial structural proteins of the skin and oral cavity: function in health and disease. Crit Rev Oral Biol Med 11(4):383–408

Purno RT, Namrata JM (2023) Oral microbial profile and their connection with systemic diseases: a review. BRAC University, Dhaka

Pussinen PJ, Paju S, Mantyla P, Sorsa T (2007) Serum microbial-and host-derived markers of periodontal diseases: a review. Curr Med Chem 14(22):2402–2412

Pussinen PJ, Könönen E, Paju S, Hyvärinen K, Gursoy UK, Huumonen S, Knuuttila M, Suominen AL (2011) Periodontal pathogen carriage, rather than periodontitis, determines the serum antibody levels. J Clin Periodontol 38(5):405–411

Rahman B, Al-Marzooq F, Saad H, Benzina D, Al Kawas S (2023) Dysbiosis of the subgingival microbiome and relation to periodontal disease in association with obesity and overweight. Nutrients 15(4):826

Ramseier CA, Woelber JP, Kitzmann J, Detzen L, Carra MC, Bouchard P (2020) Impact of risk factor control interventions for smoking cessation and promotion of healthy lifestyles in patients with periodontitis: a systematic review. J Clin Periodontol 47:90–106

Ray RR (2023) Periodontitis: an oral disease with severe consequences. Appl Biochem Biotechnol 195(1): 17–32

Rogers AP, Mileto SJ, Lyras D (2023) Impact of enteric bacterial infections at and beyond the epithelial barrier. Nat Rev Microbiol 21(4):260–274

Rosier B, Marsh P, Mira A (2018) Resilience of the oral microbiota in health: mechanisms that prevent dysbiosis. J Dent Res 97(4):371–380

Ryder MI, Fujitaki R, Johnson G, Hyun W (1998) Alterations of neutrophil oxidative burst by in vitro smoke exposure: implications for oral and systemic diseases. Ann Periodontol 3(1):76–87

Sabatini S, Lauritano D, Candotto V, Silvestre F, Nardi GM (2017) Oral probiotics in the management of gingivitis in diabetic patients: a double blinded randomized controlled study. J Biol Regul Homeost Agents 31(2 Suppl 1):197–202

Saklayen MG (2018) The global epidemic of the metabolic syndrome. Curr Hypertens Rep 20(2):1–8

Salminen A, Kopra KE, Hyvärinen K, Paju S, Mäntylä P, Buhlin K, Nieminen MS, Sinisalo J, Pussinen PJ (2015) Quantitative PCR analysis of salivary pathogen burden in periodontitis. Front Cell Infect Microbiol 5: 69

Santacroce L, Passarelli PC, Azzolino D, Bottalico L, Charitos IA, Cazzolla AP, Colella M, Topi S, Godoy FG, D'Addona A (2023) Oral microbiota in human health and disease: a perspective. Exp Biol Med 248(15):1288–1301

Scott DA, Krauss J (2012) Neutrophils in periodontal inflammation. Periodontal Dis 15:56–83

Shahoumi LA, Saleh MH, Meghil MM (2023) Virulence factors of the periodontal pathogens: tools to evade the host immune response and promote carcinogenesis. Microorganisms 11(1):115

Shoelson SE, Lee J, Goldfine AB (2006) Inflammation and insulin resistance. J Clin Invest 116(7):1793–1801

Socransky S, Haffajee A (2005) Periodontal microbial ecology. Periodontology 2000 38(1):135–187

Socransky S, Haffajee A, Cugini M, Smith C, Kent R Jr (1998) Microbial complexes in subgingival plaque. J Clin Periodontol 25(2):134–144

Souak D, Depayras S, Zommiti M, Connil N, Leblanc S, Orange N, Feuilloley M, Poc CD (2023) Particles and microbiota: interaction to death or resilience? In: Nanotechnology and human health. Elsevier, Amsterdam, pp 1–48

Squier CA, Kremer MJ (2001) Biology of oral mucosa and esophagus. JNCI Monogr 2001(29):7–15

Stoll LL, Denning GM, Li W-G, Rice JB, Harrelson AL, Romig SA, Gunnlaugsson ST, Miller FJ, Weintraub NL (2004a) Regulation of endotoxin-induced proinflammatory activation in human coronary artery cells: expression of functional membrane-bound CD14 by human coronary artery smooth muscle cells. J Immunol 173(2):1336–1343

Stoll LL, Denning GM, Weintraub NL (2004b) Potential role of endotoxin as a proinflammatory mediator of atherosclerosis. Arterioscler Thromb Vasc Biol 24(12):2227–2236

Su Y, Ye L, Hu C, Zhang Y, Liu J, Shao L (2023) Periodontitis as a promoting factor of T2D: current evidence and mechanisms. Int J Oral Sci 15(1):25

Szukiewicz D (2023) Molecular mechanisms for the vicious cycle between insulin resistance and the inflammatory response in obesity. Int J Mol Sci 24(12):9818

Talapko J, Erić S, Meštrović T, Stipetić MM, Juzbašić M, Katalinić D, Bekić S, Muršić D, Flam J, Belić D (2024) The impact of oral microbiome dysbiosis on the aetiology, pathogenesis, and development of oral cancer. Cancer 16(17):2997

Tan X, Wang Y, Gong T (2023) The interplay between oral microbiota, gut microbiota and systematic diseases. J Oral Microbiol 15(1):2213112

Tang V, Hamidi B, Janal MN, Barber CA, Godder B, Palomo L, Kamer AR (2023) Periodontal inflamed surface area (PISA) associates with composites of salivary cytokines. PLoS One 18(2):e0280333

Teeuw WJ, Slot DE, Susanto H, Gerdes VE, Abbas F, D'Aiuto F, Kastelein JJ, Loos BG (2014) Treatment of periodontitis improves the atherosclerotic profile: a systematic review and meta-analysis. J Clin Periodontol 41(1):70–79

Tekce M, Ince G, Gursoy H, Dirikan Ipci S, Cakar G, Kadir T, Yılmaz S (2015) Clinical and microbiological effects of probiotic lozenges in the treatment of chronic periodontitis: a 1-year follow-up study. J Clin Periodontol 42(4):363–372

Teughels W, Loozen G, Quirynen M (2011) Do probiotics offer opportunities to manipulate the periodontal oral microbiota? J Clin Periodontol 38:159–177

Tilg H, Moschen AR (2010) Visceral adipose tissue attacks beyond the liver: esophagogastric junction as a new target. Gastroenterology 139(6):1823–1826

Titiz Yurdakal S, Sami Güvenç I, Güngör S (2024) Transient bacteremia following the removal of four different types of rapid palatal expanders. J Orofac Orthop/ Fortschritte der Kieferorthopädie 85:1–9

Tomás I, Diz P, Tobías A, Scully C, Donos N (2012) Periodontal health status and bacteraemia from daily oral activities: systematic review/meta-analysis. J Clin Periodontol 39(3):213–228

Tylutka A, Walas Ł, Zembron-Lacny A (2024) Level of IL-6, TNF, and IL-1β and age-related diseases: a systematic review and meta-analysis. Front Immunol 15: 1330386

Vitkov L, Singh J, Schauer C, Minnich B, Krunić J, Oberthaler H, Gamsjaeger S, Herrmann M, Knopf J, Hannig M (2023) Breaking the gingival barrier in periodontitis. Int J Mol Sci 24(5):4544

Vlassara H, Uribarri J (2014) Advanced glycation end products (AGE) and diabetes: cause, effect, or both? Curr Diab Rep 14:1–10

Wade WG (2013) The oral microbiome in health and disease. Pharmacol Res 69(1):137–143

Waszkiewicz N, Zalewska-Szajda B, Chojnowska S, Szajda SD, Zalewska A, Konarzewska B, Szulc A, Wojtulewska-Supron A, Kępka A, Knaś M (2013) The salivary β-HEX A% Index as an excellent marker of periodontitis in smoking alcohol-dependent persons. Dis Markers 35(5):457–463

Williams RC, Offenbacher S (2000) Periodontal medicine: the emergence of a new branch of periodontology. Periodontology 2000 23(1):9–12

Woelber JP, Bremer K, Vach K, König D, Hellwig E, Ratka-Krüger P, Al-Ahmad A, Tennert C (2017) An oral health optimized diet can reduce gingival and periodontal inflammation in humans – a randomized controlled pilot study. BMC Oral Health 17:1–8

Xu W, Zhou W, Wang H, Liang S (2020) Roles of Porphyromonas gingivalis and its virulence factors in periodontitis. Adv Protein Chem Struct Biol 120:45–84

Xue C, Chen K, Gao Z, Bao T, Dong L, Zhao L, Tong X, Li X (2023) Common mechanisms underlying diabetic vascular complications: focus on the interaction of metabolic disorders, immuno-inflammation, and endothelial dysfunction. Cell Commun Signal 21(1):298

Yamazaki Y, Morita T, Nakai K, Konishi Y, Goto A, Yamamoto T, Seto M, Ozaki M, Tanaka H, Maeno M (2024) Impact of dental health intervention on cardiovascular metabolic risk: a pilot study of Japanese adults. J Hum Hypertens 38(6):500–509

Yoshizawa JM, Schafer CA, Schafer JJ, Farrell JJ, Paster BJ, Wong DT (2013) Salivary biomarkers: toward future clinical and diagnostic utilities. Clin Microbiol Rev 26(4):781–791

Zaura E, Keijser BJ, Huse SM, Crielaard W (2009) Defining the healthy "core microbiome" of oral microbial communities. BMC Microbiol 9:1–12

Zeigler CC, Persson GR, Wondimu B, Marcus C, Sobko T, Modéer T (2012) Microbiota in the oral subgingival biofilm is associated with obesity in adolescence. Obesity 20(1):157–164

Zeng X, Wang X, Guan X, Feng X, Lu R, Meng H (2024) The long-term effect of periodontitis treatment on changes in blood inflammatory markers in patients with generalized aggressive periodontitis. J Periodontal Res 59:689

Diabesity and Oral Immunity: Exploring the Interconnection

7

Amine Nehaoua, Amin Gasmi, Sadaf Noor,
and Asma Gasmi Benahmed

Abstract

Diabesity—obesity and type 2 diabetes mellitus (T2DM)—has a profound effect on oral health, and modulation of the oral immune system may help in managing it. This review aims to elucidate the complex interconnection between diabesity, oral immunity, and oral health consequences. Diabesity affects metabolic homeostasis, leading to changes in the immune system within the oral cavity, and increases the incidence of chronic low-grade inflammation and oral diseases. Alterations in the oral microbiome and the breakdown of immune homeostasis further contribute to oral health problems. The reciprocal relationship between diabesity and oral diseases and how inflammation in the oral cavity can worsen systemic metabolic diseases are also described. Apart from periodontal diseases, diabesity increases susceptibility to dental caries, xerostomia, and oral mucosal lesions. Novel treatment strategies are presented with special emphasis on nutrition, lifestyle changes, new drugs, and immunomodulating therapies. The possibility of using a personalized medicine approach in the prevention and treatment of oral diseases in diabesity patients is discussed with regard to genetic and epigenetic factors, biomarkers, and genetic profiles. A combined strategy focusing on both systemic metabolic and oral immunomodulation is recommended to advance the knowledge of diabesity in oral health and thus improve the oral health and quality of life (QoL) of patients with diabesity.

Keywords

Diabesity · Oral immunity · Periodontal diseases · Chronic inflammation · Personalized medicine

A. Nehaoua
Research Department, Precision Health International, Ain Oulmene, Algeria

Laboratory of Physical Activity Sciences and Public Health, Mohamed Lamine Debaghine Sétif 2 University, Setif, Algeria

A. Gasmi (✉)
Société Francophone de Nutrithérapie et de Nutrigénétique Appliquée, Villeurbanne, France

International Institute of Nutrition and Micronutrition Sciences, Saint-Étienne, France

S. Noor
Institute of Molecular Biology and Biotechnology, Bahauddin Zakariya University, Multan, Pakistan

A. G. Benahmed
Académie Internationale de Médecine Dentaire Intégrative, Paris, France

7.1 Introduction

Obesity and type 2 diabetes mellitus (T2DM) are global epidemics, and the use of the term "diabesity" captures the strong connection

N. Rezaei, H. Ziaei (eds.), *Oral Immunology*, Advances in Experimental Medicine and Biology 1492,
https://doi.org/10.1007/978-3-032-03176-1_7

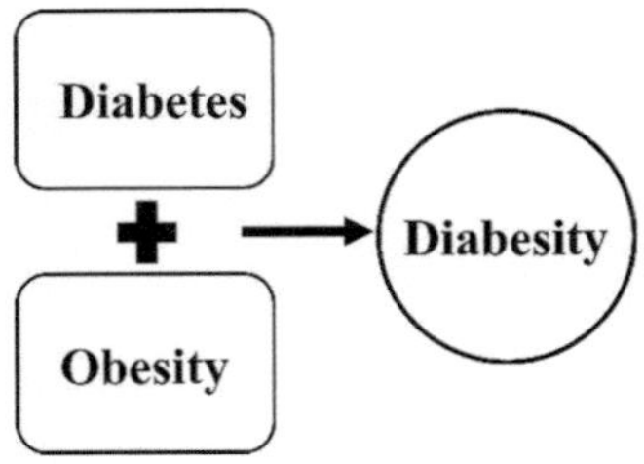

Fig. 7.1 A simple interconnection between diabetes and obesity

between the two (Fig. 7.1) (Farag and Gaballa 2011; Janez et al. 2024). Diabesity is different from other diseases because it poses a variety of risks to general health, especially to dental health and the immune system. This connection is bidirectional, and it is increasingly recognized that metabolic disorders in the context of diabesity can significantly affect the immune response in the oral cavity, while oral diseases may exacerbate systemic inflammation and metabolic disorders (Preshaw et al. 2012; Shinjo and Nishimura 2024). The oral cavity is a rather intricate environment that hosts various microorganisms and has a highly developed system of immune responses. The oral immune system is involved in the regulation of the internal environment or homeostasis as well as in the defense against pathogens and the control of inflammation. This delicate balance is especially vulnerable to systemic metabolic dysregulation, such as diabesity (Graves et al. 2019; Graves and Donaghue 2020). The oral cavity is a portal of entry and exit for substances and pathogens from the external environment because of its special anatomic structure that is in direct contact with various body systems all the time; therefore, the impact of systemic conditions like diabesity on oral health is particularly significant. Over the last decade, scientific investigations have tried to explicate the multifaceted interaction between diabesity and oral immune defense. Obesity and T2DM are characterized by insulin resistance, which has been reported to affect immune function through changes in neutrophil function, cytokine production, and adaptive immunity (Donath and Shoelson 2011; García-Macedo and de los Ángeles Fortis 2023). Furthermore, the low-grade inflammation that is a hallmark of diabesity seems to have important implications for oral health, potentially exacerbating periodontal diseases and making patients more vulnerable to oral infections (Genco et al. 2005; Isola et al. 2023).

The oral microbiome is a complex population of microorganisms living in the oral cavity, and they contribute to oral health. Recent studies have shown that diabesity may cause changes in the structure and metabolism of the oral microbial community and may increase the risk of oral diseases (Li et al. 2023; Saeb et al. 2019; Sharma et al. 2023). Although such microbial changes and their effects on oral immunity are well-recognized, it is important to elucidate these effects in therapeutic strategies. In this review, an attempt has been made to provide a detailed account of the relationship between diabesity and oral immunity. We aim to understand the connections between diabesity and oral diseases by exploring the effects of metabolic dysregulation on oral immune function and the immunopathological processes behind the oral manifestations of diabesity as well as by discussing novel treatment approaches. This understanding is imperative for developing with better approaches to address the oral health of patients with diabesity and, possibly, discover new therapeutic targets for metabolic and oral diseases.

7.2 Impact of Diabesity on Oral Immunity

7.2.1 Metabolic Dysregulation and Oral Immunity

7.2.1.1 Insulin Resistance and Immune Function

Metabolic dysfunction in diabesity significantly affects the oral immune system and involves cellular and molecular interactions. Central to these interactions is insulin resistance, a characteristic of obesity and T2DM, which has widespread immunomodulatory effects on the oral environment. Diabesity results in changes in oral immune responses due to insulin resistance, thereby making the patient more susceptible to infections and inflammation (Chandrasekaran and Weiskirchen 2024). The impaired insulin signaling impacts the immune cell population, including neutrophils, macrophages, and lymphocytes, which are indispensable for oral health (Nikolajczyk et al. 2011). The first line of defense against microbial invasion in the oral cavity is neutrophils; however, in the presence of insulin resistance, these cells lose their functional capabilities, leading to impaired chemotaxis and phagocytosis, and reduced bactericidal activity (Alba-Loureiro et al. 2007; Dhar et al. 2024). This dysfunction of neutrophils impairs the oral immune system's capacity to eliminate pathogens and thus results in recurrent infections and chronic inflammation. Another component of the innate immune system, macrophages, also exhibit functional impairment in the presence of insulin resistance (Guria et al. 2023). It has been established that macrophages in diabesity patients are in a state of chronic inflammation, thereby releasing inflammatory cytokines like tumor necrosis factor-alpha (TNF-α) and interleukin-1 beta (IL-1β) (Lumeng et al. 2007). The shift toward a pro-inflammatory state leads to chronic low-grade inflammation, characteristic of the oral tissues of diabesity patients, and fosters the development and progression of periodontal diseases. The T and B lymphocytes of the adaptive immune system are also impaired by insulin resistance (Marruganti et al. 2023). Studies have also shown that T cells in obesity and T2DM have reduced proliferative capacity and a different cytokine profile that may reduce their capability to provide an effective immune response against oral pathogens (DeFuria et al. 2013). In addition, B cells in diabesity have been identified to secrete increased quantities of pro-inflammatory cytokines and reduced quantities of regulatory cytokines that exacerbate the disruption of normal immune regulation in the oral cavity (Frasca et al. 2023; Jagannathan et al. 2010).

7.2.1.2 Adipokines and Cytokines

Adipokines and cytokines are known to be involved in the regulation of oral immune responses in the context of diabesity (Checa-Ros et al. 2024). After being regarded as just a simple energy storage depot for many years, adipose tissue is now recognized as an endocrine organ that secretes numerous signaling molecules called adipokines (Ouchi et al. 2011). In obesity, enlargement of the adipose tissue is accompanied by a change in the balance of adipokines with severe effects on oral immunity (Chaudhry and Ghafoor 2023). Leptin is a cytokine that is mainly produced in adipocytes, and it has been demonstrated that it affects immune cells in various ways. In the context of oral health, obesity-related high levels of leptin have been found to stimulate the secretion of pro-inflammatory cytokines by periodontal ligament cells and gingival fibroblasts (Checa-Ros et al. 2024; Kang et al. 2020). This may explain the fact that leptin-induced inflammation may play a role in the pathogenesis of periodontal diseases in diabesity patients. On the other hand, adiponectin, an anti-inflammatory adipokine, is usually reduced in obesity and T2DM. Its decreased level can also affect adiponectin's role in protecting periodontal tissues by inhibiting osteoclastogenesis and promoting oasteoblast differentiation (Begum et al. 2023; Yamaguchi et al. 2007).

TNF-α, IL-6, and IL-1β are highly increased in the serum and gingival crevicular fluid (GCF) of patients with diabesity (Preshaw et al. 2012). They contribute to the establishment and progression of periodontal diseases due to their effects on osteoclast activity, matrix metalloproteinases, and alveolar bone breakdown. These inflammatory mediators are constantly elevated in the oral cavity, thereby creating a vicious cycle of tissue degradation and immune system abnormalities (Radzki et al. 2024).

7.2.1.3 Oxidative Stress and Immune Dysregulation

Oxidative stress refers to the condition in which the production of reactive oxygen species (ROS) is greater than the ability of the body to neutralize these products, resulting in cellular damage. It has been implicated in the development of diabesity and oral diseases (Stavely et al. 2023). In the oral cavity, oxidative stress affects immune regulation by several processes. High ROS levels can cause direct injury to the periodontal ligament and alveolar bone and lead to immunomodulation of the leukocytes (Chapple and Matthews 2007). High glucose levels, characteristic of diabesity, affect neutrophil function and increase ROS production, while the antioxidant capacity of the body is compromised (Karima et al. 2005; Thimmappa et al. 2023). This not only weakens their bactericidal properties but also results in tissue injury and inflammation of adjacent tissues. In addition, oxidative stress has been reported to stimulate nuclear factor kappa B (NF-κB) in oral epithelial cells and immune cells, promote the synthesis of inflammatory cytokines, and thus contribute to chronic inflammation (Bullon et al. 2009; Farhad et al. 2024). As seen earlier, oxidative stress also affects the adaptive immunity of the oral cavity. It has been shown that oxidative stress can affect T cells and their cytokine production and may be a factor in the abnormal immune responses in the oral environment (Griffiths et al. 2011; Junxian et al. 2023). Furthermore, lipid peroxidation and protein and DNA oxidation in oral tissues may release more damage-associated molecular patterns (DAMPs) to stimulate the innate immune response and cause chronic inflammation (Dong et al. 2023).

7.2.2 Oral Microbiome Alterations in Diabesity

The oral microbiome refers to the complex and ever-changing microbial flora that reside in the oral cavity and are involved in maintaining oral and overall human health. It has been only in the past few years that increasing evidence has emerged regarding the effects of diabesity on the microbial population of the mouth, resulting in oral dysbiosis and alterations in oral immunity.

7.2.2.1 Changes in Oral Microbial Composition

Alterations in the oral microbial communities that occur in diabesity favor a shift toward pathogenic species. Using modern sequencing approaches, it was revealed that in obese and diabetic subjects, certain bacterial phylotypes are present in different proportions compared to their healthy counterparts (Li et al. 2023). For example, Guo et al. (2023) revealed an augmentation in the density of Gram-negative bacteria, including *Prevotella* and *Porphyromonas* species, in the subgingival biofilm of T2DM patients. These Gram-negative bacteria are known to possess lipopolysaccharide (LPS), a strong inducer of innate inflammatory reactions, which may contribute to chronic inflammation in the oral cavity. In addition, dysbiosis in diabesity is a shift not only in bacterial composition but also in fungal and viral communities (Wang et al. 2023a). *Candida* species, especially *Candida albicans*, have been identified to be more commonly isolated from the oral tissues of diabetic patients with poor glycemic control (Sardi et al. 2013). This overgrowth can lead to oral candidiasis, and, when it forms a symbiotic relationship with bacterial species, it forms mixed-species biofilms that are more resistant to both the host immune system and antimicrobial agents. It is unclear how these microbial changes relate to diabesity, but the mechanisms that promote colonization by specific microbes are probably diverse and may include alterations to the oral ecosystem. The condition of diabesity is associated with high blood glucose levels, and, thus, high glucose concentrations can be found in both saliva and GCF. These conditions may favorably influence the growth of acidogenic and aciduric organisms, namely, *Streptococcus mutans*, thereby increasing the risk of dental caries (Al-Ishaq et al. 2023; Kampoo et al. 2014).

Furthermore, long-term inflammation and oxidative stress, both characteristic features of diabesity, can affect the local gingival environment and its microbial composition. Inflammatory mediators and changes in GCF chemistry appear to produce an environment that facilitates the proliferation of periodontopathic bacteria (Isola et al. 2023). For instance, higher concentrations of inflammatory mediators promote fimbrial adhesin production in *Porphyromonas gingivalis*, a periodontal pathogen, thereby increasing its ability to adhere and colonize (Zenobia and Hajishengallis 2015). The changes in the oral microbiota that are linked to diabesity have important consequences for oral immunity. The shift in the oral microbiota toward a pathogenic one overwhelms the local immune response and results in constant inflammation and vulnerability to oral infections. Besides, dysbiosis can alter the balance between the pro- and anti-inflammatory processes in the oral cavity, which may worsen the existing periodontal diseases and contribute to the formation of new pathologies in the oral cavity (Di Stefano et al. 2023).

7.2.2.2 Biofilm Formation and Immune Evasion

Another process impacted by the metabolic disturbances of diabesity is biofilm formation, which is a key process in the colonization and persistence of oral microorganisms. Biofilms are rather intricate structures that are formed by various microorganisms and are surrounded by a matrix (Flemming et al. 2023). In the frame of diabesity, changes in biofilm formation and its characteristics can affect the oral and immune systems significantly. Hyperglycemia, which is a characteristic of diabesity, has been reported to promote biofilm formation by specific oral bacteria (Li and Ming 2024). For instance, Senpuku et al. (2019) showed that high glucose concentration enhances the production of stronger and more organized biofilms of *Streptococcus mutans*, which are widely known to be the causative agent of dental caries. These advanced biofilms exhibit increased resistance to antimicrobial compounds and enhanced evasion of the host's immune system. Diabesity also affects the biofilm dynamics of periodontal pathogens in the same way as the rest of the organisms. Some investigations have revealed that under hyperglycemic conditions, *Porphyromonas gingivalis*—one of the primary pathogens involved in periodontitis pathogenesis—forms more structurally robust and effective biofilms (Kurita-Ochiai et al. 2017). These biofilms show higher resistance to oxidative stress and an enhanced ability to invade host cells, which may explain the higher incidence and severity of periodontal diseases in individuals with diabesity.

The impact of altered biofilm formation on oral immunity is diverse. First, due to the higher structural organization and complexity of diabesity-related biofilms, it becomes difficult for immune cells and antimicrobial agents to penetrate them. This may result in repeated infections and chronic inflammation because the immune system cannot easily eliminate pathogens from the biofilm (Kurtzman et al. 2022). Biofilms act as a constant source of virulence factors and inflammatory mediators. This can lead to constant inflammation of the oral tissues and continuous stimulation of local immune reactions in the case of diabesity, in which biofilms are denser and metabolically more active (Mirghani et al. 2022). Activation of receptors within the innate immune system, for example, toll-like receptors (TLRs) by pathogen-associated molecular patterns (PAMPs) in biofilms, can trigger chronic inflammation and tissue damage (Hajishengallis 2014). Moreover, microbial communities within biofilms can alter gene expression and metabolic processes in response to environmental conditions, such as the presence of immune cells and secretion of inflammatory mediators. This adaptive capability helps biofilm-associated microorganisms avoid host immune reactions in a better manner (Zhao et al. 2023a). For instance, it is established that *P. gingivalis* modifies its lipid A composition in response to the availability of hemin, which, in turn, affects TLR4 recognition and immune activation (Al-Qutub et al. 2006).

7.3 Immunopathology of Diabesity in Oral Health

7.3.1 Chronic Low-Grade Inflammation in the Oral Cavity

Low-grade inflammation, a consequence of diabesity, has significant implications for oral health, particularly in the context of gingivitis and periodontitis. This chronic inflammatory state not only contributes to the development and exacerbation of oral diseases but also predisposes individuals to the development of new pathologies.

7.3.1.1 Mechanisms of Immune Activation

In diabesity, persistent stimulation of inflammatory signaling pathways in the oral cavity occurs due to several factors. Increased levels of the circulating pro-inflammatory cytokines TNF-α, IL-6, and IL-1β create a state of low-grade inflammation, which is now recognized to affect oral tissues (Donath and Shoelson 2011; Relvas et al. 2024). These cytokines can induce local inflammatory reactions in the gingiva and periodontal tissues, attract immune cells, and induce synthesis of other inflammatory agents. This hyperglycemic status contributes to diabesity and is also responsible for inflammation of the oral cavity (Wielento et al. 2023). High glucose levels cause the generation of advanced glycation end products (AGEs) that bind to their receptors for AGEs (RAGEs) on different cells of the oral tissues including gingival fibroblasts and periodontal ligament cells (Huang et al. 2024; Lalla et al. 2001; Shen et al. 2024). The AGE–RAGE interaction initiates intracellular signaling pathways that lead to the activation of NF-κB, which is implicated in the expression of inflammatory genes. This activation results in the production of pro-inflammatory cytokines, chemokines, and adhesion molecules and thus enhances local inflammation (Rojas et al. 2024). Furthermore, oxidative stress, which is a characteristic of diabesity, plays a major role in oral inflammation. This results in oxidative damage in oral tissues and activation of redox-sensitive transcription factors including NF-κB and activator protein 1 (AP-1) (Bullon et al. 2009; Caturano et al. 2023). These transcription factors control the production of a large number of genes related to inflammation, cytokines, chemokines, and matrix metalloproteinases (MMPs), which are involved in the tissue degradation and remodeling observed in periodontal diseases.

For diabesity patients, chronic low-grade inflammation in the oral cavity has a tremendous impact on the severity of gingivitis and periodontitis (Gupta and Jha 2023). In gingivitis, the initial stage of periodontal disease, inflammation is mostly limited to the gingival tissues (Ray 2023). However, in the presence of diabesity-related inflammation, this response may be enhanced and prolonged. The increased and prolonged inflammation causes increased permeability of blood vessels, better infiltration of immune cells, and increased levels of pro-inflammatory molecules in GCF (Preshaw et al. 2012; Salvi et al. 2023). If the inflammation process is chronic and becomes severe, then it results in periodontitis in which the periodontal ligament and alveolar bone are destroyed. Known to enhance chronic inflammation, diabesity further amplifies this process through several pathways. First, overproduction of pro-inflammatory cytokines such as TNF-α, IL-1β, and IL-6 has been shown to enhance osteoclast formation and activity, while inhibiting osteoblast differentiation, thereby accelerating alveolar bone loss (Graves et al. 2011; Hascoët et al. 2023).

7.3.1.2 Th1/Th2/Th17 Balance

In diabesity, there is a notable change in the T-helper (Th) 1/2/17 ratio, which has serious implications for oral health. Studies have established that there is a rise in the ratio of Th17 cells within the periodontal lesions of patients with obesity and T2DM (Zhao et al. 2023b; Zhu and Nikolajczyk 2014). Th17 cells are a relatively recently characterized T-cell subset that produces IL-17 and plays a multifaceted role in periodontitis. Although IL-17 is involved in host defense against extracellular pathogens, excessive or sustained Th17 responses can contribute to pathogenicity (Sun et al. 2023). The increased Th17 response in diabesity is believed to be due to several reasons, as described below. IL-6 and transforming growth factor-beta (TGF-β) levels are higher in diabesity, and these cytokines contribute to the differentiation of naive T cells into Th17 cells (Bettelli et al. 2006; Shahnoruzi et al. 2024). Furthermore, obesity and T2DM potentially affect Th17 cell differentiation through changes in the gut microbiota, which modulate the production of particular metabolites, including short-chain fatty acids (SCFAs) (Omenetti and Pizarro 2015; Sasidharan

Pillai et al. 2024). Elevation of Th17 cells and their marker cytokine, IL-17, in the periodontal tissues of individuals with diabesity, has several effects. IL-17 induces the release of neutrophil chemoattractant proteins and therefore increases neutrophil accumulation in the periodontal tissues (Irie et al. 2023). Although neutrophils are essential for host defense, their accumulation and activation are associated with tissue injury through the production of ROS and proteases (Akhter et al. 2023; Hajishengallis 2014).

Furthermore, IL-17 enhances the production of other cytokines that are pro-inflammatory, including TNF-α and IL-1β, in the oral cavity. This catabolic cooperation results in the upregulation of matrix metalloproteinases and the receptor activator of NF-κB ligand (RANKL), which are the mediators of tissue degradation and alveolar bone resorption (Awang 2014; Sadek et al. 2023). In diabesity, Th17 responses are significantly upregulated, while Th1 and Th2 responses are also affected. It has been observed that there is a relative shift toward Th1 responses in obesity and T2DM, which may adversely affect the synthesis of immunoglobulin (Ig)A antibodies against oral pathogens (Moreno et al. 2023; Surendar et al. 2011; Torres et al. 2023b). The Th1 response, which is indicated by the production of interferon gamma (IFN-γ), may vary under different conditions. In HIV-infected individuals, some studies have observed an increase in Th1 activity, while others have reported a corresponding decrease in Th17 responses, suggesting a potential inverse relationship between these T-helper cell subsets (Jagannathan et al. 2010).

7.3.2 Oral–Systemic Health Connections

7.3.2.1 Beyond Periodontitis

Although periodontitis is a well-established, diabesity-related oral manifestation, the impact of this metabolic disorder on oral health is not limited to periodontal tissues. Other oral signs and symptoms are also related to diabesity, indicating the complex interaction between general metabolic disturbances and oral tissue physiology (Marruganti et al. 2023).

Xerostomia, or dry mouth, is frequently reported by people with diabesity as well (Ayan et al. 2023). In this context, the causes of xerostomia are multifactorial, including both direct and indirect effects of metabolic disturbances on salivary glands. High blood sugar levels also cause osmotic diuresis; this means there will be systemic loss of fluids and subsequent reduction in the salivary flow (Eisbruch et al. 2003; Kim 2023). Furthermore, long-standing diabetes is characterized by autonomic neuropathy, which can impair the neural regulation of saliva-secreting glands, thereby reducing saliva production (Moore et al. 2001; Morning 2023). These three consequences of reduced salivary flow have major consequences for oral health: Saliva has several functions, which include lubrication, buffering, and antimicrobial effects, thus contributing to oral homeostasis. In diabesity-associated xerostomia, the protective salivary effects are reduced, resulting in increased susceptibility to oral infections, dental caries, and oral mucosal lesions (Plemons et al. 2014).

Moreover, saliva has a different composition in diabesity depending on the electrolyte content, pH level, and protein concentration (Liao et al. 2023). Some biochemical changes have been observed in the saliva of diabetic patients with poor glycemic control, including higher glucose concentration and total protein content, which may contribute to a shift toward cariogenic and periodontal pathogens (Alkahtani et al. 2024; Jurysta et al. 2009). The changes in the composition of saliva may also influence the formation and maturation of the acquired enamel pellicle—a smooth and continuous layer of enamel that is formed on the surface of the teeth—thus increasing the risk of dental caries and dental erosion (Enax et al. 2023; Hara and Zero 2010).

7.3.2.2 Immune Privilege and Breakdown in Oral Tissues

The immunosuppressed condition in diabesity may also contribute to an elevated caries risk, since the host's defense mechanisms may fail to adequately regulate the levels of oral bacteria. There have been changes in the composition of the oral microbiota of individuals with obesity and T2DM, resulting in a shift toward a more cariogenic microbial profile (Goodson et al. 2009; Gürsoy et al. 2024). This dysbiosis, along with the metabolic and immunological abnormalities of diabesity, creates an environment conducive to the initiation and progression of dental caries. The oral mucosa is also influenced by diabesity's systemic consequences, which include different types of mucosal lesions and diseases (Li et al. 2023). The pseudomembranous and erythematous forms of oral candidiasis are more frequently detected in patients with diabetes mellitus, particularly in those with poor glycemic control (Guggenheimer et al. 2000). The increased susceptibility to *Candida* infections is attributed to several factors, such as the effects of hyperglycemia on saliva composition, neutrophil function, and epithelial integrity (Al-Janabi 2023). Some of the other mucosal lesions seen in diabesity are lichen planus, recurrent aphthous stomatitis, and burning mouth syndrome. Although the precise mechanisms underlying these associations are still not well-studied, it is hypothesized that diabesity-induced chronic inflammation, oxidative stress, and microvascular alterations play a major role in the development and progression of these diseases (Al-MAweri et al. 2013; Sri et al. 2023; Vikas et al. 2040).

The notion of immunoprotection, which has been recognized in the central nervous system, eyes, and testes, is now gradually being applied to certain oral tissues, particularly the dental pulp and periapical tissues. Immune privilege can be described as the capacity of certain tissues to tolerate the introduction of antigens with minimal or no activation of a typical immune response, thereby preventing potentially tissue-damaging inflammation (Hargreaves and Ruparel 2016). In the setting of oral health, the dental pulp contains several features of an immune-privileged tissue. It is surrounded by a hard dental shell, which prevents it from expanding during inflammation, unlike typical connective tissues. Moreover, the pulp has a population of regulatory T cells and immunosuppressive factors that contribute to immune tolerance (Dal-Fabbro et al. 2023; Nijakowski et al. 2023). However, the immune-privileged status of certain oral tissues is seen to be compromised in diabesity, leading to increased susceptibility to infections as well as a higher likelihood of autoimmune responses. In diabesity, low-grade inflammation and metabolic changes can alter the immune homeostasis in these tissues (Kumar 2024). One potential mechanism by which diabesity affects oral immune privilege is through the disruption of the blood–dental barrier. This barrier is similar to the blood–brain barrier and controls the exchange of molecules and cells with the systemic circulation and the dental pulp. Some investigations have indicated that hyperglycemia augments the permeability of the blood–dental barrier, thereby increasing the penetration of immune cells and inflammatory agents into the pulp tissue (Leite et al. 2008).

In addition, diabesity-induced oxidative stress may upregulate the secretion of pro-inflammatory cytokines and adhesion molecules in the pulpal vasculature as well as recruitment and activation of immune cells (Bagheri et al. 2021; Sanap et al. 2024).

7.4 Innovative Therapeutic Approaches for Oral Health in Diabesity

7.4.1 Nutritional and Lifestyle Interventions

Oral healthcare in patients with diabesity should address both systemic metabolic alterations and the local oral environment. Dietary and behavioral modifications have been recognized as effective approaches to enhance oral immunity and, consequently, oral health in this group of patients.

7.4.1.1 Anti-inflammatory Diets

Anti-inflammatory diets have gained significant attention as possible therapeutic interventions in modulating the chronic low-grade inflammation in diabesity and its oral manifestations. These dietary approaches usually focus on the intake of foods that are rich in antioxidants, omega-3 fatty acids, and fibers and avoiding foods that have high inflammatory potential such as refined carbohydrates and saturated fats (Scheiber and Mank 2023). The Mediterranean diet, which is comprised of fruits, vegetables, whole grains, legumes, nuts, and olive oil, has been widely investigated for its anti-inflammatory effects (Dominguez et al. 2023). Schwingshackl and Hoffmann (2014) revealed in a systematic review and meta-analysis that the Mediterranean diet was related to a decrease in inflammation biomarkers, such as CRP and IL-6. These anti-inflammatory effects may also apply to the oral cavity and help ameliorate the exaggerated inflammation seen in periodontal diseases associated with diabesity. Some parts of the Mediterranean diet have been proven to have a direct impact on oral health. For instance, extra-virgin olive oil, a key part of this dietary pattern, contains oleocanthal, a phenolic compound with well-known anti-inflammatory properties. In vitro experiments have confirmed that oleocanthal can suppress cyclooxygenase enzymes (e.g., COX-1 and COX-2) similar to ibuprofen, which indicates that it might affect the modulation of oral inflammation (Cicerale et al. 2012). In addition, the Mediterranean diet is rich in fruits and vegetables, which serve as abundant sources of antioxidants, including vitamin C, vitamin E, and polyphenols, which may help tackle the oxidative stress of diabesity and its oral manifestations (Barber et al. 2023; Chapple et al. 2012; Laffond et al. 2023).

Another dietary regimen, effective in the management of diabesity and its manifestations in the oral cavity, is the low-glycemic-index (GI) diet. This dietary pattern involves the intake of foods that induce a slower and more gradual increase in blood glucose levels. In a randomized controlled trial carried out by Jenzsch et al. (2008), the authors showed that there were substantial improvements in the gingival and periodontal inflammation levels, when patients with T2DM adopted a low GI diet alongside improved oral hygiene practices. The authors attributed these improvements to better glycemic control and decreased oxidative stress. Among the omega-3 fatty acids, eicosapentaenoic acid (EPA) and especially docosahexaenoic acid (DHA) have been receiving a lot of attention regarding their anti-inflammatory profile in oral health. These polyunsaturated fatty acids (PUFAs) are the precursors of specialized pro-resolving mediators (SPMs); examples of these SPMs are resolvins and protectins that play an active role in resolving inflammation (Hei and Sanahanbi 2023; Saidaiah et al. 2024). A meta-analysis conducted by Elkhouli (2011) stated that omega-3 fatty acids in the diet contributed to positive changes in the clinical features of periodontal disease such as a decrease in pocket depth (PD) and gingival index scores. It is clear that the pathways through which omega-3 fatty acids control the oral immune response in diabesity are diverse. In addition to serving as precursors of SPMs, these fatty acids can also directly suppress the release of pro-inflammatory cytokines, like TNF-α and IL-1β, through peroxisome proliferator-activated receptor gamma (PPAR-γ) activation (Calder 2015; Pelczyńska et al. 2023; Salsinha et al. 2023). In addition, omega-3 polyunsaturated fatty acids (PUFAs) can also modulate the balance between pro- and anti-inflammatory lipid mediators in periodontal tissues and thus prevent the exaggerated inflammatory responses seen in diabesity-related periodontal diseases (Lee and Tribble 2023; Stańdo-Retecka et al. 2023).

Moreover, certain strains of probiotics have been found to decrease the levels of oxidized cholesterol and increase insulin receptor sensitivity, thus suggesting that probiotics may be able to modulate some of the metabolic abnormalities seen in diabesity (Bajinka et al. 2024; Wa et al. 2019). Prebiotics, which are other nondigestible food substances that encourage the growth of beneficial bacteria, may also contribute to improved oral health management in diabesity. Although there are limited studies in this area, the existing literature indicates that prebiotic compounds like fructooligosaccharides (FOSs) and galactooligosaccharides (GOSs) may promote the growth of a beneficial oral microbiota. Additionally, these compounds may enhance the production of short-chain fatty acids, which have an anti-inflammatory effect (García et al. 2023; Slomka et al. 2017; Vitetta et al. 2023).

7.4.1.2 Physical Activity and Oral Health

For a long time, exercise and physical activity have been known to be an important part of the treatment of obesity and T2DM. New findings also indicate that exercise helps in the promotion of oral health, possibly through the regulation of inflammation and immune system stimulation (Al-Mhanna et al. 2023; Syeda et al. 2023). Ferreira et al. (2019) conducted a systematic review and meta-analysis that aimed at establishing a correlation between physical activity and periodontal health; the results of their study indicated that participants with higher physical activity had a lower risk of periodontitis than those with a sedentary lifestyle. The authors also identified several mechanisms by which exercise could help prevent periodontal diseases, including improving insulin sensitivity, reducing systemic inflammation in the body, and enhancing the immune response. The anti-inflammatory effects of exercise in the context of diabesity are particularly significant for maintaining oral health. There is evidence that exercise reduces the circulating concentrations of pro-inflammatory cytokines, including TNF-α and IL-6, while upregulating anti-inflammatory molecules, including IL-10 and adiponectin (Gleeson et al. 2011; Orange et al. 2023). These systemic anti-inflammatory effects may also extend to oral tissues, thus reducing the inflammatory response observed in diabesity-associated periodontal diseases.

In addition, exercise has been proven to modulate the ratio of Th-cell subsets, which may help restore the immune balance in the oral cavity. Silveira et al. (2022) showed that moderate-intensity exercise training in obese individuals resulted in a decreased Th17/Treg ratio, which may have promising potential for regulation of the immune imbalance seen in diabesity-related oral diseases. Physical activity may also affect oral health in diabesity through moderation of glycemic control and insulin action. In a meta-analysis, Gallardo-Gómez et al. (2024) reported that exercise interventions reduced HbA1c levels by 0.86% in individuals with T2DM and that this reduction was not affected by weight loss. It has been observed that enhanced glycemic control leads to positive changes in the periodontal health status, potentially due to reduced AGEs and lower oxidative stress levels in oral tissues (Preshaw et al. 2012). Besides, it is possible to assume that increased blood flow and oxygenation of tissues can directly influence oral tissues in response to physical activity. The increased blood flow to the periodontal tissues may help in the delivery of nutrients and immune cells and, at the same time, in the removal of metabolic waste products and inflammatory mediators.

7.4.2 Pharmacological and Immunomodulatory Therapies

Diabesity is a complex metabolic condition with a close relationship with oral health, and there is a need for new pharmacological and immunomodulatory approaches for the treatment of this multifaceted metabolic disorder. New drugs for modulating systemic inflammation and metabolic dysregulation also appear to be beneficial for improving the oral health of patients with diabesity (Sugandh et al. 2023).

Among the drugs that have recently gained significant attention in the treatment of diabesity-related oral diseases are thiazolidinediones (TZDs). These insulin-sensitizing agents act as agonists of PPAR-γ, which is a nuclear receptor involved in the regulation of glucose and lipid homeostasis and immune response (Naz et al. 2023; Paniagua and Vidal-Puig 2023). In a recent randomized controlled trial (Bica et al. 2023), it has been indicated that TZDs and pioglitazone improved periodontal indices in patients with T2DM. TZDs enhance oral health in diabesity through multiple mechanisms. In addition to their role in enhancing insulin sensitization, which improves glycemic control and decreases the formation of AGEs, TZDs also possess intrinsic anti-inflammatory properties. Animal studies have shown that PPAR-γ activation can downregulate the synthesis of inflammatory mediators such as TNF-α and

IL-1β in periodontal tissues (Hasturk et al. 2007). Additionally, TZDs may be involved in regulating the differentiation and function of immune cells and helping in the recovery of the immune imbalance in the oral environment (Alexander et al. 2024). Another promising class of pharmacological agents in the management of oral health in diabesity is incretin mimetics, especially glucagon-like peptide-1 (GLP-1) receptor agonists. These drugs, which are mainly employed for glycemic management in T2DM, have emerging effects on oral health apart from glycemic regulation. Qian-yu et al. (2020), in their preclinical study, found that exendin-4, a GLP-1 receptor agonist, helped in preventing alveolar bone resorption and infiltration of inflammatory cells in a rat model of periodontitis, opening up possibilities of managing periodontal diseases in diabesity.

The possible actions by which GLP-1 receptor agonists promote oral health are through anti-inflammatory effects, modulation of bone metabolism, and stimulation of the wound-healing process. GLP-1 receptors are present in periodontal tissues, and their activation downregulates the synthesis of pro-inflammatory cytokines and upregulates osteoblast differentiation (Su et al. 2023). Moreover, GLP-1 receptor agonists may increase the synthesis of specialized pro-resolving mediators like resolvins to promote the resolution of inflammation in oral tissues (Wang et al. 2023b).

Pharmacological interventions that modulate inflammation have also been effective in treating oral health-related disorders in diabesity. One of them is anti-cytokine therapies with special emphasis on drugs that block TNF-α. Although developed for autoimmune diseases, these drugs have been found to be helpful in the context of periodontal health when general inflammation is present. Anti-TNF-α therapy is known to affect periodontal health, and a pilot study by Inchingolo et al. (2023) examined the influence of adjunctive adalimumab therapy in patients with rheumatoid arthritis and periodontitis. The authors reported significant improvement in clinical periodontal measurements as well as local inflammatory markers. Although this study did not specifically focus on diabesity, it provides preliminary evidence supporting the effectiveness of anti-cytokine treatments in managing oral inflammation in patients with systemic metabolic diseases.

The rationale for employing anti-TNF-α therapies in controlling the oral health complications related to diabesity stems from the critical role of TNF-α in both metabolic and periodontal disorders. TNF-α has been linked to insulin resistance, oxidative stress, and the heightened inflammatory reactions seen in oral diseases associated with diabesity (Cassiano et al. 2024). Since anti-TNF-α therapies specifically target this inflammatory mediator, they might be beneficial in limiting the chronic inflammation and tissue degradation processes in the oral cavity. Another approach to immunomodulation is the use of SPMs or their synthetic analogues. Resolvins, protectins, and maresins are endogenous SPMs that are synthesized from omega-3 PUFAs and that actively resolve inflammation. These molecules have been evidenced to have the capability of modulating several inflammatory disorders including periodontal diseases (Panezai and van Dyke 2023; Serhan 2014). Chiluveru et al. (2024), in their preclinical study, showed that resolvin E1 (RvE1) has the potential to prevent and treat periodontitis in rabbits. The authors stated that they observed a decrease in the infiltration of inflammatory cells, less alveolar bone resorption, and enhanced tissue repair. To the best of our knowledge, few clinical studies have been conducted in humans; however, management of oral inflammation in diabesity using SPMs or their synthetic analogues seems to be one of the potential therapeutic strategies for overcoming both e excessive inflammatory reactions and disruption of the resolution process in this pathology.

The use of SPMs in tackling the oral health complications of diabesity is not exclusive of their anti-inflammatory properties. These molecules have been described to increase insulin sensitivity, decrease oxidative stress, and polarize macrophages to an anti-inflammatory phenotype (Salsinha et al. 2023; Spite et al. 2014). Because

SPM-based therapies aim at various aspects of the pathophysiology of diabesity-associated oral diseases, the treatment of these diseases by SPMs may be more effective in maintaining oral health and immune homeostasis.

7.4.3 Personalized Medicine Approaches

Diabesity and its associated oral conditions are complex and diverse, requiring individualized treatment approaches based on genetics, the environment, and epigenetic changes. Personalized medicine is a targeted approach that has the potential to enhance the efficacy, safety, and precision of interventions aimed at managing the oral health risks associated with diabesity.

7.4.3.1 Genetic and Epigenetic Considerations

There is evidence showing that genetic predispositions are essential to the development of diabesity and its oral consequences. A genome-wide association study (GWAS) has shown that many genetic traits confer susceptibility to T2DM, obesity, and periodontal diseases (Wu et al. 2024). Knowledge of these genetic susceptibilities could help design better preventive and curative measures for treating oral diseases in individuals with diabesity. For instance, polymorphisms in the genes involved in the expression of inflammatory mediators, such as IL-1 and TNF-α, serve as predisposing factors for periodontal diseases (Cardoso et al. 2024; Santos et al. 2023). Identifying individuals with these genetic variations may offer more intensive preventive strategies or even early initiation of treatment for diabesity-related oral diseases in affected individuals. Likewise, genetic differences in taste receptors like the *TAS2R38* gene may affect dietary choices and obesity status, potentially affecting both the overall metabolic state of the body and oral health (Dotson et al. 2010; Schamarek et al. 2023). Epigenetic regulation, which involves modifications of gene expression without altering the DNA sequence, is an essential factor in the possibility of employing personalized strategies in managing oral health in diabesity. Lifestyle changes such as diet, exercise, and stress can induce epigenetic modifications that affect metabolic activity and immune responses within the oral environment.

Another study by Laberge et al. (2023) demonstrated that periodontal inflammation induces epigenetic modifications in genes governing immune response and matrix remodeling. Notably, many of these epigenetic changes were reversible following periodontal treatment, suggesting the potential for targeted interventions aimed at modulating gene expression to improve periodontal disease outcomes. In the context of diabesity, understanding individual epigenetic variations could facilitate the development of personalized dietary and lifestyle modifications as well as pharmacological treatments to enhance oral immune function and overall oral health.

7.4.3.2 Biomarker Development

Biomarkers for the early diagnosis and tracking of oral immune dysfunction in diabesity are important components of personalized medicine (Addissouky et al. 2023). Biomarkers can also provide information on the activity of the disease, the probable outcome, and the response to therapy, thus enabling early and more specific management. Noninvasive diagnostic tests such as saliva tests have been considered to be effective in evaluating the oral health status and systemic inflammation of clients with diabesity (Dongiovanni et al. 2023; Ebersole et al. 2024). One study found a panel of salivary biomarkers such as IL-1β, MMP-8, and osteoprotegerin, which may aid in differentiating the stages of periodontal disease. These biomarkers could be useful for assessing the oral health status and effectiveness of the treatment in patients with diabesity because the clinical parameters may not be accurate due to alterations in tissue reactions (Shi et al. 2023).

Another important category of biomarkers is AGEs, which may also be useful for individualized oral care in diabesity (Kamran 2024). AGEs are derived from the nonenzymatic glycation of proteins and lipids and are deposited in tissues during long-standing hyperglycemia. They have been implicated in the pathogenesis of diabetic complications, including periodontal diseases (Hadzi-Petrushev et al. 2023) (Thomas et al. 2023). Another study showed that salivary AGE levels are positively associated with periodontal disease in patients with T2DM. This suggests that AGEs could serve as potential biomarkers for assessing the oral health status of individuals with diabesity. In addition, metabolomics and proteomics analyses provide other prospects for the development of individualized approaches in the treatment of oral health disorders in patients with diabesity. These high-throughput techniques can detect all metabolites and proteins in biological samples, which can identify new biomarkers and therapeutic targets. A study by Alamri et al. (2023) using a metabolomics approach revealed changes in the salivary metabolome of individuals with periodontal disease, particularly in amino acid and lipid metabolism. Concerning diabesity, metabolomics profiling could perhaps identify the novel metabolic profiles associated with oral immune dysbiosis, potentially guiding more personalized approaches. Similarly, proteomic profiling of GCF has been used to predict the protein biomarkers related to the progression of periodontal disease in diabetes (Kido et al. 2017; Torres et al. 2023a).

The combination of genetic testing, epigenetic profiling, and biomarkers offers a promising strategy for tailoring care for oral health in patients with diabesity. Coping with individual genetic susceptibilities, external influences, and molecular signatures, physicians might be able to design more individualized preventive and therapeutic approaches. For instance, people with a high genetic predisposition to both diabetes and periodontal diseases could be advised to visit the dentist more often and receive additional information on oral hygiene and preventive measures, including prophylactic dental cleaning. People who have epigenetic markers suggesting increased inflammation could be considered for more intensive anti-inflammatory measures using specific vitamins or drugs. This information could be applied to track the effectiveness of the treatment and change the therapy regimen. For instance, persistently elevated salivary inflammatory biomarkers following conventional periodontal treatment suggest the need for host modulation therapy or more stringent glycemic control measures. Offenbacher et al. (2007) have put forward the idea of the periodontal phenotype, which can be used to incorporate these personal medicine approaches for optimizing oral health in diabesity. This model incorporates genetic predisposition and environmental and host factors to explain the ability of an individual to develop periodontal disease and its progression. When the periodontal phenotype in the context of diabesity is defined, clinicians may be able to formulate more accurate treatment approaches.

7.5 Conclusions

Diabesity and oral immunity share a bidirectional relationship, impacting both the management and research of oral health. Diabesity influences genetic and acquired immunity, modulates inflammatory responses, and alters the oral microbiota, contributing to low-grade, chronic systemic inflammation that is central to oral diseases, especially periodontal diseases. Given the fact that diabesity and oral diseases are reciprocally related, a comprehensive management strategy is required. In modern healthcare, promising strategies exist for the successful treatment of diabesity and its oral health-related complications. Anti-inflammatory diets, omega-3 supplements, weight control, and exercise can influence oral immune responses. Pharmacological and immunological therapies are now directed at the metabolic and inflammatory processes that underlie oral immune dysregulation. Personalized prevention and treatment depending on a patient's genetic, epigenetic, and biomarker profiles are expected to enhance the efficiency of oral healthcare in diabesity. An ideal model of oral healthcare management for T2DM patients should incorporate systemic metabolic control together with localized immunomodulation of oral tissues. Future oral healthcare for diabesity patients will probably use the multiple factor model that considers metabolic, immune, and oral health systems. This comprehensive approach aims to improve patient outcomes, enhance treatment satisfaction, and reduce the potential of oral diseases in patients with diabesity.

References

Addissouky T, Ali M, El Sayed IET, Wang Y (2023) Revolutionary innovations in diabetes research: from biomarkers to genomic medicine. Iran J Diabetes Obesity

Akhter S, Tasnin FM, Islam MN, Rauf A, Mitra S, Emran TB, Alhumaydhi FA, Ahmed Khalil A, Aljohani AS, Al Abdulmonem W (2023) Role of Th17 and IL-17 cytokines on inflammatory and auto-immune diseases. Curr Pharm Des 29(26):2078–2090

Alamri MM, Williams B, Le Guennec A, Mainas G, Santamaria P, Moyes DL, Nibali L (2023) Metabolomics analysis in saliva from periodontally healthy, gingivitis and periodontitis patients. J Periodontal Res 58(6):1272–1280

Alba-Loureiro TC, Munhoz CD, Martins J, Cerchiaro G, Scavone C, Curi R, Sannomiya P (2007) Neutrophil function and metabolism in individuals with diabetes mellitus. Braz J Med Biol Res 40:1037–1044

Alexander M, Cho E, Gliozheni E, Salem Y, Cheung J, Ichii H (2024) Pathology of diabetes-induced immune dysfunction. Int J Mol Sci 25(13):7105

Al-Ishaq RK, Samuel SM, Büsselberg D (2023) The influence of gut microbial species on diabetes mellitus. Int J Mol Sci 24(9):8118

Al-Janabi AAHS (2023) A positive or negative connection of diabetes mellitus to the oral microbiota. Eurasian J Med 55(1):83

Alkahtani A, Grootveld M, Bhogadia M, Baysan A (2024) Exploring salivary metabolic alterations in type 2 diabetes: implications for dental caries and potential influences of HbA1c and vitamin D levels. Meta 14(7):372

Al-MAweri SAA, Ismail NM, Ismail ARI, Abdulmlik A-G (2013) Prevalence of oral mucosal lesions in patients with type 2 diabetes attending hospital universiti sains Malaysia. MJMS 20(4):39

Al-Mhanna SB, Rocha-Rodriguesc S, Mohamed M, Batrakoulis A, Aldhahi MI, Afolabi HA, Yagin FH, Alhussain MH, Gülü M, Abubakar BD (2023) Effects of combined aerobic exercise and diet on cardiometabolic health in patients with obesity and type 2 diabetes: a systematic review and meta-analysis. BMC Sports Sci Med Rehabil 15(1):165

Al-Qutub MN, Braham PH, Karimi-Naser LM, Liu X, Genco CA, Darveau RP (2006) Hemin-dependent modulation of the lipid a structure of Porphyromonas gingivalis lipopolysaccharide. Infect Immun 74(8): 4474–4485

Awang RAR (2014) The role of IL-33 and IL-17 family cytokines in periodontal disease. University of Glasgow

Ayan G, Dayı B, Şahin Ş, Evren B (2023) Investigation of DMFT index and saliva values in morbidly obese and obese patients. Konuralp Medical Journal 15(1):91–98

Bagheri A, Ebrahimpour S, Nourbakhsh N, Talebi S, Esmaeili A (2021) Protective effect of quercetin on alteration of antioxidant genes expression and histological changes in the dental pulp of the streptozotocin-diabetic rats. Arch Oral Biol 125:105088

Bajinka O, Sylvain Dovi K, Simbilyabo L, Conteh I, Tan Y (2024) The predicted mechanisms and evidence of probiotics on type 2 diabetes mellitus (T2DM). Arch Physiol Biochem 130(4):475–490

Barber TM, Kabisch S, Pfeiffer AF, Weickert MO (2023) The effects of the Mediterranean diet on health and gut microbiota. Nutrients 15(9):2150

Begum M, Choubey M, Tirumalasetty MB, Arbee S, Mohib MM, Wahiduzzaman M, Mamun MA, Uddin MB, Mohiuddin MS (2023) Adiponectin: a promising target for the treatment of diabetes and its complications. Life 13(11):2213

Bettelli E, Carrier Y, Gao W, Korn T, Strom TB, Oukka M, Weiner HL, Kuchroo VK (2006) Reciprocal developmental pathways for the generation of pathogenic effector TH17 and regulatory T cells. Nature 441(7090):235–238

Bica I-C, Stoica RA, Salmen T, Janež A, Volčanšek Š, Popovic D, Muzurovic E, Rizzo M, Stoian AP (2023) The effects of sodium-glucose cotransporter 2-inhibitors on steatosis and fibrosis in patients with non-alcoholic fatty liver disease or steatohepatitis and type 2 diabetes: a systematic review of randomized controlled trials. Medicina 59(6):1136

Bullon P, Morillo J, Ramirez-Tortosa M, Quiles J, Newman H, Battino M (2009) Metabolic syndrome and periodontitis: is oxidative stress a common link? J Dent Res 88(6):503–518

Calder PC (2015) Marine omega-3 fatty acids and inflammatory processes: effects, mechanisms and clinical relevance. Biochim et Biophys Acta (BBA) Mol Cell Biol Lipids 1851(4):469–484

Cardoso JM, Ribeiro AC, Botelho J, Proença L, Noronha S, Alves RC (2024) The influence of genetic polymorphisms on the expression of interleukin-1beta, prostaglandin E2 and tumor necrosis factor alpha in Peri-implant Crevicular fluid: a cross-sectional study. Int J Mol Sci 25(1):651

Cassiano BA, Silveira ALPA, Kim YJ et al (2024) Role of circulating microparticles and cytokines in periodontitis associated with diabetes. Front Med 11:1394300

Caturano A, D'Angelo M, Mormone A, Russo V, Mollica MP, Salvatore T, Galiero R, Rinaldi L, Vetrano E, Marfella R (2023) Oxidative stress in type 2 diabetes: impacts from pathogenesis to lifestyle modifications. Curr Issues Mol Biol 45(8):6651–6666

Chandrasekaran P, Weiskirchen R (2024) The role of obesity in type 2 diabetes mellitus – an overview. Int J Mol Sci 25(3):1882

Chapple IL, Matthews JB (2007) The role of reactive oxygen and antioxidant species in periodontal tissue destruction. Periodontol 43(1):160

Chapple IL, Milward MR, Ling-Mountford N, Weston P, Carter K, Askey K, Dallal GE, De Spirt S, Sies H, Patel D (2012) Adjunctive daily supplementation with encapsulated fruit, vegetable and berry juice powder concentrates and clinical periodontal outcomes: a double-blind RCT. J Clin Periodontol 39(1):62–72

Chaudhry SZ, Ghafoor S (2023) Adipokines: diagnostic and prognostic markers for oral diseases. J Pak Med Assoc 73(4):858–858

Checa-Ros A, Hsueh W-C, Merck B, González-Torres H, Bermúdez V, D'Marco L (2024) Obesity and oral health: the link between Adipokines and periodontitis. *touchREVIEWS*. Endocrinology 20(1):25

Chiluveru S, Gundelly M, Pusuluri SV, Tummanepally M, Chandaka M, Koduganti RR, Sneha C, Mrunalini G, Pusuluri S, Chandaka M (2024) Resolvins in periodontitis and possible periodontal regeneration: a literature review. Cureus 16(8)

Cicerale S, Lucas L, Keast R (2012) Oleocanthal: a naturally occurring anti-inflammatory agent in virgin olive oil. In: Olive oil-constituents, quality, health properties and bioconversions, pp 357–374

Dal-Fabbro R, Swanson WB, Capalbo LC, Sasaki H, Bottino MC (2023) Next-generation biomaterials for dental pulp tissue immunomodulation. Dent Mater 39(4):333–349

DeFuria J, Belkina AC, Jagannathan-Bogdan M, Snyder-Cappione J, Carr JD, Nersesova YR, Markham D, Strissel KJ, Watkins AA, Zhu M (2013) B cells promote inflammation in obesity and type 2 diabetes through regulation of T-cell function and an inflammatory cytokine profile. Proc Natl Acad Sci 110(13): 5133–5138

Dhar AK, Anitha K, Bhatt S (2024) Immune system and diabetes mellitus. In: Biochemical immunology of diabetes and associated complications. Elsevier, pp 19–47

Di Stefano M, Santonocito S, Polizzi A, Mauceri R, Troiano G, Lo Giudice A, Romano A, Mascitti M, Isola G (2023) A reciprocal link between oral, gut microbiota during periodontitis: the potential role of probiotics in reducing dysbiosis-induced inflammation. Int J Mol Sci 24(2):1084

Dominguez LJ, Veronese N, Di Bella G, Cusumano C, Parisi A, Tagliaferri F, Ciriminna S, Barbagallo M (2023) Mediterranean diet in the management and prevention of obesity. Exp Gerontol 174:112121

Donath MY, Shoelson SE (2011) Type 2 diabetes as an inflammatory disease. Nat Rev Immunol 11(2):98–107

Dong Z, Wu L, Hong H (2023) Mitochondrial dysfunction in the pathogenesis and treatment of Oral inflammatory diseases. Int J Mol Sci 24(20):15483

Dongiovanni P, Meroni M, Casati S, Goldoni R, Thomaz DV, Kehr NS, Galimberti D, Del Fabbro M, Tartaglia GM (2023) Salivary biomarkers: novel noninvasive tools to diagnose chronic inflammation. Int J Oral Sci 15(1):27

Dotson CD, Shaw HL, Mitchell BD, Munger SD, Steinle NI (2010) Variation in the gene TAS2R38 is associated with the eating behavior disinhibition in old order Amish women. Appetite 54(1):93–99

Ebersole JL, Hasturk H, Huber M, Gellibolian R, Markaryan A, Zhang XD, Miller CS (2024) Realizing the clinical utility of saliva for monitoring oral diseases. Periodontol 95:203

Eisbruch A, Ship JA, Dawson LA, Kim HM, Bradford CR, Terrell JE, Chepeha DB, Teknos TN, Hogikyan ND, Anzai Y (2003) Salivary gland sparing and improved target irradiation by conformal and intensity modulated irradiation of head and neck cancer. World J Surg 27: 832–837

Elkhouli A (2011) The efficacy of host response modulation therapy (omega-3 plus low-dose aspirin) as an adjunctive treatment of chronic periodontitis (clinical and biochemical study). J Periodontal Res 46(2): 261–268

Enax J, Ganss B, Amaechi BT et al (2023) The composition of the dental pellicle: an updated literature review. Front Oral Health 4:1260442

Farag YM, Gaballa MR (2011) Diabesity: an overview of a rising epidemic. Nephrol Dial Trans 26(1):28–35

Farhad SZ, Karbalaeihasanesfahani A, Dadgar E, Nasiri K, Esfahaniani M, Nabi Afjadi M (2024) The role of periodontitis in cancer development, with a focus on oral cancers. Mol Biol Rep 51(1):814

Ferreira R et al (2019) Physical activity reduces the prevalence of periodontal disease: systematic review and meta-analysis. Front Physiol 10:234

Flemming H-C, van Hullebusch ED, Neu TR, Nielsen PH, Seviour T, Stoodley P, Wingender J, Wuertz S (2023) The biofilm matrix: multitasking in a shared space. Nat Rev Microbiol 21(2):70–86

Frasca D, Romero M, Diaz A, Blomberg BB (2023) Obesity accelerates age defects in B cells, and weight loss improves B cell function. Immun Ageing 20(1):35

Gallardo-Gómez D, Salazar-Martínez E, Alfonso-Rosa RM, Ramos-Munell J, del Pozo-Cruz J, del Pozo Cruz B, Álvarez-Barbosa F (2024) Optimal dose and type of physical activity to improve glycemic control in people diagnosed with type 2 diabetes: a systematic review and meta-analysis. Diabetes Care 47(2): 295–303

García G, Martínez D, Soto J, Rodríguez L, Nuez M, Domínguez N, Buchaca EF, Hernández C, Sobrino A, Pérez ER (2023) Short-chain Fructooligosaccharides improve gut microbiota composition in patients with type 2 diabetes. A randomized, open-label, controlled pilot clinical trial. J Biotechnol Biomed 6(2):244–258

García-Macedo R, de los Ángeles Fortis M (2023) The immune system and inflammation in type 2 diabetes. In: The diabetes textbook: clinical principles, patient management and public health issues. Springer, pp 171–196

Genco RJ, Grossi SG, Ho A, Nishimura F, Murayama Y (2005) A proposed model linking inflammation to obesity, diabetes, and periodontal infections. J Periodontol 76:2075–2084

Gleeson M, Bishop NC, Stensel DJ, Lindley MR, Mastana SS, Nimmo MA (2011) The anti-inflammatory effects of exercise: mechanisms and implications for the prevention and treatment of disease. Nat Rev Immunol 11(9):607–615

Goodson J, Groppo D, Halem S, Carpino E (2009) Is obesity an oral bacterial disease? J Dent Res 88(6): 519–523

Graves LE, Donaghue KC (2020) Vascular complication in adolescents with diabetes mellitus. Front Endocrinol 11:370

Graves DT, Oates T, Garlet GP (2011) Review of osteoimmunology and the host response in endodontic and periodontal lesions. J Oral Microbiol 3(1):5304

Graves DT, Corrêa JD, Silva TA (2019) The oral microbiota is modified by systemic diseases. J Dent Res 98(2):148–156

Griffiths HR, Dunston CR, Bennett SJ, Grant MM, Phillips DC, Kitas GD (2011) Free radicals and redox signalling in T-cells during chronic inflammation and ageing. Biochem Soc Trans 39(5):1273–1278. https://doi.org/10.1042/BST0391273

Guggenheimer J, Moore PA, Rossie K, Myers D, Mongelluzzo MB, Block HM, Weyant R, Orchard T (2000) Insulin-dependent diabetes mellitus and oral soft tissue pathologies. II. Prevalence and characteristics of Candida and candidal lesions. Oral Surg Oral Med Oral Pathol Oral Radiol Endod 89(5): 570–576

Guo X-J, Dai S-X, Lou J-D, Ma X-X, Hu X-J, Tu L-P, Cui J, Lu H, Jiang T, Xu J-T (2023) Distribution characteristics of oral microbiota and its relationship with intestinal microbiota in patients with type 2 diabetes mellitus. Front Endocrinol 14:1119201

Gupta A, Jha A (2023) Human microbiome and lifestyle disorders. In: Probiotics, prebiotics, Synbiotics, and Postbiotics: human microbiome and human health. Springer, pp 165–193

Guria S, Hoory A, Das S, Chattopadhyay D, Mukherjee S (2023) Adipose tissue macrophages and their role in obesity-associated insulin resistance: an overview of the complex dynamics at play. Biosci Rep 43(3): BSR20220200

Gürsoy UK, Yilmaz N, Yilmaz D, Grönroos S, Gürsoy M (2024) Type 1 diabetes mellitus and host-bacterial interactions in the oral cavity. Curr Opin Endocr Metab Res:100536

Hadzi-Petrushev N, Angelovski M, Mladenov M (2023) Advanced glycation end products and diabetes. In: Obesity, diabetes and inflammation: molecular mechanisms and clinical management. Springer, pp 99–127

Hajishengallis G (2014) The inflammophilic character of the periodontitis-associated microbiota. Mol Oral Microbiol 29(6):248–257

Hara AT, Zero DT (2010) The caries environment: saliva, pellicle, diet, and hard tissue ultrastructure. Dental Clin 54(3):455–467

Hargreaves K, Ruparel S (2016) Role of oxidized lipids and TRP channels in orofacial pain and inflammation. J Dent Res 95(10):1117–1123

Hascoët E, Blanchard F, Blin-Wakkach C, Guicheux J, Lesclous P, Cloitre A (2023) New insights into inflammatory osteoclast precursors as therapeutic targets for rheumatoid arthritis and periodontitis. Bone Research 11(1):26

Hasturk H, Kantarci A, Goguet-Surmenian E, Blackwood A, Andry C, Serhan CN, Van Dyke TE (2007) Resolvin E1 regulates inflammation at the cellular and tissue level and restores tissue homeostasis in vivo. J Immunol 179(10):7021–7029

Hei A, Sanahanbi L (2023) DHA (docosahexaenoic acid): a biomolecule with diverse roles and health benefits. In: Dietary supplements-challenges and future research. IntechOpen

Huang C-Y, Chen S-H, Lin T, Liao Y-W, Chang Y-C, Chen C-C, Yu C-C, Chen C-J (2024) Resveratrol attenuates advanced glycation end product-induced senescence and inflammation in human gingival fibroblasts. J Dental Sci 19(1):580–586

Inchingolo F, Inchingolo AM, Avantario P, Settanni V, Fatone MC, Piras F, Di Venere D, Inchingolo AD, Palermo A, Dipalma G (2023) The effects of periodontal treatment on rheumatoid arthritis and of anti-rheumatic drugs on periodontitis: a systematic review. Int J Mol Sci 24(24):17228

Irie K, Azuma T, Tomofuji T, Yamamoto T (2023) Exploring the role of IL-17a in oral dysbiosis-associated periodontitis and its correlation with systemic inflammatory disease. Dent J 11(8):194

Isola G, Santonocito S, Lupi SM, Polizzi A, Sclafani R, Patini R, Marchetti E (2023) Periodontal health and disease in the context of systemic diseases. Mediat Inflamm 2023(1):9720947

Jagannathan M, McDonnell M, Liang Y, Hasturk H, Hetzel J, Rubin D, Kantarci A, Van Dyke T, Ganley-Leal L, Nikolajczyk B (2010) Toll-like receptors regulate B cell cytokine production in patients with diabetes. Diabetologia 53:1461–1471

Janez A, Muzurovic E, Bogdanski P, Czupryniak L, Fabryova L, Fras Z, Guja C, Haluzik M, Kempler P, Lalic N (2024) Modern management of cardiometabolic continuum: from overweight/obesity to prediabetes/type 2 diabetes mellitus. Recommendations from the eastern and southern Europe diabetes and obesity expert group. Diabetes Therapy:1–28

Jenzsch A, Eick S, Rassoul F, Purschwitz R, Jentsch H (2008) Nutritional intervention in patients with periodontal disease: clinical, immunological and microbiological variables during 12 months. Br J Nutr 101(6):879–885

Junxian L, Mehrabanian M, Mivehchi H, Banakar M, Etajuri EA (2023) The homeostasis and therapeutic applications of innate and adaptive immune cells in periodontitis. Oral Dis 29(7):2552–2564

Jurysta C, Bulur N, Oguzhan B, Satman I, Yilmaz TM, Malaisse WJ, Sener A (2009) Salivary glucose concentration and excretion in normal and diabetic subjects. Biomed Res Int 2009(1):430426

Kampoo K, Teanpaisan R, Ledder RG, McBain AJ (2014) Oral bacterial communities in individuals with type 2 diabetes who live in southern Thailand. Appl Environ Microbiol 80(2):662–671

Kamran MA (2024) Salivary and crevicular fluid proinflammatory cytokines and advanced glycation end products in patients with different glycemic levels undergoing fixed orthodontic treatment. Angle Orthod 94(2):233–239

Kang K-W, Ok M, Lee S-K (2020) Leptin as a key between obesity and cardiovascular disease. J Obes Metab Syndr 29(4):248

Karima M, Kantarci A, Ohira T, Hasturk H, Jones V, Nam B, Malabanan A, Trackman P, Badwey J, Van Dyke T (2005) Enhanced superoxide release and elevated protein kinase C activity in neutrophils from diabetic patients: association with periodontitis. J Leukoc Biol 78(4):862–870

Kido D, Mizutani K, Takeda K, Mikami R, Matsuura T, Iwasaki K, Izumi Y (2017) Impact of diabetes on gingival wound healing via oxidative stress. PLoS One 12(12):e0189601

Kim Y-J (2023) Xerostomia and its cellular targets. Int J Mol Sci 24(6):5358

Kumar V (2024) Immune homeostasis: methods and protocols. Springer Nature

Kurita-Ochiai T, Hashizume-Takizawa T, Kobayashi R, Bhawal UK, Hosono A, Kinukawa N, Oguchi S (2017) Porphyromonas gingivalis promotes low-density lipoprotein oxidation and atherosclerosis. J Oral Biosc 59(1):44–49

Kurtzman GM, Horowitz RA, Johnson R, Prestiano RA, Klein BI (2022) The systemic oral health connection: biofilms. Medicine 101(46):e30517

Laberge S, Akoum D, Wlodarczyk P, Massé J-D, Fournier D, Semlali A (2023) The potential role of epigenetic modifications on different facets in the periodontal pathogenesis. Genes 14(6):1202

Laffond A, Rivera-Picón C, Rodríguez-Muñoz PM, Juárez-Vela R, Ruiz de Viñaspre-Hernández R, Navas-Echazarreta N, Sánchez-González JL (2023) Mediterranean diet for primary and secondary prevention of cardiovascular disease and mortality: an updated systematic review. Nutrients 15(15):3356

Lalla E, Lamster IB, Stern DM, Schmidt AM (2001) Receptor for advanced glycation end products, inflammation, and accelerated periodontal disease in diabetes: mechanisms and insights into therapeutic modalities. Ann Periodontol 6(1):113–118

Lee C-T, Tribble GD (2023) Roles of specialized pro-resolving mediators and omega-3 polyunsaturated fatty acids in periodontal inflammation and impact on oral microbiota. Frontiers in Oral Health 4:1217088

Leite MF, Ganzerla E, Marques MM, Nicolau J (2008) Diabetes induces metabolic alterations in dental pulp. J Endodont 34(10):1211–1214

Li F, Ming J (2024) Mulberry polyphenols restored both small and large intestinal microflora in db/db mice, potentially alleviating type 2 diabetes. Food Funct 15(16):8521–8543

Li Y, Qian F, Cheng X, Wang D, Wang Y, Pan Y, Chen L, Wang W, Tian Y (2023) Dysbiosis of oral microbiota and metabolite profiles associated with type 2 diabetes mellitus. Microbiol Spectr 11(1):e03796–e03722

Liao C, Chen X, Fu Y (2023) Salivary analysis: an emerging paradigm for non-invasive healthcare diagnosis and monitoring. Interdiscip Med 1(3):e20230009

Lumeng CN, DeYoung SM, Bodzin JL, Saltiel AR (2007) Increased inflammatory properties of adipose tissue macrophages recruited during diet-induced obesity. Diabetes 56(1):16–23

Marruganti C, Suvan JE, D'Aiuto F (2023) Periodontitis and metabolic diseases (diabetes and obesity): tackling multimorbidity. Periodontol

Mirghani R, Saba T, Khaliq H, Mitchell J, Do L, Chambi L, Diaz K, Kennedy T, Alkassab K, Huynh T (2022) Biofilms: formation, drug resistance and alternatives to conventional approaches. AIMS Microbiol 8(3):239

Moore PA, Guggenheimer J, Etzel KR, Weyant RJ, Orchard T (2001) Type 1 diabetes mellitus, xerostomia, and salivary flow rates. Oral Surg Oral Med Oral Pathol Oral Radiol Endod 92(3):281–291

Moreno CM, Boeree E, Freitas CMT, Weber KS (2023) Immunomodulatory role of oral microbiota in inflammatory diseases and allergic conditions. Front Allergy 4:1067483

Morning LF (2023) Characterization of the saliva proteome of patients with Oral lichen planus. Universitätsbibliothek der Johannes Gutenberg-Universität Mainz

Naz R, Saqib F, Awadallah S, Wahid M, Latif MF, Iqbal I, Mubarak MS (2023) Food polyphenols and type II diabetes mellitus: pharmacology and mechanisms. Molecules 28(10):3996

Nijakowski K, Ortarzewska M, Jankowski J, Lehmann A, Surdacka A (2023) The role of cellular metabolism in maintaining the function of the dentine-pulp complex: a narrative review. Meta 13(4):520

Nikolajczyk B, Jagannathan-Bogdan M, Shin H, Gyurko R (2011) State of the union between metabolism and the immune system in type 2 diabetes. Genes Immun 12(4):239–250

Offenbacher S, Barros S, Singer R, Moss K, Williams R, Beck J (2007) Periodontal disease at the biofilm–gingival interface. J Periodontol 78(10):1911–1925

Omenetti S, Pizarro TT (2015) The Treg/Th17 axis: a dynamic balance regulated by the gut microbiome. Front Immunol 6:639

Orange ST, Leslie J, Ross M, Mann DA, Wackerhage H (2023) The exercise IL-6 enigma in cancer. Trends Endocrinol Metab 34:749

Ouchi N, Parker JL, Lugus JJ, Walsh K (2011) Adipokines in inflammation and metabolic disease. Nat Rev Immunol 11(2):85–97

Panezai J, van Dyke T (2023) Polyunsaturated fatty acids and their immunomodulatory actions in periodontal disease. Nutrients 15(4):821

Paniagua JA, Vidal-Puig A (2023) Obesity in the pathophysiology of diabetes. In: The diabetes textbook: clinical principles, patient management and public health issues. Springer, pp 217–245

Pelczyńska M, Moszak M, Wesołek A, Bogdański P (2023) The preventive mechanisms of bioactive food compounds against obesity-induced inflammation. Antioxidants 12(6):1232

Plemons JM, Al-Hashimi I, Marek CL (2014) Managing xerostomia and salivary gland hypofunction: executive summary of a report from the American dental association council on scientific affairs. J Am Dent Assoc 145(8):867–873

Preshaw PM, Alba AL, Herrera D, Jepsen S, Konstantinidis A, Makrilakis K, Taylor R (2012) Periodontitis and diabetes: a two-way relationship. Diabetologia 55:21–31

Qian-yu L, Ling-Qian D, Rui Z, Tian D, Shao-hua G (2020) Effects of Exenatide-4 on proliferation, migration and osteogenic differentiation of human periodontal ligament stem cells. Shanghai J Stomatol 29(3):225

Radzki D, Negri A, Kusiak A, Obuchowski M (2024) Matrix Metalloproteinases in the periodontium – vital in tissue turnover and unfortunate in periodontitis. Int J Mol Sci 25(5):2763

Ray RR (2023) Periodontitis: an oral disease with severe consequences. Appl Biochem Biotechnol 195(1):17–32

Relvas M, Mendes-Frias A, Gonçalves M, Salazar F, López-Jarana P, Silvestre R, Viana da Costa A (2024) Salivary IL-1β, IL-6, and IL-10 are key biomarkers of periodontitis severity. Int J Mol Sci 25(15):8401

Rojas A, Lindner C, Schneider I, Gonzalez I, Uribarri J (2024) The RAGE axis: a relevant inflammatory hub in human diseases. Biomolecules 14(4):412

Sadek KM, El Moshy S, Radwan IA, Rady D, Abbass MM, El-Rashidy AA, Dörfer CE, Fawzy El-Sayed KM (2023) Molecular basis beyond interrelated bone resorption/regeneration in periodontal diseases: a concise review. Int J Mol Sci 24(5):4599

Saeb AT, Al-Rubeaan KA, Aldosary K, Raja GU, Mani B, Abouelhoda M, Tayeb HT (2019) Relative reduction of biological and phylogenetic diversity of the oral microbiota of diabetes and pre-diabetes patients. Microb Pathog 128:215–229

Saidaiah P, Banu Z, Khan AA, Geetha A, Somraj B (2024) A comprehensive review of Omega-3 fatty acids: sources, industrial applications, and health benefits. Ann Phytomed 13(1):209–225

Salsinha AS, Socodato R, Relvas JB, Pintado M (2023) The pro-and antiinflammatory activity of fatty acids. In: Bioactive lipids. Elsevier, pp 51–75

Salvi GE, Roccuzzo A, Imber JC, Stähli A, Klinge B, Lang NP (2023) Clinical periodontal diagnosis. Periodontol

Sanap A, Joshi K, Kheur S, Bhonde R (2024) Stem cell secretome restore the adipo-osteo differentiation imbalance in diabetic dental pulp-derived mesenchymal stem cells. Chronic Dis Transl Med 10:340

Santos RT, Santos BP, Júnior LD, Almeida HCR, Silveira MMF, Sobral APV (2023) Influence of interleukin genetic polymorphism on apical periodontitis in human patients: a comprehensive review. Genet Mol Res 22(31):10–4238

Sardi J, Scorzoni L, Bernardi T, Fusco-Almeida A, Mendes Giannini MJS (2013) Candida species: current epidemiology, pathogenicity, biofilm formation, natural antifungal products and new therapeutic options. J Med Microbiol 62(1):10–24

Sasidharan Pillai S, Gagnon CA, Foster C, Ashraf AP (2024) Exploring the gut microbiota: key insights into its role in obesity, metabolic syndrome, and type 2 diabetes. J Clin Endocrinol Metabol *109:2709*

Schamarek I, Anders L, Chakaroun RM, Kovacs P, Rohde-Zimmermann K (2023) The role of the oral microbiome in obesity and metabolic disease: potential systemic implications and effects on taste perception. Nutr J 22(1):28

Scheiber A, Mank V (2023) Anti-inflammatory diets

Schwingshackl L, Hoffmann G (2014) Mediterranean dietary pattern, inflammation and endothelial function: a systematic review and meta-analysis of intervention trials. Nutr Metab Cardiovasc Dis 24(9):929–939

Senpuku H, Tuna EB, Nagasawa R, Nakao R, Ohnishi M (2019) The inhibitory effects of polypyrrole on the biofilm formation of Streptococcus mutans. PLoS One 14(11):e0225584

Serhan CN (2014) Pro-resolving lipid mediators are leads for resolution physiology. Nature 510(7503):92–101

Shahnoruzi N, Salehi M, Ganji A, Mosayebi G, Ghazavi A (2024) Increase in TGF-β and CD4+ CD25+ regulatory T cells in Iranian patients with type 2 diabetes. J Kerman Univ Med Sci 31(1):40–45

Sharma AK, Sharma A, Lal S, Kumar A, Yadav NK, Tabassum F, Akhtar MS, Imam MT, Almalki ZS, Mukherjee M (2023) Dysbiosis versus diabesity: pathological signaling and promising therapeutic strategies. Drug Discov Today 28(5):103558

Shen C-Y, Lu C-H, Cheng C-F, Li K-J, Kuo Y-M, Wu C-H, Liu C-H, Hsieh S-C, Tsai C-Y, Yu C-L (2024) Advanced glycation end-products acting as Immunomodulators for chronic inflammation, Inflammaging and carcinogenesis in patients with diabetes and immune-related diseases. Biomedicine 12(8):1699

Shi F, Liu W, Yao Y, Zhang Q, Chen Z, Xian Y, Sujanamulk B (2023) Predictive salivary biomarkers for early diagnosis of periodontal diseases–current and future developments. Turkish J Biochem 48(4): 335–344

Shinjo T, Nishimura F (2024) The bidirectional association between diabetes and periodontitis, from basic to clinical. Jpn Dental Sci Rev 60:15–21

Silveira EA, Mendonça CR, Delpino FM, Souza GVE, de Souza Rosa LP, de Oliveira C, Noll M (2022) Sedentary behavior, physical inactivity, abdominal obesity and obesity in adults and older adults: a systematic review and meta-analysis. Clin Nutrit ESPEN 50:63–73

Slomka V, Hernandez-Sanabria E, Herrero ER, Zaidel L, Bernaerts K, Boon N, Quirynen M, Teughels W (2017) Nutritional stimulation of commensal oral bacteria suppresses pathogens: the prebiotic concept. J Clin Periodontol 44(4):344–352

Spite M, Claria J, Serhan CN (2014) Resolvins, specialized proresolving lipid mediators, and their potential roles in metabolic diseases. Cell Metab 19(1):21–36

Sri KB, Supriya AN, Manyam R, Pasupuleti S, Ramesh T, Ramesh MV (2023) Prevalence of oral mucosal lesions among geriatric patients: a cross-sectional study. WORLD 14(4):341

Stańdo-Retecka M, Piatek P, Namiecinska M, Bonikowski R, Lewkowicz P, Lewkowicz N (2023) Clinical and microbiological outcomes of subgingival instrumentation supplemented with high-dose omega-3 polyunsaturated fatty acids in periodontal treatment–a randomized clinical trial. BMC Oral Health 23(1):290

Stavely R, Ott LC, Rashidi N, Sakkal S, Nurgali K (2023) The oxidative stress and nervous distress connection in gastrointestinal disorders. Biomolecules 13(11):1586

Su Y, Ye L, Hu C, Zhang Y, Liu J, Shao L (2023) Periodontitis as a promoting factor of T2D: current evidence and mechanisms. Int J Oral Sci 15(1):25

Sugandh F, Chandio M, Raveena F, Kumar L, Karishma F, Khuwaja S, Memon UA, Bai K, Kashif M, Varrassi G (2023) Advances in the management of diabetes mellitus: a focus on personalized medicine. Cureus 15(8)

Sun C-Y, Yang N, Zheng Z-L, Liu D, Xu Q-L (2023) T helper 17 (Th17) cell responses to the gut microbiota in human diseases. Biomed Pharmacother 161:114483

Surendar J, Mohan V, Rao MM, Babu S, Aravindhan V (2011) Increased levels of both Th1 and Th2 cytokines in subjects with metabolic syndrome (CURES-103). Diabetes Technol Ther 13(4):477–482

Syeda UA, Battillo D, Visaria A, Malin SK (2023) The importance of exercise for glycemic control in type 2 diabetes. Am J Med Open 9:100031

Thimmappa PY, Vasishta S, Ganesh K, Nair AS, Joshi MB (2023) Neutrophil (dys) function due to altered immuno-metabolic axis in type 2 diabetes: implications in combating infections. Hum Cell 36(4):1265–1282

Thomas JT, Joseph B, Sorsa T, Mauramo M, Anil S, Waltimo T (2023) Expression of advanced glycation end products and their receptors in diabetic periodontitis patients. Oral Dis 30:2784

Torres A, Michea MA, Végvári Á, Arce M, Morales A, Lanyon E, Alcota M, Fuentes C, Vernal R, Budini M (2023a) Proteomic profile of human gingival crevicular fluid reveals specific biological and molecular processes during clinical progression of periodontitis. J Periodontal Res 58(5):1061–1081

Torres L, Miranda MCG, Martins VD, Caixeta F, de Almeida Oliveira M, Trindade LM, de Assis HC, Nascimento V, Rosa NP, Gomes E (2023b) Obesity-induced hyperglycemia impairs oral tolerance induction and aggravates food allergy. Mucosal Immunol 16(4):513–526

Vikas KT, Choudhary SS, Chahar U (2040) A hospital based prospective study to assessed the correlation between glycemic control, lipid profile and oral leukoplakia in diabetes patients. Diabetes 1

Vitetta L, Gorgani NN, Vitetta G, Henson JD (2023) Prebiotics progress shifts in the intestinal microbiome that benefits patients with type 2 diabetes mellitus. Biomolecules 13(9):1307

Wa Y, Yin B, He Y, Xi W, Huang Y, Wang C, Guo F, Gu R (2019) Effects of single probiotic-and combined probiotic-fermented milk on lipid metabolism in hyperlipidemic rats. Front Microbiol 10:1312

Wang L, Zhang K, Zeng Y, Luo Y, Peng J, Zhang J, Kuang T, Fan G (2023a) Gut mycobiome and metabolic diseases: the known, the unknown, and the future. Pharmacol Res 193:106807

Wang Q, Guo H, Mao W, Qian X, Liu Y (2023b) The oral delivery system of modified GLP-1 by probiotics for T2DM. Pharmaceutics 15(4):1202

Wielento A, Lagosz-Cwik K, Potempa J, Grabiec A (2023) The role of gingival fibroblasts in the pathogenesis of periodontitis. J Dent Res 102(5):489–496

Wu KCH, Liu L, Xu A, Chan YH, Cheung BMY (2024) Shared genetic architecture between periodontal disease and type 2 diabetes: a large scale genome-wide cross-trait analysis. Endocrine 85:1–10

Yamaguchi N, Kukita T, Li Y-J, Martinez Argueta JG, Saito T, Hanazawa S, Yamashita Y (2007) Adiponectin inhibits osteoclast formation stimulated by lipopolysaccharide from Actinobacillus actinomycetemcomitans. FEMS Immunol Med Microbiol 49(1):28–34

Zenobia C, Hajishengallis G (2015) Porphyromonas gingivalis virulence factors involved in subversion of leukocytes and microbial dysbiosis. Virulence 6(3): 236–243

Zhao A, Sun J, Liu Y (2023a) Understanding bacterial biofilms: from definition to treatment strategies. Front Cell Infect Microbiol 13:1137947

Zhao M, Xie Y, Gao W, Li C, Ye Q, Li Y (2023b) Diabetes mellitus promotes susceptibility to periodontitis – novel insight into the molecular mechanisms. Front Endocrinol 14:1192625

Zhu M, Nikolajczyk B (2014) Immune cells link obesity-associated type 2 diabetes and periodontitis. J Dent Res 93(4):346–352

8 Endodontic Inflammatory Disease: Systemic Consequences

José Luis Álvarez-Vásquez

Abstract

Endodontic inflammatory diseases (EIDs), such as pulpitis and apical periodontitis (AP), are common chronic oral inflammatory conditions, which are most commonly caused by untreated dental caries. Over the past two decades, several studies have reported positive bidirectional associations between AP and various systemic conditions, including cardiovascular disease (CVD), diabetes, hepatic disease, inflammatory bowel disease (IBD), hematological disease, and adverse pregnancy outcomes (APOs). These associations indicate that chronic oral inflammation may contribute to systemic low-grade inflammation, potentially influencing overall health. Although current data support strong correlations, no definitive causal relationship has been established to date. Recent studies have suggested that successful endodontic treatment can reduce the level of several systemic inflammatory markers, emphasizing the broader role of oral healthcare in systemic disease prevention. As the field of endodontic medicine continues to evolve, further high-quality longitudinal studies are needed to determine causality and explore the biological mechanisms underlying these associations, highlighting the importance of integrated oral–systemic healthcare.

Keywords

Pulpitis · Apical periodontitis · Chronic inflammation · Systemic disease · Public health

J. L. Álvarez-Vásquez (✉)
Discipline of Endodontics, Faculty of Dentistry, University of Cuenca, Cuenca, Ecuador
e-mail: jose.alvarezv@ucuenca.edu.ec

8.1 Introduction

Oral diseases negatively impact nearly half of the world's population, affecting daily activities, physical health, and mental health (Guarnizo-Herreño et al. 2024). Dental caries, driven by bacterial infection, is the most prevalent disease in the world (GBD 2016 Disease and Injury Incidence and Prevalence Collaborators 2017). Moreover, it is the main cause of endodontic inflammatory diseases (EIDs), such as pulpitis and apical periodontitis (AP), both of which are frequently occurring chronic oral inflammatory conditions (Lin and Huang 2020). Indeed, the most common reason for referring patients to a dentist is pulpal and periapical diseases (Raoof et al. 2019). Untreated or poorly treated tooth decay can eventually progress to irreversible pulpitis and pulp necrosis. AP, a highly prevalent condition worldwide (Tibúrcio-Machado et al. 2021), results from endodontic infection and requires root canal treatment (RCT). Its pathobiology involves neuro–immune–vascular

N. Rezaei, H. Ziaei (eds.), *Oral Immunology*, Advances in Experimental Medicine and Biology 1492,
https://doi.org/10.1007/978-3-032-03176-1_8

interactions that orchestrate the formation of the periapical bone lesion (Lin and Huang 2020). Furthermore, since AP is often asymptomatic and primarily diagnosed through radiographic examination (Lin and Huang 2020), the true burden of this endodontic disease may be underestimated. Over the past two decades, multiple studies have reported positive associations between AP and various systemic conditions (Cintra et al. 2018, 2021; Segura-Egea et al. 2015). These associations clearly indicate the possibility that endodontic inflammatory diseases may have a systemic impact on overall health—this is an impact that could potentially be mitigated through nonsurgical endodontic therapies, such as conventional root canal treatments. All these findings should prompt a paradigm shift in the field of endodontics, drawing the attention of health policymakers and professionals to the global burden of endodontic inflammatory diseases and their connection to systemic health.

8.2 Pulpitis and Systemic Conditions: Is There a Link?

Dental caries, driven by bacterial infection, has been recorded as the most prevalent disease in the world—a trend that has not changed in recent decades (GBD 2016 Disease and Injury Incidence and Prevalence Collaborators 2017). It is also the main cause of endodontic inflammatory diseases, i.e., pulpitis and AP, both being frequent chronic oral inflammatory conditions (Sebring et al. 2024). Progression of caries and the resulting carious lesions inherently lead to pulp inflammation, with pulpitis being the most common inflammatory disease in mammals (Wang et al. 2013). While these conditions are related, they are separate pathological events. The assumption that lesion progression is inevitable is now outdated (Schwendicke et al. 2019). Indeed, invasive/restorative interventions alone do not always provide optimal management of the caries process and associated lesions. In this regard, noninvasive/nonrestorative strategies should be adopted globally; however, this is not happening (Innes and Schwendicke 2017). All in all, such approaches may contribute to the maintenance of dental pulp health; for example, selective carious tissue removal was found to result in fewer pulp complications compared to nonselective removal and/or stepwise removal (Barros et al. 2020). These strategies intrinsically contribute to the prevention of AP, which will be addressed in the following sections.

While root canal treatment (RCT) remains the standard treatment for irreversible pulpitis (Hosseinzadehfard et al. 2024), less invasive treatment approaches designed to preserve the partial vitality of the pulp, collectively known as vital pulp therapies (VPTs), are attracting the attention of researchers and clinicians (European Society of Endodontology (ESE) 2019; Asgary et al. 2024; Pandya et al. 2024), especially considering the confirmed healing potential of dental pulp (Álvarez-Vásquez and Castañeda-Alvarado 2022; Duncan 2022; Galler et al. 2021). Because inflammation may be limited to the coronal pulp, leaving the radicular pulp viable (Wolters et al. 2017; Ricucci et al. 2014), pulpotomy allows the radicular pulp tissue to remain healthy and recover (Cushley et al. 2019; Duncan 2022). Thus, VPTs contribute to the maintenance of dental pulp health and eventually prevent AP. Furthermore, cutting-edge treatments like cell-based regenerative therapies have also been implemented with the same objective, demonstrating promising results in preclinical models (Quigley et al. 2024; Sabeti et al. 2024).

Considering the high global prevalence of caries and pulpitis, and the plethora of cytokines and other pro-inflammatory mediators and biomarkers released during pulpitis (Rechenberg et al. 2016; Karrar et al. 2023; Duncan 2022; Hirsch et al. 2017; Zanini et al. 2017; Kritikou et al. 2021), a possible link between pulpitis and systemic conditions could be inferred. These studies underscore the complexity of pulp inflammation, which is intricately interconnected with tissue degradation and oxidative stress (Conrads and About 2018). Levels of interleukin (IL)-1β, IL-2, IL-6,

IL-8, and tumor necrosis factor-alpha (TNF-α) are increased in irreversible pulpitis compared to healthy controls (Karrar et al. 2023; Duncan 2022). In addition, upregulation of IL-17, superoxide dismutase-3, osteocalcin, and transforming growth factor- β1 (TGF-β1) has also been observed (Kritikou et al. 2021).

On the other hand, a possible spatiotemporal association between pulpitis and systemic conditions cannot be ruled out, especially when considering the role of chronic inflammation in disease etiology across the lifespan (Furman et al. 2019). Furthermore, it could be inferred that pulpitis may potentially increase the risk of systemic diseases if not treated in a timely manner. In a recent study (Sebring et al. 2024), a higher DMFT score—reflecting the number of decayed, missing, and filled teeth—has been found to increase the risk of cardiovascular-related events. Root-filled teeth (RFT) may indicate this risk, potentially due to prior pulpitis or AP. Importantly, pain is not always a reliable indicator of pulpitis (Cabrera-Abad et al. 2024), suggesting that the true burden and implications of this endodontic disease may be underestimated. Furthermore, in the early stages of the caries process, even before direct bacterial invasion, immune cells are recruited to the pulp tissue (Rechenberg et al. 2016).

Notably, multiplatform data integration models have facilitated the identification of genes involved in the molecular mechanisms underlying pulpitis (e.g., the important role of noncoding RNAs in the epigenetic regulation of pulpitis) (Liu et al. 2021; Chen et al. 2020). Thus, in the near future, these approaches could help decipher the possible links between pulpitis and systemic conditions, especially when aligned with predictive analytic models powered by artificial intelligence (Richert et al. 2022). However, no studies to date have confirmed such an association. Importantly, a recent systematic review and meta-analysis has revealed a significant association between dental pulp stones (PSs) and cardiovascular diseases (CVDs) in patients over 40 years of age (Almadhoon et al. 2022). PSs are usually asymptomatic, and, although their exact pathogenesis is still unknown, they often result from pulpal tissue self-repair in response to noxious stimuli (Goga et al. 2008).

8.3 Apical Periodontitis and Endodontic Infection: The Cause–Effect Relationship

If left untreated, pulpitis will increase in severity, eventually leading to pulp necrosis—with its inherent bacterial infection—and even to the development of AP (Schmalz and Smith 2014). While caries is the primary cause of pulpal and/or periapical disease, periodontal disease (PD) and dental trauma can also cause pulpal infection. Ultimately, AP is a chronic inflammatory condition with high global prevalence (Tibúrcio-Machado et al. 2021), and its incidence increases with patient age (Lin and Huang 2020). Effective management typically requires root canal treatment (RCT), though root canal retreatment (RCR) may sometimes also be necessary. In some cases, when RCR fails, persistent AP may require additional apical surgery, for example, in the presence of calcified extraradicular deposits on the apical root surface of the tooth (Álvarez-Vásquez 2023).

Infection of the pulp tissue, whether due to caries or other factors such as dental trauma or periodontal disease, is the primary etiology of AP. Intra- and/or extraradicular biofilms are the primary bacterial contributors to its pathogenesis. AP pathobiology involves neuro–immune–vascular interactions that orchestrate the formation of the periapical bone lesion (Fig. 8.1) (Lin and Huang 2020). The resulting periapical inflammation is a defensive response designed to limit the spread of infection into surrounding tissues (Segura-Egea et al. 2015). All in all, once the endodontic infection is eliminated via RCT, RCR, or apical surgery, the periapical lesion resolves, reinforcing a direct cause–effect relationship (Lin and Huang 2020; Kirkevang and Vaeth 2020).

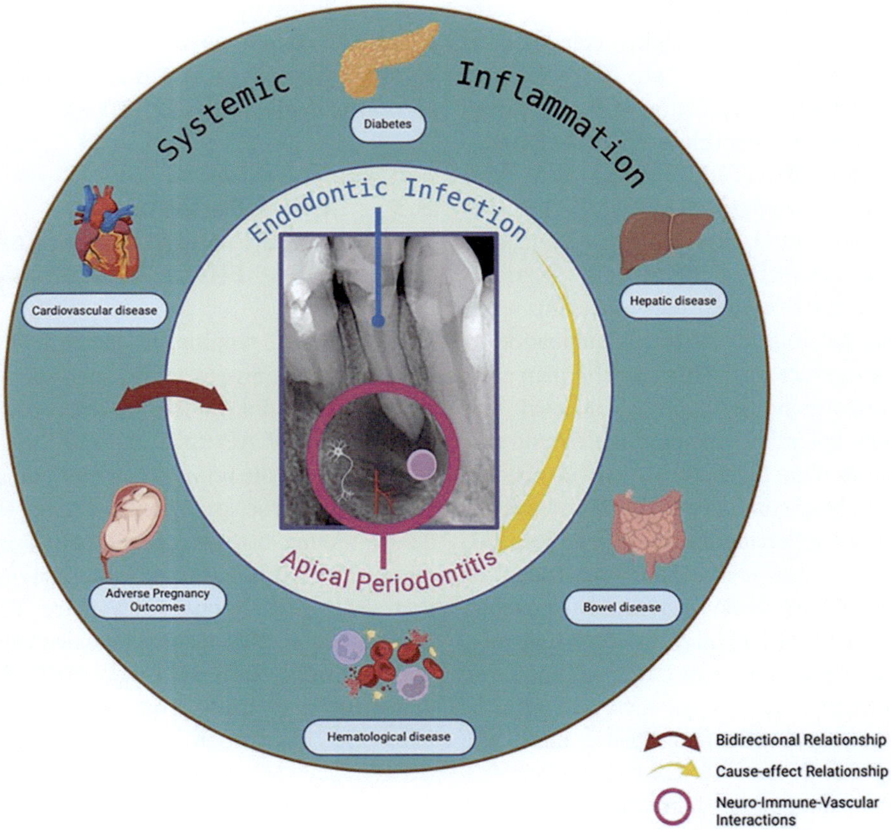

Fig. 8.1 A schematic representation of the bidirectional relationship between apical periodontitis (AP) and chronic systemic disease, depicting positive associations between AP and various systemic conditions, including cardiovascular disease (CVD), diabetes, hepatic disease, inflammatory bowel disease, hematological disease, and adverse pregnancy outcomes. The pathobiology of AP involves neuro–immune–vascular interactions, which orchestrate the formation of the periapical bone lesion, involving a massive release of immune cells and a plethora of cytokines and other pro-inflammatory mediators. Endodontic infection (EI) caused by caries or other factors, such as trauma and periodontitis, is the primary etiology of AP. Once the EI is eliminated via RCT, RCR, or apical surgery, the periapical lesion resolves, clearly denoting a cause–effect relationship between EI and AP. Contributors: José Luis Álvarez-Vásquez (figure design coordination) and Eduardo Moreno (figure design). (This figure was created using BioRender.com. *RCT* root canal treatment, *RCR* root canal retreatment)

8.4 Apical Periodontitis and Chronic Systemic Disease: A Bidirectional Relationship

Innate and adaptive immune responses, along with neurogenic inflammation, are key to AP pathogenesis, involving a massive release of immune cells and a plethora of cytokines and other pro-inflammatory mediators (Fig. 8.1) (Lin and Huang 2020). As a largely asymptomatic chronic inflammatory condition, AP can persist for long periods of time without treatment. The sustained release and presence of inflammatory mediators allows them to reach other tissues, potentially affecting existing pathological processes (Cintra et al. 2014a, b, 2016). Such mediators can induce or perpetuate a state of chronic systemic inflammation (Caplan et al. 2006).

AP is typically asymptomatic, although it can become symptomatic in certain cases, particularly when there is a direct invasion of the periapical tissues by bacteria originating from endodontic infection (Lin and Huang 2020). Being asymptomatic in nature, and primarily diagnosed through radiographic examination, the true burden of AP may be underestimated, even among patients. However, studies indicate that nearly half of the global population has at least one tooth affected by AP (Tibúrcio-Machado et al. 2021)..

With these implications in mind, recent studies have shown that AP and systemic disorders may influence each other: AP may contribute to systemic disease pathogenesis (Cintra et al. 2018, 2021; Segura-Egea et al. 2015), and systemic problems can alter blood cell and inflammatory mediator levels, leading to heightened immune responses during endodontic infection (Arya et al. 2017; Samuel et al. 2019). In addition, systemic factors, such as age, nutrition, stress, hormones, and smoking habits, as well as underlying systemic diseases may affect the healing process of periapical lesions after RCT (Segura-Egea et al. 2023). These findings affirm a bidirectional relationship between AP and systemic diseases (Fig. 8.1) and also that, evidently, AP is not merely a localized condition but one with potential systemic implications.

8.5 Apical Periodontitis and Systemic Conditions

Oral diseases are widespread, affecting a significant portion of the global population, influencing their daily activities, physical health, and mental well-being (Guarnizo-Herreño et al. 2024). Historically, various theories have speculated on the possible systemic effects of AP (Hunter 1900; Murray 1918). However, these initial theories, particularly the "focal infection theory," were discredited when it became evident that diseases attributed to dental infections were also common in individuals with healthy teeth (Easlick 1951). On the other hand, as mentioned above, recent studies have suggested that the presence of AP can indeed influence the pathogenesis of several systemic disorders (Cintra et al. 2018, 2021; Segura-Egea et al. 2015).

It is important to highlight the chronic nature of AP, which may then affect patients over extended periods. Persistent, systemic inflammation causes collateral damage and is a health-damaging phenotype linked to aging and influenced by social, environmental, and lifestyle factors (Furman et al. 2019). Importantly, a recent study of host–microbiome interactions in AP has found different inflammatory mediator concentrations in asymptomatic versus symptomatic patients (Georgiou et al. 2023).

In recent years, research on the relationship between systemic conditions and AP (endodontic medicine) has increased significantly (Cintra et al. 2018, 2021; Segura-Egea et al. 2015, 2019; Nagendrababu et al. 2020). Multiple studies have reported a positive correlation between AP and various systemic conditions, including CVD, diabetes, hepatic disease, inflammatory bowel disease (IBD), hematological disease, and adverse pregnancy outcomes (APOs) (Fig. 8.1). These associations clearly indicate that AP may have a systemic impact on overall health—an impact that might potentially be mitigated through nonsurgical endodontic therapy (RCT and RCR). All these bidirectional associations will be discussed in the following sections. While these associations highlight the potential systemic impact of endodontic infections, at this point, it must be emphasized that it has not yet been possible to establish a causal relationship (Aminoshariae et al. 2018; Cintra et al. 2018, 2021; Segura-Egea et al. 2015). Additionally, it must be pointed out that with increasing overall life expectancy, the prevalence of chronic oral and systemic diseases has risen in our aging society, affecting people's quality of life, emphasizing the importance of further research into their interplay and its implications for improving health and quality of life (Chung et al. 2024).

8.5.1 Cardiovascular Diseases

Globally, CVDs are the leading cause of death, with atherosclerosis being the primary driver of the most common types (WHO 2012). Evidence suggests that there are three potential pathways linking endodontic diseases to atherosclerotic CVD (Cotti et al. 2011a; Gomes et al. 2013): (a) transient bacteremia resulting from endodontic infection, allowing bacteria to enter the bloodstream; (b) vascular dissemination of bacterial toxins or metabolic by-products from endodontic infection; and (c) chronic low-grade systemic inflammation originating from AP. These pathways can lead to pathogenic mechanisms such as endothelial dysfunction (Cotti et al. 2011b; Nakano et al. 2006; Kebschull et al. 2010).

The severity of systemic inflammation in AP appears to correlate with the number of affected teeth in animal models, suggesting a dose-dependent effect (Cintra et al. 2016; Samuel et al. 2018). A recent study using Wistar rats has found that AP exacerbates atherosclerosis, elevates triglyceride levels, and increases carotid intima thickness and periapical bone resorption (Conti et al. 2020). However, it is important to note that while animal models provide valuable insights, their findings may not always predict human responses (Cintra et al. 2021).

While some human clinical studies have rejected the association between AP and CVD (Frisk et al. 2003; Cowan et al. 2020; Segura-Egea et al. 2010), the majority of them have confirmed a positive correlation (Pasqualini et al. 2012; Willershausen et al. 2009, 2014; Cotti et al. 2011b; Liljestrand et al. 2016; An et al. 2016; Caplan et al. 2009; Costa et al. 2014; Sebring et al. 2022, 2023). One study assessed the self-reported history of endodontic treatment (SRHET) and its association with CVD, evidencing a 21% increased risk of coronary heart disease in these patients (Caplan et al. 2009). However, a recent prospective study, with 8 years of follow-up, has shown that having one or more primary AP lesions at baseline is paradoxically associated with a reduced risk of experiencing a first CVD-related event. The authors suggested that this effect may be due to subsequent endodontic treatment of these teeth, such as RCT, which could mitigate systemic inflammatory effects and protect the potential effect of primary AP on the risk of future CVD-related events. In this regard, the potential impact of dental interventions on the incidence of CVD warrants future research (Sebring et al. 2024). Importantly, a recent study has evidenced that a weighted composite measure of endodontic inflammatory disease is linked to a higher risk of myocardial infarction (Sebring et al. 2023).

Patients who had completed RCT presented 84% and 49% reductions in their risk of developing CVD and cardiovascular-related death, respectively (Meurman et al. 2017). In addition, one cohort study showed that unfinished RCT was associated with an increased risk of hospitalization due to CVD (Lin et al. 2015). A recent study, after adjustment for potential covariates, such as sex, age, myocardial infarction, PD, smoking, diabetes, education, marital status, and a family history of CVD, has found evidence that greater tooth retention, a lower DMFT score, and fewer RFT are associated with a lower risk of both nonfatal and fatal cardiovascular events as well as CVD-specific events (Sebring et al. 2024).

Finally, recent systematic reviews have assessed the association between CVD and AP, suggesting only a moderate-to-weak association, with the level of evidence remaining low (Khalighinejad et al. 2016; Berlin-Broner et al. 2017; Aminoshariae et al. 2018). Similarly, two recent umbrella reviews (Pinto et al. 2023; Jakovljevic et al. 2020) have also confirmed an association between AP and CVD, although the evidence was of limited quality. Establishing a causal relationship between these two conditions has not yet been possible, due to the absence of prospective interventional studies. In this regard, further well-designed longitudinal research is necessary to prove this causality claim (Aminoshariae et al. 2018), especially given the long asymptomatic period of CVD, which presents an opportunity for early preventive strategies (Chauhan et al. 2023).

8.5.2 Diabetes

Diabetes mellitus (DM) is the 24th most prevalent disease worldwide, having a substantial impact on national healthcare systems, with men having a higher age-standardized incidence rate (GBD 2016 Disease and Injury Incidence and Prevalence Collaborators 2017). The immune phenotype induced by DM leads to oral complications, such as xerostomia, delayed wound healing, candidiasis, and pulp–periapical infections (Little et al. 1997).

Several studies have reported a higher prevalence and larger periapical lesions, delayed periapical healing, increased likelihood of asymptomatic infections, and worse prognoses for RFT in DM patients (Segura-Egea et al. 2015). Indeed, a recent umbrella review (Nagendrababu et al. 2020) and a systematic review and meta-analysis (León-López et al. 2023) have further supported the associations between DM and endodontic treatment outcomes, identifying DM as a main preoperative prognostic factor. There are three main biological mechanisms by which DM can influence AP (Segura-Egea et al. 2015): (a) impaired innate immunity, characterized by increased production of pro-inflammatory cytokines; (b) advanced glycated end products (AGEs), provoked by hyperglycemia, which bind to collagen and disrupt bone formation and metabolism; and (c) hyperglycemic-induced apoptosis of osteoblasts and fibroblasts, leading to reduced collagen production, as well as inhibition of osteoblastic cell proliferation and differentiation. In this regard, DM predisposes people to chronic inflammation and infection and impairs tissue repair (Segura-Egea et al. 2015).

Conversely, AP may also affect the glycemic control in diabetics through different mechanisms (Segura-Egea et al. 2015): (a) activation of the broad axis of innate immunity by AP; (b) upregulation of pro-inflammatory cytokines by bacterial lipopolysaccharides, contributing to the chronic pro-inflammatory systemic status of diabetic patients; (c) increased overall insulin resistance, since the activation of these inflammatory pathways may disrupt metabolic control in patients with both AP and DM; and (d) AGE–RAGE signaling, where AGEs bind to their receptors (RAGEs) in macrophages, activating the nuclear factor kappa B (NF-κB) signaling pathway and thus upregulating pro-inflammatory cytokines.

Many studies have confirmed a positive correlation between AP prevalence and DM. Two studies (López-López et al. 2011; Marotta et al. 2012) found significantly higher AP rates in diabetic patients compared to healthy individuals. One study (Arya et al. 2017) found that DM may negatively impact periapical healing after RCT. Another study (Sisli 2019) showed that the prevalence of AP and severe periapical bone destruction was significantly higher in DM patients compared to nondiabetic patients. In a recent cohort study with 9.8 years of follow-up, Heikkilä et al. (2022) have suggested that AP is a potential risk factor for DM incidence. It was shown that RFT with AP was more prevalent in individuals with type 1 DM than they were in nondiabetic individuals (Limeira et al. 2020). A recent observational study has shown that RCT is associated with reduced serum cytokine levels (IL-1, IL-6, and IL-8) and improved insulin responses in AP patients, suggesting that RCT may help mitigate the systemic inflammatory state in vivo and restore the in vitro insulin response (Multari et al. 2024).

A cross-sectional case–control study based on a large hospital network showed a highly significant association between type 2 DM and poorly controlled glycemia with AP (Yip et al. 2021). Importantly, four systematic reviews and meta-analyses (Segura-Egea et al. 2016; Gupta et al. 2020; Cabanillas-Balsera et al. 2019; Pérez-Losada et al. 2020) and a recent umbrella review (Pinto et al. 2023) have also confirmed associations between AP and DM, demonstrating that DM is significantly linked to the presence of periapical lesions in RFT. These studies highlight the notion that DM not only predisposes individuals to AP but also affects the healing process and prognosis of endodontic treatments.

In comparison, one study (Britto et al. 2003) found no significant association between AP and DM. A recent Mendelian randomization study has found no strong evidence to support a causal link between pulp and periapical disease and type 2 DM in either direction. Although no significant pleiotropy was detected after the removal of outliers, this model has some inherent limitations (Wang et al. 2024).

All in all, studies suggesting an association between the prevalence of AP and DM are inconclusive, warranting prospective epidemiological studies to further clarify the association. Currently, no scientific evidence supports a causal relationship between AP and metabolic control in DM (Cintra et al. 2018, 2021; Segura-Egea et al. 2015, 2019). It is important to note that, due to ethical reasons, conducting randomized controlled trials—the gold standard in the literature for evidence-based findings—to examine the causal association between AP and DM is both impermissible and impractical (Wang et al. 2024). However, dentists should be aware of this association because DM is one of the most common metabolic disorders among dental patients (Nagendrababu et al. 2020).

8.5.3 Liver Diseases

Liver diseases rank among the most common health conditions worldwide (GBD 2016 Disease and Injury Incidence and Prevalence Collaborators 2017). Research in both animal models (Furusho et al. 2013) and human studies (Nakahara et al. 2018) have detected oral microorganisms in the liver; such microorganisms may represent a risk factor for the pathogenesis and progression of liver disease.

A liver fibrosis model in rats with induced AP evidenced greater amounts of pro-inflammatory cytokines (IL-1β, IL-6, and TNF-α) and increased periapical bone resorption (Cantiga-Silva et al. 2021). A recent study in rats has demonstrated that chronic exposure to alcohol and nicotine disrupts multiple metabolic markers, compromises liver and kidney function, elevates systemic pro-inflammatory mediators, and enhances inflammation and bone destruction in AP. Furthermore, simultaneous consumption of alcohol and nicotine—both major public health concerns linked to several systemic disorders—leads to significant effects compared to either substance alone. In addition, both substances reduce production of calcitonin, a hormone inhibiting osteoclastic bone resorption (Pinto et al. 2024). On the other hand, a study in rats by Brasil et al. (2021) suggested that a high-fat diet (HFD) can influence the progression and severity of AP. In the HFD group, AP lesions were found to be larger in radiography and showed more severe inflammatory infiltrates histologically compared to controls (Brasil et al. 2021).

A recent mouse model has evidenced that gut microbiota dysbiosis links AP to liver fibrosis in nonalcoholic fatty liver disease (NAFLD), showing reduced diversity and distinct taxonomic enrichment in the AP group as well as upregulation of *LPIN1* expression (Gan et al. 2024). LPIN1 (lipin-1) is an enzyme that playa an important role in lipid metabolism, and its upregulation in the context of liver diseases indicates a dysregulation of this metabolism that can contribute to inflammation and tissue damage. On the other hand, a recent study in rats has shown that oxidative stress is implicated in AP-induced liver inflammation and that hyperlipidemia exacerbates both local and systemic inflammation (Xiao et al. 2023). Importantly, an investigation in rats by Cantiga-Silva et al. (2024) showed that liver fibrosis supports the inflammation and contamination of microorganisms in AP, triggering the expression of toll-like receptors 2 and 4, thereby modulating the innate immunity response in AP.

There are only a few clinical human studies analyzing the association between AP and liver disease. A recent investigation has evaluated the potential association of AP with NAFLD, in otherwise healthy individuals with and without AP diagnosis. The study showed that AP is associated with higher serum hepatic transaminases (alanine aminotransferase and aspartate aminotransferase), potentially contributing to NAFLD physiopathology in young adults (Bordagaray et al. 2024). Importantly, NAFLD is the most common type of liver disease, characterized by fat deposits in the liver, which is not attributable to alcohol consumption (Younossi et al. 2018). Two cross-sectional studies have also confirmed this association. A study by Castellanos-Cosano et al. (2013a) found a significant connection between AP and liver transplantation candidates, while another study (Grønkjær et al. 2016) showed that patients with AP demonstrated significantly higher incidence of cirrhosis in comparison with individuals without AP (46% vs. 27%). In comparison, an investigation by Braga Diniz et al. (2019) rejected this association because it showed that liver damage did not impair the periapical immune response.

A systematic review conducted by Khalighinejad et al. (2016) found a positive association between chronic liver disease (CLD) and AP, although there was a high risk of bias. The impaired immunological response in CLD patients is likely influenced by increased levels of endotoxins, nitric oxide, and cytokines (Tuchendler et al. 2018), highlighting the need for future studies on these factors (Pinto et al. 2023). On the other hand, a recent systematic review has highlighted the limited knowledge concerning the association between AP and gastrointestinal diseases (GIDs), noting the lack of information on the prevalence of AP in liver diseases. However, a positive correlation, albeit based on limited evidence, was observed between RFT, AP, and both inflammatory bowel disease and chronic liver disease (Jakovljevic et al. 2023).

Concerning the pathophysiological mechanisms connecting AP and liver diseases, the animal studies mentioned in this section have provided some potential fundamentals. Importantly, it has been proposed that liver inflammation may be initiated by an imbalance in the oral–gut–liver axis (Jakovljevic et al. 2023; Gan et al. 2024).

Notably, microorganisms and their virulence factors from AP can translocate (Bordagaray et al. 2021) and impair intestinal barrier function. Gut dysbiosis is pivotal in this process and can be induced through different routes (Jakovljevic et al. 2023). The blood from the dysbiotic gut circulates through the portal vein into the liver, exacerbating systemic inflammation and promoting the development of liver disease (Albuquerque-Souza and Sahingur 2022; Åberg and Helenius-Hietala 2022). Given the high prevalence of both AP and GIDs in the general adult population, efforts should be made to prevent the progression of periapical infections (Jakovljevic et al. 2023).

8.5.4 Bowel Diseases

Inflammatory bowel diseases (IBDs), such as Crohn's disease and ulcerative colitis, are conditions in which the pro-inflammatory status plays an important role. Given the inflammatory nature of both conditions, the association between IBDs and AP is highly plausible (Segura-Sampedro et al. 2022). Furthermore, IBDs are chronic idiopathic disorders that, in the last decade, have emerged as a global public health challenge, being recognized as complex and costly diseases (Ng et al. 2017). A recent systematic review by Guerrero-Gironés et al. (2021) has reported a positive association between AP and three autoimmune diseases (type I DM, rheumatoid arthritis, and IBD), although most studies reported statistically nonsignificant associations. On the other hand, a recent systematic review (Jakovljevic et al. 2023) has shown that three case–control studies (Piras et al. 2017; Poyato-Borrego et al. 2020; Segura-Sampedro et al. 2022) found no significant differences in the prevalence of AP between patients with IBD and the control groups. Nonetheless, one of these studies (Segura-Sampedro et al. 2022) found that patients with IBD had a higher prevalence of RFT and a greater percentage of RFT with periapical lesions, when compared with the control group. In addition, Poyato-Borrego et al. (2020) showed that patients with IBD were more likely to present AP when compared with the control group. The increased numbers of RFT in these patients can be linked to factors like an increased prevalence of dental caries (Marruganti et al. 2021) and persistent AP (Siqueira Jr et al. 2020). In addition, the study by Piras et al. (2017) showed that patients with IBD have a higher prevalence of periapical lesions, with significantly higher periapical index scores compared to the control group.

A longitudinal case–control study with 24 months of follow-up (Cotti et al. 2018) evidenced that IBD did not have a negative impact on the outcome of RCT. However, toward the final stage of the experiment, it was observed that the rate of healing of periapical lesions in patients with IBD was significantly higher compared to that in the group without IBD. The authors inferred that the potent effect of the immune modulators would enhance the healing process of AP, following a properly conducted RCT.

Finally, the pathophysiological mechanisms that correlate AP with IBD remain poorly elucidated (Lira-Junior and Figueredo 2016). Importantly, it must be considered that abnormal inflammatory and immune responses, host genetic factors, gut microbiota, and environmental factors are all involved in the pathogenesis of IBD (Guan 2019).

8.5.5 Hematological Diseases

Hematological diseases, such as anemia and bleeding disorders, are less common systemic conditions but have been investigated in relation to AP, with only three studies on this topic available in the literature (Cintra et al. 2021). A cross-sectional study found a significant association between AP and inherited coagulation disorders (ICDs), such as hemophilia A, hemophilia B, and Von Willebrand disease. At the same time, the study showed no significant differences between RCT frequency and ICD (Castellanos-Cosano et al. 2013b). A cohort study by Costa et al. (2013) showed a significant correlation between sickle cell anemia (SCA) and pulp necrosis. Finally, a controlled clinical trial by Ferreira et al. (2015) found a significant pro-inflammatory profile (periapical interstitial fluid) in pulp necrosis in patients with SCA compared to the control group (with and without AP). It should be pointed out that patients with SCA have an increased susceptibility to infections (Rees et al. 2010), which can lead to the development of pulp necrosis in otherwise clinically healthy permanent teeth (Costa et al. 2013). Notably, SCA is a common genetic disease worldwide; the number of people living with this disease globally increased by 41.4% from 2020 to 2021. Indeed, SCA has a large and growing global public health significance (GBD 2021 Sickle Cell Disease Collaborators 2023).

8.5.6 Adverse Pregnancy Outcomes

High-, low-, and middle-income countries share many constraints and inequities that limit access to care during pregnancy and the neonatal period (Roberts et al. 2024). Therefore, detecting potential predictive factors before and during prenatal care is crucial for preventing and controlling adverse pregnancy outcomes (APOs) (Khalighinejad et al. 2017). One of the most common APOs is preeclampsia (PE), which is among the leading causes of maternal mortality (Mammaro et al. 2009). In addition, PE causes nearly 15–20% of all preterm births (PTBs) (Backes et al. 2011).

A study by Conde-Agudelo et al. (2008) evidenced an association between systemic inflammation and APOs, such as PE. In this regard, a case–control study by Khalighinejad et al. (2017) was the first to show that maternal AP could be a strong independent predictor of PE. The authors found that pregnant women with undiagnosed/untreated AP were almost two times more likely to develop PE than those without AP. Given the high prevalence of PE, especially in developing countries (Ayele and Tilahun 2022), identifying and addressing any source of inflammation, including AP, before pregnancy is pivotal (Khalighinejad et al. 2017).

Maternal infections have also been linked to intrauterine growth restriction (IUGR) and PTB (Harjunmaa et al. 2015). Annually, 32 million infants are born underweight due to PTB or IUGR (Lee et al. 2013). Harjunmaa et al. (2015) also found a positive correlation between maternal AP and both PTB and IUGR in low-income settings, suggesting that AP treatment may help improve birth outcomes. The mechanisms through which AP affects pregnancy may be similar to those of periodontal disease (Madianos et al. 2013). Endodontic infection may potentially enter the bloodstream, leading to infection and a localized immune response in the fetoplacental unit, which may trigger labor (Harjunmaa et al. 2015). In this regard, *Fusobacterium nucleatum* is frequently detected in the amniotic fluid of women experiencing preterm labor (Han and Wang 2013; Doyle et al. 2014). Notably, a recent systematic review has shown a positive association between maternal AP and APO. However, these results were based on a limited amount of evidence, and the quality of the evidence was rated as "fair" to "good" (Jakovljevic et al. 2021).

A study by Bain et al. (2009) showed that rats with induced AP developed insulin resistance, which notably impacted their pregnancy outcomes. In rats, maternal AP is linked to offspring insulin resistance and altered signaling/inflammation pathways in adulthood (Tsosura et al. 2021a, b). Maintaining maternal oral health is therefore crucial for offspring health. On the other hand, a recent study in rats has investigated the effects of maternal true endodontic–periodontal lesions (EPLs) on insulin resistance, insulin signaling, and inflammatory pathways within the gastrocnemius muscle of adult offspring. The offspring of rats with EPLs presented with more pronounced metabolic alterations compared to those of the offspring of rats with only AP or PD (Belardi et al. 2023).

Taking the above studies into consideration, it is extremely important to detect and treat AP before pregnancy, as AP represents a source of systemic inflammation. Nonetheless, several studies have reported that most women receive insufficient dental care in the United States (Gaffield et al. 2001; Marchi et al. 2010) and others have highlighted limited dental care provided during pregnancy, along with the lack of guidance regarding oral health from health providers during and after pregnancy (Cagetti et al. 2024).

8.6 Impact of Root Canal Treatment in Systemically Healthy Human Adults

Research in healthy populations has provided no evidence of a causal relationship between AP and systemic conditions (Inchingolo et al. 2013; Poornima et al. 2021). However, the constant release of multiple mediators from AP can lead to imbalances in other preexisting inflammatory processes (Alim et al. 2021; Bakhsh et al. 2022). A recent prospective study in otherwise healthy young men presenting with AP has measured vascular changes via values of flow-mediated dilatation (FMD) and carotid intima–media thickness (c-IMT). The results indicated that RCT led to improved FMD and a reduction in c-IMT, suggesting that treating AP may help reverse the early stages of CVD (Chauhan et al. 2023).

In addition, two prospective interventional studies in systemically healthy young adults presenting with AP showed that RCT can reduce serum high-sensitivity C-reactive protein (hsCRP) levels. This result suggests that AP potentially leads to low-grade systemic inflammation. Importantly, quantifying serum CRP could play a potential role in assessing the CVD risk status (Poornima et al. 2021). Supporting the findings of this study, a recent investigation among a population of clinically healthy, young AP individuals at risk of cardiovascular conditions has shown that hsCRP levels were reduced after RCT (Garrido et al. 2024).

On the other hand, another study in otherwise systemically healthy patients with AP found that RCT may restore the systemic oxidative balance by analyzing serum glutathione levels (Bains et al. 2021). Similarly, a prospective clinical trial showed that RCT could restore a disturbed oxidative imbalance (Inchingolo et al. 2013). Other studies have also evidenced that RCT reverses the systemic impact of AP (Alim et al. 2021). These studies suggest that RCT might have a positive impact on systemic inflammation and oxidative stress associated with AP. In animal models, particularly in systemically healthy rats with local AP, an imbalance in cytokines such as TNF-α and interleukins (IL-6, IL-17, and IL-23) was shown; additionally, imbalances in cells, such as leukocytes, were shown as well as alterations in oxidative stress. A correlation was observed in which a higher number of involved teeth led to more cells and mediators being released (Cintra et al. 2016; Barcelos et al. 2020). This emphasizes the systemic effect of local AP, further supporting the importance of effective endodontic treatment in mitigating inflammation and oxidative stress.

Finally, although the above studies provide some insight into the impact of RCT in systemically healthy human adults, further research with larger sample sizes and longer follow-up periods is warranted to assess the impact.

8.7 Endodontic Medicine: A Paradigm Shift in the Field of Endodontics

As mentioned above, the main goal of endodontic medicine is to assess the interrelationship between systemic conditions and AP (Cintra et al. 2018, 2021; Segura-Egea et al. 2015, 2019). In parallel, periodontal medicine—which is ahead of endodontic medicine—has evolved in the past three decades, with multiple studies affirming the association between PD and systemic conditions (Herrera et al. 2023; Kapila 2021). Importantly, chronic endodontic and periodontal inflammatory processes have three crucial similarities: both are chronic oral infections that involve multiple microorganisms, sharing similar microbial communities, and both conditions can lead to increased cytokine production, which may be released into the bloodstream systemically during their acute and chronic phases (Segura-Egea et al. 2015).

While bidirectional relationships between these two conditions are well-recognized, the potential for multiple comorbidities and interconnected factors (multimodal relationships) warrants consideration. Genetic and environmental mechanisms likely mediate these complex interactions (Kapila 2021). Clinicians and patients should be aware of these associations, connections, and all the related implications for our systemic health. However, a recent study, regrettably, has reported that—despite the substantial body of scientific evidence on periodontal medicine—most patients remain unaware of the associations between PD and systemic health (Shimpi et al. 2020). A similar situation may exist in the field of endodontic medicine as well. It is up to academics and clinicians to take responsibility and change this scenario.

Evidently, the field of endodontics is evolving, with emerging research consistently revealing that endodontic inflammatory diseases influence systemic health, and vice versa. All these findings should prompt a paradigm shift in the field of endodontics, drawing the attention of health policymakers and professionals to the global burden of endodontic inflammatory diseases and their connection to systemic health. Notably, the recognition of oral medicine as a dental specialty in 2020 (American Dental Association 2024) will contribute to a better understanding of this fascinating world, by unifying findings from different dental specialties.

8.8 Summary and Prospects

Oral diseases affect nearly half of the world's population, from young children to older adults. These conditions lead to a wide range of negative effects on daily activities, physical health, and mental well-being of people (Guarnizo-Herreño et al. 2024). The results of investigations carried out both in humans and in animal models are inconclusive; they suggest bidirectional associations between AP and systemic conditions. These associations clearly indicate that endodontic inflammatory diseases may have a systemic impact on overall health. However, to date, there is no scientific evidence supporting a causal effect between endodontic inflammatory diseases and systemic conditions. There is a direct need for well-conducted randomized controlled trials and longitudinal cohort studies, which can establish cause–effect relationships. The possible link between pulpitis and systemic conditions requires much more investigation and attention than it is currently receiving.

On the other hand, high-throughput sequencing technologies are providing insights into the largely unexplored areas of both the oral microbiome (Kapila 2021) and the root canal microbiome (Ordinola-Zapata et al. 2024). These insights have contributed to a better description and classification of oral microbial dysbiosis (Kapila 2021) and could contribute to developing a new paradigm of personalized medicine (Jia et al. 2018). These technologies could also help us better understand the bidirectional relationship between AP and chronic systemic disease as well as the cause–effect relationship between AP and endodontic infection. In addition, considering the pivotal role of biofilms in the etiopathogenesis of AP and other oral chronic inflammatory diseases of bacterial origin, next-generation antibacterial strategies (Leveque et al. 2024; Toledano-Osorio et al. 2024) and novel intracanal medicaments (Siu et al. 2024) hold promising efficacy for biofilm eradication. Novel local drug delivery systems for the prevention and treatment of infections and infection-related manifestations in the oral cavity could also offer new avenues in the near future (Şenel et al. 2021). Importantly, given the connection between chronic oral inflammation and systemic diseases, along with the presence of oral microbial DNA in distant sites of the body, the ability of nucleic acid sensors to recognize this DNA may be an emerging link between oral and systemic diseases (Crump and Sahingur 2016). In addition, multilevel modeling, which incorporates both tooth- and patient-level predictors, provides a robust and efficient approach for analyzing AP data, which could offer valuable insights into disease patterns and progression in the future (Hussein et al. 2016; Buligon et al. 2023).

Importantly, although RCT, RCR, and apical surgery are traditional therapies used to treat AP, cutting-edge treatments like cell-based regenerative therapies have been also implemented in a move toward biologically based management of AP (Yan et al. 2023; Brizuela et al. 2024). Such therapies could revolutionize the treatment and prognosis of AP. Moreover, preventive endodontics should pave the way to reduce pulpitis and AP and, thus, their association with systemic conditions. Indeed, the 2021 WHO resolution on oral health advocates for a shift toward prevention in global oral health strategies, reorienting its agenda toward public health and challenging the prevailing focus on individual behavior and technology-based clinical care (Benzian et al. 2021). Many common chronic diseases are largely preventable, as demonstrated by preclinical, epidemiological, and clinical data. A balanced approach to healthcare, emphasizing both preventive measures and personalized treatment, is crucial for reducing health disparities (Fontana et al. 2021). Furthermore, the role of dental general practitioners, dental specialists, and family doctors in addressing noncommunicable diseases (NCDs) and promoting healthy lifestyles should be emphasized globally (Herrera et al. 2023).

References

Åberg F, Helenius-Hietala J (2022) Oral health and liver disease: bidirectional associations-a narrative review. Dentis J 10(2):16. https://doi.org/10.3390/dj10020016

Albuquerque-Souza E, Sahingur SE (2022) Periodontitis, chronic liver diseases, and the emerging oral-gut-liver axis. Periodontol 89(1):125–141. https://doi.org/10.1111/prd.12427

Alim BA, Canturk E, Koksal C (2021) The effect of treated apical periodontitis before heart valve surgery on C-reactive protein levels. Oral Dis 27(3):632–638. https://doi.org/10.1111/odi.13554

Almadhoon HW, Al-Kafarna M, Asla MM, Gbreel MI, Abd Allah MAE, Almotairy N (2022) The Association of Dental Pulp Stones to cardiovascular and renal diseases: a systematic review and meta-analysis. J Endodont 48(7):845–854. https://doi.org/10.1016/j.joen.2022.02.010

Álvarez-Vásquez JL (2023) The biomineralization conundrum of calcified extraradicular deposits on the apical root surface of teeth: a case report. Open Dent J 17: e187421062309191. https://doi.org/10.2174/18742106-v17-e230922-2023-70

Álvarez-Vásquez JL, Castañeda-Alvarado CP (2022) Dental pulp fibroblast: a star cell. J Endodont 48(8): 1005–1019. https://doi.org/10.1016/j.joen.2022.05.004

American Dental Association (2024, July 15) ADA specialties. https://ncrdscb.ada.org/recognized-dental-specialties

Aminoshariae A, Kulild JC, Fouad AF (2018) The impact of endodontic infections on the pathogenesis of cardiovascular disease(s): a systematic review with meta-analysis using GRADE. J Endodont 44(9): 1361–1366. https://doi.org/10.1016/j.joen.2018.06.011

An GK, Morse DE, Kunin M, Goldberger RS, Psoter WJ (2016) Association of radiographically diagnosed apical periodontitis and cardiovascular disease: a hospital records-based study. J Endodont 42(6):916–920. https://doi.org/10.1016/j.joen.2016.03.011

Arya S, Duhan J, Tewari S, Sangwan P, Ghalaut V, Aggarwal S (2017) Healing of apical periodontitis after nonsurgical treatment in patients with type 2 diabetes. J Endodont 43(10):1623–1627. https://doi.org/10.1016/j.joen.2017.05.013

Asgary S, Roghanizadeh L, Eghbal MJ, Akbarzadeh Baghban A, Aminoshariae A, Nosrat A (2024) Outcomes and predictive factors of vital pulp therapy in a large-scale retrospective cohort study over 10 years. Sci Rep 14(1):2063. https://doi.org/10.1038/s41598-024-52654-8

Ayele AD, Tilahun ZA (2022) Determinants of pre-eclampsia among women attending delivery services in public health institutions of Debre Tabor town: a case-control study. Reprod Health 19(1):157. https://doi.org/10.1186/s12978-022-01463-1

Backes CH, Markham K, Moorehead P, Cordero L, Nankervis CA, Giannone PJ (2011) Maternal preeclampsia and neonatal outcomes. J Pregnancy 2011: 214365. https://doi.org/10.1155/2011/214365

Bain JL, Lester SR, Henry WD, Naftel JP, Johnson RB (2009) Effects of induced periapical abscesses on rat pregnancy outcomes. Arch Oral Biol 54(2):162–171. https://doi.org/10.1016/j.archoralbio.2008.09.016

Bains R, Tikku AP, Chandra A, Ali W, Verma P (2021) Effect of endodontic treatment on serum glutathione (GSH) levels in patients with chronic apical periodontitis: a clinical pilot study. J Adv Oral Res 12(1):81–85. https://doi.org/10.1177/2320206820958239

Bakhsh A, Moyes D, Proctor G, Mannocci F, Niazi SA (2022) The impact of apical periodontitis, non-surgical root canal retreatment and periapical surgery on serum inflammatory biomarkers. Int Endod J 55(9):923–937. https://doi.org/10.1111/iej.13786

Barcelos RCS, Rosa HZ, Roversi K, Tibúrcio-Machado CDS, Inchaki PT, Burger ME, Bier CAS (2020) Apical periodontitis induces changes on oxidative stress parameters and increases Na+/K+-ATPase activity in adult rats. Arch Oral Biol 118:104849. https://doi.org/10.1016/j.archoralbio.2020.104849

Barros MMAF, De Queiroz Rodrigues MI, Muniz FWMG, Rodrigues LKA (2020) Selective, stepwise, or nonselective removal of carious tissue: which technique offers lower risk for the treatment of dental caries in permanent teeth? A systematic review and meta-analysis. Clin Oral Investig 24(2):521–532. https://doi.org/10.1007/s00784-019-03114-5

Belardi BE, Mattera MSLC, Tsosura TVS, Dos Santos RM et al (2023) Effect of maternal true combined endodontic-periodontal lesion on insulin and inflammatory pathway in adult offspring. J Periodontol 94(4): 487–497. https://doi.org/10.1002/JPER.22-0139

Benzian H, Guarnizo-Herreño CC, Kearns C, Muriithi MW, Watt RG (2021) The WHO global strategy for oral health: an opportunity for bold action. Lancet (London, England) 398(10296):192–194. https://doi.org/10.1016/S0140-6736(21)01404-5

Berlin-Broner Y, Febbraio M, Levin L (2017) Association between apical periodontitis and cardiovascular diseases: a systematic review of the literature. Int Endod J 50(9):847–859. https://doi.org/10.1111/iej.12710

Bordagaray MJ, Fernández A, Garrido M, Astorga J, Hoare A, Hernández M (2021) Systemic and extraradicular bacterial translocation in apical periodontitis. Front Cell Infect Microbiol 11:649925. https://doi.org/10.3389/fcimb.2021.649925

Bordagaray MJ, Pellegrini E, Garrido M, Hernández-Ríos P, Villalobos T, Fernández A, Hernández M (2024) Elevated serum hepatic transaminases in apical periodontitis individuals. Int Endod J 57:1395

Braga Diniz JM, Espaladori MC et al (2019) Immunological profile of teeth with inflammatory periapical disease from chronic liver disease patients. Int Endod J 52(2):149–157. https://doi.org/10.1111/iej.12992

Brasil SC, Santos RMM, Fernandes A, Lima RS, Costa CAS, Pinto KMMDC, Pires FR, Santos MD, Siqueira JF Jr, Armada L (2021) Influence of a high-fat diet in the progression of apical periodontitis. J Endodont 47(4):600–605. https://doi.org/10.1016/j.joen.2020.12.015

Britto LR, Katz J, Guelmann M, Heft M (2003) Periradicular radiographic assessment in diabetic and control individuals. Oral Surg Oral Med Oral Pathol Oral Radiol Endod 96(4):449–452. https://doi.org/10.1016/s1079-2104(03)00034-9

Brizuela C, Meza G, Khoury M (2024) Revolutionizing endodontics: innovative approaches for treating mature teeth with closed apices and apical lesions: a report of two cases. J Endodont 50(5):596–601. https://doi.org/10.1016/j.joen.2024.02.012

Buligon MP, Marin JA, Wolle CFB, Liedke GS, Sfreddo CS, Bier CAS, Moreira CHC, Morgental RD (2023) Apical periodontitis and associated factors in a rural population of southern Brazil: a multilevel analysis. Clin Oral Investig 27(6):2887–2897. https://doi.org/10.1007/s00784-023-04886-7

Cabanillas-Balsera D, Martín-González J, Montero-Miralles P, Sánchez-Domínguez B, Jiménez-Sánchez MC, Segura-Egea JJ (2019) Association between diabetes and nonretention of root filled teeth: a systematic review and meta-analysis. Int Endod J 52(3):297–306. https://doi.org/10.1111/iej.13011

Cabrera-Abad D, Jara-Vergara V, Álvarez-Vásquez JL (2024) Exploring the painless nature and potential mechanisms of asymptomatic irreversible pulpitis: a narrative review. Open Dent J 18:e18742106281444. https://doi.org/10.2174/0118742106281444240219050149

Cagetti MG, Salerno C, Ionescu AC, La Rocca S, Camoni N, Cirio S, Campus G (2024) Knowledge and attitudes on oral health of women during pregnancy and their children: an online survey. BMC Oral Health 24(1):85. https://doi.org/10.1186/s12903-023-03732-2

Cantiga-Silva C, Estrela C, Segura-Egea JJ, Azevedo JP, de Oliveira PHC, Cardoso CBM, Pinheiro TN, Ervolino E, Sivieri-Araújo G, Cintra LTA (2021) Inflammatory profile of apical periodontitis associated with liver fibrosis in rats: histological and immunohistochemical analysis. Int Endod J 54(8):1353–1361. https://doi.org/10.1111/iej.13519

Cantiga-Silva C, de Oliveira PHC, Faria FD, Justo MP, Sivieri-Araújo G, Ervolino E, Pinheiro TN, Segura-Egea JJ, Cintra LTA (2024) Increase toll-like receptors 2 and 4 in apical periodontitis of rats with chronic liver disease. Odontology 113:296

Caplan DJ, Chasen JB, Krall EA, Cai J, Kang S, Garcia RI, Offenbacher S, Beck JD (2006) Lesions of endodontic origin and risk of coronary heart disease. J Dent Res 85(11):996–1000. https://doi.org/10.1177/154405910608501104

Caplan DJ, Pankow JS, Cai J, Offenbacher S, Beck JD (2009) The relationship between self-reported history of endodontic therapy and coronary heart disease in the atherosclerosis risk in communities study. J Am Dent Assoc 140(8):1004–1012. https://doi.org/10.14219/jada.archive.2009.0311

Castellanos-Cosano L, Machuca-Portillo G, Segura-Sampedro JJ, Torres-Lagares D, López-López J, Velasco-Ortega E, Segura-Egea JJ (2013a) Prevalence of apical periodontitis and frequency of root canal treatments in liver transplant candidates. Med Oral Patol Oral 18(5):e773–e779. https://doi.org/10.4317/medoral.19148

Castellanos-Cosano L, Machuca-Portillo G, Sánchez-Domínguez B, Torrés-Lagares D, López-López J, Segura-Egea JJ (2013b) High prevalence of radiolucent periapical lesions amongst patients with inherited coagulation disorders. Haemophilia 19(3):e110–e115. https://doi.org/10.1111/hae.12089

Chauhan N, Mittal S, Tewari S, Sen J, Laller K (2023) Effect of endodontic treatment on endothelial dysfunction and subclinical atherosclerosis-a prospective intervention study. Clin Oral Investig 27(9):5617–5625. https://doi.org/10.1007/s00784-023-05183-z

Chen M, Zeng J, Yang Y, Wu B (2020) Diagnostic biomarker candidates for pulpitis revealed by bioinformatics analysis of merged microarray gene expression datasets. BMC Oral Health 20(1):279. https://doi.org/10.1186/s12903-020-01266-5

Chung YL, Lee JJ, Chien HH, Chang MC, Jeng JH (2024) Interplay between diabetes mellitus and periodontal/pulpal-periapical diseases. J Dental Sci 19(3):1338–1347. https://doi.org/10.1016/j.jds.2024.03.021

Cintra LT, da Silva Facundo AC, Prieto AK, Sumida DH, Narciso LG, Mogami Bomfim SR et al (2014a) Blood profile and histology in oral infections associated with diabetes. J Endodont 40(8):1139–1144. https://doi.org/10.1016/j.joen.2014.01.034

Cintra LT, Samuel RO, Azuma MM, Ribeiro CP, Narciso LG, de Lima VM, Sumida DH, Coclete GA, Dezan-Júnior E, Gomes-Filho JE (2014b) Apical periodontitis and periodontal disease increase serum IL-17 levels in normoglycemic and diabetic rats. Clin Oral Investig 18(9):2123–2128. https://doi.org/10.1007/s00784-014-1192-7

Cintra LT, Samuel RO, Azuma MM, de Queiróz AO, Ervolino E, Sumida DH, de Lima VM, Gomes-Filho JE (2016) Multiple apical periodontitis influences serum levels of cytokines and nitric oxide. J Endodont 42(5):747–751. https://doi.org/10.1016/j.joen.2016.01.022

Cintra LTA, Estrela C, Azuma MM, Queiroz ÍOA, Kawai T, Gomes-Filho JE (2018) Endodontic medicine: interrelationships among apical periodontitis, systemic disorders, and tissue responses of dental materials. Braz Oral Res 32(Suppl 1):e68. https://doi.org/10.1590/1807-3107bor-2018.vol32.0068

Cintra LTA, Gomes MS, da Silva CC, Faria FD, Benetti F, Cosme-Silva L, Samuel RO, Pinheiro TN, Estrela C, González AC, Segura-Egea JJ (2021) Evolution of endodontic medicine: a critical narrative review of the interrelationship between endodontics and systemic pathological conditions. Odontology 109(4):741–769. https://doi.org/10.1007/s10266-021-00636-x

Conde-Agudelo A, Villar J, Lindheimer M (2008) Maternal infection and risk of preeclampsia: systematic review and metaanalysis. Am J Obstet Gynecol 198(1):7–22. https://doi.org/10.1016/j.ajog.2007.07.040

Conrads G, About I (2018) Pathophysiology of dental caries. Monogr Oral Sci 27:1–10. https://doi.org/10.1159/000487826

Conti LC, Segura-Egea JJ, Cardoso CBM, Benetti F, Azuma MM, Oliveira PHC, Bomfim SRM, Cintra LTA (2020) Relationship between apical periodontitis and atherosclerosis in rats: lipid profile and histological study. Int Endod J 53(10):1387–1397. https://doi.org/10.1111/iej.13350

Costa CP, Thomaz EB et al (2013) Association between sickle cell anemia and pulp necrosis. J Endodont 39(2):177–181. https://doi.org/10.1016/j.joen.2012.10.024

Costa TH, de Figueiredo Neto JA, de Oliveira AE et al (2014) Association between chronic apical periodontitis and coronary artery disease. J Endodont 40(2):164–167. https://doi.org/10.1016/j.joen.2013.10.026

Cotti E, Dessì C, Piras A, Mercuro G (2011a) Can a chronic dental infection be considered a cause of cardiovascular disease? A review of the literature. Int J Cardiol 148(1):4–10. https://doi.org/10.1016/j.ijcard.2010.08.011

Cotti E, Dessì C, Piras A, Flore G, Deidda M, Madeddu C, Zedda A, Longu G, Mercuro G (2011b) Association of endodontic infection with detection of an initial lesion to the cardiovascular system. J Endodont 37(12):1624–1629. https://doi.org/10.1016/j.joen.2011.09.006

Cotti E, Mezzena S, Schirru E, Ottonello O, Mura M, Ideo F, Susnik M, Usai P (2018) Healing of apical periodontitis in patients with inflammatory bowel diseases and under anti-tumor necrosis factor alpha therapy. J Endodont 44(12):1777–1782. https://doi.org/10.1016/j.joen.2018.09.004

Cowan LT, Lakshminarayan K, Lutsey PL, Beck J, Offenbacher S, Pankow JS (2020) Endodontic therapy and incident cardiovascular disease: the atherosclerosis risk in communities (ARIC) study. J Public Health Dent 80(1):79–91. https://doi.org/10.1111/jphd.12353

Crump KE, Sahingur SE (2016) Microbial nucleic acid sensing in oral and systemic diseases. J Dent Res 95(1):17–25. https://doi.org/10.1177/0022034515609062

Cushley S, Duncan HF, Lappin MJ, Tomson PL, Lundy FT, Cooper P, Clarke M, El Karim IA (2019) Pulpotomy for mature carious teeth with symptoms of irreversible pulpitis: a systematic review. J Dent 88:103158. https://doi.org/10.1016/j.jdent.2019.06.005

Doyle RM, Alber DG, Jones HE, Harris K, Fitzgerald F, Peebles D, Klein N (2014) Term and preterm labour are associated with distinct microbial community structures in placental membranes which are independent of mode of delivery. Placenta 35(12):1099–1101. https://doi.org/10.1016/j.placenta.2014.10.007

Duncan HF (2022) Present status and future directions-vital pulp treatment and pulp preservation strategies. Int Endod J 55(Suppl 3):497–511. https://doi.org/10.1111/iej.13688

Easlick KA (1951) An evaluation of the effect of dental foci of infection on health. J Am Dent Assoc 42(6):615–697

European Society of Endodontology (ESE) et al (2019) European society of endodontology position statement: management of deep caries and the exposed pulp. Int Endod J 52(7):923–934. https://doi.org/10.1111/iej.13080

Ferreira SB, de Brito LC, Oliveira MP, Maciel KF, Martelli Júnior H, Vieira LQ, Sobrinho AP (2015) Periapical cytokine expression in sickle cell disease. J Endodont 41(3):358–362. https://doi.org/10.1016/j.joen.2014.11.016

Fontana L, Fasano A, Chong YS, Vineis P, Willett WC (2021) Transdisciplinary research and clinical priorities for better health. PLoS Med 18(7):e1003699. https://doi.org/10.1371/journal.pmed.1003699

Frisk F, Hakeberg M, Ahlqwist M, Bengtsson C (2003) Endodontic variables and coronary heart disease. Acta Odontol Scand 61(5):257–262. https://doi.org/10.1080/00016350310005510

Furman D, Campisi J, Verdin E, Carrera-Bastos P, Targ S, Franceschi C, Ferrucci L, Gilroy DW, Fasano A, Miller GW, Miller AH, Mantovani A, Weyand CM, Barzilai N, Goronzy JJ, Rando TA, Effros RB, Lucia A, Kleinstreuer N, Slavich GM (2019) Chronic inflammation in the etiology of disease across the life span. Nat Med 25(12):1822–1832. https://doi.org/10.1038/s41591-019-0675-0

Furusho H, Miyauchi M, Hyogo H, Inubushi T, Ao M, Ouhara K, Hisatune J, Kurihara H, Sugai M, Hayes CN, Nakahara T, Aikata H, Takahashi S, Chayama K, Takata T (2013) Dental infection of Porphyromonas gingivalis exacerbates high fat diet-induced steatohepatitis in mice. J Gastroenterol 48(11):1259–1270. https://doi.org/10.1007/s00535-012-0738-1

Gaffield ML, Gilbert BJ, Malvitz DM, Romaguera R (2001) Oral health during pregnancy: an analysis of information collected by the pregnancy risk assessment monitoring system. J Am Dental Assoc 132(7):1009–1016. https://doi.org/10.14219/jada.archive.2001.0306

Galler KM, Weber M, Korkmaz Y, Widbiller M, Feuerer M (2021) Inflammatory response mechanisms of the dentine-pulp complex and the periapical tissues. Int J Mol Sci 22(3):1480. https://doi.org/10.3390/ijms22031480

Gan G, Luo Y, Zeng Y, Lin S, Lu B, Zhang R, Chen S, Lei H, Cai Z, Huang X (2024) Gut microbiota dysbiosis links chronic apical periodontitis to liver fibrosis in nonalcoholic fatty liver disease: insights from a mouse model. Int Endod J 57:1608

Garrido M, Bordagaray MJ, Schweitzer C, Lucero-Mora J, Reyes M, Pellegrini E, Hernández-Ríos P, Fernández A, Hernández M (2024) Reduced C-reactive protein levels after root canal treatment in clinically healthy young apical periodontitis individuals at cardiovascular risk. A prospective study. Int Endod J 57(4):406–415. https://doi.org/10.1111/iej.14029

GBD 2016 Disease and Injury Incidence and Prevalence Collaborators (2017) Global, regional, and national incidence, prevalence, and years lived with disability for 328 diseases and injuries for 195 countries, 1990-2016: a systematic analysis for the global burden of disease study 2016. Lancet (London, England) 390(10100):1211–1259. https://doi.org/10.1016/S0140-6736(17)32154-2

GBD 2021 Sickle Cell Disease Collaborators (2023) Global, regional, and national prevalence and mortality burden of sickle cell disease, 2000-2021: a systematic analysis from the global burden of disease study 2021. Lancet Haematol 10(8):e585–e599. https://doi.org/10.1016/S2352-3026(23)00118-7

Georgiou AC, van der Waal SV, Buijs MJ, Crielaard W, Zaura E, Brandt BW (2023) Host-microbiome interactions in apical periodontitis: the endodontic microbiome in relation to circulatory immunologic markers. Int Endod J 56(6):748–764. https://doi.org/10.1111/iej.13912

Goga R, Chandler NP, Oginni AO (2008) Pulp stones: a review. Int Endod J 41(6):457–468. https://doi.org/10.1111/j.1365-2591.2008.01374.x

Gomes MS, Blattner TC, Sant'Ana Filho M, Grecca FS, Hugo FN, Fouad AF, Reynolds MA (2013) Can apical periodontitis modify systemic levels of inflammatory markers? A systematic review and meta-analysis. J Endodont 39(10):1205–1217. https://doi.org/10.1016/j.joen.2013.06.014

Grønkjær LL, Holmstrup P, Schou S, Schwartz K, Kongstad J, Jepsen P, Vilstrup H (2016) Presence and consequence of tooth periapical radiolucency in patients with cirrhosis. Hepatic Med Evid Research 8: 97–103. https://doi.org/10.2147/HMER.S113485

Guan Q (2019) A comprehensive review and update on the pathogenesis of inflammatory bowel disease. J Immunol Res 2019:7247238. https://doi.org/10.1155/2019/7247238

Guarnizo-Herreño CC, Celeste RK, Peres MA (2024) The ongoing fight for population oral health. Lancet (London, England) 404(10453):635–638. https://doi.org/10.1016/S0140-6736(24)00536-1

Guerrero-Gironés J, Ros-Valverde A, Pecci-Lloret MP, Rodríguez-Lozano FJ, Pecci-Lloret MR (2021) Association between pulpal-periapical pathology and autoimmune diseases: a systematic review. J Clin Med 10(21):4886. https://doi.org/10.3390/jcm10214886

Gupta A, Aggarwal V, Mehta N, Abraham D, Singh A (2020) Diabetes mellitus and the healing of periapical lesions in root filled teeth: a systematic review and meta-analysis. Int Endod J 53(11):1472–1484. https://doi.org/10.1111/iej.13366

Han YW, Wang X (2013) Mobile microbiome: oral bacteria in extra-oral infections and inflammation. J Dent Res 92(6):485–491. https://doi.org/10.1177/0022034513487559

Harjunmaa U, Järnstedt J, Alho L, Dewey KG, Cheung YB, Deitchler M, Ashorn U, Maleta K, Klein NJ, Ashorn P (2015) Association between maternal dental periapical infections and pregnancy outcomes: results from a cross-sectional study in Malawi. Trop Med Int Health 20(11):1549–1558. https://doi.org/10.1111/tmi.12579

Heikkilä P, Niskanen L, But A, Sorsa T, Haukka J (2022) Oral health associated with incident diabetes but not other chronic diseases: a register-based cohort study. Front Oral Health 3:956072. https://doi.org/10.3389/froh.2022.956072

Herrera D, Sanz M, Shapira L, Brotons C, Chapple I, Frese T, Graziani F, Hobbs FDR, Huck O, Hummers E, Jepsen S, Kravtchenko O, Madianos P, Molina A, Ungan M, Vilaseca J, Windak A, Vinker S (2023) Association between periodontal diseases and cardiovascular diseases, diabetes and respiratory diseases: consensus report of the joint workshop by the European Federation of Periodontology (EFP) and the European arm of the World Organization of Family Doctors (WONCA Europe). J Clin Periodontol 50(6): 819–841. https://doi.org/10.1111/jcpe.13807

Hirsch V, Wolgin M, Mitronin AV, Kielbassa AM (2017) Inflammatory cytokines in normal and irreversibly inflamed pulps: a systematic review. Arch Oral Biol 82:38–46. https://doi.org/10.1016/j.archoralbio.2017.05.008

Hosseinzadehfard P, Skučaitė N, Maciulskiene-Visockiene V, Lodiene G (2024) Blood pH changes in dental pulp of patients with pulpitis. Diagnostics (Basel, Switzerland) 14(11):1128. https://doi.org/10.3390/diagnostics14111128

Hunter W (1900) Oral sepsis as a cause of disease. Br Med J 2(2065):215–216. https://doi.org/10.1136/bmj.2.2065.215

Hussein FE, Liew AK, Ramlee RA, Abdullah D, Chong BS (2016) Factors associated with apical periodontitis: a multilevel analysis. J Endodont 42(10):1441–1445. https://doi.org/10.1016/j.joen.2016.07.009

Inchingolo F, Marrelli M, Annibali S, Cristalli MP, Dipalma G, Inchingolo AD, Palladino A, Inchingolo AM, Gargari M, Tatullo M (2013) Influence of endodontic treatment on systemic oxidative stress. Int J Med Sci 11(1):1–6. https://doi.org/10.7150/ijms.6663

Innes NPT, Schwendicke F (2017) Restorative thresholds for carious lesions: systematic review and meta-analysis. J Dent Res 96(5):501–508. https://doi.org/10.1177/0022034517693605

Jakovljevic A, Duncan HF, Nagendrababu V, Jacimovic J, Milasin J, Dummer PMH (2020) Association between cardiovascular diseases and apical periodontitis: an umbrella review. Int Endod J 53(10):1374–1386. https://doi.org/10.1111/iej.13364

Jakovljevic A, Sljivancanin Jakovljevic T, Duncan HF, Nagendrababu V, Jacimovic J, Aminoshariae A, Milasin J, Dummer PMH (2021) The association between apical periodontitis and adverse pregnancy outcomes: a systematic review. Int Endod J 54(9): 1527–1537. https://doi.org/10.1111/iej.13538

Jakovljevic A, Ideo F, Jacimovic J, Aminoshariae A, Nagendrababu V, Azarpazhooh A, Cotti E (2023) The link between apical periodontitis and gastrointestinal diseases-a systematic review. J Endodont 49(11): 1421–1431. https://doi.org/10.1016/j.joen.2023.07.024

Jia G, Zhi A, Lai PFH, Wang G, Xia Y, Xiong Z, Zhang H, Che N, Ai L (2018) The oral microbiota – a mechanistic role for systemic diseases. Br Dent J 224(6): 447–455. https://doi.org/10.1038/sj.bdj.2018.217

Kapila YL (2021) Oral health's inextricable connection to systemic health: special populations bring to bear multimodal relationships and factors connecting periodontal disease to systemic diseases and conditions. Periodontol 87(1):11–16. https://doi.org/10.1111/prd.12398

Karrar RN, Cushley S, Duncan HF, Lundy FT, Abushouk SA, Clarke M, El-Karim IA (2023) Molecular biomarkers for objective assessment of symptomatic pulpitis: a systematic review and meta-analysis. Int Endod J 56(10):1160–1177. https://doi.org/10.1111/iej.13950

Kebschull M, Demmer RT, Papapanou PN (2010) Gum bug, leave my heart alone!-epidemiologic and mechanistic evidence linking periodontal infections and atherosclerosis. J Dent Res 89(9):879–902. https://doi.org/10.1177/0022034510375281

Khalighinejad N, Aminoshariae MR, Aminoshariae A, Kulild JC, Mickel A, Fouad AF (2016) Association between systemic diseases and apical periodontitis. J Endodont 42(10):1427–1434. https://doi.org/10.1016/j.joen.2016.07.007

Khalighinejad N, Aminoshariae A, Kulild JC, Mickel A (2017) Apical periodontitis, a predictor variable for preeclampsia: a case-control study. J Endodont 43(10):1611–1614. https://doi.org/10.1016/j.joen.2017.05.021

Kirkevang LL, Vaeth M (2020) Epidemiology, treatment outcome, and risk factors for apical periodontitis. In: Ørstavik D (ed) Essential Endodontology. Prevention and treatment of apical periodontitis, 3rd edn. Willey-Blackwell, pp 143–178

Kritikou K, Imre M, Tanase M, Vinereanu A, Totan AR et al (2021) Biochemical mapping of the inflamed human dental pulp. Appl Sci 11(21):10395. https://doi.org/10.3390/app112110395

Lee AC, Katz J, Blencowe H, Cousens S, Kozuki N, Vogel JP, Adair L, Baqui AH, Bhutta ZA, Caulfield LE, Christian P, Clarke SE, Ezzati M, Fawzi W, Gonzalez R, Huybregts L, Kariuki S, Kolsteren P, Lusingu J, Marchant T et al (2013) National and regional estimates of term and preterm babies born small for gestational age in 138 low-income and middle-income countries in 2010. Lancet Glob Health 1(1):e26–e36. https://doi.org/10.1016/S2214-109X(13)70006-8

León-López M, Cabanillas-Balsera D, Martín-González J, Díaz-Flores V, Areal-Quecuty V, Crespo-Gallardo I, Montero-Miralles P, Segura-Egea JJ (2023) Prevalence of root canal treatments among diabetic patients: systematic review and meta-analysis. Appl Sci 13(10): 5957. https://doi.org/10.3390/app13105957

Leveque M, Guittat M, Thivichon-Prince B, Reuzeau A, Eveillard M, Faure M, Farges JC, Richert R, Bekhouche M, Ducret M (2024) Next generation antibacterial strategies for regenerative endodontic procedures: a scoping review. Int Endod J 57(7): 804–814. https://doi.org/10.1111/iej.13958

Liljestrand JM, Mäntylä P, Paju S, Buhlin K, Kopra KA, Persson GR, Hernandez M, Nieminen MS, Sinisalo J, Tjäderhane L, Pussinen PJ (2016) Association of Endodontic Lesions with coronary artery disease. J Dent Res 95(12):1358–1365. https://doi.org/10.1177/0022034516660509

Limeira FIR, Arantes DC, de Souza Oliveira C, de Melo DP, Magalhães CS, Bento PM (2020) Root canal treatment and apical periodontitis in a Brazilian population with type 1 diabetes mellitus: a cross-sectional paired study. J Endodont 46(6):756–762. https://doi.org/10.1016/j.joen.2020.02.010

Lin L, Huang GT (2020) Pathobiology of apical periodontitis. In: Berman L, Hargreaves K (eds) Cohen's pathways of the pulp, 11th edn. Elsevier, pp 630–659

Lin PY, Chien KL, Chang HJ, Chi LY (2015) Unfinished root canal treatments and the risk of cardiovascular disease. J Endodont 41(12):1991–1996. https://doi.org/10.1016/j.joen.2015.08.031

Lira-Junior R, Figueredo CM (2016) Periodontal and inflammatory bowel diseases: is there evidence of complex pathogenic interactions? World J Gastroenterol 22(35):7963–7972. https://doi.org/10.3748/wjg.v22.i35.7963

Little JW, Falace DA, Miller CS, Rhodus N (1997) Diabetes. In: Little JW (ed) Dental management of the medically compromised patient, 5th edn. Mosby, pp 387–409

Liu L, Wang T, Huang D, Song D (2021) Comprehensive analysis of differentially expressed genes in clinically diagnosed irreversible pulpitis by multiplatform data integration using a robust rank aggregation approach. J Endodont 47(9):1365–1375. https://doi.org/10.1016/j.joen.2021.07.007

López-López J, Jané-Salas E, Estrugo-Devesa A, Velasco-Ortega E, Martín-González J, Segura-Egea JJ (2011) Periapical and endodontic status of type 2 diabetic patients in Catalonia, Spain: a cross-sectional study. J Endodont 37(5):598–601. https://doi.org/10.1016/j.joen.2011.01.002

Madianos PN, Bobetsis YA, Offenbacher S (2013) Adverse pregnancy outcomes (APOs) and periodontal disease: pathogenic mechanisms. J Periodontol 84(4 Suppl):S170–S180. https://doi.org/10.1902/jop.2013.1340015

Mammaro A, Carrara S, Cavaliere A, Ermito S, Dinatale A, Pappalardo EM, Militello M, Pedata R (2009) Hypertensive disorders of pregnancy. J Prenat Med 3(1):1–5

Marchi KS, Fisher-Owens SA, Weintraub JA, Yu Z, Braveman PA (2010) Most pregnant women in California do not receive dental care: findings from a population-based study. Public Health Rep 125(6): 831–842

Marotta PS, Fontes TV, Armada L, Lima KC, Rôças IN, Siqueira JF Jr (2012) Type 2 diabetes mellitus and the prevalence of apical periodontitis and endodontic treatment in an adult Brazilian population. J Endodont 38(3):297–300. https://doi.org/10.1016/j.joen.2011.11.001

Marruganti C, Discepoli N, Gaeta C, Franciosi G, Ferrari M, Grandini S (2021) Dental caries occurrence in inflammatory bowel disease patients: a systematic review and meta-analysis. Caries Res 55(5):485–495. https://doi.org/10.1159/000519170

Meurman JH, Janket SJ, Surakka M, Jackson EA, Ackerson LK, Fakhri HR, Chogle S, Walls A (2017) Lower risk for cardiovascular mortality for patients with root filled teeth in a Finnish population. Int Endod J 50(12):1158–1168. https://doi.org/10.1111/iej.12772

Multari S, Bergandi L, Alovisi M, Comba A, Scotti N, Charrier L, Silvagno F, Baima G, Berutti E, Pasqualini D (2024) Endodontic treatment of chronic apical periodontitis ameliorates systemic inflammation and restores impaired cellular responses to insulin in an in vitro model. J Endodont S0099-2399(24):

00337–00336. https://doi.org/10.1016/j.joen.2024.05.013

Murray BA (1918) Mouth infection as a source of systemic disease. Can Med Assoc J 8(11):988–992

Nagendrababu V, Segura-Egea JJ, Fouad AF, Pulikkotil SJ, Dummer PMH (2020) Association between diabetes and the outcome of root canal treatment in adults: an umbrella review. Int Endod J 53(4):455–466. https://doi.org/10.1111/iej.13253

Nakahara T, Hyogo H, Ono A, Nagaoki Y, Kawaoka T, Miki D, Tsuge M, Hiraga N, Hayes CN, Hiramatsu A, Imamura M, Kawakami Y, Aikata H, Ochi H, Abe-Chayama H, Furusho H, Shintani T, Kurihara H, Miyauchi M, Takata T et al (2018) Involvement of Porphyromonas gingivalis in the progression of non-alcoholic fatty liver disease. J Gastroenterol 53(2):269–280. https://doi.org/10.1007/s00535-017-1368-4

Nakano K, Inaba H, Nomura R, Nemoto H, Takeda M, Yoshioka H, Matsue H, Takahashi T, Taniguchi K, Amano A, Ooshima T (2006) Detection of cariogenic Streptococcus mutans in extirpated heart valve and atheromatous plaque specimens. J Clin Microbiol 44(9):3313–3317. https://doi.org/10.1128/JCM.00377-06

Ng SC, Shi HY, Hamidi N, Underwood FE, Tang W, Benchimol EI, Panaccione R, Ghosh S, Wu JCY, Chan FKL, Sung JJY, Kaplan GG (2017) Worldwide incidence and prevalence of inflammatory bowel disease in the 21st century: a systematic review of population-based studies. Lancet (London, England) 390(10114):2769–2778. https://doi.org/10.1016/S0140-6736(17)32448-0

Ordinola-Zapata R, Costalonga M, Dietz M, Lima BP, Staley C (2024) The root canal microbiome diversity and function. A whole-metagenome shotgun analysis. Int Endod J 57(7):872–884. https://doi.org/10.1111/iej.13911

Pandya JK, Wheatley J, Bailey O, Taylor G, Geddis-Regan A, Edwards D (2024) Exploring deep caries management and barriers to the use of vital pulp treatments by primary care dental practitioners. Int Endod J 57:1453

Pasqualini D, Bergandi L, Palumbo L, Borraccino A, Dambra V, Alovisi M, Migliaretti G, Ferraro G, Ghigo D, Bergerone S, Scotti N, Aimetti M, Berutti E (2012) Association among oral health, apical periodontitis, CD14 polymorphisms, and coronary heart disease in middle-aged adults. J Endodont 38(12):1570–1577. https://doi.org/10.1016/j.joen.2012.08.013

Perez-Losada FDL, Estrugo-Devesa A, Castellanos-Cosano L, Segura-Egea JJ, López-López J, & Velasco-Ortega E (2020) Apical Periodontitis and Diabetes Mellitus Type 2: A Systematic Review and Meta-Analysis. J Clin Med. 9(2):540. https://doi.org/10.3390/jcm9020540

Pinto KP, Serrão G, Alves Ferreira CM, Sassone LM, Fidalgo TKDS, Silva EJNL (2023) Association between apical periodontitis and chronic diseases: an umbrella review. Iran Endod J 18(3):134–144. https://doi.org/10.22037/iej.v18i3.42560

Pinto KP, Fidalgo TKDS, de Lima CO, Lopes RT, Freitas-Fernandes LB, Valente AP, Sassone LM, Silva EJNL (2024) Chronic alcohol and nicotine consumption as catalyst for systemic inflammatory storm and bone destruction in apical periodontitis. Int Endod J 57(2): 178–194. https://doi.org/10.1111/iej.13994

Piras V, Usai P, Mezzena S, Susnik M, Ideo F, Schirru E, Cotti E (2017) Prevalence of apical periodontitis in patients with inflammatory bowel diseases: a retrospective clinical study. J Endodont 43(3):389–394. https://doi.org/10.1016/j.joen.2016.11.004

Poornima L, Ravishankar P, Abbott PV, Subbiya A, PradeepKumar AR (2021) Impact of root canal treatment on high-sensitivity C-reactive protein levels in systemically healthy adults with apical periodontitis – a preliminary prospective, longitudinal interventional study. Int Endod J 54(4):501–508. https://doi.org/10.1111/iej.13444

Poyato-Borrego M, Segura-Sampedro JJ, Martín-González J, Torres-Domínguez Y, Velasco-Ortega E, Segura-Egea JJ (2020) High prevalence of apical periodontitis in patients with inflammatory bowel disease: an age- and gender- matched case-control study. Inflamm Bowel Dis 26(2):273–279. https://doi.org/10.1093/ibd/izz128

Quigley RM, Kearney M, Kennedy OD, Duncan HF (2024) Tissue engineering approaches for dental pulp regeneration: the development of novel bioactive materials using pharmacological epigenetic inhibitors. Bioact Mater 40:182–211. https://doi.org/10.1016/j.bioactmat.2024.06.012

Raoof M, Khaleghi M, Siasar N, Mohannadalizadeh S, Haghani J, Amanpour S (2019) Antimicrobial activity of Methanolic extracts of Myrtus Communis L. and eucalyptus Galbie and their combination with calcium hydroxide powder against enterococcus Faecalis. J Dentis (Shiraz, Iran) 20(3):195–202. https://doi.org/10.30476/DENTJODS.2019.44898

Rechenberg DK, Galicia JC, Peters OA (2016) Biological markers for pulpal inflammation: a systematic review. PLoS One 11(11):e0167289. https://doi.org/10.1371/journal.pone.0167289

Rees DC, Williams TN, Gladwin MT (2010) Sickle-cell disease. Lancet (London, England) 376(9757): 2018–2031. https://doi.org/10.1016/S0140-6736(10)61029-X

Richert R, Ducret M, Alliot-Licht B, Bekhouche M, Gobert S, Farges JC (2022) A critical analysis of research methods and experimental models to study pulpitis. Int Endod J 55(Suppl 1):14–36. https://doi.org/10.1111/iej.13683

Ricucci D, Loghin S, Siqueira JF Jr (2014) Correlation between clinical and histologic pulp diagnoses. J Endodont 40(12):1932–1939. https://doi.org/10.1016/j.joen.2014.08.010

Roberts J, Abimbola S, Bale T, Barros A, Bhutta Z, Browne J, Celi A, Dube P, Graves C, Hollestelle M,

Hopkins S, Khashan A, Koi-Larbi K, Lackritz E, Myatt L, Redman C, Tunçalp Ö, Vermund S, Gravett M (2024) Global inequities in adverse pregnancy outcomes: what can we do? AJOG Global Rep 4(3): 100385. https://doi.org/10.1016/j.xagr.2024.100385c

Sabeti MA, Saqib Ihsan M, Adami D, Hassani SN, Moushekhian S, Shafieian R, Salari Sedigh H, Ghoddusi J (2024) Cell-based regenerative endodontics for the treatment of irreversible pulpitis: an in vivo investigation. J Endodont 50(3):344–350. https://doi.org/10.1016/j.joen.2023.11.014

Samuel RO, Gomes-Filho JE, Azuma MM, Sumida DH, de Oliveira SHP, Chiba FY, Bomfim SRM, Ciarlini PC, Narciso LG, Cintra LTA (2018) Endodontic infections increase leukocyte and lymphocyte levels in the blood. Clin Oral Investig 22(3):1395–1401. https://doi.org/10.1007/s00784-017-2222-z

Samuel RO, Ervolino E, de Azevedo Queiroz ÍO, Azuma MM, Ferreira GT, Cintra LTA (2019) Th1/Th2/Th17/Treg balance in apical periodontitis of Normoglycemic and diabetic rats. J Endodont 45(8):1009–1015. https://doi.org/10.1016/j.joen.2019.05.003

Schmalz G, Smith AJ (2014) Pulp development, repair, and regeneration: challenges of the transition from traditional dentistry to biologically based therapies. J Endodont 40(4 Suppl):S2–S5. https://doi.org/10.1016/j.joen.2014.01.018

Schwendicke F, Splieth C, Breschi L, Banerjee A, Fontana M, Paris S, Burrow MF, Crombie F, Page LF, Gatón-Hernández P, Giacaman R, Gugnani N, Hickel R, Jordan RA, Leal S, Lo E, Tassery H, Thomson WM, Manton DJ (2019) When to intervene in the caries process? An expert Delphi consensus statement. Clin Oral Investig 23(10):3691–3703. https://doi.org/10.1007/s00784-019-03058-w

Sebring D, Buhlin K, Norhammar A, Rydén L, Jonasson P et al (2022) Endodontic inflammatory disease: a risk indicator for a first myocardial infarction. Int Endod J 55(1):6–17. https://doi.org/10.1111/iej.13634

Sebring D, Pehrsson NG, Buhlin K, Jonasson P, Lund H, Kvist T (2023) A weighted composite of endodontic inflammatory disease is linked to a first myocardial infarction. Oral Health Prev Dent 21:375–382. https://doi.org/10.3290/j.ohpd.b4586815

Sebring D, Buhlin K, Lund H, Norhammar A, Rydén L et al (2024) Endodontic inflammatory disease and future cardiovascular events and mortality: a report from the PAROKRANK study. J Endodont 50(8): 1073–1081.e3. https://doi.org/10.1016/j.joen.2024.05.003

Segura-Egea JJ, Jimenez-Moreno E, Calvo-Monroy C, Ríos-Santos JV, Velasco-Ortega E, Sánchez-Domínguez B, Castellanos-Cosano L, Llamas-Carreras JM (2010) Hypertension and dental periapical condition. J Endodont 36(11):1800–1804. https://doi.org/10.1016/j.joen.2010.08.004

Segura-Egea JJ, Martín-González J, Castellanos-Cosano L (2015) Endodontic medicine: connections between apical periodontitis and systemic diseases. Int Endod J 48(10):933–951. https://doi.org/10.1111/iej.12507

Segura-Egea JJ, Martín-González J, Cabanillas-Balsera D, Fouad AF, Velasco-Ortega E, López-López J (2016) Association between diabetes and the prevalence of radiolucent periapical lesions in root-filled teeth: systematic review and meta-analysis. Clin Oral Investig 20(6):1133–1141. https://doi.org/10.1007/s00784-016-1805-4

Segura-Egea JJ, Cabanillas-Balsera D, Jiménez-Sánchez MC, Martín-González J (2019) Endodontics and diabetes: association versus causation. Int Endod J 52(6): 790–802. https://doi.org/10.1111/iej.13079

Segura-Egea JJ, Cabanillas-Balsera D, Martín-González J, Cintra LTA (2023) Impact of systemic health on treatment outcomes in endodontics. Int Endod J 56(Suppl 2):219–235. https://doi.org/10.1111/iej.13789

Segura-Sampedro JJ, Jiménez-Giménez C, Jane-Salas E, Cabanillas-Balsera D, Martín-González J, Segura-Egea JJ, López-López J (2022) Periapical and endodontic status of patients with inflammatory bowel disease: age- and sex-matched case-control study. Int Endod J 55(7):748–757. https://doi.org/10.1111/iej.13747

Şenel S, Özdoğan AI, Akca G (2021) Current status and future of delivery systems for prevention and treatment of infections in the oral cavity. Drug Deliv Transl Res 11(4):1703–1734. https://doi.org/10.1007/s13346-021-00961-2

Shimpi N, Glurich I, Schroeder D, Katrak C, Chyou PH, Acharya A (2020) Patient awareness of association of diabetes and periodontal disease. Health Promot Pract 21(3):464–472. https://doi.org/10.1177/1524839918801909

Siqueira JF Jr, Antunes HS, Pérez AR, Alves FRF, Mdala I, Silva EJNL, Belladonna FG, Rôças IN (2020) The apical root canal system of teeth with posttreatment apical periodontitis: correlating microbiologic, tomographic, and histopathologic findings. J Endodont 46(9):1195–1203. https://doi.org/10.1016/j.joen.2020.05.020

Sisli SN (2019) Evaluation of the relationship between type II diabetes mellitus and the prevalence of apical periodontitis in root-filled teeth using cone beam computed tomography: an observational cross-sectional study. Med Princ Pract 28(6):533–538. https://doi.org/10.1159/000500472

Siu SY, Pudipeddi A, Vishwanath V, Cheng Lee AH, Tin Cheung AW, Pan Cheung GS, Neelakantan P (2024) Effect of novel and traditional Intracanal medicaments on biofilm viability and composition. J Endodont S0099-2399(24):00398–00394. https://doi.org/10.1016/j.joen.2024.07.003

Tibúrcio-Machado CS, Michelon C, Zanatta FB, Gomes MS, Marin JA, Bier CA (2021) The global prevalence of apical periodontitis: a systematic review and meta-analysis. Int Endod J 54(5):712–735. https://doi.org/10.1111/iej.13467

Toledano-Osorio M, Osorio R, Bueno J, Vallecillo C, Vallecillo-Rivas M, Sanz M (2024) Next-generation antibacterial nanopolymers for treating oral chronic inflammatory diseases of bacterial origin. Int Endod J 57(7):787–803. https://doi.org/10.1111/iej.14040

Tsosura TVS, Mattera MSLC, Chiba FY, Carnevali ACN, Belardi BE, Dos Santos RM, Cintra LTA, Lopes FL, Scaramele NF, Matsushita DH (2021a) Effect of maternal apical periodontitis on the final step of insulin signalling and inflammatory pathway in the adult male offspring of rats. Int Endod J 54(11): 2113–2124. https://doi.org/10.1111/iej.13610

Tsosura TVS, Dos Santos RM, Chaves Neto AH, Chiba FY, Carnevali ACN, Mattera MSLC, Belardi BE, Cintra LTÂ, da Silva Machado NE, Matsushita DH (2021b) Maternal apical periodontitis increases insulin resistance and modulates the antioxidant defense system in the gastrocnemius muscle of adult offspring. J Endodont 47(7):1126–1131. https://doi.org/10.1016/j.joen.2021.04.003

Tuchendler E, Tuchendler PK, Madej G (2018) Immunodeficiency caused by cirrhosis. Clin Exp Hepatol 4(3): 158–164. https://doi.org/10.5114/ceh.2018.78119

Wang Y, Zhai S, Wang H, Jia Q, Jiang W, Zhang X, Zhang A, Liu J, Ni L (2013) Absent in melanoma 2 (AIM2) in rat dental pulp mediates the inflammatory response during pulpitis. J Endodont 39(11): 1390–1394. https://doi.org/10.1016/j.joen.2013.07.003

Wang Y, Zhu J, Tang Y, Huang C (2024) Association between pulp and periapical disease with type 2 diabetes: a bidirectional Mendelian randomization. Int Endod J 57(5):566–575. https://doi.org/10.1111/iej.14034

WHO (2012) Prevention and control of noncommunicable diseases: guidelines for primary health care in low resource settings. World Health Organization

Willershausen B, Kasaj A, Willershausen I, Zahorka D, Briseño B, Blettner M, Genth-Zotz S, Münzel T (2009) Association between chronic dental infection and acute myocardial infarction. J Endodont 35(5):626–630. https://doi.org/10.1016/j.joen.2009.01.012

Willershausen I, Weyer V, Peter M, Weichert C, Kasaj A, Münzel T, Willershausen B (2014) Association between chronic periodontal and apical inflammation and acute myocardial infarction. Odontology 102(2): 297–302. https://doi.org/10.1007/s10266-013-0112-7

Wolters WJ, Duncan HF, Tomson PL, Karim IE, McKenna G, Dorri M, Stangvaltaite L, van der Sluis LWM (2017) Minimally invasive endodontics: a new diagnostic system for assessing pulpitis and subsequent treatment needs. Int Endod J 50(9):825–829. https://doi.org/10.1111/iej.12793

Xiao S, Lei H, Li P, Ma D, Chen S, Huang X (2023) Is oxidative stress involved in the hepatic inflammatory response to apical periodontitis? A comparative study in normal and hyperlipidaemic rat. Int Endod J 56(6): 722–733. https://doi.org/10.1111/iej.13907

Yan H, De Deus G, Kristoffersen IM, Wiig E, Reseland JE, Johnsen GF, Silva EJNL, Haugen HJ (2023) Regenerative endodontics by cell homing: a review of recent clinical trials. J Endodont 49(1):4–17. https://doi.org/10.1016/j.joen.2022.09.008

Yip N, Liu C, Wu D, Fouad AF (2021) The association of apical periodontitis and type 2 diabetes mellitus: a large hospital network cross-sectional case-controlled study. J Am Dent Assoc 152(6):434–443. https://doi.org/10.1016/j.adaj.2021.01.005

Younossi Z, Anstee QM, Marietti M, Hardy T, Henry L, Eslam M, George J, Bugianesi E (2018) Global burden of NAFLD and NASH: trends, predictions, risk factors and prevention. Nat Rev Gastroenterol Hepatol 15(1): 11–20. https://doi.org/10.1038/nrgastro.2017.109

Zanini M, Meyer E, Simon S (2017) Pulp inflammation diagnosis from clinical to inflammatory mediators: a systematic review. J Endodont 43(7):1033–1051. https://doi.org/10.1016/j.joen.2017.02.009

Periodontal Disease and Osteoporosis: A Bidirectional Relationship

9

Maria Contaldo

Abstract

Osteoporosis (OP) and periodontitis (PD) are two prevalent diseases with significant public health implications, particularly among aging populations. Osteoporosis is a condition that can affect any bone, characterized by decreased bone mineral density (BMD) and deterioration of the bone architecture, resulting in weaker bones and a higher risk of fractures. Periodontitis is a chronic inflammatory condition affecting the supporting structures of the teeth, such as the gingiva, periodontal ligament, and alveolar bone, often leading to tooth loss if left untreated.

Recent studies have suggested an interrelationship between these conditions, rooted in shared biological mechanisms. Both diseases involve an imbalance in bone remodeling processes, with excessive activity of osteoclasts over osteoblasts, leading to bone resorption. In osteoporosis, this imbalance is systemic, while in periodontitis, it is localized to the jawbone.

Shared risk factors, such as aging, hormonal changes (e.g., postmenopausal estrogen deficiency), and lifestyle factors, like smoking and poor nutrition, further strengthen this association. Additionally, osteoporosis may exacerbate periodontitis by reducing alveolar bone density, while periodontal inflammation can elevate the expression of systemic biomarkers of bone resorption, influencing osteoporosis progression. Furthermore, dysbiosis of the gut and oral cavity microbial communities plays a critical role in both conditions. Gut dysbiosis can influence systemic inflammation and bone metabolism, exacerbating osteoporosis, while oral dysbiosis, marked by pathogenic shifts in the microbiota, triggers and perpetuates periodontal inflammation and alveolar bone loss. Understanding these interrelationships emphasizes the need for an integrated approach to preventing and managing these interconnected conditions.

Keywords

Osteoporosis · Periodontitis · Microbiome · Microbiota · Dysbiosis

9.1 Osteoporosis: An Overview

Osteoporosis (OP) is a systemic metabolic skeletal disorder characterized by a generalized and progressive loss of bone mineral density (BMD) as well as deterioration of the micro-architecture of the bone tissue, thus leading to an increase in bone fragility and risk of fracture, which can easily occur during daily activities or even

M. Contaldo (✉)
Multidisciplinary Department of Medical-Surgical and Dental Specialties, University of Campania L. Vanvitelli, Naples, Italy
e-mail: maria.contaldo@unicampania.it

N. Rezaei, H. Ziaei (eds.), *Oral Immunology*, Advances in Experimental Medicine and Biology 1492,
https://doi.org/10.1007/978-3-032-03176-1_9

minimal traumas ("NIH Consensus Development Panel on Osteoporosis Prevention, Diagnosis, and Therapy, March 7-29, 2000: Highlights of the Conference," 2001; Sozen et al. 2017).

Based on World Health Organization (WHO) criteria, osteoporosis is diagnosed using dual X-ray absorptiometry (DXA) of the hip and spine, calculating a fraction called "T-score," which is the difference between the patient's BMD and the average BMD of healthy young adults (Kanis et al. 2008; Kanis 2007; Sozen et al. 2017). Osteoporosis is diagnosed when the T-score is ≤ -2.5 standard deviation (SD), while T-score values between -1 and -2.5 SD indicate osteopenia, an early indicator and less severe stage of osteoporosis, characterized by lower-than-normal bone density. Conversely, severe or "established" osteoporosis is diagnosed in the presence of one or more documented fragility fractures (Kanis 2007; Sozen et al. 2017).

The global prevalence of osteoporosis is 19.7%, while osteopenia affects 40.4% of the population. It has been reported that 44.8% of males suffer from osteopenia and 10.6% from osteoporosis. Regarding females, 39.4–42.1% suffer from osteopenia and 24.8–27.4% from osteoporosis (Xiao et al. 2022). Worldwide, osteoporosis is estimated to affect 200 million women. In the European Union, approximately 22 million women and 5.5 million men have osteoporosis, leading to approximately 3.5 million fractures each year (Hernlund et al. 2013; Johnell and Kanis 2006; Xiao et al. 2022).

The main risk factors associated with OP are advanced age, body mass index (BMI) <18.5, and smoking habits (Xiao et al. 2022).

Furthermore, osteoporosis may be classified as "primary" or "secondary." Primary osteoporosis may be related to postmenopausal deficiency of estrogens or aging-related bone mass loss (Sozen et al. 2017). Secondary osteoporosis is due to medications (mainly corticosteroids, immunosuppressants, and anticoagulants), underlying diseases (such as rheumatic, neoplastic, and endocrine disorders, including hyperparathyroidism), and lifestyle factors (such as insufficient sun exposure, vitamin D and calcium intake deficiencies, smoking, excessive alcohol consumption, physical inactivity) (Sozen et al. 2017).

9.2 Periodontitis: An Overview

Periodontitis (PD) is a long-lasting inflammatory oral disease that causes progressive and irreversible destruction of the structures surrounding and supporting the teeth. These structures include the connective periodontal attachment, dental root cementum, and alveolar bone. As a result, tooth loss occurs predominantly in adults over the age of 50 years (Slot et al. 2020).

The global prevalence of PD is approximately 61.5%, but this figure increases to 79.3% for the population aged 65 and older, with no significant differences between genders (Trindade et al. 2023). In 2010, the prevalence of severe periodontitis was reported at 10.8%, affecting around 743 million people worldwide, making it the sixth most prevalent health condition (Kassebaum et al. 2014).

Periodontitis is a multifactorial disease influenced by both genetic and environmental factors. However, the critical factor initiating and perpetuating local inflammatory damage is the presence of periodontopathogenic bacteria and fungi that colonize and proliferate in the gingival pockets of susceptible individuals (Canabarro et al. 2013; Di Spirito et al. 2022; Mombelli 2018).

The persistence of Gram-negative facultative anaerobic bacteria, particularly *Treponema denticola* and *Porphyromonas gingivalis*, in the supra- and subgingival oral plaque biofilm leads to the release of various pro-inflammatory cytokines (Loesche 1996). This process enhances local inflammation, contributing to the ongoing destruction of periodontal structures (Ballini et al. 2015).

In the absence of oral hygiene, healthy individuals and those suffering from periodontitis harbor different levels of periodontopathogenic bacteria in their supra- and subgingival areas. Researchers have found that these bacterial levels are significantly higher in individuals with periodontitis at baseline and tend to reach or exceed baseline values after 2 days of plaque removal and professional dental cleaning (Uzel et al. 2011).

As dental plaque mass increases, the total number of organisms in the biofilm also rises, thereby worsening the periodontal health of affected individuals (Haffajee et al. 2009).

Beyond the quantity of bacteria, quality also plays a crucial role in periodontal damage. Haffajee et al. identified specific microbial complexes within the supra- and subgingival plaques that interactively contribute to periodontal structures' destruction (Haffajee et al. 2008). Before this, Socransky et al. (1998) examined the periodontopathogenic potential of various bacterial complexes found in mature gingival plaque, their interrelationships, and their timing of colonization.

In summary, early colonizers from the green complex, such as *Capnocytophaga* spp. and *Actinomyces actinomycetemcomitans* serotype A, as well as those from the yellow complex, like *Streptococcus* spp., adhere to the tooth surface with their fimbriae. These initial microbial communities establish a foundation that facilitates later colonization of periodontopathogenic bacteria. Subsequently, the bridge species within the orange complex—primarily *Fusobacterium nucleatum*, *Prevotella intermedia*, *Prevotella nigrescens*, and *Campylobacter* spp.—produce toxins and enzymes that progressively damage the periodontal attachment, leading to increased pocket depth. This deepening creates a favorable environment in the gingival sulcus or pocket for colonizing bacteria from the red complex, which includes the most significant periodontopathogenic bacterial species. These Gram-negative, anaerobic bacteria, such as *Porphyromonas gingivalis*, *Tannerella forsythia*, and *Treponema denticola*, penetrate and aggressively destroy periodontal attachment and cause alveolar bone loss (Contaldo et al. 2021).

Periodontal diseases are closely linked to the onset or worsening of a variety of systemic conditions, including diabetes, cardiovascular diseases, rheumatoid arthritis, and complications during pregnancy (Agueda et al. 2008; Holmstrup et al. 2017; Ionas et al. 2017; Manjunath et al. 2011). Recent evidence has suggested that both the chronic low-grade systemic inflammation associated with periodontitis and the presence of periodontopathogenic bacteria may contribute to the onset of cancer (Narayanan et al. 2023).

Periodontitis has recently been reclassified into three main groups: necrotizing periodontitis, periodontitis as a manifestation of systemic disease, and a generalized category of periodontitis, which combines two former distinct categories known as "chronic" and "aggressive" (Caton et al. 2018). This reclassification highlights that periodontal diseases can manifest as part of systemic diseases, and, conversely, some systemic diseases can adversely affect periodontal health (Jepsen et al. 2018a, b).

Periodontitis is diagnosed and staged based on a combination of parameters, including clinical attachment loss (CAL) and radiographic bone loss assessment (Albandar et al. 2018; Tonetti et al. 2018; Yu and Wang 2022)..

Periodontitis shares several risk factors with osteoporosis, including age, smoking and/or alcohol consumption, body mass index (BMI), and menopause. Both conditions significantly impact the quality of life of affected individuals and have similar prevalence rates in the population (Pepelassi et al. 2012; Pestana de Vasconcelos et al. 2024).

In 2017, Jepsen et al. highlighted the conflicting evidence regarding the relationship between osteoporosis and periodontitis, as noted by the working group on periodontal diseases and systemic conditions. According to their report, postmenopausal women with osteopenia or osteoporosis have a more significant loss of periodontal attachment, with a modest but statistically significant difference, compared to women with normal bone mineral density (Jepsen et al. 2018a, b; Penoni et al. 2017). Subsequent studies have reinforced the considerable association between osteoporosis and periodontitis, suggesting that this association is particularly robust in postmenopausal women (Qi et al. 2023).

This chapter explores the two-way relationship between periodontitis and osteoporosis, examining how each condition can influence the other and the mechanisms involved. The discussion will address the following questions:

- Can osteoporosis affect the progression of periodontitis?
- Can periodontitis influence the course of osteoporosis?
- What mechanisms are involved in this interrelationship?

9.3 Can Osteoporosis Affect the Progression of Periodontitis?

Osteoporosis is recognized globally as an independent risk factor for periodontitis. A meta-analysis by Xu et al. (2021) demonstrated that individuals with osteoporosis are 1.96 times more likely to develop periodontitis than those without osteoporosis.

In a US cross-sectional survey of almost 6000 US adults, osteoporosis was significantly associated with a 1.66 increased risk of developing periodontitis in the whole population. The risk rose to 9.66 in the subgroup of menopausal women, who experienced severe periodontitis more frequently (Peng et al. 2023). This evidence has been globally confirmed in a recent meta-analysis, which included 28 studies and more than 19,000 patients, confirming the worse periodontal indices in postmenopausal osteoporotic women than in those without osteoporosis (Qi et al. 2023).

Regardless of sex, adults under the age of 45 who have osteoporosis show a 1.2 times higher risk of developing periodontal disease (Chou et al. 2021). Individuals of either sex with decreased bone mineral density are also more likely to experience tooth loss due to periodontitis (Khunthananithi et al. 2023). Additionally, both alveolar bone loss and loss of periodontal attachment correlate with low bone mineral density (Yu and Wang 2022).

In a screening program involving more than 15,000 subjects aged 35–44 in Taiwan, Xu et al. (2021) reported that periodontal disease prevalence was 39.9% among those with osteoporosis and 33.3% among those with osteopenia. The authors also found that low bone mineral density, including both osteopenia and osteoporosis, was associated with a 13% increased risk of developing periodontitis. This risk further escalated to 21% in the presence of additional risk factors such as smoking, age, gender, education level, overweight status, and diabetes. Furthermore, osteoporosis is closely related to a twofold increased risk of future periodontitis (Al Habashneh et al. 2010; Xu et al. 2021).

However, after adjusting for age and sex, a significant association between periodontitis and osteoporosis is more frequent and significant among postmenopausal women (Gomes-Filho et al. 2013; Lin et al. 2015).

In postmenopausal women with osteoporosis or osteopenia, there is evidence of a modest but statistically significant increase in periodontal attachment loss compared to women with normal BMD. Specifically, studies on postmenopausal women suffering from generalized chronic periodontitis found that those with osteoporosis or osteopenia exhibited poorer periodontal health indicators. For instance, osteoporotic women with periodontitis showed higher average scores in the gingival index, increased bleeding upon probing, more significant gingival recession, and greater clinical attachment loss compared to women with periodontitis but without osteoporosis (Pepelassi et al. 2012; Qi et al. 2023). These differences were noted regardless of smoking habits and dental hygiene practices (Gil-Montoya et al. 2021).

Additional studies have indicated that menopausal women with osteoporosis have a ninefold increased risk of developing severe periodontitis compared to the general population (Peng et al. 2023). This heightened risk may be associated with decreased ovarian estrogen production and increased testosterone levels that occur during menopause (Minkin 2019). Estrogen deprivation during menopause reduces osteogenic differentiation and induces osteoclast death (Fischer and Haffner-Luntzer 2022), leading to an imbalance in bone homeostasis that favors bone resorption. Furthermore, a decrease in estrogen is linked to a reduction in anti-inflammatory cytokines, such as osteoprotegerin (OPG) and interleukin (IL)-10, and increased levels of pro-inflammatory ones, like IL-1, IL-6, and tumor necrosis factor (TNF). This imbalance contributes to bone mineral density loss and increases susceptibility to periodontal damage (Luo et al. 2014).

Based on this understanding, menopausal women with periodontitis may benefit from hormone replacement therapy, which has been shown to improve mandibular bone density, reduce gingival bleeding, and prevent tooth loss (Grodstein et al. 1996; Qi et al. 2023).

Osteoporosis hormonal treatments have been proven to reduce the risk of periodontitis onset in older women (Penoni et al. 2016) and ameliorate periodontal indices in those already suffering from periodontitis. A strong, consistent piece of literature confirmed that women treated with estrogen for postmenopausal osteoporosis have a lower prevalence of severe periodontitis than women not receiving treatment. For example, Passos-Soares et al. reported that postmenopausal osteoporotic women with periodontitis who received systemic estrogen—with and without progestin—and calcium and vitamin D supplements for at least 6 months showed a 44% lower prevalence of severe periodontitis. They also exhibited reduced probing depth of periodontal pockets, less clinical attachment loss, and less gingival bleeding than women not receiving treatment for osteoporosis (de Passos-Soares et al. 2017).

Similarly, a recent systematic review (Xu et al. 2024) has reported that zoledronic acid—a bisphosphonates drug indicated to prevent bone fractures in osteoporotic subjects (Black et al. 2007)—positively impacts periodontal health, reduces serum inflammation, and enhances bone metabolism in postmenopausal women with osteoporosis and periodontitis. In detail, the authors established the clinical efficacy and advantages of zoledronic acid therapy over conventional therapy on periodontal health in 113 postmenopausal women aged between 48 and 65 years suffering from osteoporosis and periodontitis. For these purposes, 52 subjects were treated with daily calcium carbonate and vitamin D, while 61 patients received a single intravenous infusion of zoledronic acid. After 3 months, compared to baseline values, the group receiving zoledronic acid reported a significantly higher treatment efficacy—demonstrated by lower serum inflammatory markers, such as C-reactive protein, IL-1, and TNF-α—and better periodontal health improvement than those treated with vitamin D and calcium, including reduced loss of attachment, probing depth, and gingival bleeding index. In summary, the treatment efficacy rate was 88.52% in the zoledronic acid group and 65.38% in the calcium/vitamin D group (Xu et al. 2024).

The periodontal benefits of osteoporotic treatments are confirmed in another study by the same research group that replicated the experiment by comparing the outcome in terms of better inflammatory and periodontal parameters between osteoporotic patients with periodontitis receiving calcium carbonate, vitamin D, and zoledronic acid and those with an additional promising Epimedium treatment (Guo et al. 2024). Epimedium is a traditional Chinese herb rich in flavonoids, which has been reported to inhibit osteoclast activity and enhance osteoblast differentiation (Cui et al. 2024). Six months after the baseline, Guo et al. reported that the group receiving Epimedium in addition to conventional therapy showed significantly higher improvements in both periodontal health indices and inflammatory markers compared to those receiving standard treatment alone with treatment efficacy rates of 91.7% for periodontitis and 95% for osteoporosis, with no significant adverse effects reported (Cui et al. 2024).

9.4 Can Periodontitis Affect the Course of Osteoporosis?

Conversely, periodontitis is recognized as a risk factor for osteoporosis and bone fractures. This association is confirmed and firmly established by a series of epidemiological studies on different populations, which consistently report a higher occurrence of osteoporosis in subjects of both sexes with periodontitis. For example, Huang et al. (2016) reported a positive correlation between the severity of periodontitis and osteoporosis occurrence in a group of more than 85,000 Taiwanese subjects, grouped according to the presence or lack of osteoporosis. In the population studied, periodontitis patients showed a sixfold risk of osteoporosis compared to subjects without periodontitis. The risk of osteoporosis increases in proportion to the severity of periodontitis and is further exacerbated by poor oral hygiene (Huang et al. 2016).

Another large cross-sectional study conducted in Taiwan with more than 135,000 participants, both with and without periodontitis, found that individuals with periodontitis had a 2.7-fold higher risk of osteoporosis and a 1.5-fold higher risk of bone fractures compared to those without periodontal disease. The odds ratio varied by age and sex, being higher in males (2.7) and females (2.2) over the age of 53 (Hong et al. 2021).

The same conclusions were drawn in two longitudinal cohort studies with more than 500,000 South Korean subjects grouped into those with and without periodontitis. The study reported that patients with periodontitis were at an increased risk of suffering from a series of noncommunicable diseases, among which osteoporosis, with a 1.37 odds ratio, was one of the most statistically significant (Lee et al. 2017; Lee and Jeong 2020). Furthermore, in a similar study comprising more than 13,000 South Korean participants with periodontitis, and under monitoring for over 10 years, osteoporosis developed less frequently in men than in women, with a 10-year incidence of 1% and 16%, respectively (Choi et al. 2017)..

The same significantly higher prevalence of osteoporosis in periodontitis patients from the Asian region was also confirmed in a study of 680 Indian patients (Saini et al. 2024).

In Europe, a longitudinal study of more than 1000 Austrian subjects suffering from periodontitis reported a significantly higher adjusted prevalence of osteoporosis compared to the general population (5% in periodontitis patients and 2% in those without periodontitis) (Sperr et al. 2018).

Interestingly, the same associations between periodontitis and osteoporosis have also been reported in animal studies, such as moose (Hoy et al. 2024), rats (Zhou et al. 2023), and great apes (Lowenstine et al. 2016).

Further clinical evidence supports the connection between periodontal health and bone quality. For instance, the correlation between bone mineral density and gingival health indicates that the gingival index can be a predictive factor for osteoporosis. Periodontopathic patients with osteoporosis had poorer periodontal indices than those without osteoporosis, and a significant inverse correlation was observed between osteoporosis and both clinical attachment loss and tooth loss (Preda et al. 2022).

Furthermore, periodontitis patients experience bone fractures 1.54 times more frequently than those without periodontitis (Hong et al. 2021).

Lastly, periodontitis has been associated with poor bone mineral density and osteoporosis via indirect mechanisms involving periodontal pathogenic bacteria. These bacteria, particularly their lipopolysaccharides, trigger an increase in salivary pro-inflammatory cytokines, which, in turn, can spread throughout the body via the systemic circulation—see later discussion—contributing to pro-osteoporotic systemic effects (Contaldo et al. 2020b; Kido et al. 2004; Pavithra et al. 2024; Potdar et al. 2023; Tabor et al. 2021).

Similarly, just as osteoporosis therapies have been shown to improve periodontal health, periodontal therapies can improve bone mineral density and ameliorate the course of osteoporosis. For example, Sharma et al. (2024) have recently reported a significant reduction in the serum levels of IL-6 after subgingival instrumentation and oral hygiene reinforcements, an improvement associated with enhanced bone health.

9.5 What Mechanisms Are Involved in This Interrelationship?

To understand the connections between osteoporosis and periodontal disease pathogenesis, we must consider various factors, including anatomical, environmental, inflammatory, genetic, and microbiological aspects.

9.5.1 Anatomical and Environmental Factors

The loss of bone mineral density (BMD) in the alveolar bone may anatomically facilitate deeper bacterial penetration into periodontal spaces, already compromised by periodontitis. This creates a vicious cycle in which local inflammation can further exacerbate and accelerate alveolar bone resorption. Consequently, osteoporosis can speed up the process of alveolar bone loss in individuals with periodontal disease (Hernández-Vigueras et al. 2016).

Among the environmental factors, smoking and alcohol consumption are the most recurrently reported and confirmed behaviors, responsible for the onset and worsening of periodontitis (Caggiano et al. 2022; Di Spirito et al. 2022), as well as osteoporosis and bone fracture risk (Zarzour and Leslie 2024).

Concerning alcohol consumption, excessive alcohol drinking increases the risk of low BMD and impaired bone metabolism (Masuko et al. 2024). Specifically, both the frequency and quantity of alcohol intake have a dose-dependent negative effect on BMD, thus increasing the risk of bone fractures, especially at the femur, in both sexes (Lai et al. 2024).

Tobacco consumption is widely associated with worsening periodontal health (Kumar 2020). It has also been reported that smokers and ex-smokers exhibit a bone fracture risk of 1.5-fold higher than nonsmokers who experience bone fractures (Zarzour and Leslie 2024). Furthermore, passive smokers also seem to have a higher risk of developing fractures, vitamin D deficiencies, and poor bone mineral density than nonsmokers (Vergatti et al. 2024).

Dietary habits are another environmental factor involved in both diseases. Hu et al. assessed the association between nutritional mineral intake—such as magnesium, potassium, and sodium intake—and BMD in more than 10,000 US adults. The results revealed a significant association between low mineral intake and an increased risk of bone fractures, osteopenia, and osteoporosis (Hu et al. 2024b). Deficiencies in these nutrients also increase periodontal damage by indirectly affecting oxidative stress promotion, bone density reduction, and immune response imbalances (Berg et al. 2024). Conversely, healthy eating habits seem to be associated with better periodontal indices and bone health (Li et al. 2022; Yu and Wang 2022).

9.5.2 Inflammatory Factors

Osteoporosis and periodontitis share common inflammatory pathways, primarily driven by pro-inflammatory cytokines that promote osteoclastic activity. These cytokines are overexpressed in both conditions (Inchingolo et al. 2020; Lerner 2006).

The main mechanisms linking osteoporosis to periodontitis include the receptor activator of nuclear factor kappa B/RANK ligand/OPG (RANK/RANKL/OPG) pathway and pro-inflammatory cytokines such as TNF-α, IL-1β, and IL-6. Additionally, oxidative stress mechanisms, which arise from localized inflammatory responses, can have systemic effects (Qi et al. 2023; Yu and Wang 2022), connecting oral and skeletal health in both directions (Sanz et al. 2020). The chronic inflammatory state observed in periodontitis is not limited to the oral cavity; it can have systemic implications that affect bone metabolism throughout the body. Research indicates that patients with severe periodontitis may exhibit higher systemic levels of inflammatory markers, such as C-reactive protein (CRP), which may contribute to the bone loss observed in osteoporosis (Larvin et al. 2023). Conversely, systemic conditions such as osteoporosis may increase bone susceptibility to periodontal inflammation, further strengthening the bidirectional relationship between the two diseases.

In periodontitis, chronic inflammation triggered by bacterial biofilms upregulates a series of pro-inflammatory cytokines, such as TNF-α, IL-1β, and IL-6. These cytokines stimulate the expression of RANKL (receptor activator of nuclear factor kappa B ligand) by osteoblasts and periodontal ligament cells. These cells bind to RANK on the osteoclast precursors' surface, thus promoting osteoclastogenesis and increasing osteoclast activity (Graves et al. 2011). The same pathway is also active in osteoporosis, where hormonal changes, particularly postmenopausal estrogen deficiency, elevate RANKL expression, leading to increased osteoclast-mediated bone resorption (Boyce and Xing 2008; Silva and Kousteni 2015; Weitzmann and Pacifici 2006).

Oxidative stress is another contributing factor to both conditions. In periodontitis, reactive oxygen species (ROS), produced during chronic inflammation, damage periodontal tissues and activate inflammatory pathways that further promote bone resorption (Chapple and Matthews 2007; Contaldo et al. 2020a). Similarly, oxidative stress plays a significant role in osteoclast activation and bone loss in osteoporosis, with evidence suggesting that antioxidants may help mitigate bone resorption (Almeida et al. 2007).

Other than circulatory inflammatory cytokines, the scientific literature has also investigated the role of a series of salivary biomarkers, such as calprotectin, interleukins, and salivary minerals. These salivary biomarkers have been found to be significantly differentially expressed in saliva from subjects with periodontitis and are statistically linked to an increased risk of osteoporosis onset. For example, calprotectin, a zinc- and calcium-binding protein involved in the acute phase of inflammation, is reported to be highly expressed in the gingival pockets of subjects suffering from untreated severe periodontitis (Kido et al. 2004). Pavithra et al. have recently reported that in a study of 24 postmenopausal women divided equally based on the presence/absence of periodontitis, the measure of salivary calprotectin levels was significantly higher in those suffering from periodontitis. Moreover, bone mineral density declined in parallel with the increase in salivary calprotectin levels (Pavithra et al. 2024).

Furthermore, postmenopausal women with periodontitis exhibited a statistically significant increase in salivary IL-8 levels (Potdar et al. 2023), while salivary calcium level is a predictive biomarker of bone mineral density and risk of bone fractures (Tabor et al. 2021).

9.5.3 Genetic Factors

Individuals with genetic predispositions that lead to a systemic reduction in bone mineral density may also be at an increased risk of damaging the alveolar bone through similar pathophysiological mechanisms. These data mean that genetic predispositions that make people more susceptible to systemic bone loss could also increase their vulnerability to periodontal damage. Generally, the factors that downregulate IL-6 gene expression are primarily estrogen and testosterone. As a result, postmenopausal women tend to have elevated levels of IL-6, even in the absence of infection, trauma, or stress. Additionally, IL-6 gene expression varies with age, linking osteoporosis to chronic periodontal disease as part of the aging process (Ershler and Keller 2000).

Furthermore, other researchers focused on genetic studies on a series of genic polymorphisms that are potentially associated with osteoporosis and periodontitis to elucidate the genetic links between these two diseases, though findings remained inconclusive (Wu et al. 2024; Yu et al. 2024).

9.5.4 Microbiological Aspect: The Role of the Microbiota

The microbiota, formerly referred to as microbial flora, is a collection of all microorganisms—primarily bacteria but also viruses and fungi—hosted in specific ecological niches of the human body, such as the eyes, nose, genitalia, stomach, skin, and mouth (Berg et al. 2020). The gut, which is the richest and most extensively studied of these niches, accounts for 95% of the human microbiota (Baquero and Nombela 2012).

An imbalance in microbiota composition can lead to dysbiosis, responsible for a cascade of effects that can have local or systemic consequences. Regarding bone metabolism, gut and oral dysbiosis have been identified as contributing factors to increased bone resorption. Moreover, the gut and mouth are interconnected, as oral bacteria have been found in the gut and the gut biofilm, significantly impacting systemic health.

Consequently, gut dysbiosis can be responsible for systemic inflammation, thus affecting bone metabolism and exacerbating conditions like osteoporosis. Conversely, oral dysbiosis, characterized by harmful shifts in the microbiota of the oral cavity, can trigger and perpetuate periodontal inflammation and alveolar bone loss directly or exert its effects indirectly through the oral–gut axis.

9.5.4.1 The Gut Microbiota and Bone Homeostasis

Dysbiosis of the intestinal microbiota leads to alterations in intestinal permeability, triggering a series of pathogenic events that directly and indirectly affect bone metabolism. The key effects include:

1. An increase in serum pro-inflammatory lipopolysaccharide (LPS) levels, promoting inflammation (Ma et al. 2024).
2. A decrease in essential mineral absorption, including calcium, phosphate, and magnesium (Behera et al. 2020).
3. Enhanced bone resorption, driven by the chronic inflammation mediated by circulating pro-inflammatory and osteoclastogenic cytokines (Ma et al. 2024).
4. Alterations in the neurotransmitter synthesis involved in bone metabolism, such as serotonin (Gershon 2013; Spohn and Mawe 2017).

Emerging evidence has also confirmed that specific compositions of the gut microbiota may contribute to the risk of osteoporosis. These studies quantitatively assessed characteristic bacterial taxa and functional pathways linked to reduced bone mineral density. Notably, osteoporosis-associated gut microbiota alterations include an increased presence of the genera *Dialister*, *Bacteroides*, *Faecalibacterium*, and *Eisenbergiella*, a higher abundance of the *Bacteroidota* phylum, and a decrease in the *Veillonella* genus (Das et al. 2019; Ma et al. 2024; Ticinesi et al. 2024; Wei et al. 2021; Zheng et al. 2025).

9.5.5 The Oral Microbiota and Bone Homeostasis

Periodontitis is closely linked to osteoporosis, with an important role played by periodontopathogenic bacteria (Contaldo et al. 2020b). In 2013, some researchers hypothesized a connection between the oral microbiota and systemic bone health (Liang et al. 2013). A subsequent 15-year prospective cohort study found that postmenopausal women with osteoporosis were nearly twice as likely to experience oral bone mineral density loss compared to those without osteoporosis, with both systemic and oral bone mineral density loss increasing with age (Banack et al. 2018).

The scientific community has reported that certain oral periodontopathogenic bacteria, including *Porphyromonas gingivalis*, *Tannerella forsythia*, *Prevotella intermedia*, and *Campylobacter rectus*, are strongly associated with severe oral bone loss (Banack et al. 2018). Furthermore, an analysis of more than 1000 postmenopausal women revealed the recurrence of 8 subgingival bacterial species associated with oral bone loss, such as *Fusobacterium nucleatum*, *Porphyromonas gingivalis*, *Streptococcus sanguis*, *Tannerella forsythia*, *Capnocytophaga* sp., *Prevotella intermedia*, *Eubacterium saburreum*, and *Campylobacter rectus* (Brennan et al. 2007).

The key mechanisms through which oral periodontopathogenic bacteria induce local and systemic bone resorption are linked to the levels of bacterial lipopolysaccharides (LPSs) and toxins released by them. These pathogenic factors act directly within the gingival pocket and systemically via dissemination through the bloodstream and/or the gut following ingestion. These molecules are strongly responsible for the neutrophil release of a series of pro-inflammatory cytokines with osteoclastic activities (Contaldo et al. 2020b). For example, it has been reported that neutrophils are highly sensitive to the LPSs

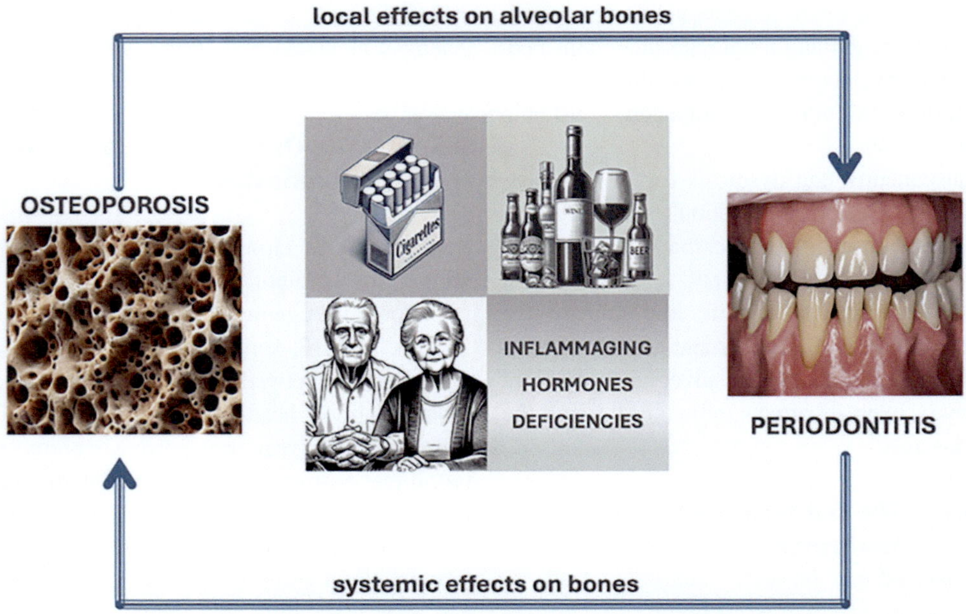

Fig. 9.1 Bidirectional interrelationships between osteoporosis and periodontitiss

secreted from *F. nucleatum*, *A. actinomycetemcomitans*, *P. intermedia*, and *P. gingivalis*. In the presence of these types of lipopolysaccharides, neutrophils release up to 16 times the baseline levels of calprotectin (Kido et al. 2004), a protein known for its pro-inflammatory and osteoclastic activity (Pavithra et al. 2024).

The effects of periodontopathogenic bacteria in exacerbating osteoporosis are indirectly supported by evidence showing that the reestablishment of proper oral hygiene positively impacts not only periodontal health indices but also bone mineral density (Huang et al. 2016).

9.6 Conclusions

There is evidence supporting a strong correlation between periodontitis and osteoporosis, primarily through shared mechanisms of inflammation and bone metabolism dysregulation. While the exact cause of this bidirectional relationship remains to be fully established, it is clear that periodontitis can exacerbate the course of osteoporosis through systemic inflammation and osteoclast activation.

Hence, the correlation between osteoporosis and periodontitis is both possible and bidirectional (Fig. 9.1). On one hand, osteoporosis can accelerate alveolar bone resorption, leading to more severe periodontitis. This occurs due to several reasons:

- Bacteria are able to penetrate deeper into the enlarged periodontal space due to the loss of bone mineral density in the alveolar bones, further exacerbating periodontal destruction.
- Both diseases have a common occurrence of overexpression of pro-inflammatory cytokines with osteoclastic activity.
- Shared risk factors, such as smoking, alcohol drinking, and hormone levels, contribute to this relationship.

On the other hand, periodontopathogenic bacteria can trigger or exacerbate systemic osteoporosis by inducing a local and systemic chronic inflammatory state. This inflammation activates osteoclasts not only at oral sites but also systemically.

Given this evidence, it is reasonable to establish screening programs at dental offices that focus on:

- Patients with known low bone mineral density (osteopenia or osteoporosis) who are at higher risks of developing periodontitis.
- Patients with periodontitis who are at a risk of developing osteopenia and osteoporosis, especially if they are smokers, alcohol drinkers, and postmenopausal women.

In each case, these individuals should be referred for screenings and investigations on bone metabolism, undergo stricter periodontal assessments and follow-ups, and maintain regular oral hygiene practices (Inagaki et al. 2005). This includes regular periodontal charting and evaluations of bone mineral density, in order to prevent the onset or progression of both diseases.

Recognizing the connection between periodontitis and osteoporosis highlights the need for an integrated healthcare approach. Dental professionals and medical practitioners should collaborate to manage both conditions. Patients diagnosed with osteoporosis should be screened for periodontitis, and, conversely, periodontal patients, particularly postmenopausal women, should be evaluated for osteoporosis risk.

Furthermore, the role of oral dysbiosis linked to the abundance of periodontopathogenic bacteria reinforces the need for in-depth studies based on omics technologies. These studies help identify the causative agents and underlying mechanisms, as well as propose beneficial therapeutic interventions, such as the use of probiotics and other noninvasive treatments, which could restore the balance of oral and intestinal microbial dysbiosis (Contaldo 2023; Hu et al. 2024a; Wang et al. 2024).

Last but not least, efforts to promote healthy lifestyle habits, such as smoking cessation, reducing alcohol consumption, and ensuring proper dietary and nutritional intake, as well as screening programs should involve active collaboration among bone specialists, dentists, nutritionists, endocrinologists, and geriatricians (Latimer et al. 2024). A multidisciplinary approach is essential to address the local and systemic pathophysiological features shared by periodontitis and osteoporosis (Shrivastava 2024).

References

Agueda A, Ramon JM, Manau C, Guerrero A, Echeverria JJ (2008) Periodontal disease as a risk factor for adverse pregnancy outcomes: a prospective cohort study. J Clin Periodontol 35(1):16–22. https://doi.org/10.1111/j.1600-051X.2007.01166.x

Al Habashneh R, Alchalabi H, Khader YS et al (2010) Association between periodontal disease and osteoporosis in postmenopausal women in Jordan. J Periodontol 81(11):1613–1621. https://doi.org/10.1902/jop.2010.100190

Albandar JM, Susin C, Hughes FJ (2018) Manifestations of systemic diseases and conditions that affect the periodontal attachment apparatus: case definitions and diagnostic considerations. J Clin Periodontol 45(S20). https://doi.org/10.1111/jcpe.12947

Almeida M, Han L, Martin-Millan M, O'Brien CA, Manolagas SC (2007) Oxidative stress antagonizes Wnt signaling in osteoblast precursors by diverting β-catenin from T cell factor- to forkhead box O-mediated transcription. J Biol Chem 282(37): 27298–27305. https://doi.org/10.1074/jbc.M702811200

Ballini A, Cantore S, Farronato D, Cirulli N, Inchingolo F, Papa F, Malcangi G, Inchingolo AD, Dipalma G, Sardaro N, De Vito D, Scacco S (2015) Periodontal disease and bone pathogenesis: the crosstalk between cytokines and porphyromonas gingivalis. J Biol Regul Homeost Agents 29(2):273–281

Banack HR, Genco RJ, LaMonte MJ, Millen AE, Buck MJ, Sun Y, Andrews CA, Hovey KM et al (2018) Cohort profile: the Buffalo OsteoPerio microbiome prospective cohort study. BMJ Open 8(12):e024263. https://doi.org/10.1136/bmjopen-2018-024263

Baquero F, Nombela C (2012) The microbiome as a human organ. Clin Microbiol Infect 18(SUPPL. 4): 2–4. https://doi.org/10.1111/j.1469-0691.2012.03916.x

Behera J, Ison J, Tyagi SC, Tyagi N (2020) The role of gut microbiota in bone homeostasis. Bone 135:115317. https://doi.org/10.1016/j.bone.2020.115317

Berg G, Rybakova D, Fischer D, Cernava T, Vergès M-CC, Charles T, Chen X, Cocolin L, Eversole K, Corral GH, Sessitsch A, Schloter M (2020) Microbiome definition re-visited: old concepts and new challenges. Microbiome 8(1). https://doi.org/10.1186/s40168-020-00875-0

Berg Y, Gabay E, Božić D, Shibli JA, Ginesin O, Asbi T, Takakura L, Mayer Y (2024) The impact of nutritional components on periodontal health: a literature review. Nutrients 16(22). https://doi.org/10.3390/nu16223901

Black DM, Delmas PD, Eastell R, Reid IR, Boonen S, Cauley JA, Cosman F, Lakatos P, Leung PC, Man Z, Mautalen C, Mesenbrink P, Hu H, Caminis J, Tong K, Rosario-Jansen T, Krasnow J, Hue TF, Sellmeyer D et al (2007) Once-yearly zoledronic acid for treatment of postmenopausal osteoporosis. N Engl J Med

356(18):1809–1822. https://doi.org/10.1056/NEJMoa067312
Boyce BF, Xing L (2008) Functions of RANKL/RANK/OPG in bone modeling and remodeling. Arch Biochem Biophys 473(2):139–146. https://doi.org/10.1016/j.abb.2008.03.018
Brennan RM, Genco RJ, Wilding GE, Hovey KM, Trevisan M, Wactawski-Wende J (2007) Bacterial species in subgingival plaque and Oral bone loss in postmenopausal women. J Periodontol 78(6):1051–1061. https://doi.org/10.1902/jop.2007.060436
Caggiano M, Gasparro R, D'Ambrosio F, Pisano M, Di Palo MP, Contaldo M (2022) Smoking cessation on periodontal and Peri-implant health status: a systematic review. Dentis J 10(9). https://doi.org/10.3390/dj10090162
Canabarro A, Valle C, Farias MR, Santos FB, Lazera M, Wanke B (2013) Association of subgingival colonization of Candida albicans and other yeasts with severity of chronic periodontitis. J Periodontal Res 48(4):428–432. https://doi.org/10.1111/jre.12022
Caton G et al (2018) A new classification scheme for periodontal and peri-implant diseases and conditions – introduction and key changes from the 1999 classification. J Clin Periodontol 45:S1–S8. https://doi.org/10.1111/jcpe.12935
Chapple ILC, Matthews JB (2007) The role of reactive oxygen and antioxidant species in periodontal tissue destruction. Periodontol 43(1):160–232. https://doi.org/10.1111/j.1600-0757.2006.00178.x
Choi J-K, Kim Y-T, Kweon H-I, Park E-C, Choi S-H, Lee J-H (2017) Effect of periodontitis on the development of osteoporosis: results from a nationwide population-based cohort study (2003-2013). BMC Womens Health 17(1):77. https://doi.org/10.1186/s12905-017-0440-9
Chou H-H, Lu S-L, Wang S-T, Huang T-H, Chen SL-S (2021) The association between bone mineral density and periodontal disease in middle-aged adults. Int J Environ Res Public Health 18(6). https://doi.org/10.3390/ijerph18063321
Contaldo M (2023) Use of probiotics for oral candidiasis: state of the art and perspective. A further step toward personalized medicine? Front Biosci (Elite Ed) 15(1). https://doi.org/10.31083/j.fbe1501006
Contaldo M, Boccellino M, Zannini G, Romano A, Sciarra A, Sacco A, Settembre G, Coppola M, Di Carlo A, D'Angelo L, Inchingolo F, Feola A, Di Domenico M (2020a) Sex hormones and inflammation role in oral cancer progression: a molecular and biological point of view. J Oncol 2020. https://doi.org/10.1155/2020/9587971
Contaldo M, Itro A, Lajolo C, Gioco G, Inchingolo F, Serpico R (2020b) Overview on osteoporosis, periodontitis and oral dysbiosis: the emerging role of oral microbiota. Appl Sci (Switzerland) 10(17). https://doi.org/10.3390/app10176000
Contaldo M, Lucchese A, Romano A, Vella FD, Di Stasio D, Serpico R, Petruzzi M (2021) Oral microbiota features in subjects with down syndrome and periodontal diseases: a systematic review. Int J Mol Sci 22(17). https://doi.org/10.3390/ijms22179251
Cui J, Lin L, Hao F, Shi Z, Gao Y, Yang T, Yang C, Wu X, Gao R, Ru Y, Li F, Xiao C, Gao Y, Wang Y (2024) Comprehensive review of the traditional uses and the potential benefits of epimedium folium. Front Pharmacol 15:1415265. https://doi.org/10.3389/fphar.2024.1415265
Das M, Cronin O, Keohane DM, Cormac EM, Nugent H, Nugent M, Molloy C, O'Toole PW, Shanahan F, Molloy MG, Jeffery IB (2019) Gut microbiota alterations associated with reduced bone mineral density in older adults. Rheumatology 58(12):2295–2304. https://doi.org/10.1093/rheumatology/kez302
de Passos-Soares J et al (2017) Association between osteoporosis treatment and severe periodontitis in postmenopausal women. Menop 24(7):789–795. https://doi.org/10.1097/GME.0000000000000830
Di Spirito F, Amato A, Romano A, Dipalma G, Xhajanka E, Baroni A, Serpico R, Inchingolo F, Contaldo M (2022) Analysis of risk factors of oral cancer and periodontitis from a sex- and gender-related perspective: gender dentistry. Appl Sci (Switzerland) 12(18):9135. https://doi.org/10.3390/app12189135
Ershler WB, Keller ET (2000) Age-associated increased interleukin-6 gene expression, late-life diseases, and frailty. Annu Rev Med 51:245–270. https://doi.org/10.1146/annurev.med.51.1.245
Fischer V, Haffner-Luntzer M (2022) Interaction between bone and immune cells: implications for postmenopausal osteoporosis. Semin Cell Dev Biol 123:14–21. https://doi.org/10.1016/j.semcdb.2021.05.014
Gershon MD (2013) 5-Hydroxytryptamine (serotonin) in the gastrointestinal tract. Curr Opin Endocrinol Diabet Obes 20(1):14–21. https://doi.org/10.1097/MED.0b013e32835bc703
Gil-Montoya JA, Garrido-Martínez M, Barrios-Rodríguez R, Ramos-García P, Lenouvel D, Montes-Castillo C, Martínez-Ramírez MJ (2021) Association between low bone mineral density and periodontitis in generally healthy perimenopausal women. J Periodontol 92(1):95–103. https://doi.org/10.1002/JPER.20-0029
Gomes-Filho IS, Oliveira TJS, Passos JS et al (2013) Effect of osteoporosis on periodontal therapy among post-menopausal women. Gerodontology 30(1):40–48. https://doi.org/10.1111/j.1741-2358.2012.00643.x
Graves DT, Oates T, Garlet GP (2011) Review of osteoimmunology and the host response in endodontic and periodontal lesions. J Oral Microbiol 3(2011). https://doi.org/10.3402/jom.v3i0.5304
Grodstein F, Colditz GA, Stampfer MJ (1996) Postmenopausal hormone use and tooth loss: a prospective study. J Am Dent Assoc 127(3):370–377. https://doi.org/10.14219/jada.archive.1996.0208
Guo Y, Ge X, Wang W, Wang R, Chen Q, Wang H (2024) Epimedium applied in the clinical treatment of

osteoporosis patients with periodontitis. Medicine 103(50):e40837. https://doi.org/10.1097/MD.0000000000040837

Haffajee AD, Socransky SS, Patel MR, Song X (2008) Microbial complexes in supragingival plaque. Oral Microbiol Immunol 23(3):196–205. https://doi.org/10.1111/j.1399-302X.2007.00411.x

Haffajee AD, Teles RP, Patel MR, Song X, Veiga N, Socransky SS (2009) Factors affecting human supragingival biofilm composition. I. Plaque mass. J Periodontal Res 44(4):511–519. https://doi.org/10.1111/j.1600-0765.2008.01154.x

Hernández-Vigueras S, Martínez-Garriga B, Sánchez MC, Sanz M, Estrugo-Devesa A, Vinuesa T, López-López J, Viñas M (2016) Oral microbiota, periodontal status, and osteoporosis in postmenopausal females. J Periodontol 87(2):124–133. https://doi.org/10.1902/jop.2015.150365

Hernlund E, Svedbom A, Ivergård M, Compston J, Cooper C, Stenmark J, McCloskey EV, Jönsson B, Kanis JA (2013) Osteoporosis in the European Union: medical management, epidemiology and economic burden. Arch Osteoporos 8(1–2):136. https://doi.org/10.1007/s11657-013-0136-1

Holmstrup P, Damgaard C, Olsen I, Klinge B, Flyvbjerg A, Nielsen CH, Hansen PR (2017) Comorbidity of periodontal disease: two sides of the same coin? An introduction for the clinician. J Oral Microbiol 9(1). https://doi.org/10.1080/20002297.2017.1332710

Hong S-J, Yang B-E, Yoo D-M, Kim S-J, Choi H-G, Byun S-H (2021) Analysis of the relationship between periodontitis and osteoporosis/fractures: a cross-sectional study. BMC Oral Health 21(1):125. https://doi.org/10.1186/s12903-021-01496-1

Hoy SR, Vucetich JA, Vucetich LM, Hindelang M, Huebner JL, Kraus VB, Peterson RO (2024) Links between three chronic and age-related diseases, osteoarthritis, periodontitis, and osteoporosis, in a wild mammal (moose) population. Osteoarthr Cartil 32(3):281–286. https://doi.org/10.1016/j.joca.2023.11.016

Hu W, Chen S, Zou X, Chen Y, Luo J, Zhong P, Ma D (2024a) Oral microbiome, periodontal disease and systemic bone-related diseases in the era of homeostatic medicine. J Adv Res 73. https://doi.org/10.1016/j.jare.2024.08.019

Hu W, Feng X, Wen C (2024b) Relationship between multi-nutrient intake and bone loss and osteoporosis in U.S. adults: findings from NHANES. Medicine 103(50):e40768. https://doi.org/10.1097/MD.0000000000040768

Huang Y-F, Chang C-T, Liu S-P, Muo C-H, Tsai C-H, Hong H-H, Shen Y-F, Wu C-Z (2016) The impact of Oral hygiene maintenance on the association between periodontitis and osteoporosis: a Nationwide population-based cross sectional study. Medicine 95(6):e2348. https://doi.org/10.1097/MD.0000000000002348

Inagaki K, Kurosu Y, Yoshinari N, Noguchi T, Krall EA, Garcia RI (2005) Efficacy of periodontal disease and tooth loss to screen for low bone mineral density in Japanese women. Calcif Tissue Int 77(1):9–14. https://doi.org/10.1007/s00223-004-0275-x

Inchingolo F, Martelli FS, Isacco CG, Borsani E, Cantore S, Corcioli F, Boddi A, Nguyễn KCD, De Vito D, Aityan SK, Dipalma G, Ballini A (2020) Chronic periodontitis and immunity, towards the implementation of a personalized medicine: a translational research on gene single nucleotide polymorphisms (SNPs) linked to chronic oral dysbiosis in 96 caucasian patients. Biomedicine 8(5). https://doi.org/10.3390/BIOMEDICINES8050115

Ionas M, Mitariu SIC, Dancila A, Ionas TH, Comaneanu RM, Botoaca O, Mitariu MMC (2017) The capacity of a specific anti-streptococcus Mutans monoclonal antibodies test to identify the diabetic patients with increased risk of periodontal disease. Rev Chim 68(11):2556–2559

Jepsen S, Caton JG, Albandar JM, Bissada NF, Bouchard P, Cortellini P, Demirel K, de Sanctis M, Ercoli C, Fan J, West NX, Yamazaki K (2018a) Periodontal manifestations of systemic diseases and developmental and acquired conditions: consensus report of workgroup 3 of the 2017 world workshop on the classification of periodontal and Peri-implant diseases and conditions. J Clin Periodontol 45:S219–S229. https://doi.org/10.1111/jcpe.12951

Jepsen S, Caton JG, Albandar JM, Bissada NF, Bouchard P, Cortellini P, Demirel K, de Sanctis M, Ercoli C, Fan J, West NX, Yamazaki K (2018b) Periodontal manifestations of systemic diseases and developmental and acquired conditions: consensus report of workgroup 3 of the 2017 world workshop on the classification of periodontal and Peri-implant diseases and conditions. J Periodontol 89:S237–S248. https://doi.org/10.1002/JPER.17-0733

Johnell O, Kanis JA (2006) An estimate of the worldwide prevalence and disability associated with osteoporotic fractures. Osteoporos Int 17(12):1726–1733. https://doi.org/10.1007/s00198-006-0172-4

Kanis JA (2007). Assessment of osteoporosis at the primary health-care level. Technical Report. World Health Organization Collaborating Centre for Metabolic Bone Diseases (Printed by the University of Sheffield., Ed.; 2007th ed.). University of Sheffield

Kanis JA, Burlet N, Cooper C, Delmas PD, Reginster J-Y, Borgstrom F, Rizzoli R (2008) European guidance for the diagnosis and management of osteoporosis in postmenopausal women. Osteoporos Int 19(4):399–428. https://doi.org/10.1007/s00198-008-0560-z

Kassebaum NJ, Bernabé E, Dahiya M, Bhandari B, Murray CJL, Marcenes W (2014) Global burden of severe periodontitis in 1990-2010: a systematic review and meta-regression. J Dent Res 93(11):1045–1053. https://doi.org/10.1177/0022034514552491

Khunthananithi P, Lertpimonchai A, Sritara C, Srithanyarat SS, Thienpramuk L, Mongkornkarn S

(2023) Decreased bone mineral density is associated with an increased number of teeth with periodontitis progression: a 5-year retrospective cohort study. Clin Oral Investig 28(1):51. https://doi.org/10.1007/s00784-023-05463-8

Kido J, Kido R et al (2004) Induction of calprotectin release by Porphyromonas gingivalis lipopolysaccharide in human neutrophils. Oral Microbiol Immunol 19(3):182–187. https://doi.org/10.1111/j.0902-0055.2004.00139.x

Kumar PS (2020) Interventions to prevent periodontal disease in tobacco-, alcohol-, and drug-dependent individuals. Periodontol 84(1):84–101. https://doi.org/10.1111/prd.12333

Lai B, Jiang H, Gao R, Zhou X (2024) Association between alcohol intake and bone mineral density: results from the NHANES 2005-2020 and two-sample Mendelian randomization. Arch Osteoporos 19(1):21. https://doi.org/10.1007/s11657-024-01382-7

Larvin H, Kang J, Aggarwal VR, Pavitt S, Wu J (2023) Periodontitis and risk of immune-mediated systemic conditions: a systematic review and meta-analysis. Community Dent Oral Epidemiol 51(5):705–717. https://doi.org/10.1111/cdoe.12812

Latimer JM, Travan S, Berkey FD, Sugai JV, Giannobile WV (2024) Physician-dentist dual referral model concept for coordinated bone anabolic therapy. J Am Dent Assoc 155(11):954–962. https://doi.org/10.1016/j.adaj.2024.08.009

Lee J-H, Jeong S-N (2020) A population-based study on the association between periodontal disease and major lifestyle-related comorbidities in South Korea: an elderly cohort study from 2002–2015. Medicina 56(11). https://doi.org/10.3390/medicina56110575

Lee J-H, Oh J-Y, Youk T-M, Jeong S-N, Kim Y-T, Choi S-H (2017) Association between periodontal disease and non-communicable diseases: a 12-year longitudinal health-examinee cohort study in South Korea. Medicine 96(26):e7398. https://doi.org/10.1097/MD.0000000000007398

Lerner UH (2006) Inflammation-induced bone remodeling in periodontal disease and the influence of post-menopausal osteoporosis. J Dent Res 85(7):596–607. https://doi.org/10.1177/154405910608500704

Li X, Wen M, Xu Y, Shen Y, Yang X (2022) The association of healthy eating index with periodontitis in NHANES 2013–2014. Front Nutr 9. https://doi.org/10.3389/fnut.2022.968073

Liang W, Li X, Li Y, Li C, Gao B, Gan H, Li S, Shen J, Kang J, Ding S, Lin X, Liao L (2013) Tongue coating microbiome regulates the changes in tongue texture and coating in patients with post-menopausal osteoporosis of Gan-shen deficiency syndrome type. Int J Mol Med 32(5):1069–1076. https://doi.org/10.3892/ijmm.2013.1485

Lin T-H, Lung C-C, Su H-P, Huang J-Y, Ko P-C, Jan S-R, Sun Y-H, Nfor ON, Tu H-P, Chang C-S, Jian Z-H, Chiang Y-C, Liaw Y-P (2015) Association between periodontal disease and osteoporosis by gender: a nationwide population-based cohort study. Medicine 94(7):e553. https://doi.org/10.1097/MD.0000000000000553

Loesche WJ (1996) In: Baron S (ed) Microbiology of dental decay and periodontal disease. University of Texas Medical Branch

Lowenstine LJ, McManamon R, Terio KA (2016) Comparative pathology of aging great apes: bonobos, chimpanzees, gorillas, and orangutans. Vet Pathol 53(2):250–276. https://doi.org/10.1177/0300985815612154

Luo K, Ma S, Guo J, Huang Y, Yan F, Xiao Y (2014) Association between postmenopausal osteoporosis and experimental periodontitis. Biomed Res Int 2014. https://doi.org/10.1155/2014/316134

Ma Z, Liu Y, Shen W, Yang J, Wang T, Li Y, Ma J, Zhang X, Wang H (2024) Osteoporosis in postmenopausal women is associated with disturbances in gut microbiota and migration of peripheral immune cells. BMC Musculoskelet Disord 25(1):791. https://doi.org/10.1186/s12891-024-07904-1

Manjunath BC, Praveen K, Chandrashekar BR, Rani RMV, Bhalla A (2011) Periodontal infections: a risk factor for various systemic diseases. Natl Med J India 24(4):214–219

Masuko K, Iwahara C, Kamiya S, Sakate S, Mizukami Y (2024) Levels of vitamin D and a bone resorption marker in the sera of young women with alcohol use disorder. J Addict Dis 42(4):447–455. https://doi.org/10.1080/10550887.2023.2264999

Minkin MJ (2019) Menopause: hormones, lifestyle, and optimizing aging. Obstet Gynecol Clin N Am 46(3):501–514. https://doi.org/10.1016/j.ogc.2019.04.008

Mombelli A (2018) Microbial colonization of the periodontal pocket and its significance for periodontal therapy. Periodontol 76(1):85–96. https://doi.org/10.1111/prd.12147

Narayanan A, Söder B, Meurman J, Lundmark A, Hu YOO, Neogi U, Yucel-Lindberg T (2023) Composition of subgingival microbiota associated with periodontitis and diagnosis of malignancy – a cross-sectional study. Front Microbiol 14. https://doi.org/10.3389/fmicb.2023.1172340

NIH Consensus Development Panel on Osteoporosis Prevention (2001) Diagnosis, and therapy, march 7-29, 2000: highlights of the conference. South Med J 94(6):569–573. https://doi.org/10.1097/00007611-200106000-00005

Pavithra J, Ramesh A, Thomas B, Kumari NS, Sharmila KP (2024) Salivary calprotectin levels and bone mineral density in post-menopausal women with and without chronic periodontitis – an evaluative study. Indian J Dental Res 35(2):140–144. https://doi.org/10.4103/ijdr.ijdr_522_21

Peng J, Chen J, Liu Y, Lyu J, Zhang B (2023) Association between periodontitis and osteoporosis in United States adults from the National Health and nutrition examination survey: a cross-sectional analysis. BMC

Oral Health 23(1). https://doi.org/10.1186/s12903-023-02990-4

Penoni DC, Torres SR, Farias MLF, Fernandes TM, Luiz RR, Leão ATT (2016) Association of osteoporosis and bone medication with the periodontal condition in elderly women. Osteoporos Int 27(5):1887–1896. https://doi.org/10.1007/s00198-015-3437-y

Penoni DC, Fidalgo TKS, Torres SR, Varela VM, Masterson D, Leão ATT, Maia LC (2017) Bone density and clinical periodontal attachment in postmenopausal women: a systematic review and meta-analysis. J Dent Res 96(3):261–269. https://doi.org/10.1177/0022034516682017

Pepelassi E, Nicopoulou-Karayianni K, Archontopoulou AD, Mitsea A, Kavadella A, Tsiklakis K, Vrotsos I, Devlin H, Horner K (2012) The relationship between osteoporosis and periodontitis in women aged 45-70years. Oral Dis 18(4):353–359. https://doi.org/10.1111/j.1601-0825.2011.01881.x

Pestana de Vasconcelos N, Martins IS, Afonso AS, Braga AC, Pina-Vaz I (2024) Osteoporosis and apical periodontitis prevalence: a systematic review. Dentis J 12(8). https://doi.org/10.3390/dj12080272

Potdar R, Ramesh A, Rajesh A (2023) Evaluation of salivary IL-8 and calcium levels in postmenopausal females with and without periodontitis-a comparative study. Indian J Dent Res 34(1):45–48. https://doi.org/10.4103/ijdr.ijdr_1192_21

Preda SA, Comanescu MC, Albulescu DM, Dascălu IT, Camen A, Cumpătă CN, Perlea P, Bugălă NM, Stoica EL, Gheorghiță LM, Diaconu OA, Tuculina MJ (2022) Correlations between periodontal indices and osteoporosis. Exp Ther Med 23(4):254. https://doi.org/10.3892/etm.2022.11179

Qi J, Chen J, Pang Y, Guo Y, Chen G, Liu Y, Wang J, Liu E (2023) Association between periodontal disease and osteoporosis in postmenopausal women: a systematic review and meta-analysis. Heliyon 9(11). https://doi.org/10.1016/j.heliyon.2023.e20922

Saini D, Dhawan S, Dhawan R, Gupta P, Kiranmayi J, Singh R (2024) Prevalence of co-morbidities among periodontitis and non-periodontitis Indian patients. Bioinformation 20(8):888–892. https://doi.org/10.6026/973206300200888

Sanz M, Marco del Castillo A, Jepsen S, Gonzalez-Juanatey JR, D'Aiuto F, Bouchard P, Chapple I, Dietrich T, Gotsman I, Graziani F, Vlachopoulos C, Wimmer G (2020) Periodontitis and cardiovascular diseases: consensus report. J Clin Periodontol 47(3):268–288. https://doi.org/10.1111/jcpe.13189

Sharma P, Sharma RK, Tewari S, Gill PS, Tanwar N, Arora R, Kaur M (2024) Impact of subgingival instrumentation on systemic inflammation and serum bone resorption marker in premenopausal women with periodontitis: a prospective interventional study. Quintessence Int 55(2):108–118. https://doi.org/10.3290/j.qi.b4867855

Shrivastava S (2024) Menopause and oral health: clinical implications and preventive strategies. J Mid-Life Health 15(3):135–141. https://doi.org/10.4103/jmh.jmh_125_24

Silva BC, Kousteni S (2015) Cellular actions of PTH: osteoblasts, osteoclasts, and osteocytes. In: The Parathyroids: basic and clinical concepts, 3rd edn. https://doi.org/10.1016/B978-0-12-397166-1.00008-4

Slot DE, Valkenburg C, Van der Weijden GA (2020) Mechanical plaque removal of periodontal maintenance patients: a systematic review and network meta-analysis. J Clin Periodontol 47(S22):107–124. https://doi.org/10.1111/jcpe.13275

Socransky SS, Haffajee AD, Cugini MA, Smith C, Kent RL Jr (1998) Microbial complexes in subgingival plaque. J Clin Periodontol 25(2):134–144. https://doi.org/10.1111/j.1600-051X.1998.tb02419.x

Sozen T, Ozisik L, Calik Basaran N (2017) An overview and management of osteoporosis. Eur J Rheumatol 4(1):46–56. https://doi.org/10.5152/eurjrheum.2016.048

Sperr M, Kundi M, Tursic V, Bristela M, Moritz A, Andrukhov O, Rausch-Fan X, Sperr WR (2018) Prevalence of comorbidities in periodontitis patients compared with the general Austrian population. J Periodontol 89(1):19–27. https://doi.org/10.1902/jop.2017.170333

Spohn SN, Mawe GM (2017) Non-conventional features of peripheral serotonin signalling – the gut and beyond. Nat Rev Gastroenterol Hepatol 14(7):412–420. https://doi.org/10.1038/nrgastro.2017.51

Tabor E, Hűpsch H, Rokicka J, Sierpińska T, Konstantynowicz J, Orywal K, Pluskiewicz W (2021) Salivary content might be associated with skeletal status in postmenopausal women: SilesiaOsteoActive study results. J Clin Densitom 24(1):14–21. https://doi.org/10.1016/j.jocd.2020.02.001

Ticinesi A, Siniscalchi C, Meschi T, Nouvenne A (2024) Gut microbiome and bone health: update on mechanisms, clinical correlations, and possible treatment strategies. Osteoporos Int 36. https://doi.org/10.1007/s00198-024-07320-0

Tonetti MS, Greenwell H, Kornman KS (2018) Staging and grading of periodontitis: framework and proposal of a new classification and case definition. J Periodontol 89:S159–S172. https://doi.org/10.1002/JPER.18-0006

Trindade D, Carvalho R, Machado V, Chambrone L, Mendes JJ, Botelho J (2023) Prevalence of periodontitis in dentate people between 2011 and 2020: a systematic review and meta-analysis of epidemiological studies. J Clin Periodontol 50(5):604–626. https://doi.org/10.1111/jcpe.13769

Uzel NG, Teles FR, Teles RP, Song XQ, Torresyap G, Socransky SS, Haffajee AD (2011) Microbial shifts during dental biofilm re-development in the absence of oral hygiene in periodontal health and disease. J Clin Periodontol 38(7):612–620. https://doi.org/10.1111/j.1600-051X.2011.01730.x

Vergatti A, Abate V, D'Elia L, De Filippo G, Piccinocchi G, Gennari L, Merlotti D, Galletti F,

Strazzullo P, Rendina D (2024) Smoking habits and osteoporosis in community-dwelling men subjected to dual-X-ray absorptiometry: a cross-sectional study. J Endocrinol Investig 47(12):3129–3135. https://doi.org/10.1007/s40618-024-02402-6

Wang F, Wei W, Liu PJ (2024) Effects of probiotic supplementation on bone health in postmenopausal women: a systematic review and meta-analysis. Front Endocrinol 15:1487998. https://doi.org/10.3389/fendo.2024.1487998

Wei M, Li C, Dai Y, Zhou H, Cui Y, Zeng Y, Huang Q, Wang Q (2021) High-throughput absolute quantification sequencing revealed osteoporosis-related gut microbiota alterations in Han Chinese elderly. Front Cell Infect Microbiol 11. https://doi.org/10.3389/fcimb.2021.630372

Weitzmann MN, Pacifici R (2006) Estrogen deficiency and bone loss: an inflammatory tale. J Clin Invest 116(5):1186–1194. https://doi.org/10.1172/JCI28550

Wu J, Yao L, Liu Y, Zhang S, Wang K (2024) Periodontitis and osteoporosis: a two-sample Mendelian randomization analysis. Braz J Med Biol Res 57:e12951. https://doi.org/10.1590/1414-431X2024e12951

Xiao P-L, Cui A-Y, Hsu C-J, Peng R, Jiang N, Xu X-H, Ma Y-G, Liu D, Lu H-D (2022) Global, regional prevalence, and risk factors of osteoporosis according to the World Health Organization diagnostic criteria: a systematic review and meta-analysis. Osteoporos Int 33(10):2137–2153. https://doi.org/10.1007/s00198-022-06454-3

Xu S, Zhang G, Guo J-F, Tan Y-H (2021) Associations between osteoporosis and risk of periodontitis: a pooled analysis of observational studies. Oral Dis 27(2):357–369. https://doi.org/10.1111/odi.13531

Xu G, Guo Y, Wang W, Yu W-Q, Chen Q-M, Wang H (2024) Zoledronic acid improves periodontal health, reduces serum inflammation, and enhances bone metabolism in postmenopausal osteoporosis complicated by periodontitis. Am J Transl Res 16(11): 6519–6527. https://doi.org/10.62347/GXEB5902

Yu B, Wang C (2022) Osteoporosis and periodontal diseases—an update on their association and mechanistic links. Periodontol 89(1):99–113. https://doi.org/10.1111/prd.12422

Yu Y, Chu T, Dong J, Deng H, Pan Y, Wang Y (2024) A Mendelian randomization study on the association of bone mineral density with periodontitis. Oral Dis 30(3):1488–1496. https://doi.org/10.1111/odi.14582

Zarzour F, Leslie WD (2024) The effect of smoking cessation versus current smoking on fracture risk: the Manitoba BMD registry. J Clin Densitom 27(4): 101523. https://doi.org/10.1016/j.jocd.2024.101523

Zheng X-Q, Wang D-B, Jiang Y-R, Song C-L (2025) Gut microbiota and microbial metabolites for osteoporosis. Gut Microbes 17(1):2437247. https://doi.org/10.1080/19490976.2024.2437247

Zhou Y, Zhu Y, Jin X, Zhang Y, Song J, Wu Z, Li Y, Yi J, Wang D, Hu M (2023) Chroogomphus rutilus regulates bone metabolism to prevent periodontal bone loss during orthodontic tooth movement in osteoporotic rats. Nutrients 15(23). https://doi.org/10.3390/nu15234906

Immunological Diagnosis and Treatments of Periodontitis in Patients with Type 2 Diabetes Mellitus

10

Koki Yoshida, Kengo Iwasaki, Osamu Uehara, Satsuki Kato, Shintaro Shimizu, Kohei Sato, Nodoka Sugiyama, Yukichi Okada, Yoshiki Fujimoto, Sho Hirose, Norihiro Nakamoto, Yuko Matsuki, Takashi Kado, Yukie Murata, Katsuya Kawanishi, Hirofumi Matsuoka, Hiroko Miura, Keiji Nagano, Yoshihiro Abiko, Yasushi Furuichi, and Toshiyuki Nagasawa

Abstract

The oral mucosa is a key barrier and entry point for microorganisms into the gastrointestinal tract. Interleukin-17 (IL-17) and T helper 17 (Th17) cells contribute to mucosal defense against bacterial and fungal pathogens. Gingival fibroblasts recruit neutrophils, construct a mechanical barrier of gingival tissue, and produce osteoprotegerin, which prevents bone resorption. In periodontitis, inflammatory cytokines induce RANKL expression by osteoblasts and periodontal ligament fibroblasts, resulting in alveolar bone resorption. In diabetes mellitus, excessive IL-17 activity increases the risk of periodontal tissue destruction and fungal infections. In patients with diabetes, periodontitis worsens insulin resistance and glycemic control while introducing periodontopathic bacteria and Th17 cells to the gut, exacerbating intestinal

K. Yoshida · Y. Abiko
Division of Oral Medicine and Pathology, Department of Human Biology and Pathophysiology, School of Dentistry, Health Sciences University of Hokkaido, Kanazawa, Hokkaido, Japan
e-mail: denty@hoku-iryo-u.ac.jp; yoshi-ab@hoku-iryo-u.ac.jp

K. Iwasaki
Division of Creative and Integrated Medicine, Advanced Medical Research Center, Translational Research Institute for Medical Innovation (TRIMI), Osaka Dental University, Hirakata, Osaka, Japan
e-mail: iwasaki-k@cc.osaka-dent.ac.jp

O. Uehara · Y. Murata · H. Matsuoka · H. Miura
Division of Disease Control & Molecular Epidemiology Department of Oral Growth & Development, School of Dentistry, Health Sciences University of Hokkaido, Kanazawa, Hokkaido, Japan
e-mail: osamu@hoku-iryo-u.ac.jp; y-murata@hoku-iryo-u.ac.jp; mazun@hoku-iryo-u.ac.jp; hmiura@hoku-iryo-u.ac.jp

S. Kato · S. Shimizu · K. Sato · N. Sugiyama · Y. Okada · Y. Fujimoto · S. Hirose · N. Nakamoto
Division of Periodontology and Endodontology, Department of Oral Rehabilitation, School of Dentistry, Health Sciences University of Hokkaido, Kanazawa, Hokkaido, Japan
e-mail: satsuki@hoku-iryo-u.ac.jp; sshimizu@hoku-iryo-u.ac.jp; kohey@hoku-iryo-u.ac.jp; nsugiyama@hoku-iryo-u.ac.jp; oka0331@hoku-iryo-u.ac.jp; yoshiki@hoku-iryo-u.ac.jp; h-sho@hoku-iryo-u.ac.jp; nakamoto@hoku-iryo-u.ac.jp

Y. Matsuki · K. Kawanishi
Division of Advanced Clinical Education, Department of Integrated Dental Education, School of Dentistry, Health Sciences University of Hokkaido, Kanazawa, Hokkaido, Japan
e-mail: matsuki@hoku-iryo-u.ac.jp; kawanisi@hoku-iryo-u.ac.jp

N. Rezaei, H. Ziaei (eds.), *Oral Immunology*, Advances in Experimental Medicine and Biology 1492,
https://doi.org/10.1007/978-3-032-03176-1_10

inflammation. Infection with periodontopathic bacteria sharing homology with β2 glycoprotein I (β2GPI) may induce antibodies against β2GPI, increasing blood clot risk. Thus, in patients with diabetes and periodontitis, symptoms vary due to localized and systemic immune responses to bacteria and fungi, necessitating a comprehensive understanding for accurate diagnosis and treatment. The biopsychosocial model aids in addressing these complexities, while the common risk approach and cognitive-behavioral therapy address social and psychological challenges. Therefore, comprehensive treatment strategies are necessary for managing periodontitis in patients with diabetes, considering both microbial factors and immune responses.

Keywords

Oral mucosa · IL-17 · Periodontitis · Diabetes · Immune response

10.1 Introduction

Diabetes mellitus can be classified into two main types: type 1, which results from the autoimmune destruction of β-cells in the pancreas, and type 2, which arises from progressive β-cell insulin deficiency in the context of insulin resistance. Other forms of diabetes include gestational diabetes mellitus and specific types of diabetes triggered by various factors (American Diabetes 2019).

A reciprocal relationship exists between periodontal disease and diabetes. Diabetes worsens periodontal disease by impairing the body's ability to fight infections and heal tissues. Conversely, periodontal disease can exacerbate insulin resistance, further complicating blood sugar control and worsening diabetes (Nagasawa et al. 2010; Shinjo and Nishimura 2024). This chapter focuses on oral immunity in type 2 diabetes, where a bidirectional relationship between diabetes and periodontal disease is observed.

Non-communicable diseases (NCDs) are medical conditions that are chronic and result from the long-term effects of several risk factors, including genetic, physiological, behavioral, and environmental factors (Budreviciute et al. 2020). Both periodontitis and diabetes are NCDs that share risk factors (Herrera et al. 2023). According to data from the Global Burden of Disease database, there were 1.1 billion cases of severe periodontitis globally in 2019 (Chen et al. 2009). The global prevalence of diabetes was 537 million people in 2021 (Sun et al. 2022). Closer collaboration between oral health professionals and family doctors is important for the early detection and management of NCDs. Information about the reported association should be made available to family doctors, oral healthcare funders, patients, and the general population (Herrera et al. 2023). Therefore, understanding the mechanisms underlying both periodontitis and diabetes is indispensable for researching, diagnosing, and treating diabetes-associated periodontitis.

T. Kado
Division of Periodontology and Endodontology, Department of Oral Rehabilitation, School of Dentistry, Health Sciences University of Hokkaido, Kanazawa, Hokkaido, Japan

Department of Integrated Dental Education Division of Dental Education Development School of Dentistry, Health Sciences University of Hokkaido, Kanazawa, Hokkaido, Japan
e-mail: kado@hoku-iryo-u.ac.jp

K. Nagano
Division of Microbiology, Department of Oral Biology, School of Dentistry, Health Sciences University of Hokkaido, Kanazawa, Hokkaido, Japan
e-mail: knagano@hoku-iryo-u.ac.jp

Y. Furuichi
Department of Integrated Dental Education Division of Dental Education Development School of Dentistry, Health Sciences University of Hokkaido, Kanazawa, Hokkaido, Japan
e-mail: furuichi@hoku-iryo-u.ac.jp

T. Nagasawa (✉)
Division of Periodontology and Endodontology, Department of Oral Rehabilitation, School of Dentistry, Health Sciences University of Hokkaido, Kanazawa, Hokkaido, Japan

Division of Advanced Clinical Education, Department of Integrated Dental Education, School of Dentistry, Health Sciences University of Hokkaido, Kanazawa, Hokkaido, Japan
e-mail: nagasawa@hoku-iryo-u.ac.jp

10.2 Mechanisms of Periodontal Tissue Destruction

10.2.1 Features of Oral Immunity

Unlike the stomach and intestines, the oral cavity, located at the entrance of the digestive tract, contains hard tissues such as teeth and alveolar bone that physically crush food and mix it with saliva, facilitating swallowing and digestion. Serving as the gateway for microorganisms, food, and airborne particles, the oral cavity directs these substances into the digestive tract, with the oral mucosa acting as the primary line of defense (Ikeuchi and Moutsopoulos 2022; Ishigame et al. 2009).

Naïve T cells express specific transcription factors in response to a particular combination of cytokines and differentiate into effector T cells, such as T helper 1 (Th1), Th2, Th17, regulatory T cells (Treg), and T follicular helper (Tfh) cells. Differentiated T cells exhibit plasticity and can occasionally undergo cross-differentiation into other T-cell types, depending on the specific combination of cytokines present (Fig. 10.1a). In T-cell-mediated defense, Th1 cells primarily protect against intracellular pathogens, while Th2 cells are involved in defense against extracellular parasites. Th17 cells confer protection against extracellular bacteria and fungi by promoting neutrophil migration to mucosal surfaces and enhancing the expression of epithelial defensins (Curtis and Way 2009).

Periodontal tissues include the gingiva, periodontal ligament, cementum, and alveolar bone. Gingival epithelial cells produce β-defensins, which prevent the outgrowth of potentially harmful bacteria and fungi (Greer et al. 2013). Although β-defensins exhibit potent antibacterial activity against aerobic bacteria, their effectiveness against anaerobic bacteria varies, depending on the specific susceptibility of the bacterial strains (Joly et al. 2004). Neutrophils produce α-defensins and LL-37, both of which exert antibacterial effects against *Porphyromonas gingivalis* and *Aggregatibacter actinomycetemcomitans* (Greer et al. 2013). The junctional epithelium, located at the interface with the tooth, does not express β-defensins (Dale et al. 2001). Under normal conditions, neutrophils migrate through this epithelium into the gingival sulcus, where they engulf bacteria and defend against infection by producing reactive oxygen species, α-defensins, and LL-37 (Greer et al. 2013) (Fig. 10.1b).

A specific distance is maintained between the epithelium and the alveolar bone, with a width of approximately 1 mm for the attachment between the epithelium and the tooth, and another 1 mm for the attachment between the subepithelial connective tissue and cementum. This structure, known as the supracrestal tissue attachment, is preserved in healthy periodontal tissue (Jepsen et al. 2018) (Fig. 10.1b). This 1-mm-wide gingival connective tissue on the alveolar bone is an anatomical (biological) line of defense, preventing direct interactions between oral bacteria and the immune cells. This barrier protects the alveolar bone supporting the teeth from inflammatory reactions that could lead to tissue destruction.

Importantly, fibroblasts in the lamina propria of gingival connective tissue secrete and organize discrete collagen networks, which physically protect the underlying tissue (Smith et al. 2019). These networks connect the cementum overlying the root surface with the gingival lamina propria and the alveolar bone with the lamina propria (Smith et al. 2019). Soluble factors, including heparin and transforming growth factor-β 1 (TGF-β1), stimulate fibroblasts to differentiate into myofibroblasts (Desmouliere et al. 1992, 1993). Myofibroblasts are characterized by the de novo expression of alpha-smooth muscle actin (α-SMA), as well as by enhanced adhesions to the extracellular matrix and an increased capacity to contract and remodel it (Desmouliere et al. 1993).

As a biochemical mechanism by which gingival connective tissue protects against alveolar bone resorption, gingival fibroblasts produce osteoprotegerin (OPG), which acts as a decoy for

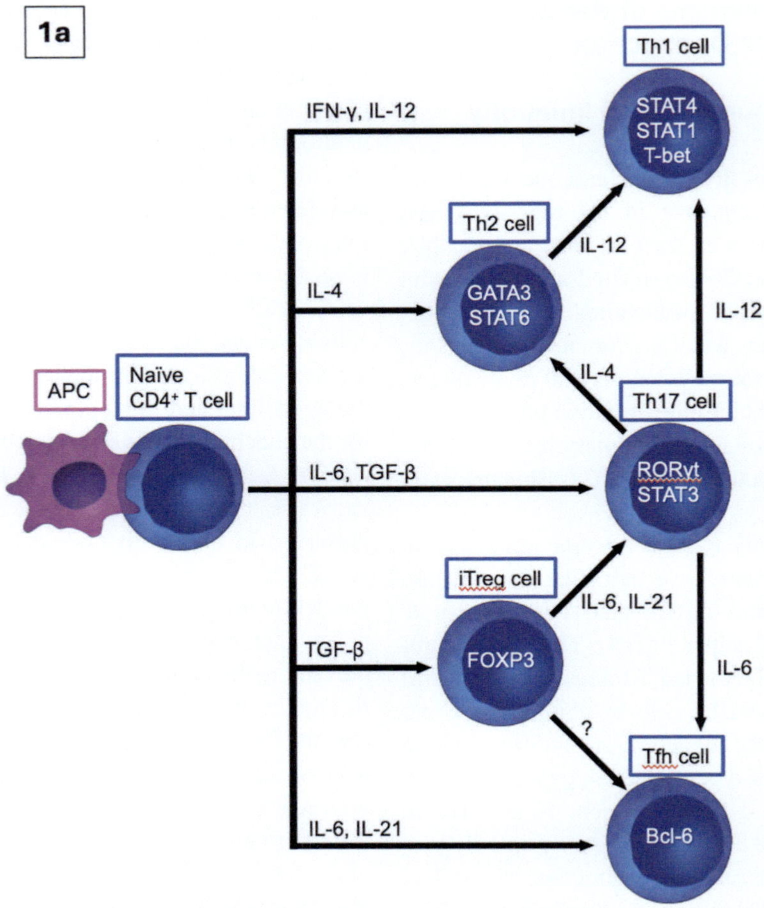

Fig. 10.1 (**a-d**) T-cell plasticity and immune mechanisms in gingival health, periodontitis, and diabetes. (**a**) Differentiated T cells exhibit plasticity and can occasionally undergo cross-differentiation into other T-cell types, depending on the specific combination of cytokines present. (**b**) A schematic representation of healthy gingival tissue, highlighting neutrophil migration into the gingival sulcus for bacterial defense, the supracrestal tissue attachment maintaining epithelial and connective tissue adherence, and key protective factors that preserve periodontal integrity and prevent alveolar bone loss. (**c**) In periodontitis, inflammatory cells infiltrate beyond the supracrestal tissue attachment, releasing cytokines and PGE2, stimulating RANKL expression in osteoblasts and periodontal ligament cells, promoting osteoclastogenesis and leading to alveolar bone resorption. (**d**) Periodontitis and diabetes mellitus. Th17 cells and AGE/RAGE, elevated in diabetes, enhance oral microflora pathogenicity, promoting periodontitis

receptor activator of nuclear factor-κB ligand (RANKL), thereby inhibiting osteoclastogenesis (Nagasawa et al. 2002). This effect is also enhanced by interleukin-1 (IL-1) and prostaglandin E_2 (PGE_2) stimulation (Hormdee et al. 2005; Kiji et al. 2007). Thus, gingival fibroblasts may suppress osteoclast differentiation and bone resorption by producing large amounts of OPG, even in the presence of inflammatory cell infiltration. However, PGE_2 stimulates RANKL production by osteoblasts (Liu et al. 2005) and periodontal ligament cells (Nukaga et al. 2004). Moreover, inflammatory cytokines, including IL-1, IL-6, IL-11, IL-17, and tumor necrosis factor (TNF), increase the expression of RANKL while decreasing that of OPG in osteoblasts (Nakashima et al. 2000). When inflammatory cells migrate beyond the supracrestal tissue

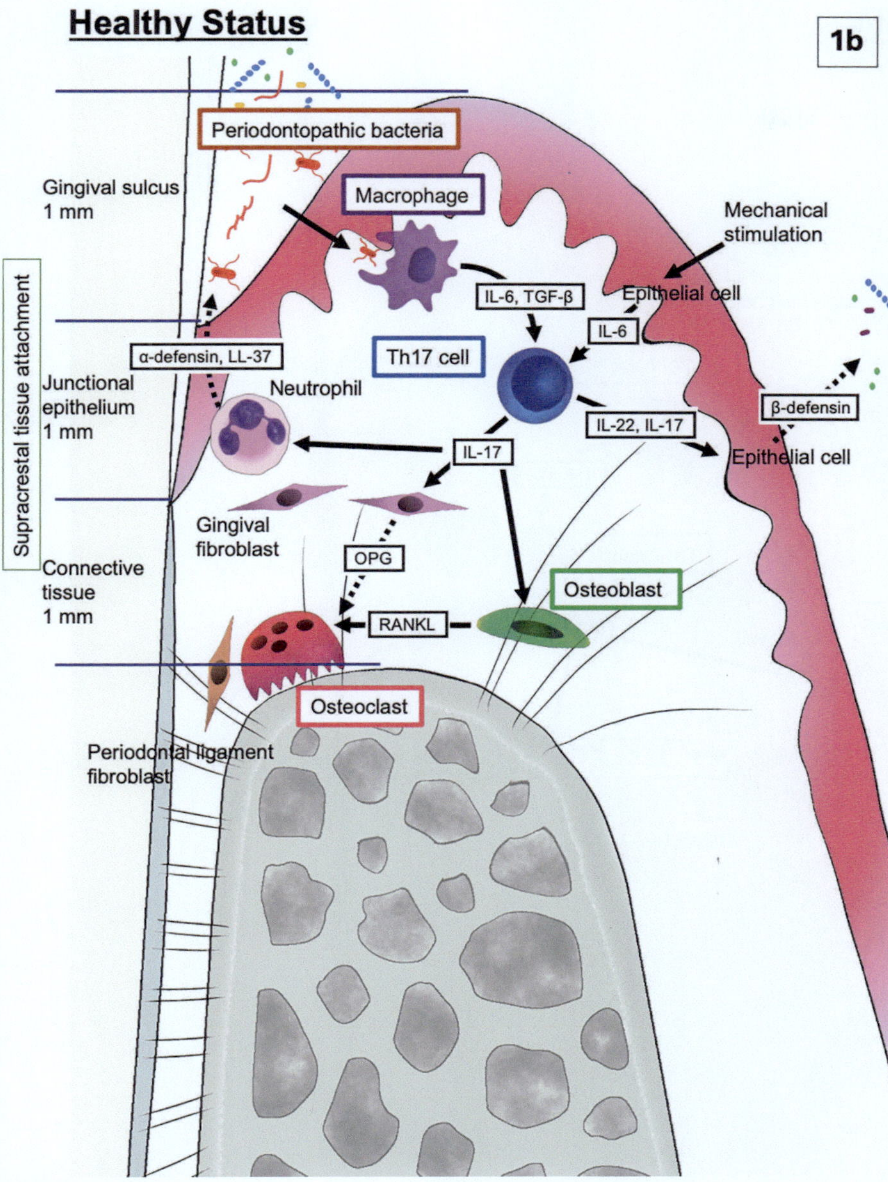

Fig. 10.1 (continued)

attachment and approach the alveolar bone, they release inflammatory cytokines and PGE_2. These molecules act on osteoblasts and periodontal ligament cells, stimulating the expression of RANKL and promoting osteoclastogenesis, which ultimately leads to alveolar bone resorption (Nagasawa et al. 2007) (Fig. 10.1c).

In the gastrointestinal tract, Th17 cells are induced as part of the immune response against bacteria and fungi (Aggor et al. 2020). In the oral cavity, Th17 cells are reportedly activated by the physical stimulus of mastication (Dutzan et al. 2017). This was demonstrated in a study analyzing gingival lymphocytes in sterile mice, which were divided into two groups: one group was fed a normal solid diet with physical stimulation, and the other was fed powdered food without physical stimulation. Gingival epithelial cells from the physically stimulated group produced IL-6, which promoted an increase in the number

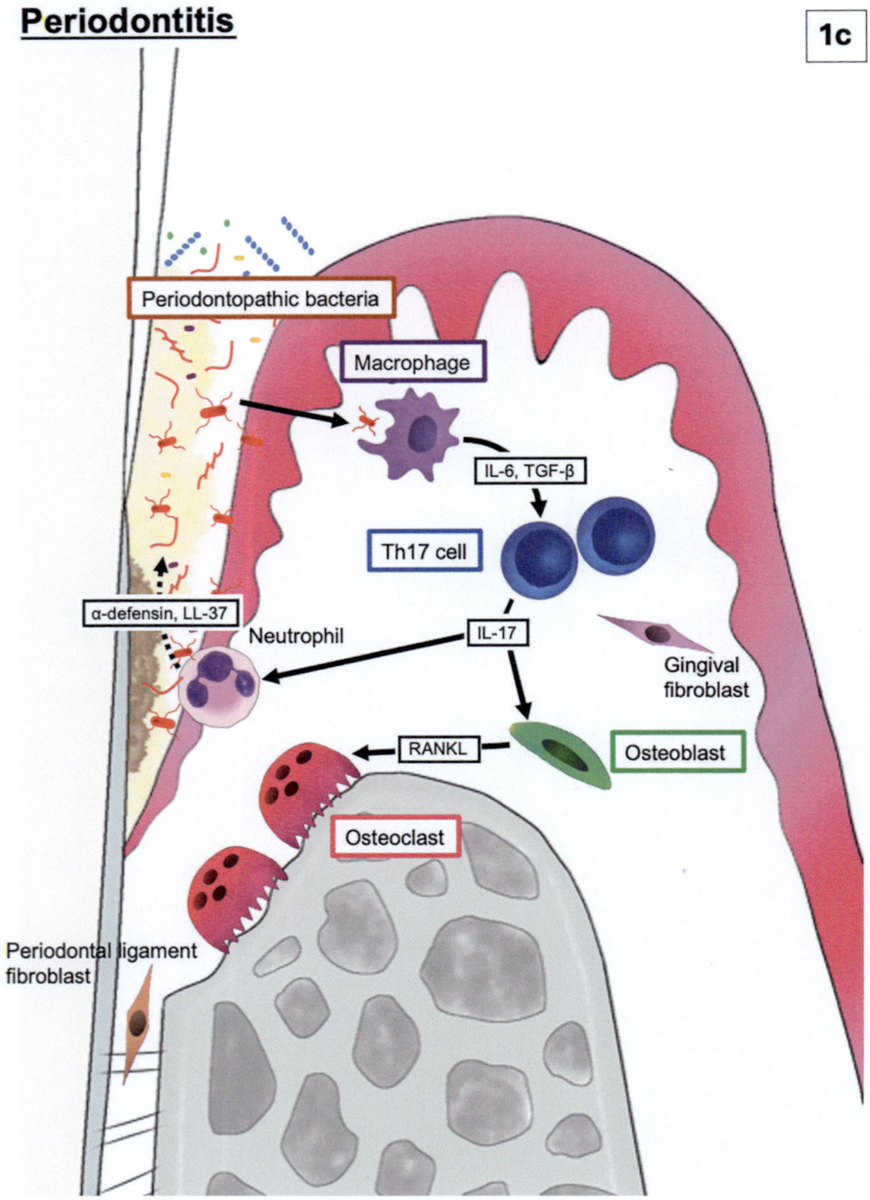

Fig. 10.1 (continued)

of gingival Th17 cells. These findings suggest that both bacterial and physical stimuli associated with mastication contribute to the activation of the gingival Th17 response (Dutzan et al. 2017). Physical stimulation is also necessary to maintain the parotid gland; powder feeding leads to parotid atrophy, which can be restored by switching to a hard diet from powder food (Takahashi et al. 2022). Therefore, Th17 cells are induced in the mucosa in response to physical stimulation by mastication, preventing bacterial invasion by secreting antimicrobial substances from the epithelium. Furthermore, chewing promotes saliva secretion, which aids in digestion and promotes the repair of damaged epithelium through the action of epidermal growth factor.

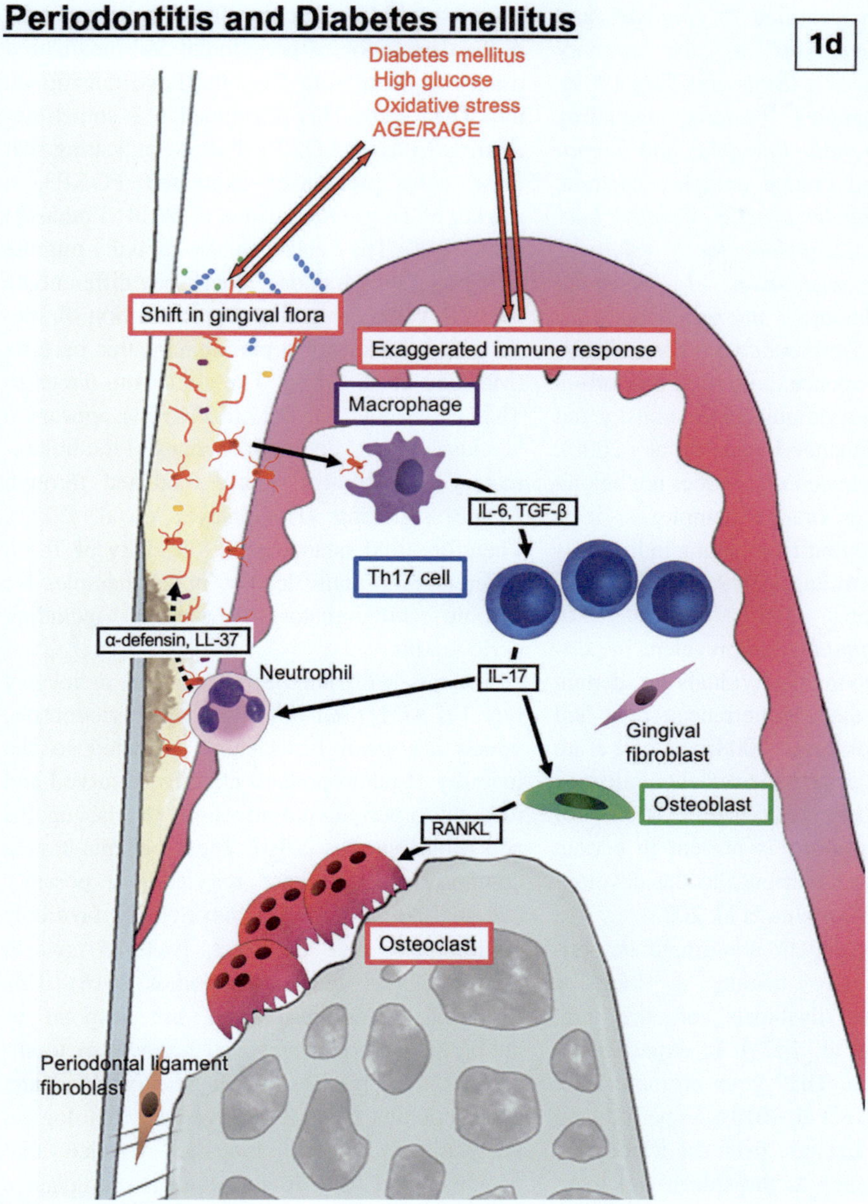

Fig. 10.1 (continued)

10.2.2 Immune Mechanisms of Periodontitis and Diabetes Mellitus

Neutrophils serve as the primary defense in the junctional epithelium, where they interact with periodontopathic bacteria. Genetic abnormalities in neutrophil function can lead to the rapid destruction of periodontal tissue, beginning in childhood. A systematic review of the treatment of periodontitis related to abnormal neutrophil function suggests that conventional therapies have limited effectiveness (Halai et al. 2020).

Gingivitis develops as a result of plaque accumulation, which promotes inflammation in the gingival tissue (Loe et al. 1965). While earlier theories suggested that periodontitis was not caused by specific bacteria (Theilade 1986), subsequent research identified *P. gingivalis* and *A. actinomycetemcomitans* as the primary periodontopathic bacteria (Slots and Ting 1999). Moreover, red complex bacteria, including *P. gingivalis, Tannerella forsythia*, and *Treponema denticola*, and orange complex bacteria, including *Campylobacter gracilis, Campylobacter rectus, Campylobacter showae, Fusobacterium nucleatum, Prevotella intermedia, Parvimonas micra, Prevotella nigrescens*, and *Streptococcus constellatus*, exhibit increased virulence, with their proportion increasing as plaque accumulates (Socransky and Haffajee 2002; Ximenez-Fyvie et al. 2000). While *A. actinomycetemcomitans* does not belong to either the red or orange complexes, it is associated with periodontitis in young individuals (Slots et al. 1990; Armitage 1999). Notably, particularly virulent JP2 strains of *A. actinomycetemcomitans* are prevalent in periodontitis among young individuals in certain regions, including the Mediterranean area and West Africa (Haubek et al. 2008; Haubek et al. 2007). Moreover, a systematic review confirmed that a highly pathogenic JP2 strain of *A. actinomycetemcomitans* is present in certain populations and may contribute to the development of periodontitis (Mehta et al. 2023).

In the absence of genetic abnormalities, periodontal tissue remains healthy as long as neutrophils prevent dysbiosis of the oral microbiome (Kim et al. 2023). In experimental periodontitis in mice, Th17 cells promote bone resorption (Tsukasaki et al. 2018). As periodontal tissue is destroyed, bacteria from the lesions are detected in organs such as the spleen and liver. Notably, once the teeth are completely lost, the bacteria also disappear, highlighting the dual role of Th17 cells in both eliminating bacteria and contributing to tissue damage (Tsukasaki et al. 2018). Under normal conditions, the Th17 response is regulated by Tregs, and the balance between these two cell types is believed to play a critical role in the extent of tissue destruction (Cafferata et al. 2019). However, helper T cells may differentiate into different cell types after initially differentiating into Tregs or Th17 cells (Geginat et al. 2014; Zhou et al. 2009) (Fig. 10.1a). In particular, Th17 cells involved in the destruction of periodontal and rheumatoid tissues originate from Tregs that have transitioned into Th17 cells. This phenomenon is sometimes referred to as ex-FOXP3 T cells, indicating that these cells previously expressed FOXP3, a marker of Tregs (Komatsu et al. 2014; Tsukasaki et al. 2018). Th17 cells can also partially migrate to regional lymph nodes, where they differentiate into Tfh cells, promoting the production of specific antibodies against periodontopathic bacteria (McClure et al. 2024). The shift from Tregs to Th17 cells, and from Th17 to Tfh cells, appears to be closely associated with changes in the inflammatory environment, likely mediated through NF-κB signaling (Hovelmeyer et al. 2022). Therefore, understanding the plasticity of T-cell differentiation could lead to novel therapies for chronic inflammatory diseases, including periodontitis.

Patients with leukocyte adhesion deficiency type 1 (LAD1) often develop severe periodontitis, which has been historically attributed to the inability of neutrophils to effectively surveil and respond to periodontal infections (Hajishengallis and Moutsopoulos 2016). These patients have a compositionally unique microbiome, possibly due to LAD fostering a nutritionally favorable environment that promotes bacterial growth (Hajishengallis and Moutsopoulos 2016). This form of periodontitis does not respond to antibiotics and/or mechanical removal of tooth-associated biofilm (Hajishengallis and Moutsopoulos 2016). Moreover, microbiological analysis of gingival biopsies revealed that patients with LAD1 do not have a tissue-invasive infection, but lipopolysaccharides can translocate into the lesions (Hajishengallis and Moutsopoulos 2016). Animal models of LAD1 have demonstrated that the fundamental cause of periodontitis in LAD1 involves IL-23 and IL-17, which drive inflammatory bone loss. IL-23 is a maturation factor for Th17 cells, and neutralizing antibodies against IL-23 have been used to suppress Th17 cells (Tang et al. 2012). Importantly, the administration of a monoclonal antibody targeting the p40 subunit of IL-23 inhibits IL-17 activity and reduces gingival inflammation in patients with LAD1 and severe periodontitis (Moutsopoulos et al. 2017). These findings further highlight the role of Th17 cells and interleukins in periodontitis, with the suppression of the Th17 cell response more directly related to symptom resolution than the microbial suppression.

Abnormal expression of LL-37 in neutrophils also cause severe periodontitis in individuals with Kostmann syndrome and Papillon-Lefevre syndrome (de Haar et al. 2006). This form of periodontitis is often caused by *A. actinomycetemcomitans* (de Haar et al. 2006), and eradication of *A. actinomycetemcomitans* can resolve severe periodontitis in Papillon-Lefevre syndrome (Boutsi et al. 1997). This suggests that periodontitis is not solely explained by an inflammatory response to accumulated plaque in individuals with poor oral hygiene, but that some molecular abnormalities may cause the overgrowth of pathogenic bacteria.

Notably, diabetes is known to increase the risk of periodontitis (Huang et al. 2020). For example, increased abundances of the families *Porphyromonas, Prevotella*, and *Fusobacterium* have been reported in patients with diabetes, even in the absence of periodontitis, compared to healthy participants (Ganesan et al. 2017; Long et al. 2017). Moreover, the families Fusobacteria and Actinobacteria are more abundant in patients with both type 2 diabetes and periodontitis compared to those with only periodontitis (Matsha et al. 2020). However, it remains unclear whether changes in the oral microbiome in patients with diabetes and periodontitis are driven by fluctuations in blood glucose or influenced by the presence of periodontitis, necessitating further research (Silva et al. 2022).

In a study where oral bacteria from diabetic mice were transplanted into germ-free mice, the recipient mice developed periodontitis. However, when these oral bacteria were transplanted into mice treated with anti-IL-17 antibodies, only mild bone resorption occurred after transplantation (Xiao et al. 2017). This study indicates that Th17 cells, which are elevated in diabetes, play a key role in enhancing the pathogenicity of oral microflora (Fig. 10.1d). Moreover, in the hyperglycemic state associated with diabetes, non-enzymatic glycation and oxidation of proteins and lipids occur, leading to the accumulation of advanced glycation end products (AGEs) in the plasma and tissues of patients with diabetes (Lalla et al. 2001). Increased levels of AGEs are observed in patients with both diabetes and periodontitis compared to periodontitis patients without diabetes (Thomas et al. 2024). Furthermore, the levels of AGEs and the receptors for AGEs (RAGE) are associated with the severity of periodontitis (Thomas et al. 2024). These results suggest that an enhanced AGE-RAGE response due to diabetes promotes periodontal tissue destruction. RAGE belong to the immunoglobulin superfamily and are expressed in various cells, including endothelial cells and macrophages (Lalla et al. 2001). The binding of AGEs on RAGE activates the NFκB pathway, triggering an inflammatory response. In macrophage lineages, AGEs induce the M1 phenotype (inflammatory macrophages) while inhibiting the polarization of the M2 phenotype (anti-inflammatory). Furthermore, M1 macrophages promote Th17 cell differentiation (Cavalla and Hernandez 2022). Collectively, these results suggest that the accumulation of AGEs in periodontal tissue induces Th17 cells in diabetes.

High insulin resistance due to diabetes impairs insulin-induced activation of the PI3K/Akt pathway, leading to impaired wound healing (Kominato et al. 2022). Furthermore, impairment of the PI3K/Akt pathway reduces the insulin-induced expression of CXCL1 in gingival fibroblasts (Shinjo et al. 2023). This suggests that diabetes increases the risk of periodontitis by preventing neutrophil migration, which is essential for immune defense.

Periodontitis is exacerbated by an immune response to bacteria, primarily Th17 cells. Diabetes worsens the inflammatory response through the accumulation of AGEs and other factors. The immune response to fungi differs from that to bacteria and should be considered separately.

10.2.3 Oral Immunity to Candida and Diabetes

The IL-17/Th17 response plays a vital role in maintaining the epithelial barrier by regulating tight junction proteins, antimicrobial peptides, and neutrophil migration (Abusleme and Moutsopoulos 2017). This response is essential for preventing infections caused by *Candida albicans*, an indigenous oral fungus (Gaffen and Moutsopoulos 2020) (Fig. 10.2a).

The response to oropharyngeal candidiasis (OPC), an infection caused by *C. albicans*, is regulated by IL-17 and IL-22, both produced by Th17 cells. IL-22 acts on the basal epithelial layer to promote epithelial cell proliferation, while IL-17 targets the superficial epithelial layer to combat infection (Aggor et al. 2020) (Fig. 10.2b).

Autosomal-dominant hyper IgE syndrome (AD-HIES) is characterized by a defect in the Th17/IL-17 response due to STAT3 deficiency. In a study comparing fungi and bacteria of the oral mucosa of individuals with AD-HIES and healthy volunteers, *C. albicans* was predominant in active candidiasis in AD-HIES. Moreover, *Streptococcus oralis* and *Streptococcus mutans* were also increased in AD-HIES, indicating an interaction between C. albicans and oral streptococci. Individuals with AD-HIES are highly susceptible to dental caries, and the increase in *S. mutans* may contribute to this susceptibility. These results suggest that the STAT3/Th17 response suppresses *C. albicans* as an oral commensal and contributes to the formation of normal fungal and bacterial communities (Abusleme et al. 2018).

Autoimmune polyendocrinopathy-candidiasis-ectodermal syndrome (APECED) is an autoimmune disease characterized by susceptibility to mucosal fungal infections due to a deficiency in the *AIRE* gene (Nagamine et al. 1997). In APECED, excessive interferon-γ (IFN-γ) disrupts the epithelial barrier function via signal transducer and activator of transcription 1 (STAT1). As a result, the inhibition of IFN-γ and Janus kinase (JAK)-STAT signaling ameliorates mucosal mycosis in patients with APECED (Oikonomou et al. 2024). These findings suggest that excessive Th1-mediated mucosal inflammation promotes susceptibility to mucosal fungal infection (Break et al. 2021).

Overall, these previous studies on rare immune disorders, including AD-HIES and APECED, have highlighted the critical role of Th17 responses in combating fungal infections. In contrast, individuals with type 2 diabetes also face an elevated risk of *C. albicans* infections, albeit for distinct reasons (Mohammed et al. 2021). Diabetes disrupts normal metabolic processes and impairs neutrophil function, which are key immune cells essential for defense. Specifically, patients with diabetes are vulnerable to *Candida* infections owing to reduced saliva flow, increased salivary glucose, and impaired neutrophil function, all of which contribute to the increased prevalence of oral *Candida* infections and mucosal lesions (Jhugroo et al. 2019) (Fig. 10.2c). Therefore, while the mechanisms differ, both conditions lead to increased vulnerability to fungal infections.

10.3 Impact of Periodontal Disease on Diabetes

10.3.1 Insulin Resistance

One possible mechanism by which periodontal disease worsens glycemic control is through the induction of insulin resistance. The US National Health and Nutrition Examination Survey (NHANES) reported a significant association between deeper periodontal pockets and increased insulin resistance. However, no such relationship was observed with attachment loss, suggesting that the periodontal infection and inflammation present at the time of measurement may play a more critical role in insulin resistance than the degree of periodontal tissue destruction (Demmer et al. 2012). In the Insulin Resistance Atherosclerosis Study, which followed 906 patients with prediabetes for an average of 5.2 years, individuals with high insulin resistance had higher C-reactive protein (CRP) levels (2.88 mg/L compared with 1.68 mg/L) than those without insulin resistance. Notably, the

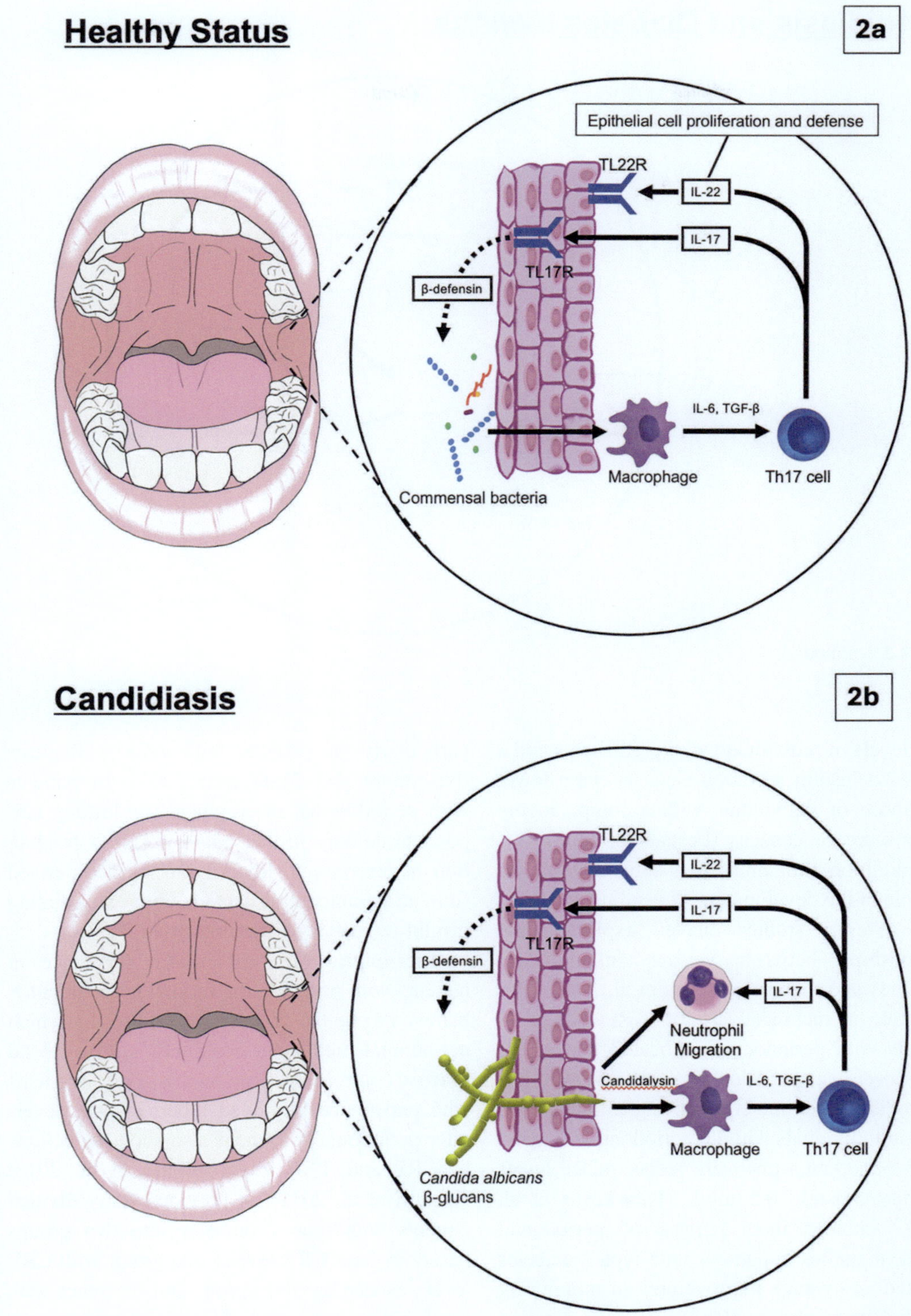

Fig. 10.2 Oral immunity to Candida in health, Candidiasis, and Candidiasis with diabetes mellitus. In a healthy state, IL-17/Th17 responses maintain the epithelial barrier, preventing Candida albicans colonization (**a**), whereas in oropharyngeal candidiasis, IL-17 and IL-22 regulate immune responses by targeting different epithelial layers to control infection (**b**), but in diabetes, metabolic dysregulation and impaired neutrophil function exacerbate susceptibility to Candida infections, leading to increased fungal overgrowth and mucosal lesions (**c**)

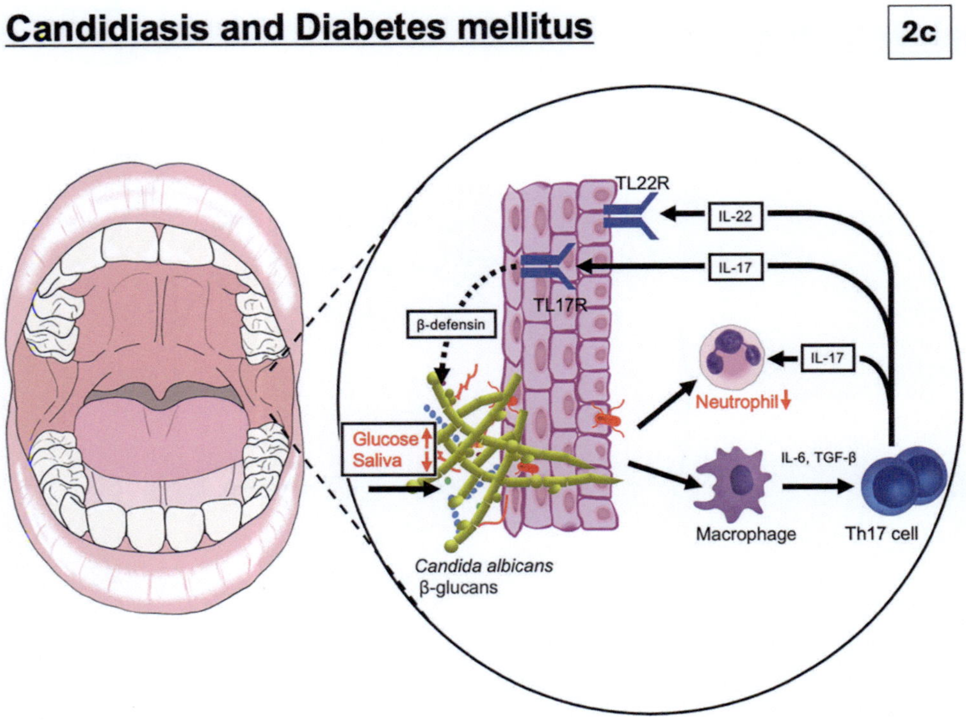

Fig. 10.2 (continued)

CRP levels of individuals with prediabetes and a decreased insulin secretion capacity were lower than those of individuals with a comparatively higher secretion capacity (Festa et al. 2003).

In addition, inflammation is believed to play a key role in the development of insulin resistance, and several studies have explored the relationships between various inflammatory cytokines and diabetes (C. Liu et al. 2016). For example, a meta-analysis of CRP levels in patients with periodontitis indicated that serum CRP levels in periodontitis patients averaged 2.1 mg/L, approximately 1.56 mg/L higher than those in individuals without periodontitis. Importantly, periodontal treatment reduced CRP levels by approximately 0.5 mg/L (Paraskevas et al. 2008). Another meta-analysis of periodontal treatment studies in patients with type 2 diabetes reported an average improvement in monitoring glycated hemoglobin (HbA1c) levels ranging from 0.4 to 0.5% (Simpson et al. 2022). Moreover, TNF-α produced by adipose tissue is another key driver of insulin resistance, particularly in patients with type 2 diabetes (Nishimura and Murayama 2001). In patients with periodontitis, macrophages infiltrating adipose tissue may further amplify TNF-α production in response to lipopolysaccharides derived from periodontopathic bacteria, thus exacerbating insulin resistance (Yamashita et al. 2007).

The improvement of insulin resistance in patients with periodontitis following treatment is believed to be the primary mechanism by which periodontal treatment positively affects blood glucose control in patients with diabetes. Patients who experience improved blood glucose levels after periodontal treatment also show reductions in CRP and TNF levels (Bharti et al. 2013; Katagiri et al. 2012). A previous study divided patients with type 2 diabetes into two groups based on their CRP levels: one group with CRP levels exceeding 500 ng/mL and the other with levels below 500 ng/mL. Notably, improvements in HbA1c levels were observed only in the high CRP group (Munenaga et al. 2013). Conversely, another study involving patients with type

2 diabetes with significant obesity indicated no improvement in glycemic control after periodontal treatment (Engebretson et al. 2013). This suggests that, in individuals with severe obesity, the increase in insulin resistance due to obesity outweighs the impact of periodontal disease. Therefore, the relative contribution of periodontal disease to overall insulin resistance is likely less significant in these patients (Fig. 10.3a).

10.3.2 Impact of Periodontopathic Bacteria on Gut Bacteria in Diabetes

Mucosal tissues, such as the oral cavity and intestine, are exposed to various exogenous antigens, including bacteria and food, and the common mucosal immune system plays a key role in protecting the body against them. Selective induction of secretory IgA antibodies to mucosal surfaces is one of the characteristics of the common mucosal immune system (Mestecky 1987). Salivary IgA protects mucosal surfaces, controls the microbiome in mucosal tissue (Brandtzaeg 2013a, b), and suppresses periodontopathic bacteria (Costalonga and Herzberg 2014). Patients with selective IgA deficiency have increased risks for type 1 diabetes and gastrointestinal infection (Odineal and Gershwin 2020). Oral tolerance, another major feature of the mucosal immune system, is a response that suppresses the immune response to orally administered antigens. Notably, oral tolerance is not induced in germ-free mice, and IgE antibodies are known to be induced against food antigens (Tanaka and Ishikawa 2004). Moreover, oral tolerance is associated with the gut microbiome and is believed to play an important role in preventing food allergy (Yang et al. 2021). The microbiome of the gastrointestinal tract and the functioning of the organism are closely related. Accordingly, microbiomes are increasingly being regarded as an essential component

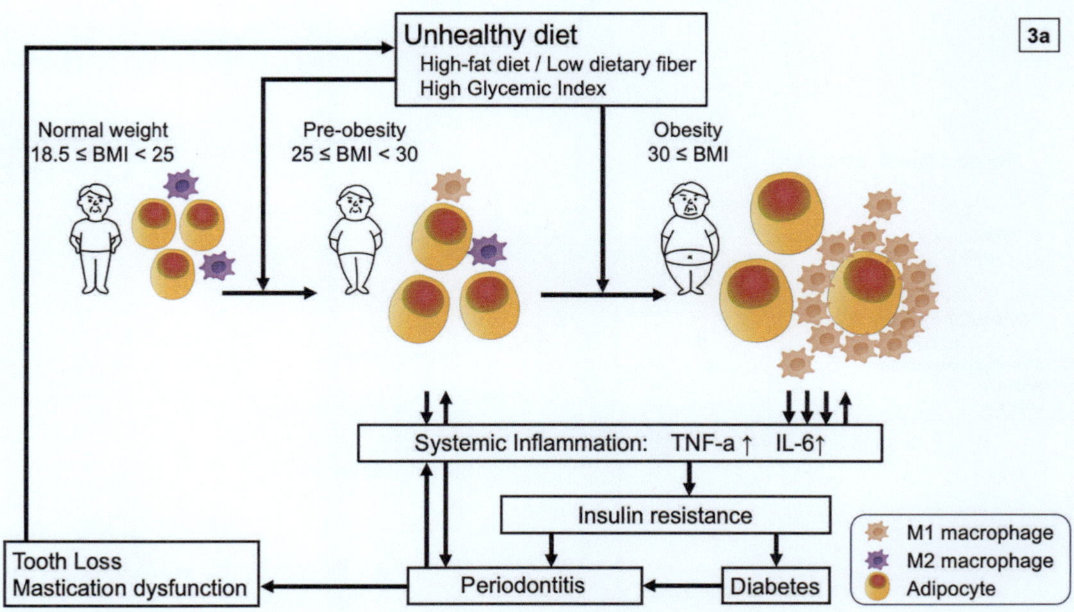

Fig. 10.3 Systemic effects of periodontitis on metabolic and immune dysregulation. (**a**) Periodontal treatment improves insulin resistance and glycemic control by reducing systemic inflammation, though its impact is less pronounced in individuals with severe obesity. (**b**) Periodontitis exacerbates gut inflammation by promoting the translocation of periodontopathic bacteria to the gut and inducing Th17 cell migration, contributing to colitis. (**c**) Periodontitis may contribute to vascular complications in diabetes by triggering antiphospholipid antibodies through molecular mimicry with periodontopathic bacteria

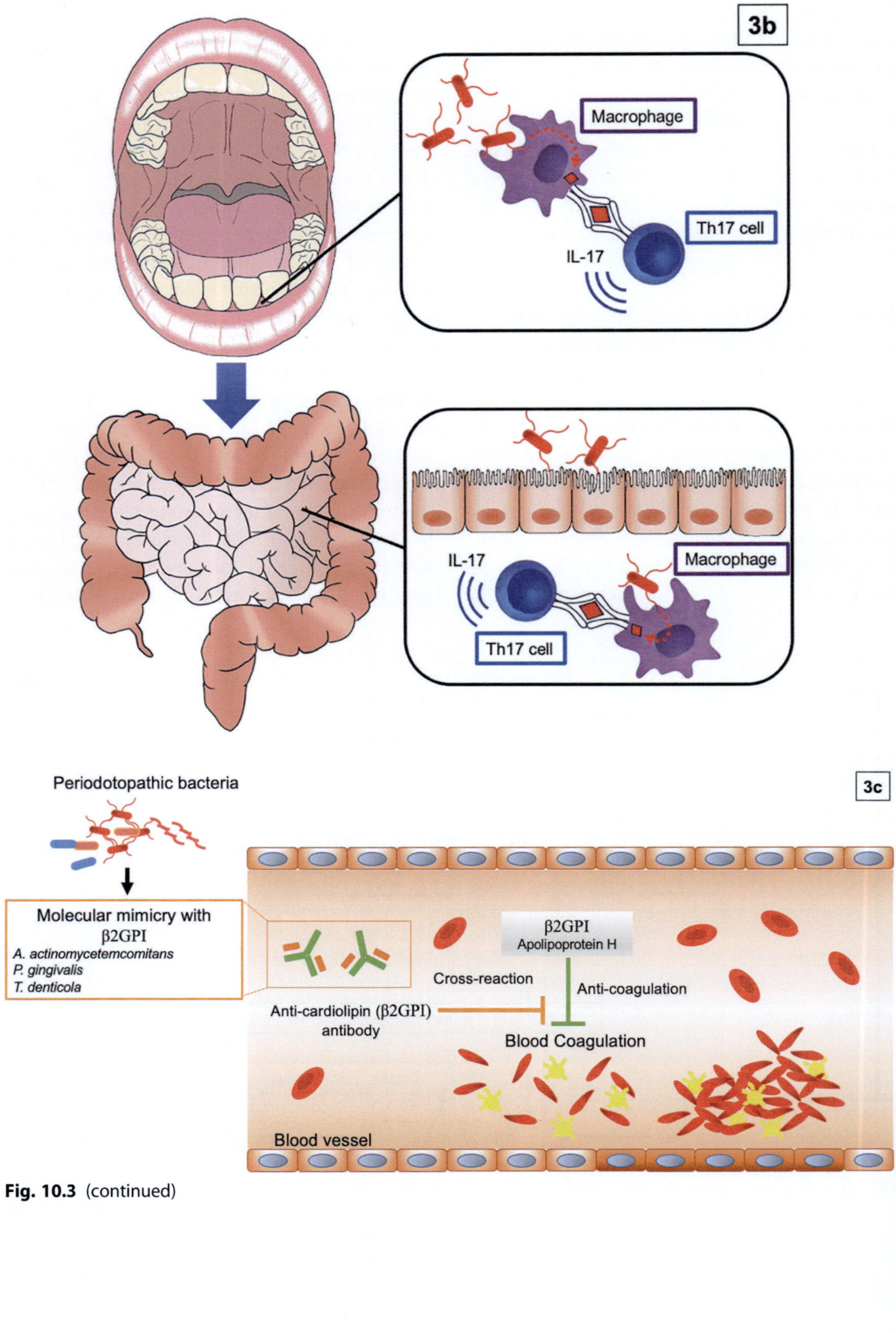

Fig. 10.3 (continued)

of the body. Changes in the microbiome are now recognized as crucial for understanding various diseases (Ma et al. 2024).

A study comparing the gut microbiota of individuals with diabetes or prediabetes demonstrated decreased abundances of *Faecalibacterium prausnitzii*, *Roseburia*, *Dialister*, *Flavonifractor*, *Alistipes*, *Haemophilus*, and *Akkermansia muciniphila*, accompanied by increased abundances of *Lactobacillus*, *Streptococcus*, *Escherichia*, *Veillonella*, and *Collinsella* in patients with diabetes (Letchumanan et al. 2022). Notably, *Lactobacillus* is positively correlated with fasting blood glucose levels, HbA1c levels, and the homeostasis model assessment of insulin resistance (HOMA-IR) score (Letchumanan et al. 2022). These findings suggest that gut bacteria may influence the pathogenesis of diabetes and that diabetes treatment can, in turn, affect the composition of gut bacteria (Zhang et al. 2024).

Similarly, in the microbiome of the oral cavity, periodontitis promotes the growth of pathogens such as *Klebsiella* spp. and *Enterobacter* spp. These pathogens are ingested and migrate to the gut, where they activate inflammasomes in colonic mononuclear phagocytes, causing inflammation. In parallel, periodontitis generates Th17 cells specific for periodontopathic bacteria. These Th17 cells migrate to the gut, where they contribute to intestinal inflammation. Th17 cells originating in the oral cavity are activated by periodontopathic bacteria that have colonized the gut, contributing to the development of colitis. However, these Th17 cells are not activated by microorganisms native to the gut. Therefore, oral inflammation, such as periodontitis, is believed to exacerbate gut inflammation by supplying both periodontopathic bacteria and Th17 cells to the gut (Kitamoto et al. 2020) (Fig. 10.3b).

10.3.3 Impact of Periodontitis on Vascular Complications in Diabetes

A systematic review demonstrated that periodontal treatment improves endothelial function and reduces the expression of biomarkers of atherosclerotic disease, especially in individuals with cardiovascular disease (CVD) and/or diabetes (Teeuw et al. 2014). A cross-sectional study indicated a correlation between periodontal disease and CVD, although the precise mechanism by which periodontitis contributes to CVD remains unclear (Lockhart et al. 2012). One proposed mechanism is that atherosclerosis caused by bacterial infection may result from molecular mimicry, in which the immune response to periodontopathic bacteria mistakenly attacks the body's cells due to homology between bacterial and host molecules, particularly heat shock proteins (Lockhart et al. 2012). These proteins have evolved from bacteria to vertebrates while maintaining molecular similarity, making them a key focus of research into autoimmune diseases triggered by molecular mimicry. Antiphospholipid antibodies, which have been linked to periodontitis, are also of interest in this context. Antiphospholipid antibody syndrome, which causes elevated anti-cardiolipin antibodies, can lead to blood clots and recurrent miscarriages (Garcia and Erkan 2018). In addition to its associations with diabetes and CVD, periodontitis is associated with systemic effects, such as preterm birth and low fetal birth weight (Beck et al. 2019). Some of the systemic effects of periodontitis may be explained by the presence of antiphospholipid antibodies, as they target β2-glycoprotein I (β2GPI) (Garcia and Erkan 2018). While the exact cause of antiphospholipid antibodies is not well understood, infection has been suggested as a potential trigger. Periodontopathic bacteria, including *A. actinomycetemcomitans, P. gingivalis*, and *T. denticola*, exhibit molecular mimicry with β2GPI, and infection with these bacteria might induce anti-β2GPI antibodies (Chen et al. 2009; Wang et al. 2008). Elevated anti-cardiolipin antibodies have been reported in patients with periodontitis (Schenkein et al. 2003), and periodontal treatment can reduce these antibody levels (Gunupati et al. 2011; Kiany and Hedayati 2015). Moreover, elevated anti-cardiolipin antibodies associated with periodontitis have also been observed in systemic lupus erythematosus (Wang et al. 2015). Notably, elevated antiphospholipid antibodies have also been detected in type 2 diabetes, suggesting a potential role in vascular complications in diabetes (Ferrannini et al. 2020). These results suggest that periodontitis, through the inflammatory response it triggers, might contribute to the elevation of antiphospholipid antibodies, which could, in turn, worsen the vascular complications often observed in diabetes.

Heparin is administered to prevent blood clots; however, a rare complication, heparin-induced thrombocytopenia (HIT) can occur following its administration (K. Nishimura et al. 2014). HIT is caused by antibodies against the complex of platelet factor 4 (PF4) and heparin, leading to thrombocytopenia and thrombosis (Greinacher et al. 2011). Notably, periodontitis is associated with an increased risk of HIT (Nishimura et al. 2014). Although the exact mechanism underlying this association is unclear, periodontal infection may trigger the formation of autoantibodies against the PF4-heparin complex (Greinacher et al. 2011).

10.4 Diagnosis and Treatment of Periodontal Disease

10.4.1 Diagnosis of Periodontal Disease in Patients with Diabetes Mellitus

The primary goal of diabetes treatment is to reduce complications such as microvascular disease, macrovascular disease, and periodontal disease. This strategy aims at enabling patients to achieve a quality of life and life expectancy comparable to those of individuals without diabetes. As described in previous sections, diabetes and periodontal disease are interrelated and present with diverse clinical manifestations, requiring comprehensive diagnostic approaches.

To properly diagnose periodontal disease in patients with diabetes, it is crucial to evaluate both systemic and local factors. Clinical evaluations should include periodontal probing depths, assessment of bleeding on probing (BOP), and radiographic evaluation of the degree of bone loss. Moreover, blood biomarker analysis provides valuable insight into the bidirectional relationship between periodontal inflammation and diabetes control. Specifically, monitoring HbA1c and inflammatory markers, such as IL-6 and CRP, helps identify how periodontal health status influences glycemic control and vice versa. The measurement of glycated albumin and 1,5-anhydroglucitol (1,5-AG) is useful to monitor blood glucose changes over 1 to 2 weeks, and the measurement of salivary glycated albumin (Aihara et al. 2023) or 1,5-AG (Xu et al. 2024) is useful for diabetes screening. Additionally, measuring HOMA-IR together with BMI might help in understanding insulin resistance and obesity, which may suggest the efficacy of periodontal treatments on HbA1c.

For proper immunological diagnosis and treatment, it is necessary to understand both the illness and disease of the patient. While "disease" refers to a defined pathological state (Emson 1987), "illness" is a subjectively experienced state that may not always align with biological symptoms (Emson 1987). In some cases, a patient's illness-related concerns may extend beyond the typical symptoms of recognized diseases, including diabetes and periodontal disease. Therefore, to fully understand a patient's illness, it is advisable to assess the patient's background according to the biopsychosocial model to understand their illness (Bahall and Bailey 2022). Subsequently, it is important to evaluate the potential for improvement through treatment and consider how to address issues that are not expected to improve (Table 10.1).

An increasing body of research highlights the significant impact of diabetes on oral health. A study assessing the glycemic control and oral symptoms of patients with prediabetes found that untreated patients had significantly higher rates of xerostomia and burning mouth sensations than those of both healthy individuals and patients with prediabetes but with good glycemic control (Javed et al. 2014). Another study using the Geriatric Oral Health Assessment Index (GOHAI) to measure dental-related quality of life indicated that patients with diabetes reported issues such as xerostomia, mucosal disease, candidiasis, dental caries, and periodontal disease

Table 10.1 Differentiating disease and illness in patients with diabetes and periodontitis using the biopsychosocial model

Diagnosis of the diseases	Diabetes with periodontitis	
Cause of illness	Illness caused by periodontitis and/or diabetes	Illness not caused by periodontitis and/or diabetes
	Biopsychosocial model and GOHAI might help the understanding patient's illness Great care must be taken to ensure that no other disease is missed	
Interaction	Along with improvement in objective parameters of periodontal disease and diabetes, amelioration in illness is expected	The illness is not expected to improve by the periodontal and/or diabetes treatment For social and psychological problems, common risk approach and cognitive behavioral therapy are effective

Understanding both disease (pathological state) and illness (subjective experience) is essential for proper immunological diagnosis and treatment. While illness may be caused by diabetes and/or periodontitis, other underlying conditions should be carefully evaluated. The biopsychosocial model and GOHAI framework aid in assessing the patient's illness perception. Treatment is expected to improve illness when linked to diabetes and periodontitis, but when the illness is unrelated, social and psychological interventions, such as cognitive behavioral therapy and a common risk approach, may be more effective

(Azogui-Levy et al. 2018). This study suggests that GOHAI might be a useful tool to screen illness in patients with both periodontitis and diabetes.

In addition to oral symptoms, diabetes also imposes psychosocial burdens. Cross-sectional studies revealed that individuals with diabetes are more prone to psychological issues, anxiety disorders, and depression associated with managing their chronic condition, often referred to as "diabetes distress," and that these challenges are frequently overlooked by healthcare professionals (Fisher et al. 2008; Peyrot et al. 2005). Cognitive-behavioral therapy (CBT) mitigates the psychological burden of diabetes, alleviates anxiety and depression, and lowers HbA1c levels (Li et al. 2022). CBT is also a valuable treatment option for oral symptoms that have a psychological component, such as dental anxiety and burning mouth syndrome (Matsuoka et al. 2017).

Dental clinics also offer unique opportunities for diabetes screening in patients with periodontitis. Undiagnosed diabetes can significantly impact an individual's health. A study measuring random blood glucose levels in 717 patients at a Japanese dental clinic found that approximately 10% of patients had hyperglycemia ($\geq$200 mg/dL), of whom one-third had no history of diabetes (Harase et al. 2015). The prevalence of diabetes in this study was 23.6%, considerably higher than the national prevalence of diabetes in Japan, which is 7.6%. This study suggests that individuals with dental disease may have an increased likelihood of undiagnosed diabetes, making dental clinics suitable environments for diabetes screening (Harase et al. 2015). While systematic blood tests are not yet standard practice in dentistry, screening through appropriate questionnaires, checklists, and referrals to internal medicine may be beneficial (American Diabetes 2019). In addition, simple diagnostic tools that use blood samples from the fingertips or gums are effective in detecting diabetes (Terada et al. 2024). It is hoped that such devices will be widely adopted in dental clinics in the future, allowing for accurate identification of undiagnosed patients with diabetes.

In summary, periodontal disease and diabetes are interrelated, and it is essential to know the status of diabetes for a correct diagnosis. Along with HbA1c and blood glucose levels, diagnostic parameters, such as glycated albumin and 1,5-AG, are being developed using saliva samples. Moreover, an accurate diagnosis requires a comprehensive understanding of the patient's illness or disease, including psychosocial aspects. This paper did not address risk factors common to both periodontal disease and diabetes, such as smoking and sugar intake, but from the perspective of NCDs, attention should be paid to these risk factors as well. Notably, the common risk factor approach addresses risk factors common to many chronic conditions within the context of the wider socio-environmental milieu. As this strategy is essential for health promotion and addressing the underlying social determinants of disease (Watt 2005), it should be applied to the management of periodontitis in patients with diabetes.

10.4.2 Control of the Progression of Periodontal Disease in Patients with Diabetes

Optimal control of diabetes is essential for managing the progression of periodontal disease, and regular visits to a family doctor are highly recommended. Patients without a family doctor should be encouraged to establish a relationship with one. Impaired healing of periodontal tissues due to poor glycemic control can improve with effective diabetes treatment (Kominato et al. 2022). Therefore, restoring periodontal health requires collaboration with internal medicine to optimize blood glucose control (Mizutani et al. 2024). Patients with prediabetes or diabetes should be regularly screened for periodontal disease and educated about the increased risk of periodontitis, with attention given to signs of inflammation (e.g., BOP), blood sugar control, and the impact on complications (Herrera et al. 2023).

The treatment of periodontal disease is based on its severity, following the 2018 classification guidelines (Caton et al. 2018) (Table 10.2). If gingivitis is diagnosed, it should be treated as a primary prevention strategy for periodontitis, while in patients with periodontitis, treatment should follow established guidelines (Herrera et al. 2022; Sanz et al. 2020a) and the patient's family physician should be informed of the diagnosis. In addition, patients with diabetes should be educated about their increased risk of peri-implantitis. Ideally, periodontal treatment for patients with diabetes or prediabetes (non-diabetic hyperglycaemia, NDH) should aim for periodontal pockets no deeper than 4 mm with no bleeding on probing or deep pockets (>6 mm) (Sanz et al. 2020a). To prevent bone resorption due to inflammatory cell infiltration, it is necessary to maintain epithelial and connective tissue attachments between the base of the gingival sulcus and the alveolar bone, each approximately 1 mm wide (Fig. 10.1b). This is particularly important when fixed restorations are involved. Maintaining root parallelism and preventing vertical bone loss are crucial for ensuring this attachment in all teeth. Periodontal regenerative therapy has proven to be effective in patients whose diabetes is well-controlled (Mikami et al. 2022). When necessary, periodontal surgery should be performed under appropriate medical supervision. Patients with stable periodontal tissues should receive maintenance care at intervals tailored to their individual needs.

In summary, the maintenance of periodontal care requires customized support that considers each patient's unique risk factors and specific needs. Ongoing professional monitoring and maintenance therapy are crucial for sustaining the stability of periodontal tissues and promoting long-term oral health and overall systemic well-being.

10.4.3 Prevention of the Progression of Microvascular and Macrovascular Complications in Patients with Periodontitis and Type 2 Diabetes

The effective prevention and management of microvascular and macrovascular complications in periodontitis patients with type 2 diabetes require a comprehensive approach that integrates glycemic control, periodontal health, and systemic risk factors.

Glycemic control plays a vital role in reducing the risk of microvascular complications such as

Table 10.2 Integrated diagnosis and management of periodontitis and diabetes

Diabetes diagnosis		Pre-diabetes (NDH)/diabetes		
Periodontitis diagnosis (2018 classification)	Undiagnosed pre-diabetes/NDH and diabetes	Gingival health	Gingivitis	Periodontitis
Periodontal findings and clinical definition	Multiple periodontal abscesses in untreated patients, and/or a poor response to optimal periodontal treatment	Gingival health is defined as $<10\%$ bleeding sites with probing depths ≤ 3 mm	Gingivitis is defined as $\geq 10\%$ bleeding sites with probing depths ≤ 3 mm	Interdental clinical attachment level is detectable at ≥ 2 non-adjacent teeth, or buccal/oral clinical attachment level of ≥ 3 mm with pocketing of >3 mm is detectable at ≥ 2 teeth and the observed CAL cannot be ascribed to non-periodontal causes such as: 1. Gingival recession of traumatic origin 2. Dental caries extending in the cervical area of the tooth 3. The presence of CAL on the distal aspect of a second molar and associated with malposition or extraction of a third molar 4. An endodontic lesion draining through the marginal periodontium 5. The occurrence of a vertical root fracture
Interaction	Undertaking risk assessment using a questionnaire validated for that population and onward referral to a family doctor for those with elevated risk (one-step protocol)	Primary prevention measures for periodontitis should be undertaken (Chapple et al. 2015)	It should be treated as a primary prevention strategy for periodontitis (Chapple et al. 2015). The need for optimal control of diabetes should also be emphasized, and regular visits to the family physician should be encouraged	It should treat the patient according to the recent EFP treatment guidelines (Herrera et al. 2022; Sanz et al. 2020a) and inform the family doctor of the diagnosis of periodontitis. If the patient does not have a family doctor, they should be advised to register with one

This table classifies patients based on diabetes status and periodontal health, outlining clinical definitions, diagnostic criteria, risk assessment protocols, and interdisciplinary management strategies to optimize both systemic and oral health retinopathy, nephropathy, and neuropathy in patients with diabetes. Specifically, for every 1% reduction in HbA1c levels, microvascular damage decreases by approximately 35% (Stratton et al. 2000). The UK Prospective Diabetes Study (UKPDS 33) examined the effect of strict glycemic control on cardiovascular events in 3867 patients with newly diagnosed type 2 diabetes mellitus. The study included an intensive treatment group using sulfonylureas and insulin and a diet-only control group over 10 years. The UKPDS 33 trial demonstrated that strict glycemic control reduces microvascular damage by 12%, although it does not significantly reduce macrovascular damage. However, a 10-year follow-up of the UKPDS 33 trial revealed that strict glycemic control also reduced macrovascular damage, confirming the long-term benefits of controlling blood sugar levels (legacy effect) (Holman et al. 2008). The Action to Control Cardiovascular Risk in Diabetes Study (ACCORD) compared high-risk patients with diabetes receiving intensive glycemic control with those receiving standard treatment. Despite achieving lower HbA1c levels (6.4% compared with 7.5% in the standard treatment group), the intensive treatment group had a higher mortality rate than anticipated, leading to the early termination of the trial after 17 months (Action to Control Cardiovascular Risk in Diabetes Study et al. 2008). These findings highlight the need to consider not only HbA1c but also rapid postprandial blood glucose spikes and hypoglycemia when managing macrovascular complications. Notably, periodontal treatment reportedly reduces HbA1c levels by approximately 0.5%, and it may help mitigate the progression of diabetic complications (Simpson et al. 2022). A previous study, which investigated the relationship between periodontal treatment and health promotion using data from the National Nutrition Survey, suggested that periodontal treatment can reduce the risk of nephropathy, retinopathy, and neuropathy by up to 20.5%, 18.4%, and 17.7%, respectively, potentially leading to reduced healthcare costs (Choi et al. 2020). Although there is no direct evidence that periodontal interventions reduce macrovascular complications, many studies have demonstrated positive effects on secondary endpoints, such as improved vascular endothelial function, following periodontal treatment (Masi et al. 2018).

In addition, patient awareness of maintaining periodontal health is crucial; it is important to explain to periodontitis patients with diabetes that they are at an increased risk of cardiovascular diseases such as myocardial infarction and stroke. Active management of cardiovascular risk factors—such as smoking cessation, regular exercise, weight control, blood pressure management, lipid control, blood glucose control, adequate periodontal treatment, and periodontal maintenance—is recommended. Moreover, while periodontal treatment can improve insulin resistance by reducing inflammation, its effect may be less pronounced in patients with severe obesity; therefore, dietary guidance to promote healthy eating habits and restore chewing function is also crucial (Nagasawa et al. 2010). Oral rehabilitation to restore effective mastication is particularly important for patients with cardiovascular disease who have lost numerous teeth.

In managing macrovascular disorders, various risk factors, including age, sex, blood pressure, and cholesterol, are used to predict events such as myocardial infarction. (Graham et al. 2021; K. Nishimura et al. 2014; Wilson et al. 1998). Risk management should be tailored to each patient's risk profile, which may vary depending on geographic location or regional health considerations (K. Nishimura et al. 2014).

For patients with periodontitis complicated by cardiovascular disease, it is essential to guide appropriate dental treatments, maintenance, and preventive strategies. These patients should receive oral hygiene education, and individualized oral hygiene management plans should be implemented (Sanz et al. 2020b). This typically includes twice-daily brushing, interdental cleaning, and, when necessary, the use of chemical plaque control. These measures should be implemented as soon as periodontitis is diagnosed and when the patient's cardiovascular condition allows (Table 10.3). During periodontal treatment, performing full-mouth therapy in a single session, rather than multiple separate

Table 10.3 Oral hygiene management for patients with cardiovascular disease (CVD). Patients with a diagnosis of CVD with periodontitis should be informed that they may be at high risk for subsequent CVD complications and should regularly adhere to recommended dental treatment, maintenance, and preventive measures

CVD Patients	CVD without periodontitis	CVD with periodontitis
Management	Implementation of preventive oral care	Should be managed as quickly as the condition of the vascular system permits
	Periodic (at least once a year) monitoring of changes in periodontal status is necessary	
	Should receive comprehensive periodontal evaluation Individualized oral hygiene, including twice-daily brushing, interdental cleaning, and possibly chemical plaque control Non-surgical treatment should be provided in several 30–45 min sessions If periodontal surgical treatment or implant treatment is required, use the same methods as for patients without CVD	

Oral hygiene management for patients with CVD
Patients with a diagnosis of CVD with periodontitis should be informed that they may be at high risk for subsequent CVD complications and should regularly adhere to recommended dental treatment, maintenance, and preventive measures

sessions, may lead to a considerable rise in serum IL-6 levels and a spike in acute systemic inflammation (Izumi et al. 2009). Therefore, for patients with cardiovascular disease, non-surgical periodontal treatment is recommended over several 30–45 min sessions to minimize systemic inflammation (Herrera et al. 2023). If surgical periodontal treatment or implant procedures are required, they should be performed in the same manner as for patients without cardiovascular disease (Herrera et al. 2023).

Given that the use of antihypertensive, antiplatelet, and anticoagulant drugs may affect dental treatment, a thorough risk assessment of hemorrhagic events is necessary, particularly in patients who have undergone coronary revascularization (Natsuaki et al. 2018). Blood pressure should be measured before any surgical procedure, and if hypertension is present (e.g., >180/100 mmHg), the surgery should be postponed until the patient's blood pressure is stabilized. In terms of antiplatelet and anticoagulant medication, the risk of bleeding during periodontal and implant surgical procedures is generally low to moderate, and in most cases, dentists should not alter the patient's medication regimen. However, local management of potential bleeding complications should be considered, and it is strongly recommended that dentists consult with the patient's physicians. If a periodontal procedure with a moderate bleeding risk is planned, it should be performed in full consultation with the health professional responsible for prescribing and managing the anticoagulant therapy. Finally, patients at the highest risk of thrombosis and ischemia—such as those with chronic atrial fibrillation, recent myocardial infarction, or recent coronary stenting—who are taking a combination of antiplatelet and anticoagulant medications should consult with their physician to determine whether their medication regimen should be adjusted for periodontal treatment (Steffel et al. 2018).

In summary, glycemic control to prevent vascular damage in diabetes requires not only lowering HbA1c but also preventing blood glucose spikes. Since periodontal disease and diabetes are interrelated and have common risk factors, comprehensive diagnosis and treatment are necessary. Maintaining periodontal care requires customized support based on each patient's risk factors and special needs. Regular professional monitoring and maintenance therapy are essential for preserving the stability of periodontal tissues and maintaining oral health and overall systemic well-being in the long term.

Competing Interests The authors declare no competing interests.

References

Abusleme L, Moutsopoulos NM (2017) IL-17: overview and role in oral immunity and microbiome. Oral Dis 23(7):854–865. https://doi.org/10.1111/odi.12598

Abusleme L, Diaz PI, Freeman AF, Greenwell-Wild T, Brenchley L, Desai JV et al (2018) Human defects in STAT3 promote oral mucosal fungal and bacterial dysbiosis. JCI Insight 3(17). https://doi.org/10.1172/jci.insight.122061

Action to Control Cardiovascular Risk in Diabetes Study, G, Gerstein HC, Miller ME, Byington RP, Goff DC Jr, Bigger JT et al (2008) Effects of intensive glucose lowering in type 2 diabetes. N Engl J Med 358(24):2545–2559. https://doi.org/10.1056/NEJMoa0802743

Aggor FEY, Break TJ, Trevejo-Nunez G, Whibley N, Coleman BM, Bailey RD et al (2020) Oral epithelial IL-22/STAT3 signaling licenses IL-17-mediated immunity to oral mucosal candidiasis. Sci Immunol 5(48). https://doi.org/10.1126/sciimmunol.aba0570

Aihara M, Jinnouchi H, Yoshida A, Ijima H, Sakurai Y, Hayashi T et al (2023) Evaluation of glycated albumin levels in tears and saliva as a marker in patients with diabetes mellitus. Diabetes Res Clin Pract 199:110637. https://doi.org/10.1016/j.diabres.2023.110637

American Diabetes A (2019) 2. Classification and diagnosis of diabetes: standards of medical care in diabetes-2019. Diabetes Care 42(Suppl 1):S13–S28. https://doi.org/10.2337/dc19-S002

Armitage GC (1999) Development of a classification system for periodontal diseases and conditions. Ann Periodontol 4(1):1–6. https://doi.org/10.1902/annals.1999.4.1.1

Azogui-Levy S, Dray-Spira R, Attal S, Hartemann A, Anagnostou F, Azerad J (2018) Factors associated with oral health-related quality of life in patients with diabetes. Aust Dent J 63(2):163–169. https://doi.org/10.1111/adj.12577

Bahall M, Bailey H (2022) The impact of chronic disease and accompanying bio-psycho-social factors on health-related quality of life. J Family Med Prim Care 11(8):4694–4704. https://doi.org/10.4103/jfmpc.jfmpc_2399_21

Beck JD, Papapanou PN, Philips KH, Offenbacher S (2019) Periodontal medicine: 100 years of progress. J Dent Res 98(10):1053–1062. https://doi.org/10.1177/0022034519846113

Bharti P, Katagiri S, Nitta H, Nagasawa T, Kobayashi H, Takeuchi Y et al (2013) Periodontal treatment with topical antibiotics improves glycemic control in association with elevated serum adiponectin in patients with type 2 diabetes mellitus. Obes Res Clin Pract 7(2):e129–e138. https://doi.org/10.1016/j.orcp.2011.11.005

Boutsi EA, Umeda M, Nagasawa T, Laosrisin N, Ishikawa I (1997) Follow-up of two cases of Papillon-Lefevre syndrome and presentation of two new cases. Int J Periodontics Restorative Dent 17(4):334–347. Retrieved from https://www.ncbi.nlm.nih.gov/pubmed/9497724

Brandtzaeg P (2013a) Secretory IgA: designed for antimicrobial defense. Front Immunol 4:222. https://doi.org/10.3389/fimmu.2013.00222

Brandtzaeg P (2013b) Secretory immunity with special reference to the oral cavity. J Oral Microbiol 5. https://doi.org/10.3402/jom.v5i0.20401

Break TJ, Oikonomou V, Dutzan N, Desai JV, Swidergall M, Freiwald T et al (2021) Aberrant type 1 immunity drives susceptibility to mucosal fungal infections. Science 371(6526). https://doi.org/10.1126/science.aay5731

Budreviciute A, Damiati S, Sabir DK, Onder K, Schuller-Goetzburg P, Plakys G, Kodzius R (2020) Management and Prevention Strategies for Non-communicable Diseases (NCDs) and Their Risk Factors. Front Public Health 8:574111. https://doi.org/10.3389/fpubh.2020.574111

Cafferata EA, Jerez A, Vernal R, Monasterio G, Pandis N, Faggion CM Jr (2019) The therapeutic potential of regulatory T lymphocytes in periodontitis: a systematic review. J Periodontal Res 54(3):207–217. https://doi.org/10.1111/jre.12629

Caton JG, Armitage G, Berglundh T, Chapple ILC, Jepsen S, Kornman KS et al (2018) A new classification scheme for periodontal and peri-implant diseases and conditions – introduction and key changes from the 1999 classification. J Clin Periodontol 45(Suppl 20):S1–S8. https://doi.org/10.1111/jcpe.12935

Cavalla F, Hernandez M (2022) Polarization profiles of T lymphocytes and macrophages responses in periodontitis. Adv Exp Med Biol 1373:195–208. https://doi.org/10.1007/978-3-030-96881-6_10

Chapple IL, Van der Weijden F, Doerfer C, Herrera D, Shapira L, Polak D, Graziani F (2015) Primary prevention of periodontitis: managing gingivitis. J Clin Periodontol 42(16)71–76. https://doi.org/10.1111/jcpe.12366

Chen YW, Nagasawa T, Wara-Aswapati N, Ushida Y, Wang D, Takeuchi Y et al (2009) Association between periodontitis and anti-cardiolipin antibodies in Buerger disease. J Clin Periodontol 36(10):830–835. https://doi.org/10.1111/j.1600-051X.2009.01467.x

Choi SE, Sima C, Pandya A (2020) Impact of treating oral disease on preventing vascular diseases: a model-based cost-effectiveness analysis of periodontal treatment among patients with type 2 diabetes. Diabetes Care 43(3):563–571. https://doi.org/10.2337/dc19-1201

Costalonga M, Herzberg MC (2014) The oral microbiome and the immunobiology of periodontal disease and caries. Immunol Lett 162(2 Pt A):22–38. https://doi.org/10.1016/j.imlet.2014.08.017

Curtis MM, Way SS (2009) Interleukin-17 in host defence against bacterial, mycobacterial and fungal pathogens. Immunology 126(2):177–185. https://doi.org/10.1111/j.1365-2567.2008.03017.x

Dale BA, Kimball JR, Krisanaprakornkit S, Roberts F, Robinovitch M, O'Neal R et al (2001) Localized

antimicrobial peptide expression in human gingiva. J Periodontal Res 36(5):285–294. https://doi.org/10.1034/j.1600-0765.2001.360503.x
de Haar SF, Hiemstra PS, van Steenbergen MT, Everts V, Beertsen W (2006) Role of polymorphonuclear leukocyte-derived serine proteinases in defense against Actinobacillus actinomycetemcomitans. Infect Immun 74(9):5284–5291. https://doi.org/10.1128/IAI.02016-05
Demmer RT, Squillaro A, Papapanou PN, Rosenbaum M, Friedewald WT, Jacobs DR Jr, Desvarieux M (2012) Periodontal infection, systemic inflammation, and insulin resistance: results from the continuous National Health and Nutrition Examination Survey (NHANES) 1999-2004. Diabetes Care 35(11):2235–2242. https://doi.org/10.2337/dc12-0072
Desmouliere A, Rubbia-Brandt L, Grau G, Gabbiani G (1992) Heparin induces alpha-smooth muscle actin expression in cultured fibroblasts and in granulation tissue myofibroblasts. Lab Investig 67(6):716–726. Retrieved from https://www.ncbi.nlm.nih.gov/pubmed/1460862
Desmouliere A, Geinoz A, Gabbiani F, Gabbiani G (1993) Transforming growth factor-beta 1 induces alpha-smooth muscle actin expression in granulation tissue myofibroblasts and in quiescent and growing cultured fibroblasts. J Cell Biol 122(1):103–111. https://doi.org/10.1083/jcb.122.1.103
Dutzan N, Abusleme L, Bridgeman H, Greenwell-Wild T, Zangerle-Murray T, Fife ME et al (2017) On-going mechanical damage from mastication drives homeostatic Th17 cell responses at the Oral barrier. Immunity 46(1):133–147. https://doi.org/10.1016/j.immuni.2016.12.010
Emson HE (1987) Health, disease and illness: matters for definition. CMAJ 136(8):811–813. Retrieved from https://www.ncbi.nlm.nih.gov/pubmed/3567788
Engebretson SP, Hyman LG, Michalowicz BS, Schoenfeld ER, Gelato MC, Hou W et al (2013) The effect of nonsurgical periodontal therapy on hemoglobin A1c levels in persons with type 2 diabetes and chronic periodontitis: a randomized clinical trial. JAMA 310(23):2523–2532. https://doi.org/10.1001/jama.2013.282431
Ferrannini G, Svenungsson E, Kjellstrom B, Elvin K, Grosso G, Nasman P et al (2020) Antiphospholipid antibodies in patients with dysglycaemia: a neglected cardiovascular risk factor? Diab Vasc Dis Res 17(3):1479164120922123. https://doi.org/10.1177/1479164120922123
Festa A, Hanley AJ, Tracy RP, D'Agostino R Jr, Haffner SM (2003) Inflammation in the prediabetic state is related to increased insulin resistance rather than decreased insulin secretion. Circulation 108(15):1822–1830. https://doi.org/10.1161/01.CIR.0000091339.70120.53
Fisher L, Skaff MM, Mullan JT, Arean P, Glasgow R, Masharani U (2008) A longitudinal study of affective and anxiety disorders, depressive affect and diabetes distress in adults with type 2 diabetes. Diabet Med 25(9):1096–1101. https://doi.org/10.1111/j.1464-5491.2008.02533.x
Gaffen SL, Moutsopoulos NM (2020) Regulation of host-microbe interactions at oral mucosal barriers by type 17 immunity. Sci Immunol 5(43). https://doi.org/10.1126/sciimmunol.aau4594
Ganesan SM, Joshi V, Fellows M, Dabdoub SM, Nagaraja HN, O'Donnell B et al (2017) A tale of two risks: smoking, diabetes and the subgingival microbiome. ISME J 11(9):2075–2089. https://doi.org/10.1038/ismej.2017.73
Garcia D, Erkan D (2018) Diagnosis and Management of the Antiphospholipid Syndrome. N Engl J Med 378(21):2010–2021. https://doi.org/10.1056/NEJMra1705454
Geginat J, Paroni M, Maglie S, Alfen JS, Kastirr I, Gruarin P et al (2014) Plasticity of human CD4 T cell subsets. Front Immunol 5:630. https://doi.org/10.3389/fimmu.2014.00630
Graham IM, Di Angelantonio E, Visseren F, De Bacquer D, Ference BA, Timmis A et al (2021) Systematic coronary risk evaluation (SCORE): JACC focus seminar 4/8. J Am Coll Cardiol 77(24):3046–3057. https://doi.org/10.1016/j.jacc.2021.04.052
Greer A, Zenobia C, Darveau RP (2013) Defensins and LL-37: a review of function in the gingival epithelium. Periodontol 63(1):67–79. https://doi.org/10.1111/prd.12028
Greinacher A, Holtfreter B, Krauel K, Gatke D, Weber C, Ittermann T et al (2011) Association of natural anti-platelet factor 4/heparin antibodies with periodontal disease. Blood 118(5):1395–1401. https://doi.org/10.1182/blood-2011-03-342857
Gunupati S, Chava VK, Krishna BP (2011) Effect of phase I periodontal therapy on anti-cardiolipin antibodies in patients with acute myocardial infarction associated with chronic periodontitis. J Periodontol 82(12):1657–1664. https://doi.org/10.1902/jop.2011.110002
Hajishengallis G, Moutsopoulos NM (2016) Role of bacteria in leukocyte adhesion deficiency-associated periodontitis. Microb Pathog 94:21–26. https://doi.org/10.1016/j.micpath.2015.09.003
Halai H, Somani C, Donos N, Nibali L (2020) Periodontal status of children with primary immunodeficiencies: a systematic review. Clin Oral Investig 24(6):1939–1951. https://doi.org/10.1007/s00784-019-03055-z
Harase T, Nishida W, Hamakawa T, Hino S, Shigematsu K, Kobayashi S et al (2015) Clinical implication of blood glucose monitoring in general dental offices: the ehime dental diabetes study. BMJ Open Diabetes Res Care 3(1):e000151. https://doi.org/10.1136/bmjdrc-2015-000151
Haubek D, Poulsen K, Kilian M (2007) Microevolution and patterns of dissemination of the JP2 clone of Aggregatibacter (Actinobacillus) actinomycetemcomitans. Infect Immun 75(6):3080–3088. https://doi.org/10.1128/IAI.01734-06

Haubek D, Ennibi OK, Poulsen K, Vaeth M, Poulsen S, Kilian M (2008) Risk of aggressive periodontitis in adolescent carriers of the JP2 clone of Aggregatibacter (Actinobacillus) actinomycetemcomitans in Morocco: a prospective longitudinal cohort study. Lancet 371(9608):237–242. https://doi.org/10.1016/S0140-6736(08)60135-X

Herrera D, Sanz M, Kebschull M, Jepsen S, Sculean A, Berglundh T et al (2022) Treatment of stage IV periodontitis: the EFP S3 level clinical practice guideline. J Clin Periodontol 49(Suppl 24):4–71. https://doi.org/10.1111/jcpe.13639

Herrera D, Sanz M, Shapira L, Brotons C, Chapple I, Frese T et al (2023) Association between periodontal diseases and cardiovascular diseases, diabetes and respiratory diseases: consensus report of the joint workshop by the European Federation of Periodontology (EFP) and the European arm of the world Organization of Family Doctors (WONCA Europe). J Clin Periodontol 50(6):819–841. https://doi.org/10.1111/jcpe.13807

Holman RR, Paul SK, Bethel MA, Matthews DR, Neil HA (2008) 10-year follow-up of intensive glucose control in type 2 diabetes. N Engl J Med 359(15):1577–1589. https://doi.org/10.1056/NEJMoa0806470

Hormdee D, Nagasawa T, Kiji M, Yashiro R, Kobayashi H, Koshy G et al (2005) Protein kinase-A-dependent osteoprotegerin production on interleukin-1 stimulation in human gingival fibroblasts is distinct from periodontal ligament fibroblasts. Clin Exp Immunol 142(3):490–497. https://doi.org/10.1111/j.1365-2249.2005.02937.x

Hovelmeyer N, Schmidt-Supprian M, Ohnmacht C (2022) NF-kappaB in control of regulatory T cell development, identity, and function. J Mol Med (Berl) 100(7):985–995. https://doi.org/10.1007/s00109-022-02215-1

Huang Z, Pei X, Graves DT (2020) The interrelationship between diabetes, IL-17 and bone loss. Curr Osteoporos Rep 18(1):23–31. https://doi.org/10.1007/s11914-020-00559-6

Ikeuchi T, Moutsopoulos NM (2022) Osteoimmunology in periodontitis; a paradigm for Th17/IL-17 inflammatory bone loss. Bone 163:116500. https://doi.org/10.1016/j.bone.2022.116500

Ishigame H, Kakuta S, Nagai T, Kadoki M, Nambu A, Komiyama Y et al (2009) Differential roles of interleukin-17A and -17F in host defense against mucoepithelial bacterial infection and allergic responses. Immunity 30(1):108–119. https://doi.org/10.1016/j.immuni.2008.11.009

Izumi Y, Nagasawa T, Umeda M, Kobayashi H, Takeuchi Y, Yashiro R et al (2009) Periodontitis and cardiovascular diseases: the link and relevant mechanisms. Jpn Dent Sci Rev 45(2):98–108. https://doi.org/10.1016/j.jdsr.2009.06.001

Javed F, Thafeed Alghamdi AS, Mikami T, Mehmood A, Ahmed HB, Samaranayake LP, Tenenbaum HC (2014) Effect of glycemic control on self-perceived oral health, periodontal parameters, and alveolar bone loss among patients with prediabetes. J Periodontol 85(2):234–241. https://doi.org/10.1902/jop.2013.130008

Jepsen S, Caton JG, Albandar JM, Bissada NF, Bouchard P, Cortellini P et al (2018) Periodontal manifestations of systemic diseases and developmental and acquired conditions: consensus report of workgroup 3 of the 2017 world workshop on the classification of periodontal and Peri-implant diseases and conditions. J Periodontol 89(Suppl 1):S237–S248. https://doi.org/10.1002/JPER.17-0733

Jhugroo C, Divakar DD, Jhugroo P, Al-Amri SAS, Alahmari AD, Vijaykumar S, Parine NR (2019) Characterization of oral mucosa lesions and prevalence of yeasts in diabetic patients: a comparative study. Microb Pathog 126:363–367. https://doi.org/10.1016/j.micpath.2018.11.028

Joly S, Maze C, McCray PB Jr, Guthmiller JM (2004) Human beta-defensins 2 and 3 demonstrate strain-selective activity against oral microorganisms. J Clin Microbiol 42(3):1024–1029. https://doi.org/10.1128/JCM.42.3.1024-1029.2004

Katagiri S, Nagasawa T, Kobayashi H, Takamatsu H, Bharti P, Izumiyama H et al (2012) Improvement of glycemic control after periodontal treatment by resolving gingival inflammation in type 2 diabetic patients with periodontal disease. J Diabetes Investig 3(4):402–409. https://doi.org/10.1111/j.2040-1124.2012.00209.x

Kiany F, Hedayati A (2015) Evaluation of serum anti-cardiolipin antibodies after non-surgical periodontal treatment in chronic periodontitis patients. Odontology 103(2):203–209. https://doi.org/10.1007/s10266-014-0149-2

Kiji M, Nagasawa T, Hormdee D, Yashiro R, Kobayashi H, Noguchi K et al (2007) Internal prostaglandin synthesis augments osteoprotegerin production in human gingival fibroblasts stimulated by lipopolysaccharide. Clin Exp Immunol 149(2):327–334. https://doi.org/10.1111/j.1365-2249.2007.03414.x

Kim TS, Silva LM, Theofilou VI, Greenwell-Wild T, Li L, Williams DW et al (2023) Neutrophil extracellular traps and extracellular histones potentiate IL-17 inflammation in periodontitis. J Exp Med 220(9). https://doi.org/10.1084/jem.20221751

Kitamoto S, Nagao-Kitamoto H, Jiao Y, Gillilland MG 3rd, Hayashi A, Imai J et al (2020) The intermucosal connection between the mouth and gut in commensal pathobiont-driven colitis. Cell 182(2):447–462.e414. https://doi.org/10.1016/j.cell.2020.05.048

Komatsu N, Okamoto K, Sawa S, Nakashima T, Oh-hora M, Kodama T et al (2014) Pathogenic conversion of Foxp3+ T cells into TH17 cells in autoimmune arthritis. Nat Med 20(1):62–68. https://doi.org/10.1038/nm.3432

Kominato H, Takeda K, Mizutani K, Mikami R, Kido D, Buranasin P et al (2022) Metformin accelerates wound healing by Akt phosphorylation of gingival fibroblasts in insulin-resistant prediabetes mice. J Periodontol 93(2):256–268. https://doi.org/10.1002/JPER.21-0362

Lalla E, Lamster IB, Stern DM, Schmidt AM (2001) Receptor for advanced glycation end products, inflammation, and accelerated periodontal disease in diabetes: mechanisms and insights into therapeutic modalities. Ann Periodontol 6(1):113–118. https://doi.org/10.1902/annals.2001.6.1.113

Letchumanan G, Abdullah N, Marlini M, Baharom N, Lawley B, Omar MR et al (2022) Gut microbiota composition in prediabetes and newly diagnosed type 2 diabetes: a systematic review of observational studies. Front Cell Infect Microbiol 12:943427. https://doi.org/10.3389/fcimb.2022.943427

Li Y, Storch EA, Ferguson S, Li L, Buys N, Sun J (2022) The efficacy of cognitive behavioral therapy-based intervention on patients with diabetes: a meta-analysis. Diabetes Res Clin Pract 189:109965. https://doi.org/10.1016/j.diabres.2022.109965

Liu XH, Kirschenbaum A, Yao S, Levine AC (2005) Cross-talk between the interleukin-6 and prostaglandin E(2) signaling systems results in enhancement of osteoclastogenesis through effects on the osteoprotegerin/receptor activator of nuclear factor-kappaB (RANK) ligand/RANK system. Endocrinology 146(4):1991–1998. https://doi.org/10.1210/en.2004-1167

Liu C, Feng X, Li Q, Wang Y, Li Q, Hua M (2016) Adiponectin, TNF-alpha and inflammatory cytokines and risk of type 2 diabetes: a systematic review and meta-analysis. Cytokine 86:100–109. https://doi.org/10.1016/j.cyto.2016.06.028

Lockhart PB, Bolger AF, Papapanou PN, Osinbowale O, Trevisan M, Levison ME et al (2012) Periodontal disease and atherosclerotic vascular disease: does the evidence support an independent association?: A scientific statement from the American Heart Association. Circulation 125(20):2520–2544. https://doi.org/10.1161/CIR.0b013e31825719f3

Loe H, Theilade E, Jensen SB (1965) Experimental gingivitis in man. J Periodontol (1930) 36:177–187. https://doi.org/10.1902/jop.1965.36.3.177

Long J, Cai Q, Steinwandel M, Hargreaves MK, Bordenstein SR, Blot WJ et al (2017) Association of oral microbiome with type 2 diabetes risk. J Periodontal Res 52(3):636–643. https://doi.org/10.1111/jre.12432

Ma Z, Zuo T, Frey N, Rangrez AY (2024) A systematic framework for understanding the microbiome in human health and disease: from basic principles to clinical translation. Signal Transduct Target Ther 9(1):237. https://doi.org/10.1038/s41392-024-01946-6

Masi S, Orlandi M, Parkar M, Bhowruth D, Kingston I, O'Rourke C et al (2018) Mitochondrial oxidative stress, endothelial function and metabolic control in patients with type II diabetes and periodontitis: a randomised controlled clinical trial. Int J Cardiol 271: 263–268. https://doi.org/10.1016/j.ijcard.2018.05.019

Matsha TE, Prince Y, Davids S, Chikte U, Erasmus RT, Kengne AP, Davison GM (2020) Oral microbiome signatures in diabetes mellitus and periodontal disease. J Dent Res 99(6):658–665. https://doi.org/10.1177/0022034520913818

Matsuoka H, Chiba I, Sakano Y, Toyofuku A, Abiko Y (2017) Cognitive behavioral therapy for psychosomatic problems in dental settings. Biopsychosocial Med 11:18. https://doi.org/10.1186/s13030-017-0102-z

McClure FA, Wemyss K, Cox JR, Bridgeman HM, Prise IE, King JI et al (2024) Th17-to-Tfh plasticity during periodontitis limits disease pathology. J Exp Med 221(8). https://doi.org/10.1084/jem.20232015

Mehta J, Eaton C, AlAmri M, Lin GH, Nibali L (2023) The association between Aggregatibacter actinomycetemcomitans JP2 clone and periodontitis: a systematic review and meta-analysis. J Periodontal Res 58(3):465–482. https://doi.org/10.1111/jre.13102

Mestecky J (1987) The common mucosal immune system and current strategies for induction of immune responses in external secretions. J Clin Immunol 7(4): 265–276. https://doi.org/10.1007/BF00915547

Mikami R, Mizutani K, Shioyama H, Matsuura T, Aoyama N, Suda T et al (2022) Influence of aging on periodontal regenerative therapy using enamel matrix derivative: a 3-year prospective cohort study. J Clin Periodontol 49(2):123–133. https://doi.org/10.1111/jcpe.13552

Mizutani K, Minami I, Mikami R, Kido D, Takeda K, Nakagawa K et al (2024) Improvement of periodontal parameters following intensive diabetes care and supragingival dental prophylaxis in patients with type 2 diabetes: a prospective cohort study. J Clin Periodontol 51(6):733–741. https://doi.org/10.1111/jcpe.13958

Mohammed L, Jha G, Malasevskaia I, Goud HK, Hassan A (2021) The interplay between sugar and yeast infections: do diabetics have a greater predisposition to develop oral and vulvovaginal candidiasis? Cureus 13(2):e13407. https://doi.org/10.7759/cureus.13407

Moutsopoulos, N. M., Zerbe, C. S., Wild, T., Dutzan, N., Brenchley, L., DiPasquale, G., . . . Holland, S. M. (2017). Interleukin-12 and interleukin-23 blockade in leukocyte adhesion deficiency type 1. N Engl J Med, 376(12), 1141–1146. doi:https://doi.org/10.1056/NEJMoa1612197

Munenaga Y, Hiroshima Study G, Yamashina T, Tanaka J, Nishimura F (2013) Improvement of glycated hemoglobin in Japanese subjects with type 2 diabetes by resolution of periodontal inflammation using adjunct topical antibiotics: results from the Hiroshima Study. Diabetes Res Clin Pract 100(1): 53–60. https://doi.org/10.1016/j.diabres.2013.01.028

Nagamine K, Peterson P, Scott HS, Kudoh J, Minoshima S, Heino M et al (1997) Positional cloning of the APECED gene. Nat Genet 17(4):393–398. https://doi.org/10.1038/ng1297-393

Nagasawa T, Kobayashi H, Kiji M, Aramaki M, Mahanonda R, Kojima T et al (2002) LPS-stimulated human gingival fibroblasts inhibit the differentiation of monocytes into osteoclasts through the production of osteoprotegerin. Clin Exp Immunol 130(2):338–344. https://doi.org/10.1046/j.1365-2249.2002.01990.x

Nagasawa T, Kiji M, Yashiro R, Hormdee D, Lu H, Kunze M et al (2007) Roles of receptor activator of nuclear factor-kappaB ligand (RANKL) and osteoprotegerin in periodontal health and disease. Periodontology 2000(43):65–84. https://doi.org/10.1111/j.1600-0757.2006.00185.x

Nagasawa T, Noda M, Katagiri S, Takaichi M, Takahashi Y, Wara-Aswapati N et al (2010) Relationship between periodontitis and diabetes – importance of a clinical study to prove the vicious cycle. Intern Med 49(10):881–885. https://doi.org/10.2169/internalmedicine.49.3351

Nakashima T, Kobayashi Y, Yamasaki S, Kawakami A, Eguchi K, Sasaki H, Sakai H (2000) Protein expression and functional difference of membrane-bound and soluble receptor activator of NF-kappaB ligand: modulation of the expression by osteotropic factors and cytokines. Biochem Biophys Res Commun 275(3):768–775. https://doi.org/10.1006/bbrc.2000.3379

Natsuaki M, Morimoto T, Yamaji K, Watanabe H, Yoshikawa Y, Shiomi H et al (2018) Prediction of thrombotic and bleeding events after percutaneous coronary intervention: CREDO-Kyoto thrombotic and bleeding risk scores. J Am Heart Assoc 7(11). https://doi.org/10.1161/JAHA.118.008708

Nishimura F, Murayama Y (2001) Periodontal inflammation and insulin resistance–lessons from obesity. J Dent Res 80(8):1690–1694. https://doi.org/10.1177/00220345010800080201

Nishimura K, Okamura T, Watanabe M, Nakai M, Takegami M, Higashiyama A et al (2014) Predicting coronary heart disease using risk factor categories for a Japanese urban population, and comparison with the Framingham risk score: the Suita study. J Atheroscler Thromb 21(8):784–798. https://doi.org/10.5551/jat.19356

Nukaga J, Kobayashi M, Shinki T, Song H, Takada T, Takiguchi T et al (2004) Regulatory effects of interleukin-1beta and prostaglandin E2 on expression of receptor activator of nuclear factor-kappaB ligand in human periodontal ligament cells. J Periodontol 75(2):249–259. https://doi.org/10.1902/jop.2004.75.2.249

Odineal DD, Gershwin ME (2020) The epidemiology and clinical manifestations of autoimmunity in selective IgA deficiency. Clin Rev Allergy Immunol 58(1):107–133. https://doi.org/10.1007/s12016-019-08756-7

Oikonomou V, Smith G, Constantine GM, Schmitt MM, Ferre EMN, Alejo JC et al (2024) The role of interferon-gamma in autoimmune polyendocrine syndrome type 1. N Engl J Med 390(20):1873–1884. https://doi.org/10.1056/NEJMoa2312665

Paraskevas S, Huizinga JD, Loos BG (2008) A systematic review and meta-analyses on C-reactive protein in relation to periodontitis. J Clin Periodontol 35(4):277–290. https://doi.org/10.1111/j.1600-051X.2007.01173.x

Peyrot M, Rubin RR, Lauritzen T, Snoek FJ, Matthews DR, Skovlund SE (2005) Psychosocial problems and barriers to improved diabetes management: results of the cross-National Diabetes Attitudes, Wishes and Needs (DAWN) Study. Diabet Med 22(10):1379–1385. https://doi.org/10.1111/j.1464-5491.2005.01644.x

Sanz M, Herrera D, Kebschull M, Chapple I, Jepsen S, Beglundh T et al (2020a) Treatment of stage I-III periodontitis-the EFP S3 level clinical practice guideline. J Clin Periodontol 47(Suppl 22):4–60. https://doi.org/10.1111/jcpe.13290

Sanz M, Marco Del Castillo A, Jepsen S, Gonzalez-Juanatey JR, D'Aiuto F, Bouchard P et al (2020b) Periodontitis and cardiovascular diseases: consensus report. J Clin Periodontol 47(3):268–288. https://doi.org/10.1111/jcpe.13189

Schenkein HA, Berry CR, Burmeister JA, Brooks CN, Barbour SE, Best AM, Tew JG (2003) Anti-cardiolipin antibodies in sera from patients with periodontitis. J Dent Res 82(11):919–922. https://doi.org/10.1177/154405910308201114

Shinjo T, Nishimura F (2024) The bidirectional association between diabetes and periodontitis, from basic to clinical. Jpn Dent Sci Rev 60:15–21. https://doi.org/10.1016/j.jdsr.2023.12.002

Shinjo T, Onizuka S, Zaitsu Y, Ishikado A, Park K, Li Q et al (2023) Dysregulation of CXCL1 expression and neutrophil recruitment in insulin resistance and Diabetes-related periodontitis in male mice. Diabetes 72(7):986–998. https://doi.org/10.2337/db22-1014

Silva DNA, Casarin M, Monajemzadeh S, Bezerra BB, Lux R, Pirih FQ (2022) The microbiome in periodontitis and Diabetes. Front Oral Health 3:859209. https://doi.org/10.3389/froh.2022.859209

Simpson TC, Clarkson JE, Worthington HV, MacDonald L, Weldon JC, Needleman I et al (2022) Treatment of periodontitis for glycaemic control in people with diabetes mellitus. Cochrane Database Syst Rev 4(4):CD004714. https://doi.org/10.1002/14651858.CD004714.pub4

Slots J, Ting M (1999) Actinobacillus actinomycetemcomitans and Porphyromonas gingivalis in human periodontal disease: occurrence and treatment. Periodontology 2000(20):82–121. https://doi.org/10.1111/j.1600-0757.1999.tb00159.x

Slots J, Feik D, Rams TE (1990) Actinobacillus actinomycetemcomitans and Bacteroides intermedius in human periodontitis: age relationship and mutual association. J Clin Periodontol 17(9):659–662. Retrieved from https://www.ncbi.nlm.nih.gov/pubmed/2250080

Smith PC, Martinez C, Martinez J, McCulloch CA (2019) Role of fibroblast populations in periodontal wound healing and tissue remodeling. Front Physiol 10:270. https://doi.org/10.3389/fphys.2019.00270

Socransky SS, Haffajee AD (2002) Dental biofilms: difficult therapeutic targets. Periodontology 2000(28): 12–55. https://doi.org/10.1034/j.1600-0757.2002.280102.x

Steffel J, Verhamme P, Potpara TS, Albaladejo P, Antz M, Desteghe L et al (2018) The 2018 European heart rhythm association practical guide on the use of non-vitamin K antagonist oral anticoagulants in patients with atrial fibrillation: executive summary. EP Europace 20(8):1231–1242. https://doi.org/10.1093/europace/euy054

Stratton IM, Adler AI, Neil HA, Matthews DR, Manley SE, Cull CA et al (2000) Association of glycaemia with macrovascular and microvascular complications of type 2 diabetes (UKPDS 35): prospective observational study. BMJ 321(7258):405–412. https://doi.org/10.1136/bmj.321.7258.405

Sun H, Saeedi P, Karuranga S, Pinkepank M, Ogurtsova K, Duncan BB, Magliano DJ (2022) IDF Diabetes Atlas: Global, regional and country-level diabetes prevalence estimates for 2021 and projections for 2045. Diabetes Res Clin Pract 183:109119. https://doi.org/10.1016/j.diabres.2021.109119

Takahashi S, Nezu A, Tanimura A, Nakamichi Y, Yamamoto T (2022) Responses of salivary glands to intake of soft diet. J Oral Biosci 64(2):210–216. https://doi.org/10.1016/j.job.2022.03.006

Tanaka K, Ishikawa H (2004) Role of intestinal bacterial flora in oral tolerance induction. Histol Histopathol 19(3):907–914. https://doi.org/10.14670/HH-19.907

Tang C, Chen S, Qian H, Huang W (2012) Interleukin-23: as a drug target for autoimmune inflammatory diseases. Immunology 135(2):112–124. https://doi.org/10.1111/j.1365-2567.2011.03522.x

Teeuw WJ, Slot DE, Susanto H, Gerdes VE, Abbas F, D'Aiuto F et al (2014) Treatment of periodontitis improves the atherosclerotic profile: a systematic review and meta-analysis. J Clin Periodontol 41(1): 70–79. https://doi.org/10.1111/jcpe.12171

Terada Y, Watanabe H, Mori M, Tomino K, Yamamoto M, Moriya M et al (2024) Reliability and utility of blood glucose levels in the periodontal pockets of patients with type 2 diabetes mellitus: a cross-sectional study. Peer J

Theilade E (1986) The non-specific theory in microbial etiology of inflammatory periodontal diseases. J Clin Periodontol 13(10):905–911. https://doi.org/10.1111/j.1600-051x.1986.tb01425.x

Thomas JT, Joseph B, Sorsa T, Mauramo M, Anil S, Waltimo T (2024) Expression of advanced glycation end products and their receptors in diabetic periodontitis patients. Oral Dis 30(5):2784–2796. https://doi.org/10.1111/odi.14769

Tsukasaki M, Komatsu N, Nagashima K, Nitta T, Pluemsakunthai W, Shukunami C et al (2018) Host defense against oral microbiota by bone-damaging T cells. Nat Commun 9(1):701. https://doi.org/10.1038/s41467-018-03147-6

Wang D, Nagasawa T, Chen Y, Ushida Y, Kobayashi H, Takeuchi Y et al (2008) Molecular mimicry of Aggregatibacter actinomycetemcomitans with beta2 glycoprotein I. Oral Microbiol Immunol 23(5): 401–405. https://doi.org/10.1111/j.1399-302X.2008.00442.x

Wang CY, Chyuan IT, Wang YL, Kuo MY, Chang CW, Wu KJ et al (2015) beta2-glycoprotein I-dependent anti-Cardiolipin antibodies associated with periodontitis in patients with systemic lupus erythematosus. J Periodontol 86(8):995–1004. https://doi.org/10.1902/jop.2015.140664

Watt RG (2005) Strategies and approaches in oral disease prevention and health promotion. Bull World Health Organ 83(9):711–718. Retrieved from https://www.ncbi.nlm.nih.gov/pubmed/16211164

Wilson PW, D'Agostino RB, Levy D, Belanger AM, Silbershatz H, Kannel WB (1998) Prediction of coronary heart disease using risk factor categories. Circulation 97(18):1837–1847. https://doi.org/10.1161/01.cir.97.18.1837

Xiao E, Mattos M, Vieira GHA, Chen S, Correa JD, Wu Y et al (2017) Diabetes enhances IL-17 expression and alters the oral microbiome to increase its pathogenicity. Cell Host Microbe 22(1):120–128 e124. https://doi.org/10.1016/j.chom.2017.06.014

Ximenez-Fyvie LA, Haffajee AD, Socransky SS (2000) Comparison of the microbiota of supra- and subgingival plaque in health and periodontitis. J Clin Periodontol 27(9):648–657. https://doi.org/10.1034/j.1600-051x.2000.027009648.x

Xu H, Chen R, Hou X, Li N, Han Y, Ji S (2024) The clinical potential of 1,5-anhydroglucitol as biomarker in diabetes mellitus. Front Endocrinol 15:1471577. https://doi.org/10.3389/fendo.2024.1471577

Yamashita A, Soga Y, Iwamoto Y, Yoshizawa S, Iwata H, Kokeguchi S et al (2007) Macrophage-adipocyte interaction: marked interleukin-6 production by lipopolysaccharide. Obesity (Silver Spring) 15(11): 2549–2552. https://doi.org/10.1038/oby.2007.305

Yang X, Liang R, Xing Q, Ma X (2021) Fighting food allergy by inducing oral tolerance: facts and fiction. Int Arch Allergy Immunol 182(9):852–862. https://doi.org/10.1159/000515292

Zhang L, Wang P, Huang J, Xing Y, Wong FS, Suo J, Wen L (2024) Gut microbiota and therapy for obesity and type 2 diabetes. Front Endocrinol 15:1333778. https://doi.org/10.3389/fendo.2024.1333778

Zhou L, Chong MM, Littman DR (2009) Plasticity of CD4+ T cell lineage differentiation. Immunity 30(5): 646–655. https://doi.org/10.1016/j.immuni.2009.05.001

Impact of Maternal Immune Response on Oral Health and Pregnancy Outcomes 11

Caroline Amaro da Silva, Francisca Marília Cruz Brasileiro, Mariné Olmos Villagómez, Maya Fernanda Manfrin Arnez, and Francisco Wanderley Garcia de Paula-Silva

Abstract

Pregnancy is a complex physiological state characterized by a series of intricate changes across the physiological, immunological, and endocrine systems that are critical for successful fetal development. However, these profound modifications also increase the mother's susceptibility to a variety of systemic and oral health conditions. This chapter explores the nuanced relationships between systemic health issues, such as gestational diabetes mellitus, and periodontal disease, with a particular emphasis on periodontitis, which is recognized as a chronic inflammatory condition that exacerbates systemic inflammatory responses during pregnancy. The oral microbiota undergoes significant alterations due to the hormonal shifts and immunological adjustments that occur during gestation. These changes can disrupt the balance of microbial communities, potentially leading to periodontal disease, which has been linked to a range of adverse maternal and neonatal outcomes. In conjunction with the physiological changes associated with pregnancy, such as endothelial dysfunction and elevated levels of pro-inflammatory cytokines, the risk of cardiovascular, metabolic, and infectious complications is markedly increased. Notably, periodontitis is a contributing factor to systemic health issues that negatively influence maternal and neonatal health. For instance, periodontitis has been associated with adverse events such as low birth weight, preterm delivery, and fetal growth restriction. By examining these interconnections, this chapter provides a comprehensive understanding of the implications of periodontal health on maternal and fetal well-being, elucidating the underlying mechanisms that may contribute to adverse pregnancy outcomes. Through this lens, clinicians and healthcare providers can develop more effective strategies to mitigate risks associated with pregnancy-related complications.

Keywords

Pregnancy · Periodontal disease · Oral health · Preterm labor · Immune response

C. A. da Silva · F. M. C. Brasileiro · M. O. Villagómez · F. W. G. d. Paula-Silva (✉)
Developmental Biology and Child Health Research Group, Ribeirão Preto School of Dentistry, University of São Paulo, São Paulo, Brazil
e-mail: franciscogarcia@forp.usp.br

M. F. M. Arnez
Developmental Biology and Child Health Research Group, Ribeirão Preto School of Dentistry, University of São Paulo, São Paulo, Brazil

Serviço de Odontologia e Estomatologia do Hospital das Clínicas da Faculdade de Medicina de Ribeirão Preto, University of São Paulo, São Paulo, Brazil

N. Rezaei, H. Ziaei (eds.), *Oral Immunology*, Advances in Experimental Medicine and Biology 1492,
https://doi.org/10.1007/978-3-032-03176-1_11

11.1 Introduction

The gestational period is characterized by numerous physical, emotional, and psychological transformations in a woman's life. During pregnancy, both the mother and the developing fetus are susceptible to the physiological changes and various health conditions that may emerge (Ayen et al. 2024). Among the major body systems that undergo significant alterations during this period are the endocrine, circulatory, respiratory, and immune systems. The immune system, in particular, undergoes adaptive modifications to maintain fetomaternal homeostasis, which is essential for the healthy progression of pregnancy (Abu-Raya et al. 2020). As a result, these physiological changes inherent to pregnancy can lead to microbial shifts in the intestines, vagina, and oral cavity of pregnant women (Koren et al. 2023).

Consequently, regarding changes in the oral cavity, there is an increased predisposition to dysbiosis characterized by periodontopathogenic bacteria, which facilitates the development of gingival and periodontal diseases during pregnancy. This condition subsequently heightens the risks of adverse pregnancy outcomes, including preeclampsia, preterm birth, and low birth weight of the newborn (Cobb et al. 2017; Gilani et al. 2024; Wen et al. 2024).

From this perspective, preterm birth stands as the leading cause of neonatal mortality and morbidity, with a global incidence of approximately 15 million births annually. A baby is classified as preterm if born before 37 weeks of gestation. This condition is often triggered by an exaggerated inflammatory response that activates the uterus-placenta complex through cytokine release, ultimately leading to the premature expulsion of the fetus (Habelrih et al. 2024). Current understanding suggests that preterm birth results from a confluence of multiple converging factors (Mariam et al. 2024). However, certain primary risk factors are well-established, such as intrauterine infection, which is attributed to 25–40% of preterm births, along with the presence of pathogens and non-infectious by-products relevant to the placenta and fetus that elicit a fetal inflammatory response marked by cytokine release (Romero et al. 2014).

In the context of adverse pregnancy outcomes, low birth weight is defined by the World Health Organization (WHO) as a birth weight of less than 2,500 g, measured within the first 24 h after birth. Low birth weight is recognized as a significant public health concern, primarily because it serves as a critical indicator of intrauterine growth retardation. Globally, there are approximately 20 million low-birth-weight newborns each year, accounting for about 1 in 7 live births (de Freitas Rocha et al. 2024). Recent studies indicate that maintaining good oral health, adhering to a nutritious diet, managing stress, prioritizing mental health, and avoiding alcohol and tobacco during pregnancy can reduce the risk of low birth weight (Arabzadeh et al. 2024).

The relationship between oral diseases, preterm birth, and low birth weight is well-documented in the literature, although the precise mechanisms underlying this association remain the subject of ongoing studies (Meriç et al. 2023). Current hypotheses suggest that multiple factors contribute to the causative link between periodontal diseases and adverse pregnancy outcomes, including genetic predispositions, socioeconomic status, heightened levels of stress during gestation, and epigenetic influences (Butera et al. 2023). In this context, the maternal immune system plays a pivotal role as a key mechanism, providing valuable insights into how immune responses during pregnancy can influence oral health and how the immune activation in oral diseases increases the risk of adverse pregnancy outcomes.

11.2 Changes in the Immune System During Pregnancy

During pregnancy, substantial adjustments occur within the immune system to facilitate 'maternal tolerance,' a phenomenon that enables the continuation of pregnancy despite the presence of foreign fetal antigens (Fuhler 2020). However, the underlying mechanisms of this tolerance remain inadequately understood. Generally, maternal tolerance is mediated by the modulation and restriction of leukocytes that migrate across the maternal-fetal interface (Ander et al. 2019). Disruptions in maternal tolerance have been associated with adverse pregnancy outcomes such as pre-eclampsia and miscarriage (Fisher 2015).

Beyond the adaptive changes associated with maternal tolerance, there is a growing consensus regarding an overall suppression of immune function during pregnancy. Although empirical evidence supporting this claim is limited, it is recognized that pregnant women exhibit an increased susceptibility to certain diseases, suggesting a significant qualitative shift in immune responses (Kourtis et al. 2014). The subsequent sections will delve into these specific changes within the immune system.

Also, physiological changes in pregnant women, including enhanced activity of the pituitary, thyroid, and adrenal glands, alongside immunological alterations, further contribute to increased susceptibility to infections and other conditions (Silva de Araujo Figueiredo et al. 2017).

11.3 Innate Immune System

11.3.1 Complement System

The complement system, an essential component of the innate immune response, undergoes significant modifications during pregnancy. These alterations play a crucial role in balancing maternal immune tolerance towards the fetus while still providing adequate protection against pathogens.

Increased activity of the complement system has been described during pregnancy, characterized by elevated plasma levels of C3a, C4a, C5a, C4d, C3, and C9, as well as an increase in the membranous attack complex (SC5b-9) (Richani et al. 2005; Derzsy et al. 2010; He et al. 2019). To maintain homeostasis, the concentration of regulatory proteins, such as Factor H, also rises, inhibiting the alternative C3 convertase (Sarma and Ward 2010). Notably, serum complement hemolytic activity, measured by CH50, increases as pregnancy advances. While enhanced complement system activation during pregnancy is a normal adaptation, its excessive activation has been associated with conditions such as pre-eclampsia and preterm birth, indicating that a finely tuned balance in the activation of the complement system is important for a healthy pregnancy (Denny et al. 2012).

11.3.2 Granulocytes

Granulocytes, a type of white blood cell involved in the innate immune response, undergo several notable changes during pregnancy. These changes are crucial for adapting the maternal immune system to support fetal development and protect the mother from infections.

Basophil and eosinophil counts remain stable during pregnancy (Abbassi-Ghanavati et al. 2010). However, an increase in eosinophil degranulation is observed during the second and third trimesters, accompanied by elevated urinary concentrations of N-methylhistamine, which suggests a reduction in mast cell degranulation in the third trimester (Matsumoto et al. 2003). Neutrophils play a role throughout pregnancy, facilitating processes from embryo implantation into the endometrium to the remodeling of the spiral arteries, and continuing until delivery. It is essential to maintain a proper balance in neutrophil concentrations, as dysregulated neutrophil activity can lead to adverse pregnancy outcomes, including pre-eclampsia, premature delivery, and spontaneous abortion (Aslanian-Kalkhoran et al. 2024).

11.3.3 Monocytes

During pregnancy, monocytes, another key component of the innate immune system, undergo various changes that support both maternal health and fetal development. These adaptations help maintain immune tolerance while ensuring an effective and adequate immune response against infections.

Monocytes can be classified into three main subtypes based on their surface markers (CD14 and CD16) and functional characteristics. Classical monocytes ($CD14^{++}$ / $CD16^{-}$) are primarily involved in phagocytosis and the production of pro-inflammatory cytokines. They play a key role in the initial immune response to infections. Conversely, non-classical monocytes ($CD14^{+}$ /$CD16^{+}$ $^{+}$ patrol the endothelium of blood vessels and are involved in the resolution of inflammation and tissue repair. They are less phagocytic compared to classical monocytes and function primarily in immune surveillance. Intermediate monocytes ($CD14^{+}$ / $CD16^{+}$) exhibit both inflammatory and phagocytic properties and their increased numbers can indicate a state of chronic inflammation (Wong et al. 2011; Mukherjee et al. 2015).

During pregnancy, women experience an increase in monocyte concentrations, beginning in the first trimester. This rise is particularly pronounced in intermediate monocytes, while classical monocytes decrease in concentration, and non-classical monocytes remain relatively unchanged (Groen et al. 2015). The elevated levels of intermediate monocytes can be attributed to increased concentrations of interleukins (IL) and tumor necrosis factor-α (TNF-α) (Germain et al. 2007). Additionally, high levels of CD11a, CD11b, CD14, and CD64, along with increased production of Reactive Oxygen Species (ROS), contribute to the heightened pro-inflammatory activity of monocytes during pregnancy (Luppi et al. 2002).

Among monocytes, macrophages are notably reduced during pregnancy (Groen et al. 2015). Naive macrophages can differentiate into M1 macrophages, which are pro-inflammatory and activated by interferon-gamma (IFN-γ), (TNF-α), and lipopolysaccharides (LPS) from Gram-negative bacteria. These M1 macrophages secrete pro-inflammatory cytokines such as IL-6, IL-1β, TNF-α, and matrix metalloproteinase-9 (MMP-9), contributing to the inflammatory process and efforts to eliminate the infectious agents. Alternatively, macrophages can differentiate into M2 macrophages, which are anti-inflammatory and activated by IL-4, IL-13, IL-10, and transforming growth factor-β (TGF-β). M2 macrophages primarily facilitate the regeneration and healing of connective tissues (Xu and Han 2022).

11.3.4 Natural Killer Cells

Natural killer (NK) cells play a crucial role in recognizing and eliminating foreign bodies. These cells are present in the placental decidua and are vital for the successful remodeling of spiral arteries and embryo/fetal implantation during the first trimester of pregnancy (Moffett and Colucci 2014; Artis and Spits 2015). While the precise concentration of circulating NK cells during pregnancy remains to be fully determined, studies have indicated a notable decrease in their levels during the third trimester (Borzychowski et al. 2005).

11.4 Adaptive Immune System

During pregnancy, significant alterations occur within the adaptive immune system to promote maternal tolerance towards the developing fetus, which is genetically distinct from the mother, while still preserving an effective immune response against pathogens.

11.4.1 T Cells

While the total number of lymphocytes remains consistent across all three trimesters of pregnancy, the quantity of T-cells is generally lower in pregnant women compared to non-pregnant individuals (Watanabe et al. 1997). Additionally, a decrease in the concentrations of $CD4^{+}$ and $CD8^{+}$ T-cells has been detected in the third trimester. This reduction may be attributed to elevated levels of estrogen and progesterone, which induce reversible thymic involution affecting $CD4^{+}$ and $CD8^{+}$ cell populations (Zoller et al. 2007).

A prominent feature of these changes is the expansion of regulatory T cells (Tregs), which play a critical role in promoting immune tolerance and preventing maternal rejection of the fetus (Tsuda et al. 2019; Zhang and Sun 2020). This increase in Tregs leads to a corresponding decrease in pro-inflammatory T helper type 1 (Th1) responses, characterized by reduced production of cytokines such as IFN-γ. Simultaneously, there is a shift towards a Th2-dominant immune response, marked by elevated levels of cytokines such as IL-4, IL-5, and IL-13 (Huang et al. 2020). This shift favors humoral immunity and enhances antibody production, consequently mitigating inflammatory responses that could potentially harm fetal development (Leng et al. 2019; Gomez-Lopes et al. 2022).

11.4.2 B Cells

During pregnancy, B cells exhibit significant alterations that are essential for facilitating maternal-fetal tolerance while ensuring adequate immune protection against infections. B cells are directly related to maternal tolerance at the maternal-fetal interface (Hussein et al. 2009). During pregnancy, there is a decrease in the concentration of B-cells in the third trimester of pregnancy, which is known as "physiological B-cell lymphopenia." This reduction is mainly due to elevated estrogen levels, which can increase the susceptibility to infections (Muzzio et al. 2014).

In addition to changes in B cell quantity and subset distribution (Leng et al. 2019), pregnant women also exhibit altered production of immunoglobulins by B cells, particularly an increase in immunoglobulin G (IgG) (Ziegler et al. 2018). This enhancement is crucial for providing passive immunity to the fetus, as IgG can cross the placenta and provide protection against pathogens during the early postnatal period.

Additionally, pregnancy is characterized by modifications in the cytokine environment, with increased levels of anti-inflammatory cytokines, such as interleukin-10 (IL-10), which promote B-cell activation, differentiation into antibody-secreting plasma cells, and modulate the overall immune response to favor tolerance (Busse et al. 2022).

11.4.3 Immunoglobulins

During pregnancy, the concentration of immunoglobulins remains uncertain; some studies report no significant changes across all trimesters, while others indicate a decrease in the third trimester (Lima et al. 2019). This observed decrease in immunoglobulin concentration during pregnancy may be attributed to several factors, including the depression of cell-mediated immunity, protein loss in urine, the transfer of IgG from the mother to the fetus via the placenta, and elevated steroid hormone levels that influence protein synthesis (Abu-Raya et al. 2020). Notably, IgG1 levels were found to be higher in pregnant women compared to non-pregnant women, while IgG3 levels were elevated during the second trimester (Ziegler et al. 2018).

Figure 11.1 depicts the main changes in pregnant women's immune system, as well as the process of "Maternal Tolerance."

11.5 Postpartum Immune Reconstitution

Postpartum immune reconstitution refers to the complex immunological process where the maternal immune system undergoes a series of changes to return to its pre-pregnancy state after childbirth. This reconstitution is essential for the mother's ability to effectively respond to pathogens and restore homeostasis following the immunomodulatory adaptations induced by pregnancy.

During pregnancy, the immune system undergoes adaptations to tolerate the developing fetus, characterized by an increase in regulatory T cells and a shift towards a Th2-dominant immune response. This change is necessary to prevent rejection of the semi-allogeneic fetus. After delivery, there is a rapid decline in the levels of

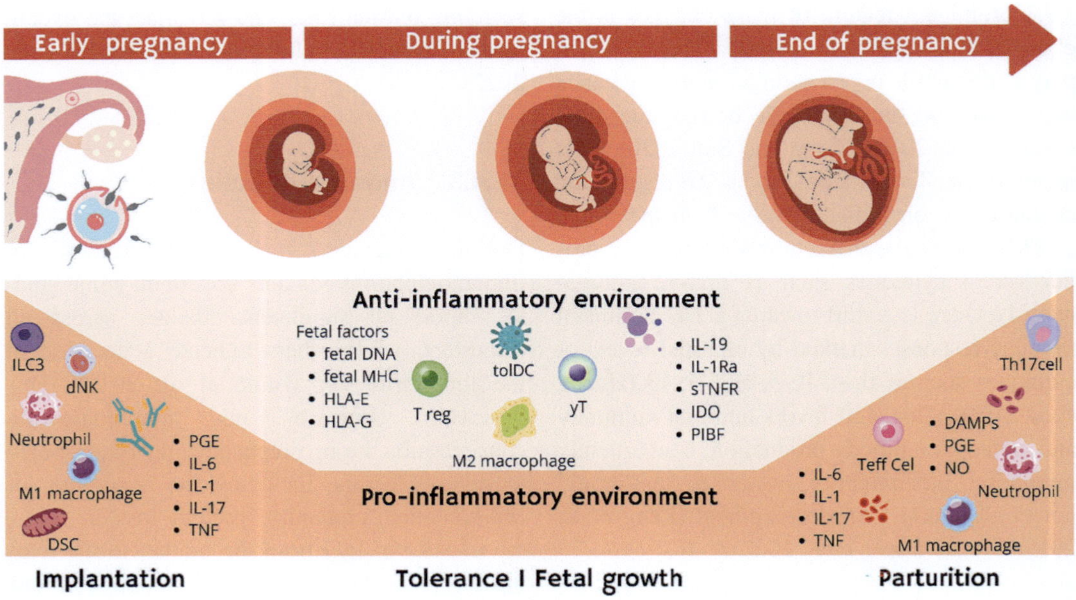

Fig. 11.1 Gestational stages: Immunomodulation mechanisms ensuring fetal retention. At the implantation stage, the presence of neutrophils, M1 macrophages (pro-inflammatory), prostaglandins (PGE), and pro-inflammatory cytokines (IL-6, IL-1, IL-17, and TNF) is noted. During the fetal tolerance development, M2 macrophages (anti-inflammatory) and anti-inflammatory cytokines (IL-19, IL-1Ra) along with regulatory T cells (Tregs) play a role. In the delivery phase, there is a resurgence of pro-inflammatory cytokines (IL-6, IL-1, IL-17, and TNF), along with M1 macrophages (pro-inflammatory), neutrophils, prostaglandins (PGE), damage-associated molecular patterns (DAMPs), Th17 cells (T helper effector lymphocytes that synthesize and secrete IL-17), immune system stem cells, and nitric oxide (NO), which is a vital component of the innate immune response

hormones such as progesterone and estrogen, which were elevated during pregnancy. This hormonal shift plays a significant role in reactivating the immune response. This hormonal shift reduces regulatory T cells and induces a resurgence of Th1-type cytokines, including IL-2, IFN-γ, and TNF-α. This cytokine profile shift is essential for re-establishing effective cellular immunity, allowing the mother to combat infections more efficiently (Ander et al. 2019; Joo et al. 2024).

The dynamics of immune cell populations undergo significant changes postpartum. Cytotoxic T lymphocytes (CTLs) and natural killer (NK) cells, both essential for the clearance of pathogens, increase in number after childbirth. This enhancement in cytotoxic activity is critical for the mother's defense against potential infections that may arise in the postpartum period. Additionally, B cell activation is marked by an increase in antibody production, which provides protection for both the mother and the newborn, particularly through breastfeeding (Singh and Perfect 2007; Mor and Cardenas 2010).

Long-term studies suggest that the immunological changes that occur during and after pregnancy may have lasting effects on a woman's susceptibility to autoimmune diseases and other chronic conditions. These insights highlight the importance of monitoring the postpartum immune landscape, as shifts in immune regulation may influence maternal health outcomes long after delivery (Orefice 2021).

11.6 Common Diseases During Pregnancy

During pregnancy, women may experience a variety of common diseases and conditions that can impact both maternal and fetal health. Although physiological changes during pregnancy are designed to protect the fetus, they may also compromise maternal health, increasing susceptibility to systemic disorders (Degasperi et al. 2021).

Gestational diabetes mellitus is one of the most prevalent complications, characterized by glucose intolerance that arises during pregnancy and typically resolves postpartum. Hypertensive disorders of pregnancy, including gestational hypertension and preeclampsia, are also common, with potential consequences such as placental abruption, preterm delivery, and fetal growth restriction. Additionally, anemia, particularly iron deficiency anemia, is a common condition due to increased maternal blood volume and nutritional demands during gestation, which can lead to fatigue and greater susceptibility to infections. Furthermore, urinary tract infections (UTIs) increase during pregnancy due to anatomical changes and hormonal fluctuations that alter urinary tract function, potentially resulting in complications such as pyelonephritis if left untreated. Other conditions, such as thyroid disorders, infections (such as listeriosis and toxoplasmosis), and exacerbations of pre-existing chronic illnesses (like asthma and autoimmune diseases), also necessitate careful monitoring and management throughout pregnancy to ensure optimal outcomes for both the mother and her developing fetus. Overall, the recognition and proper management of these common diseases during pregnancy are critical for minimizing risks and promoting maternal and fetal well-being.

Because several studies indicate that conditions such as obesity, hypertension, and diabetes mellitus are correlated with poor oral health status in pregnant women (Manrique-Corredor et al. 2019; Figueiro et al. 2020; Foratori-Junior et al. 2021; Jesuino et al. 2020), a comprehensive medical and dental monitoring during prenatal care is vital, as it enables the implementation of strategies to mitigate complications and prevent adverse outcomes for both mother and child (Degasperi et al. 2021; Paula-Silva 2023). Those diseases will be discussed further.

11.6.1 Gestational Anemia

Gestational anemia is defined as a condition characterized by a decrease in hemoglobin concentration or hematocrit levels during pregnancy, leading to a reduced capacity of the blood to carry oxygen. It is typically identified when hemoglobin levels fall below defined thresholds, which vary by gestational age. According to the World Health Organization (WHO), gestational anemia is diagnosed when hemoglobin levels decline below 11 g/dL in the first and third trimesters and drop below 10.5 g/dL in the second trimester, considering the physiological hemodilution that takes place as part of gestational physiological changes (WHO 2012).

The diagnosis of anemia during pregnancy is prevalent due to the expansion of blood volume without a proportional increase in iron levels. Both iron deficiency and iron overload can result in various pathological conditions for pregnant women. For instance, immune cells, including macrophages, T cells, and B cells, require adequate iron levels to effectively mediate their immune functions (Shah et al. 2022). Homeostasis and regulation of iron metabolism can be disrupted during infections or inflammatory diseases. Inflammatory mediators, such as interleukin-6 (IL-6) and bacterial-derived lipopolysaccharides (LPS), have been shown to induce a decrease in circulating iron levels (Nemeth et al. 2004; Armitage et al. 2011). As a result, gestational anemia can impair immune responses of pregnant women, promoting bacterial proliferation and increasing the risk of inflammation (Shah et al. 2022).

11.6.2 Hyperglycemia/Diabetes

Hyperglycemia can develop in pregnant women with pre-existing diabetes mellitus as well as in those without a prior diagnosis of the condition (Hod et al. 2015; WHO 2012). WHO guidelines define gestational diabetes mellitus as a clinical condition that can manifest during pregnancy, characterized by diabetes or glucose intolerance that may persist post-delivery (WHO 2012). The global prevalence of gestational diabetes mellitus varies widely, ranging from 1% to 28%, depending on diagnostic criteria and population characteristics. In Brazil, the prevalence within the Unified Health System (SUS) is estimated to be approximately 18% (Jesus Batista et al. 2021).

Gestational diabetes is a metabolic disorder characterized by pancreatic dysfunction, which may range from mild impairment in endocrine pancreatic function to a complete loss of insulin production capacity (Santos et al. 2021). This condition causes hyperglycemia primarily due to insulin resistance, with its pathophysiology linked to an increase in insulin counter-regulatory hormones resulting from physiological stress during pregnancy, in addition to genetic, immunological, and environmental factors (Santos et al. 2021; Mendonça et al. 2024). The main hormone associated with decreased glucose sensitivity and heightened insulin resistance is placental lactogen. However, other hormones with hyperglycemic effects, including cortisol, estrogen, progesterone, and prolactin, also contribute to the development of gestational hyperglycemia (Mendonça et al. 2024).

In cases of gestational diabetes, the fetus is exposed to elevated levels of glucose in the intrauterine environment, which can adversely affect embryonic development, leading to excessive fetal growth (fetal macrosomia), increased risk during delivery, neonatal hypoglycemia, and a higher likelihood of developing obesity and type 2 diabetes mellitus (DM2) in adulthood (Santos et al. 2021). Beyond hormonal influences, studies have shown that hyperglycemia during pregnancy affects the immune response by disrupting chemotaxis and neutrophil adhesion to vascular endothelium, which may alter the mechanisms of defense and susceptibility to pathogens, thereby increasing the risk of infections in pregnant women (Llorente et al. 2000; Barros and Offenbacher 2014; Bobetsis et al. 2020).

11.6.3 Overweight/Obesity

Obesity in pregnancy is a significant public health concern that can impact both maternal and fetal health. The prevalence of obesity among pregnant women has been rising globally, leading to a range of complications during pregnancy, labor, and postpartum recovery.

Obesity is a chronic inflammatory state influenced by genetic, metabolic, cultural, and behavioral factors. Obesity and overweight are defined in terms of body mass index (BMI). According to WHO, overweight refers to a BMI of 25.0 to 29.9, while obesity refers to a BMI of 30.0 or greater. Obesity in pregnancy is typically defined as having a body mass index (BMI) of 30 or higher, either before conception or during early pregnancy (WHO 2012).

During pregnancy, excess weight is associated with a range of adverse effects, including hypertension, pre-eclampsia, gestational diabetes mellitus, preterm birth, increased likelihood of cesarean delivery, and macrosomia, and it is also linked to metabolic and cardiovascular complications for the child (Tan et al. 2015; Gao et al. 2017). Adipose tissue produces and releases a variety of pro-inflammatory and anti-inflammatory factors, indicating its active role in inflammation and immune responses. Key mediators secreted by adipose tissue include inflammatory cytokines and adipokines such as TNF-α, IL-6, adiponectin, and leptin, which can diminish the host's immune response and contribute to a generalized inflammatory state in the body (Kim et al. 2022; Keller et al. 2015).

11.7 Oral Diseases During Pregnancy

The most prevalent oral diseases during pregnancy include dental caries, gingivitis, periodontitis, and pyogenic granuloma (Pecci-Lloret et al. 2024). Notably, dental caries is the only condition among these that is not directly associated with alterations in the immune system. Conversely, gingivitis, periodontitis, and pyogenic granuloma exhibit an immune response that is integral to their pathogenesis (Xu and Han 2022), once the infiltration of pathogens into oral tissues triggers a localized immune response, mediated by neutrophils, macrophages, complement system components, B cells, and T cells.

11.7.1 Pyogenic Granuloma

Pyogenic granuloma manifests as a red, vascularized mass, most commonly found on the gums, and is characterized by a tendency to bleed easily. It is classified as a benign lesion that can also be found on the lips, tongue, and oral mucosa (Sharma et al. 2019; Rosa et al. 2017). Despite its name, it is not a true granuloma. Instead, it represents an exaggerated proliferation of connective tissue resulting from a local inflammatory response. This condition is frequently observed during the second and third trimesters of pregnancy, a period marked by peak hormone levels, increased bleeding propensity, and potential interference with chewing (Alshuhail et al. 2023; Patrice et al. 1991; Güncü et al. 2005). The lesion's surface may become ulcerated, leading to spontaneous bleeding; however, it is generally painless unless irritated by opposing teeth during chewing. Importantly, pyogenic granuloma rarely results in alveolar bone loss (Verma et al. 2012; Bi et al. 2019).

The significant increase in progesterone and estrogen levels during pregnancy directly affects the injured endothelium. These hormonal changes inhibit inflammatory cells and fibroblasts, mechanisms that are essential for proper wound healing and new tissue formation. Moreover, modulation occurs within the body that intensifies chronic inflammatory responses. Steroid hormones can prevent the apoptosis of inflammatory cells, resulting in an intense inflammatory response. Consequently, these physiological changes during pregnancy can hinder optimal healing post-trauma, thereby increasing the likelihood of developing a pyogenic granuloma (Sharma et al. 2019). Maintaining proper oral hygiene is vital for plaque removal and preventing lesions that could lead to the formation of this pathology (Saadaqui et al. 2021; Manegold Brauer et al. 2014).

11.7.2 Gingivitis

Gingivitis is a prevalent condition during pregnancy, often referred to as pregnancy gingivitis when it occurs during this period. It is a reversible condition that can be localized or generalized, with the anterior teeth being the most commonly affected region. Currently, the most widely accepted classification is that it is plaque-induced gingivitis, with pregnancy acting as a significant modifying factor in the process (Murakami et al. 2018). During pregnancy, gingivitis typically begins in the second month of gestation and significantly increases until it reaches a peak in the eighth month. Subsequently, in some cases, there is an observation of spontaneous regression of this clinical condition after childbirth. This clinical presentation coincides with the elevated levels of progesterone that are characteristic of gestation (Díaz-Gusmán and Castellanos-Suárez 2004).

Pregnancy gingivitis is often attributed to hormonal changes. Progesterone causes an increase in exudate, affects the integrity of endothelial cells, and influences the biosynthesis of prostaglandins in the gums. This tissue can function as a target organ for these hormones, as previous studies have demonstrated the presence of estrogen and progesterone receptors in the gingival tissue (Vittek et al. 1982). In this way, the gums become inflamed, edematous, sensitive, and prone to bleeding (Díaz-Gusmán and Castellanos-Suárez 2004).

Gestational gingivitis typically manifests between the second and third months of pregnancy and may exacerbate in the following months. Common symptoms experienced by pregnant women include gum redness, inflammation, tenderness, and bleeding. Apart from changing the tissue response to pathogens, elevated hormone levels during pregnancy may also contribute to alterations in the oral microbiota, characterized by an increased proportion of both anaerobic and aerobic bacteria, such as

Bacteroides melaninogenicus, *Prevotella intermedia*, and *Porphyromonas gingivalis* (Madianos et al. 2013).

Oral pathogens, particularly periodontopathogenic bacteria, stimulate the host immune system to release pro-inflammatory cytokines, including IL-6, IL-8, and IL-1β, which induce gingival inflammation. This inflammatory response subsequently increases capillary density and vascular permeability (Bobetsis et al. 2020; Figuero et al. 2020). Additionally, progesterone improves the synthesis of prostaglandins, especially prostaglandin E_2, which can exacerbate the clinical symptoms of gingival inflammation (Markou et al. 2009).

11.7.3 Periodontitis

Epidemiological studies show that one in five pregnant women may experience severe periodontitis (Gilani et al. 2024). According to the American Academy of Periodontology (AAP) and the European Federation of Periodontology (EFP) classification, periodontitis is defined as "a multifactorial chronic inflammatory disease associated with dysbiotic biofilm, characterized by the progressive destruction of the dental attachment apparatus." Clinically, periodontitis is defined by attachment loss at two or more non-adjacent interproximal sites and ≥ 3 mm attachment loss in the buccal or lingual/palatal areas of at least two teeth. These changes should not result from traumatic gingival recession, cervical caries, second molar attachment loss due to third molar issues, endo-periodontal lesions draining through the periodontium, or cervical root fractures. Additionally, periodontitis is graded according to disease stages, which indicate the severity of the disease, reflect the evidence of disease progression, and highlight its impact on systemic health (Raza et al. 2024).

Microbial dysbiosis in periodontal tissues is characterized by a change from aerobic Gram-positive bacteria, which live in symbiosis with the oral environment, to anaerobic Gram-negative bacteria. Those bacteria are grouped as orange and red complexes, which include *Campylobacter rectus, Prevotella intermedia, Parvimonas micra, Fusobacterium nucleatum, Porphyromonas gingivalis, Tannerella forsythia,* and *Treponema denticola* (Sugiyama et al. 2023). During the gestational period, an increase in both aerobic and anaerobic bacteria in the oral environment occurs predominantly due to an increase in the concentration of estrogen and progesterone hormones, which stimulate the hypervascularization of periodontal tissues (da Silva et al. 2017).

The immunological changes associated with pregnancy also contribute to the rapid progression of periodontal disease in pregnant women, highlighting the roles of pro-inflammatory Th1 cytokines and anti-inflammatory Th2 cytokines. An elevated concentration of Th1 cytokines was found in the oral cavity of pregnant women, particularly in cases of severe periodontal disease, suggesting an enhanced inflammatory periodontal profile (Meriç et al. 2023). Pregnant women are at an increased risk of inflammation, including acute inflammatory responses in the periodontal tissues, which are associated with diminished antimicrobial activity of peripheral neutrophils (da Silva et al. 2017). The severity of periodontal disease can be further exacerbated by the altered immune response that occurs during pregnancy, characterized by a reduction in lymphocytes, suppression of T-cell activity, decreased antibody production, and impaired neutrophil function regarding phagocytosis and chemotaxis (Boyapati et al. 2021). Physiological alterations during pregnancy can also impair the turnover of periodontal connective tissue, potentially leading to an enhanced inflammatory response within the periodontium (da Silva et al. 2017).

The virulence factors of *P. gingivalis*, including surface adhesins and various enzymes, can directly damage the morphology of tissues and disrupt the microbiome of both fetal and maternal environments. This pathogen can reach the placenta through two distinct mechanisms: directly via the bloodstream or indirectly through the release of inflammatory mediators. In vitro studies demonstrate that *P. gingivalis* induces IL-8 and IFN-γ secretion and apoptosis in human extravillous trophoblasts (ATCC) via activation of extracellular signal-regulated kinase1/2 and p38 signaling pathways1 (Ren et al. 2016). Furthermore, *P. gingivalis* may enhance the production of cytokines, free radicals, and acute phase proteins within the uterine compartment, thereby elevating the risk of myometrial contractions and premature labor. Although *P. gingivalis* has been detected in the amniotic fluid and placenta in cases of obstetric complications, the specific molecular mechanisms underlying these prenatal and postnatal complications deserve further investigation (Chopra et al. 2020a, b).

A meta-analysis revealed that in 60% of the evaluated studies, there was a positive association between the presence of periodontal disease in pregnant women and preterm birth, indicating that periodontal disease increased the risk of prematurity by two-fold (Corbella et al. 2016; Manrique-Corredor et al. 2019). Additionally, another study demonstrated that the risk is increased by approximately six times, with the magnitude of the association being modified depending on the patient's glycemia (Gomes-Filho et al. 2016; Gomes-Filho et al. 2022).

It is also known that, depending on the gestational period and the severity of periodontal disease, the outcomes can differ. There is a correlation between less severe periodontal disease and preterm birth, as well as a correlation between more severe periodontal disease, beginning early in gestation, with intrauterine growth restriction and extreme prematurity (Offenbacher et al. 2006). Therefore, regardless of the mechanism, whether direct or indirect, it is already well-established that women with periodontal disease exhibit elevated plasma concentrations of mediators associated with the induction of labor (Klebanoff and Searle 2006; Beck et al. 2019; Mahapatra et al. 2021).

The mechanisms underlying adverse pregnancy outcomes associated with periodontitis may occur through two pathways (Fig. 11.2):

1. Hematogenous dissemination of bacteria (bacteremia): This pathway involves the translocation of periodontal pathogens via the bloodstream to the fetoplacental region. This can potentially lead to systemic infections and influence pregnancy outcomes. Another potential route involves the transmission and subsequent colonization of oral microbial pathogens in the vaginal microbiome, which may occur through sexual practices (Cobb et al. 2017).
2. Hematogenous dissemination of inflammatory mediators: The presence of periodontitis triggers the release of numerous inflammatory mediators, including prostaglandins (mostly PGE_2), IL-1β, IL-2, IL-6, TNF-α, and INF-γ. These mediators, produced locally in the periodontium, can reach the uterus and induce uterine contractions, endothelial dysfunction, and rupture of fetal membranes, which are critical factors contributing to preterm labor and other adverse outcomes.

These mechanisms can interact with one another, compounding the risk of poor pregnancy outcomes such as preterm birth, low birth weight, and other complications. Understanding these pathways is essential for developing effective strategies aimed at managing periodontal health during pregnancy and improving maternal and fetal outcomes.

While it may be hypothesized that supportive periodontal treatment could improve prognosis and limit the dissemination of pathogens from the oral cavity to the fetal-placental interface or amniotic fluid in pregnant women with periodontal disease, current scientific literature indicates that periodontal treatment during the second trimester does not significantly reduce the risk of adverse pregnancy outcomes (Jeffcoat et al. 2002; Papapanou 2015; Penova-Veselinovic et al. 2015).

Nonetheless, this conclusion warrants further examination due to several methodological limitations in the research conducted, including (1) Variations in the comprehensiveness of

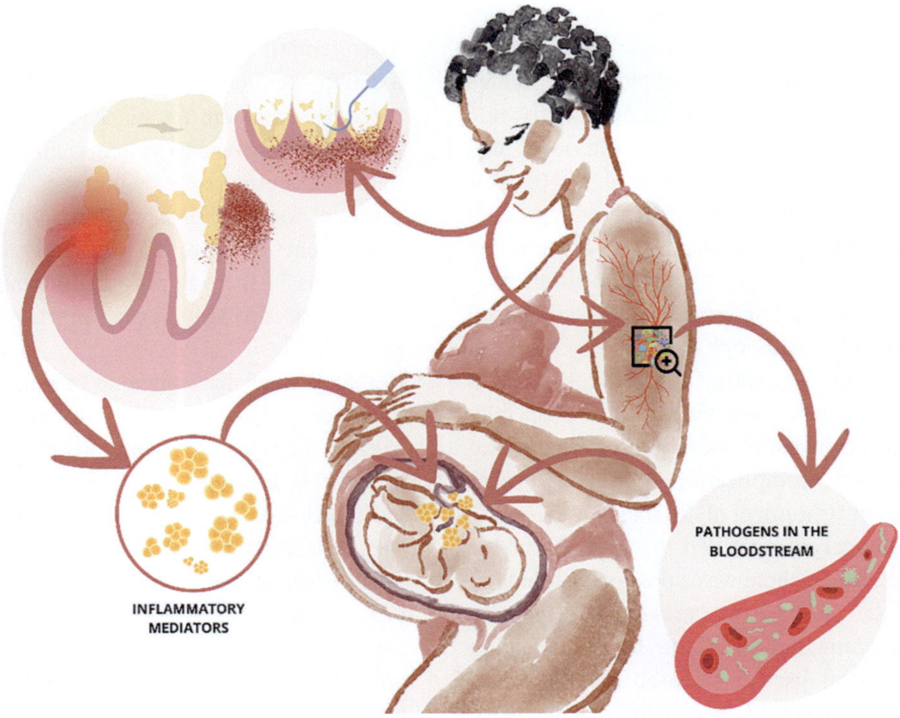

Fig. 11.2 Disruption of homeostasis—an increase in periodontal pathogens with their toxic products, together with a relatively suppressed immune system, can lead to adverse effects during pregnancy

periodontal examinations (partial versus full-mouth assessments), which can lead to inaccurate determinations of disease severity; (2) inconsistencies regarding the timing of periodontal evaluations relative to gestational age; (3) the absence of examiner blinding; (4) a lack of multivariate analyses to account for confounding variables; (5) variability in treatment protocols; (6) treatment failures that lead to insufficient reduction of periodontal inflammation; and (7) the possibility that initiating treatment in the second trimester may be too late to yield beneficial effects on pregnancy outcomes (Cobb et al. 2017).

In this context, a holistic approach to prevent adverse pregnancy outcomes and manage periodontal diseases should be adopted, as unfavorable outcomes are influenced by multiple factors.

11.8 Epigenetics, Periodontal Disease, and Adverse Pregnancy Outcomes

The relationship between epigenetics, periodontal disease, and adverse pregnancy outcomes is an emerging area of research that highlights the complex interplay between these factors.

Epigenetics studies how environmental and behavioral factors influence gene expression without altering the DNA sequence itself. These changes are heritable and can be passed down from one generation to the next. Epigenetic modifications that regulate gene expression occur through mechanisms such as DNA methylation, histone modifications, and the action of non-coding RNAs (Briceño-Pérez et al. 2023). The human genome consists of a combination of protein-coding RNAs (mRNAs) and non-protein-coding RNAs (ncRNAs). There is currently significant scientific interest in understanding the role of epigenetic mechanisms in the etiopathogenesis of various diseases (Lizárraga et al. 2023).

The inflammatory responses associated with periodontal disease can lead to epigenetic modifications in both maternal and fetal tissues. Pro-inflammatory cytokines and other mediators may induce changes in DNA methylation and histone modifications, ultimately affecting gene expression related to immune responses, inflammation, and tissue remodeling. Research conducted to elucidate the relationship between periodontal disease and pre-eclampsia identified a circulating set of nine miRNAs that were significantly associated with pre-eclampsia. The analysis aimed to pinpoint specific genes, pathways, and functional mechanisms that may be epigenetically altered, thereby mediating the biological association between periodontitis and pre-eclampsia (Zhang et al. 2022). The study of epigenetic biomarkers in relation to periodontal disease may provide insights into the mechanisms underlying adverse pregnancy outcomes.

Another study investigated the epigenetic profiles of obese and non-obese pregnant women by analyzing saliva. This research revealed changes in the expression of 20 different miRNAs and varying methylation levels of two genes linked to obesity, inflammation, and oxidative stress (Mandò et al. 2022). Epigenetic modifications during critical periods of fetal development can have long-term effects on health. For instance, changes in gene expression patterns in response to obesity may impact fetal growth, immune function, and susceptibility to diseases later in life. These epigenetic changes can influence mechanisms such as cell signaling, inflammation, and tissue development.

These findings showcase the considerable potential of epigenetics as a tool for both diagnosis and the development of therapies aimed at promoting a healthy pregnancy with minimal risk of adverse outcomes.

11.9 Conclusion

Systemic and oral diseases are intricately interconnected, demonstrating a bidirectional relationship that becomes significantly exacerbated during pregnancy. The hormonal and immunological alterations that characterize this physiological state contribute to enhanced susceptibility among pregnant women to a range of oral health conditions, particularly gingivitis and periodontitis. The inflammation and microbial dysbiosis associated with these oral diseases can, in turn, exacerbate systemic disorders, such as gestational diabetes mellitus, thereby compromising the health of both the mother and the developing fetus.

The chronic inflammatory response linked to periodontitis is particularly concerning, as it not only amplifies the severity of existing systemic conditions but also poses an increased risk for a spectrum of adverse pregnancy outcomes, including preterm birth and low birth weight. Inflammatory mediators released during periodontal disease can enter systemic circulation, leading to elevated levels of pro-inflammatory cytokines that may negatively influence placental function and fetal development.

Furthermore, the interplay between oral disease and systemic health during pregnancy underscores the importance of vigilant monitoring and management of maternal oral health. Effective interventions targeting oral diseases may thus hold significant potential for mitigating risks associated with adverse maternal and neonatal outcomes. Recognizing the implications of this bidirectional relationship is critical for healthcare providers in developing comprehensive care strategies that address the multifaceted health needs of pregnant women, ultimately promoting better outcomes for both mothers and infants. By exploring these interconnected dynamics, we can better appreciate the significance of maintaining optimal oral health as an integral part of prenatal care.

References

Abbassi-Ghanavati M, Greer LG, Cunningham FG (2010) Pregnancy and laboratory studies: a reference table for clinicians. Obstet Gynecol 115(4):868–869

Abu-Raya B, Michalski C, Sadarangani M, Lavoie PM (2020) Maternal immunological adaptation during normal pregnancy. Front Immunol 11

Alshuhail O, Alharbi AS, Alakeel N (2023) A persistent oral pyogenic granuloma: a case report with review of literature. Cureus 15(11):e49326

Ander SE, Diamond MS, Coyne CB (2019) Immune responses at the maternal-fetal interface. Sci Immunol 4(31):eaat6114

Arabzadeh H, Doosti-Irani A, Kamkari S, Farhadian M, Elyasi E, Mohammadi Y (2024) The maternal factors associated with infant low birth weight: an umbrella review. BMC Pregnancy Childbirth 24(1)

Armitage AE, Eddowes LA, Gileadi U, Cole S, Spottiswoode N, Selvakumar TA, Ho LP, Townsend AR, Drakesmith H (2011) Hepcidin regulation by innate immune and infectious stimuli. Blood 118(15): 4129–4139

Artis D, Spits H (2015) The biology of innate lymphoid cells. Nature 517(7534):293–301

Aslanian-Kalkhoran L, Mehdizadeh A, Aghebati-Maleki-L, Danaii S, Shahmohammadi-Farid S, Yousefi M (2024) Neutrophil extracellular traps (NET) in reproductive system and pregnancy complications. J Reprod Immunol:104237

Ayen SS, Kasahun AW, Zewdie A (2024) Depression during pregnancy and associated factors among women in Ethiopia: a systematic review and meta-analysis. BMC Pregnancy Childbirth 24(1)

Barros SP, Offenbacher S (2014) Modifiable risk factors in periodontal disease: epigenetic regulation of gene expression in the inflammatory response. Periodontol 64(1):95–110

Beck JD, Papapanou PN, Philips KH, Offenbacher S (2019) Periodontal medicine: 100 years of progress. J Dent Res 98(10):1053–1062

Bi J, Sun Y, Bi L, Larjava HS (2019) Large pregnancy-associated pyogenic granuloma: a case report. J Obstet Gynaecol 39(2):265–267

Bobetsis YA, Graziani F, Gürsoy M, Madianos PN (2020) Periodontal disease and adverse pregnancy outcomes. Periodontology 2000 83(1):154–174

Borzychowski A, Croy BA, Chan WL, Redman CG, Sargent I (2005) Changes in systemic type 1 and type 2 immunity in normal pregnancy and pre-eclampsia may be mediated by natural killer cells. Eur J Immunol 35(10):3054–3063

Boyapati R, Cherukuri S, Bodduru R, Kiranmaye A (2021) Influence of female sex hormones in different stages of women on periodontium. J Mid-Life Health 12(4):263

Briceño-Pérez C, Briceño-Sanabria L, Briceño-Sanabria C, Reyna-Villasmil E (2023) Early life corticosteroid overexposure: epigenetic and fetal origins of adult diseases. Int J Gynecol Obstet

Busse M, Langwisch S, Tedford K, Fischer KD, Zenclussen AC (2022) Maternal B cell signaling orchestrates fetal development in mice. Development 149(8):dev199783

Butera A, Maiorani C, Morandini A, Trombini J, Simonini M, Ogliari C, Scribante A (2023) Periodontitis in pregnant women: a possible link to adverse pregnancy outcomes. Healthcare 11(10):1372

Chopra M, Kaur N, Singh KD, Maria Jacob C, Divakar H, Babu GR, Hong Nguyen P, Bhanot A, Sabharwal M, Deb S, Baswal D, Louise Killeen S, McAuliffe FM, Hanson MA, Sethi V (2020a) Population estimates, consequences, and risk factors of obesity among pregnant and postpartum women in India: results from a national survey and policy recommendations. Int J Gynecol Obstet 151(S1):57–67

Chopra A, Radhakrishnan R, Sharma M (2020b) Porphyromonas gingivalis and adverse pregnancy outcomes: a review on its intricate pathogenic mechanisms. Crit Rev Microbiol 46(2):213–236

Cobb C, Kelly P, Williams K, Babbar S, Angolkar M, Derman R (2017) The oral microbiome and adverse pregnancy outcomes. Int J Women's Health 9:551–559

Corbella S, Taschieri S, Del Fabbro M, Francetti L, Weinstein R, Ferrazzi E (2016) Adverse pregnancy outcomes and periodontitis: a systematic review and meta-analysis exploring potential association. Quintessence Int 47(3):193–204

Da Silva MK, de Carvalho ACG, Alves EHP, da Silva FRP, Pessoa LDS, Vasconcelos DFP (2017) Genetic factors and the risk of periodontitis development: findings from a systematic review composed of 13 studies of meta-analysis with 71,531 participants. Int J Dent 2017:1–9

De Freitas Rocha AR, de Souza de Morais N, Azevedo FM, de Castro Morais D, Priore SE (2024) Maternal food insecurity increases the risk of low birth weight babies: systematic review and meta-analysis. Acta Paediatr

Degasperi JU, Dias AJW, Boleta-Ceranto DDCF (2021) Oral and systemic changes resulting from pregnancy and the importance of medical and dental prenatal care to reduce pregnancy complications. Res Soc Dev 10(3):e8810312976

Denny KJ, Woodruff TM, Taylor SM, Callaway LK (2012) Complement in pregnancy: a delicate balance. Am J Reprod Immunol 69(1):3–11

Derzsy Z, Prohászka Z, Rigó J Jr, Füst G, Molvarec A (2010) Activation of the complement system in normal pregnancy and preeclampsia. Mol Immunol 47(7–8): 1500–1506

Díaz-Guzmán LM, Castellanos-Suárez JL (2004) Lesiones de la mucosa bucal y comportamiento de la enfermedad periodontal en embarazadas. Med Oral Patol Oral Cir Bucal 9(5):430–437

Figuero E, Han YW, Furuichi Y (2020) Periodontal diseases and adverse pregnancy outcomes: mechanisms. Periodontol 83(1):175–188

Fisher SJ (2015) Why is placentation abnormal in preeclampsia? Am J Obstet Gynecol 213(4):S115–S122

Foratori-Junior GA, Missio ALT, Orenha ES, de Carvalho Sales-Peres SH (2021) Systemic condition, periodontal status, and quality of life in obese women during pregnancy and after delivery. Int Dent J 71(5):420–428

Fuhler GM (2020) The immune system and microbiome in pregnancy. Best Pract Res Clin Gastroenterol 44–45: 101671

Gao X, Yan Y, Xiang S, Zeng G, Liu S, Sha T, He Q, Li H, Tan S, Chen C, Li L, Yan Q (2017) The mutual effect of pre-pregnancy body mass index, waist circumference and gestational weight gain on obesity-related adverse pregnancy outcomes: a birth cohort study. PLoS One 12(6):e0177418

Germain SJ, Sacks GP, Soorana SR, Sargent IL, Redman CW (2007) Systemic inflammatory priming in normal pregnancy and preeclampsia: the role of circulating syncytiotrophoblast microparticles. J Immunol 178(9):5949–5956

Gilani SI, Niaz A, Afridi S (2024) Maternal periodontitis as a risk factor for preterm birth: a cross-sectional study. J Dent Res Dent Clin Dent Prospects 18(1): 72–76

Gomes-Filho IS, Pereira EC, Cruz SS, Adan LF, Vianna MI, Passos-Soares JS, Trindade SC, Oliveira EP, Oliveira MT, Cerqueira EDM, Pereira AL, Barreto ML, Seymour GJ (2016) Relationship among mothers' glycemic level, periodontitis, and birth weight. J Periodontol 87(3):238–247

Gomes-Filho IS, Trindade SC, da Cruz SS, Figueiredo ACMG, Hintz AM, Adan LFF, Cerqueira EMM, Passos-Soares JS, Vianna MIP, Coelho JMF, Alves CMC, Barreto ML, Seymour GJ, Loomer PM, Scannapieco FA (2022) Mothers' high glycemic levels and the association between periodontitis and low birth weight. J Periodontol 93(7):954–965

Gomez-Lopez N, Galaz J, Miller D, Farias-Jofre M, Liu Z, Arenas-Hernandez M, Garcia-Flores V, Shaffer Z, Greenberg JM, Theis KR, Romero R (2022) The immunobiology of preterm labor and birth: intra-amniotic inflammation or breakdown of maternal-fetal homeostasis. Reproduction 164(2):R11–R45

Groen B, van der Wijk A-E, van den Berg PP, Lefrandt JD, van den Berg G, Sollie KM, de Vos P, Links TP, Faas MM (2015) Immunological adaptations to pregnancy in women with type 1 diabetes. Sci Rep 5(1)

Güncü GN, Tözüm TF, Cağlayan F (2005) Effects of endogenous sex hormones on the periodontium-review of literature. Aust Dent J 50(3):138–145

Habelrih T, Ferri B, Côté F, Sévigny J, Augustin T-L, Sawaya K, Lubell WD, Olson DM, Girard S, Chemtob S (2024) Preventing preterm birth. Clin Perinatol

He Y, Xu B, Song D, Wang Y, Yu F, Chen Q, Zhao M (2019) Normal range of complement components during pregnancy: a prospective study. Am J Reprod Immunol 83(2)

Hod M, Kapur A, Sacks DA, Hadar E, Agarwal M, Di Renzo GC, Cabero Roura L, McIntyre HD, Morris JL, Divakar H (2015) The International Federation of Gynecology and Obstetrics (FIGO) initiative on gestational diabetes mellitus: a pragmatic guide for diagnosis, management, and care. Int J Gynaecol Obstet 131 (Suppl 3):S173–S211

Huang N, Chi H, Qiao J (2020) Role of regulatory T cells in regulating fetal-maternal immune tolerance in healthy pregnancies and reproductive diseases. Front Immunol 11:1023

Hussein MR, Abd-Elwahed AR, Abodeif ES, Abdulwahed SR (2009) Decidual immune cell infiltrate in hydatidiform mole. Cancer Investig 27(1):60–66

Jeffcoat MK, Geurs NC, Reddy MS, Cliver SP, Goldenberg RL, Hauth JC (2002) Periodontal infection and preterm birth: results of a prospective study. Obstet Gynecol Surv 57(1):5–6

Jesuino BG, Foratori-Junior GA, Missio ALT, Mascoli LS, Sales-Peres SHC (2020) Periodontal status of women with excessive gestational weight gain and the association with their newborns' health. Int Dent J 70(5):396–404

Jesus Batista MH, de Sousa LP, de Souza DMD, Silva RO, Lima EDS, Nunes TS, Schimidt CP, Rocha MA (2021) Diabetes Gestacional: Origem, Prevenção e Riscos/ gestational diabetes: origin, prevention and risks. Braz J Dev 7(1):1981–1995

Joo JS, Lee D, Hong JY (2024) Multi-layered mechanisms of immunological tolerance at the maternal-fetal interface. Immune Netw 24(4):e30

Keller A, Rohde JF, Raymond K, Heitmann BL (2015) Association between periodontal disease and overweight and obesity: a systematic review. J Periodontol 86(6):766–776

Kim CM, Lee S, Hwang W, Son E, Kim TW, Kim K, Kim YH (2022) Obesity and periodontitis: a systematic review and updated meta-analysis. Front Endocrinol 13:999455

Klebanoff M, Searle K (2006) The role of inflammation in preterm birth–focus on periodontitis. BJOG 113(Suppl 3):43–45

Koren O, Konnikova L, Brodin P, Mysorekar IU, Collado MC (2023) The maternal gut microbiome in pregnancy: implications for the developing immune system. Nat Rev Gastroenterol Hepatol

Kourtis AP, Read JS, Jamieson DJ (2014) Pregnancy and infection. N Engl J Med 370(23):2211–2218

Leng Y, Romero R, Xu Y, Galaz J, Slutsky R, Arenas-Hernandez M, Garcia-Flores V, Motomura K, Hassan SS, Reboldi A, Gomez-Lopez N (2019) Are B cells altered in the decidua of women with preterm or term labor? Am J Reprod Immunol 81(5):e13102

Lima J, Cambridge G, Vilas-Boas A, Martins C, Borrego L, Leandro M (2019) Serum markers of B-cell activation in pregnancy during late gestation, delivery, and the postpartum period. Am J Reprod Immunol 81(3):e13090

Lizárraga D, Gómez-Gil B, García-Gasca T, Ávalos-Soriano A, Casarini L, Salazar-Oroz A, García-Gasca A (2023) Gestational diabetes mellitus: genetic factors, epigenetic alterations, and microbial composition. Acta Diabetol

Llorente L, De La Fuente H, Richaud-Patin Y, Alvarado-De La Barrera C, Diaz-Borjón A, López-Ponce A, Lerman-Garber I, Jakez-Ocampo J (2000) Innate immune response mechanisms in non-insulin dependent diabetes mellitus patients assessed by flow cytoenzymology. Immunol Lett 74(3):239–244

Luppi P, Haluszczak C, Betters D, Richard CAH, Trucco M, DeLoia JA (2002) Monocytes are progressively activated in the circulation of pregnant women. J Leukoc Biol 72(5):874–884

Madianos PN, Bobetsis YA, Offenbacher S (2013) Adverse pregnancy outcomes (APOs) and periodontal disease: pathogenic mechanisms. J Clin Periodontol 40 (Suppl 14):S170–S180

Mahapatra A, Nayak R, Satpathy A, Pati BK, Mohanty R, Mohanty G, Beura R (2021) Maternal periodontal status, oral inflammatory load, and systemic inflammation are associated with low infant birth weight. J Periodontol 92(8):1107–1116

Mandò C, Abati S, Anelli GM, Favero C, Serati A, Dioni L, Zambon M, Albetti B, Bollati V, Cetin I (2022) Epigenetic profiling in the saliva of obese pregnant women. Nutrients 14(10):2122

Manegold-Brauer G, Brauer HU (2014) Oral pregnancy tumour: an update. J Obstet Gynaecol 34(2):187–188

Manrique-Corredor EJ, Orozco-Beltran D, Lopez-Pineda-A, Quesada JA, Gil-Guillen VF, Carratala-Munuera C (2019) Maternal periodontitis and preterm birth: systematic review and meta-analysis. Community Dent Oral Epidemiol 47(3):243–251

Mariam S, Hasan S, Shinde M, Gupta J, Buch SA, Rajpurohit KS, Patil V (2024) Pregnancy outcomes and maternal periodontal diseases: the unexplored connection. Cureus 16:e61697

Markou E, Eleana B, Lazaros T, Antonios K (2009) The influence of sex steroid hormones on gingiva of women. Open Dent J 3(1):114–119

Matsumoto K, Ogasawara T, Kato A, Homma T, Iida M, Akasawa A, Wakiguchi H, Saito H (2003) Eosinophil degranulation during pregnancy and after delivery by cesarean section. Int Arch Allergy Immunol 131 (Suppl. 1):34–39

Mendonça PHP, Santos HR, Martins CAL, Brasil LG (2024) Diabetes gestacional: atualização em critérios diagnósticos e terapêuticos. Revista Brasileira de Revisão de Saúde 7(2):e69035

Meriç P, Silbereisen A, Emingil G, Öztürk V-Ö, Bostanci N (2023) Clinical, oral immunological and microbiological shifts during and after pregnancy. Clin Oral Investig 28(1)

Moffett A, Colucci F (2014) Uterine NK cells: active regulators at the maternal-fetal interface. J Clin Invest 124(5):1872–1879

Mor G, Cardenas I (2010) The immune system in pregnancy: a unique complexity. Am J Reprod Immunol 63(6):425–433

Mukherjee R, Kanti Barman P, Kumar Thatoi P, Tripathy R, Kumar Das B, Ravindran B (2015) Non-classical monocytes display inflammatory features: validation in sepsis and systemic lupus erythematous. Sci Rep 5(1)

Murakami S, Mealey BL, Mariotti A, Chapple ILC (2018) Dental plaque-induced gingival conditions. J Periodontol 89(Suppl 1):S17–S27

Muzzio DO, Soldati R, Ehrhardt J, Utpatel K, Evert M, Zenclussen AC, Zygmunt M, Jensen F (2014) B cell development undergoes profound modifications and adaptations during pregnancy in mice1. Biol Reprod 91(5)

Nemeth E, Rivera S, Gabayan V, Keller C, Taudorf S, Pedersen BK, Ganz T (2004) IL-6 mediates hypoferremia of inflammation by inducing the synthesis of the iron regulatory hormone hepcidin. J Clin Invest 113(9):1271–1276

Offenbacher S, Boggess KA, Murtha AP, Jared HL, Lieff S, McKaig RG, Mauriello SM, Moss KL, Beck JD (2006) Progressive periodontal disease and risk of very preterm delivery. Obstet Gynecol 107(1):29–36

Orefice R (2021) Immunology and the immunological response in pregnancy. Best Pract Res Clin Obstet Gynaecol 76:3–12

Papapanou PN (2015) Systemic effects of periodontitis: lessons learned from research on atherosclerotic vascular disease and adverse pregnancy outcomes. Int Dent J 65(6):283–291

Patrice SJ, Wiss K, Mulliken JB (1991) Pyogenic granuloma (lobular capillary hemangioma): a clinicopathologic study of 178 cases. Pediatr Dermatol 8(4): 267–276

Paula-Silva FWG (2023) Pré-natal Odontológico. Santos Publicações

Pecci-Lloret MP, Linares-Pérez C, Pecci-Lloret MR, Rodríguez-Lozano FJ, Oñate-Sánchez RE (2024) Oral manifestations in pregnant women: a systematic review. J Clin Med 13(3):707

Penova-Veselinovic B, Keelan JA, Wang CA, Newnham JP, Pennell CE (2015) Changes in inflammatory mediators in gingival crevicular fluid following periodontal disease treatment in pregnancy: relationship to adverse pregnancy outcome. J Reprod Immunol 112: 1–10

Raza M, Abud DG, Wang J, Shariff JA (2024) Ease and practicability of the 2017 classification of periodontal diseases and conditions: a study of dental electronic health records. BMC Oral Health 24(1)

Ren H, Li Y, Jiang H, Du M (2016) Porphyromonas gingivalis induces IL-8 and INF-gamma secretion and apoptosis in human extravillous trophoblast derived HTR8/SVneo cells via activation of ERK1/2 and p38 signaling pathwas. Placenta 45:8–15

Richani K, Soto E, Romero R, Espinoza J, Chaiworapongsa T, Nien JK, Edwin S, Kim YM, Hong J-S, Mazor M (2005) Normal pregnancy is characterized by systemic activation of the clinical practice has the potential to revolutionize diagnostics and therapeutic management, enabling more personalized approaches.

12.2 Salivary Immune and Endocrine Biomarkers in Allergic Skin Conditions

12.2.1 Atopic Dermatitis (AD)

AD is a chronic inflammatory skin disorder characterized by itching and eczematous lesions (Orciani et al. 2017). Its etiology is multifactorial, involving genetic predispositions and environmental triggers, which together initiate a complex pathophysiological mechanism marked by an imbalance in type 1 helper T cell (Th1) and Th2 lymphocyte response, with a predominance of Th2 (Courtney and Su 2024). Although AD diagnosis is primarily clinical, some laboratory markers, including total serum IgE, eosinophilic cationic protein (ECP), interleukin-2 receptor (IL-2R), and thymus and activation-regulated chemokine (TARC/CCL17), can reflect disease severity (Thijs et al. 2015). However, due to the involvement of multiple biological pathways, no single biomarker reliably correlates with disease severity. Thijs et al. (2015) suggested that a combination of biomarkers might better assess AD severity and proposed alternative sampling methods, such as dried blood spots (DBS) and saliva, given the invasive nature of blood collection and its need for skilled personnel. Saliva, with its high permeability and proximity to blood capillaries, provides a non-invasive reflection of circulating blood markers. However, factors like age, sex, hydration, and flow rate can influence salivary composition, underscoring the need for standardized collection and processing methods before widespread clinical application. For instance, salivary IgA, which plays a key role in immune defense, has also been investigated as a stress marker. Some studies suggest its potential in stress assessment; however, conflicting evidence exists regarding its reliability, as environmental and handling factors may affect its concentration (Ng et al. 2003; Obayashi 2013).

In 2013, two studies investigated salivary cortisol levels in AD patients—in adults (Mizawa et al. 2013) and children (Kojima et al. 2013). Mizawa et al. (2013) identified notably higher cortisol levels in adult patients with AD, which correlated with the SCORing Atopic Dermatitis (SCORAD) index, indicating a possible hyporesponsiveness of the hypothalamic–pituitary–adrenal (HPA) axis to stress and its potential role in worsening AD symptoms. On the other hand, Kojima et al. (2013) reported no significant variation in baseline cortisol levels before stress exposure among children with different AD severities, although a negative correlation between post-stress cortisol levels and disease severity was observed. The activation of the HPA system due to stress results in cortisol secretion into the bloodstream, making serum and salivary cortisol levels reliable markers for stress evaluation (Špiljak et al. 2022, 2024). The HPA axis plays a central role in the body's acute stress response, encompassing complex interactions influenced by psychological and social dynamics. In cases of chronic psychological stress, prolonged activation of the HPA axis leads to elevated glucocorticoid levels. Persistent increases in cortisol may result in HPA axis hyperactivity and dysregulation, which has been linked to an elevated risk of various health issues, including AD (Špiljak et al. 2024).

Further research supports the use of salivary cortisol as a stress biomarker in AD patients. A study by Meštrović-Štefekov et al. (2022), involving 84 AD patients, found elevated cortisol levels in symptomatic and asymptomatic groups, noting significantly lower levels in patients with severe AD compared to those with mild or moderate forms. This suggests that while salivary cortisol may not correlate directly with disease severity, it reflects the broader psychological impact of AD, underscoring the importance of assessing psychological well-being and stress management in AD treatment plans. Topical corticosteroids are frequently used to treat AD, but their systemic absorption can impair the HPA axis, particularly in severe cases or with prolonged use. Nutan et al. (2011) demonstrated that patients with severe AD exhibited low baseline cortisol levels, which improved after topical corticosteroid treatment, suggesting that disease activity, rather than topical corticosteroid usage, primarily drives HPA axis dysfunction. Haeck et al. (2007) supported these findings, showing that low baseline cortisol values were not due to previous potent TCS use and improved with intensive AD treatment.

Other salivary biomarkers, including chromogranin A and alpha-amylase (AA), have also been investigated for their potential roles in assessing stress and AD severity. Chromogranin A (CgA), released into saliva following sympathetic stimulation, has been proposed as a reliable stress marker, especially in children (Kaneko et al. 2017; Cai et al. 2018). Additionally, sAA, a digestive enzyme and autonomic nervous system marker, shows potential as a chronic stress marker (Vineetha et al. 2014). However, studies have found no significant correlation between alpha-amylase levels and AD severity or stress (Kaneko et al. 2017). Melatonin, known for its role in sleep regulation, has also been investigated in AD, where stress or disease activity may reduce its secretion (Chang et al. 2014; Marseglia et al. 2014). Thus, lower salivary melatonin levels are associated with poorer sleep quality and greater AD severity. This body of research indicates that while salivary biomarkers present promising non-invasive tools for assessing AD and its related stress, further standardization of collection methods and more comprehensive studies are needed before routine clinical application. Table 12.1 presents clinical studies conducted on patients with atopic dermatitis over the past 10 years.

12.2.2 Allergic Contact Dermatitis (ACD)

Research on both mice and patients with ACD has shown increased levels of IL-6, IL-1, and TNF-α, either locally in skin lesions or in cultured keratinocytes (Goes et al. 2020; Li et al. 2007; Curzytek et al. 2021; Kwon et al. 2021; Galbiati et al. 2020; Sakai et al. 2019; Kondo et al. 1994). Acute skin damage from allergen exposure triggers keratinocytes and antigen-presenting cells to release these cytokines, which play a critical role in the maturation and migration of allergen-specific T cells to regional lymph nodes. Upon re-exposure to the allergen, these cytokines are essential for pro-inflammatory and chemotactic activity, guiding inflammatory cells to the skin contact site with the allergen (Silvestre et al. 2018; Peiser et al. 2012). In addition to the cytotoxic effects of allergen-specific T cells on keratinocytes, IL-6 and TNF-α further influence the skin's inflammatory process. They do this by directly affecting keratinocytes through TNF-α and by inhibiting T regulatory (Treg) cell proliferation through IL-6. Additionally, they promote the differentiation of cytotoxic T cells (CD8+) into allergen-specific T cells through both TNF-α and IL-6 (Silvestre et al. 2018).

Notably, TNF-α is part of the Th1 immune response, whereas IL-6 drives Th2 differentiation and simultaneously inhibits Th1 polarization (Diehl and Rincón 2002; Osugi et al. 1997). In mouse models, intradermal injection of TNF-α has been shown to induce dendritic cell migration to draining lymph nodes, while anti-TNF-α treatment in mice sensitized to trinitrochlorobenzene prevents the skin's inflammatory response to allergen exposure (Cumberbatch and Kimber 1992; Piguet et al. 1991). Studies on mice lacking TNF receptors further reveal that TNF-α signaling impacts dendritic cell antigen uptake and migration to lymph nodes (Becke et al. 2001; Dittmar and Schuttelaar 2017; Jia et al. 2012; Akhtar et al. 2010). In vitro studies have shown that, beyond its role in the sensitization phase, TNF-α released during the effector phase enhances the expression of adhesion molecules and cytokines, promoting the migration of allergen-specific T lymphocytes, which may sustain chronic inflammation (Dittmar and Schuttelaar 2017).

However, studies examining cytokine levels in the serum of ACD patients, especially those with hand ACD, are limited, leaving it uncertain whether ACD solely involves localized skin inflammation or, in some forms or phases, also some evidence suggesting their potential in monitoring stress responses and disease severity (Foks-Ciekalska et al. 2023). Foks-Ciekalska et al. (2023) reported that secretory IgA and CgA levels could serve as biomarkers for tracking treatment efficacy in patients with psoriasis, particularly in those receiving biological therapy.

In summary, several aspects of the "salivary signature" in psoriasis, particularly elevated levels of specific cytokines and acute-phase proteins, are being recognized as valuable tools for non-invasive disease monitoring. However, further studies with larger sample sizes and comprehensive salivary profiling are needed to validate these findings. Proteomic technologies may play a crucial role in this effort and should be applied to saliva to investigate the diagnostic and prognostic potential of cytokines such as IL-1β, TNF-α, IL-2, and IFN-γ in different clinical forms of psoriasis (Bottoni et al. 2016; Li et al. 2020).

Furthermore, Gisondi et al. (2021) observed lower salivary cortisol levels in psoriatic patients compared to controls, although Richards et al. (2005) found no significant difference. Moreover, contradictory findings regarding the HPA axis in psoriasis persist, with earlier research showing elevated salivary cortisol levels that correlate with disease severity (Brunoni et al. 2014).

12.3.2 Acne Vulgaris

Acne vulgaris is a chronic inflammatory condition of the skin that involves sebaceous gland dysfunction and is influenced by both internal and external stressors. Given that sebaceous glands are regulated by neuroendocrine mechanisms, including the HPA axis and neuroautocrine pathways, the connection between psychological stress and acne is not fully understood. Mizuno et al. (2021) investigated the relationship between psychological stress and catecholamine levels in acne lesions in a study involving 18 patients with mild to moderate acne. Utilizing the State-Trait Anxiety Inventory (STAI), the study found all participants exhibited anxiety, with six categorized as having low anxiety and 12 classified as having high anxiety. Salivary alpha-amylase (sAA) activity, a biomarker influenced by the sympatho-adrenal medullary (SAM) system, was positively associated with STAI State Anxiety (STAI-S) scores and significantly elevated in the high-anxiety group. Furthermore, normetanephrine levels within the hair follicles of acne lesions were positively correlated with STAI-S scores and were higher in individuals with high anxiety, while metanephrine levels showed no significant changes. These results imply that normetanephrine, linked to SAM system activation, may play a role in acne pathology in anxious individuals, although its levels were not associated with the severity of the acne lesions.

A link between microorganisms and immune factors, such as interleukins, has been identified in acne patients. Propionibacterium acnes plays a key role in acne pathogenesis by triggering the release of pro-inflammatory cytokines, primarily IL-1β, in the skin (Cong et al. 2019). IL-1β is released from keratinocytes, fibroblasts, and immune cells such as macrophages and mast cells, contributing to inflammatory and autoimmune responses, chemokine expression, and tissue destruction in acne patients (Rafe et al. 2019; Zindel and Kubes 2020).

In a study by Monib et al. (2022), serum IL-1β levels showed a significant positive correlation with acne severity scores, consistent with previous findings (Mochtar et al. 2018). The study also reported elevated salivary IL-1β levels in acne patients and a positive correlation between serum and salivary IL-1β levels. Other studies have identified similar correlations between salivary and serum IL-1β levels in healthy adolescent girls and individuals experiencing stress (Riis et al. 2014; La Fratta et al. 2018). However, salivary IL-1β did not significantly correlate with acne severity scores or with serum and salivary CRP levels in acne patients, likely due to IL-1β's complex role in both local salivary synthesis and systemic circulation (Seil et al. 2012). This complexity may explain the lack of significant correlation between IL-1β in saliva and acne severity. Nonetheless, the elevation of these markers in both serum and saliva, along with their correlation with disease severity, suggests they could serve as biomarkers for objectively assessing inflammation in acne patients (Monib et al. 2022).

Elevated salivary CRP has been documented in dermatology patients, particularly in psoriasis, highlighting its reflection of disease-related inflammation (Rocha-Pereira et al. 2004). Among acne patients, studies have reported significantly higher salivary CRP levels compared to controls, with salivary CRP positively correlating with disease severity as measured by GAGS scores. This correlation implies that acne may involve a systemic inflammatory response, especially in severe cases, challenging earlier assumptions that acne inflammation is strictly localized (Namazi et al. 2015). CRP, synthesized by hepatocytes during inflammation, enters saliva from the bloodstream rather than from local salivary gland production (Black et al. 2004; Megson et al. 2010). The strong correlation between serum and salivary CRP levels supports the notion that salivary CRP could serve as a less invasive inflammation biomarker in acne patients (La Fratta et al. 2018; Ouellet-Morin et al. 2011). However, currently, no effective strategies exist to make salivary CRP a reliable quantitative measure of serum CRP levels, and thus, it remains unsuitable for point-of-care (POC) testing of systemic inflammation (Pay and Shaw 2019).

Adipsin, a biomarker involved in inflammation and insulin regulation, has also been studied in acne. Albalwa et al. (2023) found that both serum and salivary adipsin levels were lower in acne patients compared to controls, with a significant negative correlation between adipsin and body mass index (BMI), GAGS scores, fasting glucose, and insulin resistance. Although adipsin's role in acne has not been extensively studied, these findings suggest its potential involvement in acne pathogenesis through its influence on insulin regulation and inflammatory responses (Kwon and Pessin 2013). Adipsin regulates the complement system, particularly the activated complement component 3 (C3a), which stimulates insulin synthesis and impacts systemic insulin sensitivity, marking it as a relevant target for further acne research (Lo et al. 2014; Lim et al. 2013).

In summary, salivary markers such as IL-1β are emerging as sensitive, non-invasive biomarkers for assessing inflammation in acne vulgaris patients. The positive correlation between serum and salivary levels of these markers, along with their association with disease severity, suggests they could be valuable for monitoring treatment efficacy and understanding acne pathogenesis. Additionally, adipsin offers a promising area for further research on acne and its metabolic connections.

12.4 Salivary Immune and Endocrine Biomarkers in Autoimmune Skin Diseases

12.4.1 Systemic Lupus Erythematosus (SLE)

Systemic lupus erythematosus (SLE) is a chronic autoimmune condition marked by widespread inflammation and the production of autoantibodies, which can affect multiple organ systems, including the salivary glands. This involvement often leads to symptoms like xerostomia (dry mouth) and a heightened risk of dental caries. Among patients with active SLE, dental caries rates can reach as high as 100%, compared to 85% in those with inactive disease. Active SLE cases are also associated with reduced salivary flow rates and lower pH levels. Additionally, saliva from SLE patients demonstrates diminished antioxidant capacity, as evidenced by decreased glutathione (GSH) levels (Loyola Rodriguez et al. 2016; Moori et al. 2016).

While serum monitoring remains the primary method for assessing autoantibody levels in SLE, advances in salivary proteomics are opening up new possibilities for saliva as a non-invasive tool for tracking disease activity. This is particularly promising for pediatric patients, given the ease of saliva collection compared to blood sampling (Jung et al. 2015). For instance, sAA and antichromatin antibodies have shown correlations with disease severity in SLE, reflecting heightened stress responses in active disease (Jung et al. 2015). To identify non-invasive diagnostic markers, researchers have also examined the presence of serum inflammatory cytokines in the saliva of SLE patients.

Stanescu et al. (2018) were among the first to monitor salivary IL-6 levels and reported significantly higher levels in SLE patients compared to controls. They found a positive correlation between salivary and serum IL-6 levels, suggesting that salivary IL-6 could serve as a reliable biomarker for assessing inflammation in SLE. Also, MCP-1 levels were elevated in the saliva of SLE patients, though no correlation was found between salivary and serum concentrations (Stanescu et al. 2018).

Further research has linked increased salivary levels of IL-6 and IL-17A in SLE patients with periodontal disease compared to healthy controls, indicating that the periodontium may be a target in SLE (Mendonça et al. 2019). However, other studies have found conflicting results regarding IL-17 salivary concentrations, with no significant differences between SLE patients with or without periodontal disease (Marques et al. 2016). Additionally, IL-6 and IL-1β levels were lower in SLE patients with periodontal disease compared to periodontal disease patients without SLE. Research suggests that SLE and periodontal disease share a similar immunological profile, increasing the risk of periodontal disease in SLE patients (Mendonça et al. 2019; Corrêa et al. 2017). A meta-analysis indicated that SLE patients have a 1.76-fold higher risk of developing periodontal disease (Rutter-Locher et al. 2017). Salivary cytokines, such as IL-1β, IL-4, and IL-33, have been proposed as potential markers to predict periodontal destruction in SLE patients. A positive correlation was found between salivary IL-1β levels and disease duration in SLE (Jung et al. 2015). Elevated levels of IL-33 in the saliva of SLE patients with periodontal disease suggest that periodontitis may contribute to increased local inflammation in SLE (Mendonça et al. 2019).

Changes in the oral microbiota in SLE patients with periodontal disease are associated with heightened local inflammation, reflected by elevated salivary IL-6, IL-17, and IL-33 levels (Corrêa et al. 2017). These findings indicate a strong connection between SLE and periodontal health, with increased salivary cytokines pointing to disease activity and progression. However, it is important to consider that saliva contains inhibitors such as mucin-like proteins, which can reduce cytokine concentrations and complicate interpretations of salivary cytokine data (Wozniak et al. 2002). A study by Ruacho et al. (2022) analyzed salivary levels of colony-stimulating factor-1 (CSF-1), TNF-α, interferon-gamma-induced protein-10 (IP-10), MCP-1, and calprotectin in SLE patients, comparing them to controls. Higher levels of these biomarkers were found in the saliva, serum, and urine of SLE patients, correlating with disease activity. Thus, CSF-1, TNF-α, IP-10, and MCP-1 in saliva and urine were particularly associated with active renal disease, suggesting that saliva may serve as a novel fluid for monitoring overall SLE activity, while urine could be more informative for renal-specific activity.

12.4.2 Bullous/Blistering Diseases

Bullous pemphigoid (BP) and pemphigus vulgaris (PV) are two acquired autoimmune bullous dermatoses characterized by involvement of the skin and/or mucous membranes. PV is characterized by acantholysis, leading to intraepidermal blisters with flaccid domes on non-inflamed skin and often involves mucous membranes, particularly the oral mucosa, in its early stages. In contrast, BP is a non-acantholytic condition characterized by sub-epidermal, tense blisters on inflamed, pruritic skin, with less common oral involvement (Corbaux and Joly 2018). The definitive diagnostic approach for both conditions involves histopathological analysis alongside direct immunofluorescence performed on skin or mucosal biopsy samples.

In PV, characteristic acantholysis (loss of cell adhesion) and intraepidermal detachment are observed. IgG is deposited against desmoglein (Dsg) 1 and/or Dsg3, which are transmembrane desmosomal proteins involved in cell adhesion (Murrell et al. 2020). Circulating autoantibodies against Dsg1 and Dsg3 can be detected in serum through indirect immunofluorescence or enzyme-linked immunosorbent assay (ELISA) techniques. Some researchers have proposed saliva as an alternative medium to detect Dsg1 and Dsg3. For instance, Andreadis et al. (2006) demonstrated that ELISA detection of Dsg1 and Dsg3 in saliva is highly sensitive and specific, showing a strong correlation with serum levels. This method has been proposed for diagnosing and monitoring PV disease activity, as well as predicting relapses.

Two studies have explored the use of salivary testing in predicting PV prognosis and mucosal severity. De et al. (2017) examined 43 histologically confirmed PV patients. A significant correlation was observed between serum and salivary levels of Dsg1 and Dsg3, though these levels did not correlate with oral mucosal disease severity as measured by the Autoimmune Bullous Skin Disorder Intensity Score (ABSIS). In contrast, Koopai et al. (2018) found a moderate correlation between salivary Dsg1 and Dsg3 levels and PV disease severity, using the Pemphigus Disease Area Index (PDAI). Mortazavi's study (2015) reported similar findings. However, Russo et al. (2017) reported discordant results between saliva and serum ELISA tests for PV, indicating that saliva may not consistently serve as a reliable medium for autoantibody detection. Despite these variations, Hallaji et al. (2010) found sensitivities of 70% and 94% for salivary Dsg1 and Dsg3, respectively, in PV diagnosis, suggesting a potential diagnostic and prognostic role for salivary testing in PV.

BP is characterized by sub-epidermal blisters filled with eosinophils and neutrophils, and immunofluorescence reveals linear deposits of IgG and complement component C3 at the basement membrane, specifically targeting the bullous pemphigoid 180 antibody (BP180) (Murrell et al. 2020). Saliva testing has also been investigated for BP; Esmaili et al. (2015) demonstrated that salivary BP180-NC16a concentrations could assist in BP diagnosis. However, the ELISA technique for BP180 has been found insufficient for detecting mucous membrane pemphigoid (MMP) in both serum and saliva (Andreadis et al. 2006).

In pemphigus foliaceus (PF), research by Rosatelli et al. (2005) found that patients exhibited decreased plasma high-molecular-weight kininogen (HK) levels and increased plasma kallikrein and kininase II activity. Elevated salivary kallikrein activity was also observed, suggesting systemic activation of the kinin system, which may contribute to blister formation. Whether this kinin system activation is a cause or consequence of blister formation requires further investigation.

In summary, while saliva shows potential for detecting Dsg1 and Dsg3 in PV and BP180 in BP, further validation is required to determine the diagnostic and prognostic utility of these techniques. Additionally, systemic markers, such as kallikrein, suggest that saliva might reflect broader inflammatory processes in bullous diseases.

12.5 Salivary Immune and Endocrine Biomarkers in Oral Conditions Associated with Dermatovenerology

12.5.1 Recurrent Aphthous Stomatitis (RAS) and Behcet's Disease

Research has demonstrated that individuals with RAS experience heightened lipid peroxidation, as indicated by elevated malondialdehyde (MDA) levels. Alterations in oxidative stress-related salivary biomarkers, such as SOD, glutathione peroxidase (GSHx), and catalase, have also been reported. These observations suggest that both enzymatic and non-enzymatic antioxidant defense mechanisms are compromised in RAS patients (Saral et al. 2005; Cimen et al. 2003; Karincaoglu et al. 2005). Interestingly, while these antioxidant enzymes are elevated in saliva, they are decreased in plasma, suggesting a localized antioxidant response to RAS, reflecting the body's capacity to direct its antioxidant defenses to affected areas. Several studies have also examined the salivary proteome in RAS patients, focusing on biomarkers such as cortisol, oxidative stress-related peptides, immunoglobulins, and cytokines.

According to literature data, RAS outbreaks are associated with significantly increased levels of immunoglobulins, specifically IgA and IgG, in saliva (Hassona and Scully 2016). Inflammatory mediators like cytokines can stimulate the production of major histocompatibility complex (MHC) class I and II antigens in epithelial cells, leading to cytotoxic T-lymphocyte responses and ulceration. Various cytokines in saliva, including TNF-α, prostaglandin E2 (PGE2), vascular endothelial growth factor (VEGF), and IL-6, have been found at elevated levels in RAS patients (Boras et al. 2006; Adisen et al. 2008; Chaudhuri et al. 2018).

Behcet's disease, a chronic systemic condition characterized by recurrent oral and genital ulcers, eye involvement, and skin lesions, is associated with increased oxidative stress. This is evidenced by enhanced superoxide production in plasma neutrophils and macrophages, which contributes to chromosomal damage. Elevated levels of lipid peroxidation products and pro-inflammatory cytokines, such as TNF-α, are additional markers of oxidative stress in Behcet's disease (Karincaoglu et al. 2005). Thiobarbituric acid-reactive substances (TBARS) are also significantly increased in affected individuals. In a pilot study by Novak et al. (2021), cytokine profiles in matched serum and saliva samples were compared across Behcet's disease patients (both with and without oral ulcers), RAS patients, and healthy controls. The study measured 12 cytokines, including IL-1β, IL-2, IL-4, IL-5, IL-6, IL-8, IL-10, IL-12p70, IL-17A, IFN-γ, TNF-α, and tumor necrosis factor-beta (TNF-β). Most cytokines were found at higher concentrations in saliva than in serum, with IL-1β, IL-2, IL-8, IL-10, and TNF-α being the most prevalent in saliva, while IL-8 dominated in serum (91.9%). Salivary levels of IL-1β, IL-6, IL-8, and TNF-α were significantly elevated in Behcet's disease patients with oral ulcers compared to healthy controls. Even those without oral ulcers exhibited markedly higher salivary IL-1β, IL-8, and TNF-α levels (Novak et al. 2021). These findings suggest that cytokine profiling in saliva could serve as a non-invasive tool for monitoring disease activity in Behcet's disease.

Higher salivary cortisol levels have been detected in RAS patients compared to healthy controls, likely reflecting stress, and anxiety often associated with this condition (Eguia-del Valle et al. 2013; Kunikullaya et al. 2017). Although total antioxidant capacity (TAC) does not appear to be directly related to RAS etiology, oxidative stress-related molecules exhibit altered levels in these patients (Karincaoglu et al. 2005; Tugrul et al. 2016; Caglayan et al. 2008).

In Behcet's disease, neutrophils play a crucial role in the disease's pathology. Neutrophils, which are white blood cells that help defend against infections, release neutrophil elastase (NE), a powerful enzyme that breaks down proteins to attack microbes. While NE is an essential component of the immune response, its excessive release can damage the body's own tissues, leading to inflammation and tissue breakdown (Novak et al. 2020). One of the hallmark symptoms of Behcet's disease is recurrent oral ulceration, which may result from an overactive neutrophil response. Excessive NE release from neutrophils in Behcet's disease can disrupt the stability of mucosal tissues, particularly when the protective balance between NE and its natural inhibitors is disrupted. Normally, NE is controlled by endogenous inhibitors like secretory leukocyte protease inhibitor (SLPI) and alpha-1 antitrypsin (α1AT), which limit NE's tissue-damaging effects. However, in Behcet's disease, reduced levels of SLPI and α1AT could mean that NE activity goes unchecked, causing more tissue damage and contributing to persistent inflammation and mucosal instability seen in Behcet's disease (Novak et al. 2020). In a study by Novak et al. (2020), levels of NE, SLPI, and α1AT were measured in saliva samples from Behcet's disease and RAS patients, along with healthy controls. Behcet's disease patients without ulcers exhibited reduced SLPI levels, while NE remained enzymatically active. The study found that treatment with azathioprine and colchicine in Behcet's disease patients helped regulate excessive NE during ulceration, suggesting a therapeutic approach to managing recurrent oral ulcerations (Novak et al. 2020).

12.5.2 Oral Lichen Planus (OLP)

Recent research on salivary biomarkers in oral lichen planus (OLP) has centered on its diagnosis and the early detection of malignant transformation. Talungchit et al. (2018) utilized proteomic analysis through two-dimensional gel electrophoresis followed by mass spectrometry on saliva samples from five OLP patients and five healthy controls. This study identified three proteins—cystatin SA, human complement component C3c (chain C), and fibrinogen fragment D (chain B)—as potential indicators for OLP. Cystatin SA, a member of the cystatin superfamily, functions as a cysteine protease inhibitor with antimicrobial properties, while fibrinogen fragments and C3c play roles in inflammatory processes. Immunofluorescence in OLP cases frequently demonstrates fibrinogen and C3 deposition (Buajeeb et al. 2015). In another study, Souza et al. (2018) identified 108 proteins differentially expressed in OLP patients, finding an absence of proteins necessary for lubrication and viscoelasticity, which correlates with xerostomia in OLP patients. Proteins such as S100A8 and S100A9, also known as myeloid-related protein 8 (MRP8) and myeloid-related protein 14 (MRP14), play roles in IL-17-mediated inflammation and cytokine production.

OLP is primarily a T-cell-mediated condition, and several cytokines released by infiltrating lymphocytes and damaged keratinocytes contribute to a self-amplifying inflammatory process (Wei et al. 2018). The imbalance in cytokine production triggers an inflammatory response and abnormal immune reaction due to local hyperproduction of cytokines (Abdel-Haq et al. 2014). Given the association of chronic inflammation with oncogenesis, the persistently high levels of salivary proinflammatory cytokines suggest their importance as mediators in cancer development. Proinflammatory cytokines activate apoptotic and non-apoptotic signaling cascades (Zhang et al. 2008; Liu et al. 2009, 2015; Kaur and Jacobs 2015).

Salivary interleukins, especially IL-6 and IL-8, have emerged as potential biomarkers. IL-6 promotes B- and T-cell differentiation and has been implicated in cancer progression by inactivating the tumor suppressor gene p53 (Hodge et al. 2005). IL-6 can also inactivate the tumor suppressor gene p53, promoting tumor cell proliferation and the development and progression of cancers such as squamous cell carcinoma (Hodge et al. 2005; Hibi et al. 1996; Tamura et al. 1995; Jinno et al. 2015). A meta-analysis by Mozaffari et al. (2018) found significantly higher IL-6 levels in OLP patients compared to controls, with salivary levels surpassing serum levels. Additionally, IL-8 is instrumental in the transition from reticular to erosive OLP and decreases with dexamethasone treatment, highlighting its potential as a therapeutic marker (Rhodus et al. 2005).

OLP patients consistently exhibit higher salivary IL-6 concentrations compared to controls (Zhang et al. 2008; Abdel-Haq et al. 2014; Lu et al. 2015; Rhodus et al. 2005; Cheng et al. 2014), particularly in erosive OLP cases. ELISA was used to measure IL-6 concentration across these studies, except in Lu et al. (2015), which applied multiple methods. This IL-6 elevation is attributed to local production by keratinocytes, monocytes, activated T lymphocytes, endothelial cells, macrophages, and fibroblasts (Hibi et al. 1996; Pachard and Libby 2008). Studies consistently report significantly elevated salivary IL-6 levels in OLP patients (Lu et al. 2015; Humberto et al. 2018; Man et al. 2004; Liu et al. 2017), with overexpression of tripartite motif-containing protein 21 (TRIM21) encoded by the TRIM21 gene, a regulator of immune processes, contributing to this increase (Wei et al. 2019). This association of IL-6 with lesion onset and advanced stages of OLP has led to its proposal as a treatment-response biomarker (Lu et al. 2015; Cheng et al. 2014). Persistent elevation of proinflammatory cytokines in saliva, such as IL-6, TNF-α, and IL-8, may indicate a heightened risk of malignant transformation in the oral mucosa (Babiuch et al. 2020).

TNF-α has transformed therapeutic approaches to dermatological inflammatory diseases. Elevated salivary TNF-α levels compared to serum suggest that salivary measurements may be more practical for monitoring treatment in OLP (Campanati et al. 2016; Radi et al. 2021). However, factors like age, OLP type, stress, alcohol use, smoking, and genetic variability complicate TNF-α interpretation (Mozaffari et al. 2017). Gavala et al. observed that alcohol inhibits serum TNF-α levels, with the highest levels found in patients with erosive/atrophic OLP (2010). In contrast, IFN-γ levels show no significant correlation with OLP severity, suggesting that IFN-γ may not be central to OLP pathogenesis (Mozaffari et al. 2019). Salivary levels of TNF-α, IL-1, IL-4, IL-6, and IL-8 have been associated with OLP diagnosis and outcomes (Humberto et al. 2018). Thus, TNF-α stimulates acute OLP inflammation by promoting other pro-inflammatory cytokines like IL-1 and IL-6, and activating T and B cells, with elevated levels observed in OLP patients (Lu et al. 2015; Humberto et al. 2018; Mozaffari et al. 2017; Sugermann et al. 1996; Thanthoni et al. 2015). IL-6, a versatile cytokine with strong pro-inflammatory properties, is secreted by various cell types and plays a central role in regulating immune and inflammatory responses (Kishimoto 2010). Similarly, IL-8, which is produced in response to tissue injury, is found at elevated levels in the saliva of OLP patients, with these levels correlating with disease severity (Humberto et al. 2018; Mozaffari et al. 2017; Ghoreishiam et al. 2017). Both IL-6 and IL-8 have been linked to OLP severity (Lu et al. 2020), though some studies suggest that salivary IL-8 may serve as a more reliable biomarker for this condition (Cheng et al. 2014; Sun et al. 2005).

Additionally, interleukins IL-1α and IL-1β, which activate T-helper lymphocytes by promoting IL-2 secretion, are also elevated in OLP patients (Lu et al. 2015; Humberto et al. 2018). While increased salivary IL-4 levels have been reported in OLP patients (Humberto et al. 2018; Mozaffari et al. 2019), the current evidence does not strongly support its utility as a biomarker for monitoring disease progression.

The etiology of OLP is largely driven by an imbalance between Th1 and Th2 lymphocytes. Proinflammatory mediators, such as IL-4, IL-10, IL-18, TNF-α, and nuclear factor kappa B (NF-κB)-related cytokines, are significantly increased in OLP patients, reflecting this immune imbalance (Hassona and Scully 2016). Treatment with immunosuppressants, such as corticosteroids and non-antibiotic macrolides, as well as alternative therapies like plant extracts, has been shown to reduce inflammation-based biomarkers (González-Moles and Scully 2010; Varoni et al. 2012; López-Jornet and Aznar-Cayuela 2016). Despite these findings, there are still no validated salivary biomarkers available to predict malignant transformation in OLP. Recent mass spectrometry-based nontargeted proteomic studies offer new insights into the etiology and diagnosis of OLP (Souza et al. 2018; Yang et al. 2006). Among protein-based biomarkers, cortisol, oxidative stress-related molecules, immunoglobulins, and cytokines are most widely investigated in the diagnosis of OLP. Altered salivary exosome profiles have also been observed in OLP; notably, a marked increase in microRNA-4484 (miR-4484) in the salivary exosomes of OLP patients suggests its utility as a biomarker for early detection, potentially enabling timely interventions to prevent malignant progression (Byun et al. 2015).

Histamine levels in saliva of OLP patients have also been studied, though differences were not statistically significant when compared to controls, suggesting further research may help elucidate the role of histamine in OLP development (Radwan-Oczko et al. 2024). Additionally, several studies have examined the relationship between salivary cortisol levels and the psychological status in OLP patients. Some case-control studies reported elevated cortisol levels in OLP patients (López-Jornet and Aznar-Cayuela 2016; Mansourian et al. 2018), whereas others found no significant differences, and a few even reported lower cortisol levels (Girardi et al. 2011; Rödström et al. 2001). Cortisol is known to reduce lymphocyte numbers and disrupt immune cell function, while HPA axis dysfunction in anxiety or depression can lead to reduced cortisol levels in both blood and saliva, contributing to immune-related diseases like OLP (González-Moles and Scully 2010). Lopez-Jornet et al. also demonstrated increased levels of adiponectin in OLP patients, along with significantly elevated levels of IgA and IgG compared to controls (López-Jornet and Aznar-Cayuela 2016). Further research conducted in 2023 showed a strong positive correlation between disease duration and salivary cortisol concentration in OLP patients, although this was not confirmed for sAA (Glavina et al. 2023).

A meta-analysis found a mean difference of 3.43 ng/ml (95% CI: 1.20–5.65) in salivary cortisol levels between OLP patients and healthy controls, confirming cortisol as a frequently quantified biomarker for psychological disorders in OLP. Other biomarkers, including dehydroepiandrosterone (DHEA), IgA, adiponectin, and sAA, were also evaluated (Pires et al. 2023). Elevated levels of cytokines, including TNF-α, granulocyte-macrophage colony-stimulating factor (GM-CSF), IL-1α, IL-1β, and IL-8, have been reported in both erosive and reticular OLP forms, with IL-13 levels being significantly lower. Salivary levels of these cytokines—especially TNF-α, GM-CSF, MIP-1α, MIP-1β, IL-1β, IL-6, and IL-8—correlate positively with the severity of OLP lesions, suggesting that an imbalance in the Th1/Th2 immune response contributes to OLP and underscores the potential of these salivary cytokines as biomarkers for diagnosis and prognosis (Zhu et al. 2022).

Several studies have investigated cortisol's role in OLP, given its influence on cytokine profiles as a biomarker of stress and anxiety (Stojanovich 2010). Some research suggests that stress triggers OLP, while the disease itself exacerbates patient stress. This dual relationship introduces complexities in assessing salivary cortisol as a biomarker in OLP, resulting in inconsistent findings across studies (Mansourian et al. 2018; Skrinjar et al. 2019; López-Jornet et al. 2019). Consequently, cortisol may not be a consistently reliable marker for OLP. Studies on salivary cortisol concentrations in OLP patients show conflicting results. Research by Lopez-Jornet et al. (2016), Koray et al. (2003), Shah et al. (2009), Nadendla et al. (2014), Karthikeyan et al. (2016), and Gaur et al. (2018) demonstrated significantly higher salivary cortisol concentrations in OLP patients compared to controls. However, Gaur A et al.'s study found no significant difference in salivary cortisol levels between erosive and non-erosive forms of OLP (Gaur et al. 2018). Contrastingly, studies by Rödström et al. (2001), Girardi et al. (2011), Nosratzehi et al. (2014), Pippi et al. (2014), and Skrinjar et al. (2019) identified no differences in salivary cortisol concentrations between OLP patients and healthy controls. One limitation of these studies is the use of various techniques to measure salivary cortisol, including ELISA, chemiluminescent microparticle immunoassay (CLIA), and radioimmunoassay (RIA).

Despite the controversies surrounding the correlation between salivary cortisol and OLP, cortisol's effect on lymphocyte numbers and other immune cells supports its role, as dysfunction of the HPA axis in patients with anxiety and depression reduces cortisol levels in both blood and saliva, contributing to immune-related diseases like OLP (Nosratzehi et al. 2014). According to research studies, elevated salivary cortisol levels in OLP patients have been observed compared to controls (Humberto et al. 2018; López-Jornet et al. 2015, 2019), suggesting a possible link between heightened cortisol and psychological stress as an OLP trigger (Shah et al. 2009). Given that stress can induce OLP recurrences, salivary cortisol has been proposed as a potential diagnostic marker for the condition (Karthikeyan and Aswath 2016). Studies on sAA activity in OLP patients are limited. A preliminary study by da Silva Simoura et al. showed higher sAA activity 30 minutes post-wake-up in OLP patients, with reductions correlating with stress, suggesting α-amylase as a promising stress biomarker (da Silva Simoura et al. 2023). However, Pippi et al.'s study found no significant variations in sAA activity in patients with inactive OLP compared to controls (Pippi et al. 2014). Additionally, in a study by Glavina et al. (2024), sAA levels showed no statistically significant difference between patients with OLP and control subjects. However, those with OLP demonstrated significantly poorer psychological well-being, with depression scores three times higher and anxiety scores twice as high as those of the controls, along with a lower reported quality of life compared to the control group.

Oxidative stress is also a well-researched aspect of OLP pathogenesis (Upadhyay et al. 2010; Sezer et al. 2007), with studies pointing to nitric oxide (NO) and reactive oxygen species (ROS) as major contributors (Tvarijonaviciute et al. 2016). Higher salivary NO levels are consistently reported in OLP patients compared to healthy controls (Humberto et al. 2018), with elevated NO linked to more severe disease progression due to mucosal lesion formation (Tvarijonaviciute et al. 2016; Ohashi et al. 1999). While ROS has been associated with cellular oxidative stress, a consensus on the correlation of salivary ROS with oxidative damage in OLP tissues is still lacking (Tvarijonaviciute et al. 2016; Darczuk et al. 2016). In addition, CRP, a common marker of inflammation, has shown higher salivary levels in OLP patients compared to controls (Tvarijonaviciute et al. 2016; Shahidi et al. 2017; Shiva et al. 2020), supporting its potential use for tracking disease progression.

Additionally, S100A8 may induce apoptosis by recruiting CD8+ and natural killer (NK) cells to affected tissues (De Carvallo et al. 2016). Further studies emphasize oxidative stress in OLP, with ROS contributing to keratinocyte apoptosis and dysfunction, perpetuated by infiltrating CD4+ T lymphocytes (Souza et al. 2018). Darczuk et al. (2016) conducted a case-control study with 62 OLP patients and 30 healthy controls, revealing significant differences in TAC, GSH, and TBARS levels. According to the results, OLP patients showed lower TAC and GSH levels and higher TBARS, especially in the erosive form, suggesting severe oxidative stress and a link with clinical severity (Abdolsamadi et al. 2014). These findings suggest potential benefits of antioxidant treatments in OLP management (Rivarola de Gutierrez et al. 2014).

A randomized, double-blind trial by Tvarijonaviciute et al. (2018) studied the effects of *Chamaemelum nobile* gel on 55 OLP patients, linking changes in salivary total antioxidant status (TAS) with increased discomfort and xerostomia, supporting the gel's potential for disease stabilization. Additionally, Wang et al. (2021) found significant decreases in TAC and uric acid levels and increases in MDA and NO in the saliva and serum/plasma of OLP patients. The same study reported elevated oxidative stress markers like 8-hydroxy-deoxyguanosine (8-OHdG) and advanced oxidation protein products (AOPP), with reduced antioxidants like vitamin C, though other antioxidants such as vitamin A, zinc, GSHx, and vitamin E showed no significant differences between OLP patients and controls. Specific markers, including SOD and 8-isoprostane, were elevated, while parameters like GSH and total protein showed dysregulation in some studies (Wang et al. 2021). Moreover, increased levels of IFN-γ and IL-33 in OLP lesions, compared to non-specific inflammatory lesions, alongside higher salivary IFN-γ levels in OLP patients, warrant further exploration of TAS, IL-6, and IL-8 in relation to neoplastic progression in OLP (Carvalho et al. 2019).

12.5.3 Oral Leukoplakia

Oral leukoplakia, particularly the homogeneous type, has been extensively studied as a representative oral potentially malignant disorder. Moreover, it is one of the primary characteristics of a genetic disorder called dyskeratosis congenita, alongside skin pigmentation abnormalities and nail dystrophy (Noto et al. 2016). Several case-control studies have employed ELISA techniques to detect and quantify inflammatory proteins in biofluids. Despite limited proteomic research on saliva for anticipating oral leukoplakia's malignant transformation, cytokines such as IL-6, IL-8, and TNF-α have been examined using ELISA in patients with oral leukoplakia (Brailo et al. 2012; Punyani and Sathawane 2013). Additionally, proteins such as complement component 4d (C4D), MDA, endothelin-1, and lactate dehydrogenase (LDH) have proven useful for differentiating between oral leukoplakia and oral squamous cell carcinoma (OSCC) (Ajona et al. 2015; Hoffmann et al. 2011). In a proteomic study using two-dimensional electrophoresis, Camisasca et al. (2017) identified 22 protein spots more abundant in oral leukoplakia patients compared to controls, including cytokeratin 10 (CK10), which was later validated through immunohistochemistry.

Deepthi et al. (2019) identified TNF-α as a potential prognostic marker for OSCC, with increased salivary TNF-α levels observed in patients with dysplasia, highlighting its utility in monitoring the malignant transformation of oral leukoplakia. Additional research has suggested that TNF-α may serve as a biomarker for the early detection of pre-oral cancer due to its elevated presence in individuals with advanced precancerous lesions (Kaur and Jacobs 2015). Moreover, polymorphisms in TNF-α have been associated with the development of precancerous oral conditions (Hsu et al. 2014).

Interleukins have also been explored as salivary markers for diagnosing and monitoring oral leukoplakia. Studies generally report increased IL-6 and IL-8 levels in oral leukoplakia patients compared to healthy individuals (Kaur and Jacobs et al. 2015; Punyani and Sathawane 2013; Juretić et al. 2013; Selvam and Sadaksharam 2015). These cytokines are associated with angiogenesis, tumor growth, and increased blood vessel density, making them promising biomarkers for early cancer detection (Chang et al. 2013). Other interleukins, such as IL-37, have also been found to be elevated in oral leukoplakia patients (Liu et al. 2017), while IL-10 levels have not shown significant associations with oral potentially malignant disorders (Hsu et al. 2014; Gonçalves et al. 2017). Contradictory findings have also been reported. Brailo et al. (2011) observed no significant differences in salivary TNF-α levels between healthy controls and individuals with oral leukoplakia or oral cancer. Similarly, Wenghoefer et al. (2010) reported no correlation between inflammation markers (IL-1β, IL-6, IL-8, IL-10, TNF-α, and cyclooxygenase-2 [COX-2]) and oral leukoplakia development, with some markers like IL-1β and IL-10 showing lower expression in oral leukoplakia patients compared to healthy controls.

In addition to cytokines, elevated salivary LDH levels, which are linked to cell necrosis, have been observed in oral leukoplakia patients, with even higher levels found in those with OSCC. This suggests that LDH could serve as a marker for evaluating the risk of malignant transformation in oral leukoplakia (Patel and Metgud 2015; Shetty et al. 2012). Other molecules, such as endothelins and growth factors like TGF-β and epidermal growth factor (EGF), have been studied, but no significant associations with OL diagnosis or prognosis have been found (Hoffmann et al. 2011; Gonçalves et al. 2017; Jaeger et al. 2015).

A single study comparing salivary IL-6 levels in proliferative verrucous leukoplakia (PVL) and control samples found significantly increased IL-6 levels in PVL cases, which is noteworthy as PVL often begins as homogeneous leukoplakia and carries a high risk of malignant transformation (Bagan et al. 2016). Overall, while some promising salivary biomarkers have been identified, the data on their usefulness in oral leukoplakia remains inconclusive. Further research is necessary to validate current findings and identify new biomolecules that could serve as reliable diagnostic and prognostic markers. Table 12.2 shows clinical studies conducted on patients with oral leukoplakia over the past 10 years.

12.6 Future Perspectives and Conclusion

Salivary diagnostics have emerged as a promising, non-invasive tool for understanding the pathophysiology of skin and mucocutaneous diseases. Over the past few decades, significant advancements have been made in identifying immune and endocrine biomarkers in saliva that correlate with various dermatological conditions. These biomarkers have opened new avenues for both diagnosing and monitoring the progression of diseases such as AD, psoriasis, and

Table 12.2 Clinical studies conducted on patients with oral leukoplakia over the past 10 years

Oral leukoplakia				
Author, Year	Analyzed factors	Methods and examinees/patients	Results	Conclusions
Deepthi et al. (2019)	Salivary TNF-α as a biomarker in oral leukoplakia and OSCC	90 participants divided into three groups (30 patients in each group): OSCC, leukoplakia, and controls. Salivary TNF-α levels measured using ELISA	Salivary TNF-α levels were significantly higher in OSCC compared to leukoplakia and controls, with a positive correlation between TNF-α levels and histological grade in OSCC and leukoplakia	Salivary TNF-α shows potential as a prognostic biomarker for OSCC and for monitoring leukoplakia's progression to malignancy
Kaur and Jacobs (2015)	Salivary and serum levels of IL-6, IL-8, and TNF-α in oral conditions and controls	54 oral lichen planus, 50 leukoplakia, 51 submucous fibrosis patients, and 50 controls; cytokines measured via enzyme-linked immunoassays	Elevated salivary and serum IL-6, IL-8, and TNF-α levels were noted in oral precancerous conditions compared to controls ($P < 0.05$). Strong correlations among cytokine levels were observed across groups ($r = 0.72$–0.82)	IL-6, IL-8, and TNF-α in saliva and serum are potential diagnostic markers for oral precancerous lesions
Hsu et al. (2014)	Cytokine gene polymorphisms in oral precancerous lesions	42 OPCL patients and 128 controls; polymorphisms in five cytokine genes analyzed using PCR with sequence-specific primers	TNF-α (−308) polymorphism was associated with a 3.63-fold increased risk for OPCLs. IL-6 and TGF-β1 polymorphisms also increased risk (20- to 35-fold for specific genotypes). No significant differences were found for IL-10 and IFN-γ polymorphisms	TNF-α, IL-6, and TGF-β1 polymorphisms appear to be linked to OPCL development
Selvam and Sadaksharam (2015)	Salivary IL-6 as a molecular marker in leukoplakia and OSCC	75 participants divided into three groups: 25 leukoplakia, 25 OSCC, 25 controls; salivary IL-6 measured using ELISA	IL-6 levels were significantly higher in OSCC than in leukoplakia and controls. Differences were also noted across OSCC stages, particularly between stages II and IV	Elevated IL-6 levels may reflect local production and serve as a marker for disease progression from leukoplakia to OSCC
Gonçalves et al. (2017)	Expression of immunomodulatory mediators in oral leukoplakia and OSCC	80 leukoplakia, 20 OSCC, and 20 control samples; tissue and saliva analyzed via immunohistochemistry and ELISA	Dysplasia was observed in 18.7% of leukoplakia samplesImmunosuppressive mediators like HLA-G, IL-10, and TGF-β were overexpressed in leukoplakia and OSCC but did not distinguish between the two in saliva	Immunosuppressive mediator overexpression in leukoplakia may indicate immune evasion potential, independent of cytological changes
Lin et al. (2016)	IL-37 expression in oral leukoplakia and OSCC	70 OSCC, 55 leukoplakia, and 10 controls; immunohistochemical analysis on tissue microarrays	IL-37 levels increased from normal mucosa to leukoplakia and OSCC. Dysplastic leukoplakia exhibited significantly higher IL-37 than non-dysplastic lesions ($P < 0.001$)	IL-37 could be a biomarker for early oral tumorigenesis and risk assessment of malignant transformation

(continued)

Table 12.2 (continued)

Oral leukoplakia				
Author, Year	Analyzed factors	Methods and examinees/patients	Results	Conclusions
Patel and Metgud (2015)	Salivary lactate dehydrogenase (LDH) in oral leukoplakia and OSCC	75 participants: 25 controls, 25 leukoplakia, 25 OSCC; LDH levels measured biochemically	Salivary LDH levels increased progressively from controls to leukoplakia and OSCC, with further increases seen in poorly differentiated OSCC	Salivary LDH may indicate cellular necrosis and serve as a diagnostic marker for oral lesions
Jaeger et al. (2015)	Salivary and tissue EGF levels in oral leukoplakia and controls	32 leukoplakia and 32 control participants; EGF and EGFR analyzed in saliva and tissue via immunohistochemistry	Salivary EGF levels were similar between leukoplakia and controls, but absence of tissue EGF correlated with multiple lesions. Dysplastic lesions showed trends toward elevated EGF	Salivary EGF cannot yet be considered a reliable biomarker for oral leukoplakia
Metgud and Patel (2014)	Serum and salivary albumin levels in oral pre-malignancy and malignancy	45 participants: 15 controls, 15 pre-malignancy, 15 malignancy; albumin measured using bromocresol green method	Serum albumin decreased in pre-malignancy and malignancy, while salivary albumin increased in these cases	Albumin alterations in saliva and serum may aid in early diagnosis of oral pre-malignant and malignant conditions

autoimmune diseases like SLE, providing clinicians with accessible tools for patient care. However, despite these advances, several challenges remain, highlighting areas for future research.

A critical area for further investigation is the validation of specific biomarkers for early detection and disease monitoring. In conditions like allergic skin diseases, including AD, CSU, and ACD, cytokines such as IL-6 and IL-8, alongside stress-related hormones like cortisol, have been widely studied. However, these findings often show inconsistencies across populations, likely influenced by external factors such as stress, diet, and individual variability. Multi-center studies with larger, diverse cohorts are needed to standardize the roles of these biomarkers, particularly to support personalized therapeutic approaches for patients. In inflammatory skin conditions such as psoriasis and acne vulgaris, immune biomarkers like TNF-α, IL-17, and oxidative stress markers have shown potential in reflecting disease severity and treatment response.

Future research should aim to integrate these biomarkers into clinical practice, enabling early intervention and improved disease management. For instance, regular monitoring of salivary cytokines in psoriasis patients could help reduce disease flares, enhancing patient quality of life. Additionally, salivary biomarker panels could be developed to predict treatment responses to biologics or other targeted therapies in patients with acne vulgaris, another autoimmune disease requiring long-term management. In autoimmune skin diseases such as SLE, pemphigus, and pemphigoid, salivary biomarkers like anti-desmoglein antibodies, stress markers, and inflammatory cytokines are of particular interest, as these conditions often involve both cutaneous and mucosal manifestations. Saliva, therefore, serves as an ideal medium for monitoring systemic immune activity. Longitudinal studies tracking salivary biomarkers during disease flare-ups and remission periods could yield valuable insights into their prognostic value, potentially allowing for more precise treatment adjustments.

Oral conditions associated with dermatovenerology, such as RAS, Behcet's disease, OLP, and oral leukoplakia, present unique challenges. These conditions share overlapping immunopathological mechanisms with systemic dermatological diseases, and salivary biomarkers offer a non-invasive way to track disease progression and potentially even malignant transformation. For example, the role of cytokines like IL-6, IL-8, and TNF-α in OLP and oral leukoplakia is well-documented, yet further research is necessary to establish these biomarkers as predictors of malignant potential. Proteomic approaches, such as those used in the identification of exosomal markers like microRNA-4484 (miR-4484) in OLP, illustrate the future of salivary diagnostics for these conditions.

Continued research into such novel biomarkers could revolutionize the early detection and treatment of oral malignancies. Despite recent progress, a significant challenge for integrating salivary diagnostics into routine dermatological practice remains: the need for cost-effective, accessible, and standardized testing methods. Current techniques, including ELISA, mass spectrometry, and immunoassays, are effective but often costly and complex. Developing point-of-care devices capable of real-time saliva sample analysis would significantly enhance the clinical utility of these biomarkers. Collaboration between academic institutions, clinical laboratories, and technology developers is essential to bring such innovations to clinical settings.

In conclusion, while salivary immune and endocrine biomarkers hold immense promise for diagnosing and managing skin and mucocutaneous diseases, further research is needed to standardize their use and integrate them into clinical practice. With continued advancements in salivary proteomics, genomics, and exosomal analysis, the future of non-invasive dermatological diagnostics looks promising. Ultimately, salivary biomarkers may not only aid in diagnosis but also play a pivotal role in personalized medicine, improving outcomes for patients with both common and complex dermatological and oral medicine conditions.

References

Abdel-Haq A, Kusnierz-Cabala B, Darczuk D, Sobuta E, Dumnicka P, Wojas-Pelc A, Chomyszyn-Gajewska M (2014) Interleukin-6 and neopterin levels in the serum and saliva of patients with lichen planus and oral lichen planus. J Oral Pathol Med 43(10):734–739

Abdolsamadi H, Rafieian N, Goodarzi MT, Feradmal J, Davoodi P, Jazayeri M, Taghavi Z, Hoseyni SM, Ahmadi-Motamayel F (2014) Levels of salivary antioxidant vitamins and lipid peroxidation in patients with oral lichen planus and healthy individuals. Chonnam Med J 50:58–62

Adisen E, Aral A, Aybay C, Gurer MA (2008) Salivary epidermal growth factor levels in Behçet's disease and recurrent aphthous stomatitis. Dermatology 217(3): 235–240

Ajona D, Pajares MJ, Chiara MD et al (2015) Complement activation product C4d in oral and oropharyngeal squamous cell carcinoma. Oral Dis 21(7):899–904

Akhtar N, Satyam A, Anand V, Verma KK, Khatri R, Sharma A (2010) Dysregulation of TH type cytokines in the patients of Parthenium induced contact dermatitis. Clin Chem Lab Med 411(23–24):2024–2028

Albalwa K, Kenawy M, El-Fallah AA, Salem RM (2023) Serum and salivary adipsin levels and its association with insulin resistance in acne vulgaris patients. J Cosmet Dermatol 22(4):1354–1360

Andreadis D, Lorenzini G, Drakoulakos D, Belazi M, Mihailidou E, Velkos G, Mourellou-Tsatsou O, Antoniades D (2006) Detection of pemphigus desmoglein 1 and desmoglein 3 autoantibodies and pemphigoid BP180 autoantibodies in saliva and comparison with serum values. Eur J Oral Sci 114:374–380

Asa'ad F, Fiore M, Alfieri A, Pigatto PDM, Franchi C, Berti E, Maiorana C, Damiani G (2018) Saliva as a future field in psoriasis research. Biomed Res Int 2018: 7290913

Babiuch K, Kuśnierz-Cabala B, Kęsek B, Okoń K, Darczuk D, Chomyszyn-Gajewska M (2020) Evaluation of proinflammatory, NF-kappaB dependent cytokines: IL-1α, IL-6, IL-8, and TNF-α in tissue specimens and saliva of patients with oral squamous cell carcinoma and oral potentially malignant disorders. J Clin Med 9(3):867

Bagan L, Sáez GT, Tormos MC, Labaig-Rueda C, Murillo-Cortes J, Bagan JV (2016) Salivary and serum interleukin-6 levels in proliferative verrucous leukoplakia. Clin Oral Investig 20:737–743

Bahramian A, Falsafi P, Abbasi T, Ghanizadeh M, Abedini M, Kavoosi F, Kouhsoltani M, Noorbakhsh F, Dabbaghi Tabriz F, Rajaeih S et al (2018) Comparing serum and salivary levels of vitamin D in patients with recurrent aphthous stomatitis and healthy individuals. J Dent 19:295–300

Bardazzi F, Magnano M, Campanati A, Loconsole F, Carpentieri A, Potenza C, Bernardini N, Di Lernia V, Carrera C, Raone B et al (2017) Biologic therapies in HIV-infected patients with psoriasis: an Italian experience. Acta Derm Venereol 97:989–990

Becke FM, Hehlgans T, Brockhoff G, Männel DN (2001) Development of allergic contact dermatitis requires activation of both tumor necrosis factor receptors. Eur Cytokine Netw 12(1):45–50

Belstrøm D, Eiberg JM, Enevold C, Grande MA, Jensen CAJ, Skov L, Hansen PR (2020) Salivary microbiota and inflammation-related proteins in patients with psoriasis. Oral Dis 26:677–687

Bešlić I, Vrtarić A, Bešlić A, Škrinjar I, Crnković D, Lugović-Mihić L (2023) Salivary melatonin values significantly correlate with reduced quality of life in chronic spontaneous urticaria patients: A pilot study. Acta Clin Croat 62(2):373–377

Black S, Kushner I, Samols D (2004) C-reactive protein. J Biol Chem 279(47):48487–48490

Boras VV, Lukac J, Brailo V, Picek P, Kordic D, Zilic IA (2006) Salivary interleukin-6 and tumor necrosis factor-alpha in patients with recurrent aphthous ulceration. J Oral Pathol Med 35(4):241–243

Bottoni U, Tiriolo R, Pullano SA, Dastoli S, Amoruso GF, Nisticò SP, Fiorillo AS (2016) Infrared saliva analysis of psoriatic and diabetic patients: similarities in protein components. IEEE Trans Biomed Eng 63:379–384

Brailo V, Vucicevic-Boras V, Lukac J, Biocina-Lukenda-D, Alajbeg I, Milenovic A, Balija M (2011) Salivary and serum interleukin 1 beta, interleukin 6, and tumor necrosis factor alpha in patients with leukoplakia and oral cancer. Med Oral Patol Oral Cir Bucal 17:e10–e15

Brailo V, Vucicevic-Boras V, Lukac J et al (2012) Salivary and serum interleukin 1 beta, interleukin 6 and tumor necrosis factor alpha in patients with leukoplakia and oral cancer. Med Oral Patol Oral Cir Bucal 17(e10–e5)

Brunoni A, Santos I, Sabbag C, Lotufo P, Benseñor I (2014) Psoriasis severity and hypothalamic-pituitary-adrenal axis function: results from the CALIPSO study. Braz J Med Biol Res 47:1102–1106

Buajeeb W, Okuma N, Thanakun S, Laothumthut T (2015) Direct immunofluorescence in oral lichen planus. J Clin Diagn Res 9:ZC34–ZC37

Byun JS, Hong SH, Choi JK, Jung JK, Lee HJ (2015) Diagnostic profiling of salivary exosomal microRNAs in oral lichen planus patients. Oral Dis 21(8):987–993

Caglayan F, Miloglu O, Altun E, Yilmaz AB (2008) Oxidative stress and myeloperoxidase levels in saliva of patients with recurrent aphthous stomatitis. Oral Dis 14(8):700–704

Cai L, Kaneko S, Morita E (2018) Changes in salivary chromogranin A levels in adults with atopic dermatitis are correlated with changes in their condition. J Dermatol 45(5):554–559

Camisasca DR, da Ros Gonçalves L, Soares MR et al (2017) A proteomic approach to compare saliva from individuals with and without oral leukoplakia. J Proteome 151:43–52

Campanati A, Moroncini G, Ganzetti G, Pozniak KN, Goteri G, Giuliano A, Martina E, Liberati G, Ricotti F, Gabrielli A et al (2013) Adalimumab

modulates angiogenesis in psoriatic skin. European J Inflamm 11:489–498
Campanati A, Ganzetti G, Giuliodori K, Molinelli E, Offidani A (2016) Biologic therapy in psoriasis: safety profile. Curr Drug Saf 11(1):4–11
Carvalho MFMS, Cavalieri D, Do Nascimento S, Lourenço TGB, Ramos DVR, Pasqualin DDC, Martins LAL, Rocha FA, Heller D, Marti L (2019) Cytokines levels and salivary microbiome play a potential role in oral lichen planus diagnosis. Sci Rep 9(1):18137
Chang KP, Kao HK, Wu CC, Fang KH, Chang YL, Huang YC, Liu SC, Cheng MH (2013) Pretreatment interleukin-6 serum levels are associated with patient survival for oral cavity squamous cell carcinoma. Otolaryngol Head Neck Surg 148:786–791
Chang YS, Chou YT, Lee JH, Lee PL, Dai YS, Sun C, Lin YT, Wang LC, Yu HH, Yang YH, Chen CA, Wan KS, Chiang BL (2014) Atopic dermatitis, melatonin, and sleep disturbance. Pediatrics 134(2):e397–e405
Chaudhuri K, Nair KK, Ashok L (2018) Salivary levels of TNF-α in patients with recurrent aphthous stomatitis: A cross-sectional study. J Dental Res Dental Clin Dental Prosp 12(1):45–48
Cheng Y-SL, Jordan L, Gorugantula LM, Schneiderman E, Chen H-S, Rees T (2014) Salivary interleukin-6 and -8 in patients with oral cancer and patients with chronic oral inflammatory diseases. J Periodontol 85:956–965
Cimen MYB, Kaya TI, Eskandari G, Tursen U, Ikizoglu G, Atik U (2003) Oxidant/antioxidant status in patients with recurrent aphthous stomatitis. Clin Exp Dermatol 28:647–650
Cong TX, Hao D, Wen X, Li XH, He G, Jiang X (2019) From pathogenesis of acne vulgaris to anti-acne agents. Arch Dermatol Res 311(5):337–349
Corbaux C, Joly P (2018) Bullous diseases. In: Puig L, Gulliver W (eds) Adverse reactions to biologics, vol 53. Karger, pp 64–69
Corrêa JD, Calderaro DC, Ferreira GA, Mendonça SM, Fernandes GR, Xiao E, Teixeira AL, Leys EJ, Graves DT, Silva TA (2017) Subgingival microbiota dysbiosis in systemic lupus erythematosus: association with periodontal status. Microbiome 5(1):34
Courtney A, Su JC (2024) The psychology of atopic dermatitis. J Clin Med 13(6):1602
Cumberbatch M, Kimber I (1992) Dermal tumor necrosis factor-alpha induces dendritic cell migration to draining lymph nodes, and possibly provides one stimulus for Langerhans' cell migration. Immunology 75(2):257–263
Curzytek K, Maes M, Kubera M (2021) Immune-regulatory and molecular effects of antidepressants on the inflamed human keratinocyte HaCaT cell line. Neurotox Res 39(4):1211–1226
Darczuk D, Krzysciak W, Vyhouskaya P, Kesek B, Galecka-Wanatowicz D, Lipska W, Kaczmarzyk T, Gluch-Lutwin M, Mordyl B, Chomyszyn-Gajewska M (2016) Salivary oxidative status in patients with oral lichen planus. J Phys Pharmacol: Off J Polish Physiol Soc 67:885–894
Das SN, Jyoti K, Rath R, Mohapatra D, Behera S, Mandal K, Patyal N, Kumar S (2023) Role of saliva in diagnosis and estimation of disease severity in pemphigus vulgaris using autoantibody levels: A comparative study between serum and saliva. Int J Sci Stud 10(11):86–93
De Carvalho GC, Domingues R, de Sousa Nogueira MA, Calvielli Castelo Branco AC, Gomes Manfrere KC, Pereira NV, Aoki V, Sotto MN, Da Silva Duarte AJ, Sato MN (2016) Up-regulation of proinflammatory genes and cytokines induced by S100A8 in CD8+ T cells in lichen planus. Acta Derm Venereol 96:485–489
De D, Khullar G, Handa S, Joshi N, Saikia B, Minz RW (2017) Correlation between salivary and serum anti-desmoglein 1 and 3 antibody titres using ELISA and between anti-desmoglein levels and disease severity in pemphigus vulgaris. Clin Exp Dermatol 42:648–650
Deepthi G, Nandan SRK, Kulkarni PG (2019) Salivary tumour necrosis factor-α as a biomarker in Oral leukoplakia and Oral squamous cell carcinoma. Asian Pac J Cancer Prev 20:2087–2093
Diehl S, Rincón M (2002) The two faces of IL-6 on Th1/Th2 differentiation. Mol Immunol 39(9):531–536
Dittmar D, Schuttelaar ML (2017) Immunology and genetics of tumour necrosis factor in allergic contact dermatitis. Contact Derm 76(5):257–271
Dzobo W, Eren E, Robson S, Shute JK (2024) Salivary IL-8 and sTREM-1 in chronic urticaria: A diagnostic test accuracy study. Clin Exp Allergy 54(5):356–358
Eguia-del Valle A, Martinez-Conde-Llamosas R, Lopez Vicente J, Uribarri-Etxebarria A, Aguirre-Urizar JM (2013) Salivary cortisol determination in patients from the Basque Country with recurrent aphthous stomatitis: a pilot study. Medicina Oral Patología Oral y Cirugía Bucal 18:e207–e211
Esmaili N, Mortazavi H, Kamyab-Hesari K, Aghazadeh N, Daneshpazhooh M, Khani S et al (2015) Diagnostic accuracy of BP180 NC16a and BP230-C3 ELISA in serum and saliva of patients with bullous pemphigoid. Clin Exp Dermatol 40:324–330
Fadel HT, Flytström I, Calander AM, Bergbrant IM, Heijl L, Birkhed D (2013) Profiles of dental caries and periodontal disease in individuals with or without psoriasis. J Periodontol 84:477–485
Fania L, Moretta G, Antonelli F, Scala E, Abeni D, Albanesi C, Madonna S (2022) Multiple roles for cytokines in atopic dermatitis: from pathogenic mediators to endotype-specific biomarkers to therapeutic targets. Int J Mol Sci 23(5):2684
Ferreira BR, Pio-Abreu JL, Figueiredo A, Misery L (2021) Pruritus, allergy and autoimmunity: paving the way for an integrated understanding of psychodermatological diseases? Frontiers in Allergy 2:688999
Foks-Ciekalska A, Jarząb J, Hadas E, Świętochowska E, Gumieniak K, Ciekalski W, Bożek A (2023) The effect of biological treatment on stress parameters determined

in saliva in patients with severe psoriasis. Medicina (Kaunas) 59(4):692

Galbiati V, Marinovich M, Corsini E (2020) Mechanistic understanding of dendritic cell activation in skin sensitization: additional evidences to support potency classification. Toxicol Lett 322:507

Ganzetti G, Campanati A, Santarelli A, Pozzi V, Molinelli E, Minnetti I, Brisigotti V, Procaccini M, Emanuelli M, Offidani A (2014) Periodontal disease: an oral manifestation of psoriasis or an occasional finding? Drug Dev Res 75(Suppl. 1):S46–S49

Ganzetti G, Campanati A, Santarelli A, Pozzi V, Molinelli E, Minnetti I, Brisigotti V, Procaccini M, Emanuelli M, Offidani A (2015) Involvement of the oral cavity in psoriasis: results of a clinical study. Br J Dermatol 172:282–285

Ganzetti G, Campanati A, Santarelli A et al (2016) Salivary interleukin-1β: Oral inflammatory biomarker in patients with psoriasis. J Int Med Res 44(1 Suppl):10–14

Gaur A, Kumar CA, Kumar JS, Nair GKR, Agrawal VM (2018) Estimation of salivary cortisol level and psychological assessment in patients with oral lichen planus. J Indian Acad Oral Med Radiol 30(1):14–17

Gavala A, Venetsanou K, Kittas C, Manolis E, Yiambides A, Myrianthefs P, Baltopoulos G (2010) Decreased whole blood TNFα production capacity after acute alcohol exposure and LPS stimulation ex vivo. Crit Care 14:P13

Gessner A, Mohrs K, Mohrs M (2005) Mast cells, basophils, and eosinophils acquire constitutive IL-4 and IL-13 transcripts during lineage differentiation that are sufficient for rapid cytokine production. J Immunol 174(2):1063–1072

Ghoreishian FS, Tavangar A, Ghalayani P, Boroujeni MA (2017) Salivary levels of interleukin-8 in oral lichen planus and diabetic patients: a biochemical study. Dental Res J 14:209–214

Giannoni M, Consales V, Campanati A, Ganzetti G, Giuliodori K, Postacchini V, Liberati G, Azzaretto L, Vichi S, Guanciarossa F et al (2015) Homocysteine plasma levels in psoriasis patients: our experience and review of the literature. J Eur Acad Dermatol Venereol 29:1781–1785

Giménez-Arnau AM, DeMontojoye L, Asero R, Cugno M, Kulthanan K, Yanase Y et al (2021) The pathogenesis of chronic spontaneous urticaria: the role of infiltrating cells. J Allergy Clin Immunol Pract 9(6):2195–2208

Girardi C, Luz C, Cherubini K, de Figueiredo MAZ, Nunes MLT, Salum FG (2011) Salivary cortisol and dehydroepiandrosterone (DHEA) levels, psychological factors in patients with oral lichen planus. Arch Oral Biol 56(9):864–868

Gisondi P, Geat D, Bellinato F, Spiazzi L, Danese E, Montagnana M, Lippi G, Girolomoni G (2021) Psychological stress and salivary cortisol levels in patients with plaque psoriasis. J Personal Med 11(11):1069

Glavina A, Lugović-Mihić L, Martinović D, Cigić L, Tandara L, Lukenda M, Biočina-Lukenda D, Šupe-Domić D (2023) Association between salivary cortisol and α-amylase with the psychological profile of patients with oral lichen planus and burning mouth syndrome: A case-control study. Biomedicine 11(8): 2182

Glavina A, Zoranić A, Tadin A, Cigić L, Šupe-Domić D, Lugović-Mihić L (2024) Is salivary α-amylase a reliable indicator of psychological status and quality of life in patients with oral lichen planus: A case-control study. Cell Physiol Biochem 58(4):311–321

Godse K, Rajagopalan M, Girdhar M, Kandhari S, Shah B, Chhajed PN et al (2016) Position statement for the use of omalizumab in the management of chronic spontaneous urticaria in Indian patients. Indian Dermatol Online J 7(1):6–11

Goes HFDO, Virgens AR, de Carvalho GC, Pietrobon AJ, Branco ACCC, Oliveira LMDS (2020) Proinflammatory and regulatory mechanisms in allergic contact dermatitis caused by methylchloroisothiazolinone and methylisothiazolinone. Exp Dermatol 29(5):490–498

Gonçalves AS, Mosconi C, Jaeger F, Wastowski I, Aguiar MCF, Silva TA, Ribeiro-Rotta R, Costa NL, Batista AC (2017) Overexpression of immunomodulatory mediators in oral precancerous lesions. Hum Immunol 78:752–757

González-Moles MA, Scully C (2010) HPA-suppressive effects of aqueous clobetasol propionate in the treatment of patients with oral lichen planus. J Eur Acad Dermatol Venereol 24(9):1055–1059

Granger DA, Johnson SB, Szanton SL, Out D, Schumann LL (2012) Incorporating salivary biomarkers into nursing research: an overview and review of best practices. Biol Res Nurs 14(4):347–356

Grieco T, Porzia A, Paolino G, Chello C, Sernicola A, Faina V et al (2020) IFN-γ/IL-6 and related cytokines in chronic spontaneous urticaria: evaluation of their pathogenetic role and changes during omalizumab therapy. Int J Dermatol 59(5):590–594

Haeck IM, Timmer-de Mik L, Lentjes EG, Buskens E, Hijnen DJ, Guikers C, Bruijnzeel-Koomen CA, de Bruin-Weller MS (2007) Low basal serum cortisol in patients with severe atopic dermatitis: potent topical corticosteroids wrongfully accused. Br J Dermatol 156(5):979–985

Hallaji Z, Mortazavi H, Lajevardi V et al (2010) Serum and salivary desmoglein 1 and 3 enzyme-linked immunosorbent assay in pemphigus vulgaris: correlation with phenotype and severity. J Eur Acad Dermatol Venereol 24(3):275–280

Hassona Y, Scully C (2016) Salivary changes in oral mucosal diseases. Periodontol 70(1):111–127

Hibi M, Nakajima K, Hirano T (1996) IL-6 cytokine family and signal transduction: A model of the cytokine system. J Mol Med 74(1):1–12

Hodge DR, Peng B, Cherry JC, Hurt EM, Fox SD, Kelley JA, Munroe DJ, Farrar WL (2005) Interleukin 6 supports the maintenance of p53 tumor suppressor gene promoter methylation. Cancer Res 65(11): 4673–4682

Hoffmann RR, Yurgel LS, Campos MM (2011) Evaluation of salivary endothelin-1 levels in oral squamous cell carcinoma and oral leukoplakia. Regul Pept 166(1–3):55–58

Hsu HJ, Yang YH, Shieh TY, Chen CH, Kao YH, Yang CF, Ko E (2014) Role of cytokine gene (interferon-γ, transforming growth factor-β1, tumor necrosis factor-α, interleukin-6, and interleukin-10) polymorphisms in the risk of oral precancerous lesions in Taiwanese. Kaohsiung J Med Sci 30:551–558

Humberto JSM, Pavanin JV, Da Rocha MJA, Motta ACF (2018) Cytokines, cortisol, and nitric oxide as salivary biomarkers in oral lichen planus: a systematic review. Braz Oral Res 32

Jaeger F, Assunção AC, Caldeira PC, Queiroz-Junior CM, Bernardes VF, Aguiar MCF (2015) Is salivary epidermal growth factor a biomarker for oral leukoplakia? A preliminary study. Oral Surg Oral Med Oral Pathol Oral Radiol 119:451–458

Jia Q, Zang D, Yi J, Dong H, Niu Y, Zhai Q et al (2012) Cytokine expression in trichloroethylene-induced hypersensitivity dermatitis: an in vivo and in vitro study. Toxicol Lett 215(1):31–39

Jinno T, Kawano S, Maruse Y, Matsubara R, Goto Y, Sakamoto T, Hashiguchi Y, Kaneko N, Tanaka H, Kitamura R, Toyoshima T, Jinno A, Moriyama M, Oobu K, Kiyoshima T, Nakamura S (2015) Increased expression of interleukin-6 predicts poor response to chemoradiotherapy and unfavorable prognosis in oral squamous cell carcinoma. Oncol Rep 33(5): 2161–2168

Jung J-Y, Nam J-Y, Kim H-A, Suh C-H (2015) Elevated salivary alpha-amylase level, association between depression and disease activity, and stress as a predictor of disease flare in systemic lupus erythematosus: a prospective case–control study. Medicine (Baltimore) 94(30):e1184

Juretić M, Cerović R, Belušić-Gobić M, Pršo IB, Kqiku L, Špalj S, Pezelj-Ribarić S (2013) Salivary levels of TNF-α and IL-6 in patients with oral premalignant and malignant lesions. Folia Biol 59:99–102

Kaneko S, Liu L, Kakamu T, Minami-Hori M, Morita E (2017) Salivary chromogranin A levels correlate with disease severity but do not reflect anxiety or personality of adult patients with atopic dermatitis. J Dermatol 44(8):920–926

Karincaoglu Y, Batcioglu K, Erdem T, Esrefoglu M, Genc M (2005) The levels of plasma and salivary antioxidants in patients with recurrent aphthous stomatitis. J Oral Pathol Med 34:7–12

Karthikeyan P, Aswath N (2016) Stress as an etiologic co-factor in recurrent aphthous ulcers and oral lichen planus. J Oral Sci 58(2):237–240

Kashyap B, Hyvärinen E, Laitinen I et al (2024) Salivary metabolomics in patients with oral lichen planus: A preliminary study based on NMR spectroscopy. Clin Oral Investig 28:103

Kaur J, Jacobs R (2015) Proinflammatory cytokine levels in oral lichen planus, oral leukoplakia, and oral submucous fibrosis. J Korean Assoc Oral Maxillofac Surg 41(4):171–175

Kimber I, Cumberbatch M, Humphreys M, Hopkins SJ (1990) Contact hypersensitivity induces plasma interleukin 6. Int Arch Allergy Appl Immunol 92(1):97–99

Kishimoto T (2010) IL-6: from its discovery to clinical applications. Int Immunol 22:347–352

Kojima R, Matsuda A, Nomura I, Matsubara O, Nonoyama S, Ohya Y, Saito H, Matsumoto K (2013) Salivary cortisol response to stress in young children with atopic dermatitis. Pediatr Dermatol 30(1):17–22

Kondo S, Pastore S, Shivji GM, McKenzie RC, Sauder DN (1994) Characterization of epidermal cytokine profiles in sensitization and elicitation phases of allergic contact dermatitis as well as irritant contact dermatitis in mouse skin. Lymphokine Cytokine Res 13(6): 367–375

Koopai M, Mortazavi H, Khatami A, Khodashenas Z (2018) Salivary and serum anti-desmoglein 1 and 3 ELISA and indirect immunofluorescence in pemphigus vulgaris: correlations with serum ELISA, indirect immunofluorescence, and disease severity. Acta Dermatovenerol Croat 26:91–99

Koray M, Dülger O, Ak G, Horasanli S, Uçok A, Tanyeri H, Badur S (2003) The evaluation of anxiety and salivary cortisol levels in patients with oral lichen planus. Oral Dis 9(6):298–301

Kuna M (2024) The association of IL-6 concentrations in saliva and serum with clinical disease indicators in patients with chronic spontaneous urticaria (doctoral dissertation). University of Zagreb, School of Dental Medicine. Retrieved from https://urn.nsk.hr/urn:nbn:hr:127:675871

Kunikullaya UK, Kumar MA, Ananthakrishnan V, Jaisri G (2017) Stress as a cause of recurrent aphthous stomatitis and its correlation with salivary stress markers. Chin J Physiol 60(4):226–230

Kwon H, Pessin JE (2013) Adipokines mediate inflammation and insulin resistance. Front Endocrinol (Lausanne) 12(4):71

Kwon B, Hong S, Kim EY, Kim JH, Kim M, Park JH et al (2021) Effect of cone of *Pinus densiflora* on DNCB-induced allergic contact dermatitis-like skin lesions in BALB/c mice. Nutrients 13(3):1–16

La Fratta I, Tatangelo R, Campagna G, Rizzuto A, Franceschelli S, Ferrone A, Patruno A, Speranza L, De Lutiis MA, Felaco M, Grilli A, Pesce M (2018) The plasmatic and salivary levels of IL-1β, IL-18, and IL-6 are associated with emotional difference during stress in young males. Sci Rep 8(1):3031

Lee JH, Bae YJ, Lee SH, Kim SC, Lee HY, Ban GY, Shin YS, Park HS, Kratzsch J, Ye YM (2019) Adaptation and validation of the Korean version of the Urticaria control test and its correlation with salivary cortisone. Allergy, Asthma Immunol Res 11(1):55–67

Li LF, Liu G, Wang J (2007) Prevalence and 1-year outcome of facial allergic contact dermatitis in patients

patch tested in a university hospital. Contact Derm 57(3):187–190
Li Y, Lin P, Wang S, Li S, Wang R, Yang L, Wang H (2020) Quantitative analysis of differentially expressed proteins in psoriasis vulgaris using tandem mass tags and parallel reaction monitoring. Clin Proteomics 17: 30
Lim J, Iyer A, Suen JY et al (2013) C5aR and C3aR antagonists each inhibit diet-induced obesity, metabolic dysfunction, and adipocyte and macrophage signaling. FASEB J 27(2):822–831
Lin L, Wang J, Liu N, Liu S, Xu H, Ji N, Zhou M, Zeng X, Zhang D, Li J et al (2016) Interleukin-37 expression and its potential role in oral leukoplakia and oral squamous cell carcinoma. Sci Rep 6:26757
Liu W, Dan H, Wang Z, Jiang L, Zhou Y, Zhao M, Chen Q, Zeng X (2009) IFNgamma and IL-4 in saliva of patients with oral lichen planus: A study in an ethnic Chinese population. Inflammation 32(3):176–181
Liu J, Shi Q, Yang S, Wang Q, Xu J, Guo B (2017) The relationship between levels of salivary and serum interleukin-6 and oral lichen planus. J Am Dent Assoc 148:743–749.e9
Lo JC, Ljubicic S, Leibiger B et al (2014) Adipsin is an adipokine that improves β cell function in diabetes. Cell 158(1):41–53
López-Jornet P, Aznar-Cayuela C (2016) Efficacy of topical chamomile management vs. placebo in patients with oral lichen planus: a randomized double-blind study. J Eur Acad Dermatol Venereol 30(10): 1783–1786
Lopez-Jornet P, Cayuela CA, Tvarijonaviciute A, Escribano D, Cerón J, Parra-Perez F (2015) Oral lichen planus: salivary biomarkers cortisol, immunoglobulin A, adiponectin. J Oral Pathol Med 45:211–217
López-Jornet P, Zavattaro E, Mozaffari HR, Ramezani M, Sadeghi M (2019) Evaluation of the salivary level of cortisol in patients with oral lichen planus: A meta-analysis. Medicina 55:213
Loyola Rodriguez JP, Galvan Torres LJ, Martinez Martinez RE, Abud Mendoza C, Medina Solis CE, Ramos Coronel S, Garcia Cortes JO, Domínguez Pérez RA (2016) Frequency of dental caries in active and inactive systemic lupus erythematosus patients: salivary and bacterial factors. Lupus 25(12): 1349–1356
Lu R, Zhang J, Sun W, Du G, Zhou G (2015) Inflammation-related cytokines in oral lichen planus: an overview. J Oral Pathol Med 44(1):1–14
Luchetti MM, Benfaremo D, Campanati A, Molinelli E, Ciferri M, Cataldi S, Capeci W, Di Carlo M, Offidani AM, Salaffi F et al (2018) Clinical outcomes and feasibility of the multidisciplinary management of patients with psoriatic arthritis: two-year clinical experience of a dermo-rheumatologic clinic. Clin Rheumatol 37:2741–2749
Maheshwari A, Garg T, Chander R, Debnath E (2022) Chronic spontaneous urticaria: an etiopathogenic study. Indian J Dermatol 67(1):92
Man GG, Martin MD, Darveau RP, Truelove E, Epstein J (2004) Oral and serum IL-6 levels in oral lichen planus patients. Oral Sur Oral Med Oral Pathol Oral Radiol Endodon 98:673–678
Mansourian A, Najafi S, Nojoumi N, Parhami P, Moosavi MS (2018) Salivary cortisol and salivary flow rate in clinical types of oral lichen planus. Skinmed 16:19–22
Marques CPC, Victor EC, Franco MM, Fernandes JM, Maor Y, de Andrade MS, Rodrigues VP, Benatti BB (2016) Salivary levels of inflammatory cytokines and their association to periodontal disease in systemic lupus erythematosus patients: a case-control study. Cytokine 85:165–170
Marseglia L, D'Angelo G, Manti S, Salpietro C, Arrigo T, Barberi I, Reiter RJ, Gitto E (2014) Melatonin and atopy: role in atopic dermatitis and asthma. Int J Mol Sci 15(8):13482–13493
Martina E, Campanati A, Diotallevi F, Offidani A (2020) Saliva and oral diseases. J Clin Med 9:466
Megson E, Fitzsimmons T, Dharmapatni K, Bartold PM (2010) C-reactive protein in gingival crevicular fluid may be indicative of systemic inflammation. J Clin Periodontol 37(9):797–804
Mendonça SMS, Corrêa JD, Souza AF, Travassos DV, Calderaro DC, Rocha NP, Vieira ÉLM, Teixeira AL, Ferreira GA, Silva TA (2019) Immunological signatures in saliva of systemic lupus erythematosus patients: influence of periodontal condition. Clin Exp Rheumatol 37(2):208–214
Meštrović-Štefekov J, Lugović-Mihić L, Hanžek M, Bešlić I, Japundžić I, Karlović D (2022) Salivary cortisol values and personality features of atopic dermatitis patients: A prospective study. Dermatitis 33(4): 341–348
Metgud R, Patel S (2014) Serum and salivary levels of albumin as diagnostic tools for oral pre-malignancy and oral malignancy. Biotech Histochem 89:8–13
Mizawa M, Yamaguchi M, Ueda C, Makino T, Shimizu T (2013) Stress evaluation in adult patients with atopic dermatitis using salivary cortisol. Biomed Res Int 2013:138027
Mizuno K, Sakaue H, Kohsaka K, Takeda H, Hayashi N, Sato T (2021) An increase in normetanephrine in hair follicles of acne lesions through the sympatho-adrenal medullary system in acne patients with anxiety. J Dermatol 48(8):1281–1285
Mochtar M, Murasmita A, Irawanto ME, Julianto I, Kariosentono H, Waskito F (2018) The difference in interleukin-19 serum on degrees of acne vulgaris severity. Int J Inflamm 2018:4141579
Mohamed RW, Fathy A, El-Sayed AE (2003) Increased circulating FcepsilonRII-bearing B lymphocytes and serum levels of IL-4 in non-autoreactive chronic idiopathic urticaria. Egyptian J Immunol 10(2):9–18

Monib KME, El-Fallah AA, Salem RM (2022) Inflammatory markers in acne vulgaris: saliva as a novel diagnostic fluid. J Cosmet Dermatol 21(3):1280–1285

Moori M, Ghafoori H, Sariri R (2016) Nonenzymatic antioxidants in saliva of patients with systemic lupus erythematosus. Lupus 25:265–271

Mortazavi H, Khatami A, Seyedin Z, Vasheghani Farahani I, Daneshpazhooh M (2015) Salivary desmoglein enzyme-linked immunosorbent assay for diagnosis of pemphigus vulgaris: A noninvasive alternative test to serum assessment. Biomed Res Int 2015: 698310

Mozaffari HR, Ramezani M, Mahmoudiahmadabadi M, Omidpanah N, Sadeghi M (2017) Salivary and serum levels of tumor necrosis factor-alpha in oral lichen planus: a systematic review and meta-analysis study. Oral Surg Oral Med Oral Pathol Oral Radiol 124: e183–e189

Mozaffari HR, Sharifi R, Mirbahari S, Montazerian S, Sadeghi M, Rostami S (2018) A systematic review and meta-analysis study of salivary and serum interleukin-8 levels in oral lichen planus. Adv Dermatol Allergol 35:599–604

Mozaffari HR, Zavattaro E, Saeedi M, Lopez-Jornet P, Sadeghi M, Safaei M, Imani MM, Nourbakhsh R, Moradpoor H, Golshah A et al (2019) Serum and salivary interleukin-4 levels in patients with oral lichen planus: A systematic review and meta-analysis. Oral Surg Oral Med Oral Pathol Oral Radiol 128:123–131

Murrell DF, Peña S, Joly P, Marinovic B, Hashimoto T, Diaz LA, Sinha AA, Payne AS, Daneshpazhooh M, Eming R, Jonkman MF, Mimouni D, Borradori L, Kim SC, Yamagami J, Lehman JS, Saleh MA, Culton DA, Czernik A, Zone JJ et al (2020) Diagnosis and management of pemphigus: recommendations of an international panel of experts. J Am Acad Dermatol 82(3): 575–585.e1

Nadendla LK, Meduri V, Paramkusam G, Pachava KR (2014) Association of salivary cortisol and anxiety levels in lichen planus patients. J Clin Diagn Res 8(12):ZC01–ZC03

Namazi MR, Parhizkar AR, Jowkar F (2015) Serum levels of hypersensitive C-reactive protein in moderate and severe acne. Indian Dermatol Online J 6(4):253–257

Ng V, Koh D, Fu Q, Chia SE (2003) Effects of storage time on stability of salivary immunoglobulin A and lysozyme. Clin Chem Lab Med 338(2):131–134

Nosratzehi T, Arbabi-Kalati F, Salimi S, Honarmand E (2014) The evaluation of psychological factors and salivary cortisol and IgA levels in patients with oral lichen planus. Zahedan J Res Med Sci 16(7):31–34

Noto Z, Tomihara K, Furukawa K, Noguchi M (2016) Dyskeratosis congenita associated with leukoplakia of the tongue. Int J Oral Maxillofac Surg 45(6):760–763

Novak T, Fortune F, Bergmeier L, Khan I, Hagi-Pavli E (2020) Neutrophil elastase and endogenous inhibitors in Behçet's disease saliva. Clin Exp Immunol 202(1): 93–105

Novak T, Hamedi M, Bergmeier LA, Fortune F, Hagi-Pavli E (2021) Saliva and serum cytokine profiles during oral ulceration in Behçet's disease. Front Immunol 12:724900

Nutan, Kanwar AJ, Bhansali A, Parsad D (2011) Evaluation of hypothalamic-pituitary-adrenal axis in patients with atopic dermatitis. Indian J Dermatol Venereol Leprol 77(3):288–293

Obayashi K (2013) Salivary mental stress proteins. Clin Chem Lab Med 425(2):196–201

Ohashi M, Iwase M, Nagumo M (1999) Elevated production of salivary nitric oxide in oral mucosal diseases. J Oral Pathol Med 28:355–359

Orciani M, Campanati A, Caffarini M, Ganzetti G, Consales V, Lucarini G, Offidani A, Di Primio R (2017) T helper (Th)1, Th17 and Th2 imbalance in mesenchymal stem cells of adult patients with atopic dermatitis: at the origin of the problem. Br J Dermatol 176(6):1569–1576

Osugi Y, Hara J, Tagawa S, Takai K, Hosoi G, Matsuda Y et al (1997) Cytokine production regulating Th1 and Th2 cytokines in hemophagocytic lymphohistiocytosis. Blood 89(11):4100–4103

Ouellet-Morin I, Danese A, Williams B, Arseneault L (2011) Validation of a high-sensitivity assay for C-reactive protein in human saliva. Brain Behav Immun 25(4):640–646

Packard RRS, Libby P (2008) Inflammation in atherosclerosis: from vascular biology to biomarker discovery and risk prediction. Clin Chem 54(1):24–38

Patel S, Metgud R (2015) Estimation of salivary lactate dehydrogenase in oral leukoplakia and oral squamous cell carcinoma: A biochemical study. J Cancer Res Ther 11:119–123

Pay JB, Shaw AM (2019) Towards salivary C-reactive protein as a viable biomarker of systemic inflammation. Clin Biochem 68:1–8

Peiser M, Tralau T, Heidler J, Api AM, Arts JHE, Basketter DA et al (2012) Allergic contact dermatitis: epidemiology, molecular mechanisms, in vitro methods and regulatory aspects. Cell Mol Life Sci 69(5):763–781

Pfaffe T, Cooper-White J, Beyerlein P, Kostner K, Punyadeera C (2011) Diagnostic potential of saliva: current state and future applications. Clin Chem 57(5):675–687

Piguet PF, Grau GE, Hauser C, Vassalli P (1991) Tumor necrosis factor is a critical mediator in hapten-induced irritant and contact hypersensitivity reactions. J Exp Med 173(3):673–679

Pippi R, Patini R, Ghiciuc CM, Sandu RB, Pasquali V, Scaccianoce S et al (2014) Diurnal trajectories of salivary cortisol, salivary α-amylase and psychological profiles in oral lichen planus patients. J Biol Regul Homeost Agents 28(1):147–156

Pires ALPV, Alves LDB, da Silva AM, Arsati F, Lima-Arsati YBO, Dos Santos JN, Calumby RT, Freitas VS (2023) Salivary biomarkers to evaluate psychological disorders in oral lichen planus: A systematic review with meta-analysis. Oral Dis 29(7):2734–2746

Pondeljak N, Lugović-Mihić L (2020) Stress-induced interaction of skin immune cells, hormones, and neurotransmitters. Clin Ther 42(5):757–770

Pondeljak N, Tomić L, Lazić Mosler E, Šitum M, Karlović D, Lugović-Mihić L (2023) Salivary cortisol and perceived psychological stress in patients with chronic contact hand dermatitis. Contact Derm 89(5): 393–395

Punyani SR, Sathawane RS (2013) Salivary level of interleukin-8 in oral precancer and oral squamous cell carcinoma. Clin Oral Investig 17(2):517–524

Radi G, Campanati A, Diotallevi F, Bianchelli T, Offidani A (2021) Novel therapeutic approaches and targets for treatment of psoriasis. Curr Pharm Biotechnol 22(1): 7–31

Radwan-Oczko M, Rybińska A, Mierzwicka A, DuśIlnicka I (2024) Salivary histamine levels in patients with oral lichen planus lesions. Medicina 60(7):1038

Rafe T, Shawon PA, Salem L, Chowdhury NI, Kabir F, Bin Zahur SM, Akhter R, Noor HB, Mohib MM, Sagor MAT (2019) Preventive role of resveratrol against inflammatory cytokines and related diseases. Curr Pharm Des 25(12):1345–1371

Rajcani J, Solarikova P, Buzgoova K, Brezina I, Jezova D (2019) Patients with atopy exhibit reduced cortisol awakening response but not cortisol concentrations during the rest of the day. Immunol Lett 67(2–3): 176–181

Rhodus NL, Cheng B, Myers S, Bowles W, Ho V, Ondrey F (2005) A comparison of the pro-inflammatory, NF-kappaB-dependent cytokines: TNF-alpha, IL-1-alpha, IL-6, and IL-8 in different oral fluids from oral lichen planus patients. Clin Immunol 114(3):278–283

Richards H, Ray D, Kirby B, Mason D, Plant D, Main C, Fortune D, Griffiths C (2005) Response of the hypothalamic-pituitary-adrenal axis to psychological stress in patients with psoriasis. Br J Dermatol 153: 1114–1120

Riis JL, Out D, Dorn LD, Beal SJ, Denson LA, Pabst S, Jaedicke K, Granger DA (2014) Salivary cytokines in healthy adolescent girls: Intercorrelations, stability, and associations with serum cytokines, age, and pubertal stage. Dev Psychobiol 56(4):797–811

Rivarola de Gutierrez E, Di Fabio A, Salomón S, Lanfranchi H (2014) Topical treatment of oral lichen planus with anthocyanins. Medicina Oral Patología Oral y Cirugía Bucal 19:e459–e466

Rocha-Pereira P, Santos-Silva A, Rebelo I, Figueiredo A, Quintanilha A, Teixeira F (2004) The inflammatory response in mild and severe psoriasis. Br J Dermatol 150(5):917–928

Rödström PO, Jontell M, Hakeberg M, Berggren U, Lindstedt G (2001) Erosive oral lichen planus and salivary cortisol. J Oral Pathol Med 30(5):257–263

Rosatelli TB, Roselino AM, Dellalibera-Joviliano R, Reis ML, Donadi EA (2005) Increased activity of plasma and tissue kallikreins, plasma kininase II, and salivary kallikrein in pemphigus foliaceus (fogo selvagem). Br J Dermatol 152(4):650–657

Ruacho G, Lira-Junior R, Gunnarsson I, Svenungsson E, Boström EA (2022) Inflammatory markers in saliva and urine reflect disease activity in patients with systemic lupus erythematosus. Lupus Sci Med 9(1): e000607

Rusiñol L, Puig L (2024) A narrative review of the IL-18 and IL-37 implications in the pathogenesis of atopic dermatitis and psoriasis: prospective treatment targets. Int J Mol Sci 25(15):8437

Russo I, Saponeri A, Michelotto A, Alaibac M (2017) Salivary samples for the diagnosis of pemphigus vulgaris using the BIOCHIP approach: A pilot study. In Vivo 31:97–99

Rutter-Locher Z, Smith TO, Giles I, Sofat N (2017) Association between systemic lupus erythematosus and periodontitis: A systematic review and meta-analysis. Front Immunol 8:112

Sakai H, Ishida T, Sato K, Mandokoro K, Yabe S, Sato F et al (2019) Interference of skin scratching attenuates accumulation of neutrophils in a murine allergic contact dermatitis model. Inflammation 42(5):2226–2235

Saral Y, Coskun BK, Ozturk P, Karatas F, Ayar A (2005) Assessment of salivary and serum antioxidant vitamins and lipid peroxidation in patients with recurrent aphthous ulceration. Tohoku J Exp Med 206:305–312

Schmetzer O, Lakin E, Topal FA, Preusse P, Freier D, Church MK et al (2018) IL-24 is a common and specific autoantigen of IgE in patients with chronic spontaneous urticaria. J Allergy Clin Immunol 142(3): 876–882

Seil M, El Ouaaliti M, Abdou Foumekoye S, Pochet S, Dehaye JP (2012) Distinct regulation by lipopolysaccharides of the expression of interleukin-1β by murine macrophages and salivary glands. Innate Immun 18(1):14–24

Selvam NP, Sadaksharam J (2015) Salivary interleukin-6 in the detection of oral cancer and precancer. Asia Pac J Clin Oncol 11:236–241

Sezer E, Ozugurlu F, Ozyurt H, Sahin S, Etikan I (2007) Lipid peroxidation and antioxidant status in lichen planus. Clin Exp Dermatol 32:430–434

Shah B, Ashok L, Sujatha GP (2009) Evaluation of salivary cortisol and psychological factors in patients with oral lichen planus. Indian J Dent Res 20(3):288–292

Shahidi M, Jafari S, Barati M, Mahdipour M, Gholami MS (2017) Predictive value of salivary microRNA-320a, vascular endothelial growth factor receptor 2, CRP and IL-6 in oral lichen planus progression. Inflammopharmacology 25:577–583

Shefler I, Pasmanik-Chor M, Kidron D, Mekori YA, Hershko AY (2014) T cell-derived microvesicles induce mast cell production of IL-24: relevance to inflammatory skin diseases. J Allergy Clin Immunol 133(1):217–224

Shetty SR, Chadha R, Babu S, Kumari S, Bhat S, Achalli S (2012) Salivary lactate dehydrogenase levels in oral leukoplakia and oral squamous cell carcinoma: A biochemical and clinicopathological study. J Cancer Res Ther 8:123

Shiva A, Arab S, Mousavi SJ, Zamanian A, Maboudi A (2020) Serum and salivary level of nitric oxide (NOx) and CRP in oral lichen planus (OLP) patients. Journal of Dentistry, Shiraz 21:6–11

Sieminska I, Pieniawska M, Grzywa TM (2024) The immunology of psoriasis—current concepts in pathogenesis. Clin Rev Allergy Immunol 66:164–191

Silvestre MC, Sato MN, Reis VMSD (2018) Innate immunity and effector and regulatory mechanisms involved in allergic contact dermatitis. An Bras Dermatol 93(2): 242–250

Simoura JA d S, Pires ALPV, Alves LDB, Arsati F, Lima-Arsati YB d O, Dos Santos JN et al (2023) Psychological profile and α-amylase levels in oral lichen planus patients: a case-control preliminary study. Oral Dis 29(3):1242–1249

Skrinjar I, Vidranski V, Brzak BL, Vidovic Juras D, Andabak Rogulj A, Brailo V, Vucicevic Boras V (2019) Salivary cortisol levels in patients with oral lichen planus—A pilot case-control study. Dentistry J 7:59

Skutnik-Radziszewska A, Maciejczyk M, Fejfer K, Krahel J, Flisiak I, Kołodziej U, Zalewska A (2020a) Salivary antioxidants and oxidative stress in psoriatic patients: can salivary total oxidant status and oxidative status index be a plaque psoriasis biomarker? Oxidative Med Cell Longev 2020:9086024

Skutnik-Radziszewska A, Maciejczyk M, Flisiak I, Kołodziej JKU, Kotowska-Rodziewicz A, Klimiuk A, Zalewska A (2020b) Enhanced inflammation and nitrosative stress in the saliva and plasma of patients with plaque psoriasis. J Clin Med 9:745

Soudan RA, Daoud SA, Mashlah AM (2011) Study of some salivary changes in cutaneous psoriatic patients. Saudi Med J 32:386–389

Souza MM, Florezi GP, Nico M, de Paula F, Paula FM, Lourenço SV (2018) Salivary proteomics in lichen planus: A relationship with pathogenesis? Oral Dis 24:784–792

Spiljak B, Vilibić M, Glavina A, Crnković M, Šešerko A, Lugović-Mihić L (2022) A review of psychological stress among students and its assessment using salivary biomarkers. Behavioral Sci 12(10):400

Spiljak B, Simunović L, Vilibić M, Hanžek M, Crnković D, Lugović-Mihić L (2024) Perceived stress, salivary cortisol, and temperament traits among students of dental medicine: A prospective and interventional study. Behavioral Sci 14(4):289

Stanescu I-I, Calenic B, Dima A et al (2018) Salivary biomarkers of inflammation in systemic lupus erythematosus. Anat Anz 219:89–93

Stojanovich L (2010) Stress and autoimmunity. Autoimmun Rev 9(6):A271–A276

Sugermann PB, Savage NW, Seymour G, Walsh LJ (1996) Is there a role for tumor necrosis factor-alpha (TNF-alpha) in oral lichen planus? J Oral Pathol Med 25:219–224

Sun A, Wang J, Chia J-S, Chiang C-P (2005) Serum interleukin-8 level is a more sensitive marker than serum interleukin-6 level in monitoring the disease activity of oral lichen planus. Br J Dermatol 152: 1187–1192

Talungchit S, Buajeeb W, Lerdtripop C, Surarit R, Chairatvit K, Roytrakul S, Kobayashi H, Izumi Y, Khovidhunkit SP (2018) Putative salivary protein biomarkers for the diagnosis of oral lichen planus: A case-control study. BMC Oral Health 18:42

Tamura S, Ouchi KF, Mori K, Endo M, Matsumoto T, Eda H, Tanaka Y, Ishitsuka H, Tokita H, Yamaguchi K (1995) Involvement of human interleukin 6 in experimental cachexia induced by a human uterine cervical carcinoma xenograft. Clin Cancer Res 1(11): 1353–1358

Thanthoni M, Sathasivasubramanian S (2015) Quantitative analysis of salivary TNF-α in oral lichen planus patients. Int J Dentistry 2015:1–5

Thijs JL, van Seggelen W, Bruijnzeel-Koomen C, de Bruin-Weller M, Hijnen D (2015) New developments in biomarkers for atopic dermatitis. J Clin Med 4(3): 479–487

Tugrul S, Kocyigit A, Dogan R et al (2016) Total antioxidant status and oxidative stress in recurrent aphthous stomatitis. Int J Dermatol 55(3):e130–e135

Tvarijonaviciute A, Aznar-Cayuela C, Rubio CP, Ceron JJ, López-Jornet P, Asta T, Cristina AC, Camila PR, Joaquin CJ (2016) Evaluation of salivary oxidative stress biomarkers, nitric oxide and C-reactive protein in patients with oral lichen planus and burning mouth syndrome. J Oral Pathol Med 46:387–392

Tvarijonaviciute A, Aznar-Cayuela C, Rubio CP, Tecles F, Ceron JJ, López-Jornet P (2018) Salivary antioxidant status in patients with oral lichen planus: correlation with clinical signs and evolution during treatment with Chamaemelum nobile. Biomed Res Int 2018:5187549

Upadhyay RB, Carnelio S, Shenoy R, Gyawali P, Mukherjee M (2010) Oxidative stress and antioxidant defense in oral lichen planus and oral lichenoid reaction. Scand J Clin Lab Invest 70:225–228

Varoni EM, Lodi G, Sardella A, Carrassi A, Iriti M (2012) Plant polyphenols and oral health: old phytochemicals for new fields. Curr Med Chem 19(11):1706–1720

Vineetha R, Pai KM, Vengal M, Gopalakrishna K, Narayanakurup D (2014) Usefulness of salivary alpha amylase as a biomarker of chronic stress and stress-related oral mucosal changes: A pilot study. J Clin Exp Dent 6(2):e132–e137

Wang J, Yang J, Wang C, Zhao Z, Fan Y (2021) Systematic review and meta-analysis of oxidative stress and antioxidant markers in oral lichen planus. Oxidative Med Cell Longev 2021:9914652

Wei W, Sun Q, Deng Y, Wang Y, Du G, Song C, Li C, Zhu M, Chen G, Tang G (2018) Mixed and inhomogeneous expression profile of Th1/Th2 related cytokines detected by cytometric bead array in the saliva of patients with oral lichen planus. Oral Surg Oral Med Oral Pathol Oral Radiol 126(2):142–151

Wei W, Wang Y, Sun Q, Jiang C, Zhu M, Song C, Li C, Du G, Deng Y, Nie H et al (2019) Enhanced T-cell proliferation and IL-6 secretion mediated by overexpression of TRIM21 in oral lesions of patients with oral lichen planus. J Oral Pathol Med 49:350–356

Wenghoefer M, Pantelis A, Najafi T, Deschner J, Allam J, Novak N, Reich R, Martini M, Berge S, Fischer H et al (2010) Gene expression of oncogenes, antimicrobial peptides, and cytokines in the development of oral leukoplakia. Oral Surg Oral Med Oral Pathol Oral Radiol Endod 110:351–356

Wozniak KL, Arribas A, Leigh JE, Fidel PL (2002) Inhibitory effects of whole and parotid saliva on immunomodulators. Oral Microbiol Immunol 17(2): 100–107

Wu KJ, Tu CC, Hu JX, Chu PH, Ma KS, Chiu HY, Kuo MY, Tsai TF, Chen YW (2022) Severity of periodontitis and salivary interleukin-1β are associated with psoriasis involvement. J Formos Med Assoc 121(10): 1908–1916

Wu J, Liu G, Jia R, Guo J (2023) Salivary extracellular vesicles: biomarkers and beyond in human diseases. Int J Mol Sci 24(24):17328

Yanase Y, Takahagi S, Ozawa K, Hide M (2021) The role of coagulation and complement factors for mast cell activation in the pathogenesis of chronic spontaneous urticaria. Cells 10(7):1759

Yang L-L, Liu X-Q, Liu W, Cheng B, Li M-T (2006) Comparative analysis of whole saliva proteomes for the screening of biomarkers for oral lichen planus. Inflamm Res 55(10):405–407

Yu L, Li L (2022) Potential biomarkers of atopic dermatitis. Front Med (Lausanne, Switzerland) 9:1028694

Zhang Y, Lin M, Zhang S, Wang Z, Jiang L, Shen J et al (2008) NF-κB-dependent cytokines in saliva and serum from patients with oral lichen planus: A study in an ethnic Chinese population. Cytokine 41(2): 144–149

Zhang Y, Zhang H, Du S, Yan S, Zeng J (2021) Advanced biomarkers: therapeutic and diagnostic targets in urticaria. Int Arch Allergy Immunol 182(10):917–931

Zhu ZD, Ren XM, Zhou MM, Chen QM, Hua H, Li CL (2022) Salivary cytokine profile in patients with oral lichen planus. J Dental Sci 17(1):100–105

Zian Z, Bouhoudan A, Mourabit N, Azizi G, Bennani Mechita M (2021) Salivary cytokines as potential diagnostic biomarkers for systemic lupus erythematosus disease. Mediat Inflamm 2021:8847557

Zindel J, Kubes P (2020) DAMPs, PAMPs, and LAMPs in immunity and sterile inflammation. Annu Rev Pathol 15:493–518

Impact of Oral Immunity on Neuroinflammation and Neurodegenerative Diseases

13

Mehrshad Ashford, Heliya Ziaei, and Nima Rezaei

Abstract

Oral health is increasingly recognized as an important factor in the regulation of systemic immune responses and neurological health. Disruption of immune homeostasis in oral tissues, particularly in chronic conditions such as periodontitis, has been linked to the development and progression of neurodegenerative diseases, such as Alzheimer's disease (AD), Parkinson's disease (PD), and multiple sclerosis (MS). This chapter explores how chronic inflammatory conditions in the oral cavity, oral dysbiosis, and the presence of specific periodontopathogens contribute to systemic immune activation, microglial priming, and blood–brain barrier dysfunction. The discussion addresses multiple mechanistic pathways underlying these relationships, such as hematogenous spread of microbial components with subsequent systemic immune stimulation, neural signaling via trigeminal and vagal pathways, and alterations in gut-brain communication through the integrated oral–gut–brain axis. The chapter also reviews findings from clinical and experimental studies demonstrating correlations between oral inflammation and cognitive impairment, structural brain alterations, and elevated inflammatory and neurodegenerative biomarkers. Oral health disturbances are considered not only as secondary outcomes of neurological disease but also as a potential contributing factor in their pathogenesis. The emphasis is on the importance of incorporating oral health assessment and maintenance into preventive and therapeutic strategies for neurodegenerative conditions, while highlighting the necessity for well-designed mechanistic investigations to further explain these complex relationships.

Keywords

Neuroinflammation · Neurodegenerative disease · Oral immunity · Alzheimer's disease · Neurodegeneration

M. Ashford · H. Ziaei
Universal Scientific Education and Research Network (USERN), Los Angeles, CA, USA

N. Rezaei (✉)
Universal Scientific Education and Research Network (USERN), Göteborg, Sweden

Network of Immunity in Infection, Malignancy and Autoimmunity (NIIMA), Universal Scientific Education and Research Network (USERN), Tehran, Iran

Research Center for Immunodeficiencies, Children's Medical Center, Tehran University of Medical Sciences, Tehran, Iran

Department of Immunology, School of Medicine, Tehran University of Medical Sciences, Tehran, Iran
e-mail: Rezaei_nima@tums.ac.ir

13.1 Introduction

The bidirectional relationship between oral health and systemic health has become a rapidly

N. Rezaei, H. Ziaei (eds.), *Oral Immunology*, Advances in Experimental Medicine and Biology 1492,
https://doi.org/10.1007/978-3-032-03176-1_13

growing area of focus in interdisciplinary research in recent years (Kapila 2021). The oral cavity has diverse tissues, which are immunologically active and continuously exposed to microbial communities that interact with both innate and adaptive immune systems (Liu et al. 2022). Current evidence demonstrates that chronic inflammatory conditions in the oral cavity and oral dysbiosis lead to disruptions in oral immune homeostasis, which can significantly impact overall health and potentially cause systemic conditions such as type 2 diabetes, atherosclerosis, hypertension, rheumatoid arthritis, adverse pregnancy outcomes, and neurodegenerative disorders. (Murray et al. 2024; Rajasekaran et al. 2024; Xiaojing et al. 2000).

Neurodegenerative diseases such as Alzheimer's disease (AD), multiple sclerosis (MS), and Parkinson's disease (PD) represent major health challenges in aging populations worldwide (Lamptey et al. 2022). According to the World Health Organization's statements in 2021, more than 55 million people worldwide live with dementia, with approximately 10 million newly diagnosed cases annually. These neurodegenerative conditions share key pathological features, such as progressive neuronal loss, chronic neuroinflammation, and the accumulation of abnormal protein aggregates, and these processes potentially support each other. Recent research suggests that neuroinflammation may not simply result from protein aggregation but might precede and stimulate this process (Wu et al. 2024; Zhang et al. 2023). Research indicates that acute and chronic low-grade inflammation impair cognitive performance, likely due to peripheral inflammation disrupting the blood–brain barrier (BBB), triggering neuroinflammatory responses, and other mechanisms (Mekhora et al. 2024). This evolving perspective emphasizes the critical importance of identifying potential initial stimulators of neuroinflammation to uncover the development of neurodegenerative diseases.

Oral health status and chronic dysregulation in oral inflammatory responses have been identified as significant factors contributing to systemic inflammation, thereby promoting neuroinflammatory risk (Konkel et al. 2019; Pruntel et al. 2024). The oral cavity is an immunologically dynamic site hosting one of the body's most diverse microbial communities and maintaining a complex network of innate and adaptive immune responses, which are connected to the systemic immune system (Kleinstein et al. 2020). These responses are essential for microbial balance, tissue integrity, and mucosal defense. When this environment is compromised through aging, inadequate oral hygiene, or microbial dysbiosis, chronic inflammatory conditions such as periodontitis can develop (Abdulkareem et al. 2023; Liu et al. 2022). Periodontitis, which is characterized by the progressive destruction of the supporting structures of the teeth, involves activation of host immune responses, sustained microbial challenge, and the release of pro-inflammatory mediators. This initially localized inflammatory activity can become systemic, as secreted pro-inflammatory cytokines enter the circulation and affect distant organs, including the nervous system (Sankowski et al. 2015). Chronic systemic inflammation originating in the oral cavity can compromise the blood–brain barrier (BBB), activate microglia, and stimulate astrocytes toward pro-inflammatory phenotypes, establishing a vicious cycle of neuroinflammation (Kwon and Koh 2020; Mayer and Fischer 2025).

Therefore, while neurodegenerative diseases can impair oral health through reduced motor control and cognitive function, emerging evidence indicates that this relationship is bidirectional. Oral inflammation, especially chronic conditions such as periodontitis, may precede, contribute to, and exacerbate neurodegenerative processes through different mechanisms. Clinical, epidemiological, and experimental research has identified periodontitis as a significant risk factor for cognitive decline (Asher et al. 2022; Sun and Li 2025).

The oral–brain connection is mediated through multiple pathways, such as systemic dissemination through the bloodstream, direct transport through peripheral nerve endings such as the trigeminal nerve, and indirect communication through immune-neural interfaces, including the vagus nerve, transmitting peripheral inflammatory signals to the brain (Sansores-España et al.

2021). The gut microbiome also plays an important role, since oral dysbiosis can alter the gut microbial composition, compromise intestinal permeability, and stimulate systemic inflammatory responses, potentially contributing to disrupted immune tolerance in the brain (Shen et al. 2025; Zhang et al. 2025). These interconnected mechanisms comprise what is now recognized as the "oral–gut–brain axis."

Identifying the oral cavity as a potential initiator of systemic and neural inflammation represents an important conceptual shift in understanding neurodegenerative disease etiology (Hajishengallis and Chavakis 2021). This suggests that improving oral health, especially managing chronic periodontitis, previously considered primarily a dental issue, could be a modifiable factor in preventing or mitigating neurodegenerative diseases (Soiniemi et al. 2024). Preventive strategies that support periodontal health and diminish microbial dysbiosis, thereby reducing systemic inflammatory burden may offer novel approaches for reducing the risk or progression of neurodegeneration (Amato et al. 2025; Hajishengallis and Chavakis 2021). Given the global burden of dementia and the high prevalence of chronic periodontitis in aging populations, this perspective has significant clinical implications.

This chapter investigates the immunological, microbiological, and neuropathological connections between oral immunity and neurodegenerative disorders. The aim is to explain how chronic oral inflammation and systemic immune activation may collectively contribute to the potential initiation or deterioration of neuroinflammation and neuronal degeneration, highlighting oral health not only as a reflection of systemic well-being but also as a potentially modifiable factor for nervous system health.

13.2 Overview of the Oral Immune Environment

The oral cavity is a complex immunological site in the body, containing various types of immune cells, including neutrophils, macrophages, dendritic cells, B cells, and T cells, which collectively work together to maintain immune surveillance and normal microbial balance while defending against pathogenic invasion. In addition to these cellular components, immunoglobulins such as secretory IgA (sIgA) play a significant role in mucosal immunity by facilitating immune tolerance and neutralizing pathogens at the epithelial surface without triggering overt inflammation (Carreto-Binaghi et al. 2024). The gingival crevicular fluid (GCF), found in the gingival sulcus, is enriched with diverse immune cells and secreted immune mediators, supporting rapid local immune responses and maintaining tissue integrity in the periodontal regions surrounding the teeth (Subbarao et al. 2019).

Equally important to the role of innate and adaptive immune system components in maintaining oral tissue health is the role of the oral microbiota in modulating immune homeostasis. Commensal microbes, such as *Streptococcus* and *Lactobacillus* species, help regulate host immune balance. However, microbial dysbiosis often results from poor oral hygiene, diet, lifestyle choices, or systemic diseases, and it favors the dominance of pathogenic species such as *Porphyromonas gingivalis* and *Fusobacterium nucleatum*, which trigger chronic inflammation in oral tissues (Liu et al. 2022; Radaic and Kapila 2021; Santacroce et al. 2023).

Major periodontal pathogens like *Porphyromonas gingivalis* and *Treponema denticola* effectively evade host defense mechanisms by influencing immune signaling pathways and impairing neutrophil clearance. These alterations may potentially lead to persistent and chronic low-grade inflammation (Mysak et al. 2014; Olsen et al. 2017). Their virulence factors, including lipopolysaccharide (LPS) and gingipains, stimulate the secretion of pro-inflammatory mediators, such as IL-1β, IL-6, TNF-α, and prostaglandin E2, from gingival epithelial cells and immune cells, which increase tissue destruction contributing to the development and progression of periodontitis (de Andrade et al. 2019; Tanabe et al. 2023).

These inflammatory processes extend beyond the oral cavity. Pro-inflammatory cytokines and bacterial products can enter the bloodstream and trigger systemic immune responses. Research has shown associations between chronic periodontitis and increased circulating levels of IL-6, TNF-α, and C-reactive protein (CRP), which are also implicated in systemic conditions, including metabolic disorders and vascular inflammation (Bansal et al. 2014; Marcaccini et al. 2009; Matsuda et al. 2024). These systemic effects highlight how disruptions in oral immune homeostasis can influence immune regulation throughout the body, emphasizing the importance of oral health in the broader context of systemic inflammatory burden.

13.3 Neuroinflammation in Neurodegenerative Diseases

Neuroinflammation is increasingly recognized as a critical factor in the pathogenesis and development of neurodegenerative diseases such as AD, PD, and MS (Adamu et al. 2024). It is characterized by the activation of glial cells in the central nervous system (CNS), particularly microglia, which are resident immune cells in the brain. Neuroinflammation may have protective roles in the short term against infections, trauma, autoimmune diseases, etc., but may potentially become damaging when chronically sustained (Adamu et al. 2024; Ransohoff et al. 2015). This persistent, low-grade inflammatory condition is involved in synaptic loss, neuronal death, and progressive cognitive decline.

Microglia are the primary mediators of neuroinflammatory responses. Under normal conditions, microglia continuously assess the neural environment and contribute to homeostatic processes such as cellular debris clearance and synaptic pruning (Gao et al. 2023; Hoogland et al. 2015). However, upon activation, which can also be initiated due to systemic immune activation and inflammatory responses, microglia shift toward a pro-inflammatory phenotype, characterized by the secretion of cytokines like IL-1β, TNF-α, and IL-6 (Hoogland et al. 2015). These cytokines disrupt neuronal signaling and can induce apoptosis.

Astrocytes, previously known as structural cells supporting neurons, are now also known as active contributors in neuroimmune regulation. In neuroinflammation, astrocytes can become reactive (a neurotoxic reactive phenotype, which is commonly referred to as A1 astrocytes), secreting cytokines and chemokines that increase microglial activity and accelerate neurodegeneration (Ding et al. 2021; Lana et al. 2021).

13.3.1 Peripheral Inflammation as a Driver of CNS Immune Responses

Several studies suggested that systemic inflammation may contribute to CNS neuroinflammation through various mechanisms (Sun et al. 2022). Inflammatory diseases such as periodontitis and type 2 diabetes have been associated with an increased risk of neurodegenerative diseases, partially due to their ability to stimulate pro-inflammatory cytokine production, which may enter the brain via the bloodstream (Plachokova et al. 2024; Szablewski 2025). Moreover, circulating levels of C-reactive protein (CRP) and IL-6, key markers of systemic inflammation, are commonly elevated in individuals with periodontitis and have been strongly linked to an increased risk of cognitive decline and neurodegenerative diseases, such as AD and PD (Gani et al. 2009; Lyra et al. 2022; Wu et al. 2023).

Peripheral cytokines can influence the CNS through various mechanisms. They can cross a compromised BBB or interact with brain regions where the BBB has higher permeability, such as the circumventricular organs (CVOs). They can also affect the CNS through neural-immune communication involving the vagus nerve and leptomeningeal cells (Dantzer 2018; Yang et al. 2022). These cytokines, upon reaching the CNS, can initiate microglial and astrocytic activation, leading to the release of additional pro-inflammatory mediators, increasing neuroinflammatory processes and maintaining a neurotoxic microenvironment (Adamu et al. 2024; Yang et al. 2022).

13.3.2 Cytokines and Related Signaling Pathways

Among different systemic cytokines involved in neuroinflammation, IL-6, TNF-α, and IL-1β are particularly important. These cytokines activate CNS immune cells and also contribute to BBB disruption, which further enables the infiltration of inflammatory mediators. Research demonstrates that chronic systemic inflammation impairs mitochondrial function, elevates reactive oxygen species (ROS) production, and increases oxidative stress, all of which are pathological features commonly present in neurodegenerative disorders (Peggion et al. 2024; Singh et al. 2019).

At the molecular level, several signaling pathways are activated during neuroinflammatory processes, such as the NF-κB pathway, which regulates the transcription of pro-inflammatory cytokines and adhesion molecules, MAPK (mitogen-activated protein kinase) pathway, which mediates cellular stress responses and cytokine production, and PI3K/Akt pathway, participating in the regulation of inflammatory processes. Collectively, the activation of these molecular pathways may promote neuronal damage and accelerate the progression of neuroinflammation and neurodegenerative diseases (Kim and Choi 2010; Wong et al. 2023).

13.3.3 The Role of Oral-Derived Inflammation

Emerging evidence increasingly connects oral inflammation, particularly inflammation originating from chronic periodontitis, as a contributor to systemic inflammation and neuroinflammation. Virulence factors, including gingipains, LPS, dentilisin, etc. secreted by pathogens such as *P. gingivalis* and *T. denticola* can enter the bloodstream and reach distant organs, including the brain (Jia et al. 2019; Poole et al. 2017). In AD patients, *P. gingivalis* DNA and gingipains have been detected in post-mortem brain tissue, colocalizing with amyloid-β (Aβ) deposits and tau pathology, suggesting a potential role in neurodegeneration (Poole et al. 2017; Ryder 2020). These results support the hypothesis that oral inflammation may initiate or exacerbate neuropathological conditions. In-vivo studies provide further mechanistic support. In a study by Ilievski et al., oral administration of *P. gingivalis* in wild-type mice led to Aβ deposition and hyperphosphorylated tau in the brain, hallmarks of AD, along with increased microglial activation and neuroinflammation (Ilievski et al. 2018). These results reinforce the hypothesis that oral-derived inflammation can initiate or exacerbate neuropathological changes.

Several studies have also emphasized systemic inflammation as a potential initiating factor for neuroinflammation. Neuroinflammation, induced by genetic susceptibility or chronic peripheral immune activation, can promote the aggregation of proteins such as Aβ and α-synuclein, thereby accelerating disease progression (Zhang et al. 2023). This insight suggests that neuroinflammation might be an early, possibly preclinical, driver of neurodegeneration, emphasizing the critical need for early evaluation and therapeutic interventions. Taken together, these findings support a growing body of research about chronic systemic inflammation playing a crucial mechanistic role in the onset and progression of neuroinflammatory and neurodegenerative processes.

13.4 Mechanistic Pathways Connecting Oral Immunity to the Brain

The recognition of oral health as a modulator of systemic immunity has resulted in emerging investigations into mechanisms by which local inflammation in the oral cavity can influence CNS function. Recent evidence suggests that inflammatory mediators and microbial byproducts, originating from chronic periodontal lesions, may reach the CNS through different mechanisms, such as hematogenous dissemination, direct transportation through neurons such as the trigeminal and vagus nerves, and the oral–gut–brain axis, which is more indirect but has been increasingly recognized (Bonaz et al. 2017; Miklossy 2011; Yamazaki 2023).

13.4.1 Hematogenous Dissemination

In oral inflammatory conditions such as chronic periodontitis, bacteria and their byproducts can enter the systemic circulation. Periodontal pathogens such as *P. gingivalis*, *T. denticola*, and *F. nucleatum* have been associated with this process. During bacteremia episodes, these microorganisms and their virulence factors, such as LPS, gingipains, and outer membrane vesicles (OMV), can enter the bloodstream and have systemic effects (Emery et al. 2021; Zhang et al. 2021b). When these components enter the circulation, they may induce systemic inflammation by stimulating immune cells to release pro-inflammatory cytokines such as IL-1β, IL-6, and TNF-α. These cytokines are known to compromise the integrity and permeability of the BBB, a semipermeable and highly selective barrier that typically protects the brain from harmful substances (Díaz-Zúñiga et al. 2020). Once the BBB is compromised, these bacteria themselves, microbial byproducts, and also inflammatory mediators such as cytokines may enter the CNS and reach brain tissues, initiating a pro-inflammatory loop that contributes to neurodegeneration (Galea 2021). Experimental models have demonstrated that oral infections involving microbial dysbiosis and high levels of *P. gingivalis* lead to the accumulation of amyloid-β (Aβ), tau hyperphosphorylation, microglial activation, and cognitive deficits in mice (Díaz-Zúñiga et al. 2020; Li et al. 2024a). These findings support the hypothesis that bacteremia and subsequent BBB disruption facilitate the translocation of neurotoxic mediators into the brain, contributing to the development of neurodegenerative diseases.

13.4.2 Neural Pathways

Another potential mechanism connecting oral health to brain inflammation is direct neural transport via cranial nerves, especially the trigeminal and vagus nerves. The trigeminal nerve, which provides sensory innervation to the orofacial region, may function as a route for microbial migration and the transportation of their toxic byproducts from periodontal tissues to the CNS (R. Li et al. 2024b). Based on previous literature, DNA from oral spirochetes, such as *T. denticola*, has been detected in the trigeminal ganglia and brain tissues of Alzheimer's disease patients. These bacteria can undergo retrograde transport from the oral cavity to the trigeminal ganglia, residing there for long periods of time (Pisani et al. 2022; Riviere et al. 2002). This retrograde axonal transport may enable oral pathogens or their components to circumvent the BBB and directly lead to localized neuroinflammation. Bacterial components such as LPS and gingipains may travel through nerve fibers and enter neural ganglia, thereby triggering immune responses that may extend to adjacent brain regions (Tran et al. 2022).

Oral inflammation may affect brain health and neuroinflammation indirectly through the vagus nerve as well. The vagus nerve does not directly innervate oral or periodontal tissues; therefore, it does not provide a direct route for oral pathogens and their toxic byproducts to enter the CNS (Bowland and Weyrich 2022). Rather, its role in transmitting inflammatory signals from the oral environment is indirect and facilitated by systemic circulation and gut-mediated inflammatory processes; the mechanisms behind this connection will be discussed in detail in Sect. 13.4.3 on the oral–gut–brain axis (Sansores-España et al. 2021).

These findings highlight the importance of direct and indirect neural–immune communication pathways in linking chronic oral inflammation to CNS health and the development of neuroinflammatory processes.

13.4.3 Oral–Gut–Brain Axis

The oral–gut–brain axis characterizes a model explaining how oral inflammation indirectly affects neurological function in the CNS through systemic effects on the immune system or neural pathways. This model is based on the continuous communication between the oral and gut microbiome through the regular ingestion of oral bacteria. Chronic oral dysbiosis can result in the swallowing of pathogenic oral bacteria, which may colonize in the gastrointestinal tract, consequently disrupting intestinal microbiome homeostasis (Adnan et al. 2025; Malan-Müller et al. 2024).

There are two main mechanisms by which oral–gut communication affects CNS health. The first one is described as the oral–gut–systemic circulation–CNS axis. In this axis, gut dysbiosis resulting from oral dysbiosis leads to systemic inflammation, which can impact CNS health (Adil et al. 2025; Singhrao and Olsen 2018). Research demonstrates that oral administration of *P. gingivalis* in murine models induces gut dysbiosis, intestinal barrier dysfunction (i.e., "leaky gut"), and elevated inflammatory cytokine levels in both systemic circulation and brain tissue. Leaky gut also permits bacterial endotoxins and microbial metabolites to enter the bloodstream, promoting systemic inflammation and further compromising BBB integrity (Doifode et al. 2021; Feng et al. 2020). Additionally, *P. gingivalis* exposure in the CNS may compromise glymphatic system function, a brain-wide waste clearance network crucial for removing metabolic byproducts, including amyloid-β, thereby increasing the susceptibility to neuroinflammation and neurodegeneration (Chi et al. 2021).

Another mechanism is through the oral–gut–vagus nerve–CNS route. Gut dysbiosis, as a result of oral microbiome dysregulation, can also influence the brain via the vagus nerve. The vagus nerve, which is part of the parasympathetic nervous system, can transmit peripheral inflammatory mediators, including those originating from systemic dissemination of oral inflammatory mediators to brain regions such as the nucleus tractus solitarius (NTS) (Breit et al. 2018). Through this mechanism, pro-inflammatory cytokines secreted in the gut can stimulate afferent fibers of the vagus nerve, which transmit immune and metabolic signals to the NTS in the brainstem. This can initiate CNS immune responses despite the absence of direct neural connections between oral tissues and vagal fibers. Based on recent studies, vagal afferent signaling has a significant impact on microglial activity and the modulation of hypothalamic–pituitary–adrenal (HPA) axis function, thereby affecting both neuroendocrine regulation and neuroinflammatory processes (Faraji et al. 2025). Supporting this concept, studies have shown that experimental vagotomy significantly decreases CNS inflammatory responses to peripheral LPS (Schweighöfer et al. 2016).

Together, these mechanisms demonstrate that oral inflammation has the potential to impact CNS homeostasis through different indirect mechanisms, highlighting the oral cavity's crucial role as a peripheral driver of neuroinflammation.

13.5 Oral Pathogens Implicated in Neurodegeneration

The role of oral pathogens, particularly those associated with periodontitis, in neurodegenerative diseases has gained increasing attention. While earlier sections discussed potential mechanisms by which oral inflammation can influence the brain, this section focuses on specific oral pathogens that have been directly associated with the pathogenesis of neurodegenerative diseases. Several microbial species, including bacteria and viruses commonly found in the oral cavity, have been correlated with AD, PD, and other neurodegenerative conditions (Nicholson and Landry 2022). These pathogens may either directly invade neural tissues or indirectly induce immune responses that promote neurodegeneration.

13.5.1 *Porphyromonas gingivalis*

P. gingivalis is one of the most studied periodontal pathogens and has been linked to the pathophysiology of AD and other neurodegenerative conditions (Huang et al. 2025). In addition to its well-known ability to stimulate systemic inflammation, activate microglia, and enhance Aβ aggregation, recent research has identified additional mechanisms that emphasize its broader neurotoxic potential.

As briefly discussed in Sect. 13.3, postmortem evaluations have shown that *P. gingivalis* DNA and gingipain proteases are present in brains of AD patients, co-localizing with tau tangles and ubiquitin-positive deposits, key histopathological markers of neurodegeneration (Dominy et al. 2025). This suggests a direct microbial contribution to tau pathology. Moreover, gingipains may directly cleave tau protein, which potentially modifies its structure and promotes its aggregation, or may indirectly exacerbate its hyperphosphorylation through stimulating inflammatory signaling pathways, thereby promoting neurofibrillary tangle formation characteristic of Alzheimer's disease (Dominy et al. 2025; Huang et al. 2025).

Recent research indicates that *P. gingivalis* contributes to systemic vascular dysfunction through specific pathophysiological mechanisms. It can induce hypercoagulability via fibrinogen degradation and modification of clot architecture, thereby potentially compromising cerebral perfusion, a hemodynamic abnormality increasingly known as a potential factor in both AD and vascular dementia etiology (Adams et al. 2019; Liu et al. 2021; Nunes et al. 2020; Xie et al. 2020; Zhang et al. 2021a). These observations suggest another mechanism by which *P. gingivalis* may contribute to neurodegeneration through vasculopathies parallel to other potential mechanisms described.

The pathogenic significance of *P. gingivalis* extends beyond AD. Experimental multiple sclerosis (MS) models have established a connection between exposure to *P. gingivalis* and aggravating demyelination processes and T-lymphocyte proliferation, suggesting broader neuroimmunological effects (Liu et al. 2021; Polak et al. 2018). In these experimental systems, *P. gingivalis* resulted in systemic immune activation and BBB hyperpermeability, which was associated with elevated CNS inflammatory markers, suggesting that *P. gingivalis* contributes to the development of MS by affecting the systemic immune system (Dai et al. 2025).

Collectively, these findings suggest that *P. gingivalis* is an important periodontopathogen with systemic effects that extend beyond the oral cavity, capable of disrupting vascular integrity, inducing systemic immune responses, and initiating or accelerating neurodegenerative processes. Recent research indicates that targeting *P. gingivalis* and its gingipains with therapeutic inhibitors might be a promising approach for reducing AD progression and neuroinflammation (Raha et al. 2020).

13.5.2 Herpes Simplex Virus (HSV)

Other than bacterial pathogens, certain viral infections, such as herpes simplex virus type 1 (HSV-1), have been found to play a significant role in developing neurodegenerative processes, particularly in the context of AD (Wu et al. 2020). HSV-1 is a neurotropic virus that establishes a persistent latency in both the PNS and CNS. In conditions such as inflammation or immunosenescence, latent HSV-1 may reactivate, triggering neuroimmune responses that may contribute to progressive neuronal damage (Duarte et al. 2019).

A mechanism by which HSV-1 contributes to neurodegenerative processes is through its interaction with TLR4, expressed on host immune cells (Feng et al. 2023). Viral reactivation stimulates TLR4-mediated signal transduction, contributing to the synthesis and accumulation of Aβ peptides as components of the innate immune response. Within this context, Aβ functions as an antimicrobial peptide with specific effects on sequestering viral particles. While this response initially serves protective functions, recurrent or persistent viral reactivation results in sustained Aβ production and its parenchymal deposition, thereby initiating and exacerbating neuroinflammatory cascades that significantly contribute to AD pathogenesis (Itzhaki et al. 2018; Fulop et al. 2018).

One key mechanism by which HSV-1 influences neurodegeneration involves its interaction with Toll-like receptor 4 (TLR4) on host immune cells. Reactivation of HSV-1 has been shown to stimulate TLR4 signaling, leading to the accumulation of Aβ, as part of an innate immune response, where Aβ acts as an antimicrobial peptide designed to sequester viral particles. While this response may initially serve a protective function, chronic or repeated reactivation results in sustained Aβ production and deposition, promoting neuroinflammatory cascades and contributing to AD pathology (Feng et al. 2023; Fulop et al. 2018).

This association is further supported by epidemiological data. Several clinical studies have shown that individuals seropositive for HSV-1 have an increased risk of developing AD or dementia (Warren-Gash et al. 2019). Moreover, this risk appears to be modulated by genetic susceptibility; carriers of the APOE4 allele not only demonstrate higher rates of HSV-1 reactivation and proliferation in the brain but may also be more susceptible to its neurotoxic effects, potentially due to decreased viral clearance or increased inflammatory signaling (Liu et al. 2025; Wainberg et al. 2021).

Importantly, preclinical studies suggest that antiviral therapy may have neuroprotective benefits. Antiviral agents such as acyclovir and valacyclovir, which inhibit HSV replication, have potential effects on reducing Aβ accumulation and improving neuronal survival in in vitro and in vivo models (Iqbal et al. 2020). Ongoing clinical trials are assessing whether long-term antiviral treatment can slow cognitive decline in HSV-1-positive individuals at risk for AD.

Collectively, these data suggest a potential link between HSV-1 and neurodegenerative disorders, representing a compelling example of how chronic latent viral infections and immune reactivation may serve as modifiable contributors to AD and other neurodegenerative disease pathogenesis, especially in genetically predisposed individuals.

13.5.3 Other Pathogens

Beyond *P gingivalis* and HSV-1, other oral microbial species have also shown associations with neurodegenerative diseases through various direct or indirect mechanisms.

Treponema denticola: This spirochetal microorganism, mainly associated with periodontal disease, has been detected in cerebral tissues of dementia patients, which raised significant questions about its neuroinvasive potential. *T. denticola* can also stimulate neural apoptosis by inducing tau hyperphosphorylation, increasing Aβ accumulation, and stimulating neuroinflammation, based on in vivo studies (Cichońska et al. 2024; Wu et al. 2022b). This pathogen may also directly reach the brain, since studies have shown its presence is significantly higher in the brain tissue of AD patients compared to cognitively healthy individuals (Li et al. 2024a).

Fusobacterium nucleatum: This species is primarily known for its role in periodontal infections and colorectal carcinogenesis; however, it has recently shown potential association with neurodegenerative conditions. This pathogen stimulates systemic inflammatory responses contributing to neuroinflammation and cognitive dysfunction (Wu et al. 2022a). Experimental animal models suggest that *F. nucleatum* increases BBB permeability, thereby enhancing CNS exposure to inflammatory mediators and neurotoxic factors (Fan et al. 2022; Wu et al. 2022a). These findings support the significant role of microbial involvement in neurological disease pathogenesis.

13.6 Systemic Immune Dysregulation as a Mediator

As previously discussed, chronic immune dysregulation is considered a critical link between oral inflammatory processes and the pathogenesis of neurodegenerative disorders. This section addresses the broader systemic immune imbalances that maintain neuroinflammatory states. In chronic inflammatory states, immune dysregulation may lead to persistent hyperactivation of immune responses, resulting in continuous secretion of pro-inflammatory mediators and compromised immune resolution (L. Chen et al. 2015).

Regarding resolution pathway impairment, continuous stimulation by bacterial virulence factors, including gingipains and LPS, during chronic periodontitis stimulates sustained cytokine production that extends beyond the oral environment. This sustained and unresolved inflammatory state contributes to systemic immune sensitization, an established risk factor for neuroinflammatory conditions (Hajishengallis and Lamont 2021; Konkel et al. 2019; Sun et al. 2022).

Chronic oral inflammation can compromise the integrity and permeability of critical biological barriers, including the gingival tissues, intestinal epithelium, and the BBB. This disruption occurs either through the direct effects of pathogenic microorganisms and their byproducts or indirectly via the upregulation of systemic inflammatory mediators (Takahashi et al. 2019; Wang et al. 2019). These processes not only facilitate the entrance of these mediators to the systemic circulation via compromised gingival tissue but also contribute to BBB breakdown, enabling their access to the CNS. These barriers normally serve as essential physical and immunological boundaries that protect the body against microbial invasion and excessive immune activation. Chronic inflammation impairs tight junction integrity, allowing cytokines, immune cells, and bacterial byproducts to reach normally restricted regions like the CNS (Gryka-Marton et al. 2025; Takata et al. 2021).

Moreover, periodontal inflammation modifies systemic T cell phenotypes; It favors pro-inflammatory Th1 and Th17 responses, capable of crossing the compromised BBB and supporting neuroinflammatory processes (Dutzan et al. 2017; Martínez-García and Hernández-Lemus 2021). Th1 and Th17 infiltration to the CNS is strongly associated with the pathogenesis of MS. In experimental MS models, *P. gingivalis* exposure enhances T cell infiltration into the CNS and exacerbates demyelination (Dai et al. 2025; Murphy et al. 2010).

Furthermore, circulating pro-inflammatory cytokines such as IL-6 and TNF-α, and IL-1β, elevated due to chronic oral inflammation, can activate microglial cells in the CNS. This leads to perpetuating the cycle of cytokine release by activated glial cells, subsequently preserving neuroinflammation and contributing to progressive neuronal damage in the CNS (Kennedy and Silver 2016; Qin et al. 2023).

13.7 Evidence from Clinical Studies

Several clinical studies demonstrate a strong association between poor oral health, particularly periodontitis and chronic oral inflammation, and an increased risk of neurodegenerative conditions, such as AD, mild cognitive impairment (MCI), cognitive decline, MS, PD, etc.

Kulkarni et al. performed one of the largest clinical cohort studies based on the TriNetX database to evaluate the association between poor oral health and AD. This study included over 30 million patients, demonstrating a 2.13-fold increased risk of developing AD in individuals with poor oral health conditions, supporting large-scale epidemiological evidence for this association (Kulkarni et al. 2023). Moreover, another study by Asher et al. combined results from 47 longitudinal studies, revealing a significantly higher risk of cognitive decline and dementia in individuals with poor periodontal status or tooth loss (Asher et al. 2022). Also, another retrospective cohort study using Korean national health data from 2002 to 2018 evaluated over 1.1 million dementia patients aged over 65. The results showed that patients who received periodontal treatment had significantly higher survival rates compared to untreated individuals. After adjusting for demographic and health variables, the untreated group had an 83% higher risk of mortality, suggesting that periodontal treatment may reduce mortality risk in dementia patients (Cho et al. 2024).

Another cross-sectional study of more than 2000 older adults demonstrated that higher serum antibody levels against *P. gingivalis* were significantly correlated with lower cognitive function scores. These findings indeed highlight a potential link between systemic exposure to periodontal pathogens and cognitive decline in elderly populations (Noble et al. 2009).

Furthermore, another study involving more than 1800 elderly participants revealed that patients with severe periodontitis demonstrated significantly lower cognitive performance and elevated serum concentrations of methylmalonic acid (MMA), an established biomarker of mitochondrial dysfunction. These findings suggest that the relationship between periodontal disease severity and reduced cognitive function is partially mediated by MMA levels. This highlights that mitochondrial dysfunction may be a potential mechanistic pathway linking oral inflammation to neurocognitive decline (Li et al. 2022).

Mayer et al. in the PAROBRAIN study identified a correlation between poor periodontal health and reduced white matter integrity, by evaluating over 2000 individuals. These data suggest that chronic oral inflammation may adversely impact brain microstructure (Mayer et al. 2023). Moreover, a two-year cohort study found that patients with periodontitis showed elevated plasma levels of phosphorylated tau and Aβ, suggesting a potential mechanism through which oral inflammation may contribute to neurodegenerative disorders such as AD (Carballo et al. 2023).

Regarding PD, although in vivo studies confirmed the correlation between periodontitis and periodontal pathogens with motor dysfunction in PD animal models, clinical studies have been controversial. A few clinical studies found an association between poor oral health and PD (Jeong et al. 2021; Verhoeff et al. 2023; Yilmaz et al. 2023); however, some studies have not found this association (Chen et al. 2023).

Collectively, clinical evidence demonstrates that oral inflammation and poor oral health conditions, specifically chronic periodontitis, represent not simply a secondary manifestation of neurocognitive and motor function decline but may be potential modifiable predictors contributing to the initiation and progression of neurodegenerative pathology. These epidemiological and clinical data emphasize the necessity of incorporating oral health examinations and on-time treatment into comprehensive approaches for the detection, risk management, and therapeutic approaches of neurodegenerative disorders.

13.8 Conclusion

This chapter highlights the potential role of oral inflammatory conditions in the pathogenesis and development of neurodegenerative diseases and neuroinflammatory processes (summarized in Fig. 13.1). Chronic oral diseases such as periodontitis are no longer considered only as local pathologies but as stimulators of the systemic immune system capable of influencing the CNS. Oral pathogens and their harmful byproducts can enter circulation, disrupt the BBB permeability, activate glial cells, and contribute to neurodegeneration through hematogenous, neural, or gut-mediated mechanisms.

Clinical and epidemiological studies increasingly support these associations, linking poor oral health to cognitive decline, structural brain alterations, and elevated biomarkers, hallmarks of diseases such as AD. Early intervention to address oral inflammatory conditions such as periodontitis also suggests that improving periodontal health may offer cognitive benefits, highlighting oral health care as a potential preventive strategy.

Although growing evidence has suggested a strong association between oral and neurocognitive health, establishing a definitive causal relationship needs further well-designed in vivo investigations and large-scale clinical studies among various neurocognitive conditions. Additional research is also needed to clarify the immune mechanisms underlying the

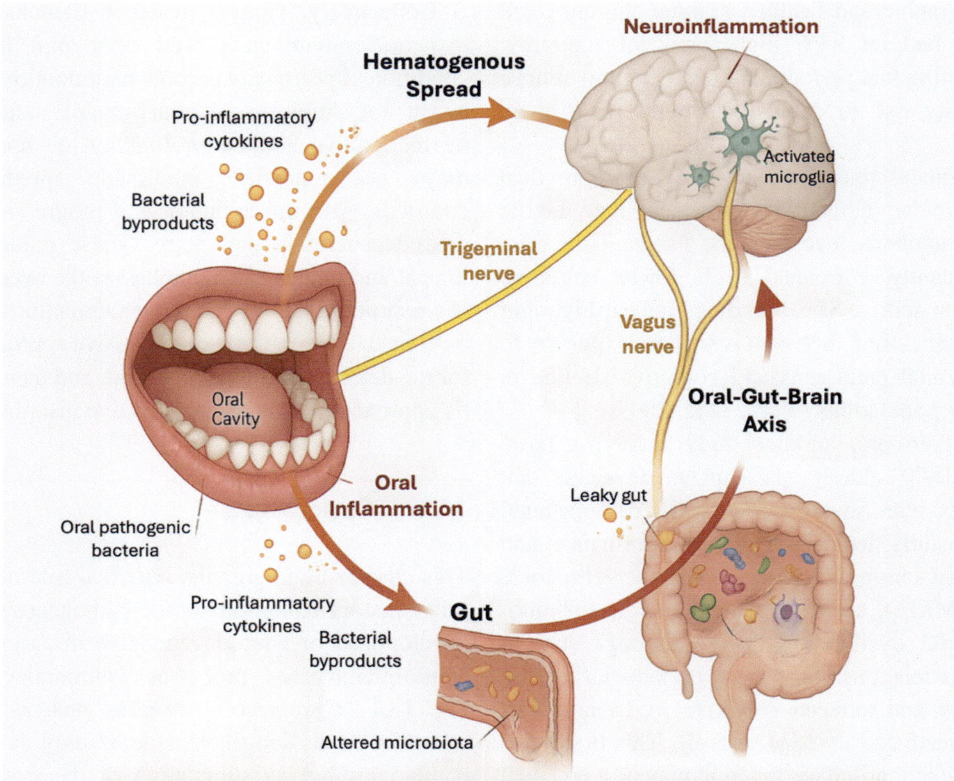

Fig. 13.1 Mechanistic pathways linking oral inflammation to neuroinflammation. This graphical abstract illustrates the interconnected pathways by which chronic oral inflammation, particularly periodontitis, may contribute to central nervous system (CNS) inflammation and neurodegeneration. Oral pathogenic bacteria and their byproducts initiate local immune responses and release pro-inflammatory cytokines. These mediators can enter the bloodstream (hematogenous spread), pass the compromised blood–brain barrier (BBB), and activate microglia in the brain, leading to inflammation. Additionally, oral bacteria can alter gut microbiota composition, increasing intestinal permeability ("leaky gut") and thereby triggering systemic inflammation, further contributing to BBB disruption and neuroinflammation. The trigeminal nerve may facilitate retrograde transport of microbial byproducts, inflammatory mediators, or microorganisms themselves from the oral cavity directly to the brain, while the vagus nerve transmits inflammatory signals from the gut to the CNS. Collectively, these pathways, including hematogenous dissemination, neural transport, and the oral–gut–brain axis, highlight oral health as a modifiable factor in the pathogenesis of neurodegenerative diseases

pathophysiology of neurodegeneration linked to oral inflammation and to evaluate the effectiveness of oral health interventions in these contexts. However, current evidence strongly supports the view that oral health is a modifiable factor affecting CNS health. Integrating routine dental care into prevention strategies for neurodegenerative disorders may offer a potential approach to reducing disease burden and improving long-term neurological outcomes.

References

Abdulkareem A, Al-Taweel F, Al-Sharqi A, Gul S, Sha A, Chapple I (2023) Current concepts in the pathogenesis of periodontitis: from symbiosis to dysbiosis. J Oral Microbiol 15:2197779. https://doi.org/10.1080/20002297.2023.2197779

Adams B, Nunes JM, Page MJ, Roberts T, Carr J, Nell TA, Kell DB, Pretorius E (2019) Parkinson's disease: a systemic inflammatory disease accompanied by bacterial Inflammagens. Front Aging Neurosci:11–2019. https://www.frontiersin.org/journals/aging-neuroscience/articles/10.3389/fnagi.2019.00210

Adamu A, Li S, Gao F, Xue G (2024) The role of neuroinflammation in neurodegenerative diseases: current understanding and future therapeutic targets. Front Aging Neurosci 16. https://www.frontiersin.org/journals/aging-neuroscience/articles/10.3389/fnagi.2024.1347987

Adil NA, Omo-Erigbe C, Yadav H, Jain S (2025) The oral–gut microbiome–brain axis in cognition. Microorganisms 13(4). https://doi.org/10.3390/microorganisms13040814

Adnan D, Engen PA, Villanueva M, Raeisi S, Ramirez V, Naqib A, Green SJ, Bishehsari F, Barnes LL, Keshavarzian A, Dhana K, Voigt RM (2025) Oral microbiome brain axis and cognitive performance in older adults. Npj Dementia 1(1):1. https://doi.org/10.1038/s44400-025-00004-4

Amato M, Polizzi A, Viglianisi G, Leonforte F, Mascitti M, Isola G (2025) Impact of periodontitis and oral dysbiosis metabolites in the modulation of accelerating ageing and human senescence. Meta 15(1). https://doi.org/10.3390/metabo15010035

Asher S, Stephen R, Mäntylä P, Suominen A, Solomon A (2022) Periodontal health, cognitive decline, and dementia: a systematic review and meta-analysis of longitudinal studies. J Am Geriatr Soc 70. https://doi.org/10.1111/jgs.17978

Bansal T, Pandey ADD, Asthana AK (2014) C-reactive protein (CRP) and its association with periodontal disease: a brief review. J Clin Diagn Res JCDR 8(7): ZE21-4. https://api.semanticscholar.org/CorpusID:28126813

Bonaz B, Sinniger V, Pellissier S (2017) The Vagus nerve in the neuro-immune axis: implications in the pathology of the gastrointestinal tract. Front Immunol 8. https://www.frontiersin.org/journals/immunology/articles/10.3389/fimmu.2017.01452

Bowland G, Weyrich L (2022) The oral-microbiome-brain axis and neuropsychiatric disorders: an anthropological perspective. Front Psych 13. https://doi.org/10.3389/fpsyt.2022.810008

Breit S, Kupferberg A, Rogler G, Hasler G (2018) Vagus nerve as modulator of the brain–gut axis in psychiatric and inflammatory disorders. Front Psych 9. https://www.frontiersin.org/journals/psychiatry/articles/10.3389/fpsyt.2018.00044

Carballo Á, López-Dequidt I, Custodia A, Botelho J, Aramburu-Núñez M, Machado V, Pías-Peleteiro JM, Ouro A, Romaus-Sanjurjo D, Vázquez-Vázquez L, Jiménez-Martín I, Aguiar P, Rodríguez-Yáñez M, Aldrey JM, Blanco J, Castillo J, Sobrino T, Leira Y (2023) Association of periodontitis with cognitive decline and its progression: contribution of blood-based biomarkers of Alzheimer's disease to this relationship. J Clin Periodontol 50(11):1444–1454. https://doi.org/10.1111/jcpe.13861

Carreto-Binaghi LE, Sztein MB, Booth JS (2024) Role of cellular effectors in the induction and maintenance of IgA responses leading to protective immunity against enteric bacterial pathogens. Front Immunol 15. https://www.frontiersin.org/journals/immunology/articles/10.3389/fimmu.2024.1446072

Chen L, Deng H, Cui H, Fang J, Zuo Z, Deng J, Li Y, Wang X, Zhao L (2015) Inflammatory responses and inflammation-associated diseases in organs. Oncotarget 9:10.18632/oncotarget.23208

Chen Y, Jin Y, Li K, Qiu H, Jiang Z, Zhu J, Chen S, Xie W, Chen G, Yang D (2023) Is there an association between Parkinson's disease and periodontitis? A systematic review and meta-analysis. J Parkinsons Dis 13: 1107–1125. https://doi.org/10.3233/JPD-230059

Chi L, Cheng X, Lishan L, Yang T, Sun J, Feng Y, Liang F, Pei Z, Teng W (2021) Porphyromonas gingivalis-induced cognitive impairment is associated with gut dysbiosis, neuroinflammation, and glymphatic dysfunction. Front Cell Infect Microbiol 11. https://doi.org/10.3389/fcimb.2021.755925

Cho H-A, Kim B-R, Shin H (2024) Association of periodontal disease treatment with mortality in patients with dementia: a population-based retrospective cohort study (2002–2018). Sci Rep 14(1):5243. https://doi.org/10.1038/s41598-024-55272-6

Cichońska D, Mazuś M, Kusiak A (2024) Recent aspects of periodontitis and Alzheimer's disease – a narrative review. Int J Mol Sci 25(5). https://doi.org/10.3390/ijms25052612

Dai D, Cao G, Huang S, Xu M, Wang J, Han X, Ma Q, Lin J (2025) Porphyromonas gingivalis exacerbates experimental autoimmune encephalomyelitis by driving Th1 differentiation via ZAP70/NF-κB signaling. Front Immunol 16. https://doi.org/10.3389/fimmu.2025.1549102

Dantzer R (2018) Neuroimmune interactions: from the brain to the immune system and vice versa. Physiol Rev 98:477–504. https://doi.org/10.1152/physrev.00039.2016

de Andrade KQ, Almeida-da-Silva CLC, Coutinho-Silva R (2019) Immunological pathways triggered by Porphyromonas gingivalis and Fusobacterium nucleatum: therapeutic possibilities? Mediat Inflamm 2019(1):7241312. https://doi.org/10.1155/2019/7241312

Díaz-Zúñiga J, More J, Melgar-Rodríguez S, Jiménez-Unión M, Villalobos-Orchard F, Muñoz-Manríquez C, Monasterio G, Valdes J, Vernal R, Paula-Lima A (2020) Alzheimer's disease-like pathology triggered by Porphyromonas gingivalis in wild type rats is serotype dependent. Front Immunol 11. https://doi.org/10.3389/fimmu.2020.588036

Ding Z-B, Song L-J, Wang Q, Kumar G, Yan Y-Q, Ma C-G (2021) Astrocytes: a double-edged sword in neurodegenerative diseases. Neural Regen Res 16:1702. https://doi.org/10.4103/1673-5374.306064

Doifode T, Giridharan VV, Generoso JS, Bhatti G, Collodel A, Schulz PE, Forlenza OV, Barichello T (2021) The impact of the microbiota-gut-brain axis on Alzheimer's disease pathophysiology. Pharmacol Res 164:105314. https://doi.org/10.1016/j.phrs.2020.105314

Dominy SS, Lynch C, Ermini F, Benedyk M, Marczyk A, Konradi A, Nguyen M, Haditsch U, Raha D, Griffin C, Holsinger LJ, Arastu-Kapur S, Kaba S, Lee A, Ryder MI, Potempa B, Mydel P, Hellvard A, Adamowicz K et al (2025) Porphyromonas gingivalis in Alzheimer's disease brains: evidence for disease causation and treatment with small-molecule inhibitors. Sci Adv 5(1):eaau3333. https://doi.org/10.1126/sciadv.aau3333

Duarte LF, Farías MA, Álvarez DM, Bueno SM, Riedel CA, González PA (2019) Herpes simplex virus type 1 infection of the central nervous system: insights into proposed interrelationships with neurodegenerative disorders. Front Cell Neurosci 13. https://www.frontiersin.org/journals/cellular-neuroscience/articles/10.3389/fncel.2019.00046

Dutzan N, Abusleme L, Bridgeman H, Greenwell-Wild T, Zangerle-Murray T, Fife ME, Bouladoux N, Linley H, Brenchley L, Wemyss K, Calderon G, Hong B-Y, Break TJ, Bowdish DME, Lionakis MS, Jones SA, Trinchieri G, Diaz PI, Belkaid Y et al (2017) On-going mechanical damage from mastication drives homeostatic Th17 cell responses at the oral barrier. Immunity 46(1):133–147. https://doi.org/10.1016/j.immuni.2016.12.010

Emery D, Seong J, Davies M, Paterson A, Allen S, West N (2021) Comparison of blood bacterial communities in periodontal health and periodontal disease. Front Cell Infect Microbiol 10. https://doi.org/10.3389/fcimb.2020.577485

Fan Z, Tang P, Li C, Yang Q, Xu Y, Su C, Li L (2022) Fusobacterium nucleatum and its associated systemic diseases: epidemiologic studies and possible mechanisms. J Oral Microbiol 15:2145729. https://doi.org/10.1080/20002297.2022.2145729

Faraji N, Payami B, Ebadpour N, Gorji A (2025) Vagus nerve stimulation and gut microbiota interactions: a novel therapeutic avenue for neuropsychiatric disorders. Neurosci Biobehav Rev 169:105990. https://doi.org/10.1016/j.neubiorev.2024.105990

Feng Y-K, Wu Q-L, Peng Y-W, Liang F-Y, You H-J, Feng Y-W, Li G, Li X-J, Liu S-H, Li Y-C, Zhang Y, Pei Z (2020) Oral P. gingivalis impairs gut permeability and mediates immune responses associated with neurodegeneration in LRRK2 R1441G mice. J Neuroinflammation 17(1):347. https://doi.org/10.1186/s12974-020-02027-5

Feng S, Liu Y, Zhou Y, Shu Z, Cheng Z, Brenner C, Feng P (2023) Mechanistic insights into the role of herpes simplex virus 1 in Alzheimer's disease. Front Aging Neurosci 15. https://doi.org/10.3389/fnagi.2023.1245904

Fulop T, Witkowski J, Bourgade K, Khalil A, Zerif E, Larbi A, Hirokawa K, Bocti C, Lacombe G, Dupuis G, Frost E (2018) Can an infection hypothesis explain the beta amyloid hypothesis of Alzheimer's disease? Front Aging Neurosci 10. https://doi.org/10.3389/fnagi.2018.00224

Galea I (2021) The blood–brain barrier in systemic infection and inflammation. Cell Mol Immunol 18. https://doi.org/10.1038/s41423-021-00757-x

Gani D, Lakshmi D, Krishnan R, Emmadi P (2009) Evaluation of C-reactive protein and interleukin-6 in the peripheral blood of patients with chronic periodontitis. J Indian Soc Periodontol 13:69–74. https://doi.org/10.4103/0972-124X.55840

Gao C, Jiang J, Tan Y, Chen S (2023) Microglia in neurodegenerative diseases: mechanism and potential therapeutic targets. Signal Transduct Target Ther 8(1): 359. https://doi.org/10.1038/s41392-023-01588-0

Gryka-Marton M, Grabowska AD, Szukiewicz D (2025) Breaking the barrier: the role of Proinflammatory cytokines in BBB dysfunction. Int J Mol Sci 26(8). https://doi.org/10.3390/ijms26083532

Hajishengallis G, Chavakis T (2021) Local and systemic mechanisms linking periodontal disease and inflammatory comorbidities. Nat Rev Immunol 21(7):426–440. https://doi.org/10.1038/s41577-020-00488-6

Hajishengallis G, Lamont R (2021) Polymicrobial communities in periodontal disease: their quasi-organismal nature and dialogue with the host. Periodontology 2000:86. https://doi.org/10.1111/prd.12371

Hoogland ICM, Houbolt C, van Westerloo DJ, van Gool WA, van de Beek D (2015) Systemic inflammation and microglial activation: systematic review of animal experiments. J Neuroinflammation 12(1):114. https://doi.org/10.1186/s12974-015-0332-6

Huang Z, Hao M, Shi N, Wang X, Yuan L, Yuan H, Wang X (2025) Porphyromonas gingivalis: a potential trigger of neurodegenerative disease. Front Immunol 16. https://www.frontiersin.org/journals/immunology/articles/10.3389/fimmu.2025.1482033

Ilievski V, Zuchowska PK, Green SJ, Toth PT, Ragozzino ME, Le K, Aljewari HW, O'Brien-Simpson NM, Reynolds EC, Watanabe K (2018) Chronic oral application of a periodontal pathogen results in brain inflammation, neurodegeneration and amyloid beta production in wild type mice. PLoS One 13(10): e0204941. https://doi.org/10.1371/journal.pone.0204941

Itzhaki RF (2018) Corroboration of a Major Role for Herpes Simplex Virus Type 1 in Alzheimer's Disease. Front Aging Neurosci 10:324. https://doi.org/10.3389/fnagi.2018.00324

Iqbal U, Zeng E, Pasinetti G (2020) The use of antimicrobial and antiviral drugs in Alzheimer's disease. Int J Mol Sci 21. https://doi.org/10.3390/ijms21144920

Jeong E, Park J-B, Park Y-G (2021) Evaluation of the association between periodontitis and risk of Parkinson's disease: a nationwide retrospective cohort study. Sci Rep 11(1):16594. https://doi.org/10.1038/s41598-021-96147-4

Jia L, Han N, Du J, Guo L, Luo Z, Liu Y (2019) Pathogenesis of important virulence factors of Porphyromonas gingivalis via toll-like receptors.

Front Cell Infect Microbiol 9. https://doi.org/10.3389/fcimb.2019.00262

Kapila Y (2021) Oral health's inextricable connection to systemic health: special populations bring to bear multimodal relationships and factors connecting periodontal disease to systemic diseases and conditions. Periodontology 2000(87):11–16. https://doi.org/10.1111/prd.12398

Kennedy RH, Silver R (2016) Neuroimmune signaling: cytokines and the CNS. In: Pfaff DW, Volkow ND (eds) Neuroscience in the 21st century. Springer, New York, pp 1–41. https://doi.org/10.1007/978-1-4614-6434-1_174-1

Kim EK, Choi E-J (2010) Pathological roles of MAPK signaling pathways in human diseases. Biochim Biophys Acta 1802(4):396–405. https://doi.org/10.1016/j.bbadis.2009.12.009

Kleinstein S, Nelson KE, Freire M (2020) Inflammatory networks linking oral microbiome with systemic health and disease. J Dent Res 99:002203452092612. https://doi.org/10.1177/0022034520926126

Konkel JE, O'Boyle C, Krishnan S (2019) Distal consequences of oral inflammation. Front Immunol 10. https://www.frontiersin.org/journals/immunology/articles/10.3389/fimmu.2019.01403

Kulkarni MS, Miller BC, Mahani M, Mhaskar R, Tsalatsanis A, Jain S, Yadav H (2023) Poor oral health linked with higher risk of Alzheimer's disease. Brain Sci 13(11). https://doi.org/10.3390/brainsci13111555

Kwon HS, Koh S-H (2020) Neuroinflammation in neurodegenerative disorders: the roles of microglia and astrocytes. Transl Neurodegener 9(1):42. https://doi.org/10.1186/s40035-020-00221-2

Lamptey RNL, Chaulagain B, Trivedi R, Gothwal A, Layek B, Singh J (2022) A review of the common neurodegenerative disorders: current therapeutic approaches and the potential role of nanotherapeutics. Int J Mol Sci 23(3). https://doi.org/10.3390/ijms23031851

Lana D, Ugolini F, Nosi D, Wenk G, Giovannini M (2021) The emerging role of the interplay among astrocytes, microglia, and neurons in the hippocampus in health and disease. Front Aging Neurosci 13:651973. https://doi.org/10.3389/fnagi.2021.651973

Li, A., Du, M., Chen, Y., Marks LAM, Visser, A., Xu, S., & Tjakkes, G.-H. (2022). Periodontitis and cognitive impairment in older adults: the mediating role of mitochondrial dysfunction. J Periodontol https://doi.org/10.1002/JPER.21-0620

Li R, Wang J, Xiong W, Luo Y, Feng H, Zhou H, Peng Y, He Y, Ye Q (2024a) The oral-brain axis: can periodontal pathogens trigger the onset and progression of Alzheimer's disease? Front Microbiol 15. https://doi.org/10.3389/fmicb.2024.1358179

Li R, Wang J, Xiong W, Luo Y, Feng H, Zhou H, Peng Y, He Y, Ye Q (2024b) The oral-brain axis: can periodontal pathogens trigger the onset and progression of Alzheimer's disease? Front Microbiol 15. https://www.frontiersin.org/journals/microbiology/articles/10.3389/fmicb.2024.1358179

Liu X-J, Che P, Xing M, Tian X-B, Gao C, Li X, Zhang N (2021) Cerebral hemodynamics and carotid atherosclerosis in patients with subcortical ischemic vascular dementia. Front Aging Neurosci 13. https://www.frontiersin.org/journals/aging-neuroscience/articles/10.3389/fnagi.2021.741881

Liu Y, Qv W, Ma Y, Zhang Y, Ding C, Chu M, Chen F (2022) The interplay between oral microbes and immune responses. Front Microbiol 13. https://www.frontiersin.org/journals/microbiology/articles/10.3389/fmicb.2022.1009018

Liu L, Bano F, Conca DV, Thorsteinsson K, Jayaweera SW, Avinens D, Pace H, Lövheim H, Olofsson A, Bally M (2025) Recruitment of apolipoprotein E facilitates herpes simplex virus 1 attachment and release. Npj Viruses 3(1):13. https://doi.org/10.1038/s44298-025-00099-9

Lyra P, Botelho J, Machado V, Rota S, Walker R, Staunton J, Proença L, Ray Chaudhuri K, Mendes JJ (2022) Self-reported periodontitis and C-reactive protein in Parkinson's disease: a cross-sectional study of two American cohorts. Npj Parkinsons Disease 8. https://doi.org/10.1038/s41531-022-00302-1

Malan-Müller S, Vidal R, O'Shea E, Montero E, Figuero E, Zorrilla I, de Diego-Adeliño J, Cano M, García-Portilla MP, González-Pinto A, Leza JC (2024) Probing the oral-brain connection: oral microbiome patterns in a large community cohort with anxiety, depression, and trauma symptoms, and periodontal outcomes. Transl Psychiatry 14(1):419. https://doi.org/10.1038/s41398-024-03122-4

Marcaccini A, Meschiari C, Sorgi C, Pereira Saraiva M, Souza A, Faccioli LH, Tanus-Santos J, Novaes A Jr, Gerlach R (2009) Circulating Interleukin-6 and high-sensitivity C-reactive protein decrease after periodontal therapy in otherwise healthy subjects. J Periodontol 80: 594–602. https://doi.org/10.1902/jop.2009.080561

Martínez-García M, Hernández-Lemus E (2021) Periodontal inflammation and systemic diseases: an overview. Front Physiol 12:709438. https://doi.org/10.3389/fphys.2021.709438

Matsuda S, Shintani T, Miyagawa T, Yumoto H, Komatsu Y, Dewake N, Iwata T, Nagano T, Morozumi T, Goto R, Kato S, Kitamura M, Shin K, Sekino S, Yamashita A, Yamashita K, Yoshimura A, Sugaya T, Takashiba S et al (2024) Effect of periodontal treatment on reducing chronic inflammation in systemically healthy patients with periodontal disease. Am J Med 137(3):273–279.e2. https://doi.org/10.1016/j.amjmed.2023.11.001

Mayer MG, Fischer T (2025) Shared mechanisms of blood-brain barrier dysfunction and neuroinflammation in COVID-19 and Alzheimer's disease. Am J Pathol. https://doi.org/10.1016/j.ajpath.2025.03.011

Mayer C, Walther C, Borof K, Nägele F, Petersen M, Schell M, Gerloff C, Kühn S, Heydecke G, Beikler T,

Cheng B, Thomalla G, Aarabi G (2023) Association between periodontal disease and microstructural brain alterations in the Hamburg City health study. J Clin Periodontol 51. https://doi.org/10.1111/jcpe.13828

Mekhora C, Lamport DJ, Spencer JPE (2024) An overview of the relationship between inflammation and cognitive function in humans, molecular pathways and the impact of nutraceuticals. Neurochem Int 181: 105900. https://doi.org/10.1016/j.neuint.2024.105900

Miklossy J (2011) Emerging roles of pathogens in AD. Expert Rev Mol Med 13:e30. https://doi.org/10.1017/S1462399411002006

Murphy ÁC, Lalor SJ, Lynch MA, Mills KHG (2010) Infiltration of Th1 and Th17 cells and activation of microglia in the CNS during the course of experimental autoimmune encephalomyelitis. Brain Behav Immun 24(4):641–651. https://doi.org/10.1016/j.bbi.2010.01.014

Murray P, Coffman J, Garcia-Godoy F (2024) Oral pathogens' substantial burden on cancer, cardiovascular diseases, Alzheimer's, diabetes, and other systemic diseases: a public health crisis – a comprehensive review. Pathogens 13:1084. https://doi.org/10.3390/pathogens13121084

Mysak J, Podzimek S, Sommerova P, Lyuya-Mi Y, Bártová J, Janatova T, Procházková J, Duskova J (2014) Porphyromonas gingivalis: major periodontopathic pathogen overview. J Immunol Res 2014:476068. https://doi.org/10.1155/2014/476068

Nicholson J, Landry K (2022) Oral dysbiosis and neurodegenerative diseases: correlations and potential causations. Microorganisms 10. https://doi.org/10.3390/microorganisms10071326

Noble JM, Borrell LN, Papapanou PN, Elkind MSV, Scarmeas N, Wright CB (2009) Periodontitis is associated with cognitive impairment among older adults: analysis of NHANES-III. J Neurol Neurosurg Psychiatry 80(11):1206. https://doi.org/10.1136/jnnp.2009.174029

Nunes M, Fillis T, Page M, Venter C, Lancry O, Kell D, Windberger U, Pretorius E (2020) Gingipain R1 and lipopolysaccharide from Porphyromonas gingivalis have major effects on blood clot morphology and mechanics. Front Immunol 11:1551. https://doi.org/10.3389/fimmu.2020.01551

Olsen I, John DL, Hajishengallis G (2017) Porphyromonas gingivalis disturbs host–commensal homeostasis by changing complement function. J Oral Microbiol 9(1):1340085. https://doi.org/10.1080/20002297.2017.1340085

Peggion C, Calì T, Brini M (2024) Mitochondria dysfunction and neuroinflammation in neurodegeneration: who comes first? Antioxidants 13(2). https://doi.org/10.3390/antiox13020240

Pisani F, Pisani V, Arcangeli F, Harding A, Singhrao S (2022) The mechanistic pathways of periodontal pathogens entering the brain: the potential role of Treponema denticola in tracing Alzheimer's disease pathology. Int J Environ Res Public Health 19:9386. https://doi.org/10.3390/ijerph19159386

Plachokova AS, Gjaltema J, Hagens ERC, Hashemi Z, Knüppe TBA, Kootstra TJM, Visser A, Bloem BR (2024) Periodontitis: a plausible modifiable risk factor for neurodegenerative diseases? A comprehensive review. Int J Mol Sci 25(8). https://doi.org/10.3390/ijms25084504

Polak D, Shmueli A, Brenner T, Shapira L (2018) Oral infection with P. gingivalis exacerbates autoimmune encephalomyelitis. J Periodontol 89. https://doi.org/10.1002/jper.17-0531

Poole S, Singhrao S, Kesavalu L, Curtis M, Crean S (2017) Determining the presence of Periodontopathic virulence factors in short-term postmortem Alzheimer's disease brain tissue. Adv Alzheimer's Dis 5:105–117. https://doi.org/10.3233/978-1-61499-706-1-105

Pruntel SM, van Munster BC, de Vries JJ, Vissink A, Visser A (2024) Oral health as a risk factor for Alzheimer disease. J Prev Alzheimers Dis 11(1): 249–258. https://doi.org/10.14283/jpad.2023.82

Qin J, Ma Z, Chen X, Shu S (2023) Microglia activation in central nervous system disorders: a review of recent mechanistic investigations and development efforts. Front Neurol 14. https://www.frontiersin.org/journals/neurology/articles/10.3389/fneur.2023.1103416

Radaic A, Kapila YL (2021) The oralome and its dysbiosis: new insights into oral microbiome-host interactions. Comput Struct Biotechnol J 19:1335–1360. https://doi.org/10.1016/j.csbj.2021.02.010

Raha D, Broce S, Haditsch U, Rodriguez L, Ermini F, Detke M, Kapur S, Hennings D, Roth T, Nguyen M, Holsinger L, Lynch C, Dominy S (2020) COR388, a novel gingipain inhibitor, decreases fragmentation of APOE in the central nervous system of Alzheimer's disease patients: human/human trials: other. Alzheimers Dement 16. https://doi.org/10.1002/alz.040578

Rajasekaran JJ, Krishnamurthy HK, Bosco J, Jayaraman V, Krishna K, Wang T, Bei K (2024) Oral microbiome: a review of its impact on oral and systemic health. Microorganisms 12(9). https://doi.org/10.3390/microorganisms12091797

Ransohoff RM, Schafer D, Vincent A, Blachère NE, Bar-Or A (2015) Neuroinflammation: ways in which the immune system affects the brain. Neurotherapeutics 12(4):896–909. https://doi.org/10.1007/s13311-015-0385-3

Riviere GR, Riviere KH, Smith KS (2002) Molecular and immunological evidence of oral Treponema in the human brain and their association with Alzheimer's disease. Oral Microbiol Immunol 17(2):113–118. https://doi.org/10.1046/j.0902-0055.2001.00100.x

Ryder M (2020) Porphyromonas gingivalis and Alzheimer's disease: recent findings and potential therapies. J Periodontol 91. https://doi.org/10.1002/jper.20-0104

Sankowski R, Mader S, Valdés-Ferrer SI (2015) Systemic inflammation and the brain: novel roles of genetic, molecular, and environmental cues as drivers of neurodegeneration. Front Cell Neurosci 9. https://api.semanticscholar.org/CorpusID:6337174

Sansores-España LD, Melgar-Rodríguez S, Olivares-Sagredo K, Cafferata EA, Martínez-Aguilar VM, Vernal R, Paula-Lima AC, Díaz-Zúñiga J (2021) Oral-gut-brain Axis in experimental models of periodontitis: associating gut dysbiosis with neurodegenerative diseases. Frontiers in Aging 2. https://www.frontiersin.org/journals/aging/articles/10.3389/fragi.2021.781582

Santacroce L, Passarelli PC, Azzolino D, Bottalico L, Charitos I, Cazzolla A, Colella M, Topi S, Garcia-Godoy F, D'Addona A (2023) Oral microbiota in human health and disease: a perspective. Exp Biol Med (Maywood) 248:15353702231187644. https://doi.org/10.1177/15353702231187645

Schweighöfer H, Rummel C, Roth J, Rosengarten B (2016) Modulatory effects of vagal stimulation on neurophysiological parameters and the cellular immune response in the rat brain during systemic inflammation. Intensive Care Med Exp 4(1):19. https://doi.org/10.1186/s40635-016-0091-4

Shen Y, Fan N, Ma S, Cheng X, Yang X, Wang G (2025) Gut microbiota Dysbiosis: pathogenesis, diseases, prevention, and therapy. MedComm 6. https://doi.org/10.1002/mco2.70168

Singh A, Kukreti R, Saso L, Kukreti S (2019) Oxidative stress: a key modulator in neurodegenerative diseases. Molecules 24(8). https://doi.org/10.3390/molecules24081583

Singhrao SK, Olsen I (2018) Are Porphyromonas gingivalis outer membrane vesicles microbullets for sporadic Alzheimer's disease manifestation? J Alzheimer's Dise Rep 2(1):219–228. https://doi.org/10.3233/ADR-180080

Soiniemi L, Solje E, Suominen AL, Kanninen KM, Kullaa AM (2024) The association between oral diseases and neurodegenerative disorders. J Alzheimer's Dis 102(3):577–586. https://doi.org/10.1177/13872877241289548

Subbarao K, Nattuthurai G, Sundararajan S, Sujith I, Joseph J, Syedshah Y (2019) Gingival Crevicular fluid: an overview. J Pharm Bioallied Sci 11:135. https://doi.org/10.4103/JPBS.JPBS_56_19

Sun Q, Li M (2025) Association between periodontitis and cognitive impairment in older adults: a cross-sectional study of the National Health and Nutrition Examination Survey. Clin Epidemiol Glob Health 102020. https://doi.org/10.1016/j.cegh.2025.102020

Sun Y, Koyama Y, Shimada S (2022) Inflammation from peripheral organs to the brain: how does systemic inflammation cause neuroinflammation? Front Aging Neurosci 14:903455. https://doi.org/10.3389/fnagi.2022.903455

Szablewski L (2025) Associations between diabetes mellitus and neurodegenerative diseases. Int J Mol Sci 26(2). https://doi.org/10.3390/ijms26020542

Takahashi N, Benso S, Miki Y-H, Takahiro T, Koichi T, Yamazaki K (2019) Gingival epithelial barrier: regulation by beneficial and harmful microbes. Tissue Barriers 7(3):e1651158. https://doi.org/10.1080/21688370.2019.1651158

Takata F, Nakagawa S, Matsumoto J, Dohgu S (2021) Blood-brain barrier dysfunction amplifies the development of Neuroinflammation: understanding of cellular events in brain microvascular endothelial cells for prevention and treatment of BBB dysfunction. Front Cell Neurosci 15. https://www.frontiersin.org/journals/cellular-neuroscience/articles/10.3389/fncel.2021.661838

Tanabe N, Tomita K, Manaka S, Ichikawa R, Takayama T, Kawato T, Ono M, Masai Y, Utsu A, Suzuki N, Sato S (2023) Co-stimulation of AGEs and LPS induces inflammatory mediators through PLCγ1/JNK/NF-κB pathway in MC3T3-E1 cells. Cells 12:1383. https://doi.org/10.3390/cells12101383

Tran V, Lee L, Cho H (2022) Neuroinflammation in neurodegeneration via microbial infections. Front Immunol 13:907804. https://doi.org/10.3389/fimmu.2022.907804

Verhoeff MC, Eikenboom D, Koutris M, de Vries R, Berendse HW, van Dijk KD, Lobbezoo F (2023) Parkinson's disease and oral health: a systematic review. Arch Oral Biol 151:105712. https://doi.org/10.1016/j.archoralbio.2023.105712

Wainberg M, Luquez T, Koelle DM, Readhead B, Johnston C, Darvas M, Funk CC (2021) The viral hypothesis: how herpesviruses may contribute to Alzheimer's disease. Mol Psychiatry 26(10):5476–5480. https://doi.org/10.1038/s41380-021-01138-6

Wang RP-H, Ho Y-S, Leung WK, Goto T, Chang RC-C (2019) Systemic inflammation linking chronic periodontitis to cognitive decline. Brain Behav Immun 81:63–73. https://doi.org/10.1016/j.bbi.2019.07.002

Warren-Gash C, Forbes HJ, Williamson E, Breuer J, Hayward AC, Mavrodaris A, Ridha BH, Rossor MN, Thomas SL, Smeeth L (2019) Human herpesvirus infections and dementia or mild cognitive impairment: a systematic review and meta-analysis. Sci Rep 9(1):4743. https://doi.org/10.1038/s41598-019-41218-w

Wong T-S, Li G, Li S, Gao W, Chen G, Gan S, Zhang M, Li H, Wu S, Du Y (2023) G protein-coupled receptors in neurodegenerative diseases and psychiatric disorders. Signal Transduct Target Ther 8(1):177. https://doi.org/10.1038/s41392-023-01427-2

Wu D, Wang C, Pang P, Kong H, Lin Z, Wang H, Chen X, Zhao J, Hao Z, Zhang T, Guo X (2020) The association between herpes simplex virus type 1 infection and Alzheimer's disease. J Clin Neurosci 82:63–70. https://doi.org/10.1016/j.jocn.2020.10.044

Wu H, Qiu W, Zhu X, Li X, Xie Z, Carreras I, Dedeoglu A, Van Dyke T, Han YW, Karimbux N, Tu Q, Cheng L, Chen J (2022a) The periodontal pathogen Fusobacterium nucleatum exacerbates Alzheimer's pathogenesis via specific pathways. Front Aging Neurosci 14. https://www.frontiersin.org/journals/aging-neuroscience/articles/10.3389/fnagi.2022.912709

Wu L, Su X, Tang Z, Jian L, Zhu H, Cheng X, Wu H (2022b) Treponema denticola induces neuronal apoptosis by promoting amyloid-β accumulation in mice. Pathogens 11(10). https://doi.org/10.3390/pathogens11101150

Wu DT, Cho YW, Spalti MD, Bishara M, Nguyen TT (2023) The link between periodontitis and Alzheimer's disease – emerging clinical evidence. Dent Rev 3(1): 100062. https://doi.org/10.1016/j.dentre.2022.100062

Wu J, Wu J, Chen T, Cai J, Ren R (2024) Protein aggregation and its affecting mechanisms in neurodegenerative diseases. Neurochem Int 180:105880. https://doi.org/10.1016/j.neuint.2024.105880

Xiaojing L, Kolltveit KM, Leif T, Ingar O (2000) Systemic diseases caused by oral infection. Clin Microbiol Rev 13(4):547–558. https://doi.org/10.1128/cmr.13.4.547

Xie M, Tang Q, Yu S, Sun J, Mei F, Zhao J, Chen L (2020) Porphyromonas gingivalis disrupts vascular endothelial homeostasis in a TLR-NF-κB axis dependent manner. Int J Oral Sci 12(1):28. https://doi.org/10.1038/s41368-020-00096-z

Yamazaki K (2023) Oral-gut axis as a novel biological mechanism linking periodontal disease and systemic diseases: a review. Jpn Dent Sci Rev 59:273–280. https://doi.org/10.1016/j.jdsr.2023.08.003

Yang J, Ran M, Li H, Lin Y, Ma K, Yang Y, Fu X, Yang S (2022) New insight into neurological degeneration: inflammatory cytokines and blood–brain barrier. Front Mol Neurosci 15. https://www.frontiersin.org/journals/molecular-neuroscience/articles/10.3389/fnmol.2022.1013933

Yilmaz M, Yay E, Balci N, Toygar H, Kılıc BB, Zirh A, Rivas CA, Kantarci A (2023) Parkinson's disease is positively associated with periodontal inflammation. J Periodontol 94(12):1425–1435. https://doi.org/10.1002/JPER.23-0274

Zhang Z, Liu D, Liu S, Zhang S, Pan Y (2021a) The role of Porphyromonas gingivalis outer membrane vesicles in periodontal disease and related systemic diseases. Front Cell Infect Microbiol 10. https://doi.org/10.3389/fcimb.2020.585917

Zhang J, Xie M, Huang X, Chen G, Yin Y, Lu X, Feng G, Yu R, Chen L (2021b) The effects of Porphyromonas gingivalis on atherosclerosis-related cells. Front Immunol 12. https://doi.org/10.3389/fimmu.2021.766560

Zhang W, Xiao D, Mao Q, Xia H (2023) Role of neuroinflammation in neurodegeneration development. Signal Transduct Target Ther 8(1):267. https://doi.org/10.1038/s41392-023-01486-5

Zhang R, Ding N, Feng X, Liao W (2025) The gut microbiome, immune modulation, and cognitive decline: insights on the gut-brain axis. Front Immunol 16. https://doi.org/10.3389/fimmu.2025.1529958

Oral Manifestations of Autoimmune Diseases

14

Afagh Tavasoli, Heliya Ziaei, and Nima Rezaei

Abstract

Autoimmune diseases arise from a complex interplay between genetic predisposition and environmental factors, including chemical exposures, infections, and psychological stress. These conditions occur when the immune system mistakenly targets the body's own healthy cells, resulting in tissue and organ damage. Both humoral and cellular immune mechanisms contribute to autoimmune disorders, leading to chronic inflammation and progressive tissue injury.

Autoimmune diseases have diverse and often overlapping symptoms, significantly impacting patients' quality of life. Early diagnosis is crucial for effective disease management; however, the clinical similarity among autoimmune disorders and the lack of definitive diagnostic tests in many cases pose major challenges to timely identification.

Notably, in many autoimmune conditions, oral manifestations are among the earliest or, in some cases, only clinical signs, as the oral epithelium is a frequent target of autoimmune responses. These manifestations often present as erosions and ulcers, causing pain and discomfort. Identifying oral lesions at an early stage can facilitate early diagnosis and improve patient outcomes. This chapter provides a comprehensive overview of autoimmune conditions and their oral manifestations.

Since immune dysregulation is central to autoimmune disease pathogenesis, local and systemic corticosteroids are commonly prescribed to control symptoms, including oral lesions. In severe or refractory cases, immunosuppressive therapies may be essential to prevent disease progression and tissue destruction.

Keywords

Autoimmune disease · Oral lesions · Pemphigus · Autoimmune disorders · Pemphigoid

A. Tavasoli
Universal Scientific Education and Research Network (USERN), Tehran, Iran

H. Ziaei
Universal Scientific Education and Research Network (USERN), Los Angeles, CA, USA

N. Rezaei (✉)
Universal Scientific Education and Research Network (USERN), Göteborg, Sweden

Network of Immunity in Infection, Malignancy and Autoimmunity (NIIMA), Universal Scientific Education and Research Network (USERN), Tehran, Iran

Research Center for Immunodeficiencies, Children's Medical Center, Tehran University of Medical Sciences, Tehran, Iran

Department of Immunology, School of Medicine, Tehran University of Medical Sciences, Tehran, Iran
e-mail: Rezaei_nima@tums.ac.ir

N. Rezaei, H. Ziaei (eds.), *Oral Immunology*, Advances in Experimental Medicine and Biology 1492,
https://doi.org/10.1007/978-3-032-03176-1_14

14.1 Introduction

Autoimmune diseases are chronic conditions with multiple etiological factors. These disorders primarily affect different organs such as skin, joints, mucous membranes, and connective tissue. Autoimmune reactions occur when the immune system mistakenly recognizes self-antigens as foreign, initiating an immune response and leading to tissue damage (Lin et al. 2023). In recent years, the incidence of autoimmune conditions has increased, and in many cases, oral manifestations may present before some systemic manifestations and are early indicators of these diseases. Thus, early diagnosis of oral presentations may lead to timely diagnosis and improved treatment outcomes (Saccucci et al. 2018). Since many autoimmune diseases can have overlapping oral presentations, distinguishing among them might be challenging. Therefore, familiarity with different oral manifestations of these diseases as well as their overlapping features is crucial for accurate early diagnosis and management (Lin et al. 2023). This chapter provides a concise yet comprehensive overview of the pathophysiology, diagnosis, treatment, and oral manifestations of autoimmune diseases. By highlighting the clinical relevance of oral manifestations in these conditions, this chapter emphasizes their role in early detection, interdisciplinary collaboration, and optimal patient care.

14.2 Pemphigus

Pemphigus is an autoimmune disease affecting epithelial tissue that negatively influences cell adhesion. In this condition, autoantibodies (IgG) directly target the specific molecules of cell membrane proteins, specifically Desmoglein (Dsg) 1 and 3. Therefore, loss of adhesion between keratinocytes leads to intraepithelial blisters that cause erosions on both skin and oral mucosa. Pemphigus disease has different subtypes such as vulgaris (PV), foliaceus (PF), paraneoplastic (PNP), and IgA pemphigus. Also, it has some variations related to these subtypes. The immune response in PV and PF is primarily humoral, while PNP involves both humoral and cellular immune mechanisms. Among these subtypes, PV is the most common type, predominantly affecting females between the ages of 50 and 60 years. In contrast, PNP is extremely rare, and only 500 cases of PNP have been reported worldwide. PNP is dominant in females as well, and it can affect both children and adults. PF mainly affects individuals in their 50 s and 60 s, but an endemic form of this disease, known as Fogo selvage, has been reported in North Africa, affecting children and young adults (Costan et al. 2021; Kasperkiewicz et al. 2017). Moreover, PNP is associated with underlying malignancies such as non-Hodgkin's lymphoma and also Castleman's disease. It is important to notice that while mucosal involvement is a hallmark of PV, it is rare in PF and IgA subtypes (Costan et al. 2021).

14.2.1 Pathophysiology

In PV, autoantibodies primarily target Dsg1 and Dsg3, although some other autoantibodies have been recognized against other structural proteins such as desmocolins. These autoantibodies can have a synergic effect on anti-Dsg autoantibodies. Dsg1 mostly exists in the superficial epidermis, while type 3 can be found mainly in deeper layers, such as the basal layer. Therefore, in the mucosal-dominant variant of PV, autoantibodies mainly target Dsg3, while in the mucocutaneous type, both Dsg1 and Dsg3 are involved. In PF, autoantibodies are against Dsg1, which exists only in the skin; therefore, unlike PV, mucosal lesions are uncommon in PF (Costan et al. 2021).

The precise pathophysiology of PNP has not been fully studied, but it seems that both autoantibody-mediated and cell-mediated immune responses are involved in this process. Autoantibodies in PNP are mainly IgG, but in rare cases, IgA autoantibodies have been reported as well. Plakin families, such as periplakin, envoplakin, desmoplakin l, and desmoplakin ll, are the dominant target of these autoantibodies in the PNP variant (Costan et al. 2021).

The development of acantholysis, the hallmark of pemphigus, is associated with dysregulation in various inflammatory mediators, while their exact mechanism of action has not been clarified. The increased level of serum concentration of some mediators in PV patients has been detected, such as IL-4, IL-6, and IL-10. However, the decreased level of IL-2 and interferon-gamma has been detected in these patients as well. Besides, serum levels of IL-17a, IL-21, and IL-27 have been increased, with IL-17a specifically detected in blister fluids of PV patients. In addition, it is revealed that TNF-α has a significant expression in skin lesions of PV patients, and the serum concentrations of this molecule correlate with disease activity and levels of circulating autoantibodies in the blood. Moreover, it is suggested that in PV, keratinocytes may release cytokines as a response to IgG autoantibody binding, contributing to disease pathogenesis (Didona et al. 2019).

14.2.2 Diagnosis

Diagnosis of pemphigus typically starts with the evaluation of oral mucosa and skin lesions by the Nikolsky sign, which is an important diagnostic indicator. This sign is positive when the normal-appearing mucosa or skin separates and/or peels away in response to gentle mechanical rubbing or pressure. Following clinical examination, the steps for diagnosis are tissue biopsy for histopathological assessment and utilization of direct immunofluorescence (DIF). In addition, indirect immunofluorescence (IIF) is commonly used for the detection of circulating autoantibodies in the blood. Also, the level of autoantibodies in the blood is correlated with the severity and extent of the disease, making serological assessments a valuable tool for monitoring disease progression (Costan et al. 2021; Didona et al. 2019).

The histopathological features of pemphigus are different among its subtypes. In PV, because of deep intraepithelial acantholysis, a suprabasal cleft can be seen above the basal cell layer. Thus, basal cells remain attached to the basal membrane, creating a tombstone-like histological appearance. In PF, acantholysis is detectable in more superficial epidermal layers; therefore, the presentation of deep clefts is not common in this subtype. Moreover, PNP exhibits variable histological features based on its clinical conditions, which consist of suprabasal acantholysis, dyskeratosis, and lichenoid interface dermatitis. Also, the main feature of the histological evaluation of IgA pemphigus is the infiltration of neutrophils. Unlike other types of pemphigus, acantholysis is not always detected in IgA pemphigus (Didona et al. 2019).

DIF in all types of pemphigus disease illustrates the binding of IgG autoantibodies to the surface of keratinocytes in patients' tissue samples. Also, IIF evaluation of this disease features a characteristic reticular (net-like) pattern of IgG deposition on the monkey's esophagus epithelium tissue as a commonly used substrate (Didona et al. 2019).

14.2.3 Treatment

There is no definite treatment for pemphigus disease, but it can be effectively managed by specific drugs such as systemic corticosteroids, aiming at controlling autoantibody production. The first-line therapy is the combination of systemic corticosteroids (prednisolone) and immunosuppressive medications such as azathioprine. Moreover, Rituximab (RTX) is a type ⍰ monoclonal anti-CD20 antibody that can be used as the first-line treatment in pemphigus. Reducing autoantibodies can substantially ameliorate lesions. Most of the patients have to take immunosuppressive medications for even years in their maintenance therapy period to prevent relapses. Although corticosteroids are effective anti-inflammatory mediators and are involved in some other mechanisms of disease remission, long-term usage of them can be responsible for adverse effects such as predisposition to infections, osteoporosis, adrenal suppression, etc. (Costan et al. 2021; Didona et al. 2019). Systemic treatments will not only help control systemic and cutaneous symptoms but also mitigate oral manifestations.

14.2.4 Oral Manifestations

PV is categorized into three subtypes: mucosal, mucocutaneous, and cutaneous (Kasperkiewicz et al. 2017). In most cases, oral lesions are the first manifestations of pemphigus disease, often appearing even 1 year before other symptoms. Generally, there are no specific sites for pemphigus lesions in oral mucosa; however, tissues with a higher incidence of frictional traumas are more susceptible (Sankar and Noujeim 2017). In the oral cavity, buccal and labial mucosa, palate, and tongue are involved in most cases (Magliocca and Fitzpatrick 2017). Also, salivary dysfunctions, such as sialorrhea, have been reported in these patients (Di Lernia et al. 2020). In the early stages of the disease, oral lesions can be similar to some ulcerative conditions such as recurrent aphthae, herpetic gingivostomatitis, and erosive lichen planus (Costan et al. 2021). In addition, blisters in the oral cavity usually break easily and lead to painful erosions. In these patients, any light pressure on the epithelial tissue can cause the formation of a blister, which is called the Nikolsky sign (Saccucci et al. 2018). In both PV and PF, this test might have positive results during the active phase of the disease. Mucosal involvement is not limited to the oropharyngeal area, and it can be detected in other mucosal tissues, such as the larynx or conjunctiva with less probability. In contrast to PV, PF does not typically affect the oral mucosa, since Dsg1, which is the primary autoantibody target, is minimally expressed in the oral mucosa. PNP presents more severe oral lesions with characteristic clinical features such as hemorrhagic stomatitis, which is resistant to treatment. Lesions mainly occur in the vermillion borders of the lips and the borders of the tongue. In comparison to PV, PNP presents more extensive oral mucosal lesions, and it has a higher likelihood of extraoral mucosal involvement (Kasperkiewicz et al. 2017).

14.3 Mucous Membrane Pemphigoid (MMP)

Pemphigoid is an autoimmune disease that has two main types of Bullous Pemphigoid (BP) and Mucous Membrane Pemphigoid (MMP), or Cicatricial Pemphigoid. BP is the most common type of this disease (Pratasava et al. 2021), and MMP is more common in females aged over 50 years (Ptasiewicz et al. 2022). The difference between these two types is that MMP is more dominant in mucosal tissue, and it leaves scars. Also, in these two types of pemphigoid disease, autoantibodies are produced against components of the basal membrane, while targeting different antigens. In BP, autoantibodies target BP230 and BP180, both of which are essential structural proteins of the basement membrane. Moreover, MMP itself has three subtypes, including classic, ocular, and anti-laminin 332. In the classic subtype, the target is the C-terminal domain of BP180, while in the ocular subtype, autoantibodies are against a part of $\alpha_6\beta_4$ integrin, a crucial component of hemi-desmosomes (Pratasava et al. 2021). Anti-laminin 332 is a rare subtype, associated with autoantibodies targeting laminin 332, which is an important protein for basement membrane adhesion.

14.3.1 Pathophysiology

In MMP, autoantibodies are produced against structural proteins of the basal membrane of epithelial or epidermal tissues and lead to detachment from underlying connective tissue (Sankar and Noujeim 2017). These targets are mainly BP180, BP230, laminin332, laminin311, and $\alpha_6\beta_4$ integrin. In classic MMP, autoantibodies are mainly IgG and IgA, which predominantly target BP180, a transmembrane protein that extends into the lamina densa of the basement membrane. BP180 autoantibodies are present in around 70 percent of MMP patients. However, in cases with ocular involvement, anti-β_4 integrin antibodies are more dominant. Overall, these mechanisms cause the loss of attachment of hemi-desmosomes to the basement membrane, which leads to the formation of subepithelial clefts (Kuten-Shorrer et al. 2020; Ptasiewicz et al. 2022). In addition, some studies have suggested the role of cellular immunity in the pathogenesis of MMP. In cases with oral involvement, an increased level of T cells has been reported, indicating a possible role of cell-mediated immune response in disease pathogenesis and manifestations (Petruzzi 2012).

14.3.2 Diagnosis

The diagnostic procedure of pemphigoid starts with clinical evaluations and is completed with histologic and immunological assessments. Tissue biopsy from the lesions is necessary for routine histological evaluation, and also perilesional tissue sampling is needed for DIF. Histopathologic evaluation in both MMP and BP shows sub-epidermal blisters. In BP, eosinophilic infiltration is detectable, while in MMP, neutrophilic and lymphocytic infiltrations are more dominant. A distinct feature that leads to ruling out other sub-epidermal bullous diseases is that, in MMP, lamellar fibrosis can be detected surrounding the lesions. However, this feature does not always exist in all cases. Moreover, DIF is the gold standard for diagnosing pemphigoid disease, showing the linear deposition of IgG and complement proteins in the basement membrane. Although serologic examinations for MMP are not generally available, some laboratory tests, such as ELISA, can detect autoantibodies targeting the c-terminal domain of BP180 or laminin332, aiding in diagnosis when other diagnostic tests are inconclusive (Pratasava et al. 2021).

14.3.3 Treatment

Generally, the treatment approach of MMP depends on the extension and severity of the lesions. In mild to moderate conditions, topical medications such as topical steroids are used, while in more severe or refractory cases, systemic therapies such as systemic corticosteroids and immunosuppressants are prescribed. Cases with oral involvement with or without skin lesions are interpreted as a low-risk presentation and can be managed with more conservative treatments. For gingival lesions, topical corticosteroids applied with a trey or gauze can be utilized to reduce inflammation. In treatment-resistant cases, intravenous immunoglobulins (IVIG) and rituximab can be used to control the disease progression and relapses (Kuten-Shorrer et al. 2020; Pratasava et al. 2021).

14.3.4 Oral Manifestations

The clinical presentation of MMP starts gradually with exacerbation and remission periods. In most of the MMP cases, the oral cavity is the initial site of disease onset (Ptasiewicz et al. 2022). Unlike BP, where skin lesions are so common, MMP mostly affects mucosal surfaces, with the oral cavity being the most commonly affected region, followed by the conjunctiva as the second most common site. Thus, oral lesions can be detected in around 85 percent of MMP cases. Also, skin lesions occur in MMP patients with a predilection to the head, neck, and upper body regions (Pratasava et al. 2021). The characteristic feature of MMP is subepithelial bullous lesions leading to scars in extra-oral sites, while the oral tissue remains safe from scarring. The most common site of involvement in the oral cavity is the gingiva, presenting as desquamative gingivitis (Kuten-Shorrer et al. 2020). However, lips, buccal mucosa, and tongue are not common sites of these lesions. Moreover, this type of gingival involvement is the most common manifestation of MMP, and sometimes it may be the only presenting sign of the disease (Ptasiewicz et al. 2022). In addition, lesions in the oral mucosa can present as erosions and erythematous regions, as intact vesicles or bullous lesions break quickly (Kuten-Shorrer et al. 2020). However, in comparison to pemphigus, MMP vesicles last longer (up to 48 hours) as they are less fragile (Saccucci et al. 2018). This is because of the different locations of tissue separations between pemphigus (intra-epithelial) and MMP (subepithelial). Also, MMP patients may report various symptoms such as a burning sensation, pain, or difficulties during mastication, particularly when lesions are present in the gingiva or palate (Saccucci et al. 2018).

14.4 Linear IgA Disease (LAD)

LAD is a rare autoimmune disease in which autoreactive IgA antibodies cause sub-epithelial blisters. Early onset of the disease is called chronic bullous disease of childhood (CBDC), and it can also be seen in individuals over 60 years old (Khan et al. 2023). LAD affects females twice as much as males (Alrashdan et al. 2024). Also, the clinical course of the disease differs by age; in children, it is self-limiting, while in adulthood it tends to follow a more chronic relapsing course (Rashid et al. 2019). Although there is not often a specific initiating factor, some conditions such as drugs (e.g., vancomycin), malignancies, gastrointestinal disease, and trauma are assumed to be potential trigger factors. In addition to cutaneous lesions, 60 to 80 percent of patients present mucosal involvement, often presenting as vesicles or blisters (Khan et al. 2023).

14.4.1 Pathophysiology

Generally, in this disease, autoreactive IgA antibodies target different antigens in the basement membrane zone, such as BP180, BP230, LAD285, and collagen VII (Khan et al. 2023). Also, the process of epitope spreading, occurring in LAD, leads to heterogeneity of involved antigens during the pathogenesis (Fortuna and Marinkovich 2012). In drug-induced LAD, autoantibodies target a drug–protein complex, where the drug acts as a hapten, modifying cutaneous proteins and triggering an immune response (Khan et al. 2023). Humoral immunity is responsible for producing autoreactive IgA antibodies, which mainly target fragments of BP180 antigen. In addition, cellular immunity contributes to disease progression. The activation of CD4+ T helper cells, which are the major lymphocyte cells in LAD disease, leads to the recruitment of neutrophils and eosinophils, resulting in further inflammation and tissue damage. Also, several cytokines, such as IL-2 and IL-5, produced by Th1 and Th2 cells, play an important role in basement membrane disruption. Furthermore, blisters are a consequence of the contribution of three important factors, including plasminogen, neutrophil-derived products, and mast cell granules. Overall, plasminogen can be activated by two main plasminogen activators (PA), including urokinase-type PA and tissue-type PA. The final product, called plasmin, is responsible for the activation of matrix metalloproteinase-9 (MMP-9) and neutrophil elastase, both of which can degrade basement membrane components. Moreover, plasmin contributes to BP180 fragmentation, generating particles that can be trigger factors for further autoimmune activation. Eventually, blisters are formed due to epithelial-dermal separation (Fortuna and Marinkovich 2012).

14.4.2 Diagnosis

Histopathologic features of LAD include subepithelial blisters, which are not a specific feature of the disease; thus, further immunological assessments are essential for accurate diagnosis. Moreover, DIF evaluation shows linear deposition of IgA autoantibodies in the basement membrane zone (Khan et al. 2023). Also, approximately 70 percent of cases show anti-basement membrane IgA antibodies in IIF assessment (Ptasiewicz et al. 2022).

14.4.3 Treatment

It is important to note that in medication-induced LAD, the trigger drug should be discontinued if possible. In addition, a treatment plan for LAD depends on the severity of the disease. In mild cases, a topical steroid such as clobetasol propionate 0.05% is prescribed. Severe cases require systemic therapy, with dapsone with a dosage of 50–150 mg/day for adults and 0.5–2 mg/kg/day for children being the first-line treatment. Moreover, if patients have G6PD deficiency, dapsone is contraindicated due to the risk of hemolysis. In such cases, only oral corticosteroids are prescribed (Khan et al. 2023).

14.4.4 Oral Manifestations

Oral involvement occurs in more than 50 percent of LAD cases (Ptasiewicz et al. 2022). The most common sites of the lesions in the oral cavity include the tongue, buccal mucosa, gingival tissue, and palatal region (Alrashdan et al. 2024). Oral lesions in LAD consist of erosion, ulceration, and desquamative gingivitis. In some cases, desquamative gingivitis may be the sole oral manifestation of the disease (Ptasiewicz et al. 2022; Rashid et al. 2019). In addition, these clinical presentations of LAD-related oral lesions are very similar to MMP, making differential diagnosis essential (Ptasiewicz et al. 2022). Moreover, LAD oral lesions may precede cutaneous manifestations, emphasizing the importance of early detection of these lesions in the timely diagnosis and management (Fortuna and Marinkovich 2012).

14.5 Lichen Planus (LP)

Lichen planus (LP) is a mucocutaneous autoimmune disease, and its oral involvement is more common than cutaneous lesions. Although skin lesions do not have a gender predisposition, oral lesions are more frequently observed in females. A key feature of this disease is Wickham's striae, a white reticular pattern on the surface of the lesions (Vičić et al. 2023). Furthermore, this disease has six subtypes, including reticular, erosive/ulcerative, plaque-like, atrophic, papular, and bullous. Among all variants of LP, the classic type can manifest as a localized or generalized condition. In the localized form, the disease is limited to extremities, while in the generalized condition, the whole body and also mucosal surfaces, such as the oral cavity, are involved. In addition, LP can affect other mucous membranes such as the larynx, conjunctiva, and genital area. The two most common types of LP are reticular and erosive forms, respectively. The erosive type, in particular, can transform into squamous cell carcinoma. In more than half of the cases, oral lichen planus (OLP) is accompanied by skin lesions, while in approximately 30 percent of the cases, OLP is the sole manifestation of the disease (Vičić et al. 2023).

14.5.1 Pathophysiology

Generally, the combination of genetic, environmental, and immunological factors may lead to LP. Genetic susceptibility can define the patient's reactions to other environmental triggers such as infections, trauma, and stress. Also, some viral infections have been introduced as possible causes of LP, such as human papillomavirus (HPV), hepatitis B virus (HBV), hepatitis C virus (HCV), Epstein-Barr virus (EBV), varicella-zoster virus (VZV), and human herpes virus-7 (HHV-7) (Kuten-Shorrer et al. 2020; Vičić et al. 2023). In addition, the salivary composition of patients has shown significant alteration of fungal and bacterial factors, such as *Candida*, *Solobacterium*, and *Fusobacterium*. Moreover, some local irritants in the oral cavity, such as dental filling materials, can cause OLP (Vičić et al. 2023).

The pathogenesis of LP is primarily driven by cellular immunity, with T lymphocytes playing a central role. Overall, by releasing damage-associated molecular patterns (DAMPs), Toll-like receptors of antigen-presenting cells (APC) such as dendritic cells (DC) are activated. Then, plasmacytoid DCs produce interferon-α (IFN-α) and Keratinocytes release interleukin1β (IL-1β) and tumor necrosis factor-α (TNFα), promoting inflammation. As inflammation progresses, IFN-α causes the activation of dermal DCs. These activated DCs migrate to regional lymph nodes, where they present antigens to naïve T cells and lead to their differentiation into T-helper 1 (Th1) and cytotoxic T cells. Eventually, these interactions result in an increase in the production of pro-inflammatory cytokines such as interferon-γ by Th1 cells, which is necessary for the pathogenesis of LP. In the final stage, apoptosis of basal Keratinocytes occurs due to the activation of the caspase cascade, a key pathologic feature of LP disease (Manchanda et al. 2024; Vičić et al. 2023). This keratinocyte destruction leads to epithelial integrity disruptions, eventually resulting in the ulcerations and erosions observed in OLP lesions.

Besides the significant role of cellular immune response in the pathogenesis of LP, humoral immunity might also play a significant role, specifically in the erosive subtype of the disease, with the detection of autoantibodies targeting Dsg1 and Dsg2 in some patients. Also, in affected oral mucosa and skin lesions of LP patients, B lymphocytes have been detected. Thus, humoral immunity has a minor, yet notable, contribution to LP pathogenesis (Vičić et al. 2023).

14.5.2 Diagnosis

Typical forms of OLP can be diagnosed clinically, while other cases need tissue biopsy for histopathologic and DIF evaluations. In histopathologic assessment, band-like infiltration of lymphocytes in the upper dermis and degeneration of basal layer cells are detectable. Although IIF is not used as a common diagnostic test for OLP, DIF is a suitable method, especially in cases of desquamative gingivitis (Manchanda et al. 2024). DIF evaluation illustrates a linear pattern of antifibrogen at the basement membrane zone and also IgM deposition on cytoid-like bodies (Alrashdan et al. 2016).

For the diagnosis of oral lichenoid reactions, patch testing can be performed. In patients who are suspected of sensitivity to dental materials, a patch test is applied on the normal skin of the upper back. After 72 hours, the region is evaluated for any probable erythema or bullous reactions (Manchanda et al. 2024).

14.5.3 Treatment

Topical corticosteroids such as clobetasol propionate, triamcinolone, betamethasone, fluocinolone, and fluticasone are the preferred treatment for localized OLP. Also, all forms of topical application, such as ointment, mouthwash, and spray, are highly effective. Patients who have refractory lesions and do not respond well to topical treatments may receive systemic corticosteroids. For example, cases with erosive lesions affecting multiple sites can benefit from systemic corticosteroid therapy. However, this systemic treatment should be prescribed for a short period of time to avoid related side effects. The preferred systemic corticosteroids for OLP management are prednisolone and methylprednisolone (Manchanda et al. 2024).

14.5.4 Oral Manifestations

The classic manifestation of OLP is symmetrical and bilateral involvement that mostly affects the buccal mucosa. Additionally, lesions can be detected on the tongue, gingiva, and labial mucosa. The most common type of OLP is the reticular form, which features papules with Wickham's striae, while bullous and papular variants are rare in the oral cavity. In addition, erosive LP presents erythematous and ulcerative oral lesions that are surrounded by reticular striae. If erosive LP affects the gingival tissue, it will cause desquamative gingivitis. Moreover, gingival lesions can be the sole manifestation of OLP in around 10 percent of cases. Plaque-like LP primarily affects the tongue and buccal mucosa in the oral cavity, and it has a similar manifestation to leukoplakia lesions. It appears as multifocal, elevated, homogenous, and white lesions. Among all subtypes, atrophic and erosive forms of OLP can lead to patients' complaints and discomfort. Unlike OLP, the oral lichenoid reaction has a definitive underlying cause; however, histological and clinical features of oral lichenoid reactions are similar to those of OLP, making the differentiation essential. Because of the different treatment approaches for these two conditions, they should be accurately diagnosed for proper management (Alrashdan et al. 2016).

14.6 Systemic Lupus Erythematosus

Systemic lupus erythematosus (SLE) is an autoimmune disease affecting many organs, such as joints, skin, and kidneys, with a high rate of mortality and morbidity. SLE mostly affects females of reproductive age, and it is accompanied by exacerbation and remission periods. In this disease, autoantibodies target different autogenous antigens, leading to the formation of immune complexes. Deposition of these complexes in tissues triggers inflammatory responses and causes tissue and organ damage (Tsang and Bultink 2021). It is important to note that oral lesions are assumed to be early manifestations of this disease, and awareness of possible signs and symptoms may be helpful in early diagnosis, which is valuable for improving the prognosis of the disease. SLE is categorized into three groups—chronic, acute, and subacute—based on its cutaneous features. Although oral lesions are rare in subacute SLE, they are frequently detectable as blisters or ulcers in the acute subtype. In chronic SLE, oral lesions do not have a symmetrical distribution in the oral cavity, and they can be seen as well-defined, atrophic, or ulcerative lesions (García-Ríos et al. 2022).

14.6.1 Pathophysiology

In the pathogenesis of SLE, both innate and adaptive immunity are involved. Neutrophils in SLE show functional impairments that contribute to reduced phagocytosis ability. Also, the accumulation of unremoved apoptotic cells contributes to providing a source of self-antigens, triggering autoimmune activation. Moreover, plasmacytoid dendritic cells produce a high level of type-I IFNs in response to circulating nuclear components such as dsDNA. Type-I IFNs lead to the activation of both B cells and T cells, promoting further activation of the autoimmune response (Accapezzato et al. 2023).

In adaptive immunity, the interaction between B cells and CD4+ T cells plays a critical role in disease progression. In SLE, B cells produce antinuclear antibodies. The target antigens for autoreactive B cells in this disease are dsDNA and RNA-associated molecules. These interactions are accompanied by abnormal engagement of Toll-like receptor (TLR)7 and TLR9. These autoantibodies are detectable in the blood circulation of SLE patients even years before the presence of clinical manifestations of the disease. Therefore, it raised questions of whether they serve as biomarkers or direct pathogenic factors. However, strong evidence suggests the pathogenic role of these autoantibodies in tissue damage. In addition, B cells in SLE can induce the formation of autoreactive T cells, since they can act as antigen-presenting cells (APC) (Accapezzato et al. 2023).

Autoreactive T cells are a principal part of SLE pathogenesis. Among all types of T cells, Th1 plays a significant role in producing a key pro-inflammatory cytokine, IFN-ɣ. On the other hand, it is revealed that the levels of Th2 and IL-4, produced by these cells, are reduced in SLE patients, suggesting the protective role of Th2 cells. Therefore, the ratio between the level of IFN-ɣ, and IL-4 can be considered as an indicator for SLE activity. Moreover, Th17 and its related pro-inflammatory cytokine (IL-17) play a part in SLE development by recruiting neutrophils, increasing B cell activities, and enhancing innate immunity response. In addition, CD8+ T cells in SLE show functional defects with impaired cytolytic capacity and reduced production of Granzyme and Perforin (Accapezzato et al. 2023).

14.6.2 Diagnosis

Based on criteria that are introduced by the American College of Rheumatology, there are 11 major components in 7 categories introduced for the diagnosis of SLE. Mucocutaneous manifestations (oral lesions, malar rash, discoid rash, and photosensitivity), hematologic manifestations (e.g., hemolytic anemia, thrombocytopenia $<100{,}000/mm3$, leukopenia$<4000/mm3$ or lymphopenia $<1000/mm3$), neurologic manifestations (e.g., seizures, psychosis), serositis (e.g., pleurisy and pericarditis), joints manifestations (e.g., non-erosive arthritis and synovitis), renal manifestations (e.g., urine protein/creatinine or 24-hr urine protein, >0.5 g), and serological manifestations (e.g. anti-nuclear antibodies, anti-Smith antibodies, anti-dsDNA antibodies, antiphospholipid antibodies, low complement levels, positive direct Coombs test). The patient must meet at least 4 out of these 11 criteria for SLE diagnosis (Benli et al. 2021).

14.6.3 Treatment

Hydroxychloroquine (HCQ) has been considered as a safe first-line treatment with a maximum dose of 5 mg/kg for all SLE patients, according to its multiple benefits and low cost. This drug can limit disease progression during the active phase and reduce cardiovascular events. Also, corticosteroid therapy should be limited to less than 7.5 mg/day to minimize long-term side effects, and discontinuation is recommended in possible circumstances. In addition, in more severe SLE cases, immunosuppressive drugs such as azathioprine, methotrexate, mycophenolate mofetil, and cyclophosphamide are prescribed to suppress the autoimmune response (Accapezzato et al. 2023).

14.6.4 Oral Manifestations

Around 40 percent of SLE cases present oral lesions (Benli et al. 2021). The most common oral manifestation of SLE is oral ulcers, such as aphthous-like lesions, which are part of the diagnostic criteria for SLE. The hard palate, soft palate, and lower lip vermilion are the common regions of these ulcers (García-Ríos et al. 2022). The classic feature of SLE oral involvement is a plaque-like lesion with an erythematous center that is surrounded by a keratotic border. However, oral lesions can be present in different clinical appearances, such as petechiae and purpura, elevated keratotic plaques, lichenoid lesions, and erythema (Benli et al. 2021). Although oral ulcers are the most common symptom of SLE, decreased salivary flow and hyperpigmentation can be present with a lower probability. With aging, hyposalivation may be exacerbated and lead to xerostomia in these patients. On the other hand, decreased salivary flow can facilitate some secondary oral conditions, such as glossodynia, fissured tongue, gingivitis and periodontal disease, angular cheilitis, and fungal infections such as candidiasis.

Additionally, as a result of the immunosuppressive medications received by SLE patients, they are susceptible to candidiasis and other oral infections. Furthermore, SLE can be categorized into three groups based on the onset of the disease, including juvenile (≤18 years), adult (19–50 years), and late (over 50 years). Among these groups, patients with juvenile-onset SLE show a higher prevalence of oral ulcerations compared to others (García-Ríos et al. 2022). Also, persistent oral ulcers in SLE patients may undergo malignant transformations that lead to squamous cell carcinoma (Benli et al. 2021).

14.7 Behcet's Disease

Behcet is a systemic vasculitis with a high prevalence among Asian populations, especially in Turkey. This syndrome typically involves both genders in their fourth decade of life. Also, this autoimmune disease has a strong genetic association with HLA-B51 (Lavalle et al. 2024; Saccucci et al. 2018). Behcet's disease can affect multiple organs. Typical manifestations of the disease are recurrent aphthous ulcers in the oral cavity, genital ulcers, and uveitis (Saccucci et al. 2018). Recurrent oral ulcers are present in around 95 percent of the cases, making them a key diagnostic feature of Behcet's disease (Lavalle et al. 2024). In addition, in most cases, mucocutaneous involvement is the first sign of the disease. Also, skin lesions can be detected on the trunk or limbs, presenting as papular or pustular eruptions (Saccucci et al. 2018). Other potential manifestations of this disease include musculoskeletal, gastrointestinal, and neurological involvement (Lavalle et al. 2024).

14.7.1 Pathophysiology

Behcet's disease is a consequence of an abnormal immune response to environmental stimuli in individuals with a genetic predisposition.

Streptococcus sanguis, herpes simplex virus, smoking, stress, and even foods are possible trigger factors. Moreover, it is known that both innate and adaptive immunity are involved in Behcet's disease. Neutrophils and monocytes produce a higher level of pro-inflammatory cytokines, such as IL-1β, IL-6, and TNF-α. Also, some specific immune mediators, secreted by innate immune cells, such as IL-1β and IL-23, are responsible for the activation of adaptive immune cells, particularly Th-17. Recent studies have shown that Th-17 and its produced mediators, such as IL-17 and IL-22, play a significant role in the pathogenesis of Behcet's disease. In addition, a higher level of these two mediators has been detected in serum and tissue samples of oral and skin lesions. Moreover, impaired Tregs contribute to autoimmune reactions and excessive immune activation (Lavalle et al. 2024).

14.7.2 Diagnosis

Diagnosis of Behcet's disease is a challenging process due to the lack of a specific diagnostic test. Instead, a diagnosis is made by excluding other possible clinical conditions. To confirm the diagnosis of Behcet's disease, a patient must have recurrent oral ulcers and at least two other conditions from the following features: recurrent genital ulcers, skin lesions, ocular lesions, and a positive pathergy test. It is important to note that not only are oral aphthous lesions common among the population, but they also occur in some other conditions, such as sarcoidosis, SLE, or Crohn's disease (Lavalle et al. 2024; Saccucci et al. 2018).

14.7.3 Treatment

Generally, the main goals of the treatment approach in this disease are relieving the patient's symptoms, reducing the exacerbation periods, and controlling possible complications. Localized or systemic corticosteroids are prescribed to reduce inflammation. In severe cases, immunosuppressive drugs such as azathioprine are administered (Lavalle et al. 2024).

14.7.4 Oral Manifestations

Recurrent oral ulcers are present in most Behcet's disease cases, and they may be the first clinical signs of the disease (Lavalle et al. 2024). Oral ulcers frequently appear in the buccal mucosa, lips, tongue, and soft palate (Saccucci et al. 2018). If lesions extend to other areas such as the tonsil or pharynx, they may lead to considerable discomfort (Nakamura et al. 2020). Behcet's oral ulcers are similar to conventional aphthous lesions, with variable sizes (Saccucci et al. 2018). Compared to the lesions in recurrent aphthous stomatitis (RAS), major oral ulcers in Behcet's disease not only tend to affect more extensive areas of the oral mucosa, but they are also more common (Nakamura et al. 2020). These cyclic lesions start as an erythematous area that subsequently develops into painful ulcers (Saccucci et al. 2018). The clinical appearance of the lesions can be defined as a center of pseudo-membranous tissue, which is surrounded by a narrow erythematous halo in the periphery. Ulcers in the oral cavity are categorized into two groups, minor (<1 cm) and major (>1 cm), based on their size. These two types have a similar appearance, while major ulcers are deeper and more painful, and they may be healed with scarring (Nakamura et al. 2020).

14.8 Epidermolysis Bullosa Acquisita (EBA)

Epidermolysis bullosa acquisita (EBA) is a chronic autoimmune disease that involves both skin and mucous membranes (Ptasiewicz et al. 2022). It can be seen in both genders with equal probability and can develop at any age (Kridin et al. 2019). Moreover, EBA can be associated with other inflammatory conditions, including Crohn's disease, ulcerative colitis, psoriasis, thyroiditis, diabetes mellitus, rheumatoid arthritis, and hepatitis C infection. In EBA, autoantibodies are produced against type VII collagen, a key component of the anchoring fibers in the basement membrane (Ptasiewicz et al. 2022). EBA can be divided into two main types, classical (mechanobullous) and inflammatory, which is more common. Blisters in the classical EBA are surrounded by non-inflamed tissue, while non-mechanobullous-type lesions have an erythematous halo in the periphery. In addition, healing with scarring and milia formation is attributed to the classical type of disease (Kridin et al. 2019). Furthermore, the inflammatory type of EBA is further categorized into bullous pemphigoid-like (BP) EBA, mucosal pemphigoid-like (MMP) EBA, and linear IgA bullous dermatosis (LABD) EBA. Approximately 50 percent of cases with BP-like EBA present oral manifestations, such as oral vesicles or erosions. In addition, MMP-like EBA dominantly affects mucous membranes, especially the oral mucosa (Ptasiewicz et al. 2022).

14.8.1 Pathophysiology

In this disease, autoantibodies are produced against the NC1 domain of type VII collagen, with less frequent involvement of the NC2 domain. Type VII collagen is a major structural part of the anchoring fibrils that can be detected in hemidesmosomes, essential for dermal-epidermal adhesion. Moreover, autoreactive antibodies in the serum of EBA patients are mainly IgG, while IgA, IgE, and IgM can be detected as well. IgG targets the dermal-epidermal junction (DEJ) of both skin and mucosa, which results in leukocyte aggregation. The amount of IgG depositions is directly correlated with dermal–epidermal disjunction, contributing to blister formation. Also, the activation of the complement system is initiated by the interaction of autoantibodies with type VII collagen, which leads to the production of pro-inflammatory cytokines. Neutrophil recruitment occurs as a consequence of two main processes: pro-inflammatory cytokine signaling and activation of natural killer and γδ-T cells due to immune complex formation. Eventually, blisters are formed by reactive oxygen species (ROS) and proteolytic enzymes, which degrade anchoring fibrils and basement membrane proteins and are released by Neutrophils (Kridin et al. 2019).

14.8.2 Diagnosis

The diagnostic process starts with clinical evaluation and is followed by tissue biopsy and laboratory assessments. The histopathology feature of EBA disease presents sub-epidermal blisters with minimal or absent inflammation. Also, as the histological appearance of the inflammatory type of EBA has similarities to bullous pemphigoid, further immunological assessments are needed (Kridin et al. 2019). DIF reveals the deposition of IgG in the basement membrane zone. However, IgM, IgA, and complement C3 may be present as well (Ptasiewicz et al. 2022). The binding of autoantibodies against their target antigen (type VII collagen) in DIF evaluation forms a u-serrated pattern that discriminates EBA from other bullous diseases such as bullous pemphigoid, mucous membrane pemphigoid, p200 pemphigoid, and anti-laminin 332 pemphigoid. In addition, the deposition of autoreactive antibodies in the dermal side of the basement membrane is beneficial to exclude EBA from other bullous diseases in IIF assessment. However, in some conditions, such as anti-laminin 332 and anti-p200 pemphigoid, autoantibodies attach to the similar side of the basement membrane as EBA, necessitating other serological tests for differentiating and accurate diagnosis (Kridin et al. 2019).

14.8.3 Treatment

The treatment approach of EBA includes patient education on lesion care and management and prevention of additional traumas to limit the formation of new lesions. In addition, systemic administration of corticosteroids is considered the first-line treatment choice, especially in the inflammatory type of EBA. The dosage of corticosteroids varies from 0.5 to 1.5 mg/kg/day, based on the severity of the disease. Also, due to possible adverse effects of long-term corticosteroid therapy, immunosuppressant drugs can be prescribed, as they have a corticosteroid-sparing effect. Dapsone (DDS) at a dose of 25–150 mg/day, colchicine at 1–2 mg/day, and Azathioprine (AZA) at a dose of 2–3 mg/kg/day can be administered in patients with EBA disease (Miyamoto et al. 2022).

14.8.4 Oral Manifestations

The most commonly affected mucosal site in EBA is the oral cavity (Kridin et al. 2019). EBA oral lesions might present as desquamative gingivitis, while this feature can be detected in other autoimmune diseases, such as oral lichen planus, autoimmune blistering diseases, lupus erythematosus, and erythema multiforme (Parlatescu et al. 2024). Also, oral lesions can be seen as vesicles, which progress to erosive tissue. In some cases, these lesions heal with crust formation or scarring (Ptasiewicz et al. 2022).

14.9 Erythema Multiforme (EM)

Erythema multiforme is an immune-mediated disease that can affect both skin and mucous membranes, such as oral tissues. This disease can affect both genders at any age; however, it is more common among young adults (Alrashdan et al. 2023). EM presents variable clinical features, with target-like or iris skin lesions being its hallmark. Moreover, this disease results from a hypersensitivity reaction to a wide range of antigens (Samim et al. 2013). This condition can be categorized into three groups, including minor, major, and oral EM. The minor type affects a single mucosal site and presents target lesions on the skin, while the major type involves at least two mucosal surfaces and presents more severe skin lesions, which might be even life-threatening. Oral EM involves non-keratinized mucosa of the oral cavity with no target lesions on the skin (Lucchese 2018). Generally, trigger factors of EM can be divided into two main groups: drugs and infections. Among viral infections, the most common factor is Herpes Simplex Virus (HSV), while Cytomegalovirus (CMV) and Epstein-Barr virus (EBV) have been suggested with less incidence. Also, recent studies have suggested that both COVID-19 (coronavirus disease 2019) and its vaccines can induce EM. However, in child-onset EM, other infections contribute to the induction of the disease as well, such as *Mycoplasma pneumoniae*. Moreover, medications such as anticonvulsants, antipsychotics, antibiotics, NSAIDs, and allopurinol can induce EM. Furthermore, some autoimmune conditions may increase the risk of EM, such as systemic lupus erythematosus or rheumatoid arthritis (Alrashdan et al. 2023).

14.9.1 Pathophysiology

The main pathogenic factor of EM is HSV-1, and with less probability, HSV-2. Studies suggest that autoimmune cross-reactivity is responsible for the pathogenesis of EM, which is induced by HSV1 and HSV2. This mechanism occurs because many peptides are shared between HSV-1 and HSV-2 and human oral mucosa proteins, such as human collagen alpha-1 (XVII) chain protein. Therefore, immune reactions due to HSV infection, especially HSV-1, can initiate autoimmune cross-reactions with different molecules in the oral mucosa and cause EM erythematosus symptoms (Lucchese 2018). Moreover, intra- and subepithelial blisters are caused by T cell activities against the keratinocytes, which present HSV antigens (Lucchese 2018). The activation of Th1 leads to the production of IFN-ɣ, which contributes to multiple consequences, such as the recruitment of autoreactive T cells, natural killer cells, and monocytes by increasing the level of cytokines and chemokines. Moreover, in medication-induced EM, CD8+ T cells are dominant, and they cause direct tissue damage via TNF-α, perforin, and granzyme B (Samim et al. 2013).

14.9.2 Diagnosis

The diagnostic procedure of EM starts with clinical evaluations and is followed by histopathologic assessment. A combination of patients' reports, including previous viral infection or medication intake with clinical features of the disease, supports the diagnosis of EM (Samim et al. 2013). In addition, histologic features of this condition consist of lymphocytes that surround necrotic keratinocytes. Also, lymphocytic and histiocytic infiltration in the dermal-epidermal junction is detectable (Alrashdan et al. 2023).

14.9.3 Treatment

The treatment approach in EM depends on the extension and severity of the disease. In minor EM, the treatment goal to alleviate patients' discomfort by the administration of topical medications. In cases of minor EM with mucosal involvement, viscous lidocaine or diphenhydramine-containing mouthwashes may relieve pain. In patients with major EM, in addition to discontinuation of the responsible medication, intensive supportive therapy is needed. In HSV-induced cases, antiviral medications such as acyclovir are prescribed to reduce recurrence (Alrashdan et al. 2023).

14.9.4 Oral Manifestations

Oral manifestations can be seen in around 70 percent of EM cases and may present with variable clinical features, ranging from diffuse erythema to ulcerations. Moreover, the most common sites of involvement in the oral cavity are the buccal mucosa, palate, and tongue, while other regions can be affected as well. Also, oral lesions heal without scarring (Samim et al. 2013). Although mucosal involvement in minor EM is not prevalent, when it occurs, it commonly affects the oral mucosa. In this condition, small erosive and ulcerative areas with a classical feature of lip crusting are detectable. Major EM, in contrast, causes extensive mucosal tissue damage and causes erosions, ulcerations, and even sloughing (Alrashdan et al. 2023).

14.10 Sjögren's Syndrome (SS)

This autoimmune disorder is caused by lymphocytic infiltration in exocrine glands, such as salivary and lacrimal glands. The manifestations of this disease vary from dryness in the eyes (xerophthalmia) and mouth (xerostomia) to systemic involvement (Hou et al. 2024). Generally, SS can be categorized into two types, primary and secondary. Secondary SS occurs in association with other autoimmune diseases, such as SLE. In terms of gender predilection, it is 10 times more common among females who are more than 55 years old (Tian et al. 2021).

14.10.1 Pathophysiology

In SS, both innate and adaptive immunity are responsible (Negrini et al. 2022). In the pathogenesis of SS, besides immune system involvement, other factors contribute, such as environmental, neuroendocrine, and genetic parameters. In terms of genetic predisposition, both IRF5 and STAT4 are considered genes that make the body susceptible to SS by inducing inflammatory mediators and autoimmune responses. Viral factors are the other possible aspect of SS pathogenesis. Studies suggest that some viral proteins have 5 to 6 amino acid sequences in common with the Ro-60 antigen, leading to molecular mimicry and triggering an autoimmune reaction as a consequence of this similarity (Tian et al. 2021). Also, it is revealed that both Epstein-Barr virus (EBV) and human T-cell leukemia virus type 1 (HTLV-1) may be attributed to the pathogenesis of SS (Hou et al. 2024). Furthermore, the salivary gland epithelial cells (SGECs) have an important role in SS pathogenesis. These cells in SS present CD80 and CD86 markers, which make them behave as antigen-presenting cells, causing T lymphocyte activation and sustained inflammation (Tian et al. 2021). The activation of Th17 and Th1 cells participates in SS pathogenesis by increasing levels of **IFN-γ, IL-17, and other pro-inflammatory cytokines** (Negrini et al. 2022). SGECs also contribute to both activation and differentiation of B lymphocytes. Autoreactive B cells promote disease progression by releasing anti-SSA and anti-SSB autoantibodies (Tian et al. 2021).

14.10.2 Diagnosis

Based on the latest diagnostic criteria, meeting a score equal to 4 or higher in patients, confirms SS diagnosis. These criteria include five key factors, and each of them gets a score. Lymphocytic sialadenitis in labial salivary glands with a focus score $\geq$ 1 is scored 3. Similarly, positive anti-SSA/Ro gets a score of 3. Other factors share an equal score of 1 including abnormal results of ocular or saliva tests such as an ocular staining score $\geq$ 5 or van Bijsterveld score $\geq$ 4, Schirmer's test $\leq$5 mm/5 min, and the unstimulated salivary flow rate of $\leq$0.1 mL/min (Negrini et al. 2022).

14.10.3 Treatment

The treatment goal in this condition is to relieve the patient's symptoms and manage systemic complications. Immunosuppressive medications and corticosteroids are administered in SS. Also, severe cases can benefit from B lymphocyte-targeted therapy. Moreover, to alleviate xerostomia, it is important to exclude all possible factors that negatively affect the salivary flow, such as alcohol or caffeine-maintaining drinks, smoking, and certain medications. Also, as mouth breathing causes more sense of dryness and discomfort for the patients, the elimination of possible trigger factors such as sinusitis should be considered. Moreover, patients are instructed to drink, sip, or gargle water frequently, which helps hydrate mucosal surfaces. For patients who have functional residual salivary glands, salivary stimulation is recommended. Chewing gum or gustatory stimulants may cause the secretion of saliva. Furthermore, some medications such as pilocarpine and cevimeline can be prescribed (Negrini et al. 2022).

14.10.4 Oral Manifestations

The most common symptom of Sjögren's syndrome is xerostomia. Reduced salivary flow leads to multiple oral problems such as infections, tooth decay, oral ulcers, and patient discomfort (Hou et al. 2024). Mucosal irritability, pain, and burning sensation are other possible symptoms (Sankar and Noujeim 2017). Moreover, one of the most common oral manifestations of SS is angular cheilitis, which is followed by candidiasis, ulcers, and glossitis (Serrano et al. 2018). Also, gingivitis is a common finding in the oral cavity of the SS cases (Saccucci et al. 2018). These patients may show some salivary gland complications, such as infection or benign enlargement. Also, as they are at risk of lymphoma, any enlarged salivary gland should be assessed to rule out malignancies (Sankar and Noujeim 2017).

14.11 Conclusion and Future Directions

Oral mucosa is the common site of autoimmune reactions, making accurate oral examinations inevitable. Also, in most autoimmune disorders, the oral cavity is involved prior to other mucosal tissues. Therefore, early detection of oral mucosal manifestations of autoimmune diseases can lead to a timely diagnosis of the condition, which may improve the prognosis.

A close collaboration between dentists and medical doctors is recommended, as it can significantly promote treatment outcomes. Accurate and comprehensive oral tissue examination by dentists and early referral of suspicious cases to medical doctors can lead to timely diagnosis and management of autoimmune diseases.

References

Accapezzato D, Caccavale R, Paroli MP, Gioia C, Nguyen BL, Spadea L, Paroli M (2023) Advances in the pathogenesis and treatment of systemic lupus erythematosus. Int J Mol Sci 24(7). https://doi.org/10.3390/ijms24076578

Alrashdan MS, Andreadis D, Zisis V, Hassona Y (2024) Immune-mediated oral mucosal pathology: a comprehensive review and update for clinicians – part II. Ital J Dermatol Venerol 159(1):11–22. https://doi.org/10.23736/s2784-8671.23.07690-9

Alrashdan MS, Arzmi MH, Ahmad Kamil WN, Al Kawas S, Leao JC (2023) Immune-mediated oral mucosal pathology: a comprehensive review and update for clinicians. Part I. Ital J Dermatol Venerol 158(5):408–418. https://doi.org/10.23736/s2784-8671.23.07676-4

Alrashdan MS, Cirillo N, McCullough M (2016) Oral lichen planus: a literature review and update. Arch Dermatol Res 308(8):539–551. https://doi.org/10.1007/s00403-016-1667-2

Benli M, Batool F, Stutz C, Petit C, Jung S, Huck O (2021) Orofacial manifestations and dental management of systemic lupus erythematosus: a review. Oral Dis 27(2):151–167. https://doi.org/10.1111/odi.13271

Costan VV, Popa C, Hâncu MF, Porumb-Andrese E, Toader MP (2021) Comprehensive review on the pathophysiology, clinical variants and management of pemphigus (review). Exp Ther Med 22(5):1335. https://doi.org/10.3892/etm.2021.10770

Di Lernia V, Casanova DM, Goldust M, Ricci C (2020) Pemphigus vulgaris and bullous pemphigoid: update on diagnosis and treatment. Dermatol Pract Concept 10(3):e2020050. https://doi.org/10.5826/dpc.1003a50

Didona D, Maglie R, Eming R, Hertl M (2019) Pemphigus: current and future therapeutic strategies. Front Immunol 10:1418. https://doi.org/10.3389/fimmu.2019.01418

Fortuna G, Marinkovich MP (2012) Linear immunoglobulin A bullous dermatosis. Clin Dermatol 30(1):38–50

García-Ríos P, Pecci-Lloret MP, Oñate-Sánchez RE (2022) Oral manifestations of systemic lupus erythematosus: a systematic review. Int J Environ Res Public Health 19(19). https://doi.org/10.3390/ijerph191911910

Hou J, Feng Y, Yang Z, Ding Y, Cheng D, Shi Z et al (2024) Primary Sjögren's syndrome: new perspectives on salivary gland epithelial cells. Eur J Med Res 29(1): 371. https://doi.org/10.1186/s40001-024-01967-5

Kasperkiewicz M, Ellebrecht CT, Takahashi H, Yamagami J, Zillikens D, Payne AS, Amagai M (2017) Pemphigus. *Nat Rev Dis Primers* 3:17026. https://doi.org/10.1038/nrdp.2017.26

Khan M, Park L, Skopit S (2023) Management options for linear immunoglobulin a (IgA) bullous dermatosis: a literature review. Cureus 15(3):e36481. https://doi.org/10.7759/cureus.36481

Kridin K, Kneiber D, Kowalski EH, Valdebran M, Amber KT (2019) Epidermolysis bullosa acquisita: a comprehensive review. Autoimmun Rev 18(8):786–795. https://doi.org/10.1016/j.autrev.2019.06.007

Kuten-Shorrer M, Menon RS, Lerman MA (2020) Mucocutaneous diseases. Dent Clin N Am 64(1):139–162. https://doi.org/10.1016/j.cden.2019.08.009

Lavalle S, Caruso S, Foti R, Gagliano C, Cocuzza S, La Via L et al (2024) Behçet's disease, pathogenesis, clinical features, and treatment approaches: a comprehensive review. Medicina (Kaunas) 60(4). https://doi.org/10.3390/medicina60040562

Lin S-Y, Huang Y-W, Błochowiak K (2023) Autoimmune diseases and their various manifestations in the oral cavity–a systematic review. European J Clin Exp Med 3:627–638

Lucchese A (2018) From HSV infection to erythema multiforme through autoimmune crossreactivity. Autoimmun Rev 17(6):576–581. https://doi.org/10.1016/j.autrev.2017.12.009

Magliocca KR, Fitzpatrick SG (2017) Autoimmune disease manifestations in the Oral cavity. Surg Pathol Clin 10(1):57–88. https://doi.org/10.1016/j.path.2016.11.001

Manchanda Y, Rathi SK, Joshi A, Das S (2024) Oral lichen planus: an updated review of Etiopathogenesis, clinical presentation, and management. Indian Dermatol Online J 15(1):8–23. https://doi.org/10.4103/idoj.idoj_652_22

Miyamoto D, Gordilho JO, Santi CG, Porro AM (2022) Epidermolysis bullosa acquisita. An Bras Dermatol 97(4):409–423. https://doi.org/10.1016/j.abd.2021.09.010

Nakamura K, Tsunemi Y, Kaneko F, Alpsoy E (2020) Mucocutaneous manifestations of Behçet's disease. Front Med (Lausanne) 7:613432. https://doi.org/10.3389/fmed.2020.613432

Negrini S, Emmi G, Greco M, Borro M, Sardanelli F, Murdaca G et al (2022) Sjögren's syndrome: a systemic autoimmune disease. Clin Exp Med 22(1):9–25. https://doi.org/10.1007/s10238-021-00728-6

Parlatescu I, Tovaru S, Tofan C, Perlea P, Milanesi E, Dobre M, Mihai LL (2024) Gingival manifestations in Oral chronic autoimmune bullous diseases: a retrospective study. Medicina (Kaunas) 60(1). https://doi.org/10.3390/medicina60010167

Petruzzi M (2012) Mucous membrane pemphigoid affecting the oral cavity: short review on etiopathogenesis, diagnosis and treatment. Immunopharmacol Immunotoxicol 34(3):363–367. https://doi.org/10.3109/08923973.2011.608684

Pratasava V, Sahni VN, Suresh A, Huang S, Are A, Hsu S, Motaparthi K (2021) Bullous pemphigoid and other pemphigoid dermatoses. Medicina (Kaunas) 57(10). https://doi.org/10.3390/medicina57101061

Ptasiewicz M, Bębnowska D, Małkowska P, Sierawska O, Poniewierska-Baran A, Hrynkiewicz R et al (2022) Immunoglobulin disorders and the Oral cavity: a narrative review. J Clin Med 11(16). https://doi.org/10.3390/jcm11164873

Rashid H, Lamberts A, Diercks GFH, Pas HH, Meijer JM, Bolling MC, Horváth B (2019) Oral lesions in autoimmune bullous diseases: an overview of clinical characteristics and diagnostic algorithm. Am J Clin Dermatol 20(6):847–861. https://doi.org/10.1007/s40257-019-00461-7

Saccucci M, Di Carlo G, Bossù M, Giovarruscio F, Salucci A, Polimeni A (2018) Autoimmune diseases and their manifestations on Oral cavity: diagnosis and clinical management. J Immunol Res 2018:6061825. https://doi.org/10.1155/2018/6061825

Samim F, Auluck A, Zed C, Williams PM (2013) Erythema multiforme: a review of epidemiology, pathogenesis, clinical features, and treatment. Dent Clin N Am 57(4):583–596. https://doi.org/10.1016/j.cden.2013.07.001

Sankar V, Noujeim M (2017) Oral manifestations of autoimmune and connective tissue disorders. Atlas Oral Maxillofac Surg Clin North Am 25(2):113–126. https://doi.org/10.1016/j.cxom.2017.04.004

Serrano J, Lopez-Pintor RM, Gonzalez-Serrano J, Fernandez-Castro M, Casanas E, Hernandez G (2018) Oral lesions in Sjogren's syndrome: a systematic review. Med Oral Patol Oral Cir Bucal 23(4):e391–e400. https://doi.org/10.4317/medoral.22286

Tian Y, Yang H, Liu N, Li Y, Chen J (2021) Advances in pathogenesis of Sjögren's syndrome. J Immunol Res 2021:5928232. https://doi.org/10.1155/2021/5928232

Tsang ASMWP, Bultink IEM (2021) New developments in systemic lupus erythematosus. Rheumatology (Oxford) 60(Suppl 6):vi21–vi28. https://doi.org/10.1093/rheumatology/keab498

Vičić M, Hlača N, Kaštelan M, Brajac I, Sotošek V, Prpić Massari L (2023) Comprehensive in into lichen planus Immunopathogenesis. Int J Mol Sci 24(3). https://doi.org/10.3390/ijms24033038

15 Recent Insights on Dental Caries Microbiota in Pediatric Patients with Inborn Errors of Immunity

Gökhan Kars, Bushra Lutf Ahmed Alkebsi, Sevgi Keleş, Halenur Altan, Hazal Özer, Can Holyavkin, and Mehmet Ali Karaselek

Abstract

Inborn errors of immunity (IEIs) are genetic disorders that impair immune defense and regulation, increasing susceptibility to infections, including those in the oral cavity. The oral microbiota plays a vital role in maintaining oral health, and in pediatric patients with IEIs, disruptions in this balance can lead to dental caries and other oral diseases. This chapter provides a comprehensive analysis of the relationship between immune deficiencies and oral microbiota dysbiosis, focusing on dental caries in children with IEIs. Omics technologies, particularly metagenomics, have enhanced understanding of the microbial diversity and metabolic activities within the oral microbiota of the patients. Key findings reveal that compromised immune responses in children with IEIs disrupt the balance of oral bacteria, making them more prone to dental caries. The chapter highlights the importance of an interdisciplinary approach, integrating microbiology, immunology, dentistry, and bioinformatics, to uncover the complex interactions between the oral microbiome and the immune system. The insights gained from this research will contribute to the development of personalized therapeutic strategies, improving the dental and overall health of pediatric patients with IEIs.

Keywords

Dental caries · Inborn errors of immunity (IEIs) · Metagenomics · Microbiota dysbiosis · Oral microbiota · Pediatric immunology

15.1 Introduction

15.1.1 Overview of the Inborn Errors of Immunity (IEIs)

Inborn errors of immunity (IEI) are disorders resulting from germline mutations in individual genes that play crucial roles in immune defense and regulation. Genetic mutations that result in haploinsufficiency, dominant negative (DN), loss of function (LOF), or gain of function (GOF) activity can lead to completely different immunologic and clinical phenotypes (Bousfiha et al. 2022; Tangye et al. 2022). According to the

G. Kars (✉) · B. L. A. Alkebsi
Faculty of Science, Department of Molecular Biology and Genetics, Necmettin Erbakan University, Konya, Türkiye
e-mail: gkars@erbakan.edu.tr

S. Keleş · M. A. Karaselek
Faculty of Medicine, Division of Pediatric Allergy and Immunology, Necmettin Erbakan University, Konya, Türkiye

H. Altan · H. Özer
Faculty of Dentistry, Department of Pediatric Dentistry, Necmettin Erbakan University, Konya, Türkiye

C. Holyavkin
Gen Era Diagnostics, İstanbul, Türkiye

N. Rezaei, H. Ziaei (eds.), *Oral Immunology*, Advances in Experimental Medicine and Biology 1492,
https://doi.org/10.1007/978-3-032-03176-1_15

Table 15.1 Current IEI classification according to the IUIS

Classification of IEIs
1. Combined immunodeficiencies.
2. Combined immunodeficiencies with syndromic features.
3. Predominantly antibody deficiencies.
4. Diseases of immune dysregulation.
5. Congenital defects of phagocytes.
6. Defects in intrinsic and innate immunity.
7. Autoinflammatory diseases.
8. Complement deficiencies.
9. Bone marrow failure.
10. Phenocopies of inborn errors of immunity.

Tangye et al. (2022)

2022 classification of the International Union of Immunological Societies (IUIS) Expert Committee (Table 15.1) (Tangye et al. 2022).

Although IEIs have previously been considered rare, recent epidemiological studies suggest that many IEIs have not yet been diagnosed. Based on a study conducted in 2022, 499 distinct disorders have been documented, with 485 identified gene defects affecting the development and/or function of the immune system (Bousfiha et al. 2022). Middle Eastern countries show a higher prevalence of IEIs than other countries in East Asia because of the higher percentage of parental consanguinity (Modell et al. 2018). The prevalence of IEI varies depending on geographic area and the use of neonatal screening programs, with an estimated incidence of approximately 1:10.000 to 1:50.000 worldwide (Abolhassani et al. 2020; Tangye et al. 2022). Recently, the Jeffrey Modell Foundation (JMF) established a global network of specialized centers aimed at improving the identification, diagnosis, treatment, and monitoring of IEI. This network comprises 792 physicians across 358 institutions in 277 cities across 86 countries. The study found that IEIs may occur more frequently than previously estimated worldwide. They suggest that 1% of the population may be affected by IEIs when all types and varieties are considered. (Modell et al. 2018). All of these results show that there are many IEIs in the world that have not yet been identified in children and adults.

Globally, 58% of patients with IEI are male, whereas 42% are female. In the United States, the distribution of male to female patients was similar, with 56% of patients being male and 44% being female. In terms of age, 64% of patients worldwide are 19 years old or younger, whereas 36% are 20 years old or older (Modell et al. 2018). Kilic et al. also evaluated the characteristics of 1435 patients with IEIs from two centers in Turkey. In this study, 94% of the patients were aged between 0 and 18 years, with a mean age of 9.2 ± 6 years (Kilic et al. 2013). Although IEIs were previously known as childhood diseases, advances in the field of genetics have revealed that these diseases are very common in adults. Although IEIs are generally diagnosed at an early age, the number of cases genetically diagnosed with IEI during adulthood has recently increased (Karabiber and Barıs 2024). Moreover, 50% of newly identified cases worldwide are reported in patients over 25 years of age (Bousfiha et al. 2013). Although existing data provide insights into the distribution of IEIs in adulthood, detailed studies on the progression of the disease during this period remain insufficient (Karabiber and Barıs 2024).

15.1.2 Significance of Oral Health in IEI Patients

The oral flora includes bacteria, fungi, and viruses (Dewhirst et al. 2010). A balanced oral microbiota in healthy individuals helps prevent the growth of pathogenic microorganisms, thus contributing to overall health (Dewhirst et al. 2010; Napiórkowska-Baran et al. 2024). Moreover, healthy teeth are essential for balanced nutrition in children and play a significant role in the development of speech skills. A healthy smile also positively affects children's social interactions and self-confidence. Dental caries arise from pH imbalances caused by the metabolic activity of the dental biofilm. Approximately 700 phylotypes of microorganisms are found in the oral cavity (Peacock et al. 2017). These microorganisms are linked to various oral diseases, such as dental caries, gingivitis, periodontitis, candidiasis, endodontic infections, orthodontic infections, and peri-implantitis. Studies on biofilm, utilizing bacterial DNA sequencing, have identified that over 40 bacteria are actively involved in the development of dental caries, and this number continues to rise (Peacock et al. 2017). The most common bacterium implicated in dental caries is *Streptococcus mutans* (Lin et al. 2021).

In patients with IEI, maintaining oral health is of greater importance due to their immune system, which makes them more susceptible to infections, including oral cavity infections (Al-Kebsi et al. 2023; Tangye et al. 2022; Yin et al. 2001). Untreated dental caries or periodontal diseases can serve as reservoirs for chronic infections, which can spread to other parts of the body, further complicating the clinical management of IEI patients (Peacock et al. 2017). Immunodeficiency increases the susceptibility to oral health issues, with various oral findings potentially impacting oral health in individuals with congenital or acquired immunodeficiencies, such as gingivitis, dental caries, periodontitis, dental calculus, mucosal ulcers, oral candidiasis, enamel defects, and missing teeth (Chaiboonyarak et al. 2024; Al-Kebsi et al. 2023; Peacock et al. 2017). These symptoms highlight the complex relationship between immune deficiencies and oral health, underscoring the need for specialized dental care for patients with IEI.

15.1.3 Purpose and Scope of the Study

The main objective of this chapter was to provide a multi-dimensional insight into the microbiota of dental caries in pediatric patients with IEIs. In line with this purpose, this chapter addresses the technical and medical viewpoints of molecular biology, bioinformatics, pediatric dentistry, and pediatric immunology. The first focus will be on the descriptive and clinical characteristics of IEIs, subsequently addressing the pathophysiological mechanisms that connect IEIs with oral health. The healthy oral microbiota and any dysbiosis caused by IEIs will be discussed next, followed by the identification of important microbial taxa linked to caries in children with IEIs. Recent advances in microbiome research, particularly omics technologies, will be discussed from a molecular biology perspective. Bioinformatics tools and techniques for interpretation of metagenomic data will then be discussed. Following this, significant discoveries from recent studies on the microbiota of dental caries in pediatric patients with IEIs will be highlighted. Finally, clinical implications and future directions will be addressed, with recommendations for dentists and healthcare providers based on the latest microbiome research. In addition, key questions and hypotheses for future investigations into the oral microbiome are proposed. Figure 15.1

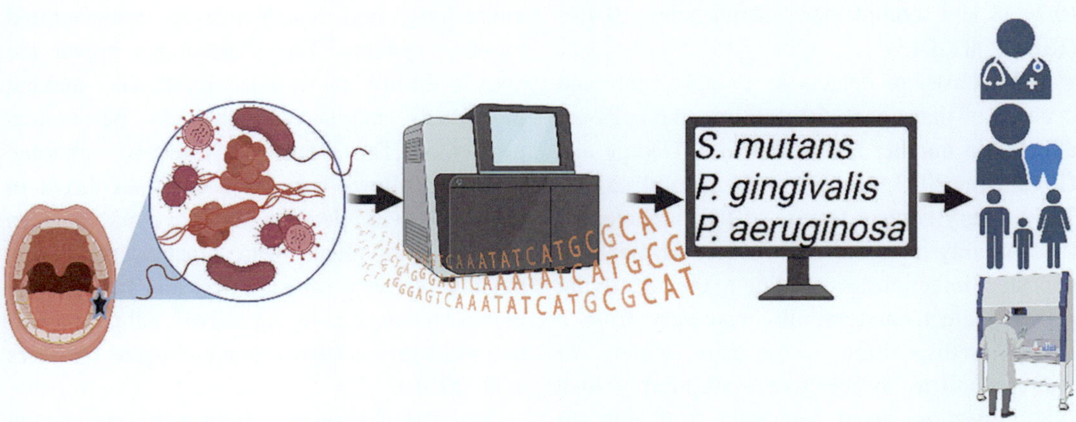

Fig. 15.1 An illustration showing the workflow of oral microbiome analysis. Oral samples are sequenced to identify bacterial species such as *Streptococcus mutans*, *Porphyromonas gingivalis*, and *Pseudomonas aeruginosa*, providing insights for both clinical and research applications. (This figure is created with BioRender)

provides a summary of the key concepts discussed in this chapter.

15.2 IEIs and Oral Health

15.2.1 Common IEIs, Clinical Features, and Systemic Impacts

Predominantly antibody deficiencies (PAD) are the most frequent IEI with a ratio of 40–57% followed by combined immunodeficiencies (CID) with associated syndromic features (12–16%) and unspecified or other deficiencies (10–17%). Selective IgA deficiency is the most prevalent PAD, followed by unspecified hypogammaglobulinemia and hyper-IgM syndrome, which are common variable immunodeficiency (CVID) according to Model's study (Modell et al. 2018). This study also evaluated the prevalence of IEIs between 2013 and 2018. They found that the ratio of IEIs in 2018 was similar to those in 2013. Another study from Turkey similarly demonstrated that predominantly antibody immunodeficiency (73.9%) was the most common IEI, followed by autoinflammatory disorders (13.3%), other well-defined immunodeficiencies (5.5%), congenital phagocyte defects (3.5%), combined T and B cell immunodeficiencies (2%), innate immunity defects (1%), immune dysregulation disorders (0.7%), and complement deficiencies (0.4%) (Kilic et al. 2013).

IEIs consist of defects in a single gene, and because of these gene defects, immune cells are affected in number and/or function (George and Govindaraj 2023; von Hardenberg et al. 2024). The primary clinical feature of IEIs is increased susceptibility to recurrent, severe, atypical, or persistent infections. According to the underlying genetic defect, susceptibility can vary from a strong predisposition to a large variety of microorganisms to selective susceptibility to a single infectious agent. Infections at all sites are the main presentation, including acute lower respiratory infections, followed by sinusitis, otitis media, and gastrointestinal infections (Pac et al. 2017). Central nervous system infections, such as meningitis, meningoencephalitis, progressive multifocal leukoencephalopathy, and chronic lymphocytic meningitis, which are usually caused by viruses, are the findings encountered in IEIs and can cause serious mortality or morbidity in patients (George and Govindaraj 2023). Patients with severe T cell defects or Mendelian susceptibility to mycobacterial infections may also present with persistent or treatment-resistant mucocutaneous infections or disseminated BCG infections (Pieniawska-Śmiech et al. 2022; Biggs et al. 2017). The clinical spectrum of IEIs is not limited to infections. The diagnosis includes a large range of clinical findings associated with immune dysregulation, such as allergy, lymphoproliferation, malignancies, and autoimmunity/auto-inflammation (Bousfiha et al. 2022). Vasculitis, autoimmune hemolytic anemia, and autoimmune thrombocytopenia are the most common manifestations of immune dysregulation (Tangye et al. 2022). As a result of these large variabilities in clinical findings, the term "primary immune deficiencies" appears to not account for congenital immune system defects and instead uses the terminology of "IEIs" (Tangye et al. 2022).

15.2.2 Oral Manifestations of IEIs

Patients with IEIs experience unique obstacles in maintaining oral health due to compromised immune systems. These conditions impair the body's ability to combat infections, making individuals more vulnerable to periodontal problems. The lack of effective immune responses allows harmful bacteria to thrive in the oral cavity, leading to decay and cavities even in individuals who practice good oral hygiene. Salivary function may also be impaired in these patients, reducing saliva's natural protective mechanisms against oral pathogens (Al-Herz et al. 2014).

Periodontal disease is frequently seen among IEI patients (del Rosario Rioboo-Crespo et al. 2005). Chronic inflammation and periodontal infection can escalate quickly due to the immune system's inability to respond adequately. This can result in severe gum disease, characterized by deep pockets between the teeth and periodontal tissues, loss of attachment, and, in advanced cases, tooth mobility, or loss (Tozum et al. 2003).

Oral candidiasis is another prevalent issue in IEI patients (Meighani et al. 2011). The compromised immune system allows for the overgrowth of *Candida* species in the oral cavity, leading to painful lesions, discomfort, and difficulty in eating or swallowing (Sampaio et al. 2012). Moreover, patients with IEIs may experience delayed wound healing after dental procedures. Their reduced ability to heal can lead to prolonged recovery times and an increased risk of complications, such as infections at the site of extraction or surgical interventions. Herpetic lesions and bacterial infections are also common in IEI patients. The interplay between oral and systemic conditions underscores the importance of interdisciplinary care.

The development of dental caries is influenced by various factors, including plaque microorganisms, dietary habits, and oral hygiene (Jagathrakshakan et al. 2015). Dental caries is caused by many different factors, such as plaque microflora, the immune system, carbohydrate intake, and oral hygiene (Jagathrakshakan et al. 2015).

In addition to dental caries, patients with IEIs experience a range of other oral manifestations that can significantly impact their overall health. Recurrent oral ulcers, severe gingivitis, and advanced periodontitis are commonly observed in these individuals due to chronic inflammation and impaired immune responses. Viral lesions, such as those caused by herpes simplex virus, are more frequent and can be particularly painful, affecting the patient's quality of life. Fungal infections, especially oral candidiasis, often present as persistent and recurrent issues, with painful lesions and difficulty in eating or swallowing. Delayed wound healing following dental procedures is another hallmark in IEI patients, increasing the risk of secondary infections and systemic complications originating from oral infections. These conditions not only exacerbate oral health problems but may also contribute to systemic health risks, such as bacteremia or sepsis in severely immunocompromised individuals. This underscores the critical importance of comprehensive oral health management and interdisciplinary care to mitigate the broader health implications of these oral manifestations (Napiórkowska-Baran et al. 2024).

The immune response can cause structural and functional changes in local oral biofilms. These changes can alter the stability of the balance between the host and resident microbiota, thereby triggering the initiation and progression of periodontitis and dental caries (Belstrøm et al. 2017). For instance, research has found that the prevalence of dental caries in children with IEIs is alarmingly high, reaching 60–80%, whereas the prevalence in healthy children typically ranges from 20% to 40%. Additionally, the severity of caries in IEI patients is often more significant, with studies showing higher rates of untreated caries and increased incidences of severe forms, such as rampant caries (Klingberg et al. 2002). These children may experience more extensive dental decay that often requires restorative procedures at an earlier age than healthy children. Dahlén et al. demonstrated that hypogammaglobulinemia and IgG subclass deficiencies had no significant impact on dental health (Dahlén et al. 1993). Furthermore, the early onset of dental caries in IEI patients can be linked to frequent hospitalizations and disruptions to regular dental care. The prevalence of dental caries and oral mucosal lesions (OML) among adolescents living with HIV (ALHIV) in Johannesburg is high. The reported prevalence was associated with high HIV RNA viral loads, shorter duration on antiretroviral treatment, and high WHO staging of HIV disease on crude analysis. Additionally, the experience of dental caries contributed to the prevalence of OML (Malele Kolisa et al. 2019). In summary, the data clearly indicate that children with primary immune deficiencies have a significantly higher prevalence and severity of dental caries than their healthy counterparts. This underscores the crucial need for targeted preventive measures and ongoing dental care for this vulnerable population, which we, as healthcare professionals, must prioritize and act upon.

15.2.3 Pathophysiological Mechanisms Linking IEIs and Oral Health

The immune system consists of an innate and adaptive immune system. Innate immune system defects such as chronic granulomatous disease (CGD) and neutrophil function compromise, resulting in an inability to effectively kill bacteria and fungi. This leads to chronic infections in the oral cavity, particularly with pathogens like *Candida albicans* and oral bacteria. Neutrophil dysfunction allows pathogens to persist in gingival tissues, resulting in chronic inflammation, gingivitis, and, eventually, periodontal destruction (Szczawinska-Poplonyk et al. 2009; Metcalfe et al. 2022). Persistent infections and delayed healing contribute to progressive gum disease and tooth loss (Casanova and Abel 2007). IEIs can also affect adaptive immune responses, particularly through T-helper (Th) cell dysfunction. Th17 cells play a crucial role in maintaining mucosal immunity, including oral tissues, by coordinating responses against bacterial and fungal infections. In cases where Th17 function is impaired (e.g., hyper IgE syndrome), the body's inability to control oral microbial communities is diminished, leading to microbial dysbiosis (Moutsopoulos and Konkel 2018). This imbalance allows pathogenic bacteria to overgrow and persist, leading to chronic inflammation of the gingiva, periodontal tissues, and systemic circulation. Furthermore, excessive production of pro-inflammatory cytokines, such as interleukin-1 (IL-1) and tumor necrosis factor-alpha (TNF-α), in IEI patients exacerbates tissue destruction. These cytokines promote the breakdown of periodontal ligaments and alveolar bone, accelerating the progression of periodontal disease (Graves and Cochran 2003). IEIs may also impact salivary gland function, leading to xerostomia. Saliva plays a critical role in maintaining oral health by providing enzymes that break down food particles, neutralize acids, and inhibit bacterial growth. Under conditions in which salivary flow is diminished, oral pathogens such as *Streptococcus mutans* thrive, promoting dental caries (Humphrey and Williamson 2001). Saliva also contains key antimicrobial proteins (e.g., lysozyme, lactoferrin) and IgA, which help control oral microbial populations. Reduced salivary output compromises this protective mechanism, allowing for unchecked bacterial growth and increasing the risk of caries and other oral infections (Cunha-Cruz et al. 2013).

Macrophages play a dual role in the formation or prevention of tooth and periodontal diseases through various polarized activities. Pro-inflammatory M1 macrophages cause inflammation and alveolar bone loss by secreting pro-inflammatory cytokines (IL-1β, IL-1α, IL-6, TNFα, IL-11) and matrix metalloproteinases (MMPs) (Metcalfe et al. 2022). M2 macrophages, which are activated via alternative pathways, play a role in suppressing inflammation and tissue regeneration (Lam et al. 2014).

Recent studies involving patients with specific gene disruptions have highlighted the importance of IL-17 signaling in oral immune surveillance and the risk of inflammatory disease. Defective production of Th17 cells in patients with STAT3 mutations, gain-of-function STAT1 mutations, or impaired IL-17/IL-22 signaling—due to mutations in the IL-17 receptor or ligand, or the presence of neutralizing autoantibodies against IL-17/IL-22—is associated with increased susceptibility to mucocutaneous candidiasis, including oral infections. Alterations in the IL-17/IL-23 axis are significantly associated with increased susceptibility to mucosal fungal infections, whereas excessive activation of these pathways is correlated with inflammatory periodontitis. Under normal conditions, neutrophils in tissue regulate the production of IL-23 by antigen-presenting cells (APCs) and Th17 cell proliferation. In conditions such as leukocyte adhesion deficiency type I (LAD-I), neutrophil migration is impaired and the IL-23/Th17 axis is disrupted. As a result, excessive IL-17 is produced, and inflammation increases excessively. Increased IL-17 levels cause tissue breakdown in the inflammatory region and tissue fragments formed because of inflammation serve as nutrients for microorganisms. Proliferating pathogenic

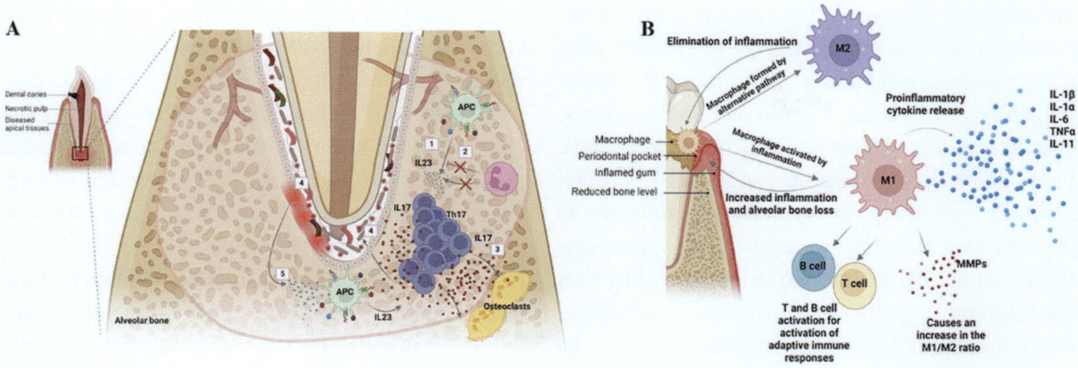

Fig. 15.2 (**a**) Th17 cell-mediated immune response in the oral cavity and the mechanism of periodontitis formation (*1*) (*2*) excessive IL-17 production and inflammation (*3*) tissue breakdown in the inflammatory region (*4*) tissue fragments formed because of inflammation, which serve as nutrients for microorganisms (*5*) periodontitis. (**b**) Macrophages in the oral cavity and their roles (*APC* antigen-presenting cells, *M1* type 1 macrophage, *M2* type 2 macrophage, *MMPs* matrix metalloproteinases). (This figure created by BioRender)

microorganisms cause excessive IL-23 production by APCs. Periodontitis occurs through this mechanism (Fig. 15.2) (Moutsopoulos and Konkel 2018).

15.3 Oral Microbiome and Dental Caries

15.3.1 Healthy Oral Microbiome

Healthy oral microbiome, a complex community of microorganisms, primarily bacteria, that inhabit the oral cavity, plays a key role in disease prevention. A typical composition includes diverse genera, such as *Streptococcus*, *Actinomyces*, *Veillonella*, *Fusobacterium*, and *Prevotella*. These microorganisms coexist in a balanced state and play essential roles in maintaining oral health, thereby empowering individuals to take proactive steps to prevent disease (Sedghi et al. 2021). One of the primary functions of the oral microbiota is to act as a protective shield against pathogenic bacterial overgrowth. Beneficial bacteria, such as *Streptococcus sanguinis*, compete for resources and space, effectively limiting the colonization of harmful pathogens that cause infections, such as dental caries and periodontal disease. This reassures us of the robustness of our oral health system (Rajasekaran et al. 2024). The oral microbiome also plays a crucial role in modulating host immune responses. A balanced microbiome can promote a healthy inflammatory response and enhance the mucosal barrier against infections, thereby providing a sense of security in oral health (Cho et al. 2021).

Several factors influence the oral microbiota balance, with diet, hygiene practices, and genetics being the most significant. The diet plays a critical role in shaping the oral microbiome. High sugar intake can increase acid-producing bacteria, promoting a shift towards a cariogenic environment. In contrast, a diet rich in fiber and nutrients can support the growth of beneficial bacteria, enhancing overall microbial diversity and health. Oral hygiene practices are also crucial for maintaining microbial balance (Deo and Deshmukh 2019). Regular brushing and flossing help remove plaque and prevent harmful bacterial accumulation. Inadequate oral hygiene can lead to dysbiosis, a condition where there is an imbalance in the oral microbiota, with harmful bacteria proliferating and beneficial bacteria decreasing, resulting in oral diseases like gingivitis and periodontitis. Genetics can influence an individual's susceptibility to certain oral conditions and their overall microbial composition. Genetic factors may affect salivary flow, immune responses, and the body's inflammatory reactions, all of which affect the stability and diversity of the oral microbiome (Cabay 2014). Additionally, environmental factors, such as exposure to antibiotics, mouthwashes containing antimicrobial agents like chlorhexidine, and certain oral care products, can disrupt the balance of oral microorganisms. Frequent use of these agents can reduce microbial diversity, allowing opportunistic pathogens to flourish and leading to oral health problems (Kilian et al. 2016). In conclusion, a healthy oral microbiome comprises diverse beneficial microorganisms that work together to maintain oral health. Various factors, including diet, oral hygiene practices, and genetics, influence the balance of the microbiome. Understanding these dynamics is essential for developing effective strategies to promote oral health and prevent diseases.

15.3.2 Microbiota Dysbiosis in Patients with IEIs

A major factor contributing to dysbiosis in IEI patients is the altered salivary function. Saliva, a vital component of maintaining oral health, provides antimicrobial agents and aids in the removal of food particles and bacteria. In individuals with IEIs, reduced saliva flow diminishes these protective mechanisms, leading to the proliferation of harmful microorganisms (Vila et al. 2019). This overgrowth can lead to conditions such as dental caries and periodontal disease, underscoring the urgency of addressing this issue (Kilian et al. 2016; van't Hof et al. 2014). Furthermore, impaired immune response in IEI patients affects their ability to combat infections. As beneficial bacteria decline, opportunistic pathogens such as *Candida* species and certain bacteria associated with periodontal disease can flourish. This shift exacerbates oral health issues and can lead to serious systemic complications because oral pathogens may enter the bloodstream, highlighting the gravity of the situation (Padron and Hernandez-Trujillo 2023). Additionally, IEI patients often undergo frequent medical treatments, including antibiotics and immunosuppressive therapies, which can further disrupt their oral microbiome. Although antibiotics target harmful bacteria, they can also eliminate beneficial strains, fostering an environment conducive to dysbiosis. In summary, primary immune deficiencies lead to significant changes in the oral microbiota owing to impaired immune responses, altered saliva production, and the impact of medical treatments. These changes increase the risk of dysbiosis, increase the likelihood of oral health issues, and complicate overall health management in these patients. This underscores the importance of a holistic approach to health management in IEI patients, where every aspect of their health, including oral health, is carefully considered and managed (Rezaei et al. 2017).

Children with IEI have distinct oral microbiota due to immune dysfunction, making them more susceptible to microbial imbalances that increase the risk of oral diseases like dental caries (Costalonga and Herzberg 2014). A comparative analysis of the oral microbiota between healthy children and those with IEIs revealed several key differences. Studies have shown that healthy children typically have a more diverse and balanced oral microbiota, with beneficial bacteria like *Streptococcus sanguinis* and *Streptococcus cristatus* playing a protective role in maintaining oral health (Chaiboonyarak et al. 2024). In contrast, children with IEIs often exhibit a less diverse microbiota, with an overrepresentation of pathogenic species such as *Streptococcus mutans* and *Lactobacillus* species, both of which are well-known contributors to dental caries (Naafs 2018; Santacroce et al. 2023). The reduced microbial diversity observed in IEI patients can be attributed to their compromised immune systems, which are less effective at controlling bacterial populations and maintaining homeostasis (Abusleme et al. 2018).

Further differences in the oral microbiota between these two groups were observed in the prevalence of certain keystone pathogens. Children with IEIs, particularly those with periodontal diseases, are more likely to harbor bacteria such as *Porphyromonas gingivalis*, *Fusobacterium*, and *Prevotella*, which are associated with periodontitis and other oral infections (Nibali et al. 2021; Al-kebsi et al. 2023). These pathogenic bacteria are less commonly found in the oral cavities of healthy children, highlighting a significant shift in the microbial landscape in IEI patients (Moyer et al. 1983; Li et al. 2022).

Age and the nature of the immune deficiency also affect the oral microbiota of children with IEIs. For instance, children with antibody deficiencies, such as X-linked agammaglobulinemia (XLA), often exhibit higher levels of *Lactobacillus* and yeast, which are associated with caries development and oral infections (Kıykım et al. 2013). In contrast, healthy children tend to have lower levels of these microorganisms, which underscores the impact of immune function on microbial composition.

15.3.3 Microbiota and Cariogenesis

Alterations in the oral microbiota can significantly contribute to developing dental caries through several interconnected mechanisms. A healthy oral microbiome is characterized by diverse beneficial bacteria that help maintain balance and protect against pathogenic organisms. When this balance is often disrupted by factors such as poor oral hygiene, high sugar intake, or immune deficiencies, dysbiosis can occur, favoring the growth of cariogenic bacteria (Szczawinska-Poplonyk et al. 2009). One of the primary contributors to dental caries is the overgrowth of acid-producing bacteria, particularly *Streptococcus mutans* and *Lactobacillus* species. These bacteria thrive in environments with high sugar availability and ferment carbohydrates to produce organic acids. The accumulation of these acids leads to a drop in the pH of the dental biofilm, creating an acidic environment that demineralizes tooth enamel. As enamel erosion progresses, cavities are created and further bacterial colonization occurs. In a dysbiotic state, the reduced abundance of beneficial bacteria diminishes the mouth's natural defenses (Luo et al. 2024). Beneficial microbes, such as *Streptococcus sanguinis*, play a crucial role by producing antimicrobial substances and competing for resources, effectively inhibiting the growth of harmful bacteria, and stressing the importance of maintaining a diverse oral microbiome. When these beneficial strains are diminished, the risk of cariogenic bacteria thriving increases, leading to a higher incidence of dental caries (Zheng et al. 2017). Additionally, dysbiosis can compromise the host's immune response; a weakened immune system, whether due to systemic conditions or altered microbial communities, can reduce the body's ability to combat infections and inflammation in the oral cavity. This creates a vicious cycle in which cariogenic bacteria not only lead to dental caries but also hamper the recovery of a balanced and healthy microbiome.

Pediatric patients with IEIs may exhibit unique alterations in their oral microbiota, which may influence the progression of dental caries (Moyer et al. 1983). For example, patients with common variable immunodeficiency (CVID) often display a higher prevalence of pathogenic bacteria, such as *Streptococcus mutans*, and a reduction in beneficial commensal species (Ghosal et al. 2024). Additionally, children with IgA deficiency and hypogammaglobulinemia typically do not exhibit severe oral pathology or craniofacial abnormalities. However, defects in phagocytic cells are associated with mucosal lesions or rapidly progressing periodontal disease. Children with T-cell and combined immunodeficiency disorders experience a more pronounced shift in their oral microbiota, further exacerbating their susceptibility to dental caries (Meighani et al. 2011; Szczawinska-Poplonyk et al. 2009).

Recent studies using high-throughput sequencing and metagenomic analyses have revealed microbial signatures associated with dental caries in pediatric IEI patients (Nibali et al. 2021; Jung et al. 2020). For instance, children with IEIs were found to have a distinct oral microbial profile characterized by an overrepresentation of cariogenic bacteria and a decrease in microbial diversity (Halai et al. 2020). The functional potential of the oral microbiota in these patients was skewed toward carbohydrate metabolism and acid production, contributing to an environment conducive to caries formation (Meighani et al. 2011).

A comparative analysis of the microbiota of healthy children versus those with IEIs, particularly autosomal dominant hyper-IgE syndrome (AD-HIES), revealed significant differences. In patients with AD-HIES, a marked reduction in microbial diversity was observed, especially in cases of oral candidiasis. Fungal communities in healthy individuals were predominantly balanced by *Malassezia*, whereas patients with AD-HIES showed severe dysbiosis characterized by an overabundance of *Candida albicans* associated with recurrent oral candidiasis (Abusleme et al. 2018). Additionally, the bacterial composition in children with IEIs showed a significant decrease in common commensal bacteria like *Streptococcus*, which are typically enriched in healthy individuals.

Key pathogens involved in dental caries in children with IEIs include *Streptococcus mutans*, a well-established contributor to dental caries. In children with neutrophil defects or CVID, pathogenic bacteria such as *Pseudomonas aeruginosa*, *Fusobacterium* spp., and *Prevotella* spp. are often associated with severe forms of periodontal disease and necrotizing ulcerative periodontitis, conditions that may also exacerbate caries risk (Szczawinska-Poplonyk et al. 2009). Furthermore, studies have shown that emergent pathobionts, including *Lachnospiraceae* spp., *Streptococcus mitis*, and *Streptococcus pneumoniae*, play a significant role in the carious process in immunocompromised patients (Chaiboonyarak et al. 2024). The presence of acidogenic and aciduric bacteria, such as *Veillonella*, *Neisseria*, and *Granulicatella*, which thrive in acidic environments, further exacerbates caries development in these children (Kıykım et al. 2013).

15.4 Recent Research Insights

15.4.1 Advances in Microbiome Research Technologies

This section of the chapter discusses advanced sequencing technologies that have profoundly transformed biological research, including microbiological techniques. Three generations of sequencing technologies have evolved since 1977 (Satam et al. 2023). Walter Gilbert and Allan Maxam devised a sequencing technique based on chemical degradation, and a significant advancement came with Fredrick Sanger's development of the chain termination sequencing method. Sanger's technique, a first-generation sequencing technology, continues to be employed to sequence several hundred bases, but complicated biological and clinical problems that require processing of large data sets can only be studied by second- and third-generation methods. Next-generation sequencing (NGS) enables rapid and high-throughput sequencing at reduced costs because of its massively parallel processing of DNA and RNA molecules in a single run. Metagenomic research can identify hundreds of microbial species in any region without the necessity of microbiological culture procedures, facilitated by advanced sequencing technology. In rRNA and Internal Transcribed Spacer (ITS)-based metagenomic approaches, relatively short reads are sufficient to obtain an operational taxonomic unit, or OTU as the basic unit of taxonomy. Therefore, all NGS platforms can be used to classify microorganisms, including bacteria, archaea, and fungi. For whole-genome sequencing, population genomics, metagenomics, and transcriptome studies, relatively long-read sequencing technologies may be advantageous (Table 15.2). Illumina has emerged as the predominant force in the second-generation sequencing landscape, setting the standard for high-throughput, accurate, and cost-effective genomic analysis. Its platforms support a variety of applications such as metagenomics, RNA Sequencing, exome and targeted sequencing, whole genome sequencing, ChIP-seq, and methylation sequencing, establishing Illumina as a key player in both research and clinical settings. Illumina sequencing platforms exhibit varying output capacities and runtimes tailored to different research needs. The MiniSeq system provides an output range of 1.8–7.5 Gb with a runtime of 4–24 h, while the MiSeq series offers 0.3–15 Gb and runs for 4–55 h. The NextSeq 500 generates 20–120 Gb in 11–29 h, and the HiSeq 4000

Table 15.2 Overview of sequencing technologies

Platform and instrument	Sequencing principle	Read/run	Read length (base)	Sequencing capacity (Gb)	Applications	References
1st gen.						
ABI Sanger	Chain termination-based sequencing using dideoxy nucleotides to terminate DNA synthesis during replication	96	1000	0.00069–0.0021	Short DNA sequencing	Sanger et al. (1977)
2nd gen.						
454 Life Sciences GS FLX	Detection of chemiluminescence from pyrophosphate released during the incorporation of dNTPs into DNA	1×10^6	400	0.5	Whole-genome sequencing and metagenomic	Slatko et al. (2018), Voelkerding et al. (2009), and Gilles et al. (2011)
Illumina NovaSeq 6000	Sequencing by synthesis using fluorescently labeled reversible terminators enabling the massively parallel sequencing of billions of DNA fragments	2×10^{10}	150	6000	Whole-genome sequencing, whole exome sequencing, metagenomics and transcriptomics	Peng et al. (2023), Blaustein et al. (2023), Huang et al. (2024), and NovaSeq (6000) System (2024)
Ion Torrent GeneStudio S5	Sequencing by synthesis using semiconductors which measure hydrogen ions released after dNTP incorporation to a DNA polymer	1.3×10^8	200	50	Whole-genome sequencing, whole-exome sequencing, targeted sequencing, metagenomics and transcriptomics	Meslier et al. (2022), Rawat et al. (2022), Ari et al. (2019), and Iontorrent application note (2024)
3rd gen.						
PacBio Revivo	Single-molecule, real-time (SMRT) sequencing detects fluorescently labeled nucleotides incorporated into circularized DNA molecules within zero-mode waveguides (ZMWs)	6×10^6/ SMRT cell	2×10^4	90/SMRT cell	Whole-genome sequencing, targeted sequencing, metagenomics and transcriptomics	*PacBio revio system* (2024)
Oxford Nanopore PromethION	Individual DNA or RNA molecules traverse nanopores, resulting in perturbations in ionic current that are monitored in real-time to ascertain the nucleotide sequence, facilitating ultra-long read lengths	400 bases/s	4×10^6	200/ PromethION flow cell	Whole-genome sequencing, population genomics, metagenomics and whole transcriptome	PromethION (2024) and Yamaguchi et al. (2023)

produces 125–1500 Gb in less than 1–3.5 days. The NovaSeq series, with an output range of 167–6000 Gb, processes data in 19–40 h, making it suitable for large-scale genomic projects (NovaSeq 6000 System 2024). The Oxford Nanopore PromethION sequencing technology is capable of generating ultra-long read lengths, often exceeding hundreds of kilobases, with the potential to reach up to four million bases, making it particularly well-suited for population genomics, whole-transcriptome analyses, and a wide range of other applications.

NGS has emerged as the primary driving force behind omics studies such as genomics, transcriptomics, and metagenomics. Its application has provided deeper insights into the diversity, function, and complex interactions of microbial communities. Metagenomic studies, in particular, have revealed the composition of microbiota in various body parts, including the oral cavity, dental caries, gut, and skin, while also enabling the identification of previously unculturable bacterial species. Furthermore, NGS has shed light on unique bacterial associations, advancing our understanding of microbial ecosystems and their roles in health and disease (Al-Kebsi et al. 2023). Transcriptomics, the analysis of RNA transcripts within a sample, uncovers the active gene expression profiles, offering valuable insights into metabolic processes occurring within the given environment. Proteomics and metabolomics elucidate the protein and metabolite profiles of a sample, thereby complementing and enhancing the insights gained from metagenomic and transcriptomic analyses (Su et al. 2023; Macowan et al. 2024). As omics technologies continue to evolve, they will provide deeper insights into complex biological challenges, potentially leading to the development of more targeted therapeutic strategies.

15.4.2 Metagenomic Approaches to Study Oral Microbiome

The oral cavity harbors one of the most diverse microbial populations in the human body, which is essential for systemic and oral health. Recent developments in bioinformatics and sequencing technologies have enhanced our ability to investigate this complex ecology. With an emphasis on two main sequencing methods and the resulting data analysis methodologies, this section of the chapter examines the metagenomic approaches used to study oral microbiota. The specifics of these methods, their use in oral health research, and difficulties in interpreting the results will be discussed.

15.4.2.1 16S rRNA Gene-Based Metagenomic Sequencing

In investigations of microbial ecology, especially those focused on the oral microbiome, the 16S rRNA gene-based metagenomic sequencing approach has emerged as a key component. This method uses variable yet globally conserved regions of the 16S rRNA gene to identify and categorize bacterial species. A mixed microbial sample's 16S rRNA gene is amplified in particular areas, and these amplicons are then sequenced at high throughput. For taxonomic classification, the generated sequences are further matched to databases of known bacterial 16S rRNA sequences.

This approach has a number of benefits for research on the oral microbiome. First and foremost, it offers a thorough understanding of the variety of bacteria, including some that are common in the oral cavity but not cultured (Aas et al. 2005). A study involving over 3900 oral samples identified 56,213 metagenome-assembled genomes (MAGs), with more than 64% containing no publicly available genomes, highlighting the richness of previously unidentified taxa within the oral microbiome (Zhu et al. 2022). This is especially crucial because many oral bacteria are difficult or impossible to cultivate using conventional techniques. By detecting previously unidentified bacteria and identifying rare species, the 16S rRNA method adds to our growing understanding of the diversity of oral microbes. Additionally, the method can process a lot of samples quickly and is relatively cost-effective (ranging from 30 to 70 USD per sample), making it appropriate for extensive epidemiological research. This has made it possible for scientists to study the oral microbiome in a range of demographics and health and illness conditions.

However, the 16S rRNA gene-based strategy does have several limitations. The technique is only specific in identifying bacteria, and it cannot identify other species, such as viruses or fungi, which are also significant components of the oral microbiome. Furthermore, some bacterial groups exhibit poor species-level identification resolution because of the high degree of similarity in their 16S rRNA sequences (Clarridge 2004). A study by Johnson et al. found that 16S rRNA sequencing could only reliably differentiate between bacterial species with less than 97% sequence similarity, potentially missing important distinctions within closely related species groups (Johnson et al. 2019). When researching closely related species that may have distinct effects on oral health, this may be difficult.

15.4.2.2 Shotgun Sequencing Approach

The shotgun sequencing technique provides a more thorough examination of the oral microbiome than the 16S rRNA method. This technique generates a picture of the complete genetic makeup of the microbial population by fragmenting all genomic material from a sample and sequencing the resulting pieces. Shotgun sequencing essentially records taxonomic and functional genetic data, providing information about the microbiome's metabolic capacity (Durazzi et al. 2021).

In shotgun sequencing, an oral sample's entire DNA is extracted, broken up, and then the fragments are sequenced. Advanced bioinformatics methods are then used to assemble and evaluate the resulting sequences. Compared to 16S rRNA gene-based techniques, this strategy has a number of advantages. It makes it possible to identify a wider variety of microorganisms, such as viruses, fungi, bacteria, and archaea, providing a more comprehensive picture of the oral microbiome.

Understanding the composition of the oral microbiome requires the ability to identify species and strains at greater resolution, which is the goal of shotgun sequencing. When researching the function of certain bacterial strains in oral health and illness, this level of precision might be especially crucial (Zhu et al. 2022). Additionally, shotgun sequencing makes functional analysis possible, providing insights into the microbial community's virulence factors, metabolic pathways, and antibiotic resistance genes (Hillmann et al. 2018). Understanding how the oral microbiota interacts with the host and reacts to environmental changes or therapeutic interventions is made much easier by this functional insight.

Despite its benefits, shotgun sequencing is typically more expensive and computationally demanding than 16S rRNA sequencing. More advanced bioinformatics analysis is required to analyze the large amount of data generated. Its application in contexts with low computational resources or large-scale studies may be restricted by these considerations.

15.4.3 Data Analysis and Interpretation

15.4.3.1 Bioinformatic Tools and Techniques

The examination and interpretation of metagenomic data from the oral microbiome research significantly depend on bioinformatics tools and methodologies. These computational methods are essential for processing large volumes of sequence data generated by both 16S rRNA and shotgun sequencing approaches. Bioinformatics pipelines typically involve several steps, including quality control of raw sequences, removal of host DNA contamination, assembly of shotgun sequencing reads (for shotgun data), taxonomic classification, and functional annotation (Navgire et al. 2022).

Popular bioinformatics tools for metagenomic analysis include MetaPhlAn and Kraken for shotgun sequencing data, and QIIME2 and MOTHUR for 16S rRNA data. These programs apply taxonomic labels, assess the abundance of diverse microbial species, and cluster comparable sequences using a variety of techniques. Tools like HUMAnN and KEGG are frequently utilized for the functional analysis of shotgun sequencing data to forecast metabolic pathways and functional capacities in the microbiome (Caporaso et al. 2010; Schloss et al. 2009).

The analysis of oral microbiome data has made use of recent advances in artificial intelligence and machine learning. These methods, including random forest algorithms and neural networks, can forecast the relationships between microbial communities and oral health outcomes and identify intricate patterns in microbiome data (Yan et al. 2022). These methods are especially helpful for managing high-dimensional and frequently noisy data that microbiome research produces.

15.4.3.2 Interpreting Microbiome Data in the Health and Disease Context

Integrating metagenomic results with clinical and environmental data is necessary to overcome the complicated challenge of interpreting microbiome data in relation to oral health and illness. Sample collection techniques, sequencing depth, and potential biases induced during sample processing and data analysis are all important considerations for researchers. Understanding the relationships between the human immune system and microbiome, identifying important taxa or functional pathways linked to certain oral diseases, and comparing the compositions of microbial communities in healthy and pathological states are common steps in the interpretation process. Differentiating between correlation and causation is a major challenge in the interpretation of oral microbiota data. Causal associations are not always implied when specific bacterial species are found in disease states. Establishing causal relationships between changes in the microbiome and oral health outcomes frequently requires longitudinal research and experimental paradigms.

Understanding the idea of the core microbiome (the collection of microbial species or genes that are regularly found in healthy people) is another crucial component of data interpretation. Determining variations in this fundamental microbiome can shed light on dysbiosis and its possible connection to oral illnesses (Zaura et al. 2009). Additionally, the changing nature of oral ecology must be considered when interpreting microbiome data. The composition of the oral microbiome can be influenced by several factors, including systemic health, dental hygiene habits, and food. A more thorough understanding of the interactions between the microbiome and host in oral health and disease can be obtained by incorporating these variables into the data analysis.

Our knowledge of the oral microbiome has greatly increased because of metagenomic techniques, which have previously provided unheard-of insights into its composition and function. Shotgun sequencing and 16S rRNA gene-based sequencing individually have advantages and disadvantages, but when combined, they offer a complete arsenal for examining the intricacy of oral microbial communities. Our capacity to analyze this large volume of data is being improved by the continuous development of bioinformatics tools and analytical methods, which will open the door to new understandings of oral health and illness. Metatranscriptomics and metabolomics will be essential for a more comprehensive understanding of the oral microbiome as the field develops by combining metagenomic data with other omics techniques. Furthermore, the use of machine learning algorithms and sophisticated statistical techniques can reveal more nuanced trends in microbiome data. The future of oral microbiome research lies in translating discoveries into practical applications, such as targeted therapies, personalized treatments, and microbiome-based diagnostics, which could revolutionize oral healthcare with more individualized and preventive approaches.

15.4.4 Key Findings from Recent Studies on the Oral Microbiota and Dental Caries in Pediatric IEI Patients

Pediatric IEI patients are highly prone to dental caries due to immune system deficiencies that alter their oral microbiota, increasing their vulnerability to opportunistic pathogens. Recent studies have provided insights into these microbial changes and dental caries risks. In a foundational study by Meighani et al. (2011), pediatric patients with Primary Antibody Deficiencies (PADs) were shown to have a higher risk of dental caries and other oral manifestations, such as aphthous ulcers, candidiasis, herpes sores, and enamel hypoplasia. These children exhibited significantly higher Decayed, Missing, and Filled Teeth DMFT scores than healthy controls, reflecting a greater prevalence of dental complications in this population. The study highlighted a distinct microbial profile in IEI patients, characterized by a higher abundance of opportunistic pathogens, likely due to decreased immunoglobulin levels. For example, children with Common Variable Immunodeficiency (CVID) demonstrated elevated, although not statistically significant, DMFT means when compared to other IEI groups. The microbial imbalance observed in IEI patients contrasts with the balanced composition of beneficial bacteria observed in healthy children, emphasizing the critical need for regular oral checkups and targeted hygiene interventions, such as the use of probiotics (Meighani et al. 2011).

A more recent study by Al-kebsi et al. expanded on previous research by comparing the oral microbiota of 15 pediatric IEI patients with dental caries with those of 15 healthy control children (Al-kebsi et al. 2023). Despite comparable DMFT scores and similar salivary buffering capacities between the groups, detailed microbial analyses revealed significant differences in bacterial composition. This study identified more than 2400 bacterial species across 1327 genera. While genera such as *Streptococcus*, *Prevotella*, *Actinomyces*, and *Veillonella* were prominent in both groups, specific opportunistic pathogens, including *Prevotella melaninogenica* ($p = 0.002$) and *Prevotella salivae* ($p = 0.004$), were significantly more prevalent in the IEI group, suggesting a potential association between these species and the immune disorder.

Further investigation of microbial diversity revealed no significant differences in alpha diversity between the two groups, as measured using the Shannon-Wiener and Simpson indices. However, beta diversity analysis, conducted using Principal Coordinate Analysis (PCoA), demonstrated a clustering of microbial profiles, particularly at the genus level. This clustering indicated a subtle yet notable shift in microbial composition between the IEI patients and healthy controls. The increased prevalence of *Prevotella* species in the IEI group underscores the presence of oral dysbiosis in these patients, reflecting an altered microbial environment that may contribute to their heightened susceptibility to dental caries and other oral health issues.

Correlation analysis in microbiota studies often uncovers significant associations between species. In this context, the Spearman test was applied to species with >3% abundance in the IEI subgroups. The analysis revealed that *P. oris* was positively correlated with both *S. wiggsiae* and *Saccharibacteria genera incertae sedis*. Notably, the correlation between *S. wiggsiae* and *Saccharibacteria* alone was even more significant (Al-kebsi et al. 2023). Additionally, studies have shown that members of *Saccharibacteria*, such as TM7x, tend to form close associations with other species, like *Actinomyces odontolyticus*, in the oral cavity. These findings suggest that interactions involving *Saccharibacteria* in patients with immunodeficiency warrant further investigation (Bor et al. 2018).

The oral microbiota of healthy children is typically balanced by beneficial bacterial species that promote oral health and prevent the overgrowth of pathogenic organisms. In contrast, pediatric IEI patients exhibit significant alterations in both microbial diversity and abundance, contributing to their increased risk of dental caries and other oral health issues. Meighani et al. also reported similar findings, where IEI patients had an oral microbial composition dominated by opportunistic pathogens, such as *Streptococcus mutans*, and a reduction in beneficial species, leading to an environment conducive to dental caries formation (Meighani et al. 2011). In contrast, healthy pediatric populations demonstrated a more balanced microbial community that included *Streptococcus*, *Lactobacillus*, and *Veillonella*, which are essential for maintaining oral health and preventing caries.

Although significant progress has been made in understanding the oral microbiota of pediatric IEI patients, several areas remain under-researched. First, the specific role of *Saccharibacteria*, particularly *TM7x*, in the progression of dental caries in IEI patients requires further investigation. Bor et al. (2018) identified close associations between *Saccharibacteria* and other oral species, such as *Actinomyces odontolyticus*, suggesting that these interactions may contribute to the progression of oral disease in IEI patients (Bor et al. 2018). More research is needed to elucidate the exact mechanisms through which *Saccharibacteria* and other emergent pathogens influence the oral health of these patients.

Additionally, future studies should focus on the long-term impact of oral microbial dysbiosis in IEI patients, particularly on the progression from early caries development to more severe periodontal disease. Halai et al. and Abusleme et al. showed that children with IEIs are at higher risk of developing periodontal diseases due to microbial imbalances. However, the specific trajectories and risk factors leading to severe oral health complications remain unclear (Halai et al. 2020; Abusleme et al. 2018).

The therapeutic potential of probiotics and synbiotics in maintaining oral health in IEI patients is a promising area for future research. Studies have indicated that specific bacterial strains, such as *Lactobacillus* and *Weissella*, may help restore microbial balance and prevent caries formation. Regular dental checkups, meticulous oral hygiene practices, and the use of fluoride and antimicrobial agents should be prioritized to manage these risks effectively. However, more clinical trials are needed to confirm the efficacy of these interventions in pediatric IEI populations (Kozak and Pawlik 2023; Berbers et al. 2020). In conclusion, although recent studies have significantly advanced our understanding of the oral microbiota in pediatric IEI patients, continued research is necessary to fill existing gaps, particularly in the areas of microbial interactions, disease progression, and potential therapeutic interventions.

15.5 Clinical Implications and Future Directions

15.5.1 Impact on Clinical Practice

Children with IEI face significant oral health challenges because of compromised immune systems, affecting both dental care and overall health. A key issue is the increased susceptibility to oral infections. A healthy immune system inhibits the growth of harmful microorganisms in the oral cavity; however, children with IEI are at risk for ulcerations, tissue necrosis, aggressive gingivitis, periodontitis, oral candidiasis, and herpetic infections (Chaiboonyarak et al. 2024; Peacock et al. 2017).

Al-kebsi et al. and Kıykım et al. found that children with IEI presented clinical features related to dental caries comparable to those of healthy children (Al-kebsi et al. 2023; Kıykım et al. 2013). There were no significant differences in the frequencies of dental caries, intraoral lesions, plaque, and gingival index values between children with IEI and healthy controls. Al-kebsi et al. found that the levels of cariogenic bacteria, such as *Streptococcus mutants* and *Lactobacillus,* were comparable between the groups (Al-kebsi et al. 2023). *Lactobacillus* levels were higher in patients with X-linked agammaglobulinemia (XLA) than in those with selective immunoglobulin A (IgA) deficiency. Except for minor differences, the numbers and species in the oral flora are quite similar between healthy children and children with IEI, and children with IEI are not at a greater disadvantage compared to healthy children in terms of oral pathogens. Periodontal destruction was increased 27 times in children who received daily G-CSF (Chaiboonyarak et al. 2024). G-CSF is a lifelong treatment administered daily or on alternate days to patients with severe congenital neutropenia (SCN) to stimulate neutrophil production in the bone marrow and prevent the risk of bacterial infections (Ye et al. 2011). In children with IEI, immune system suppression and poor oral hygiene due to medication, hospitalization, malnutrition, and mouth sores may increase the risk of dental caries. The likelihood of dental caries can decrease with the use of proper oral care, proper nutrition, and supportive preventive practices by the dentist.

15.5.2 Emerging Therapeutic Approaches

Routine dental procedures that are typically simple in healthy children can pose significant risks for IEI patients. The use of invasive procedures, such as tooth extraction and curettage, can lead to complications, such as prolonged bleeding, delayed healing, and increased risk of bacteremia. Patients receiving monthly immunoglobulin or weekly SC Ig therapy should ensure that their blood Ig level is above 600 mg/dL before dental procedures. If the Ig level is below 600 mg/dL, patients should receive a loading dose before administration. Both dentists and oral healthcare providers should be skeptical about diagnosing IEIs when developmental anomalies, periodontal diseases, and oral lesions are seen in children (Liebhaber and Dart 2013).

Recent advances in understanding the interplay between the oral microbiome and the immune system have led to novel therapeutic strategies. Probiotics, which are beneficial bacteria, are being explored to balance the oral microbiota and reduce the incidence of dental caries and periodontal diseases in IEI patients. Studies have shown that strains like L*actobacillus reuteri* and *Streptococcus salivarius* can inhibit pathogenic bacteria such as *Streptococcus mutans* (Yu et al. 2024).

In addition, antimicrobial peptides (AMPs) have emerged as a promising therapy targeting pathogenic oral bacteria while preserving beneficial microbes (Dale et al. 2006; Khurshid et al. 2016). AMPs are small naturally occurring molecules that can disrupt bacterial membranes and lead to the death of harmful bacteria without inducing significant resistance. Research is ongoing to develop more stable and effective synthetic AMPs for use in dental care.

Potential future treatments may include more sophisticated methods of microbiome engineering and immune modulation. One exciting avenue is gene editing technologies, such as CRISPR-Cas9, to alter the genetic makeup of oral bacteria precisely (Chavez-Granados et al. 2022; Mousavinasab et al. 2023). This technology can remove or disable genes in pathogenic bacteria that contribute to caries formation or periodontal disease, thereby reducing their virulence and ability to cause disease. Similarly, CRISPR can enhance beneficial bacteria, promoting a healthier and more stable oral microbiome.

Another future treatment may involve personalized medicine approaches, in which treatments are tailored to the individual's unique oral microbiome and immune profile (Mousavinasab et al. 2023). This step may involve analyzing the patient's oral microbiome and immune function in detail and creating a customized probiotic or immune-modulating therapy designed to correct any imbalance or deficiency.

15.5.3 Future Research Directions

A multi-omic approach offers valuable insights into microorganism-host interactions in children with IEIs, enabling the investigation of the central hypothesis that oral microbiota dysbiosis occurs as a consequence of IEIs. Several key questions must be addressed to validate this hypothesis. Are children with IEIs more susceptible to dental caries? Which specific types of immunodeficiency impair the immune system's ability to combat cariogenic species? Additionally, is there an association between altered caries microbiota and other systemic diseases? To address these key questions, strong interdisciplinary collaboration between microbiologists, pediatric immunologists, pediatric dentists, and bioinformaticians is essential. The ultimate goal is to translate microbiome research findings into practical therapeutic strategies for managing dental health in children with IEIs.

15.6 Conclusion

15.6.1 Summary of Main Points

In this chapter, the relationship between immune deficiencies and caries microbiota dysbiosis in children with IEIs is examined. NGS and other omics technologies enhance our understanding of the microbial diversity in patients and metabolic processes occurring within the microbiota. Key findings reveal that children with IEIs exhibit a greater susceptibility to dental caries owing to compromised immune responses that disrupt the delicate balance of the oral microbiota. In this specialized field, a multi-omic approach is a powerful tool for revealing the complex interactions between microbiota and the host, facilitated by interdisciplinary research that integrates microbiology, immunology, dental science, and bioinformatics. The findings of this research will pave the way for developing personalized therapeutic strategies and pharmaceuticals for individuals with IEIs.

15.6.2 Importance of Integrating Oral and Systemic Health

Infections are a common feature of immunodeficiencies. Pathogenic or weakly virulent, multiple infectious agents are usually responsible for infections in immunodeficiency diseases (Napiórkowska-Baran et al. 2024). Therefore, learning the history of previous infections is important for rectifying the diagnosis and treatment. Infections in IEI patients often appear in chronic, recurrent, and severe forms, with a complicated course and a poor response to treatment. In a child with a normal immune system within the first 10 years of life, it is typical to experience six upper respiratory tract infections per year, three to four episodes of otitis media, and two gastroenteritis attacks per year until the age of 2–3 years.

The oral cavity is important for general health, including digestive, respiratory, articulatory, and protective functions (Peacock et al. 2017). In the context of respiratory disorders, such as nasal obstruction, it represents the initial point of contact with pathogens. Changes in the oral cavity, such as inflammation, ulcerations, and oral lesions, may result in local pathology and contribute to the development of more serious infections. Oral diseases encompass a broad group of conditions that can affect both the soft and hard tissues of the oral cavity. These diseases are associated with craniofacial abnormalities, injuries, infectious diseases, and congenital defects (Jin et al. 2016; Peres et al. 2019).

15.6.3 Final Thoughts

Recent breakthroughs in the research of dental caries microbiota in pediatric patients with IEIs have greatly improved our understanding of how immune deficiencies lead to microbiota dysbiosis and enhance dental caries susceptibility. Multi-omic technologies enhance researchers' ability to uncover intricate interactions between the host immune system and oral microorganisms through interdisciplinary collaboration among microbiologists, immunologists, clinicians, and bioinformaticians. The outcomes of ongoing research will ultimately improve the overall health of patients with IEIs.

References

Aas JA, Paster BJ, Stokes LN, Olsen I, Dewhirst FE (2005) Defining the normal bacterial flora of the oral cavity. J Clin Microbiol 43(11):5721–5732. https://doi.org/10.1128/JCM.43.11.5721-5732.2005

Abolhassani H, Azizi G, Sharifi L, Yazdani R, Mohsenzadegan M, Delavari S, Sohani M, Shirmast P, Chavoshzadeh Z, Mahdaviani SA, Kalantari A, Tavakol M, Jabbari-Azad F, Ahanchian H, Momen T, Sherkat R, Sadeghi-Shabestarij M, Aleyasin S, Esmaeilzadeh H, Al-Herz W, Bousfiha AA, Condino-Neto A, Seppänen M, Sullivan KE, Hammarström L, Modell V, Modell F, Quinn J, Orange JS, Aghamohammadi A (2020) Global systematic review of primary immunodeficiency registries. Expert Rev Clin Immunol 16(7):717–732. https://doi.org/10.1080/1744666X.2020.1801422

Abusleme L, Diaz PI, Freeman AF, Greenwell-Wild T, Brenchley L, Desai JV, Ng WI, Holland SM, Lionakis MS, Segre JA, Kong HH, Moutsopoulos NM (2018) Human defects in STAT3 promote oral mucosal fungal and bacterial dysbiosis. JCI Insight 3(17). https://doi.org/10.1172/jci.insight.122061

Al-Herz W, Bousfiha A, Casanova JL, Chatila T, Conley ME, Cunningham-Rundles C, Etzioni A, Franco JL, Gaspar HB, Holland SM, Klein C, Nonoyama S, Ochs HD, Oksenhendler E, Picard C, Puck JM, Sullivan K, Tang ML (2014) Primary immunodeficiency diseases: an update on the classification from the International Union of Immunological Societies expert committee for primary immunodeficiency. Front Immunol 5:162. https://doi.org/10.3389/fimmu.2014.00162

Al-Kebsi BLA, Kars G, Özer H, Güner ŞN (2023) Unraveling the impact of primary immunodeficiency disorders on the microbiota of dental caries in children through 16S rRNA gene-based metagenomic analysis. Turk J Med Sci 53(5):1512–1522. https://doi.org/10.55730/1300-0144.5719

Ari O, Karabudak S, Kalcioglu MT, Gunduz AY, Durmaz R (2019) The bacteriome of otitis media with effusion: does it originate from the adenoid? Int J Pediatr Otorhinolaryngol 126:109624. https://doi.org/10.1016/j.ijporl.2019.109624

Belstrøm D, Holmstrup P, Fiehn NE, Kirkby N, Kokaras A, Paster BJ, Bardow A (2017) Salivary microbiota in individuals with different levels of caries experience. J Oral Microbiol 9(1):1270614. https://doi.org/10.1080/20002297.2016.1270614

Berbers RM, Mohamed Hoesein FAA, Ellerbroek PM, van Montfrans JM, Dalm VASH, van Hagen PM, Paganelli FL, Viveen MC, Rogers MRC, de Jong PA, Uh HW, Willems RJL, Leavis HL (2020) Low IgA associated with oropharyngeal microbiota changes and lung disease in primary antibody deficiency. Front Immunol 11:1245. https://doi.org/10.3389/fimmu.2020.01245

Biggs CM, Keles S, Chatila TA (2017) DOCK8 deficiency: insights into pathophysiology, clinical features and management. Clin Immunol 181:75–82. https://doi.org/10.1016/j.clim.2017.06.003

Blaustein RA, Shen Z, Kashaf SS, Lee-Lin S, Conlan S, Thomas J, Mullikin J, Young A, Bouffard G, Barnabas B, Brooks S, Buchter C, Crawford J, Han J, Ho S, Legaspi R, Maduro Q, Marfani H, Montemayor C, Schandler K, Schmidt B, Sison C, Stantripop M, Black S, Dekhtyar M, Masiello C, McDowell J, Park M, Thomas P, Vemulapalli M, Bosticardo M, Delmonte OM, Holmes CJ, Taylor ME, Banania G, Nagao K, Dimitrova D, Kanakry JA, Su H, Holland SM, Bergerson JRE, Freeman AF, Notarangelo LD, Kong HH, Segre JA (2023) Expanded microbiome niches of RAG-deficient patients. Cell Rep Med 4(10):101205. https://doi.org/10.1016/j.xcrm.2023.101205

Bor B, McLean JS, Foster KR, Cen L, To TT, Serrato-Guillen A, Dewhirst FE, Shi W, He X (2018) Rapid evolution of decreased host susceptibility drives a stable relationship between ultrasmall parasite TM7x and its bacterial host. Proc Natl Acad Sci 115(48): 12277–12282. https://doi.org/10.1073/pnas.1810625115

Bousfiha AA, Jeddane L, Ailal F, Benhsaien I, Mahlaoui N, Casanova JL, Abel L (2013) Primary immunodeficiency diseases worldwide: more common than generally thought. J Clin Immunol 33(1):1–7. https://doi.org/10.1007/s10875-012-9751-7

Bousfiha A, Moundir A, Tangye SG, Picard C, Jeddane L, Al-Herz W, Rundless CC, Franco JL, Holland SM, Klein C, Morio T, Oksenhendler E, Puel A, Puck J, Seppanen MRJ, Somech R, Su HC, Sullivan KE, Torgerson TE, Meyts I (2022) The 2022 update of IUIS phenotypical classification for human inborn errors of immunity. J Clin Immunol 42(7): 1508–1520. https://doi.org/10.1007/s10875-022-01352-z

Cabay RJ (2014) An overview of molecular and genetic alterations in selected benign odontogenic disorders. Arch Pathol Lab Med 138(6):754–758. https://doi.org/10.5858/arpa.2013-0057-SA

Caporaso JG, Kuczynski J, Stombaugh J, Bittinger K, Bushman FD, Costello EK, Fierer N, González Peña A, Goodrich JK, Gordon JI, Huttley GA, Kelley ST, Knights D, Koenig JE, Ley RE, Lozupone CA, McDonald D, Muegge BD, Pirrung M, Reeder J, Sevinsky JR, Turnbaugh PJ, Walters WA, Widmann J, Yatsunenko T, Zaneveld J, Knight R (2010) QIIME allows analysis of high-throughput community sequencing data. Nat Methods 7(5): 335–336. https://doi.org/10.1038/nmeth.f.303

Casanova JL, Abel L (2007) Primary immunodeficiencies: a field in its infancy. Science 317(5838):617–619. https://doi.org/10.1126/science.1142963

Chaiboonyarak T, Chantarangsu S, Gavila P, Lao-Araya M, Suratannon N, Porntaveetus T (2024) Orodental health status of patients with inborn errors of immunity. Int J Paediatr Dent 34(4):453–463. https://doi.org/10.1111/ipd.13146

Chavez-Granados PA, Manisekaran R, Acosta-Torres LS, Garcia-Contreras R (2022) CRISPR/Cas gene-editing technology and its advances in dentistry. Biochimie 194:96–107. https://doi.org/10.1016/j.biochi.2021.12.012

Cho Y-D, Kim K-H, Lee Y-M, Ku Y, Seol Y-J (2021) Oral microbiome and host health: review on current advances in genome-wide analysis. Appl Sci 11(9): 4050. https://doi.org/10.3390/app11094050

Clarridge JE (2004) Impact of 16S rRNA gene sequence analysis for identification of bacteria on clinical microbiology and infectious diseases. Clin Microbiol Rev 17(4):840–862. https://doi.org/10.1128/CMR.17.4.840-862.2004

Costalonga M, Herzberg MC (2014) The oral microbiome and the immunobiology of periodontal disease and caries. Immunol Lett 162(2 Pt A):22–38. https://doi.org/10.1016/j.imlet.2014.08.017

Cunha-Cruz J, Scott J, Rothen M, Mancl L, Lawhorn T, Brossel K, Berg J (2013) Salivary characteristics and dental caries: evidence from general dental practices. J Am Dent Assoc 144(5):e31–e40. https://doi.org/10.14219/jada.archive.2013.0159

Dahlén G, Björkander J, Gahnberg L, Slots J, Hanson LA (1993) Periodontal disease and dental caries in relation to primary IgG subclass and other humoral immunodeficiencies. J Clin Periodontol 20(1):7–13. https://doi.org/10.1111/j.1600-051x.1993.tb01752.x

Dale BA, Tao R, Kimball JR, Jurevic RJ (2006) Oral antimicrobial peptides and biological control of caries. BMC Oral Health 6:1–7. https://doi.org/10.1186/1472-6831-6-S1-S13

del Rosario Rioboo-Crespo M, Planells-del Pozo P, Rioboo-García R (2005) Epidemiology of the most common oral mucosal diseases in children. Med Oral Patol Oral Cir Bucal 10(5):376–387

Deo PN, Deshmukh R (2019) Oral microbiome: unveiling the fundamentals. J Oral Maxillofacial Pathol 23(1): 122–128. https://doi.org/10.4103/jomfp.JOMFP_304_18

Dewhirst FE, Chen T, Izard J, Paster BJ, Tanner ACR, Yu W-H, Lakshmanan A, Wade WG (2010) The human oral microbiome. J Bacteriol 192(19):5002–5017. https://doi.org/10.1128/JB.00542-10

Durazzi F, Sala C, Castellani G, Manfreda G, Remondini D, Cesare A (2021) Comparison between 16S rRNA and shotgun sequencing data for the taxonomic characterization of the gut microbiota. Sci Rep 11:Article 82726. https://doi.org/10.1038/s41598-021-82726-y

George K, Govindaraj G (2023) Infections in inborn errors of immunity with combined immune deficiency: a review. Pathogens (Basel, Switzerland) 12(2):272. https://doi.org/10.3390/pathogens12020272

Ghosal R, Gayen S, Mahmud S, Das S, Pal M (2024) A brief review on immunology of dental caries and caries vaccine. Int J Innovat Res Med Sci 9(7):408–414. https://doi.org/10.23958/ijirms/vol09-i07/1922

Gilles A, Meglécz E, Pech N, Ferreira S, Malausa T, Martin JF (2011) Accuracy and quality assessment of 454 GS-FLX titanium pyrosequencing. BMC Genomics 12:245. https://doi.org/10.1186/1471-2164-12-245

Graves DT, Cochran D (2003) The contribution of interleukin-1 and tumor necrosis factor to periodontal tissue destruction. J Periodontol 74(3):391–401. https://doi.org/10.1902/jop.2003.74.3.391

Halai H, Somani C, Donos N, Nibali L (2020) Periodontal status of children with primary immunodeficiencies: A systematic review. Clin Oral Investig 24(6): 1939–1951. https://doi.org/10.1007/s00784-019-03055-z

Hillmann B, Al-Ghalith G, Shields-Cutler R, Zhu Q, Gohl D, Beckman K, Knight R, Knights D (2018) Evaluating the information content of shallow shotgun metagenomics. mSystems 3(6):e00069-18. https://doi.org/10.1128/mSystems.00069-18

Huang YH, Qiu YR, Zhang QL, Cai MC, Yu H, Zhang JM, Jiang L, Ji MM, Xu PP, Wang L, Cheng S, Zhao WL (2024) Genomic and transcriptomic profiling of peripheral T cell lymphoma reveals distinct molecular and microenvironment subtypes. Cell Rep Med 5(2): 101416. https://doi.org/10.1016/j.xcrm.2024.101416

Humphrey SP, Williamson RT (2001) A review of saliva: normal composition, flow, and function. J Prosthet Dent 85(2):162–169. https://doi.org/10.1067/mpr.2001.113778

Illumina (2024) NovaSeq 6000 system. Illumina, October 6. https://emea.illumina.com/systems/sequencing-platforms/novaseq.html

Jagathrakshakan SN, Sethumadhava RJ, Mehta DT, Ramanathan A (2015) 16S rRNA gene-based metagenomic analysis identifies a novel bacterial co-prevalence pattern in dental caries. Eur J Dent 9(1):127–132. https://doi.org/10.4103/1305-7456.149661

Jin LJ, Lamster IB, Greenspan JS, Pitts NB, Scully C, Warnakulasuriya S (2016) Global burden of oral diseases: emerging concepts, management and interplay with systemic health. Oral Dis 22(7):609–619. https://doi.org/10.1111/odi.12428

Johnson JS, Spakowicz DJ, Hong B-Y, Petersen LM, Demkowicz P, Chen L, Leopold SR, Hanson BM, Agresta HO, Gerstein M, Sodergren E, Weinstock GM (2019) Evaluation of 16S rRNA gene sequencing for species and strain-level microbiome analysis. Nat Commun 10(1):5029. https://doi.org/10.1038/s41467-019-13036-1

Jung S, Gies V, Korganow AS, Guffroy A (2020) Primary immunodeficiencies with defects in innate immunity: focus on orofacial manifestations. Front Immunol 11: 1065. https://doi.org/10.3389/fimmu.2020.01065

Karabiber E, Barıs S (2024) Delineating the clinical and immunologic characteristics: a comparative study of inborn errors of immunity in adult versus pediatric diagnosed. Int Arch Allergy Immunol 185(11): 1123–1135. https://doi.org/10.1159/000540538

Khurshid Z, Naseem M, Sheikh Z, Najeeb S, Shahab S, Zafar MS (2016) Oral antimicrobial peptides: types and role in the oral cavity. Saudi Pharm J 24(5): 515–524. https://doi.org/10.1016/j.jsps.2015.02.015

Kilian M, Chapple I, Hannig M, Marsh P, Meuric V, Pedersen AML, Tonetti MS, Twetman S, Wade WG (2016) The oral microbiome—an update for oral healthcare professionals. Br Dent J 221(10):657–666. https://doi.org/10.1038/sj.bdj.2016.865

Kilic SS, Ozel M, Hafizoglu D, Karaca NE, Aksu G, Kutukculer N (2013) The prevalences and patient characteristics of primary immunodeficiency diseases in Turkey—two centers study. J Clin Immunol 33(1): 74–83. https://doi.org/10.1007/s10875-012-9763-3

Kıykım AA, Topçuoğlu N, Külekçi G, Çokuğraş H, Akçakaya N, Camcıoğlu Y (2013) Comparison of oral microflora in selective IgA deficiency and X-linked agammaglobulinemia cases with control group. Turk Arch Pediatr 48(3):204–209. https://doi.org/10.4274/tpa.438

Klingberg G, Oskarsdóttir S, Johannesson EL, Norén JG (2002) Oral manifestations in 22q11 deletion syndrome. Int J Paediatr Dent 12(1):14–23

Kozak M, Pawlik A (2023) The role of the oral microbiome in the development of diseases. Int J Mol Sci 24(6):5231. https://doi.org/10.3390/ijms24065231

Lam RS, O'Brien-Simpson NM, Lenzo JC, Holden JA, Brammar GC, Walsh KA, McNaughtan JE, Rowler DK, Van Rooijen N, Reynolds EC (2014) Macrophage depletion abates Porphyromonas gingivalis-induced alveolar bone resorption in mice. J Immunol (Baltimore, Md. : 1950) 193(5):2349–2362. https://doi.org/10.4049/jimmunol.1400853

Li C, Yu R, Ding Y (2022) Association between *Porphyromonas Gingivalis* and systemic diseases: focus on T cells-mediated adaptive immunity. Front Cell Infect Microbiol 12:1026457. https://doi.org/10.3389/fcimb.2022.1026457

Liebhaber M, Dart D (2013) Dental problems and immune deficient patients. IG Living. https://www.igliving.com/magazine/articles/IGL_2013-06_AR_Dental-Problems-and-Immune-Deficient-Patients.pdf

Lin D, Yang L, Wen L, Lu H, Chen Q, Wang Z (2021) Crosstalk between the oral microbiota, mucosal immunity, and the epithelial barrier regulates oral mucosal disease pathogenesis. Mucosal Immunol 14:1247–1258. https://doi.org/10.1038/s41385-021-00413-7

Luo SC, Wei SM, Luo XT, Yang QQ, Wong KH, Cheung PCK, Zhang BB (2024) How probiotics, prebiotics, synbiotics, and postbiotics prevent dental caries: an oral microbiota perspective. NPJ Biofilms Microbiomes 10(1):14. https://doi.org/10.1038/s41522-024-00488-7

Macowan M, Pattaroni C, Bonner K, Chatzis R, Daunt C, Gore M, Custovic A, Shields MD, Power UF, Grigg J, Roberts G, Ghazal P, Schwarze J, Turner S, Bush A, Saglani S, Lloyd CM, Marsland BJ (2024) Deep multiomic profiling reveals molecular signatures that underpin preschool wheeze and asthma. J Allergy Clin Immunol, Advance online publication. https://doi.org/10.1016/j.jaci.2024.08.017

Malele Kolisa Y, Yengopal V, Shumba K, Igumbor J (2019) The burden of oral conditions among adolescents living with HIV at a clinic in Johannesburg, South Africa. PLoS One 14(10): e0222568. https://doi.org/10.1371/journal.pone.0222568

Meighani G, Aghamohammadi A, Javanbakht H, Abolhassani H, Nikayin S, Jafari SM, Ghandehari Motlagh M, Shamshiri AR, Rezaei N (2011) Oral and dental health status in patients with primary antibody deficiencies. Iran J Allergy Asthma Immunol 10(4): 289–293

Meslier V, Quinquis B, Da Silva K, Oñate FP, Pons N, Roume H, Podar M, Almeida M (2022) Benchmarking second and third-generation sequencing platforms for microbial metagenomics. Sci Data 9:694. https://doi.org/10.1038/s41597-022-01762-z

Metcalfe S, Anselmi N, Escobar A, Visser MB, Kay JG (2022) Innate phagocyte polarization in the oral cavity. Front Immunol 12:768479. https://doi.org/10.3389/fimmu.2021.768479

Modell V, Orange JS, Quinn J, Modell F (2018) Global report on primary immunodeficiencies: 2018 update from the Jeffrey Modell Centers Network on disease classification, regional trends, treatment modalities, and physician-reported outcomes. Immunol Res 66(3):367–380. https://doi.org/10.1007/s12026-018-8996-5

Mousavinasab F, Karimi R, Taheri S, Ahmadvand F, Sanaaee S, Najafi S, Soltani Halvaii M, Haghgoo A, Zamany M, Majidpoor J, Khosravifar M, Baniasadi M, Talebi M, Movafagh A, Aghaei-Zarch SM, Khorram N, Farnia P, Kalhor K (2023) Microbiome modulation in inflammatory diseases: progress to microbiome genetic engineering. Cancer Cell Int 23(1):271. https://doi.org/10.1186/s12935-023-03095-2

Moutsopoulos NM, Konkel JE (2018) Tissue-specific immunity at the oral mucosal barrier. Trends Immunol 39(4):276–287. https://doi.org/10.1016/j.it.2017.08.005

Moyer IN, Kobayashi RH, Cannon ML, Simon JF, Cooley RO, Rich K (1983) Dental treatment of children with severe combined immunodeficiency. Pediatr Dent 5(1):79–82

Naafs MAB (2018) Oral mucosal immune suppression, tolerance, and silencing: a mini-review. Mod Approaches Dent Oral Health Care 1(2):25–36. https://doi.org/10.32474/MADOHC.2018.01.000109

Napiórkowska-Baran K, Darwish S, Kaczor J, Treichel P, Szymczak B, Szota M, Bartuzi Z (2024) Oral diseases as a manifestation of inborn errors of immunity. J Clin Med 13(17):5079. https://doi.org/10.3390/jcm13175079

Navgire GS, Goel N, Sawhney G, Sharma M, Kaushik P, Mohanta YK, Mohanta TK, Al-Harrasi A (2022) Analysis and interpretation of metagenomics data: an approach. Biol Proced Online 24(1):18. https://doi.org/10.1186/s12575-022-00179-7

Nibali L, Bayliss-Chapman J, Halai H, Somani C, Davies J, Ancliff P, Donos N (2021) Periodontal status in children with primary immunodeficiencies. J Periodontal Res 56(4):819–827. https://doi.org/10.1111/jre.12880

Oxford Nanopore Technologies (2024) PromethION. Oxford Nanopore Technologies, October 1. https://nanoporetech.com/products/sequence/promethion-24-48

Pac M, Bernatowska EA, Kierkuś J, Ryżko JP, Cielecka-Kuszyk J, Jackowska T, Mikołuć B (2017) Gastrointestinal disorders next to respiratory infections as leading symptoms of X-linked agammaglobulinemia in children – 34-year experience of a single center. Arch

Med Sci 13(2):412–417. https://doi.org/10.5114/aoms.2016.60338

PacBio (2024) PacBio Revio system. Pacific Biosciences, October 6. https://www.pacb.com/wp-content/uploads/Revio-brochure.pdf

Padron GT, Hernandez-Trujillo VP (2023) Autoimmunity in primary immunodeficiencies (PID). Clin Rev Allergy Immunol 65(1):1–18. https://doi.org/10.1007/s12016-022-08942-0

Peacock ME, Arce RM, Cutler CW (2017) Periodontal and other oral manifestations of immunodeficiency diseases. Oral Dis 23(7):866–888. https://doi.org/10.1111/odi.12584

Peng XP, Al-Ddafari MS, Caballero-Oteyza A, El Mezouar C, Mrovecova P, Dib SE, Massen Z, Smahi MC, Faiza A, Hassaïne RT, Lefranc G, Aribi M, Grimbacher B (2023) Next-generation sequencing (NGS)-based approach to diagnosing Algerian patients with suspected inborn errors of immunity (IEIs). Clin Immunol 256:109758. https://doi.org/10.1016/j.clim.2023.109758

Peres MA, Macpherson LMD, Weyant RJ, Daly B, Venturelli R, Mathur MR, Listl S, Celeste RK, Guarnizo-Herreño CC, Kearns C, Benzian H, Allison P, Watt RG (2019) Oral diseases: a global public health challenge. Lancet (Lond, Engl) 394(10194):249–260. https://doi.org/10.1016/S0140-6736(19)31146-8

Pieniawska-Śmiech K, Pasternak G, Lewandowicz-Uszyńska A, Jutel M (2022) Diagnostic challenges in patients with inborn errors of immunity with different manifestations of immune dysregulation. J Clin Med 11(14):Article 4220. https://doi.org/10.3390/jcm11144220

Rajasekaran JJ, Krishnamurthy HK, Bosco J, Jayaraman V, Krishna K, Wang T, Bei K (2024) Oral microbiome: a review of its impact on Oral and systemic health. Microorganisms 12(9):1797. https://doi.org/10.3390/microorganisms12091797

Rawat A, Sharma M, Vignesh P, Jindal AK, Suri D, Das J, Joshi V, Tyagi R, Sharma J, Kaur G, Lau YL, Imai K, Nonoyama S, Lenardo M, Singh S (2022) Utility of targeted next-generation sequencing for inborn errors of immunity at a tertiary care centre in North India. Sci Rep 12:10416. https://doi.org/10.1038/s41598-022-14522-1

Rezaei N, Bonilla FA, Seppänen M, de Vries E, Bousfiha AA, Puck J, Orange JD (2017) Introduction on primary immunodeficiency diseases. In: Rezaei N, Aghamohammadi A, Notarangelo L (eds) Primary immunodeficiency diseases. Springer, Berlin/Heidelberg. https://doi.org/10.1007/978-3-662-52909-6_1

Sampaio EP, Bax HI, Hsu AP, Kristosturyan E, Pechacek J, Chandrasekaran P, Paulson ML, Dias DL, Spalding C, Uzel G, Ding L, McFarland E, Holland SM (2012) A novel STAT1 mutation associated with disseminated mycobacterial disease. J Clin Immunol 32(4):681–689. https://doi.org/10.1007/s10875-012-9659-2

Sanger F, Nicklen S, Coulson AR (1977) DNA sequencing with chain-terminating inhibitors. Proc Natl Acad Sci USA 74(12):5463–5467. https://doi.org/10.1073/pnas.74.12.5463

Santacroce L, Passarelli PC, Azzolino D, Bottalico L, Charitos IA, Cazzolla AP, Colella M, Topi S, Godoy FG, D'Addona A (2023) Oral microbiota in human health and disease: A perspective. Exp Biol Med (Maywood) 248(15):1288–1301. https://doi.org/10.1177/15353702231187645

Satam H, Joshi K, Mangrolia U, Waghoo S, Zaidi G, Rawool S, Thakare RP, Banday S, Mishra AK, Das G, Malonia SK (2023) Next-generation sequencing technology: current trends and advancements. Biology 12(7):997. https://doi.org/10.3390/biology12070997

Schloss PD, Westcott SL, Ryabin T, Hall JR, Hartmann M, Hollister EB, Lesniewski RA, Oakley BB, Parks DH, Robinson CJ, Sahl JW, Stres B, Thallinger GG, Van Horn DJ, Weber CF (2009) Introducing mothur: open-source, platform-independent, community-supported software for describing and comparing microbial communities. Appl Environ Microbiol 75(23):7537–7541. https://doi.org/10.1128/AEM.01541-09

Sedghi L, DiMassa V, Harrington A, Lynch SV, Kapila YL (2021) The oral microbiome: role of key organisms and complex networks in oral health and disease. Periodontol 87(1):107–131. https://doi.org/10.1111/prd.12393

Slatko BE, Gardner AF, Ausubel FM (2018) Overview of next-generation sequencing technologies. Curr Protoc Mol Biol 122(1):e59. https://doi.org/10.1002/cpmb.59

Su S, Qi T, Wang W, Salama ES, Li Y (2023) Investigation of the oral microbiome of children associated with dental caries: a systematic study. Arch Oral Biol 154:105776. https://doi.org/10.1016/j.archoralbio.2023.105776

Szczawinska-Poplonyk A, Gerreth K, Breborowicz A, Borysewicz-Lewicka M (2009) Oral manifestations of primary immune deficiencies in children. Oral Surg Oral Med Oral Pathol Oral Radiol Endod 108(3):e9–e20. https://doi.org/10.1016/j.tripleo.2009.03.049

Tangye SG, Al-Herz W, Bousfiha A, Chatila T, Cunningham-Rundles C, Etzioni A, Franco JL, Holland SM, Klein C, Morio T, Oksenhendler E, Picard C, Puel A, Puck J, Seppanen MRJ, Somech R, Su HC, Sullivan KE, Torgerson TR, Meyts I (2022) Human inborn errors of immunity: 2022 update on the classification from the International Union of Immunological Societies Expert Committee. J Clin Immunol 42(7):1473–1507. https://doi.org/10.1007/s10875-022-01289-3

Thermo Fisher Scientific (2024) Iontorrent specification sheet. Thermo Fisher, October 6. https://assets.thermofisher.com/TFS-Assets/CSD/Specification-Sheets/PG1720-PJT2769-COL22253-P-on-GeneStudio-S5-Spec-Sheet-Global-FLR.pdf

Tozum TF, Berker E, Ersoy F, Tezcan I, Sanal O (2003) The relationship between periodontal status and peripheral levels of neutrophils in two consanguineous siblings with severe congenital neutropenia: case reports. Quintessence Int 34(3):221–226

van't Hof W, Veerman EC, Nieuw Amerongen AV, Ligtenberg AJ (2014) Antimicrobial defense systems in saliva. Monogr Oral Sci 24:40–51. https://doi.org/10.1159/000358783

Vila T, Rizk AM, Sultan AS, Jabra-Rizk MA (2019) The power of saliva: antimicrobial and beyond. PLoS Pathog 15(11):e1008058. https://doi.org/10.1371/journal.ppat.1008058

Voelkerding KV, Dames SA, Durtschi JD (2009) Next-generation sequencing: from basic research to diagnostics. Clin Chem 55(4):641–658. https://doi.org/10.1373/clinchem.2008.112789

von Hardenberg S, Klefenz I, Steinemann D, Di Donato N, Baumann U, Auber B, Klemann C (2024) Current genetic diagnostics in inborn errors of immunity. Front Pediatr 12. https://doi.org/10.3389/fped.2024.1279112

Yamaguchi M, Horiba K, Haruta K, Takeuchi S, Suzuki T, Torii Y, Kawabe S, Wada S, Ikeyama T, Ito Y, Ogi T, Kawada JI (2023) Utility of nanopore sequencing for detecting pathogens in bronchoalveolar lavage fluid from pediatric patients with respiratory failure. J Clin Virol Plus 3(2):100154. https://doi.org/10.1016/j.jcvp.2023.100154

Yan Y, Bao X, Chen B, Li Y, Yin J, Zhu G, Li Q (2022) Interpretable machine learning framework reveals microbiome features of oral disease. Microbiol Res 265:127198. https://doi.org/10.1016/j.micres.2022.127198

Ye Y, Carlsson G, Wondimu B, Fahlén A, Karlsson-Sjöberg J, Andersson M, Engstrand L, Yucel-Lindberg T, Modéer T, Pütsep K (2011) Mutations in the ELANE gene are associated with development of periodontitis in patients with severe congenital neutropenia. J Clin Immunol 31(6):936–945. https://doi.org/10.1007/s10875-011-9572-0

Yin EZ, Frush DP, Donnelly LF, Buckley RH (2001) Primary immunodeficiency disorders in pediatric patients: clinical features and imaging findings. AJR Am J Roentgenol 176(6):1541–1552. https://doi.org/10.2214/ajr.176.6.1761541

Yu X, Devine DA, Vernon JJ (2024) Manipulating the diseased oral microbiome: the power of probiotics and prebiotics. J Oral Microbiol 16(1):2307416. https://doi.org/10.1080/20002297.2024.2307416

Zaura E, Keijser BJF, Huse SM, Crielaard W (2009) Defining the healthy "core microbiome" of oral microbial communities. BMC Microbiol 9:259. https://doi.org/10.1186/1471-2180-9-259

Zheng X, He J, Wang L, Zhou S, Peng X, Huang S, Zheng L, Cheng L, Hao Y, Li J, Xu J, Xu X, Zhou X (2017) Ecological effect of arginine on oral microbiota. Sci Rep 7:7206. https://doi.org/10.1038/s41598-017-07042-w

Zhu J, Tian L, Chen P, Han M, Song L, Tong X, Sun X, Yang F, Lin Z, Liu X, Liu C, Wang X, Lin Y, Cai K, Hou Y, Xu X, Yang H, Wang J, Kristiansen K, Xiao L, Zhang T, Jia H, Jie Z (2022) Over 50,000 metagenomically assembled draft genomes for the human oral microbiome reveal new taxa. Genomics Proteomics Bioinformatics 20(2):246–259

Part III

Immunopathology of Oral Diseases and Cancer

Inflammasomes in Oral Pathologies: From Periodontal Disease to Oral Cancer

16

Gaetano Isola and Alessandro Polizzi

Abstract

Inflammasomes are multiprotein complexes that play a crucial role in the innate immune system, particularly in the regulation of inflammatory responses, and play a critical role in different oral pathologies, influencing the development and progression of conditions like periodontal disease and oral cancer by regulating inflammatory responses that contribute to tissue damage and tumorigenesis. The multiprotein inflammasome complexes have a role in the body's innate immune response by detecting pathogens, tissue damage, and other stress signals, leading to the activation of inflammatory cytokines, primarily interleukin-1β (IL-1β) and IL-18. In the context of periodontitis, a chronic inflammatory disease affecting the tissues supporting the teeth, inflammasomes are integral in mediating the inflammatory response that drives the disease progression. Their activation is pivotal in oral health and the progression of various oral diseases, ranging from periodontal diseases to oral cancer. This review explores the involvement of inflammasomes in oral pathologies, with a focus on their role in periodontal diseases, oral potentially malignant disorders (OPMDs), and oral cancer. Specifically, we examine how dysregulation of inflammasome activity contributes to inflammatory conditions like gingivitis and periodontitis, and how this can transition into more severe oral diseases, such as OPMDs and cancer. This chapter will also highlight the role of epigenetic changes in inflammasome regulation, providing insight into potential diagnostic and therapeutic strategies for oral health management. This comprehensive overview underscores the importance of inflammasome research in improving our understanding and treatment of oral pathologies, pointing to future directions for clinical applications.

Keywords

Periodontitis · Periodontal disease · Inflammasome · Inflammation · Host response Immunity

G. Isola (✉)
Department of General Surgery and Surgical-Medical Specialties, School of Dentistry, University of Catania, Catania, Italy

International Research Center on Periodontal and Systemic Health "PerioHealth", University of Catania, Catania, Italy
e-mail: gaetano.isola@unict.it

A. Polizzi
Department of General Surgery and Surgical-Medical Specialties, School of Dentistry, University of Catania, Catania, Italy

N. Rezaei, H. Ziaei (eds.), *Oral Immunology*, Advances in Experimental Medicine and Biology 1492,
https://doi.org/10.1007/978-3-032-03176-1_16

16.1 Introduction

The oral cavity, a complex ecosystem, is constantly exposed to a variety of microorganisms, food particles, and environmental pollutants, making it a primary site of immune system activity (Takahashi 2005). Maintaining oral health is crucial, as diseases affecting the oral cavity can have systemic implications, influencing general well-being and contributing to conditions such as cardiovascular disease, diabetes, and respiratory infections (Gurenlian 2009). Within this context, the innate immune system plays a vital role in preserving oral health by responding rapidly to pathogens and harmful stimuli (Akira et al. 2006). Central to this innate immune defense are inflammasomes, multiprotein complexes that regulate the body's inflammatory responses (He et al. 2016). In recent years, research has increasingly focused on inflammasomes due to their critical involvement in mediating immune responses and their implications in a variety of inflammatory diseases, including those affecting the oral cavity (Schieffer et al. 2022).

Inflammasomes are mediators linked with pathogen-associated molecular patterns (PAMPs) and damage-associated molecular patterns (DAMPs). When activated, these complexes trigger the production of pro-inflammatory cytokines, such as interleukin-1β (IL-1β) and interleukin-18 (IL-18), which are essential in initiating and sustaining the inflammatory response. While this process is a protective mechanism designed to eliminate pathogens and repair tissue damage, its dysregulation can lead to chronic inflammation and tissue destruction. This duality makes inflammasomes a key factor in both maintaining oral health and contributing to oral pathologies when aberrantly activated (Patel et al. 2017).

The role of inflammasomes in oral diseases is a rapidly evolving area of research, with studies highlighting their involvement in conditions ranging from gingivitis and periodontitis to more severe oral diseases like oral potentially malignant disorders (OPMDs) and oral cancer. Understanding the mechanisms by which inflammasomes contribute to these diseases is critical for developing targeted therapies aimed at modulating their activity to restore balance in the immune response. In healthy individuals, inflammasomes serve as gatekeepers, ensuring that immune responses remain proportional to the threat posed by pathogens. However, when the regulatory mechanisms controlling inflammasome activation fail, the resulting chronic inflammation can exacerbate tissue damage and contribute to disease progression (Karaaslan 2023).

Periodontal disease, one of the most common chronic inflammatory conditions worldwide, offers a prime example of how inflammasomes function in the oral cavity. This disease, which includes both gingivitis and periodontitis, is driven by microbial infection but significantly influenced by the host's immune response. Inflammasomes, and their related nucleotide-binding domain, leucine-rich repeat (NLR) family pyrin domain-containing 3 (NLRP3) inflammasome, are crucial for the immune response in periodontal tissues. The activation of the NLRP3 inflammasome in response to bacterial colonization in the gingival sulcus leads to the production of IL-1β and IL-18, and immunoglobulin antibodies contributing to the characteristic inflammation seen in gingivitis and periodontitis. However, when inflammasome activity becomes excessive or prolonged, it drives the chronic inflammation and tissue destruction that characterizes advanced periodontitis (Chen et al. 2021; Isola et al. 2022; Okano et al. 2018).

Beyond periodontal diseases, inflammasomes are also implicated in the development of OPMDs, such as, among others, leukoplakia, erythroplakia, and oral lichen planus (OLP). OPMDs are characterized by chronic inflammation and an increased risk of progression to oral cancer (Warnakulasuriya et al. 2021). The precise role of inflammasomes in OPMDs is still under investigation, but emerging evidence suggests that inflammasome dysregulation may contribute to the chronic inflammatory environment that fosters malignant transformation. Additionally, epigenetic changes in inflammasome-related genes have been identified in OPMDs, indicating that inflammasomes may not only be involved in disease progression but also serve as potential biomarkers for early detection and risk assessment (Thi Do et al. 2018).

Oral cancer, particularly oral squamous cell carcinoma (OSCC), is another area where inflammasomes have been found to play a pivotal role. Chronic inflammation is a well-established risk factor for cancer development, and inflammasome activation is a key driver of this inflammation in the oral mucosa (Stasiewicz and Karpiński 2022). The overproduction of IL-1β and IL-18 in response to persistent inflammasome activation can promote a pro-tumorigenic environment by fostering cell proliferation, angiogenesis, and immune evasion. Additionally, the inflammasome-mediated inflammatory milieu in the oral cavity can enhance the invasion and metastasis of malignant cells, contributing to the progression of oral cancer (Wang et al. 2018).

In recent years, the potential for targeting inflammasomes in oral disease has gained significant attention. Given their central role in modulating the immune response and inflammation, inflammasomes represent a promising therapeutic target for both the prevention and treatment of oral diseases. Novel therapeutic strategies aimed at inhibiting inflammasome activation or modulating their downstream effects are currently being explored (Marchesan et al. 2020). These approaches have the potential to not only reduce inflammation and tissue destruction in conditions like periodontitis but also to interrupt the inflammatory pathways that contribute to the development of OPMDs and oral cancer (Bullon et al. 2018).

Therefore, inflammasomes are key regulators of the immune response in the oral cavity, balancing the need for protection against pathogens with the risk of chronic inflammation and tissue damage. Their dysregulation is implicated in a range of oral diseases, from periodontal disease to oral cancer, highlighting their importance in both health and disease conditions. A deeper understanding of how inflammasomes contribute to these conditions will be crucial for developing new diagnostic tools and therapeutic strategies aimed at improving oral health outcomes. As research continues to uncover the complex roles of inflammasomes in oral pathologies, they are likely to emerge as critical targets for interventions that could revolutionize the management of oral diseases.

16.2 Inflammasome and Innate Immune System in Oral Pathologies

The oral cavity serves as the body's first line of defense against various external threats, including pathogenic microorganisms, environmental pollutants, and dietary substances. To combat these threats, the body relies on the innate immune system, the non-specific, rapid-response arm of the immune system that responds to invaders and tissue damage (Zasloff 2002). Within the framework of the innate immune system, inflammasomes play a pivotal role in initiating and regulating inflammatory responses. Inflammasomes are cytosolic multiprotein complexes that detect the presence of pathogen-associated molecular patterns (PAMPs) and damage-associated molecular patterns (DAMPs) (Neudorf et al. 2019). Upon activation, they initiate the maturation and secretion of pro-inflammatory cytokines such as IL-1β and IL-18, which are critical in mediating the inflammatory process (Porritt et al. 2021).

In oral pathologies, inflammasomes are implicated in both the protective and pathogenic roles of the immune response. In healthy tissues, inflammasomes are essential in maintaining immune homeostasis and ensuring rapid responses to oral pathogens. However, when dysregulated, inflammasome activity can lead to chronic inflammation, tissue destruction, and the progression of oral diseases, such as periodontal diseases, OPMDs, and oral cancer (Karaaslan 2023). This section explores the relationship between inflammasomes and the innate immune system, providing a foundational understanding of their roles in oral health and pathology.

16.2.1 Innate Immune Response in Oral Health

The innate immune system in the oral cavity comprises various components, including epithelial cells, resident immune cells (e.g., macrophages, dendritic cells), and chemical barriers (e.g., antimicrobial peptides and enzymes present in saliva). Together, these elements provide an initial line of defense against oral pathogens. The epithelial lining of the oral mucosa acts as a physical barrier, preventing microbial invasion, linked with pattern recognition receptors (PRRs) such as Toll-like receptors (TLRs) and nucleotide-binding oligomerization domain (NOD)-like receptors (NLRs) (Marchesan et al. 2020).

Among the NLR family, the NLRP3 inflammasome (Fig. 16.1) is particularly well-studied in the context of oral pathologies. It acts as a sensor of both microbial and endogenous danger signals. When activated, the NLRP3 inflammasome facilitates the assembly of an inflammasome complex, which leads to the activation of caspase-1. Caspase-1, in turn, cleaves pro-IL-1β and pro-IL-18 into their active forms, IL-1β and IL-18, respectively (Fig. 16.2). These cytokines are crucial in recruiting additional immune cells to the site of infection, where they work to contain and eliminate pathogens (Koca-Ünsal et al. 2022).

In addition to IL-1β and IL-18, the inflammasome activation process can trigger pyroptosis, a form of programmed cell death that acts as a defense mechanism to eliminate infected or damaged cells. Pyroptosis also contributes to the release of additional DAMPs, amplifying the immune response and facilitating the clearance of pathogens. This process is vital in preventing the spread of infection and maintaining tissue integrity (Toldo and Abbate 2024).

In healthy oral tissues, the activation of inflammasomes is tightly regulated to avoid excessive inflammation that could lead to tissue damage. When properly controlled, inflammasome activation ensures that the immune response is proportionate to the threat posed by oral pathogens, maintaining a state of immune homeostasis that protects against both infection and chronic inflammation (Ebersole et al. 2016; Yilmaz and Lee 2015).

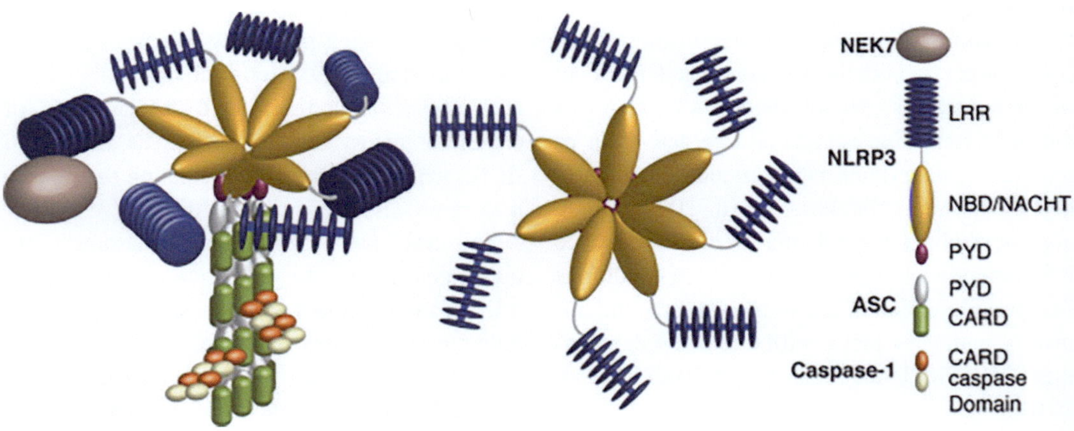

Fig. 16.1 NLRP3 inflammasome model. NLRP3 oligomerizes in response to cell stress after priming. By binding to the leucine-rich repeat of NLRP3, NEK7 stabilizes the NLRP3 oligomer and creates a platform of pyrin domains that interact with one another to cause the development of ASC filaments. Through CARD-CARD interactions, several ASC molecules facilitate filament building and caspase-1 binding. Abbreviations: NEK7, NIMA-related kinase 7; LRR, leucine-rich repeat; NBD, nucleotide-binding domain; PYD, pyrin domain; CARD, caspase activation and recruitment domain. (From (Marchesan et al. 2020) under the terms of the Creative Commons CC BY license)

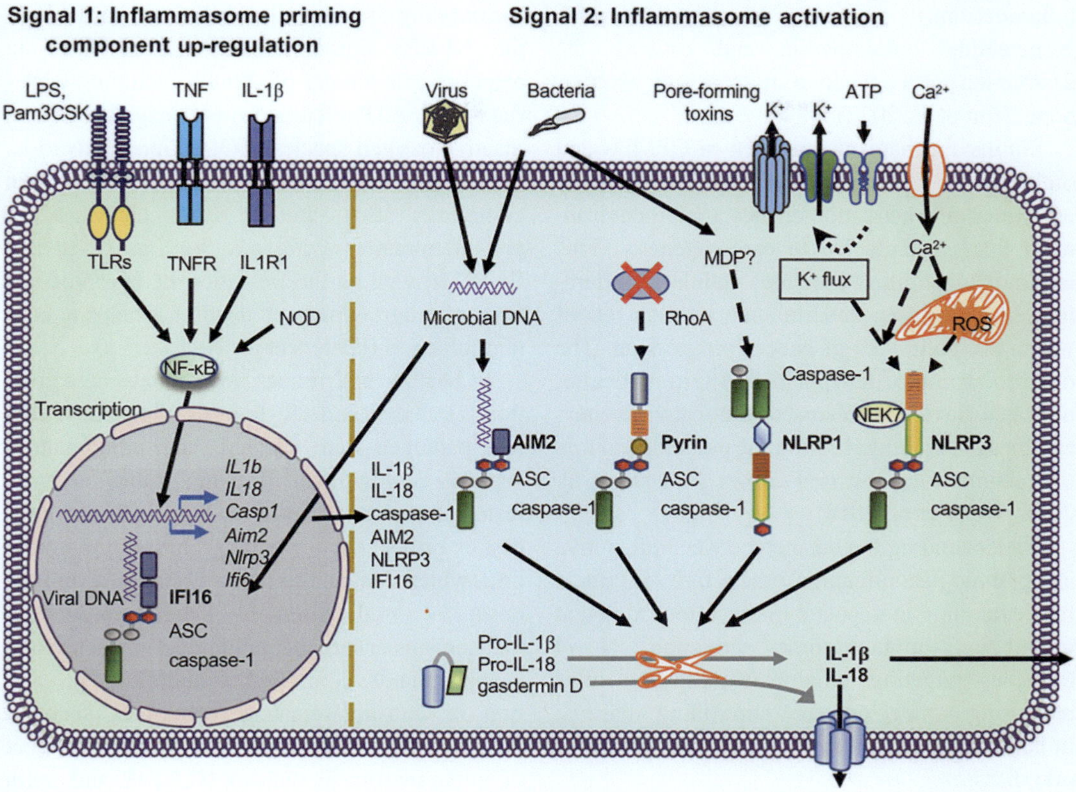

Fig. 16.2 Inflammasome activation and priming. Prior to activation (signal 2), inflammasomes need to be primed (signal 1). First, heightened expression of inflammasome components (IL1B, IL18, CAS1 AIM2, NLRP3, IFI16) is induced by a nuclear factor-κB-activating stimulus (lipopolysaccharide, tumor necrosis factor-β), which in turn causes increased expression of these proteins. A second, distinct signal is needed for inflammasome activation following priming in order to activate each individual inflammasome and cause the development of distinct inflammasome complexes. Inflammasome sensors multimerize in response to an activating stimulus, attracting ASC and pro-caspase-1 to aid in caspase-1 autoactivation. Proproteins known as interleukin-1β and interleukin-18 are produced, activated by caspase-1, and released by cells. (From (Marchesan et al. 2020) under the terms of the Creative Commons CC BY license)

16.2.2 Dysregulation of Inflammasomes in Periodontitis and Related Diseases of the Oral Cavity

Although inflammasomes are essential for defending the body against oral pathogens, their dysregulation can result in pathological outcomes. Excessive or prolonged inflammasome activation can lead to chronic inflammation, which is a common underlying feature of many oral diseases. Chronic inflammation occurs when the immune system fails to resolve the initial immune response, leading to persistent tissue damage and an increased risk of disease progression (Gurung et al. 2015).

In periodontal diseases, for example, dysregulated inflammasome activity contributes to the chronic inflammatory state that drives tissue destruction. Periodontal disease is a polymicrobial infection that triggers an immune response, characterized by the infiltration of immune cells into the gingival tissues (Isola et al. 2021, 2023a, b, c, 2024). These cells, in turn, activate inflammasomes in response to bacterial PAMPs and tissue DAMPs, leading to the secretion of IL-1β and IL-18. However, in cases of persistent infection or immune dysfunction, inflammasome activity remains elevated, perpetuating inflammation and causing the destruction of connective tissues and alveolar bone (Zhu et al. 2022).

Similar mechanisms are seen in OPMDs and oral cancer. Chronic inflammation driven by inflammasome activation creates a microenvironment that is conducive to carcinogenesis. This inflammatory milieu promotes cellular proliferation, angiogenesis, and immune evasion, all of which are hallmarks of cancer progression. The overproduction of IL-1β and IL-18, in particular, has been linked to enhanced tumor invasiveness and metastasis, underscoring the pathological role of inflammasomes in oral cancer (Thi Do et al. 2018; Wang et al. 2018).

Understanding the balance between protective and pathological inflammasome activity is critical for developing therapeutic interventions aimed at modulating inflammasome activation. Novel therapies targeting inflammasomes may offer promising strategies for controlling chronic inflammation and preventing the progression of oral diseases.

16.2.3 Role of Inflammasome in Healthy Mouth

In a healthy oral cavity, the inflammasome acts as a crucial sentinel of the innate immune system, playing a pivotal role in maintaining immune homeostasis. The primary function of inflammasomes in the healthy mouth is to ensure a balanced immune response that protects against microbial invasion without causing excessive tissue damage or chronic inflammation. This balance is achieved through the finely tuned regulation of inflammasome activation, allowing for a rapid yet controlled response to microbial and environmental challenges.

16.2.3.1 Inflammasomes as Regulators of Oral Immune Homeostasis

One of the key functions of inflammasomes in the healthy mouth is to regulate immune homeostasis by detecting and responding to pathogens in a manner that limits collateral damage to the surrounding tissues. Inflammasomes, particularly the NLRP3 inflammasome, are activated in response to a variety of stimuli, including bacterial toxins, ATP released from damaged cells, and reactive oxygen species (ROS). Upon activation, inflammasomes assemble into multiprotein complexes that initiate the cleavage of pro-inflammatory cytokines, such as IL-1β and IL-18, as well as the induction of pyroptosis, a form of programmed cell death that helps to contain infection (Peeters et al. 2015).

In healthy oral tissues, inflammasome activation is transient and self-limiting. Once the invading pathogens are cleared, anti-inflammatory signals downregulate inflammasome activity, restoring tissue homeostasis. This regulatory process is critical for preventing chronic inflammation, which can lead to tissue destruction and the onset of oral diseases. Several regulatory mechanisms ensure that inflammasome activation is appropriately controlled, including the production of anti-inflammatory cytokines, such as IL-10, and the expression of negative regulators of inflammasomes, such as NLRP12 and pyrin (Gurung et al. 2015).

16.2.3.2 Oral Microbiota and Inflammasome Activation

The healthy mouth harbors a diverse microbiota that plays an important role in oral health. This microbiota is composed of a variety of bacterial species, many of which exist in a symbiotic relationship with the host. The oral microbiota contributes to the maintenance of immune homeostasis by preventing the overgrowth of pathogenic species and promoting immune tolerance to commensal organisms (Ptasiewicz et al. 2022).

Inflammasomes are integral to maintaining the balance between the host and the oral microbiota. In the presence of a healthy microbiota, inflammasomes are minimally activated, allowing the immune system to tolerate commensal bacteria without eliciting a strong inflammatory response. However, when the balance of the microbiota is disrupted, as in the case of dysbiosis, inflammasomes become activated in response to the overgrowth of pathogenic bacteria. This activation triggers an immune response aimed at restoring microbial balance and protecting the host from infection (Cugini et al. 2021; Marchesan et al. 2020; Shibata 2018).

In a healthy mouth, inflammasomes help maintain the integrity of the oral mucosa by promoting the clearance of harmful microbes while minimizing inflammation. This balance is crucial for preventing the development of oral diseases associated with dysbiosis, such as periodontal diseases and dental caries (Bullon et al. 2018).

16.2.3.3 Role of Inflammasomes in Tissue Repair

Another important function of inflammasomes in the healthy mouth is their role in tissue repair and regeneration. In response to tissue damage caused by mechanical injury or infection, inflammasomes are activated to support the clearance of damaged cells and facilitate the repair process. IL-1β and IL-18, the pro-inflammatory cytokines produced by inflammasome activation, play a dual role in this process. While they initiate the inflammatory response necessary for clearing infection, they also promote tissue repair by triggering the proliferation of epithelial cells and the production of extracellular matrix components (Rathinam and Chan 2018).

In this context, inflammasomes act as mediators of both inflammation and tissue regeneration, ensuring that tissue damage is repaired in a timely manner. However, the repair process is dependent on the resolution of inflammation. Prolonged inflammasome activation can impair tissue repair, leading to chronic inflammation and the development of fibrotic tissue, which can compromise the function of the affected tissues (Josiane Bazzo de Alencar et al. 2020).

16.3 Inflammasome in Periodontal Diseases

Periodontal diseases, including gingivitis and periodontitis, are among the most prevalent chronic inflammatory conditions affecting humans (Khabadze et al. 2023). Characterized by the inflammation of the supporting structures of the teeth, these diseases arise primarily from microbial infection and the host's immune response. While gingivitis is generally a reversible inflammation confined to the gingiva, periodontitis is a more severe and progressive condition that leads to the destruction of periodontal ligament, alveolar bone, and tooth loss if untreated (Abusleme et al. 2021). The immune response in periodontal diseases is a critical factor in their progression, and recent research has highlighted the role of inflammasomes, particularly the NLRP3 inflammasome, in modulating this response (Isola et al. 2022).

This section delves into the role of inflammasomes in both gingivitis and periodontitis, examining how their activation contributes to disease progression.

16.3.1 Role of Inflammasome in Gingivitis

Gingivitis, the mildest form of periodontal disease, is characterized by inflammation of the gingival tissues without the destruction of the supporting structures of the teeth. It is initiated by the accumulation of dental plaque, a biofilm of bacteria that forms on the surfaces of teeth. In response to this microbial challenge, the host immune system activates, recruiting immune cells to the gingival tissues to combat the bacterial invasion (Chapple et al. 2018). Inflammasomes play a key role in orchestrating this immune response.

16.3.1.1 Bacterial Triggers and Inflammasome Activation in Gingivitis

The accumulation of dental plaque in gingivitis creates an environment rich in microbial PAMPs, such as lipopolysaccharides (LPS) from Gram-negative bacteria. These PAMPs are recognized by pattern recognition receptors (PRRs) on immune cells, such as Toll-like receptors (TLRs) and NOD-like receptors (NLRs) (Okano et al. 2018). The NLRP3 inflammasome, one of the most well-characterized inflammasomes, is activated in response to these microbial signals.

Upon activation, the NLRP3 inflammasome facilitates the cleavage of pro-caspase-1 into its active form, caspase-1, which in turn cleaves pro-IL-1β and pro-IL-18 into their active cytokine forms (Marchesan et al. 2020).

IL-1β is a major pro-inflammatory cytokine that plays a critical role in gingivitis by recruiting neutrophils and other immune cells to the site of infection. These immune cells help contain the microbial infection and prevent it from spreading deeper into the periodontal tissues (Sawada et al. 2013). IL-18, another cytokine activated by the inflammasome, also contributes to the inflammatory response in gingival tissue by promoting the production of interferon-gamma (IFN-γ), which activates macrophages and enhances the immune response against the microbial challenge (Orozco et al. 2006).

While inflammasome activation is essential for mounting an effective immune response in gingivitis, excessive or prolonged inflammasome activity can lead to chronic inflammation. The continuous production of IL-1β and IL-18 can exacerbate the inflammatory process, causing collateral damage to the gingival tissues. This highlights the dual role of inflammasomes in gingivitis: they are necessary for controlling bacterial infections, but their dysregulation can contribute to the persistence of inflammation.

16.3.1.2 Resolution of Inflammation in Gingivitis

In a healthy immune response, once the microbial threat is eliminated, anti-inflammatory signals are activated to resolve the inflammation and promote tissue healing. In gingivitis, the resolution of inflammation involves the downregulation of inflammasome activity and the production of anti-inflammatory cytokines such as IL-10 and transforming growth factor-beta (TGF-β). These cytokines suppress inflammasome activation and promote the clearance of apoptotic immune cells, facilitating tissue repair (Marchesan et al. 2020).

Failure to resolve inflammation in gingivitis can lead to its progression to periodontitis, where inflammasome activity becomes more pronounced and destructive. Understanding the mechanisms that regulate inflammasome activation and resolution in gingivitis is crucial for developing strategies to prevent the transition to more severe periodontal disease.

16.3.2 Role of Inflammasome in Periodontitis

Periodontitis represents the advanced stage of periodontal disease, characterized by the destruction of the connective tissue and bone that support the teeth. Unlike gingivitis, periodontitis is not limited to the gingival tissues but affects the deeper periodontal structures, leading to irreversible damage if left untreated (Papapanou et al. 2018). The progression from gingivitis to periodontitis is driven by a dysregulated immune response to the bacterial biofilm, with inflammasomes playing a central role in this process.

16.3.2.1 Inflammasome Activation in Periodontitis

In periodontitis, the bacterial biofilm becomes more complex, and the immune response shifts from an acute inflammatory response to a chronic, dysregulated state. PAMPs from periodontopathogenic bacteria, such as *Porphyromonas gingivalis*, *Tannerella forsythia*, *Aggregatibacter actinomycetemcomitans,* and *Treponema denticola*, linked to the activation of NLRP3 in immune host cells such as macrophages and dendritic cells. In addition to PAMPs, DAMPs released from damaged periodontal tissues further activate the inflammasome, perpetuating the inflammatory cycle (C. Cugini et al. 2021; J. B. de Alencar et al. 2020; Kelk et al. 2022).

The activation of the NLRP3 inflammasome in periodontitis leads to the persistent production of IL-1β and IL-18, which amplify the inflammatory response. IL-1β, in particular, is a potent mediator of tissue destruction in periodontitis. It stimulates the production of matrix metalloproteinases (MMPs), enzymes that degrade the extracellular matrix, and promotes osteoclast differentiation, leading to bone resorption. This results in the characteristic alveolar bone loss seen in periodontitis (Bostanci et al. 2009; Strowig et al. 2012).

Moreover, IL-1β and IL-18 contribute to the recruitment of additional immune cells, including neutrophils, macrophages, and T cells, to the site of infection. While these cells are essential for controlling bacterial infection, their prolonged presence, and activity in the periodontal tissues can lead to the destruction of both microbial invaders and host tissues. The chronic activation of the inflammasome in periodontitis thus creates a self-perpetuating cycle of inflammation and tissue destruction (Marchesan et al. 2020; Peeters et al. 2015).

16.3.2.2 Role of Inflammasomes in Bone Resorption

One of the most significant consequences of inflammasome activation in periodontitis is the loss of alveolar bone, which supports the teeth. Bone resorption in periodontitis is driven by an imbalance between bone-forming osteoblasts and bone-resorbing osteoclasts. IL-1β, produced by inflammasome activation, plays a critical role in promoting osteoclast differentiation and activity. This cytokine enhances the expression of receptor activator of nuclear factor kappa-B ligand (RANKL), a key mediator of osteoclastogenesis, while simultaneously inhibiting the production of osteoprotegerin (OPG), a natural inhibitor of RANKL. The resulting increase in RANKL/OPG ratio leads to excessive bone resorption (He et al. 2020; Jandinski 1988).

Inflammasomes also contribute to bone resorption through the production of other TNF-α and IL-6, which in turn promotes osteoclastogenesis and bone destruction. This dysregulated inflammatory environment ultimately results in the irreversible loss of alveolar bone, compromising tooth stability and function (de Molon et al. 2018).

16.3.2.3 Epigenetic Regulation of Inflammasome Activity in Periodontitis

Emerging research suggests that epigenetic modifications, such as DNA methylation, histone modifications, and non-coding RNA regulation, may play a role in the dysregulation of inflammasome activity in periodontitis. Epigenetic changes can alter the expression of inflammasome components, such as NLRP3 and pro-IL-1β, leading to increased inflammasome activation and cytokine production (Poli et al. 2020).

MicroRNAs (miRNAs), small non-coding RNAs that regulate gene expression, have also been implicated in the regulation of inflammasome activity in periodontitis. Certain miRNAs, such as miR-223, act as negative regulators of the NLRP3 inflammasome, and their downregulation in periodontitis has been linked to increased inflammasome activation and inflammation (Xia et al. 2021). Understanding the epigenetic regulation of inflammasomes in periodontitis may provide new insights into the mechanisms driving chronic inflammation and tissue destruction, as well as potential targets for therapeutic intervention.

16.3.2.4 Therapeutic Targeting of Inflammasomes in Periodontitis

Given the central role of inflammasomes in the pathogenesis of periodontitis, targeting inflammasome activation presents a promising therapeutic strategy for controlling inflammation and preventing tissue destruction. Several approaches are being explored to modulate inflammasome activity that could be applied in the future for periodontitis, including:

- ***Direct Inhibition of NLRP3***: Small-molecule inhibitors of the NLRP3 inflammasome, such as MCC950, have shown promise in preclinical models of inflammatory diseases. By preventing the assembly and activation of the NLRP3 inflammasome, these inhibitors reduce the production of IL-1β and IL-18, attenuating inflammation and tissue damage (Corcoran et al. 2021).
- ***Caspase-1 Inhibitors***: Caspase-1 is a key enzyme in the inflammasome pathway responsible for cleaving pro-IL-1β and pro-IL-18 into their active forms. Inhibitors of caspase-1, such as VX-765, have been shown to reduce

inflammation in animal models of periodontitis by blocking the production of these pro-inflammatory cytokines (Marchesan et al. 2020).
- ***IL-1 Receptor Antagonists***: Since IL-1β is a major driver of inflammation and tissue destruction in periodontitis, blocking its receptor with IL-1 receptor antagonists, such as anakinra, may help mitigate the inflammatory response. Anakinra has been successfully used in other inflammatory conditions and is being investigated for its potential use in periodontal disease (Cavalli and Dinarello 2018; Papathanasiou et al. 2020).
- ***Targeting Epigenetic Modifications***: Modulating epigenetic changes that contribute to inflammasome dysregulation, such as DNA methylation and miRNA expression, may offer new therapeutic avenues for controlling inflammation in periodontitis. Epigenetic therapies, including inhibitors of DNA methyltransferases (DNMTs) and histone deacetylases (HDACs), are being explored in the context of inflammatory diseases and may have potential in periodontal therapy (Shanmugam and Sethi 2013; Shao and Gudjonsson 2020).

Future research into the epigenetic regulation of inflammasomes and the development of targeted therapies holds promise for improving the management of periodontal diseases and preserving oral health.

16.4 Inflammasome in Oral Diseases

Oral diseases encompass a broad spectrum of conditions, ranging from inflammatory disorders such as gingivitis and periodontitis to more severe pathological conditions, including potentially malignant disorders and oral cancer. The role of the immune system, particularly the inflammasome pathway, in modulating the onset, progression, and severity of these diseases is increasingly recognized. Inflammasomes, as key components of the innate immune system, serve to detect harmful stimuli such as pathogens and cellular damage, initiating a cascade of inflammatory responses aimed at protecting the host. However, dysregulation of inflammasome activity can lead to chronic inflammation and tissue damage, potentially contributing to the development of malignancies (Bullon et al. 2018).

In the context of OPMDs, such as leukoplakia and oral submucous fibrosis (OSMF), and their progression to oral cancer, understanding the contribution of inflammasomes offers valuable insights into disease mechanisms and possible therapeutic targets. Additionally, recent research has shed light on the role of epigenetic changes in regulating inflammasome activity in OPMDs, further highlighting the complexity of these pathways in disease progression. This section will explore the role of inflammasomes in OPMDs and examine the epigenetic mechanisms that modulate inflammasome function in these disorders.

16.4.1 Role of Inflammasome in Oral Potentially Malignant Disorders

OPMDs are oral mucosa lesions with important risk development in OSCC. Common examples of OPMDs include leukoplakia, erythroplakia, OLP, and OSMF. The transformation of these lesions into malignant conditions is influenced by various factors, including chronic inflammation, genetic mutations, and epigenetic alterations (Warnakulasuriya et al. 2021). Recent studies have suggested that the inflammasome pathway, particularly the NLRP3 inflammasome, plays a pivotal role in the pathogenesis and progression of OPMDs.

16.4.1.1 Chronic Inflammation and Inflammasome Activation in OPMDs

Chronic inflammation is a well-established risk factor for the development of cancer, and inflammasomes are central players in regulating inflammatory responses. In OPMDs, the persistent activation of inflammasomes can contribute to the chronic inflammatory environment that predisposes tissues to malignant transformation. For example, in oral leukoplakia, a common OPMD characterized by white patches on the oral mucosa, inflammasome activation has been associated with elevated levels of pro-inflammatory cytokines such as IL-1β and IL-18. These cytokines promote the recruitment of immune cells to the site of the lesion, sustaining a state of chronic inflammation that can induce DNA damage and cellular dysplasia, both of which are hallmarks of carcinogenesis (Jain et al. 2024; Lopez-Labady et al. 2021).

Similarly, in OLP, a chronic inflammatory condition affecting the oral mucosa, inflammasomes have been implicated in maintaining the inflammatory microenvironment. OLP is characterized by a T-cell-mediated immune response, which is thought to be driven in part by inflammasome activation (Polizzi et al. 2023; Talley et al. 2023). The NLRP3 inflammasome, in particular, has been shown to be activated in response to cellular stress and damage in OLP, leading to the production of IL-1β and other pro-inflammatory mediators. These cytokines contribute to the chronic inflammation observed in OLP and may play a role in the malignant transformation of OLP lesions into OSCC.

16.4.1.2 Role of Inflammasome in OSMF

OSMF is another OPMD that has a well-documented link to the development of oral cancer. OSMF is a chronic, progressive condition characterized by the fibrosis (scarring) of the oral mucosa, which leads to restricted mouth opening and significant morbidity. The pathogenesis of OSMF involves chronic exposure to areca nut, a common chewing habit in certain populations, which induces an inflammatory response and subsequent fibrotic changes in the oral tissues (Shih et al. 2019).

Recent studies have implicated the potential role of inflammasomes in the pathogenesis of OSMF. The NLRP3 inflammasome is activated in response to oxidative stress and tissue damage caused by the metabolites of areca nut, such as arecoline. This activation results in the release of IL-1β and IL-18, which promote the fibrotic response by stimulating the production of collagen and other extracellular matrix components. The sustained activation of inflammasomes in OSMF creates a chronic inflammatory and fibrotic environment, which not only leads to the clinical symptoms of the disease but also increases the risk of malignant transformation to OSCC (You et al. 2019).

Moreover, the activation of inflammasomes in OSMF is thought to contribute to the immune evasion mechanisms that allow malignant cells to proliferate unchecked. By promoting chronic inflammation and fibrosis, inflammasomes may create a microenvironment that supports tumor growth and progression. Therefore, targeting inflammasome activation in OSMF may represent a potential therapeutic strategy for preventing the progression of this condition to oral cancer (Zeng et al. 2024).

16.4.1.3 Inflammasomes and Immune Surveillance in OPMDs

While inflammasomes are primarily involved in initiating and sustaining inflammatory responses, they also play a role in modulating immune surveillance, the process by which the immune system detects and eliminates emerging malignant cells. In OPMDs, dysregulated inflammasome activity can impair immune surveillance, allowing dysplastic or pre-malignant cells to escape immune detection and proliferate. For example, chronic inflammasome activation in the oral mucosa may lead to the exhaustion of cytotoxic T cells and natural killer (NK) cells, both of which are critical for controlling tumor growth (Huang et al. 2017).

The balance between inflammasome activation and immune suppression is a critical factor in the progression of OPMDs to oral cancer. Excessive inflammasome activation can create an immunosuppressive environment, while inadequate activation may fail to initiate an effective immune response against pre-malignant cells. Understanding this balance is essential for developing therapeutic interventions that target inflammasomes to prevent malignant transformation in OPMDs.

16.4.2 Epigenetics of Oral Potentially Malignant Disorders

In the context of oral potentially malignant disorders, epigenetic changes can influence the expression and activity of inflammasome components, contributing to the dysregulation of inflammation and immune responses (Kumari et al. 2022). This section explores the epigenetic mechanisms that modulate inflammasome activity in OPMDs and their implications for disease progression and therapeutic interventions.

16.4.2.1 DNA Methylation and Inflammasome Regulation in OPMDs

DNA methylation involves several mechanisms typically leading to the suppression of gene expression. In OPMDs, aberrant DNA methylation patterns have been observed in genes related to inflammasome activation, suggesting that epigenetic changes may contribute to the dysregulated inflammatory response in these disorders (Herreros-Pomares et al. 2021).

Specifically, the NLRP3 gene promoter hypermethylation has strictly related to the NLRP3 inflammasome in certain OPMDs, leading to impaired inflammasome activation and a decreased immune response. This suppression of inflammasome activity may allow pre-malignant cells to evade immune surveillance, increasing the risk of malignant transformation. Conversely, hypomethylation of inflammasome-related genes may result in excessive inflammasome activation, contributing to chronic inflammation and tissue damage in OPMDs such as leukoplakia (Fu et al. 2018).

Targeting DNA methylation patterns to restore normal inflammasome activity presents a potential therapeutic strategy for controlling inflammation and preventing the progression of OPMDs to oral cancer. Epigenetic therapies, such as DNA methyltransferase inhibitors (DNMTis), are currently being explored in the context of cancer treatment and may have applications in managing OPMDs (Ushio et al. 2022).

16.4.2.2 Histone Modifications and Inflammasome Activation in OPMDs

Histone modifications, such as acetylation, methylation, and phosphorylation, regulate gene expression by altering the structure of chromatin, making it more or less accessible to transcriptional machinery. In OPMDs, dysregulated histone modifications have been implicated in the abnormal expression of inflammasome components, leading to either excessive or insufficient inflammasome activation (Mancuso et al. 2009).

For instance, histone acetylation of the NLRP3 promoter region may enhance the expression of the NLRP3 inflammasome, contributing to chronic inflammation in OPMDs such as OSMF. (Sharma et al. 2020).

Histone deacetylase inhibitors (HDACis) are a class of drugs that target histone modifications and are currently being investigated for their potential in treating cancer and inflammatory diseases (Ushio et al. 2022). Specifically, it has been found that HDAC6 inhibitors reduce inflammation by suppressing IL-1β release, but this in not depend on disrupting NLRP3 inflammasome assembly, suggesting HDAC6 plays a role in the upstream priming phase of inflammasome activation, likely through interactions with TLR signaling rather than directly assembling the inflammasome (Bockstiegel et al. 2023).

16.4.2.3 Non-coding RNAs and Inflammasome Regulation in OPMDs

Non-coding RNAs, and long non-coding RNAs (lncRNAs), a family of microRNAs (miRNAs), are key mediators of gene expression and activities in modulating inflammasome activity. In OPMDs, dysregulated expression of non-coding RNAs has been linked to aberrant inflammasome activation and chronic inflammation (Polizzi et al. 2024).

For example, certain miRNAs, such as miR-223, act as negative regulators of the NLRP3 inflammasome (Yan et al. 2019).

Downregulation of these miRNAs in OPMDs may lead to excessive inflammasome activation, promoting chronic inflammation and increasing the risk of malignant transformation. Conversely, the upregulation of miRNAs that target inflammasome components may suppress the immune response, allowing pre-malignant cells to evade immune detection (Jain et al. 2024).

Targeting non-coding RNAs to modulate inflammasome activity is an emerging area of research in the field of cancer and inflammatory diseases. Developing therapeutic strategies that restore the normal expression of these non-coding RNAs may provide new opportunities for managing OPMDs and preventing their progression to oral cancer.

16.5 Inflammasome in Oral Cancer

Oral cancer, OSCC, is one of the most prevalent malignancies worldwide, with a high morbidity and mortality rate due to its aggressive nature and tendency for late diagnosis. Chronic inflammation plays a significant role in the initiation, promotion, and progression of oral cancer, and emerging evidence suggests that the inflammasome pathway is a critical mediator in this process. The NLRP3 inflammasome, in particular, has been extensively studied in the context of oral cancer, with research revealing its dual role in both tumor suppression and tumor promotion, depending on the stage of cancer and the microenvironment (Chamoli et al. 2021).

16.5.1 Inflammasome Activation in Oral Cancer Microenvironment

Inflammasomes, especially NLRP3, play a central role in modulating the inflammatory response within the tumor microenvironment (TME). The TME is composed of cancer cells, immune cells, fibroblasts, and extracellular matrix components, all of which contribute to the progression of cancer through complex interactions. In the early stages of carcinogenesis, inflammasomes are activated in response to cellular stress, DNA damage, and exposure to carcinogens such as tobacco smoke, alcohol, and areca nut. This activation leads to the production of IL-1β and IL-18, which recruit immune cells to the site of the tumor and initiate an anti-tumor immune response. At this stage, inflammasomes play a protective role by promoting immune surveillance and the clearance of pre-malignant and dysplastic cells (Huang et al. 2017).

However, as the tumor progresses, the chronic activation of inflammasomes can contribute to a pro-tumorigenic environment. Sustained inflammasome activation leads to the overproduction of pro-inflammatory cytokines, creating a persistent inflammatory state that promotes tumor growth, angiogenesis, and immune evasion. IL-1β, in particular, has been shown to enhance vascular endothelial growth factor (VEGF), which promotes angiogenesis and supports the supply of nutrients to the growing tumor (Kamatani et al. 2013). Additionally, chronic inflammasome activation can induce immunosuppressive cell recruitment, such as myeloid-derived suppressor cells (MDSCs) and regulatory T cells (Tregs), related to immune response in several types of cancer (Chu et al. 2012).

16.5.2 Inflammasome-Mediated Immune Evasion in Oral Cancer

As oral cancer progresses, tumor cells can exploit inflammasome-mediated pathways to evade immune surveillance. Chronic inflammasome activation within the TME leads to the exhaustion of cytotoxic T lymphocytes (CTLs) and natural killer (NK) cells, both of which are essential for recognizing and eliminating cancer cells (Lin et al. 2021). Additionally, inflammasome-induced production of IL-18 can stimulate the production of pro-tumorigenic cytokines such as IL-10, which suppress the activity of anti-tumor immune cells and promote an immunosuppressive environment (Mirlekar 2022).

Moreover, recent studies suggest that inflammasomes may contribute to the epithelial-to-mesenchymal transition (EMT), a process by which epithelial cells acquire mesenchymal traits that enhance their migratory and invasive capabilities. EMT is a key step in cancer metastasis, and inflammasome-mediated inflammation has been shown to promote EMT in oral cancer cells, facilitating their invasion into surrounding tissues and distant organs (Marandi et al. 2021; Wang et al. 2018).

16.5.3 Therapeutic Targeting of Inflammasomes in Oral Cancer

Given the dual role of inflammasomes in both tumor suppression and tumor promotion, targeting inflammasome pathways presents a promising strategy for oral cancer therapy. Inhibitors of inflammasome components, such as caspase-1 inhibitors, IL-1β antagonists, and NLRP3 inhibitors, have shown potential in preclinical studies for reducing inflammation, slowing tumor growth, and enhancing the efficacy of immunotherapies. Additionally, targeting downstream effectors of inflammasome activation, such as VEGF or immunosuppressive cytokines, may help to reprogram the TME and restore anti-tumor immunity (Missiroli et al. 2021).

Furthermore, combination therapies that include inflammasome inhibitors alongside treatments such as chemotherapy and radiation therapy, with certain outcomes in patients with oral cancer. By reducing the chronic inflammatory state within the TME, these therapies could prevent immune evasion and improve the effectiveness of anti-tumor immune responses (Feng et al. 2017).

Despite the promising preclinical data, clinical trials investigating the efficacy of inflammasome-targeting therapies in oral cancer are still in the early stages. Future research is needed to further elucidate the complex role of inflammasomes in oral cancer progression and to develop targeted therapies that can selectively modulate inflammasome activity in the TME.

16.6 Conclusions and Future Perspectives

The role of inflammasomes in oral pathologies, from periodontal disease to oral cancer, highlights their critical function in modulating inflammation and immune responses. While inflammasomes are essential for maintaining homeostasis and protecting against infections and cellular damage, their dysregulation can lead to chronic inflammation, tissue destruction, and an increased risk of malignant transformation. In periodontal diseases, the NLRP3 inflammasome is a key mediator of the immune response to bacterial infection, driving the progression from gingivitis to periodontitis. In oral potentially malignant disorders, such as leukoplakia and oral submucous fibrosis, chronic inflammasome activation contributes to the persistent inflammatory environment that promotes carcinogenesis. In oral cancer, inflammasomes play a dual role, initially protecting against tumor development but eventually facilitating tumor growth, immune evasion, and metastasis.

The emerging understanding of the inflammasome's role in oral diseases offers new opportunities for therapeutic intervention. Targeting the inflammasome pathway, either by inhibiting its activation or modulating its downstream effects, presents a promising approach for controlling inflammation, preventing disease progression, and improving patient outcomes. Therapies that target inflammasome components, such as caspase-1 inhibitors, IL-1β antagonists, and NLRP3 inhibitors, have shown potential in preclinical studies and may offer new treatment options for patients with periodontal diseases, OPMDs, and oral cancer.

Looking to the future, several key areas of research will be critical for advancing our understanding of inflammasome regulation in oral health and disease. First, further studies are needed to clarify the mechanisms by which inflammasomes contribute to immune surveillance and immune evasion in oral cancer. Understanding how inflammasome activity influences the balance between anti-tumor immunity and immunosuppression could lead to the development of more effective immunotherapies. Second, exploring the epigenetic regulation of inflammasomes in oral potentially malignant disorders may reveal new targets for preventing malignant transformation. Epigenetic therapies that restore normal inflammasome activity could offer novel approaches for managing OPMDs and reducing the risk of oral cancer.

Lastly, clinical trials are needed to evaluate the safety and efficacy of inflammasome-targeting therapies in patients with oral diseases. While preclinical studies have demonstrated the potential of these therapies, translating these findings into clinical practice will require careful evaluation of their effects on the immune system, tumor microenvironment, and overall patient health.

In conclusion, the inflammasome represents a critical link between inflammation, immune regulation, and oral disease pathogenesis. By targeting inflammasome pathways, future therapies may offer new hope for preventing and treating a wide range of oral diseases, from periodontal disease to oral cancer.

References

Abusleme L, Hoare A, Hong BY, Diaz PI (2021) Microbial signatures of health, gingivitis, and periodontitis. Periodontol 86(1):57–78

Akira S, Uematsu S, Takeuchi O (2006) Pathogen recognition and innate immunity. Cell 124(4):783–801

Bockstiegel J, Wurnig SL, Engelhardt J, Enns J, Hansen FK, Weindl G (2023) Pharmacological inhibition of HDAC6 suppresses NLRP3 inflammasome-mediated IL-1beta release. Biochem Pharmacol 215:115693. https://doi.org/10.1016/j.bcp.2023.115693

Bostanci N, Emingil G, Saygan B, Turkoglu O, Atilla G, Curtis MA, Belibasakis GN (2009) Expression and regulation of the NALP3 inflammasome complex in periodontal diseases. Clin Exp Immunol 157(3): 415–422. https://doi.org/10.1111/j.1365-2249.2009.03972.x

Bullon P, Pavillard LE, de la Torre-Torres R (2018) Inflammasome and oral diseases. In: Inflammasomes: clinical and therapeutic implications, pp 153–176

Cavalli G, Dinarello CA (2018) Anakinra therapy for non-cancer inflammatory diseases. Front Pharmacol 9:1157. https://doi.org/10.3389/fphar.2018.01157

Chamoli A, Gosavi AS, Shirwadkar UP, Wangdale KV, Behera SK, Kurrey NK, Kalia K, Mandoli A (2021) Overview of oral cavity squamous cell carcinoma: risk factors, mechanisms, and diagnostics. Oral Oncol 121: 105451. https://doi.org/10.1016/j.oraloncology.2021.105451

Chapple IL, Mealey BL, Van Dyke TE, Bartold PM, Dommisch H, Eickholz P, Geisinger ML, Genco RJ, Glogauer M, Goldstein M (2018) Periodontal health and gingival diseases and conditions on an intact and a reduced periodontium: consensus report of workgroup 1 of the 2017 world workshop on the classification of periodontal and Peri-implant diseases and conditions. J Periodontol 89:S74–S84

Chen Y, Yang Q, Lv C, Chen Y, Zhao W, Li W, Chen H, Wang H, Sun W, Yuan H (2021) NLRP3 regulates alveolar bone loss in ligature-induced periodontitis by promoting osteoclastic differentiation. Cell Prolif 54(2):e12973

Chu M, Su YX, Wang L, Zhang TH, Liang YJ, Liang LZ, Liao GQ (2012) Myeloid-derived suppressor cells contribute to oral cancer progression in 4NQO-treated mice. Oral Dis 18(1):67–73. https://doi.org/10.1111/j.1601-0825.2011.01846.x

Corcoran SE, Halai R, Cooper MA (2021) Pharmacological inhibition of the nod-like receptor family pyrin domain containing 3 Inflammasome with MCC950. Pharmacol Rev 73(3):968–1000. https://doi.org/10.1124/pharmrev.120.000171

Cugini C, Ramasubbu N, Tsiagbe VK, Fine DH (2021) Dysbiosis from a microbial and host perspective relative to Oral health and disease. Front Microbiol 12: 617485. https://doi.org/10.3389/fmicb.2021.617485

de Alencar JB, Zacarias JMV, Tsuneto PY, Souza VH, Silva COE, Visentainer JEL, Sell AM (2020) Influence of inflammasome NLRP3, and IL1B and IL2 gene polymorphisms in periodontitis susceptibility. PLoS One 15(1):e0227905. https://doi.org/10.1371/journal.pone.0227905

de Molon RS, Park CH, Jin Q, Sugai J, Cirelli JA (2018) Characterization of ligature-induced experimental periodontitis. Microsc Res Tech 81(12):1412–1421. https://doi.org/10.1002/jemt.23101

Ebersole JL, Kirakodu S, Novak MJ, Exposto CR, Stromberg AJ, Shen S, Orraca L, Gonzalez-Martinez J, Gonzalez OA (2016) Effects of aging in the expression of NOD-like receptors and inflammasome-related genes in oral mucosa. Mol Oral Microbiol 31(1):18–32

Feng X, Luo Q, Zhang H, Wang H, Chen W, Meng G, Chen F (2017) The role of NLRP3 inflammasome in 5-fluorouracil resistance of oral squamous cell carcinoma. J Exp Clin Cancer Res 36(1):81. https://doi.org/10.1186/s13046-017-0553-x

Fu J, Su Y, Liu Y, Zhang XY (2018) Comparative analysis of methylation profiles in tissues of oral leukoplakia and oral squamous cell carcinoma. Zhonghua Kou Qiang Yi Xue Za Zhi 53(4):248–253. https://doi.org/10.3760/cma.j.issn.1002-0098.2018.04.007

Gurenlian JR (2009) The relationship between Oral health and systemic disease. The Dental Assistance

Gurung P, Li B, Subbarao Malireddi R, Lamkanfi M, Geiger TL, Kanneganti T-D (2015) Chronic TLR

stimulation controls NLRP3 inflammasome activation through IL-10 mediated regulation of NLRP3 expression and caspase-8 activation. Sci Rep 5(1):14488
He Y, Hara H, Núñez G (2016) Mechanism and regulation of NLRP3 inflammasome activation. Trends Biochem Sci 41(12):1012–1021
He Z, Sun Y, Wu J, Xiong Z, Zhang S, Liu J, Liu Y, Li H, Jin T, Yang Y, Yang S (2020) Evaluation of genetic variants in IL-1B and its interaction with the predisposition of osteoporosis in the northwestern Chinese Han population. J Gene Med 22(10):e3214. https://doi.org/10.1002/jgm.3214
Herreros-Pomares A, Llorens C, Soriano B, Bagan L, Moreno A, Calabuig-Fariñas S, Jantus-Lewintre E, Bagan J (2021) Differentially methylated genes in proliferative verrucous leukoplakia reveal potential malignant biomarkers for oral squamous cell carcinoma. Oral Oncol 116:105191. https://doi.org/10.1016/j.oraloncology.2021.105191
Huang CF, Chen L, Li YC, Wu L, Yu GT, Zhang WF, Sun ZJ (2017) NLRP3 inflammasome activation promotes inflammation-induced carcinogenesis in head and neck squamous cell carcinoma. J Exp Clin Cancer Res 36(1):116. https://doi.org/10.1186/s13046-017-0589-y
Isola G, Polizzi A, Alibrandi A, Williams RC, Lo Giudice A (2021) Analysis of galectin-3 levels as a source of coronary heart disease risk during periodontitis. J Periodontal Res 56(3):597–605. https://doi.org/10.1111/jre.12860
Isola G, Polizzi A, Santonocito S, Alibrandi A, Williams RC (2022) Periodontitis activates the NLRP3 inflammasome in serum and saliva. J Periodontol 93(1):135–145
Isola G, Polizzi A, Mascitti M, Santonocito S, Ronsivalle V, Cicciu M, Pesce P (2023a) Impact of periodontitis on circulating cell-free DNA levels as a measure of cardiovascular disease. Clin Oral Investig 27(11):6855–6863. https://doi.org/10.1007/s00784-023-05300-y
Isola G, Tartaglia GM, Santonocito S, Chaurasia A, Marya A, Lo Giudice A (2023b) Growth differentiation factor-15 and circulating biomarkers as predictors of periodontal treatment effects in patients with periodontitis: a randomized-controlled clinical trial. BMC Oral Health 23(1):582. https://doi.org/10.1186/s12903-023-03237-y
Isola G, Tartaglia GM, Santonocito S, Polizzi A, Williams RC, Iorio-Siciliano V (2023c) Impact of N-terminal pro-B-type natriuretic peptide and related inflammatory biomarkers on periodontal treatment outcomes in patients with periodontitis: an explorative human randomized-controlled clinical trial. J Periodontol 94(12):1414–1424. https://doi.org/10.1002/JPER.23-0063
Isola G, Polizzi A, Santonocito S, Alibrandi A, Pesce P, Kocher T (2024) Effect of quadrantwise versus full-mouth subgingival instrumentation on clinical and microbiological parameters in periodontitis patients: A randomized clinical trial. J Periodontal Res 59. https://doi.org/10.1111/jre.13279
Jain T, Chandra A, Mishra SP, Khairnar M, Rajoria S, Maheswari R, Keerthika R, Tiwari S, Agrawal R (2024) Unravelling the significance of NLRP3 and IL-β1 in Oral squamous cell carcinoma and potentially malignant Oral disorders: A diagnostic and prognostic exploration. Head Neck Pathol 18(1):77. https://doi.org/10.1007/s12105-024-01685-8
Jandinski JJ (1988) Osteoclast activating factor is now interleukin-1 beta: historical perspective and biological implications. J Oral Pathol 17(4):145–152. https://doi.org/10.1111/j.1600-0714.1988.tb01515.x
Kamatani T, Shiogama S, Yoshihama Y, Kondo S, Shirota T, Shintani S (2013) Interleukin-1 beta in unstimulated whole saliva is a potential biomarker for oral squamous cell carcinoma. Cytokine 64(2):497–502. https://doi.org/10.1016/j.cyto.2013.08.011
Karaaslan S (2023) Relationship of NLRP3 Inflammasome with Oral health. Curr Oral Health Rep 10(4):196–202
Kelk P, Moghbel NS, Hirschfeld J, Johansson A (2022) Aggregatibacter actinomycetemcomitans Leukotoxin activates the NLRP3 Inflammasome and cell-to-cell communication. Pathogens 11(2). https://doi.org/10.3390/pathogens11020159
Khabadze Z, Kulikova A, Generalova Y, Abdulkerimova S, Dashtieva M, Bakaev Y, Shainyan A, Yusupov M, Golovina V, Karapetov K (2023) The prevalence of inflammatory periodontal diseases (gingivitis, periodontitis) among the population. J Int Dental Med Res 16(4):1830–1835
Koca-Ünsal RB, Şehirli AÖ, Sayıner S, Aksoy U (2022) Relationship of NLRP3 inflammasome with periodontal, endodontic and related systemic diseases. Mol Biol Rep 49(11):11123–11132
Kumari P, Debta P, Dixit A (2022) Oral potentially malignant disorders: etiology, pathogenesis, and transformation into Oral cancer. Front Pharmacol 13:825266. https://doi.org/10.3389/fphar.2022.825266
Lin T-Y, Tsai M-C, Tu W, Yeh H-C, Wang S-C, Huang S-P, Li C-Y (2021) Role of the NLRP3 inflammasome: insights into cancer hallmarks. Front Immunol 11:610492
Lopez-Labady J, Bologna-Molina R, Villarroel-Dorrego M (2021) Expression of interleukin-1ß and Interleukin-8 in Oral potentially malignant disorders and carcinomas. Front Oral Health 2:649406. https://doi.org/10.3389/froh.2021.649406
Mancuso M, Matassa DS, Conte M, Colella G, Rana G, Fucci L, Piscopo M (2009) H3K4 histone methylation in oral squamous cell carcinoma. Acta Biochim Pol 56(3):405–410
Marandi Y, Hashemzade S, Tayebinia H, Karimi J, Zamani A, Khodadadi I (2021) NLRP3-inflammasome activation is associated with epithelial-mesenchymal transition and progression of colorectal cancer. Iran J Basic Med Sci 24(4):483

Marchesan JT, Girnary MS, Moss K, Monaghan ET, Egnatz GJ, Jiao Y, Zhang S, Beck J, Swanson KV (2020) Role of inflammasomes in the pathogenesis of periodontal disease and therapeutics. Periodontol 82(1):93–114

Mirlekar B (2022) Tumor promoting roles of IL-10, TGF-β, IL-4, and IL-35: its implications in cancer immunotherapy. SAGE Open Med 10: 20503121211069012

Missiroli S, Perrone M, Boncompagni C, Borghi C, Campagnaro A, Marchetti F, Anania G, Greco P, Fiorica F, Pinton P, Giorgi C (2021) Targeting the NLRP3 Inflammasome as a new therapeutic option for overcoming cancer. Cancers (Basel) 13(10). https://doi.org/10.3390/cancers13102297

Neudorf H, Durrer C, Myette-Cote E, Makins C, O'Malley T, Little JP (2019) Oral ketone supplementation acutely increases markers of NLRP3 inflammasome activation in human monocytes. Mol Nutr Food Res 63(11):1801171

Okano T, Ashida H, Suzuki S, Shoji M, Nakayama K, Suzuki T (2018) Porphyromonas gingivalis triggers NLRP3-mediated inflammasome activation in macrophages in a bacterial gingipains-independent manner. Eur J Immunol 48(12):1965–1974

Orozco A, Gemmell E, Bickel M, Seymour GJ (2006) Interleukin-1beta, interleukin-12 and interleukin-18 levels in gingival fluid and serum of patients with gingivitis and periodontitis. Oral Microbiol Immunol 21(4):256–260. https://doi.org/10.1111/j.1399-302X.2006.00292.x

Papapanou PN, Sanz M, Buduneli N, Dietrich T, Feres M, Fine DH, Flemmig TF, Garcia R, Giannobile WV, Graziani F, Greenwell H, Herrera D, Kao RT, Kebschull M, Kinane DF, Kirkwood KL, Kocher T, Kornman KS, Kumar P, S.,...Tonetti, M. S. (2018) Periodontitis: consensus report of workgroup 2 of the 2017 world workshop on the classification of periodontal and Peri-implant diseases and conditions. J Periodontol 89(Suppl 1):S173–s182. https://doi.org/10.1002/jper.17-0721

Papathanasiou E, Conti P, Carinci F, Lauritano D, Theoharides TC (2020) IL-1 superfamily members and periodontal diseases. J Dent Res 99(13): 1425–1434. https://doi.org/10.1177/0022034520945209

Patel MN, Carroll RG, Galván-Peña S, Mills EL, Olden R, Triantafilou M, Wolf AI, Bryant CE, Triantafilou K, Masters SL (2017) Inflammasome priming in sterile inflammatory disease. Trends Mol Med 23(2):165–180

Peeters PM, Wouters EF, Reynaert NL (2015) Immune homeostasis in epithelial cells: evidence and role of inflammasome signaling reviewed. J Immunol Res 2015(1):828264

Poli G, Fabi C, Bellet MM, Costantini C, Nunziangeli L, Romani L, Brancorsini S (2020) Epigenetic mechanisms of Inflammasome regulation. Int J Mol Sci 21(16). https://doi.org/10.3390/ijms21165758

Polizzi A, Santonocito S, Lo Giudice A, Alibrandi A, De Pasquale R, Isola G (2023) Analysis of the response to two pharmacological protocols in patients with oral lichen planus: A randomized clinical trial. Oral Dis 29(2):755–763. https://doi.org/10.1111/odi.13960

Polizzi A, Santonocito S, Distefano A, De Pasquale R, Alibrandi A, Alanazi AM, Li Volti G, Isola G (2024) Analysis of oral lichen planus severity on micro-RNA linked with malignant transformation risks. Oral Dis 30(5):2918–2928. https://doi.org/10.1111/odi.14758

Porritt RA, Zemmour D, Abe M, Lee Y, Narayanan M, Carvalho TT, Gomez AC, Martinon D, Santiskulvong C, Fishbein MC (2021) NLRP3 inflammasome mediates immune-stromal interactions in vasculitis. Circ Res 129(9):e183–e200

Ptasiewicz M, Grywalska E, Mertowska P, Korona-Głowniak I, Poniewierska-Baran A, Niedźwiedzka-Rystwej P, Chałas R (2022) Armed to the teeth—the oral mucosa immunity system and microbiota. Int J Mol Sci 23(2):882

Rathinam VA, Chan FK-M (2018) Inflammasome, inflammation, and tissue homeostasis. Trends Mol Med 24(3):304–318

Sawada S, Chosa N, Ishisaki A, Naruishi K (2013) Enhancement of gingival inflammation induced by synergism of IL-1β and IL-6. Biomed Res 34(1):31–40

Schieffer L, Manzl C, Schatz C, Haybaeck J, Crismani A (2022) Nrf2 in the field of dentistry with special attention to NLRP3. Antioxidants 11(1):149

Shanmugam MK, Sethi G (2013) Role of epigenetics in inflammation-associated diseases. Subcell Biochem 61:627–657. https://doi.org/10.1007/978-94-007-4525-4_27

Shao S, Gudjonsson JE (2020) Epigenetics of psoriasis. Adv Exp Med Biol 1253:209–221. https://doi.org/10.1007/978-981-15-3449-2_8

Sharma M, Fonseca FP, Hunter KD, Radhakrishnan R (2020) Loss of oral mucosal stem cell markers in oral submucous fibrosis and their reactivation in malignant transformation. Int J Oral Sci 12(1):23. https://doi.org/10.1038/s41368-020-00090-5

Shibata K-i (2018) Historical aspects of studies on roles of the inflammasome in the pathogenesis of periodontal diseases. Mol Oral Microbiol 33(3):203–211

Shih YH, Wang TH, Shieh TM, Tseng YH (2019) Oral submucous fibrosis: A review on Etiopathogenesis, diagnosis, and therapy. Int J Mol Sci 20(12). https://doi.org/10.3390/ijms20122940

Stasiewicz M, Karpiński TM (2022) The oral microbiota and its role in carcinogenesis. Semin Cancer Biol 86: 633

Strowig T, Henao-Mejia J, Elinav E, Flavell R (2012) Inflammasomes in health and disease. Nature 481(7381):278–286. https://doi.org/10.1038/nature10759

Takahashi N (2005) Microbial ecosystem in the oral cavity: metabolic diversity in an ecological niche and its relationship with oral diseases. Int Congr Ser 1284:103

Talley S, Rademacher DJ, Campbell EM (2023) Inflammasome activation occurs in CD4(+) and CD8 (+) T cells during graft-versus-host disease. Cell Death Dis 14(9):632. https://doi.org/10.1038/s41419-023-06138-8

Thi Do T, Phoomak C, Champattanachai V, Silsirivanit A, Chaiyarit P (2018) New evidence of connections between increased O-GlcNAcylation and inflammasome in the oral mucosa of patients with oral lichen planus. Clin Exp Immunol 192(1):129–137

Toldo S, Abbate A (2024) The role of the NLRP3 inflammasome and pyroptosis in cardiovascular diseases. Nat Rev Cardiol 21(4):219–237

Ushio R, Hiroi M, Matsumoto A, Mori K, Yamamoto N, Ohmori Y (2022) Enhanced cytotoxic effects in human Oral squamous cell carcinoma cells treated with combined methyltransferase inhibitors and histone deacetylase inhibitors. Biomedicine 10(4). https://doi.org/10.3390/biomedicines10040763

Wang H, Luo Q, Feng X, Zhang R, Li J, Chen F (2018) NLRP3 promotes tumor growth and metastasis in human oral squamous cell carcinoma. BMC Cancer 18:1–10

Warnakulasuriya S, Kujan O, Aguirre-Urizar JM, Bagan JV, González-Moles MÁ, Kerr AR, Lodi G, Mello FW, Monteiro L, Ogden GR (2021) Oral potentially malignant disorders: A consensus report from an international seminar on nomenclature and classification, convened by the WHO collaborating Centre for Oral Cancer. Oral Dis 27(8):1862–1880

Xia Y, Zhou K, Sun M, Shu R, Qian J, Xie Y (2021) The miR-223-3p regulates Pyroptosis through NLRP3-caspase 1-GSDMD signal Axis in periodontitis. Inflammation 44(6):2531–2542. https://doi.org/10.1007/s10753-021-01522-y

Yan Y, Lu K, Ye T, Zhang Z (2019) MicroRNA-223 attenuates LPS-induced inflammation in an acute lung injury model via the NLRP3 inflammasome and TLR4/NF-κB signaling pathway via RHOB. Int J Mol Med 43(3):1467–1477. https://doi.org/10.3892/ijmm.2019.4075

Yilmaz Ö, Lee KL (2015) The inflammasome and danger molecule signaling: at the crossroads of inflammation and pathogen persistence in the oral cavity. Periodontol 69(1):83–95

You Y, Huang Y, Wang D, Li Y, Wang G, Jin S, Zhu X, Wu B, Du X, Li X (2019) Angiotensin (1-7) inhibits arecoline-induced migration and collagen synthesis in human oral myofibroblasts via inhibiting NLRP3 inflammasome activation. J Cell Physiol 234(4):4668–4680. https://doi.org/10.1002/jcp.27267

Zasloff M (2002) Innate immunity, antimicrobial peptides, and protection of the oral cavity. Lancet 360(9340):1116–1117

Zeng Y, Luo M, Yao Z, Xiao X (2024) Adiponectin inhibits ROS/NLRP3 inflammatory pathway through FOXO3A to ameliorate oral submucosal fibrosis. Odontology 112(3):811–825. https://doi.org/10.1007/s10266-023-00891-0

Zhu X, Huang H, Zhao L (2022) PAMPs and DAMPs as the bridge between periodontitis and atherosclerosis: the potential therapeutic targets. Front Cell Develop Biol 10:856118

The Role of Proinflammatory Biomarkers in Oral Cancer Pathophysiology

17

Busra Yilmaz and Bulent Ozpolat

Abstract

Oral cancer (OC) is one of the most commonly diagnosed head and neck cancers and includes malignancies arising from the tongue, lips, floor of the mouth, buccal mucosa, hard palate, alveolar ridge, gingiva, and oropharynx. The most common histological type is squamous cell carcinoma, which originates from squamous cells lining the mouth and throat and accounts for over 90% of all OC cases. Notably, OC, with a five-year survival rate of approximately 68% represents a significant health burden worldwide. Biomarkers are used as molecular signatures, enabling the objective differentiation between normal and pathological conditions as well as quantifiable characteristics of disease processes, pharmacological responses to therapeutic interventions, prediction of clinically relevant endpoints, and development of personalized therapies. In the intricate landscape of OC progression, inflammation within the tumor microenvironment emerges as a central player. Predictive biomarkers include tumor necrosis factor-alpha (TNF-α), interleukin (IL)-17, and interferon-gamma (IFN-γ), which modulate tumor growth and affect response to cancer therapy. Prognostic biomarkers, such as IL-6, IL-8, IL-1α, IL-1β, cyclooxygenase (COX)-2, C-reactive protein (CRP), and hematological parameters, such as neutrophil-to-lymphocyte ratio, platelet-to-lymphocyte ratio, lymphocyte-to-monocyte ratio, and monocyte-to- lymphocyte ratio provide insights into clinical outcomes. Diagnostic biomarkers, such as IL-1β, IL-8, TNF-α, and COX-2, assist in predicting carcinogenesis or determining cancer subtypes. Monitoring biomarkers, including salivary cytokines (IL-1β, IL-8, and IL-6) and TNF-α, help evaluate changes in disease status. Overall, emerging data suggest that biomarkers of inflammation provide invaluable tools for the evaluation of cancer progression, prognosis, and clinical outcomes.

Keywords

Oral squamous cell carcinoma · Biomarkers · Inflammation mediators · Cytokines · Tumor microenvironment

B. Yilmaz (✉)
Department of Oral Medicine and Maxillofacial Radiology, School of Dental Medicine, Istinye University, Istanbul, Turkey

Department of Nanomedicine, Houston Methodist Research Institute, Houston, TX, USA
e-mail: byilmaz@houstonmethodist.org

B. Ozpolat (✉)
Department of Nanomedicine, Houston Methodist Research Institute, Houston, TX, USA
e-mail: bozpolat@houstonmethodist.org

N. Rezaei, H. Ziaei (eds.), *Oral Immunology*, Advances in Experimental Medicine and Biology 1492,
https://doi.org/10.1007/978-3-032-03176-1_17

17.1 Introduction

17.1.1 Overview of Oral Cancer

17.1.1.1 Definition and Epidemiology

Oral cancer (OC), one of the most common types of head and neck cancers, is a broad term representing cancers originating from the tissues of the mouth. It frequently involves malignancies arising in the tongue, lips, floor of the mouth, buccal mucosa, hard palate, alveolar ridge, gingiva, and oropharynx. The most common histological type is squamous cell carcinoma (SCC), which originates from the flat, squamous cells lining the mouth and throat, accounting for over 90% of all OC cases (Cancer.Net 2023). Based on global estimates, approximately 476,100 cases were diagnosed with oral or oropharyngeal cancer in 2020 (Cancer.Net 2023), and regrettably, OC represents a significant worldwide health burden with around 389,900 new cases and 188,400 deaths annually (Ferlay et al. 2024).

The epidemiology of OC involves a multifaceted examination, covering parameters such as incidence, prevalence, distribution, and temporal trends. The incidence rates of OC exhibit significant geographic variation, with a more than 20-fold difference observed worldwide. Asia holds the highest prevalence rate of OC at 62.4%, largely due to regional factors and lifestyle practices. Europe is accompanied by a prevalence rate of 18.5%, reflecting various risk factors and healthcare practices. Prevalence rates in the Americas vary, with North America at 9.7% and Latin America and the Caribbean at 5.1%. Africa shows a prevalence rate of 3.1%, while Oceania has the lowest prevalence at 1.2% (Ferlay et al. 2024).

Recent epidemiological studies report that the incidence of OC is approximately two or three times higher in males than females, making it the eighth most common cancer among men worldwide (Louredo et al. 2022; Stepan et al. 2023). Furthermore, despite the average age of diagnosis reported to be 64, there has been a significant increase in incidence rates among younger age groups in recent years (Cancer.Net 2023). In particular, oral SCC (OSCC) tends to occur in individuals under the age of 45 who do not smoke or consume alcohol, distinguishing them from conventional head and neck SCC (HNSCC) patients. These trends underscore the changing demographic patterns and etiological factors contributing to the epidemiology of OC (Ng et al. 2017; Tran et al. 2023). In addition to demographic factors, socioeconomic status, access to healthcare, and environmental exposures also influence the epidemiology of OC. According to the systematic review by Conway et al., low socioeconomic status and lifestyle are potentially significant factors in the development of OC (Conway et al. 2008). Moreover, ancestral and ethnic disparities significantly influence OC incidence and mortality, often attributed to diverse cultural practices and risk behaviors. Regions like South and East Asia, and Melanesia, where betel quid chewing is prevalent, show higher OC rates (Ramanathan and Lakshimi 1974). In the United States, social and environmental factors play a pivotal role in the higher rates observed among black males (Morse and Kerr 2006). Understanding these epidemiological patterns is crucial for informing preventive strategies, screening programs, and healthcare policies aimed at reducing the burden of OC on a global scale.

17.1.1.2 Risk Factors and Common Etiological Agents

Tobacco use, whether in the form of smoking or smokeless products, is a major causal factor for oral and pharyngeal cancers, supported by numerous epidemiological studies. Smokers face a significantly heightened risk, with heavy smokers particularly vulnerable. Continuing to smoke post-treatment for OC worsens the prognosis and raises the risk of secondary cancers (GBD 2019 Lip, Oral, and Pharyngeal Cancer Collaborators et al. 2023). The primary carcinogens in tobacco use, such as benzo(a)-pyrene (B(a)P) and tobacco-specific nitrosamines (TSNAs), induce DNA damage. TSNAs like N′--nitrosonornicotine (NNN) and N-nitroso-4-(methylnitrosamino)-1-(3-pyridyl)-1-butanone

(NNK) trigger tumors in the oral carcinogenesis (Warnakulasuriya et al. 1999). NNK, NNN, and their metabolites form covalent bonds with the DNA of keratinocyte stem cells, leading to the formation of DNA adducts responsible for critical mutations in DNA replication (Warnakulasuriya et al. 1999). The metabolism of these carcinogens involves oxygenation by P450 enzymes in cytochromes and conjugation by glutathione-S-transferase. The genetic predisposition to tobacco-related OCs may be linked to polymorphisms in the genes encoding these enzymes (Yesupatham et al. 2023).

The influence of alcohol on oral carcinogenesis depends on the interactive effect that emerges from its combined use with tobacco. This phenomenon, accounting for approximately 75% of OCs, arises from alcohol's ability to increase the permeability of the oral mucosa, trigger epithelial atrophy, and facilitate the infiltration of carcinogens into the oral mucosa (Marziliano et al. 2020). The main metabolite of alcohol produced by the alcohol dehydrogenase (ADH) enzyme is acetaldehyde, which is oxidized to acetate via aldehyde dehydrogenase (ALDH). Acetaldehyde not only interferes with DNA synthesis and repair, causing DNA damage but also induces sister chromatid exchanges and specific gene mutations. Another effect of acetaldehyde is its inhibition of the 6-methylguanine transferase enzyme responsible for repairing injuries caused by alkylating agents (Lee et al. 2019). Hence, the accumulation of acetaldehyde, with all its adverse effects, is acknowledged to induce carcinogenesis. Additionally, the decrease in ALDH enzyme activity leads to the accumulation of acetaldehyde, while ADH type-3 genotypes accelerate the oxidation of acetaldehyde from alcohol, thereby increasing the risk of OC (Murti et al. 1985).

Human papillomavirus (HPV) infection, a well-established biological risk factor for oral carcinomas, also serves as a biomarker for such malignancies. Specifically, it has emerged as a significant risk factor strongly associated with the development of OSCC, particularly for subtypes like HPV-16 (Radaic et al. 2023). In contrast to traditional risk factors like smoking and alcohol abuse, HPV infection categorizes OSCC tumors, impacting prognosis, and treatment response (Lambert et al. 2011). HPV-positive (HPV+) OSCC patients generally experience better outcomes, including higher survival rates and reduced risk of disease progression and recurrence compared to HPV-negative (HPV-) individuals (Fakhry and D'Souza 2013). The presence of p16, a prognostic biomarker, further enhances the favorable prognosis for HPV-positive (HPV+) OSCC patients (Wang et al. 2017). Additionally, the interaction between p53 and HPV-encoded E6 protein is pivotal in OSCC pathogenesis, with HPV+ tumors typically displaying wild-type p53 levels associated with a better prognosis than HPV- tumors characterized by elevated p53 mutations (Singh et al. 2017).

Diet and nutrition significantly influence OC risk, as confirmed by the International Agency for Research on Cancer (IARC). Low intake of fruits and vegetables increases the risk, while a diet rich in processed meats, cakes, butter, eggs, and red meat also may raise this risk (Jeng et al. 2001). The protective effects of fruits and vegetables stem from their rich content of vitamins, antioxidants, and bioactive compounds, such as vitamins A, C, and E, β-carotene, and selenium. These micronutrients act as antioxidants, combating free radical damage to DNA and cellular membranes. Although establishing direct cause-effect relationships through large-scale trials is challenging, evidence suggests that a diet abundant in nutrient-dense foods can significantly lower the incidence of OSCCs, highlighting the importance of dietary modifications in cancer prevention (Tan et al. 2023).

17.1.2 Importance of Biomarkers in Cancer Research

17.1.2.1 Definition and Types of Biomarkers

A biomarker, functioning as a quantifiable characteristic, signifies normal biological functions, disease processes, or pharmacological responses to therapeutic interventions, while also serving as a biological observation that can substitute for and ideally predict clinically relevant endpoints or intermediate outcomes (Aronson and Ferner 2017). This term goes beyond simple substances, encompassing a wide range of indicators such as radiological findings, laboratory data from bodily fluids or tissues, and genetic or genomic information derived from DNA, RNA, peptides, and proteins. Biomarkers serve as molecular signatures, enabling the objective differentiation between normal and pathological conditions and contributing to the comprehension of the body's response to treatments. As stated by the National Cancer Institute (NCI) of the National Institutes of Health (NIH), biomarkers are crucial for monitoring the effectiveness of disease treatments, advancing medical diagnostics, and developing personalized therapies (Biomarkers Definitions Working Group 2001).

Biomarkers are categorized into predictive, prognostic, diagnostic, monitoring, pharmacodynamic/response, susceptibility/risk, safety, and digital biomarkers by the Food and Drug Administration (FDA) (FDA-NIH Biomarker Working Group 2016). Over the past two decades, advancements in molecular and cellular cancer research have driven the emergence of numerous biomarkers, which have become indispensable tools for early cancer detection and the discovery of novel targeted therapies. These biomarkers, essential in cancer diagnosis and screening, originate from non-invasive sources like saliva, plasma, serum, cyst fluids, cerebrospinal fluid, and urine. The primary aim in clinical settings is for these markers to accurately reflect the presence or absence of the disease, as well as responses to treatment (Diamandis 2010).

Predictive biomarkers enhance therapeutic precision and aid in developing personalized treatment strategies by evaluating the likelihood of a patient's response to cancer management. Prognostic biomarkers aid treatment planning and patient management by offering insights into potential cancer outcomes, whether with or without intervention (Ballman 2015). Diagnostic biomarkers facilitate early detection and precise diagnosis by anticipating cancer occurrence or categorizing individuals by disease subtype. Risk biomarkers provide valuable guidance for preventive measures in clinical settings by evaluating the probability of an individual developing an illness or health condition. Monitoring biomarkers detects changes in disease status, including occurrences, recurrences, progressions, and other clinically significant developments. Response biomarkers reflect the physiological response of an individual to a therapeutic intervention or environmental stimulus. Safety biomarkers assess the probability and severity of toxicity as an adverse reaction, both before and after exposure to a therapeutic drug or environmental substance (Al-Tashi et al. 2023).

17.1.2.2 Role of Biomarkers in Oral Cancer Diagnosis, Prognosis, and Treatment

The early detection of OC commonly involves visual and tactile examination, enabling the identification of abnormalities in more than 10% of cases (Bouquot 1986). While leukoplakia is common, biopsies of lesions with persistent ulcers or erythroplakia have a higher potential for OSCC development (Shafer and Waldron 1975). Dysplasia may also be present in normal mucosa (Thomson 2002). In this context, additional diagnostic tools include vital staining (e.g., toluidine blue), cytologic testing, autofluorescence (e.g., VELscope), salivary adjuncts (e.g., MOP genetic OC screening), and tissue reflectance (e.g., ViziLite Plus). The American Dental Association (ADA) emphasizes an immediate biopsy for suspected malignant lesions; however, cytological adjuncts may provide useful information if a biopsy is refused (Lingen et al. 2017). After diagnosis, in the management of OC, a personalized treatment plan is typically developed based on the assessment of clinical examination, radiology, and pathology findings. The initial step involves determining whether curative or palliative treatment modalities should be presented to the patient. Surgical intervention is preferred in cases of OSCC with resectable lesions and limited primary involvement. Potential therapeutic approaches encompass surgical, radiotherapy, or chemoradiotherapy modalities. Furthermore, when encountering distant metastasis and surgical inaccessibility of regional lymph nodes, palliative treatment is recommended for OSCC (Wong and Wiesenfeld 2018).

Despite treatment advancements, the 5-year relative survival rate stands at approximately 64% (Howlader et al. 2013). Additionally, the stage-dependent nature of OSCC survival, coupled with a decrease in survival rates 5 years post-diagnosis following regional and distant metastases, and the late-stage diagnosis of about 70% of new OSCC cases, underscores the importance of early diagnosis and preventive strategies (Siegel et al. 2017). Hence, biomarkers derived from genetics, proteins, metabolites, autoantibodies, and the microbiome play a pivotal role in oncology. These biomarkers may predict OSCC development and/or progression, aiding early detection, treatment decision-making, and response prediction. Moreover, reliable biomarkers for OSCC detection, including those predicting oral dysplasia progression to cancer, promise to enhance OSCC management as non-invasive tools (Radaic et al. 2023). In this context, detailed data exist on the role of various biomarkers in detecting OSCC. Nucleic acid-based biomarkers, such as circulating tumor DNA (ctDNA) and HPV DNA, are considered reliable indicators for identifying the presence and characteristics of OSCC (Payne et al. 2018). Additionally, metabolite-based markers, particularly trimethylamine N-oxide (TMAO) (Supawat et al. 2021), lactate, and glutamine, reflect the activation of cancer-related metabolic pathways, providing potential diagnostic indicators (Paul et al. 2020). The presence of tumor-associated autoantibodies is also significant for early diagnosis and prognosis assessment. Increased levels of matrix metalloproteinases (MMPs) along with changes in other protein levels, facilitate cancer cell invasion (Hsueh et al. 2022). As potential oral microbiome-based biomarkers, heightened levels of bacteria like *Porphyromonas* spp. and *Fusobacterium* spp., alongside a notable increase in fungi such as *Candida albicans*, may play pivotal roles in disease advancement and inflammation processes associated with cancer (Binder Gallimidi et al. 2015).

In recent years, mounting evidence has underscored the pivotal role of inflammation and cell-mediated immunity in oral carcinogenesis. Studies have highlighted the potential efficacy of various proinflammatory cytokines such as interleukin (IL)-6, IL-8, tumor necrosis factor-α (TNF-α), and transforming growth factor (TGF)-β1 (Schiegnitz et al. 2018). Additionally, cytokines with the C-X-C motif including CXCL1, CXCL3, CXCL2, and CXCL10 (Goertzen et al. 2018), along with inflammatory proteins implicated in OSCC pathophysiology such as C-reactive protein (CRP) (Chang et al. 2013a, b), MMP-2, −9 (Patel et al. 2007), play a crucial role in the intricate development and progression of OC. As a result, this section is devoted to uncovering the potential of proinflammatory biomarkers for the exploration and evidence-based prediction, diagnosis, evaluation, and strategic management planning of OC (Table 17.1).

17.2 Inflammation and Cancer

Inflammation represents the body's reaction to internal or external stimuli, aimed at shielding it from harmful agents and toxic insults. Initially, it presents as a pyrogenic or acute inflammatory response, possessing therapeutic qualities. However, should this condition persist or become chronic, proinflammatory cytokines prompt immune cells to release reactive oxygen species (ROS) and reactive nitrogen species (RNS) (Coussens and Werb 2002). The cytotoxic impact of these substances not only aids in eliminating invaders but also leads to the demise of host cells or triggers epigenetic changes and mutations (Singh et al. 2019). When the immune system fails to function effectively and healing remains incomplete, inflammation can transition from acute to chronic, resulting in recurring injuries and delayed wound healing, potentially creating an environment that supports cancer initiation (Coussens and Werb 2002).

In fact, in the nineteenth century, Rudolf Virchow proposed the presence of leukocytes in tumors as the first evidence of a potential link between inflammation and cancer, establishing the association between inflammation and cancer (Karin 2006). Definitive evidence for the role of

Table 17.1 Proinflammatory biomarkers and their functions and applications in OC

Biomarker	Function	Clinical application	Research application	Relationship with oral cancer
IL-1α (Bae et al. 2014)	Inflammation and immune response	Early diagnosis	Identifying new therapeutic targets	Increased levels have been linked to enhanced tumor proliferation and metastatic spread
IL-1β (Lee et al. 2015)	Inflammation and immune response	Diagnosis, prognosis	Tumor microenvironment analysis	High concentrations indicate invasive tumor behavior and are linked to unfavorable clinical outcomes
IL-6 (Pradhan et al. 2024)	Inflammation and immune response	Early diagnosis, prognosis	Identifying new therapeutic targets	Increased levels correlate with advanced disease stages, aggressive tumor progression, and reduced survival rates
IL-8 (Ralli et al. 2020)	Chemotaxis and cell migration	Diagnosis, tumor invasion	Tumor microenvironment analysis	Augmented levels signify an increased potential for tumor cell migration and metastasis
TNF-α (Niklander 2021)	Cell death and inflammation	Tumor growth, inflammation	Immunotherapy research	Raised levels contribute to chronic inflammation, stimulate tumor proliferation, and are indicative of poor clinical outcomes
TGF-β1 (Magnussen et al. 2017)	Cell growth and differentiation	Tumor progression	Tumor suppression mechanisms	Enhanced concentrations support tumor growth, facilitate immune evasion, and promote disease progression
CXCL1 (Goertzen et al. 2018; Lee et al. 2015)	Chemotaxis and inflammation	Tumor growth, angiogenesis	Angiogenesis inhibitors	High levels drive angiogenesis and tumor expansion, supporting a pro-tumorigenic environment
CXCL2 (Goertzen et al. 2018; Lee et al. 2015)	Inflammation and immune response	Inflammation	Investigating inflammatory response	Increased levels foster chronic inflammation and enhance tumor invasiveness
CXCL3 (Goertzen et al. 2018; Lee et al. 2015)	Chemotaxis and cell migration	Tumor growth, metastasis	Metastasis mechanisms	High concentrations enhance tumor cell migration, contributing to invasion and metastatic potential
CXCL10 (Goertzen et al. 2018; Lee et al. 2015)	Antiviral response and chemotaxis	Antiviral response, chemotaxis	Immunotherapy strategies	Increased levels modulate immune cell recruitment, fostering a tumor-supportive microenvironment and facilitating progression
CRP (Liu et al. 2023; Salema et al. 2024)	Inflammation marker	Inflammation	Monitoring inflammatory response	Heightened levels are indicative of systemic inflammation and correlate with tumor advancement
MMP-2 (Chang et al. 2013a, b)	Extracellular matrix degradation	Preventing metastasis	Invasion	Raised levels contribute to extracellular matrix breakdown, facilitating invasion and metastatic spread
MMP-9 (Chang et al., 2013a, b)	Extracellular matrix degradation	Preventing metastasis	Invasion	Enhanced concentrations promote tissue remodeling, aiding tumor infiltration and distant metastases
PGE-2 (Nasry et al. 2018)	Inflammation and pain	Pain management, inflammation	Modulating pro-inflammatory response	Increased levels are pivotal in linking inflammation to tumor development and progression
COX-2 (Nasry et al. 2018)	Inflammation and pain	Pain management, inflammation	Modulating pro-inflammatory response	High concentrations are essential in sustaining inflammatory responses and driving tumor expansion
NLR (Magdum et al. 2024; Mariani et al. 2022)	Inflammation and immune response	Prognostic marker	Modulating immune response	High levels reflect systemic inflammation, correlating with advanced disease stages and unfavorable clinical outcomes.

(continued)

Table 17.1 (continued)

Biomarker	Function	Clinical application	Research application	Relationship with oral cancer
LMR (Ding et al. 2021)	Inflammation and immune response	Prognostic marker	Modulating immune response	Low levels indicate suppressed anti-tumor immunity and are linked to enhanced tumor progression
PLR (Kumarasamy et al. 2021)	Inflammation and immune response	Prognostic marker	Modulating immune response	High levels signify heightened inflammatory activity, supporting tumor growth and metastatic potential

Abbreviations: *CRP* C-reactive protein, *COX-2* Cyclooxygenase-2, *CXCL* cytokines harboring the C-X-C motif, *IL* interleukin, *MMP* Matrix metalloproteinase, *NLR* neutrophil-to-lymphocyte ratio, *LMR* lymphocyte-to-monocyte ratio, *PLR* platelet-to-lymphocyte ratio, *PGE-2* Prostaglandin E2, *TNF-α* tumor necrosis factor-α, *TGF-β1* transforming growth factor-β1

inflammation in carcinogenesis was later demonstrated through studies showing that the use of nonsteroidal anti-inflammatory drugs (NSAIDs) (Hawk and Levin 2005) or aspirin to reduce inflammation significantly decreased the incidence of various cancers, including those of the brain, esophagus, lung, and stomach (Rothwell et al. 2011).

17.2.1 The Link Between Inflammation and Cancer

Inflammation contributing to cancer development is driven by the recruitment of immune cells, including macrophages and neutrophils, which serve as key producers of ROS and RNS. With repeated tissue damage, these reactive species induce mutations in DNA and RNA and significantly alter the structures of lipids, glycans, and proteins. Within the tumor microenvironment, inflammatory cytokines, including TNF-α, are released by activated macrophages. These cytokines stimulate ROS production in neighboring cells, thereby perpetuating inflammation that promotes tumor growth. Moreover, TNF-α initiates the activation of nuclear factor-κB (NF-κB), a vital transcription factor within the inflammatory pathway, which is commonly activated across various cancer types. In most cancers, extracellular proteases like MMPs, cytokines such as IL-6 and IL-8, and the activation of signal transducer and activator of transcription 3 (STAT3) synergistically contribute to cellular proliferation, angiogenesis, and survival, thereby supporting tumor progression (Ben-Neriah and Karin 2011).

Carcinogenesis progresses through, initiation, promotion, malignant conversion, and tumor progression driven by cumulative genetic and epigenetic alterations resulting from mutations in proto-oncogenes and tumor suppressor genes, ultimately leading to malignancy. Cells in an initiated state, signifying a premalignant phase, persist in tissues, encountering diverse promotional stimuli, and acquiring distinct traits known as hallmarks of cancer, facilitating tumor growth and metastasis (Hanahan and Weinberg 2011). Cancer and inflammation are linked through two pathways, intrinsic and extrinsic, and the types of inflammation tied to a heightened risk of cancer development, as well as the inflammatory signals, vary depending on the nature of the initial damage. The intrinsic pathway is triggered by genetic alterations, including oncogene activation through mutations, chromosomal rearrangements, or gene amplifications, alongside the inactivation of tumor suppressor genes. These genetic changes drive the production of inflammatory mediators, fostering an inflammatory tumor microenvironment that supports neoplastic development. In contrast, the "extrinsic pathway" is augmented by inflammatory or infectious conditions, increasing the risk of cancer development at specific anatomical sites. These two pathways converge through the activation of pivotal transcription factors within tumor cells, such as NF-κB, STAT3, and hypoxia-inducible factor-

1α (HIF-1α). These transcriptional regulators orchestrate the production of cytokines and chemokines, stimulating diverse leukocyte populations and fostering a pro-inflammatory microenvironment that supports cancer progression (Mantovani et al. 2008).

17.2.1.1 Chronic Inflammation and Carcinogenesis

Chronic inflammation influences approximately a quarter of all cancer cases globally (Mantovani et al. 2008). For example, inflammatory bowel diseases like chronic ulcerative colitis and Crohn's disease, are associated with an increased risk of colon cancer, hepatitis C infection is linked to liver cancer, and chronic *Helicobacter pylori* infection may lead to stomach cancer. This connection is elucidated by the DNA damage caused by persistent infections and inflammatory processes, predisposing cells to neoplastic transformation (Ernst and Gold 2000).

Driven by persistent infections within the host, chronic inflammation perpetuates continuous tissue damage and regeneration. Throughout this process, high RNS and ROS released by inflammatory cells interact with DNA in proliferating epithelial cells, leading to enduring genomic alterations. Consequently, this heightens DNA damage and fosters the accumulation of potential oncogenic mutations. The release of the macrophage migration inhibitory factor from macrophages and T lymphocytes inhibits p53 function, thereby weakening the cellular response to DNA damage (Hudson et al. 1999). This inhibition of p53's regulatory functions due to chronic inflammation promotes cell proliferation and creates an environment with an insufficient response to DNA damage, thereby fostering a microenvironment conducive to cancer development (Coussens and Werb 2002). Inflammation affects other cells through messenger molecules such as cytokines, prostaglandins, chemokines, and angiogenic factors (Segal 2006). Inflammatory cytokines activate transcription factors like NF-κB, which upregulate genes involved in cell proliferation, apoptosis, and carcinogenesis, while also promoting the production of pro-inflammatory cytokines that amplify the inflammatory response (Khansari et al. 2009). Additionally, oxygen-derived free radicals damage cellular components and recruit more inflammatory cells, further exacerbating the damage. Ultimately, the continuous DNA damage and inflammatory signals generated by the chronic inflammatory microenvironment induce carcinogenesis, initiating its proliferation and spread.

During the progression of oral carcinogenesis, extending from non-dysplastic hyperkeratosis to OSCC, there is a gradual rise noted in both the quality and concentration of the inflammatory infiltration (Goertzen et al. 2018). Oral inflammations such as submucous fibrosis, oral discoid lupus, and oral lichen planus, by harboring active cytokines, chemokines, prostaglandins, ROS, and transcription factors in their microenvironments, increase susceptibility to OSCC development and underscore the strong link between chronic inflammation and oral carcinogenesis (Speight et al. 2018). These mediators possess the ability to induce proliferation, epithelial-to-mesenchymal transition (EMT), and invasion in lesions bearing mutations in tumor-suppressor genes and/or oncogenes, thereby promoting tumor progression (Lee et al. 2015). Chronic inflammation remains a hallmark feature in established OSCC, fueling oncogenesis, invasion, and metastasis (Goertzen et al. 2018). Therefore, the potential of various inflammatory molecules as both prognostic markers and therapeutic targets for OC has gained significant attention.

17.2.1.2 Molecular Mechanisms Linking Inflammation to Oral Cancer

Inflammation within the tumor microenvironment (TME) plays a central role in the progression of OC. Beyond the neoplastic cells, the TME hosts a dynamic interplay of various stromal cells. These diverse constituents engage in a complex dialogue with tumor cells, driving transformative alterations in the extracellular matrix that promote critical processes including proliferation, invasion, angiogenesis, and metastasis (D'Arcangelo et al. 2020). Characterized by genetic instability, malignant oral keratinocytes serve as instigators, unleashing a cascade of events by activating transcription factors like activator protein (AP)-1 and NF-κB. These molecular orchestrators, in turn, awaken oncogenes that govern crucial cellular functions such as apoptosis, angiogenesis, and cell growth (Rao et al. 2010). Additionally, these rogue cells trigger the expression of inflammatory genes, leading to the sustained secretion of inflammatory mediators like IL-1, IL-6, IL-8, TNF-α, and TGF-β. This creates a self-propagating cycle that further amplifies inflammation and drives tumor progression (Rhodus et al. 2005).

An ensemble of immune cells, including tumor-associated macrophages, T lymphocytes, neutrophils, and mast cells, significantly shape the inflammatory milieu within OSCC (Goertzen et al. 2018). Tumor-associated macrophages, known for their role in promoting proliferation and invasion, are key players in disease progression and poor prognoses (Hu et al. 2016; Mohapatra et al. 2021). Serving as reservoirs of cytokines, metalloproteinases, and growth factors, they play a crucial role in shaping the dynamics of the TME. Similarly, T lymphocytes, constituting the majority of inflammatory infiltrates in OSCC, act as a double-edged sword, either promoting or inhibiting tumor progression based on their secretory profiles (Greten and Grivennikov 2019). Meanwhile, neutrophils promote OSCC invasion via matrix degradation and invadopodia formation, while mast cells, involved in the process of angiogenesis, release an abundance of cytokines and chemokines (Goertzen et al. 2018).

Cancer-associated fibroblasts dominate the TME landscape, responding to signals like TGF-β and actively contributing to tumor progression by secreting vascular endothelial growth factor (VEGF), IL-6, IL-8, Prostaglandin E2 (PGE2), and TGF-β (Prime et al. 2017). Furthermore, cancer-associated fibroblasts orchestrate the activation of NF-κB, a key driver of poor prognostic outcomes and bone invasion in OSCC (Elmusrati et al. 2017). Senescent cells, which are most prevalent in aging individuals and cancer contexts, have a significant influence over chronic inflammation through the senescent-associated secretory phenotype (SASP). Characterized by a surge of pro-oncogenic cytokines like IL-1, IL-6, and IL-8, SASP drives EMT, invasion, and metastasis (Davalos et al. 2010). Within the chronic inflammatory microenvironment, persistent DNA damage and inflammatory signals combine to promote the process of carcinogenesis in OC, in terms of its initiation, proliferation, and dissemination.

17.2.2 Proinflammatory Biomarkers

17.2.2.1 Definition and Classification

Proinflammatory biomarkers are specific molecules, primarily proteins, crucial in regulating and signaling the body's inflammatory response. They facilitate communication among immune cells, orchestrate the body's defense mechanisms against pathogens, and contribute to tissue repair processes. In clinical practice, proinflammatory biomarkers are indispensable tools for early disease detection, assessing disease progression, and evaluating the effectiveness of therapeutic interventions. These specific markers, known for their regulatory role in immune response and diagnostic utility in identifying inflammation-related diseases, are measurable through non-invasive or minimally invasive procedures (Tampa et al. 2018). With reasonable half-lives and stability under various conditions, they ensure accurate measurement. Clinical validation is essential to confirm their reliability in predicting and monitoring disease states (Biomarkers Definitions Working Group 2001).

Proinflammatory biomarkers can be categorized based on their specific clinical roles and relevance in inflammation-related conditions (Table 17.1). Predictive biomarkers include factors like TNF-α, IL-17, and interferon-gamma (IFN-γ), which modulate tumor growth and affect response to cancer therapy. Prognostic biomarkers, such as IL-6, IL-8, and IL-1 family members (IL-1α, IL-1β), Cyclooxygenase-2 (COX-2), CRP and hematological parameters such as neutrophil-to-lymphocyte ratio (NLR), platelet-to-lymphocyte ratio (PLR), lymphocyte-to-monocyte ratio (LMR), and monocyte-to-lymphocyte ratio (MLR) provide insights into cancer outcomes. Diagnostic biomarkers like IL-1β, IL-8, TNF-α, and COX-2 assist in predicting carcinogenesis or categorizing individuals by disease subtype. Monitoring biomarkers, including salivary cytokines (IL-1β, IL-8, IL-6) and TNF-α, detect changes in disease status. Additionally, other biomarkers such as NF-κB, PGE2, and chemokines play significant roles in inflammation and cancer progression.

17.2.2.2 Common Proinflammatory Biomarkers in Oral Cancer

Common proinflammatory biomarkers in OC include IL-6 and IL-8. IL-6 plays a pivotal role in immune response modulation, hematopoiesis, and oncogenesis (Choudhary et al. 2016; Surcel et al. 2017). Elevated IL-6 levels maintain chronic inflammation by recruiting monocytes through monocyte chemoattractant protein (MCP)-1 secretion and are correlated with larger tumor sizes, advanced disease stages, and chemotherapy resistance in OSCCs (Gabay 2006). Additionally, IL-8, which is triggered by NF-κB and influenced by stimuli like hypoxia, binds to CRCX-1 and CRCX-2 receptors on inflammatory and tumor cells (Sahibzada et al. 2017). Studies demonstrated significantly higher IL-8 levels in OSCC patients compared to those with premalignant lesions or healthy individuals (Ralli et al. 2020). High levels of chemokines, such as CXCL8 and CCL2, along with their receptors, indicate aggressive tumor behavior, metastasis, and poor prognosis in OSCC by regulating immune cell recruitment and modulating the TME (Chen et al. 2024; Shan et al. 2021). Similarly, heightened expression of CXCL9 in OSCC tissues and serum serves as an indicator of advanced disease (Chang et al. 2013a, b).

CRP is another significant marker of systemic inflammation linked to oral carcinogenesis. Studies of premalignant lesions (Metgud and Bajaj 2016) and OSCCs with buccal mucosa involvement and lymph node invasion consistently highlight elevated CRP levels, emphasizing its potential as a prognostic marker (Chen et al. 2011). Additionally, TNF-α is crucial in OSCC development, promoting invasion, angiogenesis, and EMT. It upregulates genes linked to neutrophil recruitment and invasion, reducing overall and disease-free survival. Elevated TNF-α receptor-1 signaling enhances MMP-2 and MMP-9 production, contributing to metastasis via NF-κB, AKT, and PI3K pathways. Furthermore, TNF-α-induced EMT is associated with cancer stem cells, therapy resistance, and poor prognosis. Targeting TNF-α may offer therapeutic benefits for OSCC (Niklander 2021).

Chang and colleagues confirmed the predictive ability of E-selectin and Serum Amyloid A (SAA), in addition to IL-6 and CRP, among inflammatory markers in OSCC treatment (Chang et al. 2013a, b). Moreover, PGE2 contributes to OSCC progression by inducing oncogenic transcription factors through ROS production, suppressing anti-tumor immune responses, promoting angiogenesis, and synergizing with IL-6 to enhance cell transformation and malignancy progression via STAT3 activation (Nasry et al. 2018). Epithelial COX-2 induces oral carcinogenesis through the increased production of PGE2 (Husvik et al. 2009).

MMPs, known for signaling invasion potential, offer valuable prognostic insights into OC. Increases in total serum and plasma levels of MMP-2 and MMP-9 before and after the development of OC can predict advanced stages of disease and metastasis (Patel et al. 2007). Periodontal inflammation has been linked to the progression to OSCC, driven by factors like IL-1β, CXCL8, MMP12, and MMP13. This inflammation can promote OSCC through pathways associated with inflammation, such as the IL-17 signaling pathway (Chen et al. 2024).

Recent studies have shown that the systemic inflammatory response indicates tissue hypoxia, damage, and necrosis (Yilmaz et al. 2024). This effect in oral carcinogenesis is reflected by CRP and white blood cells (Tampa et al. 2018). NLR is a significant proinflammatory biomarker for OSCC because elevated NLR is associated with chronic inflammation and has been identified as a negative prognostic indicator. High NLR levels are linked to advanced and aggressive cancer types, lower survival rates, and increased recurrence risks (Charles et al. 2016). Low levels of LMR, another marker of systemic inflammatory response, promote inflammation and indicate advanced stages of OSCC (Ding et al. 2021). Monocytes counteracting the effects of lymphocytes promote tumor progression by enhancing angiogenesis and immunosuppression (Kumarasamy et al. 2021). Higher monocyte counts have a detrimental effect on patient survival. Platelet count, morphology, and functions change in OSCC, with platelet activation playing a crucial role in this cancer (Tampa et al. 2018). Consequently, low LMR and increased NLR, PLR, and MLR may aid in completing OSCC diagnosis and evaluating treatment.

NF-κB's potential proinflammatory marker role in OC is implicated in triggering inflammatory processes and its involvement in tumor development (Tampa et al. 2018). Additionally, IFN-γ, a proinflammatory cytokine involved in Th1-driven immune responses, may contribute to OSCC development through abnormal methylation of its promoter (Tian et al. 2017).

17.3 Proinflammatory Biomarkers in Oral Cancer

17.3.1 Cytokines and Chemokines

17.3.1.1 Role of Cytokines

Cytokines, once classified as lymphokines or monokines based on their cell of origin, are currently recognized as crucial immune response modulators secreted primarily by helper T cells, macrophages, and other nucleated cells (Zhang and An 2007). In OC, key proinflammatory biomarkers include IL-1, IL-6, IL-8, TNF-α, and TGF-β (Tampa et al. 2018). The pathology of OC involves alterations in immune response, inflammation, angiogenesis, and metabolism (Vissink et al. 2008). NF-kB-dependent cytokines serve as biomarkers for these changes, promoting the expression of proinflammatory, immunoregulatory, and proangiogenic cytokines crucial in carcinogenesis (Sahibzada et al. 2017). NF-kB activation in OSCC results in elevated levels of IL-1α, IL-6, IL-8, and TNF-α across tissues, tissue-infiltrating lymphocytes, serum, and cell lines (Rhodus et al. 2005). The expression of these proinflammatory cytokines has been reported in tissue samples not only from OSCC but also from potential malignant oral disorders such as oral leukoplakia and oral lichen planus (Babiuch et al. 2020).

IL-1α becomes a crucial factor in oral carcinogenesis by modulating inflammatory responses, promoting cellular proliferation, and preventing apoptosis. Released from OSCC cells, IL-1α stimulates carcinoma-associated fibroblasts (CAFs) to secrete chemokines such as IL-8, CC chemokine ligand (CCL)7, and CXCL1 (Bae et al. 2014). These chemokines facilitate cancer cell invasion and metastasis by remodeling the extracellular matrix and creating a supportive microenvironment for tumor growth. Within the TME, IL-1α recruits immune cells to the tumor site initially as part of the body's defense mechanism, but the tumor can exploit this process to foster inflammation and promote cancer progression. Moreover, IL-1α exhibits anti-apoptotic effects by preventing programmed cell death, allowing cancer cells to evade the body's natural defense mechanisms. This anti-apoptotic signaling supports uncontrolled cellular proliferation and tumor growth (Sakamoto et al. 2010). Ultimately, IL-1α-induced stimulation of CAFs and promotion of other pro-inflammatory cytokine secretion increase tumor aggressiveness and potential metastasis to distant regions. Bancroft's study provided additional insights into IL-1α's role in HNSCC. The study showed that elevated IL-1α in HNSCC triggers the activation of transcription factors NF-κB and AP-1, enhancing IL-8 expression (Bancroft et al. 2002). This mechanism further elucidates IL-1α's involvement in inflammation and tumor progression, reinforcing IL-1α's impact on cancer progression.

In OC development and progression, IL-1β, another proinflammatory cytokine, modifies the TME and supports tumor growth and progression through autocrine signaling. Cancer cells and surrounding cells produce pro-IL-1β in response to inflammatory signals. Activated inflammasomes then convert pro-IL-1β to active IL-1β via caspase-1. Released IL-1β promotes chronic inflammation, alters the TME, enhances cell proliferation and survival, facilitates invasion and metastasis, and increases chemotherapy resistance (Lee et al. 2015; Schroder and Tschopp 2010).

IL-6 is produced during acute inflammatory responses, contributing to immune function, and regulating oncogenesis, hematopoiesis, and inflammation (Choudhary et al. 2016; Tanaka et al. 2014). By enhancing MCP-1 secretion, IL-6 recruits monocytes to the site of inflammation, triggering the transition from acute to chronic inflammation (Gabay 2006). Renowned for its multifaceted role in growth promotion and apoptosis inhibition, IL-6 modulates the activation of Janus kinases (JAK) and STAT proteins, which crucially activate mitogen-activated protein kinase (MAPK), supporting cancer progression (Colotta et al. 2009; Sahibzada et al. 2017) (Fig. 17.1). Furthermore, IL-6 promotes the activation of oncogenes Ras and c-erbB2 through increased secretion of the transcription factor AP-2 in the context of cancer. Additionally, it supports hypermethylation of the promoter region of the tumor suppressor gene p53 (Hodge et al. 2005; Oyama et al. 1999). This dual action suppresses apoptosis in cancer cells, ultimately contributing to unchecked cancer cell growth. Moreover, IL-6 has been implicated in the progression of OC by enhancing cancer stem cell growth, proliferation, metastasis, and angiogenesis. According to a recent study by Pradhan and colleagues, IL-6 may enhance the growth, proliferation, metastasis, and angiogenesis of cancer stem cells in OC. The research showed that IL-6 increases PD-L1 expression through the JAK2/STAT3 pathway, contributing significantly to OC cell development and spread (Pradhan et al. 2024). In the same way, Wei et al. investigated IL-6-induced STAT3 signaling pathways in 111 OSCC samples, finding that IL-6 enhances tumor cell survival and progression via STAT3, and serves as a stage-independent predictor of survival in OCs (Wei et al. 2022) These results underscore IL-6 as a proinflammatory biomarker closely linked to the mechanism of OC, particularly in processes like metastasis and angiogenesis.

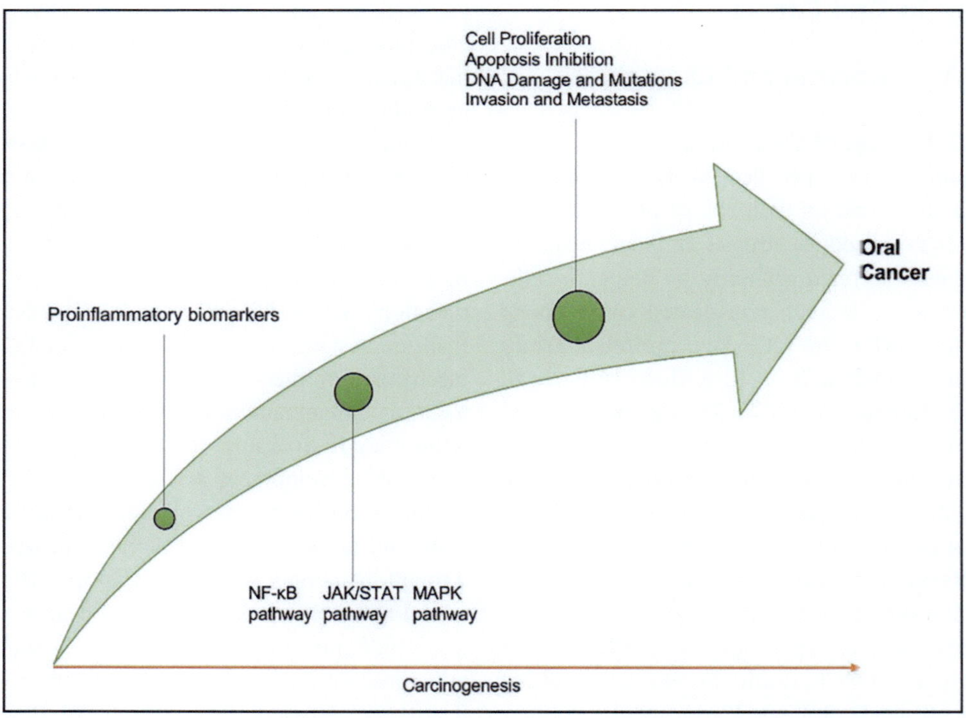

Fig. 17.1 Proinflammatory biomarkers related to the signaling pathways of oral cancer development

IL-8, the prototype molecule within the chemokine family, stands out as a proinflammatory cytokine responsible for prolonged effects in acute inflammation and inflammatory sites (Remick 2005). IL-8's activity in carcinogenesis resides in its capacity to recruit neutrophils, stimulate angiogenesis, enhance cell proliferation and survival, and inhibit apoptosis. NF-κB triggers IL-8 release from macrophages and neutrophils in response to stimuli like hypoxia and inflammation, which then interacts with CXC chemokine receptors (CXCR1 and CXCR2) receptors on tumor-associated inflammatory and tumor cells, promoting tumor progression (Sahibzada et al. 2017; Tampa et al. 2018). Evidence that IL-8 is a biomarker in OC was provided by Babiuch et al., who compared normal oral mucosa with OSCC samples. They reported that IL-8 was present in all layers of the epithelium in 65% of OSCC cases, while no IL-8 was detected in any of the normal oral mucosa samples. This finding suggests that IL-8 is a significant biological marker in the malignant transformation process of the oral mucosa ($p = 0.021$) (Babiuch et al. 2020).

TNF-α acts in OC pathogenesis by regulating inflammation, angiogenesis, programmed cell death, and cell proliferation through the TNF-TNF receptor signaling system (Krishnan et al. 2014). Moreover, TNF-α can lead to malignancy by inducing DNA damage in cells through the induction of ROS (Yan et al. 2006). It enhances OC invasion and metastasis by promoting cancer invadopodia formation, recruiting neutrophils to the TME, and supporting autocrine and paracrine signaling between cancer cells and stromal cells (Glogauer et al. 2015; Goertzen et al. 2018). Salvo and colleagues clinically reported that TNF-α overexpression in OCs, through Schwann cell activation, supports cancer progression and pain. They suggested that TNF-α inhibition could reduce OC progression and cancer-related neuropathic pain (Salvo et al. 2021).

17.3.1.2 Chemokines and Their Receptors

Chemokines are a subset of cytokines crucial for immune response, regulating the migration and activation of immune cells in response to pro-inflammatory signals like IL-1 family members, TNFs, and IFN-γ. In OSCC, chemokines and their receptors intricately orchestrate pathogenesis and progression (Dikova et al. 2021).

MCP-1, or CCL2, governs monocyte migration and is implicated in the aggressive behavior of squamous cell carcinomas (Tsou et al. 2007). Elevated MCP-1 levels correlate with tumor cells in HNSCCs, including OSCC, and are linked to metastatic lymph nodes and invasion (Ferreira et al. 2008; Ohta et al. 2002). Hemofiltrate C-C motif chemokine 1 (HCC-1), also known as CCL14, acts as a homeostatic chemokine promoting angiogenesis and OSCC progression (Cai et al. 2020). Its role in oral carcinogenesis is highlighted by its impact on blood vessel formation and tumor growth via related non-coding RNAs in OSCC patients (Feng et al. 2017). Platelet factor 4 (PF-4), also known as CXCL14, another homeostatic chemokine, influences immune cell recruitment and epithelial cell mobility (Giacobbi et al. 2024). Dysregulated PF-4 in OSCC, with elevated saliva levels, correlates with poor prognosis, affecting immune regulation and epithelial function (Ozawa et al. 2006).

The relationship between chemokines and their receptors in OSCC is multifaceted. These molecules contribute not only to inflammation but also to cancer progression. For example, MCP-1 enhances tumor invasion, while HCC-1 promotes tumor progression and angiogenesis. PF-4, through its dysregulation, affects immune regulation and patient outcomes (Dikova et al. 2021). Therefore, understanding how chemokines affect monocyte migration, angiogenesis, and immune cell recruitment in OSCC can guide potential therapeutic strategies for tumor progression and metastasis as a biomarker.

17.3.2 Acute Phase Proteins

17.3.2.1 C-Reactive Protein

CRP, an acute-phase protein of the pentraxin family, serves as a key biomarker of inflammation within the TME. Chronic inflammation triggers excessive cell proliferation, activating diverse cellular functions and fostering malignant alterations in DNA (Murata 2018). Although numerous studies have shown increased CRP levels associated with cancer, the direct connection between CRP and carcinogenesis remains unclear (Liu et al. 2023; Knittelfelder et al. 2020; Tampa et al. 2018). Nevertheless, insights from the current literature provide a basis for the potential hypothesis regarding the association between CRP and OC.

Various factors affecting healthy oral mucosa, such as smoking, alcohol consumption, viral infections, poor oral hygiene, and periodontal diseases, can lead to chronic inflammation (Abati et al. 2020). This persistent condition results in the continuous activation of inflammatory cells and cytokines. In response to chronic inflammation, CRP levels, produced by the liver as an indicator of inflammation, are increased (Asegaonkar et al. 2015). CRP functions as part of the inflammatory response, promoting the activation of inflammatory cells and the release of inflammatory mediators. Proinflammatory cytokines like IL-1, IL-6, and TNF-α activate transcription factors such as NF-κB. NF-κB regulates cell proliferation, anti-apoptotic responses, and the expression of inflammatory genes, thereby facilitating continuous cell growth (Colotta et al. 2009). CRP contributes to the accumulation of inflammatory cells and mediators in the oral mucosa and supports the maintenance of ongoing inflammatory processes. During this process, CRP binds to fibronectin, enhancing monocyte adhesion and reinforcing the inflammatory microenvironment (Ramos et al. 2016).

Recently, Salema and colleagues demonstrated that pre-treatment CRP levels of 5 mg/L in oral pre-malignant and malignant lesions were statistically significant ($p < 0.001$) in indicating malignancy, supporting the histopathological findings and suggesting CRP levels as a biomarker for early OC diagnosis (Salema et al. 2024). Similarly, Liu et al. found in their study among the European population that elevated serum CRP levels were markedly correlated with an increased risk of oral and oropharyngeal cancer (OR 1.52, 95% CI 1.14 to 2.02) (Liu et al. 2023). These findings provide robust evidence supporting the role of elevated serum CRP levels as a proinflammatory biomarker in predicting OC risk.

17.3.2.2 Serum Amyloid A

SAA, a protein produced by the liver, significantly increases during the acute-phase response to inflammation. In the context of oral carcinogenesis, elevated SAA levels reflect the proinflammatory environment, thereby serving as a valuable biomarker for assessing inflammation associated with tumor presence and progression. High SAA levels are associated with increased tumor aggressiveness and poorer prognosis in OC patients. Together with CRP and IL-6 values, elevated SAA levels offer a comprehensive insight into the patient's inflammatory and tumor status. The synergistic interplay between SAA, CRP, and IL-6 underscores their collective role in oral cancer-associated inflammation. IL-6 drives the production of SAA and CRP, amplifying the acute-phase response, while SAA and CRP provide systemic insights into the tumor's inflammatory environment. Together, these biomarkers enhance the precision of monitoring tumor progression and recurrence risk. Monitoring SAA levels facilitates early detection of disease recurrence and assessment of treatment response in OC patients (Chang et al. 2013a, b).

Similarly,, Shinriki et al. demonstrated significantly increased serum SAA levels in HNSCC patients compared to healthy individuals ($p < 0.0001$). They also noted elevated SAA1 gene and protein expression in HNSCC tissues relative to normal oral mucosa. Furthermore, inhibition of the STAT3 signaling pathway reduced SAA1 expression in HNSCC cell lines, indicating a potential regulatory role via the Janus

JAK-STAT3 pathway (Shinriki et al. 2010). However, while these data highlight the potential of SAA in detecting and assessing OC, more data are needed to fully understand its predictive and prognostic properties within the OC mechanism.

17.3.3 Other Proinflammatory Molecules

17.3.3.1 Prostaglandins

In addition to cytokines, other inflammatory mediators, such as prostaglandins, COX enzymes, and MMPs, can induce genetic and epigenetic alterations. These changes lead to the suppression of tumor suppressor gene function through mechanisms such as DNA methylation and post-translational modifications (Munn 2017). The lipid mediator originating from arachidonic acid in inflammation is PGE2. Arachidonic acid undergoes conversion into various bioactive eicosanoid types through two main pathways including the COX pathway, producing prostaglandins such as PGE2, and the lipoxygenase pathway, contributing to the synthesis of leukotrienes alongside anti-inflammatory lipoxins (Wang and Dubois 2010). While products from these processes regulate inflammation, PGE2 is actively directing in the transition from inflammation to oral carcinogenesis. Hence, the data showing that aspirin intake diminishes the risk of several cancers, including gastroesophageal, lung, breast, colon, and prostate cancers, is particularly relevant to understanding the role of inflammation, PGE2, and cancer (Lloyd et al. 2022). Aspirin's ability to inhibit COX enzymes, particularly COX-2, reduces PGE2 production, a key mediator in the transition from inflammation to carcinogenesis. This mechanism may also have implications for oral cancers, where PGE2-driven pathways contribute to tumor progression and metastasis. Additionally, St John noted that PGE2 is overexpressed in OSCC and suggested that treatments targeting the COX pathway might decrease tumor metastasis in HNSCCs by preventing the PGE2-driven induction of E-cadherin transcriptional repressors (St John 2015).

In OSCC, factors such as tobacco, alcohol, and HPV infections lead to the overexpression of COX-2 and an increase in the production of PGE2 (Nasry et al. 2018). PGE2 binds to its receptors (EP1–4) on various cell types, triggering a series of inflammatory responses (Nakanishi and Rosenberg 2013). These responses result in chronic inflammation, causing oxidative damage and changes in gene expression. Chronic inflammation induced by PGE2 increases the expression of inflammation-associated transcription factors, initiating pro-tumorigenic processes such as cellular proliferation and migration. Additionally, PGE2 activates signaling pathways such as the epidermal growth factor receptor (EGFR) pathway, HIF-1α, and VEGF, promoting tumor growth and angiogenesis. This activation enhances the proliferation of tumor cells and their ability to form new blood vessels, supporting tumor nutrition and growth. Furthermore, PGE2 increases the activity of MMPs and CD147, facilitating the invasion and metastasis of tumor cells to surrounding tissues. It also modulates the TME by recruiting immune cells that support tumor growth and suppress anti-tumor immunity, creating a favorable environment for cancer progression (Nasry et al. 2018). Ultimately, PGE2 promotes cellular proliferation, migration, and invasion through its receptors, and all these processes collectively contribute to the progression of OSCC.

17.3.3.2 Matrix Metalloproteinases

MMP enzymes belong to the zinc-dependent family, comprising over 20 members with proteolytic activity capable of degrading nearly all types of extracellular matrix components. They play roles in various physiological processes such as tissue remodeling or regeneration. However, they are also associated with abnormal activation in pathological conditions such as inflammation and tumorigenesis (Tampa et al. 2018; Yadav et al. 2014). MMPs participate in cell migration and regulate cytokine levels at inflammation sites. Inflammatory cells release MMPs, which in turn affect non-matrix substrates, thereby either promoting or inhibiting the inflammatory response (Nissinen and Kähäri 2014).

Key mechanisms supporting MMP involvement in carcinogenesis include regulation of tumor cell growth and angiogenesis, modulation of apoptosis, degradation of the extracellular matrix facilitating tumor invasion, and EMT (Yadav et al. 2014). According to a systematic review and meta-analysis by Deng et al., overexpression of MMP-9 and -2 serves as prognostic biomarkers for OC (Deng et al. 2019). Additionally, Chang et al. found that saliva MMP-1 levels significantly distinguished OSCC patients from non-cancerous groups ($p < 0.001$), proposing MMP-1 as a novel biomarker for detecting and monitoring OSCC (Chang et al. 2020). Furthermore, another study identified MMP-2, MMP-9, CRP, TGF-β1, and E-selectin as reliable inflammation markers in oral leukoplakia and OSCC patients, highlighting MMP-9's strongest correlation with progression from normal tissue to neoplasm (80% sensitivity for oral malignancy, with a specificity of 90%) (Chang et al. 2013a, b). Kadeh et al. suggested MMP-10 as a potential indicator of malignant transformation of normal mucosa to OSCC (Kadeh et al. 2015). Moreover, MMP-7 expression correlates with higher metastatic risk and increased local invasion in early-stage oral tongue SCC, indicating its potential as a prognostic biomarker. Ren et al. (2020) demonstrated that MMP-7 expression is markedly higher in metastatic OSCC tumors compared to nonmetastatic ones, highlighting its critical role in driving tumor invasiveness. The upregulation of MMP-7, potentially mediated by COX-2 activation, emphasizes its involvement in extracellular matrix degradation and metastatic progression (Ren et al. 2020).

17.3.4 Hematologic Proinflammatory Biomarkers

17.3.4.1 Neutrophil-to-Lymphocyte Ratio

Neutrophils and macrophages initiate tumor formation by releasing cytokines and chemokines, leading to heightened cell proliferation and reduced cell death. Inflammation also promotes neo-angiogenesis and metastatic progression (Yang et al. 2018). The TME is composed of innate immune cells such as neutrophils and adaptive immune cells like lymphocytes, which modulate tumor growth by secreting cytokines and chemokines (de Visser et al. 2006; Lin and Karin 2007). Neutrophils, under the influence of TGF-β, can either promote or inhibit tumors and suppress T cells via integrin macrophage-1 antigen (Mac-1) and hydrogen peroxide (Fridlender et al. 2009; Pillay et al. 2012). T lymphocytes counteract tumor cell proliferation and metastasis by triggering cytotoxic cell death and producing cytokines. A decrease in lymphocyte count compromises the immune response (Yilmaz et al. 2023a). Thus, an elevated NLR indicates a poor prognosis in OSCC (Yang et al. 2018). In addition, systematic reviews and meta-analyses have confirmed that an elevated NLR serves as a predictive and prognostic biomarker in HNSCCs (Mariani et al. 2022; Mascarella et al. 2018).

Smoking or tobacco consumption can weaken host immune responses in oral potentially malignant disorders, resulting in ROS production that impairs the vitality and phagocytic activity of oral neutrophils (Hasnis et al. 2004). As a natural response to inflammation provoked by oral potentially malignant disorders, neutrophilia, and lymphopenia can be observed (Do et al. 2023). The reason for lymphopenia is the decline in saliva's antioxidant capacity due to chronic smoking and the generation of free radicals and RNS emitted from tobacco products, which diminish the survival rate of oral lymphocytes (Do et al. 2023). Therefore, the increase in NLR observed in oral saliva samples as a consequence of neutrophilia and lymphopenia may be proposed as a more reliable predictive indicator of OC compared to blood samples (Domnich et al. 2020). Addressing this context, Magdum and colleagues identified that NLR values of 4 or higher, in the presence of oral potentially malignant disorders, exhibited 92% sensitivity and specificity in predicting OC (Magdum et al. 2024). In a study conducted by Singh et al., three groups were compared: OC, oral potentially premalignant lesions (including leukoplakia, erythroplakia, oral lichen planus, and oral submucous fibrosis), and healthy controls. NLR values were found to be significantly higher in OC and potential malignant lesions compared to healthy controls ($P < 0.001$) (Singh et al. 2021). Therefore, due to being a non-invasiveness process to obtain samples and practicality, saliva NLR represents a promising pro-inflammatory biomarker option increasingly poised for utilization in the diagnosis and treatment of OC.

17.3.4.2 Lymphocyte-to-Monocyte Ratio

The inhibitory effect of lymphocytes on tumor growth, contrasted with the supportive role of monocytes in tumor proliferation through the production of pro-inflammatory cytokines and promotion of angiogenesis, highlights the prognostic value of the LMR in solid tumors such as HNSCC (Kano et al. 2017). In this line, it has been indicated that LMR levels may mirror the immune response of the patient within the TME in advanced-stage OCs, potentially inhibiting or promoting tumor growth, and a low LMR has been identified as an indicator of poor prognosis in this patient group (Supanimitjaroenporn et al. 2022). Chen et al. proposed LMR for prognostic prediction in OC, noting that a high pre-treatment LMR correlates with better overall survival. (HR = 0,65; %95 GA = 0,44-0,98) (Chen et al. 2021). Hence, clinically, LMR may function as a prognostic biomarker in OC patients.

17.3.4.3 Platelet-to-Lymphocyte Ratio

PLR is a significant biomarker in the context of OC, reflecting the interplay between inflammation and immunity. Platelets, beyond their traditional role in hemostasis and thrombosis, participate in a variety of pathophysiological processes (Yilmaz et al. 2023b), including cancer progression. Tumor cells can influence platelets in multiple ways, both numerically and functionally. This includes the secretion of mediators and cytokines that activate platelets, a process that can be initiated through various platelet receptors or mediated by cytokines. Once activated, platelets release numerous growth factors from their dense α-granules and peroxisomes, which can directly promote tumor growth, adhesion, and metastasis. Additionally, tumor microvesicles can stimulate platelets, leading to the release of platelet microparticles that further support thrombosis and tumor growth (Goubran et al. 2014). This interaction creates a self-perpetuating cycle of activation where platelets facilitate tumor growth and protect metastatic cells from immune surveillance, thereby enhancing the potential for oral malignancy. Kumarasamy and colleagues' meta-analysis demonstrated a significant association between higher PLR values and poorer overall survival in head and neck cancer, reporting a pooled effect estimate of 1.461 (95% CI 1.329–1.674) (Kumarasamy et al. 2021). PLR may also serve as a potent prognostic marker in OC; however, further large-scale studies are necessary to validate its clinical relevance.

17.4 Mechanisms of Proinflammatory Biomarkers in Oral Cancer Pathophysiology

17.4.1 Tumor Microenvironment and Inflammation

17.4.1.1 Interaction Between Tumor Cells and Immune Cells

TME in OSCC is densely infiltrated by various immune cells, including tumor-associated macrophages, T lymphocytes, and dendritic cells. These cells facilitate tumor growth and metastasis through the secretion of cytokines and growth factors such as VEGF and MMPs, promoting angiogenesis and the degradation of the extracellular matrix. Additionally, proinflammatory biomarkers play a significant role in these processes. Among these immune cells, tumor-associated macrophages are crucial players within the TME, promoting tumor progression by releasing cytokines, growth factors, and MMPs. These elements aid in tumor cell proliferation, invasion, and angiogenesis. Consequently, tumor-associated macrophages are often linked with poor prognosis in OSCC because they modulate the immune response and enhance the inflammatory environment, thus supporting tumor growth and metastasis (Nasry et al. 2018). For instance, TNF-α, a proinflammatory cytokine, is often secreted by tumor-associated macrophages and can promote tumor growth and inflammation. In addition to tumor-associated macrophages, T lymphocytes play a significant role in the TME and can exhibit both anti-tumor and tumor-promoting effects. T helper type 17 (Th17) cells produce IL-17, which induces pro-inflammatory cytokines and promotes tumor growth. Conversely, cytotoxic T lymphocytes directly target and kill tumor cells. However, tumor cells often evade immune surveillance by expressing immune checkpoint molecules like PD-L1, leading to immune suppression. Furthermore, regulatory T cells (Tregs) recruited by tumor cells suppress the anti-tumor immune response, enabling tumor cell proliferation and metastasis (Gonda et al. 2009). Neutrophils also support tumor invasion by releasing enzymes that degrade the extracellular matrix, while mast cells contribute to the inflammatory microenvironment within the TME by releasing cytokines and chemokines, further supporting angiogenesis and immune modulation (Nasry et al. 2018). Proinflammatory biomarkers such as IL-6 and IL-8 are fundamental in maintaining chronic inflammation and driving tumor progression.

17.4.1.2 Role of the Extracellular Matrix

The extracellular matrix in the TME is not merely a structural scaffold but an active participant in cancer progression. It is continuously remodeled by MMPs and other enzymes secreted by both tumor and stromal cells. MMPs, for instance, are enzymes capable of degrading various components of the extracellular matrix. In OSCC, overexpression of MMPs such as MMP-2 and MMP-9 is common and is associated with tumor invasion and metastasis (Patel et al. 2007). By breaking down extracellular matrix barriers, these enzymes allow tumor cells to invade neighboring tissues and spread to distant sites.

The extracellular matrix remodeling involves not only the degradation but also the deposition of new matrix components, creating a dynamic environment that supports tumor growth and dissemination. The extracellular matrix interacts with tumor cells through various signaling pathways, influencing cell behavior, gene expression, and resistance to apoptosis. NF-κB, a transcription factor activated by various proinflammatory signals, is a key regulator of these processes, driving the expression of genes involved in inflammation and extracellular matrix remodeling. Additionally, CAFs are another key component of the TME. These cells secrete cytokines, chemokines, and extracellular matrix components that support tumor growth and invasion (Niklander 2021). By producing collagen and fibronectin, CAFs contribute to extracellular matrix remodeling, creating a scaffold for tumor cell migration. Proinflammatory cytokines like IL-1 and IL-6 secreted by CAFs further enhance the inflammatory environment and support tumor progression. The rapid growth of tumor cells often leads to hypoxic conditions within the TME, inducing HIFs that stimulate angiogenesis through the upregulation of VEGF and other pro-angiogenic factors. However, the newly formed blood vessels are often abnormal and leaky, which further contributes to an inflammatory and immunosuppressive TME (Nasry et al. 2018; Niklander 2021).

17.4.2 Signal Transduction Pathways

17.4.2.1 NF-κB Pathway

As a mediator of inflammation and immune responses, the activation of NF-κB in the pathophysiology of OC can be induced by various stimuli such as tobacco, alcohol, pathogens, and environmental and proinflammatory factors. Upon activation, NF-κB translocates to the nucleus, where it promotes the transcription of genes involved in cell proliferation, survival, and the production of inflammatory cytokines and chemokines, contributing to a microenvironment that supports cancer cell growth and invasion (Feller et al. 2013).

NF-κB is activated through Toll-like receptor (TLR)-MyD88 intracellular signaling pathways in epithelial cells and immuno-inflammatory cells of the innate immune system. This activation leads to the production of inflammatory factors such as TNF-α, IL-1β, COX-2, and PGE-2 that further activate NF-κB, thereby promoting cell survival by upregulating anti-apoptotic genes and genes regulating cell cycle checkpoints, such as c-Myc, Mc-1, cyclin-D, Bcl-2, c-Flip, and survivin. This process results in a state of cellular genomic instability, which is conducive to cancer progression (Colotta et al. 2009; Feller et al. 2013; Mantovani et al. 2008).

The dynamic interaction between cancer cells and the surrounding non-cancerous stromal cells plays a crucial role in tumor progression. Cancer cells stimulate stromal immuno-inflammatory cells to release inflammatory mediators, such as IL-6, IL-10, and MMP-9. These mediators, in turn, promote cancer cell proliferation, enhance survival, and drive angiogenesis, processes fundamental to tumor development (Feller et al. 2013; Hanahan and Weinberg 2011). This interaction explains why rapidly progressing cancers often exhibit an intense inflammatory infiltrate microscopically.

17.4.2.2 JAK/STAT Pathway

The Janus kinase (JAK)/STAT pathway is another crucial signaling cascade involved in OC. Cytokines such as IL-6 and TNF-α can activate this pathway. Upon activation, STAT3 translocates to the nucleus, where it induces the expression of genes that drive cell proliferation and survival.

Activated STAT3 underpins the modulation of the immune response and the development of an immunosuppressive TME, enabling tumor evasion from immune surveillance. This pathway also involves ROS production and RNS, which can cause direct DNA damage and dysregulate mechanisms of DNA repair, cell-cycle checkpoint control, and apoptosis. This genomic instability favors the evolution of random mutations and the development of cancer (Colotta et al. 2009; Hanahan and Weinberg 2011). Additionally, the JAK/STAT pathway contributes to the production of inflammatory mediators that further promote cancer cell proliferation and survival. The sustained activation of this pathway creates a feedback loop that perpetuates the inflammatory microenvironment, supporting tumor progression and metastasis.

17.4.3 Angiogenesis and Metastasis

17.4.3.1 Role of Proinflammatory Biomarkers in Promoting Angiogenesis

Angiogenesis, involving the development of new blood vessels from existing ones, fuels tumor growth and metastasis. In OC, as in other malignancies, angiogenesis provides the necessary oxygen and nutrients to proliferating tumor cells, facilitating tumor progression and the spread of cancer cells to distant sites. One of the primary mediators of angiogenesis is the VEGF family, particularly VEGF-A, which binds to its receptors VEGFR1 and VEGFR2 on endothelial cells, initiating a cascade of downstream signaling events. These events encompass the activation of the MAPK and phosphatidylinositol 3-kinase (PI3K)/protein kinase B (PKB/AKT) pathways, resulting in endothelial cell proliferation, migration, and neovascularization (Alessandrini et al. 2023; Vassilakopoulou et al. 2015).

Hypoxia, a common feature of the TME, upregulates VEGF expression through the stabilization of HIF-1α. Other factors such as IL-8, epidermal growth factor (EGF), and platelet-derived growth factor (PDGF) also contribute to the angiogenic process in OC. These factors can be produced by tumor cells, stromal cells, or infiltrating immune cells, creating a pro-angiogenic milieu that supports tumor progression and dissemination. Beyond VEGF, other proinflammatory biomarkers PGE2 are also crucial in driving angiogenesis. PGE2 enhances the expression of VEGF and other angiogenic factors, facilitating the formation of new blood vessels. The binding of PGE2 to its receptors (EP1–4) on endothelial cells and other target cells triggers downstream signaling events that promote endothelial cell proliferation, migration, and tube formation, essential steps in angiogenesis (Nasry et al. 2018; Vassilakopoulou et al. 2015).

Macrophages and neutrophils attracted to the tumor site produce proinflammatory markers, enhancing angiogenesis by increasing the availability of VEGF and other mediators. This interplay between proinflammatory biomarkers and angiogenic factors fosters a favorable environment for tumor development (Kim and Bae 2016).

Anti-angiogenic agents targeting angiogenesis in HNSCCs, particularly VEGF monoclonal antibodies, and tyrosine kinase inhibitors, show promise. These agents can normalize tumor vasculature and enhance the efficacy of chemotherapy and radiation (Vassilakopoulou et al. 2015). However, research is focused on identifying predictive biomarkers and optimizing treatment regimens to determine the clinical benefits of these therapies in OC.

17.4.3.2 Mechanisms of Metastasis Influenced by Inflammation

Proinflammatory cytokines and other mediators promote the expression of MMPs, which degrade the extracellular matrix and basement membrane, facilitating cancer cell invasion and migration. For instance, TNF-α and IL-1 can upregulate MMPs, enabling tumor cells to breach tissue barriers and enter the bloodstream or lymphatic system (Karnezis et al. 2012; Nasry et al. 2018). PGE2 modulates the TME, increasing cancer cell invasiveness by upregulating intercellular adhesion molecule (ICAM)-1, which aids interaction with endothelial cells and promotes intravasation (Yang et al. 2010). Additionally, PGE2 induces EMT, enhancing cell motility and invasiveness.

Inflammation-induced hypoxia within the TME further promotes metastasis by stabilizing HIF-1α, which upregulates genes involved in angiogenesis, invasion, and survival (Lee et al. 2018). Inflammatory cells, such as macrophages and neutrophils, facilitate metastasis. These cells secrete additional proinflammatory cytokines and growth factors, creating a feedback loop that sustains chronic inflammation and supports metastasis. Proinflammatory biomarkers in OC accelerate angiogenesis and metastasis by modifying tumor cells, degrading the extracellular matrix, and increasing cancer cell invasiveness, which can reveal therapeutic targets to inhibit tumor progression and metastasis.

17.5 Clinical Implications

Non-invasive techniques for detecting biomarkers in saliva, serum, and other biological fluids have markedly advanced the early detection of OC. For example, biomarkers like interleukins IL-6 and IL-8 found in saliva are indicative of OC's presence and progression, which supports earlier diagnosis. The advent of advanced diagnostic technologies, including next-generation sequencing and liquid biopsy, has revolutionized the landscape of OC detection and management. The next-generation sequencing allows for comprehensive genetic profiling of tumors, enabling the formulation of personalized treatment regimens, while liquid biopsy permits the non-invasive monitoring of tumor dynamics and therapeutic response through the analysis of biomarkers in blood samples (Menditti et al. 2023).

Prognostic biomarkers are equally critical in forecasting the trajectory and outcomes of OC. Elevated levels of IL-6 and TNF-α, along with mutations in p53 and the overexpression of EGFR, are indicative of aggressive tumor phenotypes and poor clinical prognoses. These biomarkers facilitate the stratification of patients according to risk profiles, thereby informing and optimizing therapeutic strategies (Dikova et al. 2021). Chronic inflammation is recognized as a significant contributor to oral carcinogenesis, making the targeting of proinflammatory pathways a promising therapeutic approach. Inhibitors of COX-2 and specific cytokine inhibitors can mitigate the inflammatory milieu within the TME, thereby impeding tumor growth and metastasis (Nasry et al. 2018). Furthermore, advancements in immunotherapy, particularly those aimed at activating the immune system to target cancer cells, offer new avenues for the treatment of OC.

17.6 Future Directions and Research

17.6.1 Emerging Biomarkers

The identification of novel biomarkers represents only the initial step. Equally critical is their validation to ensure clinical applicability. This process requires extensive research to enhance these biomarkers' sensitivity, specificity, and reproducibility across diverse patient populations. Large-scale, multicenter studies are essential to validate potential biomarkers, ensuring their reliability in predicting disease presence, progression, and response to treatment. Moreover, integrating bioinformatics and machine learning processes may assist in analyzing complex datasets to validate biomarkers effectively. Alongside proinflammatory biomarkers, employing a combination of multiple indicators may strengthen diagnostic and prognostic efficacy. Multi-biomarker frameworks that integrate genetic, proteomic, and metabolomic factors can provide a more accurate risk stratification, facilitate early detection, and contribute to a unified understanding of OC.

17.6.2 Personalized Medicine

Personalized medicine, tailored to an individual's genetic and molecular profile, offers significant potential for improving OC outcomes. Research efforts are increasingly focused on elucidating how emerging biomarkers can guide the development of personalized treatment strategies. Personalized medicine can reduce adverse effects and enhance the overall survival rates of patients by aligning treatment strategies to the unique characteristics of each patient's cancer. However, the integration of biomarker research into clinical practice requires the establishment of standardized protocols for biomarker testing and the development of advanced clinical decision support systems to guide treatment selections.

17.6.3 Challenges and Limitations

Despite extensive research in biomarkers, several challenges and limitations need to be addressed. The heterogeneity of OC and the variability in biomarker expression among patients pose significant obstacles. Additionally, the cost and accessibility of advanced diagnostic technologies can limit the widespread adoption of biomarker-based approaches. Cost-effective and widely accessible biomarker tests may overcome these barriers.

Variability in proinflammatory biomarker levels due to genetic, environmental, and lifestyle factors may impact their clinical utility. Studies on how factors such as age, gender, smoking, and diet influence biomarker levels and their association with OC are likely to advance the reliability of proinflammatory biomarker-based diagnostics. Standardizing proinflammatory biomarker analyses in diagnostic workflows demands setting reference ranges for biomarker measurements, ensuring test reproducibility, and implementing robust quality control measures. International collaboration and consensus are necessary to establish globally applicable guidelines, ensuring that biomarker-based diagnostics and treatments are both consistent and reliable.

Conflicts of Interest There is no conflict of interest to declare.

References

Abati S, Bramati C, Bondi S, Lissoni A, Trimarchi M (2020) Oral Cancer and Precancer: A narrative review on the relevance of early diagnosis. Int J Environ Res Public Health 17(24):9160. https://doi.org/10.3390/ijerph17249160

Alessandrini L, Astolfi L, Daloiso A, Sbaraglia M, Mondello T, Zanoletti E, Franz L, Marioni G (2023) Diagnostic, prognostic, and therapeutic role for angiogenesis markers in head and neck squamous cell carcinoma: A narrative review. Int J Mol Sci 24(13):10733. https://doi.org/10.3390/ijms241310733

Al-Tashi Q, Saad MB, Muneer A, Qureshi R, Mirjalili S, Sheshadri A, Le X, Vokes NI, Zhang J, Wu J (2023)

Machine learning models for the identification of prognostic and predictive Cancer biomarkers: A systematic review. Int J Mol Sci 24(9):7781. https://doi.org/10.3390/ijms24097781

Aronson JK, Ferner RE (2017) Biomarkers-A General Review. Current Protocols Pharmacol 76: 9.23.1–9.23.17. https://doi.org/10.1002/cpph.19

Asegaonkar SB, Asegaonkar BN, Takalkar UV, Advani S, Thorat AP (2015) C-reactive protein and breast Cancer: new insights from old molecule. Int J Breast Cancer 2015:145647. https://doi.org/10.1155/2015/145647

Babiuch K, Kuśnierz-Cabala B, Kęsek B, Okoń K, Darczuk D, Chomyszyn-Gajewska M (2020) Evaluation of Proinflammatory, NF-kappaB dependent cytokines: IL-1α, IL-6, IL-8, and TNF-α in tissue specimens and saliva of patients with Oral squamous cell carcinoma and Oral potentially malignant disorders. J Clin Med 9(3):867. https://doi.org/10.3390/jcm9030867

Bae JY, Kim EK, Yang DH, Zhang X, Park YJ, Lee DY, Che CM, Kim J (2014) Reciprocal interaction between carcinoma-associated fibroblasts and squamous carcinoma cells through interleukin-1α induces cancer progression. Neoplasia (New York) 16(11):928–938. https://doi.org/10.1016/j.neo.2014.09.003

Ballman KV (2015) Biomarker: predictive or prognostic? J Clin Oncol Off J Am Soc Clin Oncol 33(33): 3968–3971. https://doi.org/10.1200/JCO.2015.63.3651

Bancroft CC, Chen Z, Yeh J, Sunwoo JB, Yeh NT, Jackson S, Jackson C, Van Waes C (2002) Effects of pharmacologic antagonists of epidermal growth factor receptor, PI3K and MEK signal kinases on NF-kappaB and AP-1 activation and IL-8 and VEGF expression in human head and neck squamous cell carcinoma lines. Int J Cancer 99(4):538–548. https://doi.org/10.1002/ijc.10398

Ben-Neriah Y, Karin M (2011) Inflammation meets cancer, with NF-κB as the matchmaker. Nat Immunol 12(8):715–723. https://doi.org/10.1038/ni.2060

Binder Gallimidi A, Fischman S, Revach B, Bulvik R, Maliutina A, Rubinstein AM, Nussbaum G, Elkin M (2015) Periodontal pathogens Porphyromonas gingivalis and Fusobacterium nucleatum promote tumor progression in an oral-specific chemical carcinogenesis model. Oncotarget 6(26):22613–22623. https://doi.org/10.18632/oncotarget.4209

Biomarkers Definitions Working Group (2001) Biomarkers and surrogate endpoints: preferred definitions and conceptual framework. Clin Pharmacol Ther 69(3):89–95. https://doi.org/10.1067/mcp.2001.113989

Bouquot JE (1986) Common oral lesions found during a mass screening examination. J Amer Dental Assoc (1939) *112*(1):50–57. https://doi.org/10.14219/jada.archive.1986.0007

Cai Y, Ling Y, Huang L, Huang H, Chen X, Xiao Y, Zhu Z, Chen J (2020) C-C motif chemokine 14 as a novel potential biomarker for predicting the prognosis of epithelial ovarian cancer. Oncol Lett 19(4): 2875–2883. https://doi.org/10.3892/ol.2020.11378

Cancer.Net (2023) Oral and Oropharyngeal Cancer: Introduction. https://www.cancer.net/cancer-types/oral-and-oropharyngeal-cancer/statistics. Accessed 27 May 2024

Charles KA, Harris BD, Haddad CR, Clarke SJ, Guminski A, Stevens M, Dodds T, Gill AJ, Back M, Veivers D, Eade T (2016) Systemic inflammation is an independent predictive marker of clinical outcomes in mucosal squamous cell carcinoma of the head and neck in oropharyngeal and non-oropharyngeal patients. BMC Cancer 16:124. https://doi.org/10.1186/s12885-016-2089-4

Choudhary MM, France TJ, Teknos TN, Kumar P (2016) Interleukin-6 role in head and neck squamous cell carcinoma progression. World J Otorhinolaryngol – Head Neck Surg 2(2):90–97. https://doi.org/10.1016/j.wjorl.2016.05.002

Conway DI, Petticrew M, Marlborough H, Berthiller J, Hashibe M, Macpherson LM (2008) Socioeconomic inequalities and oral cancer risk: a systematic review and meta-analysis of case-control studies. Int J Cancer 122(12):2811–2819. https://doi.org/10.1002/ijc.23430

Chang KP, Wu CC, Fang KH, Tsai CY, Chang YL, Liu SC, Kao HK (2013a) Serum levels of chemokine (C-X-C motif) ligand 9 (CXCL9) are associated with tumor progression and treatment outcome in patients with oral cavity squamous cell carcinoma. Oral Oncol 49(8):802–807. https://doi.org/10.1016/j.oraloncology.2013.05.006

Chang PY, Kuo YB, Wu TL, Liao CT, Sun YC, Yen TC, Chan EC (2013b) Association and prognostic value of serum inflammation markers in patients with leukoplakia and oral cavity cancer. Clin Chem Lab Med 51(6): 1291–1300. https://doi.org/10.1515/cclm-2012-0504

Chang YT, Chu LJ, Liu YC, Chen CJ, Wu SF, Chen CH, Chang IY, Wang JS, Wu TY, Dash S, Chiang WF, Chiu SF, Gou SB, Chien CY, Chang KP, Yu JS (2020) Verification of saliva matrix Metalloproteinase-1 as a strong diagnostic marker of Oral cavity Cancer. Cancer 12(8):2273. https://doi.org/10.3390/cancers12082273

Chen HH, Chen IH, Liao CT, Wei FC, Lee LY, Huang SF (2011) Preoperative circulating C-reactive protein levels predict pathological aggressiveness in oral squamous cell carcinoma: a retrospective clinical study. Clinical Otolaryngol: Off J ENT-UK; Off J Netherlands Soc Oto-Rhino-Laryngology Cervico-Facial Sur 36(2):147–153. https://doi.org/10.1111/j.1749-4486.2011.02274.x

Chen L, Qian J, Lin L, Lin J, Chen Q, Zhuang Z, Hong Y, Wang J, Qiu Y, Pan L, Shi B, Wang J, Liu F, Cai L, Hu Z, He B, Chen F (2021) Prognostic value of preoperative lymphocyte-to-monocyte ratio in oral cancer patients and establishment of a dynamic nomogram. Oral Dis 27(5):1127–1136. https://doi.org/10.1111/odi.13629

Chen X, Lei H, Cheng Y, Fang S, Sun W, Zhang X, Jin Z (2024) CXCL8, MMP12, and MMP13 are common biomarkers of periodontitis and oral squamous cell carcinoma. Oral Dis 30(2):390–407. https://doi.org/10.1111/odi.14419

Colotta F, Allavena P, Sica A, Garlanda C, Mantovani A (2009) Cancer-related inflammation, the seventh hallmark of cancer: links to genetic instability. Carcinogenesis 30(7):1073–1081. https://doi.org/10.1093/carcin/bgp127

Coussens LM, Werb Z (2002) Inflammation and cancer. Nature 420(6917):860–867. https://doi.org/10.1038/nature01322

D'Arcangelo E, Wu NC, Cadavid JL, McGuigan AP (2020) The life cycle of cancer-associated fibroblasts within the tumour stroma and its importance in disease outcome. Br J Cancer 122(7):931–942. https://doi.org/10.1038/s41416-019-0705-1

Davalos AR, Coppe JP, Campisi J, Desprez PY (2010) Senescent cells as a source of inflammatory factors for tumor progression. Cancer Metastasis Rev 29(2):273–283. https://doi.org/10.1007/s10555-010-9220-9

Deng W, Peng W, Wang T, Chen J, Zhu S (2019) Overexpression of MMPs functions as a prognostic biomarker for Oral Cancer patients: A systematic review and meta-analysis. Oral Health Prev Dent 17(6):505–514. https://doi.org/10.3290/j.ohpd.a43636

de Visser KE, Eichten A, Coussens LM (2006) Paradoxical roles of the immune system during cancer development. Nat Rev Cancer 6(1):24–37. https://doi.org/10.1038/nrc1782

Diamandis EP (2010) Cancer biomarkers: can we turn recent failures into success? J Natl Cancer Inst 102(19):1462–1467. https://doi.org/10.1093/jnci/djq306

Dikova V, Jantus-Lewintre E, Bagan J (2021) Potential non-invasive biomarkers for early diagnosis of Oral squamous cell carcinoma. J Clin Med 10(8):1658. https://doi.org/10.3390/jcm10081658

Ding M, Song Y, Jing J, Tian M, Ding L, Li Q, Zhou C, Dong H, Ni Y, Mou Y (2021) The ratio of preoperative serum biomarkers predicts prognosis in patients with Oral squamous cell carcinoma. Front Oncol 11:719513. https://doi.org/10.3389/fonc.2021.719513

Do HT, Nguyen TT, Vo TL, Huynh NC, Nguyen ATK (2023) The influence of smoking on oral neutrophils and matrix metalloproteinase-8 in periodontitis patients before and after nonsurgical treatment. J Oral Biol Craniofacial Res 13(3):442–447. https://doi.org/10.1016/j.jobcr.2023.05.004

Domnich M, Riedesel J, Pylaeva E, Kürten CHL, Buer J, Lang S, Jablonska J (2020) Oral neutrophils: underestimated players in Oral Cancer. Front Immunol 11:565683. https://doi.org/10.3389/fimmu.2020.565683

Elmusrati AA, Pilborough AE, Khurram SA, Lambert DW (2017) Cancer-associated fibroblasts promote bone invasion in oral squamous cell carcinoma. Br J Cancer 117(6):867–875. https://doi.org/10.1038/bjc.2017.239

Ernst PB, Gold BD (2000) The disease spectrum of helicobacter pylori: the immunopathogenesis of gastroduodenal ulcer and gastric cancer. Ann Rev Microbiol 54:615–640. https://doi.org/10.1146/annurev.micro.54.1.615

Fakhry C, D'Souza G (2013) Discussing the diagnosis of HPV-OSCC: common questions and answers. Oral Oncol 49(9):863–871. https://doi.org/10.1016/j.oraloncology.2013.06.002

FDA-NIH Biomarker Working Group (2016) BEST (Biomarkers, Endpoint S, and other Tools) Resource. Silver Spring (MD): Food and Drug Administration (US); Bethesda (MD): National Institutes of Health (US). https://www.ncbi.nlm.nih.gov/books/NBK326791/. Accessed 29 May 2024

Feller L, Altini M, Lemmer J (2013) Inflammation in the context of oral cancer. Oral Oncol 49(9):887–892. https://doi.org/10.1016/j.oraloncology.2013.07.003

Feng L, Houck JR, Lohavanichbutr P, Chen C (2017) Transcriptome analysis reveals differentially expressed lnc RNAs between oral squamous cell carcinoma and healthy oral mucosa. Oncotarget 8(19):31521–31531. https://doi.org/10.18632/oncotarget.16358

Ferlay J, Ervik M, Lam F, Laversanne M, Colombet M, Mery L, Piñeros M, Znaor A, Soerjomataram I, Bray F (2024) Global Cancer observatory: Cancer today. International Agency for Research on Cancer, Lyon. https://gco.iarc.who.int/today. Accessed 27 May 2024

Ferreira FO, Ribeiro FL, Batista AC, Leles CR, de Cássia Gonçalves Alencar R, Silva TA (2008) Association of CCL2 with lymph node metastasis and macrophage infiltration in oral cavity and lip squamous cell carcinoma. Tumour biol: J Int Soc Oncodevelop Biol Med 29(2):114–121. https://doi.org/10.1159/000137669

Fridlender ZG, Sun J, Kim S, Kapoor V, Cheng G, Ling L, Worthen GS, Albelda SM (2009) Polarization of tumor-associated neutrophil phenotype by TGF-beta: "N1" versus "N2" TAN. Cancer Cell 16(3):183–194. https://doi.org/10.1016/j.ccr.2009.06.017

Gabay C (2006) Interleukin-6 and chronic inflammation. Arthritis Res Ther *8*(Suppl 2):S3. https://doi.org/10.1186/ar1917

GBD 2019 Lip, Oral, and Pharyngeal Cancer Collaborators, Cunha ARD, Compton K, Xu R, Mishra R, Drangsholt MT, Antunes JLF, Kerr AR, Acheson AR, Lu D, Wallace LE, Kocarnik JM, Fu W, Dean FE, Pennini A, Henrikson HJ, Alam T, Ababneh E, Abd-Elsalam S, Abdoun M et al (2023) The global, regional, and National Burden of adult Lip, Oral, and Pharyngeal Cancer in 204 countries and territories: A systematic analysis for the global burden of disease study 2019. JAMA Oncol 9(10):1401–1416. https://doi.org/10.1001/jamaoncol.2023.2960

Giacobbi NS, Mullapudi S, Nabors H, Pyeon D (2024) The chemokine CXCL14 as a potential immunotherapeutic agent for Cancer therapy. Viruses 16(2):302. https://doi.org/10.3390/v16020302

Glogauer JE, Sun CX, Bradley G, Magalhaes MA (2015) Neutrophils increase Oral squamous cell carcinoma invasion through an Invadopodia-dependent pathway. Cancer Immunol Res 3(11):1218–1226. https://doi.org/10.1158/2326-6066.CIR-15-0017

Goertzen C, Mahdi H, Laliberte C, Meirson T, Eymael D, Gil-Henn H, Magalhaes M (2018) Oral inflammation promotes oral squamous cell carcinoma invasion. Oncotarget 9(49):29047–29063. https://doi.org/10.18632/oncotarget.25540

Gonda TA, Tu S, Wang TC (2009) Chronic inflammation, the tumor microenvironment and carcinogenesis. Cell cycle (Georgetown, Tex) 8(13):2005–2013. https://doi.org/10.4161/cc.8.13.8985

Goubran HA, Stakiw J, Radosevic M, Burnouf T (2014) Platelet-cancer interactions. Semin Thromb Hemost 40(3):296–305. https://doi.org/10.1055/s-0034-1370767

Greten FR, Grivennikov SI (2019) Inflammation and Cancer: triggers, mechanisms, and consequences. Immunity 51(1):27–41. https://doi.org/10.1016/j.immuni.2019.06.025

Hanahan D, Weinberg RA (2011) Hallmarks of cancer: the next generation. Cell 144(5):646–674. https://doi.org/10.1016/j.cell.2011.02.013

Hasnis E, Reznick AZ, Pollack S, Klein Y, Nagler RM (2004) Synergistic effect of cigarette smoke and saliva on lymphocytes–the mediatory role of volatile aldehydes and redox active iron and the possible implications for oral cancer. Int J Biochem Cell Biol 36(5):826–839. https://doi.org/10.1016/j.biocel.2003.09.003

Hawk ET, Levin B (2005) Colorectal cancer prevention. J Clin Oncol Off J Am Soc Clin Oncol 23(2):378–391. https://doi.org/10.1200/JCO.2005.08.097

Hodge DR, Peng B, Cherry JC, Hurt EM, Fox SD, Kelley JA, Munroe DJ, Farrar WL (2005) Interleukin 6 supports the maintenance of p53 tumor suppressor gene promoter methylation. Cancer Res 65(11):4673–4682. https://doi.org/10.1158/0008-5472.CAN-04-3589

Howlader N, Noone AM, Krapcho M, Garshell J, Neyman N, Altekruse SF, Kosary CL, Yu M, Ruhl J, Tatalovich Z, Cho H, Mariotto A, Lewis DR, Chen HS, Feuer EJ, Cronin KA (eds) (2013) SEER Cancer Statistics Review, 1975–2010. National Cancer Institute, Bethesda. https://seer.cancer.gov/archive/csr/1975_2010/. Accessed 1 June 2024

Hsueh PC, Chang KP, Liu HP, Chiang WF, Chan XY, Hung CM, Chu LJ, Wu CC (2022) Development of a salivary autoantibody biomarker panel for diagnosis of oral cavity squamous cell carcinoma. Front Oncol 12:968570. https://doi.org/10.3389/fonc.2022.968570

Hu Y, He MY, Zhu LF, Yang CC, Zhou ML, Wang Q, Zhang W, Zheng YY, Wang DM, Xu ZQ, Wu YN, Liu LK (2016) Tumor-associated macrophages correlate with the clinicopathological features and poor outcomes via inducing epithelial to mesenchymal transition in oral squamous cell carcinoma. J Exp Clin Cancer Res: CR 35:12. https://doi.org/10.1186/s13046-015-0281-z

Hudson JD, Shoaibi MA, Maestro R, Carnero A, Hannon GJ, Beach DH (1999) A proinflammatory cytokine inhibits p53 tumor suppressor activity. J Exp Med 190(10):1375–1382. https://doi.org/10.1084/jem.190.10.1375

Husvik C, Khuu C, Bryne M, Halstensen TS (2009) PGE2 production in oral cancer cell lines is COX-2-dependent. J Dent Res 88(2):164–169. https://doi.org/10.1177/0022034508329519

Jeng JH, Chang MC, Hahn LJ (2001) Role of areca nut in betel quid-associated chemical carcinogenesis: current awareness and future perspectives. Oral Oncol 37(6):477–492. https://doi.org/10.1016/s1368-8375(01)00003-3

Kadeh H, Saravani S, Heydari F, Keikha M, Rigi V (2015) Expression of matrix Metalloproteinase-10 at invasive front of squamous cell carcinoma and verrucous carcinoma in the Oral cavity. Asian Pacific J Cancer Prevention: APJCP 16(15):6609–6613. https://doi.org/10.7314/apjcp.2015.16.15.6609

Kano S, Homma A, Hatakeyama H, Mizumachi T, Sakashita T, Kakizaki T, Fukuda S (2017) Pretreatment lymphocyte-to-monocyte ratio as an independent prognostic factor for head and neck cancer. Head Neck 39(2):247–253. https://doi.org/10.1002/hed.24576

Karin M (2006) Nuclear factor-kappaB in cancer development and progression. Nature 441(7092):431–436. https://doi.org/10.1038/nature04870

Karnezis T, Shayan R, Fox S, Achen MG, Stacker SA (2012) The connection between lymphangiogenic signalling and prostaglandin biology: a missing link in the metastatic pathway. Oncotarget 3(8):893–906. https://doi.org/10.18632/oncotarget.593

Khansari N, Shakiba Y, Mahmoudi M (2009) Chronic inflammation and oxidative stress as a major cause of age-related diseases and cancer. Recent Patents Inflamm Allergy Drug Discov 3(1):73–80. https://doi.org/10.2174/187221309787158371

Kim J, Bae JS (2016) Tumor-associated macrophages and neutrophils in tumor microenvironment. Mediat Inflamm 2016:6058147. https://doi.org/10.1155/2016/6058147

Knittelfelder O, Delago D, Jakse G, Lukasiak K, Thurner EM, Thurnher D, Pichler M, Renner W, Stranzl-Lawatsch H, Langsenlehner T (2020) The pre-treatment C-reactive protein represents a prognostic factor in patients with Oral and oropharyngeal Cancer treated with radiotherapy. Cancer 12(3):626. https://doi.org/10.3390/cancers12030626

Krishnan R, Thayalan DK, Padmanaban R, Ramadas R, Annasamy RK, Anandan N (2014) Association of serum and salivary tumor necrosis factor-α with histological grading in oral cancer and its role in differentiating premalignant and malignant oral disease. Asian Pacific J Cancer Prevention: APJCP 15(17):7141–7148. https://doi.org/10.7314/apjcp.2014.15.17.7141

Kumarasamy C, Tiwary V, Sunil K, Suresh D, Shetty S, Muthukaliannan GK, Baxi S, Jayaraj R (2021) Prognostic utility of platelet-lymphocyte ratio, neutrophil-lymphocyte ratio and monocyte-lymphocyte ratio in head and neck cancers: A detailed PRISMA compliant systematic review and meta-analysis. Cancer 13(16): 4166. https://doi.org/10.3390/cancers13164166

Lambert R, Sauvaget C, de Camargo Cancela M, Sankaranarayanan R (2011) Epidemiology of cancer from the oral cavity and oropharynx. Eur J Gastroenterol Hepatol 23(8):633–641. https://doi.org/10.1097/MEG.0b013e3283484795

Lee CH, Chang JS, Syu SH, Wong TS, Chan JY, Tang YC, Yang ZP, Yang WC, Chen CT, Lu SC, Tang PH, Yang TC, Chu PY, Hsiao JR, Liu KJ (2015) IL-1β promotes malignant transformation and tumor aggressiveness in oral cancer. J Cell Physiol 230(4): 875–884. https://doi.org/10.1002/jcp.24816

Lee LT, Wong YK, Chan MY, Chang KW, Chen SC, Chang CT, Wang J (2018) The correlation between HIF-1 alpha and VEGF in oral squamous cell carcinomas: expression patterns and quantitative immunohistochemical analysis. J Chinese Med Assoc: JCMA 81(4):370–375. https://doi.org/10.1016/j.jcma.2017.06.025

Lin WW, Karin M (2007) A cytokine-mediated link between innate immunity, inflammation, and cancer. J Clin Invest 117(5):1175–1183. https://doi.org/10.1172/JCI31537

Lee YA, Li S, Chen Y, Li Q, Chen CJ, Hsu WL, Lou PJ, Zhu C, Pan J, Shen H, Ma H, Cai L, He B, Wang Y, Zhou X, Ji Q, Zhou B, Wu W, Ma J, Boffetta P et al (2019) Tobacco smoking, alcohol drinking, betel quid chewing, and the risk of head and neck cancer in an east Asian population. Head Neck 41(1):92–102. https://doi.org/10.1002/hed.25383

Lingen MW, Abt E, Agrawal N, Chaturvedi AK, Cohen E, D'Souza G, Gurenlian J, Kalmar JR, Kerr AR, Lambert PM, Patton LL, Sollecito TP, Truelove E, Tampi MP, Urquhart O, Banfield L, Carrasco-Labra A (2017) Evidence-based clinical practice guideline for the evaluation of potentially malignant disorders in the oral cavity: A report of the American dental association. J Amer Dental Assoc (1939) 148(10):712–727.e10. https://doi.org/10.1016/j.adaj.2017.07.032

Lloyd KE, Hall LH, King N, Thorneloe RJ, Rodriguez-Lopez R, Ziegler L, Taylor DG, MacKenzie M, Smith SG, AsCaP Group (2022) Aspirin use for cancer prevention: A systematic review of public, patient and healthcare provider attitudes and adherence behaviours. Prev Med 154:106872. https://doi.org/10.1016/j.ypmed.2021.106872

Louredo BV, Vargas PA, Pérez-de-Oliveira ME, Lopes MA, Kowalski LP, Curado MP (2022) Epidemiology and survival outcomes of lip, oral cavity, and oropharyngeal squamous cell carcinoma in a southeast Brazilian population. Medicina Oral Patologia Oral Cirugia Bucal 27(3):e274–e284. https://doi.org/10.4317/medoral.25147

Magdum DB, Kulkarni NA, Kavle PG, Paraye S, Pohankar PS, Giram AV (2024) Salivary neutrophil-to-lymphocyte ratio as a prognostic predictor of Oral premalignant and malignant disorders: a prospective study. Cureus 16(3):e56273. https://doi.org/10.7759/cureus.56273

Magnussen SN, Hadler-Olsen E, Costea DE, Berg E, Jacobsen CC, Mortensen B, Salo T, Martinez-Zubiaurre I, Winberg JO, Uhlin-Hansen L, Svineng G (2017) Cleavage of the urokinase receptor (uPAR) on oral cancer cells: regulation by transforming growth factor - β1 (TGF-β1) and potential effects on migration and invasion. BMC Cancer 17(1):350. https://doi.org/10.1186/s12885-017-3349-7

Mantovani A, Allavena P, Sica A, Balkwill F (2008) Cancer-related inflammation. Nature 454(7203): 436–444. https://doi.org/10.1038/nature07205

Mariani P, Russo D, Maisto M, Troiano G, Caponio VCA, Annunziata M, Laino L (2022) Pre-treatment neutrophil-to-lymphocyte ratio is an independent prognostic factor in head and neck squamous cell carcinoma: meta-analysis and trial sequential analysis. J Oral Pathol Med: Off Pub Int Assoc Oral Pathol Amer Acad Oral Pathol 51(1):39–51. https://doi.org/10.1111/jop.13264

Marziliano A, Teckie S, Diefenbach MA (2020) Alcohol-related head and neck cancer: summary of the literature. Head Neck 42(4):732–738. https://doi.org/10.1002/hed.26023

Mascarella MA, Mannard E, Silva SD, Zeitouni A (2018) Neutrophil-to-lymphocyte ratio in head and neck cancer prognosis: A systematic review and meta-analysis. Head Neck 40(5):1091–1100. https://doi.org/10.1002/hed.25075

Metgud R, Bajaj S (2016) Altered serum and salivary C-reactive protein levels in patients with oral premalignant lesions and oral squamous cell carcinoma. Biotechnic Histochemistry: Off Publ Biological Stain Comm 91(2):96–101. https://doi.org/10.3109/10520295.2015.1077393

Mohapatra S, Pioppini C, Ozpolat B, Calin GA (2021) Non-coding RNAs regulation of macrophage polarization in cancer. Mol Cancer 20(1):24. https://doi.org/10.1186/s12943-021-01313-x

Morse DE, Kerr AR (2006) Disparities in oral and pharyngeal cancer incidence, mortality and survival among black and white Americans. J Ame Dental Assoc (1939) *137*(2):203–212. https://doi.org/10.14219/jada.archive.2006.0146

Munn LL (2017) Cancer and inflammation. Wiley interdisciplinary reviews. Sys Biology Med 9(2). https://doi.org/10.1002/wsbm.1370

Murata M (2018) Inflammation and cancer. Environ Health Prev Med 23(1):50. https://doi.org/10.1186/s12199-018-0740-1

Murti PR, Bhonsle RB, Pindborg JJ, Daftary DK, Gupta PC, Mehta FS (1985) Malignant transformation rate in oral submucous fibrosis over a 17-year period. Community Dent Oral Epidemiol 13(6):340–341. https://doi.org/10.1111/j.1600-0528.1985.tb00468.x

Nakanishi M, Rosenberg DW (2013) Multifaceted roles of PGE2 in inflammation and cancer. Semin

Immunopathol 35(2):123–137. https://doi.org/10.1007/s00281-012-0342-8
Nasry WHS, Rodriguez-Lecompte JC, Martin CK (2018) Role of COX-2/PGE2 mediated inflammation in Oral squamous cell carcinoma. Cancer 10(10):348. https://doi.org/10.3390/cancers10100348
Niklander SE (2021) Inflammatory mediators in Oral Cancer: pathogenic mechanisms and diagnostic potential. Front Oral Health 2:642238. https://doi.org/10.3389/froh.2021.642238
Nissinen L, Kähäri VM (2014) Matrix metalloproteinases in inflammation. Biochim Biophys Acta 1840(8): 2571–2580. https://doi.org/10.1016/j.bbagen.2014.03.007
Ng JH, Iyer NG, Tan MH, Edgren G (2017) Changing epidemiology of oral squamous cell carcinoma of the tongue: A global study. Head Neck 39(2):297–304. https://doi.org/10.1002/hed.24589
Ohta M, Kitadai Y, Tanaka S, Yoshihara M, Yasui W, Mukaida N, Haruma K, Chayama K (2002) Monocyte chemoattractant protein-1 expression correlates with macrophage infiltration and tumor vascularity in human esophageal squamous cell carcinomas. Int J Cancer 102(3):220–224. https://doi.org/10.1002/ijc.10705
Oyama N, Iwatsuki K, Homma Y, Kaneko F (1999) Induction of transcription factor AP-2 by inflammatory cytokines in human keratinocytes. J Invest Dermatol 113(4):600–606. https://doi.org/10.1046/j.1523-1747.1999.00734.x
Ozawa S, Kato Y, Komori R, Maehata Y, Kubota E, Hata R (2006) BRAK/CXCL14 expression suppresses tumor growth in vivo in human oral carcinoma cells. Biochem Biophys Res Commun 348(2):406–412. https://doi.org/10.1016/j.bbrc.2006.07.070
Paul A, Srivastava S, Roy R, Anand A, Gaurav K, Husain N, Jain S, Sonkar AA (2020) Malignancy prediction among tissues from Oral SCC patients including neck invasions: a ^{1}H HRMAS NMR based metabolomic study. Metabol Off J Metabol Soc 16(3):38. https://doi.org/10.1007/s11306-020-01660-8
Payne K, Spruce R, Beggs A, Sharma N, Kong A, Martin T, Parmar S, Praveen P, Nankivell P, Mehanna H (2018) Circulating tumor DNA as a biomarker and liquid biopsy in head and neck squamous cell carcinoma. Head Neck 40(7):1598–1604. https://doi.org/10.1002/hed.25140
Patel BP, Shah SV, Shukla SN, Shah PM, Patel PS (2007) Clinical significance of MMP-2 and MMP-9 in patients with oral cancer. Head Neck 29(6):564–572. https://doi.org/10.1002/hed.20561
Pillay J, Kamp VM, van Hoffen E, Visser T, Tak T, Lammers JW, Ulfman LH, Leenen LP, Pickkers P, Koenderman L (2012) A subset of neutrophils in human systemic inflammation inhibits T cell responses through Mac-1. J Clin Invest 122(1):327–336. https://doi.org/10.1172/JCI57990
Pradhan R, Paul S, Acharya SS, Sinha S, Dash SR, Kundu CN (2024) Nano formulated resveratrol inhibits PD-L1 in oral cancer cells by deregulating the association between tumor associated macrophages and cancer associated fibroblasts through IL-6/JAK2/STAT3 signaling axis. J Nutr Biochem 125:109568. https://doi.org/10.1016/j.jnutbio.2024.109568
Prime SS, Cirillo N, Hassona Y, Lambert DW, Paterson IC, Mellone M, Thomas GJ, James EN, Parkinson EK (2017) Fibroblast activation and senescence in oral cancer. J Oral Pathol Med Off Publ Int Assoc Oral Pathol Amer Acad Oral Pathol 46(2):82–88. https://doi.org/10.1111/jop.12456
Radaic A, Kamarajan P, Cho A, Wang S, Hung GC, Najarzadegan F, Wong DT, Ton-That H, Wang CY, Kapila YL (2023) Biological biomarkers of oral cancer. Periodontology 2000. https://doi.org/10.1111/prd.12542
Ralli M, Grasso M, Gilardi A, Ceccanti M, Messina MP, Tirassa P, Fiore M, Altissimi G, Salzano A, F., De Vincentiis, M., & Greco, A. (2020) The role of cytokines in head and neck squamous cell carcinoma: A review. Clin Ter 171(3):e268–e274. https://doi.org/10.7417/CT.2020.2225
Ramanathan K, Lakshimi S (1974) Oral carcinoma in Malaysian Indian males. Singap Dent J 13(2):5–11
Ramos de GO, Bernardi L, Lauxen I, Sant'Ana Filho M, Horwitz AR, Lamers ML (2016) Fibronectin modulates cell adhesion and signaling to promote single cell migration of highly invasive Oral squamous cell carcinoma. PLoS One 11(3):e0151338. https://doi.org/10.1371/journal.pone.0151338
Rao SK, Pavicevic Z, Du Z, Kim JG, Fan M, Jiao Y, Rosebush M, Samant S, Gu W, Pfeffer LM, Nosrat CA (2010) Pro-inflammatory genes as biomarkers and therapeutic targets in oral squamous cell carcinoma. J Biol Chem 285(42):32512–32521. https://doi.org/10.1074/jbc.M110.150490
Remick DG (2005) Interleukin-8. Crit Care Med 33(12 Suppl):S466–S467. https://doi.org/10.1097/01.ccm.0000186783.34908.18
Ren ZH, Wu K, Yang R, Liu ZQ, Cao W (2020) Differential expression of matrix metalloproteinases and miRNAs in the metastasis of oral squamous cell carcinoma. BMC Oral Health 20(1):24. https://doi.org/10.1186/s12903-020-1013-0
Rhodus NL, Ho V, Miller CS, Myers S, Ondrey F (2005) NF-kappaB dependent cytokine levels in saliva of patients with oral preneoplastic lesions and oral squamous cell carcinoma. Cancer Detect Prev 29(1):42–45. https://doi.org/10.1016/j.cdp.2004.10.003
Rothwell PM, Fowkes FG, Belch JF, Ogawa H, Warlow CP, Meade TW (2011) Effect of daily aspirin on long-term risk of death due to cancer: analysis of individual patient data from randomised trials. Lancet (London, England) 377(9759):31–41. https://doi.org/10.1016/S0140-6736(10)62110-1
Sahibzada HA, Khurshid Z, Khan RS, Naseem M, Siddique KM, Mali M, Zafar MS (2017) Salivary IL-8, IL-6 and TNF-α as potential diagnostic biomarkers for Oral Cancer. Diagnostics (Basel, Switzerland) 7(2):21. https://doi.org/10.3390/diagnostics7020021
Sakamoto K, Hikiba Y, Nakagawa H, Hayakawa Y, Yanai A, Akanuma M, Ogura K, Hirata Y, Kaestner

KH, Omata M, Maeda S (2010) Inhibitor of kappaB kinase beta regulates gastric carcinogenesis via interleukin-1alpha expression. Gastroenterology 139(1):226–38.e6. https://doi.org/10.1053/j.gastro.2010.03.047

Salvo E, Tu NH, Scheff NN, Dubeykovskaya ZA, Chavan SA, Aouizerat BE, Ye Y (2021) TNFα promotes oral cancer growth, pain, and Schwann cell activation. Sci Rep 11(1):1840. https://doi.org/10.1038/s41598-021-81500-4

Schiegnitz E, Kämmerer PW, Schön H, Blatt S, Berres M, Sagheb K, Al-Nawas B (2018) Proinflammatory cytokines as serum biomarker in oral carcinoma-A prospective multi-biomarker approach. J Oral Pathol Medicine Off Publ Int Assoc Oral Pathol Amer Acad Oral Pathol 47(3):268–274. https://doi.org/10.1111/jop.12670

Segal AW (2006) How superoxide production by neutrophil leukocytes kills microbes. Novartis Found Symp 279:92–219

Shafer WG, Waldron CA (1975) Erythroplakia of the oral cavity. Cancer 36(3):1021–1028. https://doi.org/10.1002/1097-0142(197509)36:3<1021::aid-cncr2820360327>3.0.co;2-w

Shan Q, Takabatake K, Omori H, Kawai H, Oo MW, Nakano K, Ibaragi S, Sasaki A, Nagatsuka H (2021) Stromal cells in the tumor microenvironment promote the progression of oral squamous cell carcinoma. Int J Oncol 59(3):72. https://doi.org/10.3892/ijo.2021.5252

Shinriki S, Ueda M, Ota K, Nakamura M, Kudo M, Ibusuki M, Kim J, Yoshitake Y, Fukuma D, Jono H, Kuratsu J, Shinohara M, Ando Y (2010) Aberrant expression of serum amyloid A in head and neck squamous cell carcinoma. J Oral Pathol Med Off Publ Int Assoc Oral Pathol Amer Acad Oral Pathol 39(1):41–47. https://doi.org/10.1111/j.1600-0714.2009.00777.x

Siegel RL, Miller KD, Jemal A (2017) Cancer statistics, 2017. CA Cancer J Clin 67(1):7–30. https://doi.org/10.3322/caac.21387

Singh N, Baby D, Rajguru JP, Patil PB, Thakkannavar SS, Pujari VB (2019) Inflammation and cancer. Ann Afr Med 18(3):121–126. https://doi.org/10.4103/aam.aam_56_18

Singh S, Singh J, Ganguly R, Chandra S, Samadi FM, Suhail S (2021) Diagnostic efficacy of neutrophil to lymphocyte ratio (NLR) in oral potentially malignant disorders and oral cancer. Indian J Pathol Microbiol 64(2):243–249. https://doi.org/10.4103/IJPM.IJPM_114_20

Singh V, Husain N, Akhtar N, Khan MY, Sonkar AA, Kumar V (2017) p16 and p53 in HPV-positive versus HPV-negative oral squamous cell carcinoma: do pathways differ? Journal Oral Pathol Med Off Pub Int Assoc Oral Pathol Amer Acad Oral Pathol 46(9):744–751. https://doi.org/10.1111/jop.12562

Speight PM, Khurram SA, Kujan O (2018) Oral potentially malignant disorders: risk of progression to malignancy. Oral Surg Oral Med Oral Pathol Oral Radiol 125(6):612–627. https://doi.org/10.1016/j.oooo.2017.12.011

Stepan KO, Mazul AL, Larson J, Shah P, Jackson RS, Pipkorn P, Kang SY, Puram SV (2023) Changing epidemiology of Oral cavity Cancer in the United States. Otolaryngol Head Neck Sur Off J Amer Acad Otolaryngol-Head Neck Sur 168(4):761–768. https://doi.org/10.1177/01945998221098011

St John MA (2015) Inflammatory mediators drive metastasis and drug resistance in head and neck squamous cell carcinoma. Laryngoscope 125(Suppl 3):S1–S11. https://doi.org/10.1002/lary.24998

Surcel M, Constantin C, Caruntu C, Zurac S, Neagu M (2017) Inflammatory cytokine pattern is sex-dependent in mouse cutaneous melanoma experimental model. J Immunol Res 2017:9212134. https://doi.org/10.1155/2017/9212134

Supanimitjaroenporn P, Kirtsreesakul V, Tangthongkum M, Leelasawatsuk P, Prapaisit U (2022) Prognostic value of pretreatment lymphocyte-to-monocyte ratio in patients with advanced oral cavity cancer. Laryngoscope Invest Otolaryngol 7(3):740–745. https://doi.org/10.1002/lio2.790

Supawat B, Aye KT, Ritpanja J, Nueangwong W, Kothan S, Pan J, Tungjai M (2021) Differences in spectroscopic properties of saliva taken from Normal subjects and Oral Cancer patients: comparison studies. J Fluoresc 31(3):747–754. https://doi.org/10.1007/s10895-021-02707-2

Tan Y, Wang Z, Xu M, Li B, Huang Z, Qin S, Nice EC, Tang J, Huang C (2023) Oral squamous cell carcinomas: state of the field and emerging directions. Int J Oral Sci 15(1):44. https://doi.org/10.1038/s41368-023-00249-w

Tanaka T, Narazaki M, Kishimoto T (2014) IL-6 in inflammation, immunity, and disease. Cold Spring Harb Perspect Biol 6(10):a016295. https://doi.org/10.1101/cshperspect.a016295

Tampa M, Mitran MI, Mitran CI, Sarbu MI, Matei C, Nicolae I, Caruntu A, Tocut SM, Popa MI, Caruntu C, Georgescu SR (2018) Mediators of inflammation – A potential source of biomarkers in Oral squamous cell carcinoma. J Immunol Res 2018:1061780. https://doi.org/10.1155/2018/1061780

Thomson PJ (2002) Field change and oral cancer: new evidence for widespread carcinogenesis? Int J Oral Maxillofac Surg 31(3):262–266. https://doi.org/10.1054/ijom.2002.0220

Tian S, Jiang C, Liu X, Xu S, Zhang Z, Chen H, Zhang Y, Liu Y, Ma D (2017) Hypermethylation of IFN-γ in oral cancer tissues. Clin Oral Investig 21(8):2535–2542. https://doi.org/10.1007/s00784-017-2052-z

Tran Q, Maddineni S, Arnaud EH, Divi V, Megwalu UC, Topf MC, Sunwoo JB (2023) Oral cavity cancer in young, non-smoking, and non-drinking patients: A contemporary review. Crit Rev Oncol Hematol 190:104112. https://doi.org/10.1016/j.critrevonc.2023.104112

Tsou CL, Peters W, Si Y, Slaymaker S, Aslanian AM, Weisberg SP, Mack M, Charo IF (2007) Critical roles for CCR2 and MCP-3 in monocyte mobilization from bone marrow and recruitment to inflammatory sites. J

Clin Invest 117(4):902–909. https://doi.org/10.1172/JCI29919

Vassilakopoulou M, Psyrri A, Argiris A (2015) Targeting angiogenesis in head and neck cancer. Oral Oncol 51(5):409–415. https://doi.org/10.1016/j.oraloncology.2015.01.006

Vissink A, Wolff A, Veerman ECI (2008) Saliva collectors. In: Wong DT (ed) Salivary diagnostics. Wiley-Blackwell, Ames, IA, pp 37–59

Wang D, Dubois RN (2010) Eicosanoids and cancer. Nat Rev Cancer 10(3):181–193. https://doi.org/10.1038/nrc2809

Wang F, Zhang H, Xue Y, Wen J, Zhou J, Yang X, Wei J (2017) A systematic investigation of the association between HPV and the clinicopathological parameters and prognosis of oral and oropharyngeal squamous cell carcinomas. Cancer Med 6(5):910–917. https://doi.org/10.1002/cam4.1045

Warnakulasuriya KA, Johnson NW, Linklater KM, Bell J (1999) Cancer of mouth, pharynx and nasopharynx in Asian and Chinese immigrants resident in Thames regions. Oral Oncol 35(5):471–475. https://doi.org/10.1016/s1368-8375(99)00019-6

Wei LY, Lin HC, Tsai FC, Ko JY, Kok SH, Cheng SJ, Lee JJ, Chia JS (2022) Effects of Interleukin-6 on STAT3-regulated signaling in oral cancer and as a prognosticator of patient survival. Oral Oncol 124: 105665. https://doi.org/10.1016/j.oraloncology.2021.105665

Wong T, Wiesenfeld D (2018) Oral Cancer. Aust Dent J 63(Suppl 1):S91–S99. https://doi.org/10.1111/adj.12594

Yadav L, Puri N, Rastogi V, Satpute P, Ahmad R, Kaur G (2014) Matrix metalloproteinases and cancer – roles in threat and therapy. Asian Pacific J Cancer Prevention: APJCP 15(3):1085–1091. https://doi.org/10.7314/apjcp.2014.15.3.1085

Yang Y, Liu R, Ren F, Guo R, Zhang P (2018) Prognostic and clinicopathological significance of neutrophil-to-lymphocyte ratio in patients with oral cancer. Biosci Rep 38(6):BSR20181550. https://doi.org/10.1042/BSR20181550

Yesupatham ST, Dayanand CD, Azeem Mohiyuddin SM, Harendra Kumar ML (2023) An insight into Survivin in relevance to hematological, biochemical and genetic characteristics in tobacco chewers with Oral squamous cell carcinoma. Cells 12(10):1444. https://doi.org/10.3390/cells12101444

Yan B, Wang H, Rabbani ZN, Zhao Y, Li W, Yuan Y, Li F, Dewhirst MW, Li CY (2006) Tumor necrosis factor-alpha is a potent endogenous mutagen that promotes cellular transformation. Cancer Res 66(24): 11565–11570. https://doi.org/10.1158/0008-5472.CAN-06-2540

Yang SF, Chen MK, Hsieh YS, Chung TT, Hsieh YH, Lin CW, Su JL, Tsai MH, Tang CH (2010) Prostaglandin E2/EP1 signaling pathway enhances intercellular adhesion molecule 1 (ICAM-1) expression and cell motility in oral cancer cells. J Biol Chem 285(39): 29808–29816. https://doi.org/10.1074/jbc.M110.108183

Yilmaz B, Somay E, Topkan E, Kucuk A, Pehlivan B, Selek U (2023a) Utility of pre-chemoradiotherapy Pan-immune-inflammation-value for predicting the osteoradionecrosis rates in locally advanced nasopharyngeal cancers. Strahlenther Onkol 199(10):910–921. https://doi.org/10.1007/s00066-023-02119-0

Yilmaz B, Somay E, Topkan E, Kucuk A, Pehlivan B, Selek U (2023b) The predictive value of pretreatment hemoglobin-to-platelet ratio on osteoradionecrosis incidence rates of locally advanced nasopharyngeal cancer patients managed with concurrent chemoradiotherapy. BMC Oral Health 23(1):231. https://doi.org/10.1186/s12903-023-02937-9

Yilmaz B, Topkan E, Besen AA, Mertsoylu H, Selek U (2024) The predictive power of biomarkers in osteoradionecrosis. In: Interdisciplinary Cancer Research. Springer, Cham. https://doi.org/10.1007/16833_2024_266

Zhang JM, An J (2007) Cytokines, inflammation, and pain. Int Anesthesiol Clin 45(2):27–37. https://doi.org/10.1097/AIA.0b013e318034194e

Kate, Schroder Jurg, Tschopp (2010) The Inflammasomes Cell 140(6):821–832. https://doi.org/10.1016/j.cell.2010.01.040

Chia Huei, Lee Jeffrey Shu-Ming, Chang Shih-Han, Syu Thian-Sze, Wong Jimmy Yu-Wai, Chan Ya-Chu, Tang ZhiPing, Yang Wen-Chan, Yang Chiung-Tong, Chen Shao-Chun, Lu Pei-Hua, Tang Tzu-Ching, Yang Pei-Yi, Chu Jenn-Ren, Hsiao Ko-Jiunn, Liu (2015) IL-1β Promotes Malignant Transformation and Tumor Aggressiveness in Oral Cancer Journal of Cellular Physiology 230(4):875–884. https://doi.org/10.1002/jcp.v230.4 10.1002/jcp.24816

Hamza, Salema Samir, Joshi Sudhir, Pawar Vivek S, Nair Vedangi V, Deo Manali M, Sanghai Evaluation of the Role of C-reactive Protein as a Prognostic Indicator in Oral Pre-malignant and Malignant Lesions Cureus https://doi.org/10.7759/cureus.60812

Weixing, Liu Yue, Liu Pei, Li Zhiyuan, Wang Jia, Chen Hui, Liu Jin, Ye (2023) Causal association of serum biomarkers with oral cavity and oropharyngeal cancer: a mendelian randomization study Abstract BMC Oral Health 23(1). https://doi.org/10.1186/s12903-023-03729-x

Dardo, Menditti Mario, Santagata Gianmaria, Imola Samuel, Staglianò Rita, Vitagliano Ciro Emiliano, Boschetti Angelo Michele, Inchingolo (2023) Personalized Medicine in Oral Oncology: Imaging Methods and Biological Markers to Support Diagnosis of Oral Squamous Cell Carcinoma (OSCC): A Narrative Literature Review Journal of Personalized Medicine 13(9):1397. https://doi.org/10.3390/jpm13091397

Characterizing the Head and Neck Cancer Microenvironment: Implications for Immunotherapy and Prognosis

18

Shabnam Varmazyari, Hossein Borhani, Zohreh Moradipour, Heliya Ziaei, and Nima Rezaei

Abstract

Head and neck cancers (HNCs), predominantly squamous cell carcinomas, exhibit high relapse and poor survival rates despite multimodal treatment approaches. One of the main challenges in HNC is the tumor microenvironment (TME), which is composed of both immune and non-immune cells, as well as non-cellular components such as the extracellular matrix, which collectively foster an immunosuppressive environment that enables tumor growth, invasion, and therapeutic resistance. This chapter explores the intricate nature of the TME in HNC and its pivotal role in influencing treatment outcomes, particularly with immunotherapy. It delves into the characteristics and functions of key cellular components of the TME, including cancer-associated fibroblasts, dendritic cells, T cells, B cells, tumor-associated macrophages, and neutrophils, each presenting both challenges and therapeutic opportunities. Furthermore, it explores novel therapeutic strategies that target the TME, such as immune checkpoint inhibitors, adoptive cell transfer, and cancer vaccines. Additionally, it discusses predictive and prognostic factors related to immune, genetic, and microbial signatures in HNC, offering insights into their potential role in improving treatment outcomes. By exploring both current and emerging therapies, the present chapter underscores the significance of modulating the TME to enhance the efficacy of cancer treatments.

Keywords

Head and neck cancer · Oral squamous cell carcinoma · Immunotherapy · Cancer immunotherapy · Oral cancer

S. Varmazyari
Laser Research Center of Dentistry, Dentistry Research Institute, Tehran University of Medical Sciences, Tehran, Iran

H. Borhani
Dental Students' Scientific Research Center, School of Dentistry, Tehran University of Medical Sciences, Tehran, Iran

Z. Moradipour
School of Dentistry, Tehran University of Medical Sciences, Tehran, Iran

H. Ziaei
Universal Scientific Education and Research Network (USERN), Los Angeles, CA, USA

N. Rezaei (✉)
Universal Scientific Education and Research Network (USERN), Göteborg, Sweden

Network of Immunity in Infection, Malignancy and Autoimmunity (NIIMA), Universal Scientific Education and Research Network (USERN), Tehran, Iran

Research Center for Immunodeficiencies, Children's Medical Center, Tehran University of Medical Sciences, Tehran, Iran

Department of Immunology, School of Medicine, Tehran University of Medical Sciences, Tehran, Iran
e-mail: Rezaei_nima@tums.ac.ir

N. Rezaei, H. Ziaei (eds.), *Oral Immunology*, Advances in Experimental Medicine and Biology 1492,
https://doi.org/10.1007/978-3-032-03176-1_18

18.1 Introduction

Head and neck cancers (HNCs) are lethal malignancies with poor prognosis, high relapse rate, and frequent metastases, with head and neck squamous cell carcinomas (HNSCCs) comprising 90% of cases (Chen et al. 2020; Elmusrati et al. 2021). HNSCC is frequently detected at advanced stages with local or distant metastases, and it presents significant treatment complications due to the complex maxillofacial anatomy, surgical challenges, and the need to preserve the patient's functionality and aesthetics (Elmusrati et al. 2021; Mazilu et al. 2021). As a result, treatment typically requires a collaborative and multidisciplinary strategy, including surgery, radiotherapy, chemotherapy, and immunotherapy (Wang et al. 2019). Despite these efforts, about 40–60% of locally aggressive HNSCC tumors eventually progress or become recurrent (Bhat et al. 2021), with up to 30% of patients experiencing relapse (Canning et al. 2019; Chen et al. 2020). In addition, despite the multidisciplinary treatment regimens, the prognosis of HNSCC remains poor, with a 40–50% 5-year survival rate that has witnessed little improvement over time (Marur and Forastiere 2008; Mazilu et al. 2021; Wondergem et al. 2020). Given the considerable rates of treatment failure, new strategies are required for managing HNSCC.

Cetuximab, the first FDA-approved immunotherapy for HNSCC, initially showed promise but achieved only 10–20% efficacy as monotherapy (Jie et al. 2015; Moy et al. 2017; Srivastava et al. 2013; Vermorken et al. 2008a). Although the drug improved treatment outcomes in combination with chemotherapy, survival rates of HNC patients remained limited with a median of around 10 months (Vermorken et al. 2008b). Later implementation of immune checkpoint blockade agents such as pembrolizumab and nivolumab further enhanced outcomes for metastatic HNSCC patients, yet only 15–20% of patients experienced a durable response (Chow et al. 2016; Cramer et al. 2019; Ferris et al. 2016; Mazilu et al. 2021; Moy et al. 2017; Ruffin et al. 2023; Wang et al. 2019). This limited effectiveness suggests the involvement of other immunosuppressive mechanisms within the tumor microenvironment (TME) (Wondergem et al. 2020).

Latest advancements in immunotherapy have highlighted the critical functions of the TME and its significance in treatment outcomes (Economopoulou et al. 2020; Wondergem et al. 2020). The TME is not merely a collection of malignant cells but a complicated ecosystem of immune cells, non-immune cells, and extracellular components (Wondergem et al. 2020). In HNSCC, tumor cells exploit immune constituents of the TME to foster tumor growth, angiogenesis, and metastasis while creating an immunosuppressive environment (Elmusrati et al. 2021; Qin et al. 2021). This environment not only supports tumor progression but also contributes to treatment resistance and immune system evasion (Qin et al. 2021).

Understanding tumor-TME interactions is key to overcoming immunotherapy resistance and improving treatment outcomes in HNC. Therefore, this chapter explores the mechanisms of immunosurveillance, immune escape, and immune evasion in the TME. It details the characteristics, functions, and roles of each TME constituent, as well as their therapeutic implications. Additionally, it covers novel therapeutic approaches that indirectly affect the TME and provides a brief overview of immunological, microbial, and genetic prognostic factors influencing HNC outcomes.

18.2 Immunosurveillance and Immunoediting

Burnet's theory of "cancer immunosurveillance" suggests that tumor cells express unique antigens that allow the immune system's innate and adaptive branches to work together to identify and eliminate cancerous cells, and prevent these cells from becoming invasive (Buchwald and Schmitt 2021; Burnet 1967; Economopoulou et al. 2020; Sim et al. 2019). The idea of "immunoediting" was an elaboration of the concept of immunosurveillance (Sim et al. 2019). Immunoediting is an evolving process consisting of three main stages of elimination, equilibrium, and escape (Mohan et al. 2019; Sim et al. 2019). During elimination, both branches of the immune system work together to identify and destroy emerging cancer cells before they become clinically detectable (Mohan et al. 2019; Sim et al. 2019). Ideally, this stage would result in the complete eradication of all tumor cells (Mohan et al. 2019; Sim et al. 2019). However, if any cancer cells survive, the process moves into equilibrium. In this stage, the immune system continues to suppress tumor growth, primarily through adaptive immunity, maintaining a stable but subclinical state (Sim et al. 2019). Genetic changes within the tumor may occur during this time, contributing to its persistence (Mohan et al. 2019; Sim et al. 2019). If tumor cells evade the immune system at this stage, they progress to the escape stage and adopt further strategies to avoid detection (Mohan et al. 2019; Sim et al. 2019). A number of these escape strategies by which tumors establish an immunosuppressive microenvironment are mentioned in the following text.

18.2.1 Disruption in Antigen Presentation and Processing

Tumors manipulate antigen presentation to evade immune recognition. Under-expression and dysfunction of the human leukocyte antigen (HLA) complex impairs tumor antigen handling and displaying to T cells (Wondergem et al. 2020). Disruption in antigen-processing machinery (APM) genes further inhibits the handling and displaying of tumor-associated antigens (Wondergem et al. 2020). Mutations in MHC class I molecules allow tumor cells to evade identification and elimination by Natural killer (NK) cells (Moy et al. 2017). Overexpression of epidermal growth factor receptor (EGFR) downregulates HLA through pathways such as SHP2/STAT1, PI3K/AKT, and MAPK, promoting immune evasion (Wondergem et al. 2020). Another key mechanism involves the reduction in neoantigen production which limits the identification and targeting of tumor cells (Moy et al. 2017; Wondergem et al. 2020).

There are potential treatment strategies to address disruption in antigen presentation in tumor cells. HLA class I and APM deficiencies in HNSCC can be addressed by targeting SHP2 or enhancing STAT1 signaling through IFN-γ, which may restore antigen presentation and improve tumor recognition (Buchwald and Schmitt 2021). Additionally, EGFR antagonists can be used to restore HLA class I expression and boost proinflammatory cytokine production, thereby enhancing anti-tumor immune responses (Buchwald and Schmitt 2021).

18.2.2 Exclusion of Immune Cells and Induction of Immunosuppressive Environments

This strategy that impairs anti-tumor immunity includes inducing T cell dysfunction, anergy, and apoptosis alongside recruiting M2 macrophages, regulatory T cells (Tregs), and Myeloid-Derived Suppressor Cells (MDSCs) to the TME (Wondergem et al. 2020). These alterations contribute to suppressing effective anti-tumor immunity.

18.2.3 Upregulation of Inhibitory Receptors

Upregulation of Inhibitory Receptors is a key immune evasion strategy employed by tumors to suppress antitumor immune responses.

18.2.3.1 Cytotoxic T-Lymphocyte Associated Protein-4 (CTLA-4)

Regarding the upregulation of inhibitory receptors, CTLA-4, displayed on T cells after T Cell Receptor (TCR) activation, counteracts CD28 signals to prevent excessive immune responses (Honeychurch et al. 2015; Moy et al. 2017; Rudd et al. 2009; Schwartz 1992). While CTLA-4 expression is transient in CD8+ T cells, where it helps regulate function and memory, it remains highly expressed in CD4+ regulatory T cells (Tregs), where it plays a key role in immune suppression (Alegre et al. 1996; Chambers et al. 1999; Egen and Allison 2002; Moy et al. 2017; Takahashi et al. 2000; Wing et al. 2008). With higher affinity than CD28, CTLA-4 competes for APC B7 ligands (CD80/CD86) with other receptors, blocking the second signal required for T cell activation (Linsley et al. 1994). It also promotes B7 ligand internalization and degradation, further suppressing T cell activation (Moy et al. 2017; Oderup et al. 2006; Qureshi et al. 2011).

18.2.3.2 PD/PDL-1 Pathway

PD-L1 is displayed in 50–60% of HNSCCs, aiding immune evasion and tumor growth (Moy et al. 2017). PD-1, found on activated immune cells, binds to PD-L1 and PD-L2 (Xu 2020). While PD-L1, present on various cells including tumors, suppresses T cell activity, PD-L2, mainly on antigen-presenting cells (APCs), regulates T cell function in inflammation (Moy et al. 2017). Both ligands are upregulated by IFN-γ and TNF-α, indicating a feedback inhibition mechanism (Freeman et al. 2000; Latchman et al. 2001; Liang et al. 2003). Tumors can overproduce PD-L1 through signaling pathways such as PI3K-AKT and EGFR-JAK2/STAT1 (Butte et al. 2007; Eppihimer et al. 2002). Increased expression of PD-1 on cytotoxic T cells indicates immune exhaustion, while increased display of PD-L1 on APCs supports Treg-mediated immune suppression in tumors (Concha-Benavente et al. 2016; Parsa et al. 2007; Xu 2020).

18.2.3.3 LAG-3

LAG3 is an inhibitory receptor that promotes Tregs while suppressing cytotoxic T cells (Ribas et al. 2016). It interacts with MHC class II molecules (Xu 2020), which are also present on dendritic cells (DCs), further influencing tumor suppression (Zaretsky et al. 2016).

18.2.3.4 TIM-3 (HAVcr2)

TIM-3 is predominantly found on IFN-γ-producing T cells, NK cells, and Tregs, playing a significant role in immune suppression and identifying dysfunctional cytotoxic T cells and NK cells (Moy et al. 2017). As TIM-3 expression is mostly restricted to tumor tissues, targeting it has a lower likelihood of autoimmune side effects compared to other checkpoint inhibitors (Moy et al. 2017).

18.2.3.5 Killer Immunoglobulin-Like Receptors (KIR)

Finally, KIRs are expressed on NK cells and interact with HLA class I molecules, particularly HLA-C, on target cells to regulate NK cell activity (Moy et al. 2017). Inhibitory KIRs bind to autologous HLA-C molecules, suppressing NK cell activation and protecting healthy cells. Tumor cells exploit this mechanism by maintaining HLA class I expression, evading immune detection. Activating KIRs recognize altered or non-self HLA molecules or stress-induced ligands, triggering NK cell cytotoxicity and promoting tumor elimination (Nishimura et al. 1999; Tivol et al. 1995).

To therapeutically target inhibitory receptors, preclinical studies have shown that CTLA-4-targeting antibodies can eliminate cancer cells and support immune memory (Leach et al. 1996). While CTLA-4 blockade alone was less effective, it significantly improved outcomes when combined with tumor-antigen vaccines, radiation, and chemotherapy (Hurwitz et al. 2000; Hurwitz et al. 1998; Moy et al. 2017; van Elsas et al. 1999). This resulted in the creation of ipilimumab, which has been FDA-approved for metastatic melanoma, and tremelimumab, which is in phase III clinical trials (Gao and Chen 2016; Hodi et al. 2010; Honeychurch et al. 2015; Ribas et al. 2013). Both agents are being tested in combination with other HNSCC treatments (Moy et al. 2017). Similarly, blocking PD-1 with specific antibodies can enhance immune response and tumor clearance (Wherry 2011). Pembrolizumab, nivolumab, and durvalumab are effective anti-PD-L1 treatments for HNSCC, with pembrolizumab being FDA-approved and durvalumab showing promising survival rates (Okazaki et al. 2001; Terme et al. 2011). Combination therapies including PD-1/PD-L1 blockers with CTLA-4 inhibitors have shown increased effectiveness, with ongoing trials exploring the effectiveness of these combinations as well as novel biomarkers such as PD-L2 (Pai et al. 2016; Segal et al. 2016).

Moreover, LAG-3 and PD-1 are frequently co-expressed on exhausted T cells, and targeting both PD-1 and LAG-3 can also help reverse T cell dysfunction (Goldberg and Drake 2011; Xu 2020). Anti-LAG-3 therapies are being evaluated with nivolumab in clinical trials for HNSCC (NCT01968109) (Moy et al. 2017). Additionally, blocking TIM-3 can help counteract T cell exhaustion and rejuvenate NK cell functionality (Moy et al. 2017). Combining TIM-3 inhibition with PD-1 blockade has shown superior tumor regression compared to targeting either pathway alone (Grosso et al. 2009; Jie et al. 2013). Early stage clinical trials are evaluating monoclonal antibodies against TIM-3, such as MBG-453 and TSR-022, as monotherapy and also in combination with PD-1 inhibitors (NCT02608268 and NCT02817633) (Moy et al. 2017). Finally, targeting inhibitory KIRs is another potential approach to enhance NK cell-mediated tumor elimination. Anti-KIR monoclonal antibody (mAb) lirilumab is in early-phase clinical trials aimed at enhancing NK cell activity by blocking inhibitory KIRs (NCT01714739 and NCT01750580) (Moy et al. 2017).

18.2.4 Use of Costimulatory Receptors

Costimulatory receptors play an important role in improving immune activation and reinforcing anti-tumor immune response. Downregulation of these receptors in the TME leads to immune suppression, while therapeutic activation of them can restore immune function and reduce tumor progression. When it comes to downregulating costimulatory receptors in tumors, one mechanism is targeting CD137, which is expressed on T cells, DCs, and NK cells (Moy et al. 2017). Its interaction with its ligand, CD137-L, boosts NK cell cytotoxicity, promotes T cell differentiation, and inhibits Tregs (Ruggeri et al. 2007; Verheyden et al. 2004). Another option is OX40, a receptor present on T cells, which is crucial in enhancing T cell proliferation, releasing antitumor cytokines, and generating memory T cells (Fulop et al. 2019; Moy et al. 2017). However, it has limited effectiveness due to the low expression of its ligand, OX40-L, within tumors (Srivastava et al. 2017). In addition, CD40 is displayed on APCs and a number of tumor cells (Moy et al. 2017). It plays a dual role, as its activation enhances T cell immunity, boosts macrophage tumor-killing, and stimulates apoptosis in CD40+ tumor cells. On the other hand, it also has the capacity to promote tumor growth and angiogenesis in tumor-associated blood vessels (Moy et al. 2017).

A number of therapeutic implications exist for costimulatory receptors. CD137 agonists in HNSCC are being tested coupled with Cetuximab and tremelimumab for improved efficacy (Bauman and Grandis 2014; Lucido et al. 2014). Furthermore, enhancing OX40 signaling could potentially boost tumor-specific immune responses and improve patient outcomes (Moy et al. 2017). Research indicates that activating OX40 can increase disease-free survival in various solid tumors (Vetto et al. 1997). Agonistic OX40 monoclonal antibodies (mAbs) are under investigation for their potential to enhance other immunotherapies, such as CD137 agonists, and to improve outcomes following surgery or radiation (Moy et al. 2017). Additionally, advancing agonistic CD40 mAbs and recombinant ligands have shown limited response rates, at around 20% in HNSCC cases, with some patients experiencing thromboembolic events, cytokine release syndrome, and elevated tumor growth (Sathawane et al. 2013; von Leoprechting et al. 1999; Vonderheide et al. 2007). To improve efficacy, combinations with other treatments are being explored (Moy et al. 2017).

18.3 Tumor Microenvironment (TME)

Historically, cancer was seen merely as a mass of undifferentiated tumor cells, overlooking the crucial role of the surrounding stromal cells in the TME (Bhat et al. 2021; Mazilu et al. 2021).

However, in HNC, the TME is now recognized as a complex and dynamic ecosystem, characterized by constant interactions and evolution (Canning et al. 2019; Elmusrati et al. 2021; Forster and Devlin 2018; Horton et al. 2019; Koontongkaew 2013; Markwell and Weed 2015; Qin et al. 2021; Santuray et al. 2018; Wondergem et al. 2020). This intricate environment regulates abnormal tissue function, playing a crucial role in helping HNC cells evade immune detection, suppress immune responses, diminish anti-cancer immunity, and progress into more aggressive forms (Canning et al. 2019; Chen et al. 2015, 2020; Curry et al. 2014; Giancotti 2014; Mazilu et al. 2021; Nisar et al. 2021; Wondergem et al. 2020). The response to immunotherapy in HNC is significantly influenced by the TME's pre-existing immune conditions, which also contributes to these tumors' increased incidence of relapse and metastasis (Mazilu et al. 2021; Wondergem et al. 2020). The TME consists of non-cellular components and diverse cell groups that infiltrate the tumor, interacting with both tumor cells and each other through complex signaling networks (Chen et al. 2020).

18.3.1 Non-cellular Elements of the TME

The non-cellular elements of the TME include extracellular matrix (ECM), cytokines, growth factors, exosomes, pH levels, oxygen concentration, and interstitial pressure (Bhat et al. 2021; Qin et al. 2021; Quezada et al. 2011). ECM, which comprises proteins such as collagen, fibronectin, elastin, laminin, and tenascin, influences proliferation, invasion, metastasis, survival, immune evasion, and resistance to therapy (Dobrenis et al. 2015; Hu et al. 2015; Mazilu et al. 2021; Quezada et al. 2011; Wang et al. 2019).

Fibronectin, the major glycoprotein in the ECM, mediates interactions with collagen and fibrin, facilitating tumor cell adhesion and migration (Wang et al. 2019). Matrix metalloproteinases (MMPs), produced primarily by the ECM, degrade ECM components, facilitating cancer cell migration and activating factors like fibroblast growth factor (bFGF), vascular endothelial growth factor (VEGF), and TGF-β to promote angiogenesis and tumor progression (Egeblad and Werb 2002). Additionally, other adhesion molecules within ECM also contribute to cell proliferation and tumor development (Bergers et al. 2000; Tatti et al. 2008; Wang et al. 2019).

18.3.2 Cellular Elements of the TME

18.3.2.1 Non-immune TME Cells

These cells include endothelial cells, cancer-associated fibroblasts (CAFs), adipocytes, mast cells, neuroendocrine cells, and blood and lymphatic vascular cells (Bhat et al. 2021; Cristina et al. 2019; Nisar et al. 2021).

18.3.2.1.1 CAFs

CAFs are crucial components of the TME in HNSCC (Economopoulou et al. 2020), playing diverse roles due to their varied origins from mesenchymal stem cells, fibrocytes, pericytes, endothelial cells, resting fibroblasts, and epithelial cells (Elmusrati et al. 2021; Petrelli et al. 2023; Wang et al. 2019). Their formation is driven by processes such as mesothelial- and epithelial-mesenchymal transitions (EMT) and TGFβ pathways, which are further amplified by interleukin (IL) 1β (Elmusrati et al. 2021; Wang et al. 2019). Moreover, cancer cells secrete FAP and platelet-derived growth factor (PDGF-A), likely contributing to the enhanced accumulation of CAFs within the TME (Elmusrati et al. 2021; Liu et al. 2019; Qin et al. 2018).

CAFs are typically identified by indicators such as fibroblast activation protein (FAP), alpha-smooth muscle actin (α-SMA), vimentin, and PDGFR-β (Economopoulou et al. 2020; Sasidharan Nair and Elkord 2018). However, these markers are not entirely specific to CAFs due to their wide genetic variations (Elmusrati et al. 2021). To better characterize CAF subtypes, one study employed single-cell RNA sequencing to classify CAFs into three main categories based on gene expression profiles: classical CAFs, normally activated fibroblasts, and elastic fibroblasts (Kürten et al. 2021; Ruffin et al. 2023). Classical CAFs express fibroblast activation protein (FAP), PDGFRA, LOX, and MMPs, contributing to tumor progression and ECM remodeling, while elastic fibroblasts express *ELN*, *FBLN1*, and *MFAP4*, associated with tissue fibrosis and elasticity (Kürten et al. 2021; Ruffin et al. 2023). The wide variation in CAF phenotypes across human solid tumors results in diverse functional roles in tumor progression (Kürten et al. 2021; Ruffin et al. 2023).

CAFs are highly active components of the TME in HNC, promoting tumor angiogenesis, proliferation, invasion, and spread, as well as TME shaping (Qin et al. 2021; Ruffin et al. 2023; Wang et al. 2019). They produce collagen and other structural proteins of the ECM and maintain tissue stability (Qin et al. 2021; Ruffin et al. 2023). A critical function of CAFs is the regulation of MMPs, which break down and remodel the ECM, resulting in enhanced cancer cell invasiveness and facilitated cell movement within tissues (Qin et al. 2021; Ruffin et al. 2023; Wang et al. 2019). This ECM remodeling creates a dense stromal barrier that supports tumor growth and progression, while also hindering proper drug delivery and impairing the effectiveness of cancer therapies (Economopoulou et al. 2020; Wang et al. 2019).

CAFs also correlate with cancer stem cell formation, which is often resistant to treatment (Wang et al. 2019). This is mediated through mechanisms such as the secretion of periostin, a key pro-stemness factor that promotes a phenotype in cancer cells that resembles stem cells (Álvarez-Teijeiro et al. 2018; Fiori et al. 2019; Wang et al. 2019). Additionally, CAFs collaborate with tumor cells to establish an immunosuppressive network that helps tumors evade immune responses (Qin et al. 2021). They suppress T cell proliferation, impair T cell function via PD-1 and PD-L1 checkpoint pathways, induce T cell apoptosis, and promote the expansion of Tregs through secreting VEGF, TGF-β, IL-6, TNF, and CXCL-8 (Elmusrati et al. 2021; Qin et al. 2021; Sasidharan Nair and Elkord 2018; Takahashi et al. 2015). They also recruit M2 macrophages and MDSCs to enhance immune evasion (Birbrair 2020; Qin et al. 2021). Finally, CAFs affect the metabolism of tumor cells, primarily through secreting hepatocyte growth factor (HGF), thus enhancing glycolysis in tumor cells and driving HNSCC progression (Elmusrati et al. 2021; Ruffin et al. 2023). In response, HNSCC cells release basic bFGF which further boosts the secretion of HGF by CAFs. This reciprocal interaction with the fibroblast growth factor receptor (FGFR) pathway enhances oxidative phosphorylation, tumor cell proliferation, and migration (Elmusrati et al. 2021; Ruffin et al. 2023).

Emerging therapeutic strategies in targeting CAFs within the TME offer new avenues for cancer treatment, although challenges remain. One such strategy is inhibiting the pathways through which CAFs promote cancer stem cells, potentially reducing treatment resistance and leading to more effective tumor suppression (Wang et al. 2019). Another potential strategy is inhibiting c-Met and FGF receptors, disrupting the feedback loop between CAFs and tumor cells, which potentially reduces tumor growth (Wang et al. 2019). Other novel approaches such as depleting CAFs, functionally modifying them, or utilizing them as drug-delivery vehicles are under investigation (Economopoulou et al. 2020).

However, CAF targeting in cancer treatment has potential challenges and complexities. For example, targeting α-SMA has paradoxically resulted in more aggressive tumors in other cancer types (Chen and Song 2019; Economopoulou et al. 2020; Özdemir et al. 2014). In addition, while preclinical models have shown promise in targeting the FAP, translating this success into clinical trials has been challenging, with limited efficacy observed so far (Economopoulou et al. 2020).

In addition, inhibiting the IL-6 and JAK/STAT3 pathways has been useful as a potential therapeutic approach in HNSCC. IL-6, known for its association with aggressive HNSCC behavior, activates STAT3 to promote tumor growth (Economopoulou et al. 2020). However, despite promising preclinical results, Siltuximab, an IL-6 monoclonal antibody, failed to demonstrate efficacy in HNSCC clinical trials (Angevin et al. 2014; Economopoulou et al. 2020; Johnson et al. 2018). Moreover, C188–9, a small molecule that inhibits STAT3, has been effective in causing tumor regression in mouse models and is currently being tested for HNSCC (Bharadwaj et al. 2016; Economopoulou et al. 2020; Wong et al. 2017). In addition, JAK2 inhibitors have shown tumor-suppressive effects in patient-derived HNSCC xenografts, and ruxolitinib, a JAK1/2 inhibitor, is under evaluation for patients with operable HNSCC (Sen et al. 2015). Moreover, efforts to block MMPs, enzymes that help break down the extracellular matrix have had mixed results in clinical trials, showing the challenge of targeting the ECM in HNSCC treatment (Economopoulou et al. 2020). An example of such efforts is S-3304, an oral MMP inhibitor that was evaluated in two HNSCC patients and led to disease stabilization in only one of them, showing limited clinical success in MMP-targeting therapies (Chiappori et al. 2007).

18.3.2.1.2 DCs

DCs link innate and adaptive immunity by displaying tumor antigens via MHC classes I and II, initiating the activation of naive T cells (Qin et al. 2021; Ruffin et al. 2023). They also enhance cytotoxic T lymphocyte functions through helper T (Th) cells (Qin et al. 2021). In addition, mature DCs attract and stimulate NK cells through CXCL8, CXCL10, CX3CL1, and IL-12p70 and contribute to overall anti-tumor responses (Ruffin et al. 2023).

Regarding treatment strategies, targeting DCs in HNSCC treatment has emerged as a promising strategy. Cetuximab therapy has proven to improve treatment outcomes by activating NK cells, which enhances the differentiation of DCs (Fulop et al. 2019; Gerosa et al. 2002; Vujanovic et al. 2010; Wculek et al. 2020). In parallel, DC-based vaccines trigger a targeted immune response through patient-derived monocytes cultured with tumor antigens and cytokines (Schuler et al. 2014). Examples of such vaccines include a DC vaccine infused with p53 peptides was found to be safe for HNSCC after going through phase 1 clinical trials for HNSCC (Schuler et al. 2014; Tan et al. 2018). Additionally, cancer stem cell-DC vaccines have shown preclinical success in improving survival rate and disease management in SCC (Hu et al. 2016; Tan et al. 2018).

18.3.2.2 Immune TME Cells

18.3.2.2.1 T Cells

T cells, a crucial component of the adaptive immune system, are categorized into two main groups of CD4+ and CD8+ T cells (Elmusrati et al. 2021; Shurin et al. 2014). CD8+ T cells, or cytotoxic T cells, identify and directly eradicate tumor cells by secreting perforin, granzymes, and pro-inflammatory cytokines (Chen et al. 2020; Elmusrati et al. 2021; Fulop et al. 2019; Ruffin et al. 2023). In HNSCC, reductions in cytotoxic T cells in the blood and TME indicate impaired anti-tumor immune response (Ferrandino et al. 2020; Fulop et al. 2019; Senovilla et al. 2012). CD4+ T cells play a crucial role in shaping the TME by differentiating into effector subtypes and secreting cytokines that modulate immune responses (Fulop et al. 2019; Ruffin et al. 2023). Within the CD4+ category, distinct subgroups include Th cells and Treg cells (Elmusrati et al. 2021; Gao and Chen 2016).

Th cells assist CD8+ T cells in targeting and eliminating tumor cells (Ruffin et al. 2023). This category includes TH1, TH17, and T follicular helper (TFH) cells. Each of these cells plays distinct and specialized roles in cancer immunity (17th International Congress of Immunology 2019; Ruffin et al. 2023).

TH1 cells help initiate antitumor responses by releasing IFN-γ and TNF (Fulop et al. 2019; Ruffin et al. 2023). These cytokines boost the presence of MHC class I and II in the TME which in turn enhances the identification of tumor cells (Gao and Chen 2016; Joyce and Pollard 2009; Ruffin et al. 2023).

Meanwhile, TH17 cells increase in number in the blood and lymph nodes surrounding the tumor and demonstrate both pro- and anti-tumor effects (Ruffin et al. 2023; Shurin et al. 2014; Tesmer et al. 2008; Ye et al. 2013). In HNSCC, they can suppress tumor growth and reduce the expression of angiogenesis-related proteins in vitro, suggesting a potential protective role in certain contexts (Kesselring et al. 2010; Li et al. 2011). Additionally, TFH cells, characterized by CXCR5 and PD-1 expression, play an important role in modulating adaptive immune response within the TME. TFH cells produce CXCL13 to recruit B and CD8+ T cells, thereby enhancing anti-tumor immunity (Ruffin et al. 2023; Shurin et al. 2014). They activate B cells via IL-21, Inducible T cell CoStimulator (ICOS), and CD40L, therefore coordinating B cell and cytotoxic T cell activity within the TME (Ruffin et al. 2023; Shurin et al. 2014; Tahiliani et al. 2017; Zotos et al. 2010).

Additionally, Tregs are crucial for the immune balance by suppressing excessive immune responses and preventing autoimmunity (17th International Congress of Immunology 2019; Boucek et al. 2010; Echarti et al. 2019; Peltanova et al. 2019). However, in HNSCC, they significantly contribute to the immunosuppressive TME. These cells are found in both the blood and the TME of HNSCC patients (Jie et al. 2013; Ruffin et al. 2023; Strauss et al. 2007), where they exhibit enhanced immunosuppressive activity compared to circulating Tregs, primarily due to their higher expression of immune checkpoint molecules (Jie et al. 2013; Wang et al. 2019). Tregs display CD4, CD25, and FOXP3 (Fulop et al. 2019; Qin et al. 2021). Their attraction and accumulation within the TME are driven by chemokine signaling pathways including CXCL12-CXCR4, CCR8-CCL1, CCL28-CCR10, CCR4-CCL17/22, and CXCR3-CCL10, along with elevated TGF-β levels. This increase in TGF-β further enhances the immunosuppression by Tregs in the TME (Chen et al. 2020; de Medeiros et al. 2017; Facciabene et al. 2011; Jie et al. 2015; Lahl et al. 2007; Maggioni et al. 2017; Wang et al. 2019; Yu et al. 2018; Zou et al. 2004).

In HNSCC, Tregs display checkpoint inhibitors such as CTLA-4, PD-1, and PD-L1, which intensify immune suppression and contribute to cytotoxic T cell dysfunction (Dorta-Estremera et al. 2019; Kesselring et al. 2010; Liu et al. 2018c; Ohue and Nishikawa 2019). Moreover, Tregs upregulate CTLA-4 in response to cetuximab, suppressing NK cell activity. They also express the inhibitory receptor NRP1, which is associated with poor prognosis and can be therapeutically targeted by modifying TCR signaling (Chuckran et al. 2021; Ruffin et al. 2023). The TIM-3+ subtype of Tregs inhibits cytotoxic T cell proliferation, further weakening the immune response against tumors (Liu et al. 2018b, c; Ruffin et al. 2023). Tregs also reduce local IL-2 levels due to their CD25 expression, leading to diminished T cell function (Buoncervello et al. 2019). They also secrete IL-10, IL-35, TGF-β and produce granzyme B and perforin, which can directly eliminate APCs or effector immune cells (Dadey et al. 2020; Fulop et al. 2019; Jarnicki et al. 2006; Jie et al. 2013; Sawant et al. 2019; Setoguchi et al. 2005; Togashi et al. 2019; Wang et al. 2019). In addition, using CD39 and CD73, Tregs turn ATP into adenosine, thus suppressing T cells through the adenosine receptor 2A (A2AR) while also delivering negative signals to both T cells and APCs (Deaglio et al. 2007; Jie et al. 2013; Kuksin et al. 2021; Liu et al. 2024; Maj et al. 2017; Wang et al. 2019). Tregs interact with MHC class II through LAG-3, decreasing T cell activity (Camisaschi et al. 2010; Xu 2020) and also reducing tryptophan levels in DCs, thus inhibiting T cells (Acovic et al. 2018; Liu et al. 2024). Lastly, Tregs inhibit DC function by downregulating CD80 through CTLA-4 interactions, impairing APC maturation and contributing to the immunosuppressive TME in HNSCC (Qureshi et al. 2011).

Effector T cells can experience exhaustion and dysfunction due to the chronic stimulation and persistent inflammation induced by cancer (17th International Congress of Immunology 2019; Ma et al. 2019; Qin et al. 2021). This dysfunction, commonly observed in HNSCC, is characterized by decreased proliferation, reduced production of IL-2, TNF, and IFN-γ, and increased PD-1, TIM-3, LAG-3, CTLA-4, and TIGIT in tumor-infiltrating lymphocytes (TILs) (Chen et al. 2020; Ruffin et al. 2023). This exhaustion induces T cell apoptosis and a reduction in antigen-specific TILs (Qin et al. 2021). Several mechanisms lead to T cell dysfunction and exhaustion in the TME. These mechanisms include increased expression of PD-1, LAG-3, TIM-3, CTLA-4, and 2B4, induced CD8+ T cell apoptosis by Fas-L+ micro-vesicles released by the tumor, increased number of Treg and Th17 cells, driven by IL-23, IL-6, and TGF-β secretion, reduced MHC I expression on tumor cells, and reduced mitochondrial mass, indicating metabolic insufficiency (Chen et al. 2020; Fulop et al. 2019; Grandis et al. 2000; Scharping et al. 2016).

There are several therapeutic strategies that involve targeting T cells in the TME. One is using immune checkpoint inhibitors, which block inhibitory receptors that suppress T cell activation. This prevents T cell anergy and apoptosis while enhancing immune cytotoxicity against tumor cells (Ferris 2015; Mei et al. 2020; Moy et al. 2017). Commonly targeted pathways include CTLA-4, PD-1, PD-L1, LAG-3, and Tim-3 (Ferris 2015; Mei et al. 2020; Moy et al. 2017). Another therapeutic strategy is adoptive cell transfer (ACT), which uses T cells to target and eliminate tumor cells, with larger transfers of these cells boosting the body's anti-tumor response (Menon et al. 2016). This approach can involve either tumor-infiltrating lymphocytes (TILs) or genetically engineered T cells, offering a potentially curative option (Met et al. 2019). The TIL approach includes isolating T cells from tumors, developing them in culture, and then reintroducing them to the patient (Sahu and Suryawanshi 2021). The engineered TCR therapy involves introducing specific antigen receptors into T cells using genetic engineering (Ti et al. 2018). TCR-modified T cells identify antigens in a manner dependent on the MHC (Ellis et al. 2021). These modified T cells can then be mass-produced for patient treatment (Ellis et al. 2021). Chimeric antigen receptors (CARs) are also engineered structures delivered to the surface of T cells via a viral vector (Mo and Mamonkin 2020). Equipped with both antigen-binding and intracellular signaling components, CARs enable T cells to recognize antigens independently of APCs and trigger cellular activation (Menon et al. 2016).

Another therapeutic approach involves enhancing STAT1 activity, which has reversed T cell exhaustion in HNSCC mouse models by boosting T cell responses and limiting suppressive myeloid cells, thereby potentially improving treatment efficacy and outcomes (Ryan et al. 2020). Anti-CD73 antibody treatments can also be useful, as in an HNSCC mouse model, elevated CD73 on T cells was linked to T cell exhaustion. Anti-CD73 antibody treatment reduced tumor growth and reversed T cell exhaustion by lowering PD-1 and CTLA-4 levels (Deng et al. 2018). Additionally, in HNSCC mice, anti-TIGIT treatment improved CD8+ T cell function, reduced Tregs, and countered MDSCs, leading to delayed tumor growth (Wu et al. 2019). Finally, restoring PPAR-gamma coactivator 1α (PGC1α) expression, a crucial regulator of mitochondrial biogenesis, in tumor-infiltrating T cells can reprogram them, enhancing their metabolism and functionality (Scharping et al. 2016).

To target Tregs as another therapeutic approach, anti-CTLA-4, PD-1, and PD-L1 approaches are being explored as potential treatments for HNSCC (Dorta-Estremera et al.

2019). In addition, anti-TIGIT treatments can activate CD8+ T cell effector functions and reduce Treg cell numbers, significantly delaying tumor growth in transgenic HNSCC mice (Qin et al. 2021; Wu et al. 2019). Targeting Notch1 signaling pathway may also lead to reducing the number of MDSCs and Tregs, downregulating PD-1 and CTLA-4, and modulating CD47 expression. Blocking CD47 stimulates cytotoxic T cells and improves the immunosuppressive environment (Mao et al. 2018; Qin et al. 2021). Inhibiting A2AR, which is positively associated with HIF-1α, CD73, and Foxp3, and contributes to Treg expansion, can also be useful by leading to a reduction in Foxp3+ Tregs and increasing CD8+ T cells' ability to kill cancerous cells (Ma et al. 2017a; Qin et al. 2021). Inhibiting A2AR so significantly improves anti-tumor immunity and reduces Treg cell numbers, alleviating tumor cell toxicity (Liu et al. 2018b; Qin et al. 2021). Furthermore, targeting the inhibitory receptor NRP1, which is expressed on Tregs and is associated with poor prognosis, represents a potential therapeutic strategy to counteract immune suppression. Modifying TCR signaling to disrupt NRP1 can help in reducing Treg-mediated immune suppression, improve CD8+ T cell responses, and restore effective antitumor immunity (Chuckran et al. 2021; Ruffin et al. 2023).

18.3.2.2.2 B Cells

Tumor-infiltrating B cells are a key element of the TME in HNC. They are often observed in high numbers at the invasive margins, within the tumor itself, or as part of tertiary lymphoid structures (TLS) (Gavrielatou et al. 2021). TLSs are clusters of immune cells that form as a reaction to persistent inflammation or infection (Neyt et al. 2012; Pitzalis et al. 2014; Ruffin et al. 2021). These structures are organized into specific areas for T cells, B cell follicles, and mature DCs (Ruffin et al. 2021; Sautès-Fridman et al. 2016, 2019). TLSs can attract immune cells into solid tumors and often contain germinal centers (GCs), which create B cells with high affinities, leading to the production of memory B cells and lasting humoral immunity (Budi and Farhood 2023; Ruffin et al. 2021).

Distinct B cell phenotypes are identified more commonly in the TME compared to healthy tissues. Some examples include increased levels of CD86+ B cells in tumor tissue compared to healthy mucosa and isolated Peripheral Blood Mononuclear Cells (PBMCs). Additionally, memory B cells with the IgD−/CD27+ phenotype are elevated in both tumor tissue and peripheral blood, compared to those in healthy donors, indicating a shift in adaptive immune response. Another notable B cell phenotype is higher levels of CD27+/CD38hi/CD138hi/CD20− plasma cells in tumor tissues compared to PBMCs (Gavrielatou et al. 2021; Lechner et al. 2019).

B cells' functions in cancer remain incompletely understood (17th International Congress of Immunology 2019; Budi and Farhood 2023; Shurin et al. 2014). Some of their roles in cancer immunity include antigen presentation to T cells, aiding T cells' expansion and memory formation, and assisting in cross-presentation to other APCs (Gao and Chen 2016; Nelson 2010; Rivera et al. 2001). B cells also contribute to tumor immune responses through antibody-dependent cell-mediated Fytotoxicity (ADCC); B cells produce tumor-antigen-specific IgG antibodies that bind to tumor cells and engage Fc receptors on NK cells, monocytes, and neutrophils, triggering cancer cell destruction through non-phagocytic mechanisms (Gavrielatou et al. 2021; Zahavi et al. 2018). B cells also demonstrate direct cytotoxic abilities, primarily through granzyme B production, which induces apoptosis (Budi and Farhood 2023; Gavrielatou et al. 2021). Additionally, B cells can express FasL, enabling them to kill Fas + tumor cells (Gavrielatou et al. 2021). The presence of B cells may help prevent secondary malignancies as well, as their depletion following Rituximab treatment is linked to a higher risk of developing additional cancers (Tarella et al. 2011).

B cells in the TME of HNC can also contribute to cancer progression through various mechanisms. For example, B cells promote tumorigenesis in HPV16+ premalignant tissues via immune complex formation (Andreu et al. 2010; de Visser et al. 2005; Gavrielatou et al. 2021). Moreover, removing B cells with anti-CD20 mAb enhances the effectiveness of chemotherapy in mouse models (Affara et al. 2014; Gavrielatou et al. 2021). In addition to tumorigenesis, different subpopulations of regulatory B cells (Bregs) within the TME contribute to immunosuppression (Gavrielatou et al. 2021). These include CD24hi/CD38hi/CD19+, adenosine-producing, and CD19 + IL-10+ Bregs, each of which exerts distinct effects based on phenotype and localization (Gavrielatou et al. 2021). Their pro-tumor activities involve inducing T cell exhaustion, inhibiting effector B cells, expanding Tregs, and promoting tumor growth by producing inhibitory cytokines (Gavrielatou et al. 2021).

Treatment strategies involving B cells should aim to stimulate the formation of TLS, enhance B cell maturation, and target specific B cell subsets (Ruffin et al. 2023). The often-impaired function of B cells in cancer, often due to the overexpression of PD-1, PD-L1, and CTLA-4, indicates that their activity could be modified by Immune Checkpoint Blockade (ICB) therapies (Budi and Farhood 2023; Wherry and Kurachi 2015). Additionally, ADCC-based therapies remain a pivotal component of anti-cancer therapies involving monoclonal antibodies (Gavrielatou et al. 2021; Zahavi et al. 2018).

Regarding B regs, phenotypes such as PD-L1+ B regs in metastatic melanoma can suppress T cell activity via PD-1/PD-L1 pathways, making them principal candidates for ICB therapy (Wu et al. 2020; Xiao et al. 2016). Furthermore, B reg-targeting agents such as resveratrol and MK886 have shown preclinical promise in reducing tumor growth and metastasis, suggesting their potential role in future cancer immunotherapy strategies (Lee-Chang et al. 2013; Wejksza et al. 2013).

18.3.2.2.3 MDSCs

MDSCs are a diverse group of highly immunosuppressive innate immune cells that expand under disease conditions like cancer (Gabrilovich et al. 2012; Ruffin et al. 2023; Veglia et al. 2021). MDSCs recruitment and accumulation in the TME is driven by VEGF, IL-6, macrophage colony-stimulating factor (GM-CSF), CXCL1, and IL-8, with migration guided by receptors such as CXCR2 and CSFR1 (Economopoulou et al. 2020; Elmusrati et al. 2021).

MDSCs are classified into monocytic MDSCs (M-MDSCs), marked by CD14 + CD11b + HLA-DRlo/–CD15−, and polymorphonuclear MDSCs (PMN-MDSCs), identified by CD15 + CD11b + CD14− (Ansell et al. 2024; Ruffin et al. 2023). A less mature subset, early-stage MDSCs, includes both M-MDSCs and PMN-MDSCs, lacks mature myeloid markers, and resembles neutrophils, yet still contributes to immune suppression as they develop (Elmusrati et al. 2021; Ribechini et al. 2010; Wang et al. 2019).

MDSCs contribute to tumor growth and spread by suppressing NK cells, DCs, macrophages, and T cells through the production of ROS, NO, arginase 1, and PGE2 (Ruffin et al. 2023; Chen et al. 2020). MDSCs deplete amino acids, such as L-tryptophan, L-arginine, and L-cysteine (Wang et al. 2019), downregulate CD3/CD28 expression, and thus, inhibit T cells in the TME (Corzo et al. 2010; Solito et al. 2011). In addition, these cells promote IL-10 and TGF-β production, leading to Treg development and activation, further suppressing anti-tumor response (Maggioni et al. 2017; Yu et al. 2018). These cells also promote the activation of Th2 cells (Elmusrati et al. 2021) and increase IL-1 and IL-6 levels, leading to impaired APC maturation and indirectly promoting tumor growth (Qin et al. 2021). Moreover, MDSCs compete with APCs, such as DCs and TAM, leading to cysteine depletion, a critical factor for T lymphocyte activation (Elmusrati et al. 2021). In some cases, they suppress NK cell activity, diminishing the effectiveness of NK cell-based immunotherapies in HNSCC (Elmusrati et al. 2021). Under tumor hypoxia, MDSCs exhibit increased PD-L1 levels, which bind to PD-1 on T cells and impair T cell-driven immunity (Elmusrati et al. 2021). Moreover, elevated levels of CD38, induced by IFN-γ and

TNF-α, promote tumor growth and expand M-MDSCs, further contributing to immune suppression (Blacher et al. 2015; Chen et al. 2018).

A number of therapeutic strategies target MDSCs to impede tumor progression and enhance immune responses. These approaches include delaying tumor progression by using Anti-CD38 monoclonal antibodies and suppressing JAK2/STAT3 signaling to reduce MDSC levels and inhibit angiogenesis (Chen et al. 2020; Liu et al. 2018a). Additionally, disabling MDSCs through SEMA4D targeting and inhibiting STAT4 has shown promise in limiting MDSC-mediated immune suppression in tumors (Anderson et al. 2019; Younis et al. 2016).

Another strategy includes modulating CD24 expression, which negatively regulates MDSC number and function (Fugle et al. 2016). To counteract MDSC-mediated immune suppression, it can also be beneficial to target the TIGIT–CD155 pathway, to improve PD-L1 inhibition, or inhibit Indoleamine 2,3-dioxygenase 1 (IDO1) with agents, such as epacadostat, navoximod, and BMS-986205 (Economopoulou et al. 2020; Mao et al. 2021). Blocking CXCR1 and CXCR2, using agents such as SX-682, helps to eliminate tumor-infiltrating MDSCs and enhance NK cell performance (Fulop et al. 2019; Greene et al. 2020; Sun et al. 2019). Furthermore, inhibiting PDE5 using agents such as tadalafil can reduce MDSC accumulation and enhance antitumor immunity (Califano et al. 2015).

18.3.2.2.4 Tumor-Associated Macrophages (TAMs)

TAMs are a significant, prevalent constituent of the TME (Qin et al. 2021). They are recruited by IL-10, macrophage colony-stimulating factor (M-CSF), PDGF, VEGF, and monocyte chemotactic protein 1 (MCP-1) (Economopoulou et al. 2020; Liu et al. 2024). Specifically, elevated levels of MCP1/CCL2 and MIP3α/CCL20 further indicate the increased presence of TAMs in HNSCC, highlighting their role in immunosuppression and tumor progression (Marcus et al. 2004; Ruffin et al. 2023). In the TME of HNSCC, TAMs are often found in hypoxic regions within the tumor and exhibit higher infiltration compared to normal mucosa, correlating with recurrent or metastatic disease (Economopoulou et al. 2020; Ruffin et al. 2023).

TAMs promote cancer cell motility and angiogenesis via breaking down and restructuring the ECM of the TME (Birbrair 2020; Elmusrati et al. 2021). They also secrete TNF-α, IL-1β, and macrophage migration inhibitory factor (MIF) (Chen et al. 2020; Fulop et al. 2019; Shurin et al. 2014). MIF facilitates neutrophil adhesion and recruitment, thereby worsening tumor invasiveness (Chen et al. 2020). Moreover, TAMs in HNSCC are significant contributors to PDL1 expression, which is crucial for interaction with CD8+ T cells and suppressing their effective immune responses (Kürten et al. 2021; Ruffin et al. 2023).

TAMs can be classified as M1 and M2 phenotypes (Qin et al. 2021). These phenotypes are functionally distinct, with M1 macrophages promoting a Th1 immune response and exerting anti-tumor effects, while M2 macrophages foster a Th2 response, linked to tumor development and spread (Kim and Bae 2016). M1 macrophages, or classically activated macrophages, are involved in pro-inflammatory responses and tumor suppression (Fulop et al. 2019; Qin et al. 2021). They produce reactive oxygen species (ROS) and release TNF-α, IL-1, IL-6, and IL-12, which help Th1 attraction and anti-tumor responses (Qin et al. 2021). However, within the TME, TAMs often adopt the M2 tumor-promoting phenotype (Economopoulou et al. 2020). M2 macrophages, associated with immunosuppression and tumor progression, are induced by cytokines like IL-4 and IL-10 (Qin et al. 2021). They produce chemokines such as CCL24 and IL-10, which suppress M1 macrophages and promote tissue remodeling (Qin et al. 2021). M2 macrophages also secrete tumor growth factor (TGF-β), create an immunosuppressive environment, and release VEGF and EGF to facilitate angiogenesis, invasion, and metastasis (Economopoulou et al. 2020; Petrelli et al. 2023).

Strategies for targeting TAMs generally fall into one of the two categories of inhibiting TAM recruitment or repolarizing TAMs to the M1 phenotype (Economopoulou et al. 2020).

CCL2, a chemokine crucial for TAM recruitment, has been associated with metastasis in OSCC, and its inhibition by curcumin has demonstrated promising therapeutic potential (Ferreira et al. 2008; Gao et al. 2016). The effectiveness of CCL2 inhibitors such as carlumab and PF04136309 in HNSCC remains uncertain and requires further clinical validation (Brana et al. 2015; Nywening et al. 2016). Similarly, blocking CSF1R, a key regulator of TAM migration, has been investigated as another therapeutic strategy. Using CSF1R inhibitor, RG7155 successfully reduced TAMs and increased T cell infiltration in animal models, but clinical success was limited (Ries et al. 2014). Currently, pexidartinib, another CSF1R inhibitor, is being evaluated in advanced solid tumors (NCT02734433) (Economopoulou et al. 2020).

Given the plasticity of TAMs, repolarizing immunosuppressive M2 TAMs to pro-inflammatory M1s enhances anti-tumor responses (Qin et al. 2021). Repolarization approaches include the use of TLR agonists, CD40 agonists, and PI3K inhibitors (Economopoulou et al. 2020). TLR7/9 agonists, combined with PD-1 blockade, improve the M1/M2 ratio and induce tumor regression (Sasidharan Nair and Elkord 2018). Specific TLR agonists like imiquimod and SD-101 are currently being evaluated for squamous cell carcinoma and HNSCC (NCT03370406, NCT0232319) (Economopoulou et al. 2020). In addition, CD40 agonists, such as selicrelumab and CDX-1140, aim to convert M2 macrophages to a M1 phenotype (Economopoulou et al. 2020; Sasidharan Nair and Elkord 2018). Moreover, PI3K inhibitors, including duvelisib and IPI-549, reverse TAM polarization and boost anti-tumor responses, with IPI-549 being tested with nivolumab in resectable HNSCC (NCT03795610) (Economopoulou et al. 2020; Michaeli et al. 2017; Shaul and Fridlender 2019; Valero et al. 2017).

18.3.2.2.5 NK Cells

NK cells are critical elements of the innate immune system that directly eliminate tumor cells by cytotoxic mechanisms and cytokine release (Elmusrati et al. 2021; Qin et al. 2021). NK cells are often not detectable near cancer cells, despite their accumulation in the stromal TME (Elmusrati et al. 2021; Qin et al. 2021; Terrén et al. 2019).

NK cells are categorized into two primary subsets, based on CD56 and CD16 expression. The $CD56^{bright}CD16^{-}$ NK cells, which primarily produce cytokines like IFN-γ and TNF-α (Wang et al. 2019), and $CD56^{dim}CD16^{+}$ NK Cells, which are known for their potent cytotoxic capabilities (Ruffin et al. 2023). Recent studies also suggest the existence of a third subset, which is particularly proficient in IFN-γ production and has a unique chemokine receptor profile, aiding its enhanced migration to tumors (Vujanovic et al. 2012; Vujanovic et al. 2019). HNSCC shows the highest median infiltration of this subtype of NK cells among tumors (Fulop et al. 2019; Mandal et al. 2016). Moreover, NK cell infiltration is notably higher in HPV-positive tumors compared to HPV-negative ones, suggesting that HPV-driven tumors might be more responsive to NK-cell mediated immune response (Vujanovic et al. 2012; Wagner et al. 2016).

NK cells exert anti-tumor effects through various mechanisms, primarily by inducing apoptosis via the secretion of perforin, granzymes, and TRAIL. NK cells also engage in ADCC, which eventually leads to targeted lysis of antibody-coated cancer cells (17th International Congress of Immunology 2019; Wang et al. 2019). The NKG2D receptors on NK cells and other T cells help identify tumor cells by binding to NKG2D ligands, expressed on tumor cells. However, high levels of shed NKG2D ligands in HNSCC have been correlated with lowered NK cell function and tumor progression (Wang et al. 2019; Weil et al. 2017). NK cells also enhance adaptive immunity by rapidly producing IFN-γ, which activates Th1 cells and promotes MDSC infiltration, impacting the overall immune response (Elmusrati et al. 2021; Morvan and Lanier

2016). NK cells also enhance the adaptive immune system by producing cytokines such as IL-2, IL-12, IL-15, IFN-γ, and TNF-α, which support overall immune regulation (Elmusrati et al. 2021; Mandal et al. 2016; Morvan and Lanier 2016; Santos et al. 2019).

Key NK cell-based immunotherapeutic approaches in cancer treatment include targeting tumor-specific ligands, enhancing NK cell infiltration into tumors, and activating multiple NK cell receptors (Ben-Shmuel et al. 2020; Qin et al. 2021; Ruffin et al. 2023). An example of receptor activation is the use of cetuximab, an antibody targeting EGFR, which enhances tumor elimination by increasing CD16/FcγRIII$^+$ NK cell-mediated ADCC (Patterson et al. 2024). However, its effectiveness in primary tumors may be limited due to a reduced population of CD16$^+$ NK cells within the TME (Patterson et al. 2024). Despite this limitation, Cetuximab might still be beneficial in recurrent or metastatic cases (Kurai et al. 2007; López-Albaitero et al. 2009; Ruffin et al. 2023; Srivastava et al. 2013).

18.3.2.2.6 Tumor-Associated Neutrophils (TANs)

TANs take part in adaptive immunity (Elmusrati et al. 2021; Petrelli et al. 2023; Wang et al. 2019) and exhibit pro- or anti-tumor functions, determined by the stage of tumor development (Elmusrati et al. 2021). These cells exert pro-tumor functions by releasing pro-inflammatory substances that facilitate the breakdown of basement membrane and permit cancer cell invasion, increasing metastasis in HNSCC (Singh and Gautam 2023). They also increase cancer cell clustering in the bloodstream, enhancing these cells' survival and metastasis, leading to lower survival rates in patients (Singh and Gautam 2023). Moreover, TANs can accelerate the transition of metastasized cells to a more invasive phenotype, creating favorable conditions for tumor growth (Singh and Gautam 2023).

A number of TANs' anti-tumor functions include exerting direct cytotoxicity on tumor cells, limiting tumor growth and metastasis and activating DCs and T cells (Singh and Gautam 2023).

Neutrophils are split into two subsets based on their anti- or pro-tumorigenic functions: N1 (anti-tumorigenic) neutrophils and N2 (pro-tumorigenic) neutrophils (Singh and Gautam 2023). The shift into a pro-tumorigenic phenotype can occur due to the action of factors within the TME (Wang et al. 2019; Zhang et al. 2016). N1 neutrophils boost cytotoxic T cell proliferation, trigger apoptosis of tumor cells, and recruit other anti-tumor cells through releasing antimicrobial and cytotoxic granules, as well as immune mediators (Elmusrati et al. 2021; Wang et al. 2019). N2 neutrophils, characterized by CXCR4, VEGF, and MMP-9 markers, contribute to cancer development by accelerating tumor cell growth, breaking down the TME, fostering new blood vessel formation, and facilitating metastasis. These cells enhance tumorigenesis, tumor expansion, and immune suppression (Elmusrati et al. 2021; Fridlender et al. 2009; Shaul and Fridlender 2019; Wang et al. 2019).

Targeting pro-tumor neutrophils, neutrophil extracellular traps, and key cytokines like TGF-β1 and IL-17A could reduce metastasis and improve outcomes (Singh and Gautam 2023). Additionally, combining these treatment strategies with existing cancer treatments and identifying neutrophil-related biomarkers may further optimize patient responses (Singh and Gautam 2023).

18.4 Novel Therapeutic Approaches Indirectly Targeting the TME

18.4.1 Vaccines

HNSCC vaccines are broadly categorized into peptide-based, nucleic acid-based, pathogen vector-based, and cell-based vaccines (Tan et al. 2018). To be effective, vaccines should deliver antigens with optimal 3D density, including a balanced combination of Th and cytotoxic T cell epitopes (Bot et al. 2011), and ensure both antigen and adjuvant co-delivery to the same cells to reduce antigen-specific tolerance in APCs (Bot et al. 2011). Moreover, they must target lymph nodes efficiently, while maintaining safety with minimal side effects. Each vaccine approach has its own strengths, but no single strategy fully accomplishes all these goals.

18.4.1.1 Peptide-Based Vaccines

Peptide-based vaccines are relatively straightforward to produce and are generally safe and well-tolerated. However, their application is restricted because they bind exclusively to certain HLAs, making them less broadly applicable (Bot et al. 2011; Tan et al. 2018). Challenges with these vaccines include low intracellular delivery efficiency and difficulty in controlling antigen 3-dimensional density (Bullock et al. 2003; Kenter et al. 2009; Tan et al. 2018).

18.4.1.2 Nucleic Acid–Based Vaccines

Nucleic acid-based cancer vaccines, including INO-3112 (INO-9012) (Devaraja 2019), employ plasmid DNA to generate neoantigens, that activate pattern recognition receptors (PRRs) such as STING and TLRs (Tan et al. 2018). This activation stimulates the immune system to target cancer cells, with the vaccine's framework also serving as an adjuvant to amplify the immune response (Delaloye et al. 2009).

18.4.1.3 Pathogen-Vector Vaccines

Bacteria and viruses are used as delivery systems in this type of vaccine, because they efficiently activate the innate immune system (Tan et al. 2018; Wallecha et al. 2012).

18.4.1.4 Cell-Based Vaccines

These vaccines are among advanced immunological approaches that leverage modified cells to trigger anti-tumor immune responses. An example of these vaccines is MVX-ONCO-1, which delivers irradiated tumor cells encapsulated with an allogeneic cell line that secretes granulocyte-GM-CSFs (Tan et al. 2018). In a phase 1 clinical trial, this approach showed desirable safety and positive immune responses in over 50% of solid-tumor patients (Vernet et al. 2024). In addition, vaccines using DCs to target cancer stem-like cells proved most effective when administered post-surgery, reducing tumor recurrence and boosting survival rates in SCC mouse models, particularly when combined with anti-PD-L1 (Hu et al. 2016). Moreover, RTX-240, an autologous genetically modified red blood cell vaccine, expanded T cells and NK cells in melanoma animal models, thus controlling cancer with minimal organ damage due to its limited systemic distribution (McArdel et al. 2021; Mohsin 2024).

18.4.2 Oncolytic Viruses

Oncolytic viruses have shown promise in cancer treatment with both intravenous and intra-tumoral injections and have proven to be safe options (Garcia-Carbonero et al. 2017; Old et al. 2016; Samson et al. 2022; Todo et al. 2022). Intra-tumoral injections are particularly well-suited for HNC due to the visibility and accessibility of tumors (Goradel et al. 2021; Old et al. 2016). Several viruses, including measles, vaccinia, reovirus, adenovirus, and herpes, have demonstrated varying degrees of effectiveness in cancer treatment (Goradel et al. 2021; Old et al. 2016). These viruses specifically infect, replicate within, and lyse tumor cells, while also triggering anti-tumor responses. Talimogene laherparepvec (T-VEC), an engineered oncolytic HSV-1 expressing GM-CSF has also been successful in treating HNSCC (Goradel et al. 2021; Old et al. 2016). In China, H101, along with several other oncolytic viruses, have received approval for use in advanced HNC cases (Goradel et al. 2021; Old et al. 2016).

18.4.2.1 Oncolytic Herpes Viruses

Genetically modified oncolytic herpes simplex viruses replicate exclusively in tumor tissues, sparing normal cells by lacking essential replication genes, like ICP 34.5 (Goradel et al. 2021; Old et al. 2016). While clinical trials have confirmed their safety, further improvements are needed to achieve significant therapeutic efficacy (Hamdan et al. 2023; Lin et al. 2023). A number of examples include three viruses created from HSV-1, that are currently being tested for safety and efficacy in patients with HNC (Hamdan et al. 2023). Furthermore, clinical trials are underway for HF10, a naturally occurring HSV-1 mutant (Lin et al. 2023). Another promising candidate, HSV1716, an oncolytic herpes simplex virus, has shown a favorable safety profile in patients with different types of solid cancers and is currently being tested in clinical trials as for testing their efficacy in cancer treatment (Lin et al. 2023).

18.4.2.2 Oncolytic Poxviruses

The vaccinia viruses can target both primary and metastatic tumors when delivered intravenously, as most individuals lack pre-existing antibodies against it. The safety of an example of such vaccine, GL-ONCI, which is administered intravenously was tested in a phase I trial in 19 patients with locally advanced HNSCC, with or without distant metastases (Hamada and Yura 2020; Mell et al. 2017; Schwab 2011). The only significant adverse effect observed was a systemic rash. Over a 30-month follow-up period, the trial reported a 2-year overall and progression-free survival rates of 64.1% and 69.2%, respectively (Mell et al. 2017). Additionally, preclinical studies found that vvDD, a vaccinia virus with double deletions, and vvDD-DAI, which overexpresses the DAI receptor, were more effective against mouse melanoma models (Fatima et al. 2023). Moreover, the myxomavirus (MYXV) was engineered by deleting the M0111L gene, making brain tumor cells more vulnerable to virus-induced apoptosis and extending the survival rate in glioblastoma mouse models (Fatima et al. 2023). The MVA-EL vaccine was also created using this modified virus to target Epstein-Barr virus (EBV). Intradermal MVA-EL vaccination administered in patients with nasopharyngeal cancer in phase I clinical trial studies resulted in enhanced immunity against at least one EBV antigen in the majority of participants (Fatima et al. 2023).

18.4.2.3 Oncolytic Reovirus

Phase I studies on intratumoral reovirus injection indicated favorable safety and tolerable side effects (Old et al. 2016). Later studies combining reovirus with platinum-taxane treatments for incurable or relapsed/metastatic HNC showed notable efficacy and extended survival rates (Hamada and Yura 2020; Old et al. 2016).

In a phase III clinical trial study (REO 018; NCT01166542) involving platinum-refractory, taxane-naïve relapsed or metastatic HNSCC patients, adding reovirus to paclitaxel and carboplatin significantly enhanced disease-free survival and overall survival rates (Old et al. 2016; Sze et al. 2013).

18.4.2.4 Oncolytic Adenoviruses

Adenoviruses are small DNA viruses with genomes of 26 to 45 kb that have been extensively modified for cancer therapy by the integration of foreign genes for creating various mutant strains (Cervantes-García et al. 2008). Phase II clinical trials of the gene-deleted adenovirus ONYX-015 showed promising results in cancers with p53 deficiency (Cervantes-García et al. 2008). In stages III–IV hypopharyngeal carcinoma patients, combining Ad-p53, a replication-deficient adenovirus with p53, with surgery and CRT improved survival rates (Engeland and Bell 2020). In recurrent HNSCC patients receiving re-irradiation, AdGV.EGR.TNF.11D, a non-replicating adenovirus engineered to express TNF-α, resulted in a high response rate and improved survival rates (Hamada and Yura 2020; Runge et al. 2022). Within paramyxoviridae, the measles virus MV-NIS, which produces the sodium/iodide symporter (NIS), exhibited improved anticancer effects in HNSCC cells, especially when combined with radioisotopes (Wang et al. 2023).

18.4.2.5 Oncolytic Rhabdoviruses

An example of these viruses is VSV-IFN, an engineered vesicular stomatitis virus (VSV) expressing interferon beta, which selectively targets tumor cells over normal tissue. Preclinical studies have shown strong antitumor effects when combined with immune checkpoint inhibitors (ICIs) (Shilpa et al. 2014).

18.5 Prediction and Treatment Prognosis

18.5.1 Immunological Factors

To predict tumor reaction to immunotherapy, recent studies have concentrated on determining the immunological profile of tumors classified as hot (inflamed) or cool (non-inflamed) (Smussi et al. 2023). Inflamed tumors demonstrate abundant dysfunctional CD8+ cytotoxic T cells, along with macrophages, B cells, myeloid cells, Tregs, IFN-γ, various cytokines, and upregulated PD-L1 and IDO (Economopoulou et al. 2020; Sim et al. 2019). This suggests the presence of an initial anti-tumor response blocked by an immunosuppressive TME. This immune resistance develops once T cells infiltrate the tumor (Economopoulou et al. 2020; Sim et al. 2019).

Non-inflamed TMEs, categorized as immune-excluded or immune-desert, are characterized by a lack of T cells, innate immune effectors, chemokines, CTL infiltration, IFN-γ signatures, and active inhibitory pathways (Davis et al. 2016; Economopoulou et al. 2020; Sim et al. 2019). Instead, these TMEs are dominated by immunosuppressive components, such as TAMs and Tregs (Chen and Mellman 2017; Economopoulou et al. 2020). In immune-excluded tumors, immune cells are restricted to the stroma, which reduces their effectiveness and results in poor treatment outcomes (Economopoulou et al. 2020). Defining HNSCC's immune classification is challenging due to its varied immune cell and cytokine distribution, but the disease often presents with an inflamed TME (Economopoulou et al. 2020; Wang et al. 2019).

18.5.1.1 Prognostic Potential of TAMs

An increase in TAMs in HNSCC is strongly linked to poor survival outcomes (Ruffin et al. 2023; Wu et al. 2019). Elevated M2 TAMs, particularly CD163+ macrophages, are strongly linked to poor prognosis in HNSCC, including nasopharyngeal carcinoma (Balermpas et al. 2014; Fu et al. 2020; Ugel et al. 2015; Wu et al. 2019). However, M1 macrophages are generally correlated with better prognosis in certain cancers, and a higher M1-to-M2 TAM ratio is correlated with improved outcomes in HNSCC, especially in HPV-positive tumors (Balermpas et al. 2014; Fu et al. 2020; Petitprez et al. 2020; Qin et al. 2021).

18.5.1.2 Prognostic Potential of DCs

In HNSCC, the presence and increased number of DCs are linked to higher infiltration of T cells and NK cells, reduced tumor progression, and improved treatment outcomes (Jardim et al. 2018; Reichert et al. 2001).

18.5.1.3 Prognostic Potential of NK Cells

In HNSCC, higher NK cell counts associate with improved prognosis and survival rates (Mandal et al. 2016; Ruffin et al. 2023).

18.5.1.4 Prognostic Potential of CAFs

The presence of a CAF gene signature is associated with worse survival outcomes in HPV-positive HNSCC, whereas its impact is less pronounced in HPV-negative cases (Mandal et al. 2016). Elevated levels of CAFs, particularly α-SMA-positive CAFs, EMT-positive CAFs, MHC class II-expressing CAFs, and myofibroblasts are correlated with lower survival rates, more advanced disease stages, greater tumor aggressiveness and invasion, recurrence, treatment resistance, and overall poorer prognosis (Economopoulou et al. 2020; Mandal et al. 2016).

18.5.1.5 Prognostic Potential of Tregs

High concentrations of FoxP3+ Treg in HNSCC have been linked to conflicting outcomes. While high Treg infiltration has been linked to longer relapse-free survival and improved overall survival, some other studies suggest that higher Treg levels is associated with poor prognosis

(Economopoulou et al. 2020; Saito et al. 2016; Seminerio et al. 2019).

18.5.1.6 Prognostic Potential of TANs

Elevated neutrophil levels, both in circulation and within the TME, are linked to worse patient outcomes (Sionov et al. 2015). The neutrophil-to-lymphocyte ratio (NLR), which compares circulating neutrophils to leukocytes, serves as a prognostic indicator in HNSCC, with higher NLR associated with poor prognosis (Ferrandino et al. 2020; Yasumatsu et al. 2019).

18.5.1.7 Prognostic Potential of MDSCs

Elevated levels of MDSCs in HNSCCs are linked to advanced clinical stage of the tumor, poor pathological grade, tumor recurrence, and cancer progression, primarily through suppressing T cell functions (Gabrilovich 2017; Lang et al. 2018; Liu et al. 2024; Ma et al. 2017b; Pang et al. 2020; Wang et al. 2019; Weed et al. 2015). Higher levels of CD11b+/CD16+ PMN-MDSCs in HNSCC are associated with adverse clinical outcomes including lower survival rates. Moreover, higher frequencies of the CD66b+/CD11b+/CD16+ subset correlate with elevated arginase I activity, which suppresses T cell proliferation and cytokine release, contributing more strongly to immune evasion than other MDSCs (Lang et al. 2018; Wang et al. 2019).

18.5.1.8 Prognostic Potential of B Cells

CD20+ B cells have been associated with extended recurrence-free survival and enhanced condition-specific survival in HNSCC cases (Pretscher et al. 2009; Suárez-Sánchez et al. 2021). A higher density of CD20+ B cells has been linked to a lower T-stage in HNSCC and enhanced responses to ICI therapy, although the clinical evidence is conflicting (Gavrielatou et al. 2021; Karpathiou et al. 2017; Ruffin et al. 2023).

18.5.1.9 Prognostic Potential of T Cells

In HNSCC, increased levels of CD4+ and CD8+ TILs, particularly CD4+ TILs, are linked to higher overall and relapse-free survival rates (Nguyen et al. 2016). Similarly, in oropharyngeal cancer, high CD8+ T cell levels are linked to better outcomes, and in oral SCC, a higher frequency of peritumoral CD8+ TILs correlate with favorable clinical outcomes, including smaller tumor dimensions and decreased lymph node metastasis (Cho et al. 2011; Sasidharan Nair and Elkord 2018).

Additionally, in HPV-positive HNSCC, TFH cells are linked to better progression-free survival, and in HPV-negative HNSCC, despite a decrease in CD8+ T cells, similar outcomes to HPV-positive cases indicate the essential function of tissue-localized CD8+ T cells in antitumor immunity (Huang et al. 2006; Noy and Pollard 2014; Ruffin et al. 2023).

18.5.2 Genetical Prognostic Factors

Neo-epitopes from non-synonymous tumor mutations are crucial for immune recognition and T cell priming in cancer (Payne et al. 2019). Tumors with more missense mutations tend to have higher CD8+ T cell infiltration and better prognoses, though only a small share of these mutations produce neo-antigens that trigger T cell responses (Payne et al. 2019). Additionally, recent RNA sequencing studies across various tumor types indicate that immunogenic mutations, rather than overall tumor mutational burden (TMB) alone, are more strongly associated with improved survival outcomes (Rodrigo et al. 2024). These mutations elevate the expression of immune exhaustion markers and CD8A, creating conditions favorable for immunotherapy response (Rodrigo et al. 2024). Additionally, higher TMB levels in HNC, particularly in HPV-positive or EBV-positive patients have brought better outcomes in anti-PD-1/PD-L1 therapies (Liu et al. 2024; Rodrigo et al. 2024). Particularly among HPV-negative/EBV-negative responders, mutations in genes such as *NOTCH1* and *SMARCA4* have been found to be more common in responders than in non-responders (Baysal et al. 2022). Moreover,

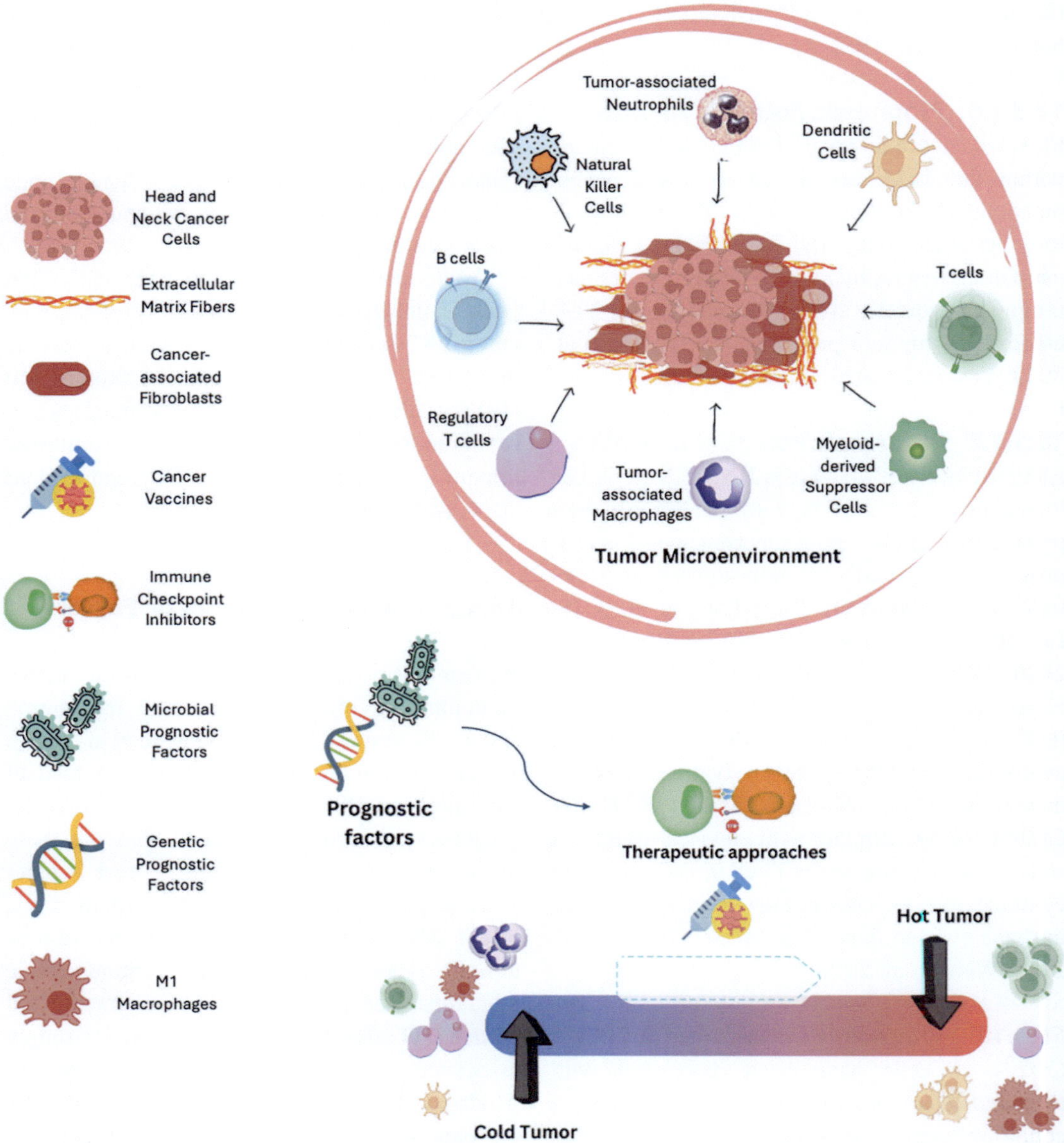

Fig. 18.1 The tumor microenvironment (TME) in head and neck cancer (HNC) consists of immune and non-immune cells, along with extracellular matrix components, creating an immunosuppressive niche that promotes tumor growth, invasion, and treatment resistance. Key cellular components—cancer-associated fibroblasts, dendritic cells, T cells, B cells, tumor-associated macrophages, and neutrophils—play dual roles in tumor progression and therapy response. Targeting the TME through immune checkpoint inhibitors, adoptive cell transfer, and cancer vaccines offers promising therapeutic strategies. These approaches aim to convert "cold" tumors with low immune activity into "hot" tumors with enhanced immune infiltration, improving response to immunotherapy. Additionally, immune, genetic, and microbial prognostic markers provide insights for personalized treatment approaches

the IFN-γ signature has been correlated with higher response rates and progression-free survival rates, potentially predicting which patients should be excluded from immunotherapy and may not benefit from this approach (Gavrielatou et al. 2020; Seiwert et al. 2016).

18.5.3 Microbial Factors

Oral microbes are being explored as potential biomarkers for pre-malignant lesions, namely leukoplakia and erythroplakia, which can progress into HNSCC. Identifying microbial signatures specific to these early stages could enable earlier therapeutic interventions and improve outcomes (Burcher et al. 2022). Research has highlighted differences in the oral microbiome between OPMD, healthy, and HNSCC samples(Burcher et al. 2022). Increased prevalence of *Fusobacterium*, *Prevotella*, and *Porphyromonas* is noted in **oral potentially malignant disorders** (OPMD) and HNSCC cases, while bacteria such as *Cloacibacillus* and *Gemmiger* are more abundant in OPMD. These changes might indicate a progression from OPMD to HNSCC (Burcher et al. 2022). Moreover, the oral microbiome in HNSCC often shows reduced diversity, leading to dysbiosis, which may influence disease progression and treatment response(Burcher et al. 2022). Specific genera, including *Fusobacterium*, *Peptostreptococcus*, and *Capnocytophaga*, are more common in HNSCC, though findings are inconsistent (Burcher et al. 2022). In addition, certain microbiota, such as *F. nucleatum* in HNSCC have been associated with better outcomes, potentially affecting survival rates and disease progression (Burcher et al. 2022). However, results are mixed, and further research is necessary to establish reliable prognostic markers (Burcher et al. 2022). *Porphyromonas gingivalis* has also been associated with oral cancer, potentially aiding cancer development by evading immune responses and increasing proMMP9 levels through various signaling pathways (Orlandi et al. 2019). These findings suggest that the oral microbiome might serve as a prognostic tool for HNSCC, but further studies are required to confirm its clinical relevance.

18.6 Conclusions

This chapter highlighted the pivotal role of the TME in HNC, affecting disease progression, treatment resistance, and therapeutic outcomes. The complex interactions between tumor cells and various TME constituents, including immune cells, non-immune cells, and non-cellular components, create an immunosuppressive environment that diminishes the efficacy of current treatments. To counteract their effects, each TME component can be directly targeted in cancer treatments. Moreover, therapies that indirectly modulate the TME, including oncolytic viruses and vaccines, show promise in overcoming immune evasion and improving treatment responses. Additionally, emerging predictive and prognostic factors, such as immune cell infiltration, genetic mutations, and microbial signatures, can enhance the understanding of tumor behavior and guide therapeutic strategies (Fig. 18.1). Future therapeutic approaches must focus on both direct and indirect TME modulation to achieve more durable and effective outcomes in HNC.

References

17th International Congress of Immunology (2019) 17th international congress of immunology, 19-23 October 2019, Beijing, China. Eur J Immunol 3(Suppl 3): 1–2223. https://doi.org/10.1002/eji.201970400

Acovic A, Simovic Markovic B, Gazdic M, Arsenijevic A, Jovicic N, Gajovic N, Jovanovic M, Zdravkovic N, Kanjevac T, Harrell CR, Fellabaum C, Dolicanin Z, Djonov V, Arsenijevic N, Lukic ML, Volarevic V (2018) Indoleamine 2,3-dioxygenase-dependent expansion of T-regulatory cells maintains mucosal healing in ulcerative colitis. Ther Adv Gastroenterol 11:1756284818793558. https://doi.org/10.1177/1756284818793558

Affara NI, Ruffell B, Medler TR, Gunderson AJ, Johansson M, Bornstein S, Bergsland E, Steinhoff M, Li Y, Gong Q, Ma Y, Wiesen JF, Wong MH, Kulesz-Martin M, Irving B, Coussens LM (2014) B cells regulate macrophage phenotype and response to chemotherapy in squamous carcinomas. Cancer Cell 25(6):809–821. https://doi.org/10.1016/j.ccr.2014.04.026

Alegre ML, Noel PJ, Eisfelder BJ, Chuang E, Clark MR, Reiner SL, Thompson CB (1996) Regulation of surface and intracellular expression of CTLA4 on mouse T cells. J Immunol 157(11):4762–4770

Álvarez-Teijeiro S, García-Inclán C, Villaronga M, Casado P, Hermida-Prado F, Granda-Díaz R, Rodrigo JP, Calvo F, Del-Río-Ibisate N, Gandarillas A, Morís F, Hermsen M, Cutillas P, García-Pedrero JM (2018) Factors secreted by cancer-associated fibroblasts that sustain cancer stem properties in head and neck squamous carcinoma cells as potential therapeutic targets. Cancers (Basel) 10(9). https://doi.org/10.3390/cancers10090334

Anderson K, Ryan N, Volpedo G, Varikuti S, Satoskar AR, Oghumu S (2019) Immune suppression mediated by STAT4 deficiency promotes lymphatic metastasis in HNSCC. Front Immunol 10:3095. https://doi.org/10.3389/fimmu.2019.03095

Andreu P, Johansson M, Affara NI, Pucci F, Tan T, Junankar S, Korets L, Lam J, Tawfik D, DeNardo DG, Naldini L, de Visser KE, De Palma M, Coussens LM (2010) FcRgamma activation regulates inflammation-associated squamous carcinogenesis. Cancer Cell 17(2):121–134. https://doi.org/10.1016/j.ccr.2009.12.019

Angevin E, Tabernero J, Elez E, Cohen SJ, Bahleda R, van Laethem JL, Ottensmeier C, Lopez-Martin JA, Clive S, Joly F, Ray-Coquard I, Dirix L, Machiels JP, Steven N, Reddy M, Hall B, Puchalski TA, Bandekar R et al (2014) A phase I/II, multiple-dose, dose-escalation study of siltuximab, an anti-interleukin-6 monoclonal antibody, in patients with advanced solid tumors. Clin Cancer Res 20(8):2192–2204. https://doi.org/10.1158/1078-0432.Ccr-13-2200

Ansell S, Bhardwaj V, Yang Z-Z, Jalali S, Villasboas J, Mudappathi R, Wang J, Mukherjee P, Paludo J, Tang X, Kim H, Krull J, Wenzl K, Novak A, Mondello P (2024) Expanded tumor-associated polymorphonuclear Myeloid-derived suppressor cells in Waldenstrom macroglobulinemia display immune suppressive activity. Blood Cancer J 14(1):217

Balermpas P, Rödel F, Liberz R, Oppermann J, Wagenblast J, Ghanaati S, Harter PN, Mittelbronn M, Weiss C, Rödel C, Fokas E (2014) Head and neck cancer relapse after chemoradiotherapy correlates with CD163+ macrophages in primary tumour and CD11b+ myeloid cells in recurrences. Br J Cancer 111(8):1509–1518. https://doi.org/10.1038/bjc.2014.446

Bauman JE, Grandis JR (2014) Targeting secondary immune responses to cetuximab: CD137 and the outside story. J Clin Invest 124(6):2371–2375. https://doi.org/10.1172/jci76264

Baysal H, Siozopoulou V, Zaryouh H, Hermans C, Lau HW, Lambrechts H, Fransen E, De Pauw I, Jacobs J, Peeters M, Pauwels P, Vermorken JB, Smits E, Lardon F, De Waele J, Wouters A (2022) The prognostic impact of the immune signature in head and neck squamous cell carcinoma. Front Immunol 13:1001161. https://doi.org/10.3389/fimmu.2022.1001161

Ben-Shmuel A, Biber G, Barda-Saad M (2020) Unleashing natural killer cells in the tumor microenvironment-the next generation of immunotherapy? Front Immunol 11:275. https://doi.org/10.3389/fimmu.2020.00275

Bergers G, Brekken R, McMahon G, Vu TH, Itoh T, Tamaki K, Tanzawa K, Thorpe P, Itohara S, Werb Z, Hanahan D (2000) Matrix metalloproteinase-9 triggers the angiogenic switch during carcinogenesis. Nat Cell Biol 2(10):737–744. https://doi.org/10.1038/35036374

Bharadwaj U, Eckols TK, Xu X, Kasembeli MM, Chen Y, Adachi M, Song Y, Mo Q, Lai SY, Tweardy DJ (2016) Small-molecule inhibition of STAT3 in radioresistant head and neck squamous cell carcinoma. Oncotarget 7(18):26307–26330. https://doi.org/10.18632/oncotarget.8368

Bhat AA, Yousuf P, Wani NA, Rizwan A, Chauhan SS, Siddiqi MA, Bedognetti D, El-Rifai W, Frenneaux MP, Batra SK, Haris M, Macha MA (2021) Tumor microenvironment: an evil nexus promoting aggressive head and neck squamous cell carcinoma and avenue for targeted therapy. Signal Transduct Target Ther 6(1):12. https://doi.org/10.1038/s41392-020-00419-w

Birbrair A (2020) tumor microenvironments in organs from the brain to the skin – part A: from the brain to the skin – part A. https://doi.org/10.1007/978-3-030-36214-0

Blacher E, Levy A, Baruch BB, Green KD, Garneau-Tsodikova S, Fridman M, Stein R (2015) Targeting CD38 in the tumor microenvironment; a novel approach to treat glioma. Cancer Cell Microenviron 2

Bot A, Obrocea M, Marincola F, Farsaci B (2011) Cancer vaccines: from research to clinical practice. CRC Press

Boucek J, Mrkvan T, Chovanec M, Kuchar M, Betka J, Boucek V, Hladikova M, Betka J, Eckschlager T, Rihova B (2010) Regulatory T cells and their prognostic value for patients with squamous cell carcinoma of the head and neck. J Cell Mol Med 14(1–2):426–433. https://doi.org/10.1111/j.1582-4934.2008.00650.x

Brana I, Calles A, LoRusso PM, Yee LK, Puchalski TA, Seetharam S, Zhong B, de Boer CJ, Tabernero J, Calvo E (2015) Carlumab, an anti-C-C chemokine ligand 2 monoclonal antibody, in combination with four chemotherapy regimens for the treatment of patients with solid tumors: an open-label, multicenter phase 1b study. Target Oncol 10(1):111–123. https://doi.org/10.1007/s11523-014-0320-2

Buchwald ZS, Schmitt NC (2021) Immunotherapeutic strategies for head and neck cancer. Otolaryngol Clin N Am 54(4):729–742. https://doi.org/10.1016/j.otc.2021.04.004

Budi HS, Farhood B (2023) Targeting oral tumor microenvironment for effective therapy. Cancer Cell Int

23(1):101. https://doi.org/10.1186/s12935-023-02943-5

Bullock TN, Mullins DW, Engelhard VH (2003) Antigen density presented by dendritic cells in vivo differentially affects the number and avidity of primary, memory, and recall CD8+ T cells. J Immunol 170(4):1822–1829. https://doi.org/10.4049/jimmunol.170.4.1822

Buoncervello M, Gabriele L, Toschi E (2019) The Janus face of tumor microenvironment targeted by immunotherapy. Int J Mol Sci 20(17). https://doi.org/10.3390/ijms20174320

Burcher KM, Burcher JT, Inscore L, Bloomer CH, Furdui CM, Porosnicu M (2022) A review of the role of oral microbiome in the development, detection, and management of head and neck squamous cell cancers. Cancers 14(17)

Burnet FM (1967) Immunological aspects of malignant DiseasE. Lancet 289(7501):1171–1174. https://doi.org/10.1016/S0140-6736(67)92837-1

Butte MJ, Keir ME, Phamduy TB, Sharpe AH, Freeman GJ (2007) Programmed death-1 ligand 1 interacts specifically with the B7-1 costimulatory molecule to inhibit T cell responses. Immunity 27(1):111–122. https://doi.org/10.1016/j.immuni.2007.05.016

Califano JA, Khan Z, Noonan KA, Rudraraju L, Zhang Z, Wang H, Goodman S, Gourin CG, Ha PK, Fakhry C, Saunders J, Levine M, Tang M, Neuner G, Richmon JD, Blanco R, Agrawal N, Koch WM et al (2015) Tadalafil augments tumor specific immunity in patients with head and neck squamous cell carcinoma. Clin Cancer Res 21(1):30–38. https://doi.org/10.1158/1078-0432.Ccr-14-1716

Camisaschi C, Casati C, Rini F, Perego M, De Filippo A, Triebel F, Parmiani G, Belli F, Rivoltini L, Castelli C (2010) LAG-3 expression defines a subset of CD4(+)CD25(high)Foxp3(+) regulatory T cells that are expanded at tumor sites. J Immunol 184(11):6545–6551. https://doi.org/10.4049/jimmunol.0903879

Canning M, Guo G, Yu M, Myint C, Groves MW, Byrd JK, Cui Y (2019) Heterogeneity of the head and neck squamous cell carcinoma immune landscape and its impact on immunotherapy. Front Cell Dev Biol 7:52. https://doi.org/10.3389/fcell.2019.00052

Cervantes-García D, Ortiz-López R, Mayek-Pérez N, Rojas-Martínez A (2008) Oncolytic virotherapy. Ann Hepatol 7(1):34–45

Chambers CA, Kuhns MS, Allison JP (1999) Cytotoxic T lymphocyte antigen-4 (CTLA-4) regulates primary and secondary peptide-specific CD4(+) T cell responses. Proc Natl Acad Sci USA 96(15):8603–8608. https://doi.org/10.1073/pnas.96.15.8603

Chen DS, Mellman I (2017) Elements of cancer immunity and the cancer-immune set point. Nature 541(7637):321–330. https://doi.org/10.1038/nature21349

Chen X, Song E (2019) Turning foes to friends: targeting cancer-associated fibroblasts. Nat Rev Drug Discov 18(2):99–115. https://doi.org/10.1038/s41573-018-0004-1

Chen F, Zhuang X, Lin L, Yu P, Wang Y, Shi Y, Hu G, Sun Y (2015) New horizons in tumor microenvironment biology: challenges and opportunities. BMC Med 13:45. https://doi.org/10.1186/s12916-015-0278-7

Chen L, Diao L, Yang Y, Yi X, Rodriguez BL, Li Y, Villalobos PA, Cascone T, Liu X, Tan L, Lorenzi PL, Huang A, Zhao Q, Peng D, Fradette JJ, Peng DH, Ungewiss C, Roybal J et al (2018) CD38-mediated immunosuppression as a mechanism of tumor cell escape from PD-1/PD-L1 blockade. Cancer Discov 8(9):1156–1175. https://doi.org/10.1158/2159-8290.Cd-17-1033

Chen SMY, Krinsky AL, Woolaver RA, Wang X, Chen Z, Wang JH (2020) Tumor immune microenvironment in head and neck cancers. Mol Carcinog 59(7):766–774. https://doi.org/10.1002/mc.23162

Chiappori AA, Eckhardt SG, Bukowski R, Sullivan DM, Ikeda M, Yano Y, Yamada-Sawada T, Kambayashi Y, Tanaka K, Javle MM, Mekhail T, O'Bryant C et al (2007) A phase I pharmacokinetic and pharmacodynamic study of s-3304, a novel matrix metalloproteinase inhibitor, in patients with advanced and refractory solid tumors. Clin Cancer Res 13(7):2091–2099. https://doi.org/10.1158/1078-0432.Ccr-06-1586

Cho YA, Yoon HJ, Lee JI, Hong SP, Hong SD (2011) Relationship between the expressions of PD-L1 and tumor-infiltrating lymphocytes in oral squamous cell carcinoma. Oral Oncol 47(12):1148–1153. https://doi.org/10.1016/j.oraloncology.2011.08.007

Chow LQM, Haddad R, Gupta S, Mahipal A, Mehra R, Tahara M, Berger R, Eder JP, Burtness B, Lee SH, Keam B, Kang H, Muro K, Weiss J, Geva R, Lin CC, Chung HC, Meister A et al (2016) Antitumor activity of Pembrolizumab in biomarker-unselected patients with recurrent and/or metastatic head and neck squamous cell carcinoma: results from the phase Ib KEYNOTE-012 expansion cohort. J Clin Oncol 34(32):3838–3845. https://doi.org/10.1200/jco.2016.68.1478

Chuckran CA, Cillo AR, Moskovitz J, Overacre-Delgoffe-A, Somasundaram AS, Shan F, Magnon GC, Kunning SR, Abecassis I, Zureikat AH, Luketich J, Pennathur A, Sembrat J, Rojas M, Merrick DT, Taylor SE, Orr B, Modugno F et al (2021) Prevalence of intratumoral regulatory T cells expressing neuropilin-1 is associated with poorer outcomes in patients with cancer. Sci Transl Med 13(623):eabf8495. https://doi.org/10.1126/scitranslmed.abf8495

Concha-Benavente F, Srivastava RM, Trivedi S, Lei Y, Chandran U, Seethala RR, Freeman GJ, Ferris RL (2016) Identification of the cell-intrinsic and -extrinsic pathways downstream of EGFR and IFNγ that induce PD-L1 expression in head and neck cancer. Cancer Res 76(5):1031–1043. https://doi.org/10.1158/0008-5472.Can-15-2001

Corzo CA, Condamine T, Lu L, Cotter MJ, Youn JI, Cheng P, Cho HI, Celis E, Quiceno DG, Padhya T, McCaffrey TV, McCaffrey JC, Gabrilovich DI (2010) HIF-1α regulates function and differentiation of myeloid-derived suppressor cells in the tumor microenvironment. J Exp Med 207(11):2439–2453. https://doi.org/10.1084/jem.20100587

Cramer JD, Burtness B, Ferris RL (2019) Immunotherapy for head and neck cancer: recent advances and future directions. Oral Oncol 99:104460. https://doi.org/10.1016/j.oraloncology.2019.104460

Cristina V, Herrera-Gómez RG, Szturz P, Espeli V, Siano M (2019) Immunotherapies and future combination strategies for head and neck squamous cell carcinoma. Int J Mol Sci 20(21)

Curry JM, Sprandio J, Cognetti D, Luginbuhl A, Bar-ad V, Pribitkin E, Tuluc M (2014) Tumor microenvironment in head and neck squamous cell carcinoma. Semin Oncol 41(2):217–234. https://doi.org/10.1053/j.seminoncol.2014.03.003

Dadey RE, Workman CJ, Vignali DAA (2020) Regulatory T cells in the tumor microenvironment. Adv Exp Med Biol 1273:105–134. https://doi.org/10.1007/978-3-030-49270-0_6

Davis RJ, Ferris RL, Schmitt NC (2016) Costimulatory and coinhibitory immune checkpoint receptors in head and neck cancer: unleashing immune responses through therapeutic combinations. Cancers Head Neck 1:12. https://doi.org/10.1186/s41199-016-0013-x

de Medeiros MC, Banerjee R, Liu M, Anovazzi G, D'Silva NJ, Junior CR (2017) HNSCC subverts PBMCs to secrete soluble products that promote tumor cell proliferation. Oncotarget 8(37):60860–60874. https://doi.org/10.18632/oncotarget.18486

de Visser KE, Korets LV, Coussens LM (2005) De novo carcinogenesis promoted by chronic inflammation is B lymphocyte dependent. Cancer Cell 7(5):411–423. https://doi.org/10.1016/j.ccr.2005.04.014

Deaglio S, Dwyer KM, Gao W, Friedman D, Usheva A, Erat A, Chen JF, Enjyoji K, Linden J, Oukka M, Kuchroo VK, Strom TB, Robson SC (2007) Adenosine generation catalyzed by CD39 and CD73 expressed on regulatory T cells mediates immune suppression. J Exp Med 204(6):1257–1265. https://doi.org/10.1084/jem.20062512

Delaloye J, Roger T, Steiner-Tardivel QG, Le Roy D, Knaup Reymond M, Akira S, Petrilli V, Gomez CE, Perdiguero B, Tschopp J, Pantaleo G, Esteban M, Calandra T (2009) Innate immune sensing of modified vaccinia virus Ankara (MVA) is mediated by TLR2-TLR6, MDA-5 and the NALP3 inflammasome. PLoS Pathog 5(6):e1000480. https://doi.org/10.1371/journal.ppat.1000480

Deng W-W, Li Y-C, Ma S-R, Mao L, Yu G-T, Bu L-L, Kulkarni AB, Zhang W-F, Sun Z-J (2018) Specific blockade CD73 alters the "exhausted" phenotype of T cells in head and neck squamous cell carcinoma. Int J Cancer 143(6):1494–1504. https://doi.org/10.1002/ijc.31534

Devaraja K (2019) Current prospects of molecular therapeutics in head and neck squamous cell carcinoma. Pharmaceut Med 33(4):269–289. https://doi.org/10.1007/s40290-019-00288-x

Dobrenis K, Gauthier LR, Barroca V, Magnon C (2015) Granulocyte colony-stimulating factor off-target effect on nerve outgrowth promotes prostate cancer development. Int J Cancer 136(4):982–988. https://doi.org/10.1002/ijc.29046

Dorta-Estremera S, Hegde VL, Slay RB, Sun R, Yanamandra AV, Nicholas C, Nookala S, Sierra G, Curran MA, Sastry KJ (2019) Targeting interferon signaling and CTLA-4 enhance the therapeutic efficacy of anti-PD-1 immunotherapy in preclinical model of HPV(+) oral cancer. J Immunother Cancer 7(1):252. https://doi.org/10.1186/s40425-019-0728-4

Echarti A, Hecht M, Büttner-Herold M, Haderlein M, Hartmann A, Fietkau R, Distel L (2019) CD8+ and regulatory T cells differentiate tumor immune phenotypes and predict survival in locally advanced head and neck cancer. Cancers (Basel) 11(9). https://doi.org/10.3390/cancers11091398

Economopoulou P, Kotsantis I, Psyrri A (2020) Tumor microenvironment and immunotherapy response in head and neck cancer. Cancers (Basel) 12(11). https://doi.org/10.3390/cancers12113377

Egeblad M, Werb Z (2002) New functions for the matrix metalloproteinases in cancer progression. Nat Rev Cancer 2(3):161–174. https://doi.org/10.1038/nrc745

Egen JG, Allison JP (2002) Cytotoxic T lymphocyte antigen-4 accumulation in the immunological synapse is regulated by TCR signal strength. Immunity 16(1):23–35. https://doi.org/10.1016/s1074-7613(01)00259-x

Ellis GI, Sheppard NC, Riley JL (2021) Genetic engineering of T cells for immunotherapy. Nat Rev Genet 22(7):427–447. https://doi.org/10.1038/s41576-021-00329-9

Elmusrati A, Wang J, Wang CY (2021) Tumor microenvironment and immune evasion in head and neck squamous cell carcinoma. Int J Oral Sci 13(1):24. https://doi.org/10.1038/s41368-021-00131-7

Engeland CE, Bell JC (2020) Introduction to oncolytic Virotherapy. Methods Mol Biol 2058:1–6. https://doi.org/10.1007/978-1-4939-9794-7_1

Eppihimer MJ, Gunn J, Freeman GJ, Greenfield EA, Chernova T, Erickson J, Leonard JP (2002) Expression and regulation of the PD-L1 immunoinhibitory molecule on microvascular endothelial cells. Microcirculation 9(2):133–145. https://doi.org/10.1038/sj/mn/7800123

Facciabene A, Peng X, Hagemann IS, Balint K, Barchetti A, Wang L-P, Gimotty PA, Gilks CB, Lal P, Zhang L, Coukos G (2011) Tumour hypoxia promotes tolerance and angiogenesis via CCL28 and Treg cells. Nature 475(7355):226–230. https://doi.org/10.1038/nature10169

Fatima M, Amraiz D, Navid MT (2023) Oncolytic Virotherapy. Cancer Treat Res 185:105–126. https://doi.org/10.1007/978-3-031-27156-4_7

Ferrandino RM, Roof S, Garneau J, Haidar Y, Bates SE, Park YA, Bauml JM, Genden EM, Miles B, Sigel K (2020) Neutrophil-to-lymphocyte ratio as a prognostic indicator for overall and cancer-specific survival in squamous cell carcinoma of the head and neck. Head Neck 42(10):2830–2840. https://doi.org/10.1002/hed.26329

Ferreira FO, Ribeiro FL, Batista AC, Leles CR et al (2008) Association of CCL2 with lymph node metastasis and macrophage infiltration in oral cavity and lip squamous cell carcinoma. Tumour Biol 29(2):114–121. https://doi.org/10.1159/000137669

Ferris RL (2015) Immunology and immunotherapy of head and neck cancer. J Clin Oncol 33(29):3293–3304. https://doi.org/10.1200/jco.2015.61.1509

Ferris R, Blumenschein G, Fayette J, Guigay J, Colevas A, Licitra L, Harrington K, Kasper S, Vokes E, Even C, Worden F, Haddad R, Kiyota N, Tahara M, Monga M, Lynch M, Geese W, Kopit J, Shaw J, Gillison M (2016) Further evaluations of nivolumab (nivo) versus investigator's choice (IC) chemotherapy for recurrent or metastatic (R/M) squamous cell carcinoma of the head and neck (SCCHN): CheckMate 141. J Clin Oncol 34:6009–6009. https://doi.org/10.1200/JCO.2016.34.15_suppl.6009

Fiori ME, Di Franco S, Villanova L, Bianca P, Stassi G, De Maria R (2019) Cancer-associated fibroblasts as abettors of tumor progression at the crossroads of EMT and therapy resistance. Mol Cancer 18(1):70. https://doi.org/10.1186/s12943-019-0994-2

Forster MD, Devlin MJ (2018) Immune checkpoint inhibition in head and neck cancer. Front Oncol 8:310. https://doi.org/10.3389/fonc.2018.00310

Freeman GJ, Long AJ, Iwai Y, Bourque K, Chernova T, Nishimura H, Fitz LJ, Malenkovich N, Okazaki T, Byrne MC, Horton HF, Fouser L, Carter L, Ling V, Bowman MR, Carreno BM, Collins M, Wood CR, Honjo T (2000) Engagement of the PD-1 immunoinhibitory receptor by a novel B7 family member leads to negative regulation of lymphocyte activation. J Exp Med 192(7):1027–1034. https://doi.org/10.1084/jem.192.7.1027

Fridlender ZG, Sun J, Kim S, Kapoor V, Cheng G, Ling L, Worthen GS, Albelda SM (2009) Polarization of tumor-associated neutrophil phenotype by TGF-beta: "N1" versus "N2" TAN. Cancer Cell 16(3):183–194. https://doi.org/10.1016/j.ccr.2009.06.017

Fu E, Liu T, Yu S, Chen X, Song L, Lou H, Ma F, Zhang S, Hussain S, Guo J, Sun J, Yu P, Mao X, Wei L (2020) M2 macrophages reduce the radiosensitivity of head and neck cancer by releasing HB-EGF. Oncol Rep 44(2):698–710. https://doi.org/10.3892/or.2020.7628

Fugle CW, Zhang Y, Hong F, Sun S, Westwater C, Rachidi S, Yu H, Garret-Mayer E, Kirkwood K, Liu B, Li Z (2016) CD24 blunts oral squamous cancer development and dampens the functional expansion of myeloid-derived suppressor cells. Onco Targets Ther 5(10):e1226719. https://doi.org/10.1080/2162402x.2016.1226719

Fulop T, Franceschi C, Hirokawa K, Pawelec G (2019) Handbook of immunosenescence basic understanding and clinical implications: basic understanding and clinical implications. https://doi.org/10.1007/978-3-319-99375-1

Gabrilovich DI (2017) Myeloid-derived suppressor cells. Cancer Immunol Res 5(1):3–8. https://doi.org/10.1158/2326-6066.Cir-16-0297

Gabrilovich DI, Ostrand-Rosenberg S, Bronte V (2012) Coordinated regulation of myeloid cells by tumours. Nat Rev Immunol 12(4):253–268. https://doi.org/10.1038/nri3175

Gao XH, Chen HD (2016) Practical immunodermatology. https://doi.org/10.1007/978-94-024-0902-4

Gao L, Wang FQ, Li HM, Yang JG, Ren JG, He KF, Liu B, Zhang W, Zhao YF (2016) CCL2/EGF positive feedback loop between cancer cells and macrophages promotes cell migration and invasion in head and neck squamous cell carcinoma. Oncotarget 7(52):87037–87051. https://doi.org/10.18632/oncotarget.13523

Garcia-Carbonero R, Salazar R, Duran I, Osman-Garcia I, Paz-Ares L, Bozada JM, Boni V, Blanc C, Seymour L, Beadle J, Alvis S, Champion B, Calvo E, Fisher K (2017) Phase 1 study of intravenous administration of the chimeric adenovirus enadenotucirev in patients undergoing primary tumor resection. J Immunother Cancer 5(1):71. https://doi.org/10.1186/s40425-017-0277-7

Gavrielatou N, Doumas S, Economopoulou P, Foukas PG, Psyrri A (2020) Biomarkers for immunotherapy response in head and neck cancer. Cancer Treat Rev 84:101977. https://doi.org/10.1016/j.ctrv.2020.101977

Gavrielatou N, Vathiotis I, Economopoulou P, Psyrri A (2021) The role of B cells in head and neck cancer. Cancers (Basel) 13(21). https://doi.org/10.3390/cancers13215383

Gerosa F, Baldani-Guerra B, Nisii C, Marchesini V, Carra G, Trinchieri G (2002) Reciprocal activating interaction between natural killer cells and dendritic cells. J Exp Med 195(3):327–333. https://doi.org/10.1084/jem.20010938

Giancotti FG (2014) Deregulation of cell signaling in cancer. FEBS Lett 588(16):2558–2570. https://doi.org/10.1016/j.febslet.2014.02.005

Goldberg MV, Drake CG (2011) LAG-3 in cancer immunotherapy. Curr Top Microbiol Immunol 344:269–278. https://doi.org/10.1007/82_2010_114

Goradel NH, Baker AT, Arashkia A, Ebrahimi N, Ghorghanlu S, Negahdari B (2021) Oncolytic virotherapy: challenges and solutions. Curr Probl Cancer 45(1):100639. https://doi.org/10.1016/j.currproblcancer.2020.100639

Grandis JR, Falkner DM, Melhem MF, Gooding WE, Drenning SD, Morel PA (2000) Human leukocyte antigen class I allelic and haplotype loss in squamous cell carcinoma of the head and neck: clinical and immunogenetic consequences. Clin Cancer Res 6(7): 2794–2802

Greene S, Robbins Y, Mydlarz WK, Huynh AP, Schmitt NC, Friedman J, Horn LA, Palena C, Schlom J, Maeda DY, Zebala JA, Clavijo PE, Allen C (2020) Inhibition of MDSC trafficking with SX-682, a CXCR1/2 inhibitor, enhances NK-cell immunotherapy in head and neck cancer models. Clin Cancer Res 26(6): 1420–1431. https://doi.org/10.1158/1078-0432.Ccr-19-2625

Grosso JF, Goldberg MV, Getnet D, Bruno TC, Yen HR, Pyle KJ, Hipkiss E, Vignali DA, Pardoll DM, Drake CG (2009) Functionally distinct LAG-3 and PD-1 subsets on activated and chronically stimulated CD8 T cells. J Immunol 182(11):6659–6669. https://doi.org/10.4049/jimmunol.0804211

Hamada M, Yura Y (2020) Efficient delivery and replication of oncolytic virus for successful treatment of head and neck cancer. Int J Mol Sci 21(19). https://doi.org/10.3390/ijms21197073

Hamdan F, Fusciello M, Cerullo V (2023) Personalizing oncolytic Virotherapy. Hum Gene Ther 34(17–18): 870–877. https://doi.org/10.1089/hum.2023.122

Hodi FS, O'Day SJ, McDermott DF, Weber RW, Sosman JA, Haanen JB, Gonzalez R, Robert C, Schadendorf D, Hassel JC, Akerley W, van den Eertwegh AJ, Lutzky J, Lorigan P, Vaubel JM, Linette GP, Hogg D, Ottensmeier CH et al (2010) Improved survival with ipilimumab in patients with metastatic melanoma. N Engl J Med 363(8):711–723. https://doi.org/10.1056/NEJMoa1003466

Honeychurch J, Cheadle EJ, Dovedi SJ, Illidge TM (2015) Immuno-regulatory antibodies for the treatment of cancer. Expert Opin Biol Ther 15(6):787–801. https://doi.org/10.1517/14712598.2015.1036737

Horton JD, Knochelmann HM, Day TA, Paulos CM, Neskey DM (2019) Immune evasion by head and neck cancer: foundations for combination therapy. Trends Cancer 5(4):208–232. https://doi.org/10.1016/j.trecan.2019.02.007

Hu P, Wang G, Shen M, Zhang P, Zhang J, Du J, Liu Q (2015) Intratumoral polymorphonuclear granulocyte is associated with poor prognosis in squamous esophageal cancer by promoting epithelial-mesenchymal transition. Future Oncol 11(5):771–783. https://doi.org/10.2217/fon.14.306

Hu Y, Lu L, Xia Y, Chen X, Chang AE, Hollingsworth RE, Hurt E, Owen J, Moyer JS, Prince ME, Dai F, Bao Y, Wang Y, Whitfield J, Xia JC, Huang S, Wicha MS, Li Q (2016) Therapeutic efficacy of cancer stem cell vaccines in the adjuvant setting. Cancer Res 76(16):4661–4672. https://doi.org/10.1158/0008-5472.Can-15-2664

Huang B, Pan PY, Li Q, Sato AI, Levy DE, Bromberg J, Divino CM, Chen SH (2006) Gr-1+CD115+ immature myeloid suppressor cells mediate the development of tumor-induced T regulatory cells and T-cell anergy in tumor-bearing host. Cancer Res 66(2):1123–1131. https://doi.org/10.1158/0008-5472.Can-05-1299

Hurwitz AA, Yu TF, Leach DR, Allison JP (1998) CTLA-4 blockade synergizes with tumor-derived granulocyte-macrophage colony-stimulating factor for treatment of an experimental mammary carcinoma. Proc Natl Acad Sci USA 95(17):10067–10071. https://doi.org/10.1073/pnas.95.17.10067

Hurwitz AA, Foster BA, Kwon ED, Truong T, Choi EM, Greenberg NM, Burg MB, Allison JP (2000) Combination immunotherapy of primary prostate cancer in a transgenic mouse model using CTLA-4 blockade. Cancer Res 60(9):2444–2448

Jardim JF, Gondak R, Galvis MM, Pinto CAL, Kowalski LP (2018) A decreased peritumoral CD1a+ cell number predicts a worse prognosis in oral squamous cell carcinoma. Histopathology 72(6):905–913. https://doi.org/10.1111/his.13415

Jarnicki AG, Lysaght J, Todryk S, Mills KHG (2006) Suppression of antitumor immunity by IL-10 and TGF-β-producing T cells infiltrating the growing tumor: influence of tumor environment on the induction of CD4+ and CD8+ regulatory T Cells1. J Immunol 177(2):896–904. https://doi.org/10.4049/jimmunol.177.2.896

Jie HB, Gildener-Leapman N, Li J, Srivastava RM, Gibson SP, Whiteside TL, Ferris RL (2013) Intratumoral regulatory T cells upregulate immunosuppressive molecules in head and neck cancer patients. Br J Cancer 109(10):2629–2635. https://doi.org/10.1038/bjc.2013.645

Jie HB, Schuler PJ, Lee SC, Srivastava RM, Argiris A, Ferrone S, Whiteside TL, Ferris RL (2015) CTLA-4$^+$ regulatory T cells increased in Cetuximab-treated head and neck cancer patients suppress NK cell cytotoxicity and correlate with poor prognosis. Cancer Res 75(11): 2200–2210. https://doi.org/10.1158/0008-5472.Can-14-2788

Johnson DE, O'Keefe RA, Grandis JR (2018) Targeting the IL-6/JAK/STAT3 signalling axis in cancer. Nat Rev Clin Oncol 15(4):234–248. https://doi.org/10.1038/nrclinonc.2018.8

Joyce JA, Pollard JW (2009) Microenvironmental regulation of metastasis. Nat Rev Cancer 9(4):239–252. https://doi.org/10.1038/nrc2618

Karpathiou G, Casteillo F, Giroult JB, Forest F, Fournel P, Monaya A, Froudarakis M, Dumollard JM, Prades JM, Peoc'h M (2017) Prognostic impact of immune microenvironment in laryngeal and pharyngeal squamous cell carcinoma: immune cell subtypes, immunosuppressive pathways and clinicopathologic characteristics. Oncotarget 8(12):19310–19322. https://doi.org/10.18632/oncotarget.14242

Kenter GG, Welters MJ, Valentijn AR, Lowik MJ, Berends-van der Meer DM, Vloon AP, Essahsah F, Fathers LM, Offringa R, Drijfhout JW, Wafelman AR, Oostendorp J, Fleuren GJ, van der Burg SH, Melief CJ (2009) Vaccination against HPV-16 oncoproteins for vulvar intraepithelial neoplasia. N

Engl J Med 361(19):1838–1847. https://doi.org/10.1056/NEJMoa0810097

Kesselring R, Thiel A, Pries R, Trenkle T, Wollenberg B (2010) Human Th17 cells can be induced through head and neck cancer and have a functional impact on HNSCC development. Br J Cancer 103(8):1245–1254. https://doi.org/10.1038/sj.bjc.6605891

Kim J, Bae JS (2016) Tumor-associated macrophages and neutrophils in tumor microenvironment. Mediat Inflamm 2016:6058147. https://doi.org/10.1155/2016/6058147

Koontongkaew S (2013) The tumor microenvironment contribution to development, growth, invasion and metastasis of head and neck squamous cell carcinomas. J Cancer 4(1):66–83. https://doi.org/10.7150/jca.5112

Kuksin M, Morel D, Aglave M, Danlos FX, Marabelle A, Zinovyev A, Gautheret D, Verlingue L (2021) Applications of single-cell and bulk RNA sequencing in onco-immunology. Eur J Cancer 149:193–210. https://doi.org/10.1016/j.ejca.2021.03.005

Kurai J, Chikumi H, Hashimoto K, Yamaguchi K, Yamasaki A, Sako T, Touge H, Makino H, Takata M, Miyata M, Nakamoto M, Burioka N, Shimizu E (2007) Antibody-dependent cellular cytotoxicity mediated by cetuximab against lung cancer cell lines. Clin Cancer Res 13(5):1552–1561. https://doi.org/10.1158/1078-0432.Ccr-06-1726

Kürten CHL, Kulkarni A, Cillo AR, Santos PM, Roble AK, Onkar S, Reeder C, Lang S, Chen X, Duvvuri U, Kim S, Liu A, Tabib T, Lafyatis R, Feng J, Gao SJ, Bruno TC, Vignali DAA et al (2021) Investigating immune and non-immune cell interactions in head and neck tumors by single-cell RNA sequencing. Nat Commun 12(1):7338. https://doi.org/10.1038/s41467-021-27619-4

Lahl K, Loddenkemper C, Drouin C, Freyer J, Arnason J, Eberl G et al (2007) Selective depletion of Foxp3+ regulatory T cells induces a scurfy-like disease. J Exp Med 204(1):57–63. https://doi.org/10.1084/jem.20061852

Lang S, Bruderek K, Kaspar C, Höing B, Kanaan O, Dominas N, Hussain T, Droege F, Eyth C, Hadaschik B, Brandau S (2018) Clinical relevance and suppressive capacity of human myeloid-derived suppressor cell subsets. Clin Cancer Res 24(19):4834–4844. https://doi.org/10.1158/1078-0432.Ccr-17-3726

Latchman Y, Wood CR, Chernova T, Chaudhary D, Borde M, Chernova I, Iwai Y, Long AJ, Brown JA, Nunes R, Greenfield EA, Bourque K, Boussiotis VA, Carter LL, Carreno BM, Malenkovich N, Nishimura H, Okazaki T et al (2001) PD-L2 is a second ligand for PD-1 and inhibits T cell activation. Nat Immunol 2(3):261–268. https://doi.org/10.1038/85330

Leach DR, Krummel MF, Allison JP (1996) Enhancement of antitumor immunity by CTLA-4 blockade. Science 271(5256):1734–1736. https://doi.org/10.1126/science.271.5256.1734

Lechner A, Schlößer HA, Thelen M, Wennhold K, Rothschild SI, Gilles R, Quaas A, Siefer OG, Huebbers CU, Cukuroglu E, Göke J, Hillmer A, Gathof B, Meyer MF, Klussmann JP, Shimabukuro-Vornhagen A, Theurich S, Beutner D, von Bergwelt-Baildon M (2019) Tumor-associated B cells and humoral immune response in head and neck squamous cell carcinoma. Onco Targets Ther 8(3):1535293. https://doi.org/10.1080/2162402x.2018.1535293

Lee-Chang C, Bodogai M, Martin-Montalvo A, Wejksza K, Sanghvi M, Moaddel R, de Cabo R, Biragyn A (2013) Inhibition of breast cancer metastasis by resveratrol-mediated inactivation of tumor-evoked regulatory B cells. J Immunol 191(8):4141–4151. https://doi.org/10.4049/jimmunol.1300606

Li C, Zhao Y, Zhang W, Zhang W (2011) Increased prevalence of T(H)17 cells in the peripheral blood of patients with head and neck squamous cell carcinoma. Oral Surg Oral Med Oral Pathol Oral Radiol Endod 112(1):81–89. https://doi.org/10.1016/j.tripleo.2010.11.032

Liang SC, Latchman YE, Buhlmann JE, Tomczak MF, Horwitz BH, Freeman GJ, Sharpe AH (2003) Regulation of PD-1, PD-L1, and PD-L2 expression during normal and autoimmune responses. Eur J Immunol 33(10):2706–2716. https://doi.org/10.1002/eji.200324228

Lin D, Shen Y, Liang T (2023) Oncolytic virotherapy: basic principles, recent advances and future directions. Signal Transduct Target Ther 8(1):156. https://doi.org/10.1038/s41392-023-01407-6

Linsley PS, Greene JL, Brady W, Bajorath J, Ledbetter JA, Peach R (1994) Human B7-1 (CD80) and B7-2 (CD86) bind with similar avidities but distinct kinetics to CD28 and CTLA-4 receptors. Immunity 1(9):793–801. https://doi.org/10.1016/s1074-7613(94)80021-9

Liu JF, Deng WW, Chen L, Li YC, Wu L, Ma SR, Zhang WF, Bu LL, Sun ZJ (2018a) Inhibition of JAK2/STAT3 reduces tumor-induced angiogenesis and myeloid-derived suppressor cells in head and neck cancer. Mol Carcinog 57(3):429–439. https://doi.org/10.1002/mc.22767

Liu JF, Wu L, Yang LL, Deng WW, Mao L, Wu H, Zhang WF, Sun ZJ (2018b) Blockade of TIM3 relieves immunosuppression through reducing regulatory T cells in head and neck cancer. J Exp Clin Cancer Res 37(1):44. https://doi.org/10.1186/s13046-018-0713-7

Liu Z, McMichael EL, Shayan G, Li J, Chen K, Srivastava R, Kane LP, Lu B, Ferris RL (2018c) Novel effector phenotype of Tim-3(+) regulatory T cells leads to enhanced suppressive function in head and neck cancer patients. Clin Cancer Res 24(18):4529–4538. https://doi.org/10.1158/1078-0432.Ccr-17-1350

Liu T, Zhou L, Li D, Andl T, Zhang Y (2019) Cancer-associated fibroblasts build and secure the tumor microenvironment. Front Cell Dev Biol 7:60. https://doi.org/10.3389/fcell.2019.00060

Liu S, Wang R, Fang J (2024) Exploring the frontiers: tumor immune microenvironment and immunotherapy in head and neck squamous cell carcinoma. Discov Oncol 15(1):22. https://doi.org/10.1007/s12672-024-00870-z

López-Albaitero A, Lee SC, Morgan S, Grandis JR, Gooding WE, Ferrone S, Ferris RL (2009) Role of polymorphic fc gamma receptor IIIa and EGFR expression level in cetuximab mediated, NK cell dependent in vitro cytotoxicity of head and neck squamous cell carcinoma cells. Cancer Immunol Immunother 58(11): 1853–1864. https://doi.org/10.1007/s00262-009-0697-4

Lucido CT, Vermeer PD, Wieking BG, Vermeer DW, Lee JH (2014) CD137 enhancement of HPV positive head and neck squamous cell carcinoma tumor clearance. Vaccines (Basel) 2(4):841–853. https://doi.org/10.3390/vaccines2040841

Ma SR, Deng WW, Liu JF, Mao L, Yu GT, Bu LL, Kulkarni AB, Zhang WF, Sun ZJ (2017a) Blockade of adenosine A2A receptor enhances CD8(+) T cells response and decreases regulatory T cells in head and neck squamous cell carcinoma. Mol Cancer 16(1):99. https://doi.org/10.1186/s12943-017-0665-0

Ma X, Sheng S, Wu J, Jiang Y, Gao X, Cen X, Wu J, Wang S, Tang Y, Tang Y, Liang X (2017b) LncRNAs as an intermediate in HPV16 promoting myeloid-derived suppressor cell recruitment of head and neck squamous cell carcinoma. Oncotarget 8(26): 42061–42075. https://doi.org/10.18632/oncotarget.14939

Ma X, Bi E, Lu Y, Su P, Huang C, Liu L, Wang Q, Yang M, Kalady MF, Qian J, Zhang A, Gupte AA, Hamilton DJ, Zheng C, Yi Q (2019) Cholesterol induces CD8(+) T cell exhaustion in the tumor microenvironment. Cell Metab 30(1):143–156.e145. https://doi.org/10.1016/j.cmet.2019.04.002

Maggioni D, Pignataro L, Garavello W (2017) T-helper and T-regulatory cells modulation in head and neck squamous cell carcinoma. Onco Targets Ther 6(7): e1325066. https://doi.org/10.1080/2162402x.2017.1325066

Maj T, Wang W, Crespo J, Zhang H, Wang W, Wei S, Zhao L, Vatan L, Shao I, Szeliga W, Lyssiotis C, Liu JR, Kryczek I, Zou W (2017) Oxidative stress controls regulatory T cell apoptosis and suppressor activity and PD-L1-blockade resistance in tumor. Nat Immunol 18(12):1332–1341. https://doi.org/10.1038/ni.3868

Mandal R, Şenbabaoğlu Y, Desrichard A, Havel JJ, Dalin MG, Riaz N, Lee KW, Ganly I, Hakimi AA, Chan TA, Morris LG (2016) The head and neck cancer immune landscape and its immunotherapeutic implications. JCI Insight 1(17):e89829. https://doi.org/10.1172/jci.insight.89829

Mao L, Zhao ZL, Yu GT, Wu L, Deng WW, Li YC, Liu JF, Bu LL, Liu B, Kulkarni AB, Zhang WF, Zhang L, Sun ZJ (2018) γ-Secretase inhibitor reduces immunosuppressive cells and enhances tumour immunity in head and neck squamous cell carcinoma. Int J Cancer 142(5):999–1009. https://doi.org/10.1002/ijc.31115

Mao L, Xiao Y, Yang QC, Yang SC, Yang LL, Sun ZJ (2021) TIGIT/CD155 blockade enhances anti-PD-L1 therapy in head and neck squamous cell carcinoma by targeting myeloid-derived suppressor cells. Oral Oncol 121:105472. https://doi.org/10.1016/j.oraloncology.2021.105472

Marcus B, Arenberg D, Lee J, Kleer C, Chepeha DB, Schmalbach CE, Islam M, Paul S, Pan Q, Hanash S, Kuick R, Merajver SD, Teknos TN (2004) Prognostic factors in oral cavity and oropharyngeal squamous cell carcinoma. Cancer 101(12):2779–2787. https://doi.org/10.1002/cncr.20701

Markwell SM, Weed SA (2015) Tumor and stromal-based contributions to head and neck squamous cell carcinoma invasion. Cancers (Basel) 7(1):382–406. https://doi.org/10.3390/cancers7010382

Marur S, Forastiere AA (2008) Head and neck cancer: changing epidemiology, diagnosis, and treatment. Mayo Clin Proc 83(4):489–501. https://doi.org/10.4065/83.4.489

Mazilu L, Suceveanu AI, Stanculeanu DL, Gheorghe AD, Fricatel G, Negru SM (2021) Tumor microenvironment is not an 'innocent bystander' in the resistance to treatment of head and neck cancers (review). Exp Ther Med 22(4):1128. https://doi.org/10.3892/etm.2021.10562

McArdel SL, Dugast AS, Hoover ME, Bollampalli A, Hong E, Castano Z, Leonard SC, Pawar S, Mellen J, Muriuki K, McLaughlin DC, Bayhi N, Carpenter CL, Turka LA, Wickham TJ, Elloul S (2021) Anti-tumor effects of RTX-240: an engineered red blood cell expressing 4-1BB ligand and interleukin-15. Cancer Immunol Immunother 70(9):2701–2719. https://doi.org/10.1007/s00262-021-03001-7

Mei Z, Huang J, Qiao B, Lam AK (2020) Immune checkpoint pathways in immunotherapy for head and neck squamous cell carcinoma. Int J Oral Sci 12(1):16. https://doi.org/10.1038/s41368-020-0084-8

Mell LK, Brumund KT, Daniels GA, Advani SJ, Zakeri K, Wright ME, Onyeama SJ, Weisman RA, Sanghvi PR, Martin PJ, Szalay AA (2017) Phase I trial of intravenous oncolytic vaccinia virus (GL-ONC1) with cisplatin and radiotherapy in patients with Locoregionally advanced head and neck carcinoma. Clin Cancer Res 23(19):5696–5702. https://doi.org/10.1158/1078-0432.Ccr-16-3232

Menon S, Shin S, Dy G (2016) Advances in cancer immunotherapy in solid tumors. Cancers (Basel) 8(12). https://doi.org/10.3390/cancers8120106

Met Ö, Jensen KM, Chamberlain CA, Donia M, Svane IM (2019) Principles of adoptive T cell therapy in cancer. Semin Immunopathol 41(1):49–58. https://doi.org/10.1007/s00281-018-0703-z

Michaeli J, Shaul ME, Mishalian I, Hovav AH, Levy L, Zolotriov L, Granot Z, Fridlender ZG (2017) Tumor-associated neutrophils induce apoptosis of non-activated CD8 T-cells in a TNFα and

NO-dependent mechanism, promoting a tumor-supportive environment. Onco Targets Ther 6(11): e1356965. https://doi.org/10.1080/2162402x.2017.1356965

Mo F, Mamonkin M (2020) Generation of chimeric antigen receptor T cells using Gammaretroviral vectors. Methods Mol Biol 2086:119–130. https://doi.org/10.1007/978-1-0716-0146-4_8

Mohan SP, Bhaskaran MK, George AL, Thirutheri A, Somasundaran M, Pavithran A (2019) Immunotherapy in oral cancer. J Pharm Bioallied Sci 11(Suppl 2): S107–s111. https://doi.org/10.4103/jpbs.Jpbs_31_19

Mohsin SF (2024) Vaccine a promising immunotherapy option for head and neck cancer patients. Pak J Med Sci 40(7):1578–1583. https://doi.org/10.12669/pjms.40.7.8791

Morvan MG, Lanier LL (2016) NK cells and cancer: you can teach innate cells new tricks. Nat Rev Cancer 16(1):7–19. https://doi.org/10.1038/nrc.2015.5

Moy JD, Moskovitz JM, Ferris RL (2017) Biological mechanisms of immune escape and implications for immunotherapy in head and neck squamous cell carcinoma. Eur J Cancer 76:152–166. https://doi.org/10.1016/j.ejca.2016.12.035

Nelson BH (2010) CD20+ B cells: the other tumor-infiltrating lymphocytes. J Immunol 185(9): 4977–4982. https://doi.org/10.4049/jimmunol.1001323

Neyt K, Perros F, GeurtsvanKessel CH, Hammad H, Lambrecht BN (2012) Tertiary lymphoid organs in infection and autoimmunity. Trends Immunol 33(6): 297–305. https://doi.org/10.1016/j.it.2012.04.006

Nguyen N, Bellile E, Thomas D, McHugh J, Rozek L, Virani S, Peterson L, Carey TE, Walline H, Moyer J, Spector M, Perim D, Prince M, McLean S, Bradford CR, Taylor JM, Wolf GT (2016) Tumor infiltrating lymphocytes and survival in patients with head and neck squamous cell carcinoma. Head Neck 38(7): 1074–1084. https://doi.org/10.1002/hed.24406

Nisar S, Yousuf P, Masoodi T, Wani NA, Hashem S, Singh M, Sageena G, Mishra D, Kumar R, Haris M, Bhat AA, Macha MA (2021) Chemokine-cytokine networks in the head and neck tumor microenvironment. Int J Mol Sci 22(9). https://doi.org/10.3390/ijms22094584

Nishimura H, Nose M, Hiai H, Minato N, Honjo T (1999) Development of lupus-like autoimmune diseases by disruption of the PD-1 gene encoding an ITIM motif-carrying immunoreceptor. Immunity 11(2):141–151. https://doi.org/10.1016/s1074-7613(00)80089-8

Noy R, Pollard JW (2014) Tumor-associated macrophages: from mechanisms to therapy. Immunity 41(1):49–61. https://doi.org/10.1016/j.immuni.2014.06.010

Nywening TM, Wang-Gillam A, Sanford DE, Belt BA, Panni RZ, Cusworth BM, Toriola AT, Nieman RK, Worley LA, Yano M, Fowler KJ, Lockhart AC, Suresh R, Tan BR, Lim K-H, Fields RC, Strasberg SM, Hawkins WG, DeNardo D et al (2016) Targeting tumour-associated macrophages with CCR2 inhibition in combination with FOLFIRINOX in patients with borderline resectable and locally advanced pancreatic cancer: a single-Centre, open-label, dose-finding, non-randomised, phase 1b trial. Lancet Oncol 17(5): 651–662. https://doi.org/10.1016/S1470-2045(16)00078-4

Oderup C, Cederbom L, Makowska A, Cilio CM, Ivars F (2006) Cytotoxic T lymphocyte antigen-4-dependent down-modulation of costimulatory molecules on dendritic cells in CD4+ CD25+ regulatory T-cell-mediated suppression. Immunology 118(2):240–249. https://doi.org/10.1111/j.1365-2567.2006.02362.x

Ohue Y, Nishikawa H (2019) Regulatory T (Treg) cells in cancer: can Treg cells be a new therapeutic target? Cancer Sci 110(7):2080–2089. https://doi.org/10.1111/cas.14069

Okazaki T, Maeda A, Nishimura H, Kurosaki T, Honjo T (2001) PD-1 immunoreceptor inhibits B cell receptor-mediated signaling by recruiting src homology 2-domain-containing tyrosine phosphatase 2 to phosphotyrosine. Proc Natl Acad Sci USA 98(24): 13866–13871. https://doi.org/10.1073/pnas.231486598

Old MO, Wise-Draper T, Wright CL, Kaur B, Teknos T (2016) The current status of oncolytic viral therapy for head and neck cancer. World J Otorhinolaryngol Head Neck Surg 2(2):84–89. https://doi.org/10.1016/j.wjorl.2016.05.009

Orlandi E, Iacovelli NA, Tombolini V, Rancati T, Polimeni A, De Cecco L, Valdagni R, De Felice F (2019) Potential role of microbiome in oncogenesis, outcome prediction and therapeutic targeting for head and neck cancer. Oral Oncol 99:104453. https://doi.org/10.1016/j.oraloncology.2019.104453

Özdemir BC, Pentcheva-Hoang T, Carstens JL, Zheng X, Wu CC, Simpson TR, Laklai H, Sugimoto H, Kahlert C, Novitskiy SV, De Jesus-Acosta A, Sharma P, Heidari P, Mahmood U, Chin L, Moses HL, Weaver VM, Maitra A, Allison J et al (2014) Depletion of carcinoma-associated fibroblasts and fibrosis induces immunosuppression and accelerates pancreas cancer with reduced survival. Cancer Cell 25(6):719–734. https://doi.org/10.1016/j.ccr.2014.04.005

Pai SI, Zandberg DP, Strome SE (2016) The role of antagonists of the PD-1:PD-L1/PD-L2 axis in head and neck cancer treatment. Oral Oncol 61:152–158. https://doi.org/10.1016/j.oraloncology.2016.08.001

Pang X, Fan HY, Tang YL, Wang SS, Cao MX, Wang HF, Dai LL, Wang K, Yu XH, Wu JB, Tang YJ, Liang XH (2020) Myeloid derived suppressor cells contribute to the malignant progression of oral squamous cell carcinoma. PLoS One 15(2):e0229089. https://doi.org/10.1371/journal.pone.0229089

Parsa AT, Waldron JS, Panner A, Crane CA, Parney IF, Barry JJ, Cachola KE, Murray JC, Tihan T, Jensen MC, Mischel PS, Stokoe D, Pieper RO (2007) Loss of tumor suppressor PTEN function increases B7-H1

expression and immunoresistance in glioma. Nat Med 13(1):84–88. https://doi.org/10.1038/nm1517
Patterson C, Hazime KS, Zelenay S, Davis DM (2024) Prostaglandin E2 impacts multiple stages of the natural killer cell antitumor immune response. Eur J Immunol 54(2):e2350635. https://doi.org/10.1002/eji.202350635
Payne K, Brooks J, Spruce R, Batis N, Taylor G, Nankivell P, Mehanna H (2019) Circulating tumour cell biomarkers in head and neck cancer: current Progress and future prospects. Cancers (Basel) 11(8). https://doi.org/10.3390/cancers11081115
Peltanova B, Raudenska M, Masarik M (2019) Effect of tumor microenvironment on pathogenesis of the head and neck squamous cell carcinoma: a systematic review. Mol Cancer 18(1):63. https://doi.org/10.1186/s12943-019-0983-5
Petitprez F, de Reyniès A, Keung EZ, Chen TW, Sun CM, Calderaro J, Jeng YM, Hsiao LP, Lacroix L, Bougoüin A, Moreira M, Lacroix G, Natario I, Adam J, Lucchesi C, Laizet YH, Toulmonde M, Burgess MA et al (2020) B cells are associated with survival and immunotherapy response in sarcoma. Nature 577(7791):556–560. https://doi.org/10.1038/s41586-019-1906-8
Petrelli F, Malaspina L, Piane RM, Ambrosio MR (2023) The evolving concept of viruses and immune system interaction in head and neck neoplasms. Springer International Publishing, pp 1–28. https://doi.org/10.1007/16833_2023_157
Pitzalis C, Jones GW, Bombardieri M, Jones SA (2014) Ectopic lymphoid-like structures in infection, cancer and autoimmunity. Nat Rev Immunol 14(7):447–462. https://doi.org/10.1038/nri3700
Pretscher D, Distel LV, Grabenbauer GG, Wittlinger M, Buettner M, Niedobitek G (2009) Distribution of immune cells in head and neck cancer: CD8+ T-cells and CD20+ B-cells in metastatic lymph nodes are associated with favourable outcome in patients with oro- and hypopharyngeal carcinoma. BMC Cancer 9:292. https://doi.org/10.1186/1471-2407-9-292
Qin X, Yan M, Wang X, Xu Q, Wang X, Zhu X, Shi J, Li Z, Zhang J, Chen W (2018) Cancer-associated fibroblast-derived IL-6 promotes head and neck cancer progression via the Osteopontin-NF-kappa B signaling pathway. Theranostics 8(4):921–940. https://doi.org/10.7150/thno.22182
Qin Y, Zheng X, Gao W, Wang B, Wu Y (2021) Tumor microenvironment and immune-related therapies of head and neck squamous cell carcinoma. Mol Ther Oncolytics 20:342–351. https://doi.org/10.1016/j.omto.2021.01.011
Quezada SA, Peggs KS, Simpson TR, Allison JP (2011) Shifting the equilibrium in cancer immunoediting: from tumor tolerance to eradication. Immunol Rev 241(1):104–118. https://doi.org/10.1111/j.1600-065X.2011.01007.x
Qureshi OS, Zheng Y, Nakamura K, Attridge K, Manzotti C, Schmidt EM, Baker J, Jeffery LE, Kaur S, Briggs Z, Hou TZ, Futter CE, Anderson G, Walker LS, Sansom DM (2011) Trans-endocytosis of CD80 and CD86: a molecular basis for the cell-extrinsic function of CTLA-4. Science 332(6029):600–603. https://doi.org/10.1126/science.1202947
Reichert TE, Scheuer C, Day R, Wagner W, Whiteside TL (2001) The number of intratumoral dendritic cells and zeta-chain expression in T cells as prognostic and survival biomarkers in patients with oral carcinoma. Cancer 91(11):2136–2147
Ribas A, Kefford R, Marshall MA, Punt CJ, Haanen JB, Marmol M, Garbe C, Gogas H, Schachter J, Linette G, Lorigan P, Kendra KL, Maio M, Trefzer U, Smylie M, McArthur GA, Dreno B, Nathan PD et al (2013) Phase III randomized clinical trial comparing tremelimumab with standard-of-care chemotherapy in patients with advanced melanoma. J Clin Oncol 31(5):616–622. https://doi.org/10.1200/jco.2012.44.6112
Ribas A, Hamid O, Daud A, Hodi FS, Wolchok JD, Kefford R, Joshua AM, Patnaik A, Hwu WJ, Weber JS, Gangadhar TC, Hersey P, Dronca R, Joseph RW, Zarour H, Chmielowski B, Lawrence DP, Algazi A, Rizvi N et al (2016) Association of Pembrolizumab with tumor response and survival among patients with advanced Melanoma. JAMA 315(15):1600–1609. https://doi.org/10.1001/jama.2016.4059
Ribechini E, Greifenberg V, Sandwick S, Lutz MB (2010) Subsets, expansion and activation of myeloid-derived suppressor cells. Med Microbiol Immunol 199(3):273–281. https://doi.org/10.1007/s00430-010-0151-4
Ries CH, Cannarile MA, Hoves S, Benz J, Wartha K, Runza V, Rey-Giraud F, Pradel LP, Feuerhake F, Klaman I, Jones T, Jucknischke U, Scheiblich S, Kaluza K, Gorr IH, Walz A, Abiraj K, Cassier PA et al (2014) Targeting tumor-associated macrophages with anti-CSF-1R antibody reveals a strategy for cancer therapy. Cancer Cell 25(6):846–859. https://doi.org/10.1016/j.ccr.2014.05.016
Rivera A, Chen CC, Ron N, Dougherty JP, Ron Y (2001) Role of B cells as antigen-presenting cells in vivo revisited: antigen-specific B cells are essential for T cell expansion in lymph nodes and for systemic T cell responses to low antigen concentrations. Int Immunol 13(12):1583–1593. https://doi.org/10.1093/intimm/13.12.1583
Rodrigo JP, Sánchez-Canteli M, Otero-Rosales M, Martínez-Camblor P, Hermida-Prado F, García-Pedrero JM (2024) Tumor mutational burden predictability in head and neck squamous cell carcinoma patients treated with immunotherapy: systematic review and meta-analysis. J Transl Med 22(1):135. https://doi.org/10.1186/s12967-024-04937-x
Rudd CE, Taylor A, Schneider H (2009) CD28 and CTLA-4 coreceptor expression and signal transduction. Immunol Rev 229(1):12–26. https://doi.org/10.1111/j.1600-065X.2009.00770.x
Ruffin AT, Cillo AR, Tabib T, Liu A, Onkar S, Kunning SR, Lampenfeld C, Atiya HI, Abecassis I, Kürten CHL, Qi Z, Soose R, Duvvuri U, Kim S,

Oesterrich S, Lafyatis R, Coffman LG, Ferris RL, Vignali DAA, Bruno TC (2021) B cell signatures and tertiary lymphoid structures contribute to outcome in head and neck squamous cell carcinoma. Nat Commun 12(1):3349. https://doi.org/10.1038/s41467-021-23355-x

Ruffin AT, Li H, Vujanovic L, Zandberg DP, Ferris RL, Bruno TC (2023) Improving head and neck cancer therapies by immunomodulation of the tumour microenvironment. Nat Rev Cancer 23(3):173–188. https://doi.org/10.1038/s41568-022-00531-9

Ruggeri L, Mancusi A, Capanni M, Urbani E, Carotti A, Aloisi T, Stern M, Pende D, Perruccio K, Burchielli E, Topini F, Bianchi E, Aversa F, Martelli MF, Velardi A (2007) Donor natural killer cell allorecognition of missing self in haploidentical hematopoietic transplantation for acute myeloid leukemia: challenging its predictive value. Blood 110(1):433–440. https://doi.org/10.1182/blood-2006-07-038687

Runge A, Petersson M, Riechelmann H (2022) Oncolytic virotherapy in head and neck cancer. Laryngorhinootologie 101(10):787–796

Ryan N, Anderson K, Volpedo G, Hamza O, Varikuti S, Satoskar AR, Oghumu S (2020) STAT1 inhibits T-cell exhaustion and myeloid derived suppressor cell accumulation to promote antitumor immune responses in head and neck squamous cell carcinoma. Int J Cancer 146(6):1717–1729. https://doi.org/10.1002/ijc.32781

Sahu M, Suryawanshi H (2021) Immunotherapy: the future of cancer treatment. J Oral Maxillofac Pathol 25(2):371. https://doi.org/10.4103/0973-029x.325257

Saito T, Nishikawa H, Wada H, Nagano Y, Sugiyama D, Atarashi K, Maeda Y, Hamaguchi M, Ohkura N, Sato E, Nagase H, Nishimura J, Yamamoto H, Takiguchi S, Tanoue T, Suda W, Morita H, Hattori M et al (2016) Two FOXP3(+)CD4(+) T cell subpopulations distinctly control the prognosis of colorectal cancers. Nat Med 22(6):679–684. https://doi.org/10.1038/nm.4086

Samson A, West EJ, Carmichael J, Scott KJ, Turnbull S, Kuszlewicz B, Dave RV, Peckham-Cooper A, Tidswell E, Kingston J, Johnpulle M, da Silva B, Jennings VA, Bendjama K, Stojkowitz N, Lusky M, Prasad KR, Toogood GJ et al (2022) Neoadjuvant intravenous oncolytic vaccinia virus therapy promotes anticancer immunity in patients. Cancer Immunol Res 10(6):745–756. https://doi.org/10.1158/2326-6066.Cir-21-0171

Santos EM, Rodrigues de Matos F, Freitas de Morais E, Galvão HC, de Almeida Freitas R (2019) Evaluation of Cd8+ and natural killer cells defense in oral and oropharyngeal squamous cell carcinoma. J Craniomaxillofac Surg 47(4):676–681. https://doi.org/10.1016/j.jcms.2019.01.036

Santuray RT, Johnson DE, Grandis JR (2018) New therapies in head and neck cancer. Trends Cancer 4(5):385–396. https://doi.org/10.1016/j.trecan.2018.03.006

Sasidharan Nair V, Elkord E (2018) Immune checkpoint inhibitors in cancer therapy: a focus on T-regulatory cells. Immunol Cell Biol 96(1):21–33. https://doi.org/10.1111/imcb.1003

Sathawane D, Kharat RS, Halder S, Roy S, Swami R, Patel R, Saha B (2013) Monocyte CD40 expression in head and neck squamous cell carcinoma (HNSCC). Hum Immunol 74(1):1–5. https://doi.org/10.1016/j.humimm.2012.09.004

Sautès-Fridman C, Lawand M, Giraldo NA, Kaplon H, Germain C, Fridman WH, Dieu-Nosjean MC (2016) Tertiary lymphoid structures in cancers: prognostic value, regulation, and manipulation for therapeutic intervention. Front Immunol 7:407. https://doi.org/10.3389/fimmu.2016.00407

Sautès-Fridman C, Petitprez F, Calderaro J, Fridman WH (2019) Tertiary lymphoid structures in the era of cancer immunotherapy. Nat Rev Cancer 19(6):307–325. https://doi.org/10.1038/s41568-019-0144-6

Sawant DV, Yano H, Chikina M, Zhang Q, Liao M, Liu C, Callahan DJ, Sun Z, Sun T, Tabib T, Pennathur A, Corry DB, Luketich JD, Lafyatis R, Chen W, Poholek AC, Bruno TC, Workman CJ, Vignali DAA (2019) Adaptive plasticity of IL-10(+) and IL-35(+) T(reg) cells cooperatively promotes tumor T cell exhaustion. Nat Immunol 20(6):724–735. https://doi.org/10.1038/s41590-019-0346-9

Scharping NE, Menk AV, Moreci RS, Whetstone RD, Dadey RE, Watkins SC, Ferris RL, Delgoffe GM (2016) The tumor microenvironment represses T cell mitochondrial biogenesis to drive Intratumoral T cell metabolic insufficiency and dysfunction. Immunity 45(2):374–388. https://doi.org/10.1016/j.immuni.2016.07.009

Schuler PJ, Harasymczuk M, Visus C, Deleo A, Trivedi S, Lei Y, Argiris A, Gooding W, Butterfield LH, Whiteside TL, Ferris RL (2014) Phase I dendritic cell p53 peptide vaccine for head and neck cancer. Clin Cancer Res 20(9):2433–2444. https://doi.org/10.1158/1078-0432.Ccr-13-2617

Schwab M (2011) Encyclopedia of cancer. https://doi.org/10.1007/978-3-642-16483-5

Schwartz RH (1992) Costimulation of T lymphocytes: the role of CD28, CTLA-4, and B7/BB1 in interleukin-2 production and immunotherapy. Cell 71(7):1065–1068. https://doi.org/10.1016/s0092-8674(05)80055-8

Segal NH, Ou S-H, Balmanoukian AS, Massarelli E, Brahmer J, Weiss J, Schoffski P, Antonia SJ, Massard C, Zandberg DP, Maher C, Khleif S, Jin X, Rebelatto M, Steele K, Antal J, Gupta A, Spreafico A (2016) Updated safety and efficacy of durvalumab (MEDI4736), an anti-PD-L 1 antibody, in patients from a squamous cell carcinoma of the head and neck (SCCHN) expansion cohort. Jpn Soc Med Oncol 27. https://doi.org/10.1093/annonc/mdw376.01

Seiwert TY, Burtness B, Mehra R, Weiss J, Berger R, Eder JP, Heath K, McClanahan T, Lunceford J, Gause C, Cheng JD, Chow LQ (2016) Safety and clinical activity of pembrolizumab for treatment of recurrent or metastatic squamous cell carcinoma of the head and neck (KEYNOTE-012): an open-label, multicentre, phase 1b trial. Lancet Oncol 17(7):956–965. https://doi.org/10.1016/S1470-2045(16)30066-3

Seminerio I, Descamps G, Dupont S, de Marrez L, Laigle JA, Lechien JR, Kindt N, Journe F, Saussez S (2019) Infiltration of FoxP3+ regulatory T cells is a strong and independent prognostic factor in head and neck squamous cell carcinoma. Cancers (Basel) 11(2). https://doi.org/10.3390/cancers11020227

Sen M, Pollock NI, Black J, DeGrave KA, Wheeler S, Freilino ML, Joyce S, Lui VW, Zeng Y, Chiosea SI, Grandis JR (2015) JAK kinase inhibition abrogates STAT3 activation and head and neck squamous cell carcinoma tumor growth. Neoplasia 17(3):256–264. https://doi.org/10.1016/j.neo.2015.01.003

Senovilla L, Vacchelli E, Galon J, Adjemian S, Eggermont A, Fridman WH, Sautès-Fridman C, Ma Y, Tartour E, Zitvogel L, Kroemer G, Galluzzi L (2012) Trial watch: prognostic and predictive value of the immune infiltrate in cancer. Onco Targets Ther 1(8):1323–1343. https://doi.org/10.4161/onci.22009

Setoguchi R, Hori S, Takahashi T, Sakaguchi S (2005) Homeostatic maintenance of natural Foxp3(+) CD25 (+) CD4(+) regulatory T cells by interleukin (IL)-2 and induction of autoimmune disease by IL-2 neutralization. J Exp Med 201(5):723–735. https://doi.org/10.1084/jem.20041982

Shaul ME, Fridlender ZG (2019) Tumour-associated neutrophils in patients with cancer. Nat Rev Clin Oncol 16(10):601–620. https://doi.org/10.1038/s41571-019-0222-4

Shilpa P, Kaul R, Bhat S, Sultana N, Pandeshwar P (2014) Oncolytic viruses in head and neck cancer: a new ray of hope in the management protocol. Ann Med Health Sci Res 4(Suppl 3):S178–S184. https://doi.org/10.4103/2141-9248.141953

Shurin M, Umansk V, Malyguine A (2014) The tumor immunoenvironment. https://doi.org/10.1007/978-94-007-6217-6

Sim F, Leidner R, Bell RB (2019) Immunotherapy for head and neck cancer. Oral Maxillofac Surg Clin North Am 31(1):85–100. https://doi.org/10.1016/j.coms.2018.09.002

Singh K, Gautam PK (2023) Emulating the role of neutrophils in head and neck cancer microenvironment: prognostic role and therapeutic strategies. J Cancer Immunol 5:61

Sionov RV, Fridlender ZG, Granot Z (2015) The multifaceted roles neutrophils play in the tumor microenvironment. Cancer Microenviron 8(3):125–158. https://doi.org/10.1007/s12307-014-0147-5

Smussi D, Mattavelli D, Paderno A, Gurizzan C, Lorini L, Romani C, Bignotti E, Grammatica A, Ravanelli M, Bossi P (2023) Revisiting the concept of neoadjuvant and induction therapy in head and neck cancer with the advent of immunotherapy. Cancer Treat Rev 121: 102644. https://doi.org/10.1016/j.ctrv.2023.102644

Solito S, Bronte V, Mandruzzato S (2011) Antigen specificity of immune suppression by myeloid-derived suppressor cells. J Leukoc Biol 90(1):31–36. https://doi.org/10.1189/jlb.0111021

Srivastava RM, Lee SC, Andrade Filho PA, Lord CA, Jie HB, Davidson HC, López-Albaitero A, Gibson SP, Gooding WE, Ferrone S, Ferris RL (2013) Cetuximab-activated natural killer and dendritic cells collaborate to trigger tumor antigen-specific T-cell immunity in head and neck cancer patients. Clin Cancer Res 19(7):1858–1872. https://doi.org/10.1158/1078-0432.Ccr-12-2426

Srivastava RM, Trivedi S, Concha-Benavente F, Gibson SP, Reeder C, Ferrone S, Ferris RL (2017) CD137 stimulation enhances Cetuximab-induced natural killer: dendritic cell priming of antitumor T-cell immunity in patients with head and neck cancer. Clin Cancer Res 23(3):707–716. https://doi.org/10.1158/1078-0432.Ccr-16-0879

Strauss L, Bergmann C, Gooding W, Johnson JT, Whiteside TL (2007) The frequency and suppressor function of CD4+CD25highFoxp3+ T cells in the circulation of patients with squamous cell carcinoma of the head and neck. Clin Cancer Res 13(21): 6301–6311. https://doi.org/10.1158/1078-0432.Ccr-07-1403

Suárez-Sánchez FJ, Lequerica-Fernández P, Rodrigo JP, Hermida-Prado F, Suárez-Canto J, Rodríguez-Santamarta T, Domínguez-Iglesias F, García-Pedrero JM, de Vicente JC (2021) Tumor-infiltrating CD20(+) B lymphocytes: significance and prognostic implications in Oral cancer microenvironment. Cancers (Basel) 13(3). https://doi.org/10.3390/cancers13030395

Sun L, Clavijo PE, Robbins Y, Patel P, Friedman J, Greene S, Das R, Silvin C, Van Waes C, Horn LA, Schlom J, Palena C, Maeda D, Zebala J, Allen CT (2019) Inhibiting myeloid-derived suppressor cell trafficking enhances T cell immunotherapy. JCI Insight 4(7). https://doi.org/10.1172/jci.insight.126853

Sze DY, Reid TR, Rose SC (2013) Oncolytic virotherapy. J Vasc Interv Radiol 24(8):1115–1122. https://doi.org/10.1016/j.jvir.2013.05.040

Tahiliani V, Hutchinson TE, Abboud G, Croft M, Salek-Ardakani S (2017) OX40 cooperates with ICOS to amplify follicular Th cell development and germinal center reactions during infection. J Immunol 198(1): 218–228. https://doi.org/10.4049/jimmunol.1601356

Takahashi T, Tagami T, Yamazaki S, Uede T, Shimizu J, Sakaguchi N, Mak TW, Sakaguchi S (2000) Immunologic self-tolerance maintained by CD25(+)CD4(+) regulatory T cells constitutively expressing cytotoxic T lymphocyte-associated antigen 4. J Exp Med 192(2): 303–310. https://doi.org/10.1084/jem.192.2.303

Takahashi H, Sakakura K, Kawabata-Iwakawa R, Rokudai S, Toyoda M, Nishiyama M, Chikamatsu K (2015) Immunosuppressive activity of cancer-associated fibroblasts in head and neck squamous cell carcinoma. Cancer Immunol Immunother 64(11):

1407–1417. https://doi.org/10.1007/s00262-015-1742-0
Tan YS, Sansanaphongpricha K, Prince MEP, Sun D, Wolf GT, Lei YL (2018) Engineering vaccines to reprogram immunity against head and neck cancer. J Dent Res 97(6):627–634. https://doi.org/10.1177/0022034518764416
Tarella C, Passera R, Magni M, Benedetti F, Rossi A, Gueli A, Patti C, Parvis G, Ciceri F, Gallamini A, Cortelazzo S, Zoli V, Corradini P, Carobbio A, Mulé A, Bosa M, Barbui A, Di Nicola M et al (2011) Risk factors for the development of secondary malignancy after high-dose chemotherapy and autograft, with or without rituximab: a 20-year retrospective follow-up study in patients with lymphoma. J Clin Oncol 29(7):814–824. https://doi.org/10.1200/jco.2010.28.9777
Tatti O, Vehviläinen P, Lehti K, Keski-Oja J (2008) MT1-MMP releases latent TGF-beta1 from endothelial cell extracellular matrix via proteolytic processing of LTBP-1. Exp Cell Res 314(13):2501–2514. https://doi.org/10.1016/j.yexcr.2008.05.018
Terme M, Ullrich E, Aymeric L, Meinhardt K, Desbois M, Delahaye N, Viaud S, Ryffel B, Yagita H, Kaplanski G, Prévost-Blondel A, Kato M, Schultze JL, Tartour E, Kroemer G, Chaput N, Zitvogel L (2011) IL-18 induces PD-1-dependent immunosuppression in cancer. Cancer Res 71(16):5393–5399. https://doi.org/10.1158/0008-5472.Can-11-0993
Terrén I, Orrantia A, Vitallé J, Zenarruzabeitia O, Borrego F (2019) NK cell metabolism and tumor microenvironment. Front Immunol 10:2278. https://doi.org/10.3389/fimmu.2019.02278
Tesmer LA, Lundy SK, Sarkar S, Fox DA (2008) Th17 cells in human disease. Immunol Rev 223:87–113. https://doi.org/10.1111/j.1600-065X.2008.00628.x
Ti D, Niu Y, Wu Z, Fu X, Han W (2018) Genetic engineering of T cells with chimeric antigen receptors for hematological malignancy immunotherapy. Sci China Life Sci 61(11):1320–1332. https://doi.org/10.1007/s11427-018-9411-4
Tivol EA, Borriello F, Schweitzer AN, Lynch WP, Bluestone JA, Sharpe AH (1995) Loss of CTLA-4 leads to massive lymphoproliferation and fatal multiorgan tissue destruction, revealing a critical negative regulatory role of CTLA-4. Immunity 3(5):541–547. https://doi.org/10.1016/1074-7613(95)90125-6
Todo T, Ito H, Ino Y, Ohtsu H, Ota Y, Shibahara J, Tanaka M (2022) Intratumoral oncolytic herpes virus G47Δ for residual or recurrent glioblastoma: a phase 2 trial. Nat Med 28(8):1630–1639. https://doi.org/10.1038/s41591-022-01897-x
Togashi Y, Shitara K, Nishikawa H (2019) Regulatory T cells in cancer immunosuppression – implications for anticancer therapy. Nat Rev Clin Oncol 16(6):356–371. https://doi.org/10.1038/s41571-019-0175-7
Ugel S, De Sanctis F, Mandruzzato S, Bronte V (2015) Tumor-induced myeloid deviation: when myeloid-derived suppressor cells meet tumor-associated macrophages. J Clin Invest 125(9):3365–3376. https://doi.org/10.1172/jci80006
Valero C, Pardo L, López M, García J, Camacho M, Quer M, León X (2017) Pretreatment count of peripheral neutrophils, monocytes, and lymphocytes as independent prognostic factor in patients with head and neck cancer. Head Neck 39(2):219–226. https://doi.org/10.1002/hed.24561
van Elsas A, Hurwitz AA, Allison JP (1999) Combination immunotherapy of B16 melanoma using anti-cytotoxic T lymphocyte-associated antigen 4 (CTLA-4) and granulocyte/macrophage colony-stimulating factor (GM-CSF)-producing vaccines induces rejection of subcutaneous and metastatic tumors accompanied by autoimmune depigmentation. J Exp Med 190(3):355–366. https://doi.org/10.1084/jem.190.3.355
Veglia F, Sanseviero E, Gabrilovich DI (2021) Myeloid-derived suppressor cells in the era of increasing myeloid cell diversity. Nat Rev Immunol 21(8):485–498. https://doi.org/10.1038/s41577-020-00490-y
Verheyden S, Bernier M, Demanet C (2004) Identification of natural killer cell receptor phenotypes associated with leukemia. Leukemia 18(12):2002–2007. https://doi.org/10.1038/sj.leu.2403525
Vermorken JB, Herbst RS, Leon X, Amellal N, Baselga J (2008a) Overview of the efficacy of cetuximab in recurrent and/or metastatic squamous cell carcinoma of the head and neck in patients who previously failed platinum-based therapies. Cancer 112(12):2710–2719. https://doi.org/10.1002/cncr.23442
Vermorken JB, Mesia R, Rivera F, Remenar E, Kawecki A, Rottey S, Erfan J, Zabolotnyy D, Kienzer HR, Cupissol D, Peyrade F, Benasso M, Vynnychenko I, De Raucourt D, Bokemeyer C, Schueler A, Amellal N, Hitt R (2008b) Platinum-based chemotherapy plus cetuximab in head and neck cancer. N Engl J Med 359(11):1116–1127. https://doi.org/10.1056/NEJMoa0802656
Vernet R, Fernandez E, Migliorini D, Ancrenaz V, Charrier E, Belkouch M-C, Von Rohr O, Urwyler M, De Vito C, Renaux J, Villard J, Rubin O, Grogg J, Mach N (2024) A first-in-human phase I clinical study with MVX-ONCO-1, a personalized active immunotherapy, in patients with advanced solid tumors. Cancer Res Commun 4(8):2089–2100. https://doi.org/10.1158/2767-9764.CRC-24-0150
Vetto JT, Lum S, Morris A, Sicotte M, Davis J, Lemon M, Weinberg A (1997) Presence of the T-cell activation marker OX-40 on tumor infiltrating lymphocytes and draining lymph node cells from patients with melanoma and head and neck cancers. Am J Surg 174(3):258–265. https://doi.org/10.1016/s0002-9610(97)00139-6
von Leoprechting A, van der Bruggen P, Pahl HL, Aruffo A, Simon JC (1999) Stimulation of CD40 on immunogenic human malignant melanomas augments their cytotoxic T lymphocyte-mediated lysis and induces apoptosis. Cancer Res 59(6):1287–1294

Vonderheide RH, Flaherty KT, Khalil M, Stumacher MS, Bajor DL, Hutnick NA, Sullivan P, Mahany JJ, Gallagher M, Kramer A, Green SJ, O'Dwyer PJ, Running KL, Huhn RD, Antonia SJ (2007) Clinical activity and immune modulation in cancer patients treated with CP-870,893, a novel CD40 agonist monoclonal antibody. J Clin Oncol 25(7):876–883. https://doi.org/10.1200/jco.2006.08.3311

Vujanovic L, Szymkowski DE, Alber S, Watkins SC, Vujanovic NL, Butterfield LH (2010) Virally infected and matured human dendritic cells activate natural killer cells via cooperative activity of plasma membrane-bound TNF and IL-15. Blood 116(4): 575–583. https://doi.org/10.1182/blood-2009-08-240325

Vujanovic L, Ballard W, Thorne SH, Vujanovic NL, Butterfield LH (2012) Adenovirus-engineered human dendritic cells induce natural killer cell chemotaxis via CXCL8/IL-8 and CXCL10/IP-10. Onco Targets Ther 1(4):448–457. https://doi.org/10.4161/onci.19788

Vujanovic L, Chuckran C, Lin Y, Ding F, Sander CA, Santos PM, Lohr J, Mashadi-Hossein A, Warren S, White A, Huang A, Kirkwood JM, Butterfield LH (2019) CD56(dim) CD16(−) natural killer cell profiling in melanoma patients receiving a cancer vaccine and interferon-α. Front Immunol 10:14. https://doi.org/10.3389/fimmu.2019.00014

Wagner S, Wittekindt C, Reuschenbach M, Hennig B, Thevarajah M, Würdemann N, Prigge ES, von Knebel Doeberitz M, Dreyer T, Gattenlöhner S, Klussmann JP (2016) CD56-positive lymphocyte infiltration in relation to human papillomavirus association and prognostic significance in oropharyngeal squamous cell carcinoma. Int J Cancer 138(9):2263–2273. https://doi.org/10.1002/ijc.29962

Wallecha A, French C, Petit R, Singh R, Amin A, Rothman J (2012) Lm-LLO-based immunotherapies and HPV-associated disease. J Oncol 2012:542851. https://doi.org/10.1155/2012/542851

Wang HC, Chan LP, Cho SF (2019) Targeting the immune microenvironment in the treatment of head and neck squamous cell carcinoma. Front Oncol 9:1084. https://doi.org/10.3389/fonc.2019.01084

Wang Z, Sun P, Li Z, Xiao S (2023) Clinical advances and future directions of oncolytic Virotherapy for head and neck cancer. Cancers (Basel) 15(21). https://doi.org/10.3390/cancers15215291

Wculek SK, Cueto FJ, Mujal AM, Melero I, Krummel MF, Sancho D (2020) Dendritic cells in cancer immunology and immunotherapy. Nat Rev Immunol 20(1):7–24. https://doi.org/10.1038/s41577-019-0210-z

Weed DT, Vella JL, Reis IM, De la Fuente AC, Gomez C, Sargi Z, Nazarian R, Califano J, Borrello I, Serafini P (2015) Tadalafil reduces myeloid-derived suppressor cells and regulatory T cells and promotes tumor immunity in patients with head and neck squamous cell carcinoma. Clin Cancer Res 21(1):39–48. https://doi.org/10.1158/1078-0432.Ccr-14-1711

Weil S, Memmer S, Lechner A, Huppert V, Giannattasio A, Becker T, Müller-Runte A, Lampe K, Beutner D, Quaas A, Schubert R, Herrmann E, Steinle A, Koehl U, Walter L, von Bergwelt-Baildon MS, Koch J (2017) Natural killer group 2D ligand depletion reconstitutes natural killer cell Immunosurveillance of head and neck squamous cell carcinoma. Front Immunol 8:387. https://doi.org/10.3389/fimmu.2017.00387

Wejksza K, Lee-Chang C, Bodogai M, Bonzo J, Gonzalez FJ, Lehrmann E, Becker K, Biragyn A (2013) Cancer-produced metabolites of 5-lipoxygenase induce tumor-evoked regulatory B cells via peroxisome proliferator-activated receptor α. J Immunol 190(6):2575–2584. https://doi.org/10.4049/jimmunol.1201920

Wherry EJ (2011) T cell exhaustion. Nat Immunol 12(6): 492–499. https://doi.org/10.1038/ni.2035

Wherry EJ, Kurachi M (2015) Molecular and cellular insights into T cell exhaustion. Nat Rev Immunol 15(8):486–499. https://doi.org/10.1038/nri3862

Wing K, Onishi Y, Prieto-Martin P, Yamaguchi T, Miyara M, Fehervari Z, Nomura T, Sakaguchi S (2008) CTLA-4 control over Foxp3+ regulatory T cell function. Science 322(5899):271–275. https://doi.org/10.1126/science.1160062

Wondergem NE, Nauta IH, Muijlwijk T, Leemans CR, van de Ven R (2020) The immune microenvironment in head and neck squamous cell carcinoma: on subsets and subsites. Curr Oncol Rep 22(8):81. https://doi.org/10.1007/s11912-020-00938-3

Wong ALA, Hirpara JL, Pervaiz S, Eu JQ, Sethi G, Goh BC (2017) Do STAT3 inhibitors have potential in the future for cancer therapy? Expert Opin Investig Drugs 26(8):883–887. https://doi.org/10.1080/13543784.2017.1351941

Wu L, Mao L, Liu J-F, Chen L, Yu G-T, Yang L-L, Wu H, Bu L-L, Kulkarni AB, Zhang W-F, Sun Z-J (2019) Blockade of TIGIT/CD155 signaling reverses T-cell exhaustion and enhances antitumor capability in head and neck squamous cell carcinoma. Cancer Immunol Res 7(10):1700–1713. https://doi.org/10.1158/2326-6066.CIR-18-0725

Wu H, Xia L, Jia D, Zou H, Jin G, Qian W, Xu H, Li T (2020) PD-L1(+) regulatory B cells act as a T cell suppressor in a PD-L1-dependent manner in melanoma patients with bone metastasis. Mol Immunol 119:83–91. https://doi.org/10.1016/j.molimm.2020.01.008

Xiao X, Lao XM, Chen MM, Liu RX, Wei Y, Ouyang FZ, Chen DP, Zhao XY, Zhao Q, Li XF, Liu CL, Zheng L, Kuang DM (2016) PD-1hi identifies a novel regulatory B-cell population in human Hepatoma that promotes disease progression. Cancer Discov 6(5):546–559. https://doi.org/10.1158/2159-8290.Cd-15-1408

Xu J (2020) Regulation of cancer immune checkpoints. Adva Exp Med Biol 1248. https://doi.org/10.1007/978-981-15-3266-5

Yasumatsu R, Wakasaki T, Hashimoto K, Nakashima K, Manako T, Taura M, Matsuo M, Nakagawa T (2019) Monitoring the neutrophil-to-lymphocyte ratio may be useful for predicting the anticancer effect of nivolumab in recurrent or metastatic head and neck cancer. Head Neck 41(8):2610–2618. https://doi.org/10.1002/hed.25737

Ye J, Livergood RS, Peng G (2013) The role and regulation of human Th17 cells in tumor immunity. Am J Pathol 182(1):10–20. https://doi.org/10.1016/j.ajpath.2012.08.041

Younis RH, Han KL, Webb TJ (2016) Human head and neck squamous cell carcinoma-associated Semaphorin 4D induces expansion of myeloid-derived suppressor cells. J Immunol 196(3):1419–1429. https://doi.org/10.4049/jimmunol.1501293

Yu L, Yang F, Zhang F, Guo D, Li L, Wang X, Liang T, Wang J, Cai Z, Jin H (2018) CD69 enhances immunosuppressive function of regulatory T-cells and attenuates colitis by prompting IL-10 production. Cell Death Dis 9(9):905. https://doi.org/10.1038/s41419-018-0927-9

Zahavi D, AlDeghaither D, O'Connell A, Weiner LM (2018) Enhancing antibody-dependent cell-mediated cytotoxicity: a strategy for improving antibody-based immunotherapy. Antib Ther 1(1):7–12. https://doi.org/10.1093/abt/tby002

Zaretsky JM, Garcia-Diaz A, Shin DS, Escuin-Ordinas H, Hugo W, Hu-Lieskovan S, Torrejon DY, Abril-Rodriguez G, Sandoval S, Barthly L, Saco J, Homet Moreno B, Mezzadra R, Chmielowski B, Ruchalski K, Shintaku IP, Sanchez PJ, Puig-Saus C et al (2016) Mutations associated with acquired resistance to PD-1 blockade in melanoma. N Engl J Med 375(9):819–829. https://doi.org/10.1056/NEJMoa1604958

Zhang X, Zhang W, Yuan X, Fu M, Qian H, Xu W (2016) Neutrophils in cancer development and progression: roles, mechanisms, and implications (review). Int J Oncol 49(3):857–867. https://doi.org/10.3892/ijo.2016.3616

Zotos D, Coquet JM, Zhang Y, Light A, D'Costa K, Kallies A, Corcoran LM, Godfrey DI, Toellner KM, Smyth MJ, Nutt SL, Tarlinton DM (2010) IL-21 regulates germinal center B cell differentiation and proliferation through a B cell-intrinsic mechanism. J Exp Med 207(2):365–378. https://doi.org/10.1084/jem.20091777

Zou L, Barnett B, Safah H, LaRussa VF, Evdemon-Hogan M, Mottram P, Wei S, David O, Curiel TJ, Zou W (2004) Bone marrow is a reservoir for CD4+CD25+ regulatory T cells that traffic through CXCL12/CXCR4 signals. Cancer Res 64(22):8451–8455. https://doi.org/10.1158/0008-5472.CAN-04-1987

In Silico Analysis of Squamous Cell Carcinoma

19

Snežana M. Jovičić

Abstract

In silico methodology is discussed as a comprehensive approach to studying squamous cell carcinoma (SCC) and oral squamous cell carcinoma (OSCC). The advancement of computational methodologies has enabled the discovery of novel biomarkers by enhancing our understanding of tumor biology. In this chapter, we explore how bioinformatics tools improve our knowledge of the molecular mechanisms underlying these diseases and contribute to potential therapeutic advancements.

Cancer research has been revolutionized through the integration of computational techniques, facilitating the identification of biomarkers and improving our understanding of tumor progression. By combining genomic, transcriptomic, and proteomic datasets, bioinformaticians enable researchers to uncover key pathways involved in disease development, drug resistance, and interactions within the tumor microenvironment.

This chapter discusses various computational techniques, including gene expression analysis, network-based modeling, and machine learning algorithms, which aid in identifying prognostic and diagnostic markers. Additionally, we highlight the significance of publicly available datasets and high-throughput sequencing technologies in advancing in silico analysis.

The utilization of bioinformatics bridges the gap between basic research and clinical applications, paving the way for more personalized treatment strategies. By emphasizing the role of in silico methodologies, this chapter demonstrates how these approaches contribute to precision medicine. Advancing our understanding of the molecular mechanisms of SCC and OSCC facilitates the development of targeted therapeutics, ultimately leading to improved patient outcomes.

Keywords

In silico analysis · Oral squamous cell carcinoma (OSCC) · Biomarker discovery · Bioinformatics · Precision medicine

Abbreviations

AdaBoost	Adaptive Boosting
AKT	Protein Kinase B
c-Myc	Myelocytomatosis Oncogene
CIBERSORT	Cell-type Identification By Estimating Relative Subsets Of RNA Transcripts

S. M. Jovičić (✉)
Department of Genetics, Faculty of Biology, University of Belgrade, Belgrade, Serbia
e-mail: b3008_2014@stud.bio.bg.ac.rs

N. Rezaei, H. Ziaei (eds.), *Oral Immunology*, Advances in Experimental Medicine and Biology 1492,
https://doi.org/10.1007/978-3-032-03176-1_19

CK17	Cytokeratin 17
CMap	Connectivity Map
CNN	Convolutional Neural Networks
CTLA-4	Cytotoxic T-Lymphocyte-Associated Protein 4
CTNNB1	Catenin Beta 1
CDKN2A	Cyclin-Dependent Kinase Inhibitor 2A
DAVID	Database for Annotation, Visualization, and Integrated Discovery
DEG	Differentially Expressed Genes
DGE	Differential Gene Expression
EGFR	Epidermal Growth Factor Receptor
EMT	Epithelial-Mesenchymal Transition
ERK	Extracellular Signal-Regulated Kinase
FDA	Food and Drug Administration
FN1	Fibronectin 1
GEO	Gene Expression Omnibus
GRN	Gene Regulatory Network
GSEA	Gene Set Enrichment Analysis
HPA	Human Protein Atlas
HPV	Human Papillomavirus
IDH1/2	Isocitrate Dehydrogenase 1/2
IL-6	Interleukin-6
ImmPort	Immunology Database and Analysis Portal
IMvigor210	Immunogenomic Data Set for Cancer Immunotherapy Response
IPA	Ingenuity Pathway Analysis
JAK/STAT	Janus Kinase/Signal Transducer and Activator of Transcription signaling pathway
KEGG	Kyoto Encyclopedia of Genes and Genomes
LINCS	Library of Integrated Network-Based Cellular Signatures
MMP	Matrix Metalloproteinase
MMP9	Matrix Metallopeptidase 9
ML	Machine Learning;
MOFA	Multi-Omics Factor Analysis
mTOR	Mechanistic Target of Rapamycin
miR	MicroRNA
miRBase	MicroRNA Database
NGS	Next-Generation Sequencing
NF-κB	Nuclear Factor Kappa-light-chain-enhancer of activated B cells
NN	Neural Networks
NOTCH1	Neurogenic Locus Notch Homolog Protein 1
OSCC	Oral Squamous Cell Carcinoma
PD-L1	Programmed Death-Ligand 1
PI3K	Phosphoinositide 3-Kinase
PI3K/AKT/mTOR	Phosphoinositide 3-Kinase/Protein Kinase B/Mammalian Target of Rapamycin signaling pathway
PIK3CA	Phosphatidylinositol-4,5-Bisphosphate 3-Kinase Catalytic Subunit Alpha
PPI	Protein-Protein Interaction
PharmGKB	Pharmacogenomics Knowledgebase
PTEN	Phosphatase and Tensin Homolog
PTM	Post-Translational Modifications
RAS	Rat Sarcoma
RAF	Rapidly Accelerated Fibrosarcoma
RF	Random Forest
S100A7	S100 Calcium Binding Protein A7
SCC	Squamous Cell Carcinoma
SNAI2	Snail Family Transcriptional Repressor 2
SOX2	SRY-Box Transcription Factor 2
SNP	Single Nucleotide Polymorphism
STRING	Search Tool for the Retrieval of Interacting Genes/Proteins
STAT3	Signal Transducer and Activator of Transcription 3
SVM	Support Vector Machine
TCGA	The Cancer Genome Atlas
TCA cycle	Tricarboxylic Acid Cycle (Krebs Cycle)
TME	Tumor Microenvironment
TP53	Tumor Protein 53
VEGFA	Vascular Endothelial Growth Factor A
XGBoost	eXtreme Gradient Boosting

19.1 Introduction to In Silico Analysis in Cancer Research and Oral Immunity

In silico analysis represents a computational method to simulate biological processes and analyze data, enabling the handling of datasets generated through high-throughput technologies. This enables the identification of genetic mutations and gene expression patterns in disease (Barh et al. 2014). Since the completion of the Human Genome Project in 2003, cancer genomic research has been advanced significantly, enabling the detection of key genomic alterations, for example, *TP53* mutation, which are related to various malignancies and diseases (Olivier et al. 2010). Integration of bioinformatic data enables a holistic understanding of the squamous cell carcinoma (SCC) disease (Supplitt et al. 2021). Identification of pathways involved in tumorigenesis depends on RNA sequencing analysis (Supplitt et al. 2021). This chapter aims to provide an understanding of in silico analysis, which focuses on bioinformatic and machine learning (ML) techniques in oral immunity and identifies novel biomarkers.

19.1.1 Overview of SCC and Oral Squamous Cell Carcinoma (OSCC)

Squamous epithelial cells are in diverse organs such as the skin, mucosa, lungs, and esophagus (Sánchez-Danés and Blanpain 2018). Malignant tumor neoplasms derived from the squamous epithelial cells represent SCC (Sánchez-Danés and Blanpain 2018). SCC is the second most common type of skin cancer, with a 20% prevalence in regions with high sun exposure, following basal cell carcinoma (Fania et al. 2021). SCC can affect the respiratory and digestive systems (Jin 2013). OSCC accounts for more than 90% of oral cancers and is the sixth most common cancer worldwide due to the environmental influence and genetic factors (Badwelan et al. 2023). Usually, tobacco and alcohol consumption with oncogenic virus exposure like human papillomavirus (HPV) leads to the development of OSCC, representing the highest incidence rate of all cancers in Southeast Asia and Europe (Yang et al. 2023; Fatima et al. 2024). In OSCC, the oral cavity affects the tongue, mouth floor, and the inner cheek, often presenting as non-healing ulcers (Fitzpatrick et al. 2019). OSCC contributes to more than 300,000 new cases and 145,000 deaths annually (Blatt et al. 2022). Etiology and risk factors involve environmental and genetic factors like smoking and alcohol (Pérez Jardón et al. 2023). Smokers are 3 to 7 times more likely to develop OSCC compared to non-smokers (Pérez Jardón et al. 2023). Infection with HPV-16 leads to the development of OSCC, with unknown molecular mechanisms (Pérez Jardón et al. 2023). In OSCC, mutations of tumor suppressor genes like ***TP53*** affect apoptosis and cell proliferation along with signaling pathways (e.g. NOTCH1, EGFR, and PI3K/AKT) (Pérez Jardón et al. 2023).

19.1.2 Clinical and Pathological Features of OSCC

OSCC often presents as red or white lesions or as an abnormal growth or mass in the oral cavity (Pérez Jardón et al. 2023; Tan et al. 2023). Early stage OSCC may be asymptomatic or cause minimal discomfort, leading to delayed diagnosis (Pérez Jardón et al. 2023; Tan et al. 2023). As the tumor progresses, patients experience pain, difficulty in swallowing, and speech impairments (Pérez Jardón et al. 2023; Tan et al. 2023). Advanced OSCC can invade nearby structures, including the jawbone, and metastasize to cervical lymph nodes (Pérez Jardón et al. 2023; Tan et al. 2023).

Pathological characteristics of OSCC include changes in histopathology, molecular signaling, and tumor microenvironment (Barh et al. 2014; Olivier et al. 2010; Supplitt et al. 2021; Sánchez-Danés and Blanpain 2018; Fania et al. 2021; Jin 2013; Badwelan et al. 2023; Yang et al. 2023; Fatima et al. 2024; Blatt et al. 2022; Pérez Jardón et al. 2023; Tan et al. 2023). Histologically,

OSCC is characterized by abnormal squamous cells that invade underlying tissues, forming keratin pearls (Barh et al. 2014; Olivier et al. 2010; Supplitt et al. 2021; Sánchez-Danés and Blanpain 2018; Fania et al. 2021; Jin 2013; Badwelan et al. 2023; Yang et al. 2023; Fatima et al. 2024; Blatt et al. 2022; Pérez Jardón et al. 2023; Tan et al. 2023). Poorly differentiated OSCC is more aggressive with the lack of cellular organization and proper cellular differentiation, often leading to a higher chance of invasion and metastasis, even in less advanced forms (Pérez Jardón et al. 2023; Tan et al. 2023). Activation of oncogenic pathways and dysregulation of epidermal growth factor receptor (EGFR) leads to the development of OSCC due to genetic mutations in tumor suppressor genes and epigenetic modifications (Barh et al. 2014; Olivier et al. 2010; Supplitt et al. 2021; Sánchez-Danés and Blanpain 2018; Fania et al. 2021; Jin 2013; Badwelan et al. 2023; Yang et al. 2023; Fatima et al. 2024; Blatt et al. 2022; Pérez Jardón et al. 2023; Tan et al. 2023). Cell cycle arrest and apoptosis in response to DNA damage are affected by the TP53 gene mutation in more than 50% of the cases (Barh et al. 2014; Olivier et al. 2010; Supplitt et al. 2021; Sánchez-Danés and Blanpain 2018; Fania et al. 2021; Jin 2013; Badwelan et al. 2023; Yang et al. 2023; Fatima et al. 2024; Blatt et al. 2022; Pérez Jardón et al. 2023; Tan et al. 2023).

Moreover, overexpression of EGFR increases cell proliferation and survival. Among targeted therapies, the most efficient one in treating OSCC is an EGFR inhibitor, cetuximab (Halder et al. 2023). Changes in the PI3K/AKT/mTOR signaling pathway cause uncontrolled cell proliferation and apoptosis resistance (Halder et al. 2023).

Fibroblasts, immune cells, and extracellular matrix components of the tumor microenvironment (TME) promote angiogenesis, invasion, and immune evasion in OSCC (Halder et al. 2023). Diagnosis, prognosis, and treatment remain challenging for OSCC, since late visibility of the asymptomatic disease due to delayed diagnosis contributes to poor prognosis (Wang et al. 2023; González-Moles et al. 2022). So far, noninvasive diagnostic techniques include brush biopsy and salivary biomarkers, where treatment includes surgery, radiation therapy, and chemotherapy (Wang et al. 2023; González-Moles et al. 2022). Immunotherapy represents a promising approach to disease treatment (Wang et al. 2023; González-Moles et al. 2022). The survival rate of OSCC patients is low, due to late diagnosis and high recurrence rate (Wang et al. 2023; González-Moles et al. 2022).

Tumor stage at the time of diagnosis, HPV status, and genetic alteration affect OSCC prognosis (Powell et al. 2021; Sharkey Ochoa et al. 2021). Better response to radiation therapy and prognosis occur in HPV-positive OSCC cases, compared to HPV-negative tumors associated with tobacco and alcohol use, which are aggressive (Powell et al. 2021; Sharkey Ochoa et al. 2021).

Clinical and pathological features of OSCC represent the essential foundation for in silico investigations in OSCC, highlighting the significance of bioinformatics in researching the molecular mechanisms of the disease. This approach has facilitated revealing novel biomarkers such as ***ANXA1*** and ***SPINK5*** (Arora et al. 2023; Wu et al. 2021a, b; Li et al. 2023a, b). The following sections will enable a deeper exploration of the epidemiological, clinical, and pathological features of the OSCC, and the role of in silico analysis in revolutionizing cancer research along with a multi-omics approach, making a high impact on personalized medicine (Arora et al. 2023; Wu et al. 2021a, b; Li et al. 2023a, b).

19.2 The Role of In Silico Analysis in Cancer Research and Oral Immunity

19.2.1 Introduction to In Silico Analysis and Its Importance in Modern Cancer Research and Oral Immunity

In silico analysis includes computational models, simulations, and bioinformatics techniques to analyze and interpret data from biological sources (Jovičić 2024; Barh et al. 2014). This approach is significant in cancer research, providing scientists with efficient analysis of large datasets, predicting molecular interactions, and identifying potential biomarkers or therapeutic targets while reducing the reliance on extensive and expensive in vivo or in vitro experiments (Jovičić 2024; Barh et al. 2014). Complex biological systems in SCC and OSCC are investigated faster and more cost-effective with in silico methodology, revolutionizing our understanding of cancer biology (Tseng et al. 2022; Makarev et al. 2017). Next-generation sequencing (NGS) and high-throughput data are easy to access, manage, analyze, and interpret with in silico methodology (Tseng et al. 2022; Makarev et al. 2017). Researchers can easily identify mutations, copy number variations, and single-nucleotide polymorphisms (SNPs) associated with the disease (Ahuja et al. 2023; Satam et al. 2023). Potential biomarkers are revealed by predictive modeling, where computational drug design enables fast testing, saving time and resources in drug development (Ahuja et al. 2023; Satam et al. 2023). In silico analysis in oral immunity is significant in OSCC to understand the immune microenvironment, tumor progression, and therapeutic efficacy (Venkatesiah et al. 2022; Zhao et al. 2022). Researchers will better understand infiltrated immune cells to determine the impact on tumor progression and evading immune detection (Venkatesiah et al. 2022; Zhao et al. 2022).

19.2.2 Bioinformatics Tools and Databases in SCC Research

Powerful in silico methodology enables the interpretation of diverse datasets from multi-omic analysis, allowing researchers to uncover the molecular mechanisms underlying SCC and discover potential therapeutic targets (Jovičić 2024; Wan et al. 2022). In SCC and OSCC research, key bioinformatic tools and databases include the Cancer Genome Atlas (TCGA), Gene Expression Omnibus (GEO), Ensemble, and UCSC Genome Browser (Zou et al. 2019; Shen et al. 2019). One of the biggest databases for cancer genomics is the TCGA database to analyze genes and mutations in OSCC, like ***TP53, PIK3CA***, and ***NOTCH1*** genes (Saidak et al. 2021; Nakagaki et al. 2017; Pansare et al. 2019). Gene expression profile comparison with microarray and RNA-seq data is enabled with the GEO public repository (Jovičić 2024; Rao et al. 2018; Kim et al. 2024; Zhao 2015; Genome Browser 2024; Van Riet et al. 2022). Ensembl and UCSC Genome Browser platforms access genomic sequences and annotations, which are critical for exploring mutations, structural variations, and gene regulatory elements in SCC (Rao et al. 2018; Kim et al. 2024; Zhao 2015; Genome Browser 2024; Van Riet et al. 2022). Bioinformatic in silico analysis enables the identification of cancer driver mutations (Jovičić 2024; Rao et al. 2018; Kim et al. 2024; Zhao 2015; Genome Browser 2024; Van Riet et al. 2022). Search Tool for the Retrieval of Interacting Genes/Proteins (STRING), Kyoto Encyclopedia of Genes and Genomes (KEGG), and Database for Annotation, Visualization, and Integrated Discovery (DAVID) database present tools for proteomic and pathway analysis (Jovičić 2024; Rao et al. 2018; Kim et al. 2024; Zhao 2015; Genome Browser 2024; Van Riet et al. 2022). Protein-protein interaction (PPI) is enabled by the STRING database to map interactions between overexpressed and mutated proteins in OSCC (Rao et al. 2018; Kim et al. 2024; Zhao 2015; Genome Browser 2024; Van Riet et al. 2022). KEGG maps molecular pathways and helps elucidate the functional significance of genes and proteins, to investigate altered signaling pathways in OSCC, like PI3K/AKT/mTOR, EGFR, and MAPK pathways in OSCC that are usually impaired (Jovičić 2024; Rao et al. 2018; Kim et al. 2024; Zhao 2015; Genome Browser 2024; Van Riet et al. 2022). Overexpression of differentially expressed genes (DEG) occurs with DAVID database functional annotation and enrichment analysis (Rao et al. 2018; Kim et al. 2024; Zhao 2015; Genome Browser 2024; Van Riet et al. 2022).

Moreover, database platforms such as cBioPortal, PharmGKB, and DrugBank provide valuable sources for OSCC research. cBioPortal enables the interpretation of mutations and drug interactions by integrating data from tools like TCGA in OSCC (Belizário et al. 2016; Gao et al. 2013; Brlek et al. 2020). It also includes survival analysis tools, enabling users to correlate specific genetic alterations with clinical outcomes in OSCC (Jovičić 2024; Belizário et al. 2016; Gao et al. 2013; Brlek et al. 2020). PharmGKB platform links genetic variation to drug responses, providing an understanding of patient SCC therapy response and resistance to standard chemotherapy or radiation in OSCC (Belizário et al. 2016; Gao et al. 2013; Brlek et al. 2020). PharmGKB enables linking genetic variants to drug efficacy (Jovičić 2024; Belizário et al. 2016; Gao et al. 2013; Brlek et al. 2020). DrugBank provides detailed information on drug-target relationships and pharmacological data to detect potential drugs targeting specific proteins involved in SCC or OSCC progression, facilitating drug repurposing efforts or new targeted therapies (Belizário et al. 2016; Gao et al. 2013; Brlek et al. 2020).

Tools for Immunogenomic Analysis include silico methodologies like CIBERSORT, ImmPort, and IMvigor210 Dataset (Chen et al. 2018; Finotello and Trajanoski 2018; Wang et al. 2020; Racle et al. 2017; Hoffman-Censits et al. 2016). CIBERSORT enables the quantification of immune cell populations from transcriptomic data and can detect the immune landscape of tumors, providing insights into the immune cells that influence tumor progression and response to immunotherapy in OSCC (Finotello and Trajanoski 2018; Wang et al. 2020). The ImmPort database provides immunological data, where immune evasion mechanisms play a critical role in disease progression in OSCC (Racle et al. 2017). IMvigor210 Dataset includes genomic and immune data to evaluate how OSCC patients respond to immune checkpoint inhibitors (Hoffman-Censits et al. 2016).

ML tools for predictive modeling include XGBoost and AdaBoost in cancer research for optimizing therapeutic strategies and revealing novel biomarkers with the integration of multiomic (Al-Tashi et al. 2023; Ma et al. 2020; Wei et al. 2023; Lynch et al. 2023). In silico approaches accelerate the discovery rate of novel therapeutics and the development of personalized treatment strategies, improving patient outcomes in OSCC (Al-Tashi et al. 2023; Ma et al. 2020; Wei et al. 2023; Lynch et al. 2023).

19.3 Genomic and Transcriptomic Analysis

High-throughput technology like NGS enables fast and reliable detection of changes in the genome (Tseng et al. 2022). Mutation in the ***NOTCG1*** gene in OSCC highlights their role as a potential therapeutic target with diagnostic potential via TEL2/SERPINE1 axis (Salameti et al. 2019). Bioinformatic tools like DESeq2 or edgeR enable the identification of downregulated and upregulated genes involved in inflammation and immune response such as IL-6 and TNF, suggesting a correlation between inflammation and tumor environment (Jovičić 2024). Moreover, tools like gene ontology (GO) and KEGG provide insight into the cell cycle regulation and apoptosis pathway enrichment, providing insight into the molecular mechanisms of OSCC (Jovičić 2024). Integrating genomic and transcriptomic data enables the combination of data from whole exome sequencing with RNA profiling to identify changes in expression patterns. This integrative approach enables better clinical decisions by enabling targeted and personalized therapeutic strategies (Jovičić 2024; Salameti et al. 2019).

19.3.1 Utilizing NGS Data to Reveal Genetic Mutations in OSCC

NGS analysis provides fast analysis of the entire genome to detect somatic and copy number variations, fusions, and other significant genetic alterations (Jovičić 2024; Tseng et al. 2022). NGS in OSCC enables insight into the disease genomic landscape with TCGA, enabling the identification of mutations and genomic alterations that drive the disease (Tseng et al. 2022). In 60–80% of cases of OSCC mutation in TP53 occur, changing the function of the p53 protein, affecting cell proliferation and tumorigenesis (Chen et al. 2022). Cell growth, survival, and metabolism are affected by the mutation targeting PI3CA genes involved in the PI3K/AKT/mTOR signaling (Peng et al. 2022). Moreover, cell differentiation and growth are affected by mutations in ***NOTCH1,*** leading to poor disease prognosis. In OSCC, mutations or deletions in ***CDKN2A*** disrupt the regulation of the cell cycle, contributing to uncontrolled cell division and tumor growth, associated with HPV-negative OSCC, exhibiting more aggressive behavior (Salameti et al. 2019; Chen et al. 2022; Peng et al. 2022). In HPV-positive OSCC, ***TP53*** gene mutations are less common; instead, E6 and E7 viral oncogenes inactivate p53 and retinoblastoma (Rb) proteins, effectively inactivating anti-tumor effects of these genes. Despite this, HPV-positive OSCC is associated with better clinical outcomes (Samara et al. 2024).

19.3.2 DEG Analysis Identifies Significant Genes of Tumorigenesis and Immunity

Identification of upregulated and downregulated genes in cells derived from tumor and normal tissue is possible with powerful DEG analysis (Jovičić 2024). In OSCC, DGE analysis revealed numerous genes in tumor growth, invasion, metastasis, and immune evasion (Mohtasham et al. 2024). Genes such as ***EGFR*** and ***MMP*** are overexpressed, while tumor suppressor genes like ***CDKN2A***, ***E-cadherin***, as well as immune-related genes including ***PD-L1***, ***CTLA-4***, and ***IL-6,*** are downregulated, promoting cell proliferation, survival, and angiogenesis, facilitating tumor metastasis and invasion (Mohtasham et al. 2024). Cell cycle progression is unchecked with ***CDKN2A*** gene downregulation (Shi et al. 2022). Tumor invasiveness and metastatic potential are promoted by reduced ***E-cadherin*** expression and EMT (Shi et al. 2022). In OSCC, overexpression of ***PD-L1*** correlates with immune evasion, inhibiting the activity of T cells and enabling tumor cells to escape immune detection, along with ***CTLA-4*** upregulation. ***IL-6*** promotes tumor growth and immune evasion (Shi et al. 2022). KEGG and Reactome tools identify dysregulated pathways. Providing deeper insights into the underlying mechanism of OSCC (Badoiu et al. 2023; Wan et al. 2015; Ghafouri et al. 2022). Overexpression of ***PIK3CA*** and ***EGFR*** due to a mutation in the components of the PI3K/AKT/mTOR pathway promotes cell survival, growth, and invasion, while dysregulation of immune-related pathways such as NF-κB and JAK/STAT pathways promotes inflammation and immune evasion (Badoiu et al. 2023; Wan et al. 2015; Ghafouri-Fard et al. 2022). The molecular landscape of OSCC is more well understood by NGS and DGE methods, providing insight into mutational changes in OSCC, uncovering significantly altered genes, enhancing our knowledge, and improving patient outcomes (Mohtasham et al. 2024; Shi et al. 2022; Badoiu et al. 2023; Wan et al. 2015; Ghafouri-Fard et al. 2022).

19.4 Proteomics and Metabolomics

Comparing protein and metabolomic expression levels of expression in cancerous and healthy tissues provides the identification of novel biomarkers like SOX2 in head and neck SCC. Sox2 has been involved in cancer stem cell characteristics, particularly through the PI3K/Akt/mTOR pathway (Acharya et al. 2024). Identification of profile changes in proteomics and metabolomics have a significant clinical implication, providing valuable insight for diagnosis and therapeutic decision-making (Jovičić 2024).

19.4.1 Understanding Protein Expression Patterns and Metabolic Changes in OSCC

Cell function and metabolism insight are enabled by proteomic and metabolomic analysis and revealing distinct expression patterns in cancer cells (Gao and Chen 2023; Khowal et al. 2021). In the case of OSCC, proteomics identifies and quantifies expressed proteins in normal and cancerous tissues with mass spectrometry and two-dimensional gel electrophoresis (Gao and Chen 2023; Khowal et al. 2021). Based on the applicability of OSCC techniques, new findings include information about the overexpression of oncogenic proteins like EGFR and S100 proteins, dysregulation of tumor suppressor proteins, and post-translational modifications (Pillai et al. 2021). In OSCC, oncogenic proteins, such as EGFR and S100 protein upregulate, affecting tumor initiation, growth, and metastasis like EGFR and S100 protein (Pillai et al. 2021). Several tumor suppressor proteins show reduced expression in OSCC like p53 (Lindemann et al. 2018). Low levels of p53 cause unchecked cell growth and genomic instability, contributing to the aggressive nature of OSCC (Lindemann et al. 2018). Post-translational modifications (PTMs), such as phosphorylation and glycosylation, alter protein function, where phosphorylation of AKT and STAT3 enhances oncogenic activity in OSCC, providing insights into the protein regulation in cancer, and highlights new therapeutic targets (Li et al. 2023a, b). Metabolomic changes in OSCC include dysregulated pathways that support cell growth, proliferation, and survival. Key metabolomic changes include the Warburg effect, amino acid and lipid metabolism, TCA cycle, and oxidative phosphorylation (Gupta et al. 2024; Schiliro and Firestein 2021; Luo et al. 2024). OSCC produces energy primarily through aerobic glycolysis in the presence of oxygen (Gupta et al. 2024; Schiliro and Firestein 2021; Luo et al. 2024). Glycoliytic metabolites such as lactate and pyruvate provide energy for tumor growth (Gupta et al. 2024; Schiliro and Firestein 2021; Luo et al. 2024). Additionlly, OSCC exhibits the upregulation of glutamine metabolism, a nitrogen and carbon source for anabolic processes in OSCC (Gupta et al. 2024; Schiliro and Firestein 2021; Luo et al. 2024). Moreover, dysregulation of fatty acid metabolism and beta-oxidation has been implicated in OSCC (Gupta et al. 2024; Schiliro and Firestein 2021; Luo et al. 2024). When glycolysis is upregulated, metabolomics data support cellular bioenergetics and tumor adaptation (Gupta et al. 2024; Schiliro and Firestein 2021; Luo et al. 2024).

19.4.2 Integrating Multi-omics Data in Tumor Biology

Multi-omics data combine genomic, transcriptomic, proteomic, and metabolomic analyses, providing a comprehensive view of OSCC by capturing molecular changes across different levels of regulation, allowing correlation of genetic mutations with changes in protein expression and metabolic pathways, offering insights into the molecular mechanisms of OSCC (Jovičić 2024; Wan et al. 2022). Challenges in multi-omics integration exist due to the dataset complexity, variations in measurement techniques, and scale differences, making the efficient integration and analysis of datasets crucial in OSCC research (Jovičić 2024; Wan et al. 2022). Tools for bioinformatic multi-omic integration include Ingenuity Pathway Analysis (IPA) and Multi-Omics Factor Analysis (MOFA). Enrichment and network analysis enable IPA to detect significant genes and signaling pathways contributing to the disease (Wan et al. 2022; Athieniti and Spyrou 2023; Cervantes-Gracia et al. 2022). MOFA as a statistical framework enables the identification of shared and unique patterns across different omics layers, providing insights into how genomic alterations impact transcriptomics, proteomics, and metabolomics (Athieniti and Spyrou 2023; Cervantes-Gracia et al. 2022). Recent studies apply multi-omics approaches to OSCC to uncover novel insights into the molecular complexity of the disease (Athieniti and Spyrou 2023; Cervantes-Gracia et al. 2022). In OSCC, mutations in genes such as ***TP53***, ***PIK3CA***, and ***NOTCH1*** correlate with downstream changes in protein expression, confirming the functional impact of these mutations in cancer development (Zheng et al. 2024). Expression of AKT and downstream targets are closely associated with mutations in PI3K/AKT/mTOR pathways and integration of multi-omics data integration reveals that mutations in *IDH1/2* correlate to dysregulated TCA cycle metabolism contributing to cancer progression (Zheng et al. 2024; Glaviano et al. 2023).

The integration of multi-omics data is significant for personalized medicine in OSCC and by understanding the molecular alterations at multiple levels, clinicians can stratify patients based on their molecular profiles and identify personalized therapeutic targets such as PI3K and EGFR inhibitors (Jayawickrama et al. 2024; Diao et al. 2024). Profiling TME enables the development of checkpoint inhibitors (Jayawickrama et al. 2024; Diao et al. 2024). Proteomic and metabolomic research provides insight into the diversity of molecular changes within the cell, enabling the allocation of multi-omic research into in silico analysis. This approach improves the development of targeted therapeutic strategies in OSCC patients (Jayawickrama et al. 2024; Diao et al. 2024).

19.5 Identification of Novel Biomarkers and Therapeutic Targets

Bioinformatic platforms TCGA and GEO enable access to large-scale datasets for biomarker discovery in OSCC, such as CD44 which has been identified as a prognostic marker (Jovičić 2024; Matsuoka and Yashiro 2024; Arora et al. 2023; Hou et al. 2022). ML techniques predict the outcome based on gene expression profiles, demonstrating potential algorithms in personalized medicine Matsuoka et al. Matsuoka and Yashiro 2024; Arora et al. 2023; Hou et al. 2022). By analyzing pathways and networks, *EGFR* mutations contribute to SCC, indicating the development of potential inhibitors as therapeutics (Matsuoka and Yashiro 2024; Arora et al. 2023; Hou et al. 2022). Gene regulatory network development shows gene interaction with regulatory elements where Cytoscape visualizes these networks involving transcription factors governing gene expression in OSCC (Matsuoka and Yashiro 2024; Arora et al. 2023; Hou et al. 2022). Therapeutic implications derived from pathway analysis may lead to drug design and contribute to improved treatment outcomes (Matsuoka and Yashiro 2024; Arora et al. 2023; Hou et al. 2022). CRISPR gene editing and single-cell RNA sequencing (scRNA-seq) enable new avenues for future research, leading to an understanding of tumor heterogeneity and treatment (Jovičić 2024).

19.5.1 Bioinformatic Tools Identify Early Detection and Prognostic Biomarkers of OSCC

The physiological state of the organism and pathological conditions arise from identified biomarkers in specific biological samples (Jovičić 2024). Biomarkers apply in cancer research, for early detection, prognosis, and treatment monitoring (Passaro et al. 2024; Das et al. 2023; Liang et al. 2024). Personalized medicine is a significant weapon for early diagnosis and disease prognosis, impacting the prediction of disease outcomes (Jovičić 2024; Passaro et al. 2024; Das et al. 2023; Liang et al. 2024). Bioinformatics tools for biomarker discovery play a role in analyzing the vast amount of omics data generated by high-throughput technologies (Passaro et al. 2024; Das et al. 2023; Liang et al. 2024). Bioinformatics tools and databases like oncomine, TCGA, Human Protein Atlas (HPA), and miRBase apply for OSCC biomarker research (Jovičić 2024; Passaro et al. 2024; Das et al. 2023; Liang et al. 2024). A comprehensive cancer microarray database enables the comparison of gene expression profiles across normal and cancerous tissues to identify DEG as potential biomarkers (Jovičić 2024). TCGA provides insight into genomic and transcriptomic data, identifying genetic mutations, copy number variations, and gene expression changes linked to OSCC (Passaro et al. 2024; Das et al. 2023; Liang et al. 2024).

In order to make an appropriate comparison between healthy and OSCC tissue, HPA methodology integrates proteomic and transcriptomic data to validate protein expression (Digre and Lindskog 2020). miRBase is involved in post-transcriptional gene regulation and helps identify dysregulated miRNAs in OSCC that could serve as diagnostic or prognostic biomarkers (Park et al. 2024; Kampel et al. 2023; Zhou et al. 2024). Diverse biomarkers apply for early detection of

OSCC with computational methodologies like circulating tumor DNA (ctDNA), microRNAs (miRNAs), proteins, and glycoproteins, and prognostic biomarkers like ***TP53*** mutations and ***EGFR*** overexpression (Park et al. 2024; Kampel et al. 2023; Zhou et al. 2024). In OSCC ctDNA is released into the bloodstream by tumor cells, whereas in silico analysis of liquid biopsy detects genetic mutations or methylation patterns specific to OSCC, like *TP53* and *PIK3CA* detected in ctDNA for noninvasive cancer screening (Zavarykina et al. 2023; Yang et al. 2024). Moreover, miRNAs are significant biomarkers, and tools like miRNet and miRDB identify miRNA signatures associated with the OSCC, such as miR-21, miR-31, and miR-146a (Kaunein et al. 2021; Lin et al. 2022; Liu et al. 2023). As a marker of tumor progression and metastasis Cytokeratin 17 (CK17), matrix metalloproteinases (MMPs), and S100A7 proteins are upregulated in OSCC, as other markers of tumor progression and metastasis (Sanguansin et al. 2021; Hu et al. 2023). Poor prognosis and shorter survival time of the disease correlate to mutations in the TP53 gene (Ali et al. 2017; Tarle et al. 2023). EGFR overexpression is linked to poor survival in OSCC, correlating with increased tumor growth and aggressiveness, making it both a prognostic marker and a therapeutic target (Ali et al. 2017; Tarle et al. 2023).

19.5.2 Utilizing Computational Approaches to Discover New Therapeutic Targets and Drug Repurposing Opportunities

Computational tools enable the identification of novel treatments by utilizing novel molecules that lead to tumor growth (Jovičić 2024). In silico analysis offers advantages, like screening thousands of potential targets simultaneously and predicting the efficacy of drugs based on molecular interactions (Jovičić 2024). Bioinformatic platforms like STRING Database, CMap, Molecular Docking, and Virtual Screening are applied to identify novel therapeutic targets in OSCC (Singh et al. 2024). STRING Database focuses on protein-protein interactions (PPI) and enables visualization of the interaction networks of significant proteins in OSCC, serving as potential drug targets like PI3K/AKT/mTOR and Wnt/β-catenin signaling pathways (Singh et al. 2024; Wu et al. 2024). CMap platform matches gene expression signatures from OSCC cells with drug-induced gene expression profiles, allowing researchers to identify compounds that reverse the gene expression changes associated with OSCC, facilitating drug-repurposing efforts (Musa et al. 2018). CMap suggests the potential repositioning of drugs like metformin and celecoxib for OSCC treatment (Li et al. 2024). AutoDock and Schrodinger Glide tools enable the prediction of the binding affinity between a drug and target molecule by computational analysis (Niazi and Mariam 2023). In silico methodologies facilitate the identification of significant signaling pathways, leading to the development of drug-targeted therapies like everolimus and rapamycin (Vittorio et al. 2021; Palve et al. 2020). Additionally, immune checkpoint inhibitors, nivolumab, and pembrolizumab, apply in the case of upregulation of ***PD-L1*** and inhibitor ***LGK974*** as a Wnt signaling pathway inhibitor in OSCC treatment (Yin et al. 2023; Velcheti et al. 2020). For OSCC, several drugs could be repurposed like Metformin, Celecoxib, and Statins, meaning that they have another purpose, but can be used for OSCC treatment as well, based on the in silico findings (Li et al. 2024; Tripathi et al. 2024; Yoshida et al. 2019). Downregulation of the mTOR pathway in OSCC reduces cell proliferation with metformin, a drug primarily designed for diabetes (Hua et al. 2023). Inhibition of the COX-2 enzyme is a powerful therapeutic approach in OSCC treatment with celecoxib, a nonsteroidal anti-inflammatory drug (NSAID) (Gao et al. 2021; Shaker Nasry et al. 2018). Moreover, statins are commonly used to lower cholesterol. Statin inhibits HMG-CoA reductase, reducing OSCC cell proliferation and inducing apoptosis by modulating the mevalonate pathway (Tripathi et al. 2024; Ricco and Kron 2023). In silico methodology enables the identification of novel biomarkers and therapeutic targets in OSCC by leveraging large-scale data analysis and enabling the usage of existing drugs for treatment (Lin et al. 2022).

19.6 Pathway and Network Analysis

OSCC progression and development are influenced by the immune microenvironment, leading to poor clinical outcomes (Kondoh et al. 2019; Shi et al. 2024). Dysregulation of the immune checkpoint protein is associated with poor prognosis, highlighting the need for immunotherapeutic approaches for improving clinical outcomes (Kondoh et al. 2019; Shi et al. 2024). Several mechanisms enable immune detection, such as downregulation of antigen presentation, immune suppression, or cytokine production (Kondoh et al. 2019; Shi et al. 2024). The downregulation of ***MHC*** affects the ability of immune cells to recognize cancer cells, while immunosuppressive cells may inhibit anti-tumor immune response, contributing to OSCC immune evasion (Cornel et al. 2020; Tie and Tang 2022). Cytokines may also promote immune suppression and facilitate tumor growth (Cornel et al. 2020; Tie and Tang 2022). It is significant to understand immune interaction with the tumor microenvironment to indicate a favorable prognosis, underscoring the importance of immune interactions (Cornel et al. 2020; Tie and Tang 2022). To enhance anti-tumor immunity in OSCC, scientists apply immune checkpoint inhibitors, cancer vaccines, and combination therapies. Patients' survival rate improves with inhibitors targeting ***PD-1/PD-L1*** or ***CTLA-4*** molecules. Combining immunotherapies with chemotherapy or radiation enhances anti-tumor response and improves clinical outcomes. Future research should focus on biomarker identification to understand tumor heterogeneity and patient stratification for an effective treatment response (Wojtukiewicz et al. 2021; Ravi and Babu 2024).

19.6.1 Analysis of Signaling Pathways and Gene Regulatory Networks Implicated in OSCC

Pathway and network analysis are bioinformatics tools for understanding the molecular mechanisms of OSCC (Kumari et al. 2023; Pomella et al. 2023). Dysregulation of signaling pathways in cancer affects critical cellular functions, contributing to tumor initiation and progression, identification of key molecules within PI3K/AKT/mTOR, EGFR, Wnt/β-Catenin, NF-κB, and p53 signaling pathways, enhances our understanding of the OSCC disease, pathogenesis, progression, and therapy (Kumari et al. 2023; Pomella et al. 2023). PI3K/AKT/mTOR pathway aberrations in OSCC promote cell survival, growth, and resistance to apoptosis (Kumari et al. 2023; Pomella et al. 2023). Tumor aggressiveness and resistance depend on biomarkers like PIK3CA and PTEN; therefore, mutations in these genes serve as therapeutic targets, with inhibitors like cetuximab, an EGFR inhibitor, which targets RAS/RAF/MEK/ERK and PI3K/AKT signaling pathways (Xu et al. 2017; Davis et al. 2014). Dysregulation of the Wnt/β-Catenin pathway is associated with tumor growth and metastasis, affecting its downstream CTNNB1 (β-catenin) (Paluszczak 2020). Upstream mutations or environmental factors like smoking, promote tumor growth and resistance to apoptosis in OSCC, as a chronic activation of NF-κB where chemoradiotherapy reduces the disease size (Qian et al. 2022). Network analysis shows how mutated p53 interacts with proteins to evade cell death. Gene regulatory networks (GRNs) involve the complex interplay between transcription factors, regulatory RNAs, and genes that control gene expression (Pilley et al. 2021). Overexpression of transcriptional factors and noncoding RNA molecules, such as ***SOX2, STAT3***, and ***c-Myc***, causes molecular aberrations that support tumor proliferation and immune evasion in OSCC (Vastrad et al. 2023). Moreover, the upregulation of miR-21 causes tumor growth by inhibiting the expression of tumor suppressor genes (Rhim et al. 2022).

19.6.2 Identification of Drivers and Hubs in Oncogenic Networks

Network analysis maps the interactions between genes, proteins, and other molecules, which help identify key oncogenic drivers of OSCC. These mutations and altered gene expressions cause tumorigenesis, providing insights into potential therapeutic targets (Creixell et al. 2015). Common driver mutations include ***TP53, PIK3CA, CDKN2A***, and ***NOTCH1*** (Pandey et al. 2021). Gene prediction algorithms like Oncodrive distinguish driver mutations from passenger mutations, which have no direct impact on cancer progression (Nourbakhsh et al. 2024). In oncogenic networks, hub genes or proteins maintain the stability of the network (Nourbakhsh et al. 2024). Hub analysis identifies critical regulators in OSCC networks whose targeting disrupts multiple cancer-promoting pathways, providing valuable insights into potential therapeutic targets with widespread effects on tumor growth (Nourbakhsh et al. 2024). Pathway enrichment and functional analysis reveal biological processes or pathways most affected by the dysregulation of genes in OSCC, using GSEA, KEGG, and Reactome, identifying immune responses and providing insights into the broader biological context of these changes (Das et al. 2024). Once pathways and network drivers are identified, computational methods prioritize therapeutic targets that play a role in cancer networks, and these can be targeted by small molecules or biologics like EGFR and PI3K and network-based drug repurposing (Alharbi and Vakanski 2023). We can conclude that pathway and network analysis is significant in understanding OSCC molecular mechanisms. Researchers can reveal potential therapeutic targets and improve personalized medical treatment by integrating computational analysis.

19.7 Machine Learning and Predictive Modeling

ML analyzes large datasets to make predictions by identifying pattern differences in the analyzed data, enabling prediction of tumor type based on gene expression profiles (Jovičić 2024; Lavanya 2024; Gupta et al. 2022; Albaradei et al. 2020; Vollmer et al. 2024; Varalakshmi et al. 2024). Data cleaning and preprocessing were performed to identify relevant changes and optimize the model-making process, eventually enhancing the model's accuracy (Jovičić 2024). Model development and validation are performed by splitting the dataset into training and testing sets to evaluate model performance, thereby demonstrating clinical utility (Jovičić 2024). Predictive models for treatment arise from ML, where it is possible to predict the effectiveness of immunotherapy based on tumor mutation burden (Jovičić 2024; Lavanya 2024; Gupta et al. 2022; Albaradei et al. 2020; Vollmer et al. 2024; Varalakshmi et al. 2024). Deep learning and transfer learning methods represent future trends in ML methodology, improving prediction and facilitating the analysis (Jovičić 2024; Lavanya 2024; Gupta et al. 2022; Albaradei et al. 2020; Vollmer et al. 2024; Varalakshmi et al. 2024). ML has applications to predict outcomes, identify tumor subtypes, and drug response prediction (Jovičić 2024). Clinical data and molecular profiles predict survival rates in OSCC patients, stratifying and tailoring therapeutic approaches (Lavanya 2024; Gupta et al. 2022; Albaradei et al. 2020; Vollmer et al. 2024; Varalakshmi et al. 2024). Clustering methods enable the identification of molecular subtypes in OSCC, where unique genomic organization guides personalized treatment (Lavanya 2024; Gupta et al. 2022; Albaradei et al. 2020; Vollmer et al. 2024; Varalakshmi et al. 2024). Using the OSCC cell line dataset, supporting vectors predict the sensitivity of chemotherapy and enable the selection of optimal therapy (Lavanya 2024; Gupta et al. 2022; Albaradei et al. 2020; Vollmer et al. 2024; Varalakshmi et al. 2024). Challenges in ML applications include limitations of data quality and quantity, issues with data interpretability, and generalizability, which impact the reliability of predictions in clinical settings (Jovičić 2024). Model accuracy is usually hindered in OSCC research due to the available datasets (Lavanya 2024; Gupta et al. 2022; Albaradei et al. 2020; Vollmer et al. 2024; Varalakshmi et al. 2024). The clinical decision-making process can be complicated to predict gene expression (Lavanya 2024; Gupta et al. 2022; Albaradei et al. 2020; Vollmer et al. 2024; Varalakshmi et al. 2024).

Some datasets lack generality, leading to poor model performance (Jovičić 2024; Lavanya 2024; Gupta et al. 2022; Albaradei et al. 2020; Vollmer et al. 2024; Varalakshmi et al. 2024). Future research in the direction of ML includes integration of multi-omics data, development of explainable AI, and real-world utilization of data, enhancing predictability and accuracy, facilitating understanding of model prediction, and enhancing validation and applicability of the results in clinical settings.

19.7.1 Application of Machine Learning Algorithms to Predict Disease Progression and Patient Outcomes

ML enables the analysis of large, complex datasets to predict disease progression, patient outcomes, and treatment responses (Jovičić 2024). By learning patterns from multi-omics data, histopathological images, and clinical data, ML algorithms improve our understanding of disease mechanisms and identify potential biomarkers for early diagnosis and personalized therapy in OSCC (Dixit et al. 2023; Zhang et al. 2023a, b). Several machine learning algorithms like Random Forest (RF), Support Vector Machines (SVM), Neural Networks (NN), XGBoost, and AdaBoost in OSCC research predict outcomes like tumor progression, recurrence, and patient survival (Alabi et al. 2021; Wu et al. 2021a, b; Varlamova et al. 2023; Noh et al. 2022). RF constructs multiple decision trees to make accurate predictions to identify gene signatures from transcriptomic data that predict OSCC aggressiveness and metastatic potential, assessing the importance of individual genes or features in these predictions (Meirelles et al. 2022). Molecular profile, gene expression data, and imaging enable prediction of the likelihood of tumor recurrence and response to therapy with SVM as a learning algorithm (Jovičić 2024). NN-like deep learning models can process large, high-dimensional datasets such as histopathology images (Camalan et al. 2021). Convolutional neural networks (CNNs) analyze tissue slides to automatically detect OSCC lesions, grade tumor severity, and predict patient prognosis, integrating multi-omics data for more comprehensive predictions (Alabi et al. 2023). XGBoost is a gradient-boosted decision tree algorithm highly effective for tabular data analysis, clinical features, or genetic markers, known for its accuracy and speed, used to build predictive models for OSCC outcomes, including survival rates, tumor recurrence, and treatment responses (Yu and Wang 2022). Combining gene expression data, histopathology, and clinical and radiological data enables ML disease model prediction and patient outcome. Utilizing the transcriptomic dataset, the ML model identifies gene expression patterns of OSCC (Lee 2023; Butt et al. 2024). Histopathology images are analyzed using CNN to grade the severity of the tumor, predict recurrence, and facilitate faster diagnosis (Zhang et al. 2023a, b). ML algorithms apply clinical features like demographics, tumor stage, and radiological imaging data to predict survival outcomes, response to radiotherapy, or the likelihood of tumor recurrence. Patient outcome prediction is enabled by the integration of data with these tools to determine the disease course along with DEG and clinical features (Menditti et al. 2023).

19.7.2 Development of Predictive Models for Personalized Medicine in OSCC

Personalized medicine tailors treatments based on the molecular and clinical profile of individual patients (Jovičić 2024). Personalized approaches are significant in heterogeneous diseases like OSCC the disease, where each patient exhibits different genetic mutations, immune responses, and drug sensitivities (Melo-Alvim et al. 2022). ML develops predictive models leading to effective personalized therapy combining multi-omics data, treatment histories, and patient-specific factors (Jovičić 2024). ML models predict OSCC patient response to different treatments, such as chemotherapy, radiotherapy, or immunotherapy (Wu et al. 2020). These models are trained on data from past patients, including molecular profiles and treatment outcomes, allowing the algorithm to predict therapies that are the most effective for individual patients (Jovičić 2024). ML analyzes gene expression or mutation data to identify molecular signatures that predict sensitivity or resistance to chemotherapy drugs commonly used in OSCC, such as cisplatin (Zhang and Wang 2022). Analysis of immune-related genes is efficient due to ML models and knowledge of the tumor microenvironment. Radiotherapy is a key treatment for OSCC, but patient sensitivity varies. ML models link genomic features with radiological imaging, to predict patients' sensitivity to radiation to optimize the dosage of radiation (Dixit et al. 2023; Lin et al. 2022). ML algorithms identify existing repurposing drugs for OSCC treatment by comparing gene expression profiles of OSCC cells with drug-induced gene expression changes from publicly available databases, such as LINCS and CMap. By analyzing these patterns, ML models predict drugs that can reverse oncogenic expression patterns in OSCC disease (Lin et al. 2022; Liu et al. 2023; Liu & Mei 2023). Drugs like metformin or statins are potential candidates for OSCC treatment using ML-driven approaches. ML models integrate multi-omics data to provide a comprehensive overview of a patient's tumor biology, enabling the development of highly accurate predictive models for personalized therapy (Ideta et al. 2021). By researching genomic mutations and copy number variations, ML predicts effective targeted therapies (Jovičić 2024). Transcriptomic profiles reveal dysregulated pathways in OSCC, where ML predicts the response to pathway-specific inhibitors (Pham et al. 2024). Protein expression patterns and metabolic changes in OSCC cells inform predictive models for treatment response, where overexpression or altered metabolic pathways help predict the efficacy of targeted therapies (Pham et al. 2024).

ML and predictive modeling enable efficient diagnosis, monitoring, and treatment of OSCC by applying advanced algorithms to multi-omics and clinical data, tailoring personalized medicine approaches to each patient, improving treatment outcomes, and minimizing side effects (Almansoori et al. 2022).

19.8 Case Studies and Applications

In silico research raises several ethical considerations, including data privacy and informed consent. Adhering to regulations like GDPR when handling personal health information is essential, and consent from patients involved in the study should be obtained, making patients aware of the data usage (Jovičić 2024). In silico analysis includes limitations regarding data quality and generalizability, depending on the completeness of the data included in the research, making them not applicable to a broader population (Jovičić 2024). By utilizing bioinformatic tools, researchers should prioritize transparency and peer review, reporting methodology and findings, which enables reproducibility of the analysis and trust in the outcomes, and allows rigorous evaluation and validation (Jovičić 2024). Future research should engage other stakeholders and develop ethical guidance when involving patients and community representatives in the discussion of research practices. Ethical consideration will help maintain integrity and trust in the scientific field.

19.8.1 Presentation of Case Studies Demonstrating Successful Applications of In Silico Analysis in OSCC Research

Case Study 1 Identification of prognostic biomarkers using transcriptomic data analysis of DEG from large transcriptomic data is possible with in silico methods and publicly available datasets from TCGA and GEO (Kaddoura et al. 2023; Quintero-Fabián et al. 2019). Upregulation of genes, such as ***MMP9, IL-6,*** and ***VEGFA,*** is observed in tumor samples of OSCC, revealing their role in cell processes correlated to inflammation, angiogenesis, and extracellular matrix remodeling (de Paula Souza et al. 2023). Through ML models and survival analysis, this gene panel was validated as a potential prognostic biomarker for OSCC (Ribeiro Cavalcante et al. 2022). Increased gene expression leads to worse outcomes, demonstrating the strength of in silico methodology in personalized medicine (Jovičić 2024).

Case Study 2 Computational screening of drug repurposing of existing drugs as treatments is another notable application of in silico analysis (Lin et al. 2022; Lin & Chen 2023). Computational screening of drug databases identifies drugs that target pathways or genes frequently altered in OSCC (Shamsi et al. 2024). By leveraging structure-based drug design and molecular docking techniques, the researchers identified several FDA-approved drugs, such as metformin and dasatinib, that exhibit a high affinity for OSCC-related molecular targets, such as ***EGFR*** and PI3K (Li et al. 2024).

In vitro validation of these drug candidates shows promising results, with metformin reducing OSCC cell proliferation, illustrating the power of in silico analysis in streamlining the drug discovery process, by repurposing existing drugs, saving time, and reducing costs compared to traditional drug development methods (Ge et al. 2019).

Case Study 3: Network-Based Analysis to Identify Key Drivers of OSCC Metastasis In a third case study, researchers used network-based analysis to study the mechanisms underlying OSCC metastasis by PPI networks and gene co-expression networks, where they identified several key hub proteins involved in the metastatic spread of OSCC (Shaikh et al. 2023; Park et al. 2024; Li et al. 2017; Ghantous et al. 2024). Proteins such as FN1 and SNAI2, from EMT, were highlighted as central nodes in metastatic networks (Huynh et al. 2021; Shetty et al. 2020). Network simulation and perturbation analysis show that these proteins play a pivotal role in OSCC metastasis, suggesting that targeting EMT-related pathways could suppress metastatic spread, providing novel insights into the molecular drivers of OSCC metastasis and efficient therapy (Huynh et al. 2021; Shetty et al. 2020; Mohr et al. 2024).

19.8.2 Discussion of Challenges and Future Directions in the Field

Challenges of in silico analysis for OSCC research include data quality and standardization, integration of multi-omics data, validation of computational predictions, and limited availability of OSCC-specific databases (Makarev et al. 2017; Gupta et al. 2024; Nwosu and Piccolo 2021). One of the primary challenges in the application of in silico analysis to OSCC research is the variability in data quality (Makarev et al. 2017). Publicly available datasets from GEO or TCGA arise from diverse patient populations and experimental conditions, leading to inconsistencies in data interpretation (Liu et al. 2019; Gunardi et al. 2023). Differences in sample handling, sequencing technologies, and bioinformatics pipelines introduce analysis noise, making it difficult to derive robust and consistent conclusions (Jovičić 2024). Future efforts focus on improving standardization in data collection and processing to ensure the reliability of results (Jovičić 2024). Integration of multi-omics data represents a secondary challenge in silico bioinformatic analysis (Jovičić 2024). Although researchers collect various omics data, dataset integration remains a significant challenge, since combining them to generate a holistic view of tumor biology is complex (Jovičić 2024). Advanced algorithms and computational frameworks for multi-omics integration will be crucial for advancing the field (Jovičić 2024). Validation of computational predictions generates valuable hypotheses; however, experimental validation is necessary to confirm the biological relevance (Jovičić 2024). Bridging the gap between in silico predictions and in vitro or in vivo validation is resource-intensive, requiring substantial time and funding (Jovičić 2024). The future guiding stone is more reliable in silico validation techniques, such as virtual cell models and

AI-based simulations, which could accelerate the validation process. Limited availability of OSCC-specific databases exists, limiting the ability to apply in silico methods to OSCC research on a large scale. Future research should create dedicated OSCC databases with comprehensive genomic, transcriptomic, proteomic, and clinical data. Collaborative initiatives between research institutions, hospitals, and industry help expand data availability. Future directions of in silico analysis for OSCC include AI and deep learning integration, personalized medicine applications, advances in network biology and drug repurposing using big data (Wang et al. 2023; Jovičić 2024; Makarev et al. 2017; Suri et al. 2024; Jensen et al. 2019; Bebek et al. 2012).

As AI and deep learning continue to evolve, their integration into in silico cancer research becomes significantly promising (Menditti et al. 2023). AI algorithms can process vast amounts of data and identify complex patterns that traditional methods overlook (Menditti et al. 2023). AI approaches improve the accuracy of biomarker identification, predict patient response to therapies, and assist in drug discovery by simulating drug-target interactions with greater precision in OSCC (Menditti et al. 2023). By integrating data from an individual's genomic and transcriptomic profile, computational models predict effective therapies (Jovičić 2024). Personalized predictive models and machine learning help clinician's select optimal treatment strategies based on a patient's unique molecular characteristics (Jovičić 2024; Menditti et al. 2023).

Advances in network biology are enabled by the increasing availability of high-throughput data (Panditrao et al. 2022; Loscalzo 2022). Future developments in network-based algorithms provide insights into the molecular interactions driving OSCC by focusing on entire molecular networks rather than individual genes or proteins, to identify more effective therapeutic targets and uncover novel treatments (Panditrao et al. 2022; Loscalzo 2022).

Drug repurposing is a promising area in cancer research since in silico analysis offers a rapid and cost-effective way to identify existing drugs in OSCC treatment. Future research will leverage big data and AI to screen massive databases of approved drugs, uncovering new therapeutic applications for these agents in OSCC (Lin et al. 2022; Randhawa and Acharya 2015; Shi et al. 2023).

The case studies presented above demonstrate the significant potential of in silico analysis in advancing OSCC research from the identification of novel biomarkers to repurposing drugs. However, as computational methods evolve with the integration of AI and multi-omics data, the future of in silico analysis in OSCC holds the potential for even greater precision.

19.9 Conclusion and Future Perspectives

In this chapter, we explored the pivotal role of in silico analysis in advancing our understanding of OSCC. Key findings from various sections illustrate how computational methodologies have revolutionized cancer research by enabling a comprehensive understanding of tumor biology, identification of biomarkers, targeted therapeutics, and drug repurposing, machine learning applications, network analysis, and system biology (Fig. 19.1 (Graphical abstract)).

19.9.1 Summary of Key Findings and Their Implications for OSCC Research and Treatment

In silico approaches, including genomic and transcriptomic analyses, present the complex molecular underpinnings of OSCC by identifying critical genetic mutations, gene expression patterns, and protein interactions, providing valuable insights into the biological processes driving tumorigenesis and metastasis.

The integration of bioinformatics tools facilitates the novel biomarker discovery for early detection and prognosis of OSCC disease. Identifying specific gene expression profiles associated with disease outcomes leads to the development of targeted screening methods and

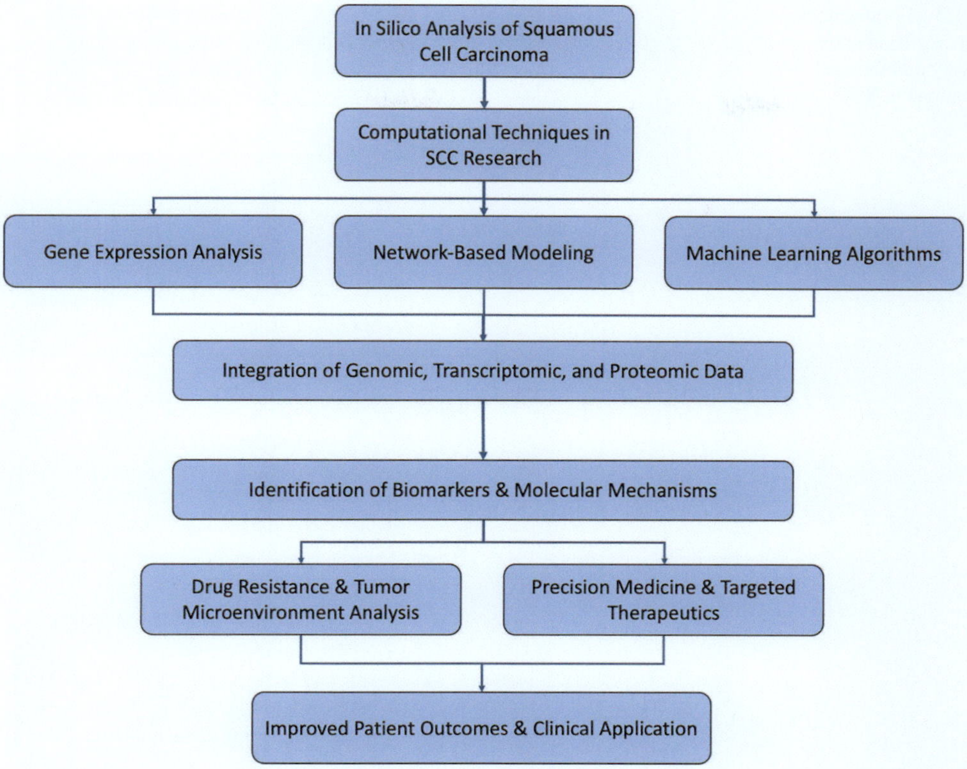

Fig. 19.1 (Graphical abstract). Visual representation of the chapter's key points, showing how computational techniques contribute to SCC research, biomarker identification, and clinical advancements

personalized treatment strategies. Targeted therapeutics and drug repurposing lead to novel biomarker discoveries in OSCC. Computational predictions, confirmed by experimental validations, enable easier application of FDA-approved drugs for OSCC therapy. The use of ML is powerful in predicting disease progression and patient outcomes, enabling clinicians to modify and tailor treatment plans to individual patient profiles and enhancing the efficacy of therapeutic interventions. Network analysis and system biology explore significant drivers and hubs in OSCC, where understanding these networks is critical for developing targeted therapies that disrupt the pathways responsible for cancer progression and resistance. Eventually, we can actually conclude that the implications of these findings are profound, emphasizing the potential role of in silico analysis in OSCC to improve diagnostic accuracy and lead to the development of novel therapeutic strategies.

19.9.2 Future Trends in In Silico Analysis and Its Potential Impact on Cancer Research

The personalized medicine approach in OSCC treatment for every patient depends on multi-omics data integration with in silico analysis, AI, and ML in the future. Combining this information will enable the identification of complex interactions and molecular mechanisms underlying the disease, leading to the application of more effective treatments. This will increase the predictive model capacity and uncover novel patterns and relationships within large datasets. Emerging technology developments, like wearable devices and health applications, incorporate real-time patient data into computational models to monitor disease dynamics and treatment responses, facilitating personalized patient-centric approaches. Patient-sensitive information should be handled with high ethical standards, ensuring

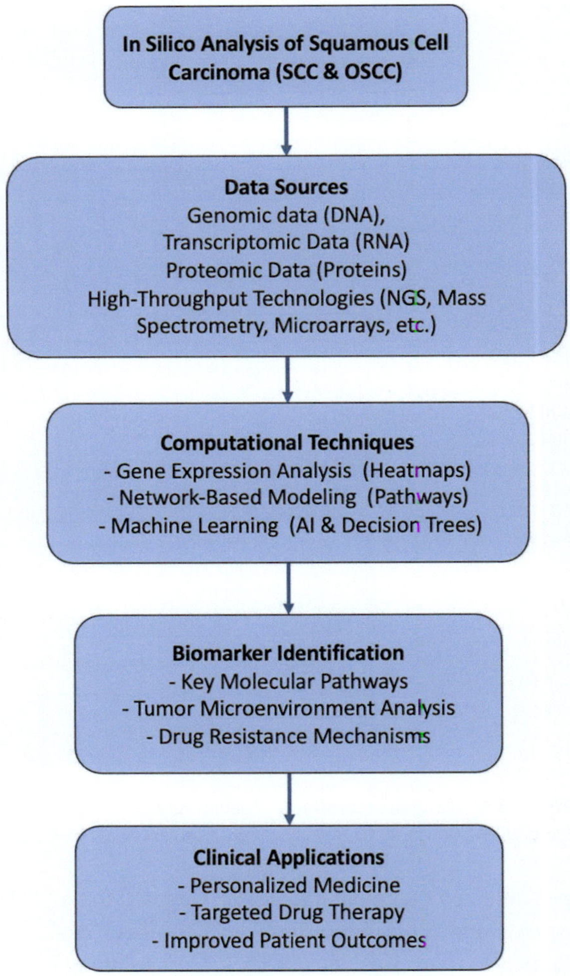

Fig. 19.2 Bioinformatics Pipeline for Biomarker Discovery and Clinical Applications in SCC & OSCC

data privacy and maintaining confidentiality. Collaboration with experts in diverse fields and institutions will enable research development, providing insights into disease biology and identifying novel biomarkers in OSCC, where technology advancement improves patient outcomes and therapy. Key findings and future trends enable the summarization of significant information regarding current and future research trends in cancer biology and OSCC disease, highlighting the transformative role of in silico methodology and being a valuable resource for researchers.

This book chapter illustrates the significance of the integration of bioinformatic and ML algorithms in understanding OSCC, where genetic and proteomic biomarkers enable making better treatment strategies, identifying therapeutic targets, enhancing the precision of disease management, and identifying the critical role of the immune microenvironment in OSCC progression and treatment response (Fig. 19.2). Future research should focus on multi-omics data integration and personalized medicine to understand the holistic understanding and improve clinical outcomes. Collaboration among researchers and institutions is essential for providing a multidisciplinary approach to enhance ssknowledge in the field and foster innovation in disease treatment. Each section of this chapter offers insight and examples to illustrate key concepts.

Acknowledgments The author thanks the Faculty of Biology, University of Belgrade, for their continued support and access to research and publication resources.

Dedication This chapter is dedicated to my lovely family, my dear mom and dad. Thank you so much for providing me with everything in life and guiding me on the right path. Love you so much.

Preface This chapter was written to bridge the gap between traditional cancer research methodologies and cutting-edge computational techniques, emphasizing the role of bioinformatics in decoding the complexities of SCC and oral immunity. It aims to be a resource for both clinicians and bioinformaticians working towards improved therapeutic outcomes in SCC patients.

Glossary

A short list of specific terminology in bioinformatics and molecular biology.

1. **Bioinformatics**: The tool involving a computer that analyzes data from biological sources.
2. **Differential Expression**: A statistical analysis utilized to identify upregulated and downregulated genes that compares gene expression levels between sample groups in a specific condition.
3. **Inhibitors**: Drugs that block specific molecules to prevent normal or abnormal system functioning.
4. **Machine Learning**: Computational tools with algorithms for predictive modeling.
5. **RNA-Sequencing**: Molecular biology technique to determine gene expression levels.
6. **Proteomics**: The research biology area are investigating functions and structures of proteins to identify biomarkers and therapeutic targets.
7. **Single Nucleotide Polymorphism (SNP)**: A variation in a single nucleotide at a specific position in the genome.
8. **Pathway Analysis**: A method for biological pathway identification and analysis.
9. **Survival Analysis**: A statistical method that analyzes time-to-event data to estimate the patient's survival rates.
10. **Cohort Study**: Following up observational study over time to assess outcome.

Appendix

Data Sources

The data source for OSCC analysis with bioinformatics tools include:

1. **TCGA:** contains multi-dimensional maps of the genomic, transcriptomic, epigenomic, and clinical data changes in various cancers, enabling exploration of the potential biomarkers and heterogeneity in OSCC.
2. **GEO:** stores high-throughput gene data from microarray and RNA-sequencing to compare expression profiles between disease and healthy samples to identify genes in OSCC progression and resistance to therapy.
3. **STRING:** Represent protein-protein interaction database with integrated experimental data, computational predictions, and public databases useful for the biological pathways, implicated in OSCC.
4. **cBioPortal:** provides visualization and analysis of multidimensional cancer genomics data, allowing researchers to explore the genetic alterations associated with OSCC and their clinical implications.
5. **The Human Protein Atlas:** contains information about the tissue-specific expression and localization of proteins in human cells, aiding in diagnostic and therapeutic applications of OSCC.

Software and Tools

A list of bioinformatics software and ML in OSCC research include:

1. **R:** a programming language containing packages such as Bioconductor and DESeq2 package in R to analyze RNA-seq data and identify changes in OSCC.

2. **Python**: programming language containing libraries such as NumPy, pandas, and SciPy and machine learning libraries like sci-kit-learn and TensorFlow for data analysis to predict patient outcomes based on genomic data from OSCC patients for model building and evaluation.
3. **Cytoscape**: open source platform for visualizing complex networks and integrating data in OSCC.
4. **GenePattern:** a genomic analysis that performs gene expression analysis, clustering, and machine learning to analyze microarray data from OSCC studies.
5. **SPSS**: a software package for interactive or batched statistical analysis, offering tools for data management, statistical analysis, and reporting to analyze clinical correlation data between patient demographics, treatment regimens, and survival outcomes.

References

Acharya SK, Shai S, Choon YF (2024) Cancer stem cells in oral squamous cell carcinoma: a narrative review on experimental characteristics and methodological challenges. Biomedicine 12(9):2111

Ahuja SK, Shrimankar DD, Durge AR (2023) A study and analysis of disease identification using genomic sequence processing models: an empirical review. Curr Genomics 24(4):207–235

Alabi RO, Youssef O, Pirinen M et al (2021) Machine learning in oral squamous cell carcinoma: current status, clinical concerns and prospects for future-a systematic review. Artif Intell Med 115:102060

Alabi RO, Elmusrati M, Leivo I et al (2023) Machine learning explainability in nasopharyngeal cancer survival using LIME and SHAP. Sci Rep 13:13

Albaradei S, Thafar M, Alsaedi A et al (2020) Machine learning and deep learning methods that use omics data for metastasis prediction. Comput Struct Biotechnol J 19:5008–5018

Alharbi F, Vakanski A (2023) Machine learning methods for cancer classification using gene expression data: a review. Bioengineering 10(2)

Ali SA, Khan HA, Irfan O et al (2017) Correlation of TP53 overexpression and clinical parameters with five-year survival in oral squamous cell carcinoma patients. Cureus. 9(6)

Almansoori A, Bhamidimarri PM, Bendardaf R et al (2022) In silico analysis of publicly available transcriptomics data identifies putative prognostic and therapeutic molecular targets for papillary thyroid carcinoma. Int J General Med 15:3097–3120

Al-Tashi Q, Saad MB, Muneer A (2023) Machine learning models for the identification of prognostic and predictive cancer biomarkers: a systematic review. Int J Mol Sci 24(9)

Arora R, Cao C, Kumar M et al (2023) Spatial transcriptomics reveals distinct and conserved tumor core and edge architectures that predict survival and targeted therapy response. Nat Commun 14(1):1–19

Athieniti E, Spyrou GM (2023) A guide to multi-omics data collection and integration for translational medicine. Comput Struct Biotechnol J 21:134–149

Badoiu SC, Greabu M, Miricescu D et al (2023) PI3K/AKT/mTOR dysregulation and reprogramming metabolic pathways in renal cancer: crosstalk with the VHL/HIF Axis. Int J Mol Sci 24(9)

Badwelan M, Muaddi H, Ahmed A et al (2023) Oral squamous cell carcinoma and concomitant primary tumors, what do we know? A review of the literature. Curr Oncol 30(4):3721–3734

Barh D, Chaitankar V, Yiannakopoulou EC et al (2014) In Silico models: from simple networks to complex diseases. Anim Biotechnol:385–404

Bebek G, Koyutürk M, Price ND et al (2012) Network biology methods integrating biological data for translational science. Brief Bioinform 13(4):446–459

Belizário JE, Sangiuliano BA, Perez-Sosa M et al (2016) Using Pharmacogenomic databases for discovering patient-target genes and small molecule candidates to cancer therapy. Front Pharmacol 7

Blatt S, Krüger M, Sagheb K et al (2022) Tumor recurrence and follow-up intervals in oral squamous cell carcinoma. J Clin Med 11(23)

Brlek P, Kafka A, Bukovac A (2020) Integrative cBioPortal analysis revealed molecular mechanisms that regulate EGFR-PI3K-AKT-mTOR pathway in diffuse gliomas of the brain. Cancer 13(13):3247

Butt MA, Kaleem MF, Bilal M et al (2024) Using multilabel ensemble CNN classifiers to mitigate labelling inconsistencies in patch-level Gleason grading. PLoS One 19(7):e0304847

Camalan S, Mahmood H, Binol H et al (2021) Convolutional neural network-based clinical predictors of oral dysplasia: class activation map analysis of deep learning results. Cancer 13(6)

Cervantes-Gracia K, Chahwan R, Husi H (2022) Integrative OMICS data-driven procedure using a derivatized meta-analysis approach. Front Genet 13

Chen B, Khodadoust MS, Liu CL et al (2018) Profiling tumor infiltrating immune cells with CIBERSORT. In: Methods in molecular biology, vol 1711, Clifton, p 243

Chen X, Zhang T, Su W et al (2022) Mutant p53 in cancer: from molecular mechanism to therapeutic modulation. Cell Death Dis 13(11):974

Cornel AM, Mimpen IL, Nierkens S (2020) MHC class I downregulation in cancer: underlying mechanisms and potential targets for cancer immunotherapy. Cancer 12(7)

Creixell P, Reimand J, Haider S et al (2015) Pathway and network analysis of cancer genomes. Nat Methods 12(7):615

Das S, Dey MK, Devireddy R et al (2023) Biomarkers in cancer detection, diagnosis, and prognosis. Sensors 24(1):37

Das D, Maitra A, Panda CK et al (2024) Breakthroughs in AI and multi-omics for cancer drug discovery: a review. Eur J Med Chem 280:116925

Davis NM, Sokolosky M, Stadelman K et al (2014) Deregulation of the EGFR/PI3K/PTEN/Akt/mTORC1 pathway in breast cancer: possibilities for therapeutic intervention. Oncotarget 5(13):4603–4650

de Paula Souza DPS et al (2023) Identification of potential biomarkers and survival analysis for oral squamous cell carcinoma: a transcriptomic study. Oral Dis 29(7):2658–2666

Diao P, Dai Y, Wang A et al (2024) Integrative multiomics analyses identify molecular subtypes of head and neck squamous cell carcinoma with distinct therapeutic vulnerabilities. Cancer Res 84(18):3101–3117

Digre A, Lindskog C (2020) The human protein atlas – spatial localization of the human proteome in health and disease. Protein Sci 30(1):218–233

Dixit S, Kumar A, Srinivasan K (2023) A current review of machine learning and deep learning models in oral cancer diagnosis: recent technologies, open challenges, and future research directions. Diagnostics 13(7)

Fania L, Didona D, Di Pietro F et al (2021) Cutaneous squamous cell carcinoma: from pathophysiology to novel therapeutic approaches. Biomedicine 9(2)

Fatima J, Fatima E, Mehmood F et al (2024) Comprehensive analysis of oral squamous cell carcinomas: clinical, epidemiological, and histopathological insights with a focus on prognostic factors and survival time. Cureus 16(2)

Finotello F, Trajanoski Z (2018) Quantifying tumor-infiltrating immune cells from transcriptomics data. Cancer Immunol Immunother 67(7):1031–1040

Fitzpatrick SG, Cohen DM, Clark AN (2019) Ulcerated lesions of the oral mucosa: clinical and histologic review. Head Neck Pathol 13(1):91–102

Gao L, Chen Y (2023) A metabolomic and proteomic study to elucidate the molecular mechanisms of immunotherapy resistance in patients with oesophageal squamous cell carcinoma. Biomed Rep 18(5):36

Gao J, Aksoy BA, Dogrusoz U et al (2013) Integrative analysis of complex cancer genomics and clinical profiles using the cBioPortal. Sci Signal 6(269):l1

Gao L, Wang TH, Chen CP et al (2021) Targeting COX-2 potently inhibits proliferation of cancer cells *in vivo* but not *in vitro* in cutaneous squamous cell carcinoma. Transl Cancer Res 10(5):2219

Ge Y, Li W, Ni Q et al (2019) Weighted gene co-expression network analysis identifies hub genes associated with occurrence and prognosis of oral squamous cell carcinoma. Med Sci Monit 25:7272–7288

Genome Browser (2024). Available from: https://genome.ucsc.edu/. Accessed 07 Oct 2024

Ghafouri-Fard S, Alamdari AN, Alamdari YN et al (2022) Role of PI3K/AKT pathway in squamous cell carcinoma with an especial focus on head and neck cancers. Cancer Cell Int 22:254

Ghantous Y, Mozalbat S, Nashef A et al (2024) EMT dynamics in lymph node metastasis of oral squamous cell carcinoma. Cancer 16(6)

Glaviano AC, Foo AS, Lam HY et al (2023) PI3K/AKT/mTOR signaling transduction pathway and targeted therapies in cancer. Mol Cancer 22:138

González-Moles MÁ, Aguilar-Ruiz M, Ramos-García P (2022) Challenges in the early diagnosis of oral cancer, evidence gaps and strategies for improvement: a scoping review of systematic reviews. Cancer 14(19)

Gunardi I, Sufiawati I, Goenawan H et al (2023) Research trends in molecular biological studies on oral squamous cell carcinoma: a bibliometric analysis. Oncol Rev 17:17

Gupta S, Gupta MK, Shabaz M et al (2022) Deep learning techniques for cancer classification using microarray gene expression data. Front Physiol 13

Gupta I, Badrzadeh F, Tsentalovich Y (2024) Connecting the dots: investigating the link between environmental, genetic, and epigenetic influences in metabolomic alterations in oral squamous cell carcinoma. J Exp Clin Cancer Res 43:43

Halder S, Basu S, Lal S et al (2023) Targeting the EGFR signaling pathway in cancer therapy: what's new in 2023? Expert Opin Ther Targets 27(4–5):305

Hoffman-Censits JH, Grivas P, Heijden MS et al (2016) IMvigor 210, a phase II trial of atezolizumab (MPDL3280A) in platinum-treated locally advanced or metastatic urothelial carcinoma (mUC). JCO 34:355

Hou W, Kong L, Hou Z et al (2022) CD44 is a prognostic biomarker and correlated with immune infiltrates in gastric cancer. BMC Med Genet 15

Hu Y, Han Y, He M et al (2023) S100 proteins in head and neck squamous cell carcinoma (review). Oncol Lett 26(2):362

Hua Y, Zheng Y, Yao Y et al (2023) Metformin and cancer hallmarks: shedding new lights on therapeutic repurposing. J Transl Med 21:403

Huynh NC, Huang T, Nguyen CT et al (2021) Comprehensive integrated single-cell whole transcriptome analysis revealed the p-EMT tumor cells – CAFs communication in oral squamous cell carcinoma. Int J Mol Sci 23(12):6470

Ideta Y, Tagawa T, Hayashi Y et al (2021) Transcriptomic profiling predicts multiple pathways and molecules associated with the metastatic phenotype of oral cancer cells. Cancer Genomics Proteomics 18(1):17–27

Jayawickrama SM, Ranaweera PM, Gunaratnege Roshan Pradeep RG et al (2024) Developments and future prospects of personalized medicine in head and neck squamous cell carcinoma diagnoses and treatments. Cancer Rep 7(3):e2045

Jensen JS, Jakobsen KK, Mirian C et al (2019) The Copenhagen oral cavity squamous cell carcinoma database: protocol and report on establishing a

comprehensive oral cavity cancer database. Clin Epidemiol 11:733–741
Jin M (2013) Gastric and duodenal squamous cell carcinoma: metastatic or primary? World J Surg Oncol 11: 204
Jovičić SM (2024) In silico analysis of precursor messenger RNA as a potential biomarker of myasthenia gravis thymoma. Eur J Inflamm 22
Kaddoura R, Alqutami F, Asbaita M et al (2023) In Silico analysis of publicly available transcriptomic data for the identification of triple-negative breast cancer-specific biomarkers. Life. 13(2)
Kampel L, Feldstein S, Tsuriel S et al (2023) Mutated TP53 in circulating tumor DNA as a risk level biomarker in head and neck squamous cell carcinoma patients. Biomolecules. 13(9):1418
Kaunein N, Ramani RS, Koo K et al (2021) A systematic review of MicroRNA signatures associated with the progression of leukoplakia with and without epithelial dysplasia. Biomolecules 11(12)
Khowal S, Monga S, Naqvi SH et al (2021) Molecular winnowing, expressional analyses and interactome scrutiny of cellular proteomes of oral squamous cell carcinoma. Adv Cancer Biol Metast 2:100003
Kim WJ, Choi BR, Noh JJ et al (2024) Comparison of RNA-Seq and microarray in the prediction of protein expression and survival prediction. Front Genet 15: 1342021
Kondoh N, Mizuno-Kamiya M, Umemura N et al (2019) Immunomodulatory aspects in the progression and treatment of oral malignancy. Jpa Dental Sci Rev 55(1):113–120
Kumari M, Chhikara BS, Singh P et al (2023) Signaling and molecular pathways implicated in oral cancer: a concise review. Chem Biol Lett 11(1):652
Lavanya JMS (2024) Innovative approach towards early prediction of ovarian cancer: machine learning-enabled XAI techniques. Heliyon 10(9):e29197
Lee M (2023) Deep learning techniques with genomic data in cancer prognosis: a comprehensive review of the 2021–2023 literature. Biology 12(7):893
Li S, Chen X, Liu X et al (2017) Complex integrated analysis of lncRNAs-miRNAs-mRNAs in oral squamous cell carcinoma. Oral Oncol 73:1–9
Li Y, Zhang R, Hei H (2023a) Advances in post-translational modifications of proteins and cancer immunotherapy. Front Immunol 14
Li Z, Fu R, Wen X et al (2023b) The significant clinical correlation of the intratumor oral microbiome in oral squamous cell carcinoma based on tissue-derived sequencing. Front Physiol 13:1089539
Li JH, Hsin PY, Hsiao YC et al (2024) A narrative review: repurposing metformin as a potential therapeutic agent for oral cancer. Cancers (Basel) 16(17):3017
Liang L, Liang X, Yu X et al (2024) Bioinformatic analyses and integrated machine learning to predict prognosis and therapeutic response based on E3 ligase-related genes in colon cancer. J Cancer 15(16): 5376–5395
Lin YC, Chen BS (2022) Identifying drug targets of oral squamous cell carcinoma through a systems biology method and genome-wide microarray data for drug discovery by deep learning and drug design specifications. Int J Mol Sci 23(18):10409
Lin X, Wu W, Ying Y et al (2022) MicroRNA-31: a pivotal oncogenic factor in oral squamous cell carcinoma. Cell Death Discovery. 8:8
Lindemann A, Takahashi H, Patel AA et al (2018) Targeting the DNA damage response in OSCC with TP53 mutations. J Dent Res 97(6):635–644
Liu Y, Mei Y (2023) Prediction of drug sensitivity based on multi-omics data using deep learning and similarity network fusion approaches. Front Bioeng Biotechnol 11:11
Liu X, Wang J, Chen M et al (2019) Combining data from TCGA and GEO databases and reverse transcription quantitative PCR validation to identify gene prognostic markers in lung cancer. Onco Targets Ther 12:709–720
Liu T, Yu C, Lu Y et al (2023) MiR-146a participates in the regulation of cancer stemness of oral carcinoma cells. J Dental Sci 18(2):503–509
Loscalzo J (2022) Molecular interaction networks and drug development: novel approach to drug target identification and drug repositioning. FASEB J 37(1)
Luo G, Wang S, Lu W et al (2024) Application of metabolomics in oral squamous cell carcinoma. Oral Dis 30(6):3719–3731
Lynch SM, Heeran AB, Burke C (2023) Precision oncology, artificial intelligence, and novel therapeutic advancements in the diagnosis, prevention, and treatment of cancer: highlights from the 59th Irish Association for Cancer Research (IACR) annual conference. Cancer 16(11):1989
Ma B, Meng F, Yan G et al (2020) Diagnostic classification of cancers using extreme gradient boosting algorithm and multi-omics data. Comput Biol Med 121: 103761
Makarev E, Schubert AD, Kanherkar RR et al (2017) In silico analysis of pathways activation landscape in oral squamous cell carcinoma and oral leukoplakia. Cell Death Discovery 3:17022
Matsuoka T, Yashiro M (2024) Bioinformatics analysis and validation of potential markers associated with prediction and prognosis of gastric cancer. Int J Mol Sci 25(11)
Meirelles AL, Kurc T, Kong J et al (2022) Building efficient CNN architectures for histopathology images analysis: a case-study in tumor-infiltrating lymphocytes classification. Front Med 9:9
Melo-Alvim C, Neves ME, Santos JL et al (2022) Radiotherapy, chemotherapy and immunotherapy – current practice and future perspectives for recurrent/metastatic oral cavity squamous cell carcinoma. Diagnostics 13(1)
Menditti D, Santagata M, Imola G et al (2023) Personalized medicine in oral oncology: imaging methods and biological markers to support diagnosis

of oral squamous cell carcinoma (OSCC): a narrative literature review. J Pers Med 13(9)

Mohr AEPC, Whisner CM, Jasbi P (2024) Navigating challenges and opportunities in multi-omics integration for personalized healthcare. Biomedicine 12(7):1496

Mohtasham N, Zarepoor M, Shooshtari Z et al (2024) Genes involved in metastasis in oral squamous cell carcinoma: a systematic review. Health Sci Rep 7(4): e1977

Musa A, Ghoraie LS, Zhang D et al (2018) A review of connectivity map and computational approaches in pharmacogenomics. Brief Bioinform 19(3):506–523

Nakagaki T, Tamura M, Kobashi K et al (2017) Profiling cancer-related gene mutations in oral squamous cell carcinoma from Japanese patients by targeted amplicon sequencing. Oncotarget 8(35):59113–59122

Niazi SK, Mariam Z (2023) Computer-aided drug design and drug discovery: a prospective analysis. Pharmaceuticals 17(1)

Noh JK, Woo SR, Kong M et al (2022) Gene signature predicting recurrence in oral squamous cell carcinoma is characterized by increased oxidative phosphorylation. Mol Oncol 17(1):134–149

Nourbakhsh M, Degn K, Saksager A et al (2024) Prediction of cancer driver genes and mutations: the potential of integrative computational frameworks. Brief Bioinform 25(2)

Nwosu IO, Piccolo SR (2021) A systematic review of datasets that can help elucidate relationships among gene expression, race, and immunohistochemistry-defined subtypes in breast cancer. Cancer Biol Ther 22(7–9):417–429

Olivier M, Hollstein M, Hainaut P (2010) TP53 mutations in human cancers: origins, consequences, and clinical use. Cold Spring Harb Perspect Biol 2(1):a001008

Paluszczak J (2020) The significance of the dysregulation of canonical Wnt signaling in head and neck squamous cell carcinomas. Cells 9(3)

Palve V, Liao Y, Remsing Rix LL et al (2020) Turning liabilities into opportunities: off-target based drug repurposing in cancer. Semin Cancer Biol 68:209

Pandey R, Johnson N, Cooke L et al (2021) TP53 mutations as a driver of metastasis signaling in advanced cancer patients. Cancer 13(4)

Panditrao G, Bhowmick R, Meena C et al (2022) Emerging landscape of molecular interaction networks: opportunities, challenges and prospects. J Biosci 47(2)

Pansare K, Gardi N, Kamat S et al (2019) Establishment and genomic characterization of gingivobuccal carcinoma cell lines with smokeless tobacco associated genetic alterations and oncogenic PIK3CA mutation. Sci Rep 9(1):1–10

Park N, Ryu K, Ju Y (2024) The potential MicroRNA diagnostic biomarkers in oral squamous cell carcinoma of the tongue. Curr Issues Mol Biol 46(7):6746–6756

Passaro A, Bakir MA, Hamilton EG et al (2024) Cancer biomarkers – emerging trends and clinical implications for personalized treatment. Cell 187(7):1617

Peng Y, Wang Y, Zhou C (2022) PI3K/AKT/mTOR pathway and its role in cancer therapeutics: are we making headway? Front Oncol 12:819128

Pérez Jardón A, Arroyo E, Gándara Vila P et al (2023) Risk factors of oral cancer: identification and mechanism of prevention. In: Rezaei N (ed) Handbook of cancer and immunology. Springer, Cham

Pham TD, Teh M, Chatzopoulou D et al (2024) Artificial intelligence in head and neck cancer: innovations, applications, and future directions. Curr Oncol 31(9): 5255–5290

Pillai J, Chincholkar T, Dixit R et al (2021) A systematic review of proteomic biomarkers in oral squamous cell cancer. World J Surg Oncol 19:315

Pilley S, Rodriguez TA, Vousden KH (2021) Mutant p53 in cell-cell interactions. Genes Dev 35(7–8):433–448

Pomella S, Melaiu O, Cifaldi L et al (2023) Biomarkers identification in the microenvironment of oral squamous cell carcinoma: a systematic review of proteomic studies. Int J Mol Sci 25(16):8929

Powell SF, Vu L, Spanos WC et al (2021) The key differences between human papillomavirus-positive and -negative head and neck cancers: biological and clinical implications. Cancer 13(20)

Qian Y, Wang W, Chen D et al (2022) Cigarette smoking induces the activation of RIP2/caspase-12/NF-κB axis in oral squamous cell carcinoma. PeerJ 10:10

Quintero-Fabián S, Arreola R, Becerril-Villanueva E et al (2019) Role of matrix Metalloproteinases in angiogenesis and cancer. Front Oncol 9:9

Racle J, Jonge K, Baumgaertner P et al (2017) Simultaneous enumeration of cancer and immune cell types from bulk tumor gene expression data. elife 6:e26476

Randhawa V, Acharya V (2015) Integrated network analysis and logistic regression modeling identify stage-specific genes in oral squamous cell carcinoma. BMC Med Genet 8:39

Rao MS, Van Vleet TR, Ciurlionis R et al (2018) Comparison of RNA-Seq and microarray gene expression platforms for the Toxicogenomic evaluation of liver from short-term rat toxicity studies. Front Genet 9

Ravi P, Babu S (2024) Emerging immune checkpoint inhibitors for the treatment of oropharyngeal squamous cell carcinoma. Oral Oncol Rep 12:100650

Rhim J, Baek W, Seo Y (2022) From molecular mechanisms to therapeutics: understanding MicroRNA-21 in cancer. Cells 11(18)

Ribeiro Cavalcante BR, Freitas RD, Rocha GV et al (2022) In silico approaches for drug repurposing in oncology: protocol for a scoping review of existing evidence. PLoS One 17(7)

Ricco N, Kron SJ (2023) Statins in cancer prevention and therapy. Cancer 15(15)

Saidak Z, Lailler C, Testelin S et al (2021) Contribution of genomics to the surgical management and study of oral cancer. Ann Surg Oncol 28(11):5842–5854

Salameti V, Bhosale PG, Ames-Draycott A et al (2019) NOTCH1 signaling in oral squamous cell carcinoma

via a TEL2/SERPINE1 axis. Oncotarget 10(63): 6791–6804
Samara P, Athanasopoulos M, Mastronikolis S et al (2024) The role of oncogenic viruses in head and neck cancers: epidemiology, pathogenesis, and advancements in detection methods. Microorganisms 12(7)
Sánchez-Danés A, Blanpain C (2018) Deciphering the cells of origin of squamous cell carcinomas. Nat Rev Cancer 18(9):549
Sanguansin S, Kosanwat T, Juengsomjit R et al (2021) Diagnostic value of cytokeratin 17 during oral carcinogenesis: an Immunohistochemical study. Int J Dent 2021:1
Satam H, Joshi K, Mangrolia U et al (2023) Next-generation sequencing technology: current trends and advancements. Biology 12(7)
Schiliro C, Firestein BL (2021) Mechanisms of metabolic reprogramming in cancer cells supporting enhanced growth and proliferation. Cells 10(5)
Shaikh S, Yadav DK, Bhadresha K et al (2023) Integrated computational screening and liquid biopsy approach to uncover the role of biomarkers for oral cancer lymph node metastasis. Sci Rep 13:13
Shaker Nasry WH, Rodriguez-Lecompte JC, Martin CK (2018) Role of COX-2/PGE2 mediated inflammation in oral squamous cell carcinoma. Cancer 10(10)
Shamsi A, Khan MS, Yadav DK et al (2024) Structure-based screening of FDA-approved drugs identifies potential histone deacetylase 3 repurposed inhibitor: molecular docking and molecular dynamic simulation approaches. Front Pharmacol 15:1424175
Sharkey Ochoa I, Toner M, Kay E et al (2021) The role of HPV in determining treatment, survival, and prognosis of head and neck squamous cell carcinoma. Cancer 14(17):4321
Shen Y, Liu J, Zhang L et al (2019) Identification of potential biomarkers and survival analysis for head and neck squamous cell carcinoma using bioinformatics strategy: a study based on TCGA and GEO datasets. Biomed Res Int 2019:1
Shetty SS, Sharma M, Fonseca FP et al (2020) Signaling pathways promoting epithelial mesenchymal transition in oral submucous fibrosis and oral squamous cell carcinoma. Jpn Dental Sci Rev 56(1):97–108
Shi K, Li H, Bai S (2022) The cell cycle-associated protein CDKN2A may promotes colorectal cancer cell metastasis by inducing epithelial-mesenchymal transition. Front Oncol 12:12
Shi S, Luo H, Ji Y et al (2023) Repurposing Dihydroartemisinin to combat oral squamous cell carcinoma, associated with mitochondrial dysfunction and oxidative stress. Oxidative Med Cell Longev 2023:1
Shi X, Wang X, Yao W et al (2024) Mechanism insights and therapeutic intervention of tumor metastasis: latest developments and perspectives. Signal Transduct Target Ther 9(1):1–46
Singh P, Patel M, Bhowmmik D et al (2024) Identification of common biomarkers affecting patient survibal in cancer. World Acad Sci J 6
Supplitt S, Karpinski P, Sasiadek M et al (2021) Current achievements and applications of Transcriptomics in personalized cancer medicine. Int J Mol Sci 22(3):1422
Suri S, Boora GS, Kaur R et al (2024) Recent advances in minimally invasive biomarkers of OSCC: from generalized to personalized approach. Front Oral Health 5:1426507
Tan Y, Wang Z, Xu M et al (2023) Oral squamous cell carcinomas: state of the field and emerging directions. International. J Oral Sci 15:15
Tarle M, Raguž M, Muller D et al (2023) Nuclear epidermal growth factor receptor overexpression as a survival predictor in oral squamous cell carcinoma. Int J Mol Sci 24(6)
Tie Y, Tang F (2022) Immunosuppressive cells in cancer: mechanisms and potential therapeutic targets. J Hematol Oncol 15:15
Tripathi S, Gupta E, Galande S (2024) Statins as anti-tumor agents: a paradigm for repurposed drugs. Cancer Reports 7(5):e2078
Tseng C, Tsou H, Huang Y et al (2022) Using next-generation sequencing and Bioinformatic methods to predict new genes that may be regulated by CD47 in oral squamous cell carcinoma. Curr Issues Mol Biol 44(5):2243–2256
Van Riet J, Saha C, Strepis N et al (2022) CRISPRs in the human genome are differentially expressed between malignant and normal adjacent to tumor tissue. Commun Biol 5(1):1–13
Varalakshmi D, Tharaheswari M, Anand T et al (2024) Transforming oral cancer care: the promise of deep learning in diagnosis. Oral Oncol Rep 10:100482
Varlamova EV, Butakova MA, Semyonova VV et al (2023) Mach Learn Meets Cancer 16(6):1100
Vastrad SJ, Ritesh GVSS, Saraswathy GR et al (2023) Panoramic view of key cross-talks underpinning the oral squamous cell carcinoma stemness – unearthing the future opportunities. Front Oncol 13:13
Velcheti V, Chandwani S, Chen X et al (2020) Pembrolizumab for previously treated, PD-L1-expressing advanced NSCLC: real-world time on treatment and overall survival. Clin Lung Cancer 21(5): e445–e455
Venkatesiah SS, Augustine D, Mishra D et al (2022) Immunology of oral squamous cell carcinoma – a comprehensive insight with recent concepts. Life 12(11)
Vittorio S, Gitto R, Adornato I et al (2021) In Silico strategy for targeting the mTOR kinase at rapamycin binding site by small molecules. Molecules 26(4)
Vollmer A, Hartmann S, Vollmer M et al (2024) Multimodal artificial intelligence-based pathogenomics improves survival prediction in oral squamous cell carcinoma. Sci Rep 14(1):1–14
Wan X, Li X, Yang J et al (2015) Genetic association between PIK3CA gene and oral squamous cell carcinoma: a case control study conducted in Chongqing, China. Int J Clin Exp Pathol 8(10):13360–13366
Wan Z, Xiong H, Tan X et al (2022) Integrative multi-omics analysis reveals candidate biomarkers for oral squamous cell carcinoma. Front Oncol 11:794146

Wang L, Yang Z, Cao Y (2020) Regulatory T cell and activated natural killer cell infiltration in hepatocellular carcinoma: immune cell profiling using the CIBERSORT. Ann Transl Med 8(22):1483

Wang S, Yang M, Li R et al (2023) Current advances in noninvasive methods for the diagnosis of oral squamous cell carcinoma: a review. Eur J Med Res 28:53

Wei L et al (2023) Artificial intelligence (AI) and machine learning (ML) in precision oncology: a review on enhancing discoverability through multiomics integration. Br J Radiol 96(1150)

Wojtukiewicz MZ, Rek MM, Karpowicz K et al (2021) Inhibitors of immune checkpoints – PD-1, PD-L1, CTLA-4 – new opportunities for cancer patients and a new challenge for internists and general practitioners. Cancer Metastasis Rev 40(3):949–982

Wu H, Chen W, Li G et al (2020) Analysis of the differentially expressed genes induced by cisplatin resistance in oral squamous cell carcinomas and their interaction. Front Genet 10:504709

Wu HT, Chen WT, Chen WJ et al (2021a) Bioinformatics analysis reveals that ANXA1 and SPINK5 are novel tumor suppressor genes in patients with oral squamous cell carcinoma. Transl Cancer Res 10(4):1761

Wu X, Yao Y, Dai Y et al (2021b) Identification of diagnostic and prognostic signatures derived from preoperative blood parameters for oral squamous cell carcinoma. Ann Transl Med 9(15):1220

Wu Y, Wang Y, Liu H et al (2024) Mechanism of apoptosis in oral squamous cell carcinoma promoted by cardamonin through PI3K/AKT signaling pathway. Sci Rep 14(1):1–14

Xu M, Wang Y, Wang L et al (2017) PIK3CA mutations contribute to acquired cetuximab resistance in metastatic colorectal cancer patients. Clin Cancer Res 23(16):4602

Yang Z, Sun P, Dahlstrom KR et al (2023) Joint effect of human papillomavirus exposure, smoking and alcohol on risk of oral squamous cell carcinoma. BMC Cancer 23:457

Yang J, Lin N, Niu M et al (2024) Circulating tumor DNA mutation analysis: advances in its application for early diagnosis of hepatocellular carcinoma and therapeutic efficacy monitoring. Aging (Albany NY) 16(14): 11460–11474

Yin S, Chen Z, Chen D et al (2023) Strategies targeting PD-L1 expression and associated opportunities for cancer combination therapy. Theranostics 13(5): 1520–1544

Yoshida H, Yoshimura H, Matsuda S et al (2019) Celecoxib suppresses lipopolysaccharide-stimulated oral squamous cell carcinoma proliferation in vitro and in vivo. Oncol Lett 18(6):5793–5800

Yu C, Wang J (2022) Data mining and mathematical models in cancer prognosis and prediction. Med Rev 2(3):285–307

Zavarykina TM, Lomskova PK, Pronina IV et al (2023) Circulating tumor DNA is a variant of liquid biopsy with predictive and prognostic clinical value in breast cancer patients. Int J Mol Sci 24(23)

Zhang L, Wang X (2022) An immune-related gene signature can predict clinical outcomes and immunotherapeutic response in oral squamous cell carcinoma. Front Genet 13

Zhang B, Shi H, Wang H (2023a) Machine learning and AI in cancer prognosis, prediction, and treatment selection: a critical approach. J Multidiscip Healthc 16: 1779–1791

Zhang C, Xu J, Tang R et al (2023b) Novel research and future prospects of artificial intelligence in cancer diagnosis and treatment. J Hematol Oncol 16:114

Zhao S (2015) Zhang BA comprehensive evaluation of ensembl, RefSeq, and UCSC annotations in the context of RNA-seq read mapping and gene quantification. BMC Genomics 16:97

Zhao Y, Chen D, Yin J et al (2022) Comprehensive analysis of tumor immune microenvironment characteristics for the prognostic prediction and immunotherapy of oral squamous cell carcinoma. Front Genet 13

Zheng S, He S, Liang Y et al (2024) Understanding PI3K/Akt/mTOR signaling in squamous cell carcinoma: mutated PIK3CA as an example. Mol Biomed 5:13

Zhou Y, Tao L, Qiu J et al (2024) Tumor biomarkers for diagnosis, prognosis and targeted therapy. Signal Transduct Target Ther 9(1):1–86

Zou B, Li J, Xu K et al (2019) Identification of key candidate genes and pathways in oral squamous cell carcinoma by integrated bioinformatics analysis. Exp Ther Med 17(5):4089–4099

Infectious/Inflammatory Oral Diseases and Medication-Related Osteonecrosis of the Jaw (MRONJ)

20

Luciano Tavares Angelo Cintra, Leonardo Perez Faverani, Carolina Sayuri Wajima, Estêvão Lopes Pereira, Luan Felipe Toro, João Martins de Mello-Neto, Tiburtino José de Lima Neto, and Edilson Ervolino

Abstract

Medication-related osteonecrosis of the jaw (MRONJ) is a rare and potentially debilitating condition affecting patients treated with medications that inhibit bone resorption. It is characterized by exposed bone, usually in the mandible or maxilla, with clinical signs of inflammation, pain, and local infection. MRONJ is often triggered by invasive procedures in the oral cavity, such as dental extractions or trauma, and is associated with a predisposition to secondary infections. The interaction between MRONJ and oral infectious/inflammatory diseases, such as periodontal disease, apical periodontitis, and peri-implantitis, is complex, as oral infections lead to bone infection, exacerbating the bone necrosis characteristic of MRONJ. Furthermore, the inflammatory process of oral infections can worsen the patient's immune response and hinder the healing of bone lesions. The clinical management of patients with MRONJ resulting from oral infections requires a careful approach, with emphasis on strict control of the infection and preservation of periodontal health. Prevention and control of oral infections are essential to minimize the risk of complications and improve the patient's quality of life. This chapter addresses the relationship between MRONJ and different oral infections, presenting current aspects of this relationship and special care for its treatment.

Keywords

Osteonecrosis · Oral health-related quality of life · Periodontal Medicine · Endodontic Medicine · Preventive dentistry

L. T. A. Cintra (✉) · C. S. Wajima
Department of Restorative Dentistry, Araçatuba School of Dentistry, São Paulo State University (UNESP), Araçatuba, São Paulo, Brazil
e-mail: luciano.cintra@unesp.br; cs.wajima@unesp.br

L. P. Faverani · T. J. de Lima Neto
Department of Oral Diagnosis, Piracicaba Dental School, University of Campinas (UNICAMP), Piracicaba, Brazil
e-mail: faverani@unicamp.br

E. L. Pereira · E. Ervolino
Department of Basic Sciences, Araçatuba School of Dentistry, São Paulo State University (UNESP), Araçatuba, São Paulo, Brazil
e-mail: estevao.pereira@unesp.br; e.ervolino@unesp.br

L. F. Toro
Marília Medical School (FAMEMA), Marília, São Paulo, Brazil
e-mail: luan.toro@unesp.br

J. M. de Mello-Neto
College of Medicine and Dentistry, James Cook University, Cairns, QLD, Australia
e-mail: joao.martinsdemelloneto@jcu.edu.au

20.1 Introduction

The medication-related osteonecrosis of the jaw (MRONJ) is characterized by the progressive destruction and death of bone tissue in the oral

N. Rezaei, H. Ziaei (eds.), *Oral Immunology*, Advances in Experimental Medicine and Biology 1492,
https://doi.org/10.1007/978-3-032-03176-1_20

and maxillofacial regions. This condition is often linked to the use of antiresorptive drugs, such as bisphosphonates, RANKL inhibitors, and/or antiangiogenic drugs (Ruggiero et al. 2022; Campisi et al. 2020). The occurrence of MRONJ is usually associated with at least one systemic risk factor, i.e., the use of the antiresorptive drug, and one or more local risk factors. Local conditions that predispose individuals to MRONJ include invasive dental procedures, such as tooth extractions, and chronic inflammatory-infectious processes, such as periodontitis (Khan et al. 2017; Kuroshima et al. 2019). Several studies indicate that a significant number of patients who developed MRONJ following tooth extraction had their teeth removed due to severe periodontal or periapical disease (Ruggiero et al. 2022; Ficarra et al. 2005).

Given the increase in foci of infection, MRONJ has emerged as a crucial factor in the bilateral aggravation and triggering of this pathology (Kang et al. 2013; Aguirre et al. 2021; Khan et al. 2017; Song et al. 2016). Among the main oral risk factors associated with the development of MRONJ, invasive dental procedures account for 50–70% of cases (Vardas et al. 2014; Yazdi and Schiodt 2015; Schiodt et al. 2018). However, chronic conditions of an inflammatory and infectious nature, such as periodontitis and periapical lesions, are also significant risk factors (Khan et al. 2017; McGowan et al. 2018; Kuroshima et al. 2019). Studies suggest that the presence of dental or periodontal infections, particularly before tooth extraction, may represent the main local risk factor for MRONJ (Barasch et al. 2011; Bolette et al. 2019; Nicolatou-Galitis et al. 2015; Song et al. 2016; Otto et al. 2017; Kim et al. 2018; Soutome et al. 2018; Hadaya et al. 2019; Hasegawa et al. 2019).

It is known that most tooth extractions are performed to resolve preexisting pathological lesions, including periodontal or periapical diseases (Ficarra et al. 2005; Boonyapakorn et al. 2008; Ruggiero et al. 2014). Thus, it is hypothesized that preexisting local inflammatory lesions initiated by bacterial infection may be related to the early stages of MRONJ (Yarom et al. 2010) and have already predisposed the affected area to develop MRONJ after tooth extraction (Mawardi et al. 2009; López-Jornet et al. 2011; Aguirre et al. 2012; Kang et al. 2013; Kim et al. 2018).

Many clinical studies suggest that periodontal or periapical infection of the teeth is the leading cause of most tooth extractions and significantly increases the risk of MRONJ (Hasegawa et al. 2019; Epstein et al. 2010; Dimopoulos et al. 2009). Tooth extraction is the precipitating event that causes MRONJ in 78% of cases (Marx et al. 2005) and can increase the risk of osteonecrosis by 33 times (Ruggiero et al. 2014; Martins et al. 2017). As these risk factors are associated with inflammation and/or infection, they induce the local production of pro-inflammatory cytokines, such as TNF-α and IL-1, which stimulate inflammation and the death of osteocytes in the alveolar bone, exacerbating MRONJ lesions (Khan et al. 2017; Song et al. 2016). In addition to the inflammatory effect, studies have shown that pathogenic bacteria colonize tooth extraction sites in animals and humans treated with bisphosphonates, exacerbating the development of MRONJ (Katsarelis et al. 2015).

Another factor that has been investigated to verify its correlation with MRONJ is the osseointegrated implant placement in patients with a history or current use of antiresorptive drugs. Prolonged use of these drugs orally and the higher potency of intravenous drugs have been associated with a greater predisposition to MRONJ. The main questions raised to mitigate the occurrence of MRONJ include whether implant placement after a drug suspension period (known as a drug holiday) would reduce the risk, whether patients with periodontitis prior to implant treatment are more predisposed to peri-implantitis, and whether the presence of peri-implantitis negatively impacts treatment outcomes in these cases. Although MRONJ is often related to tooth extractions, recent cases have shown its occurrence around dental implants, referred to as peri-implantitis-like medication-related osteonecrosis of the jaw (PI-MRONJ). This condition presents similar symptoms to peri-implantitis, including inflammation and bone resorption, complicating its diagnosis and delaying appropriate treatment (Tempesta et al. 2022).

This chapter is divided into three parts, considering the specificity of periodontal disease, apical periodontitis, and peri-implantitis, so that the subject can be presented in a specific way, considering the nature of each of these infectious/inflammatory oral diseases.

20.2 Periodontitis and MRONJ

Periodontitis is considered an important local risk factor related to MRONJ, since there is involvement of the alveolar bone, which is also the target of antiresorptive therapy (Fig. 20.1). This section will address the importance of maintaining periodontal health throughout therapy with antiresorptive drugs, the relationship between periodontitis and MRONJ, and the care that should be taken in the treatment of periodontitis patients with this condition.

Periodontitis is a chronic multifactorial inflammatory disease affecting the teeth' supporting tissues, including the cementum, the periodontal ligament, and the alveolar bone (Papapanou et al. 2018). This condition arises from a harmful interaction between pathogenic microorganisms and the immune response of a susceptible host, leading to tissue destruction and potentially resulting in tooth loss (Loos and Van Dyke 2020). Diabetes mellitus, smoking, and genetic predispositions can worsen this condition, making it a systemic concern due to its links to various comorbidities (Chapple 2022).

Understanding the link between periodontitis and MRONJ is essential due to the increasing number of periodontitis patients taking these medications. Additionally, it is important to consider how oral health may compromise the effectiveness of pharmacological treatments.

Experimental studies indicate that periodontitis may enhance the susceptibility and occurrence of MRONJ. Animal models have demonstrated the development of osteonecrotic-like lesions in rodents that were treated with potent bisphosphonates and had induced periodontitis, even without tooth extractions or other surgical or invasive procedures (Hadaya et al. 2024; Gkouveris et al. 2022; Li et al. 2016; Aguirre et al. 2012; Aghaloo et al. 2011). These experimental studies have also been employed to

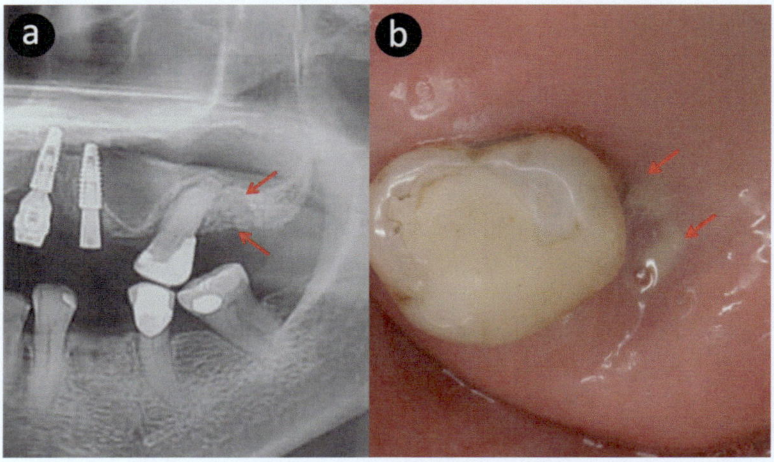

Fig. 20.1 Patient, 53 years old with a history of using bisphosphonates for 8 years. Radiographic appearance (**a**) and clinical appearance (**b**) of the left upper first molar (#26). The radiographic appearance shows extensive thickening of the periodontal ligament and the presence of an associated radiolucent lesion. Laterally to the tooth, there are areas of lesions characteristic of bone sequestration, as indicated by the red arrows. In the clinical aspect, in the mucosal region corresponding to the area of bone sequestration there is purulent secretion and an area of necrosis, as indicated by the red arrows

understand the role of the inflammatory response in the pathophysiology of MRONJ, as its etiopathogenesis remains poorly understood.

Clinical studies have also shown a strong association between periodontitis and the development of MRONJ. Nicolatou-Galitis et al. (2015) conducted a series of clinical case reports that identified various changes in periodontal tissues in cancer patients undergoing potent bisphosphonate treatment. These changes included pain, bleeding, fistulas, purulent secretion, tooth mobility, and the presence of periodontal pockets. Notably, these symptoms occurred before the development of osteonecrotic lesions. Supporting clinical studies indicate that periodontitis is one of the primary dental comorbidities associated with the development of MRONJ (Olofsson et al. 2024; Otsuru et al. 2024; Manfredi et al. 2017; Marx et al. 2005). On the other hand, Vahtsevanos et al. (2009) conducted an observational study that found no connection between periodontal disease, smoking, or endodontic treatment and an increased risk of MRONJ in patients receiving bisphosphonate treatment.

It is important to note that while tooth extractions are recognized as the primary local risk factor for the development of MRONJ (Ruggiero et al. 2022); extractions are mainly performed when inflammatory-infectious conditions affect the tooth and its supporting tissues. Thus, recent experimental studies evaluating management strategies for MRONJ have included periodontitis before tooth extraction to better mimic the clinical conditions in which MRONJ develops (Ervolino et al. 2019, 2022; Toro et al. 2019; Statkievicz et al. 2018).

The presence of infection and heightened inflammation in periodontal tissues seems to play a critical role in the mechanisms underlying MRONJ. In MRONJ experimental models, an increase in pro-inflammatory cytokines such as TNF-α, IL-1β, and IL-6 has been strongly associated with the disease development (Soma et al. 2021; Funayama et al. 2019; Morita et al. 2017). Morita et al. (2017) demonstrated that knockout mice lacking some of the key pro-inflammatory cytokines became resistant to osteonecrosis development, even when they received high doses of bisphosphonates. Soma et al. (2021) also proposed using medications such as Etanercept, a TNF-α inhibitor, or neutralizing antibodies to IL-6 as strategies capable of preventing MRONJ by reducing local inflammation. Further highlighting the participation of the inflammatory processes in the development of osteonecrotic lesions, a clinical study conducted by Nakamura et al. (2024) concluded that extracting teeth with local infections in cancer patients is more effective in preventing MRONJ than the maintenance of teeth with periodontal and/or periapical involvement.

At the cellular level, it has been proposed that bisphosphonates do not act exclusively on osteoclasts, although their activity on osteoclasts is considered this class of drugs' main mechanism of action (Cremers et al. 2020). Recent studies suggest that bisphosphonates also have secondary effects on cells of the immune system, causing dysfunctions in innate and acquired responses. Consequently, these effects alter the expression of cytokines with pro- and anti-inflammatory activity. Under these conditions, antiresorptive agents appear to stimulate the polarization and activation of M1 macrophages, phagocytes that are related to pro-inflammatory events and the production of cytokines and mediators such as TNF-α, IL-1β, IL-6, IL-12, reactive oxygen species (ROS), and nitric oxide. Furthermore, these drugs inhibit the polarization and activation of M2 macrophages, which play a crucial role in anti-inflammatory and tissue repair processes (Wehrhan et al. 2017; Zhang et al. 2021). Other defense cells are also positively influenced by bisphosphonates, such as T lymphocytes and polymorphonuclear neutrophils (de Barros Silva et al. 2016; Clézardin and Massaia 2010). Thus, the result of the action of bisphosphonates on the cells of the immune system is the exacerbation and prolongation of the local inflammatory state, which becomes greatly enhanced in the presence of other inflammatory disorders, such as periodontitis, which favors the occurrence of tissue damage and makes it difficult to resolve the process.

Under these clinical conditions, a significant increase in pro-inflammatory cytokines in periodontal and adjacent tissues seems to activate molecular signaling pathways, leading to programmed cell death in nearby osteocytes. In patients receiving bisphosphonate treatment, the strong suppression of osteoclast activity slows or even inhibits the removal of non-vital bone tissue. As a result, osteocytes undergoing apoptosis or in an advanced state of autophagy are not effectively removed and remain in the lacuno-canalicular network of the bone matrix for an extended period, thereby increasing the likelihood of secondary necrosis (Aguirre et al. 2021).

In addition to osteocyte apoptosis caused by heightened inflammation, another type of cell death associated with MRONJ is osteocyte necroptosis. Unlike apoptosis, necroptosis arises from pre-programmed intracellular pathways but leads to cell death characterized by accidental features, similar to necrosis. This process involves rupturing the plasma membrane and releasing cytoplasmic contents into the extracellular space (Souza et al. 2024; Aguirre et al. 2021; Dhuriya and Sharma 2018). Various factors have been identified as triggers for necroptosis in osteocytes, including TNF-α, lipopolysaccharides from Gram-negative bacteria, and pathogen-associated molecular patterns (PAMPs). All these molecules can be found in tissues affected by periodontitis, further strengthening the link between periodontal disease and the development of MRONJ (Grootjans et al. 2017; Pasparakis and Vandenabeele 2015). The death of osteocytes, often associated with intensified local inflammation, is the main histopathological indicator of MRONJ (Franco-Pretto et al. 2014).

Although there is still no definitive consensus that MRONJ is triggered or sustained by an infectious process, it is a fact that the presence of bacteria is a frequent histopathological finding in biopsies of these lesions (Boff et al. 2014; Kim et al. 2024). The accumulation of bacterial biofilm on the tooth surface, along with dysbiosis of the resident oral microbiota during periodontal disease, facilitates the rapid proliferation of pathogenic microorganisms, allowing them to invade deeper tissues. Bacterial endotoxins, mainly LPS from the cell wall of periodontopathogenic Gram-negative bacteria, are capable of stimulating the differentiation of monocytes and the activation of M1 macrophages, generating cellular responses that lead to the synthesis of more and more pro-inflammatory cytokines in the affected region. Furthermore, some bacterial exotoxins may be capable of activating inflammasomes or acting as superantigens, resulting in the release of IL-1β, IL-18, and other cytokines by T lymphocytes (Cavaillon 2018). Therefore, it can be inferred that the presence of infection stimulates the maintenance and exacerbation of the inflammatory process, generating self-amplifying responses that delay tissue repair and favor the triggering and progression of secondary lesions, such as osteonecrosis.

Despite the increasing prevalence of MRONJ, there is currently no well-established and fully effective treatment protocol. Preventive strategies have been emphasized due to the strong link between periodontitis and MRONJ. Periodontal health must be closely monitored throughout and after treatment with antiresorptive drugs, especially when administered at oncological doses (Ruggiero et al. 2022; Campisi et al. 2020; Mücke et al. 2016; Vandone et al. 2012). In this situation, the treatment of periodontitis should be a minimally invasive approach, such as scaling and root planing. Surgical treatment of periodontitis is not recommended in this situation. There are reports in the literature that even the use of scaling and root planing in the management of periodontitis in patients undergoing treatment with antiresorptive drugs was a precipitating factor for MRONJ (Diniz-Freitas et al. 2018; Braun and Iacono 2006). In view of this, the use of appropriate adjuvant therapies alongside scaling and root planing can enhance the efficacy and safety of periodontal treatment. Among the numerous existing adjuvant therapies, selection should be based on their ability to promote antimicrobial action, modulate immune responses, and stimulate tissue repair (Li et al. 2024).

20.3 Apical Periodontitis and MRONJ

Apical periodontitis (AP) is an inflammatory disease of infectious origin that can lead to systemic imbalance. Previous studies have linked AP to systemic inflammation, with its severity depending on the number of affected teeth. This condition has been shown to elevate levels of inflammatory cells (Azuma et al. 2017a; Samuel et al. 2018), pro-inflammatory mediators such as interleukin (IL)-6, IL-17, IL-23, tumor necrosis factor α (TNF-α), and nitric oxide (Cintra et al. 2016). Additionally, it has been demonstrated in animal models that AP has been associated with decreased insulin signaling in blood and skeletal muscle (Astolphi et al. 2013, 2015) and increased insulin resistance in these tissues (Pereira et al. 2016, 2017). These changes were also observed in the offspring of rats with apical lesions (Tsosura et al. 2019, 2021). On the other hand, patients undergoing therapy with antiresorptive drugs experience impaired healing (Hellstein et al. 2011) and a negative impact on their quality of life (Winter et al. 2022). Additionally, MRONJ has been linked to the presence of oral infections. This topic addresses the current aspects of the possible relationship between endodontic infection and MRONJ.

Endodontic infection can be described as the bacterial colonization of root canals due to pulp exposure, caused mainly by caries or dental trauma (Cintra et al. 2018). When bacteria enter the root canal system, the pulp reacts through a localized inflammatory process (Smith 2002), damaging the pulp cells and impairing pulp vitality (Blake 2023). AP is the result of this persistent bacterial contamination of the root canal system, characterized by inflammation of the periradicular tissues (Kakehashi et al. 1965; Howait et al. 2019) and degradation of the periapical bone (Cintra et al. 2021). This inflammation is limited to the periodontal ligament, where neutrophils release leukotrienes and prostaglandins and attract macrophages, which release cytokines (Nair 2004). If this initial inflammation is not resolved, it will become chronic, dominated by macrophages encapsulated in connective tissue (Nair 2004), with the presence of periapical bone degradation that occurs through the recruitment of inflammatory cells and the dynamics of osteoclasts, osteoblasts, and transcription factors, such as receptor activator of NF-κB ligand (RANKL) and receptor activator of NF-κB (RANK) (Liu and Peng 2013). In addition, the recruitment of inflammatory cells occurs (Xiong et al. 2009; Colic et al. 2010), along with the release of pro-inflammatory substances, such as interleukins and tumor necrosis factors. These mediators can lead to unfavorable results, since their release and activation initiate or potentiate the pathogenesis of various diseases, driven by the same mediators and inflammatory mechanisms (Khalighinejad et al. 2016; Cintra et al. 2018; Segura-Egea et al. 2019; Cantiga-Silva et al. 2024). Studies show that the incidence of endodontic infection in compromised individuals is higher when compared to the population without systemic disorders (Cintra et al. 2018; Katz et al. 2021). This effect on systemic health is associated with the number of foci of infection (Cintra et al. 2016; Samuel et al. 2018), making this disease a major public health problem, due to its high global prevalence, with 52% of the world's adult population having at least one tooth with AP (Tibúrcio-Machado et al. 2021).

The relationship between AP and systemic disorders or systemic conditions has been the subject of several studies, especially in relation to diabetes (Cintra et al. 2013, 2014a, b, c, 2017 Sánchez-Domínguez et al. 2015; Azuma et al. 2017b; Ferreira et al. 2017, Prieto et al. 2017 Samuel et al. 2019;), cardiovascular disease (Caplan et al. 2006; Cotti et al. 2011, 2020), liver disease (Castellanos-Cosano et al. 2013; Grønkjær et al. 2016; Cantiga-Silva et al. 2021), osteoporosis (López-López et al. 2015; Gomes-Filho et al. 2015a, 2015b), kidney disease (Stenvinkel et al. 2005; Khalighinejad et al. 2017), and pregnancy (Bain et al. 2009; Tsosura et al. 2019). On the other hand, few studies have analyzed the effect of systemic treatments on the development and severity of AP.

Few studies have explored the relationship between AP and MRONJ, making it difficult to establish a link between them. Since AP has a high prevalence worldwide and is an inflammatory and infectious lesion, which can introduce bacteria into the alveolar bone through the root canal system (Rao et al. 2017), it is hypothesized that this condition may contribute to the development of osteonecrosis. Marques-Ferreira et al. (2023) conducted an experimental animal study to investigate whether experimentally induced periapical lesions contribute to the development of MRONJ after administration of zoledronic acid (ZOL) (0.1 mg/kg). The study consisted of two groups: the control group (control animals) and the experimental group (animals that received ZOL). The animals in the experimental group received three intraperitoneal injections per week for 3 weeks. The animals in the control group followed the same protocol, but received the saline solution instead. One week after the last injection, the animals were subjected to pulp exposure for the development of AP for 21 days. After this period, the animals received an intravenous injection of 40 ± 17 MBq of 99mTechnetium hydroxymethylene diphosphonate (99mTc-HMDP) for scintigraphic analysis. Three hours after administration of the tracer, bone scintigraphy was performed in a GE Millennium gamma camera. Radiographic and histological analyses of the mandibles with periapical lesions were also performed. The results showed that periapical lesions in animals treated with ZOL were significantly smaller compared to the control group. These findings were corroborated by their radiological and nuclear medicine results, which indicated wider periapical lesions in the control group compared to the ZOL-treated group. In addition, there was an increase in radionuclide uptake in the periapical lesions, suggesting that ZOL is involved in the infection of the alveolar bone and limits the area of periapical disease.

In the study by Kang et al. (2013), an experimental model was used with mice, in which a ZOL dose higher than that used in cancer therapies was administered (200 μg/kg, approximately three times higher than the cancer dose of 66 μg/kg). The justification for this was to increase the incidence of MRONJ in the animals, considering that there seems to be a dose-dependent effect of bisphosphonates on the incidence of MRONJ in humans. The animals received intraperitoneal injections of either VEH (endotoxin-free water) or ZOL three times a week for 1 week before the periapical lesion induction. Injections continued for another 7 weeks, following the same three-times-a-week protocol. At the end of the protocol, the animals received a total of 24 ZOL injections, which corresponds to 2 years of treatment in humans with cancer (Ibrahim et al. 2003). This protocol was adopted because the average time for the onset of ONJ in patients treated with ZOL is approximately 18 months (Maerevoet et al. 2005). The mandibles were analyzed by MicroCT and by histological and immunohistochemical analysis for tartrate-resistant acid phosphatase (TRAP), a marker for osteoclast activity. The results showed that the animals treated with bisphosphonates presented less widening of the periodontal ligament space and less volumetric bone loss when compared to the group treated with VEH. It was also observed that treatment with bisphosphonates caused a reduction in the periodontal ligament space and an increase in the thickness of the lamina dura in the furcation of healthy teeth. Histological analysis showed that osteonecrosis was present only at the perforated site of the animals treated with bisphosphonates, while no osteonecrosis was observed in any of the VEH animals. On the other hand, the number of empty bone gaps was not different as a function of bisphosphonate administration. Immunostaining for TRAP showed a significant increase in the number of TRAP-positive cells in the periapex of the perforated teeth compared to healthy teeth for both animals treated with VEH or ZOL. However, there was no difference between animals treated with VEH or ZOL.

In a study with ovariectomized animals (OVX), Wayama et al. (2015) also showed regression of periapical lesions caused by the use of ZOL (0.04 mg/body). This study consisted of groups of control and ovariectomized female Wistar rats treated with vehicle or ZOL. The substances were administered intravenously once a week for 4 weeks. The animals had their dental pulps exposed to the oral environment to induce periapical lesions. The analysis periods were 7 and 30 days. The jaws were examined by MicroCT and histopathological, histometric and immunohistochemical analyses. MicroCT analysis showed that the ovariectomized group experienced greater bone loss, which was mitigated by ZOL treatment at both time points. Similarly, the ovariectomized group exhibited a higher number of TRAP-labeled cells compared to the ovariectomized group treated with ZOL. According to the histopathology, the inflammation remained relatively similar between these experimental groups, with an accentuated inflammatory response in the ovariectomized group without ZOL treatment. Thus, it was verified that the low concentration of hormones influences the exacerbation of periapical lesions, and ZOL can help contain periapical bone destruction. The authors report that their results suggest ZOL acts similarly to estrogen, inhibiting the differentiation and resorption activity of osteoclasts, inducing their apoptosis, and consequently reducing the number of TRAP-positive cells.

This decrease in periapical bone resorption in animals that received ZOL was also found in the experimental study by Rao et al. (2019) that used OVX mice with the administration of ZOL (66 μg/kg). Despite this, another study by Rao et al. (2017), also using an OVX mouse model, found that the dosage of ZOL (66 μg/kg) did not have a significant effect on reducing periapical bone loss in the presence of periapical disease, but did substantially reduce periodontal bone loss. Although one study showed a decrease in periapical bone resorption while the other did not, both had similar histological results. Both studies observed more visible and extensive areas of lacunae in the bone, along with loss of osteocytes and an increase in their death, indicating the presence of micronecrosis. Due to the antiresorptive effect of bisphosphonates, the degradation of non-viable osteocytes may be decreased, leading to their accumulation (Ikebe 2013), which can lead to bone necrosis and the development of MRONJ. This strengthens the hypothesis that bisphosphonates exert a direct toxic effect on osteocytes and cause their death (Rao et al. 2017).

These aforementioned results, observing that ZOL can inhibit alveolar bone resorption in the presence of AP, are expected. This is because bisphosphonates act by interrupting the mevalonate pathway, which has a direct impact on bone metabolism. They interact with the enzyme farnesyl diphosphate synthase, blocking the formation of farnesyl diphosphate and geranylgeranyl diphosphate, which are crucial for the prenylation of small GTPases such as Ras, Rho, and Rac, the main targets of bisphosphonates. The lack of these proteins interferes with osteoclastogenesis, inducing early apoptosis in active osteoclasts and causing significant changes in the cytoskeleton dynamics of these cells. This leads to impaired vesicular trafficking and defective formation of the clear and pleated zones, both of which are essential for the interaction of osteoclasts with the bone matrix, the creation of the resorptive microenvironment, and effective bone resorption (Soares et al. 2016; Ebetino et al. 2011; Rogers et al. 2011). Despite the gradual decrease in bone resorption, studies indicate that the bone tissue in the affected area remains avascular and non-vital (Ervolino et al. 2022; Hadad et al. 2022; Ristow et al. 2023; Nicolatou-Galitis et al. 2015). This means that although the bone still has greater volume, it is composed of tissue prone to microfractures, which can result in areas of non-vital bone, favoring infections and aggravating bone necrosis. In addition, the strong antiangiogenic effect of bisphosphonates can contribute to necrosis (Kuroshima et al. 2019; Aghaloo et al. 2015; Ruggiero et al. 2014; Badel et al. 2013).

As demonstrated, osteocytes play important roles in biological and pathological mechanisms related to MRONJ (Aguirre et al. 2021). Osteocytes are directly affected by drugs such as bisphosphonates and angiogenesis inhibitors, which reduce bone-remodeling capacity and, consequently, compromise the vitality of bone tissue. The suppression of osteocyte activity leads to reduced bone remodeling and microdamage accumulation, providing an environment favorable to bone necrosis. In addition, in situations of bone stress, such as periapical infections, the osteocytes affected by drug therapy are less able to respond adequately, intensifying the risk of MRONJ (Aguirre et al. 2021).

In a review study by de Aguirre et al. (2021), preclinical models simulating MRONJ in animals were explored, especially those involving drugs such as bisphosphonates, antiresorptive and antiangiogenic agents, associated with inflammatory dental diseases such as periapical infections. It has been reported that periapical infection acts as an inflammatory trigger that exacerbates the local bone response, contributing to tissue necrosis. It also causes an uncontrollable inflammatory response and prevents bone repair. This mechanism occurs due to the interaction between the inflammation process caused by the infection and the effects of the drug in suppressing bone remodeling, resulting in the formation of necrotic areas that can progress to MRONJ. Therefore, the article highlights that periapical infections should be carefully evaluated in patients using these drugs, as they represent an important risk for osteonecrosis of the jaws, especially in prolonged and intensive treatments.

Regarding the use of antibiotics in patients with AP who use bisphosphonates, many gaps still need to be studied. In the study by Du et al. (2022), a mouse model with long-term antibiotic treatment and combined induction of AP and subsequent MRONJ by tooth extraction was used. The findings demonstrated that animals that had AP induction with subsequent tooth extraction and treatment with ZOL had unrepaired alveoli and a significant increase in bone necrosis, as determined by empty lacunae and necrotic bone. This study revealed that long-term overtreatment with antibiotics in teeth with AP undergoing tooth extraction can further aggravate bone necrosis by increasing the number of osteoclasts. On the other hand, a study by López-Jornet et al. (2011) showed that antibiotic prophylaxis significantly reduced the development of MRONJ in rats. Prophylactic antibiotic treatment has been recommended in the clinic when tooth extraction is indicated in patients taking antiresorptive agents.

Since antibiotic therapy is a treatment recommended by the American Association of Oral and Maxillofacial Surgeons for Bisphosphonate-Related Osteonecrosis of the Jaw (Ruggiero et al. 2009), more in-depth studies should be performed in the presence of previous infections, such as AP, since it is not yet fully known how the etiopathogenesis of AP affects MRONJ.

Although there are currently few studies aimed at understanding the mechanisms by which AP may favor the development of MRONJ. There are indications that the infectious and inflammatory processes of AP may contribute to this role, since AP is a risk factor for this pathology (Fig. 20.2). Despite the conflicting results in the scientific literature, which show that bisphosphonates cause regression of periapical bone loss, the role of osteocytes can lead to the development of MRONJ. The inactivity of osteocytes in periapical lesions caused by ZOL is associated with areas of micronecrosis, and the lack of bone remodeling further makes bone tissue more susceptible to infections and necrotic processes. However, further research should be conducted to better understand the relationship between AP and the effect of using bisphosphonates, which can contribute to understanding the etiopathogenesis of MRONJ.

20.4 Peri-Implantitis and MRONJ

The discussion about the interrelationship between implants and osteonecrosis of the jaws (ONJ) is recent. The question lies in whether surgery is the cause of the pathology's development or if the pathology develops due to biomechanical processes associated with the implant/prosthesis. It is known that both local and systemic factors are associated with the development of ONJ, with systemic factors such as diabetes, hypertension, and hormonal changes, and local factors such as surgical, prosthetic, and periodontal trauma being the most related to the pathology's development (Yong et al. 2024).

Implant presence-triggered osteonecrosis is a distinct condition from MRONG but shares similarities in presentation and pathogenesis. It occurs around dental implants in patients

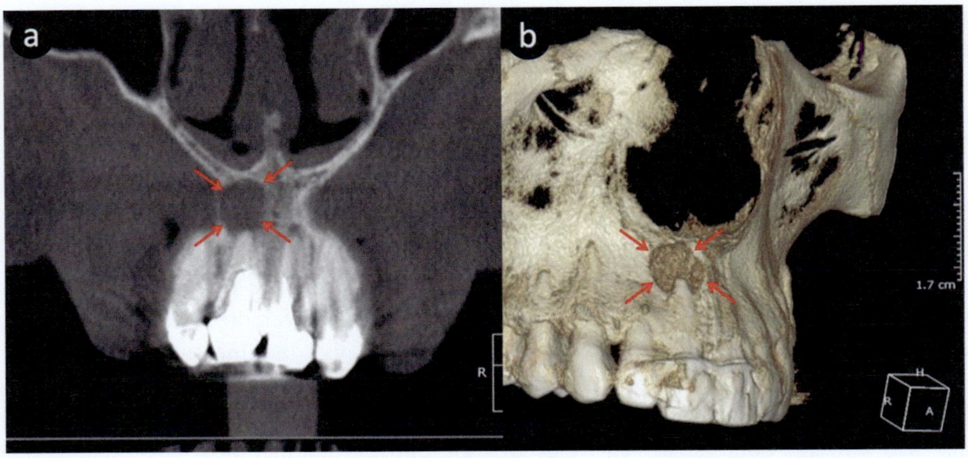

Fig. 20.2 Patient, 47 years old, with a history of using bisphosphonates for 6 years. In the tomographic appearance of the anterior region of the maxilla (**a**, **b**), there is the presence of a well-defined, circular radiolucent lesion in the region of the apex of the right central incisor (#11), characteristic of an apical periodontitis lesion, as indicated by the red arrows. The tooth corresponding to the lesion has artifacts that are consistent with the presence of a crown and a metal intraradicular post, associated with unsatisfactory endodontic treatment

undergoing antiresorptive or antiangiogenic therapy for conditions like osteoporosis and bone metastases. Unlike MRONJ, implant-triggered osteonecrosis is linked specifically to the biomechanical impact of implants (Yong et al. 2024).

Yong et al. (2024) published a scoping review looking for evidence on implant presence-triggered osteonecrosis (IPTO), focusing on its etiopathogenesis, clinical features, and management. This systematic review searched databases including PubMed, EMBASE, CINAHL, and Cochrane Library using terms such as "implant presence-triggered osteonecrosis" and "peri-implant medication-related osteonecrosis." Eligible studies included cases where implants were placed before initiating antiresorptive or antiangiogenic therapy. The studies included 108 patients, primarily postmenopausal women with a mean age of 54–95 years. The majority of cases were linked to therapies such as bisphosphonates and Denosumab, used for osteoporosis or cancer-related conditions. IPTO was diagnosed based on criteria adapted from MRONJ guidelines, with modifications. These include the presence of osseointegrated implants placed at least 6 months before the initiation of therapy and the onset of non-healing peri-implant lesions. Symptoms such as mucosal inflammation, bone resorption, and infection are prevalent. Most studies reported that IPTO occurred more frequently in the mandible (72.8%) than in the maxilla (27.2%). Common clinical signs included erythema, swelling, pain, bone exposure, and purulent discharge. Radiographic features commonly observed in ITPO cases included peri-implant osteolysis, bone sclerosis, and sequestration. Regarding the timeline, IPTO typically developed years after implant placement, with a mean time of 68 months from therapy initiation. In conclusion, this scoping review addressed that early recognition and timely surgical intervention are critical for favorable outcomes. However, more research is necessary to develop preventive strategies and improve understanding of this emerging clinical problem.

The relationship between ONJ and dental implants evolving into peri-implantitis can be understood in two ways. The first is due to surgical trauma during implant placement, which includes surgical access, osteotomy, and subsequent compression of the bone walls by the implant during placement. The second is due to biomechanical overload during the secondary stability of the implant, with the use of the prosthesis, which can alter periodontal and bone physiology (Escobedo et al. 2020; Degidi et al. 2005). These factors increase the transmission of forces to the peri-implant bone tissue, raising the metabolic demand. Due to the antiresorptive and antiangiogenic actions of these drugs, a deficiency in osteoclast response and reduced vascular supply may explain the increased occurrence of ONJ (Lorenzo-Pouso et al. 2020). This condition is still hypothesized to be a result of a combination of mechanical stress and infection. Implants lack the periodontal ligament, concentrating occlusal forces on the peri-implant bone. Antiresorptive agents impair bone remodeling, leading to microfractures that accumulate over time. These factors, combined with secondary infections, drive disease progression (Yong et al. 2024).

Although the pathophysiology of ONJ is not fully understood, especially regarding peri-implantitis, the strongest correlation involves the interaction of local factors, as previously explained, and systemic changes, such as diabetes, hypertension, hypothyroidism, adrenal system diseases, and hypovitaminosis that disrupt bone metabolism. Implants can induce local stress, triggering an inflammatory process that compromises the adjacent bone, leading to necrosis and worsening the inflammatory process, thus fueling the pathological cycle (Lorenzo-Pouso et al. 2020). This pathological cycle is intensified in peri-implantitis, as the oral cavity is highly predisposed to bacterial contamination (Seki and Hagiwara 2022; Kwon et al. 2014).

The literature classifies implant-associated osteonecrosis into those related to the consequences of the surgical procedure and those that are spontaneous. Six months before or after surgery, was defined as a key point, as this is when bone tissue maturation occurs during the osseointegration process. After this period, any complications are no longer associated with the surgical procedure (Yong et al. 2024; Seki and Hagiwara 2022; Tempesta et al. 2022). It is clear that, when correlating with factors that may increase predisposition to ONJ, spontaneity may not be the best term to describe these cases. Early cases are directly related to surgical timing and osseointegration, while late cases occur after the achievement of secondary stability.

The clinical characteristics of ONJ include changes such as edema, pain, and infection, which may lead to bone exposure around the implant, resulting in implant failure (Fig. 20.3). In some cases, the adjacent bone is also compromised, which can significantly affect the surrounding structures and severely impact the patient's quality of life (Kwon et al. 2014).

In radiographic or tomographic examination, osteolytic areas are observed in the peri-implant region, sometimes extending to the adjacent bone and with areas of detachment from the basal bone, known as bone sequestrum.

In order to accurately assess the relationship between MRONJ and peri-implantitis, the following criteria should be considered:

1. Current or prior treatment with antiresorptive or antiangiogenic agents
2. Persistent intra or extraoral bone exposure exceeding 8 weeks, or refractory peri-implant disease accompanied by clinical or radiographic evidence of compromised peri-implant bone viability
3. No history of radiation therapy or metastatic jaw disease
4. Dental implants placed at least 6 months before antiresorptive therapy, assessed and classified as osseointegrated (Yong et al. 2024)

Tempesta et al. 2022 addressed peri-implantitis-like medication-related osteonecrosis of the jaw (PI-MRONJ), a condition increasingly reported among patients undergoing treatment with antiresorptive or antiangiogenic drugs. The primary aim of the study was to explore cases of PI-MRONJ in 19 osteoporotic patients, all previously misdiagnosed with peri-implantitis, through a combination of clinical observations and histological evaluations. The study included 19 postmenopausal osteoporotic female patients treated with antiresorptive drugs. Each patient had dental implants placed at least 6 months before beginning antiresorptive therapy. These patients presented with peri-implantitis symptoms that persisted despite nonsurgical treatments.

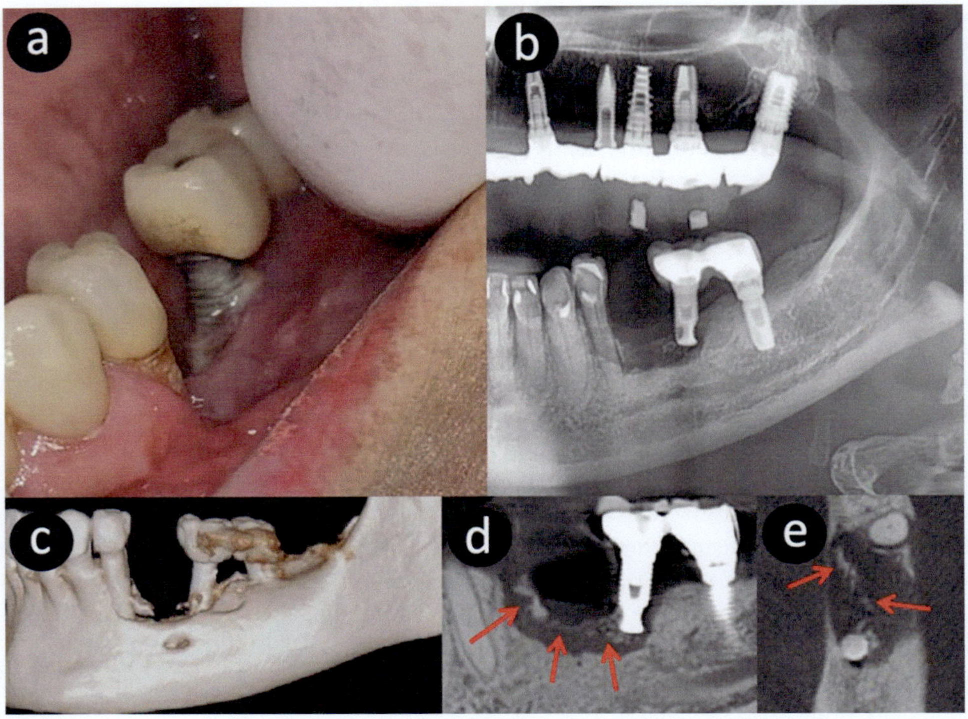

Fig. 20.3 Patient, 65 years old, with a history of osteoporosis and using zoledronic acid for 4 years to treat osteoporosis. Clinical aspect (**a**) of peri-implantitis in the lower left first molar (#36) with exposure of bone tissue 3 months after implant placement. The patient presented with pain, infection with pus drainage and implant mobility. Radiographic (**b**) and tomographic (**c**, **d**, **e**) appearance of the lesions, with the presence of bone loss involving the implant that extends to the lower first premolar (#34). In this area, bone sequestration is evident, as indicated by the red arrows

Diagnostic protocols involved clinical examinations, radiological assessments (e.g., X-rays and computed tomography), and periodontal chart reviews. In cases where peri-implantitis signs did not resolve after five nonsurgical treatments, surgical intervention was pursued following a mandatory three-month drug holiday. The surgical procedure included the removal of implants and necrotic bone, followed by osteoplasty and the application of wound-healing agents. The study observed persistent signs of peri-implantitis, such as bleeding on probing and increased probing depths, with an average of 7 mm. Radiological examinations revealed bone defects characteristic of peri-implantitis. Despite nonsurgical treatments, none of the cases achieved complete healing. Therefore, surgical removal of implants and surrounding bone was performed. Histopathological analysis of the surgical specimens revealed features typical of MRONJ, including avascular bone necrosis, inflammatory infiltrates in Haversian canals, and bacterial colonies within necrotic debris. Confocal laser scanning microscopy (CLSM) provided detailed imaging, highlighting internal osteoclastic resorption and confirming the diagnosis of PI-MRONJ. All lesions healed without recurrence during a 20-month follow-up period. This study underscores the need for heightened vigilance in diagnosing and managing PI-MRONJ, particularly in patients undergoing antiresorptive or antiangiogenic therapy. Peri-implantitis-like symptoms in such patients should prompt consideration of PI-MRONJ, even in the absence of traditional MRONJ signs. Accurate diagnosis through histopathological examination and timely surgical intervention are critical for achieving complete healing. Future studies are essential to establish standardized guidelines for PI-MRONJ prevention, early diagnosis, and management.

The management of implant-associated osteonecrosis presents a complex challenge. Conservative treatment approaches, incorporating antibiotic therapy and supportive therapies like debridement and implantoplasty, lack consensus in the literature. Conversely, surgical intervention tends to yield more favorable outcomes, though it is often delayed due to challenges in early diagnosis and suboptimal patient compliance. This challenge is further complicated by the fact that surgical treatment frequently results in implant loss and a reversion to edentulism, making it a less desirable option for patients. (Vanpoecke et al. 2020).

Initial conservative treatments, such as antibiotics and non-surgical peri-implant cleaning, are generally ineffective (Junquera et al. 2011; Marín-Fernández et al. 2015; Ueda et al. 2022). Surgical intervention, including implant removal and bone debridement, is the most effective approach. While surgical outcomes are favorable, they often lead to loss of functional bone stock, complicating oral rehabilitation (Yong et al. 2024).

Regardless of the chosen treatment for peri-implantitis associated with MRONJ, it is important to keep in mind that the soft tissue and bone are compromised. Adjuvant therapies, even though they lack clinical consensus, can be considered as options for tissue optimization (Ervolino et al. 2019, 2022; Ruggiero et al. 2022). When patients are eligible for hyperbaric oxygen therapy sessions, the use of drugs such as pentoxifylline and tocopherol, along with other therapies with antimicrobial and biomodulatory effects, such as antimicrobial photodynamic therapy, low-level laser therapy, and ozone therapy, may contribute to stabilizing tissues compromised by MRONJ (Colapinto et al. 2023; Kimura et al. 2024; Silva et al. 2024).

Therefore, professional and patient awareness represent the initial step in preventing complications, as certain situations necessitate withholding dental implant treatment. In instances where patients have undergone dental implant rehabilitation and subsequently require antiresorptive or antiangiogenic therapy, meticulous consideration must be given to occlusal force distribution, peri-implant keratinized tissue presence, prosthetic connection stability, and regular follow-up appointments (Lorenzo-Pouso et al. 2020). According to Yong et al. (2024), dentists are supposed to monitor patients undergoing antiresorptive therapy more closely, especially those with persistent peri-implantitis unresponsive to standard treatments. Enhanced awareness can facilitate early intervention, reducing morbidity.

References

Aghaloo TL, Kang B, Sung EC, Shoff M, Ronconi M, Gotcher JE, Bezouglaia O, Dry SM, Tetradis S (2011) Periodontal disease and bisphosphonates induce osteonecrosis of the jaws in the rat. J Bone Miner Res 26(8):1871–1882

Aghaloo T, Hazboun R, Tetradis S (2015) Pathophysiology of osteonecrosis of the jaws. Oral Maxillofac Surg Clin North Am 27(4):489–496

Aguirre JI, Akhter MP, Kimmel DB, Pingel JE, Williams A, Jorgensen M, Kesavalu L, Wronski TJ (2012) Oncologic doses of zoledronic acid induce osteonecrosis of the jaw-like lesions in rice rats (Oryzomys palustris) with periodontitis. J Bone Miner Res 27(10):2130–2143

Aguirre JI, Castillo EJ, Kimmel DB (2021) Biologic and pathologic aspects of osteocytes in the setting of medication-related osteonecrosis of the jaw (MRONJ). Bone 153:116168

Astolphi RD, Curbete MM, Colombo NH, Shirakashi DJ, Chiba FY, Prieto AK, Cintra LT, Bomfim SR, Ervolino E, Sumida DH (2013) Periapical lesions decrease insulin signal and cause insulin resistance. J Endodont 39(5):648–652

Astolphi RD, Curbete MM, Chiba FY, Cintra LT, Ervolino E, da Mota MS, Antoniali C, Garbin CA, Sumida DH (2015) Periapical lesions decrease insulin signaling in rat skeletal muscle. J Endodont 41(8): 1305–1310

Azuma MM, Gomes-Filho JE, Ervolino E, Pipa CB, Cardoso CBM, Andrada AC, Kawai T, Cintra LTA (2017a) Omega 3 fatty acids reduce bone resorption while promoting bone generation in rat apical periodontitis. J Endodont 43(6):970–976

Azuma MM, Gomes-Filho JE, Prieto AKC, Samuel RO, de Lima VMF, Sumida DH, Ervolino E, Cintra LTA (2017b) Diabetes increases interleukin-17 levels in periapical, hepatic, and renal tissues in rats. Arch Oral Biol 83:230–235

Badel T, Pavicin IS, Carek AJ, Rosin-Grget K, Grbesa D (2013) Pathophysiology of osteonecrosis of the jaw in patients treated with bisphosphonate. Coll Antropol 37(2):645–651

Bain JL, Lester SR, Henry WD, Naftel JP, Johnson RB (2009) Effects of induced periapical abscesses on rat pregnancy outcomes. Arch Oral Biol 54(2):162–171

Barasch A, Cunha-Cruz J, Curro FA, Hujoel P, Sung AH, Vena D, Voinea-Griffin AE, Beadnell S, Craig RG, DeRouen T, Desaranayake A, Gilbert A, Gilbert GH, Goldberg K, Hauley R, Hashimoto M, Holmes J, Latzke B, Leroux B, Yin W (2011) Risk factors for osteonecrosis of the jaws: a case-control study from the CONDOR dental PBRN. J Dent Res 90(4):439–444

Blake A, Tuttle T, & McKinney R (2023) Apical Periodontitis. In StatPearls. StatPearls Publishing

Boff RC, Salum FG, Figueiredo MA, Cherubini K (2014) Important aspects regarding the role of microorganisms in bisphosphonate-related osteonecrosis of the jaws. Arch Oral Biol 59(8):790–799

Bolette A, Lecloux G, Rompen E, Albert A, Kerckhofs G, Lambert F (2019) Influence of induced infection in medication-related osteonecrosis of the jaw development after tooth extraction: a study in rats. J Craniomaxillofac Surg 47(2):349–356

Boonyapakorn T, Schirmer I, Reichart PA, Sturm I, Massenkeil G (2008) Bisphosphonate-induced osteonecrosis of the jaws: prospective study of 80 patients with multiple myeloma and other malignancies. Oral Oncol 44(9):857–869

Braun E, Iacono VJ (2006) Bisphosphonates: case report of nonsurgical periodontal therapy and osteochemonecrosis. Int J Periodontics Restorative Dent 26(4):315–319

Campisi G, Mauceri R, Bertoldo F, Bettini G, Biasotto M, Colella G, Consolo U, Di Fede O, Favia G, Fusco V, Gabriele M, Lo Casto A, Lo Muzio L, Marcianò A, Mascitti M, Meleti M, Mignogna MD, Oteri G, Panzarella V, Romeo U et al (2020) Medication-related osteonecrosis of jaws (MRONJ) prevention and diagnosis: Italian consensus update 2020. Int J Environ Res Public Health 17(16):5998

Cantiga-Silva C, Estrela C, Segura-Egea JJ, Azevedo JP, de Oliveira PHC, Cardoso CBM, Pinheiro TN, Ervolino E, Sivieri-Araújo G, Cintra LTA (2021) Inflammatory profile of apical periodontitis associated with liver fibrosis in rats: histological and immunohistochemical analysis. Int Endod J 54(8):1353–1361

Cantiga-Silva C, de Oliveira PHC, Faria FD, Justo MP, Sivieri-Araújo G, Ervolino E, Pinheiro TN, Segura-Egea JJ, Cintra LTA (2024) Increase toll-like receptors 2 and 4 in apical periodontitis of rats with chronic liver disease. Odontology *113:296*

Caplan DJ, Chasen JB, Krall EA, Cai J, Kang S, Garcia RI, Offenbacher S, Beck JD (2006) Lesions of endodontic origin and risk of coronary heart disease. J Dent Res 85(11):996–1000

Castellanos-Cosano L, Machuca-Portillo G, Segura-Sampedro JJ, Torres-Lagares D, López-López J, Velasco-Ortega E, Segura-Egea JJ (2013) Prevalence of apical periodontitis and frequency of root canal treatments in liver transplant candidates. Med Oral 18(5):e773–e779

Cavaillon JM (2018) Exotoxins and endotoxins: inducers of inflammatory cytokines. Toxicon 149:45–53

Chapple I (2022) Time to take gum disease seriously. Br Dent J 232:360–361

Cintra LT, da Silva Facundo AC, Azuma MM, Sumida DH, Astolphi RD, Bomfim SR, Narciso LG, Gomes-Filho JE (2013) Pulpal and periodontal diseases increase triglyceride levels in diabetic rats. Clin Oral Investig 17(6):1595–1599

Cintra LT, da Silva Facundo AC, Prieto AK, Sumida DH, Narciso LG, Mogami Bomfim SR et al (2014a) Blood profile and histology in oral infections associated with diabetes. J Endodont 40(8):1139–1144

Cintra LT, Samuel RO, Azuma MM, Ribeiro CP, Narciso LG, de Lima VM, Sumida DH, Coclete GA, Dezan-Júnior E, Gomes-Filho JE (2014b) Apical periodontitis and periodontal disease increase serum IL-17 levels in normoglycemic and diabetic rats. Clin Oral Investig 18(9):2123–2128

Cintra LT, Samuel RO, Facundo AC, Prieto AK, Sumida DH, Bomfim SR, Souza JC, Dezan-Júnior E, Gomes-Filho JE (2014c) Relationships between oral infections and blood glucose concentrations or HbA1c levels in normal and diabetic rats. Int Endod J 47(3):228–237

Cintra LT, Samuel RO, Azuma MM, de Queiróz AO, Ervolino E, Sumida DH, de Lima VM, Gomes-Filho JE (2016) Multiple apical periodontitis influences serum levels of cytokines and nitric oxide. J Endodont 42(5):747–751

Cintra LT, Samuel RO, Prieto AK, Sumida DH, Dezan-Júnior E, Gomes-Filho JE (2017) Oral health, diabetes, and body weight. Arch Oral Biol 73:94–99

Cintra LTA, Estrela C, Azuma MM, Queiroz ÍOA, Kawai T, Gomes-Filho JE (2018) Endodontic medicine: interrelationships among apical periodontitis, systemic disorders, and tissue responses of dental materials. Braz Oral Res 32(suppl 1):e68

Cintra LTA, Gomes MS, da Silva CC, Faria FD, Benetti F, Cosme-Silva L, Samuel RO, Pinheiro TN, Estrela C, González AC, Segura-Egea JJ (2021) Evolution of endodontic medicine: a critical narrative review of the interrelationship between endodontics and systemic pathological conditions. Odontology 109(4):741–769

Clézardin P, Massaia M (2010) Nitrogen-containing bisphosphonates and cancer immunotherapy. Curr Pharm Des 16(27):3007–2014

Colapinto G, Goker F, Nocini R, Albanese M, Nocini PF, Sembronio S, Argenta F, Robiony M, Del Fabbro M (2023) Outcomes of a pharmacological protocol with Pentoxifylline and tocopherol for the Management of Medication-Related Osteonecrosis of the jaws (MRONJ): a randomized study on 202 osteoporosis patients. J Clin Med 12(14):4662

Colic M, Gazivoda D, Vasilijic S, Vucevic D, Lukic A (2010) Production of IL-10 and IL-12 by antigen-presenting cells in periapical lesions. J Oral Pathol Med 39(9):690–696

Conti LC, Segura-Egea JJ, Cardoso CBM, Benetti F, Azuma MM, Oliveira PHC, Bomfim SRM, Cintra LTA (2020) Relationship between apical periodontitis and atherosclerosis in rats: lipid profile and histological study. Int Endod J 53(10):1387–1397

Cotti E, Dessì C, Piras A, Mercuro G (2011) Can a chronic dental infection be considered a cause of cardiovascular disease? A review of the literature. Int J Cardiol 148(1):4–10

Cremers S, Ebetino FH, Phipps R (2020) On the pharmacological evaluation of bisphosphonates in humans. Bone 139:115501

de Barros Silva PG, de Oliveira CC, Brizeno L, Wong D, Lima Júnior R, Gonçalves RP, Sousa FB, Mota M, de Albuquerque Ribeiro R, Alves A (2016) Immune cellular profile of bisphosphonate-related osteonecrosis of the jaw. Oral Dis 22(7):649–657

Degidi M, Scarano A, Piattelli M, Perrotti V, Piattelli A (2005) Bone remodeling in immediately loaded and unloaded titanium dental implants: a histologic and histomorphometric study in humans. J Oral Implantol 31(1):18–24

Dhuriya YK, Sharma D (2018) Necroptosis: a regulated inflammatory mode of cell death. J Neuroinflammation 15(1):199

Dimopoulos MA, Kastritis E, Bamia C, Melakopoulos I, Gika D, Roussou M, Migkou M, Eleftherakis-Papaiakovou E, Christoulas D, Terpos E, Bamias A (2009) Reduction of osteonecrosis of the jaw (ONJ) after implementation of preventive measures in patients with multiple myeloma treated with zoledronic acid. Ann Oncol 20(1):117–120

Diniz-Freitas M, Fernández-Feijoo J, Diz Dios P, Pousa X, Limeres J (2018) Denosumab-related osteonecrosis of the jaw following non-surgical periodontal therapy: a case report. J Clin Periodontol 45(5):570–577

Du W, Yang M, Kim T, Kim S, Williams DW, Esmaeili M, Hong C, Shin KH, Kang MK, Park NH, Kim RH (2022) Indigenous microbiota protects development of medication-related osteonecrosis induced by periapical disease in mice. Int J Oral Sci 14(1):16

Ebetino FH, Hogan AM, Sun S, Tsoumpra MK, Duan X, Triffitt JT, Kwaasi AA, Dunford JE, Barnett BL, Oppermann U, Lundy MW, Boyde A, Kashemirov BA, McKenna CE, Russell RG (2011) The relationship between the chemistry and biological activity of the bisphosphonates. Bone 49(1):20–33

Epstein MS, Wicknick FW, Epstein JB, Berenson JR, Gorsky M (2010) Management of bisphosphonate-associated osteonecrosis: pentoxifylline and tocopherol in addition to antimicrobial therapy. An initial case series. Oral Surg Oral Med Oral Pathol Oral Radiol Endod 110(5):593–596

Ervolino E, Statkievicz C, Toro LF, de Mello-Neto JM, Cavazana TP, Issa JPM, Dornelles RCM, de Almeida JM, Nagata MJH, Okamoto R, Casatti CA, Garcia VG, Theodoro LH (2019) Antimicrobial photodynamic therapy improves the alveolar repair process and prevents the occurrence of osteonecrosis of the jaws after tooth extraction in senile rats treated with zoledronate. Bone 120:101–113

Ervolino E, Olivo MB, Toro LF, Freire JOA, Ganzaroli VF, Guiati IZ, Nuernberg MAA, Franciscon JPS, Ângelo Cintra LT, Garcia VG, Wainwright M, Theodoro LH (2022) Effectiveness of antimicrobial photodynamic therapy mediated by butyl toluidine blue in preventing medication-related osteonecrosis of the jaws in rats. Photodiagn Photodyn Ther 40:103172

Escobedo MF, Cobo JL, Junquera S, Milla J, Olay S, Junquera LM (2020) Medication-related osteonecrosis of the jaw. Implant presence-triggered osteonecrosis: case series and literature review. J Stomatol Oral Maxillofac Surg 121(1):40–48

Ferreira LL, Gomes JE, Filho S et al (2017) Diabetic rats present high mean platelet count in the presence of oral infections. Braz Dent J 28(5):548–551

Ficarra G, Beninati F, Rubino I, Vannucchi A, Longo G, Tonelli P, Pini Prato G (2005) Osteonecrosis of the jaws in periodontal patients with a history of bisphosphonates treatment. J Clin Periodontol 32(11): 1123–1128

Franco-Pretto E, Pacheco M, Moreno A, Messa O, Gnecco J (2014) Bisphosphonate-induced osteonecrosis of the jaws: clinical, imaging, and histopathology findings. Oral Surg Oral Med Oral Pathol Oral Radiol 118(4):408–417

Funayama H, Tashima I, Okada S, Ogawa T, Yagi H, Tada H, Wakita R, Asada Y, Endo Y (2019) Effects of Zoledronate on local and systemic production of IL-1β, IL-18, and TNF-α in mice and augmentation by lipopolysaccharide. Biol Pharm Bull 42(6):929–936

Gkouveris I, Hadaya D, Elzakra N, Soundia A, Bezouglaia O, Dry SM, Pirih F, Aghaloo T, Tetradis S (2022) Inhibition of HMGB1/RAGE signaling reduces the incidence of medication-related osteonecrosis of the jaw (MRONJ) in mice. J Bone Miner Res 37(9):1775–1786

Gomes-Filho JE, Wayama MT, Dornelles RC, Ervolino E, Yamanari GH, Lodi CS, Sivieri-Araújo G, Dezan-Júnior E, Cintra LT (2015a) Raloxifene modulates regulators of osteoclastogenesis and angiogenesis in an oestrogen deficiency periapical lesion model. Int Endod J 48(11):1059–1068

Gomes-Filho JE, Wayama MT, Dornelles RC, Ervolino E, Coclete GA, Duarte PC, Yamanri GH, Lodi CS, Dezan-Júnior E, Cintra LT (2015b) Effect of raloxifene on periapical lesions in ovariectomized rats. J Endodont 41(5):671–675

Grønkjær LL, Holmstrup P, Schou S, Schwartz K, Kongstad J, Jepsen P, Vilstrup H (2016) Presence and consequence of tooth periapical radiolucency in patients with cirrhosis. Hepat Med 8:97–103

Grootjans S, Vanden Berghe T, Vandenabeele P (2017) Initiation and execution mechanisms of necroptosis: an overview. Cell Death Differ 24(7):1184–1195

Hadad H, Kawamata de Jesus L, Piquera Santos AF, Rinaldi Matheus H, de Souza Rodrigues LG, Paolo Poli P et al (2022) Beta tricalcium phosphate, either alone or in combination with antimicrobial photodynamic therapy or doxycycline, prevents medication-related osteonecrosis of the jaw. Sci Rep 12(1):16510

Hadaya D, Soundia A, Gkouveris I, Dry SM, Aghaloo TL, Tetradis S (2019) Development of medication-related osteonecrosis of the jaw after extraction of teeth with experimental periapical disease. J Oral Maxillofac Surg 77(1):71–86

Hadaya D, Soundia A, Bezouglaia O, Pirih F, Aghaloo TL, Tetradis S (2024) The incidence and severity of medication related osteonecrosis of the jaws is similar in male and female mice. J Oral Maxillofac Surg 82(9): 1139–1146

Hasegawa T, Hayashida S, Kondo E, Takeda Y, Miyamoto H, Kawaoka Y, Ueda N, Iwata E, Nakahara H, Kobayashi M, Soutome S, Yamada SI, Tojyo I, Kojima Y, Umeda M, Fujita S, Kurita H, Shibuya Y, Kirita T, Komori T et al (2019) Medication-related osteonecrosis of the jaw after tooth extraction in cancer patients: a multicenter retrospective study. Osteoporos Int 30(1):231–239

Hellstein JW, Adler RA, Edwards B, Jacobsen PL, Kalmar JR, Koka S, Migliorati CA, Ristic H et al (2011) Managing the care of patients receiving antiresorptive therapy for prevention and treatment of osteoporosis: executive summary of recommendations from the American dental association council on scientific affairs. J Am Dent Assoc 142(11):1243–1251

Howait M, Albassam A, Yamada C, Sasaki H, Bahammam L, Azuma MM, Cintra LTA, Satoskar AR, Yamada S, White R, Kawai T, Movila A (2019) Elevated expression of macrophage migration inhibitory factor promotes inflammatory bone resorption induced in a mouse model of Periradicular periodontitis. J Immunol 202(7):2035–2043

Ibrahim A, Scher N, Williams G, Sridhara R, Li N, Chen G, Leighton J, Booth B, Gobburu JV, Rahman A, Hsieh Y, Wood R, Vause D, Pazdur R (2003) Approval summary for zoledronic acid for treatment of multiple myeloma and cancer bone metastases. Clin Cancer Res 9(7):2394–2399

Ikebe T (2013) Pathophysiology of BRONJ: drug-related osteoclastic disease of the jaw. Oral Science International 10(1):1–8

Junquera L, Gallego L, Pelaz A (2011) Multiple myeloma and bisphosphonate-related osteonecrosis of the mandible associated with dental implants. Case Rep Dent 2011:568246

Kakehashi S, Stanley HR, Fitzgerald RJ (1965) The effects of surgical exposures of dental pulps in germ-free and conventional laboratory rats. Oral Surg Oral Med Oral Pathol 20:340–349

Katz J, & Rotstein I (2021) Prevalence of Periapical Lesions in Patients with Osteoporosis. J Endod 47(2): 234–238. https://doi.org/10.1016/j.joen.2020.10.019

Kang B, Cheong S, Chaichanasakul T, Bezouglaia O, Atti E, Dry SM, Pirih FQ, Aghaloo TL, Tetradis S (2013) Periapical disease and bisphosphonates induce osteonecrosis of the jaws in mice. J Bone Miner Res 28(7):1631–1640

Katsarelis H, Shah NP, Dhariwal DK, Pazianas M (2015) Infection and medication-related osteonecrosis of the jaw. J Dent Res 94(4):534–539

Khalighinejad N, Aminoshariae MR, Aminoshariae A, Kulild JC, Mickel A, Fouad AF (2016) Association between systemic diseases and apical periodontitis. J Endodont 42(10):1427–1434

Khalighinejad N, Aminoshariae A, Kulild JC, Sahly K, Mickel A (2017) Association of end-stage renal disease with radiographically and clinically diagnosed apical periodontitis: a hospital-based study. J Endodont 43(9):1438–1441

Khan AA, Morrison A, Kendler DL, Rizzoli R, Hanley DA, Felsenberg D, McCauley LK, O'Ryan F, Reid IR, Ruggiero SL, Taguchi A, Tetradis S, Watts NB, Brandi ML, Peters E, Guise T, Eastell R, Cheung AM, Morin SN, Masri B et al (2017) Case-based review of osteonecrosis of the jaw (ONJ) and application of the international recommendations for management from the international task force on ONJ. J Clin Densitom 20(1):8–24

Kim T, Kim S, Song M, Lee C, Yagita H, Williams DW, Sung EC, Hong C, Shin KH, Kang MK, Park NH, Kim RH (2018) Removal of pre-existing periodontal inflammatory condition before tooth extraction ameliorates medication-related osteonecrosis of the jaw-like lesion in mice. Am J Pathol 188(10): 2318–2327

Kim HY, Jung YS, Park W, Choi YJ, Kim JY (2024) Can medication-related osteonecrosis of the jaw be attributed to specific microorganisms through oral microbiota analyses? A preliminary study. BMC Oral Health 24(1):160

Kimura T, Kusano K, Sakata KI, Sato J, Kitagawa Y (2024) The effectiveness of hyperbaric oxygen therapy on older patients with medication-related osteonecrosis of the jaws: a case series. Cureus 16(9):e69226

Kuroshima S, Sasaki M, Sawase T (2019) Medication-related osteonecrosis of the jaw: a literature review. J Oral Biosci 61(2):99–104

Kwon T, Lee C, Park J, Choi S, Rijal G, Shin H (2014) Osteonecrosis associated with dental implants in patients undergoing bisphosphonate treatment. Clin Oral Implants Res 25(5):632–640

Li CL, Lu WW, Seneviratne CJ, Leung WK, Zwahlen RA, Zheng LW (2016) Role of periodontal disease in bisphosphonate-related osteonecrosis of the jaws in ovariectomized rats. Clin Oral Implants Res 27(1):1–6

Li J, Wang Y, Tang M, Zhang C, Fei Y, Li M, Li M, Gui S, Guo J (2024) New insights into nanotherapeutics for periodontitis: a triple concerto of antimicrobial activity, immunomodulation and periodontium regeneration. J Nanobiotechnol 22(1):19

Liu L, Peng B (2013) The expression of macrophage migration inhibitory factor is correlated with receptor activator of nuclear factor kappa B ligand in induced rat periapical lesions. J Endodont 39(8):984–989

Loos BG, Van Dyke TE (2020) The role of inflammation and genetics in periodontal disease. Periodontol 83(1): 26–39

López-Jornet P, Camacho-Alonso F, Martínez-Canovas A, Molina-Miñano F, Gómez-García F, Vicente-Ortega V (2011) Perioperative antibiotic regimen in rats treated with pamidronate plus dexamethasone and subjected to dental extraction: a study of the changes in the jaws. J Oral Maxillofac Surg 69(10):2488–2493

López-López J, Castellanos-Cosano L, Estrugo-Devesa A, Gómez-Vaquero C, Velasco-Ortega E, Segura-Egea JJ (2015) Radiolucent periapical lesions and bone mineral density in post-menopausal women. Gerodontology 32(3):195–201

Lorenzo-Pouso AI, Pérez-Sayáns M, Chamorro-Petronacci C, Gándara-Vila P, López-Jornet P, Carballo J, García-García A (2020) Association between periodontitis and medication-related osteonecrosis of the jaw: a systematic review and meta-analysis. J Oral Pathol Med 49(3):190–200

Maerevoet M, Martin C, Duck L (2005) Osteonecrosis of the jaw and bisphosphonates. N Engl J Med 353(1): 99–102

Manfredi M, Mergoni G, Goldoni M, Salvagni S, Merigo E, Meleti M, Vescovi P (2017) A 5-year retrospective longitudinal study on the incidence and the risk factors of osteonecrosis of the jaws in patients treated with zoledronic acid for bone metastases from solid tumors. Med Oral 22(3):e342–e348

Marín-Fernández AB, García Medina B, Aguilar-Salvatierra A, Jiménez-Burkhardt A, Gómez-Moreno G (2015) Jaw osteonecrosis management around a dental implant inserted 2 years before starting treatment with zoledronic acid. J Clin Exp Dent 7(3):e444–e446

Marques-Ferreira M, Abrantes AM, Paula A, Laranjo M, Pires AS, Caramelo F, Segura-Egea JJ, Brito A, Carvalho L, Botelho MF, Carrilho E, Marto CM, Paulo S (2023) The role of apical periodontitis disease in the development of bisphosphonate-related osteonecrosis of the jaw: an animal study. Dent J 11(7):168

Martins AS, Correia JA, Salvado F, Caldas C, Santos N, Capelo A, Palmela P (2017) Relevant factors for treatment outcome and time to healing in medication-related osteonecrosis of the jaws – a retrospective cohort study. J Craniomaxillofac Surg 45(10): 1736–1742

Marx RE, Sawatari Y, Fortin M, Broumand V (2005) Bisphosphonate-induced exposed bone (osteonecrosis/osteopetrosis) of the jaws: risk factors, recognition, prevention, and treatment. J Oral Maxillofac 63(11):1567–1575

Mawardi H, Treister N, Richardson P, Anderson K, Munshi N, Faiella RA, Woo SB (2009) Sinus tracts – an early sign of bisphosphonate-associated osteonecrosis of the jaws? J Oral Maxillofac Surg 67(3):593–601

McGowan K, McGowan T, Ivanovski S (2018) Risk factors for medication-related osteonecrosis of the jaws: a systematic review. Oral Dis 24(4):527–536

Morita M, Iwasaki R, Sato Y, Kobayashi T, Watanabe R, Oike T, Nakamura S, Keneko Y, Miyamoto K, Ishihara K, Iwakura Y, Ishii K, Matsumoto M, Nakamura M, Kawana H, Nakagawa T, Miyamoto T (2017) Elevation of pro-inflammatory cytokine levels following anti-resorptive drug treatment is required for osteonecrosis development in infectious osteomyelitis. Sci Rep 7:46322

Mücke T, Deppe H, Hein J, Wolff KD, Mitchell DA, Kesting MR, Retz M, Gschwend JE, Thalgott M (2016) Prevention of bisphosphonate-related osteonecrosis of the jaws in patients with prostate cancer treated with zoledronic acid – a prospective study over 6 years. J Craniomaxillofac Surg 44(10): 1689–1693

Nair PN (2004) Pathogenesis of apical periodontitis and the causes of endodontic failures. Crit Rev Oral Biol Med 15(6):348–381

Nakamura N, Otsuru M, Miyoshi T, Suyama K, Omori K, Morishita K, Soutome S, Rokutanda S, Miura KI, Umeda M (2024) Risk factors for the development of medication-related osteonecrosis of the jaw and effects of tooth extraction with local infection. J Dental Sci 19(3):1770–1782

Nicolatou-Galitis O, Razis E, Galiti D, Galitis E, Labropoulos S, Tsimpidakis A, Sgouros J, Karampeazis A, Migliorati C (2015) Periodontal disease preceding osteonecrosis of the jaw (ONJ) in cancer patients receiving antiresorptives alone or combined with targeted therapies: report of 5 cases and literature review. Oral Surg Oral Med Oral Pathol Oral Radiol 120(6):699–706

Olofsson R, Korytowska M, Almhöjd U, Almståhl A, Cevik-Aras H (2024) Oral health, dental treatment, and medication related osteonecrosis of the jaw in multiple myeloma – a longitudinal cohort study. BMC Oral Health 24(1):184

Otsuru M, Fujiki Y, Soutome S, Nakamura N, Miyoshi T, Naruse T, Ohnuma M, Hotokezaka Y, Rokutanda S, Umeda M (2024) Risk factors for dental findings of the development of medication-related osteonecrosis of the jaw: investigation of 3734 teeth in cancer patients receiving high dose antiresorptive agents. J Dental Sci 19(1):203–210

Otto S, Pautke C, Martin Jurado O, Nehrbass D, Stoddart MJ, Ehrenfeld M, Zeiter S (2017) Further development of the MRONJ minipig large animal model. J Craniomaxillofac Surg 45(9):1503–1514

Papapanou PN, Sanz M, Buduneli N, Dietrich T, Feres M, Fine DH, Flemmig TF, Garcia R, Giannobile WV, Graziani F, Greenwell H, Herrera D, Kao RT, Kebschull M, Kinane DF, Kirkwood KL, Kocher T, Kornman KS, Kumar PS, Loos BG et al (2018) Periodontitis: consensus report of workgroup 2 of the 2017 world workshop on the classification of periodontal and Peri-implant diseases and conditions. J Periodontol 89(Suppl 1):S173–S182

Pasparakis M, Vandenabeele P (2015) Necroptosis and its role in inflammation. Nature 517(7534):311–320

Pereira RF et al (2016) Periapical lesions decrease Akt serine phosphorylation and plasma membrane GLUT4 content in rat skeletal muscle. Clin Oral Investig 20(7):1625–1630

Pereira RF, Cintra LTA, Tessarin GWL, Chiba FY et al (2017) Periapical lesions increase macrophage infiltration and inflammatory signaling in muscle tissue of rats. J Endodont 43(6):982–988

Prieto AKC, Gomes-Filho JE, Azuma MM, Sivieri-Araújo G, Narciso LG, Souza JC, Ciarlini PC, Cintra LTA (2017) Influence of apical periodontitis on stress oxidative parameters in diabetic rats. J Endodont 43(10):1651–1656

Rao NJ, Wang JY, Yu RQ, Leung YY, Zheng LW (2017) Role of periapical diseases in medication-related osteonecrosis of the jaws. Biomed Res Int 2017: 1560175

Rao NJ, Yu RQ, Wang JY, Helm A, Zheng LW (2019) Effect of periapical diseases in development of MRONJ in immunocompromised mouse model. Biomed Res Int 2019:1271492

Ristow O, Schnug G, Smielowksi M, Moratin J, Pilz M, Engel M, Freudlsperger C, Hoffmann J, Rückschloß T (2023) Diagnostic accuracy comparing OPT and CBCT in the detection of non-vital bone changes before tooth extractions in patients with antiresorptive intake. Oral Dis 29(3):1039–1049

Rogers MJ, Crockett JC, Coxon FP, Mönkkönen J (2011) Biochemical and molecular mechanisms of action of bisphosphonates. Bone 49(1):34–41

Ruggiero SL, Dodson TB, Assael LA, Landesberg R, Marx RE, Mehrotra B et al (2009) American Association of oral and maxillofacial surgeons position paper on bisphosphonate-related osteonecrosis of the jaw – 2009 update. Aust Endod J 35(3):119–130

Ruggiero SL, Dodson TB, Fantasia J, Goodday R, Aghaloo T, Mehrotra B, O'Ryan F et al (2014) American Association of oral and maxillofacial surgeons position paper on medication-related osteonecrosis of the jaw – 2014 update. J Oral Maxillofac Surg 72(10): 1938–1956

Ruggiero SL, Dodson TB, Aghaloo T, Carlson ER, Ward BB, Kademani D (2022) American Association of Oral and Maxillofacial Surgeons' position paper on medication-related osteonecrosis of the Jaws-2022 update. J Oral Maxillofac Surg 80(5):920–943

Samuel RO, Gomes-Filho JE, Azuma MM, Sumida DH, de Oliveira SHP, Chiba FY, Bomfim SRM, Ciarlini PC, Narciso LG, Cintra LTA (2018) Endodontic infections increase leukocyte and lymphocyte levels in the blood. Clin Oral Investig 22(3):1395–1401

Samuel RO, Ervolino E, de Azevedo Queiroz ÍO, Azuma MM, Ferreira GT, Cintra LTA (2019) Th1/Th2/Th17/Treg balance in apical periodontitis of Normoglycemic and diabetic rats. J Endodont 45(8):1009–1015

Sánchez-Domínguez B, López-López J, Jané-Salas E, Castellanos-Cosano L, Velasco-Ortega E, Segura-Egea JJ (2015) Glycated hemoglobin levels and prevalence of apical periodontitis in type 2 diabetic patients. J Endodont 41(5):601–606

Schiodt M, Vadhan-Raj S, Chambers MS, Nicolatou-Galitis O, Politis C, Coropciuc R, Fedele S, Jandial D, Zhang J, Ma H, Saunders DP (2018) A multicenter case registry study on medication-related osteonecrosis of the jaw in patients with advanced cancer. Support Care Cancer 26(6):1905–1915

Segura-Egea JJ, Cabanillas-Balsera D, Jiménez-Sánchez MC, Martín-González J (2019) Endodontics and diabetes: association versus causation. Int Endod J 52(6): 790–802

Seki K, Hagiwara Y (2022) Peri-implantitis-induced medication-related osteonecrosis of the jaw: a case report. J Dental Sci 17(1):576–577

Silva MC, Delamura IF, de Sá Simon ME, Barbosa S, Ting DT, Bechara K, Shibli JA, Mourão CF, Bassi APF, Ervolino E, Faverani LP (2024) Is there an ideal concentration of ozonized oil for the prevention and modulation of Zoledronate-induced mandibular osteonecrosis? A study on senescent rats. J Funct Biomater 15(12):353

Smith AJ (2002) Pulpal responses to caries and dental repair. Caries Res 36(4):223–232. https://doi.org/10.1159/000063930

Soares AP et al (2016) Bisphosphonates: pharmacokinetics, bioavailability, mechanisms of action, clinical applications in children, and effects on tooth development. Environ Toxicol Pharmacol 42:212–217

Soma T, Iwasaki R, Sato Y, Kobayashi T, Nakamura S, Kaneko Y, Ito E, Okada H, Watanabe H, Miyamoto K, Matsumoto M, Nakamura M, Asoda S, Kawana H, Nakagawa T, Miyamoto T (2021) Tooth extraction in mice administered zoledronate increases inflammatory cytokine levels and promotes osteonecrosis of the jaw. J Bone Miner Metab 39(3):372–384

Song M, Alshaikh A, Kim T, Kim S, Dang M, Mehrazarin S, Shin KH, Kang M, Park NH, Kim RH (2016) Preexisting periapical inflammatory condition exacerbates tooth extraction-induced bisphosphonate-related osteonecrosis of the jaw lesions in mice. J Endodont 42(11):1641–1646

Soutome S, Hayashida S, Funahara M, Sakamoto Y, Kojima Y, Yanamoto S, Umeda M (2018) Factors affecting development of medication-related osteonecrosis of the jaw in cancer patients receiving high-dose bisphosphonate or denosumab therapy: is tooth extraction a risk factor? PLoS One 13(7): e0201343

Souza EQM, Felipe Toro L, Franzão Ganzaroli V et al (2024) Peri-implantitis increases the risk of medication-related osteonecrosis of the jaws associated with osseointegrated implants in rats treated with zoledronate. Sci Rep 14(1):627

Statkievicz C, Toro LF, de Mello-Neto JM, de Sá DP, Casatti CA, Issa JPM, Cintra LTA, de Almeida JM, Nagata MJH, Garcia VG, Theodoro LH, Ervolino E (2018) Photomodulation multiple sessions as a promising preventive therapy for medication-related osteonecrosis of the jaws after tooth extraction in rats. J Photochem Photobiol B 184:7–17

Stenvinkel P, Ketteler M, Johnson RJ, Lindholm B, Pecoits-Filho R, Riella M, Heimbürger O, Cederholm T, Girndt M (2005) IL-10, IL-6, and TNF-alpha: central factors in the altered cytokine network of uremia – the good, the bad, and the ugly. Kidney Int 67(4):1216–1233

Tempesta A, Capodiferro S, Mauceri R, Lauritano D, Maiorano E, Favia G, Limongelli L (2022) Peri-implantitis-like medication-related osteonecrosis of the jaw: clinical considerations and histological evaluation with confocal laser scanning microscope. Oral Dis 28(6):1603–1609

Tibúrcio-Machado CS, Michelon C, Zanatta FB, Gomes MS, Marin JA, Bier CA (2021) The global prevalence of apical periodontitis: a systematic review and meta-analysis. Int Endod J 54(5):712–735

Toro LF, de Mello-Neto JM, Santos FFVD, Ferreira LC, Statkievicz C, Cintra LTÂ, Issa JPM, Dornelles RCM, de Almeida JM, Nagata MJH, Garcia VG, Theodoro LH, Casatti CA, Ervolino E (2019) Application of autologous platelet-rich plasma on tooth extraction site prevents Occurence of medication-related osteonecrosis of the jaws in rats. Sci Rep 9(1):22

Tsosura TVS, Chiba FY, Mattera MSLC, Pereira RF, Cintra LTA, Conti LC, Santos RMD, Mateus JHP, Garbin CAS, Sumida DH (2019) Maternal apical periodontitis is associated with insulin resistance in adult offspring. Int Endod J 52(7):1040–1050

Tsosura TVS, Dos Santos RM, Chaves Neto AH, Chiba FY, Carnevali ACN, Mattera MSLC, Belardi BE, Cintra LTÂ, da Silva Machado NE, Matsushita DH (2021) Maternal apical periodontitis increases insulin resistance and modulates the antioxidant defense system in the gastrocnemius muscle of adult offspring. J Endodont 47(7):1126–1131

Ueda N, Imada M, Kato Y, Okuda N, Nakaue K, Horita S, Kinoshita S, Kasahara K, Kirita T (2022) Bevacizumab-associated implant presence-triggered osteonecrosis: a case report and literature review. J Oral Implantol 48(4):325–331

Vahtsevanos K, Kyrgidis A, Verrou E, Katodritou E, Triaridis S, Andreadis CG, Boukovinas I, Koloutsos GE, Teleioudis Z, Kitikidou K, Paraskevopoulos P, Zervas K, Antoniades K (2009) Longitudinal cohort study of risk factors in cancer patients of bisphosphonate-related osteonecrosis of the jaw. J Clin Oncol 27(32):5356–5362

Vandone AM, Donadio M, Mozzati M, Ardine M, Polimeni MA, Beatrice S, Ciuffreda L, Scoletta M (2012) Impact of dental care in the prevention of bisphosphonate-associated osteonecrosis of the jaw: a single-center clinical experience. Ann Oncol 23(1): 193–200

Vanpoecke J, Verstraete L, Smeets M, Ferri J, Nicot R, Politis C (2020) Medication-related osteonecrosis of the jaw (MRONJ) stage III: conservative and conservative surgical approaches versus an aggressive surgical intervention: a systematic review. J Craniofac Surg 48(4):435–443

Vardas E, Coward T, Papadopoulou E, Nicolatou-Galitis O (2014) Dental extractions as the major local risk factor of bisphosphonates related jaw osteonecrosis in cancer patients receiving intravenous bisphosphonates therapy: a systematic review. Mediterr Oncol J 1:26–33

Wayama MT, Yoshimura H, Ohba S, Yoshida H, Matsuda S, Kobayashi J, Kobayashi M, Gomes Filho JE, Sano K (2015) Diminished progression of periapical lesions with Zoledronic acid in Ovariectomized rats. J Endodont 41(12):2002–2007

Wehrhan F, Moebius P, Amann K, Ries J, Preidl R, Neukam FW, Weber M (2017) Macrophage and osteoclast polarization in bisphosphonate associated necrosis and osteoradionecrosis. J Craniomaxillofac Surg 45(6):944–953

Winter A, Schulz SM, Schmitter M, Brands RC, Straub A, Kübler A, Borgmann A, Hartmann S (2022) Oral-health-related quality of life in patients with medication-related osteonecrosis of the jaw: a prospective clinical study. Int J Environ Res Public Health 19(18):11709

Xiong H, Wei L, Peng B (2009) Immunohistochemical localization of IL-17 in induced rat periapical lesions. J Endodont 35(2):216–220

Yarom N, Fedele S, Lazarovici TS, Elad S (2010) Is exposure of the jawbone mandatory for establishing the diagnosis of bisphosphonate-related osteonecrosis of the jaw? J Oral Maxillofac Surg 68(3):705

Yazdi PM, Schiodt M (2015) Dentoalveolar trauma and minor trauma as precipitating factors for medication-related osteonecrosis of the jaw (ONJ): a retrospective study of 149 consecutive patients from the Copenhagen ONJ cohort. Oral Surg Oral Med Oral Pathol Oral Radiol 119(4):416–422

Yong CW, Sng TJH, Choo SHJ, Chew JRJ, Islam I (2024) Implant presence-triggered osteonecrosis: a scoping review. J Stomatol Oral Maxillofac Surg 125(5): 101759

Zhang W, Gao L, Ren W, Li S, Zheng J, Li S, Jiang C, Yang S, Zhi K (2021) The role of the immune response in the development of medication-related osteonecrosis of the jaw. Front Immunol 12:606043

Inflammatory Mediators in Peri-Implantitis: From Pathophysiology to Therapeutic Targets

21

Akira Hasuike and Shuichi Sato

Abstract

Peri-implantitis is a complex inflammatory disease affecting dental implants, characterized by microbial involvement, titanium particle release, and host inflammatory reactions. Conventional diagnostic approaches are limited, prompting exploration of biomarkers in saliva and peri-implant crevicular fluid. These include proinflammatory, anti-inflammatory, and bone metabolism-related markers. Advanced perspectives focus on gene polymorphism, proteomics, and single-cell RNA sequencing, which have illuminated detailed cellular interactions and inflammatory pathways unique to peri-implantitis. The future of peri-implantitis management trends toward precision medicine, leveraging genomic data and computational tools for personalized treatment strategies. A comprehensive understanding of peri-implantitis etiology and pathogenesis is crucial for developing effective targeted therapies to benefit patients long-term.

Keywords

Peri-implantitis · Biomarkers · Inflammation · Therapeutic targets

21.1 Introduction

In 2017, the World Workshop on the Classification of Periodontal and Peri-implant Diseases and Conditions defined peri-implantitis as "A plaque-associated pathological condition occurring in the tissues around dental implants, characterized by inflammation in the peri-implant mucosa and subsequent progressive loss of supporting bone (Berglundh et al. 2018)."

This definition highlights the primary role of plaque association in peri-implant diseases, distinguishing them from cases caused by technical errors or overload. Although plaque accumulation on the implant/abutment surface is considered the predominant cause, various other risk factors also contribute to this condition (Fu and Wang 2020).

The host response to plaque and the associated risk factors remains an area of ongoing research and evolving understanding. Just as Arsène Lupin, Maurice Leblanc's beloved gentleman burglar, captivated readers with his skillful masquerades and ability to effortlessly switch between various personas, peri-implantitis presents itself in various forms and manifestations, reflecting an uncertain etiology

A. Hasuike (✉) · S. Sato
Department of Periodontology, Nihon University School of Dentistry, Tokyo, Japan

Division of Applied Oral Sciences, Nihon University Graduate School of Dentistry, Tokyo, Japan
e-mail: hasuike.akira@nihon-u.ac.jp

N. Rezaei, H. Ziaei (eds.), *Oral Immunology*, Advances in Experimental Medicine and Biology 1492,
https://doi.org/10.1007/978-3-032-03176-1_21

and various risk factors. Peri-implantitis can be described as infectious, vascular, fibroblastic, foreign-body-reactive, and iatrogenic, underscoring its dynamic and multifaceted nature (Fragkioudakis et al. 2021; Wilson Jr. et al. 2015; Yamamoto et al. 2021). The accurate word to describe this elusive adversary is "inflammatory." Inflammation is a fundamental protective response to harmful stimuli, such as pathogens or damaged cells, to remove injurious stimuli and initiate healing. However, disturbances in regulating inflammation can lead to chronic inflammatory diseases. During peri-implantitis progression, the inflammatory cascade in peri-implant tissues is orchestrated through a complex interplay of cytokines and chemokines, which are generated following the recruitment and activation of various immune cells (Duarte et al. 2016; Park et al. 2024). This chapter discusses the role of inflammatory mediators in peri-implantitis from the etiopathogenesis, diagnostic, and therapeutic targets, and explores future perspectives.

21.2 Etiopathogenesis of Peri-Implantitis

21.2.1 Microbial Etiopathogenesis

The oral cavity is regularly exposed to a wide variety of bacteria from the external environment, and approximately 700 bacterial strains are considered indigenous to this ecosystem (Aas et al. 2005). These bacterial communities vary significantly among individuals and fluctuate within the same individual, influenced by factors such as oral hygiene and overall health (Palmer Jr. 2014). Titanium implants have unique surface characteristics that promote the formation of distinct biofilms and microbial communities that differ considerably from those found in natural teeth. Peri-implantitis, a condition associated with these implants, is associated with a complex and diverse microbiota. A recent systematic review demonstrated notable differences in the bacterial profiles between peri-implantitis and implants affected solely by mucositis (Carvalho et al. 2023). The review identified several bacteria potentially associated with peri-implantitis, such as *S. epidermidis*, *F. nucleatum*, *T. denticola*, *T. forsythia*, *P. intermedia*, and *P. gingivalis*. Additionally, studies suggest that the Epstein–Barr virus exacerbates peri-implantitis in the presence of *P. gingivalis* (Al-Zawawi 2020). Furthermore, *Candida albicans* has been implicated in the etiology of peri-implantitis, particularly in the presence of periopathogens, potentially contributing to increased soft tissue inflammation and crestal bone loss around the implants (Al-Zawawi 2020). Similar to periodontitis, the microbial role in peri-implantitis may involve either the direct activation of virulence factors or the induction of chronic low-grade immune activation. However, the submucosal microbiota in peri-implantitis lesions remains underexplored using culture-independent methods, leaving the microbial profile of peri-implantitis incomplete (Schwarz et al. 2018).

21.2.2 The Role of Titanium Particles in Peri-Implantitis

Peri-implantitis involves a multifaceted interaction with the microbiota, yet attributing its etiology exclusively to microbial factors provides an incomplete picture. Recent literature emphasizes that titanium implants, in cases of peri-implantitis, may no longer remain inert within the body (Kandaswamy et al. 2024). Titanium implants undergo corrosion and wear, leading to the release of titanium particles and ions into the adjacent tissues. These titanium byproducts are considered critical for the onset and progression of peri-implantitis (Kheder et al. 2021). Numerous studies have identified the presence of titanium particles in peri-implant tissues, both in the intercellular spaces and within cells, particularly in macrophages (Fretwurst et al. 2016; Olmedo et al. 2013; Wilson Jr. et al. 2015). The concentration of titanium in peri-implant plaques has been reported to be higher in cases of peri-implantitis than in healthy implants (Daubert et al. 2018). Notably, these titanium particles can provoke inflammatory responses in surrounding tissues, stimulating the release of proinflammatory cytokines, and potentially contributing to bone resorption by activating osteoclasts (Pettersson et al. 2017; Wakuda et al. 2024). This titanium-induced inflammation may account for the typically rapid and aggressive bone loss observed in peri-implantitis, in contrast to the generally slower progression of periodontitis. Additionally, considering that titanium particles can affect the peri-implant microbiome is essential (Daubert et al. 2018; Kotsakis and Olmedo 2021). Higher titanium concentrations in plaques have been linked to reduced microbial diversity and changes in community structure (Daubert et al. 2018). This indirect effect of titanium on the microbial environment highlights a key distinction between peri-implantitis from periodontitis.

21.3 Biological Characteristics of Peri-Implantitis

21.3.1 Histopathological Characteristics of Peri-Implantitis

More than a decade ago, Carcuac and Berglundh conducted a comparative study on peri-implantitis and periodontitis lesions, investigating their cellular properties and histopathological distinctions (Carcuac and Berglundh 2014). Histological samples from 40 patients with severe peri-implantitis were compared with those from patients with severe chronic periodontitis in this study. The findings revealed that peri-implantitis lesions were more than twice the size of periodontitis lesions ($3.48 \pm 2.54\,mm^2$ vs. $1.49 \pm 1.05\,mm^2$). Additionally, peri-implantitis lesions exhibited considerably larger proportions, higher counts, and greater densities of plasma cells, macrophages, and neutrophils than periodontitis lesions. Moreover, peri-implantitis lesions extended apically to the pocket epithelium and were not surrounded by non-infiltrated connective tissue, unlike periodontal lesions. A further noteworthy distinction was observed in the vascular structures, as peri-implantitis lesions had significantly higher densities of vascular structures in the connective tissue lateral to the infiltrated tissue compared to within the infiltrate itself. This study suggested that the absence of an epithelial barrier between the lesion and biofilm in the pocket, coupled with increased peripheral vascular density in peri-implantitis, might account for the prominence of neutrophil granulocytes in these lesions.

A more recent systematic review of 33 studies examining the histopathological features of peri-implant diseases found that peri-implant diseases are characterized by increased inflammatory infiltrates and vascular proliferation (Shafizadeh et al. 2021). Peri-implantitis lesions exhibit more intense inflammatory infiltrates, primarily composed of plasma cells extending into the apical region. The surface area of the infiltrated connective tissue was notably larger in the peri-implantitis group than in the peri-implant mucositis and periodontitis groups. Most studies have reported a more aggressive inflammatory infiltrate in peri-implantitis, characterized by cell infiltration and vascular proliferation being more pronounced in the apical portion of the pocket and central regions of the infiltrated connective tissue. Additionally, titanium particles are commonly observed in peri-implantitis lesions (in $>90\%$ of cases), and their presence is linked to an increased inflammatory response.

21.3.2 Transcriptomic Insights into the Inflammatory Mechanisms of Peri-Implantitis

Recent transcriptomic studies have revealed a complex and intense inflammatory response in peri-implantitis that distinguishes it from periodontitis. These findings offer crucial insights into the molecular and cellular mechanisms driving the aggressive nature of peri-implant inflammation.

Martin et al. used a transcriptome-wide microarray profiling workflow for human peri-implantitis samples, which cover more than 20,000 well-annotated genes, to characterize the inflammatory landscape of peri-implantitis (Martin et al. 2022). Their findings highlighted a significant upregulation of genes involved in the endosomal-lysosomal pathway (WASH1, BLOC1S4, and DNAJC28), particularly those related to actin polymerization, in diseased tissues. This may suggest a heightened intracellular activity in response to bacteria and titanium particles. Moreover, pathways linked to cellular respiration and oxidative stress were highly transcribed across all peri-implant samples, indicating that implant-specific factors may induce persistent oxidative stress. Interestingly, despite the assumption that proinflammatory cytokines like interleukin (IL)-1β play a crucial role in peri-implantitis, the study found no significant differential expression of IL-1β between diseased and healthy tissues. This may be attributed to the predominance of other inflammation-related genes involved in the immune responses to peri-implantitis. The authors proposed that post-translational modifications of IL-1β, particularly its regulation via inflammasomes, might reveal differences between healthy and diseased states upstream of the cytokine's activity.

Wang et al. provided additional evidence of an increased inflammatory state in peri-implantitis using bulk RNA sequencing (RNA-seq) and machine learning techniques to evaluate immune cell profiles in human tissue samples from peri-implantitis patients (Wang et al. 2021). This study focused on a cohort of patients undergoing regenerative therapy in which peri-implant granulation tissue was analyzed using bulk RNA-seq. A robust outlier-resistant machine-learning algorithm was employed for immune deconvolution, allowing the classification of disease sites into three risk categories (low, intermediate, and high) based on immune cell composition. These classifications correlate with clinical outcomes following regenerative therapy. The low-risk group exhibited higher M1/M2 macrophage ratios, an increased presence of CD4+ and regulatory T cells, and reduced B cell levels. Furthermore, this group displayed enhanced complement signaling and elevated Th1 and Th17 cytokine levels. In contrast, the high-risk group showed elevated levels of pathogens, such as *Fusobacterium nucleatum* and *Prevotella intermedia*. Notably, post-treatment, the low-risk group sustained lower levels of pathogens, while the high-risk group exhibited a more rapid recolonization of pathogens. The authors concluded that the peri-implant immune environment influences the microbial composition and affects regenerative treatment outcomes in peri-implantitis. They highlighted the potential of immune signatures to improve risk assessment and optimize treatment planning for peri-implantitis.

The major limitation of microarray and bulk RNA-seq is that they profile the average gene expression across all cell types in a tissue sample. This averaging effect prevents differentiation between the gene expression profiles of individual cell types and the capture of specific cellular details such as mutations. Single-cell RNA sequencing (scRNA-seq) allows for the examination of gene expression at the individual cell level, providing a more detailed understanding of cellular diversity and the specific characteristics of different cell types (Tang et al. 2009). The scRNA-seq workflow comprises five essential steps. First, individual cells are isolated and tagged with unique nucleotide barcodes for identification. Following cell lysis and mRNA capture, cDNA libraries are constructed and sequenced. The final step involves computational analysis of the sequencing data. scRNA-seq has recently been applied to peri-implantitis tissue samples, offering a granular view of the cellular players involved in peri-implantitis inflammation. Mo et al. observed a quantitative increase in neutrophil subtypes, particularly CXCR2+ ones, in peri-implantitis compared to periodontitis and healthy individuals, indicating a shift toward a more aggressive inflammatory phenotype (Mo et al. 2024). A key finding of this study was the identification of a proinflammatory fibroblast subpopulation that was more prevalent in peri-implantitis. Fibroblasts, characterized by high expression of chemokines and cytokines (CXCL1, CXCL2, CXCL6, CXCL8, CXCL13, IL24), appear to play a central role in orchestrating inflammatory responses. They also observed enhanced CXCL8$^+$ fibroblast–CXCR2$^+$ neutrophil interactions, which play a key role in the mechanism of sustained and amplified inflammation in peri-implantitis.

Another single-cell RNA-seq study elucidated the inflammatory dynamics in peri-implantitis by analyzing cell–cell communication networks (Li et al. 2024). They found a marked reduction in stromal cells, particularly fibroblasts, accompanied by an increase in immune cells such as neutrophils and monocytes. Researchers have identified a specific subpopulation of fibroblasts characterized by CXCL13 expression, which plays a crucial role in recruiting immune cells to the site of inflammation. One of the key findings was that neutrophils showed a more proinflammatory profile, with increased activation of the NF-kappa B signaling pathway. Additionally, monocytes are more likely to differentiate into osteoclasts, potentially explaining rapid and severe bone loss. This study also highlighted the importance of cell–cell communication in the peri-implantitis microenvironment. CXCL13+ fibroblasts recruit both neutrophils and monocytes through specific chemokine–receptor interactions.

Collectively, these studies depict peri-implantitis as a condition characterized by a complex and intensified inflammatory response. Transcriptomic analyses revealed unique gene expression patterns and increased activation of pathways related to bacterial and titanium stimulation. Notably, proinflammatory fibroblasts and activated neutrophil subtypes play crucial roles in sustaining inflammation, as revealed by scRNA-seq, which highlights dynamic cellular interactions contributing to rapid tissue degradation. These findings underscore the need for targeted therapeutic strategies that focus on specific inflammatory mechanisms in peri-implantitis to enhance treatment outcomes.

21.4 Diagnostic and Monitoring Applications of Biomarkers for Peri-Implantitis

21.4.1 Need for Biomarkers in Peri-Implantitis Diagnosis and Monitoring

The standard clinical approach for evaluating peri-implant conditions adopts methods long used for natural teeth, employing a periodontal probe to measure pocket depths and observe bleeding during probing. According to the 2017 workshop, the case definition of peri-implantitis is defined as follows: (1) bleeding (BoP) and/or suppuration on gentle probing, (2) increased probing pocket depth (PD), and (3) bone loss beyond crestal bone-level changes resulting from initial bone remodeling (Berglundh et al. 2018). In the absence of longitudinal data, a secondary case definition was proposed, which includes bleeding on probing and/or suppuration on probing at ≥ 1 site, probing depth ≥ 6 mm, and bone level ≥ 3 mm. However, the conventional diagnostic approach has limitations when applied to implants because these metrics alone do not reliably predict peri-implant disease activity, future bone loss, or implant failure (Dukka et al. 2021). The absence of a periodontal ligament around the implants, combined with the structural complexity of the prosthetic components, complicates both the execution and interpretation of probing depth measurements. Additionally, the mucosal seal around implants offers less resistance to probing than natural teeth, potentially causing mechanical bleeding, even in healthy implants. To mitigate excessive probing depths, the application of a low probing pressure (approximately 0.25 N) is recommended, considering the fragile nature of the peri-implant mucosal attachment (Dukka et al. 2021). Furthermore, large, over-contoured implant structures can lead to traumatic bleeding on probing due to restricted probing directions. This underscores the need for noninvasive diagnostic and monitoring tools such as biomarker analysis. Biomarkers are objectively measurable indicators of normal biological processes, pathological states, and therapeutic responses. Their evaluation may facilitate the early detection of peri-implantitis, which is a persistent challenge.

21.4.2 Properties of Saliva and Peri-Implant Crevicular Fluid (PICF) in Peri-Implantitis

Saliva and peri-implant crevicular fluid (PICF) are crucial biological fluids with distinct properties used to diagnose and monitor peri-implantitis (Haririan et al. 2021; Pliavga et al. 2023). Saliva offers a non-invasive, accessible, and cost-effective collection, making it suitable for routine screening. Secretions from the major salivary glands (parotid, submandibular, and sublingual) contain numerous proteins and peptides and maintain the integrity of the oral cavity. Additionally, secreted saliva may play a significant role in the establishment and progression of periodontal diseases, including peri-implantitis, because of its importance in oral biofilm formation and host defense. However, because saliva composition reflects the entire oral environment, it may obscure localized conditions.

Contrastingly, PICF collected from the peri-implant sulcus provides specific, localized information directly related to implant health. PICF is both a physiological fluid and an inflammatory exudate derived from the gingival plexus in the peri-implant mucosal corium, located beneath the epithelial lining of the implant–mucosal interface. It traverses the basement membrane and the junctional epithelium to enter the crevice. The composition of PICF allows the detection of subclinical changes in tissue metabolism, inflammatory cell recruitment, and connective tissue remodeling. Cytokines and enzymes present in the peri-implant tissues contribute to connective tissue collagen degradation and alveolar bone resorption, originating from the host's immune response to the plaque biofilm. Both fluids are valuable for managing peri-implantitis, with saliva offering convenient collection, and PICF providing precise localized data. Researchers employ various techniques to analyze the constituents of PICF, targeting the diagnosis of peri-implantitis (Alassy et al. 2019). The standard collection method involves gently placing paper strips into the crevicular space for 30 s to absorb the fluid (Griffiths 2003). The absorbed fluid is then eluted from the strips into a suitable solution for further processing. These samples are analyzed using advanced detection methods such as enzyme-linked immunosorbent assay, flow cytometry, Luminex, and spectrophotometry, providing detailed insights into the molecular composition of PICF. Although PICF collection requires a clinical setting, it provides highly accurate insights into the peri-implant inflammatory processes. Understanding the distinct characteristics of these fluids is essential for effective clinical application in the diagnosis and monitoring of peri-implantitis.

21.5 Potential Biomarkers for Peri-Implantitis Detection

21.5.1 Proinflammatory Biomarkers

Cytokines, primarily secreted by leukocytes, either function as communication proteins that can promote or inhibit inflammation. In peri-implantitis, the balance between proinflammatory and anti-inflammatory cytokines is disrupted, resulting in a shift toward a proinflammatory state (Duarte et al. 2016; Oliveira et al. 2023).

IL-1, a key proinflammatory cytokine, is predominantly produced by macrophages but is also secreted by neutrophilic granulocytes and other cells (Dinarello 2013). IL-1 plays a critical role in the innate immune response and is a central regulator of inflammation in various human diseases. The IL-1 family consists of 11 cytokines, including seven agonistic ligands: IL-1α, IL-1β, IL-18, IL-33, IL-36α, IL-36β, and IL-36γ (Fields et al. 2019). Among these, IL-1β has been extensively studied as a biomarker for peri-implantitis (Cardoso et al. 2022). Numerous studies have demonstrated significantly higher levels of IL-1β in individuals with peri-implantitis compared to those with healthy implants (Al-Askar et al. 2018; Bhavsar et al. 2019; Che et al. 2017; Gürlek et al.

2017; Milinkovic et al. 2021; Song et al. 2022; Yakar et al. 2019). However, whether peri-implantitis is associated with higher IL-1β levels than peri-implant mucositis remains controversial.

IL-6 is also a key inflammatory cytokine, and numerous studies have reported significantly elevated levels in peri-implantitis samples compared with healthy implants (Abduljabbar et al. 2019; Al-Askar et al. 2018; Jiang et al. 2021; Milinkovic et al. 2021). Moreover, a significant difference was observed between peri-implantitis and peri-implant mucositis (Milinkovic et al. 2021).

IL-8 and IL-17 are also classified as proinflammatory cytokines in general (Meniailo et al. 2018; Zenobia and Hajishengallis 2015). Although IL-8 levels are often elevated in peri-implantitis, some studies have reported comparable levels in healthy and peri-implant conditions (Ata-Ali et al. 2015). In patients with herpes viral infection, the control group had higher IL-8 levels in the PICF than those with peri-implantitis, although the underlying reason for this phenomenon remains undetermined (Sahoo et al. 2021). Furthermore, a positive correlation between probing depth and IL-17 levels has been identified (Darabi et al. 2013). When comparing IL-17 concentrations between mucositis and peri-implantitis, the mucositis group exhibited the highest mean concentrations, with both the mucositis and peri-implantitis groups showing significantly elevated levels compared to healthy patients (Farhad et al. 2019).

The proinflammatory cytokine tumor necrosis factor (TNF) and its associated superfamily (TNFSF) represent key inflammatory mediators. Several studies consistently reported significantly elevated levels of cytokines such as TNF-α (Abduljabbar et al. 2019; Jiang et al. 2021), TNFSF12 (Jansson et al. 2021; Yakar et al. 2019), TNFSF13B (Jansson et al. 2021), sTNF-R1 (Jansson et al. 2021), in localized samples such as saliva or PICF in the peri-implantitis group compared to healthy individuals.

Although not classified as cytokines, several matrix metalloproteinase (MMP) indicators were also evaluated in PICF, and all showed higher levels in the peri-implantitis condition than in healthy implants: MMP-2 (Che et al. 2017; Zhang et al. 2022), MMP-8 (Bhavsar et al. 2019; Hentenaar et al. 2021), and MMP-9 (Che et al. 2017).

21.5.2 Anti-Inflammatory Biomarkers

IL-10 is a widely recognized anti-inflammatory cytokine. Recent studies have increasingly reported reduced IL-10 levels in PICF, which are inversely correlated with the presence of peri-implantitis (Alassy et al. 2019). A previous study has identified a disrupted IL-1β/IL-10 ratio in these cases (Gleiznys et al. 2019). However, several studies have documented elevated IL-10 levels, often accompanied by increased concentrations of proinflammatory cytokines such as IL-1 and TNFα (Ata-Ali et al. 2015; Corrêa et al. 2019). Notably, studies have shown that IL-10 levels are higher in the tissues surrounding implants affected by peri-implantitis compared to peri-implant mucositis (Eckert et al. 2018; Farhad et al. 2019). To fully understand the simultaneous elevation of both proinflammatory and anti-inflammatory cytokines in peri-implantitis conditions, further investigations on tissue dynamics, blood flow patterns, and cytokine sensitivity are essential.

21.5.3 Biomarkers Related to Bone Metabolism

The inflammatory bone resorption in peri-implantitis is driven by the interactions between inflammatory mediators and regulators of bone resorption. Central to this mechanism is the signaling pathway involving the receptor activator of nuclear factor κB (RANK), its ligand (RANKL), and osteoprotegerin (OPG) (Udagawa et al. 2021). The binding of RANKL to RANK promotes osteoclast differentiation and regulates bone resorption. Conversely, OPG acts as a decoy receptor and inhibits RANKL from binding to RANK, thus preventing osteoclast differentiation and activity. The balance between RANK,

RANKL, and OPG is crucial in maintaining bone homeostasis and can serve as an indicator of peri-implant health (Arikan et al. 2011; Rakic et al. 2013). Although several studies have evaluated the effectiveness of OPG as a biomarker for peri-implantitis, its clinical relevance remains controversial (Gürlek et al. 2017; Hentenaar et al. 2021; Rakic et al. 2020; Yakar et al. 2019). In contrast, studies assessing RANKL as a biomarker for peri-implantitis have consistently shown its elevated levels in the peri-implantitis group compared with both groups of healthy individuals (Milinkovic et al. 2021; Rakic et al. 2020; Yakar et al. 2019) and peri-implant mucositis (Zhang et al. 2019). Additionally, RANKL levels in peri-implant mucositis are higher than those observed in healthy implants (Rakic et al. 2020).

Other bone remodeling markers, such as cathepsin K and sclerostin, contribute to this pathological process (Ashifa et al. 2021). Cathepsin K is involved in the degradation of the bone matrix, whereas sclerostin negatively regulates bone formation by reducing bone mineral density. Elevated cathepsin K levels have been associated with increased PD, BoP, and plaque index in peri-implantitis, indicating its role in disease progression (Strbac et al. 2006). A study suggested that cathepsin K expression may occur early in the disease process, potentially helping to distinguish peri-implant mucositis from peri-implantitis (Rakic et al. 2014). Furthermore, a recent finding suggested that the clinical diagnostic accuracy for peri-implantitis can be enhanced by combining traditional clinical parameters, such as PD and BoP, with biomarker levels, such as RANKL. For example, RANKL levels above a certain threshold ($>$19.9 pg/site) are indicative of peri-implant mucositis, while lower levels ($\leq$19.9 pg/site) may suggest peri-implantitis (Rakic et al. 2020).

21.5.4 MicroRNAs as Diagnostic Biomarkers in Peri-implantitis

MicroRNAs (miRNAs) are small noncoding RNA molecules, typically 18–22 base pairs in length, that play critical roles in regulating gene expression (Irwandi and Vacharaksa 2016). These endogenous molecules bind to the complementary sequences of target messenger RNA (mRNA), resulting in mRNA degradation or translational repression (Huntzinger and Izaurralde 2011). As key modulators of numerous biological processes such as immune responses and bone metabolism, miRNAs have gained considerable attention in oral health research, particularly concerning periodontal and peri-implant diseases. The potential of miRNAs as diagnostic tools for peri-implantitis is gaining traction (Asa'ad et al. 2020). Insights derived from periodontitis research and preliminary studies on peri-implantitis have indicated that miRNAs are significantly involved in the regulation of immune responses and bone metabolism, both of which are vital for peri-implant health and disease (Asa'ad et al. 2020). The advantages of miRNAs as potential biomarkers of peri-implantitis are noteworthy. These molecules are stable, easily detectable in body fluids, and capable of being transported via extracellular vesicles (EVs), making them suitable for PICF analysis (Mitchell et al. 2008). miRNAs have been shown to influence bone metabolism by promoting osteogenesis and regulating osteoclast activity (Zhao et al. 2014). Given that peri-implantitis is characterized by inflammatory bone loss around the dental implants, these miRNAs may serve as indicators of disease progression and treatment efficacy.

Chaparro et al. explored this potential by focusing on the diagnostic value of extracellular vesicles (EVs) and specific miRNAs in PICF (Chaparro et al. 2021). Researchers examined PICF samples from patients with healthy implants, peri-implant mucositis, and peri-implantitis. They found significantly higher concentrations of EVs in peri-implantitis than in healthy implants or peri-implant mucositis. Moreover, miRNA-21-3p and miRNA-150-5p were downregulated in peri-implantitis compared to both healthy implants and those with peri-implant mucositis. A model combining these two miRNAs showed high diagnostic accuracy for distinguishing peri-implantitis from other conditions.

Hamed and Abdulbaqi investigated miRNA-146a and miRNA-155 expression in human peri-implant pocket tissues (Hamed and Abdulbaqi 2024). They analyzed samples from 15 peri-implantitis patients and 15 healthy controls using qPCR. The results showed significant upregulation of both miRNAs in peri-implantitis tissues. miRNA-146a and miRNA-155 demonstrated high diagnostic performance, with a sensitivity of 87.5% for both miRNAs and a specificity of 73.3% and 66.7%, respectively. This study suggests the potential of these miRNAs as diagnostic biomarkers of peri-implantitis. However, the role of miRNAs in peri-implantitis is complex and poorly understood. Future studies are required to better understand the role of miRNAs in peri-implantitis and assess their potential as diagnostic markers.

21.6 Future Perspective of Detection and Diagnosis of Peri-implantitis

21.6.1 Assessment of Cytokine Gene Polymorphism

Most studies using PICF aimed to distinguish between peri-implantitis, peri-implant mucositis, and healthy implants. Additionally, PICF analysis is being explored to identify high-risk individuals who are predisposed to peri-implant diseases. Genetic susceptibility is a crucial risk factor for peri-implantitis (Turkmen and Firatli 2022). Specifically, genetic polymorphisms, variations in specific DNA sequences, have been implicated, with single-nucleotide polymorphisms (SNPs) being the most prevalent (Lafuente-Ibáñez de Mendoza et al. 2022; Thorisson et al. 2005). The detection of SNPs can aid in the identification of altered genes or proteins involved in disease pathogenesis. In light of this genetic predisposition, Lafuente-Ibáñez de Mendoza et al. conducted a systematic review and meta-analysis to examine the relationship between proinflammatory genetic polymorphisms and peri-implantitis (Lafuente-Ibáñez de Mendoza et al. 2022). This review included 12 studies involving 1015 patients. The meta-analysis revealed no statistically significant association between IL-1β or TNF-α polymorphisms and peri-implantitis. However, the C/C genotype of the IL-1β SNP was found to be associated with an increased risk of peri-implantitis. The associations between inflammatory polymorphisms and peri-implantitis must be interpreted with caution because of the varied methodological designs, sample sizes, and diagnostic criteria across studies. This variability limits the generalizability of these findings. Therefore, future studies should explore the relationship between specific SNPs and peri-implantitis. Addressing these challenges will be crucial for improving the treatment and prevention of peri-implantitis.

21.6.2 Application of Proteomics

The composition of the PICF, which changes in cases of peri-implantitis, peri-implant mucositis, and healthy cases, has been analyzed using antigen-based techniques. However, these targeted methods are limited to the detection of predefined proteins, leading researchers to focus mainly on a narrow range of proteins, predominantly proinflammatory cytokines. Therefore, diagnostic tools based on a single cytokine or MMPs have limited specificity for peri-implant conditions. Moreover, antigen-based assays have failed to provide a complete overview of PICF composition. Consequently, nontargeted proteomic analyses of PICF are needed. Proteomics allows the identification and quantification of numerous proteins in an untargeted manner, offering insights into post-translational modifications, proteolytic patterns, and microbial components. This approach aims to enhance our understanding of disease mechanisms and facilitate the discovery of novel biomarkers and therapeutic targets.

Esberg et al. conducted a groundbreaking untargeted proteomic analysis of PICF by analyzing samples from 25 patients before and after surgical treatment for peri-implantitis. Using label-free proteomics, they identified around 200 proteins per sample (Esberg et al. 2019). Their findings highlighted distinct proteomic clusters associated with treatment outcomes, including implant retention over 12 months or implant failure. Fifty-two proteins exhibited differential expression and were aligned with biological pathways such as wound healing, apoptosis, bacterial defense, and inflammatory and stress responses.

Subsequently, Halstenbach et al. carried out a proteomic study comparing PICF samples from healthy and diseased implants and healthy teeth, analyzing 13 cases (Halstenbach et al. 2023). Their analysis revealed a notable proteomic similarity between healthy teeth and healthy implants. However, PICF samples collected from peri-implantitis patients presented 59 upregulated and 31 downregulated proteins compared to healthy implants. Gene Ontology analysis suggested that the identified biological processes were consistent with the findings of Esberg et al., who emphasized humoral immune responses and antimicrobial activities. Notably, neutrophil-derived proteins, such as azurocidin, cathelicidin antimicrobial peptide, and myeloperoxidase, were among the upregulated proteins, underscoring the potential role of neutrophils in peri-implantitis immunology. Although neutrophil infiltration is a hallmark of peri-implantitis, their precise contribution to disease progression remains unclear. Additionally, increased protease activity in peri-implantitis was evidenced by the identification of N-terminal cleavage sites, indicating the involvement of proinflammatory proteases, such as cathepsin G, neutrophil elastase, and MMPs.

21.6.3 Application of Single Cell Approach

Advances in transcriptomics, proteomics, and metabolomics have contributed significantly to improvements in disease control and treatment through the implementation of precision medicine. This approach can categorize individuals into subgroups based on their susceptibility to the disease and responses to specific treatments. Precision medicine depends on molecular research, various biomarkers, population studies, and large-scale data analyses to model complex diseases. In the future, the accuracy, predictive power, and clinical benefits of these models, including new molecular targets, are expected to enhance the management of multifactorial and complex conditions, such as peri-implantitis. Although this methodology has been well established in oncology and genetic disorders, it is still in its early stages of development in the field of dentistry. Research on novel biomarkers in peri-implant tissues could aid in monitoring the clinical status of implants, allowing for the classification of both implants and patients based on the actual risk of developing biological complications.

scRNA-seq is a highly effective and powerful method for biofluid analysis, providing novel insights into cellular heterogeneity and function at the cellular level. This approach enables the comprehensive gene expression profiling of thousands of cells simultaneously, uncovering subtle variations in cellular states, identifying rare cell populations, and elucidating complex cellular interactions. In biofluids, scRNA-seq has demonstrated efficacy in analyzing blood and cerebrospinal fluid (CSF), yielding essential information about both health and disease processes. In hematological disorders, scRNA-seq has uncovered novel cellular subpopulations and disease-specific transcriptional patterns. scRNA-seq application to CSF has resulted in significant findings and insights, especially in neurological disorders. These applications underscore the capacity of scRNA-seq to reveal complex cellular heterogeneity and dynamics in biofluids, thereby advancing our understanding of pathological processes and opening new avenues for enhanced diagnostics and personalized medicine. Moreover, the integration of scRNA-seq with spatial techniques offers a promising approach for deepening our understanding of cellular biology in biofluids. Although biofluids lack the structured spatial organization characteristics of solid tissues, the combination of these technologies may yield significant insights. Spatial approaches can trace the origins of cells in biofluids back to their respective tissues, thereby establishing connections between circulating cells and their tissue microenvironments. This integration sheds light on the cellular changes occurring during the transition from tissues to biofluids, offering valuable insights into disease processes and cell migration. Incorporating spatial data into scRNA-seq results could enhance interpretation by providing contextual information regarding the cellular neighborhoods from which the biofluid cells originated.

However, in the context of peri-implantitis, analyzing PICF is a promising yet challenging endeavor. The extremely limited volume of PICF, typically 1–2 μL, necessitates the development of ultra-sensitive analytical methods. Potential strategies to address this limitation include optimizing sampling procedures, employing concentration or amplification techniques, developing in situ analytical methods, utilizing microfluidic devices, and leveraging advanced bioinformatic approaches. Despite current technological limitations, rapid advancements in microsample analysis suggest that effective PICF analysis in peri-implantitis may soon be feasible, potentially revolutionizing the diagnosis, treatment, and management of the condition.

21.7 Therapeutic Application of Biomarkers

21.7.1 Effect of Conventional Peri-Implantitis Therapy on Biomarkers

The prevalence of peri-implantitis is rapidly increasing among patients with dental implants, creating an urgent need for effective treatment strategies. Moaven et al. conducted a systematic review that summarized changes in inflammatory biomarkers following conventional peri-implantitis treatment (Moaven et al. 2022). Their comprehensive analysis examined various treatment modalities and their effects on different inflammatory biomarkers. Nonsurgical treatment alone led to minimal or no significant changes in biomarkers such as IL-1β, IL-6, TNF-α, monocyte chemoattractant protein-1 (MCP-1/CCL2), macrophage inflammatory protein-1 alpha (MIP-1α/CCL3), interferon-γ, MMP-8, sRANKL, OPG, and granulocyte-colony stimulating factor (G-CSF). Laser-assisted nonsurgical treatment showed a significant decrease in MMP-9 with the Er-laser, while other markers, including IL-1α, IL-1β, IL-6, IL-8, TNF-α, MMP-1, MMP-3, MMP-8, MMP-13, and C-reactive protein (CRP), showed no significant changes. In clinically stable cases, MCP-1β, vascular endothelial growth factor, and IL-6 levels decreased. Adjunctive antibiotics with nonsurgical treatment did not significantly affect IL-1β, MPO, or alkaline phosphatase (ALP) levels. Flap surgery alone led to significant reductions in MMP-8 and TNF-α levels at 12 months post-treatment. Additionally, the combination of antibiotics with flap surgery significantly reduced MMP-8 levels, though IL-1β and MIP-1α remained unchanged. The use of an enamel matrix derivative with flap surgery significantly decreased the expression of 40 proteins in PICF, although the specific proteins were not identified. The addition of Ti particles to flap surgery did not show significant differences between groups at baseline or 12 months post-treatment for biomarkers such as MMP-8, IL-6, OPG, osteocalcin, leptin, osteopontin, parathyroid hormone, TNF-α, adiponectin, and insulin.

The authors cautioned against direct comparisons owing to variations in evaluation methods and timing across studies. However, this review provides valuable insights into the effects of different peri-implantitis treatments on inflammatory biomarkers. These findings highlight the complexity of interpreting data from diverse study designs and underscore the inherent challenges of synthesizing research outcomes in this field. This has set the stage for a broader discussion on the current state of peri-implantitis research and treatment, which remains a significant challenge in dental implantology.

21.7.2 Inflammatory Mediators as Therapeutic Targets

Research on therapeutic strategies for peri-implantitis often focuses on targeting the cytokines involved in inflammatory processes. One approach involves inhibiting IL-1β, a cytokine closely linked to bone destruction. Agents such as Anakinra, an IL-1 receptor antagonist, and Canakinumab, a monoclonal antibody specific to IL-1β, are designed to reduce inflammation by blocking IL-1β activity. Similarly, TNF-α inhibitors like infliximab and adalimumab have been considered for their ability to prevent tissue degradation around implants by inhibiting TNF-α.

In addition to blocking proinflammatory cytokines, efforts are underway to induce anti-inflammatory cytokines. These methods aim to mitigate the progression of peri-implantitis by promoting a well-regulated immune response. Nevertheless, the clinical use of IL-1β and TNF-α inhibitors in peri-implantitis is not currently advisable due to significant side effects. These agents are generally reserved for the treatment of rheumatic and autoimmune diseases, where the benefits of systemic immunosuppression outweigh the risks. Contrastingly, peri-implantitis is a localized condition, and the systemic immunosuppressive effects of these drugs, including the increased risk of infection, are considered disproportionate to their potential benefits in this context. Future studies should focus on developing local treatment options that balance both efficacy and safety.

21.7.3 Precision Medicine in Peri-implantitis

scRNA-seq revealed distinct cellular responses in human tissue samples from peri-implantitis cases, offering an unprecedented resolution of cellular complexity. Similarly, spatial omics techniques have been introduced to map cell locations and interactions in periodontitis, thereby providing a spatial context for gene expression data. These technologies allow for the identification of specific cellular niches and enable detailed analyses of spatially resolved cell–cell interactions within the affected tissues. Such granular insights are critical at the preclinical level, facilitating the discovery of biomarkers and the mapping of complex cellular networks, and in clinical applications, to guide personalized treatments. Preclinically, scRNA-seq and spatial omics offer unparalleled details for understanding disease mechanisms by highlighting variations at the single-cell level and providing the spatial context of these cells within the tissue architecture. With advancements in technology, their integration into clinical practice is anticipated, and they can be used for tailored diagnostic evaluations. This involves collecting PICF, cells, and tissues from patients through both noninvasive and surgical methods to provide precise, customized treatment plans. In the future, the potential application of the Drug2Cell tool could significantly advance research on peri-implantitis. Drug2Cell is a computational tool that utilizes scRNA-seq data to identify specific cellular targets of drugs and uncover the cellular mechanisms affected by pharmacological agents (Kanemaru et al. 2023). By analyzing how drugs interact with individual cell types, this tool can identify effective therapeutic targets within the complex tissue environment of peri-implantitis. Although not yet applied in this specific field, its use in peri-implantitis could lead to more targeted and effective treatments, thereby enhancing the safety and efficacy of future pharmaceutical interventions.

21.8 Conclusion

Peri-implantitis is defined as "a plaque-associated pathological condition," yet it is a complex inflammatory disease influenced by various risk factors. Characterized by microbial involvement, titanium particle release, and host inflammatory reactions, recent transcriptomic advances, particularly scRNA-seq, have illuminated detailed cellular interactions and specific inflammatory pathways unique to peri-implantitis. Conventional diagnostic approaches to peri-implantitis

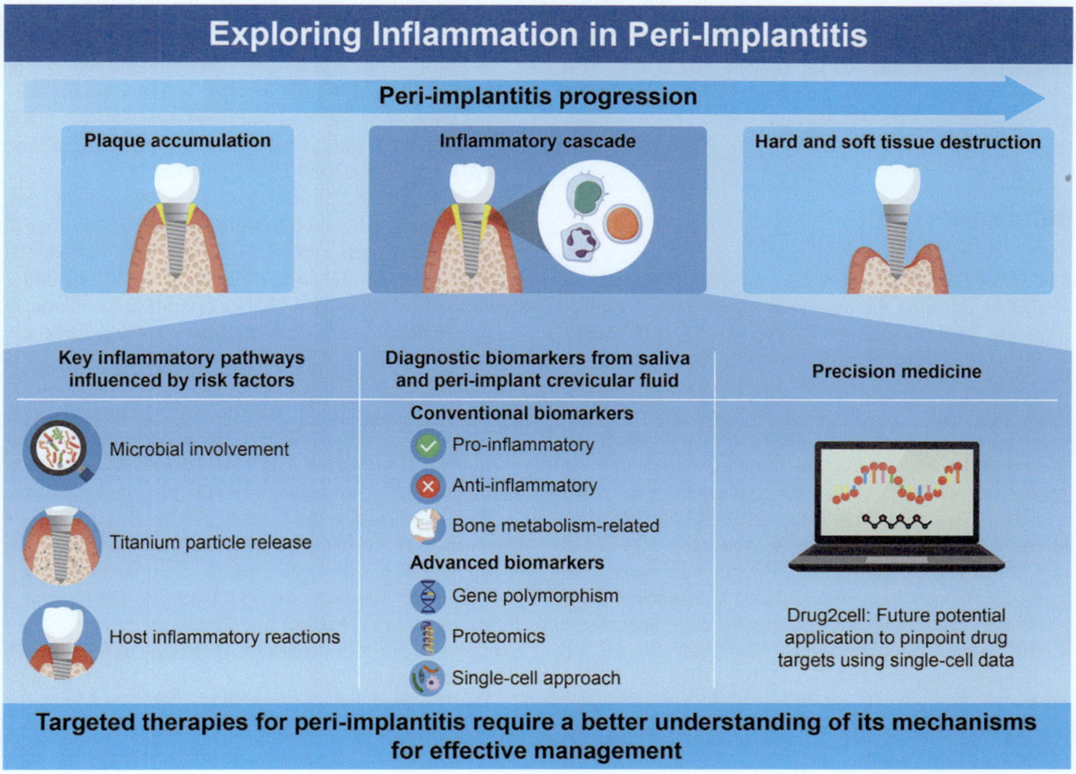

Fig. 21.1 Schematic graphical abstract. Comprehensive overview of peri-implantitis: from pathogenesis to potential precision medicine. This illustration depicts the progression of peri-implantitis from initial plaque formation to bone loss, highlighting the inflammatory cascade. Key inflammatory pathways influenced by microbial involvement, titanium particle release, and host reactions are shown. Diagnostic biomarkers from saliva and peri-implant crevicular fluid are categorized, emphasizing their pro- and anti-inflammatory roles, as well as their relation to bone metabolism. The precision medicine approach, represented by the Drug2cell method, suggests the potential for future targeted therapies based on single-cell data. This integrative view underscores the need for a deeper understanding of peri-implantitis mechanisms to develop effective, targeted treatments

often fall short, driving the need to explore biomarkers in saliva and peri-implant crevicular fluid for early detection and accurate monitoring. These biomarkers include proinflammatory, anti-inflammatory, and bone metabolism-related markers, each holds significant potential for improving diagnostic precision. The future of peri-implantitis management is moving toward precision medicine, leveraging cutting-edge genomic data and computational tools such as Drug2Cell to enable personalized and effective treatment strategies in dental implantology.

While the debate continues regarding various treatment modalities, the etiology and pathogenesis of peri-implantitis are not fully understood. Ideally, research should follow a logical sequence–elucidating disease mechanisms, developing targeted therapies, and evaluating their efficacy (Fig. 21.1). However, current practices reflect the tension between addressing immediate clinical needs and developing treatments based on a robust scientific foundation. Bridging this knowledge gap could lead to the development of more effective targeted therapies based on a comprehensive understanding of peri-implantitis. Although this approach may take longer to yield clinical application, it is more likely to result in robust long-term treatment strategies for peri-implantitis that will significantly benefit patients.

Acknowledgments This work was financially supported by the JSPS KAKENHI Grant-in-Aid for Scientific Research (C) (number 22 K10042 to A.H.). This work was also partially supported by Nihon University through a 2022 Overseas Researcher grant.

References

Aas JA, Paster BJ, Stokes LN, Olsen I, Dewhirst FE (2005) Defining the normal bacterial flora of the oral cavity. J Clin Microbiol 43(11):5721–5732. https://doi.org/10.1128/jcm.43.11.5721-5732.2005

Abduljabbar T, Vohra F, Ullah A, Alhamoudi N, Khan J, Javed F (2019) Relationship between self-rated pain and peri-implant clinical, radiographic and whole salivary inflammatory markers among patients with and without peri-implantitis. Clin Implant Dent Relat Res 21(6):1218–1224. https://doi.org/10.1111/cid.12866

Al-Askar M, Ajlan S, Alomar N, Al-Daghri NM (2018) Clinical and radiographic Peri-implant parameters and whole salivary interleukin-1β and Interleukin-6 levels among Type-2 diabetic and nondiabetic patients with and without Peri-Implantitis. Med Princ Pract 27(2): 133–138. https://doi.org/10.1159/000488032

Alassy H, Parachuru P, Wolff L (2019) Peri-implantitis diagnosis and prognosis using biomarkers in peri-implant crevicular fluid: a narrative review. Diagnostics 9(4):214

Al-Zawawi AS (2020) Contribution of fungi and viruses towards the etiopathogenesis peri-implantitis: a literature review of currently available evidence. Surg Pract Sci 2:100017. https://doi.org/10.1016/j.sipas.2020.100017

Arikan F, Buduneli N, Lappin DF (2011) C-telopeptide pyridinoline crosslinks of type I collagen, soluble RANKL, and osteoprotegerin levels in crevicular fluid of dental implants with peri-implantitis: a case-control study. Int J Oral Maxillofac Implants 26(2): 282–289

Asa'ad F, Garaicoa-Pazmiño C, Dahlin C, Larsson L (2020) Expression of microRNAs in periodontal and peri-implant diseases: a systematic review and meta-analysis. Int J Mol Sci 21(11):4147

Ashifa N, Viswanathan K, Sundaram R, Srinivasan S (2021) Sclerostin and its role as a bone modifying agent in periodontal disease. J Oral Biosci 63(2): 104–110

Ata-Ali J, Flichy-Fernández AJ, Alegre-Domingo T, Ata-Ali F, Palacio J, Peñarrocha-Diago M (2015) Clinical, microbiological, and immunological aspects of healthy versus peri-implantitis tissue in full arch reconstruction patients: a prospective cross-sectional study. BMC Oral Health 15(1):43. https://doi.org/10.1186/s12903-015-0031-9

Berglundh T, Armitage G, Araujo MG, Avila-Ortiz G, Blanco J, Camargo PM, Chen S, Cochran D, Derks J, Figuero E, Hämmerle CHF, Heitz-Mayfield LJA, Huynh-Ba G, Iacono V, Koo KT, Lambert F, McCauley L, Quirynen M et al (2018) Peri-implant diseases and conditions: consensus report of workgroup 4 of the 2017 world workshop on the classification of periodontal and Peri-implant diseases and conditions. J Clin Periodontol 45(Suppl 20):S286–s291. https://doi.org/10.1111/jcpe.12957

Bhavsar I, Miller CS, Ebersole JL, Dawson DR 3rd, Thompson KL, Al-Sabbagh M (2019) Biological response to peri-implantitis treatment. J Periodontal Res 54(6):720–728. https://doi.org/10.1111/jre.12681

Carcuac O, Berglundh T (2014) Composition of human peri-implantitis and periodontitis lesions. J Dent Res 93(11):1083–1088. https://doi.org/10.1177/0022034514551754

Cardoso JM, Duarte S, Ribeiro AC, Mascarenhas P, Noronha S, Alves RC (2022) Association between IL-1A, IL-1B and IL-1RN polymorphisms and peri-implantitis: a systematic review and meta-analysis. Appl Sci 12(14):6958

Carvalho É et al (2023) Microbiota associated with peri-implantitis-a systematic review with meta-analyses. Clin Oral Implants Res 34(11):1176–1187. https://doi.org/10.1111/clr.14153

Chaparro A, Atria P, Realini O, Monteiro LJ, Betancur D, Acuña-Gallardo S, Ramírez V, Bendek MJ, Pascual A, Nart J, Beltrán V, Sanz A (2021) Diagnostic potential of peri-implant crevicular fluid microRNA-21-3p and microRNA-150-5p and extracellular vesicles in peri-implant diseases. J Periodontol 92(6):11–21. https://doi.org/10.1002/jper.20-0372

Che C, Liu J, Ma L, Xu H, Bai N, Zhang Q (2017) LOX-1 is involved in IL-1β production and extracellular matrix breakdown in dental peri-implantitis. Int Immunopharmacol 52:127–135. https://doi.org/10.1016/j.intimp.2017.09.003

Corrêa MG, Pimentel SP, Ribeiro FV, Cirano FR, Casati MZ (2019) Host response and peri-implantitis. Braz Oral Res 33(suppl 1):e066. https://doi.org/10.1590/1807-3107bor-2019.vol33.0066

Darabi E, Kadkhoda Z, Amirzargar A (2013) Comparison of the levels of tumor necrosis factor-α and interleukin-17 in gingival crevicular fluid of patients with peri-implantitis and a control group with healthy implants. Iran J Allergy Asthma Immunol 12(1):75–80

Daubert D, Pozhitkov A, McLean J, Kotsakis G (2018) Titanium as a modifier of the peri-implant microbiome structure. Clin Implant Dent Relat Res 20(6):945–953. https://doi.org/10.1111/cid.12676

Dinarello CA (2013) Overview of the interleukin-1 family of ligands and receptors. Semin Immunol 25(6): 389–393. https://doi.org/10.1016/j.smim.2013.10.001

Duarte PM, Serrão CR, Miranda TS, Zanatta LC, Bastos MF, Faveri M, Figueiredo LC, Feres M (2016) Could cytokine levels in the peri-implant crevicular fluid be used to distinguish between healthy implants and implants with peri-implantitis? A systematic review. J Periodontal Res 51(6):689–698. https://doi.org/10.1111/jre.12354

Dukka H, Saleh MHA, Ravidà A, Greenwell H, Wang HL (2021) Is bleeding on probing a reliable clinical indicator of peri-implant diseases? J Periodontol 92(12): 1669–1674. https://doi.org/10.1002/jper.20-0890

Eckert M, Mizgalska D, Sculean A, Potempa J, Stavropoulos A, Eick S (2018) In vivo expression of proteases and protease inhibitor, a serpin, by periodontal pathogens at teeth and implants. Mol Oral Microbiol 33(3):240–248. https://doi.org/10.1111/omi.12220

Esberg A, Isehed C, Holmlund A, Lundberg P (2019) Peri-implant crevicular fluid proteome before and after adjunctive enamel matrix derivative treatment of peri-implantitis. J Clin Periodontol 46(6):669–677. https://doi.org/10.1111/jcpe.13108

Farhad SZ, Rezazadeh F, Mohammadi M (2019) Interleukin – 17 and Interleukin-10 as inflammatory and prevention biomarkers in Periimplant diseases. Int J Prev Med 10:137. https://doi.org/10.4103/ijpvm.IJPVM_27_19

Fields JK, Günther S, Sundberg EJ (2019) Structural basis of IL-1 family cytokine signaling. Front Immunol 10: 1412. https://doi.org/10.3389/fimmu.2019.01412

Fragkioudakis I, Tseleki G, Doufexi AE, Sakellari D (2021) Current concepts on the pathogenesis of Peri-implantitis: a narrative review. Eur J Dent 15(2): 379–387. https://doi.org/10.1055/s-0040-1721903

Fretwurst T, Buzanich G, Nahles S, Woelber JP, Riesemeier H, Nelson K (2016) Metal elements in tissue with dental peri-implantitis: a pilot study. Clin Oral Implants Res 27(9):1178–1186. https://doi.org/10.1111/clr.12718

Fu JH, Wang HL (2020) Breaking the wave of peri-implantitis. Periodontol 84(1):145–160. https://doi.org/10.1111/prd.12335

Gleiznys D, Gleiznys A, Abraškevičiūtė L, Vitkauskienė A, Šaferis V, Sakalauskienė J (2019) Interleukin-10 and interleukin-1β cytokines expression in leukocytes of patients with chronic Peri-Mucositis. Med Sci Monit 25:7471–7479. https://doi.org/10.12659/msm.915464

Griffiths GS (2003) Formation, collection and significance of gingival crevice fluid. Periodontol 31:32–42. https://doi.org/10.1034/j.1600-0757.2003.03103.x

Gürlek Ö, Gümüş P, Nile CJ, Lappin DF, Buduneli N (2017) Biomarkers and bacteria around implants and natural teeth in the same individuals. J Periodontol 88(8):752–761. https://doi.org/10.1902/jop.2017.160751

Halstenbach T, Nelson K, Iglhaut G, Schilling O, Fretwurst T (2023) Impact of peri-implantitis on the proteome biology of crevicular fluid: a pilot study. J Periodontol 94(7):835–847. https://doi.org/10.1002/jper.22-0461

Hamed MN, Abdulbaqi HR (2024) Expression of miRNAs (146a and 155) in human peri-implant tissue affected by peri-implantitis: a case control study. BMC Oral Health 24(1):856. https://doi.org/10.1186/s12903-024-04579-x

Haririan H, Andrukhov O, Laky M, Rausch-Fan X (2021) Saliva as a source of biomarkers for periodontitis and Periimplantitis. Front Dental Med 2:687638

Hentenaar DFM, De Waal YCM, Vissink A, Van Winkelhoff AJ, Meijer HJA, Liefers SC, Kroese FGM, Raghoebar GM (2021) Biomarker levels in peri-implant crevicular fluid of healthy implants, untreated and non-surgically treated implants with peri-implantitis. J Clin Periodontol 48(4):590–601. https://doi.org/10.1111/jcpe.13423

Huntzinger E, Izaurralde E (2011) Gene silencing by microRNAs: contributions of translational repression and mRNA decay. Nat Rev Genet 12(2):99–110

Irwandi RA, Vacharaksa A (2016) The role of microRNA in periodontal tissue: a review of the literature. Arch Oral Biol 72:66–74

Jansson L, Lundmark A, Modin C, Abadji D, Yucel-Lindberg T (2021) Intra-individual cytokine profile in peri-implantitis and periodontitis: a cross-sectional study. Clin Oral Implants Res 32(5):559–568. https://doi.org/10.1111/clr.13725

Jiang J, Gao G, Wang Q (2021) Levels of SERPIN family proteins in peri-implant crevicular fluid in patients with peri-implantitis. J Clin Lab Anal 35(9):e23926. https://doi.org/10.1002/jcla.23926

Kandaswamy E, Harsha M, Joshi VM (2024) Titanium corrosion products from dental implants and their effect on cells and cytokine release: a review. J Trace Elem Med Biol 84:127464. https://doi.org/10.1016/j.jtemb.2024.127464

Kanemaru K, Cranley J, Muraro D, Miranda AMA, Ho SY, Wilbrey-Clark A, Patrick Pett J, Polanski K, Richardson L, Litvinukova M, Kumasaka N, Qin Y, Jablonska Z, Semprich CI, Mach L, Dabrowska M, Richoz N, Bolt L et al (2023) Spatially resolved multiomics of human cardiac niches. Nature 619(7971):801–810. https://doi.org/10.1038/s41586-023-06311-1

Kheder W, Al Kawas S, Khalaf K, Samsudin AR (2021) Impact of tribocorrosion and titanium particles release on dental implant complications – a narrative review. Jpn Dent Sci Rev 57:182–189. https://doi.org/10.1016/j.jdsr.2021.09.001

Kotsakis GA, Olmedo DG (2021) Peri-implantitis is not periodontitis: scientific discoveries shed light on microbiome-biomaterial interactions that may determine disease phenotype. Periodontol 86(1):231–240. https://doi.org/10.1111/prd.12372

Lafuente-Ibáñez de Mendoza I, Setien-Olarra A, García-De la Fuente AM, Aguirre-Urizar JM, Marichalar-Mendia X (2022) Role of proinflammatory mutations in peri-implantitis: systematic review and meta-analysis. Int J Implant Dent 8(1):2. https://doi.org/10.1186/s40729-022-00400-y

Li J, Ye LJ, Dai YW, Wang HW, Gao J, Shen YH, Wang F, Dai QG, Wu YQ (2024) Single-cell analysis reveals a unique microenvironment in peri-implantitis. J Clin Periodontol 51. https://doi.org/10.1111/jcpe.13982

Martin A, Zhou P, Singh BB, Kotsakis GA (2022) Transcriptome-wide gene expression analysis in Peri-implantitis reveals candidate cellular pathways. JDR Clin Trans Res 7(4):415–424. https://doi.org/10.1177/23800844211045297

Meniailo ME, Malashchenko VV, Shmarov VA, Gazatova ND, Melashchenko OB, Goncharov AG, Seledtsova GV, Seledtsov VI (2018) Interleukin-8 favors pro-inflammatory activity of human monocytes/macrophages. Int Immunopharmacol 56:217–221. https://doi.org/10.1016/j.intimp.2018.01.036

Milinkovic I, Djinic Krasavcevic A, Nikolic N, Aleksic Z, Carkic J, Jezdic M, Jankovic S, Milasin J (2021) Notch down-regulation and inflammatory cytokines and RANKL overexpression involvement in peri-implant mucositis and peri-implantitis: a cross-sectional study. Clin Oral Implants Res 32(12):1496–1505. https://doi.org/10.1111/clr.13850

Mitchell PS, Parkin RK, Kroh EM, Fritz BR, Wyman SK, Pogosova-Agadjanyan EL, Peterson A, Noteboom J, O'Briant KC, Allen A (2008) Circulating microRNAs as stable blood-based markers for cancer detection. Proc Natl Acad Sci 105(30):10513–10518

Mo JJ, Lai YR, Huang QR, Li YR, Zhang YJ, Chen RY, Qian SJ (2024) Single-cell sequencing identifies inflammation-promoting fibroblast-neutrophil interaction in peri-implantitis. J Clin Periodontol 51(2):196–208. https://doi.org/10.1111/jcpe.13912

Moaven H, Giacaman A, Beltrán V, Sam YH, Betancur D, Mainas G, Tarjomani SA, Donos N, Sousa V (2022) Biomarker expression of Peri-implantitis lesions before and after treatment: a systematic review. Int J Environ Res Public Health 19(21). https://doi.org/10.3390/ijerph192114085

Oliveira JA, de Oliveira Alves R, Nascimento IM, Hidalgo MAR, Scarel-Caminaga RM, Cristina Pigossi S (2023) Pro-and anti-inflammatory cytokines and osteoclastogenesis-related factors in peri-implant diseases: systematic review and meta-analysis. BMC Oral Health 23(1):420

Olmedo DG, Nalli G, Verdú S, Paparella ML, Cabrini RL (2013) Exfoliative cytology and titanium dental implants: a pilot study. J Periodontol 84(1):78–83. https://doi.org/10.1902/jop.2012.110757

Palmer RJ Jr (2014) Composition and development of oral bacterial communities. Periodontol 64(1):20–39. https://doi.org/10.1111/j.1600-0757.2012.00453.x

Park SE, Park K, Kim E, Kim CY, Hwang SM, Lee JM, Suh JY, Lee Y, Kim MO, Kim YG (2024) CXCL5/CXCL8 induces neutrophilic inflammation in peri-implantitis. J Periodontal Res 59(4):698–711. https://doi.org/10.1111/jre.13230

Pettersson M, Kelk P, Belibasakis GN, Bylund D, Molin Thorén M, Johansson A (2017) Titanium ions form particles that activate and execute interleukin-1β release from lipopolysaccharide-primed macrophages. J Periodontal Res 52(1):21–32. https://doi.org/10.1111/jre.12364

Pliavga V, Peceliunaite G, Daugela P, Leketas M, Gervickas A, Juodzbalys G (2023) Peri-implantitis diagnosis and prognosis using biomarkers: a systematic literature review. Int J Oral Maxillofac Implants 38(6):1095

Rakic M, Lekovic V, Nikolic-Jakoba N, Vojvodic D, Petkovic-Curcin A, Sanz M (2013) Bone loss biomarkers associated with peri-implantitis. A cross-sectional study. Clin Oral Implants Res 24(10):1110–1116. https://doi.org/10.1111/j.1600-0501.2012.02518.x

Rakic M, Struillou X, Petkovic-Curcin A, Matic S, Canullo L, Sanz M, Vojvodic D (2014) Estimation of bone loss biomarkers as a diagnostic tool for peri-implantitis. J Periodontol 85(11):1566–1574. https://doi.org/10.1902/jop.2014.140069

Rakic M, Monje A, Radovanovic S, Petkovic-Curcin A, Vojvodic D, Tatic Z (2020) Is the personalized approach the key to improve clinical diagnosis of peri-implant conditions? The role of bone markers. J Periodontol 91(7):859–869. https://doi.org/10.1002/jper.19-0283

Sahoo SK, Jalaluddin M, Bhuyan L, Dash KC, Mishra S, Mishra P (2021) Assessment of cytokine and herpesvirus level in Peri-implantitis and healthy patients. J Pharm Bioallied Sci 13(Suppl 2):S1418–s1421. https://doi.org/10.4103/jpbs.jpbs_233_21

Schwarz F, Derks J, Monje A, Wang HL (2018) Peri-implantitis. J Clin Periodontol 45(Suppl 20):S246–s266. https://doi.org/10.1111/jcpe.12954

Shafizadeh M, Amid R, Mahmoum M, Kadkhodazadeh M (2021) Histopathological characterization of peri-implant diseases: a systematic review and meta-analysis. Arch Oral Biol 132:105288. https://doi.org/10.1016/j.archoralbio.2021.105288

Song L, Jiang J, Li J, Zhou C, Chen Y, Lu H, He F (2022) The characteristics of microbiome and cytokines in healthy implants and Peri-Implantitis of the same individuals. J Clin Med 11(19). https://doi.org/10.3390/jcm11195817

Strbac GD, Monov G, Cei S, Kandler B, Watzek G, Gruber R (2006) Cathepsin K levels in the crevicular fluid of dental implants: a pilot study. J Clin Periodontol 33(4):302–308. https://doi.org/10.1111/j.1600-051X.2006.00904.x

Tang F, Barbacioru C, Wang Y, Nordman E, Lee C, Xu N, Wang X, Bodeau J, Tuch BB, Siddiqui A, Lao K, Surani MA (2009) mRNA-Seq whole-transcriptome analysis of a single cell. Nat Methods 6(5):377–382. https://doi.org/10.1038/nmeth.1315

Thorisson GA, Smith AV, Krishnan L, Stein LD (2005) The international HapMap project web site. Genome Res 15(11):1592–1593. https://doi.org/10.1101/gr.4413105

Turkmen M, Firatli E (2022) The study of genetic predisposition on periodontitis and peri-implantitis. Niger J Clin Pract 25(11):1799–1804

Udagawa N, Koide M, Nakamura M, Nakamichi Y, Yamashita T, Uehara S, Kobayashi Y, Furuya Y,

Yasuda H, Fukuda C (2021) Osteoclast differentiation by RANKL and OPG signaling pathways. J Bone Miner Metab 39:19–26

Wakuda S, Hasuike A, Fujiwara K, Sakai R, Chaurasia A, Uchiyama T, Sato S (2024) Titanium particle-induced inflammasome in human gingival epithelial cells. J Dental Sci 20. https://doi.org/10.1016/j.jds.2024.06.013

Wang CW, Hao Y, Di Gianfilippo R, Sugai J, Li J, Gong W, Kornman KS, Wang HL, Kamada N, Xie Y, Giannobile WV, Lei YL (2021) Machine learning-assisted immune profiling stratifies peri-implantitis patients with unique microbial colonization and clinical outcomes. Theranostics 11(14): 6703–6716. https://doi.org/10.7150/thno.57775

Wilson TG Jr, Valderrama P, Burbano M, Blansett J, Levine R, Kessler H, Rodrigues DC (2015) Foreign bodies associated with peri-implantitis human biopsies. J Periodontol 86(1):9–15. https://doi.org/10.1902/jop.2014.140363

Yakar N, Guncu GN, Akman AC, Pınar A, Karabulut E, Nohutcu RM (2019) Evaluation of gingival crevicular fluid and peri-implant crevicular fluid levels of sclerostin, TWEAK, RANKL and OPG. Cytokine 113:433–439. https://doi.org/10.1016/j.cyto.2018.10.021

Yamamoto R, Amano K, Takahashi S-W, To, M, Takahashi S, Matsuo M (2021) Changes in the microcirculation in periodontal tissue due to experimental peri-implantitis. J Oral Biosci 63(2):153–160

Zenobia C, Hajishengallis G (2015) Basic biology and role of interleukin-17 in immunity and inflammation. Periodontol 69(1):142–159. https://doi.org/10.1111/prd.12083

Zhang Q, Liu J, Ma L, Bai N, Xu H (2019) LOX-1 is involved in TLR2 induced RANKL regulation in peri-implantitis. Int Immunopharmacol 77:105956. https://doi.org/10.1016/j.intimp.2019.105956

Zhang Q, Xu T, Bai N, Tan F, Zhao H, Liu J (2022) Lectin-type oxidized LDL receptor 1 modulates matrix metalloproteinase 2 production in peri-implantitis. Exp Ther Med 23(2):171. https://doi.org/10.3892/etm.2021.11094

Zhao X, Xu D, Li Y, Zhang J, Liu T, Ji Y, Wang J, Zhou G, Xie X (2014) MicroRNAs regulate bone metabolism. J Bone Miner Metab 32:221–231

22 Titanium Dental Implant Corrosion in Peri-Implantitis and Implant Failure

Heba Allah Madi, Emerita Jocelyne Feine, and Simon D. Tran

Abstract

Titanium (Ti) is an inert metallic element with the atomic number 22. It is alloyed with other elements to create strong, lightweight materials like titanium alloy Ti6Al4V, a titanium alloy composed of 90% titanium, 6% aluminum, and 4% vanadium. These alloys have been widely used in dentistry and surgery for their biocompatibility, mechanical strength, and corrosion resistance. A key feature of titanium implants is their ability to osseointegrate with bone, creating a stable connection between the implant and the surrounding bone tissue. Titanium naturally forms a protective oxide layer during implant placement, enhancing its corrosion resistance. However, continuous exposure to the oral environment poses challenges that may lead to its degradation and corrosion. This process releases metal ions, microparticles, and nanoparticles into the surrounding tissues, triggering local and systemic inflammatory reactions. Alongside bacterial infection and mechanical wear, this corrosion can exacerbate peri-implant inflammatory diseases, such as mucositis and peri-implantitis, potentially leading to implant failure.

This chapter explores the mechanisms of titanium implant corrosion, risk factors accelerating the process, the role of titanium hypersensitivity in response to corrosion by-products, as well as its effects on implant success. Furthermore, it explores how corrosion contributes to peri-implant inflammation and affects the surrounding peri-implant tissues. It also highlights strategies to minimize corrosion, including the use of innovative materials in implant coatings and surface treatments, and concludes by discussing potential future research directions.

Keywords

Titanium corrosion · Dental implants · Peri-implantitis · Implant failure · Osseointegration

22.1 Introduction

22.1.1 Overview of Dental Implants

Earlier dental implants were fabricated from stainless steel and cobalt-chromium alloys. However, these alloys had high failure rates, were prone to corrosion, and caused unfavorable tissue reactions (Steinemann 1998).

The introduction of titanium as an implant material marked a turning point in implantology. Titanium dental implants have become the standard treatment for tooth restoration, providing a

H. A. Madi (✉) · J. Feine · S. D. Tran
Faculty of Dental Medicine and Oral Health Sciences, McGill University, Montréal, QC, Canada
e-mail: heba.madi@mail.mcgill.ca

N. Rezaei, H. Ziaei (eds.), *Oral Immunology*, Advances in Experimental Medicine and Biology 1492,
https://doi.org/10.1007/978-3-032-03176-1_22

stable foundation for prosthetic teeth that closely resemble the appearance and function of natural dentition (Gaviria et al. 2014). Their widespread use is attributed to high success and survival rates, predictability, and strong patient acceptance (Kupka et al. 2024). According to the American Dental Association, 5 million implants are placed in the U.S. annually, with comparable numbers/populations globally (Chen et al. 2023b). Three million people in the United States have implants, and this number grows by 500,000 annually (Dentistry 2022).

Titanium (Ti) is an inert metallic element with the atomic number 22. It is usually combined with other elements to produce a strong, lightweight alloy. Commercially pure titanium implants (cp-Ti), particularly Ti6Al4V (90% titanium, 6% aluminum, and 4% vanadium), have been used in surgery and dentistry for the past three decades due to their excellent biocompatibility, mechanical properties, and high corrosion resistance (Balasubramanian Gayathri et al. 2022).

The success of dental implants depends mainly on their ability to integrate with the surrounding bone tissue through osseointegration. This biological fusion ensures the implant remains firmly anchored, allowing patients to restore proper function and improve aesthetics (Adell et al. 1981; Albrektsson et al. 1981).

22.1.2 Types of Titanium Alloys

Titanium alloys are available in three structural forms: α, β, or a combination of α-β titanium alloys (Table 22.1).

22.1.2.1 Alpha (α) Titanium Alloys

The structure of Alpha (α) titanium alloy is a hexagonal close-packed crystal, stabilized by adding α-phase elements such as aluminum, oxygen, nitrogen, and occasionally zirconium or tin. These elements enhance the alloy's strength and mechanical properties at both room temperature and elevated temperatures, up to 550 °C. Alpha titanium alloys also exhibit high corrosion resistance and weldability, making them suitable for aerospace and chemical processing applications, where strength and resistance to creep at high temperatures are critical. While alpha titanium alloys are biocompatible due to their inertness and excellent corrosion resistance, they are seldom used in medical and dental implants because they have lower strength at room temperature compared to beta and alpha-beta titanium alloys. Medical and dental implants require materials that can endure mechanical loading over long periods, which makes other titanium alloys more suitable for these applications (Donachie 2000; Peters et al. 2003).

22.1.2.2 Beta (β) Titanium Alloys

The structure of Beta (β) titanium alloy is a body-centered cubic crystal that is stabilized by adding β-phase stabilizer elements to the titanium, including molybdenum, vanadium, niobium, and chromium. β titanium alloys are highly formable, have a lower modulus of elasticity, better fatigue resistance, greater flexibility than alpha alloys, and can be heat-treated to improve their strength. Standard heat treatment methods include solution treatment followed by aging, which enhances strength by transforming part of the β phase into martensitic structures. While β titanium alloys typically offer lower corrosion resistance than alpha alloys, they still perform well in room and moderate temperature applications, maintaining good overall corrosion resistance. They are not suitable for very high-temperature applications, as their strength and creep resistance decrease when the temperature is higher, and they lose strength at temperatures above 400 °C. Therefore, they are highly suitable for applications requiring formability and flexibility at room temperature. They are used in aerospace components, sports equipment, and bio-medical devices, including medical and dental implants (Boyer and Briggs 2005; Donachie 2000; Peters et al. 2003).

22.1.2.3 Alpha-Beta (α-β) Titanium Alloys

The structure of alpha-beta (α-β) titanium alloy is a mix of hexagonal close-packed alpha and body-centered cubic beta phases. This dual-phase

Table 22.1 Types of titanium alloys: phase structure and composition, alloying elements, advantages, and typical applications

Titanium alloys	Composition	Advantages	Applications
Alpha (α) titanium alloys	Titanium, aluminum, oxygen, nitrogen, zirconium, tin	High corrosion resistance, high weldability	Aerospace, chemical processing
Beta (β) titanium alloys	Titanium, molybdenum, vanadium, niobium, chromium	High formability, good fatigue resistance, low modulus of elasticity	Aerospace components, sports equipment, bio-medical implants
Alpha-beta (α-β) titanium alloys	Titanium, aluminum, vanadium, molybdenum, niobium	Good balance of strength, fatigue resistance and formability	Aircraft, automotive, medical, and dental implants

Note: Specific alloy compositions and applications may vary based on industry requirements and advancements in material science

structure is stabilized by adding a combination of α-phase and β-phase stabilizer elements, including aluminum for the α-phase and molybdenum, vanadium, or niobium for the β-phase. α-β alloys balance strength, fatigue resistance, and formability, making them more flexible than alpha or beta alloys. α-β alloys can be heat-treated in a process called aging to improve their mechanical properties. During aging, β-phase stabilizers such as vanadium and molybdenum facilitate the precipitation of fine α-phase particles within the β-phase matrix, enhancing the alloy's strength and hardness. Although they do not exhibit the same strength as α alloys at high temperatures, their performance at moderate temperatures remains good, and they exhibit good creep resistance to 400 °C. These properties, along with their high corrosion resistance, make them well-suited for various applications, including aircraft, automotive, medical, and dental implants (Donachie 2000; Peters et al. 2003).

22.1.3 Osseointegration

In 1952, Per-Ingvar Brånemark, a Swedish orthopedic surgeon, conducted the first animal study to demonstrate titanium implants' biocompatibility. In this study, Brånemark placed a titanium case in an optic chamber within a rabbit's thighbone to observe the bone's vascularity. When he attempted to remove the titanium case, he found that it had fused to the bone, making extraction nearly impossible. This demonstrated that titanium implants could irreversibly integrate with bone; this phenomenon was later termed "osseointegration." Brånemark defined osseointegration as "a direct structural and functional connection between ordered, living bone, and the surface of a load-carrying implant" (Brånemark et al. 1969).

In 1965, Brånemark conducted the first human clinical trials by placing 10 mm long, root-shaped, cylindrical titanium implants into the human alveolar bone. He extensively studied the use of pure titanium screws in dental prosthetics. Later, Brånemark published the first guidelines for the successful use of endosseous implants in 1977, and by 1983, he presented the results of the 15-year follow-up of these implants. Following Brånemark's research, commercially pure titanium became the material of choice for dental implants, providing a reliable method for restoring edentulous ridges (Adell et al. 1981; Brånemark 1983; Brånemark et al. 1969, 1977).

Between 1978 and 1983, studies from different research groups replicated Brånemark's findings, further validating the results of his research. This work laid the foundation for developing implant dentistry as a field of science (Smith and Zarb 1989; Zarb and Schmitt 1990).

Further research focused on understanding osseointegration at clinical, anatomical, and histological levels, including detailed in-vivo and in-vitro studies of bone/metal interactions and how the characteristics of titanium influence implants (Albrektsson et al. 1981; Davies 1998).

Dental implant designs have evolved over the years, and modern research focuses on altering implant surfaces to maximize the surface area of bone-implant integration and reduce corrosion, enhancing osseointegration and implant stability (Le Guéhennec et al. 2007; Wennerberg and Albrektsson 2009). Osseointegrated titanium implants have demonstrated high success rates (Buser et al. 2012). The metal's resilience allowed the development of different implant forms, such as single-tooth implants, multiple teeth replacement, and full-mouth reconstructions, establishing its importance in contemporary dental practice (Jokstad et al. 2003; Pjetursson et al. 2012).

22.1.3.1 Role of the Immune System in Titanium Implant Osseointegration

Osseointegration is a complex healing process akin to fracture healing, involving overlapping phases: initial bone destruction, repair, implant integration, and stabilization with load adaptation. The healing sequence begins with hemostasis, followed by inflammatory, proliferative, and remodeling phases (Terheyden et al. 2012).

Immune and inflammatory mechanisms are triggered upon implant insertion, with neutrophils occupying the implant surface within hours, followed by monocytes and macrophages within 2–4 days. These immune cells regulate osseointegration by releasing cytokines and growth factors, promoting angiogenesis and osteogenesis, leading to collagen matrix deposition and early bone formation. Over time, woven bone forms on the implant surface, later maturing into lamellar bone. Osseointegration remains a dynamic process involving continuous bone remodeling and adaptation to functional demands (Berglundh et al. 2003; Wang et al. 2016).

Recent research recognizes osseointegration as an osteoimmune reaction rather than a classic bone healing response. The immune system recognizes the implant as a foreign object and responds by forming bone around it, with the immune system effectively shielding the implant from the host bone tissue (Albrektsson et al. 2022). The immune system participates in implant integration by modulating inflammation, bone remodeling, and tissue regeneration. Various immune cells, including macrophages, T cells, and bone cells (e.g., osteoblasts, osteoclasts, and osteocytes) orchestrate these interactions to regulate implant stability and integration (Kondo et al. 2024).

Macrophages are pivotal regulators of osseointegration. M1 macrophages dominate the initial response, promoting inflammation and osteoclastogenesis activity and secreting pro-inflammatory cytokines such as IL (interleukin)-1β, IL-6, and tumor necrosis factor-alpha (TNF-α). In contrast, M2 macrophages support tissue repair and osteogenesis by secreting bone morphogenetic protein-2 (BMP-2), transforming growth factor Beta 1 (TGF-β1), and vascular endothelial growth factor (VEGF), facilitating bone matrix deposition and vascularization (Liang et al. 2021; Fang et al. 2024). A balanced transition from M1 to M2 macrophages is essential for successful implant integration (Amengual-Peñafiel et al. 2021). Additionally, osteal macrophages, known as osteoMacs, play a vital role in immune surveillance during implant healing and bone homeostasis (Yang et al. 2022).

Bone resorption and formation are tightly regulated by the RANKL (Receptor Activator of Nuclear Factor Kappa-B Ligand)/OPG (Osteoprotegerin) axis. RANKL, secreted by osteoblasts and stromal cells, binds to RANK on osteoclast precursors, inducing their differentiation into mature osteoclasts and promoting bone resorption. Conversely, OPG, a decoy receptor for RANKL, inhibits osteoclast formation, maintaining bone homeostasis. An imbalance in RANKL/OPG expression can lead to excessive bone resorption, jeopardizing implant stability and osseointegration (Emam and Moussa 2024; Kondo et al. 2024).

The physical and chemical properties of titanium implants, particularly geometry, surface roughness, and hydrophilicity, influence the behavior of immune and bone-forming cells. Surface modifications at macro, micro, and nano levels play a crucial role in implant stability and osseointegration. Macro-level modifications, such as implant body shape and thread pattern, enhance primary fixation and long-term stability by increasing bone-implant contact and mechanical interlocking. These features optimize load distribution and create spaces for bone ingrowth (Kligman et al. 2021).

Micro-level modifications increase the implant's surface area, enhancing osteoconduction by facilitating fibrin matrix formation and supporting osteogenic cell attachment. Micrometer-range roughness fosters bone matrix deposition, accelerating healing and integration. Nano-level modifications improve hydrophilicity and surface roughness, creating optimal conditions for osteoblast adhesion, proliferation, and differentiation (Kligman et al. 2021).

These modifications can also promote an anti-inflammatory response in macrophages and modulate neutrophil activity, fostering a pro-osteogenic environment. This, in turn, enhances both contact and distance osteogenesis, where new bone forms directly on the implant surface and in the surrounding area (Kondo et al. 2024).

While a balanced immune response is essential for successful osseointegration, titanium corrosion can disrupt this process. The release of titanium particles and ions can trigger excessive inflammatory responses, macrophage polarization, and osteoclast activity, leading to peri-implant inflammation, bone loss, and implant failure. The following sections will explore how titanium implant corrosion influences these immune pathways.

22.1.3.2 Role of the Titanium Oxide Layer in Titanium Implants

Titanium's superior performance in implantology is primarily due to its high oxygen affinity, allowing it to form a stable, continuous, and self-protective oxide layer (TiO_2) when exposed to air or fluids (Textor et al. 2001). This 'passive layer' protects titanium from further oxidation and promotes a biologically favorable environment for bone cell attachment and proliferation, facilitating osseointegration. The titanium oxide is a thin layer that ranges between 1 and 20 nm with a low-crystalline oxide film (Souza et al. 2020). This thin layer exhibits thermodynamic stability and firm adherence to the titanium metal surface. Although the titanium oxide layer is chemically stable, it is not immune to degradation. The passivation model implies that the oxide layer undergoes partial dissolution and reprecipitation cycles in an aqueous environment. However, if the dissolution exceeds the reprecipitation, metal ions will gradually be released, leading to corrosion of the titanium implant surface (Hanawa 2004).

The titanium oxide (TiO_2) layer plays a pivotal role in implant biocompatibility by minimizing immune activation and supporting osseointegration. This passive layer interacts with host immune cells, reducing the recognition of titanium as a foreign body and limiting excessive inflammatory responses. However, corrosion-induced disruption of the oxide layer can expose the underlying titanium, triggering an immune response characterized by macrophage activation, cytokine secretion, and potential hypersensitivity reactions. The balance between immune tolerance and activation is crucial in maintaining implant longevity and preventing peri-implant inflammation (Kandaswamy et al. 2024; Stepanovska et al. 2020).

22.1.4 Peri-Implant Inflammatory Diseases

Peri-implant inflammatory diseases are described in two entities, including peri-implant mucositis and peri-implantitis. Peri-implant mucositis is defined as a pathologic condition in the tissue around dental implants, characterized by bleeding on probing and visual signs of inflammation and increased probing depths. Peri-implantitis is defined as a chronic pathologic condition occurring in the tissue around dental implants and characterized by bleeding on probing, visual signs of inflammation, increased probing depth, and progressive bone loss (Berglundh et al. 2018; Heitz-Mayfield 2024; Scarano et al. 2023).

While peri-implant mucositis is reversible when treated, peri-implantitis can be managed if detected early. If left untreated, it may progress in a nonlinear and accelerating pattern, leading to implant failure (Heitz-Mayfield 2024). Implant failure refers to the condition in which implants require removal or have already been lost (Misch et al. 2008). It is well documented in the literature that plaque and bacteria are the primary etiologic factors for peri-implant mucositis and peri-implantitis (Berglundh et al. 2018). In addition to bacterial infection, recent evidence suggests that tribocorrosion, along with the accumulation of microparticles and nanoparticles of titanium corrosion in the peri-implant tissues, triggers peri-implant inflammation and bone loss (Kheder et al. 2021; Souza et al. 2020).

22.2 Mechanisms of Titanium Implant Corrosion

Unlike other medical implants placed in sterile environments, dental implants exist in a complex, non-sterile oral environment, making them more susceptible to corrosion. The oxide layer may degrade, resulting in different forms of corrosion (Rodrigues et al. 2013). Understanding the mechanisms of titanium corrosion is critical to improving implant success and preventing further complications.

22.2.1 Types of Corrosion Affecting Titanium Implants

Several types of corrosion can affect the surface of titanium implants (Table 22.2) as follows:

22.2.1.1 Galvanic Corrosion

Galvanic corrosion occurs when two dissimilar metals come into contact in an electrolytic solution, such as saliva. In dental implants, titanium often interacts with another metal, typically in abutments or prosthetic frameworks, forming a galvanic cell. The more electropositive metal (e.g., titanium) acts as the cathode. The less electropositive metal (e.g., gold, cobalt-chromium, or stainless steel) acts as the anode. Saliva, which contains ions (e.g., Na^+, Cl^-, K^+, Ca^{2+}), serves as the electrolyte. The anode oxidizes, releasing metal ions into the surrounding tissue. The cathode remains protected, but the corrosion accelerates at the anode. Titanium implants connected to gold or stainless-steel prosthetic components can cause corrosion of the abutment. Released metal ions increase inflammatory responses, contributing to peri-implantitis and potential implant failure (Boffano et al. 2024; Gittens et al. 2011; Souza et al. 2015, 2020).

22.2.1.2 Crevice Corrosion

Crevice corrosion is a localized form of corrosion in areas with limited access to oral fluids and oxygen, such as contacting interfaces between implants, abutments, and prosthetic structures, such as the implant–abutment connection, micro-gaps between implant components. In areas where saliva and fluids stagnate, oxygen depletion occurs, causing the fluid to become acidic as the pH decreases. This breakdown of the protective oxide layer leads to corrosion. Crevice corrosion is particularly concerning as it often remains undetected until substantial damage has occurred (Boffano et al. 2024; Gittens et al. 2011; Souza et al. 2015, 2020).

22.2.1.3 Pitting Corrosion

Pitting corrosion occurs when the titanium oxide layer becomes vulnerable to attack by chlorides and other aggressive ions in saliva and oral fluids. Chloride ions aggressively attack weak points in the oxide layer, forming pits. Pits or holes on the implant surface can grow deeper over time, compromising the structural integrity of the implant. Pitting is particularly concerning because it can progress rapidly once initiated and is challenging to detect (Boffano et al. 2024; Gittens et al. 2011; Souza et al. 2015, Souza et al. 2020).

22.2.1.4 Fretting Corrosion

Fretting corrosion occurs when the masticatory and occlusal forces create a micro-movement between implant components or poorly fitted abutments. Cyclic mechanical loading causes titanium fatigue, degrading the protective oxide layer and exposing the underlying metal to corrosion. This type of corrosion is exacerbated in areas of

Table 22.2 Types of corrosion in titanium dental implants

Corrosion type	Description
Galvanic corrosion	Occurs when titanium interacts with another metal in an electrolytic solution, forming a galvanic cell. The anode oxidizes, releasing metal ions into surrounding tissue
Crevice corrosion	Occurs in micro-gaps between implant components due to stagnant fluids and oxygen depletion. The resulting acidic environment breaks down the protective oxide layer, leading to undetected damage
Pitting corrosion	Occurs when chloride ions aggressively attack weak points in the titanium oxide layer, leading to deep pits that compromise implant integrity
Fretting corrosion	Results from micro-movements between implant components due to masticatory forces, causing mechanical fatigue and breakdown of the protective oxide layer
Tribocorrosion	A combination of mechanical wear, chemical degradation, and microbiological interactions leads to the continuous release of titanium particles and inflammatory responses in peri-implant tissues

Note: Understanding these corrosion types is essential for improving the longevity and performance of titanium dental implants

high mechanical loading, such as posterior teeth (Boffano et al. 2024; Gittens et al. 2011; Souza et al. 2015, 2020).

22.2.1.5 Tribocorrosion

Tribocorrosion in titanium dental implants results from the synergistic effects of mechanical wear and chemical, biochemical, and microbiological processes. Mechanical wear from mastication forces, micro-movement and micro-gap, chemical corrosion, and plaque biofilm interaction contribute to the continuous release of titanium particles. Intrinsic and extrinsic factors in the oral environment can synergistically lead to wear and corrosion of the dental implant surface and the release of corrosive by-products into the surrounding environment. The accumulation of ions, microparticles, and nanoparticles triggers peri-implant inflammation and bone loss (Boffano et al. 2024; Gaur et al. 2022; Souza et al. 2015).

22.2.2 Factors Influencing Corrosion of Titanium Implants

The implant surface is continuously challenged in the dynamic oral environment by fluctuation in pH, bacterial colonization and biofilm byproducts, exposure to fluoride ions, and masticatory forces within the oral environment. These stresses compromise the integrity of the oxide layer and expose the underlying metal to corrosion. This accelerates titanium corrosion, releasing metal ions, microparticles, and nanoparticles, which trigger adverse biological reactions and inflammatory responses. As shown in Fig. 22.1, the mechanisms of titanium implant corrosion involve various factors like mechanical wear, plaque biofilm, and acidic conditions (Souza et al. 2020).

22.2.2.1 Electrochemical Factors

Electrochemical factors include an individual's health, microbiome, pH, saliva composition, glycoproteins, and temperature, which directly influence the electrochemical properties of the oral environment (Souza et al. 2020). Bacterial colonization begins shortly after implant placement, and bacteria are often found at implant–abutment and crown joints. Bacterial species (e.g., *Streptococcus mutans*, *Porphyromonas gingivalis*, and *Prevotella intermedia*) produce lactic acid, hydrogen peroxide (H_2O_2), and lipopolysaccharides (LPS), which lower pH and cause corrosion. This degradation of the TiO_2 layer releases titanium ions, microparticles, and nanoparticles, triggering an inflammatory immune response (Kotsakis and Olmedo 2021). The interaction between the titanium dioxide layer and bacteria may involve binding ions from the titanium dioxide surface to molecules on the bacterial cell wall, such as proteins or polysaccharides, leading to stable bacterial attachment to the implant surface. This facilitates bacterial adsorption and electrochemical changes on the titanium surface, causing corrosion of the implant surface (Chen et al. 2023a; Kotsakis and Olmedo 2021).

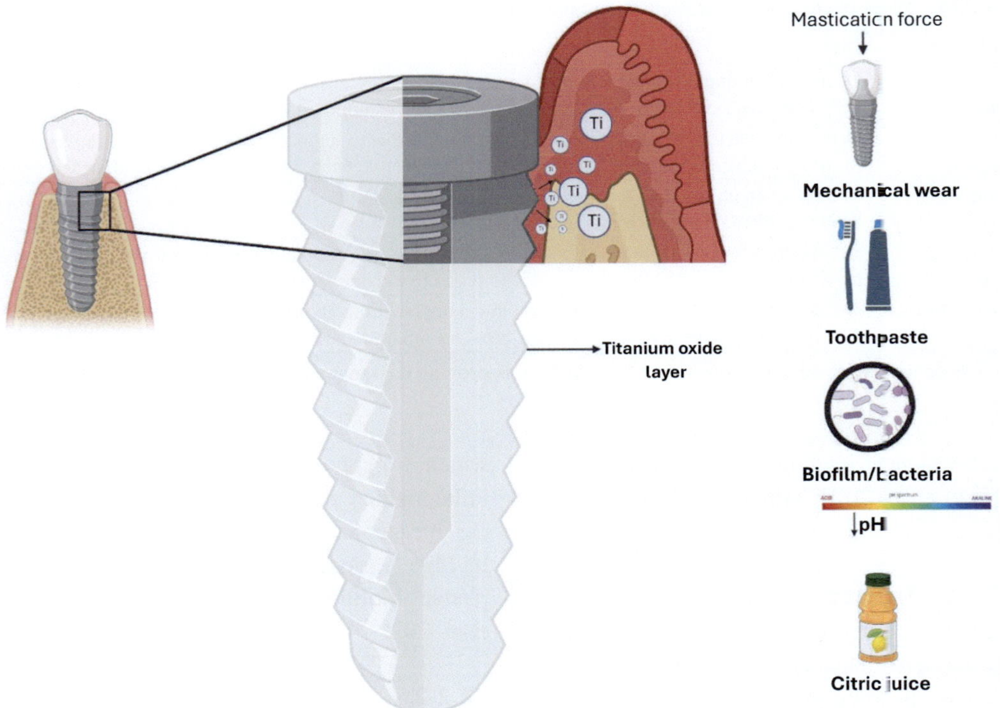

Fig. 22.1 Mechanisms of titanium implant corrosion in the oral environment: Mechanical wear from mastication, abrasives, and fluoride in toothpaste, combined with biofilm and bacteria, leads to oxygen depletion and fluid retention in the crevice area. This lowers the pH, compromising the protective titanium oxide layer, which leads to corrosion and the release of titanium ions, microparticles, and nanoparticles (Created using BioRender. Madi, HA. (2024) BioRender.com/x411403)

22.2.2.2 Environmental Factors

Environmental factors, including therapeutic substances (e.g., mouthwash), dietary habits, and poor oral hygiene, influence the corrosion of titanium implants. Hydrofluoric acids degrade the TiO_2 layer when saliva pH drops to as low as 2.0 due to acidic fluoride treatments. Mechanical stress from mastication and micromovements at the implant-abutment connection produces friction, exacerbating corrosion (Apaza-Bedoya et al. 2017; Boffano et al. 2024; Kotsakis and Olmedo 2021; Souza et al. 2020). Friction between the implant and bone surface during placement, along with mechanical, ultrasonic, and chemical cleaning during decontamination, can contribute to titanium surface corrosion (Delgado-Ruiz and Romanos 2018; Eger et al. 2017; Suárez-López Del Amo et al. 2018).

22.2.3 Distribution of Titanium Particles

Titanium particles in patients with dental implants are primarily derived from the implants themselves. Titanium particles accumulate in the peri-implant soft tissues, gingiva, alveolar bone, and submucosal plaque. They can also spread systemically, and their presence has been documented in the lymph nodes, lungs, liver, and blood (Chen et al. 2023a; Kandaswamy et al. 2024; Olmedo et al. 2002).

22.3 Immune Response to Corrosion of Titanium Implants

The inflammatory and immune responses to titanium implants are complex and involve a combination of factors, including bacterial colonization, mechanical wear, and corrosion. While all these factors contribute to peri-implant diseases, this section will focus on the immune and inflammatory reactions triggered by corrosion and the release of titanium particles and ions.

22.3.1 Cellular Response to Corrosion of Titanium Implants

Various immune cells, including lymphocytes, neutrophils, macrophages, and granulocytes, are present in the peri-implant gingival tissues (Comino-Garayoa et al. 2020; Gudima et al. 2024). The immune response to titanium corrosion by-products involves both innate and adaptive immunity (Kandaswamy et al. 2024). It begins with an innate immune response within minutes to hours, during which antigen-presenting cells, particularly macrophages, recognize these titanium particles as foreign bodies, triggering a cascade of local inflammatory responses (Gudima et al. 2024; Kheder et al. 2021). This response involves the recruitment of various immune cells to the site and triggering the release of pro-inflammatory cytokines and reactive oxygen species (ROS) (Fage et al. 2016; Gudima et al. 2024; Kheder et al. 2021). It is followed by the adaptive immune response with a Type IV hypersensitivity reaction, which occurs within 24–72 hours after exposure and involves T lymphocyte activation (Kheder et al. 2021). Exposure to titanium nanoparticles reduced cell viability and increased the frequency of mutations, leading to the formation of binucleated and multinucleated cells (Kandaswamy et al. 2024).

22.3.1.1 Neutrophils

Neutrophils play a key role in the early innate immune response to titanium corrosion by phagocytosing Ti particles smaller than 5 microns and secreting pro-inflammatory cytokines, such as IL-1β and TNF-α, which amplify local inflammation (Noronha Oliveira et al. 2018; Souza et al. 2020). Titanium particles can induce NETosis, a programmed neutrophil cell death where neutrophils release web-like NETs (Neutrophil Extracellular Traps) composed of DNA and antimicrobial proteins to capture and neutralize pathogens (Gudima et al. 2024). While NETosis typically responds to microbial infections, it can also be induced in sterile inflammatory conditions, such as exposure to titanium particles. This overactive immune response may contribute to excessive inflammation, tissue damage, and bone loss around the implant (Bressan et al. 2019; Gudima et al. 2024; Souza et al. 2020).

22.3.1.2 Oxidative Stress and Reactive Oxygen Species (ROS)

Titanium particles are known to induce oxidative stress, primarily by stimulating the production of reactive oxygen species (ROS) in neutrophils, macrophages, and other immune cells. ROS are highly reactive molecules that can damage key cellular components, including DNA, proteins, and lipids. Elevated ROS levels amplify inflammation and intensify immune response; they also recruit excessive neutrophils, which release metalloproteinases that degrade collagen fibers, contributing to tissue destruction and peri-implant bone loss (Bressan et al. 2019; Kandaswamy et al. 2024).

22.3.1.3 Macrophages

Macrophages are the first immune cells to respond to titanium particles. These foreign particles are detected through pattern recognition receptors (PRRs), such as Toll-like receptors (TLRs) found on the cell surface and NOD-like receptors (NLRs) located in the cell cytoplasm. These receptors identify titanium as a foreign material, and once recognized, macrophages engulf these particles in a process known as phagocytosis. This activation of macrophages triggers several inflammatory signaling pathways, including nuclear factor kappa B (NF-κB), which leads to the release of proinflammatory cytokines (e.g., IL-1α, IL-1β, IL-6, and TNF-α), chemokines, and growth factors. These molecules recruit additional immune cells to the site of inflammation, amplifying the host response (Gudima et al. 2024).

Macrophages typically phagocytose particles that are smaller than ten microns. However, when titanium particles are too large, macrophages fuse to form multinucleated foreign body giant cells (FBGCs), which secrete extracellular enzymes to break down these larger particles. Prolonged exposure to titanium particles can lead to the persistent formation of FBGCs, contributing to chronic inflammation and the foreign body reaction (Fage et al. 2016; Gudima et al. 2024; Kandaswamy et al. 2024; Kheder et al. 2023; Sheikh et al. 2015).

22.3.1.4 NOD-, LRR-, and Pyrin Domain-Containing Protein 3 (NLRP3) Inflammasome

Macrophages can detect titanium particles that form protein aggregates, which signal the presence of foreign material and trigger the activation of the NLRP3 inflammasome, a protein complex involved in inflammatory responses. Upon detecting these particles, the inflammasome activates pro-caspase-1, converting it into its active form, caspase-1. This, in turn, promotes the maturation of pro-inflammatory cytokines, such as pro-IL-1β, into their active form, IL-1β, which drives a strong inflammatory response. Additionally, caspase-1 can trigger a form of programmed cell death called pyroptosis, which helps eliminate infected or damaged cells. This inflammatory cascade attracts immune cells, triggering osteoclast activation and bone resorption around titanium implants. The continuous activation of the NLRP3 inflammasome by titanium particles results in ongoing tissue damage, osteoclast formation, and progressive peri-implant bone loss, contributing to peri-implantitis and eventual implant failure (Huang et al. 2023; Li et al. 2020; Pettersson et al. 2022; Poli et al. 2021).

22.3.1.5 T Lymphocytes

T lymphocytes, a key component of the adaptive immune system, play a central role in the immune response to titanium corrosion by-products. These cells are primarily involved in type IV delayed hypersensitivity reactions. Although T lymphocytes do not directly interact with the titanium surface, they are activated by macrophages that have phagocytosed titanium particles. These macrophages migrate to the lymph nodes and present the antigen via the major histocompatibility complex (MHC) to naïve T cells, triggering their activation. Once activated, T cells differentiate into two main types: T-helper (Th) cells ($CD4^+$ T cells) and cytotoxic T cells ($CD8^+$ T cells). Th1 cells stimulate macrophages by releasing cytokines (e.g., IL-1, IL-6, TNF-α, and interferon-gamma (IFN-γ)). These cytokines enhance the inflammatory response and promote osteoclast differentiation, leading to bone resorption around the implant, a key feature of peri-implantitis. Th17 cells produce IL-17, which helps recruit neutrophils and amplifies inflammation. $CD8^+$ cytotoxic T cells, although less prominent in peri-implantitis than $CD4^+$ cells, recognize and destroy damaged or infected cells. They release cytokines like IFN-γ and induce apoptosis in compromised cells, further contributing to inflammation. Regulatory T cells (Tregs) help balance the immune response by suppressing excessive inflammation. However, in peri-implantitis, the balance between pro-inflammatory T cells and anti-inflammatory Tregs is disrupted, leading to chronic inflammation and bone loss (Annunziato et al. 2015; Comino-Garayoa et al. 2020; Fage et al. 2016; Kandaswamy et al. 2024; Kheder et al. 2023; Petersen et al. 2018).

22.3.1.6 Fibroblasts and Osteoblasts

Titanium corrosion by-products, particularly titanium ions and particles, can exert cytotoxic effects on both fibroblasts and osteoblasts. An in-vitro study demonstrated that titanium particles at concentrations of 0.1 mg/mL and higher can adversely affect cell viability and proliferation, with cytotoxic effects of titanium particles on human gingival fibroblasts and human calvarial osteoblasts, leading to cellular dysfunction and inflammatory responses (Happe et al. 2019). When exposed to titanium particles, fibroblasts release pro-inflammatory cytokines such as IL-1β, IL-6, IL-10, and TNF-α, amplifying the local inflammatory reaction. This inflammation may exacerbate tissue damage surrounding the implant (Messous et al. 2021). Osteoblasts, the bone-forming cells, also react to titanium particles by increasing the secretion of cytokines like IL-6 and IL-8. These pro-inflammatory signals further contribute to inflammation and can impair the bone formation process. Additionally, titanium particles suppress collagen synthesis in osteoblasts, directly affecting the structural integrity of the bone matrix. Titanium ions influence bone remodeling by disrupting the balance between RANKL and OPG in osteoblast cells. RANKL promotes osteoclastogenesis (the formation of bone-resorbing cells), while OPG inhibits this process. Titanium particles enhance RANKL expression, increasing osteoclast differentiation and activity, which accelerates bone resorption around the implant. This imbalance between bone formation and resorption is a critical factor in peri-implant bone loss and implant failure (Choi et al. 2005; Mine et al. 2010).

22.4 Titanium Implant Hypersensitivity

Titanium implant hypersensitivity is rare and results from an exaggerated immune response to titanium ions or particles, leading to allergic-like symptoms and peri-implant inflammation. In severe cases, it may cause severe inflammation and implant failure (Delgado-Ruiz and Romanos 2018). Studies with small sample sizes have reported hypersensitivity reactions to titanium dental implants, typically identified as type IV hypersensitivity (Fage et al. 2016).

22.4.1 Testing for Hypersensitivity to Titanium Implants

Testing for hypersensitivity reactions to titanium is challenging due to the absence of a reliable, standardized diagnostic test. Various diagnostic tests are available, but it should be noted that these tests lack the sensitivity or specificity for detecting titanium ions, microparticles, and nanoparticles, making the diagnosis of titanium hypersensitivity clinically challenging. Understanding the application and limitations of such tests may help clinicians monitor and treat patients at risk of hypersensitivity reactions to titanium implant surface corrosion by-products (Chen and Kurmis 2024).

22.4.1.1 Patient History and Clinical Examination

The diagnosis of titanium hypersensitivity begins with a comprehensive patient history, including allergies to metals, dermatological conditions, and acute reactions to jewelry or implanted metals. A general clinical examination follows this to assess signs of hypersensitivity such as unexplained pain, persistent swelling or inflammation around the implant site, localized redness peri-implant tissue, and localized skin rashes near the implant site. In addition to systemic signs of hypersensitivity, such as dermatitis or generalized allergic reactions (Fage et al. 2016; Razzante et al. 2019).

22.4.1.2 Patch Test

This test is the most common diagnostic tool for evaluating hypersensitivity reactions to metals, including titanium. Titanium salt or oxide is prepared in a carrier solution and introduced into the superficial skin layers with a fine needle, followed by readings at 48 and 72 hours. A positive result typically presents as erythema or vesicles at the test site, indicating a type IV hypersensitivity reaction. The reliability of this test is questionable due to the poor absorption and solubility of titanium salts into the skin, leading to potential false-negative results. Moreover, patch testing might not detect deeper tissue hypersensitivity at the implant site, limiting its relevance to implant failure (Comino-Garayoa et al. 2020; Müller-Heupt et al. 2022).

22.4.1.3 Lymphocyte Transformation (LTT) Test

This test measures lymphocyte proliferation in response to metal ions. Lymphocytes are isolated from blood, and when sensitized to metal, they proliferate and release pro-inflammatory cytokines, indicating hypersensitivity. Interpretation requires specialized laboratories and expertise, limiting its accessibility (Fage et al. 2016; Müller-Heupt et al. 2022; Müller and Valentine-Thon 2006).

22.4.1.4 Memory Lymphocyte Immunostimulation Assay (MELISA) Test

This test is an optimized form of LTT, measuring the proliferation of memory T lymphocytes in response to metal ions, making it more specific for detecting metal-induced hypersensitivity. MELISA has a high sensitivity and can detect metal sensitization that may be missed by LTT or patch testing. However, similar to LTT, MELISA requires specialized equipment and technical expertise, limiting its availability to specialized centers (Fage et al. 2016; Müller-Heupt et al. 2022; Müller and Valentine-Thon 2006).

22.4.1.5 Cytokine Profiling

Cytokine profiling measures the levels of various cytokines such as IL-1β, IL-6, and TNF-α from blood serum, saliva, or peri-implant crevicular fluid. This test provides insight into the immune response during peri-implantitis. However, determining whether the inflammation results from mechanical irritation, bacterial infection, or titanium corrosion requires further investigation beyond cytokine profiling. While it is a valuable research tool, cytokine profiling is still being evaluated for its clinical effectiveness in distinguishing between the causes of implant-related inflammation (Ghassib et al. 2019).

22.4.1.6 Biopsy and Histological Examination Test

Biopsy and histology can definitively confirm hypersensitivity reactions to titanium dental implants. However, biopsies of peri-implant tissues, including gingiva, are invasive and typically reserved for cases where other diagnostic methods fail to provide conclusive results. Histological examinations can identify macrophages, mast cells, and eosinophils, indicating an allergic reaction, along with foreign body giant cells around titanium particles, suggesting chronic inflammation due to titanium hypersensitivity (Pettersson et al. 2017; Poli et al. 2021).

22.5 Titanium Implant Corrosion: Prevention, Alternative Materials, and Surface Modifications

Preventive strategies and future directions in implantology focus on material innovation, patient-customized treatments, and advanced surface modifications to minimize risks and extend implant longevity. Exploring new biomaterials and enhancements to implant surface treatments aims to make dental implants more biocompatible, corrosion-resistant, and better suited to patient needs.

22.5.1 Corrosion-Resistant Implant Materials

Dental implant materials are continuously refined to improve mechanical properties and corrosion resistance. Alternatives to commercially pure titanium implants include the following:

22.5.1.1 Zirconia Implants

Zirconia ceramic implants represent a promising metal-free alternative to titanium, particularly for patients with allergies or sensitivities to metals. Zirconia is highly biocompatible, corrosion-resistant, and has a modulus of elasticity closer to that of bone, reducing stress shielding. Moreover, its tooth-like appearance makes zirconia particularly suitable for anterior implant placements, where aesthetics is critical. Zirconia also exhibits low bacterial adhesion, making it less susceptible to peri-implant inflammation and peri-implantitis, which are common complications of titanium implants. Although zirconia implants perform well in terms of osseointegration and mechanical stability, ongoing research aims to improve their resistance to long-term fatigue, as cyclic mechanical loading can result in micro-crack complications and eventual implant failure (Cionca et al. 2017; Sadowsky 2024; Sivaraman et al. 2018; Webber et al. 2021).

22.5.1.2 Polyetheretherketone Implants (PEEK)

PEEK is a high-performance polymer that offers a promising alternative to metal. Its modulus of elasticity is close to that of human bone, and it is highly corrosion-resistant. This minimizes stress shielding, making PEEK ideal for load-bearing implants, particularly in patients at risk for bone resorption and metal hypersensitivity. However, PEEK implants are bioinert and do not naturally integrate with bone without surface modification. Researchers are addressing this limitation by adding bioactive coatings like hydroxyapatite, which enhances osseointegration. These modifications enhance PEEK implants' bonding to bone while maintaining their excellent mechanical properties (Moharil et al. 2023; Rahmitasari et al. 2017).

22.5.1.3 Beta (β) Titanium Alloys

β titanium alloys have emerged as one of the key advancements in implant biomaterials. β titanium alloys help prevent bone resorption by better distributing the load across the bone–implant interface, which is critical for maintaining implant stability and longevity. They remain stable in the oral environment, minimizing immune responses and peri-implant inflammation that could lead to implant failure (Chen et al. 2020; Shao et al. 2021).

22.5.1.4 Tantalum Implants

Tantalum implants are emerging as a strong alternative to titanium, offering superior biocompatibility and corrosion resistance. Tantalum's porous nature stimulates bone ingrowth, improving mechanical integration with the surrounding bone. Unlike zirconia, tantalum is highly resistant to long-term fatigue and micro-cracks under cyclic mechanical loading. Tantalum's superior mechanical properties and osseointegration capabilities make it ideal for high-stress applications with better osseointegration, fewer corrosion-related complications (Kim et al. 2013; Liu et al. 2015; Wang et al. 2021).

22.5.2 Advanced Surface Treatments and Coatings of Titanium Implant

Surface modifications of titanium implants are essential for enhancing osseointegration, biocompatibility, and corrosion resistance. These modifications also promote cell attachment to the implant surface while minimizing bacterial adhesion, thereby reducing the risk of inflammation and improving long-term implant stability (Kurup et al. 2021).

22.5.2.1 Sand Blasting and Acid Etching

The most common methods for roughening the dental implant surface are sandblasting and acid etching. These techniques increase the roughness of the implant surface and the area that contacts the bone, improving the mechanical interlocking between the implant and the bone. The sandblasting technique creates surface irregularities and microporosities by bombarding the implant surface with abrasive particles. In addition, acid etching modifies the surface using strong acids, particularly hydrofluoric, sulfuric, or nitric acid. These acids chemically roughen the surface by the selective dissolution of small amounts of titanium implant material. Acid etching uses acids (e.g., hydrochloric acid (HCl), nitric acid (HNO_3), sulfuric acid (H_2SO_4), and hydrofluoric acid (HF)) to dissolve the protective oxide layer, allowing the acids to interact with the underlying metallic titanium. Hydrogen ions penetrate the metal subsurface during this process and form titanium hydride. Once the etching is complete, a new titanium oxide layer spontaneously reforms on the surface. Dual Acid Etch technology uses double acid etching without first sandblasting. Combining both surface treatments alters the surface characteristics, allowing faster and more effective osseointegration and increasing the overall stability of the dental implant (Han et al. 2023; Kligman et al. 2021; Tuikampee et al. 2024).

22.5.2.2 Laser Surface Texturing

Laser surface texturing enhances corrosion resistance and strengthens the oxide layer by generating precise micro- and nano-scale surface patterns on titanium implants. It also strengthens the mechanical bonding between the implant and bone and reduces bacterial adhesion, making it less susceptible to infection (Bush et al. 2011; Ferraris et al. 2024; Singh et al. 2021).

22.5.2.3 Anodization

Anodization is an electrochemical process that thickens the oxide layer in titanium, thus providing a more stable and corrosion-resistant surface. This thicker oxide layer improves osteoblast attachment and accelerates bone formation. Anodized titanium implants resist bacterial colonization and corrosion, reducing the risk of peri-implant inflammation (Alipal et al. 2021; Guo et al. 2022).

22.5.2.4 Coatings for Titanium Implant Surfaces

The development of implant surface coatings has become a significant focus in implantology, as they combat biofilm formation and promote bone regeneration. Different techniques are used to deposit biomaterials and coat the titanium implant surface.

Biodegradable coatings gradually break down over time while promoting bone integration and corrosion resistance. Calcium phosphate compounds (e.g., hydroxyapatite, tricalcium phosphate, and biphasic calcium phosphate) degrade slowly while also promoting bone integration (Inchingolo et al. 2023; Zafar et al. 2019). Discrete Crystalline Deposition (DCD) is a sol–gel technique that enhances early osseointegration by modifying a double acid-etched implant surface with calcium phosphate particles ranging from 20 to 100 nm, improving bone-to-implant contact (Kligman et al. 2021). Polylactic-co-glycolic acid (PLGA) acts as a polymer-based barrier, protecting the implant surface by reducing titanium ion release and preventing corrosion (Kligman et al. 2021).

Similarly, extracellular matrix coatings (e.g., collagen, fibronectin, and hyaluronic acid) gradually degrade in the body while modulating the immune response, reducing excessive inflammation, promoting tissue regeneration, and enhancing cell attachment and proliferation (Abdulghafor et al. 2024). Likewise, bone-stimulating agents, including bone morphogenetic proteins (BMPs), accelerate bone healing and naturally degrade after fulfilling their function (Inchingolo et al. 2023; Kligman et al. 2021).

Furthermore, chitosan is a biopolymer that provides antimicrobial benefits and is biocompatible while gradually breaking down in the body (Kligman et al. 2021). Lastly, Polyhydroxyalkanoate (PHA) (Lumbikananda et al. 2024), a naturally occurring polyester produced by microorganisms, has excellent biocompatibility and biodegradability (Inchingolo et al. 2023).

Non-biodegradable coatings remain stable on the implant surface while providing long-lasting corrosion resistance. Silver nanoparticles offer antibacterial properties, preventing bacterial colonization and remaining intact over time (Inchingolo et al. 2023; Kligman et al. 2021). Fluoride coatings enhance bone-to-implant contact, promoting osseointegration while maintaining their structural integrity (Barfeie et al. 2015). Titanium nitride (TiN) coatings improve wear and corrosion resistance, creating a durable, long-lasting implant surface (Barfeie et al. 2015; Kligman et al. 2021). Zinc oxide coatings provide antimicrobial and corrosion-resistant properties (Barfeie et al. 2015; Inchingolo et al. 2023). Bioactive glass (e.g., 45S5 Bioactive Glass) has an excellent bond with the surrounding bone, maintaining a stable implant–bone interface (Zafar et al. 2019).

Antibiotic coatings (e.g., tetracycline, simvastatin) release therapeutic agents gradually, but the coating itself remains intact after drug depletion (Barfeie et al. 2015; Kligman et al. 2021). Similarly, antimicrobial peptide-based coatings (e.g., GL13K and human beta-defensins) adhere to the implant surface, preventing bacterial colonization without breaking down. Lastly, graphene-based coatings exhibit antimicrobial properties and enhance osteoblast proliferation while remaining structurally stable over time (Inchingolo et al. 2023).

In summary, the choice of materials and surface modifications for dental implants plays a crucial role in regulating immune response and minimizing inflammation. Biocompatible materials, along with specialized surface treatments and coatings, can reduce implant corrosion, modulate tissue interactions, decrease bacterial adhesion, and suppress inflammatory reactions. Furthermore, incorporating antimicrobial and anti-inflammatory/bacterial agents into implant coatings can further suppress immune activation, promoting better integration with surrounding tissues. These strategies collectively aim to mitigate adverse immune and hypersensitivity reactions, fostering a stable and biocompatible interface between the implant and host tissue.

22.6 Systematic Approach for Managing Titanium Implant Hypersensitivity

Further research is required to develop a reliable, standardized diagnostic test for diagnosing hypersensitivity reactions caused by titanium corrosion of dental implant surfaces. There are ongoing debates in the literature about the role of titanium allergy or hypersensitivity to play a major role in peri-implant disease (Mombelli et al. 2018). The following systematic approach (Figs. 22.2 and 22.3) can assist in diagnosing and managing titanium implant hypersensitivity in the absence of a standardized diagnostic test (Chen and Kurmis 2024; Chen and Thyssen 2018).

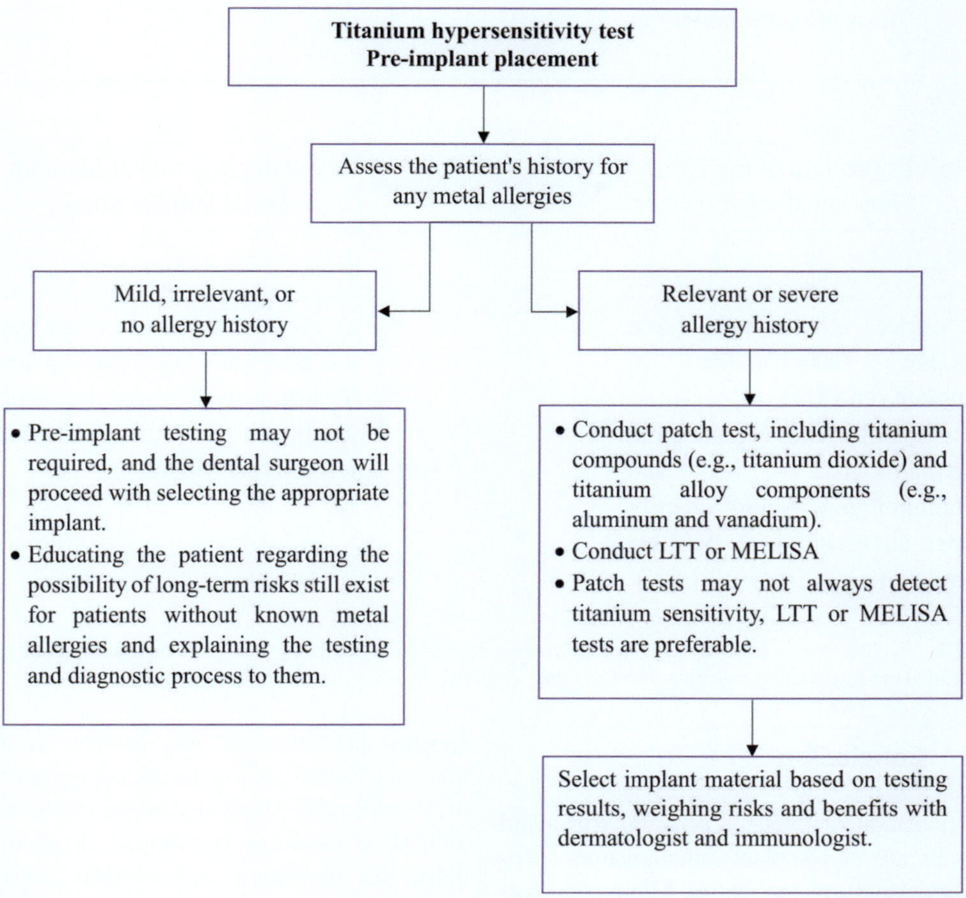

Fig. 22.2 Titanium hypersensitivity assessment flowchart before implant placement. The flowchart outlines the decision-making process for evaluating titanium hypersensitivity in patients before implant placement

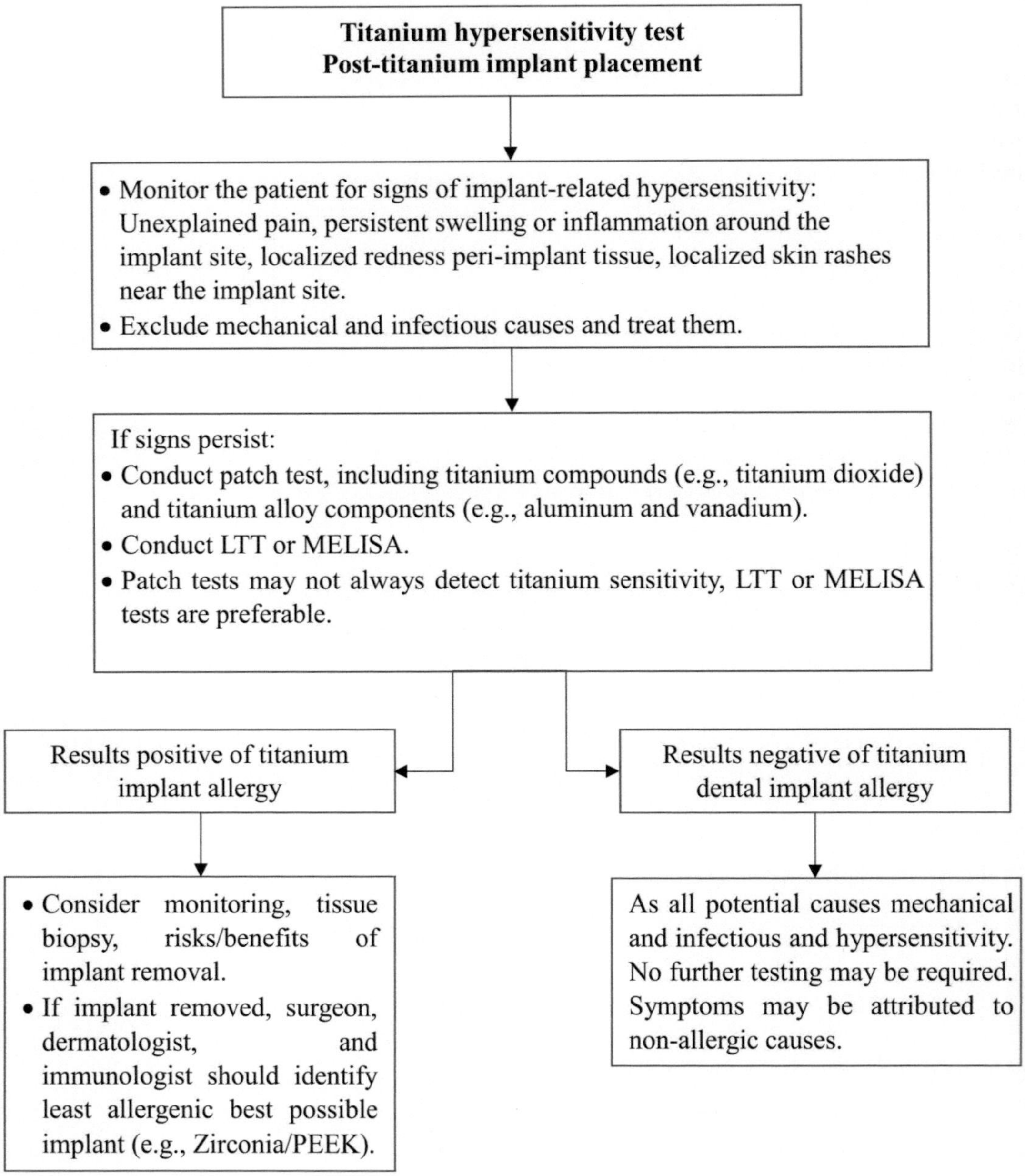

Fig. 22.3 Titanium hypersensitivity assessment flowchart after titanium implant placement. The flowchart outlines the decision-making process for evaluating titanium hypersensitivity in patients after implant placement

22.7 Conclusion

The prevention of implant corrosion will avoid the risks of peri-implant inflammation, peri-implantitis, and implant failure. Minimizing corrosion, while effectively managing bacterial infection and mechanical stress, can significantly enhance the longevity and success of dental implants. Understanding the underlying corrosion of dental implant mechanisms and implementing preventive measures is essential. It should be noted that titanium dental implants consist of various alloys and superstructures, including abutments, which may contain metals beyond titanium that could trigger hypersensitivity reactions. Different implant materials are now available, allowing clinicians to personalize dental implant treatment to meet the needs of each patient. Future research should prioritize the development of reliable diagnostic tests for titanium hypersensitivity while exploring alternative materials like zirconia and PEEK to mitigate immune responses and corrosion-related complications. The success and survival of titanium dental implants depend upon regular follow-up, enabling early detection of soft tissue inflammation that could otherwise lead to peri-implantitis, bone loss, and implant failure.

References

Abdulghafor MA, Mahmood MK, Tassery H, Tardivo D, Falguiere A, Lan R (2024) Biomimetic coatings in implant dentistry: a quick update. J Funct Biomater 15(1):15

Adell R, Lekholm U, Rockler B, Brånemark PI (1981) A 15-year study of osseointegrated implants in the treatment of the edentulous jaw. Int J Oral Surg 10(6): 387–416. https://doi.org/10.1016/s0300-9785(81)80077-4

Albrektsson T, Brånemark PI, Hansson HA, Lindström J (1981) Osseointegrated titanium implants. Requirements for ensuring a long-lasting, direct bone-to-implant anchorage in man. Acta Orthop Scand 52(2):155–170. https://doi.org/10.3109/17453678108991776

Albrektsson T, Tengvall P, Amengual L, Coli P, Kotsakis GA, Cochran D (2022) Osteoimmune regulation underlies oral implant osseointegration and its perturbation. Front Immunol 13:1056914. https://doi.org/10.3389/fimmu.2022.1056914

Alipal J, Lee TC, Koshy P, Abdullah HZ, Idris MI (2021) Evolution of anodised titanium for implant applications. Heliyon 7(7):e07408. https://doi.org/10.1016/j.heliyon.2021.e07408

Amengual-Peñafiel L, Córdova LA, Constanza Jara-Sepúlveda M, Brañes-Aroca M, Marchesani-Carrasco-F, Cartes-Velásquez R (2021) Osteoimmunology drives dental implant osseointegration: a new paradigm for implant dentistry. Jpn Dent Sci Rev 57:12–19. https://doi.org/10.1016/j.jdsr.2021.01.001

Annunziato F, Romagnani C, Romagnani S (2015) The 3 major types of innate and adaptive cell-mediated effector immunity. J Allergy Clin Immunol 135(3): 626–635. https://doi.org/10.1016/j.jaci.2014.11.001

Apaza-Bedoya K, Tarce M, Benfatti CAM, Henriques B, Mathew MT, Teughels W, Souza JCM (2017) Synergistic interactions between corrosion and wear at titanium-based dental implant connections: a scoping review. J Periodontal Res 52(6):946–954. https://doi.org/10.1111/jre.12469

Balasubramanian Gayathri YK, Kumar RL, Ramalingam VV, Priyadharshini GS, Kumar KS, Prabhu TR (2022) Additive manufacturing of Ti-6Al-4V alloy for biomedical applications. J Bio- and Tribo-Corros 8(4): 98. https://doi.org/10.1007/s40735-022-00700-1

Barfeie A, Wilson J, Rees J (2015) Implant surface characteristics and their effect on osseointegration. Br Dent J 218(5):E9–E9. https://doi.org/10.1038/sj.bdj.2015.171

Berglundh T, Abrahamsson I, Lang NP, Lindhe J (2003) De novo alveolar bone formation adjacent to endosseous implants. Clin Oral Implants Res 14(3): 251–262. https://doi.org/10.1034/j.1600-0501.2003.00972.x

Berglundh T, Armitage G, Araujo MG, Avila-Ortiz G, Blanco J, Camargo PM, Chen S, Cochran D, Derks J, Figuero E, Hämmerle CHF, Heitz-Mayfield LJA, Huynh-Ba G, Iacono V, Koo KT, Lambert F, McCauley L, Quirynen M et al (2018) Peri-implant diseases and conditions: consensus report of workgroup 4 of the 2017 world workshop on the classification of periodontal and Peri-implant diseases and conditions. J Clin Periodontol 45(Suppl 20):S286–s291. https://doi.org/10.1111/jcpe.12957

Boffano P, Brucoli M, Rocchetti V (2024) Corrosion features of titanium alloys in dental implants: a systematic review. J Maxillofac Oral Surg. https://doi.org/10.1007/s12663-024-02169-6

Boyer RR, Briggs RD (2005) The use of β titanium alloys in the aerospace industry. J Mater Eng Perform 14(6): 681–685. https://doi.org/10.1361/105994905X75448

Brånemark PI (1983) Osseointegration and its experimental background. J Prosthet Dent 50(3):399–410. https://doi.org/10.1016/s0022-3913(83)80101-2

Brånemark PI, Adell R, Breine U, Hansson BO, Lindström J, Ohlsson A (1969) Intra-osseous anchorage of dental prostheses. I. Experimental studies. Scand J Plast Reconstr Surg 3(2):81–100. https://doi.org/10.3109/02844316909036699

Brånemark PI, Hansson BO, Adell R, Breine U, Lindström J, Hallén O, Ohman A (1977) Osseointegrated implants in the treatment of the edentulous jaw. Experience from a 10-year period. Scand J Plast Reconstr Surg Suppl 16:1–132

Bressan E, Ferroni L, Gardin C, Bellin G, Sbricoli L, Sivolella S, Brunello G, Schwartz-Arad D, Mijiritsky E, Penarrocha M, Penarrocha D, Taccioli C, Tatullo M, Piattelli A, Zavan B (2019) Metal nanoparticles released from dental implant surfaces: potential contribution to chronic inflammation and Peri-implant bone loss. Materials 12(12):2036

Buser D, Janner SF, Wittneben JG, Brägger U, Ramseier CA, Salvi GE (2012) 10-year survival and success rates of 511 titanium implants with a sandblasted and acid-etched surface: a retrospective study in 303 partially edentulous patients. Clin Implant Dent Relat Res

14(6):839–851. https://doi.org/10.1111/j.1708-8208.2012.00456.x

Bush JR, Nayak BK, Nair LS, Gupta MC, Laurencin CT (2011) Improved bio-implant using ultrafast laser induced self-assembled nanotexture in titanium. J Biomed Mater Res B Appl Biomater 97(2):299–305. https://doi.org/10.1002/jbm.b.31815

Chen A, Kurmis AP (2024) Understanding immune-mediated cobalt/chromium allergy to orthopaedic implants: a meta-synthetic review. Art 6(1):1. https://doi.org/10.1186/s42836-023-00227-x

Chen J, Thyssen J (2018) Metal allergy: from dermatitis to implant and device failure. Springer. https://doi.org/10.1007/978-3-319-58503-1

Chen L-Y, Cui Y-W, Zhang L-C (2020) Recent development in Beta titanium alloys for biomedical applications. Meta 10(9):1139

Chen L, Tong Z, Luo H, Qu Y, Gu X, Si M (2023a) Titanium particles in peri-implantitis: distribution, pathogenesis and prospects. Int J Oral Sci 15(1):49. https://doi.org/10.1038/s41368-023-00256-x

Chen YC, Chen MY, Chen TY, Chan ML, Huang YY, Liu YL, Lee PT, Lin GJ, Li TF, Chen CA, Chen SL, Li KC, Abu PAR (2023b) Improving dental implant outcomes: CNN-based system accurately measures degree of Peri-Implantitis damage on periapical film. Bioengineering (Basel) 10(6). https://doi.org/10.3390/bioengineering10060640

Choi MG, Koh HS, Kluess D, O'Connor D, Mathur A, Truskey GA, Rubin J, Zhou DX, Sung KL (2005) Effects of titanium particle size on osteoblast functions in vitro and in vivo. Proc Natl Acad Sci USA 102(12):4578–4583. https://doi.org/10.1073/pnas.0500693102

Cionca N, Hashim D, Mombelli A (2017) Zirconia dental implants: where are we now, and where are we heading? Periodontol 73(1):241–258. https://doi.org/10.1111/prd.12180

Comino-Garayoa R, Cortés-Bretón Brinkmann J, Peláez J, López-Suárez C, Martínez-González JM, Suárez MJ (2020) Allergies to titanium dental implants: what do we really know about them? A scoping review. Biology (Basel) 9(11). https://doi.org/10.3390/biology9110404

Davies JE (1998) Mechanisms of endosseous integration. Int J Prosthodont 11(5):391–401

Delgado-Ruiz R, Romanos G (2018) Potential causes of titanium particle and ion release in implant dentistry: a systematic review. Int J Mol Sci 19(11). https://doi.org/10.3390/ijms19113585

Dentistry AAOI (2022) What are dental implants? Retrieved October 21, 2024, from https://aaid-implant.org/what-are-dental-implants/

Donachie MJ (2000) Titanium : a technical guide, 2nd edn. ASM International

Eger M, Sterer N, Liron T, Kohavi D, Gabet Y (2017) Scaling of titanium implants entrains inflammation-induced osteolysis. Sci Rep 7:39612. https://doi.org/10.1038/srep39612

Emam SM, Moussa N (2024) Signaling pathways of dental implants' osseointegration: a narrative review on two of the most relevant; NF-κB and Wnt pathways. BDJ Open 10(1):29. https://doi.org/10.1038/s41405-024-00211-w

Fage SW, Muris J, Jakobsen SS, Thyssen JP (2016) Titanium: a review on exposure, release, penetration, allergy, epidemiology, and clinical reactivity. Contact Derm 74(6):323–345. https://doi.org/10.1111/cod.12565

Fang X, Sun D, Li Y, Han X, Gan Y, Jiao J (2024) Macrophages in the process of osseointegration around the implant and their regulatory strategies. Connect Tissue Res 65(1):1–15. https://doi.org/10.1080/03008207.2023.2300455

Ferraris S, Cochis A, Scalia AC, Tori A, Rimondini L, Spriano S (2024) Laser surface texturing of Ti-cp and Ti6Al4V alloy for the improvement of fibroblast adhesion and alignment and the reduction of bacterial adhesion. J Mater Res Technol 29:5464–5472. https://doi.org/10.1016/j.jmrt.2024.03.033

Gaur S, Agnihotri R, Albin S (2022) Bio-Tribocorrosion of titanium dental implants and its toxicological implications: a scoping review. Scient World J 2022:4498613. https://doi.org/10.1155/2022/4498613

Gaviria L, Salcido JP, Guda T, Ong JL (2014) Current trends in dental implants. J Korean Assoc Oral Maxillofac Surg 40(2):50–60. https://doi.org/10.5125/jkaoms.2014.40.2.50

Ghassib I, Chen Z, Zhu J, Wang HL (2019) Use of IL-1 β, IL-6, TNF-α, and MMP-8 biomarkers to distinguish peri-implant diseases: a systematic review and meta-analysis. Clin Implant Dent Relat Res 21(1):190–207. https://doi.org/10.1111/cid.12694

Gittens RA, Olivares-Navarrete R, Tannenbaum R, Boyan BD, Schwartz Z (2011) Electrical implications of corrosion for Osseointegration of titanium implants. J Dent Res 90(12):1389–1397. https://doi.org/10.1177/0022034511408428

Gudima A, Hesselbarth D, Li G, Riabov V, Michel J, Liu Q, Schmuttermaier C, Jiao Z, Sticht C, Jawhar A, Obertacke U, Klüter H, Vrana NE, Kzhyshkowska J (2024) Titanium induces proinflammatory and tissue-destructive responses in primary human macrophages. J Leukoc Biol 116(4):706–725. https://doi.org/10.1093/jleuko/qiae072

Guo T, Ivanovski S, Gulati K (2022) Optimizing titanium implant nano-engineering via anodization. Mater Des 223:111110. https://doi.org/10.1016/j.matdes.2022.111110

Han X, Ma J, Tian A, Wang Y, Li Y, Dong B, Tong X, Ma X (2023) Surface modification techniques of titanium and titanium alloys for biomedical orthopaedics applications: a review. Colloids Surf B Biointerfaces 227:113339. https://doi.org/10.1016/j.colsurfb.2023.113339

Hanawa T (2004) Metal ion release from metal implants. Mater Sci Eng C 24(6):745–752. https://doi.org/10.1016/j.msec.2004.08.018

Happe A, Sielker S, Hanisch M, Jung S (2019) The biological effect of particulate titanium contaminants of dental implants on human osteoblasts and gingival fibroblasts. Int J Oral Maxillofac Implants 34(3): 673–680. https://doi.org/10.11607/jomi.6929

Heitz-Mayfield LJA (2024) Peri-implant mucositis and peri-implantitis: key features and differences. Br Dent J 236(10):791–794. https://doi.org/10.1038/s41415-024-7402-z

Huang W, Zhang Z, Qiu Y, Gao Y, Fan Y, Wang Q, Zhou Q (2023) NLRP3 inflammasome activation in response to metals. Front Immunol 14:1055788. https://doi.org/10.3389/fimmu.2023.1055788

Inchingolo AM, Malcangi G, Ferrante L, Del Vecchio G, Viapiano F, Inchingolo AD, Mancini A, Annicchiarico C, Inchingolo F, Dipalma G, Minetti E, Palermo A, Patano A (2023) Surface coatings of dental implants: a review. J Funct Biomater 14(5):287

Jokstad A, Braegger U, Brunski JB, Carr AB, Naert I, Wennerberg A (2003) Quality of dental implants. Int Dent J 53(6 Suppl 2):409–443. https://doi.org/10.1111/j.1875-595x.2003.tb00918.x

Kandaswamy E, Harsha M, Joshi VM (2024) Titanium corrosion products from dental implants and their effect on cells and cytokine release: a review. J Trace Elem Med Biol 84:127464. https://doi.org/10.1016/j.jtemb.2024.127464

Kheder W, Al Kawas S, Khalaf K, Samsudin AR (2021) Impact of tribocorrosion and titanium particles release on dental implant complications – a narrative review. Jpn Dent Sci Rev 57:182–189. https://doi.org/10.1016/j.jdsr.2021.09.001

Kheder W, Bouzid A, Venkatachalam T, Talaat IM, Elemam NM, Raju TK, Sheela S, Jayakumar MN, Maghazachi AA, Samsudin AR, Hamoudi R (2023) Titanium particles modulate lymphocyte and macrophage polarization in Peri-implant gingival tissues. Int J Mol Sci 24(14). https://doi.org/10.3390/ijms241411644

Kim DG, Huja SS, Tee BC, Larsen PE, Kennedy KS, Chien HH, Lee JW, Wen HB (2013) Bone ingrowth and initial stability of titanium and porous tantalum dental implants: a pilot canine study. Implant Dent 22(4):399–405. https://doi.org/10.1097/ID.0b013e31829b17b5

Kligman S, Ren Z, Chung CH, Perillo MA, Chang YC, Koo H, Zheng Z, Li C (2021) The impact of dental implant surface modifications on Osseointegration and biofilm formation. J Clin Med 10(8). https://doi.org/10.3390/jcm10081641

Kondo T, Yamada M, Egusa H (2024) Innate immune regulation in dental implant osseointegration. J Prosthodont Res 68(4):511–521. https://doi.org/10.2186/jpr.JPR_D_23_00198

Kotsakis GA, Olmedo DG (2021) Peri-implantitis is not periodontitis: scientific discoveries shed light on microbiome-biomaterial interactions that may determine disease phenotype. Periodontol 86(1): 231–240. https://doi.org/10.1111/prd.12372

Kupka JR, König J, Al-Nawas B, Sagheb K, Schiegnitz E (2024) How far can we go? A 20-year meta-analysis of dental implant survival rates. Clin Oral Investig 28(10):541. https://doi.org/10.1007/s00784-024-05929-3

Kurup A, Dhatrak P, Khasnis N (2021) Surface modification techniques of titanium and titanium alloys for biomedical dental applications: a review. Mater Today Proc 39:84–90. https://doi.org/10.1016/j.matpr.2020.06.163

Le Guéhennec L, Soueidan A, Layrolle P, Amouriq Y (2007) Surface treatments of titanium dental implants for rapid osseointegration. Dent Mater 23(7):844–854. https://doi.org/10.1016/j.dental.2006.06.025

Li X, Tang L, Ye Myat T, Chen D (2020) Titanium ions play a synergistic role in the activation of NLRP3 Inflammasome in Jurkat T cells. Inflammation 43(4): 1269–1278. https://doi.org/10.1007/s10753-020-01206-z

Liang B, Wang H, Wu D, Wang Z (2021)Macrophage M1/M2 polarization dynamically adapts to changes in microenvironment and modulates alveolar bone remodeling after dental implantation. J Leukoc Biol 110(3):433–447. https://doi.org/10.1002/JLB.1MA0121-001R

Liu Y, Bao C, Wismeijer D, Wu G (2015) The physicochemical/biological properties of porous tantalum and the potential surface modification techniques to improve its clinical application in dental implantology. Mater Sci Eng C Mater Biol Appl 49:323–329. https://doi.org/10.1016/j.msec.2015.01.007

Lumbikananda S, Srithanyarat SS, Mattheos N, Osathanon T (2024) Oral fluid biomarkers for Peri-Implantitis: a scoping review. Int Dent J 74(3):387–402. https://doi.org/10.1016/j.identj.2023.11.005

Messous R, Henriques B, Bousbaa H, Silva FS, Teughels W, Souza JCM (2021) Cytotoxic effects of submicron- and nano-scale titanium debris released from dental implants: an integrative review. Clin Oral Investig 25(4):1627–1640. https://doi.org/10.1007/s00784-021-03785-z

Mine Y, Makihira S, Nikawa H, Murata H, Hosokawa R, Hiyama A, Mimura S (2010) Impact of titanium ions on osteoblast-, osteoclast- and gingival epithelial-like cells. J Prosthodont Res 54(1):1–6. https://doi.org/10.1016/j.jpor.2009.07.003

Misch CE, Perel ML, Wang HL, Sammartino G, Galindo-Moreno P, Trisi P, Steigmann M, Rebaudi A, Palti A, Pikos MA, Schwartz-Arad D, Choukroun J, Gutierrez-Perez JL, Marenzi G, Valavanis DK (2008) Implant success, survival, and failure: the international congress of Oral Implantologists (ICOI) Pisa Consensus conference. Implant Dent 17(1):5–15. https://doi.org/10.1097/ID.0b013e3181676059

Moharil S, Reche A, Durge K (2023) Polyetheretherketone (PEEK) as a biomaterial: an overview. Cureus 15(8): e44307. https://doi.org/10.7759/cureus.44307

Mombelli A, Hashim D, Cionca N (2018) What is the impact of titanium particles and biocorrosion on implant survival and complications? A critical review. Clin Oral Implants Res 29(Suppl 18):37–53. https://doi.org/10.1111/clr.13305

Müller K, Valentine-Thon E (2006) Hypersensitivity to titanium: clinical and laboratory evidence. Neuro Endocrinol Lett 27(Suppl 1):31–35

Müller-Heupt LK, Schiegnitz E, Kaya S, Jacobi-Gresser E, Kämmerer PW, Al-Nawas B (2022) Diagnostic tests for titanium hypersensitivity in implant dentistry: a systematic review of the literature. Int J Implant Dent 8(1):29. https://doi.org/10.1186/s40729-022-00428-0

Noronha Oliveira M, Schunemann WVH, Mathew MT, Henriques B, Magini RS, Teughels W, Souza JCM (2018) Can degradation products released from dental implants affect peri-implant tissues? J Periodontal Res 53(1):1–11. https://doi.org/10.1111/jre.12479

Olmedo D, Guglielmotti MB, Cabrini RL (2002) An experimental study of the dissemination of titanium and zirconium in the body. J Mater Sci Mater Med 13(8):793–796. https://doi.org/10.1023/a:1016131310025

Peters M, Hemptenmacher J, Kumpfert J, Leyens C (2003) Structure and properties of titanium and titanium alloys. In: Titanium and titanium alloys, pp 1–36. https://doi.org/10.1002/3527602119.ch1

Petersen TH, Geisler C, Bonefeld CM (2018) Acquired immunity in metal allergy: T cell responses. In: Chen JK, Thyssen JP (eds) Metal allergy: from dermatitis to implant and device failure. Springer International Publishing, pp 85–95. https://doi.org/10.1007/978-3-319-58503-1_9

Pettersson M, Kelk P, Belibasakis GN, Bylund D, Molin Thorén M, Johansson A (2017) Titanium ions form particles that activate and execute interleukin-1β release from lipopolysaccharide-primed macrophages. J Periodontal Res 52(1):21–32. https://doi.org/10.1111/jre.12364

Pettersson M, Almlin S, Romanos GE, Johansson A (2022) Ti ions induce IL-1β release by activation of the NLRP3 Inflammasome in a human macrophage cell line. Inflammation 45(5):2027–2037. https://doi.org/10.1007/s10753-022-01672-7

Pjetursson BE, Thoma D, Jung R, Zwahlen M, Zembic A (2012) A systematic review of the survival and complication rates of implant-supported fixed dental prostheses (FDPs) after a mean observation period of at least 5 years. Clin Oral Implants Res 23(Suppl 6): 22–38. https://doi.org/10.1111/j.1600-0501.2012.02546.x

Poli PP, de Miranda FV, Polo TOB, Santiago Júnior JF, Lima Neto TJ, Rios BR, Assunção WG, Ervolino E, Maiorana C, Faverani LP (2021) Titanium allergy caused by dental implants: a systematic literature review and case report. Materials 14(18):5239

Rahmitasari F, Ishida Y, Kurahashi K, Matsuda T, Watanabe M, Ichikawa T (2017) PEEK with reinforced materials and modifications for dental implant applications. Dent J (Basel) 5(4). https://doi.org/10.3390/dj5040035

Razzante MC, Ehredt DJ, Clougherty CO, Kriger SJ, Menninger BA, Behan Dionisopoulos S (2019) Type IV cellmediated hypersensitivity reaction caused by titanium implant following double calcaneal osteotomy and first metatarsal-cuneiform arthrodesis: a case report and review of the literature. J Foot Ankle Surg 58(5):974–979. https://doi.org/10.1053/j.jfas.2018.11.031

Rodrigues DC, Valderrama P, Wilson TG, Palmer K, Thomas A, Sridhar S, Adapalli A, Burbano M, Wadhwani C (2013) Titanium corrosion mechanisms in the oral environment: a retrieval study. Materials 6(11):5258–5274

Sadowsky SJ (2024) Zirconia implants: a mapping review. Oral 4(1):9–22

Scarano A, Khater AGA, Gehrke SA, Serra P, Francesco I, Di Carmine M, Tari SR, Leo L, Lorusso F (2023) Current status of Peri-implant diseases: a clinical review for evidence-based decision making. J Funct Biomater 14(4). https://doi.org/10.3390/jfb14040210

Shao L, Du Y, Dai K, Wu H, Wang Q, Liu J, Tang Y, Wang L (2021) β-Ti alloys for orthopedic and dental applications: a review of progress on improvement of properties through surface modification. Coatings 11(12):1446

Sheikh Z, Brooks PJ, Barzilay O, Fine N, Glogauer M (2015) Macrophages, foreign body Giant cells and their response to implantable biomaterials. Materials (Basel) 8(9):5671–5701. https://doi.org/10.3390/ma8095269

Singh I, George SM, Tiwari A, Ramkumar J, Balani K (2021) Influence of laser surface texturing on the wettability and antibacterial properties of metallic, ceramic, and polymeric surfaces. J Mater Res 36(19): 3985–3999. https://doi.org/10.1557/s43578-021-00273-8

Sivaraman K, Chopra A, Narayan AI, Balakrishnan D (2018) Is zirconia a viable alternative to titanium for oral implant? A critical review. J Prosthodont Res 62(2):121–133. https://doi.org/10.1016/j.jpor.2017.07.003

Smith DE, Zarb GA (1989) Criteria for success of osseointegrated endosseous implants. J Prosthet Dent 62(5):567–572. https://doi.org/10.1016/0022-3913(89)90081-4

Souza JCM, Henriques M, Teughels W, Ponthiaux P, Celis J-P, Rocha LA (2015) Wear and corrosion interactions on titanium in oral environment: literature review. J Bio- Tribo-Corros 1(2):13. https://doi.org/10.1007/s40735-015-0013-0

Souza JCM, Apaza-Bedoya K, Benfatti CAM, Silva FS, Henriques B (2020) A comprehensive review on the corrosion pathways of titanium dental implants and their biological adverse effects. Meta 10(9):1272

Steinemann SG (1998) Titanium – the material of choice? Periodontol 17:7–21. https://doi.org/10.1111/j.1600-0757.1998.tb00119.x

Stepanovska J, Matejka R, Rosina J, Bacakova L, Kolarova H (2020) Treatments for enhancing the biocompatibility of titanium implants. Biomed Pap Med Fac Univ Palacky Olomouc Czech Repub 164(1): 23–33. https://doi.org/10.5507/bp.2019.062

Suárez-López Del Amo F, Garaicoa-Pazmiño C, Fretwurst T, Castilho RM, Squarize CH (2018) Dental implants-associated release of titanium particles: a systematic review. Clin Oral Implants Res 29(11): 1085–1100. https://doi.org/10.1111/clr.13372

Terheyden H, Lang NP, Bierbaum S, Stadlinger B (2012) Osseointegration – communication of cells. Clin Oral Implants Res 23(10):1127–1135. https://doi.org/10.1111/j.1600-0501.2011.02327.x

Textor M, Sittig C, Frauchiger V, Tosatti S, Brunette DM (2001) Properties and biological significance of natural oxide films on titanium and its alloys. In: Brunette DM, Tengvall P, Textor M, Thomsen P (eds) Titanium in medicine: material science, surface science, engineering, biological responses and medical applications. Springer, Berlin Heidelberg, pp 171–230. https://doi.org/10.1007/978-3-642-56486-4_7

Tuikampee S, Chaijareenont P, Rungsiyakull P, Yavirach A (2024) Titanium surface modification techniques to enhance osteoblasts and bone formation for dental implants: a narrative review on current advances. Meta 14(5):515

Wang Y, Zhang Y, Miron RJ (2016) Health, maintenance, and recovery of soft tissues around implants. Clin Implant Dent Relat Res 18(3):618–634. https://doi.org/10.1111/cid.12343

Wang X, Ning B, Pei X (2021) Tantalum and its derivatives in orthopedic and dental implants: Osteogenesis and antibacterial properties. Colloids Surf B: Biointerfaces 208:112055. https://doi.org/10.1016/j.colsurfb.2021.112055

Webber LP, Chan H-L, Wang H-L (2021) Will zirconia implants replace titanium implants? Appl Sci 11(15): 6776

Wennerberg A, Albrektsson T (2009) Effects of titanium surface topography on bone integration: a systematic review. Clin Oral Implants Res 20(Suppl 4):172–184. https://doi.org/10.1111/j.1600-0501.2009.01775.x

Yang RM, Talib HS, Miron RJ, Wiedemann TG (2022) OsteoMacs and their role in early implant failure and Osseointegration. Compend Contin Educ Dent 43(10): 698–703

Zafar MS, Farooq I, Awais M, Najeeb S, Khurshid Z, Zohaib S (2019) Chapter 11 – bioactive surface coatings for enhancing Osseointegration of dental implants. In: Kaur G (ed) Biomedical, therapeutic and clinical applications of bioactive glasses. Woodhead Publishing, pp 313–329. https://doi.org/10.1016/B978-0-08-102196-5.00011-2

Zarb GA, Schmitt A (1990) The longitudinal clinical effectiveness of osseointegrated dental implants: the Toronto study. Part I: surgical results. J Prosthet Dent 63(4):451–457. https://doi.org/10.1016/0022-3913(90)90237-7

23 Innate Immunity Gene Variants Affect Periodontitis Susceptibility

Martin Levine and Zsolt M. Lohinai

Abstract

Periodontitis is a chronic bacterial infection that destroys the epithelial attachment, forming a crevice beneath the gingival margins. The bacteria form dentally attached biofilms that enter the crevice and cause systemic inflammatory diseases (comorbidities). Susceptibility to moderate or severe (established) periodontitis is associated with smoking, diabetes, inadequate oral hygiene, and a genetic factor related to innate immunity function. One key factor in innate immunity function and its significance for periodontitis is related to the essential amino acid lysine. While lysine is absent in saliva, it is present in gingival crevicular fluid (GCF), an inflammatory exudate from the gingival crevice that permeates dental biofilms. After 16 periodontally healthy adults refrained from oral hygiene for 1 week, nine exhibited strong and seven exhibited weak GCF exudation, despite similar biofilm lysine concentrations. Strong responders, who effectively moved biofilm bacteria back into saliva, possess the adenine (A) single nucleotide polymorphism (SNP) at site −511 of the interleukin-1β (*IL1B*) gene instead of guanine (G). Weak responders possess thymidine (T) SNP at site +3877 of *IL1B* instead of cytidine (C), leading to bacterial retention in gingival crevices where lysine promotes periodontopathic biofilm development. If both *IL1B* + 3877(T) and *IL1B*-511 (A) are present, an SNP in *IL6*, *IL10,* or *CD14* determines the GCF response. These genetic profiles identify 40% of the US population as susceptible to periodontitis and comorbidities, causing bacterially driven systemic inflammation.

Keywords

Periodontitis · Innate immunity · Inflammasomes · Genetic variants · Susceptibility

M. Levine (✉)
(Volunteer) Department of Periodontology, University of Oklahoma University HSC, OK City, OK, USA

(Retired 2019) Department of Biochemistry & Molecular Biology, University of Oklahoma HSC, OK City, OK, USA
e-mail: martin-levine@ouhsc.edu

Z. M. Lohinai
Restorative Dentistry and Endodontics, Semmelweis University, Budapest, Hungary

23.1 Introduction: Periodontal Disease Susceptibility Is Inherited

Periodontitis is an inflammatory disease in which bacteria destroy the connective tissue and alveolar bone, supporting teeth in the oral cavity. The disease begins as gingival inflammation (gingivitis) during adolescence and young adulthood in

N. Rezaei, H. Ziaei (eds.), *Oral Immunology*, Advances in Experimental Medicine and Biology 1492,
https://doi.org/10.1007/978-3-032-03176-1_23

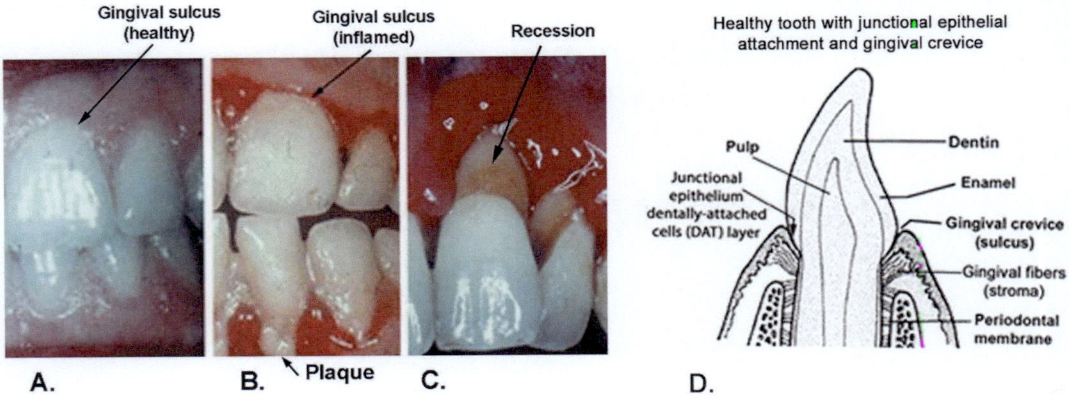

Fig. 23.1 Healthy gingiva, inflamed gingiva (gingivitis), and periodontitis in the oral cavity. (**A**) *Healthy*: The gingiva is light pink, stippled, and attached tightly to clean teeth. The gingival crevice or sulcus is indicated with an arrow (top right). (**B**). *Gingivitis*: The gingiva is red, especially at the free gingival margins (top arrow), and plaque is present (bottom arrow). Patients with established gingivitis often complain that their gums bleed on toothbrushing. (**c**) *Periodontitis*: The gingiva has receded and exposed the cementum of the tooth root (arrow). The gingiva is swollen and inflamed (red), and the sulcus is deepened into a periodontal pocket between the exposed cementum and the inflamed (red) gingiva. (**D**) *Dental attachments*: The gingival crevice or sulcus (right) is formed by dentally attached (DAT) cells of junctional epithelium (JE). These cells form the crevice base (left). Immediately below the DAT cells lie the dentally attached fibers of the connective tissue stroma, and below that, labeled on the right, are the periodontal membrane ligaments. As periodontal disease develops, the connective tissue stroma takes on the appearance shown and described in Fig. 23.1c. If the infection continues, the base of the pocket migrates apically until half or more of the periodontal ligaments and alveolar bone have disappeared. The teeth become loose and may eventually exfoliate, indicating severe periodontitis. (Sources: Figure 22.1a–c from Levine (2011), Fig. 13.1, Chap. 13, p. 232; Fig. 23.1d is from the same source, Fig. 3.5, Chap. 3, p. 37)

individuals with inadequate oral hygiene. All individuals who abstain from oral hygiene develop gingivitis beneath bacterial biofilms that adhere to their teeth at the gingival margins above the shallow crevice. Figure 23.1a–c illustrate the clinical appearance of gingival health, gingivitis, and moderate to severe (established) periodontitis. Figure 23.1d diagrams the site of the gingival crevice and the underlying gingival fibers in health. Figure 23.2 is an enlarged diagram of the gingival crevice in health and disease. On the diseased side, dentally adherent bacteria are located above the inflamed junctional epithelium (JE), where gingival crevicular fluid (GCF) appears in association with lysine depletion by decarboxylation to cadaverine and carbon dioxide, the equation above the diagram. The loss of lysine at the site of bacteria shown in Fig. 23.1B exposes the crevice base to innate immune activation by the products from the overlying bacterial biofilm (Fig. 23.2). This interaction is detailed in Section 23.5.2. Investigators have proposed that the development of periodontitis from gingivitis may depend on the host's "immune fitness" (Loos and Van Dyke 2020) or result from a host-mediated "aberrant immune response" (Hajishengallis 2022). Because periodontitis is inherited (Michalowicz et al. 2000), "immune fitness" or "aberrant response" must come at least partially from innate immunity genes individually associated with periodontitis. These genes are primarily but not exclusively *IL1A*, *IL1B*, *IL6, IL10,* and *CD14*. Certain single-nucleotide polymorphism (SNP) variants in these genes appear to act together in providing strong or weak GCF exudation traits within a week of experimental gingivitis (EG) (Lohinai et al. 2023).

As recorded by NCBI (The US National Center for Biotechnology Information), the nucleotide sequence of DNA in all genes, including human ones, is read in the 5′ to 3′ direction along the chromosome. Each chromosome contains double-stranded DNA, and every

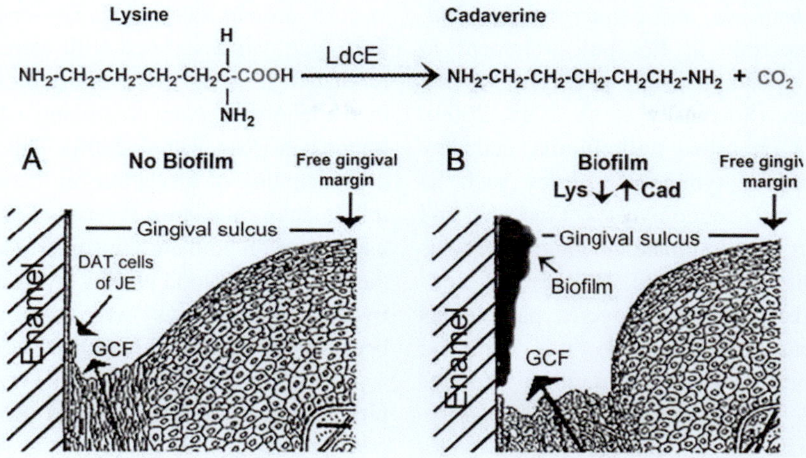

Fig. 23.2 Dentally attached cells and gingival crevicular fluid (GCF). The top panel shows the reaction catalyzed by biofilm lysine decarboxylase (LdcE). (**A**) Healthy crevice. GCF provides nutrients to maintain the integrity of the dentally attached tissue (DAT) and junctional epithelium (JE). (**B**) Infected crevice (incipient pocket). LdcE converts the amino acid lysine to cadaverine, resulting in DAT cell starvation and periodontal inflammation. (Original copyright 1977 S. Karger AG, and cited by Levine and Lohinai (2021), where it was republished by MDPI with permission under an open access Creative Common CC BY license)

individual has two copies of chromosomes, one inherited from each parent. An SNP in the DNA of a gene encoded in a chromosome can therefore be heterozygous (as a haplotype), or if present on both chromosomes, can be homozygous (as a haplotype pair or holotype). The DNA is also double-stranded, so there are two sets of nucleotides on genes, the forward (5′) strand, and the reverse (complementary) strand, either of which can possess an SNP. For genes encoded on the forward (5′) strand, SNPs are annotated directly on the gene's nucleotide sequence as displayed in the NCBI database, but for genes encoded on the complementary strand, SNPs are named with their complementary nucleotides. This means that the SNP on a gene encoded on the complementary strand—for example *IL1B* + 3954(T), reported as thymidine in the literature—is listed as *IL1B* + 3954(A), adenine on NCBI. Similarly, *ILB* + 3954(C), cytidine, is reported as *ILB* + 3954(G), guanine in the NCBI database. We followed the NCBI nucleotide naming system throughout this chapter. The plus or minus signs indicate that the site of the SNP is downstream (+) or upstream (−) from the transcription start site nucleotide of the encoded protein, and its associated number locates the nucleotide relative to the gene's encoded protein. The NCBI reference SNP (rs) number references the first discovered and usually most common nucleotide variant, followed by one or more alternative nucleotides. The reference site for SNP information is https://www.ncbi.nlm.nih.gov/snp/. Finally, we note that our published research article (Lohinai et al. 2023) incorrectly reported that the results were from *IL1B* + 3954 (rs1143634), whereas we later found that they were from *IL1B* + 3877 (rs1143633). As reiterated in Sect. 23.8.1, this article accurately cites and presents results for the correct gene, *IL1B* + 3877 (rs1143633).

23.2 Genetic and Microbial Factors in Periodontitis Susceptibility

23.2.1 Symbionts and Pathobionts

Incubating the salivary microbiome from periodontally healthy hosts in vitro in a medium resembling GCF for 3 weeks results in the appearance of the pathogenic "red" complex (Naginyte et al. 2019). This mixture of gram-negative bacteria, which includes *Porphorymonas gingivalis*,

Treponema denticola, and *Tannerella forsythia*, is barely detectible in EG but prominent in established periodontitis and referred to as the "red" complex (Socransky et al. 1998, 2000). Such bacteria are called **pathobionts**, and they gradually replace symbiotic species such as streptococci and actinomycetes (Jochum and Stecher 2020), that dominate the gingival crevice in health (Diaz et al. 2016). Besides the "red" complex, three recently discovered pathobionts may accompany or replace the "red" complex: *Filifactor alocis* (Edmisson et al. 2018), "rough" strains of *Aggregatibacter actinomycetemcomitans* (Fu et al. 2022), and butyrate-producing strains of *Fusobacteria nucleatum* (Jorth et al. 2014). Current studies have shown that biofilm from the gingival crevice in health is primarily composed of symbionts, whereas biofilms from periodontal pockets in severe periodontitis patients are almost 50% pathobionts, mostly *P. gingivalis* and *T. forsythia* (Hong et al. 2015).

23.2.2 Genes and SNPs Associated with Periodontitis Susceptibility

Comparisons of identical and non-identical twins indicate that up to 50% of the variance in periodontitis is inherited (Michalowicz et al. 1991). Although the genes involved in periodontitis were identified by 2010 (Laine et al. 2010), how SNPs in these genes relate to clinical measures of periodontitis has remained unclear. SNPs comprise only a fraction of the sequence, and one must determine not only a gene, but also each SNP within the gene sequence to identify a relationship.

Systemic inflammation increases C-reactive protein (CRP) in the blood plasma of healthy individuals (Machado et al. 2021) and indicates activated innate immunity in the body. Peripheral blood monocytes, derived from gingival crevicular fluid samples possessing certain interleukin-1B (*IL1B*) gene haplotype pairs, were previously associated with increased CRP levels. When these monocytes were examined in vitro, protein levels of IL-1β—encoded by the haplotype pair associated with increased CRP—produced 86–87% more IL-1β protein in vitro than a reference group possessing a different haplotype pair (Rogus et al. 2008). This was the first demonstration of a relationship between in vivo inflammatory mediator levels and gene promoter haplotypes on both chromosomes. Unfortunately, the use of CRP blood plasma levels as an alternative to periodontitis was later discontinued because this indicator is not disease-specific.

Genome-wide associations (GWAS) in complex, common diseases such as dental caries or periodontal disease have found few genetic associations that were both common and useful (Shungin et al. 2019). GWAS results provide a low portion of heritability explained by identified SNPs and an inability to account for underlying biological and pathological mechanisms (Tong et al. 2019). Nevertheless, studies show that using established periodontitis, i.e., moderate or severe periodontitis, as the determining factor in GWAS resulted in the discovery of an SNP within the protein-coding region of *IL1B* at residue +3954 (rs1143634). The presence of the homo- or heterozygous A SNP, but not of the homozygous G SNP, was associated with established periodontitis, as evidenced by increased IL-1β secretion and the presence of pathobionts (Ferreira Jr. et al. 2008). Using this direct GWAS approach, together with additional clinical studies, SNPs in other genes were also found to be associated with established periodontitis (Table 23.2), However, whether these associations act synergistically or independently could not be determined.

Gingival inflammation (gingivitis) precedes the onset of periodontitis (Chapple et al. 2015). EG is a procedure in which gingivitis-free adults stop all oral hygiene for 1 to 3 weeks. A follow-up study examined the *IL1B* + 3954 A and G variants reported by Ferreira et al. to be associated or not with pathobiont colonization in established periodontitis (Trombelli et al. 2010). The Ferreira study reported that the amount of IL-1β protein after EG was significantly increased in a diseased (gingivitis) quadrant compared with control (cleaned) quadrants. Nevertheless, hosts possessing homozygous *ILB* + 3954(AA), rs1143634, had significantly higher levels of IL-1β throughout the study than those possessing a homozygous *IL1B* + 3954(GG) genotype. Heterozygous *ILB* + 3954(GA) hosts had intermediate mean levels of IL-1β that did not differ significantly from either homozygous hosts (Trombelli et al. 2010). The EG study therefore supports Ferreira's findings on established periodontitis, in which *ILB* + 3954(AA+BB) hosts had higher IL1β levels, more pathobionts, and more established disease (Ferreira Jr. et al. 2008). In other words, the presence of an alternate gene (A) tends to enhance IL1β expression before the development of established periodontitis. However, the reason why more pathobionts were present in homozygous *ILB* + 3954 adenine (A) but not homozygous guanine (G) individuals is not clear. In Sect. 23.7.1, we discuss evidence for faster and slower gingivitis development associated with differences in microbial composition (Bamashmous et al. 2021; Huang et al. 2014). Some of these differences suggest an unknown host factor that may influence susceptibility to periodontitis and thereby could pave the way for more individualized approaches to prevention and therapy.

23.3 Host–Microbial Interactions and Immune Responses in Periodontitis

23.3.1 Periodontitis and Oral Hygiene

Measurements based on clinical and radiographic examinations and the number of lost teeth categorize periodontitis as not present, mild, moderate, or severe (Caton et al. 2018). When applied to the results of the US periodontitis survey (NHANES 2009–2012), Table 23.1 indicated that 62% of US adults aged 30 years and older (representing approximately 65 million adults) had mild or no periodontitis, and that 38% of the adult population had established periodontitis (Genco and Borgnakke 2020). The chronic and cumulative nature of periodontitis makes it increasingly common in older adults, and almost

Table 23.1 Periodontitis in US aged 30–65 years

Severity	Prevalence (%)
No disease	52.8
Mild	8.7
Moderate	30.0
Severe	8.5
Total	100.0

Compiled from un-tabulated information in the text of Genco and Borgnakke (2013). Source was Table 23.1 in Lohinai et al. (2023) published by MDPI under an open access Creative Common CC BY license

two-thirds of individuals ≥65 years of age were affected in the USA in 2012 (Eke et al. 2015).

Another way of looking at the study reported by Eke et al. is to consider individuals without established periodontitis. As evidenced by mild or no periodontitis in 60% of US Caucasians prior to age 65 in 2010, it can be inferred that there is a slow progression from gingivitis in most individuals (Table 23.1). This suggests that the number of protected individuals decreased at an average rate of 1.14% of individuals per year, with 40% developing established periodontitis within 35 years (ages 30–65 years). This decrease in protection persisted above age 65 at a similar rate, with an average rate of 1.20% of individuals per year. Over that later 20-year span (adults aged between 65 and 85 years), the fraction of protected individuals dropped from 60% to 36%. By age 85, just over one-third of the US population remained protected from established periodontitis (Eke et al. 2015).

Reducing susceptibility to periodontitis requires good oral hygiene, especially after dental caries (cavities), crowded or misaligned teeth are first treated (Janakiram and Dye 2020). Brushing with a commercial toothpaste and a mechanical toothbrush, followed by using an interdental brush and/or floss in the morning and evening, should be considered (Worthington et al. 2019). Smoking is a primary cause of established periodontitis and should be avoided (Apatzidou 2022); moreover, diabetes (type 1 or type 2) should be controlled (Giannobile et al. 2013; Preshaw et al. 2012). Finally, adults should attend a dentist or dental hygienist at least twice a year to remove calculus, which consists of mineralized deposits of calcium that firmly adhere to the tooth surfaces and protect the biofilm from being removed by toothbrushing. The calcium ions originate from the inflammatory GCF fluid, derived from blood plasma (serum) in the gingival crevice. Within this niche, bacteria present in the biofilm above or within the gingival crevice metabolize serum-derived proteins and amino acids to ammonia, thereby increasing the local pH to alkaline levels (Barros et al. 2016). Fair to poor oral hygiene increases the risk of periodontitis two- to five-fold, while good oral hygiene reduces the risk (Lertpimonchai et al. 2017).

23.3.2 Innate Immunity, Inflammasomes, and Periodontitis

Histories of various infections in different populations or ethnic groups have led to a high frequency of SNPs in genes that regulate innate immunity (Lazarus et al. 2002). The first gene expressed by the activation of innate immunity is interleukin1A (*IL1A*), whose encoded protein (IL-1α) is intrinsic to the nucleus and cytosol of most body cells (Dinarello 2018). Infection or injury to host cells causes IL-1α secretion with only minor differences. Secretion of IL-1α occurs because infecting bacteria secrete pathogen-activated molecular products (PAMPs), and injured host cells release damage-associated molecular patterns (DAMPs). PAMPs and DAMPs interact with cell-surface or intracellular pattern recognition receptors (PRRs), which distinguish "self" from "non-self." Typical PAMPs are cell surface lipopolysaccharides, lipoproteins, and metabolic products; typical DAMPs include heat shock proteins, uric acid, ATP, DNA, and nuclear proteins (Festjens et al. 2006; Rider et al. 2017).

PAMP- and DAMP-PRRs share domains homologous with Toll-like receptors (TLRs) or nucleotide oligomerization domains (NOD)-like receptors (NLRs), and absent in melanoma-2 (-AIM2)-like receptors (ALRs). PRRs involved in cancer or autoimmunity have other binding domains, for example, those found in retinoic acid-inducible gene-i-like receptors (RLRs) and C-type lectin receptors (CLRs). Once a microbial PRR has recognized and bound to a PAMP (or DAMP) at the gingival epithelial cell surface, the complex is endocytosed and cooperatively assembles with TLRs, NLRs, and ALRs into a multimeric platform within the cell cytosol called an NLRP3 inflammasome (Evavold and Kagan 2018). Chronic, low-grade inflammation associated with chronic inflammatory diseases such as periodontitis and its comorbidities activates inflammasomes, and the increase in systemic inflammation is typically detected by a blood test such as C-reactive protein (Rogus et al. 2008) as reported in Sect. 23.2.2.

Inflammasomes activate a caspase, a hydrolytic enzyme that cleaves a cytosolic protein, gasdermin-D (GSDMD), by freeing its hydrophobic N-terminal domain to release IL-1α (illustrated in Fig. 23.3A). Once in the interstitial fluid, IL-1α binds to its receptor, IL-1R, on surrounding cells. The binding directly activates the release of more IL-1α, and subsequently, the synthesis and release of IL-1β. These two IL-1 proteins then act through an IL-1-combination receptor on surrounding cells (Fig. 23.3B), from which mostly IL-1β is released and strongly induces inflammation by attracting and activating PMNs to the site of infected or damaged cells (Dinarello 2018). Namely, the JE at the base of the gingival crevice or pocket in both gingivitis and periodontitis (Fig. 23.2) (Caton et al. 2018). Measurements of the amount of IL-1β in GCF or other inflamed tissue exudates reflect the intensity of inflammation present clinically (Liu et al. 1996).

23.3.3 PMNs Have Multiple Roles in Inflammation

PMNs serve as the first line of defense against microbial threats by effectively controlling infections and preserving mucosal integrity through processes such as phagocytosis, degranulation, and the release of neutrophil extracellular traps (NETs). Unfortunately, correlating PMN

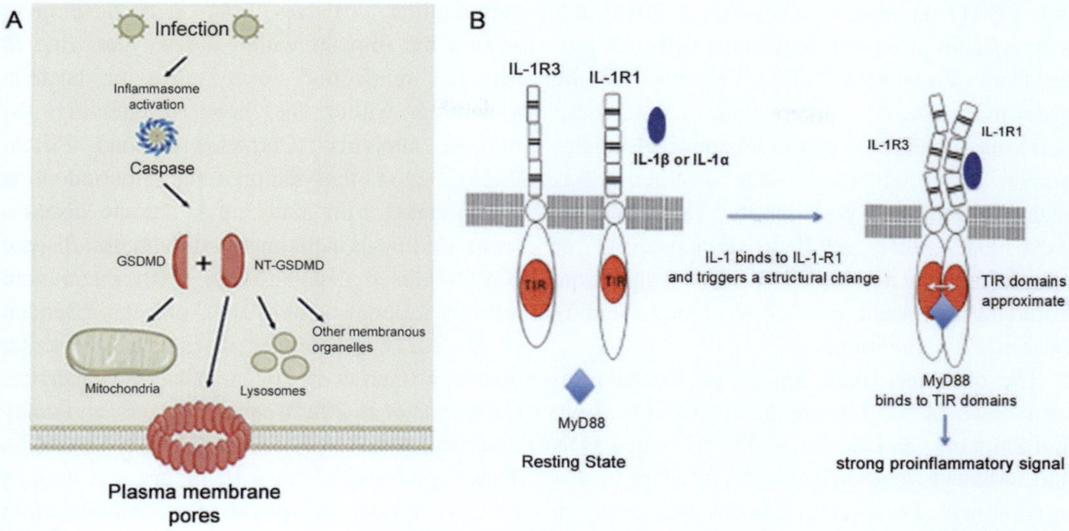

Fig. 23.3 How infection activates inflammation. (**A**) Bacterial exposure to gingival crevice basal cells triggers innate immune complexes called inflammasomes, that activate caspases to cleave the cytosolic protein gasdermin D (GSDMD), and release its large N-terminal domain (NT-GSDMD). The latter translocates to the outer region of the plasma membrane, where it forms transmembrane pores. NT-GSDMD makes similar membrane pores in other cytosolic organelles, for example, mitochondria, nucleus, and lysosomes. The cell is eventually lysed, but it first releases pre-formed interleukin-1α (IL-1α), which diffuses out from the cytosol and into the interstitial fluid between cells. Simplified from Fig. 1 in Devant & Kagan, Nature Immunology, 2023, Volume 24, Molecular mechanisms of gasdermin D pore formation, p. 1066). Copyright permission for Fig. 23.3A granted by Springer Nature publishers. Fig. 1b. (**B**) Once released, IL-1α binds to the IL1 receptor-1 (IL-1R1), an integral membrane protein present in almost all nucleated cells; it crosses the outer cell membrane so that binding of IL-1α in the outer (interstitial) fluid causes a change in the intracellular cytosolic apportionments. These changes trigger a structural change that activates a nearby second receptor, IL-1R3 on the same cell to bind to a third cytosolic protein, MyD88 (Myeloid Differentiation Primary Response Protein 88). MyD88 binds to the TIR (Toll/interleukin-1 Receptor/resistance protein) domains and becomes phosphorylated. The IL-1 receptor-activated kinases (IRAKs) then participate in a strong pro-inflammatory signal to the nucleus that activates the expression, synthesis, and secretion of IL-1β into the interstitial fluid. IL-1α and IL-1β then activate the formation of the IR1/R3 receptor in surrounding cells and more IL-1β synthesis. IL-1β is the primary agent that attracts leukocytes that have roles in tissue destruction or protection from acquired immunity around the base of the gingival crevice (Fig. 23.2B). In addition, TIR-domain proteins not only play a role in innate immunity by assisting with the formation of inflammasomes but also with the subsequent tissue destruction in established periodontitis. (Copyright permission granted by Wiley-Blackwell for Dinarello 2018, Fig. 2B, on p. 36, slightly modified, Copyright permission for the portion of Fig. 23.3A,B granted by Wiley-Blackwell

responses with the severity of periodontitis remains challenging because our inability to predict periodontitis susceptibility makes it impractical to personalize or target therapy (Kim and Moutsopoulos 2024).

PMNs activated by IL-1 induce a secondary response that can enhance or resolve inflammation, depending on the site and cause of the infection or cell damage (Rao et al. 2014). These different environmental conditions regulate PMN function so that they continue digesting the affected site or releasing chemoattracting factors, such as cathepsin G, azurocidin, and others that recruit blood monocytes and lymphocytes to activate a protective immune process called adaptive or acquired immunity (Prame Kumar et al. 2018). Monocytes differentiate into macrophages that interact with bone marrow-derived (B) lymphocytes to make antibodies or activate T lymphocytes that interact with and eliminate specific pathogens from the body. These macrophages are called alternatively activated type 2 (M2) macrophages to distinguish them from the better-known proinflammatory type 1 (M1) macrophages (Wynn et al. 2013). M2 macrophages also produce memory cells that protect from subsequent infection. Yet sometimes the PMN-monocyte interactions activate (M1) macrophages to cause uncontrolled inflammation and progressive tissue destruction—as seen in established periodontitis. Teeth become loose and either exfoliate spontaneously or require removal to stop the persistent infection. Tooth loss is a common outcome of long-standing established periodontitis.

The difference may depend on the environment, such as oral hygiene habits, teeth spacing, and smoking, and on the ability of both PMNs and monocyte/macrophage cells to adapt to that environment. For example, in the oral cavity at the tooth's gingival interface, the host immune system faces a two-way choice: (1) adaptive immunity is induced, and periodontitis is controllable, or (2) adaptive immunity is not induced, and periodontitis is uncontrollable. Phenotype modulation and functional targeting of PMNs by monocyte/macrophage activation determine whether the immune response promotes inflammation or tissue repair. Curcumin, an active component of turmeric, exhibits potent immunomodulatory effects that activate and enhance M2 macrophage activation and tissue repair (Deng et al. 2024). A novel, chemically modified curcumin (CMC2.24) is a new medication that might control human periodontitis development (Deng et al. 2023) and is discussed further in Sect. 23.9.

23.4 Periodontitis Comorbidities, a Major Health Problem in Older Adults

A major reason for preventing periodontitis from becoming established in older adults is the need to limit the risk of systemic infection by replacing the teeth with dentures or implants. Implants are functionally superior to dentures, but implant failure (implantitis) is common in patients whose teeth were extracted because of established periodontitis. This condition, resembling periodontitis, involves bone loss developing around the implant within a few years due to bacterial reinfection from saliva or bacteria remaining within the bone re-infecting the implant supports (Chmielewski and Pilloni 2023). Beyond that, the presence of periodontitis is associated with many other chronic diseases (comorbidities), including cardiovascular disease (CVD) (Genco and Van Dyke 2010), rheumatoid arthritis (Sandal et al. 2016), diabetes (Bendek et al. 2021), Alzheimer disease and vascular dementia (Guo et al. 2021), metabolic syndrome (Aizenbud et al. 2023), certain cancers, and many other inflammatory diseases (Nara et al. 2021). These comorbidities may be induced, at least in part, by a systemic spread of pathobionts from periodontal pockets around the teeth of individuals with established periodontitis (Hajishengallis 2022).

23.5 Genetic and Microbial Determinants of Periodontitis Progression

23.5.1 An *IL1* Genotype Predicts Severe Tooth Loss

Pathobionts were originally designated as indicators of refractory, chronic periodontitis (Socransky et al. 2002). Subsequently, the study reported in Sect. 23.2.2 (Ferreira Jr. et al. 2008) indicated that the presence of *IL1B* + 3954 (A) was correlated with greater pathobiont presence and greater periodontal destruction. Nevertheless, the need for over a hundred *IL1B* + 3954 (A) patients and a similar number of controls suggests the presence of unknown confounding SNP variants from the same or different genes. Other studies indicated that the adenine (A) at *IL1A*-889 also associates with several chronic inflammatory diseases (Kornman et al. 1997). These investigators found that established periodontitis often appears before age 50 in patients carrying a genotype consisting of *IL1B* + 3954 (A) and *IL1A*-889(A), which they named "PerioPredict." Conversely, PerioPredict negative patients only developed periodontitis after age 50. The PerioPredict genotype was later tested and found to be unrelated to the results of our 2023 study (Sect. 23.8.1).

In a study of 5117 US Caucasian dental patients who were 34–55 years of age, 534 exhibited tooth loss from established periodontitis, of whom 67.6% were PerioPredict positive (Giannobile et al. 2013). These patients possessed any two of the following three factors: a persistent smoking habit, adult-onset Type 2 diabetes, or the PerioPredict genotype. Aside from this limited subgroup, the recorded frequency of the PerioPredict genotype was 38%. The high prevalence of PerioPredict in the general population is confounding and indicates a need to identify genetic biomarkers of periodontal susceptibility with more general validity and clinical utility to enable personalized therapy (Kim and Moutsopoulos 2024). Giannobile's only evidence of periodontitis was tooth loss from severe periodontitis, which occurs in less than 10% of the population (Table 23.1), mostly smokers (Tomar and Asma 2000) or individuals with diabetes (Liccardo et al. 2019), who are already well-known predictors of severe periodontitis.

Below, we describe and justify a different set of genetic markers (Lohinai et al. 2023). Note that references to *IL1B* + 3954 in that article should be changed to *IL1B* + 3877 and the text re-interpreted as described below. These markers are based on two SNPs in *IL1B* alone, at −511 and + 3877, or together with a third SNP on *IL6* or *IL10* or *CD14* (Sect. 23.2.1). The presence of these markers results in either strong or weak GCF responses, which are likely to influence periodontitis susceptibility as described below through Sect. 23.8. Table 23.2 lists seven genes and ten SNPs that we tested using DNA from each of the six weak and nine strong participants in our earlier EG study (Lohinai et al. 2012). One participant had moved from Hungary and was unavailable. Table 23.3A indicates that *IL1B*-511(A) alone was significantly associated with a strong GCF response. Table 23.3B shows that *IL1B* + 3877(T) was significantly associated with a weak response when epistasis by *IL6*–1363 in the host's genome was included.

Table 23.2 Selected genes associated with periodontitis

Name and site[a]	Alleles[b]	MAFα[c]	rs number[d]	Reference article[e]
IL1A-889(G)[f]	G> A, C, T	0.285	rs1800587	Kornman et al. (1997)
IL1B + 3954(G)[f]	G > A	0.228	rs1143634	Ferreira Jr. et al. (2008))
IL1B-511 (A)	A > G	0.357	rs16944	Rogus et al. (2008)
IL1B + 3877(C)[g]	C > A, G, T	0.350	rs1143633	Wu et al. (2015)
IL6-1363(T)	G > C, T	0.076	rs2069827	Nibali et al. (2013)
IL10-592(T)	T > G	0.294	rs1800872	Wong et al. (2018)
IL10-1082(T)	T > C	0.453	rs1800896	Wong et al. (2018)
CD14-260(G)	G > A, C, T	0.485	rs2569190	Nicu et al. (2009)
COX2 + 8473(A)	A > G, T	0.349	rs5275	Prakash et al. (2015)
MMP8-799(G)	A > C, G	0.479	rs11225395	Izakovicova Holla et al. (2012)

Source was Table 23.2 in Lohinai et al. (2023) published by MDPI under an open access Creative Common CC BY license

[a]Gene name is italicized, and the location is indicated by the number of nucleotides before (−) or after (+) the translation start site

[b]Reference nucleotide is first, the alternate SNP nucleotide(s) follow the > sign, and the one used is underlined if a choice is possible

[c]Minor SNP allele frequency (MAFα) validated in a large European population

[d]Reference SNP cluster ID number

[e]Reference article

[f]PerioPredict genotype

[g]*IL1B* + 3877 and not *IL1B* + 3954 was later identified as the source of our results. *IL1B* + 3954 MAFα and Reference columns are included

Table 23.3 Genes associated with strength of GCF exudation

a. *IL1B*-511	**Wi A**	**No A**
Strong	9	0
Weak	3	3
Fisher exact test odds = ∞		
Two-tail $p = 0.044$ ($p > 0.05$)		
b. *IL1B* + 3877[a]	**Wi T**	**No T**
Strong	2	7
Weak	6	0
Fisher's exact test odds = 0		
Two-tail $p = 0.007$ ($p > 0.05$)		

Source was part of Fig. 23.5 in Lohinai et al. (2023) and published by MDPI under an open access Creative Common CC BY license

[a]*IL1B* + 3877 is adjusted for epistasis in the presence of *IL6*–1363(T)

Epistasis (Phillips 2008) refers to a genetic interaction in which the effect of one gene is modified by another. In this context, it represents a new and poorly understood conversion of a trait associated with *IL1B* + 3877, in which strong GCF exudation, normally associated with homozygous *IL1B* + 3877(CC), was converted to weak, and weak GCF exudation, normally associated with *IL1B* + 3877(TT or TC), was converted to strong. How we defined strong and weak GCF exudation traits and found their genetic associations is detailed in our 2023 publication (Lohinai et al. 2023). As noted at the beginning of this paragraph, references to *IL1B* + 3954 in that article should be changed to *IL1B* + 3877. Figures and Tables from that article are valid if *IL1B* + 3954 is replaced with *IL1B* + 3877, and the data reinterpreted accordingly (Sect. 23.8).

23.5.2 Lysine Decarboxylase from *Eikenella corrodens* (LdcE) Determines Biofilm Lysine Content After Experimental Gingivitis (EG)

Biofilms removed from teeth contain a protein that inhibits mammalian cell growth and was later identified as lysine decarboxylase (LdcE), an enzyme produced by *Eikenella corrodens* (Levine et al. 2001). LdcE decarboxylates the amino acid lysine by converting it to cadaverine and carbon dioxide (Fig. 23.2). This activity drew our attention to a role that lysine and cadaverine might play in EG. Lysine is an amino acid whose presence in the diet is essential for maintaining dentally attached JE cell turnover at the base of gingival crevices. Interstitial fluids and inflammatory exudates such as GCF contain lysine at concentrations of 0.17 μmol/mL (Lohinai et al. 2012), but lysine is present in saliva at concentrations of less than 0.01 μmol/mL (Lohinai et al. 2012; Tabi et al. 2008; Van Wuyckhuyse et al. 1995). We therefore performed an EG study in which we assessed biofilm accumulation using the plaque index (PI), measured the host response via GCF, and compared both to the lysine concentration of the biofilm (Lohinai et al. 2012). Prior to EG, we fabricated a bioguard to cover the teeth in an upper quadrant (Trombelli et al. 2004), which allowed the teeth in the other three quadrants to be cleaned normally during EG.

After 1 week without oral hygiene during which biofilm accumulation was even less than in Fig. 23.1b, we assessed PI accumulation on the labial surface of three teeth in the bioguarded quadrant, and the rate of GCF exudation per minute from the gingival crevices of the same teeth (Lohinai et al. 2012). Biofilm was then removed, pooled, and diluted with 0.07 mM saline, centrifuged, and the lysine content of the supernatant was measured (Tabi et al. 2008). Figure 23.4a shows a linear relationship between PI and lysine concentration, and Fig. 23.4b shows a quadratic (arched) relationship between GCF exudation and lysine concentration. The peak of the arch coincided with the minimal lysine concentration of healthy blood plasma, 0.11 μmol/mL (Tan and Gajra 2006). Lysine depletion below that value led to reductions in both GCF exudation and PI. Interestingly, when biofilm lysine concentrations exceeded 0.11 μmol/g, GCF decreased, even as PI increased. Figure 23.4b suggests that inhibiting biofilm LdcE in the absence of oral hygiene increases the biofilm lysine concentration but reduces the

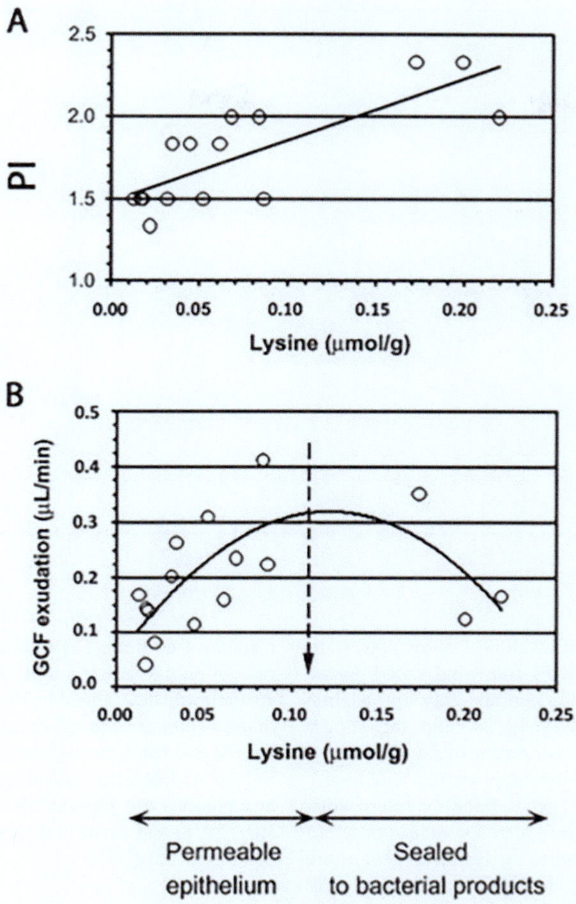

Fig. 23.4 Relationships between PI and GCF exudation rate relative to biofilm lysine content. (**A**) The Plaque Index (PI) showed a significant correlation with biofilm lysine (Lys) content following oral hygiene restriction. Spearman r2 = 0.58, $P < 0.001$. The graph shows the result of a least-squares regression whose equation was as follows: PI = 1.48 + 3.75 * Lys; $R^2 = 0.61$, df = 1,14, ANOVA, F = 24.1, $P < 0.001$. $R^2 = 0.61$, $P < 0.001$. Notably, this association remained statistically significant even after excluding the three highest lysine values (values exceeding 0.11 mg/mL) (Spearman $r^2 = 0.33$, $P = 0.042$). (**B**) GCF exudation rate was regressed on biofilm lysine content following oral hygiene restriction. The vertical arrow indicates the minimal lysine concentration of normal blood plasma (0.11 mg/mL). A quadratic relationship was significant: least-squares regression $R^2 = 0.37$, df = 2,13, ANOVA, F = 5.4, $P < 0.02$. GCF = −18.5 * Lys^2 + 4.5 * Lys + 0.05. Coefficients: Lys^2 t statistic = −3.09, $P < 0.01$; Lys t statistic = 3.27, $P < 0.01$; constant t statistic = 1.01, not significant. (Original source was Fig. 23.4 in Lohinai et al. (2012), published as a supplement to Lohinai et al. 2023)

GCF response by strengthening the host barrier to biofilm (Levine and Lohinai 2021; Lohinai et al. 2015).

We subsequently reported that the curve in Fig. 23.4b, showing the arched relationship of GCF exudation to lysine in all participants, separated into two distinct response patterns of GCF response, shown as blue and red curves in Fig. 23.5 (Levine and Lohinai 2018). Measuring GCF as dependent on the biofilm lysine concentration after 1 week of EG detected a consistent difference in host GCF responsiveness to lysine in biofilms at the gingival crevice. We investigated whether this difference in GCF

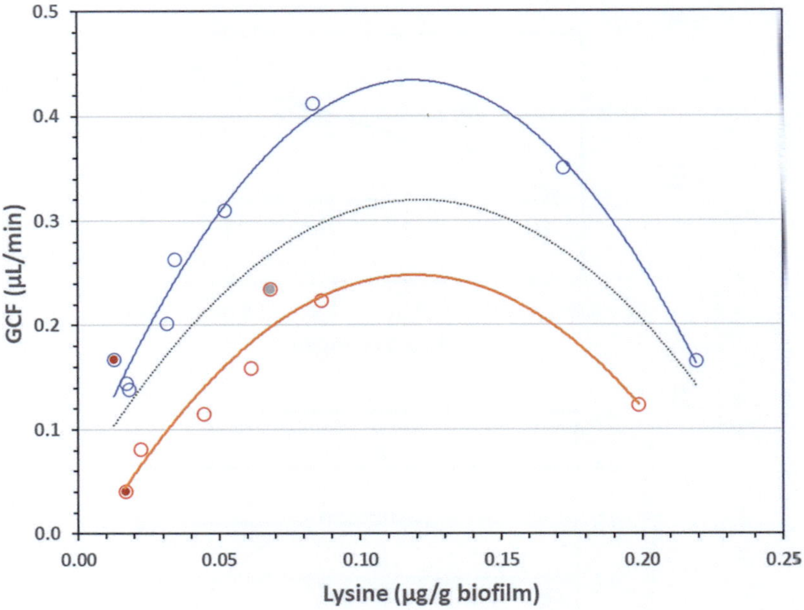

Fig. 23.5 Dual relationship of biofilm lysine content to gingival crevicular fluid (GCF) following oral hygiene restriction for one week. The central, gray curved line indicates the single quadratic (arched) relationship shown in Fig. 23.4b, but a much stronger fit of parallel strong (blue) and weak (red) GCF response curves was highly significant (F Ratio $= 41.77$, $p < 0.0001$). Interactions with biofilm lysine and lysine-squared contents were not significant. The GCF response to the lysine content of the biofilm (blue line) was about twice that of the weak response (red line). The blue and red circles possessing a red dot on the far left are the 2 hosts whose response was epistatic and determined by the presence of *IL6*–1363)GT) in their genome. The red circle with the black dot between the red and black curves was from a host who was unable to provide DNA for this study. Figure 23.5 was presented as a poster at the EuroPerio9 conference in Amsterdam in June 2018, and in an oral presentation at the IADR-PER meeting in July 2018

exudation rate resulted from innate immunity activating different SNPs in genes associated with periodontitis during the first week of EG. Was the rate of GCF exudation influenced by different SNPs in innate immunity genes associated with periodontitis? Was innate immune activation during the first week of EG influenced by the presence of a reference (more common) SNP compared to an alternate (less common) SNP in different individuals? In short, do the strong and weak inflammatory responses to the same biofilm lysine concentration reflect distinct GCF exudation traits, and if so, does each trait influence periodontitis susceptibility differently?

23.6 Physiologically and Microbiologically Informed Innate Immunity

23.6.1 GCF Exudation in the Apparent Absence of Bacteria

The JE consists of squamous epithelial cells with basal (dividing) cells attached to tooth cementum on one side and the connective tissue stroma of the gingiva on the other side (Fig. 23.1d). The central, squamous portion of the JE is interconnected by few desmosmes and gap junctions, resulting in wide intercellular spaces that allow GCF, containing inflammatory and immune cells, to transit to the crevice base (Fig. 23.2b) (Bosshardt 2018). In premolars extracted for orthodontic reasons in teenagers, the dentally attached JE always contains small numbers of polymorphonuclear leukocytes (PMNs), indicating that mild tissue damage occurs even in healthy gingiva, when there are few or no bacteria present in the crevice. GCF must continuously nourish the JE cells attached to the teeth (Figs. 23.1d and 23.2a) to maintain their attachment, and a lack of lysine—an essential amino acid supplied only by GCF—can impair this nourishment and compromise attachment integrity.

23.6.2 Chewing Traumatizes JE Cells and Releases Damage-Associated Molecular Patterns (DAMPs)

One obvious cause of persistent mild inflammation in the absence of biofilm is mastication, during which minor tooth movements can damage the JE cells, leading to the release of DAMPs. DAMPs interact with PRRs to activate inflammasomes, leading to the release of IL-1α as a downstream consequence of the caspase-cleaved N-terminal region of gasdermin D (Fig. 23.3B and Sect. 23.3.2). The released IL-1α then binds to its receptor on the surrounding uninjured JE cells and thereby enables IL-1β to be synthesized and expressed. IL-1β subsequently activates its receptor on adjacent epithelial or fibroblast cells. When sufficient IL-1β is present, it triggers nearby cells to produce IL-6, IL-23, and IL-17. IL-17, in particular, promotes the recruitment and activation of Th17 lymphocytes, which amplify the inflammatory cascade at epithelial barriers that separate bacterially colonized regions such as the gingival crevice or gastrointestinal tract from underlying tissues (Dutzan et al. 2017; Rider et al. 2017). Th17 cells also recruit and interact with monocytes, which are differentiated into macrophages that remove DAMPs and tissue debris. The minor but persistent source of damage to the JE from tooth movements during mastication makes the crevice region biofilm especially susceptible to pathobiont colonization (Sect. 23.6.3).

23.6.3 The GCF Favors Pathobiont Outgrowth

GCF is derived from defibrinated blood plasma (serum), which provides nutrients that support the growth of anaerobic bacteria (Naginyte et al. 2019), including pathobionts discussed in Sect. 23.2.2. During the first few days of oral hygiene restriction, low levels of pathobionts present in saliva can adhere to teeth surfaces within existing symbiont-dominated biofilms. However, these pathobionts grow more efficiently than symbionts on the traces of GCF found at the crevice orifices. As the pathobiont population expands, the gingival margins become inflamed (Figs. 23.1b and 23.2b). Nevertheless, there is evidence that a sufficiently rapid flow of GCF can reduce the rate of pathobiont accumulation (Brill 1959).

Periodontitis develops from gingivitis (Chapple et al. 2015) by gradually deepening and widening the gingival crevice (Fig. 23.2b), where pathogenic bacteria may become partly calcified by ammonia production (Sect. 23.3.1). The resulting rise in pH facilitates the precipitation of calcium phosphate (calculus) from GCF and protects pathobionts from being removed by toothbrushing (Velsko et al. 2019). However, if pathobiont colonization is restricted by a faster GCF flow rate, the progression from gingivitis to periodontitis may be slowed, thereby delaying periodontitis pocket development. In other words, the finding of strong and weak exudation in Fig. 23.5 should be evidence of two distinct types of host immune response profiles and two kinds of bacterial composition in gingivitis and periodontitis.

23.7 Bacterial Biofilm Composition and Host Response Variability in Gingivitis and Periodontitis

23.7.1 Two Types of Bacterial Biofilms and Host Responses in EG and Periodontitis

The GCF inflammatory exudate from the crevice is strong enough to remove sterile charcoal particles in addition to bacteria from healthy gingival crevices (Brill 1959). This suggests that GCF plays a crucial role in maintaining gingival health. However, the rate of GCF exudation after 3 weeks of EG is an indicator of gingivitis severity (Barros et al. 2016). Despite the established understanding that periodontitis develops from gingivitis (Chapple et al. 2015), how EG relates to periodontitis susceptibility remains uncertain.

An EG study using group-based trajectory modeling identified two significantly different response groups based on the progression of gingivitis over 3 weeks. The first group, "fast" responders, comprised 15 participants (36%), while the second group, "slow" responders, comprised almost twice as many individuals, with 27 participants (64%). Absolute model fit statistics indicated a correct classification assignment for both groups, with the major difference in trajectory occurring during the first 5–7 days, during which the "fast" responders achieved higher mean GI scores earlier than the "slow" responders (Nascimento et al. 2019b). These GI score differences were confirmed by finding that the "fast" responders had more activated proinflammatory chemokines in their GCF than the "slow responders" (Nascimento et al. 2019a; Silbereisen et al. 2019).

"Fast" GCF responders are not necessarily strong GCF responders that expel bacteria efficiently from gingival crevices. Instead, their rapid activation of local inflammatory cytokines also suggests a weak barrier to bacteria and/or a greater baseline presence of pathobionts. Another group also performed a three-week EG study, examining the types of bacteria and the combination of activated cytokines in 21 periodontally healthy volunteers (Bamashmous et al. 2021). Based on these criteria, three distinct clinical phenotypes were reported: high responders (strong, $n = 6$, 28.5%), low responders (weak, $n = 6$, 28.5%), or slow responders ($n = 9$, 43%). The slow-response phenotype displayed no change in biofilm accumulation and a greater relative abundance of *Streptococcus* spp. The remaining two phenotypes of strong and weak clinical response demonstrated significantly different inflammatory responses despite similar microbial succession patterns.

The most relevant EG study was conducted in China and investigated how a strong or weak gingivitis response after 3 weeks was related to a difference in microbial colonization (Huang et al. 2014). Using ribosomal RNA sequencing, the authors identified 27 microbial genera whose abundance in biofilms significantly correlated with either stronger or weaker gingivitis responses in 50 participants after 3 weeks of abstaining from oral hygiene in an EG study. Surprisingly, these different microbial patterns were also already present in naturally occurring biofilms obtained at the time of enrollment, before the participants even started the intensive oral hygiene period preceding EG. The same result was observed when pre- and post-EG samples from a second cohort of 41 additional participants were obtained as part of the same study. The same two distinct microbial patterns were identified and predicted the subsequent development of strong or weak gingivitis during EG with 95% accuracy. The authors concluded that the different preexisting microbial structures of an individual's biofilm before EG can reliably predict susceptibility to gingivitis progression and severity during and after EG. We propose that this difference may indicate an environmental influence or genetic predisposition, based on the studies described in the following sections.

23.7.2 Weak GCF Responders Seem more Likely to Develop Pathobionts

Chronic periodontitis rarely appears before age 30 (Genco and Borgnakke 2013), typically developing at least 10–15 years after the initial onset of gingivitis in adolescents and young adults whose oral hygiene is inadequate. Eventually, the gingival crevices provide an environment conducive to the overgrowth of pathobionts, including members of the "red" complex containing *P. gingivalis* and other associated species. Our evidence suggests that individuals with weak GCF responses are more prone to developing the "red complex."

Within a group of 34 non-smoking US Caucasian patients with periodontitis (mean age: 52 ± 12 years, standard deviation), *P. gingivalis* colonization in the deepest periodontal pockets reached almost 50% of total bacterial composition, 44% of patients, compared to only 12% in the remaining 56% of patients (Hong et al. 2015). Interestingly, this fraction (44%) of high *P. gingivalis* colonization in a middle-aged population with established periodontitis happens to be the same as the fraction of weak GCF responders (7 out of 16 participants) identified in our EG study (Fig. 23.5). Because GCF is a richer environmental source of bacterial nutrients than saliva, those participants with a weak GCF response in EG are more likely to promote faster development of pathobiont growth over symbionts (Naginyte et al. 2019), thereby shifting the microbiome toward a disease-associated state.

In a later study (Huang et al. 2021), rapid and consistent alterations in biofilm metabolites and cytokines in saliva were reported, emerging as early as 24–72 h after restricting oral hygiene. These alterations included a rapid, full activation of 11 salivary cytokines, accompanied by a steep, simultaneous decrease in plaque-derived betaine (Zhao et al. 2018) and *Rothia* species (Tsuzukibashi et al. 2017), both of which are health-associated. The decreases in betaine and Rothia suggest the development of a suboptimal health stage that displaces earlier anti-inflammation mechanisms associated with health-promoting symbionts in the gingiva. Unfortunately, whether their optimal or suboptimal health stage was related to their 2014 microbiome prediction of gingivitis severity or to differences in GCF flow was not reported. Based on these findings, in addition to the conclusion of Sect. 23.5.2, we addressed whether strong and weak GCF responses to biofilm-derived lysine concentrations might be determined by SNPs in genes associated with periodontitis susceptibility.

23.8 Genetic Determinants of GCF Response and Periodontitis Susceptibility

23.8.1 Associations of SNPs with Strong and Weak GCF Exudation Are Modified by Epistasis

The results in this section are based on the study by Lohinai et al. (2023), which reported findings for the *IL1B*+3877 (rs1143633) SNP, but incorrectly attributed and discussed them as relating to *IL1B*+3954 (rs1143634)—see last 2 sentences of Sect. 23.1. In that article, selected SNPs from genes associated with periodontitis (Table 23.2) were investigated in hosts who had exhibited strong or weak GCF responses to lysine after EG (Lohinai et al. 2012). Tables 23.2, 23.3, 23.4, and 23.5 in the current report have been corrected from the 2023 publication, but none of the Figures. The rows of Table 23.4 summarize all 4 possible combinations of SNPs at *IL1B*+3877 and *IL1B*-511, relating them to GCF response phenotypes, and by implication, to periodontitis susceptibility.

Row 1 summarizes data from seven individuals who possessed homozygous reference SNPs (CC) at *IL1B*+3877, with either homo- or heterozygous reference (A) alleles (AA or AG) at *IL1B*-511. All but one produced the strong GCF trait, which retards periodontitis development by removing bacteria, including trace levels of

Table 23.4 Summary of *IL1B* gene combinations, GCF trait, and susceptibility to periodontitis

Hosts	(%)	*IL1B* + 3877	*IL1B*-511	GCF trait[a]	Periodontitis[b]
7	47	*IL1B* + 3877(CC)[c]	*IL1B*-511(AG)[d]	6S+1W[e]	Mild or none[f]
3	20	*IL1B* + 3877(CT)	*IL1B*-511(GG)	3 W	Established[g]
5	33	*IL1B* + 3877(CT)[c]	*IL1B*-511(AG)	2S+1S[e]+2 W	2nd Gene[h]
0	0	*IL1B* + 3877(CC)	*IL1B*-511(GG)	Not present	Unknown

Source was part of Table 23.4 from Lohinai et al. (2023) published by MDPI under an open access Creative Common CC BY license
[a]GCF trait: S = Strong; W = Weak;
[b]Periodontitis susceptibility
[c]Epistasis which changed S to W trait or W to S trait (see Table 23.5)
[d]AG or AA
[e]Which converted S to W
[f]No disease or Mild except for epistasis
[g]Moderate or Severe
[h]A second gene's SNP (see Table 23.5) determines periodontitis susceptibility

pathobionts, from the biofilm. So, periodontitis susceptibility in this group is expected to remain low. The SNPs at *IL1B* + 3954 appear unrelated to gingivitis (Mazurek-Mochol et al. 2019), as discussed in Sect. 23.5.1. One of seven individuals was a unique case in also possessing *IL6*–1363(T), an uncommon SNP in Europeans (Table 23.2). This SNP converted the normal, strong GCF trait of *IL1B* + 3877(CC) plus *IL1B*-511(AG) to a weak trait.

Epistasis refers to an unexpected change of trait or phenotype against a fixed genetic background, normally detected only in humans with large families (Miton et al. 2021). Indeed, the fact that periodontitis develops from gingivitis (Chapple et al. 2015) highlights why the use of SNPs associated with periodontitis provided the necessary shared genetic background. Epistatic conversions exclusive to *IL1B* + 3877 converted the homozygous strong host (C) to weak, but the homo- or heterozygous weak host (T) to strong. Such reversals of trait (or phenotype) are typical of specific epistasis (Phillips 2008). The *IL6*–1363(GT) genotype is associated with aggressive (severe) chronic periodontitis (Nibali et al. 2008, 2013), and in this study, it correlates with the smallest biofilm lysine content within each GCF response group (Fig. 23.5). A study of refractory patients (Levine et al. 2017) suggests that greater LdcE activity accounts for greater lysine depletion from biofilms and is linked to a refractory response to periodontal therapy.

In **row 2** of Table 23.4, the alternate SNPs (T) at *IL1B* + 3877 and (G) at *IL1B*-511 were present in 3 hosts, all of whom had a weak GCF response that enhances pathobiont colonization and susceptibility to established periodontitis. The weak GCF flow permits pathobionts to persist longer and proliferate in GCF, which is a much better substrate than saliva (Naginyte et al. 2019), thereby enhancing susceptibility to established periodontitis. The group in **row 3** had a conflicted genotype, featuring the alternate T allele at *IL1B* + 3877 with the homozygous or heterozygous reference (A) allele at *IL1B*-511. One *IL1B*+3877(T) allele was converted from weak to strong by epistasis from *IL6*–1363(T). Two hosts exhibited a strong GCF response due to the presence of the reference SNP at *IL10*–592 (GG). In contrast, the presence of an alternate SNP (TT) at *IL10*–592 or (AA) at *CD14*–260 delivered a weak GCF response.

There was no representative for **row 4**, which would include the homozygous reference SNP pair, *IL1B* + 3877(CC), typically associated with a strong GCF response, combined with the homozygous alternate allele at *IL1B*-511(GG), which is linked to a potentially weak GCF response. On the other hand, significant associations of *IL1B* + 3877 with chronic periodontitis in

Caucasians and African Americans could be caused by three upstream SNP variants *IL1B*-511, *IL1B*-1464, and *IL1B*-3737 (Wu et al. 2015). These investigators validated that hypothesis by studying two additional racial groups: Hispanics (OR = 1.95, p = 0.04) and Asians (OR = 3.27, p = 0.01), and provided a meta-analysis of all four groups (Hispanics, Asians, US Africans, and US Caucasians). The absence of *IL1B* + 3877(CC)/*IL1B*-511(GG) genotype combination in this study is consistent with functional variability. It suggests that *IL1B*-511 (GG) may only co-occur with *IL1B* + 3877 (CC) when *IL1B*-511(A) is present on at least one chromosome in Caucasians. Conversely, SNP variants of *IL1B* + 3877(CT) do not show a consistent allelic association with either the A or G variant of *IL1B*-511, which appears to be independent and capable of interacting with IL6, IL10, and CD14 SNPs to influence periodontitis susceptibility in Caucasians. In African Americans, Hispanics, and Asians, other *IL1B* + 3877 interactions with *IL1B*-511, or perhaps with *IL1B*-1464 and/or *IL1B*-3737, may indicate periodontitis susceptibilities that should be investigated.

The two PerioPredict SNPs are IL1A-889 and IL1B + 3954 (Sect. 23.5.1). After realizing that our 2023 article had not included *IL1A*-889 and had also mis-represented our data as coming from *IL1B* + 3954, we tested the SNPs at *IL1A*-889 and re-evaluated *IL1B* + 3954, comparing the results with those from *IL1B* + 3877. The findings confirm that *IL1B* + 3877 was the SNP actually responsible for the results previously attributed to *IL1B* + 3954. In addition, *IL1B* + 3954 (T) and *IL1A*-889(T) both identified four strong responders and 1 weak responder, making 5 of the 15 hosts positive for the PerioPredict genotype, 33% compared with 38% positive in the Giannobile study. To summarize, our results show that the GCF response is a strong or weak genetically encoded trait that could determine whether an individual exhibits a strong or weak GCF response to bacteria after 1 week of EG. The strong GCF response trait facilitates bacterial clearance and prevents the growth of pathobionts responsible for periodontitis. In contrast, the weak trait resembles the effect of smoking, in which reduced and weak GCF exudation allows salivary bacteria to persist and proliferate (Molnar et al. 2015), predisposing individuals to colonization by pathobionts such as *P. gingivalis*. These data underscore the importance of integrating environmental to genetic factors that shape individual susceptibility to periodontitis.

Table 23.5 Determination of GCF trait from SNP allelic genotypes

Strong GCF trait (periodontitis protected)	**Host (%)**
IL1B + 3877(CC) + *IL1B*-511(AG)[1a]	40.0
IL1B + 3877(CT) + *IL1B*-511(AG) + ***IL6*–1363**(GT)[1b]	6.7
IL1B + 3877(CT) + *IL1B*-511(AG) + *IL10*–592(GG)[1c]	13.3
Weak GCF trait (periodontitis susceptible)	**Host (%)**
IL1B + 3877(CT) + *IL1B*-511(GG)[2a]	20.0
IL1B + 3877(CC) + *IL1B*-511(AG) + ***IL6*–1363**(GT)[2b]	6.7
IL1B + 3877(CT) + *IL1B*-511(AG) + *IL10*–592(TT)[2c]	6.7
IL1B + 3877(CT) + *IL1B*-511(AG) + *CD14*–260(AA)[2d]	6.7

Source is a simplified version of Table 23.5 from Lohinai et al. (2023) published by MDPI under an open access Creative Common CC BY license

[1a]Strong GCF trait

[2a]Weak GCF trait

[1b]Bold=gene *IL6*-1363(GT) acts on *IL1B*+3877(CT) to produce a strong GCF trait and on

[2b]*IL1B*+3877(CC) to produce a weak GCF trait

[1c]*IL10*-592(CC) or *COX2*+8473(AG) identifies a conflicted genotype as strong GCF trait, and

[2c]*IL10*-592(TT) or

[2d]*CD14*-260(AA) as weak trait

23.8.2 Determining Susceptibility to Periodontitis and Its Comorbidities at any Age

We are working on a proof-of-concept study to validate our findings to assess if they could reliably determine individual susceptibility to established periodontitis and its numerous comorbidities, which is a major aim of modern periodontal and medical research. Our approach is illustrated in Table 23.5 and suggests that the GCF trait can be determined from the SNP genotypes identified in Table 23.4. Potential genetic screening methods could include a microarray-based approach to detect the specific groups of SNPs shown in Table 23.5. In this method, the SNP sequences are pre-loaded on a glass slide (Conzone and Pantano 2004), where they serve as probes to detect matching sequences in DNA samples from patients who provide a cheek scraping. Methodological details can be found at: https://online.stat.psu.edu/stat555/node/28/#:~:text=Affymetrix%20microarrays%20are%20still%20very,SNP%20variant%20at%20a%20locus

For our 2023 study, we used the TaqMan assay (ThermoFisher Scientific, Waltham, MA, USA) to genotype the SNP alleles at each locus in each of the genes listed in Table 23.2 (Sect. 23.2.2). Each assay kit contained two primers for amplifying the sequence of interest and two allele-specific and differently labeled TaqMan minor groove binder (MGB) probes for allele detection. Alleles were ordered based on their rs number; however, the problem was that the sequence we received for *IL1B*-3954 (rs1143634) was, in fact, the gene segment for rs1143633, *IL1B*-3877 (Lohinai et al. 2023), and it was almost a year later that we discovered the error.

23.9 Conclusions: Two Genetically Based Personalized Therapies for Preventing Periodontitis

The SNPs identified in this study (Sect. 23.3.2 and 23.8.2) reflect evolutionary processes that shaped human responses to bacterial invasion and cellular damage (Lazarus et al. 2002). Evolution implies reproduction, the passing of genes with their acquired SNPs to the next generation. If one of two SNPs at the same genomic site becomes associated with more descendants, that SNP may come to dominate in that locus, becoming the reference gene. The results suggest that longer human survival is associated with the strong GCF trait in hosts carrying reference SNPs at *IL1B* + 3877 and *IL1B*-511, namely the homozygous C gene at *IL1B* + 3877, and the homo- or heterozygous A gene at *IL1B*-511. Conversely, weak responders carry alternate SNPs, respectively T or G, which may have been the original nucleotides at these sites. Epistasis or some other function of *IL6*–1363 (GT) may have been more common in the past, but this SNP is now rare in the European population (Table 23.2).

Lysine degradation to cadaverine, observed in both weak and strong responders, stabilizes the biofilm and facilitates colonization by pathobionts that catabolize GCF proteins to butyrate (Levine and Lohinai 2021). Butyrate is a mammalian cell growth inhibitor in a mixture of short-chain fatty acids produced by pathobionts that account for 15% of the growth inhibition mediated by LdcE through lysine depletion in dental biofilms (Levine 1985; Levine et al. 2001). LdcE is inhibited by zinc (Levine et al. 2021) and butyrate by stannous fluoride (Cannon et al. 2018). A dentifrice called Dual Zinc plus Arginine (DZA) was developed by Colgate-Palmolive and significantly reduced bleeding on probing compared to a control dentifrice (Delgado et al. 2018). However, another study that compared this result in individuals using stannous fluoride to inhibit butyrate production (He et al. 2021) did not replicate these findings. The only significant difference from the control dentifrice was observed in participants using the fluoride dentifrice, not the DZA dentifrice.

The mean age of participants in the Delgado study was 33.6 years, but in the He study it was 38.8 years. Because cadaverine facilitates subsequent pathobiont colonization and butyrate production, targeting these metabolic processes at an earlier age is critical. A young population is essential to prevent pathobionts from becoming established in dental biofilms after age 30 (Sect. 23.3.1). Thus, if a DZA dentifrice is used regularly after age 15, inhibition of cadaverine production by this dentifrice should prevent pathobionts from developing in the biofilm. The DZA was tested on a population significantly younger than the sodium fluoride control. The mean age of the DZA group was 33.6 years, compared to 38.8 years in the stannous fluoride group. Because stannous fluoride primarily inhibits butyrate and is effective only after pathobionts have already become established, it would be valuable to compare both toothpastes in adults under the age of 30. Even more promising would be the development of a combined dentifrice containing both DZA and stannous fluoride, which could work independently from the same dentifrice, thereby preventing periodontitis development more efficiently. Such a combination could be especially important for preventing the development of periodontitis in individuals with the weak GCF trait in this study.

Another study promotes a therapy for periodontitis by inhibiting matrix collagenases, especially matrix metalloproteinases (MMP-8, MMP-9, MMP-12, and MMP-13), which are active beneath the gingival JE. Inhibitors of these enzymes are produced by M2 macrophages induced by PMNs, previously activated by *IL1B* (Sect. 23.3.3). Surprisingly, tetracycline antibiotics and their derivatives also inhibit these MMPS by mechanisms other than their antibiotic activity (Golub and Lee 2020; Golub et al. 1998). However, repeated use of tetracycline could lead to antibiotic resistance. A promising alternative is curcumin, the major curcuminoid of turmeric in spices, which has been used for centuries to treat chronic inflammatory diseases. A chemical modification of curcumin called CMC2.24 was evaluated in canine models over 3 months. Compared to placebo controls, CMC2.24 treatment reduced gingival inflammation, gingival index, GCF exudation, pocket depth (PD), and PD $\geq$4 mm, compared with placebo. CMC2.24 also reduced levels of active MMP-9 and MMP-2 in gingival tissue, in addition to reducing alveolar bone loss and IL-1β levels in GCF (Deng et al. 2020).

A subsequent study using rat macrophages showed that CMC2.24 is a potent inhibitor of the pro-inflammatory M1 phenotype, while promoting the pro-resolving M2 phenotype (Sect. 23.3.3). In other words, it acts as a critical "switch" to reduce chronic inflammation (Deng et al. 2023), activated genetically by weak GCF exudation (Table 23.5). If CMC2.24 or curcumin can be used to supplement standard periodontal therapy for a month, two or three times a year, it may help prevent weak responders from developing established periodontitis and its comorbidities.

Acknowledgments The authors sincerely thank Dr Anna Földes, Dept of Oral Biology, Semmelweis University, for providing and helping us edit the gene sequencing information reported in this study.

References

Aizenbud I, Wilensky A, Almoznino G (2023) Periodontal disease and its association with metabolic syndrome-a comprehensive review. Int J Mol Sci 24(16). https://doi.org/10.3390/ijms241613011

Apatzidou DA (2022) The role of cigarette smoking in periodontal disease and treatment outcomes of dental implant therapy. Periodontol 90(1):45–61. https://doi.org/10.1111/prd.12449

Bamashmous S, Kotsakis GA, Kerns KA, Leroux BG, Zenobia C, Chen D, Trivedi HM, McLean JS, Darveau RP (2021) Human variation in gingival inflammation. Proc Natl Acad Sci USA 118(27). https://doi.org/10.1073/pnas.2012578118

Barros SP, Williams R, Offenbacher S, Morelli T (2016) Gingival crevicular fluid as a source of biomarkers for periodontitis. Periodontol 70(1):53–64. https://doi.org/10.1111/prd.12107

Bendek MJ, Canedo-Marroquín G, Realini O, Retamal IN, Hernández M, Hoare A, Busso D, Monteiro LJ, Illanes SE, Chaparro A (2021) Periodontitis and gestational diabetes mellitus: a potential inflammatory vicious cycle. Int J Mol Sci 22(21). https://doi.org/10.3390/ijms222111831

Bosshardt DD (2018) The periodontal pocket: pathogenesis, histopathology and consequences. Periodontol 76(1):43–50. https://doi.org/10.1111/prd.12153

Brill N (1959) Removal of particles and bacteria from gingival pockets by tissue fluid. Acta Odontol Scand 17(4):431–440. https://doi.org/10.3109/00016355908993933

Cannon M, Khambe D, Klukowska M, Ramsey DL, Miner M, Huggins T, White DJ (2018) Clinical effects of stabilized stannous fluoride dentifrice in reducing plaque microbial virulence ii: Metabonomic changes. J Clin Dent 29(1):1–12

Caton JG, Armitage G, Berglundh T, Chapple ILC, Jepsen S, Kornman KS, Mealey BL, Papapanou PN, Sanz M, Tonetti MS (2018) A new classification scheme for periodontal and peri-implant diseases and conditions – introduction and key changes from the 1999 classification. J Periodontol 89(Suppl 1):S1–s8. https://doi.org/10.1002/jper.18-0157

Chapple IL, Van der Weijden F, Doerfer C, Herrera D, Shapira L, Polak D, Madianos P, Louropoulou A, Machtei E, Donos N, Greenwell H, Van Winkelhoff AJ, Eren KB, Arweiler N, Teughels W, Aimetti M, Molina A, Montero E, Graziani F (2015) Primary prevention of periodontitis: managing gingivitis. J Clin Periodontol 42(Suppl 16):S71–S76. https://doi.org/10.1111/jcpe.12366

Chmielewski M, Pilloni A (2023) Current molecular, cellular and genetic aspects of peri-implantitis disease: a narrative review. Dent J (Basel) 11(5). https://doi.org/10.3390/dj11050134

Conzone SD, Pantano CG (2004) Glass slides to DNA microarrays. Mater Today 7(3):20–26. https://doi.org/10.1016/S1369-7021(04)00122-1

Delgado E, Garcia-Godoy F, Montero-Aguilar M, Mateo LR, Ryan M (2018) A clinical investigation of a dual zinc plus arginine dentifrice in reducing established dental plaque and gingivitis over a six-month period of product use. J Clin Dent 29:A33–A40

Deng J, Golub LM, Lee HM, Lin MC, Bhatt HD, Hong HL, Johnson F, Scaduto J, Zimmerman T, Gu Y (2020) Chemically-modified curcumin 2.24: a novel systemic therapy for natural periodontitis in dogs. J Exp Pharmacol 12:47–60. https://doi.org/10.2147/jep.S236792

Deng J, Golub LM, Lee HM, Bhatt HD, Johnson F, Xu TM, Gu Y (2023) A novel modified-curcumin 2.24 resolves inflammation by promoting m2 macrophage polarization. Sci Rep 13(1):15513. https://doi.org/10.1038/s41598-023-42848-x

Deng T, Xu J, Wang Q, Wang X, Jiao Y, Cao X, Geng Q, Zhang M, Zhao L, Xiao C (2024) Immunomodulatory effects of curcumin on macrophage polarization in rheumatoid arthritis. Front Pharmacol 15:1369337. https://doi.org/10.3389/fphar.2024.1369337

Diaz PI, Hoare A, Hong BY (2016) Subgingival microbiome shifts and community dynamics in periodontal diseases. J Calif Dent Assoc 44(7):421–435

Dinarello CA (2018) Overview of the IL-1 family in innate inflammation and acquired immunity. Immunol Rev 281(1):8–27. https://doi.org/10.1111/imr.12621

Dutzan N, Abusleme L, Bridgeman H, Greenwell-Wild T, Zangerle-Murray T, Fife ME, Bouladoux N, Linley H, Brenchley L, Wemyss K, Calderon G, Hong BY, Break TJ, Bowdish DM, Lionakis MS, Jones SA, Trinchieri G, Diaz PI, Belkaid Y et al (2017) On-going mechanical damage from mastication drives homeostatic th17 cell responses at the oral barrier. Immunity 46(1):133–147. https://doi.org/10.1016/j.immuni.2016.12.010

Edmisson JS, Tian S, Armstrong CL, Vashishta A, Klaes CK, Miralda I, Jimenez-Flores E, Le J, Wang Q, Lamont RJ, Uriarte SM (2018) Filifactor alocis modulates human neutrophil antimicrobial functional responses. Cell Microbiol 20(6):e12829. https://doi.org/10.1111/cmi.12829

Eke PI, Dye BA, Wei L, Slade GD, Thornton-Evans GO, Borgnakke WS, Taylor GW, Page RC, Beck JD, Genco RJ (2015) Update on prevalence of periodontitis in adults in the United States: Nhanes 2009 to 2012. J Periodontol 86(5):611–622. https://doi.org/10.1902/jop.2015.140520

Evavold CL, Kagan JC (2018) How inflammasomes inform adaptive immunity. J Mol Biol 430(2):217–237. https://doi.org/10.1016/j.jmb.2017.09.019

Ferreira SB Jr, Trombone AP, Repeke CE, Cardoso CR, Martins W Jr, Santos CF, Trevilatto PC, Avila-Campos MJ, Campanelli AP, Silva JS, Garlet GP (2008) An interleukin-1beta (IL-1beta) single-nucleotide polymorphism at position 3954 and red complex periodontopathogens independently and additively modulate the levels of IL-1beta in diseased periodontal tissues. Infect Immun 76(8):3725–3734. https://doi.org/10.1128/iai.00546-08

Festjens N, Vanden Berghe T, Vandenabeele P (2006) Necrosis, a well-orchestrated form of cell demise: signalling cascades, important mediators and concomitant immune response. Biochim Biophys Acta 1757(9–10):1371–1387. https://doi.org/10.1016/j.bbabio.2006.06.014

Fu Y, Maaβ S, du Teil Espina M, Wolters AHG, Gong Y, de Jong A, Raangs E, Buist G, Westra J, Becher D, van Dijl JM (2022) Connections between exoproteome heterogeneity and virulence in the oral pathogen aggregatibacter actinomycetemcomitans. mSystems 7(3):e0025422. https://doi.org/10.1128/msystems.00254-22

Genco RJ, Borgnakke WS (2013) Risk factors for periodontal disease. Periodontol 62(1):59–94. https://doi.org/10.1111/j.1600-0757.2012.00457.x

Genco RJ, Borgnakke WS (2020) Diabetes as a potential risk for periodontitis: association studies. Periodontol 83(1):40–45. https://doi.org/10.1111/prd.12270

Genco RJ, Van Dyke TE (2010) Prevention: reducing the risk of cvd in patients with periodontitis. Nat Rev Cardiol 7(9):479–480. https://doi.org/10.1038/nrcardio.2010.120

Giannobile WV, Braun TM, Caplis AK, Doucette-Stamm-L, Duff GW, Kornman KS (2013) Patient stratification for preventive care in dentistry. J Dent Res 92(8):694–701. https://doi.org/10.1177/0022034513492336

Golub LM, Lee HM (2020) Periodontal therapeutics: current host-modulation agents and future directions. Periodontol 82(1):186–204. https://doi.org/10.1111/prd.12315

Golub LM, Lee HM, Ryan ME, Giannobile WV, Payne J, Sorsa T (1998) Tetracyclines inhibit connective tissue breakdown by multiple non-antimicrobial mechanisms. Adv Dent Res 12(2):12–26. https://doi.org/10.1177/08959374980120010501

Guo H, Chang S, Pi X, Hua F, Jiang H, Liu C, Du M (2021) The effect of periodontitis on dementia and cognitive impairment: a meta-analysis. Int J Environ Res Public Health 18(13). https://doi.org/10.3390/ijerph18136823

Hajishengallis G (2022) Interconnection of periodontal disease and comorbidities: evidence, mechanisms, and implications. Periodontol 89(1):9–18. https://doi.org/10.1111/prd.12430

He T, Mazor S, Zou Y, Grender J, Farrell S, Sagel P, Biesbrock AR, Timm H, Zini A (2021) Randomized clinical trial assessing anti-gingivitis efficacy of two stannous fluoride dentifrices and zinc/arginine dentifrice. Am J Dent 34(2):110–115

Holla LI, Hrdlickova B, Vokurka J, Fassmann A (2012) Matrix metalloproteinase 8 (MMP8) gene polymorphisms in chronic periodontitis. Arch Oral Biol, 57(2):188–196. https://doi.org/10.1016/j.archoralbio.2011.08.018

Hong BY, Furtado Araujo MV, Strausbaugh LD, Terzi E, Ioannidou E, Diaz PI (2015) Microbiome profiles in periodontitis in relation to host and disease characteristics. PLoS One 10(5):e0127077. https://doi.org/10.1371/journal.pone.0127077

Huang S, Li R, Zeng X, He T, Zhao H, Chang A, Bo C, Chen J, Yang F, Knight R, Liu J, Davis C, Xu J (2014) Predictive modeling of gingivitis severity and susceptibility via oral microbiota. ISME J 8(9):1768–1780. https://doi.org/10.1038/ismej.2014.32

Huang S, He T, Yue F, Xu X, Wang L, Zhu P, Teng F, Sun Z, Liu X, Jing G, Su X, Jin L, Liu J, Xu J (2021) Longitudinal multi-omics and microbiome meta-analysis identify an asymptomatic gingival state that links gingivitis, periodontitis, and aging. MBio 12(2). https://doi.org/10.1128/mBio.03281-20

Janakiram C, Dye BA (2020) A public health approach for prevention of periodontal disease. Periodontol 84(1):202–214. https://doi.org/10.1111/prd.12337

Jochum L, Stecher B (2020) Label or concept – what is a pathobiont? Trends Microbiol 28(10):789–792. https://doi.org/10.1016/j.tim.2020.04.011

Jorth P, Turner KH, Gumus P, Nizam N, Buduneli N, Whiteley M (2014) Metatranscriptomics of the human oral microbiome during health and disease. MBio 5(2)

Kim TS, Moutsopoulos NM (2024) Neutrophils and neutrophil extracellular traps in oral health and disease. Exp Mol Med 56(5):1055–1065. https://doi.org/10.1038/s12276-024-01219-w

Kornman KS, Crane A, Wang HY, di Giovine FS, Newman MG, Pirk FW, Wilson TG Jr, Higginbottom FL, Duff GW (1997) The interleukin-1 genotype as a severity factor in adult periodontal disease. J Clin Periodontol 24(1):72–77. https://doi.org/10.1111/j.1600-051x.1997.tb01187.x

Laine ML, Loos BG, Crielaard W (2010) Gene polymorphisms in chronic periodontitis. Int J Dent 2010:324719. https://doi.org/10.1155/2010/324719

Lazarus R, Vercelli D, Palmer LJ, Klimecki WJ, Silverman EK, Richter B, Riva A, Ramoni M, Martinez FD, Weiss ST, Kwiatkowski DJ (2002) Single nucleotide polymorphisms in innate immunity genes: abundant variation and potential role in complex human disease. Immunol Rev 190:9–25. https://doi.org/10.1034/j.1600-065x.2002.19002.x

Lertpimonchai A, Rattanasiri S, Arj-Ong Vallibhakara S, Attia J, Thakkinstian A (2017) The association between oral hygiene and periodontitis: a systematic review and meta-analysis. Int Dent J 67(6):332–343. https://doi.org/10.1111/idj.12317

Levine M (1985) The role for butyrate and propionate in mediating hela-cells growth inhibition by human dental plaque fluid from adult periodontal disease. Arch Oral Biol 30(2):155–159. https://doi.org/10.1016/0003-9969(85)90108-6

Levin M, Lohinai ZM (2018) Two host types with distinct susceptibility to human experimental gingivitis. Abstracts of EuroPerio9, Amsterdam

Levine M, Lohinai ZM (2021) Resolving the contradictory functions of lysine decarboxylase and butyrate in periodontal and intestinal diseases. J Clin Med 10(11). https://doi.org/10.3390/jcm10112360

Levine M, Progulske-Fox A, Denslow ND, Farmerie WG, Smith DM, Swearingen WT, Miller FC, Liang Z, Roe BA, Pan HQ (2001) Identification of lysine decarboxylase as a mammalian cell growth inhibitor in *eikenella corrodens*: possible role in periodontal disease. Microb Pathog 30(4):179–192. https://doi.org/10.1006/mpat.2000.0421

Levine M, Lohinai Z, Teles RP (2017) Low biofilm lysine content in refractory chronic periodontitis. J Periodontol 88(2):181–189. https://doi.org/10.1902/jop.2016.160302

Levine M, Collins LM, Lohinai Z (2021) Zinc chloride inhibits lysine decarboxylase production from *eikenella corrodens in vitro* and its therapeutic implications. J Dent 104:103533. https://doi.org/10.1016/j.jdent.2020.103533

Liccardo D, Cannavo A, Spagnuolo G, Ferrara N, Cittadini A, Rengo C, Rengo G (2019) Periodontal disease: a risk factor for diabetes and cardiovascular disease. Int J Mol Sci 20(6). https://doi.org/10.3390/ijms20061414

Liu CM, Hou LT, Wong MY, Rossomando EF (1996) Relationships between clinical parameters, interleukin 1b and histopathologic findings of gingival tissue in periodontitis patients. Cytokine 8(2):161–167. https://doi.org/10.1006/cyto.1996.0023

Lohinai Z, Keremi B, Szoko E, Tabi T, Szabo C, Tulassay Z, Levine M (2012) Bacterial lysine decarboxylase influences human dental biofilm lysine content, biofilm accumulation, and subclinical gingival inflammation. J Periodontol 83(8):1048–1056. https://doi.org/10.1902/jop.2011.110474

Lohinai Z, Keremi B, Szoko E, Tabi T, Szabo C, Tulassay Z, DiCesare JC, Davis CA, Collins LM, Levine M (2015) Biofilm lysine decarboxylase, a new therapeutic target for periodontal inflammation. J Periodontol 86(10):1176–1184. https://doi.org/10.1902/jop.2015.140490

Lohinai ZM, Ruksakiet K, Földes A, Dinya E, Levine M (2023) Genetic control of gcf exudation: innate immunity genes and periodontitis susceptibility. Int J Mol Sci 24(18). https://doi.org/10.3390/ijms241814249

Loos BG, Van Dyke TE (2020) The role of inflammation and genetics in periodontal disease. Periodontol 83(1): 26–39. https://doi.org/10.1111/prd.12297

Machado V, Botelho J, Escalda C, Hussain SB, Luthra S, Mascarenhas P, Orlandi M, Mendes JJ, D'Aiuto F (2021) Serum c-reactive protein and periodontitis: a systematic review and meta-analysis. Front Immunol 12:706432. https://doi.org/10.3389/fimmu.2021.706432

Mazurek-Mochol M, Dembowska E, Malinowski D, Safranow K, Pawlik A (2019) IL-1ß rs1143634 and rs16944 polymorphisms in patients with periodontal disease. Arch Oral Biol 98:47–51. https://doi.org/10.1016/j.archoralbio.2018.11.004

Michalowicz BS, Aeppli DP, Kuba RK, Bereuter JE, Conry JP, Segal NL, Bouchard TJ Jr, Pihlstrom BL (1991) A twin study of genetic variation in proportional radiographic alveolar bone height. J Dent Res 70(11):1431–1435. https://doi.org/10.1177/00220345910700110701

Michalowicz BS, Diehl SR, Gunsolley JC, Sparks BS, Brooks CN, Koertge TE, Califano JV, Burmeister JA, Schenkein HA (2000) Evidence of a substantial genetic basis for risk of adult periodontitis. J Periodontol 71(11):1699–1707. https://doi.org/10.1902/jop.2000.71.11.1699

Miton CM, Buda K, Tokuriki N (2021) Epistasis and intramolecular networks in protein evolution. Curr Opin Struct Biol 69:160–168. https://doi.org/10.1016/j.sbi.2021.04.007

Molnar E, Lohinai Z, Demeter A, Mikecs B, Toth Z, Vag J (2015) Assessment of heat provocation tests on the human gingiva: the effect of periodontal disease and smoking. Acta Physiol Hung 102(2):176–188. https://doi.org/10.1556/036.102.2015.2.8

Naginyte M, Do T, Meade J, Devine DA, Marsh PD (2019) Enrichment of periodontal pathogens from the biofilms of healthy adults. Sci Rep 9(1):5491. https://doi.org/10.1038/s41598-019-41882-y

Nara PL, Sindelar D, Penn MS, Potempa J, Griffin WST (2021) *Porphyromonas gingivalis* outer membrane vesicles as the major driver of and explanation for neuropathogenesis, the cholinergic hypothesis, iron dyshomeostasis, and salivary lactoferrin in alzheimer's disease. J Alzheimer's Dis 82(4):1417–1450. https://doi.org/10.3233/jad-210448

Nascimento GG, Baelum V, Sorsa T, Tervahartiala T, Skottrup PD, López R (2019a) Salivary levels of mpo, mmp-8 and timp-1 are associated with gingival inflammation response patterns during experimental gingivitis. Cytokine 115:135–141. https://doi.org/10.1016/j.cyto.2018.12.002

Nascimento GG, Danielsen B, Baelum V, Lopez R (2019b) Identification of inflammatory response patterns in experimental gingivitis studies. Eur J Oral Sci 127(1):33–39. https://doi.org/10.1111/eos.12588

Nibali L, Griffiths GS, Donos N, Parkar M, D'Aiuto F, Tonetti MS, Brett PM (2008) Association between interleukin-6 promoter haplotypes and aggressive periodontitis. J Clin Periodontol 35(3):193–198. https://doi.org/10.1111/j.1600-051X.2007.01188.x

Nibali L, Pelekos G, D'Aiuto F, Chaudhary N, Habeeb R, Ready D, Parkar M, Donos N (2013) Influence of IL-6 haplotypes on clinical and inflammatory response in aggressive periodontitis. Clin Oral Investig 17(4): 1235–1242. https://doi.org/10.1007/s00784-012-0804-3

Nicu EA, Laine ML, Morré SA, Van der Velden U, Loos BG (2009) Soluble CD14 in periodontitis. Innate Immunity 15(2):121–128. https://doi.org/10.1177/1753425908101577

Phillips PC (2008) Epistasis – the essential role of gene interactions in the structure and evolution of genetic systems. Nat Rev Genet 9(11):855–867. https://doi.org/10.1038/nrg2452

Prakash G, Uma M, Ajay S, Bali D, Upadhyay R, Gupta KK, Dixit J, Mittal B (2015) COX-2 gene polymorphisms and risk of chronic periodontitis: a case-control study and meta-analysis. Oral Diseases 21(1):38–45. https://doi.org/10.1111/odi.12203

Prame Kumar K, Nicholls AJ, Wong CHY (2018) Partners in crime: neutrophils and monocytes/macrophages in inflammation and disease. Cell Tissue Res 371(3): 551–565. https://doi.org/10.1007/s00441-017-2753-2

Preshaw PM, Alba AL, Herrera D, Jepsen S, Konstantinidis A, Makrilakis K, Taylor R (2012) Periodontitis and diabetes: a two-way relationship. Diabetologia 55(1):21–31. https://doi.org/10.1007/s00125-011-2342-y

Rao X, Zhong J, Sun Q (2014) The heterogenic properties of monocytes/macrophages and neutrophils in inflammatory response in diabetes. Life Sci 116(2):59–66. https://doi.org/10.1016/j.lfs.2014.09.015

Rider P, Voronov E, Dinarello CA, Apte RN, Cohen I (2017) Alarmins: feel the stress. J Immunol 198(4): 1395–1402. https://doi.org/10.4049/jimmunol.1601342

Rogus J, Beck JD, Offenbacher S, Huttner K, Iacoviello L, Latella MC, de Gaetano M, Wang HY, Kornman KS, Duff GW (2008) Il1b gene promoter haplotype pairs predict clinical levels of interleukin-1beta and c-reactive protein. Hum Genet 123(4):387–398. https://doi.org/10.1007/s00439-008-0488-6

Sandal I, Karydis A, Luo J, Prislovsky A, Whittington KB, Rosloniec EF, Dong C, Novack DV, Mydel P, Zheng SG, Radic MZ, Brand DD (2016) Bone loss and aggravated autoimmune arthritis in hla-drbeta1-bearing humanized mice following oral challenge with *porphyromonas gingivalis*. Arthritis Res Ther 18(1): 249. https://doi.org/10.1186/s13075-016-1143-6

Shungin D, Haworth S, Divaris K, Agler CS, Kamatani Y, Keun Lee M, Grinde K, Hindy G, Alaraudanjoki V, Pesonen P, Teumer A, Holtfreter B, Sakaue S, Hirata J, Yu YH, Ridker PM, Giulianini F, Chasman DI, Magnusson PK et al (2019) Genome-wide analysis of dental caries and periodontitis combining clinical and self-reported data. Nat Commun 10(1):2773. https://doi.org/10.1038/s41467-019-10630-1

Silbereisen A, Hallak AK, Nascimento GG, Sorsa T, Belibasakis GN, Lopez R, Bostanci N (2019) Regulation of PGLYRP1 and TREM-1 during progression and resolution of gingival inflammation. JDR Clin Trans Res 2380084419844937. https://doi.org/10.1177/2380084419844937

Socransky SS, Haffajee AD, Cugini MA, Smith C, Kent RL Jr (1998) Microbial complexes in subgingival plaque. J Clin Periodontol 25(2):134–144

Socransky SS, Haffajee AD, Smith C, Duff GW (2000) Microbiological parameters associated with IL-1 gene polymorphisms in periodontitis patients. J Clin Periodontol 27(11):810–818. https://doi.org/10.1034/j.1600-051x.2000.027011810.x

Socransky SS, Smith C, Haffajee AD (2002) Subgingival microbial profiles in refractory periodontal disease. J Clin Periodontol 29(3):260–268. https://doi.org/10.1034/j.1600-051x.2002.290313.x

Tabi T, Lohinai Z, Palfi M, Levine M, Szoko E (2008) Ce-lif determination of salivary cadaverine and lysine concentration ratio as an indicator of lysine decarboxylase enzyme activity. Anal Bioanal Chem 391. https://doi.org/10.1007/s00216-008-2026-8

Tan IK, Gajra B (2006) Plasma and urine amino acid profiles in a healthy adult population of Singapore. Ann Acad Med Singap 35(7):468–475

Tomar SL, Asma S (2000) Smoking-attributable periodontitis in the United States: findings from nhanes iii. National health and nutrition examination survey. J Periodontol 71(5):743–751. https://doi.org/10.1902/jop.2000.71.5.743

Tong H, Wei Z, Yin J, Zhang B, Zhang T, Deng C, Huang Y, Zhang N (2019) Genetic susceptibility of common polymorphisms in Nin and siglec5 to chronic periodontitis. Sci Rep 9(1):2088. https://doi.org/10.1038/s41598-019-38632-5

Trombelli L, Tatakis DN, Scapoli C, Bottega S, Orlandini E, Tosi M (2004) Modulation of clinical expression of plaque-induced gingivitis. Ii. Identification of "high-responder" and "low-responder" subjects. J Clin Periodontol 31(4): 239–252. https://doi.org/10.1111/j.1600-051x.2004.00478.x

Trombelli L, Scapoli C, Carrieri A, Giovannini G, Calura G, Farina R (2010) Interleukin-1 beta levels in gingival crevicular fluid and serum under naturally occurring and experimentally induced gingivitis. J Clin Periodontol 37(8):697–704

Tsuzukibashi O, Uchibori S, Kobayashi T, Umezawa K, Mashimo C, Nambu T, Saito M, Hashizume-Takizawa T, Ochiai T (2017) Isolation and identification methods of rothia species in oral cavities. J Microbiol Methods 134:21–26. https://doi.org/10.1016/j.mimet.2017.01.005

Van Wuyckhuyse BC, Perinpanayagam HE, Bevacqua D, Raubertas RF, Billings RJ, Bowen WH, Tabak LA (1995) Association of free arginine and lysine concentrations in human parotid saliva with caries experience. J Dent Res 74(2):686–690. https://doi.org/10.1177/00220345950740021001

Velsko IM, Fellows Yates JA, Aron F, Hagan RW, Frantz LAF, Loe L, Martinez JBR, Chaves E, Gosden C, Larson G, Warinner C (2019) Microbial differences between dental plaque and historic dental calculus are related to oral biofilm maturation stage. Microbiome 7(1):102. https://doi.org/10.1186/s40168-019-0717-3

Wong HC, Ooi Y, Pulikkotil SJ, Naing C (2018) The role of three interleukin 10 gene polymorphisms (- 1082 A G, - 819 C T, - 592 A C) in the risk of chronic and aggressive periodontitis: a meta-analysis and trial sequential analysis. BMC Oral Health 18(1):171. https://doi.org/10.1186/s12903-018-0637-9

Worthington HV, MacDonald L, Poklepovic Pericic T, Sambunjak D, Johnson TM, Imai P, Clarkson JE (2019) Home use of interdental cleaning devices, in addition to toothbrushing, for preventing and controlling periodontal diseases and dental caries. Cochrane Database Syst Rev 4(4):Cd012018. https://doi.org/10.1002/14651858.CD012018.pub2

Wu X, Offenbacher S, López NJ, Chen D, Wang HY, Rogus J, Zhou J, Beck J, Jiang S, Bao X, Wilkins L, Doucette-Stamm L, Kornman K (2015) Association of interleukin-1 gene variations with moderate to severe chronic periodontitis in multiple ethnicities. J Periodontal Res 50(1):52–61. https://doi.org/10.1111/jre.12181

Wynn TA, Chawla A, Pollard JW (2013) Macrophage biology in development, homeostasis and disease. Nature 496(7446):445–455. https://doi.org/10.1038/nature12034

Zhao G, He F, Wu C, Li P, Li N, Deng J, Zhu G, Ren W, Peng Y (2018) Betaine in inflammation: mechanistic aspects and applications. Front Immunol 9:1070. https://doi.org/10.3389/fimmu.2018.01070

Part IV

Bone Metabolism, Orthodontics, and Immunological Interactions

Oral and Craniofacial Development and Immunology

24

Mina Goudarzi, Seyed Mobin Tafreshi, Sara Zare, Heliya Ziaei, and Nima Rezaei

Abstract

This chapter explores the intricate interplay between immune regulation and craniofacial development, emphasizing the critical role of the immune system in both normal and pathological conditions. The process of craniofacial development, from the formation of pharyngeal arches and facial prominences to palate formation, is outlined, highlighting the susceptibility of these structures to maternal immune disruptions.

Key immune interactions at the maternal–fetal interface are discussed, focusing on immune tolerance mechanisms that safeguard the fetus from maternal immune rejection. Dysregulation of this immune balance, whether through maternal infections, inflammation, or nutritional deficiencies, can lead to significant craniofacial defects such as cleft lip and palate. The chapter delves into the teratogenic effects of infections, such as Campylobacter rectus, rubella, and cytomegalovirus (CMV), which disrupt gene expression and interfere with crucial developmental pathways.

The chapter further examines immune-mediated craniofacial disorders, including spondyloenchondrodysplasia (SPENCD), cherubism, autoimmune polyendocrinopathy-candidiasis-ectodermal dystrophy (APECED), Langerhans Cell Histiocytosis (LCH), and juvenile idiopathic arthritis (JIA), explaining how immune dysfunction leads to abnormal bone remodeling and developmental anomalies. Finally, the chapter addresses preventive strategies, emphasizing the importance of maternal health optimization, adequate nutritional support, and early management of infections to mitigate the impact of immune dysregulation on craniofacial development. This comprehensive exploration provides a foundation for understanding the critical role of the immune system in oral and craniofacial development within the broader context of oral immunity.

M. Goudarzi · S. M. Tafreshi
University of California Los Angeles School of Dentistry, Los Angeles, CA, USA

S. Zare
Research Development Center, Sina Hospital, Tehran University of Medical Sciences, Tehran, Iran

H. Ziaei
Universal Scientific Education and Research Network (USERN), Los Angeles, CA, USA

N. Rezaei (✉)
Universal Scientific Education and Research Network (USERN), Göteborg, Sweden

Network of Immunity in Infection, Malignancy and Autoimmunity (NIIMA), Universal Scientific Education and Research Network (USERN), Tehran, Iran

Research Center for Immunodeficiencies, Children's Medical Center, Tehran University of Medical Sciences, Tehran, Iran

Department of Immunology, School of Medicine, Tehran University of Medical Sciences, Tehran, Iran
e-mail: Rezaei_nima@tums.ac.ir

N. Rezaei, H. Ziaei (eds.), *Oral Immunology*, Advances in Experimental Medicine and Biology 1492,
https://doi.org/10.1007/978-3-032-03176-1_24

Keywords

Craniofacial development · Oral immunology · Maternal Immunity · Craniofacial defects · Oral immunity

24.1 Oral and Craniofacial Development

24.1.1 Craniofacial Development Overview

Craniofacial development begins during the fourth week of embryonic development with the formation of the pharyngeal apparatus, including pharyngeal arches, pouches, and clefts. This process continues through the tenth week, during which the face, neck, and associated structures are being formed. The pharyngeal clefts are the grooves invaginating inwards on the external surface and are lined with ectodermal tissue, while pharyngeal pouches are on the interior side of the apparatus and are made up of endodermal tissue. The pharyngeal arches consist of mesoderm, contributing to muscles and vasculature, and neural crest cells, forming connective tissue, cartilage, and other craniofacial components.

There are five pharyngeal arches, but the fifth arch is not numbered since it is underdeveloped. These arches develop into different craniofacial structures like muscles, bone, nerves, arteries, and cartilage.

The layers in each arch develop into their associated specific craniofacial structures, consisting of their related cartilages, cranial nerves, and aortic arch.

By the fifth week, neural crest cells (NCC) migrate and form facial prominences (FP), including the frontonasal prominence (FNP) and the maxillary (MxP) and mandibular prominences (MP), which originate from the first pharyngeal arch. These prominences merge to form the nose, upper lip, jaws, and related structures. The palate, which separates the nasal and oral cavities, consists of the primary palate (from the FNP) and the secondary palate (from the MxPs). The primary palate forms first, followed by the secondary palate, which develops as palatal shelves grow vertically beside the tongue, elevate, reorient horizontally, before fusing at the midline by the sixth week of embryonic development.

Facial and palate formation involves the merging of mesenchymal cells and the fusion of epithelial-covered structures. This process relies on cellular proliferation, matrix production, epithelial-to-mesenchymal transformation to enable cell migration, and apoptosis. The viscerocranium (facial skeleton) and neurocranium (cranial skeleton) develop from neural crest mesenchyme and paraxial mesoderm through ossification. Postnatally, craniofacial bones undergo significant growth to accommodate the developing face and brain, with sutures and fontanelles remaining unfused at birth, allowing for expansion and skull flexibility during childbirth. These structures eventually ossify as growth ceases. Disruptions in these complex processes of craniofacial development can lead to congenital defects like cleft lip and palate, and understanding these mechanisms is essential in managing associated risks and complications (Rinkoff 2023; Roth et al. 2021).

24.2 Introduction to Role of Immune System in Oral and Craniofacial Development

24.2.1 Maternal–Fetal Immune System Interactions

The interaction between the maternal immune system and fetus starts as early as the implantation stage (Johnson et al. 1999). Since the fetus inherits half of its genetic material from the paternal origin, fetal cells are recognized as foreign antigens by the maternal immune system. Thus, for a successful implementation and fetal development, the maternal immune system must differentiate between the fetal cells and its self-antigens, avoiding an immune response against the fetus (Rackaityte and Halkias 2020). This immunological tolerance is mediated by multiple regulatory mechanisms to establish a conducive immunological environment for the developing fetus. These encompass modified antigen presentation by placental cells, the differential of maternal T cells into regulatory ones (e.g., Treg), suppression of cytotoxic T cell activity, and proliferation of immunosuppressive regulatory cells, Moreover, maternal immune system is influenced hormonal alterations such as increased progesterone, and the secretion of immunomodulatory cytokines such as interlukin-10 (IL-10), and other factors (Than et al. 2019).

24.2.2 Regulations in the Maternal Immune System in Response to the Formation of Fetus

In the female reproductive tract, the maternal immune system uniquely maintains a remarkable balance and modulates its response to simultaneously protect the developing fetus while also defending against potential pathogens. While the female genital tract encounters various foreign materials, such as spermatozoa and bacteria, the human endometrium lacks a traditional mucosal immune system. Although IgA-synthesizing plasma cells are found in adjacent tissues like the endocervix, ectocervix, vagina, and fallopian tube, they are rarely found within the endometrium itself. Instead, the endometrium relies on IgA synthesized elsewhere in distant tissues, which is transported via circulation (Johnson et al. 1999; Kutteh and Mestecky 1994).

Despite the fetus being genetically different from the mother, it is not rejected due to highly evolved tolerance mechanisms (Billingham 1964; Medawar 1948). These mechanisms include maternal immune tolerance, physical barriers like the placenta, and modulation of fetal antigen presentations (Arcellana-Panlilio and Schultz 1994; Warner and Gollnick 1993).

Antigens and regulatory molecules from paternal seminal fluid are processed by maternal dendritic cells and then presented to T-cells, fostering immune tolerance (Moldenhauer et al. 2009). Additionally, the zona pellucida surrounding the blastocyst provides a temporary protective barrier against maternal immune cells until implantation occurs (Arcellana-Panlilio and Schultz 1994; Warner and Gollnick 1993).

For successful embryo implantation, maintaining a balanced response from the innate immune pathways of the uterus is essential (Raziel et al. 2007; Zhou et al. 2008). Disruption in this balance, such as infections in the endometrium, can lead to implantation failures or spontaneous abortions (Moreno et al. 2016).

Trophoblast cells, which originate from the blastocyst shortly after fertilization, play a critical role in an important process called trophoblast invasion. During placental development, trophoblasts invade the decidualized endometrium and spiral arteries, leading to extensive arterial wall remodeling and lumen widening to accommodate increased blood flow, necessary to sustain the developing fetus (Pijnenborg et al. 2006). Disturbances in this mechanism can lead to significant complications for both the mother and fetus. Insufficient invasion disrupts the maternal-fetal interface, potentially causing implantation failure, infertility, recurrent miscarriage, or intrauterine growth retardation (IUGR) (Ochoa-Bernal and Fazleabas 2020). In preeclampsia, trophoblast invasion is particularly shallow, contributing to this condition (Peng et al. 2008).

The decidualized endometrium (decidua) is essential for establishing the main components to help create immune tolerance at the maternal–fetal interface. One of the key factors in this tolerance is the presence of leukocytes, which are distributed within the epithelium and stroma, especially in proximity to glands. The proportion of these immune cells varies dynamically during the menstrual cycle and early pregnancy, with increases in late secretory phases and early gestation (Aplin 1996; Johnson et al. 1999). The leukocyte population in the decidua includes several specialized cell types with critical roles, such as decidual natural killers (dNK), which remodel spiral arteries and promote trophoblast invasion. Decidual macrophages are another type that differentiate into M1 (pro-inflammatory) and M2 (anti-inflammatory) macrophages. They facilitate artery remodeling, trophoblast invasion, angiogenesis, tissue remodeling, and phagocytosis of apoptotic trophoblasts. The last type is regulatory T-cells (Tregs), which ensure maternal tolerance to fetal antigens and reduce inflammation to support successful embryo implantation (Balasundaram and Farhana 2024).

24.3 Effects of Maternal Immune System Dysregulation on Fetal Craniofacial Development

24.3.1 Overview of Effects on Fetus

Inflammatory responses during pregnancy can lead to significant epigenetic changes within the developing fetus's genome (Pradhan et al. 2023). These changes, such as DNA methylation and histone modifications, can be triggered by maternal infections and have profound effects on the expression of genes that are critical for fetal development (Pacis et al. 2015, 2019). The consequences of these immune-mediated epigenetic alterations may be long-lasting, potentially resulting in structural abnormalities like cleft lip, cleft palate, and other craniofacial defects (Bhagirath et al. 2021). Inflammation resulting from infections can interfere with the normal gene expression essential for the correct formation of craniofacial features. An example of this is *Zika Virus* infection. Inflammation from the Zika virus interferes with craniofacial development through direct viral neurotoxicity, inflammation-induced apoptosis, and disruption of neural crest cell migration. This leads to microcephaly, craniofacial disproportions, ocular defects, and joint contractures. (Moore et al. 2017).

24.3.2 Maternal Nutritional Deficiencies and Craniofacial Defects

Furthermore, maternal nutritional deficiencies, particularly in essential nutrients such as Iron, Magnesium, Vitamin B6, and Vitamin B12, are strongly linked to an increased risk of Neural Tube Defects (NTDs) in the developing fetus (Felkner et al. 2007; Nashaat and Mansour 2014). Infections, such as *Helicobacter pylori*, can exacerbate these deficiencies by impairing nutrient absorption in the gastrointestinal tract. This compounding effect significantly increases the risk of developmental defects. The interplay between nutrient availability and infection underscores the importance of ensuring adequate maternal nutrition to mitigate these risks and support healthy fetal development (Felkner et al. 2007; Golalipour et al. 2012).

24.3.3 Dental Anomalies and Associated Infections

In addition to craniofacial defects and neural tube issues, maternal infections have also been associated with dental anomalies in children. These anomalies may include delayed tooth eruption and structural defects, which can significantly impact a child's oral health (Paulsson et al. 2019). For example, congenital syphilis is known to cause specific dental deformities, such as Hutchinson's teeth, characterized by notched, peg-shaped incisors that are a hallmark of this infection (Bhagirath et al. 2021; Hutchinson 1861). Other maternal infections that are associated with dental anomalies are rubella virus infection, affecting enamel mineralization and leading to hypoplastic enamel, and cytomegalovirus infection, which is linked to delayed tooth eruption (Bhatia et al. 2012; Jaskoll et al. 2008). Moreover, maternal periodontal infections are linked to several congenital tooth anomalies (Blumer et al. 2015).

24.4 Maternal Viral and Bacterial Infections Impacting Fetus Craniofacial Development

24.4.1 Rubella Infection and Its Effects

Rubella infection during pregnancy is particularly dangerous during the first trimester due to its high teratogenicity. Rubella infection can cause non-inflammatory necrosis in the chorionic epithelium and endothelial cells, allowing damaged cells to enter the fetal circulation and leading to ischemic lesions in vital organs like the brain and heart, potentially resulting in craniofacial malformations. (Bouthry et al. 2014) Additionally, the maternal immune response increases the production of interferons and cytokines in fetal cells, which can hinder the proper development and differentiation of fetal tissues. This combination of direct cellular damage and immune-mediated interference can lead to congenital defects, including those affecting the craniofacial region(Adamo et al. 2008). The virus can cross the placenta and infect fetal tissues, resulting in Congenital Rubella Syndrome (CRS) (Givens et al. 1993). Major criteria of CRS include cataracts, heart defects, sensorineural deafness, and minor criteria include craniofacial issues such as microcephaly. CRS is also associated with a set of distinctive facial characteristics known as rubella facies. These features include a triangular face in which the shape of the face resembles a triangle with the forehead as the base and the chin as the apex. The nose is prominent with an increased distance between the subnasale and the pronasale. The nasal bridge is sharp and aligns in a straight line with the tip of the nose, while the forehead appears wide or broad. A receding frontal hairline creates an expanded space above the eyebrows, and a circular pattern in the hair growth near the front hairline, known as Rubella Whorl, may also be observed. In Rubella Whorl, the hair spirals around a central point determined by the direction of the hair follicles (Begum 2020; Mawson and Croft 2019; Smith and Gong 1973).

In some cases, the rubella virus can persist in fetal tissues, causing chronic infections and further disrupting development, including craniofacial structures. This viral persistence is associated with immune system deficiencies, particularly in T-cell function, which may also increase the risk of developing autoimmune disorders later in life (Bouthry et al. 2014; Rabinowe et al. 1986).

24.4.2 Human Cytomegalovirus (CMV) Infection

Human cytomegalovirus (CMV) is one of the most common congenital infections worldwide, affecting 0.2–2% of all live births. The virus is transmitted to the fetus in about 40% of maternal infections, with 0.5–1% of children acquiring it in utero. Although 90% of infected infants are asymptomatic at birth, they are at risk for developing hearing, neurological, and developmental issues. Approximately 10% of affected infants exhibit symptoms at birth or during childhood, such as central nervous system anomalies and growth restrictions like intrauterine growth retardation, hydrops fetalis, and oligohydramnios. Early childhood CMV infection is also common, especially in group daycare settings, where the children shed the virus for extended periods. Despite its prevalence and potential severity, CMV often remains undetected due to its silent or asymptomatic nature (Divya et al. 2017; Weichert et al. 2010).

In a study by Divya et al., blood samples were collected from four groups of patients: children with cleft lip/palate, children with intellectual disabilities, children who were completely deaf, and normal, healthy children (control group). These samples were tested for HCMV-specific IgG and IgM antibodies using ELISA. The presence of HCMV Immunoglobulin G (IgG) antibodies in the blood of children with cleft lip and/or palate, mental retardation, and deafness indicated that these children have been exposed to HCMV in the past or may harbor a latent CMV infection. This exposure or latent infection may have contributed to the development of these conditions (Divya et al. 2017). Other studies also support that cleft lips and palates may be associated with congenital cytomegalovirus (CMV) infection. Laboratory research in mice has shown that CMV can inhibit chondrogenesis and osteogenesis during mandibular embryonic development (Slots and Slots 2024; Weichert et al. 2010). These findings suggest that CMV may disrupt normal craniofacial development by interfering with cartilage and bone formation during critical periods of embryogenesis.

24.5 Immune System Dysregulations and Craniofacial Disorders

24.5.1 Immune Dysregulation in Spondyloenchondrodysplasia (SPENCD) and Its Effects on Craniofacial Development

Spondyloenchondrodysplasia (SPENCD) is a rare genetic condition resulting from mutations in the ACP5 gene, which encodes tartrate-resistant acid phosphatase (TRAP) (Girschick et al. 2015). TRAP plays a crucial role in bone resorption and regulates immune responses. In SPENCD patients, the deficiency in TRAP activity leads to increased levels of osteopontin, a multifunctional glycoprotein implicated in bone metabolism and immune regulation. Elevated osteopontin in turn triggers the production of type I interferon, stimulating a cascade of immune dysregulation. The increase in type I interferon production and subsequent immune dysregulations, associated with SPENCD, results in interferonopathy, a group of disorders characterized by dysregulated immune responses. This interferonopathy negatively impacts the endochondral ossification process, which is vital for the development of long bones, as well as craniofacial structures. Since endochondral ossification is essential for the development of the cranial base and facial skeleton, any disruption can result in various craniofacial anomalies (An et al. 2017).

Patients with SPENCD often exhibit distinctive craniofacial anomalies due to issues with ossification. These include delayed spheno-occipital synchondrosis, which is a critical growth site in the cranial base, and its delayed fusion can disrupt cranial base development. Another craniofacial anomaly is abnormal cranial base angles, often correlating with midface hypoplasia and retrognathia. The delay in the closure of the spheno-occipital synchondrosis, together with an obtuse cranial base angle, can cause abnormal development of facial bones, leading to conditions like retrognathia (a backward-positioned lower jaw) and overdevelopment of the sphenoidal sinuses (Alfaqeeh et al. 2013).

Beyond skeletal dysplasia, the immune dysfunction seen in SPENCD extends to significant consequences on facial growth as well, directly impacting the proper development of craniofacial structures. Elevated type I interferon levels, resulting from TRAP deficiency, alter the balance of cytokine and growth factors critical for craniofacial development. The dysregulated immune system can directly impact osteoblast and chondrocyte differentiation, leading to improper growth of facial and cranial bones. This results in conditions such as malocclusion, delayed tooth eruption, and other dental and craniofacial anomalies (Alfaqeeh et al. 2013). These findings emphasized the crucial role the immune system plays in craniofacial development and highlighted how systemic immune disorders can manifest as skeletal dysplasia and facial abnormalities (Hong et al. 2018; Stoustrup et al. 2012). Early detection of these conditions allows for timely interventions such as orthodontic treatment, while research into the molecular pathways, particularly the roles of type I interferon and osteopontin, could lead to targeted therapies for treating the skeletal and immune abnormalities associated with SPENCD.

24.5.2 Cherubism: SH3BP2 Mutations and Immune System Disruption Affecting Jawbone Remodeling

Cherubism is a genetic autoinflammatory bone disorder that manifests during childhood, characterized by abnormal bone remodeling in the jaw, leading to distinctive facial swelling and deformities. This condition is primarily caused by mutations in the SH3BP2 gene, which disrupt normal bone development and immune system regulation. The disease features bilateral and symmetrical proliferative fibro-osseous lesions limited to the mandible and maxilla (Papadaki et al. 2012). Manifestations can range from no clinical symptoms to severe overgrowth of the jawbones, resulting in respiratory, vision, speech, and swallowing difficulties. The characteristic symmetrical jaw enlargement often leads to dental anomalies such as displaced, unerupted, malformed, or missing teeth, appearing to "float" within cyst-like spaces on imaging. Other dental complications include malocclusion, premature loss of deciduous teeth, and root resorption (Jones et al. 1952).

Mutations in the SH3BP2 gene, which encodes an adaptor protein involved in adaptive and innate immune responses and interacts with c-Abl, are identified as the genetic basis of cherubism (Bell et al. 1997; Ren et al. 1993). These mutations disrupt critical signaling pathways that regulate bone remodeling, leading to abnormal bone development in the mandible and maxilla. The SH3BP2 gene mutations enhance osteoclast progenitor cell activity by increasing pro-inflammatory signaling pathways, such as TNF-α and RANKL production, leading to rapid bone resorption, which is a hallmark of cherubism, particularly in the jawbones (Reichenberger et al. 2012). Therefore, the immune system's role in cherubism lies in its regulation of osteoclast genesis. This influence is particularly significant during key stages of dental development in children, where immune dysregulation can lead to the characteristic facial deformities seen in cherubism (Reichenberger et al. 2012; Ueki et al. 2001).

The clustering of specific amino-acid mutations in SH3BP2 likely results in a gain-of-function or dominant-negative effect, further disturbing normal pathways that regulate bone and immune system activities (Reichenberger et al. 2012). The onset of craniofacial abnormalities in cherubism may be linked to disruptions in SH3BP2-mediated signal transduction pathways, particularly during critical periods of dental development, such as the eruption of secondary teeth (Kannu and Bowdin 2024).

24.5.3 Autoimmune Polyendocrinopathy-Candidiasis-Ectodermal Dystrophy (APECED): Autoimmune Destruction Impacting Craniofacial Structures

Autoimmune polyendocrinopathy-candidiasis-ectodermal dystrophy (APECED), also known as autoimmune polyglandular syndrome type 1, is a rare autosomal recessive disorder that significantly impacts the immune system. Caused by mutations in the *AIRE* (Autoimmune Regulator) gene located on chromosome 21q22.3, this condition impairs the development of immune tolerance, leading the immune system to mistakenly attack the body's own organs and tissues. APECED commonly presents with chronic mucocutaneous candidiasis, hypoparathyroidism, and adrenal insufficiency (Addison's disease). Other autoimmune disorders associated with APECED can include hypothyroidism, hypogonadism, type 1 diabetes mellitus, chronic active hepatitis, pernicious anemia, vitiligo, alopecia, biliary cirrhosis, and ectodermal dysplasia (Buzi et al. 2003).

In individuals with APECED, the immune system can falsely attack and destroy tissues in endocrine glands such as the parathyroid and adrenal glands. This autoimmune destruction can also affect ectodermal tissues, including those involved in the development of craniofacial structures such as tooth enamel and jaw bones, leading to developmental defects in these tissues (Ahonen et al. 1990; Clemente et al. 1997).

Enamel hypoplasia is a developmental dental condition characterized by the incomplete or defective formation of tooth enamel. It occurs when ameloblasts—specialized epithelial cells responsible for enamel production—experience disrupted matrix formation during the formative stage of amelogenesis. This impairment can result in pitting, grooving, or even a complete absence of enamel on the tooth surface (Via Jr 1968). Enamel hypoplasia is commonly found in the permanent teeth of APECED patients, while enamel hypomaturation and changes in deciduous teeth are less frequently reported. Some patients show both enamel hypoplasia and hypomaturation in deciduous teeth before an APECED diagnosis, suggesting that enamel defects can occur early, even before other clinical signs. These early dental abnormalities vary among patients and could serve as risk indicators for APECED, although the reasons for the variation are not well understood (Lukinmaa et al. 1996; Myllärniemi and Perheentupa 1978).

Ameloblasts originate from epithelial tissue, similar to the endocrine gland cells commonly targeted in autoimmune diseases (Ahonen et al. 1987). In APECED, the immune system mistakenly attacks ameloblasts, and chronic immune dysregulation promotes inflammation and autoantibody production, further disrupting the proteins ameloblasts secrete into the extracellular matrix (ECM). This autoimmune response can reduce the amount of enamel matrix, causing hypoplasia and altering enamel composition. These changes result in defects in enamel patterning, mineralization, and maturation, further compromising enamel quality (Lukinmaa et al. 1996; Pavlič and Waltimo-Sirén 2009).

24.5.4 Langerhans Cell Histiocytosis (LCH): Immune Dysregulation and Craniofacial Bone Implications

Langerhans cell histiocytosis (LCH) is a rare yet the most prevalent histiocytic disorder characterized by the abnormal function, differentiation, or proliferation of Langerhans cells—specialized dendritic cells involved in immune responses within the mononuclear phagocyte system. The term "histiocyte," although archaic and meaning "tissue cell," refers to phagocytic cells with mononuclear morphology (Favara et al. 1997; van de Ven et al. 2011). In LCH, Langerhans cells grow uncontrollably and form granulomatous lesions composed of langerin-positive (CD207+) histiocytes, accompanied by an inflammatory infiltrate. These lesions can develop in virtually any organ system but have a particular tendency to affect bones, skin, lungs, and the pituitary gland. The disease spectrum ranges from a mild, self-resolving condition to a severe, multi-system illness that can be fatal. Underlying causes involve both immune system dysregulation and genetic mutations, contributing to its complex and varied clinical presentation (Badalian-Very et al. 2010).

In LCH, clonal Langerhans cells often carry mutations, most notably the BRAF V600E mutation. The BRAF V600E mutation activates the mitogen-activated protein kinase (MAPK) pathway, a signaling cascade that starts with ligand binding to a receptor, activating RAS, RAF, MEK, and ERK. Activated ERK translocates into the nucleus to initiate transcription factor activity and alterations in specific gene expressions, a process critical to the progression of LCH. As a result, this mutation disrupts normal cell signaling, leading to the pathological growth of these cells (Davies et al. 2002; Hingorani et al. 2003; Michaloglou et al. 2008). In addition to genetic factors, LCH is also associated with a significant inflammatory response and immune dysregulation, particularly the upregulation of inflammatory cytokines such as interleukin-17A (Coury et al. 2008).

In patients with multisystem LCH, bone involvement is common. Bone lesions are typically lytic with beveled edges, sometimes featuring a button sequestrum in radiological examination. These lesions may be asymptomatic or present as palpable masses. Such aggressive osteolytic activity compromises the structural integrity of affected bones, which, in turn, affects normal craniofacial growth and development. Bone lesions have a significant predilection for the skull, particularly the calvarium and often the parietal bone, the Sphenoid bone, and the petrous temporal bone. Bone lesions in the temporal bone can result in destructive mastoid lesions, where their proximity to critical anatomical structures can lead to severe complications, including nerve compression and significant neurological deficits, such as sixth nerve (abducens nerve) palsy.

Facial swelling, another common clinical feature in pediatric LCH patients, often presents as rapidly progressive and can be challenging to differentiate from other conditions like osteomyelitis due to overlapping symptoms. The uncontrolled proliferation of Langerhans cells in LCH results in a cascade of immune-driven effects that directly impact craniofacial development. The resulting lesions extend into neighboring anatomical areas, including the sinuses, tympanic cavity, and orbits, further compromising craniofacial growth and leading to facial deformities and functional impairments such as difficulty chewing, visual disturbances, and hearing loss.

Craniofacial frequently affects the maxillofacial region, with the mandible being the most common site. Mandibular lesions often lead to alveolar ridge resorption, creating the characteristic "floating teeth" appearance on imaging. In addition, involvement of the maxilla or maxillary sinus can also be seen, while the hard palate is affected less, often manifesting with gingival bleeding or swelling. Orbital involvement, once considered rare, is now being reported in more patients. Patients with orbital lesions typically present with proptosis and periorbital edema, and imaging reveals destructive lesions of the lateral orbital wall that can extend into adjacent structures and erode the sphenoid bone. Although intraocular and brain involvement are rare, these manifestations emphasize the disease's potential for widespread tissue infiltration (D'Ambrosio et al. 2008; Forlino et al. 2013).

Radiological assessment is crucial for diagnosing and managing LCH. Imaging techniques, including computed tomography (CT) and magnetic resonance imaging (MRI), are crucial for assessing the extent of bone involvement, characterizing the lesions, and detecting soft tissue extension. Radiologically, LCH lesions often appear as "punched out" areas with poorly defined and beveled edges, which are characteristic of aggressive bone destruction. MRI is particularly useful in delineating the soft tissue component of these lesions and assessing any extension into adjacent structures. In many cases, these lesions present as large masses that enhance with contrast and can compress or displace surrounding structures, such as the optic nerve, resulting in complications like proptosis and facial asymmetry (D'Ambrosio et al. 2008).

Ultimately, the immune dysfunction in LCH not only disrupts craniofacial bone formation and integrity but also has a broader systemic impact, affecting other organs and exacerbating the disease's impact on overall growth and development. Effective management of LCH requires early recognition, thorough radiological evaluation, and differentiation from other conditions such as osteomyelitis and tumors, which can present with similar symptoms. Early and accurate diagnosis allows for appropriate intervention, which is essential for minimizing the impact on craniofacial development, preventing severe complications, and improving the overall prognosis for affected pediatric patients. Given the complexity and variability of LCH presentations, a multidisciplinary approach involving pediatricians, radiologists, oncologists, and surgeons is often necessary to provide comprehensive and optimal care and outcomes (D'Ambrosio et al. 2008; Forlino et al. 2013).

24.5.5 Juvenile Idiopathic Arthritis (JIA): Chronic Inflammation and Craniofacial Deformities

Juvenile idiopathic arthritis (JIA) is the most prevalent type of arthritis affecting children, characterized by persistent joint inflammation beginning before age 16 and lasting for at least 6 weeks. As a heterogeneous condition with multiple subtypes, the exact etiology of JIA remains unclear; both genetic and environmental factors play a role in prolonged synovial inflammation and joint damage. Genetic elements, including human leukocyte antigen (HLA) haplotypes such as HLA-B27 and non-HLA genes, lead to dysregulation of immune responses and overproduction of proinflammatory cytokines, such as TNF-α, IL-1, IL-6, and IL-17. These cytokines mediate cartilage degradation, bone erosion, and joint deformities by promoting osteoclast activity and disrupting the balance of bone remodeling (Huang et al. 2024; La Bella et al. 2023).

One of the major concerns in JIA is the involvement of the temporomandibular joint (TMJ), with a 40–96% prevalence rate of TMJ arthritis in affected children, which can severely affect craniofacial development if not diagnosed early and treated properly. This arthritis can significantly impact both the structure and function of the TMJ, leading to alterations in dentofacial growth (Ringold and Cron 2009; Stoll et al. 2018; Stoll et al. 2012; Weiss et al. 2008). Chronic inflammation of the TMJ in children with JIA leads to condylar destruction, micrognathia, retrognathia, posterior mandibular rotation, reduced vertical growth, malocclusions, and open bite, especially in severe or early-onset TMJ arthritis. These alterations eventually disrupt normal facial growth, often causing a "bird-face" deformity characterized by a convex profile. These deformities affect both aesthetics and function, since resulting mandibular retrognathia and posterior rotation contribute to a receding chin and restricted airway, often causing children to extend their heads for better airflow. Reduced muscle function further limits mandibular growth, and dento-alveolar compensations, such as lower incisor proclination, may arise in response to these skeletal anomalies (Arvidsson et al. 2009; Bakke et al. 2001; Hsieh et al. 2016; Peltomäki et al. 2015; Savioli et al. 2004).

As mentioned earlier, immune dysregulation plays a critical role in temporomandibular joint disorder (TMD) in JIA patients. These immune dysregulations, such as the release of pro-inflammatory cytokines and autoantibodies such as rheumatoid factor (RF) and anti-cyclic citrullinated peptide (anti-CCP), are key markers of immune dysregulation. These autoantibodies themselves can contribute to inflammatory processes by binding to their target antigens, activating immune cells, and releasing pro-inflammatory cytokines (Zaripova et al. 2021). This leads to exacerbating inflammation and disrupting bone remodeling, potentially impacting craniofacial growth and development and structural changes in TMJ, leading to functional impairments.

JIA-related TMJ inflammation often begins asymptomatically, which can delay diagnosis and treatment, exacerbating craniofacial deformities in the long run. Traditional imaging techniques like radiography are insufficient for detecting early inflammation, only revealing damage after it has occurred. Although magnetic resonance imaging (MRI) is considered the gold standard for identifying active TMJ inflammation, it also has limitations, which can delay treatment and lead to significant joint damage and permanent craniofacial deformities (Abramowicz et al. 2011; Kellenberger et al. 2018; Koos et al. 2014; Müller et al. 2009).

24.6 Orofacial Clefts and Maternal Immune Dysregulation

24.6.1 Overview of Orofacial Clefts (OFC)

Orofacial clefts (OFC) are the most common craniofacial anomalies, encompassing cleft lip (CL), cleft palate (CP), and cleft lip with palate (CLP). These malformations occur when facial structures fail to fuse properly during early pregnancy, creating visible separations in the upper lip or palate. OFCs can be either non-syndromic or syndromic, with most cases being non-syndromic. The prevalence varies geographically, influenced by factors such as genetic predispositions, maternal smoking, alcohol use, folate deficiency, and health conditions during pregnancy. Understanding these risk factors is crucial for the prevention and management of OFCs.

Pierre Robin sequence (PRS) is a condition related to OFCs involving a wide U-shaped cleft palate, small chin (micrognathia), and glossoptosis, which obstructs the airway. The sequence results from inadequate space for the tongue to descend during early development, preventing proper fusion of palatal processes. Identifying these underlying causes helps guide treatment options and preventive measures for affected individuals, ultimately improving outcomes (Bhagirath et al. 2021; Kini 2023).

24.6.2 Maternal Infections and Their Association with Orofacial Clefts

As previously mentioned, maternal immune dysregulation during pregnancy, particularly due to viral infections like cytomegalovirus (CMV), rubella, and influenza, can significantly affect fetal development, leading to congenital anomalies such as orofacial clefts (Garcia et al. 1980; Griffith et al. 1972). These infections, especially when occurring between the 26th and 56th days of gestation, can disrupt facial development, acting as potent teratogens and contributing to the formation of orofacial clefts through a combination of genetic and environmental factors (Molnárová et al. 2018).

Several studies established a link between maternal rubella infection and craniofacial defects like OFC in the neonatal stages of development. Evidence indicates that rubella symptoms may arise from liver dysfunction affecting vitamin A metabolism, resulting in hypervitaminosis A. Excess retinoids alter the signaling pathways and gene expression patterns that could lead to congenital rubella syndrome (CRS), and they could change the normal formation of facial features, including orofacial clefts (Bhagirath et al. 2021; Mawson and Croft 2019).

Maternal exposure to acute infections like influenza and chronic infections such as syphilis was linked to increased risk of cleft lip in newborns, possibly due to vascular disruptions caused by hyperthermia or salicylate use during embryonic development (Molnárová et al. 2018). Influenza during the second month of pregnancy increases OFC risk, while occurring in the third month may have protective effects (Dymanus et al. 2020). Similar associations were observed between influenza and congenital anomalies, including neural tube defects, OFC, and congenital heart defects (Mátrai et al. 2023). Maternal fever or common cold during pregnancy may also increase the risk of cleft lip and palate in offspring (Shi et al. 2023). Moreover, chronic infections like CMV and Epstein-Barr virus (EBV) can provoke sustained inflammation in the mother. This ongoing immune response can release cytokines and other inflammatory mediators, which can cross the placenta and interfere with the normal development of the fetus's craniofacial structures. This disruption can inhibit the cellular processes essential for the proper formation of these structures, thereby increasing the predisposition of orofacial clefts (Molnárová et al. 2018).

Maternal genitourinary infections, particularly those involving Chlamydia, have been linked to an increased risk of cleft lip in offspring, with the risk being higher when Chlamydia exposure is reported (Bhagirath et al. 2021). Chlamydia trachomatis can cause a persistent fallopian tube infection, leading to infertility through the production of heat shock protein (hsp60), which may induce autoimmunity. The inflammatory hsp60 protein can activate immune responses during early pregnancy, potentially causing embryo rejection. The production of anti-HSP60 antibodies that interfere with fetal tissue formation, maternal inflammation, and cellular stress affecting the developing embryo, and potential infection of placental tissues by *Chlamydia trachomatis*, allowing the pathogen to impact craniofacial development (Molnárová et al. 2006; Witkin et al. 1997).

Treponema pallidum, the bacterium that causes syphilis, can significantly disrupt the normal development and fusion of facial structures in a fetus, leading to congenital anomalies such as cleft lip or cleft palate due to improper tissue formation. This disruption is likely linked to an inflammatory response triggered by the infection (Bhagirath et al. 2021). Congenital syphilis can be transmitted at any stage of pregnancy, but is more likely to occur after the first trimester when Treponema pallidum crosses the placenta. Transmission is facilitated by placental inflammation rather than Langerhans cells. Severe infections may lead to miscarriage, stillbirth, or perinatal death, while milder cases can present as early or late congenital syphilis (Soares et al. 2004). Early signs of congenital syphilis include a mucocutaneous rash, osteochondritis, and saddle nose (Monhian et al. 2001; Siegel 1996). Late congenital syphilis is characterized by Hutchinson's triad, which includes interstitial keratitis, eighth nerve deafness, and dental abnormalities like Hutchinson's incisors and Mulberry molars. Other oral symptoms can include atrophic glossitis and cleft palate (Fiumara et al. 1978; Monhian et al. 2001; Siegel 1996).

Studies have suggested that toxoplasmosis may also be a causative factor in the development of some orofacial clefts, as mothers and children affected by these conditions had higher toxoplasma antibody levels compared to healthy controls (Gu et al. 1992; Holková et al. 1994). It has been reported that antigens related to Toxoplasma were detected in the congenital cleft lip and palate using RCEP and COA tests, indicating the presence of the infection. Additionally, antibodies against Toxoplasma were found in the mothers' serum using IHA, IFA, and RIPEGA tests. Researchers suggested that Toxoplasma infection in the mothers may have contributed to the congenital defects observed in their offspring. Moreover, a rare combination of anomalies, including anophthalmia and craniofacial clefts in newborns, has reportedly been linked to maternal toxoplasmosis (Arce-Estrada et al. 2017).

24.6.3 Immune-Related Genetic Factors Contributing to Orofacial Clefts

The *PVRL1* gene influences the immune system by interacting with the TIGIT pathway, which plays a role in immune regulation. In hepatocellular carcinoma (HCC), *PVRL1* is overexpressed, leading to the stabilization of the poliovirus receptor (PVR) and suppression of the cytotoxic T-cell response. This interaction inhibits T-cell proliferation and function, contributing to immune evasion by the tumor (Peng and Fu 2020).

The PVRL1 gene, part of the immunoglobulin superfamily, is essential for maintaining epithelial cell junctions, and its mutations are linked to orofacial clefts, particularly the cleft lip and palate/ectodermal dysplasia 1 syndrome (CLPED1). Studies have identified various PVRL1 variants, including both rare and common mutations, associated with non-syndromic cleft lip with or without cleft palate (NSCL/P). Some variants, such as S112T, T131A, and S447L, have shown associations with increased cleft risk, potentially affecting cell adhesion and embryonic facial development. The findings indicate that PVRL1 may play a minor role in NSCL/P susceptibility, but its impact appears to vary among populations. Larger studies are needed to further explore the role of these genetic variants, including possible maternal transmission and epigenetic effects (Park et al. 2024; Sözen et al. 2009; Turhani et al. 2005).

24.7 Preventive Strategies for Immune Dysregulation and Craniofacial Issues

To mitigate the risk of immune system dysregulation leading to craniofacial developmental defects, a multifaceted approach that includes preconception care, proper nutrition, infection management, immune modulation, and public health interventions is essential. Here is an expanded and rephrased discussion of these strategies:

24.7.1 Optimizing Maternal Health Before Pregnancy

Ensuring that women are in optimal health before conception is crucial for preventing complications during pregnancy that could affect fetal development. Preconception care should include comprehensive health screenings to identify and address any existing infections or chronic conditions, such as autoimmune disorders, that might contribute to immune dysregulation. Vaccinations, especially against Rubella, should be up to date to protect both the mother and the

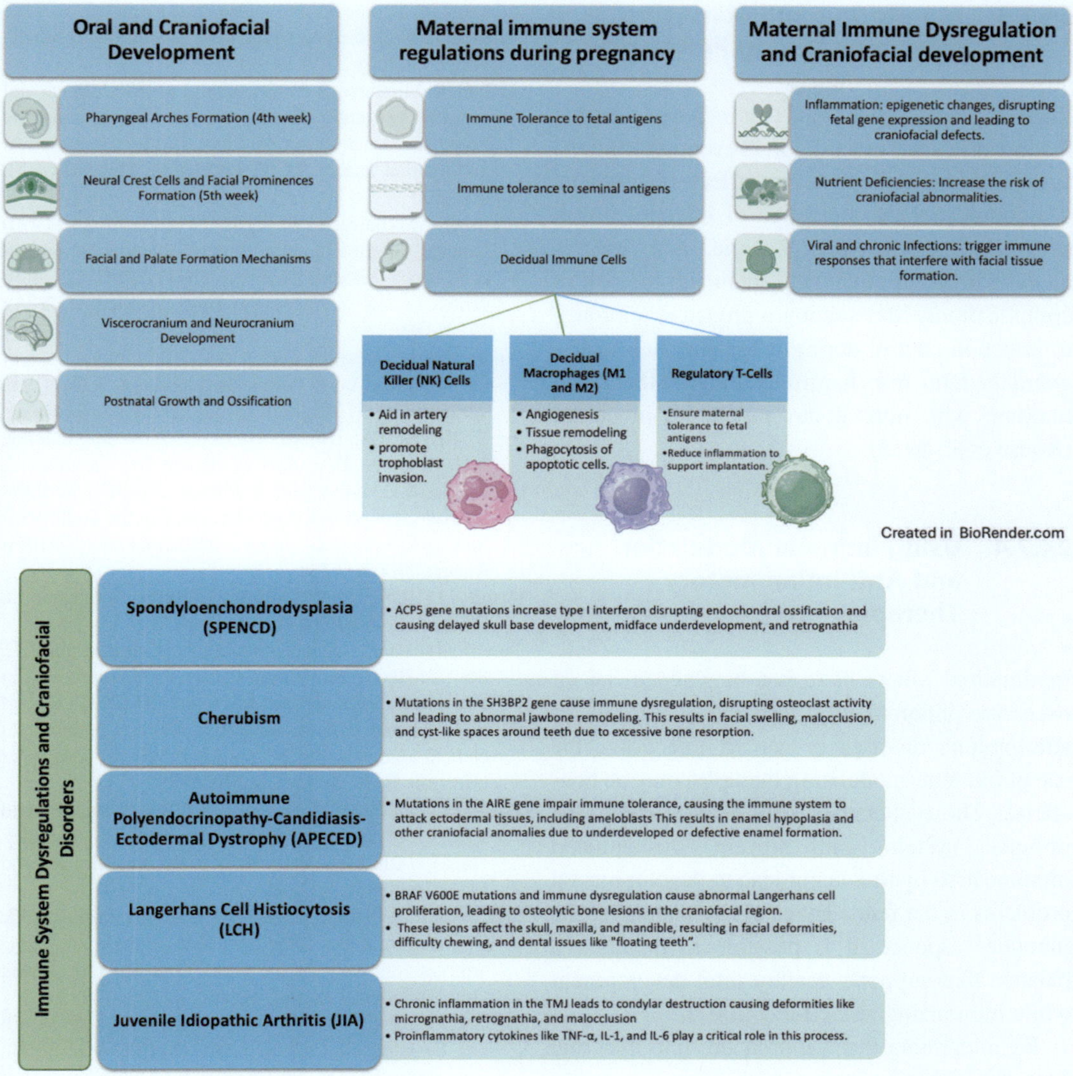

Fig. 24.1 Chart summarizes oral and craniofacial developmental stages, maternal immune system regulation during pregnancy, the potential impacts of maternal immune dysregulation on craniofacial development, and immune system dysregulations in infants, children, and juveniles with potential effects on craniofacial disorders

fetus from preventable infections that could disrupt normal development (Bernstein et al. 2023; Liu 2021).

24.7.2 Ensuring Adequate Nutritional Support During Pregnancy

Proper nutrition plays a vital role in supporting both the mother's immune system and the healthy development of the fetus. Nutritional deficiencies, particularly in essential vitamins and minerals like iron, magnesium, vitamin B6, and vitamin B12, have been linked to an increased risk of neural tube defects and other developmental abnormalities. Providing pregnant women with the necessary dietary supplements and encouraging a balanced diet can help mitigate these risks and support overall fetal development (Black et al. 2013).

24.7.3 Monitoring and Managing Infections During Pregnancy

Regular prenatal check-ups that include screening for infections are essential in preventing immune system disruptions that can affect craniofacial development. Early detection of infections such as *Helicobacter pylori*, Rubella, and syphilis allows for prompt treatment, reducing the risk of complications. Maintaining a proactive approach to infection control during pregnancy helps safeguard the fetus from harmful pathogens that could interfere with normal developmental processes (Rouse et al. 1995).

24.7.4 Using Immune Modulation and Anti-Inflammatory Therapies

In situations where there is a heightened risk of excessive inflammatory responses, such as in autoimmune diseases or certain infections, the use of immune-modulating therapies may be beneficial. These therapies can help regulate the immune system and prevent dysregulated immune activity that might lead to developmental problems in the fetus. By carefully managing the immune response, it is possible to maintain a balance that supports healthy fetal development while minimizing risks (Levy 2007).

By integrating these approaches into maternal care, the risks associated with immune system dysregulation during pregnancy can be significantly reduced, leading to better health outcomes for both mothers and their children. Regular monitoring, proper nutrition, and timely treatment of infections all contribute to preventing craniofacial developmental problems linked to immune system dysregulation (Fig. 24.1 summarizes findings of this review).

References

Abramowicz S, Cheon JE, Kim S, Bacic J, Lee EY (2011) Magnetic resonance imaging of temporomandibular joints in children with arthritis. J Oral Maxillofac Surg 69(9):2321–2328

Adamo MP, Zapata M, Frey TK (2008) Analysis of gene expression in fetal and adult cells infected with rubella virus. Virology 370(1):1–11

Ahonen P, Miettinen A, Perheentupa J (1987) Adrenal and steroidal cell antibodies in patients with autoimmune Polyglandular disease type I and risk of adrenocortical and ovarian failure. J Clin Endocrinol Metabol 64(3): 494–500

Ahonen P, Myllärniemi S, Sipilä I, Perheentupa J (1990) Clinical variation of autoimmune polyendocrinopathy–candidiasis–ectodermal dystrophy (APECED) in a series of 68 patients. N Engl J Med 322(26): 1829–1836

Alfaqeeh SA, Gaete M, Tucker AS (2013) Interactions of the tooth and bone during development. J Dent Res 92(12):1129–1135

An J, Briggs TA, Dumax-Vorzet A, Alarcón-Riquelme ME, Belot A, Beresford M, Bruce IN, Carvalho C, Chaperot L, Frostegård J, Plumas J (2017) Tartrate-resistant acid phosphatase deficiency in the predisposition to systemic lupus erythematosus. Arthrit Rheumatol 69(1):131–142

Aplin JD (1996) The cell biology of human implantation. Placenta 17(5–6):269–275

Arce-Estrada GE, Gómez-Toscano V, Cedillo-Peláez C, Sesman-Bernal AL, Bosch-Canto V, Mayorga-Butrón JL, Vargas-Villavicencio JA, Correa D (2017) Report of an unsual case of anophthalmia and craniofacial cleft in a newborn with toxoplasma gondii congenital infection. BMC Infect Dis 17(1):459

Arcellana-Panlilio MY, Schultz GA (1994) Temporal and spatial expression of major histocompatibility complex class I H-2K in the early mouse embryo. Biol Reprod 51(2):169–183

Arvidsson LZ, Flatø B, Larheim TA (2009) Radiographic TMJ abnormalities in patients with juvenile idiopathic arthritis followed for 27 years. Oral Surg Oral Med Oral Pathol Oral Radiol Endod 108(1):114–123

Badalian-Very G, Vergilio J-A, Degar BA, MacConaill LE, Brandner B, Calicchio ML, Kuo FC et al (2010) Recurrent BRAF mutations in Langerhans cell histiocytosis. Blood J Am Soc Hematol 116(11): 1919–1923

Bakke M, Zak M, Jensen BL, Pedersen FK, Kreiborg S (2001) Orofacial pain, jaw function, and temporomandibular disorders in women with a history of juvenile chronic arthritis or persistent juvenile chronic arthritis. Oral Surg Oral Med Oral Pathol Oral Radiol Endod 92(4):406–414

Balasundaram P, Farhana A (2024) Immunology at the maternal-fetal interface. In: StatPearls. StatPearls Publishing Copyright © 2024, StatPearls Publishing LLC

Begum NNF (2020) Novel facial characteristics in congenital rubella syndrome: a study of 115 cases in a cardiac hospital of Bangladesh. BMJ Paediatrics Open 4(1):e000860

Bell SM, Shaw M, Jou YS, Myers RM, Knowles MA (1997) Identification and characterization of the human homologue of SH3BP2, an SH3 binding domain protein within a common region of deletion

at 4p16.3 involved in bladder cancer. Genomics 44(2): 163–170
Bernstein M, Afshar Y, Han CS (2023) Preconception care: in the continuum of women's healthcare. In: Shoupe D (ed) Handbook of gynecology. Springer International Publishing, pp 39–61
Bhagirath AY, Medapati MR, de Jesus VC, Yadav S, Hinton M, Dakshinamurti S, Atukorallaya D (2021) Role of maternal infections and inflammatory responses on craniofacial development. Front Oral Health 2:735634
Bhatia S, Goyal A, Dubey M, Kapur A, Ritwik P (2012) Congenital rubella syndrome: dental manifestations and management in a 5 year old child. J Clin Pediatr Dent 37(1):71–75
Billingham RE (1964) Transplantation immunity and the maternal-fetal relationship. Trans Stud Coll Physicians Phila 31:187–203
Black RE, Victora CG, Walker SP, Bhutta ZA, Christian P, De Onis M, Ezzati M, Grantham-McGregor S, Katz J, Martorell R, Uauy R (2013) Maternal and child undernutrition and overweight in low-income and middle-income countries. Lancet 382(9890):427–451
Blumer S, Peretz B, Costa L (2015) Periodontal health during pregnancy and the dental health of the child. Refu'at Ha-peh Veha-shinayim 32(2):28–31
Bouthry E, Picone O, Hamdi G, Grangeot-Keros L, Ayoubi JM, Vauloup-Fellous C (2014) Rubella and pregnancy: diagnosis, management and outcomes. Prenat Diagn 34(13):1246–1253
Buzi F, Badolato R, Mazza C, Giliani S, Notarangelo LD, Radetti G, Plebani A, Notarangelo LD (2003) Autoimmune Polyendocrinopathy-candidiasis-ectodermal dystrophy syndrome: time to review diagnostic criteria? J Clin Endocrinol Metabol 88(7):3146–3148
Clemente MG, Obermayer-Straub P, Meloni A, Strassburg CP, Arangino V, Tukey RH, De Virgiliis S, Manns MP (1997) Cytochrome P450 1A2 is a hepatic autoantigen in autoimmune polyglandular syndrome type 1. J Clin Endocrinol Metabol 82(5):1353–1361
Coury F, Annels N, Rivollier A, Olsson S, Santoro A, Speziani C, Azocar O, Flacher M, Djebali S, Tebib J, Brytting M (2008) Langerhans cell histiocytosis reveals a new IL-17A–dependent pathway of dendritic cell fusion. Nat Med 14(1):81–87
D'Ambrosio N, Soohoo S, Warshall C, Johnson A, Karimi S (2008) Craniofacial and intracranial manifestations of Langerhans cell histiocytosis: report of findings in 100 patients. Am J Roentgenol 191(2):589–597
Davies H, Bignell GR, Cox C, Stephens P, Edkins S, Clegg S, Teague J et al (2002) Mutations of the BRAF gene in human cancer. Nature 417(6892): 949–954
Divya DV, Prasad MGS, Radhakrishna AN, Reddy SP, Pratyusha K, Kumar KS, Sandeep RV (2017) The serological evidence of cytomegalovirus infection as a potent aetiological factor for cleft lip/palate, mental retardation and deafness. J Clin Diagn Res 11(6):ZC51
Dymanus K, Chishom T, Moraczewski J, et al (2020) The association between influenza infection rates and the incidence of orofacial clefts in the United States. FACE 2(1):23–29. https://doi.org/10.1177/2732501620973033
Favara BE, Feller AC, Pauli M, Jaffe ES, Weiss LM, Arico M, Bucsky P, Egeler RM, Elinder G, Gadner H, Gresik M (1997) Contemporary classification of histiocytic disorders. Med Pediatr Oncol 29(3): 157–166
Felkner M, Suarez L, Liszka B, Brender JD, Canfield M (2007) Neural tube defects, micronutrient deficiencies, and helicobacter pylori: a new hypothesis. Birth Defects Res A Clin Mol Teratol 79(8):617–621
Fiumara NJ, Grande DJ, Giunta JL (1978) Papular secondary syphilis of the tongue: report of a case. Oral Surg Oral Med Oral Pathol 45(4):540–542
Forlino D, Manzone P, Gomel MC, Nicoli MB, Pedrini C (2013) Craniofacial and skull base findings in Langerhans cell histiocytosis in pediatric patients. Pediatrics 77(1):1–9
Garcia AG, Pereira JMS, Vidigal N, Lobato YY, Pegado CS, Branco JPC (1980) Intrauterine infection with mumps virus. Obstet Gynecol 56(6):756–758
Girschick H, Wolf C, Morbach H, Hertzberg C, Lee-Kirsch MA (2015) Severe immune dysregulation with neurological impairment and minor bone changes in a child with spondyloenchondrodysplasia due to two novel mutations in the ACP5 gene. Pediatr Rheumatol 13(1):37
Givens KT, Lee DA, Jones T, Ilstrup DM (1993) Congenital rubella syndrome: ophthalmic manifestations and associated systemic disorders. Br J Ophthalmol 77(6): 358–363
Golalipour MJ, Sedehi M, Qorbani M (2012) Does maternal helicobacter pylori infection increase the risk of occurrence of neural tube defects in newborns in northern Iran? Neurosci J 17(3):219–225
Griffith GW, Adelstein AM, Lambert PM, Weatherall JAC (1972) Influenza and infant mortality. Br Med J 3(5826):553–556
Gu SZ, Lu S, Ma YW (1992) Study on the relationship between congenital toxoplasmosis and monsters of cleft lip and palate accompanied by multiple malformation. Shanghai J Stomatol 1(2):66–69
Hingorani SR, Jacobetz MA, Robertson GP, Herlyn M, Tuveson DA (2003) Suppression of BRAFV599E in human melanoma abrogates transformation. Cancer Res 63(17):5198–5202
Holková R, Molnárová A, Nemec R, Brozman M, Fábry J, Horáková E (1994) Relation between toxoplasmosis and orofacial clefts in children. Bratislavske Lekarske Listy 95(2):64–67
Hong SW, Huh KH, Lee JK, Kang JH (2018) Craniofacial anomalies associated with spondyloenchondrodysplasia: two case reports. Medicine (Baltimore) 97(50):e13644
Hsieh YJ, Darvann TA, Hermann NV, Larsen P, Liao YF, Bjoern-Joergensen J, Kreiborg S (2016) Facial

morphology in children and adolescents with juvenile idiopathic arthritis and moderate to severe temporomandibular joint involvement. Am J Orthod Dentofacial Orthop 149(2):182–191

Huang HYR, Wireko AA, Miteu GD, Khan A, Roy S, Ferreira T, Garg T, Aji N, Haroon F, Zakariya F, Alshareefy Y (2024) Advancements and progress in juvenile idiopathic arthritis: a review of pathophysiology and treatment. Medicine 103(13):e37567

Hutchinson J (1861) Clinical lecture on heredito-syphilitic struma: and on the teeth as a means of diagnosis. Br Med J 1(20):515

Jaskoll T, Abichaker G, Jangaard N, Bringas P Jr, Melnick M (2008) Cytomegalovirus inhibition of embryonic mouse tooth development: a model of the human amelogenesis imperfecta phenocopy. Arch Oral Biol 53(5):405–415

Johnson PM, Christmas SE, Vince GS (1999) Immunological aspects of implantation and implantation failure. Hum Reprod 14(Suppl 2):26–36

Jones WA, Gerrie J, Pritchard J (1952) Cherubism – a familial fibrous dysplasia of the jaws. Oral Surg Oral Med Oral Pathol 5(3):292–305

Kannu P, B. B. Bowdin (1993–2024) GeneReviews® [Internet]. University of Washington, Seattle, Seattle (WA)

Kellenberger CJ, Abramowicz S, Arvidsson LZ, Kirkhus E, Tzaribachev N, Larheim TA (2018) Recommendations for a standard magnetic resonance imaging protocol of temporomandibular joints in juvenile idiopathic arthritis. J Oral Maxillofac Surg 76(12): 2463–2465

Kini U (2023) Genetics and orofacial clefts: a clinical perspective. Br Dent J 234(12):947–952

Koos B, Twilt M, Kyank U, Fischer-Brandies H, Gassling V, Tzaribachev N (2014) Reliability of clinical symptoms in diagnosing temporomandibular joint arthritis in juvenile idiopathic arthritis. J Rheumatol 41(9):1871–1877

Kutteh WH, Mestecky J (1994) Secretory immunity in the female reproductive tract. Am J Reprod Immunol 31(1):40–46

La Bella S, Rinaldi M, Di Ludovico A, Di Donato G, Di Donato G, Salpietro V, Chiarelli F, Breda L (2023) Genetic background and molecular mechanisms of juvenile idiopathic arthritis. Int J Mol Sci 24(3):1846

Levy O (2007) Innate immunity of the newborn: basic mechanisms and clinical correlates. Nat Rev Immunol 7(5):379–390

Liu JM (2021) Emphasize the significance of preconception health in achieving favorable pregnancy outcomes, highlighting the need for early health interventions to prevent complications. Matern Child Health J:515–521

Lukinmaa P-L, Waltimo J, Pirinen S (1996) Microanatomy of the dental enamel in autoimmune polyendocrinopathy-candidiasis-ectodermal dystrophy (APECED): report of three cases. J Craniofac Genet Dev Biol 16(3):174–181

Mátrai Á, Teutsch B, Pethő B, Kaposi AD, Hegyi P, Ács N (2023) Reducing the risk of birth defects associated with maternal influenza: insights from a Hungarian case – control study. J Clin Med 12(21):6934

Mawson AR, Croft AM (2019) Rubella virus infection, the congenital rubella syndrome, and the link to autism. Int J Environ Res Public Health 16(19):3543

Medawar P (1948) Immunity to homologous grafted skin. III. The fate of skin homographs transplanted to the brain, to subcutaneous tissue, and to the anterior chamber of the eye. Br J Exp Pathol 29(1):58

Michaloglou C, Vredeveld LCW, Mooi WJ, Peeper DS (2008) BRAFE600 in benign and malignant human tumours. Oncogene 27(7):877–895

Moldenhauer LM, Diener KR, Thring DM, Brown MP, Hayball JD, Robertson SA (2009) Cross-presentation of male seminal fluid antigens elicits T cell activation to initiate the female immune response to pregnancy. J Immunol 182(12):8080–8093

Molnárová A, Kováčová E, Majtán J, Fedeleš J, Bieliková E, Cvachová S, Vojtaššák J, Repiská V (2006) Chlamydia and mycoplasma infections during pregnancy and their relationships to orofacial cleft. Biologia 61(6):719–723

Molnárová A, Palenčár D, Fekiačová D et al (2018) Orofacial clefts and infections during pregnancy. Biologia 73:629–635

Monhian N, Murr AH, Mandpe AH, Baehner FL (2001) Pathology forum. Quiz case 1. Secondary syphilis. Arch Otolaryngol Head Neck Surg 127(2):217

Moore CA, Staples JE, Dobyns WB, Pessoa A, Ventura CV, Fonseca EB et al (2017) Characterizing the pattern of anomalies in congenital Zika syndrome for pediatric clinicians. JAMA Pediatr 171(3):288–295

Moreno I, Codoñer FM, Vilella F, Valbuena D, Martinez-Blanch JF, Jimenez-Almazán J, Alonso R, Alamá P, Remohí J, Pellicer A, Ramon D (2016) Evidence that the endometrial microbiota has an effect on implantation success or failure. Am J Obstet Gynecol 215(6): 684–703

Müller L, Kellenberger CJ, Cannizzaro E, Ettlin D, Schraner T, Bolt IB, Peltomäki T, Saurenmann RK (2009) Early diagnosis of temporomandibular joint involvement in juvenile idiopathic arthritis: a pilot study comparing clinical examination and ultrasound to magnetic resonance imaging. Rheumatology 48(6): 680–685

Myllärniemi S, Perheentupa J (1978) Oral findings in the autoimmune polyendocrinopathy-candidosis syndrome (APECS) and other forms of hypoparathyroidism. Oral Surg Oral Med Oral Pathol 45(5):721–729

Nashaat EH, Mansour GM (2014) Helicobacter pylori and anemia with pregnancy. Arch Gynecol Obstet 289(6): 1197–1202

Ochoa-Bernal MA, Fazleabas AT (2020) Physiologic events of embryo implantation and Decidualization in human and non-human primates. Int J Mol Sci 21(6)

Pacis A, Tailleux L, Morin AM, Lambourne J, MacIsaac JL, Yotova V et al (2015) Bacterial infection remodels the DNA methylation landscape of human dendritic cells. Genome Res 25(12):1801–1811

Pacis A, Mailhot-Léonard F, Tailleux L, Randolph HE, Yotova V, Dumaine A, Grenier JC, Barreiro LB (2019) Gene activation precedes DNA demethylation in response to infection in human dendritic cells. Proc Natl Acad Sci 116(14):6938–6943

Papadaki ME, Lietman SA, Levine MA, Olsen BR, Kaban LB, Reichenberger EJ (2012) Cherubism: best clinical practice. Orphanet J Rare Dis 7(1):S6

Park J, Kang G, Baek S-H, Kim Y (2024) Genetic associations and parent-of-origin effects of PVRL1 in nonsyndromic cleft lip with or without cleft palate across multiple ethnic populations. Epidemiol Health 46:e2024069

Paulsson L, Arvini S, Bergström N, Klingberg G, Lindh C (2019) The impact of premature birth on dental maturation in the permanent dentition. Clin Oral Investig 23(2):855–861

Pavlič A, Waltimo-Sirén J (2009) Clinical and microstructural aberrations of enamel of deciduous and permanent teeth in patients with autoimmune polyendocrinopathy-candidiasis-ectodermal dystrophy. Arch Oral Biol 54(5):424–431

Peltomäki T, Kreiborg S, Pedersen TK, Ogaard B (2015) Craniofacial growth and dento-alveolar development in juvenile idiopathic arthritis patients. Semin Orthod 21(2):84–93

Peng H, Fu YX (2020) The inhibitory PVRL1/PVR/TIGIT axis in immune therapy for hepatocellular carcinoma. Gastroenterology 159(2):434–436

Peng M, Ling Y, Ding YL, Zhou CJ (2008) Trophoblast cells invaing the placenta bed and change of spiral arteries and microvessels in pre-eclampsia. J Central South Univer Med Sci 33(2):121–129

Pijnenborg R, Vercruysse L, Hanssens M (2006) The uterine spiral arteries in human pregnancy: facts and controversies. Placenta 27(9–10):939–958

Pradhan J, Mallick S, Mishra N, Tiwari A, Negi VD (2023) Pregnancy, infection, and epigenetic regulation: a complex scenario. Biochim Biophys Acta Mol basis Dis 1869(7):166768

Rabinowe SL, George KL, Loughlin R, Soeldner JS, Eisenbarth GS (1986) Congenital rubella monoclonal antibody-defined T cell abnormalities in young adults. Am J Med 81(5):779–782

Rackaityte E, Halkias J (2020) Mechanisms of fetal T cell tolerance and immune regulation. Front Immunol 11: 588

Raziel A, Schachter M, Strassburger D, Bern O, Ron-El R, Friedler S (2007) Favorable influence of local injury to the endometrium in intracytoplasmic sperm injection patients with high-order implantation failure. Fertil Steril 87(1):198–201

Reichenberger EJ, Levine MA, Olsen BR, Papadaki ME, Lietman SA (2012) The role of SH3BP2 in the pathophysiology of cherubism. Orphanet J Rare Dis 7(1):S5

Ren R, Mayer BJ, Cicchetti P, Baltimore D (1993) Identification of a ten-amino acid proline-rich SH3 binding site. Science 259(5098):1157–1161

Ringold S, Cron RQ (2009) The temporomandibular joint in juvenile idiopathic arthritis: frequently used and frequently arthritic. Pediatr Rheumatol 7(1):11

Rinkoff S (2023) Embryology, craniofacial growth, and development. StatPearls

Roth DM, Bayona F, Baddam P, Graf D (2021) Craniofacial development: neural crest in molecular embryology. Head Neck Pathol 15(1):1–15

Rouse DJ, Andrews WW, Goldenberg RL, Owen J (1995) Screening and treatment of asymptomatic bacteriuria of pregnancy to prevent pyelonephritis: a cost-effectiveness and cost-benefit analysis. Obstet Gynecol 86(1):119–123

Savioli C, Silva CA, Lin HC, Campos LM, Prado EF, Siqueira JTT (2004) Dental and facial characteristics of patients with juvenile idiopathic arthritis. Revista do Hospital das Clinicas 59:93–98

Shi FP, Huang YY, Dai QQ, Chen YL, Jiang HY, Liang SY (2023) Maternal common cold or fever during pregnancy and the risk of orofacial clefts in the offspring: a systematic review and meta-analysis. The Cleft Palate Craniofacial Journal 60(4):446–453

Siegel MA (1996) Syphilis and gonorrhea. Dent Clin N Am 40(2):369–383

Slots J, Slots V (2024) Severe periodontitis and congenital cytomegalovirus: cleft lip and cleft palate. Periodontol 96(1):185–188

Smith DW, Gong BT (1973) Scalp hair patterning as a clue to early fetal brain development. J Pediatr 83(3): 374–380

Soares AB, Gonzaga HFS, Jorge MA, Barraviera S (2004) Oral manifestations of syphilis: a review. J Venom Anim Toxins Incl Trop Dis 10:2–9

Sözen MA, Hecht JT, Spritz RA (2009) Mutation analysis of the PVRL1 gene in caucasians with nonsyndromic cleft lip/palate. Genet Test Mol Biomarkers 13(5): 617–621

Stoll ML, Sharpe T, Beukelman T, Good J, Young D, Cron RQ (2012) Risk factors for temporomandibular joint arthritis in children with juvenile idiopathic arthritis. J Rheumatol 39(9):1880–1887

Stoll ML, Kau CH, Waite PD, Cron RQ (2018) Temporomandibular joint arthritis in juvenile idiopathic arthritis, now what? Pediatr Rheumatol 16(1):32

Stoustrup P, Kristensen KD, Verna C, Küseler A, Herlin T, Pedersen TK (2012) Orofacial symptoms related to temporomandibular joint arthritis in juvenile idiopathic arthritis: smallest detectable difference in self-reported pain intensity. J Rheumatol 39(12):2352–2358

Than NG, Hahn S, Rossi SW, Szekeres-Bartho J (2019) Editorial: fetal-maternal immune interactions in pregnancy. Front Immunol 10:2729

Turhani D, Item CB, Watzinger E, Sinko K, Watzinger F, Lauer G, Ewers R (2005) Mutation analysis of CLPTM 1 and PVRL 1 genes in patients with non-syndromic clefts of lip, alveolus and palate. J Craniofac Surg 33(5):301–306

Ueki Y et al (2001) Mutations in the gene encoding c-Abl-binding protein SH3BP2 cause cherubism. Nat Genet 28:125–126 van de Ven R, van den Hout MF, Lindenberg JJ, Sluijter BJ, van Leeuwen PA, Lougheed SM, Meijer S, van den Tol MP, Scheper RJ, de Gruijl TD (2011) Characterization of four conventional dendritic cell subsets in human skin-draining lymph nodes in relation to T-cell activation. Blood J Am Soc Hematol 118(9): 2502–2510

Via WF Jr (1968) Enamel defects induced by trauma during tooth formation. Oral Surg Oral Med Oral Pathol 25(1):49–54

Warner CM, Gollnick SO (1993) Expression of H-2K major histocompatibility antigens on preimplantation mouse embryos. Biol Reprod 48(5):1082–1087

Weichert A, Vogt M, Dudenhausen JW, Kalache KD (2010) Evidence in a human fetus of micrognathia and cleft lip as potential effects of early cytomegalovirus infection. Fetal Diagn Ther 28(4):225–228

Weiss PF, Arabshahi B, Johnson A, Bilaniuk LT, Zarnow D, Cahill AM, Feudtner C, Cron RQ (2008) High prevalence of temporomandibular joint arthritis at disease onset in children with juvenile idiopathic arthritis, as detected by magnetic resonance imaging but not by ultrasound. Arthritis Rheum 58(4): 1189–1196

Witkin SS, Neuer A, Giraldo P, Jeremias J, Tolbert V, Korneeva IL, Kneissl D, Bongiovanni AM (1997) Chlamydia trachomatis infection, immunity, and pregnancy outcome. Infect Dis Obstet Gynecol 5(2): 128–132

Zaripova LN, Midgley A, Christmas SE, Beresford MW, Baildam EM, Oldershaw RA (2021) Juvenile idiopathic arthritis: from aetiopathogenesis to therapeutic approaches. Pediatr Rheumatol 19(1):135

Zhou L, Li R, Wang R, Huang HX, Zhong K (2008) Local injury to the endometrium in controlled ovarian hyperstimulation cycles improves implantation rates. Fertil Steril 89(5):1166–1176

Immune Involvement in the Development and Maintenance of Oral and Craniofacial Bones

25

Takehito Ono

Abstract

The oral and craniofacial bones are essential for determining individual appearance and enabling mastication, respiration, and phonation. The cranium exhibits a complex architecture composed of multiple small bones. The development of this structured organization is a consequence of the well-coordinated functions of various bone cells, such as chondrocytes, osteoblasts, osteoclasts, and osteocytes. These cells contribute to both the development and maintenance of bone, known as the modeling and remodeling phases, respectively. The differentiation and function of these bone cells can be influenced by intrinsic and extrinsic stimuli, resulting in the adaptation of bone morphology and functions when stimulation occurs within physiological ranges. In contrast, excessive or abnormal stimuli, including genetic mutations, endocrine disorders, abnormal mechanical loading, tumor formation, and aberrant immune responses, can lead to bone malformations and dysfunction. The immune system, originally developed as a defense mechanism against non-self entities, can disrupt the physiological structures and functions of organs when dysregulated. Bone metabolism can be influenced by the immune system, both directly and indirectly via other biological systems, and irregular activation of the immune system can cause inflammatory bone diseases as well as bone diseases of other categories. A precise understanding of these bone diseases requires insights into the involvement of the immune system. This chapter summarizes our knowledge of how the immune system functions during the development and maintenance of oral and craniofacial bones.

Keywords

Immune system · Growth · Development · Morphology · Bone metabolism

25.1 Introduction

The structure of the oral and craniofacial regions is a determining factor of individual appearance. The morphology of this structure is largely shaped by the underlying oral and craniofacial bones, as evidenced by facial reconstruction techniques in forensic medicine and dentistry, which use cranial remains to predict the facial appearance of deceased individuals (Chowdhry et al. 2018). The human cranium is composed of 22 bones that are classified into 14 types, which are connected by sutures and two

T. Ono (✉)
Laboratory of Drug Discovery and Pharmacology, Faculty of Veterinary Medicine, Okayama University of Science, Ehime, Japan
e-mail: ta-ono@ous.ac.jp

N. Rezaei, H. Ziaei (eds.), *Oral Immunology*, Advances in Experimental Medicine and Biology 1492,
https://doi.org/10.1007/978-3-032-03176-1_25

temporomandibular joints (TMJs) (Ono and Nakashima 2022).

The process of bone development and growth, including that of the cranium, is called the bone modeling phase. After this modeling phase, bones undergo a maintenance process called the bone remodeling phase, where damaged tissue is removed and replaced with new bone (Bolamperti et al. 2022). The processes of bone formation and resorption are significantly influenced by genetic factors, the endocrine system, mechanical environment, and the immune system (Okamoto et al. 2017; Tsukasaki and Takayanagi 2019). Regulation of bone modeling and remodeling by these components modifies an individual's skeleton. Compared to other parts of the body, the shapes of oral and craniofacial bones exhibit significant individual variations, possibly due to their susceptibility to these stimuli.

The immune system, which protects the self against the non-self entities, consists of various immune cells and their derivative molecules, including cytokines. Some of these cells and molecules have been reported to directly or indirectly regulate bone development and maintenance. This chapter focuses on how the immune system functions during the development and maintenance of the oral and craniofacial bones (Fig. 25.1).

25.2 Biology of the Bone

The process of bone development is driven by a dynamic interplay of chondrocytes, osteoblasts, osteoclasts, and osteocytes. Once developed, bone functions together with its adjacent tissues, such as bone marrow, nerves, blood vessels, ligaments, and tendons, to provide structural support, protect internal organs, enable locomotion, facilitate hematopoiesis, and store minerals. To understand the immune regulation of bone development and maintenance, basic knowledge of bone biology is outlined below.

25.2.1 Chondrocytes and Cartilage of the Bone

Chondrocytes are specialized cells responsible for producing and maintaining cartilage throughout the body. In the context of skeletal development, chondrocytes form a template for long bones that is later replaced by mature bone tissue through the process of endochondral ossification (Aghajanian and Mohan 2018; Li et al. 2022). Cartilage remains in the growth plate of long bones until the completion of growth and on the articular surface throughout life. Thus, dysfunction of chondrocytes in the bone modeling phase results in skeletal deformities, such as abnormal patterning of the extremities and altered bone length, whereas disturbance of these cells during the remodeling phase is commonly associated with joint diseases.

25.2.2 Osteoblasts and Bone Formation

Bone is not merely a mass of calcium. The elaborate function of osteoblasts builds a composite structure of the extracellular matrix (ECM) mineralized with hydroxyapatite. The composite of well-oriented ECM (mostly type I collagen) and hydroxyapatite crystals endows the bone with both rigidity and resilience. The alignment of hydroxyapatite crystals, an integral factor for bone strength, is influenced by osteocalcin, another ECM protein produced by osteoblasts (Moriishi et al. 2020). Osteoblasts facilitate bone formation through two mechanisms: endochondral and intramembranous ossification. Endochondral ossification is predominantly observed in the long bones, as previously discussed, and intramembranous ossification is observed in the cranium and clavicle (Li et al. 2022). Disorders in osteoblast differentiation and activity lead to skeletal deformities and increased bone fragility. In addition to bone formation, osteoblasts regulate bone resorption by expressing macrophage colony-stimulating factor (M-CSF) and receptor activator of NF-κB ligand (RANKL). Alterations

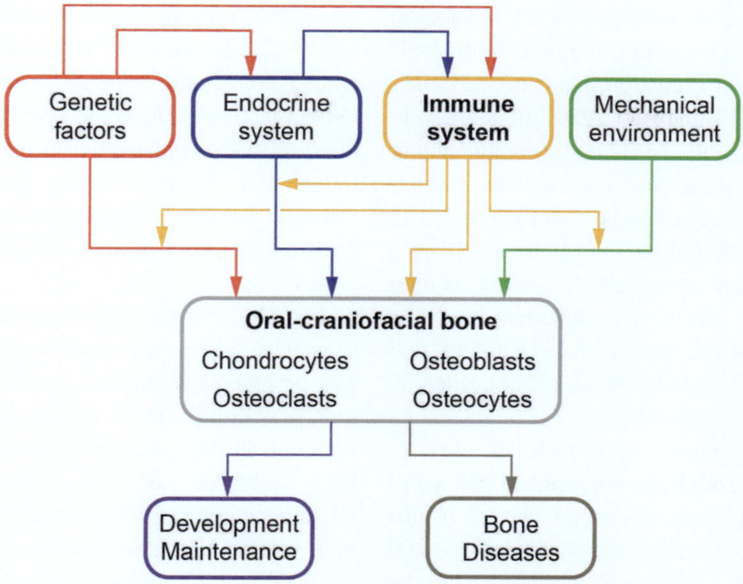

Fig. 25.1 Conceptual diagram of immune involvement in the development and maintenance of oral and craniofacial bones
The development and maintenance of oral and craniofacial bones are influenced by various biological systems. These regulatory systems can directly or indirectly affect the bone cells by interacting with other systems. The proper regulation of bone metabolism ensures the development and maintenance of physiological bone structure and function, and its dysregulation can lead to bone diseases. Dysregulation of the immune system can cause many bone diseases, both directly and indirectly, as discussed in the main text

in the expression of these cytokines are associated with various bone diseases, manifested by abnormal bone resorption.

25.2.3 Osteoclasts and Bone Resorption

For physiological bone modeling and remodeling, bone formation alone is not sufficient; coordinated bone resorption is essential as well. In the process of bone modeling, osteoclasts participate in resorptive bone modeling by degrading bone tissue to create cavities and concave structures. This process ensures the creation of spaces for internal organs and tissues, including the bone marrow, as well as pathways for blood vessels and nerves. Deficiency in osteoclastogenesis or osteoclast function leads to osteopetrosis, a condition characterized by defective bone marrow formation and constricted nerve pathways, leading to functional disturbances (Ono et al. 2020). During the remodeling phase, osteoclasts contribute to the removal of damaged bone, as observed during bone fracture healing. Osteoclasts originate from monocyte/macrophage-lineage cells and therefore share many signaling molecules with immune cells (Ono and Nakashima 2018). Mutations in the gene encoding these molecules can lead to abnormalities during both bone resorption and inflammatory processes.

25.2.4 Osteocytes and Modification of Bone Metabolism

Osteocytes are one of the terminal differentiation forms of osteoblasts. Osteoblasts undergo substantial morphological and functional changes when they become embedded within the bone matrix. A key morphological feature of osteocytes is their network of cellular processes, which radiate extensively within the bone matrix to establish connections with other osteocytes and other cell types. This connection allows the cells constituting bone to function in a concerted manner (Robling and Bonewald 2020). In addition to their function in cellular communication, osteocytes can sense the mechanical environment, detecting mechanical stimuli such as matrix strain and tissue fluid flow. In response to changes in the mechanical environment, osteocytes alter the expression levels of cytokines, such as RANKL, sclerostin, and IGF-1 (Robling and Bonewald 2020; Sasaki et al. 2021), thereby regulating bone metabolism.

Bone development and maintenance are accomplished through the coordinated actions of bone-composing cells described above. If this coordination is disturbed, the skeletal shape and balance of bone formation/resorption can be compromised, as discussed below.

25.3 Direct Regulation of Oral-Craniofacial Bone Modeling and Remodeling by the Immune System

The immune system is activated in response to tissue damage, such as infection, traumatic injury, and tumor development, facilitating pathogen eradication and tissue regeneration. In autoimmune conditions, immune responses are driven by self-reactive immune cells that recognize autoantigens. Dysregulation of inflammation developed in these contexts can disturb bone modeling and remodeling of the affected regions, establishing lesions that perturb bone morphology and function. This section focuses on diseases in which dysregulated inflammation directly influences bone metabolism.

25.3.1 Infection

Congenital infections with microorganisms, particularly those caused by TORCH pathogens (***T**oxoplasma gondii*, ***O**ther*, ***R**ubella virus*, ***C**ytomegalovirus*, ***H**erpes simplex virus*), are known to cause craniofacial malformations (Megli and Coyne 2022). In some infectious diseases, the cranial bones and teeth are reportedly affected. However, these effects are potentially due to the direct effect of pathogens on tissue and organs, rather than indirect effects mediated by the immune system, since the immune system remains immature during the fetal and perinatal stages.

Periodontitis is an infectious disease primarily caused by oral bacteria and is roughly categorized into peripheral and apical periodontitis. The former, periodontitis in the narrow sense, is the most common chronic infectious disease that is caused by dysbiosis of the oral microbiome (Hajishengallis and Chavakis 2021; Sirisereephap et al. 2022). Bacterial invasion of the gingival sulcus triggers gingival inflammation and leads to the formation of periodontal pockets. As inflammation proceeds, osteoclastic bone resorption occurs in the alveolar bone, resulting in the loss of tooth support and subsequent tooth loss. Immune cells play crucial roles in the production of pro-inflammatory cytokines and RANKL expression, promoting osteoclastic bone resorption (Settem et al. 2021; Tsukasaki et al. 2018). Loss of molars increases the mechanical burden on the incisors, which can induce labial inclination of these teeth, leading to anterior deep bite, labial protrusion, and a decrease in lower anterior face height (LAFH). Tooth loss causes further alveolar bone resorption, and in edentulous patients, complete loss of the alveolar process and alveolar part results in a severe decrease in LAFH. Thus, peripheral periodontitis significantly disrupts oral and craniofacial bone metabolism in the remodeling phase when it reaches its advanced stage, altering both morphology and function. However, it is unlikely to affect bones in the modeling phase, as peripheral periodontitis usually affects adults. On the other hand, aggressive periodontitis, a distinct form of periodontitis that manifests in adolescent patients (Bouziane et al. 2020), can impact the oral and craniofacial bones during their modeling phase.

The progression of dental caries can lead to pulpitis and apical periodontitis. Therefore, apical periodontitis can theoretically occur as early as the eruption of the lower central deciduous teeth. As bacterial infection originating from the carious lesion spreads apically, it reaches the maxillary or mandibular bone. If left untreated, the infection can expand and involve the bone marrow, leading to osteomyelitis of the jaw. Garré's osteomyelitis, a form of chronic osteomyelitis, can occur subsequent to apical periodontitis, even in pre-adolescent children whose bones are in the bone modeling phase (Nidhin Philip et al. 2021). This type of osteomyelitis induces a periosteal reaction, stimulating osteoblastic bone formation on the bone surface, causing enlargement and asymmetry of the bone near the root apex of the infected teeth. Thus, infection-induced inflammation can affect the bone modeling and remodeling phases. In orthopedics and pediatrics, osteomyelitis in the long bones is known to affect the growth plate, resulting in the impairment of bone growth (Jaramillo et al. 2017). In this context, bone destruction is induced directly by bacterial infection and indirectly by the immune system (Veis and Cassat 2021). This may also occur in the condyle, which is the site of mandibular growth. However, the incidence of osteomyelitis near the condyle is rare (Chattopadhyay et al. 2017); thus, it is unclear whether jaw osteomyelitis affects mandibular condyle growth.

25.3.2 Traumatic Injury

Traumatic injury triggers acute inflammation, which facilitates the eradication of tissue debris and subsequent tissue regeneration. During traumatic bone injuries, such as bone fractures and bone surgery procedures, blood vessels near the bone injury site are damaged, forming a hematoma. Within the hematoma, immune cells infiltrate and induce acute inflammation. Innate immune cells (macrophages, γδT cells, etc.) and their pro-inflammatory cytokines (IL-6, TNF-α, IL-17A, etc.) are observed at the site of bone injury, promoting the healing process (Duda et al. 2023; Ono and Takayanagi 2017). After the resolution of inflammation, mesenchymal cells increase at the injury site through chemoattraction and proliferation (Ono et al. 2016; Qing et al. 2024). These cells further differentiate into chondrocytes and osteoblasts to form callus tissue that bridges the fractured bone. The callus undergoes remodeling by osteoclasts and osteoblasts, restoring the fractured bone's structure and functions. Thus, the immune system directly regulates bone remodeling during fracture healing. However, prolonged or overactive inflammation can result in delayed or incomplete healing (nonunion), potentially disturbing bone growth in children, though this is extremely rare (Lipphardt et al. 2023).

25.3.3 Tumor

The immune system is deeply involved in every aspect of tumor pathology, including initiation, progression, invasion, metastasis, and eradication. Chronic inflammation, induced by infection, autoimmunity, metabolic diseases, or tissue injury, often precedes tumorigenesis. Neutrophils and macrophages can induce mutagenesis by producing reactive oxygen species (ROS) and reactive nitrogen species (RNS). Pro-inflammatory cytokines can lead to the dysregulation of the epigenetic status of cells, resulting in tumorigenesis (Greten and Grivennikov 2019). In the tumor microenvironment (TME), a niche composed of tumor cells, cancer-associated fibroblasts (CAFs), endothelial cells, immune cells, various cytokines, and growth factors is produced to enhance tumor development. Additionally, several immune cells and cytokines induce angiogenesis and ECM degradation, promoting metastasis (Mantovani et al. 2022; Zhao et al. 2021). The tumors, emerging as described above, are targeted and eliminated by the immune system. $CD8^{+}$ T cells (cytotoxic T lymphocytes, CTLs) recognize tumor antigens and are activated by antigen-presenting cells (APCs) (Mellman et al. 2023). Natural killer (NK) cells are innate lymphoid cells involved in anti-viral or anti-tumor immunity. These cells eliminate tumor cells through their cytotoxicity and the production of cytokines, such as interferon-γ (IFN-γ) (Wolf et al. 2022). Thus, the immune system plays a crucial role in tumor pathology by both tumor-suppressive and tumor-promoting impacts. Inflammation associated with tumor pathology often influences adjacent tissues, including the bone, in both primary and metastatic tumors. In tumors, various cytokines are produced by immune cells, tumors, and tumor stromal cells (H. Zhao et al. 2021). These cytokines can affect both osteoblasts and osteoclasts, leading to bone formation and resorption at the site of primary or metastatic sites.

Head and neck cancers account for approximately 5% of all cancers and cause 460,000 deaths per year (WHO 2022). Most cancers in this region exhibit histological characteristics of squamous cell carcinoma (SCC). Various types of immune cells have been reported to infiltrate the TME of head and neck SCC (HNSCC), including neutrophils, myeloid-derived suppressor cells (MDSCs), dendritic cells (DCs), NK cells, and T cells. These immune cells, as well as tumor cells, produce various types of cytokines, including IL-6 and TNF-α (Elmusrati et al. 2021). Given the proximity of the primary site of carcinoma (the oral epithelium) to the underlying jawbone, these cytokines can easily affect jawbone metabolism. Cytokines produced in tumors stimulate mesenchymal cells to express RANKL, and induce osteoclastogenesis and bone resorption. Osteoclastic bone resorption releases growth factors, such as IGF-1 and TGF-β, from the degraded bone matrix, thereby promoting tumor growth. The tumor thus enlarges produces additional cytokines, exacerbating bone resorption. This vicious cycle of tumor progression and bone resorption weakens the jawbone structure and may result in its fracture (Ono and Nakashima 2022). In some cases, osteoclast activation due to HNSCC progression also leads to tooth root resorption. Thus, HNSCC severely disturbs jawbone remodeling and oral function, partly through immune-mediated mechanisms.

Osteosarcoma is another type of malignant tumor that affects the jawbones. Although this is the most common bone tumor, its occurrence is rare in the oral-craniofacial region; it accounts for only 10% of osteosarcomas and less than 1% of malignant head and neck tumors (Weber et al. 2023). Cranial osteosarcoma differs from extracranial sarcomas (primarily found in the extremities, also known as long bone osteosarcomas) in terms of the peak age of onset, frequency of metastasis, and treatment response. These differences may stem from the distinct embryonic origins of the bone; cranial bones develop from the ectoderm, whereas most extracranial bones originate from the mesoderm. Owing to these differences, we cannot easily apply the insights obtained from extracranial osteosarcomas to cranial osteosarcomas. Additionally, the rarity of cranial osteosarcomas presents challenges in accumulating a comprehensive knowledge of their pathophysiology. Macrophages, neutrophils, MDSCs, DCs, NK cells, B cells, and T cells are known to infiltrate lesions of extracranial osteosarcoma (Bertin et al. 2020; Tian et al. 2023). These immune and tumor cells secrete various pro-inflammatory cytokines that can affect bone metabolism. Aggressive bone formation in osteosarcoma is partly driven by immune regulation. Unlike extracranial osteosarcomas, cranial osteosarcomas have a peak incidence in patients in their 30s and 40s. Therefore, these tumors are more likely to affect the bone remodeling phase rather than the bone modeling phase.

25.3.4 Juvenile Idiopathic Arthritis (JIA)

Certain types of inflammatory diseases occur in pediatric patients whose bones are in the bone-modeling phase. JIA, formerly known as Still's disease, is a chronic arthritis of unknown etiology that persists for at least 6 weeks in children under the age of 16 (Nigrovic et al. 2021). Although specific human leukocyte antigen (HLA) types have been associated with genetic predispositions to JIA, its causative genes have not been determined (For this reason, JIA is not described in the "Involvement of the Immune System in Genetic Regulation of Oral-Craniofacial Bone Modeling and Remodeling" section.). JIA consists of seven different categories, as defined by the International League of Associations for Rheumatology (ILAR): systemic arthritis, oligoarthritis, polyarthritis rheumatoid factor (RF)-positive, polyarthritis RF-negative, enthesitis-related arthritis, psoriatic arthritis (PsA), and undifferentiated arthritis (Martini et al. 2019), the first of which is a disease of autoinflammation, whereas the others are autoimmune diseases. Therefore, JIA represents a group of inflammatory bone diseases with a heterogeneous pathology.

JIA affects the joints during the growth phase; therefore, patients with JIA can suffer from growth retardation. In addition, prolonged corticosteroid treatment can result in growth suppression and osteoporosis (Zaripova et al. 2021). A substantial proportion of patients with JIA experience TMJ involvement, which leads to the impairment of mastication, malocclusion, and micrognathia. Intra-articular corticosteroid injection technique is frequently used for the treatment of TMJ inflammation in patients, but the insertion of injection needles and glucocorticoids into the TMJ region may negatively impact the mandibular growth (Ringold and Cron 2009). The use of biologic disease-modifying antirheumatic drugs (bDMARDs) can decrease the use of corticosteroids, potentially mitigating side effects on skeletal growth (Onel et al. 2022; Stoustrup et al. 2023).

25.3.5 Autoimmune Arthritis of the Adult

Inflammatory joint diseases can develop after the growth period as well. When systemic JIA develops in patients aged 16 and above, it is classified as adult-onset Still's disease (AOSD). In some cases, joint inflammation in AOSD worsens and leads to erosion (Gerfaud-Valentin et al. 2014). Although joint destruction can occur in the TMJ in these patients, such cases are rare.

Rheumatoid arthritis (RA) is a chronic autoimmune disease that causes joint inflammation and destruction. Its pathogenesis is known to be a complex combination of genetic and environmental risk factors, the accumulation of which leads to autoimmunity. Once established, synovial inflammation causes joint pain and swelling in the distal and small joints. As synovitis proceeds, osteoclastic bone resorption occurs, disrupting the joint architecture (Di Matteo et al. 2023; Weyand and Goronzy 2021). In a significant proportion of RA patients, lesions are established in the TMJs, either unilaterally or bilaterally. The most common symptoms are joint pain and condylar resorption, resulting in anterior open bite and subsequent masticatory dysfunction. Ankylosis can be observed in severe cases as well (Savtekin and Sehirli 2018).

Psoriasis, another autoimmune disease, is an inflammatory skin disease accompanied by joint inflammation. Psoriatic arthritis (PsA), which is an inflammatory arthropathy associated with psoriasis, is induced by the interaction of multiple factors, including psoriatic skin inflammation, intestinal dysbiosis, metabolic conditions, and mechanical stimuli (Griffiths et al. 2021; Schett et al. 2022). Unlike RA, PsA joints tend to exhibit bone-forming lesions associated with enthesitis, resulting in joint ankylosis. In the TMJ, pathological bone formation occurs in the condyle to form osteophytes, leading to joint pain and impaired mouth opening. However, facial deformity in PsA is relatively less severe compared to RA (Shivhare et al. 2022; Wang et al. 2014).

Taken together, the immune system, which is activated in various contexts, can directly regulate bone modeling and remodeling, possibly leading to impaired structure and function of oral-craniofacial bones.

25.4 Involvement of the Immune System in Genetic Regulation of Oral-Craniofacial Bone Modeling and Remodeling

Genomic mutations can lead to skeletal diseases. As of 2023, 771 diseases with 552 causative genes have been classified as genetic skeletal disorders (GSDs) (Unger et al. 2023). The causative genes of these disorders are associated with the differentiation of cells into the skeleton and their functions, such as collagen synthesis and mineralization. In several GSD cases, mutations affect both skeletal tissues and the immune system, suggesting a potential link between skeletal and immune disorders. Since GSD genomic mutations arise during embryonic stages, they can influence both bone modeling and remodeling phases.

25.4.1 Spondyloenchondrodysplasia (SPENCD)

SPENCD is an autosomal recessive genetic disease caused by loss-of-function mutations in *ACP5*, the gene encoding tartrate-resistant acid phosphatase (TRAP) (Briggs et al. 2016; Elhossini et al. 2023). Patients with this condition exhibit impaired endochondral ossification and develop short stature, rhizomelic micromelia, and barrel chest. Abnormalities in the oral-craniofacial region include calcification of the basal ganglia, delayed spheno-occipital synchondrosis, delayed tooth eruption, and anterior open bite (Hong et al. 2018). Due to the overactivation of DCs, the type I immune response is irregularly driven, leading to inflammation (Briggs et al. 2011; Lausch et al. 2011). TRAP is expressed in osteoclasts, macrophages, and DCs. This enzyme dephosphorylates substrates, including osteopontin. Phosphorylated osteopontin anchors osteoclasts onto bone surfaces and modulates mineralization (Ek-Rylander et al. 1994; Hayman et al. 1996). The accumulation of phosphorylated osteopontin in TRAP deficiency also results in dysregulated inflammation (Lausch et al. 2011). Thus, the *ACP5* mutations in SPENCD independently affect the skeletal tissue and immune system. Some patients with SPENCD develop arthritis (Hong et al. 2018; Kara et al. 2020), in which the immune-skeletal interactions may occur in a dysregulated inflammation-dependent manner.

25.4.2 Ehlers-Danlos Syndrome (EDS)

EDS is a group of disorders characterized by connective tissue fragility, leading to joint laxity and skin hyperextensibility. There are 13 subtypes of EDS, the causative genes of which are mostly involved in the synthesis of ECM components, including collagen and proteoglycans. An exception is periodontal EDS (pEDS), which is caused by mutations in *C1R* or *C1S* in an autosomal dominant manner in genes encoding complement system proteins rather than ECM components (Malfait et al. 2017) (Although pEDS is not categorized as a GSD, it is a monogenic skeletal disease and is included in this section for convenience).

These complement components are expressed in gingival fibroblasts. Mutant C1r has been reported to exhibit constitutive activities, and mutant C1s degrade collagen in fibroblasts (Amberger et al. 2023; Grobner et al. 2019), connecting the immune system to ECM metabolism. Patients with pEDS experience extensive gingivitis and early-onset periodontitis, with mild plaque accumulation that can lead to tooth loss (Malfait et al. 2017). Thus, pEDS disrupts alveolar bone remodeling by promoting inflammatory bone loss.

25.4.3 Cherubism

Cherubism is a rare genetic disorder characterized by bilateral swelling of the middle and lower face, resulting from abnormal bone resorption and fibrous tissue overgrowth in the jawbone. This disease is caused by gain-of-function mutations in *SH3BP2* (the gene encoding SH3 domain-binding protein 2) and is inherited in an autosomal dominant manner (Ueki et al. 2001). SH3BP2 protein functions as a scaffold for signaling molecules that integrate multiple signaling pathways, including those involved in bone metabolism and immune regulation (Mukai et al. 2019; Srivastava et al. 2024). Pathogenic *SH3BP2* mutations lead to SYK phosphorylation, increasing TNF-α production, and osteoclastogenesis of macrophages, which are responsible for inflammation and bone resorption (Ueki et al. 2007).

Inflammation-driven mechanisms are suggested to play a crucial role in cherubism pathogenesis. Enhanced inflammation is suggested to cause the proliferation of fibrous tissues in cherubism lesions. It is also suggested that the inflammation can induce the expression of M-CSF and RANKL from the stromal cells in the lesion, indirectly augmenting osteoclastogenesis. TLR signaling is involved in the pathogenesis of cherubism, indicating that bacterial infections can trigger disease onset (Fujii et al. 2022). It is reasonable for this disease to preferentially affect the jawbone because the oral mucosa provides a potential entry site for bacteria: the gingival sulcus and periodontal pocket.

25.4.4 Fibrodysplasia Ossificans Progressiva (FOP)

FOP is a rare autosomal dominant GSD with an incidence of 1 in 2,000,000 individuals. It is caused by mutations in the *ACVR1* gene (Shore et al. 2006), which encodes Activin A receptor type I (ACVR1), also known as ALK2, a type I BMP receptor. This mutation renders ALK2 responsive to activin A, a ligand that does not activate the wild-type receptor (Hatsell et al. 2015; Hino et al. 2015). Upon activation of mutant ALK2 by traumatic injury, downstream signaling pathways are triggered, leading to the characteristic pathological features of FOP: heterotopic bone formation within soft tissues, including skeletal muscles, ligaments, and tendons. This process begins in the neck, shoulders, and back, and then progresses to the trunk and limbs, severely disturbing mobility, respiration, mastication, and neurological functions (Huning and Gillessen-Kaesbach 2014; Kitterman et al. 2012; Susami et al. 2012). In the pathogenesis of FOP, heterotopic bone formation is preceded by severe inflammation at the injury site, known as a flare-up, indicating that the immune system is involved in FOP development. This was further supported by clinical evidence showing that pharmacological immunosuppression decreased FOP activity (Kaplan et al. 2007). Additionally, traumatic injury alone can induce heterotopic bone formation in skeletal muscle (myositis ossificans) via inflammatory mechanisms without *ACVR1* mutation (McCarthy and Sundaram 2005; Tu et al. 2021). Taken together, these results suggest that the immune system regulates bone modeling and remodeling in the pathogenesis of FOP. However, the direct impact of *ACVR1* mutations on immune cells remains largely unknown.

25.4.5 Chronic Infantile Neurologic Cutaneous Articular (CINCA) Syndrome

Mutations in genes related to the immune system can lead to its overactivation, leading to the disease conditions known as autoinflammatory diseases, some of which affect skeletal tissues. Cryopyrin-associated periodic syndrome (CAPS) is a group of autoinflammatory diseases caused by gain-of-function mutations in NLRP3 (cryopyrin), a component of the inflammasome (Broderick and Hoffman 2022). Hyperactive NLRP3 in CINCA patients converts pro-IL-1β into active IL-1β in a dysregulated manner, provoking severe inflammation in the skin, joints, and the nervous system. CAPS consists of familial cold autoinflammatory, Muckle-Wells, and CINCA syndromes, the latter of which is classified as a GSD. Patients with CINCA syndrome (also known as neonatal-onset multisystem inflammatory disease; NOMID) develop severe autoinflammatory symptoms soon after birth. One of the most prominent symptoms of this disease is enlarged joints with overgrowth of the epi-metaphyseal cartilage in the long bones. Recent studies using induced pluripotent stem cells (iPSCs) from patients with CINCA syndrome have shown that excess production of cartilage is mediated by the caspase 1-independent cAMP/protein kinase A/CREB pathway (Yokoyama et al. 2015), indicating that IL-1β is not directly responsible for the overgrowth of cartilage. In addition, in vitro studies have shown that IL-1β suppresses chondrogenesis (Lu et al. 2023; Wu et al. 2022), further supporting the inflammation-independent cartilage overgrowth in patients with CINCA syndrome. However, in genetically modified mice carrying the CINCA-type mutant *NLRP3*, the infiltration of inflammatory cells was observed in joint tissues (Snouwaert et al. 2016), suggesting that inflammation induced by the mutation may be involved in the inflammatory bone remodeling associated with this disease.

Thus, in several types of GSDs, the immune system can directly or indirectly regulate bone modeling and remodeling, influencing the morphology and function of oral and craniofacial bones.

25.5 Involvement of the Immune System in Endocrine Regulation of Oral-Craniofacial Bone Modeling and Remodeling

The endocrine system regulates physiological and behavioral activities by releasing hormones into the bloodstream to influence distant organs. This system consists of the hypothalamus, pituitary gland, thyroid gland, parathyroid gland, pancreas, and genital glands. Since bone is a target of endocrine regulation, disruption in the endocrine system leads to an imbalance in bone modeling and remodeling, as observed in acromegaly, hypopituitarism, and hypothyroidism. Additionally, certain endocrine disorders are involved in immune regulation, suggesting a potential interaction between the endocrine system and immune pathways in bone homeostasis.

25.5.1 Grave's Disease

The thyroid gland is an essential endocrine organ that targets almost all nucleated cells, and its dysfunction severely perturbs growth and metabolism, including that of bone. In Graves' disease (also known as Basedow's disease), autoantibodies activate thyroid-stimulating hormone receptors, enhancing the production of thyroid hormones (Taylor et al. 2018). Thyroid hormones promote both osteoblastic bone formation and osteoclastic bone resorption, resulting in secondary osteoporosis (Kim and Mohan 2013; Williams and Bassett 2018). This osteoporosis can occur in the maxilla and mandible, which may be related to an increased susceptibility to periodontitis and accelerated tooth eruption (Chandna and Bathla 2011).

25.5.2 Hashimoto's Thyroiditis

Autoimmunity does not always activate the thyroid gland. Autoantibodies against thyroid peroxidase (TPO) and thyroglobulin contribute to the destruction of thyroid tissue, leading to inflammatory hypothyroidism known as Hashimoto's thyroiditis (Kolanu et al. 2024). Patients with this disease can exhibit osteoporosis, an anterior open bite, and delayed tooth eruption (Chandna and Bathla 2011). Autoimmune thyroid diseases usually initiate at the ages of 20–30 years. Therefore, these diseases are more likely to affect the bone remodeling phase rather than the modeling phase.

25.5.3 Autoimmune Hypoparathyroidism and Parathyroid Hormone Resistance

Parathyroid hormone (PTH) upregulates serum Ca^{2+} level and is known to directly promote osteoblastogenesis and indirectly promote osteoclastogenesis via RANKL expression on mesenchymal cells (Zhou et al. 2021). Therefore, PTH dysregulation can lead to bone diseases. Hypoparathyroidism is most commonly a post-surgical complication, but it can also occur due to autoimmunity. In autoimmune hypoparathyroidism, autoantibodies activate the calcium-sensing receptor (CaSR), leading to suppression of PTH secretion and leading to hypocalcemia. The development of hypoparathyroidism leads to elevated bone mass and bone mineral density (BMD), with the balance of bone remodeling shifting towards formation (Khan et al. 2022). The generation of autoantibodies against the PTH receptor (PTH1R) results in resistance to PTH, leading to the development of a pathology similar to autoimmune hypothyroidism (Mandl et al. 2021). Because decreased PTH signaling affects bone remodeling to induce hypocalcemia, the mineralization and eruption of teeth are disturbed in some cases (Hejlesen et al. 2018). Autoimmune hypoparathyroidism can be a component of autoimmune polyglandular syndrome (APS) or autoimmune polyendocrinopathy-candidiasis-ectodermal dystrophy (APECED). Therefore, patients with these conditions may exhibit the previously described symptoms (Hejlesen et al. 2018).

25.5.4 Type I Diabetes

The pancreas has exocrine and endocrine functions, the latter of which are deeply connected to energy metabolism. Autoantibodies against insulin, glutamic acid decarboxylase (GAD), insulinoma-associated antigen-2 (IA-2), or zinc transporter 8 (ZnT8) are observed in patients with type I diabetes (Tandel et al. 2024). Due to the autoimmune destruction of β-cells, insulin production severely decreases, and these patients require replacement therapies. Patients with this disease have decreased BMD and an increased risk of fractures due to suppression of osteoblastic bone formation (Hough et al. 2016). Diabetes compromises the immune response to infection; therefore, periodontitis progresses more severely in patients with this disease (Darwitz Benjamin et al. 2024; M. Zhao et al. 2023).

Overall, the immune system regulates bone modeling and remodeling indirectly via the regulation of the endocrine system, mostly in the context of autoimmunity.

25.6 Involvement of the Immune System in the Oral-Craniofacial Bone Modeling and Remodeling Under Mechanical Stimuli

The mechanical environment of the bone influences its morphology and strength. Exercise reinforces the bone, as mechanical loading shifts the balance of bone remodeling towards formation. Mimicking exercises with a locomomimetic drug has been shown to strengthen the bone (Ono et al. 2022). In contrast, conditions such as prolonged bed rest or microgravity during spaceflight lead to decreased bone mass and BMD due to increased bone resorption (Sasaki et al. 2021). These mechanical influences are not only limited to bones of the trunk and extremities but also affect oral and craniofacial bones, which adapt their shapes and mechanical properties by modulating their bone remodeling balances.

25.6.1 Masticatory Force

Masticatory muscles exert strong forces equivalent to body weight when they function during occlusion. Masticatory forces are closely linked to facial profiles, but the mechanisms underlying this association remain largely elusive. To investigate these mechanisms, rodent models of reduced or forceful mastication have been used, demonstrating that masticatory forces are related to the shape and BMD of the mandible (Fukushima-Nakayama et al. 2017; Hichijo et al. 2014; Inoue et al. 2019). Masticatory force modulates osteocyte activity by upregulating IGF-1 and downregulating sclerostin production. Together, these induce the osteoblastogenesis of cells in the tendon of the masseter muscle, forming a stout jawbone (Inoue et al. 2019). Several pro-inflammatory cytokines and chemokines are produced in the skeletal muscles during their function (Chow et al. 2022; Graca et al. 2023), which may mediate jawbone modeling and remodeling.

25.6.2 Orthodontic Tooth Movement (OTM)

OTM is established by alveolar bone remodeling, which is induced by mechanical loading onto teeth by orthodontic appliances. Osteocytes are thought to sense the mechanical stimuli of orthodontic forces because osteocyte depletion suppresses OTM (Matsumoto et al. 2013). Osteocytes express RANKL and are involved in osteoclastic bone resorption on the compression side of the alveolar bone, towards which the teeth are driven (Shoji-Matsunaga et al. 2017). In addition, the expression of sclerostin, DMP1, and MEPE, molecules characteristic of osteocytes and related to bone formation, were reported to change during OTM, suggesting that osteocytes may respond to orthodontic force to regulate bone formation at the tension side of the alveolar bone (Gluhak-Heinrich et al. 2007; Shu et al. 2017).

Patients frequently complain of pain during orthodontic treatment. Orthodontic intervention induces inflammation in the periodontal ligament (PDL) (Klein et al. 2020). There is an immune regulation over OTM by macrophages and T cells (Wald et al. 2021; Xu et al. 2022; Yan et al. 2015), as well as by inflammatory cytokines, including IL-6 and TNF-α (Toyama et al. 2023; Yoshimatsu et al. 2006). IL-6 is produced in the PDL upon orthodontic intervention and increases the responsiveness of macrophages to M-CSF and RANKL, facilitating osteoclastic bone resorption on the compression side (Toyama et al. 2023).

Although previous studies have indicated the involvement of various immune cells and their downstream in OTM, the detailed mechanisms are not fully understood. Mechanical loading can stimulate immune cells via ECM interactions and tissue fluid dynamics (Du et al. 2023). Damage-associated molecular patterns (DAMPs), which are released during OTM-induced tissue injury, can also result in the activation of immune cells. These activated immune cells can regulate osteoclastic bone resorption and osteoblastic bone formation. Thus, the immune system participates to some extent in alveolar bone remodeling during OTM; however, further research is required to explore the molecular mechanisms involved.

Mechanical stimuli regulate the modeling and remodeling of the oral-craniofacial bone. Beyond the previously described contexts, disturbances in the mechanical environment of oral and craniofacial regions can also occur due to parafunctional oral habits, such as bruxism and tongue thrusting, and abnormal oral functions in patients with deformities, such as cleft lip and palate. Bone modeling and remodeling are affected in such situations; however, it is unclear whether the immune system is involved.

25.7 Conclusion

The connection between the immune system and bone metabolism may not seem apparent at first glance. However, as discussed in this chapter, a closer look at the mechanisms underlying diseases reveals the relationship between bone metabolism and the immune system. In certain genetic diseases, single pathogenic mutations can cause abnormalities in both the skeletal and immune systems independently without evident interactions. However, in other genetic diseases and inflammatory bone diseases, immune system disorders cause imbalances in bone metabolism (Fig. 25.1). Recent advances in molecular biology and analytical techniques have substantially contributed to uncovering the underlying mechanisms. Future advances will provide further insights into the development, maintenance, and pathogenesis of the oral and craniofacial bones, factoring in immune regulation within.

References

Aghajanian P, Mohan S (2018) The art of building bone: emerging role of chondrocyte-to-osteoblast transdifferentiation in endochondral ossification. Bone Res 6:19. https://doi.org/10.1038/s41413-018-0021-z

Amberger A, Pertoll J, Traunfellner P, Kapferer-Seebacher I, Stoiber H, Klimaschewski L et al (2023) Degradation of collagen I by activated C1s in periodontal Ehlers-Danlos syndrome. Front Immunol 14: 1157421. https://doi.org/10.3389/fimmu.2023.1157421

Bertin H, Gomez-Brouchet A, Redini F (2020) Osteosarcoma of the jaws: an overview of the pathophysiological mechanisms. Crit Rev Oncol Hematol 156:103126. https://doi.org/10.1016/j.critrevonc.2020.103126

Bolamperti S, Villa I, Rubinacci A (2022) Bone remodeling: an operational process ensuring survival and bone mechanical competence. Bone Res 10(1):48. https://doi.org/10.1038/s41413-022-00219-8

Bouziane A, Hamdoun R, Abouqal R, Ennibi O (2020) Global prevalence of aggressive periodontitis: a systematic review and meta-analysis. J Clin Periodontol 47(4):406–428. https://doi.org/10.1111/jcpe.13266

Briggs TA, Rice GI, Daly S, Urquhart J, Gornall H, Bader-Meunier B et al (2011) Tartrate-resistant acid phosphatase deficiency causes a bone dysplasia with autoimmunity and a type I interferon expression signature. Nat Genet 43(2):127–131. https://doi.org/10.1038/ng.748

Briggs TA, Rice GI, Adib N, Ades L, Barete S, Baskar K et al (2016) Spondyloenchondrodysplasia due to mutations in ACP5: a comprehensive survey. J Clin Immunol 36(3):220–234. https://doi.org/10.1007/s10875-016-0252-y

Broderick L, Hoffman HM (2022) IL-1 and autoinflammatory disease: biology, pathogenesis and therapeutic targeting. Nat Rev Rheumatol 18(8):448–463. https://doi.org/10.1038/s41584-022-00797-1

Chandna S, Bathla M (2011) Oral manifestations of thyroid disorders and its management. Indian J Endocrinol Metab 15(Suppl 2):S113–S116. https://doi.org/10.4103/2230-8210.83343

Chattopadhyay PK, Nagori SA, Menon RP, Thanneermalai B (2017) Osteomyelitis of the mandibular condyle: a report of 2 cases with review of the literature. J Oral Maxillofac Surg 75(2):322–335. https://doi.org/10.1016/j.joms.2016.08.018

Chow LS, Gerszten RE, Taylor JM, Pedersen BK, van Praag H, Trappe S et al (2022) Exerkines in health, resilience and disease. Nat Rev Endocrinol 18(5):273–289. https://doi.org/10.1038/s41574-022-00641-2

Chowdhry A, Kapoor P, Popli DB, Sircar K, Miglani R (2018) Inclusion of forensic Odontologist in team of forensic facial approximation-a proposal and technical note. J Clin Diagn Res. https://doi.org/10.7860/jcdr/2018/35863.12003

Darwitz Benjamin P, Genito Christopher J, Thurlow Lance R (2024) Triple threat: how diabetes results in worsened bacterial infections. Infect Immun 92:e00509–e00523. https://doi.org/10.1128/iai.00509-23

Di Matteo A, Bathon JM, Emery P (2023) Rheumatoid arthritis. Lancet 402(10416):2019–2033. https://doi.org/10.1016/S0140-6736(23)01525-8

Du H, Bartleson JM, Butenko S, Alonso V, Liu WF, Winer DA, Butte MJ (2023) Tuning immunity through tissue mechanotransduction. Nat Rev Immunol 23(3):174–188. https://doi.org/10.1038/s41577-022-00761-w

Duda GN, Geissler S, Checa S, Tsitsilonis S, Petersen A, Schmidt-Bleek K (2023) The decisive early phase of bone regeneration. Nat Rev Rheumatol 19(2):78–95. https://doi.org/10.1038/s41584-022-00887-0

Ek-Rylander B, Flores M, Wendel M, Heinegård D, Andersson G (1994) Dephosphorylation of osteopontin and bone sialoprotein by osteoclastic tartrate-resistant acid phosphatase. Modulation of osteoclast adhesion in vitro. J Biol Chem 269(21):14853–14856. https://doi.org/10.1016/S0021-9258(17)36541-9

Elhossini RM, Elbendary HM, Rafat K, Ghorab RM, Abdel-Hamid MS (2023) Spondyloenchondrodysplasia in five new patients: identification of three novel ACP5 variants with variable neurological presentations. Mol Gen Genomics 298(3):709–720. https://doi.org/10.1007/s00438-023-02009-1

Elmusrati A, Wang J, Wang CY (2021) Tumor microenvironment and immune evasion in head and neck squamous cell carcinoma. Int J Oral Sci 13(1):24. https://doi.org/10.1038/s41368-021-00131-7

Fujii Y, Monteiro N, Sah SK, Javaheri H, Ueki Y, Fan Z et al (2022) Tlr2/4-mediated Hyperinflammation promotes Cherubism-like jawbone expansion in Sh3bp2 (P416R) Knockin mice. JBMR Plus 6(1):e10562. https://doi.org/10.1002/jbm4.10562

Fukushima-Nakayama Y, Ono T, Hayashi M, Inoue M, Wake H, Ono T, Nakashima T (2017) Reduced mastication impairs memory function. J Dent Res 96(9):1058–1066. https://doi.org/10.1177/0022034517708771

Gerfaud-Valentin M, Maucort-Boulch D, Hot A, Iwaz J, Ninet J, Durieu I et al (2014) Adult-onset still disease: manifestations, treatment, outcome, and prognostic factors in 57 patients. Medicine (Baltimore) 93(2):91–99. https://doi.org/10.1097/MD.0000000000000021

Gluhak-Heinrich J, Pavlin D, Yang W, MacDougall M, Harris SE (2007) MEPE expression in osteocytes during orthodontic tooth movement. Arch Oral Biol 52(7):684–690. https://doi.org/10.1016/j.archoralbio.2006.12.010

Graca FA, Stephan A, Minden-Birkenmaier BA, Shirinifard A, Wang YD, Demontis F, Labelle M (2023) Platelet-derived chemokines promote skeletal muscle regeneration by guiding neutrophil recruitment to injured muscles. Nat Commun 14(1):2900. https://doi.org/10.1038/s41467-023-38624-0

Greten FR, Grivennikov SI (2019) Inflammation and cancer: triggers, mechanisms, and consequences. Immunity 51(1):27–41. https://doi.org/10.1016/j.immuni.2019.06.025

Griffiths CEM, Armstrong AW, Gudjonsson JE, Barker J (2021) Psoriasis. Lancet 397(10281):1301–1315. https://doi.org/10.1016/S0140-6736(20)32549-6

Grobner R, Kapferer-Seebacher I, Amberger A, Redolfi R, Dalonneau F, Bjorck E et al (2019) C1R mutations trigger constitutive complement 1 activation in periodontal Ehlers-Danlos syndrome. Front Immunol 10:2537. https://doi.org/10.3389/fimmu.2019.02537

Hajishengallis G, Chavakis T (2021) Local and systemic mechanisms linking periodontal disease and inflammatory comorbidities. Nat Rev Immunol 21(7):426–440. https://doi.org/10.1038/s41577-020-00488-6

Hatsell SJ, Idone V, Wolken DM, Huang L, Kim HJ, Wang L et al (2015) ACVR1R206H receptor mutation causes fibrodysplasia ossificans progressiva by imparting responsiveness to activin a. Sci Transl Med 7(303):303ra137. https://doi.org/10.1126/scitranslmed.aac4358

Hayman AR, Jones SJ, Boyde A, Foster D, Colledge WH, Carlton MB et al (1996) Mice lacking tartrate-resistant acid phosphatase (Acp 5) have disrupted endochondral ossification and mild osteopetrosis. Development 122(10):3151–3162. https://doi.org/10.1242/dev.122.10.3151

Hejlesen J, Underbjerg L, Gjorup H, Bloch-Zupan A, Sikjaer T, Rejnmark L, Haubek D (2018) Dental findings in patients with non-surgical hypoparathyroidism and Pseudohypoparathyroidism: a systematic review. Front Physiol 9:701. https://doi.org/10.3389/fphys.2018.00701

Hichijo N, Kawai N, Mori H, Sano R, Ohnuki Y, Okumura S et al (2014) Effects of the masticatory demand on the rat mandibular development. J Oral Rehabil 41(8): 581–587. https://doi.org/10.1111/joor.12171

Hino K, Ikeya M, Horigome K, Matsumoto Y, Ebise H, Nishio M et al (2015) Neofunction of ACVR1 in fibrodysplasia ossificans progressiva. Proc Natl Acad Sci USA 112(50):15438–15443. https://doi.org/10.1073/pnas.1510540112

Hong SW, Huh KH, Lee JK, Kang JH (2018) Craniofacial anomalies associated with spondyloenchondrodysplasia: two case reports. Medicine (Baltimore) 97(50):e13644. https://doi.org/10.1097/MD.0000000000013644

Hough FS, Pierroz DD, Cooper C, Ferrari SL, Bone IC, Diabetes Working G (2016) Mechanisms in endocrinology: mechanisms and evaluation of bone fragility in type 1 diabetes mellitus. Eur J Endocrinol 174(4): R127–R138. https://doi.org/10.1530/EJE-15-0820

Huning I, Gillessen-Kaesbach G (2014) Fibrodysplasia ossificans progressiva: clinical course, genetic mutations and genotype-phenotype correlation. Mol Syndromol 5(5):201–211. https://doi.org/10.1159/000365770

Inoue M, Ono T, Kameo Y, Sasaki F, Ono T, Adachi T, Nakashima T (2019) Forceful mastication activates osteocytes and builds a stout jawbone. Sci Rep 9(1): 4404. https://doi.org/10.1038/s41598-019-40463-3

Jaramillo D, Dormans JP, Delgado J, Laor T, St Geme JW (2017) Hematogenous osteomyelitis in infants and children: imaging of a changing disease. Radiology 283(3):629–643. https://doi.org/10.1148/radiol.2017151929

Kaplan FS, Glaser DL, Shore EM, Pignolo RJ, Xu M, Zhang Y et al (2007) Hematopoietic stem-cell contribution to ectopic skeletogenesis. J Bone Joint Surg Am 89(2):347–357. https://doi.org/10.2106/jbjs.F.00472

Kara B, Ekinci Z, Sahin S, Gungor M, Gunes AS, Ozturk K et al (2020) Monogenic lupus due to spondyloenchondrodysplasia with spastic paraparesis and intracranial calcification: case-based review. Rheumatol Int 40(11):1903–1910. https://doi.org/10.1007/s00296-020-04653-x

Khan AA, Bilezikian JP, Brandi ML, Clarke BL, Gittoes NJ, Pasieka JL et al (2022) Evaluation and management of hypoparathyroidism summary statement and guidelines from the second international workshop. J Bone Miner Res 37(12):2568–2585. https://doi.org/10.1002/jbmr.4691

Kim HY, Mohan S (2013) Role and mechanisms of actions of thyroid hormone on the skeletal development. Bone Res 1(2):146–161. https://doi.org/10.4248/BR201302004

Kitterman JA, Strober JB, Kan L, Rocke DM, Cali A, Peeper J et al (2012) Neurological symptoms in individuals with fibrodysplasia ossificans progressiva. J Neurol 259(12):2636–2643. https://doi.org/10.1007/s00415-012-6562-y

Klein Y, Fleissig O, Polak D, Barenholz Y, Mandelboim O, Chaushu S (2020) Immunorthodontics: in vivo gene expression of orthodontic tooth movement. Sci Rep 10(1):8172. https://doi.org/10.1038/s41598-020-65089-8

Kolanu ND, Awan NA, Butt AI, Reza T, Almadhoun M, Janoowala T et al (2024) From antibodies to artificial intelligence: a comprehensive review of diagnostic challenges in Hashimoto's thyroiditis. Cureus 16(2): e54393. https://doi.org/10.7759/cureus.54393

Lausch E, Janecke A, Bros M, Trojandt S, Alanay Y, De Laet C et al (2011) Genetic deficiency of tartrate-resistant acid phosphatase associated with skeletal dysplasia, cerebral calcifications and autoimmunity. Nat Genet 43(2):132–137. https://doi.org/10.1038/ng.749

Li Q, Xu R, Lei K, Yuan Q (2022) Insights into skeletal stem cells. Bone Res 10(1):61. https://doi.org/10.1038/s41413-022-00235-8

Lipphardt M, Zamzam M, Saleh ES (2023) Pediatric humeral shaft fracture nonunion: a case report. Cureus 15(12):e51019. https://doi.org/10.7759/cureus.51019

Lu R, Wang YG, Qu Y, Wang SX, Peng C, You H et al (2023) Dihydrocaffeic acid improves IL-1beta-induced inflammation and cartilage degradation via inhibiting NF-kappaB and MAPK signalling pathways. Bone Joint Res 12(4):259–273. https://doi.org/10.1302/2046-3758.124.BJR-2022-0384.R1

Malfait F, Francomano C, Byers P, Belmont J, Berglund B, Black J et al (2017) The 2017 international classification of the Ehlers-Danlos syndromes. Am J Med Genet C Semin Med Genet 175(1):8–26. https://doi.org/10.1002/ajmg.c.31552

Mandl A, Burbelo PD, Di Pasquale G, Tay YS, Welch J, Lionakis MS et al (2021) Parathyroid hormone resistance and autoantibodies to the PTH1 receptor. N Engl J Med 385(21):1974–1980. https://doi.org/10.1056/NEJMoa2109409

Mantovani A, Allavena P, Marchesi F, Garlanda C (2022) Macrophages as tools and targets in cancer therapy. Nat Rev Drug Discov 21(11):799–820. https://doi.org/10.1038/s41573-022-00520-5

Martini A, Ravelli A, Avcin T, Beresford MW, Burgos-Vargas R, Cuttica R et al (2019) Toward new classification criteria for juvenile idiopathic arthritis: first steps, pediatric rheumatology international trials organization international consensus. J Rheumatol 46(2): 190–197. https://doi.org/10.3899/jrheum.180168

Matsumoto T, Iimura T, Ogura K, Moriyama K, Yamaguchi A (2013) The role of osteocytes in bone resorption during orthodontic tooth movement. J Dent

Res 92(4):340–345. https://doi.org/10.1177/0022034513476037

McCarthy EF, Sundaram M (2005) Heterotopic ossification: a review. Skeletal Radiol 34(10):609–619. https://doi.org/10.1007/s00256-005-0958-z

Megli CJ, Coyne CB (2022) Infections at the maternal-fetal interface: an overview of pathogenesis and defence. Nat Rev Microbiol 20(2):67–82. https://doi.org/10.1038/s41579-021-00610-y

Mellman I, Chen DS, Powles T, Turley SJ (2023) The cancer-immunity cycle: indication, genotype, and immunotype. Immunity 56(10):2188–2205. https://doi.org/10.1016/j.immuni.2023.09.011

Moriishi T, Ozasa R, Ishimoto T, Nakano T, Hasegawa T, Miyazaki T et al (2020) Osteocalcin is necessary for the alignment of apatite crystallites, but not glucose metabolism, testosterone synthesis, or muscle mass. PLoS Genet 16(5):e1008586. https://doi.org/10.1371/journal.pgen.1008586

Mukai T, Fujita S, Morita Y (2019) Tankyrase (PARP5) inhibition induces Bone loss through accumulation of its substrate SH3BP2. Cells 8(2). https://doi.org/10.3390/cells8020195

Nidhin Philip LM, Akkara F, Khwaja T, Narayan T, Kamath AG, Jose NP (2021) Surgical management of garre's osteomyelitis in an 8-year-old child. Afr J Paediatr Surg 18(2):111–113. https://doi.org/10.4103/ajps.AJPS_66_20

Nigrovic PA, Colbert RA, Holers VM, Ozen S, Ruperto N, Thompson SD et al (2021) Biological classification of childhood arthritis: roadmap to a molecular nomenclature. Nat Rev Rheumatol 17(5):257–269. https://doi.org/10.1038/s41584-021-00590-6

Okamoto K, Nakashima T, Shinohara M, Negishi-Koga T, Komatsu N, Terashima A et al (2017) Osteoimmunology: the conceptual framework unifying the immune and skeletal systems. Physiol Rev 97(4): 1295–1349. https://doi.org/10.1152/physrev.00036.2016

Onel KB, Horton DB, Lovell DJ, Shenoi S, Cuello CA, Angeles-Han ST et al (2022) 2021 American College of Rheumatology Guideline for the treatment of juvenile idiopathic arthritis: therapeutic approaches for Oligoarthritis, temporomandibular joint arthritis, and systemic juvenile idiopathic arthritis. Arthritis Rheumatol 74(4):553–569. https://doi.org/10.1002/art.42037

Ono T, Nakashima T (2018) Recent advances in osteoclast biology. Histochem Cell Biol 149(4):325–341. https://doi.org/10.1007/s00418-018-1636-2

Ono T, Nakashima T (2022) Oral bone biology. J Oral Biosci 64(1):8–17. https://doi.org/10.1016/j.job.2022.01.008

Ono T, Takayanagi H (2017) Osteoimmunology in Bone fracture healing. Curr Osteoporos Rep 15(4):367–375. https://doi.org/10.1007/s11914-017-0381-0

Ono T, Okamoto K, Nakashima T, Nitta T, Hori S, Iwakura Y, Takayanagi H (2016) IL-17-producing gammadelta T cells enhance bone regeneration. Nat Commun 7:10928. https://doi.org/10.1038/ncomms10928

Ono T, Hayashi M, Sasaki F, Nakashima T (2020) RANKL biology: bone metabolism, the immune system, and beyond. Inflamm Regen 40:2. https://doi.org/10.1186/s41232-019-0111-3

Ono T, Denda R, Tsukahara Y, Nakamura T, Okamoto K, Takayanagi H, Nakashima T (2022) Simultaneous augmentation of muscle and bone by locomomimetism through calcium-PGC-1alpha signaling. Bone Res 10(1):52. https://doi.org/10.1038/s41413-022-00225-w

Qing Y, Ono T, Kohara Y, Watanabe A, Ogiso N, Ito M et al (2024) Emilin2 marks the target region for mesenchymal cell accumulation in bone regeneration. Inflamm Regen 44(1):27. https://doi.org/10.1186/s41232-024-00341-6

Ringold S, Cron RQ (2009) The temporomandibular joint in juvenile idiopathic arthritis: frequently used and frequently arthritic. Pediatr Rheumatol Online J 7:11. https://doi.org/10.1186/1546-0096-7-11

Robling AG, Bonewald LF (2020) The osteocyte: new insights. Annu Rev Physiol 82:485–506. https://doi.org/10.1146/annurev-physiol-021119-034332

Sasaki F, Hayashi M, Ono T, Nakashima T (2021) The regulation of RANKL by mechanical force. J Bone Miner Metab 39(1):34–44. https://doi.org/10.1007/s00774-020-01145-7

Savtekin G, Sehirli AO (2018) Rheumatoid arthritis in temporo-mandibular joint: a review. Niger J Clin Pract 21(10):1243–1246. https://doi.org/10.4103/njcp.njcp_117_18

Schett G, Rahman P, Ritchlin C, McInnes IB, Elewaut D, Scher JU (2022) Psoriatic arthritis from a mechanistic perspective. Nat Rev Rheumatol 18(6):311–325. https://doi.org/10.1038/s41584-022-00776-6

Settem RP, Honma K, Chinthamani S, Kawai T, Sharma A (2021) B-cell RANKL contributes to pathogen-induced alveolar Bone loss in an experimental periodontitis mouse model. Front Physiol 12:722859. https://doi.org/10.3389/fphys.2021.722859

Shivhare P, Ghimire AC, Karishma S et al (2022) Temporomandibular joint disorders associated with psoriatic arthritis: a report of three cases. Cureus 14(12):e32959. https://doi.org/10.7759/cureus.32959

Shoji-Matsunaga A, Ono T, Hayashi M, Takayanagi H, Moriyama K, Nakashima T (2017) Osteocyte regulation of orthodontic force-mediated tooth movement via RANKL expression. Sci Rep 7(1):8753. https://doi.org/10.1038/s41598-017-09326-7

Shore EM, Xu M, Feldman GJ, Fenstermacher DA, Cho TJ, Choi IH et al (2006) A recurrent mutation in the BMP type I receptor ACVR1 causes inherited and sporadic fibrodysplasia ossificans progressiva. Nat Genet 38(5):525–527. https://doi.org/10.1038/ng1783

Shu R, Bai D, Sheu T, He Y, Yang X, Xue C et al (2017) Sclerostin promotes bone remodeling in the process of tooth movement. PLoS One 12(1):e0167312. https://doi.org/10.1371/journal.pone.0167312

Sirisereephap K, Maekawa T, Tamura H, Hiyoshi T, Domon H, Isono T et al (2022) Osteoimmunology in periodontitis: local proteins and compounds to alleviate periodontitis. Int J Mol Sci 23(10). https://doi.org/10.3390/ijms23105540

Snouwaert JN, Nguyen M, Repenning PW, Dye R, Livingston EW, Kovarova M et al (2016) An NLRP3 mutation causes Arthropathy and osteoporosis in humanized mice. Cell Rep 17(11):3077–3088. https://doi.org/10.1016/j.celrep.2016.11.052

Srivastava T, Garola RE, Zhou J, Boinpelly VC, Rezaiekhaligh MH, Joshi T et al (2024) Scaffold protein SH3BP2 signalosome is pivotal for immune activation in nephrotic syndrome. JCI Insight 9(3). https://doi.org/10.1172/jci.insight.170055

Stoustrup P, Resnick CM, Abramowicz S, Pedersen TK, Michelotti A, Kuseler A et al (2023) Management of orofacial manifestations of juvenile idiopathic arthritis: interdisciplinary consensus-based recommendations. Arthritis Rheumatol 75(1):4–14. https://doi.org/10.1002/art.42338

Susami T, Mori Y, Tamura K, Ohkubo K, Nagahama K, Takahashi N et al (2012) Facial morphology and occlusion of a patient with fibrodysplasia ossificans progressiva (FOP): a case report. Spec Care Dentist 32(4):165–170. https://doi.org/10.1111/j.1754-4505.2012.00260.x

Tandel D, Hinton B, de Jesus Cortez F, Seftel D, Robinson P, Tsai CT (2024) Advances in risk predictive performance of pre-symptomatic type 1 diabetes via the multiplex antibody-detection-by-agglutination-PCR assay. Front Endocrinol (Lausanne) 15:1340436. https://doi.org/10.3389/fendo.2024.1340436

Taylor PN, Albrecht D, Scholz A, Gutierrez-Buey G, Lazarus JH, Dayan CM, Okosieme OE (2018) Global epidemiology of hyperthyroidism and hypothyroidism. Nat Rev Endocrinol 14(5):301–316. https://doi.org/10.1038/nrendo.2018.18

Tian H, Cao J, Li B, Nice EC, Mao H, Zhang Y, Huang C (2023) Managing the immune microenvironment of osteosarcoma: the outlook for osteosarcoma treatment. Bone Res 11(1):11. https://doi.org/10.1038/s41413-023-00246-z

Toyama N, Ono T, Ono T, Nakashima T (2023) The interleukin-6 signal regulates orthodontic tooth movement and pain. Biochem Biophys Res Commun 684:149068. https://doi.org/10.1016/j.bbrc.2023.09.096

Tsukasaki M, Takayanagi H (2019) Osteoimmunology: evolving concepts in bone-immune interactions in health and disease. Nat Rev Immunol 19(10):626–642. https://doi.org/10.1038/s41577-019-0178-8

Tsukasaki M, Komatsu N, Nagashima K, Nitta T, Pluemsakunthai W, Shukunami C et al (2018) Host defense against oral microbiota by bone-damaging T cells. Nat Commun 9(1):701. https://doi.org/10.1038/s41467-018-03147-6

Tu B, Yu B, Wang W, Li J, Yuan F, Zhu J, Fan C (2021) Inhibition of IL-17 prevents the progression of traumatic heterotopic ossification. J Cell Mol Med 25(16):7709–7719. https://doi.org/10.1111/jcmm.16617

Ueki Y, Tiziani V, Santanna C, Fukai N, Maulik C, Garfinkle J et al (2001) Mutations in the gene encoding c-Abl-binding protein SH3BP2 cause cherubism. Nat Genet 28(2):125–126. https://doi.org/10.1038/88832

Ueki Y, Lin CY, Senoo M, Ebihara T, Agata N, Onji M et al (2007) Increased myeloid cell responses to M-CSF and RANKL cause bone loss and inflammation in SH3BP2 "cherubism" mice. Cell 128(1):71–83. https://doi.org/10.1016/j.cell.2006.10.047

Unger S, Ferreira CR, Mortier GR, Ali H, Bertola DR, Calder A et al (2023) Nosology of genetic skeletal disorders: 2023 revision. Am J Med Genet A 191(5):1164–1209. https://doi.org/10.1002/ajmg.a.63132

Veis DJ, Cassat JE (2021) Infectious osteomyelitis: marrying bone biology and microbiology to shed new light on a persistent clinical challenge. J Bone Miner Res 36(4):636–643. https://doi.org/10.1002/jbmr.4279

Wald S, Leibowitz A, Aizenbud Y, Saba Y, Zubeidat K, Barel O et al (2021) γδT cells are essential for orthodontic tooth movement. J Dent Res 100(7):731–738. https://doi.org/10.1177/0022034520984774

Wang Z, Zhao Y, Ma X (2014) Ankylosis of temporomandibular joint caused by psoriatic arthritis: a report of four cases with literature review. Chin J Dent Res 17(1):49–55. https://doi.org/10.3290/j.cjdr.a32436

Weber V, Stigler R, Lutz R, Kesting M, Weber M (2023) Systematic review of craniofacial osteosarcoma regarding different clinical, therapeutic and prognostic parameters. Front Oncol 13:1006622. https://doi.org/10.3389/fonc.2023.1006622

Weyand CM, Goronzy JJ (2021) The immunology of rheumatoid arthritis. Nat Immunol 22(1):10–18. https://doi.org/10.1038/s41590-020-00816-x

WHO (2022) Estimated number of new cancer cases in 2022, lip, oral cavity, nasopharynx, oropharynx, hypopharynx and larynx. Cancer fact sheets. Retrieved from https://gco.iarc.who.int/today/en/fact-sheets-cancers

Williams GR, Bassett JHD (2018) Thyroid diseases and bone health. J Endocrinol Investig 41(1):99–109. https://doi.org/10.1007/s40618-017-0753-4

Wolf NK, Kissiov DU, Raulet DH (2022) Roles of natural killer cells in immunity to cancer, and applications to immunotherapy. Nat Rev Immunol 23(2):90–105. https://doi.org/10.1038/s41577-022-00732-1

Wu X, Zhao K, Fang X, Lu F, Cheng P, Song X et al (2022) Saikosaponin D inhibited IL-1beta induced ATDC 5 chondrocytes apoptosis in vitro and delayed articular cartilage degeneration in OA model mice in vivo. Front Pharmacol 13:845959. https://doi.org/10.3389/fphar.2022.845959

Xu H, Zhang S, Sathe AA, Jin Z, Guan J, Sun W et al (2022) CCR2(+) macrophages promote orthodontic tooth movement and alveolar Bone remodeling. Front Immunol 13:835986. https://doi.org/10.3389/fimmu.2022.835986

Yan Y, Liu F, Kou X, Liu D, Yang R, Wang X et al (2015) T cells are required for orthodontic tooth movement. J Dent Res 94(10):1463–1470. https://doi.org/10.1177/0022034515595003

Yokoyama K, Ikeya M, Umeda K, Oda H, Nodomi S, Nasu A et al (2015) Enhanced chondrogenesis of induced pluripotent stem cells from patients with neonatal-onset multisystem inflammatory disease occurs via the caspase 1-independent cAMP/protein kinase A/CREB pathway. Arthritis Rheumatol 67(1): 302–314. https://doi.org/10.1002/art.38912

Yoshimatsu M, Shibata Y, Kitaura H, Chang X, Moriishi T, Hashimoto F et al (2006) Experimental model of tooth movement by orthodontic force in mice and its application to tumor necrosis factor receptor-deficient mice. J Bone Miner Metab 24(1): 20–27. https://doi.org/10.1007/s00774-005-0641-4

Zaripova LN, Midgley A, Christmas SE, Beresford MW, Baildam EM, Oldershaw RA (2021) Juvenile idiopathic arthritis: from aetiopathogenesis to therapeutic approaches. Pediatr Rheumatol Online J 19(1): 135. https://doi.org/10.1186/s12969-021-00629-8

Zhao H, Wu L, Yan G, Chen Y, Zhou M, Wu Y, Li Y (2021) Inflammation and tumor progression: signaling pathways and targeted intervention. Signal Transduct Target Ther 6(1):263. https://doi.org/10.1038/s41392-021-00658-5

Zhao M, Xie Y, Gao W, Li C, Ye Q, Li Y (2023) Diabetes mellitus promotes susceptibility to periodontitis-novel insight into the molecular mechanisms. Front Endocrinol (Lausanne) 14:1192625. https://doi.org/10.3389/fendo.2023.1192625

Zhou R, Guo Q, Xiao Y, Guo Q, Huang Y, Li C, Luo X (2021) Endocrine role of bone in the regulation of energy metabolism. Bone Res 9(1):25. https://doi.org/10.1038/s41413-021-00142-4

Crosstalk Between Hematopoietic and Mesenchymal Cells in Bone and Tooth Immune Response

26

Larissa Sthefani Sales, Alice Corrêa Silva-Sousa, Guido Artemio Marañón-Vásquez, and Francisco Wanderley Garcia de Paula-Silva

Abstract

The immune response within dental tissues encompasses intricate interactions between hematopoietic and mesenchymal cells, which are pivotal for maintaining homeostasis and responding to pathological challenges. Hematopoietic cells, representing a diverse array of immune constituents, play an essential role in orchestrating inflammatory responses and regulating tissue repair mechanisms following injury or infection. These immune cells are instrumental in mediating inflammation, as well as directing the healing processes necessary for restoring tissue integrity. In parallel, mesenchymal cells, which include fibroblasts, odontoblasts, and osteoblasts, are indispensable during the inflammatory response and subsequent resolution of inflammation. The orchestration of these immune responses is governed by complex cellular signaling pathways involving specific cell surface receptors, as well as the secretion of various cytokines and growth factors that facilitate communication between cells. Mesenchymal cells possess not only the capacity to differentiate into multiple dental cell types but also the ability to respond to immunological signals, thus modulating inflammatory responses and actively participating in tissue regeneration and repair. A comprehensive understanding of these mechanisms and cellular interactions provides essential insights into the pathophysiology of inflammatory conditions affecting both connective and mineralized dental tissues. Such knowledge can significantly inform the development of more effective therapeutic strategies aimed at managing these conditions. This chapter delves into the cellular interactions and molecular mediators that are commonly observed in contexts of bacterial contamination of dental tissues and during orthodontic tooth movement, thereby offering a thorough biomolecular perspective on these complex processes.

Keywords

Hematopoietic cells · Mesenchymal cells · Bone remodeling · Dental tissue · Inflammation

26.1 Introduction

Teeth and their supporting periodontal tissues are susceptible to pathological stimuli that can elicit complex immunological events. Pathogen-associated molecular patterns (PAMPs) and damage-associated molecular patterns (DAMPs)

L. S. Sales · A. C. Silva-Sousa · G. A. Marañón-Vásquez · F. W. G. d. Paula-Silva (✉)
Developmental Biology and Child Health Research Group, Ribeirão Preto School of Dentistry, University of São Paulo, São Paulo, Brazil
e-mail: franciscogarcia@forp.usp.br

N. Rezaei, H. Ziaei (eds.), *Oral Immunology*, Advances in Experimental Medicine and Biology 1492,
https://doi.org/10.1007/978-3-032-03176-1_26

serve as critical triggers for the activation of innate immunity, which constitutes the host's initial line of defense against external insults (Mogensen 2009; Olive 2012; Gong et al. 2020). Innate immunity is instrumental in mediating the early vascular inflammatory response, engaging both resident and hematopoietic myeloid-derived immune cells in this process (Barreiros et al. 2018). Upon recognition of antigens, the adaptive immune response is subsequently activated, involving lymphoid-origin hematopoietic cells, which contribute to the multifaceted immune landscape.

The interaction of immune cells orchestrates a series of alterations within the local cellular microenvironment, facilitating cross talk with neighboring mesenchymal cells present in both bone and dental tissues (Fig. 26.1). Depending on the context and the specific stage of the immune response, these interactions can promote the expression of either pro-inflammatory or anti-inflammatory mediators, influencing the balance between tissue damage and repair mechanisms. In mineralized tissues, such as enamel, dentin, and bone, repair processes are predominantly anabolic, necessitating mineral deposition to restore tissue integrity (Almeida-Junior et al. 2023), whereas tissue damage is often characterized by catabolic mineral loss (Paula-Silva et al. 2020a, b).

The receptor activator of nuclear factor-kB ligand (RANKL)/receptor activator of nuclear factor-kB (RANK)/osteoprotegerin (OPG) signaling pathway stands as the principal mechanism by which inflammatory immune processes facilitate bone resorption. RANKL, a cytokine widely expressed by mesenchymal stem cells and immune cells, binds to its receptor RANK on myeloid lineage-derived pre-osteoclasts, promoting their differentiation into mature osteoclasts, thus sustaining their activation and longevity. Conversely, OPG acts as a decoy receptor, predominantly produced by osteoblasts, which competes with RANK for RANKL binding, subsequently inhibiting osteoclastogenesis and suppressing osteoclast activity (Jiang et al. 2015).

In addition to the RANKL/RANK/OPG axis, various other mechanisms have been identified that regulate bone and tooth resorption as well as repair in response to different stimuli. The following sections will delineate the primary interactions between hematopoietic and mesenchymal cells that govern these processes, particularly in the contexts of dental pulp lesions, periodontal disease, and orthodontic tooth movement (OTM). By elucidating these cellular dynamics, one can gain deeper insights into the biological responses associated with oral health and disease.

26.2 Dental Pulp Lesions: Immunological Communication Between Hematopoietic and Mesenchymal Cells

Dental pulp lesions refer to pathological conditions affecting the dental pulp, the innermost soft tissue of the tooth that contains nerves, blood vessels, and connective tissue. One of the most common etiological factors of dental pulp lesions is dental caries, in which bacteria demineralize the tooth structure, progressively invading deeper layers until they reach the pulp chamber. The resulting infection can cause pulpitis, which may present in reversible or irreversible forms depending on the severity of the tissue damage. Trauma also plays a significant role in the development of dental pulp lesions. Physical trauma from accidents or sports injuries can lead to tooth fractures or luxations, exposing the pulp to external irritants and compromising the vascular supply. Such injuries can result in a cascade of inflammatory responses, potentially resulting in pulp necrosis if left untreated. Restorative dental procedures can inadvertently contribute to pulp damage. Aggressive cavity preparation during filling or crown placement, as well as the use of thermal or chemical irritants during procedures, may generate heat or release substances that irritate the pulp, leading to inflammation (Sipert et al. 2022).

The complex interplay between hematopoietic and mesenchymal cells with tissue during injury and repair is characterized by a sophisticated network of signaling pathways, molecular mediators,

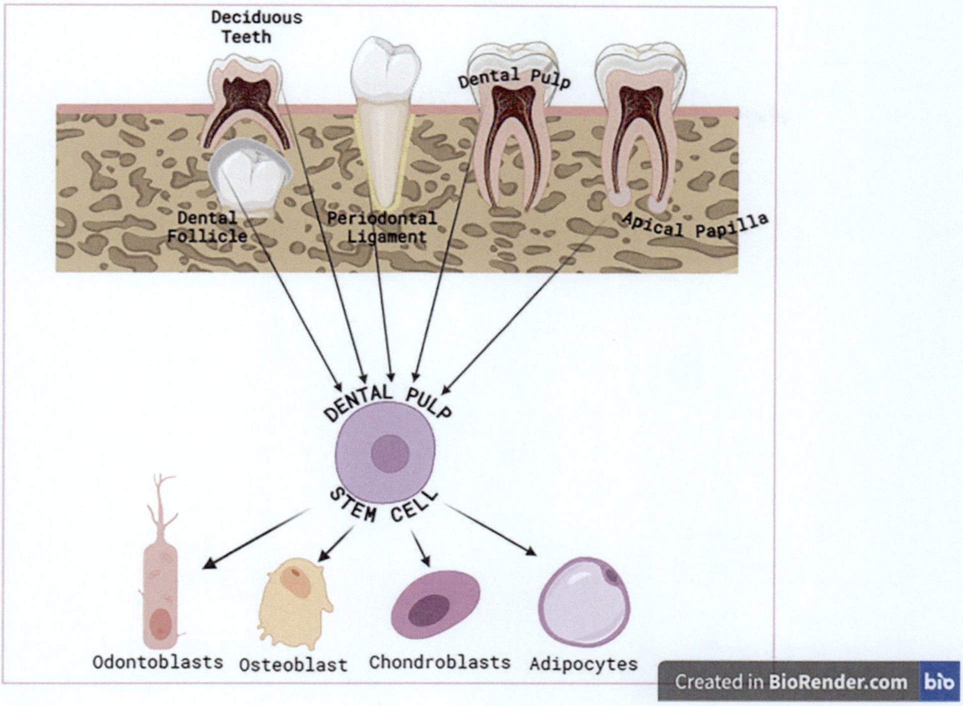

Fig. 26.1 Dental stem cells can be obtained from dental follicle, exfoliated deciduous teeth, dental pulp from permanent teeth, apical papilla from immature teeth, and periodontal ligament. These stem cells, under specific stimuli, can differentiate into odontoblasts, osteoblasts, chondroblasts, and adipocytes

and direct cell-to-cell communication. Upon dental pulp injury, hematopoietic inflammatory cells are recruited to the site of damage and proliferate, releasing a diverse array of cytokines and growth factors. These molecular signals significantly influence the behavior and functional responses of mesenchymal cells, affecting the processes of inflammation, tissue repair, and regeneration (Fig. 26.2) (Lorencetti-Silva et al. 2022a, b).

Dental pulp stem cells (DPSCs) represent a specialized subset of mesenchymal stem cells (MSCs), identifiable by specific surface markers, including CD29, CD44, CD59, CD90, CD105, CD146, and CD166 (Liu et al. 2015; Zhou et al. 2023). First identified and characterized by Gronthos et al. (2000), DPSCs have attracted substantial interest within the field of regenerative medicine due to their remarkable properties, which encompass self-renewal capacity, rapid proliferation, immunomodulatory potential, and the ability to differentiate into multiple cell types. Notably, DPSCs can be easily harvested from extracted third molars or exfoliated deciduous teeth, making them accessible for therapeutic applications (Zhou et al. 2023).

A particularly significant aspect of DPSCs is their unique capacity to replace damaged odontoblasts and produce reparative dentin. In response to stimuli from dental caries or trauma, especially in cases where native odontoblasts are absent, DPSCs can differentiate into odontoblast-like cells. This differentiation allows the production of the dentin matrix and subsequent mineralization processes, resulting in the formation of reparative dentin. Beyond their local functional contributions to dental health, DPSCs hold considerable potential for broader applications in regenerative medicine (Arana-Chavez and Massa 2004; Janebodin et al. 2011; Ma et al. 2012; Zhou et al. 2023; Sipert et al. 2022).

The intricate dynamics of pulp repair are further modulated by various biomolecules,

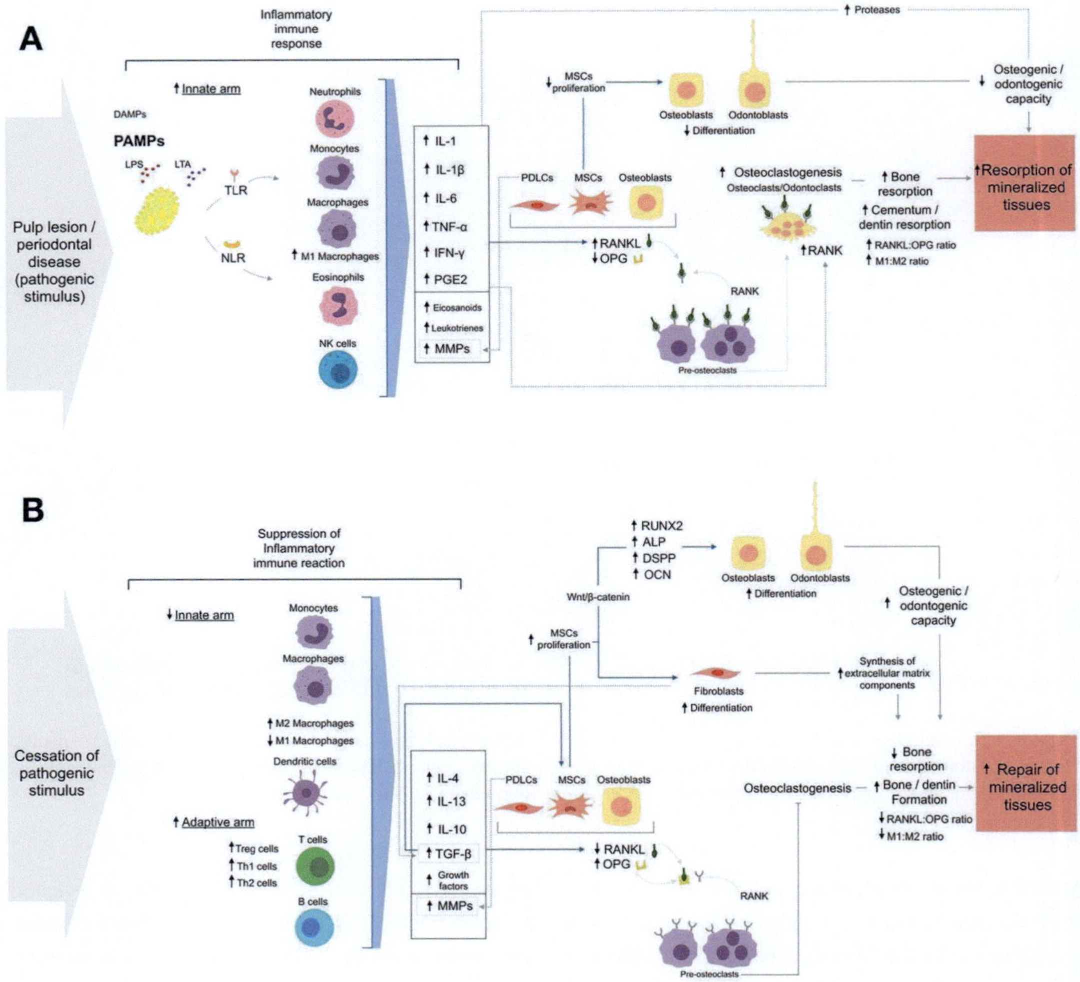

Fig. 26.2 Cross talk between mesenchymal and hematopoietic stem cells in dental pulp and periodontal lesions in two conditions: (**a**) in response to a pathogenic stimulus or (**b**) after the cessation of the stimulus

including bone morphogenetic proteins (BMPs), cytokines, matrix metalloproteinases (MMPs), and a range of bioactive proteins that collectively influence these complex reparative processes (Roberts-Clark and Smith 2000; Farges et al. 2015; Galler et al. 2016; Smith et al. 2016; Conrads and About 2018; Bjørndal et al. 2019). The involvement of immune cells, in coordination with odontoblasts, pulp fibroblasts, and DPSCs, is essential for initiating the immune responses within dental pulp tissue, particularly in the context of reparative dentin formation (Galler et al. 2021). By clarifying the roles of these biomolecules and cellular interactions, the mechanisms governing pulp repair and regeneration can be better understood. This knowledge is crucial for ultimately advancing strategies aimed at enhancing regenerative therapeutic interventions in dental medicine.

Numerous biomolecules are important players in modulating critical processes involved in pulp repair, including chemotaxis, angiogenesis, neurogenesis, mesenchymal stem cell recruitment, and mineralization (Farges et al. 2015; Zhou et al. 2023). Among these signaling molecules, transforming growth factor-beta (TGF-β) stands out as a potent mediator that significantly influences wound healing and tissue repair. TGF-β promotes the formation of granulation tissue and plays a vital role in regulating the inflammatory response through complex interactions between hematopoietic and mesenchymal cells (Farges et al. 2015; Zhou et al. 2023).

The TGF-β superfamily, which includes BMPs, mediates intracellular signaling through Smad proteins that transduce signals from the cell surface to the nucleus (Zhou et al. 2023). BMPs, in particular, are closely associated with Smad protein activation, facilitating the translocation of signaling molecules to the nucleus where they regulate gene expression (Wu et al. 2016; Zhou et al. 2023). Research demonstrates that BMP/Smad signaling pathways regulate dentin sialophosphoprotein (DSPP) both in vitro and in vivo (Chen et al. 2008, 2009; Cho et al. 2010). Notably, BMP2 has been shown to induce the differentiation of DPSCs into odontoblast-like cells, thereby promoting dentin formation (Woo et al. 2016; Hu et al. 2019, 2021).

Additionally, vascular endothelial growth factor (VEGF) regulates angiogenesis, an essential component of effective tissue repair. VEGF stimulates the formation of new blood vessels within injured pulp tissue, thereby enhancing the delivery of oxygen and nutrients required for efficient healing processes (Roberts-Clark and Smith 2000; Farges et al. 2015; Smith et al. 2016).

The Wnt signaling pathway emerges as a central regulator in mesenchymal stem cell differentiation and tissue repair, particularly within the dental pulp environment. Activation of this pathway is essential for guiding the differentiation of mesenchymal stem cells into odontoblast-like cells, which are indispensable for the formation of dentin (Zhou et al. 2023). The Wnt signaling cascade can be divided into two pathways: canonical and noncanonical (Rao and Kuhl 2010). In the canonical pathway, in the absence of Wnt signaling, a protein complex composed of adenomatous polyposis coli (APC), glycogen synthase kinase-3β (GSK-3β), Axin, and casein kinase 1 (CK1) is formed. This complex phosphorylates β-catenin, leading to its degradation and preventing signal transduction. Conversely, the presence of Wnt ligands facilitates their binding to the Frizzled receptor and low-density lipoprotein receptor-related proteins 5 and 6 (LRP5/6), leading to the disassembly of this protein complex. This disruption allows β-catenin to accumulate within the cytoplasm and its translocation into the nucleus, where it directly activates various target genes such as *Runx2*, *ALP*, *DSPP*, and *OCN*, all of which are involved in odontogenic differentiation (Zhou et al. 2023). The noncanonical Wnt pathways, mainly mediated by Wnt5a, also contribute to tissue repair and regeneration through mechanisms that do not involve β-catenin. These pathways are primarily categorized into the planar cell polarity (PCP) pathways and Wnt/Ca2+ pathways (Monroe et al. 2012; He et al. 2014).

In addition to the Wnt signaling pathway, nuclear factor kappa B (NF-κB), a canonical transcription factor involved in the inflammatory response, contributes to pulp repair mechanisms by activating the expression of pro-inflammatory cytokines and chemokines, which in turn attract immune cells to the injury site (Bjørndal et al. 2019).

Cytokines, a heterogeneous group of proteins, serve as critical regulators of inflammation, immune response, and cellular differentiation. In the context of dental pulp lesions, pro-inflammatory cytokines such as tumor necrosis factor-alpha (TNF-α), interleukin-1 beta (IL-1β), and interleukin-6 (IL-6) are secreted by activated immune cells, particularly macrophages. These cytokines play a crucial role in recruiting and activating additional immune cells, thereby initiating the inflammatory cascade essential for host defense (Liu et al. 2015; Johnston et al. 2016; Anitua et al. 2018; Hossein-Khannazer et al. 2019). While these inflammatory cytokines are vital for the immune response against pathogens, their unchecked or prolonged activity can lead to significant tissue damage, ultimately leading to detrimental effects on the dental pulp.

In parallel, eicosanoids—bioactive signaling molecules derived from arachidonic acid—play a critical role in a wide range of physiological and pathological processes, including inflammation, immune modulation, and pain sensation. The role of eicosanoids in dental pulp disease has been extensively documented in various studies (McGowan et al. 2013; Petrini et al. 2012). For example, levels of prostaglandin E2 (PGE2) are frequently reported to be elevated in inflamed dental pulp tissues. Increased PGE2 can enhance the expression of additional inflammatory mediators, creating a positive feedback loop that perpetuates inflammation within the pulp (Ohsaki et al. 2006; Wang et al. 2009; Wahyudi et al. 2023). Furthermore, eicosanoids are implicated in the pathology of apical periodontitis, a common sequela of untreated dental pulp disease. Research has demonstrated that the upregulation of specific eicosanoids in periapical tissues contributes to bone resorption and exacerbates inflammatory processes associated with apical periodontitis (Ribeiro-Santos et al. 2019; Paula-Silva et al. 2020a, b; Petean et al. 2021).

The intricate interplay between cytokines and growth factors in pulp lesions orchestrates a balanced response to injury. Inflammatory cytokines serve to initiate the inflammatory response, whereas growth factors are essential for the subsequent repair and regeneration of damaged tissues. Dysregulation of this delicate balance can lead to chronic inflammation, impaired healing processes, and ultimately, irreversible tissue damage. A deeper understanding of these molecular dynamics is vital for the development of targeted therapeutic strategies aimed at restoring homeostasis within the dental pulp and promoting effective tissue regeneration. By elucidating the roles of specific cytokines and eicosanoids in the pathogenesis of pulp diseases, researchers and clinicians can develop more effective interventions that mitigate destructive inflammatory responses while enhancing the regenerative potential of pulp tissues.

26.3 Periodontal Disease

The periodontium is defined as all the supporting tissues surrounding the tooth, comprising root cementum, periodontal ligament, alveolar bone, and gingiva (Nanci and Bosshardt 2006). Periodontitis is a chronic inflammatory disease characterized by a microbial dysbiosis resulting from the colonization and persistence of pathogenic bacteria, including *Porphyromonas gingivalis*, *Tannerella forsythia*, and *Treponema denticola*, which adhere to the subgingival biofilm. This dysbiotic environment subsequently disrupts the host immune balance, leading to alveolar bone resorption around the affected tooth, rendering periodontitis one of the most prevalent inflammatory diseases afflicting humans (Berezow and Darveau 2010; Haque et al. 2022).

Within the periodontal environment, periodontal ligament stem cells (PDLSCs) hold substantial therapeutic potential for preserving periodontal homeostasis. These cells exhibit remarkable immunomodulatory capabilities, enabling them to interact synergistically with immune cells during the inflammatory processes associated with periodontal disease (Haque et al. 2022; Arora et al. 2022). PDLSCs are recognized as a crucial cellular source for the regeneration of periodontium, facilitating tissue repair and regeneration in response to injury (Song et al. 2015).

PDLSCs exhibit self-renewal capacity and differentiation into osteoblastic and cementoblastic lineages, thereby playing a role in the maintenance of periodontal tissue homeostasis (Gronthos et al. 2000; Paula-Silva et al. 2010). PDLSCs exert their effects in a paracrine manner, significantly impacting surrounding immune cells, such as macrophages, neutrophils, eosinophils, and monocytes. These immune cells recognize extracellular signals and bacterial stimuli, thereby initiating an inflammatory cascade through the release of various pro-inflammatory mediators (Liu et al. 2019). The regulatory interactions between PDLSCs and immune cells are essential for the effective repair and regeneration of periodontal tissue. However, the chronic inflammation associated with periodontitis can lead to diminished osteogenic differentiation capacities of PDLSCs, ultimately compromising periodontal regeneration (Tang et al. 2016).

In instances of tissue damage or cellular stress, local cells release alarmins and DAMPs, which serve as danger signals to activate the immune system and initiate an immune response. Toll-like receptors (TLRs) and NOD-like receptors (NLRs) are components of the innate immune response, as they recognize PAMPs. This recognition activates a complex cascade of intracellular signaling events, resulting in the activation of NF-κB and other interferon regulatory factors, ultimately regulating the expression of genes associated with the inflammatory response (Julier et al. 2017). Pathogen-associated molecular patterns (PAMPs) are typically found on the surface of microorganisms and play a crucial role in recognition by the innate immune system. Among these components, lipopolysaccharides (LPS), which are characteristic of Gram-negative bacteria, and lipoteichoic acid (LTA), present in Gram-positive bacteria, are well-studied examples. These molecules are recognized by specialized receptors, triggering immune responses that help protect the body against infections (Han et al. 2015). Upon recognition of PAMPs, pro-inflammatory cytokines such as interleukin-1 (IL-1) and TNF-α are released by innate immune cells to attract circulating immune cells, including neutrophils, monocytes, and macrophages, to the site of inflammation. Additionally, these cytokines activate local macrophages into a pro-inflammatory M1 phenotype (Julier et al. 2017).

Dendritic cells, monocytes, and macrophages play a role in initiating the anti-inflammatory response following tissue injury, particularly within the context of periodontal disease. These immune cells are highly specialized at phagocytosing bacteria and cellular debris, thereby contributing to infection clearance and the removal of damaged tissue to promote the resolution of inflammation. Concurrently, T and B lymphocytes are essential participants in tissue regeneration processes, further elaborating the immune landscape during healing responses (Kolaczkowska and Kubes 2013).

T cells, particularly during the pro-inflammatory phase, release a multitude of cytokines that exacerbate inflammation and further recruit additional immune cells to the affected site. This phase is characterized by increased cytokine production, which facilitates the recruitment and activation of other immune effectors, including osteoclasts—cells that mediate bone resorption. This process directly influences the structural integrity of the periodontal tissues, contributing to alveolar bone (Yuce et al. 2017; Galler et al. 2021). As inflammation resolves, T cells also transit into the anti-inflammatory phase, where certain subsets, such as regulatory T cells (Tregs) and T helper 2 (Th2) cells, play critical roles. These subsets are involved in downregulating inflammatory responses, maintaining local tissue homeostasis, and promoting tissue healing. Tregs are particularly influential, as they can induce the proliferation of mesenchymal stem cells (MSCs) by releasing TGF-β and other growth factors. This interaction facilitates the differentiation of these stem cells into terminal, functional cell types that are essential for effective tissue regeneration (Xu et al. 2021; Gu and Han 2020; Julier et al. 2017).

The regeneration of the periodontium presents a complex challenge due to the sophisticated architecture and diverse functions of different types of tissues in the periodontium. In the context of periodontitis, mesenchymal stem cells emerge as critical mediators, capable of modulating the inflammatory response while concurrently facilitating tissue regeneration. Their ability to communicate with both immune cells and the local extracellular environment underscores their potential therapeutic value in periodontal rehabilitation.

Understanding the intricate interactions between stem cells, immune cells, and the inflammatory milieu is fundamental for leveraging the immunomodulatory properties of MSCs for periodontal regeneration. Such insights can inform innovative therapeutic strategies designed to enhance the regenerative potential of periodontal tissues, thereby improving clinical outcomes for patients suffering from periodontal disease. Approaches that effectively manipulate the local immune environment to promote healing and regeneration could transform periodontal therapies, shifting treatment strategies toward biologically driven regeneration rather than disease management (Wada et al. 2009; Liu et al. 2012; Wang et al. 2020). As research progresses, elucidating these cellular interactions will allow the development of advanced regenerative techniques that utilize the innate capabilities of stem cells within the periodontal microenvironment.

26.4 Orthodontic Tooth Movement

Orthodontic tooth movement refers to the biological and mechanical processes involved in the repositioning of teeth through the application of controlled forces over time, typically accomplished during orthodontic treatment. This process is essential for correcting dental malocclusions, which can lead to functional, aesthetic, and health issues. The success of OTM is reliant on a combination of mechanical principles and biological responses of the periodontal structures.

The application of mechanical force to teeth induces micro-injuries within the periodontium, initiating a sterile inflammatory response that activates a series of coordinated biological processes resulting in both bone resorption and apposition on opposing sides of the force application (Krishnan and Davidovitch 2006). This response is intricately orchestrated by hematopoietic immune cells of myeloid and lymphoid origins, which engage in complex interactions with mesenchymal-derived cells within the dental and periodontal tissues to regulate bone turnover and maintain periodontal health (Chaushu et al. 2022; Nakai et al. 2023) (Fig. 26.3).

Upon the application of mechanical stimulus (Fig. 26.3a), the acute inflammatory reaction is triggered within the periodontium, initiating a characteristic innate immune response. This response is marked by the rapid infiltration of neutrophils, which are among the first responders to sites of injury (Marcaccini et al. 2010; Klein et al. 2020; Wald et al. 2021). These neutrophils secrete various chemotactic mediators that recruit monocytes and macrophages to the affected area (Klein et al. 2020). In addition to neutrophils, the presence of other granulocytes, as well as natural killer cells, has also been observed (Yamasaki et al. 1982; Klein et al. 2020). Granulocytes play a key role in initiating inflammation by releasing histamine and promoting the synthesis of eicosanoids, which further enhance the recruitment of leukocytes and stimulate the secretion of pro-inflammatory cytokines (Groeger et al. 2020).

Interestingly, research has demonstrated that the adaptive immune system participates cooperatively during this immediate response to mechanical stress, leading to the upregulation of resident γδ T cells, B cells (CD220+), and various subsets of T cells, including CD4+ T helper cells (Th1, Th7) (Alhashimi et al. 2004; Kook et al. 2011; Li et al. 2020; Klein et al. 2020; Wald et al. 2021). γδ T cells, in particular, have been found to play an essential role in recruiting additional neutrophils, monocytes, and osteoclasts during OTM (Wald et al. 2021).

Monocytes, which are myeloid-derived precursors, are critical for differentiating into macrophages and osteoclasts, while macrophages and dendritic cells can also differentiate into osteoclasts in the presence of key osteoclastogenic factors, such as RANKL and macrophage colony-stimulating factor (M-CSF) (Boyce 2013; Wu et al. 2019). This cascade of cellular events ultimately results in an increased population of osteoclasts within the periodontal ligament following the application of mechanical force (Kook et al. 2011; Yan et al. 2015; Zeng et al. 2015).

The alteration of the periodontal microenvironment is associated with elevated levels of various cytokines that critically modulate the inflammatory response (Alikhani et al. 2015, 2018; Yan et al. 2015; Liu et al. 2017; Klein et al. 2020). Most pro-inflammatory cytokines exert their effects either through direct or indirect upregulation of RANK expression in preosteoclasts and mature osteoclasts, or by stimulating RANKL expression in periodontal ligament cells (PDLCs), osteoblasts, osteocytes, and even T lymphocytes (Suzawa et al. 2000; Tomita et al. 2002; Kitaura et al. 2004; Yamaguchi 2009; Jiang et al. 2015; Shoji-

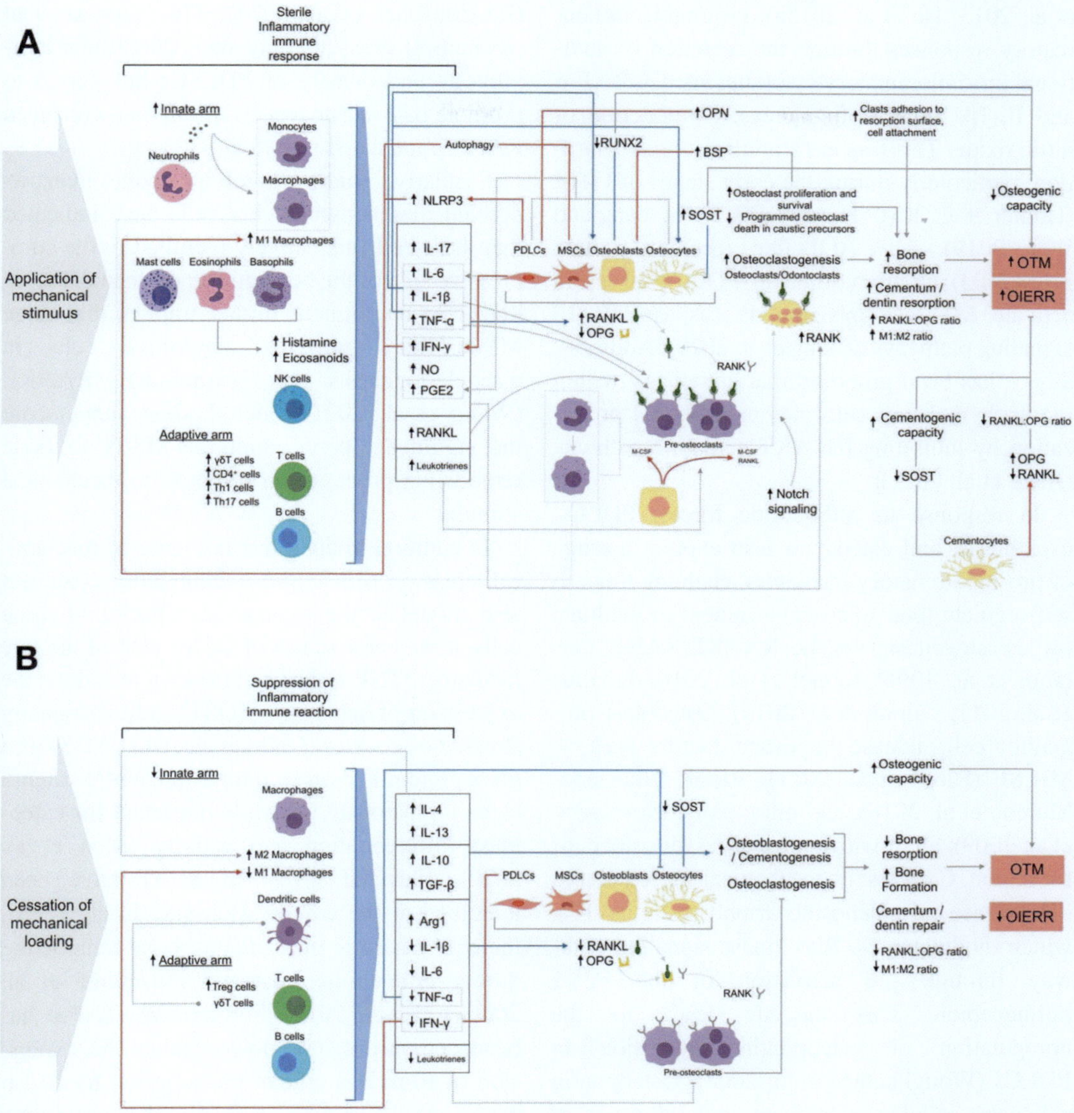

Fig. 26.3 Cross talk of mesenchymal and hematopoietic stem cells in orthodontic tooth movement (**a**) following the application of mechanical stimulus and (**b**) the cessation of mechanical loading

Matsunaga et al. 2017; Kato 2020; Wu et al. 2020; Miyazaki et al. 2021).

During OTM, key pro-inflammatory mediators such as TNF-α, IL-1β, IL-6, interleukin-17 (IL-17), interferon-gamma (IFN-γ), and PGE2 are significantly upregulated at the site of compression, contributing to the recruitment and activation of osteoclast precursors (Davidovitch et al. 1988; Alhashimi et al. 2000; Yan et al. 2015; Jayaprakash et al. 2019; Schroder et al. 2020; Wald et al. 2021). These mediators stimulate or enhance osteoclastogenesis primarily through the RANKL/RANK signaling axis, which is well-documented for its central role in bone resorption and remodeling in the literature (Udagawa et al. 1995; Hofbauer et al. 1999; Lam et al. 2000; Kanzaki et al. 2002; Palmqvist et al. 2002; Wei et al. 2005; Mayahara et al. 2012; Salla et al. 2012; Lin et al. 2014; Tanaka 2017; Schroder et al. 2020; Gong et al. 2023).

IFN-γ specifically plays a role in macrophage polarization towards the M1 phenotype (Davies et al. 2013; He et al. 2015a), promoting inflammatory responses through the secretion of additional pro-inflammatory cytokines such as TNF-α and IL-1β, along with increased production of nitric oxide. This process stimulates bone resorption, particularly during the early stages of OTM (Hunter et al. 2010; He et al. 2015a, b; Yang and Wan 2019). The NOD-like receptor protein 3 (NLRP3) inflammasome in PDLCs serves to activate M1 macrophages via caspase-1/IL-1β signaling pathways (Zhang et al. 2020). Additionally, it has been proposed that autophagy within mesenchymal stem cells may promote M1 polarization by inhibiting the AKT signaling pathway (Jiang et al. 2021).

In response to orthodontic forces, PDLCs, osteoblasts, and osteocytes also express a range of pro-inflammatory mediators, enabling them to self-regulate their microenvironment to facilitate osteoclastogenesis via the RANKL/RANK axis (Kaji et al. 1996; Kamel et al. 2010; Ignatius et al. 2011; Ullrich et al. 2019). Osteoblast progenitor cells release important factors such as M-CSF (Hioki et al. 2021), VEGF (Manzano-Moreno et al. 2018), and nitric oxide (Sosroseno et al. 2009), all of which are critical for osteoclast formation. Conversely, mechanical stress induces the release of adenosine triphosphate (ATP), which modulates the Rho kinase signaling pathway through the activation of the P2Y1 purinoceptor. This cascade leads to the upregulation of osteopontin expression in PDLCs (Wongkhantee et al. 2008). Osteopontin enhances cellular attachment and adhesion of osteoclasts to resorptive bone surfaces, thereby facilitating effective bone resorption.

MSCs are notably responsive to mechanical stress, as evidenced by several studies demonstrating their adaptive behavior under varying mechanical conditions. For instance, PDLSCs subjected to in vitro compressive forces exhibit increased cell confluence, accompanied by a significant reduction in the secretion levels of OPG, particularly in cells derived from maxillary periodontal tissues. Concurrently, there is a notable trend toward an increase in RANKL expression in these compressed cells (Radermacher et al. 2024). This suggests that mechanical stress not only alters the cellular morphology and density of PDLSCs but may also promote osteoclastogenesis through an enhanced RANKL-mediated signaling pathway.

Similarly, studies involving bone marrow-derived mesenchymal stem cells have indicated that these cells are actively recruited to the compressive side of the bone during mechanical loading. This recruitment further implies that these MSCs may play a supportive role in osteoclastogenesis by expressing RANKL (Wang et al. 2021). Such findings underscore the significant contribution of MSCs to bone remodeling processes in response to mechanical stimuli.

In addition to the direct influence of mechanical stress on MSCs, pro-inflammatory cytokines also modulate the osteogenic capacity of these cells at the early stages of OTM. One of the key mediators, TNF-α, has been shown to induce the expression of sclerostin (SOST) in mechanically stressed osteocytes (Ohori et al. 2019). SOST is a glycoprotein known to inhibit the Wnt/β-catenin signaling pathway, which is important for osteoblast differentiation and activity (Men et al. 2020). Elevated levels of SOST have been identified at the compression site during OTM, indicating its role in regulating bone remodeling during orthodontic treatment (Nishijima et al. 2006; Lu et al. 2019). Furthermore, TNF-α has been demonstrated to downregulate the expression of Runx2, a critical transcription factor for osteoblast differentiation, while also promoting apoptosis in osteoblasts (Baum and Gravallese 2014). Also, IL-1β has been identified as an inhibitory factor for both migration and differentiation of osteoblasts, further complicating the dynamics of bone remodeling during inflammation (Hengartner et al. 2013).

Emerging evidence suggests that both IL-1β and PGE2 can attenuate the secretion of OPG in PDLCs, impacting the delicate balance between RANKL and OPG in the local microenvironment (Fujisaki et al. 2007; Mayahara et al. 2012). This alteration in the RANKL/OPG ratio is critical, as it shifts the balance in favor of osteoclastogenesis over osteogenesis, thereby promoting bone resorption and influencing the overall health and integrity of periodontal tissues.

As mechanical stress diminishes, the initial inflammatory response gradually subsides, initiating a resolution phase during which the periodontal tissue seeks to restore homeostasis (Chaushu et al. 2022). However, given that the mechanical stimulus persists throughout OTM, this resolution is only partially achieved. During this phase, there is concomitant proliferation of preosteoblasts and osteoclasts (Holland et al. 2019; Li et al. 2020). Despite the progression towards resolution, a predominance of M1 macrophages continues to be evident, thereby sustaining ongoing bone resorption and facilitating the continued biomechanics of OTM (He et al. 2015a).

At this stage, while the populations of CD4+ T helper cells, including Th1 and Th17 cells, decline, there is a notable increase in the percentage of γδ T cells and Tregs (Li et al. 2020; Wald et al. 2021). Elevated levels of IL-17 persist in areas of periodontal ligament compression, predominantly produced by γδ T cells, which serve as a bridge between the innate and adaptive immune systems (Madureira et al. 2015; Wald et al. 2021). Treg cells play a critical role in modulating the inflammatory response, promoting bone formation, and selectively suppressing pro-inflammatory cytokine production associated with T-cell activity (Li et al. 2020).

As mechanical stress continues to decrease, there is a corresponding reduction in the activity of the innate immune system (Fig. 26.3b). This transition is characterized by a significant reduction in the proportions of neutrophils, monocytes, γδ T cells, and CD4+ T cells (Wald et al. 2021), while there is a marked increase in B cells and CD3+ T cells, indicating a shift toward the adaptive immune response (Wald et al. 2021). As the inflammatory milieu changes, there is a concomitant decrease in the expression of pro-inflammatory cytokines, including TNF-α, IL-1β, IL-6, and IFN-γ (Alhashimi et al. 2001; Alikhani et al. 2015; Klein et al. 2020), further favoring the downregulation of RANKL.

The reduction in mechanical stress, coupled with decreased levels of TNF-α, results in lower expression of SOST by osteocytes, thereby enhancing the osteogenic capacity while reducing both osteoclastic activity and RANK expression (Shu et al. 2017). The diminishing presence of IFN-γ also facilitates a decrease in M1 macrophages (He et al. 2015a). Interestingly, certain studies suggest that IFN-γ may exert a suppressive effect on osteoclastogenesis (Takayanagi et al. 2000). For instance, the injection of IFN-γ has been shown to reduce OTM by diminishing the number of osteoclasts on the compression side (Kohara et al. 2012).

In contrast to the observed decrease in pro-inflammatory cytokines, there is an increase in the levels of anti-inflammatory cytokines, such as interleukin 10 (IL-10) and TGF-β, which promotes osteoblastogenesis (Garlet et al. 2007; Klein et al. 2020). Following the resorption of the bone matrix by osteoclasts, matrix-associated factors, including TGF-β and insulin-like growth factors (IGFs), are released to regulate the activity of osteoblasts (Baum and Gravallese 2014). Furthermore, cytokines such as IL-4 and IL-13 are upregulated during this phase, resulting in the activation of macrophage polarization toward the M2 phenotype (Davies et al. 2013; He et al. 2015a). This M2 polarization promotes the cessation of bone resorption and initiates the tissue repair processes essential for restoring periodontal integrity. IL-4 and IL-10 are upregulated by M2 macrophages to suppress osteoclastogenesis by inhibiting the Nuclear factor of activated T cells 1 (NFATc1) (Yang and Wan 2019). T cells induce B cells to release OPG through the expression of CD40 ligand that binds to CD40 on the surface of B cells (Li et al. 2007).

All the processes previously discussed influence the delicate balance between the activities of osteoclasts and osteoblasts, ultimately determining whether bone resorption or apposition occurs. The application of mechanical stimulation, particularly on the compressive side of the periodontal ligament, enhances resorption through the upregulation of RANKL and the downregulation of OPG (Garlet et al. 2007; Castroflorio et al.

2017). Research indicates that when compressive forces in the range of 2.0 to 3.0 g/cm^2 are applied to PDLCs, there is a measurable increase in RANKL and a corresponding decrease in OPG levels, all in a force-dependent manner (Tyrovola 2015). Specifically, as the magnitude of the applied force increases, the RANKL/OPG ratio escalates, leading to enhanced osteoclastogenesis and bone resorption.

Upon cessation of mechanical stress, RANKL expression may persist (Alikhani et al. 2018; Kirschneck et al. 2020); however, the levels of OPG typically increase (Garlet et al. 2007). This rise in OPG expression contributes to the reduction of the RANKL/OPG ratio, thereby decreasing bone resorption activity and promoting bone formation in the aftermath of mechanical loading.

Orthodontically induced root resorption (OIERR) represents a pathological side effect of OTM, characterized by the concomitant resorption of cementum and/or dentin mediated by osteoclast-like cells known as odontoclasts (Wang and McCauley 2011). The underlying cellular and molecular mechanisms that govern bone resorption and repair processes similarly apply to root resorption.

The ratio of M1 to M2 macrophages has been identified as a significant indicator of OIERR progression. An elevated M1/M2 ratio correlates with increased expression of inflammatory mediators such as TNF-α, IFN-γ, and nitric oxide, which collectively amplify inflammatory responses and thereby favor the occurrence of OIERR. Conversely, a reduced M1/M2 ratio is associated with higher levels of anti-inflammatory cytokines, including IL-4, IL-10, and arginase I, which can mitigate the severity of root resorption (He et al. 2015a). Additionally, Notch signaling has been implicated in the stimulation of OIERR through the upregulation of RANKL and IL-6 in PDLCs (Kikuta et al. 2015).

The role of IL-1β in OIERR remains a subject of debate. Some studies report that the production of IL-1β, in conjunction with compressive forces, significantly induces RANKL expression in cementoblasts, suggesting that IL-1β contributes to OIERR during orthodontic treatment (Diercke et al. 2012). Conversely, other research suggests a negative correlation between IL-1β and OIERR (Viecilli et al. 2009). For instance, experiments involving IL-1β knockout (−/−) mice showed a significant increase in OIERR, indicating that the absence of IL-1β may exacerbate root resorption during orthodontic interventions (Al-Qawasmi et al. 2004).

The ratio of RANKL to OPG is significantly elevated in patients experiencing OIERR (Yamaguchi et al. 2006). Several studies have identified protective mechanisms that counteract this increase and may prevent or mitigate the occurrence of OIERR. It is hypothesized that resident MSCs reduce RANKL expression in clastic cells while simultaneously enhancing the expression of OPG (Zheng et al. 2015). Evidence supporting this hypothesis includes findings that demonstrate the transfer of MSCs to the periodontal ligament during OTM effectively supports these protective responses (Amuk et al. 2020). Notably, the transfer of MSCs during the retention phase, when mechanical loading is not applied, has proven more effective in preventing and repairing OIERR than during the active phase of orthodontic force application (Amuk et al. 2020).

Interestingly, the behavior of cementocytes in response to mechanical stimulation differs markedly from that of osteocytes. Research indicates that cementocytes can significantly upregulate OPG levels while concurrently downregulating RANKL expression, resulting in a decreased RANKL/OPG ratio. This protective mechanism helps inhibit cementum resorption (Walker et al. 2008; Xiong et al. 2015). Moreover, cementocytes have been shown to suppress *Sost* expression in cementum under mechanical stress, thereby promoting the formation of new cellular cementum. This process is facilitated through the modulation of Wnt signaling activity via SOST inhibition (Lim et al. 2014).

In summary, the interplay between various immune cell types in response to mechanical forces is critical for the modulation of inflammatory and regenerative processes within the periodontal environment. A comprehensive understanding of these dynamic interactions is essential for devising effective strategies in orthodontic treatment and periodontal regeneration, as well as for elucidating the underlying mechanisms that drive bone remodeling in response to mechanical stimuli. Enhanced insight into these biological responses may lead to the development of improved interventions that optimize clinical outcomes in both dental and orthodontic practices.

26.5 Conclusion and Future Perspectives

The signaling cascades that are activated through the crosstalk between hematopoietic and mesenchymal cells, within the context of bone and dental immune responses, demonstrate that these cells are key players in not only the inflammatory stimuli but also in orchestrating critical processes of tissue regeneration and repair.

A deeper understanding of the molecular mechanisms underlying this crosstalk may reveal potential innovative diagnostic strategies and targeted treatments in regenerative dentistry. Key implications include the identification of novel diagnostic biomarkers that could facilitate early detection and monitoring of periodontal and pulpal pathologies, as well as the host response to orthodontic tooth movement.

The incorporation of biomolecules into biomaterials that support and promote essential cellular responses represents another significant frontier in regenerative dentistry. The advancement of specific cell therapies offers an additional avenue for treatment, potentially allowing for the targeted repair of damaged tissues through the strategic application of cells. Additionally, the innovative use of gene silencing technologies presents an exciting possibility for modulating inflammatory responses and promoting tissue repair, which could be pivotal in managing chronic inflammatory conditions associated with oral and periodontal diseases.

In conclusion, the dynamic interplay between hematopoietic and mesenchymal cells in the bone and dental immune context represents a rich area of research with profound implications for regenerative therapies. Advancing this field requires continued interdisciplinary collaboration and robust ethical oversight to navigate future challenges and optimize innovative regenerative solutions.

References

Alhashimi N, Frithiof L, Brudvik P, Bakhiet M (2000) Orthodontic movement induces high numbers of cells expressing IFN-gamma at mRNA and protein levels. J Interf Cytokine Res 20(1):7–12

Alhashimi N, Frithiof L, Brudvik P, Bakhiet M (2001) Orthodontic tooth movement and de novo synthesis of proinflammatory cytokines. Am J Orthod Dentofacial Orthop 119(3):307–312

Alhashimi N, Frithiof L, Brudvik P, Bakhiet M (2004) CD40-CD40L expression during orthodontic tooth movement in rats. Angle Orthod 74(1):100–105

Alikhani M, Alyami B, Lee IS, Almoammar S, Vongthongleur T, Alikhani M, Alansari S, Sangsuwon C, Chou MY, Khoo E, Boskey A, Teixeira CC (2015) Saturation of the biological response to orthodontic forces and its effect on the rate of tooth movement. Orthod Craniofac Res 18(Suppl 1):8–17

Alikhani M, Sangsuwon C, Alansari S, Nervina JM, Teixeira CC (2018) Biphasic theory: breakthrough understanding of tooth movement. J World Fed Orthod 7(3):82–88

Almeida-Junior LA, Araujo LDC, Lamarque GCC, Arnez MFM, Kapila YL, Silva LABD, Paula-Silva FWG (2023) Reparative dentin formation following dental pulp capping is mediated by TNFR1 in vivo. J Endodont 49(10):1329–1336

Al-Qawasmi RA, Hartsfield JK, Everett ET, Weaver MR, Foroud TM, Roberts WE (2004) Root resorption associated with orthodontic force in IL-1Beta knockout mouse. J Musculoskelet Neuronal Interact 4(4): 383–385

Amuk NG, Kurt G, Karsli E, Ozcan S, Acar MB, Amuk M, Lekesizcan A, Gurgan CA (2020) Effects of mesenchymal stem cell transfer on orthodontically induced root resorption and orthodontic tooth movement during orthodontic arch expansion protocols: an experimental study in rats. Eur J Orthod 42(3): 305–316

Anitua E, Troya M, Zalduendo MAR (2018) Progress in the use of dental pulp stem cells in regenerative medicine. Cytotherapy. https://doi.org/10.1016/j.jcyt.2017.12.011

Arana-Chavez VE, Massa LF (2004) Odontoblasts: the cells forming and maintaining dentine. Int J Biochem Cell Biol 36(8):1367–1373. https://doi.org/10.1016/j.biocel.2004.01.006

Arora P, Li W, Huang X, Yu W, Huang R, Jiang Q (2022) Chen C: metabolic reconfiguration activates the staminality and immunomodulation of PDLSCs. Int J Mol Sci:23 (7)

Barreiros D, Pucinelli CM, Oliveira KMH, Paula-Silva FWG, Nelson-Filho P, Silva LAB, Küchler EC, Silva RAB (2018) Immunohistochemical and mRNA expression of RANK, RANKL, OPG, TLR2 and MyD88 during apical periodontitis progression in mice. J Appl Oral Sci 26:e20170512

Baum R, Gravallese EM (2014) Impact of inflammation on the osteoblast in rheumatic diseases. Curr Osteoporos Rep 12(1):9–16

Berezow AB, Darveau RP (2010) Microbial shift and periodontitis. Periodontol 55(1):36–47

Bjørndal L, Simon S, Tomson PL, Duncan HF (2019) Management of deep caries and the exposed pulp. Int Endod J 52(7):949–973

Boyce BF (2013) Advances in the regulation of osteoclasts and osteoclast functions. J Dent Res 92(10):860–867

Castroflorio T, Gamerro EF, Caviglia GP, Deregibus A (2017) Biochemical markers of bone metabolism during early orthodontic tooth movement with aligners. Angle Orthod 87(1):74–81

Chaushu S, Klein Y, Mandelboim O, Barenholz Y, Fleissig O (2022) Immune changes induced by orthodontic forces: a critical review. J Dent Res 101(1): 11–20

Chen S, Gluhak-Heinrich J, Martinez M, Li T, Wu Y, Chuang HH et al (2008) Bone morphogenetic protein 2 mediates dentin sialophosphoprotein expression and odontoblast differentiation via NF-Y signaling. J Biol Chem 283(28):19359–19370. https://doi.org/10.1074/jbc.M709492200

Chen S, Gluhak-Heinrich J, Wang YH, Wu YM, Chuang HH, Chen L et al (2009) Runx2, osx, and dspp in tooth development. J Dent Res 88(10):904–909. https://doi.org/10.1177/0022034509342873

Cho YD, Yoon WJ, Woo KM, Baek JH, Park JC, Ryoo HM (2010) The canonical BMP signaling pathway plays a crucial part in stimulation of dentin sialophosphoprotein expression by BMP-2. J Biol Chem 285(47):36369–36376. https://doi.org/10.1074/jbc.M110.103093

Conrads G, About I (2018) Pathophysiology of dental caries. Monogr Oral Sci 27:1–10

Davidovitch Z, Nicolay OF, Ngan PW, Shanfeld JL (1988) Neurotransmitters, cytokines, and the control of alveolar bone remodeling in orthodontics. Dent Clin N Am 32(3):411–435

Davies LC, Jenkins SJ, Allen JE, Taylor PR (2013) Tissue-resident macrophages. Nat Immunol 14(10):986–995

Diercke K, Kohl A, Lux CJ, Erber R (2012) IL-1β and compressive forces lead to a significant induction of RANKL-expression in primary human cementoblasts. J Orofac Orthop 73(5):397–412

Farges JC, Alliot-Licht B, Renard E, Ducret M, Gaudin A, Smith AJ, Cooper PR (2015) Dental pulp defence and repair mechanisms in dental caries. Mediat Inflamm 2015:230251

Fujisaki K, Tanabe N, Suzuki N, Kawato T, Takeichi O, Tsuzukibashi O, Makimura M, Ito K, Maeno M (2007) Receptor activator of NF-kappa B ligand induces the expression of carbonic anhydrase II, cathepsin K, and matrix metalloproteinase-9 in osteoclast precursor RAW264.7 cells. Life Sci 80(14):1311–1318

Galler KM, Widbiller M, Buchalla W, Eidt A, Hiller KA, Hoffer PC, Schmalz G (2016) EDTA conditioning of dentine promotes adhesion, migration and differentiation of dental pulp stem cells. Int Endod J 49(6): 581–590

Galler KM, Weber M, Korkmaz Y, Widbiller M, Feuerer M (2021) Inflammatory response mechanisms of the dentine-pulp complex and the periapical tissues. Int J Mol Sci 22(3):1480

Garlet TP, Coelho U, Silva JS, Garlet GP (2007) Cytokine expression pattern in compression and tension sides of the periodontal ligament during orthodontic tooth movement in humans. Eur J Oral Sci 115(5):355–362

Gong T, Liu L, Jiang W, Zhou R (2020) DAMP-sensing receptors in sterile inflammation and inflammatory diseases. Nat Rev Immunol 20(2):95–112

Gong X, Sun S, Yang Y, Huang X, Gao X, Jin A, Xu H, Wang X, Liu Y, Liu J, Dai Q, Jiang L (2023) Osteoblastic STAT3 is crucial for orthodontic force driving alveolar bone remodeling and tooth movement. J Bone Miner Res 38(1):214–227

Groeger M, Spanier G, Wolf M, Deschner J, Proff P, Schroder A, Kirschneck C (2020) Effects of histamine on human periodontal ligament fibroblasts under simulated orthodontic pressure. PLoS One 15(8): e0237040

Gronthos S, Mankani M, Brahim J, Robey PG, Shi S (2000) Postnatal human dental pulp stem cells (DPSCs) in vitro and in vivo. Proc Natl Acad Sci USA 97(25):13625–13630. https://doi.org/10.1073/pnas.240309797

Gu Y, Han X (2020) Toll-like receptor signaling and immune regulatory lymphocytes in periodontal disease. Int J Mol Sci 21(9):3329

Han M-X, Ding C, Kyung HM (2015) Genetic polymorphisms in pattern recognition receptors and risk of periodontitis: evidence based on 12,793 subjects. Hum Immunol 76:496–504

Haque MM, Yerex K, Kelekis-Cholakis A, Duan K (2022) Advances in novel therapeutic approaches for periodontal diseases. BMC Oral Health 22(1):492

He W, Wang Z, Zhou Z, Zhang Y, Zhu Q, Wei K et al (2014) Lipopolysaccharide enhances Wnt5a expression through toll-like receptor 4, myeloid differentiating factor 88, phosphatidylinositol 3-OH kinase/AKT and nuclear factor kappa B pathways in human dental pulp stem cells. J Endodont 40(1):69–75. https://doi.org/10.1016/j.joen.2013.09.011

He D, Kou X, Luo Q, Yang R, Liu D, Wang X, Song Y, Cao H, Zeng M, Gan Y, Zhou Y (2015a) Enhanced M1/M2 macrophage ratio promotes orthodontic root resorption. J Dent Res 94(1):129–139

He D, Kou X, Yang R, Liu D, Wang X, Luo Q, Song Y, Liu F, Yan Y, Gan Y, Zhou Y (2015b) M1-like macrophage polarization promotes orthodontic tooth movement. J Dent Res 94(9):1286–1294

Hengartner NE, Fiedler J, Ignatius A, Brenner RE (2013) IL-1β inhibits human osteoblast migration. Mol Med 19(1):36–42

Hioki T, Tokuda H, Kuroyanagi G, Kim W, Tachi J, Matsuchima-Nishiwaki R, Iida H, Kozawa O (2021) Olive polyphenols attenuate TNF-α-stimulated M-CSF and IL-6 synthesis in osteoblasts: Suppression of Akt and p44/p42 MAP kinase signaling pathways. Biomed Pharmacother 141:111816

Ho CT, Huang YW, Chen TR, Lo CH, Lo WC (2021) Discovering the ultimate limits of protein secondary structure prediction. Biomolecules 11(11):1627. https://doi.org/10.3390/biom11111627

Hofbauer LC, Lacey DL, Dunstan CR, Spelsberg TC, Riggs BL, Khosla S (1999) Interleukin-1beta and tumor necrosis factor-alpha, but not interleukin-6, stimulate osteoprotegerin ligand gene expression in human osteoblastic cells. Bone 25(3):255–259

Holland R, Bain C, Utreja A (2019) Osteoblast differentiation during orthodontic tooth movement. Orthod Craniofac Res 22(3):177–182

Hossein-Khannazer N, Hashemi SM, Namaki S, Ghanbarian H, Sattari M, Khojasteh A (2019) Study of the immunomodulatory effects of osteogenic differentiated human dental pulp stem cells. Life Sci 216:111–118

Hu X, Zhong Y, Kong Y, Chen Y, Feng J, Zheng J (2019) Lineage-specific exosomes promote the odontogenic differentiation of human dental pulp stem cells (DPSCs) through TGFβ1/smads signaling pathway via transfer of microRNAs. Stem Cell Res Ther 10(1):170. https://doi.org/10.1186/s13287-019-1278-x

Hunter M, Wang A, Parhar K, Johnston M, Van Rooijen N, Beck PL, McKay DM (2010) In vitro-derived alternatively activated macrophages reduce colonic inflammation in mice. Gastroenterology 138(4):1395–1405

Ignatius A, Schoengraf P, Kreja L, Liedert A, Recknagel S, Kandert S, Brenner RE, Schneider M, Lambris JD, Huber-Lang M (2011) Complement C3a and C5a modulate osteoclast formation and inflammatory response of osteoblasts in synergism with IL-1β. J Cell Biochem 112(9):2594–2605

Janebodin K, Horst OV, Ieronimakis N, Balasundaram G, Reesukumal K, Pratumvinit B et al (2011) Isolation and characterization of neural crest-derived stem cells from dental pulp of neonatal mice. PLoS One 6(11): e27526. https://doi.org/10.1371/journal.pone.0027526

Jayaprakash PK, Basavanna JM, Grewal H, Modi P, Sapawat P, Bohara PD (2019) Elevated levels of Interleukin (IL)-1β, IL-6, tumor necrosis factor-α, epidermal growth factor, and β2-microglobulin levels in gingival crevicular fluid during human orthodontic tooth movement (OTM). J Family Med Prim Care 8(5):1602–1606

Jiang C, Li Z, Quan H, Xiao L, Zhao J, Jiang C, Wang Y, Liu J, Gou Y, An S, Huang Y, Yu W, Zhang Y, He W, Yi Y, Chen Y, Wang J (2015) Osteoimmunology in orthodontic tooth movement. Oral Dis 21(6):694–704

Jiang N, He D, Ma Y, Su J, Wu X, Cui S, Li Z, Zhou Y, Yu H, Liu Y (2021) Force-induced autophagy in periodontal ligament stem cells modulates M1 macrophage polarization via AKT signaling. Front Cell Dev Biol 9: 666631

Johnston CJC, Smyth DJ, Dresser DW, Maizels RM (2016) TGF-b in tolerance, development and regulation of immunity. Cell Immunol 299:14–22. https://doi.org/10.1016/j.cellimm.2015.10.006

Julier Z, Park AJ, Briquez PS, Martino MM (2017) Promoting tissue regeneration by modulating the immune system. Acta Biomater 53:13–28

Kaji H, Sugimoto T, Kanatani M, Fukase M, Kumegawa M, Chihara K (1996) Prostaglandin E2 stimulates osteoclast-like cell formation and bone-resorbing activity via osteoblasts: role of cAMP-dependent protein kinase. J Bone Miner Res 11(1): 62–71

Kamel MA, Picconi JL, Lara-Castillo N, Johnson M (2010) Activation of β-catenin signaling in MLO-Y4 osteocytic cells versus 2T3 osteoblastic cells by fluid flow shear stress and PGE2: implications for the study of mechanosensation in bone. Bone 47(5):872–881

Kanzaki H, Chiba M, Shimizu Y, Mitani H (2002) Periodontal ligament cells under mechanical stress induce osteoclastogenesis by receptor activator of nuclear factor kappa B ligand up-regulation via prostaglandin E2 synthesis. J Bone Miner Res 17(2):210–220

Kato M (2020) New insights into IFN-γ in rheumatoid arthritis: role in the era of JAK inhibitors. Immunol Med 43:72–78

Kikuta J, Yamaguchi M, Shimizu M, Yoshino T, Kasai K (2015) Notch signaling induces root resorption via RANKL and IL-6 from hPDL cells. J Dent Res 94(1):140–147

Kirschneck C, Bauer M, Gubernator J, Proff P, Schroder A (2020) Comparative assessment of mouse models for experimental orthodontic tooth movement. Sci Rep 10(1):12154

Kitaura H, Sands MS, Aya K, Zhou P, Hirayama T, Uthgenannt B, Wei S, Takeshita S, Novack DV, Silva MJ, Abu-Amer Y, Ross FP, Teitelbaum SL (2004) Marrow stromal cells and osteoclast precursors differentially contribute to TNF-alpha-induced osteoclastogenesis in vivo. J Immunol 173(8): 4838–4846

Klein Y, Fleissig O, Polak D, Barenholz Y, Mandelboim O, Chaushu S (2020) Immunorthodontics: in vivo gene expression of orthodontic tooth movement. Sci Rep 10(1):8172

Kohara H, Kitaura H, Yoshimatsu M, Fujimura Y, Morita Y, Eguchi T, Yoshida N (2012) Inhibitory effect of interferon-γ on experimental tooth movement in mice. J Interf Cytokine Res 32(9):426–431

Kolaczkowska E, Kubes P (2013) Neutrophil recruitment and function in health and inflammation. Nat Rev Immunol 13:159–175

Kook SH, Jang YS, Lee JC (2011) Human periodontal ligament fibroblasts stimulate osteoclastogenesis in response to compression force through TNF-α-mediated activation of CD4+ T cells. J Cell Biochem 112(10):2891–2901

Krishnan V, Davidovitch Z (2006) Cellular, molecular, and tissue-level reactions to orthodontic force. Am J Orthod Dentofacial Orthop 129(4):469.e461–e432

Lam J, Takeshita S, Barker JE, Kanagawa O, Ross FP, Teitelbaum SL (2000) TNF-alpha induces osteoclastogenesis by direct stimulation of macrophages exposed to permissive levels of RANK ligand. J Clin Invest 106(12):1481–1488

Li Y, Toraldo G, Li A, Yang X, Zhang H, Qian W, Weitzmann MN (2007) B cells and T cells are critical for the preservation of bone homeostasis and attainment of peak bone mass in vivo. Blood 109(9): 3839–3848

Li J, Yu TT, Yan HC, Qiao YQ, Wang LC, Zhang T, Li Q, Zhou YH, Liu DW (2020) T cells participate in bone remodeling during the rapid palatal expansion. FASEB J 34(11):15327–15337

Lim WH, Liu B, Hunter DJ, Cheng D, Mah SJ, Helms J (2014) Downregulation of Wnt causes root resorption. Am J Orthod Dentofacial Orthop 146(3):337–345

Lin D, Li L, Sun Y, Wang W, Wang X, Ye Y, Chen X, Xu Y (2014) IL-17 regulates the expressions of RANKL and OPG in human periodontal ligament cells via TRAF6/TBK1-JNK/NF-κB pathways. Immunology 144(3):472–485

Liu D, Xu J, Liu O et al (2012) Mesenchymal stem cells derived from inflamed periodontal ligaments exhibit impaired immunomodulation. J Clin Periodontol 39: 1174–1182

Liu J, Yu F, Sun Y, Jiang B, Zhang W, Yang J et al (2015) Concise reviews: characteristics and potential applications of human dental tissue-derived mesenchymal stem cells. Stem Cells 33(3):627–638. https://doi.org/10.1002/stem.1909

Liu Y, Zhang T, Zhang C, Jin SS, Yang RL, Wang XD, Jiang N, Gan YH, Kou XX, Zhou YH (2017) Aspirin blocks orthodontic relapse via inhibition of CD4(+) T lymphocytes. J Dent Res 96(5):586–594

Liu J, Chen B, Bao J, Zhang Y, Lei L, Yan F (2019) Macrophage polarization in periodontal ligament stem cells enhanced periodontal regeneration. Stem Cell Res Ther 10(1):320

Lorencetti-Silva F, Sales LS, Lamarque GCC, Caixeta GA, Arnez MFM, Faccioli LH, Paula-Silva FWG (2022a) Effects of inflammation in dental pulp cell differentiation and reparative response. Front Dent Med 3:942714. https://doi.org/10.3389/fdmed.2022.942714

Lorencetti-Silva F, Arnez MFM, Thomé JPQ, Carvalho MS, Carvalho FK, Queiroz AM, Faccioli LH, Paula-Silva FWG (2022b) Leukotriene B4 loaded in microspheres inhibits osteoclast differentiation and activation. Braz Dent J 33(5):35–45. https://doi.org/10.1590/0103-6440202204827. PMID: 36287497; PMCID: PMC9645171

Lu W, Zhang X, Firth F, Mei L, Yi J, Gong C, Li H, Zheng W, Li Y (2019) Sclerostin injection enhances orthodontic tooth movement in rats. Arch Oral Biol 99: 43–50

Ma D, Gao J, Yue J, Yan W, Fang F, Wu B (2012) Changes in proliferation and osteogenic differentiation of stem cells from deep caries in vitro. J Endodont 38(6):796–802. https://doi.org/10.1016/j.joen.2012.02.014

Madureira DF, da Silva JM, Teixeira AL, Abreu MH, Pretti H, Lages EM, da Silva TA (2015) Cytokine measurements in gingival crevicular fluid and periodontal ligament: are they correlated? Am J Orthod Dentofacial Orthop 148(2):293–301

Manzano-Moreno FJ, Costela-Ruiz VJ, Melguizo-Rodríguez L, Illescas-Montes R, García-Martínez O, Ruiz C, Ramos Torrecillas J (2018) Inhibition of VEGF gene expression in osteoblast cells by different NSAIDs. Arch Oral Biol 92:75–78

Marcaccini AM, Amato PA, Leao FV, Gerlach RF, Ferreira JT (2010) Myeloperoxidase activity is increased in gingival crevicular fluid and whole saliva after fixed orthodontic appliance activation. Am J Orthod Dentofacial Orthop 138(5):613–616

Mayahara K, Yamaguchi A, Takenouchi H, Kariya T, Taguchi H, Shimizu N (2012) Osteoblasts stimulate osteoclastogenesis via RANKL expression more strongly than periodontal ligament cells do in response to PGE(2). Arch Oral Biol 57(10):1377–1384

McGowan R, Hurley C, Walsh LJ (2013) The role of eicosanoids in the pathogenesis of dental pulp disease. Mol Oral Microbiol 28(1):17–28

Men Y, Wang Y, Yi Y, Jing D, Luo W, Shen B, Stenberg W, Chai Y, Ge W, Feng JQ, Zhao H (2020) Gli1+ periodontium stem cells are regulated by osteocytes and occlusal force. Dev Cell 54(5): 639–654.e6

Miyazaki T, Kurimoto R, Chiba T, Matsuchima T, Nakamichi R, Tsutsumi H, Takada K, Yagasaki L, Kato T, Shishido K, Kobayashi Y, Matsumoto T, Moriyama K, Asahara H (2021) Mkx regulates the orthodontic tooth movement via osteoclast induction. J Bone Miner Metab 39(5):780–786

Mogensen TH (2009) Pathogen recognition and inflammatory signaling in innate immune defenses. Clin Microbiol Rev 22(2):240–273

Monroe DG, McGee-Lawrence ME, Oursler MJ, Westendorf JJ (2012) Update on Wnt signaling in bone cell biology and bone disease. Gene 492(1): 1–18. https://doi.org/10.1016/j.gene.2011.10.044

Nakai Y, Praneetpong N, Ono W, Ono N (2023) Mechanisms of osteoclastogenesis in orthodontic tooth movement and orthodontically induced tooth root resorption. J Bone Metab 30(4):297–310

Nanci A, Bosshardt DD (2006) Structure of periodontal tissues in health and disease. Periodontol 40:11–28

Nishijima Y, Yamaguchi M, Kojima T, Aihara N, Nakajima R, Kasai K (2006) Levels of RANKL and OPG in gingival crevicular fluid during orthodontic tooth movement and effect of compression force on releases from periodontal ligament cells in vitro. Orthod Craniofac Res 9(2):63–70

Ohori F, Kitaura H, Marahleh A, Kishikawa A, Ogawa S, Qi J, Shen W, Noguchi T, Nara Y, Mizoguchi I (2019) Effect of TNF-α-induced sclerostin on osteocytes during orthodontic tooth movement. J Immunol Res 2019: 9716758

Ohsaki K, Kato H, Yamaoka T (2006) Prostaglandin E2 and its receptor in human dental pulp. J Endodont 32(9):872–877

Olive C (2012) Pattern recognition receptors: sentinels in innate immunity and targets of new vaccine adjuvants. Expert Rev Vaccin 11(2):237–256

Palmqvist P, Persson E, Conaway HH, Lerner UH (2002) IL-6, leukemia inhibitory factor, and oncostatin M stimulate bone resorption and regulate the expression of receptor activator of NF-kappa B ligand, osteoprotegerin, and receptor activator of NF-kappa B in mouse calvariae. J Immunol 169(6):3353–3362

Paula-Silva FWG, Ghosh A, Arzate H et al (2010) Calcium hydroxide promotes cementogenesis and induces cementoblastic differentiation of mesenchymal periodontal ligament cells in a CEMP1- and ERK-dependent manner. Calcif Tissue Int 87:144–157

Paula-Silva FWG, Arnez MFM, Petean IBF, Almeida-Junior LA, Silva RAB, Silva LAB, Faccioli LH (2020a) Effects of 5-lipoxygenase gene disruption on inflammation, osteoclastogenesis and bone resorption in polymicrobial apical periodontitis. Arch Oral Biol 112:104670

Paula-Silva FWG, Ribeiro-Santos FR, Petean IBF, Manfrin Arnez MF, Almeida-Junior LA, Carvalho FK, Silva LABD, Faccioli LH (2020b) Root canal contamination or exposure to lipopolysaccharide differentially modulate prostaglandin E 2 and leukotriene B 4 signaling in apical periodontitis. J Appl Oral Sci 28:e20190699. https://doi.org/10.1590/1678-7757-2019-0699. Epub 2020 May 11. PMID: 32401938; PMCID: PMC7213784

Petean IBF, Almeida-Junior LA, Arnez MFM, Queiroz AM, Silva RAB, Silva LAB, Faccioli LH, Paula-Silva FWG (2021) Celecoxib treatment dampens LPS-induced periapical bone resorption in a mouse model. Int Endod J 54(8):1289–1299. https://doi.org/10.1111/iej.13472. Epub 2021 May 28. PMID: 33403674

Petrini M, Ferrante M, Ciavarelli L, Brunetti L, Vacca M, Spoto G (2012) Prostaglandin E2 to diagnose between reversible and irreversible pulpitis. Int J Immunopathol Pharmacol 25(1):157–163. https://doi.org/10.1177/039463201202500118. PMID: 22507328

Radermacher C, Craveiro RB, Jahnen-Dechent W, Beier JP, Bülow A, Wolf M, Neuss S (2024) Impact of compression forces on different mesenchymal stem cell types regarding orthodontic indication. Stem Cells Transl Med szae057

Rao TP, Kuhl M (2010) An updated overview on Wnt signaling pathways: a prelude for more. Circ Res 106(12):1798–1806. https://doi.org/10.1161/CIRCRESAHA.110.219840

Ribeiro-Santos FR, Silva GGD, Petean IBF, Arnez MFM, Silva LABD, Faccioli LH, Paula-Silva FWG (2019) Periapical bone response to bacterial lipopolysaccharide is shifted upon cyclooxygenase blockage. J Appl Oral Sci 27:e20180641. https://doi.org/10.1590/1678-7757-2018-0641. PMID: 31166414; PMCID: PMC6546268

Roberts-Clark DJ, Smith AJ (2000) Angiogenic growth factors in human dentine matrix. Arch Oral Biol 45(11):1013–1016

Salla JT, Taddei SR, Queiroz-Junior CM, Junior IA, Teixeira MM, Silva TA (2012) The effect of IL-1 receptor antagonist on orthodontic tooth movement in mice. Arch Oral Biol 57(5):519–524

Schroder A, Kappler P, Nazet U, Jantsch J, Proff P, Cieplik F, Deschner J, Kirschneck C (2020) Effects of compressive and tensile strain on macrophages during simulated orthodontic tooth movement. Mediat Inflamm 2020:2814015

Shoji-Matsunaga A, Ono T, Hayashi M, Takayanagi H, Moriyama K, Nakashima T (2017) Osteocyte regulation of orthodontic force-mediated tooth movement via RANKL expression. Sci Rep 7(1):8753

Shu R, Bai D, Sheu T, He Y, Yang X, Xue C, He Y, Zhao M, Han X (2017) Sclerostin promotes bone remodeling in the process of tooth movement. PLoS One 12(1):e0167312

Sipert CR, Perez DEC, Paula-Silva FWG, Palo RM (2022) Aspectos Biológicos na Endodontia. In: Sousa-Neto MD, Duarte MAH, Gavini G, Baratto-Filho F, Estrela C (eds) Endodontia: fundamentos científicos para a prática clínica. Editora Manole Ltda, Santana do Parnaíba, pp 108–123

Smith AJ, Duncan HF, Diogenes A, Simon S, Cooper PR (2016) Exploiting the bioactive properties of the dentin-pulp complex in regenerative endodontics. J Endodont 42(1):47–56

Song IS, Han YS, Lee JH et al (2015) Periodontal ligament stem cells for periodontal regeneration. Curr Oral Health Rep 2:236–244

Sosroseno W, Bird PS, Seymour GJ (2009) Nitric oxide production by a human osteoblast cell line stimulated with Aggregatibacter actinomycetemcomitans lipopolysaccharide. Oral Microbiol Immunol 24(1): 50–55

Suzawa T, Miyaura C, Inada M, Maruyama T, Sugimoto Y, Ushikubi F, Ichikawa A, Narumiya S, Suda T (2000) The role of prostaglandin E receptor subtypes (EP1, EP2, EP3, and EP4) in bone resorption: an analysis using specific agonists for the respective EPs. Endocrinology 141(4):1554–1559

Takayanagi H, Ogasawara K, Hida S, Chiba T, Murata S, Sato K, Takaoka A, Yokochi T, Oda H, Tanaka K, Nakamura K, Taniguchi T (2000) T-cell-mediated regulation of osteoclastogenesis by signalling cross-talk between RANKL and IFN-gamma. Nature 408(6812): 600–605

Tanaka S (2017) RANKL-independent osteoclastogenesis: a long-standing controversy. J Bone Miner Res 32:431–433
Tang HN, Xia Y, Yu Y, Wu RX, Gao LN, Chen FM (2016) Stem cells derived from “inflamed” and healthy periodontal ligament tissues and their sheet functionalities: a patient-matched comparison. J Clin Periodontol 43(1):72–84
Tomita M, Li X, Okada Y, Woodiel FN, Young RN, Pilbeam CC, Raisz LG (2002) Effects of selective prostaglandin EP4 receptor antagonist on osteoclast formation and bone resorption in vitro. Bone 30(1): 159–163
Tyrovola JB (2015) The “mechanostat theory” of frost and the OPG/RANKL/ RANK system. J Cell Biochem 116(12):2724–2729
Udagawa N, Takahashi N, Katagiri T, Tamura T, Wada S, Findlay DM, Martin TJ, Hirota H, Taga T, Kishimoto T, Suda T (1995) Interleukin (IL)-6 induction of osteoclast differentiation depends on IL-6 receptors expressed on osteoblastic cells but not on osteoclast progenitors. J Exp Med 182(5):1461–1468
Ullrich N, Schröder A, Jantsch J, Spanier G, Proff P, Kirschneck C (2019) The role of mechanotransduction versus hypoxia during simulated orthodontic compressive strain-an in vitro study of human periodontal ligament fibroblasts. Int J Oral Sci 11(4):33
Viecilli RF, Katona TR, Chen J, Hartsfield JK, Roberts WE (2009) Orthodontic mechanotransduction and the role of the P2X7 receptor. Am J Orthod Dentofacial Orthop 135(6):694.e1–694.16
Wada N, Menicanin D, Shi S, Bartold PM, Gronthos S (2009) Immunomodulatory properties of human periodontal ligament stem cells. J Cell Physiol 219:667–676
Wahyudi R, Seang S, Everts V, Osathanon T, Limjeerajarus CN (2023) Anti-inflammatory effects of the prostacyclin analogue iloprost in an in vitro model of inflamed human dental pulp cells. Aust Endod J 49(Suppl 1):330–338. https://doi.org/10.1111/aej.12736. Epub 2023 Feb 1. PMID: 36723392
Wald S, Leibowitz A, Aizenbud Y, Saba Y, Zubeidat K, Barel O, Koren N, Heyman O, Wilharm A, Sandrock I, Fleissig O, Tal Y, Prinz I, Aizenbud D, Chaushu S, Hovav AH (2021) γδT cells are essential for orthodontic tooth movement. J Dent Res 100(7):731–738
Walker CG, Ito Y, Dangaria S, Luan X, Diekwisch TG (2008) Rankl, osteopontin, and osteoclast homeostasis in a hyperocclusion mouse model. Eur J Oral Sci 116(4):312–318
Wang Z, McCauley LK (2011) Osteoclasts and odontoclasts: signaling pathways to development and disease. Oral Dis 17(2):129–142
Wang X, Wu P, Liu Z (2009) Prostaglandin E2 levels in pulpal tissues of teeth affected by irreversible pulpitis. Int Endod J 42(7):652–658
Wang M, Xie J, Wang C, Zhong D, Xie L, Fang H (2020) Immunomodulatory properties of stem cells in periodontitis: current status and future prospective. Stem Cells Int 2020:9836518
Wang J, Jiao D, Huang X, Bai Y (2021) Osteoclastic effects of mBMMSCs under compressive pressure during orthodontic tooth movement. Stem Cell Res Ther 12(1):1–13
Wei S, Kitaura H, Zhou P, Ross FP, Teitelbaum SL (2005) IL-1 mediates TNF-induced osteoclastogenesis. J Clin Invest 115(2):282–290
Wongkhantee S, Yongchaitrakul T, Pavasant P (2008) Mechanical stress induces osteopontin via ATP/P2Y1 in periodontal cells. J Dent Res 87(6):564–568
Woo SM, Kim WJ, Lim HS, Choi NK, Kim SH, Kim SM et al (2016) Combination of mineral trioxide aggregate and platelet-rich fibrin promotes the odontoblastic differentiation and mineralization of human dental pulp cells via BMP/Smad signaling pathway. J Endodont 42(1):82–88. https://doi.org/10.1016/j.joen.2015.06.019
Wu M, Chen G, Li YP (2016) TGF-beta and BMP signaling in osteoblast, skeletal development, and bone formation, homeostasis and disease. Bone Res 4:16009. https://doi.org/10.1038/boneres.2016.9
Wu L, Luo Z, Liu Y, Jia L, Jiang Y, Du J, Guo L, Bai Y, Liu Y (2019) Aspirin inhibits RANKL-induced osteoclast differentiation in dendritic cells by suppressing NF-κB and NFATc1 activation. Stem Cell Res Ther 10(1):375
Wu L, Su Y, Lin F, Zhu S, Wang J, Hou Y, Du J, Liu Y, Guo L (2020) MicroRNA-21 promotes orthodontic tooth movement by modulating the RANKL/OPG balance in T cells. Oral Dis 26(2):370–380
Xiong J, Piemontese M, Onal M, Campbell J, Goellner JJ, Dusevich V, Bonewald L, Manolagas SC, O’Brien CA (2015) Osteocytes, not osteoblasts or lining cells, are the main source of the RANKL required for osteoclast formation in remodeling bone. PLoS One 10(9): e0138189
Xu XW, Liu X, Shi C, Sun HC (2021) Roles of immune cells and mechanisms of immune responses in periodontitis. Chin J Dent Res 24(4):219–230
Yamaguchi M (2009) RANK/RANKL/OPG during orthodontic tooth movement. Orthod Craniofac Res 12:113–119
Yamaguchi M, Aihara N, Kojima T, Kasai K (2006) RANKL increase in compressed periodontal ligament cells from root resorption. J Dent Res 85(8):751–756
Yamasaki K, Shibasaki Y, Fukuhara T (1982) Behavior of mast cells in periodontal ligament associated with experimental tooth movement in rats. J Dent Res 61(12):1447–1450
Yan Y, Liu F, Kou X, Liu D, Yang R, Wang X, Song Y, He D, Gan Y, Zhou Y (2015) T cells are required for orthodontic tooth movement. J Dent Res 94(10): 1463–1470
Yang D, Wan Y (2019) Molecular determinants for the polarization of macrophage and osteoclast. Semin Immunopathol 41(5):551–563

Yuce HB, Gokturk O, Turkal A, Inanir A, Benli I, Demir O (2017) Assessment of local and systemic 25-hydroxy-vitamin D, RANKL, OPG, and TNF levels in patients with rheumatoid arthritis and periodontitis. J Oral Sci 59:397–404

Zeng M, Kou X, Yang R, Liu D, Wang X, Song Y, Zhang J, Yan Y, Liu F, He D, Gan Y, Zhou Y (2015) Orthodontic force induces systemic inflammatory monocyte responses. J Dent Res 94(9):1295–1302

Zhang J, Liu X, Wan C, Liu Y, Wang Y, Meng C, Zhang Y, Jiang C (2020) NLRP3 inflammasome mediates M1 macrophage polarization and IL-1β production in inflammatory root resorption. J Clin Periodontol 47(4):451–460

Zheng Y, Chen M, He L, Marão HF, Sun DM, Zhou J, Kim SG, Song S, Wang SL, Mao JJ (2015) Mesenchymal dental pulp cells attenuate dentin resorption in homeostasis. J Dent Res 94(6):821–827

Zhou L, Zhao S, Xing X (2023) Effects of different signaling pathways on odontogenic differentiation of dental pulp stem cells: a review. Front Physiol 14:1272764. https://doi.org/10.3389/fphys.2023.1272764. PMID: 37929208; PMCID: PMC10622672

27 The Interplay of Immune Response and Dental Enamel Formation in Health and Sickness

Juliana de Lima Gonçalves, Roberta Duarte Leme, Júlia Ingryd Targino de Sousa, Alexandra Mussolino de Queiroz, Fabrício Kitazono de Carvalho, and Francisco Wanderley Garcia de Paula-Silva

Abstract

Tooth formation is a complex and multifaceted biological process, intricately regulated by a diverse array of molecular entities and signaling pathways. Within this context, amelogenesis specifically refers to the development of enamel, an essential structure that provides protection and structural integrity to teeth. During amelogenesis, various mediators of the immune response, including transforming growth factor-β (TGF-β), tumor necrosis factor-α (TNF-α), and c-Jun N-terminal kinase (JNK), play critical roles. Disruptions occurring during the crucial secretion and biomineralization phases of enamel development can lead to a spectrum of dental enamel defects, which are often characterized by structural anomalies. The etiology of these defects is inherently complex and multifactorial, encompassing genetic predispositions, systemic influences, and local environmental factors, which may act individually or in combination to contribute to the adverse outcomes observed. Recent investigations have begun to uncover the significant influence of immune response mechanisms on amelogenesis, suggesting that these immunological factors may play a substantial role in the pathogenesis of dental enamel defects. This chapter aims to elucidate the diverse roles of specific components of the immune system in the context of dental enamel development. Furthermore, it explores the intricate interactions between various molecular mediators and signaling pathways, detailing their collective influence on amelogenesis. A thorough understanding of the interplay between immunological factors and tooth development is critical for advancing the frontiers of diagnostic and therapeutic strategies. Such insights may pave the way for innovative approaches to address enamel defects and enhance oral health outcomes in affected individuals.

Keywords

Odontogenesis · Amelogenesis · Dental enamel defects · Immune response

J. de Lima Gonçalves · R. D. Leme · J. I. T. de Sousa
A. M. de Queiroz · F. K. de Carvalho
F. W. G. de Paula-Silva (✉)
Developmental Biology and Child Health Research Group, Ribeirão Preto School of Dentistry, University of São Paulo, São Paulo, Brazil
e-mail: franciscogarcia@forp.usp.br

N. Rezaei, H. Ziaei (eds.), *Oral Immunology*, Advances in Experimental Medicine and Biology 1492,
https://doi.org/10.1007/978-3-032-03176-1_27

27.1 Introduction

Tooth development is a multifaceted process that initiates during intrauterine life and continues through a series of well-defined stages. This developmental trajectory is governed by intricate signaling pathways, with specific molecules playing pivotal roles in ensuring proper morphogenesis. The mature tooth is composed of three principal components: dental pulp, dentin, and enamel, each exhibiting unique biochemical and structural characteristics that evolve throughout their respective developmental phases (Abramyan et al. 2021).

A central aspect of this process is the dynamic interplay between epithelial and mesenchymal tissues, which collectively give rise to specialized cell types responsible for the secretion and maturation of both the dentin and enamel matrices. Notably, it has become increasingly evident that immune and inflammatory mediators are implicated in various facets of tooth development. These mediators have an influence not only during physiological formation but also in the pathogenesis of dental defects (Gonçalves et al. 2022; Resende et al. 2023; Luo et al. 2021; Ikeda et al. 2019; Lacruz et al. 2017; Unterbrink et al. 2009).

The following sections will elucidate contemporary insights regarding the influence of the immune system on dental development, with particular emphasis on its critical role in the intricate process of enamel formation. By exploring the various immune mediators and their interactions with developmental pathways, we aim to underscore the significance of understanding immune contributions to tooth morphogenesis, as well as their potential implications for the pathogenesis of enamel defects. This exploration will provide a comprehensive overview of how immunological factors may shape not only the normal development of teeth but also the manifestation of dental anomalies.

27.2 Odontogenesis and Immune Response

Tooth formation, or odontogenesis, is a highly organized biological process that initiates with the localized thickening of the dental epithelium. This thickening subsequently invaginates into the underlying oral mesenchyme, leading to the formation of a structure known as the dental lamina (Kawashima and Okiji 2016). During the bud stage of development, ectomesenchymal tissue, which is derived from the neural crest, undergoes a process of condensation surrounding the dental epithelium. As development progresses to the cap stage, the epithelial cervical loops begin to elongate and enclose the condensed mesenchyme, surrounding the developing dental papilla, a structure essential for subsequent stages of tooth development (Rothová et al. 2012).

The dental papilla is composed of ectomesenchyme-derived cells that are destined for differentiation into mature dental pulp. Within the dental pulp, odontoblasts, which are specialized columnar cells with polarized nuclei, play a crucial role in the secretion and promotion of dentin biomineralization. These cells produce a mineralized extracellular matrix that forms the dentin, thereby contributing to the structural integrity and functional capacity of the tooth (Thesleff 2003; Luo et al. 2021; Kawashima and Okiji 2016).

Although the mechanisms underlying odontoblast function, including their roles in dentinogenesis, have been extensively studied, certain aspects of odontoblast polarity remain incompletely understood. Notably, the activation of the c-Jun N-terminal kinase (JNK) signaling pathway is one such aspect that warrants further exploration (Luo et al. 2021; Chang et al. 2019). The complexities of JNK signaling may play a pivotal role in regulating odontoblast polarization and functionality, influencing their ability to secrete and mineralize under both physiological and pathological conditions.

JNK is a member of the mitogen-activated protein kinase (MAPK) family and plays a role in cell proliferation, differentiation, migration, and apoptosis. Within the immune system, one of its primary functions is to modulate the inflammatory response by inducing the production of pro-inflammatory cytokines and activating immune cells, including macrophages and T cells (Yan et al. 2024). The JNK signaling pathway is activated through the phosphorylation of MAPK proteins by upstream kinases. Upon activation, JNK translocates from the cytoplasm to the nucleus, where it activates various transcription factors that regulate gene expression (Yan et al. 2024; Johnson and Nakamura 2007).

In tooth development, the JNK pathway is involved in odontoblast differentiation by mediating the polarization of mesenchymal cells, a critical step in this process (Luo et al. 2021; Wang et al. 2013). Also, there is a connection between JNK activation and the Wnt signaling pathway, which is essential for tooth formation (Rahman et al. 2022). When JNK was inhibited in cultured cells derived from mouse tooth germs, changes in the expression of Golgi matrix protein 130 kD (GM130), a protein associated with Golgi complex organization and cytoskeletal polarization, were observed. Under normal JNK activity, GM130 was localized to the lateral sides of odontoblasts, indicating proper nuclear polarization. However, JNK inhibition resulted in reduced GM130 levels and changed the spatial expression pattern. Additionally, the absence of JNK was found to affect the expression of Ras homolog family member A (RhoA), a protein vital for maintaining cell morphology by regulating cytoskeletal dynamics (Chang et al. 2019). Specifically, JNK inhibition led to decreased RhoA expression and activity in odontoblasts compared to those in which JNK was present. Collectively, these findings indicate that the JNK pathway is important to nuclear polarization, a pivotal process in odontoblast differentiation (Luo et al. 2021).

Growth factors are essential for tooth development. Transforming growth factor-β (TGF-β) is a multifunctional cytokine that orchestrates a variety of cellular processes, including proliferation, differentiation, chemotaxis, and apoptosis, in different cell types, such as monocytes, epithelial, mesenchymal, and neuronal cells (Niwa et al. 2018; Kubiczkova et al. 2012). Molecules from this superfamily participate in various physiological processes, including embryonic development, tissue repair, and tooth development (Kobayashi-Kinoshita et al. 2016; Cho et al. 2013; Benedete et al. 2008). Specifically in teeth, TGF-β isoforms are involved in odontoblast and ameloblast differentiation and dentin and enamel biomineralization (Chang et al. 2019; Oka et al. 2007; Thyagarajan et al. 2001). All three isoforms of TGF-β—namely TGF-β1, TGF-β2, and TGF-β3—are expressed throughout the various stages of odontogenesis (Niwa et al. 2018). Notably, TGF-β1 is present from the initiation stage of tooth development until adulthood, with its expression in odontoblasts persisting throughout the lifespan (Thyagarajan et al. 2001).

Studies have demonstrated that TGF-β1 is uniformly expressed in the dental papilla and inner enamel epithelium, with an increase in its levels in the odontoblast and ameloblast layers as well as in the intermediate stratum during dentin matrix formation (Fan et al. 1998; Benedete et al. 2008). TGF-β1 typically exerts stimulatory effects on mesenchymal cells, including pre-odontoblasts and odontoblasts, promoting their differentiation and functional activity during tooth development (Unterbrink et al. 2009).

Dentin sialophosphoprotein (DSPP) is a non-collagenous protein secreted by odontoblasts, vital for dentin secretion and mineralization (Unterbrink et al. 2009; D'Souza et al. 1998). Mutations in the *DSPP* gene are associated with dentinogenesis imperfecta, a hereditary condition characterized by malformations in tooth morphology and dentin structure (Martín-Vacas et al. 2022; Isono et al. 2021). DSPP is a dentin-associated protein that plays a critical role in the nucleation of hydroxyapatite crystals within dentin. The regulation of DSPP is crucial for proper odontoblast differentiation and dentin development, with TGF-β acting as one of the key mediators in this regulatory process. Overexpression of TGF-β1 in transgenic mice has been linked to dysregulation of DSPP expression and resulted in dentin malformations (Thyagarajan et al. 2001). The effects of TGF-β1 overexpression were further elucidated using a DSPP promoter-luciferase construct. This technique involves adding a specific genetic sequence to the promoter region of DNA, which initiates the transcription process. When a designated fragment is incorporated, transcription factors can influence gene expression, either enhancing or repressing it (Fokkema and Dunnen 2021).

Oka et al. (2007) examined the impact of TGF-β receptor 2 (TGF-βR2) deficiency on odontoblast differentiation and dentin mineralization in mice. Scanning electron microscopy (SEM) images revealed that transgenic mice displayed decreased dentin thickness and malformations in the dentin tubules and matrix. Additionally, while wild-type mice exhibited regular expression of DSPP and type I collagen, transgenic mice failed to express DSPP and showed reduced levels of type I collagen, underscoring the crucial role of TGF-β signaling in these processes (Pan et al. 2023; Seo et al. 2017).

The impact of the immune system on the molecular signaling involved in dentin formation remains underexplored, as information on this interaction is still limited. However, it is important to note that immune response signaling pathways and molecules are fundamental to the development of other dental structures, including enamel secretion and biomineralization (Gonçalves et al. 2022; Luo et al. 2021; Benedete et al. 2008).

27.3 Enamel Development

Dental enamel is the hardest biological structure in the human body, serving as the protective outer layer of the tooth crown (Lacruz et al. 2017; Gil-Bona and Bidlack 2020). Ameloblasts are highly specialized cells responsible for the formation of enamel and exhibit variation in their morphology and function throughout the distinct stages of amelogenesis. The morphology of ameloblasts is indicative of the specific phase of amelogenesis in which they participate (Bartlett 2013). The pre-secretory stage is characterized by the differentiation of epithelial cells from the inner dental epithelium into pre-ameloblasts. These pre-ameloblasts exhibit a cuboidal shape and are responsible for signaling the initiation of enamel formation (Bei 2009a, b). As odontoblasts deposit the first layer of dentin, a signaling cascade activates pre-ameloblasts, prompting them to elongate, polarize their nuclei, and accumulate essential organelles for enamel production within their cytoplasm (Bei 2009a, b). In the subsequent secretory stage, ameloblasts adopt a cylindrical morphology with polarized nuclei and cytoplasmic projections. Secretory ameloblasts are responsible for releasing minerals, such as calcium (Ca) and phosphorus (P), as well as enamel matrix proteins including amelogenin, ameloblastin, and enamelin (Bei 2009a, b; Gil-Bona and Bidlack 2020). Hydroxyapatite crystals are deposited in the extracellular space, and the enamel matrix proteins form a scaffold that guides the deposition of these crystals. As amelogenesis progresses through the secretory stage, the projections of ameloblasts become increasingly developed, resulting in the formation of Tomes' process (Bartlett 2013). Tomes' process is a specialized cellular extension that is crucial not only for releasing vesicles containing enamel matrix proteins and minerals but also for establishing the patterns of rods and prisms in the enamel layer (Bei 2009a, b). The following stage of amelogenesis is the maturation stage, during which ameloblasts begin to lose Tomes' process. During this stage, cytoplasmic projections secrete proteases such as matrix metalloproteinase-20 (MMP-20) and kallikrein-related peptidase 4 (KLK-4), which degrade the enamel matrix proteins, thereby facilitating the incorporation of minerals into the hydroxyapatite crystals and enhancing the mineral content of the enamel (Bartlett 2013; Bei 2009a, b). Once enamel is fully formed, ameloblasts undergo apoptosis, leaving behind a fully mineralized enamel layer (Liu et al. 2015).

27.4 Molecular Signaling During Amelogenesis

Amelogenesis is a highly intricate biological process that is essential for the formation of dental enamel, a vital structure known for its role in protecting the underlying tooth structures from various stresses and wear (Lacruz et al. 2017). This process relies on a diverse array of molecules and signaling pathways that coordinate the differentiation, function, and activities of ameloblasts, the specialized cells responsible for enamel synthesis.

Key components of this regulatory network include inflammatory cytokines, such as TGF-β and members of the tumor necrosis factor-α (TNF-α) superfamily. These molecules, along with their corresponding ligands and receptors, are crucial for both the secretory phase and the subsequent biomineralization stage of enamel development. TGF-β, for example, is known to modulate various cellular functions, including proliferation, differentiation, and the secretion of extracellular matrix proteins by ameloblasts, thereby ensuring the proper formation of the enamel matrix (Bussaneli et al. 2019).

Similarly, TNF-α and its associated superfamily members exhibit a range of functions that have implications for enamel development. These cytokines are known for their critical roles in mediating inflammatory responses and can influence cell signaling pathways that affect ameloblast activity, enamel matrix secretion, and mineral deposition. The interaction of these cytokines with their receptors on ameloblasts is essential for the regulation of gene expression and the orchestration of enamel matrix proteins, which are fundamental for successful enamel formation (Kobayashi-Kinoshita et al. 2016; Cho et al. 2013; Gao et al. 2009).

Furthermore, the involvement of these inflammatory mediators in amelogenesis highlights a sophisticated interplay between developmental processes and immune responses. Dysregulation of inflammatory cytokines may not only impair the delicate balance required for enamel formation but could also lead to an increased risk for enamel defects, such as hypoplasia or hypomineralization, thus underscoring the importance of understanding their roles in this context.

27.4.1 TGF-β Role in Dental Enamel Matrix Biomineralization

In the context of enamel formation, TGF-β is particularly important for the regulation of organic enamel matrix secretion and biomineralization (Kobayashi-Kinoshita et al. 2016; Cho et al. 2013). Expression of TGF-β isoforms has been detected in the inner dental epithelium, as well as in secretory and mature ameloblasts, indicating their involvement throughout the whole process of amelogenesis (Okubo et al. 2019; Gao et al. 2009; Benedete et al. 2008). The pre-secretory stage is marked by the presence of epithelial cells surrounding the dental papilla, and the expression of TGF-β isoforms at this stage suggests a role for this cytokine in the proliferation and differentiation of epithelial cells destined to differentiate towards an ameloblastic phenotype (Okubo et al. 2019; Benedete et al. 2008).

During the secretion and maturation stages, the levels of all three TGF-β isoforms are elevated. Specifically, TGF-β1 is expressed in secretory ameloblasts, where its primary function is to regulate the expression of MMP-20 and KLK-4 (Okubo et al. 2019; Kobayashi-Kinoshita et al. 2016). MMP-20 is released by secretory ameloblasts, whereas KLK-4 is secreted during the later stages of enamel secretion and maturation (Cho et al. 2013; Bartlett 2013). The regulatory mechanism initiates with the release of a latent form of TGF-β1 by ameloblasts. Upon the secretion of MMP-20 into the organic enamel matrix, this metalloproteinase cleaves amelogenins and the latency-associated peptide of TGF-β1. The activated form of TGF-β1 then binds to cleavage products of amelogenin, forming a complex known as TGF-β1-amelogenin. This complex interacts with the TGF-β1 receptors on the ameloblast cell membrane, thereby activating TGF-β1 signaling. When KLK-4 is released during the mineralization stage, the activity of TGF-β1 is diminished, as KLK-4 has the capacity to cleave the TGF-β1-amelogenin complex (Kobayashi-Kinoshita et al. 2016). TGF-β knockout mice exhibit enamel with hypomineralization, which was detected using radiographs of the mandibles. Additionally, a decrease in density was observed in the enamel layer (Liu et al. 2019a, b). Analysis of the expression of essential amelogenesis mediators showed that TGF-β participates in the regulatory process of Runt-related transcription factor 2 (RUNX2) and tryptophan-aspartate repeat domain 72 (WDR72), which are directly related to the formation and mineralization of tooth enamel (Katsura et al. 2014; Chu et al. 2018; Katsura et al. 2022). RUNX2 plays a role during tooth germ formation, tooth morphogenesis, and amelogenesis, as its presence has been detected in maturation ameloblasts (Liu et al. 2018; Chu et al. 2018). The ablation of RUNX2 in mice has shown that this transcription factor can impact the expression of enamel proteins and enzymes and alter the ultrastructural characteristics of the enamel layer (Chu et al. 2018). The role of WDR72 is not fully elucidated, but studies have shown that this protein is responsible for the removal of the organic enamel matrix, an essential process (Katsura et al. 2022).

Meanwhile, the effects of TGF-β overexpression on enamel formation were investigated in transgenic mice, and DSPP expression was analyzed as a gene highly expressed during the early stages of enamel and dentin formation, and its absence can lead to defects in dental enamel (Liang et al. 2021; Park et al. 2018; Thyagarajan et al. 2001). DSPP is an essential protein for dentin deposition, participating in the odontoblast differentiation process, as a hydroxyapatite nucleator, and allowing dentin to develop properly (Thyagarajan et al. 2001). In TGF-β overexpression mice, DSPP expression was significantly reduced. Overexpression of TGF-β during the early stages of amelogenesis results in detachment of the ameloblast layer from dentin, leading to the formation of cyst-like structures within the enamel organic matrix space. Later on, TGF-β1 and bone morphogenetic protein-2 (BMP-2) promote the differentiation of pre-ameloblasts into fully functional ameloblasts, a process that occurs once the first layer of dentin has been deposited by odontoblasts (Khaddam et al. 2014; Bei 2009a, b).

During odontogenesis, following the degradation of the basal membrane that separates epithelial from ectomesenchymal cells, a direct connection is established between enamel and dentin, forming the enamel-dentin junction (EDJ) (Fukae et al. 2002). Adhesion at the EDJ is mediated by various proteins including amelogenins and tuftelin, and their expression is regulated by TGF-β (Deutsch et al. 2002). Elevated levels of TGF-β during the early secretory stage lead to the detachment of the ameloblast layer, adversely affecting the secretion and biomineralization of the enamel matrix (El-Zainy et al. 2017). Scanning electron microscopy also revealed hypoplastic areas and an aprismatic, amorphous structure within the enamel layer.

As amelogenesis transitions, approximately 25% of ameloblasts undergo apoptosis, with an additional 25% undergoing apoptosis during the maturation phase (Okubo et al. 2019; Smith and Warshawsky 1997). By the end of amelogenesis, all ameloblasts are lost (Baranova et al. 2020; Pandya and Diekwisch 2019). Apoptosis is a vital process throughout all stages of amelogenesis, playing a key role in regulating the number of ameloblasts and removing residual epithelial cells and inactive ameloblasts (Abramyan et al. 2021). TGF-β isoforms are implicated in mediating ameloblast apoptosis (Okubo et al. 2019). TGF-β1 plays a role in regulating apoptosis by inducing the expression of the pro-apoptotic Bcl-2-associated X protein (BAX) while decreasing the levels of B cell lymphoma 2 (Bcl-2), an antiapoptotic molecule that binds to BAX and inhibits its activation, thus preventing apoptosis. When the ratio of BAX to Bcl-2 becomes dysregulated, BAX is activated and integrated into the mitochondrial membrane, triggering the apoptotic pathway and leading to cell death. Notably, enamelin has also been identified as a regulator of apoptosis, exerting protective effects against cell death (Abramyan et al. 2021; Hu et al. 2018). Studies have shown that in the absence of enamelin during the secretion stage, there are significant failures in the formation of hydroxyapatite crystals, which can disrupt the signaling stimuli necessary for ameloblasts. This disruption ultimately leads to the apoptosis of these cells (Hu et al. 2018).

27.4.2 TNF-α Superfamily Synthesis During Amelogenesis

Tumor necrosis factor-α (TNF-α) is a cytokine recognized as one of the primary regulators of inflammatory responses (You et al. 2021). It is involved in various biological processes, including embryonic development, cell differentiation, and the maintenance of body homeostasis (You et al. 2021; Idriss and Naismith 2000). At least 18 human genes are classified within the tumor necrosis factor superfamily, encompassing both ligands and receptors, which are capable of modulating diverse functions within the human body (Wallach 2018; Ward-Kavanagh et al. 2016; Ohazama and Sharpe 2004).

In the context of tooth development, the role of receptors expressed in lymphoid tissues (RELT) and ectodysplasin A receptor (EDAR)—members of the TNF-α superfamily—has been the focus of several studies (Ikeda et al. 2019; Tucker et al. 2000). RELT, initially identified in lymphoid and hematopoietic tissues, can modulate inflammatory responses, as well as influence cell proliferation, apoptosis, and morphogenesis (Kim et al. 2019; Aggarwal et al. 2012). Because it does not bind to any of the 19 ligands associated with TNF signaling pathways (Resende et al. 2023; Nikolopoulos et al. 2020; Ikeda et al. 2019; Kim et al. 2019), the role of RELT during immune response is still under investigation (Cusick et al. 2023; Sica et al. 2001). Interestingly, in enamel development, RELT synthesis has been detected in pre-secretory and secretory ameloblasts (Resende et al. 2023; Ikeda et al. 2019). Mutations in the *Relt* gene have been linked to amelogenesis imperfecta in human (Ikeda et al. 2019; Kim et al. 2019) and, in mice, the absence of *Relt* gene causes hypomineralization and the formation of a rough enamel surface in incisors. Histological analyses revealed the presence of unusual spaces at the dentin–enamel junction (DEJ) (Kim et al. 2019). Although the precise molecular mechanisms by which RELT contributes to amelogenesis remain unclear, understanding its role is crucial for elucidating the complex processes involved in tooth development and enamel formation.

Ectodysplasin A (EDA) and its receptor, ectodysplasin A receptor (EDAR), are critical components in the development of ectodermal structures, including glands, hair, and teeth (Wallach 2018). During the initial stages of odontogenesis, EDAR is expressed in ectodermal signaling centers that play a pivotal role in tooth morphogenesis, influencing both early and later stages of enamel formation. Misexpression of EDAR can adversely affect tooth morphogenesis and the mineralization of the enamel matrix. However, it is important to note that while this signaling pathway contributes to enamel development, it is not strictly essential for proper enamel formation, as other compensatory pathways may mitigate the effects of reduced EDAR expression. Although EDAR is a ligand of the TNF-α superfamily, it does not participate directly in the immune response (Wallach 2018). This nuanced understanding of the role of EDA and EDAR in dental development highlights their importance in tooth morphogenesis while recognizing the existence of alternative pathways that can support enamel formation despite potential disruptions in EDAR signaling.

TNF-α is expressed throughout the entire ameloblastic layer, indicating the presence of this cytokine during amelogenesis (Fig. 27.1). Given the presence of TNF-α superfamily ligands and receptors in ameloblasts, a study was conducted to examine the TNF-α-TNFR1 signaling pathway in mice. The findings revealed that the ablation of TNF receptor 1 (TNFR1) in mice adversely affected amelogenesis by altering the molecular signaling pathways related to enamel matrix protein expression and protease activity, both of which are essential for proper enamel formation. SEM analysis demonstrated that TNFR1 knockout animals exhibited a loss of the characteristic enamel rod patterns on the enamel surface. Further analysis of the enamel layer's composition revealed an increased carbon content, suggesting a higher organic component in

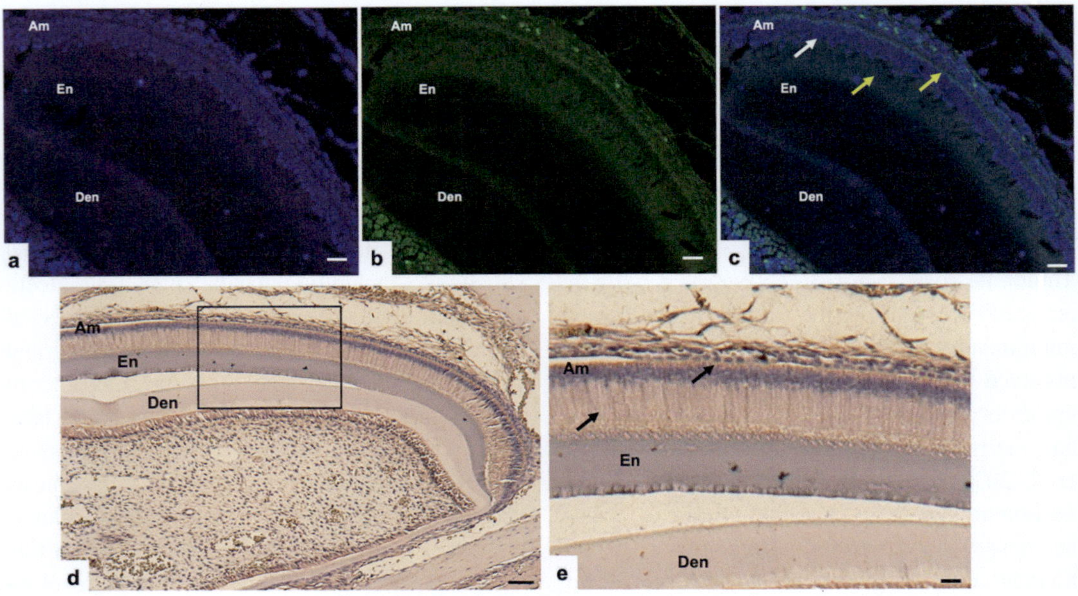

Fig. 27.1 (**a**–**c**) Indirect immunofluorescence microscopy of an 8-week-old TNFR1 knockout mouse incisors. (a) DAPI staining marking cell nuclei (blue). Scale bar = 20 μm. (**b**) FITC-labeled TNF-α expression in the ameloblast layer (green). Scale bar = 20 μm. (**c**) Overlay of (**a**) and (**b**) images showing TNF-α expression in ameloblast cytoplasm in green (yellow arrow) and the nucleus in blue (white arrow). Scale bar = 20 μm. (**d, e**) Immunohistochemistry analysis of TNF-α synthesis in ameloblasts. (**d**) TNF-α staining (brown) distributed throughout the ameloblast layer (black arrows). Scale bar = 50 μm. (**e**) Higher magnification showing TNF-α localization. Scale bar = 20 μm. *Am* ameloblasts, *En* enamel, *Den* dentin

the enamel of these teeth. This elevated organic material was associated with a reduction in the hardness of the enamel layer, highlighting the critical role of TNFR1 in enamel mineralization (Gonçalves et al. 2024b).

27.4.3 Disturbances During Amelogenesis Result in Dental Enamel Defects

Because amelogenesis is a fine-tuned process, disruptions during this period can cause developmental enamel defects (DEDs) (Lopes-Fatturi et al. 2024). These defects can typically be classified into two major categories of hypoplasia and hypomineralization. Hypoplasia is a quantitative defect resulting from disturbances in the secretory stage, and hypomineralization is caused by abnormalities during the maturation phase and overall enamel composition (Salanitri and Seow 2013). Teeth affected by hypomineralization exhibit characteristics such as reduced hardness, decreased mineral content, and increased protein content in the enamel structure (Gonçalves et al. 2024a; Gil-Bona and Bidlack 2020; Lacruz et al. 2017).

The etiology of enamel defects is considered multifactorial, involving a combination of hereditary, environmental, systemic, and local factors (Bussaneli et al. 2019; Salanitri and Seow 2013). Genetic and epigenetic influences, along with various external stressors—such as metabolic disturbances, compromised blood supply, hypoxia, immune responses, and exposure to xenobiotics—can contribute to the development of these defects (Wright 2023; Collignon et al. 2022; Brook et al. 1997). For instance, hypoxia can disrupt both enamel formation and mineralization, potentially resulting in DEDs (Cruvinel et al. 2012). Additionally, epigenetic factors may create site-specific defects by disrupting enamel formation at particular stages and locations (Alzahrani et al. 2023; Bekes 2020).

While research on epigenetics is still in the early stages, preliminary findings indicate that epigenetics may increase the risk of enamel defects through changes in gene expression, particularly concerning genes essential for normal enamel development (Wright 2023). Furthermore, investigations have focused on mutations in genes responsible for the synthesis of enamel matrix proteins and proteases, which are critical for enamel formation (Lopes-Fatturi et al. 2024).

Several environmental factors during pregnancy and early childhood have been associated with the occurrence of developmental enamel defects. Exposure to passive smoking, alcohol consumption, air pollution, and low socioeconomic status are among the risk factors that can adversely affect amelogenesis (Garot et al. 2021; Salanitri and Seow 2013; Chun et al. 2024). Additionally, immune-mediated systemic illnesses, which trigger the persistent release of pro-inflammatory molecules as part of the immune response, are prevalent during childhood and can impact enamel formation (Acosta et al. 2022; Bussaneli et al. 2019).

Gestational diabetes mellitus has been identified as a significant factor in the development of enamel defects in primary teeth. The underlying mechanisms involve disturbances caused by elevated glucose levels during pregnancy, which can adversely impact the function of ameloblasts (Xu et al. 2022; Marler et al. 2022). Elevated maternal glucose levels can disrupt metabolic processes and lead to alterations in the expression of key genes and proteins involved in amelogenesis. As a result, the enamel matrix may be inadequately formed, leading to DEDs that manifest as hypoplasia, hypomineralization, or various enamel color changes of primary teeth.

In addition to the effects of gestational diabetes mellitus, substantial evidence indicates that preterm infants are at a higher risk for developing enamel defects. This increased susceptibility is largely attributed to their compromised immune systems, which can impair their ability to withstand environmental challenges and infections during critical periods of tooth development (Xu et al. 2022). The connection between gestational diabetes mellitus, preterm birth, and enamel defects underscores the importance of prenatal care and the monitoring of maternal health conditions that may have long-term effects on dental development. It also highlights a need for increased awareness and education among healthcare providers regarding the potential dental implications of gestational diabetes and preterm birth. Identifying these mechanisms may lead to better preventive strategies and interventions for those at risk. Moreover, enhancing maternal health and nutritional support during pregnancy may serve as a critical approach to mitigate the risks of enamel defects and promote optimal dental health in children.

Respiratory diseases can be related to molar-incisor hypomineralization (MIH) through several interconnected mechanisms that may influence tooth development during critical developmental periods (Né et al. 2022). When children experience respiratory infections, particularly during early childhood, the resulting immune response triggers the release of pro-inflammatory cytokines. This inflammatory response can alter the local environment and impact the function of ameloblasts. Disruptions in amelogenesis due to inflammation can lead to inadequate enamel matrix production, resulting in defects characteristic of MIH (Rizzardi et al. 2022).

The timing of respiratory illnesses is particularly critical. Episodes occurring during the mineralization stages of primary teeth and first permanent molars can be especially detrimental. During these formative periods, disturbances in normal developmental processes caused by illness can predispose teeth to structural weaknesses, leading to hypomineralization.

Furthermore, certain respiratory diseases may involve chronic issues, such as asthma or recurrent bronchitis, which can exacerbate systemic health challenges. These chronic conditions may lead to prolonged inflammation and repeated cycles of stress on the child's immune system, further affecting enamel development (Rizzardi et al. 2022).

27.4.4 Molar Incisor Hypomineralization

Molar-incisor hypomineralization (MIH) is a qualitative enamel defect that is increasingly recognized for its high prevalence and clinical significance in pediatric dentistry. This condition is clinically characterized by the presence of enamel opacities that can vary in color from white to yellow or brown. MIH predominantly affects the first permanent molars and incisors, emerging during the critical mineralization phase of amelogenesis (Bezamat et al. 2021).

The presence of these opacities is not merely an aesthetic concern; they are often associated with increased dental enamel breakdown, leading to significant diagnostic and therapeutic challenges for clinicians (Juárez-López et al. 2023). Such challenges arise from the need to balance an effective management of the existing hypomineralized enamel with the prevention of further aggravation, highlighting the importance of a thorough understanding of the condition's etiology.

The etiology of MIH is multifactorial and remains incompletely understood, involving a complex interplay of genetic, epigenetic, and environmental factors (Crombie et al. 2009). Recent studies have elucidated how these varied factors contribute to the development of MIH. Specifically, research has shown that genetic polymorphisms in genes associated with enamel formation and immune response may interact in intricate ways, thereby influencing the overall risk of developing this condition (Lopes-Fatturi et al. 2024; Bussaneli et al. 2019; Jeremias et al. 2016).

Of particular interest is the impact of systemic conditions such as respiratory diseases, high fever, otitis media, and other infectious illnesses during the first 3 years of life—the same crucial period during which the mineralization of permanent dentition occurs (Garot et al. 2021). The high incidence of these systemic health concerns during early childhood correlates with an exacerbated immune response, which can detrimentally impact the enamel mineralization process. Immune dysregulation in this period may interfere with the normal secretory and biomineralization phases of amelogenesis, leading to insufficient mineral deposition and the development of enamel defects (Bussaneli et al. 2019; Elhennawy et al. 2017).

Fever can significantly impact tooth formation by creating a systemic environment that can adversely affect key developmental processes involved in dental mineralization. It is supposed that elevated body temperatures can disrupt normal cellular functions and metabolic processes of ameloblasts, which are important for tooth formation. Additionally, fever is often associated with inflammatory responses and the release of cytokines that can alter the local tissue environment in the oral cavity. These responses may lead to changes in the signaling pathways critical for physiological tooth development, possibly resulting in defects such as hypoplasia or hypomineralization (Simmer and Hu 2001). Some studies have analyzed children's medical records and the occurrence of high fever episodes during early childhood, and results have shown an association with MIH cases (Lopes-Fatturi et al. 2024; Garot et al. 2021; Juárez-López et al. 2023). Overall, the impact of fever on enamel development is complex, intertwining physiological changes and immune response, albeit the mechanisms involved in this process warrant further investigation.

This intricate relationship between systemic health, immune response, and enamel development underscores the need for further research into MIH. A deeper understanding of the contributing factors to this condition will be essential for developing effective preventive and therapeutic strategies. Given the potential long-term implications of MIH on oral health, early detection and targeted interventions must be prioritized to mitigate the impact of this enamel defect on affected individuals. Continued investigations into the underlying mechanisms of MIH will not only enhance clinical approaches but may also illuminate broader principles of enamel biology and its vulnerabilities during early development.

A potential association between polymorphisms in immune response-related genes and an increased risk of MIH has been identified. Bussaneli et al. (2019) conducted a study involving 101 families and 390 participants, including at least one family member diagnosed with MIH. They performed genotyping on all samples and analyzed genes associated with inflammatory responses, such as TGF-β, TNF-α, and interleukins. A polymorphism (rs10733708) in the TGF-β receptor 1 (*TGFBR1)* was associated with the severity of MIH. TGF-β1 is a molecule expressed during the secretion and early maturation phases, responsible for regulating the expression of MMP-20 and KLK-4, enzymes essential for removing protein content from enamel and incorporating calcium and phosphorus into hydroxyapatite crystals (Gonçalves et al. 2024a; Bartlett 2013). The single nucleotide polymorphism (SNP) in the *TGFBR1* gene may alter TGF-β expression and, consequently, affect the activity of MMP-20 and KLK-4. However, Bezamat et al. (2021) reported that although TGF-β alone may not directly contribute to the occurrence of MIH, its interaction with the interferon regulatory factor 6 (*IRF6)* may increase susceptibility to the condition. *Irf6* is a gene involved in craniofacial development, and its interaction with immune response genes, such as interleukins, may lead to disturbances in jawbone growth and enamel formation (Bezamat et al. 2021).

Furthermore, polymorphisms in *TNF* and *IL10* genes have also been considered risk factors for the development of MIH. TNF-α and interleukins are molecules involved in the inflammatory response during systemic diseases, such as high fever. The role of these molecules in amelogenesis is not yet fully understood, but TNF-α is a cytokine expressed in the ameloblast layer, and the absence of TNF-α receptor 1 is associated with reduced enamel hardness and mineral content (Gonçalves et al. 2024a). Regarding interleukins, the mechanism by which these cytokines act during amelogenesis remains unclear, necessitating further research (Lopes-Fatturi et al. 2024). One proposed hypothesis is that polymorphisms in immune system genes may increase angiogenesis in the dental germ, leading to increased vascularization and a higher chance of serum albumin penetration into the enamel layer. The presence of serum albumin has been detected in MIH-affected teeth, and since proteases responsible for removing protein content cannot degrade albumin, it remains trapped in the enamel matrix, interfering with the growth of hydroxyapatite crystals and contributing to DEDs (Perez et al. 2020).

These findings emphasize the importance of the interactions between immunological and genetic factors and how genetic variants can impact amelogenesis. Understanding these complex interactions is crucial for fully elucidating the etiology of MIH and developing effective diagnostic and treatment strategies. Future studies should explore these relationships in greater detail to enhance knowledge of the condition and improve clinical approaches.

27.5 Conclusion

Given that mediators of the immune response are integral to the process of enamel formation, any disruptions occurring during the critical phases of secretion and biomineralization can significantly contribute to the development of dental enamel defects. During the secretion phase, ameloblasts synthesize and secrete enamel proteins, which are essential for the formation of the enamel matrix. The activity of these cells is intricately regulated by various signaling molecules, including those involved in immune responses. Similarly, the biomineralization phase, which involves the deposition and maturation of mineral components within the enamel matrix, is also influenced by immune-related mediators that facilitate extracellular matrix mineralization. Disruptions to these phases can occur due to a variety of factors, including genetic mutations, systemic health conditions, and environmental stresses. For instance, an aberrant immune response or the presence of inflammatory mediators can interfere with the normal signaling pathways that are vital for ameloblast function, leading to inadequate enamel matrix formation or improper mineralization. Consequently, this can manifest as enamel hypoplasia, characterized by insufficient enamel quantity, or enamel hypomineralization, where the enamel lacks the necessary mineral content, resulting in structural weaknesses and increased susceptibility to caries and other dental diseases. Furthermore, emerging research suggests a bidirectional relationship between enamel formation and immune processes. Immune mediators can influence not only the developmental stages of amelogenesis but may also be modulated by the state of the enamel itself. Therefore, understanding this interplay is crucial for elucidating the etiology of enamel defects and developing targeted therapeutic approaches.

References

Abramyan J, Geetha-Loganathan P, Šulcová M, Buchtová M (2021) Role of cell death in cellular processes during odontogenesis. Front Cell DevBiol 9:671475. https://doi.org/10.3389/fcell.2021.671475

Acosta E, Cortes O, Guzman S, Catala M, Lorente M, Arense JJ (2022) Relationship between molar incisor hypomineralization, intrapartum medication, and illnesses in the first year of life. Sci Rep 12:1637. https://doi.org/10.1038/s41598-022-05628-7

Aggarwal BB, Gupta SC, Kim JH (2012) Historical perspectives on tumor necrosis factor and its superfamily: 25 years later, a golden journey. Blood 119(3): 651–665. https://doi.org/10.1182/blood-2011-04-325225

Alzahrani AY, Alamoudi NMH, El Meligy OAES (2023) Contemporary understanding of the etiology and management of molar incisor hypomineralization: a literature review. Dent J 11(7):157. https://doi.org/10.3390/dj11070157

Baranova J, Büchner D, Götz W, Schulze M, Tobiasch E (2020) Tooth formation: Are the hardest tissues of the human body hard to regenerate? Int J Mol Sci 21(11): 4031. https://doi.org/10.3390/ijms21114031

Bartlett JD (2013) Dental enamel development: Proteinases and their enamel matrix substrates. ISRN Dent 684607. https://doi.org/10.1155/2013/684607

Bei M (2009a) Molecular genetics of ameloblast cell lineage. J Exp Zool B Mol Dev Evol 312B(5):437–444. https://doi.org/10.1002/jez.b.21261

Bei M (2009b) Molecular genetics of tooth development. Curr Opin Genet Dev 19(5):504–510. https://doi.org/10.1016/j.gde.2009.09.002

Bekes K (2020) Molar incisor hypomineralization. Springer

Benedete APS, Sobral APV, Lima DMC, Kamibeppu L, Soares FA, Lourenço SV (2008) Expression of transforming growth factor-beta 1, -beta 2, and -beta 3 in human developing teeth: Immunolocalization according to the odontogenesis phases. Pediatr Dev Pathol 11(3):206–212. https://doi.org/10.2350/07-09-0333.1

Bezamat M, Souza JF, Silva FM, Corrêa EG, Fatturi AL, Brancher JA et al (2021) Gene-environment interaction in molar-incisor hypomineralization. PLoS One 16(1): e0241898. https://doi.org/10.1371/journal.pone.0241898

Brook AH, Fearne JM, Smith JM (1997) Environmental causes of enamel defects. Ciba Found Symp 205:212–221

Bussaneli DG, Restrepo M, Fragelli CMB, Santos-Pinto L, Jeremias F, Cordeiro R et al (2019) Genes regulating immune response and amelogenesis interact in increasing the susceptibility to molar-incisor hypomineralization. Caries Res 53(2):217–227. https://doi.org/10.1159/000491644

Chang B, Svoboda KKH, Liu X (2019) Cell polarization: From epithelial cells to odontoblasts. Eur J Cell Biol 98(1):1–11

Cho A, Haruyama N, Hall B, Danton MJS, Zhang L, Arany P et al (2013) TGF-β regulates enamel mineralization and maturation through KLK4 expression. PLoS One 8(11):e82267. https://doi.org/10.1371/journal.pone.0082267

Chu Q, Gao Y, Gao X, Dong Z, Song W, Xu Z et al (2018) Ablation of Runx2 in ameloblasts suppresses enamel maturation in tooth development. Sci Rep 8:9594

Chun YP, Tan C, Villanueva O, Colley ME, Quintanilla TJ, Basiouny MS et al (2024) Overexpression of ameloblastin in secretory ameloblasts results in demarcated, hypomineralized opacities in enamel. Front Physiol 14:1233391. https://doi.org/10.3389/fphys.2023.1233391

Collignon AM, Vergnes JN, Germa A, Azogui S, Breinig S, Hollande C et al (2022) Factors and mechanisms involved in acquired developmental defects of enamel: a scoping review. Front Pediatr 10: 836708. https://doi.org/10.3389/fped.2022.836708

Crombie F, Manton D, Kilpatrick N (2009) Aetiology of molar–incisor hypomineralization: a critical review. Int J Paediatr Dent 19(2):73–83. https://doi.org/10.1111/j.1365-263X.2008.00963.x

Cruvinel VR, Gravina DB, Azevedo TD, Rezende CS, Bezerra AC, Toledo OA (2012) Prevalence of enamel defects and associated risk factors in both dentitions in preterm and full term born children. J Appl Oral Sci 20: 310–317. https://doi.org/10.1590/s1678-77572012000300003

Cusick JK, Alcaide J, Shi Y (2023) The RELT family of proteins: an increasing awareness of their importance for cancer, the immune system, and development. Biomedicine 11(10):2695. https://doi.org/10.3390/biomedicines11102695

D'Souza R, Cavender A, Sood R, Tarnuzzer R, Dickinson DP, Roberts A, Letterio J (1998) Dental abnormalities in mice lacking a functional transforming growth factor-beta1 TGF-β gene indicate a role for TGF-β1 in biomineralization. J Oral Biol 23(3):119–131

Deutsch D, Leiser Y, Shay B, Fermon E, Taylor A, Rosenfeld E et al (2002) The human tuftelin gene and the expression of tuftelin in mineralizing and nonmineralizing tissues. Connect Tissue Res 43:425–434. https://doi.org/10.1080/03008200290001186

Elhennawy K, Manton DJ, Crombie F, Zaslansky P, Radlanski RJ, Jost-Brinkmann PG, Schwendicke F (2017) Structural, mechanical, and chemical evaluation of molar-incisor hypomineralization-affected enamel: a systematic review. Arch Oral Biol 83:272–281. https://doi.org/10.1016/j.archoralbio.2017.08.017

El-Zainy MA, El-Haddad KN, Obeid RF (2017) Efficacy of transforming growth factor-β on development of ameloblasts and odontoblasts in tooth germ of young albino rats. Future Dent J 3(2):39–46. https://doi.org/10.1016/j.fdj.2017.08.001

Fan MW, Bian Z, Gao YG (1998) Immunohistochemistry and in situ hybridization investigation of transforming growth factor-beta during odontoblast and ameloblast differentiation. Chin J Dent Res 1(2):17–21

Fokkema IFAC, Dunnen JT (2021) Chapter 2 – general considerations: terminology and standards. In: In translational and applied genomics. Clinical DNA variant interpretation, pp 9–27. https://doi.org/10.1016/B978-0-12-820519-8.00009-0

Fukae M, Tanabe T, Nagano T, Ando H, Yamakoshi Y, Yamada M et al (2002) Odontoblasts enhance the maturation of enamel crystals by secreting EMSP1 at the enamel-dentin junction. J Dent Res 81(10):668–672. https://doi.org/10.1177/154405910208101003

Gao Y, Li D, Han T, Sun Y, Zhang Y (2009) TGF-beta1 and TGFBR1 are expressed in ameloblasts and promote MMP20 expression. Anat Rec 292:885–890. https://doi.org/10.1002/ar.20901

Garot E, Rouas P, Somani C, Taylor GD, Wong F, Lygidakis NA (2021) An update of the aetiological factors involved in molar incisor hypomineralisation (MIH): a systematic review and meta-analysis. Eur Arch Paediatr Dent 23:23–28. https://doi.org/10.1007/s40368-021-00646-x

Gil-Bona A, Bidlack FB (2020) Tooth enamel and its dynamic protein matrix. Int J Mol Sci 21(12):4458. https://doi.org/10.3390/ijms21124458

Gonçalves JL, Duarte ACA, Almeida-Junior LA, de Carvalho FK, de Queiroz AM, Arnez MFM, Faccioli LH, Paula-Silva FWG (2022) Enamel biomineralization under the effects of indomethacin and celecoxib non-steroidal anti-inflammatory drugs. Sci Rep 12(1):15823. https://doi.org/10.1038/s41598-022-19583-w. PMID: 36138112, PMCID: PMC9500046

Gonçalves JL, Carvalho FK, Queiroz AM, Paula-Silva FWG (2024a) Implications of Histological and Ultrastructural Characteristics on the Chemical and Mechanical Properties of Hypomineralised Enamel andClinical Consequences. Monogr Oral Sci, 32:43–55. https://doi.org/10.1159/000538865. PMID: 39321779

Gonçalves, Almeida-Júnior J, Arnez L, Queiroz M, De Carvalho A, Paula-Silva F, Francisco (2024b) Impact of TNFR1 pro-inflammatory receptor ablation in amelogenesis and dental enamel biomineralization in mice. https://doi.org/10.21203/rs.3.rs-4042598/v1

Hu T, Xu Z, Huang Y, Chen J, Yang F (2018) TGF-β in amelogenesis: a mechanistic review. J Dent Res 97(10):1077–1085. https://doi.org/10.1177/0022034518799821

Idriss HT, Naismith JH (2000) TNF alpha and the TNF receptor superfamily: structure-function relationship(s). Microsc Res Tech 50(3):184–195. https://doi.org/10.1002/1097-0029(20000801)50:3<184::AID-JEMT2>3.0.CO;2-H

Ikeda A, Shahid S, Blumberg BR, Suzuki M, Bartlett JD (2019) ADAM10 is expressed by ameloblasts, cleaves the RELT TNF receptor extracellular domain and facilitates enamel development. Sci Rep 9:14086. https://doi.org/10.1038/s41598-019-50277-y

Isono K, Takahashi E, Miyoshi I, Tsuneto M, Hikosaka-Kuniishi M, Yamane T et al (2021) Simultaneous fluorescent identification of odontoblasts and ameloblasts. J Dent Res 100(5):532–541. https://doi.org/10.1177/0022034520974576

Jeremias F, Pierri RA, Souza JF, Fragelli CMB, Restrepo M, Finoti LS et al (2016) Family-based genetic association for molar-incisor hypomineralization. Caries Res 50(3):310–318. https://doi.org/10.1159/000445982

Johnson GL, Nakamura K (2007) The c-jun kinase/stress-activated pathway: regulation, function and role in human disease. Biochim Biophys Acta 1773(8):1341–1348. https://doi.org/10.1016/j.bbamcr.2006.12.009

Juárez-López MLA, Salazar-Treto LV, Hernández-Monjaraz B, Molina-Frechero N (2023) Etiological factors of molar incisor hypomineralization: a systematic review and meta-analysis. Dent J 11(5):111. https://doi.org/10.3390/dj11050111

Katsura KA, Horst JA, Chandra D, Le TQ, Nakano Y, Zhang Y et al (2014) WDR72 models of structure and function: a stage-specific regulator of enamel mineralization. Matrix Biol 38:48–58. https://doi.org/10.1016/j.matbio.2014.06.005

Katsura K, Nakano Y, Zhang Y, Shemirani R, Li W, DenBesten P (2022) WDR72 regulates vesicle trafficking in ameloblasts. Sci Rep 12:2820

Kawashima N, Okiji T (2016) Odontoblasts: Specialized hard-tissue-forming cells in the dentin-pulp complex. Congenit Anom 56(4):144–153. https://doi.org/10.1111/cga.12169

Khaddam M, Huet E, Vallée B, Bensidhoum M, Le Denmat D, Filatova A et al (2014) EMMPRIN/CD147 deficiency disturbs ameloblast-odontoblast crosstalk and delays enamel mineralization. Bone 66:256–266. https://doi.org/10.1016/j.bone.2014.06.019

Kim JW, Zhang H, Seymen F, Koruyucu M, Hu Y, Kang J (2019) Mutations in RELT cause autosomal recessive amelogenesis imperfecta. Clin Genet 95(3):375–383. https://doi.org/10.1111/cge.13487

Kobayashi-Kinoshita S, Yamakoshi Y, Onuma K, Yamamoto R, Asada Y (2016) TGF-β1 autocrine signalling and enamel matrix components. Sci Rep 6: 33644. https://doi.org/10.1038/srep33644

Kubiczkova L, Sedlarikova L, Hajek R, Sevcikova S (2012) TGF-β - an excellent servant but a bad master. J Transl Med 10:183. https://doi.org/10.1186/1479-5876-10-183

Lacruz RS, Habelitz S, Wright JT, Paine ML (2017) Dental enamel formation and implications for oral health and disease. Physiol Rev 97(3):939–993. https://doi.org/10.1152/physrev.00030.2016

Liang T, Xu Q, Zhang H, Wang S, Diekwisch TGH, Qin C, Lu Y (2021) Enamel defects associated with dentin Sialophosphoprotein mutation in mice. Front Physiol 12:724098. https://doi.org/10.3389/fphys.2021.724098

Liu C, Niu Y, Zhou X, Xu X, Yang Y, Zhang Y (2015) Cell cycle control, DNA damage repair, and apoptosis-related pathways control pre-ameloblasts differentiation during tooth development. BMC Genomics 16(1):592. https://doi.org/10.1186/s12864-015-1783-y

Liu X, Wang Y, Zhang L, Xu Z, Chu Q, Xu C et al (2018) Combination of Runx2 and Cbfβ upregulates Amelotin gene expression in ameloblasts by directly interacting with cis-enhancers during amelogenesis. Mol Med Rep 17(4):6068–6076. https://doi.org/10.3892/mmr.2018.8564

Liu X, Xu C, Tian Y, Sun Y, Zhang J, Bai J et al (2019a) RUNX2 contributes to TGF-β1-induced expression of Wdr72 in ameloblasts during enamel mineralization. Biomed Pharmacother 118:109235. https://doi.org/10.1016/j.biopha.2019.109235

Liu Z, Tian Y, Zhang X (2019b) Molecular mechanisms in amelogenesis: a comprehensive review. J Biochem Mol Biol 52(4):231–240. https://doi.org/10.5483/BMBRep.2019.52.4.032

Lopes-Fatturi A, Wambier LM, Chibinski AC, Assunção LRS, Brancher JA, Reis A, Souza JF (2024) A systematic review and meta-analysis of systemic exposure associated with molar incisor hypomineralization. Community Dent Oral Epidemiol 47(5):407–415. https://doi.org/10.1111/cdoe.12467

Luo J, Tan X, Ye L, Wang C (2021) C-Jun N-terminal kinase (JNK) pathway activation is essential for dental papilla cells polarization. PLoS One 16(3):e0233944. https://doi.org/10.1371/journal.pone.0233944

Marler J, Howland R, Kimmons LA, Mohrien K, Vandigo JE, Jones GM (2022) Safety of propofol when used for rapid sequence intubation in septic patients: a multicenter cohort study. Hosp Pharm 57(2):287–293. https://doi.org/10.1177/00185787211029547

Martín-Vacas A, Nova MJ, Sagastizabal B, García-Barbero AE, Vera-González V (2022) Morphological study of dental structure in dentinogenesis imperfecta type I with scanning electron microscopy. Healthcare (Basel) 10(8):1453

Né YGS, Frazão DR, Lopes GO, Fagundes NCF, Souza-Rodrigues RD, Paula-Silva FWG, Maia LC, Lima RR (2022) Association between respiratory diseases and molar-incisor hypomineralization: a systematic review and meta-analysis. Front Med 9:990421. https://doi.org/10.3389/fmed.2022.990421

Nikolopoulos G, Smith CEL, Brookes SJ, El Asrag ME, Brown CJ, Patel A et al (2020) New missense variants in RELT causing hypomineralised amelogenesis imperfecta. Clin Genet 97(5):688–695. https://doi.org/10.1111/2cge.13721

Niwa T, Yamakoshi Y, Yamazaki H, Karakida T, Chiba R, Hu JC, Nagano T, Yamamoto R, Simmer JP, Margolis HC, Gomi K et al (2018) The dynamics of TGF-β in dental pulp, odontoblasts and dentin. Sci Rep 8(1): 4450. https://doi.org/10.1038/s41598-018-22823-7

Ohazama A, Sharpe PT (2004) TNF signaling in tooth development. Curr Opin Genet Dev 14(5):513–519. https://doi.org/10.1016/j.gde.2004.07.008

Oka S, Oka K, Xu X, Sasaki T, Bringas-Jr P, Chai Y (2007) Cell autonomous requirement for TGF-β signaling during odontoblast differentiation and dentin matrix formation. Mech Dev 124(6):409–415. https://doi.org/10.1016/j.mod.2007.02.003

Okubo M, Chiba R, Karakida T, Yamazaki H, Yamamoto R, Kobayashi S et al (2019) Potential function of TGF-β isoforms in maturation-stage ameloblasts. J Oral Biosci 61(1):43–54. https://doi.org/10.1016/j.job.2018.12.002

Pan H, Yang Y, Xu H, Jin A, Huang X, Gao X et al (2023) The odontoblastic differentiation of dental mesenchymal stem cells: molecular regulation mechanism and related genetic syndromes. Front Cell Dev Biol 11. https://doi.org/10.3389/fcell.2023.1174579

Pandya M, Diekwisch TGH (2019) Enamel biomimetics – fiction or future of dentistry. Int J Oral Sci 11(1):8. https://doi.org/10.1038/s41368-018-0038-6

Park SJ, Lee HK, Seo YM, Son C, Bae HS, Park JC (2018) Dentin sialophosphoprotein expression in enamel is regulated by Copine-7, a preameloblast-derived factor. Arch Oral Biol 86:131–137. https://doi.org/10.1016/j.archoralbio.2017.12.004

Perez VA, Mangum JE, Hubbard MJ (2020) Pathogenesis of molar Hypomineralisation: aged albumin demarcates chalky regions of Hypomineralised enamel. Front Physiol 11:579015. https://doi.org/10.3389/fphys.2020.579015

Rahman SU, Kim WJ, Chung SH, Woo KM (2022) Nanofibrous topography-driven altered responsiveness to Wnt5a mediates the three-dimensional polarization of odontoblasts. Mater Today Bio 2(17):100479. https://doi.org/10.1016/j.mtbio.2022.100479

Resende KKM, Riou MC, Yamaguti PM, Fournier B, Rondeau S, Pacot L et al (2023) Oro-dental phenotyping and report of three families with RELT-associated amelogenesis imperfecta. Eur J Hum Genet 31:1337–1134. https://doi.org/10.1038/s41431-023-01440-7

Rizzardi KF, da Silva Toledo E, Ferraz LFC, Darrieux M, Girardello R, de Lima Marson FA, Parisotto TM (2022) Association between asthma and enamel defects in primary and young permanent teeth – a systematic review. Pediatr Pulmonol 57(1):26–37. https://doi.org/10.1002/ppul.25737

Rothová M, Peterková R, Tucker AS (2012) Fate map of the dental mesenchyme: dynamic development of the dental papilla and follicle. Dev Biol 366(2):244–254. https://doi.org/10.1016/j.ydbio.2012.03.018

Salanitri S, Seow WK (2013) Developmental enamel defects in the primary dentition: aetiology and clinical management. Aust Dent J 58(2):133–140. https://doi.org/10.1111/adj.12039

Seo YM, Park SJ, Lee HK, Park JC (2017) Copine-7 binds to the cell surface receptor, nucleolin, and regulates ciliogenesis and Dspp expression during odontoblast differentiation. Sci Rep 7:11283. https://doi.org/10.1038/s41598-017-11641-y

Sica GL, Zhu G, Tamada K, Liu D, Ni J, Chen L (2001) RELT, a new member of the tumor necrosis factor receptor superfamily, is selectively expressed in hematopoietic tissues and activates transcription factor NF-κB. Blood 97(9):2704–2707. https://doi.org/10.1182/blood.v97.9.2702

Simmer JP, Hu JC (2001) Dental enamel formation and its impact on clinical dentistry. J Dent Educ 65(9):896–905

Smith CE, Warshawsky H (1997) Quantitative analysis of cell turnover in the enamel organ of the rat incisor. Evidence for ameloblast death immediately after enamel matrix secretion. Anat Rec 187(1):63–98. https://doi.org/10.1002/ar.1091870106

Thesleff I (2003) Epithelial-mesenchymal signalling regulating tooth morphogenesis. J Cell Sci 116(9):1647–1648. https://doi.org/10.1242/jcs.00678

Thyagarajan T, Sreenath T, Cho A, Wright JT, Kulkarni AB (2001) Reduced expression of dentin Sialophosphoprotein is associated with dysplastic dentin in mice overexpressing transforming growth factor-β1 in teeth. Mol Basis Cell Dev Biol 276(14):11016-20. 11016-1102. https://doi.org/10.1074/jbc.M010502200

Unterbrink A, O'Sullivan M, Chen S, MacDougall M (2009) TGFβ-1 downregulates DMP-1 and DSPP in odontoblasts. Connect Tissue Res 43:354–358. https://doi.org/10.1080/03008200290000565

Wallach D (2018) The tumor necrosis factor family: family conventions and private idiosyncrasies. Cold Spring Harbour Perspect Biol 10(10):a028431. https://doi.org/10.1101/cshperspect.a028431

Wang C, Zhao Y, Su Y, Li R, Lin Y, Zhou X et al (2013) C-Jun N-terminal kinase (JNK) mediates Wnt5a-induced cell motility dependent or independent of RhoA pathway in human dental papilla cells. PLoS One 8(7):e69440. https://doi.org/10.1371/journal.pone.0069440

Ward-Kavanagh LK, Lin WW, Šedý JR, Ware CF (2016) The TNF receptor superfamily in co-stimulating and co-inhibitory responses. Immunity 44(5):1005–1019. https://doi.org/10.1016/j.immuni.2016.04.019

Wright JT (2023) Enamel phenotypes: genetic and environmental determinants. Genes (Basel) 14(3):545. https://doi.org/10.3390/genes14030545

Xu S, Zhao C, Jia L, Ma Z, Zhang X, Shi H (2022) Relationship between preterm, low birth weight, and development defects of enamel in the primary dentition: a meta-analysis. Front Pediatr 10:975340. https://doi.org/10.3389/fped.2022.975340

Yan H, He L, Lv D, Yang J, Yuan Z (2024) The role of the dysregulated JNK signaling pathway in the pathogenesis of human diseases and its potential therapeutic strategies: a comprehensive review. Biomolecules 14(2):243. https://doi.org/10.3390/biom14020243

You K, Gu H, Yuan Z, Xu X (2021) Tumor necrosis factor alpha signaling and organogenesis. Front Cell Dev Biol 9:727075. https://doi.org/10.3389/fcell.2021.727075

Immune Responses in Orthodontic Treatment

28

Chen Zong

Abstract

Orthodontic tooth movement (OTM) involves the application of mechanical forces, leading to bone remodeling through a coordinated immune response. This process comprises four stages—initial, lag, acceleration, and retention—each characterized by specific immune mechanisms involving cytokines, immune cells, and signaling pathways. Inflammatory mediators, such as interleukin (IL)-1β and tumor necrosis factor-alpha (TNF-α), initiate bone resorption, while regulatory cytokines like IL-10 promote tissue repair. The nuclear factor kappa-B ligand (RANKL)/osteoprotegerin (OPG) balance regulates osteoclast activity, ensuring controlled bone remodeling. Understanding these immune responses provides valuable insights into optimizing orthodontic treatment outcomes by controlling inflammation and enhancing therapeutic efficiency. This knowledge is significant for alleviating pain during orthodontic treatment, reducing excessive resorption of periodontal tissues, and preventing post-treatment relapse.

Keywords

Tooth movement · Cytokines · Inflammation · Bone remodeling · Retention

C. Zong (✉)
College of Dental Medicine, Columbia University Irving Medical Center, New York, NY, USA
e-mail: cz2891@cumc.columbia.edu

28.1 Overview of Immune Responses in Orthodontic Treatment

Orthodontic tooth movement (OTM) involves a sequence of biological events activated by the application of mechanical force to the teeth, leading to controlled bone remodeling. This process encompasses four primary stages—initial, lag, acceleration, and retention—each involving distinct immune responses that influence cellular activities in the periodontal ligament (PDL) and alveolar bone (Zong et al. 2022). The immune response in OTM includes various immune cells, cytokines, and signaling pathways, which collectively facilitate or modulate bone resorption in the compression areas and bone formation in the tension areas. Understanding these immune responses provides insight into both the biological basis of orthodontic treatment and potential strategies to optimize outcomes by modulating the inflammatory process (Chaushu et al. 2022).

28.1.1 The Initial Phase: Activation of Inflammatory Mediators

In the initial phase of OTM, the mechanical force applied to teeth compresses the PDL in the compression areas while stretching it in the tension areas. This physical stress disrupts blood flow, creating a hypoxic and acidic environment that

N. Rezaei, H. Ziaei (eds.), *Oral Immunology*, Advances in Experimental Medicine and Biology 1492,
https://doi.org/10.1007/978-3-032-03176-1_28

stimulates the production of inflammatory mediators such as prostaglandin E2 (PGE2) and various cytokines, including interleukin-1 beta (IL-1β), tumor necrosis factor-alpha (TNF-α), and interleukin-6 (IL-6). These cytokines activate local immune cells, notably macrophages, and osteoclast precursor cells. Cytokine signaling also induces the upregulation of receptor activator of nuclear factor kappa-B ligand (RANKL) on osteoblasts and fibroblasts, which is critical for osteoclast differentiation and subsequent bone (Bawyan et al. 2021).

IL-1β and TNF-α are particularly significant in the initial response due to their role in recruiting and activating immune cells like monocytes and neutrophils, which migrate to the site of compression. The production of monocyte chemoattractant protein-1 (MCP-1) aids in this process, ensuring an adequate immune cell presence in the PDL to begin tissue remodeling. These recruited cells produce further cytokines and enzymes that contribute to the local inflammatory response, initiating the process of bone resorption required for tooth movement (Zeng et al. 2015). Recent human transcriptomic data further confirmed that PDL tissues show stress responses after 7 days of orthodontic force, with inflammation and enhanced bone resorption (Zhao et al. 2024a).

28.1.2 The Lag Phase: Modulation of the Immune Response

As orthodontic force continues, the immune response transitions to a lag phase, where cellular activity slows down despite the ongoing mechanical load. This phase is marked by a balance between pro-inflammatory and anti-inflammatory signals, allowing tissues to adapt and prevent excessive inflammation that could damage periodontal structures. During this stage, a shift occurs within the macrophage population, with M1 macrophages (pro-inflammatory) giving way to M2 macrophages (anti-inflammatory) that support tissue repair and resolution of inflammation. This change is essential for maintaining tissue integrity while ensuring that bone remodeling can continue without excessive damage (Bilgic et al. 2016).

The RANKL/osteoprotegerin (OPG) ratio plays a critical role in the lag phase. OPG acts as a decoy receptor for RANKL, inhibiting osteoclast differentiation and limiting bone resorption. This balance between RANKL and OPG helps modulate the activity of osteoclasts, preventing an excessive inflammatory response and maintaining bone homeostasis. Additionally, Tregs are recruited during this phase to release interleukin-10 (IL-10) and transforming growth factor-beta (TGF-β), both of which contribute to an anti-inflammatory environment and promote tissue repair (Chaushu et al. 2022; Kikuta et al. 2020).

28.1.3 The Acceleration Phase: Heightened Bone Remodeling Activity

In the acceleration phase, bone remodeling is highly promoted with the increasing activity of osteoclasts and osteoblasts, resulting in faster bone resorption in the compression areas and bone formation in the tension areas. This phase is characterized by a second wave of immune cell recruitment, including B cells and T helper cells. These cells secrete cytokines, which could further amplify the remodeling process. TNF-α, IL-6, and interleukin-17 (IL-17) are upregulated during this phase, promoting osteoclast activity and facilitating bone resorption (Babenko et al. 2021).

In the tension areas, fibroblasts and PDL cells respond to cytokines and growth factors such as fibroblast growth factor (FGF) and bone morphogenetic protein (BMP), inducing osteoblast differentiation and subsequent osteogenesis. These cellular activities are regulated by transcription factors such as nuclear factor kappa-B (NF-κB) and hypoxia-inducible factor-1 alpha (HIF-1α), which modulate gene expression to sustain the remodeling activities necessary for tooth movement (Jeon et al. 2021; Kirschneck et al. 2020a).

28.1.4 The Retention Phase: Immune Resolution and Tissue Stabilization

The purpose of retention is to stabilize the newly positioned teeth and maintain periodontal integrity. This phase involves the resolution of inflammation and the reconstruction of tissue homeostasis. Anti-inflammatory cytokines, including IL-10 and TGF-β, play significant roles in suppressing residual inflammatory responses (Bagheri et al. 2019). Additionally, M2 macrophages act predominantly in this stage, promoting tissue repair and contributing to a reconstruction of homeostasis in the PDL and alveolar bone (Alvarez et al. 2020).

The downregulation of RANKL and upregulation of OPG levels effectively reduce osteoclast activity. This process is crucial for preserving bone density and preventing further bone resorption. During this stage, tissue-specific macrophages and fibroblasts secrete extracellular matrix components that repair the periodontal tissue, helping to secure the tooth in its new position. This phase is essential to prevent orthodontic relapse and to ensure long-term treatment success.

28.1.5 Conclusion

The four stages of OTM—initial, lag, acceleration, and retention—represent a coordinated immune response involving diverse immune cells, cytokines, and signaling pathways that collectively regulate bone resorption and formation. Each stage relies on a finely tuned balance between pro-inflammatory and anti-inflammatory mechanisms that control the extent of bone remodeling while protecting periodontal tissue integrity. Effective orthodontic treatment outcomes depend on maintaining this balance, which facilitates OTM as well as minimizes the risk of adverse effects. Understanding these immune responses offers valuable insights into optimizing orthodontic treatments and may guide the development of interventions to modulate inflammation and enhance therapeutic outcomes in orthodontics.

28.2 Inflammatory Cytokines in Orthodontic Treatment

Orthodontic treatment induces mechanical stress on the PDL, leading to inflammatory responses that play a crucial role in facilitating tooth movement. Cytokines, as signaling molecules, are central to the regulation of inflammation and bone remodeling. The inflammatory response triggered by orthodontic forces involves a range of cytokines, including pro-inflammatory cytokines such as IL-1β, TNF-α, and IL-6, which promote osteoclastogenesis and anti-inflammatory cytokines that balance this response and stabilize the tissues. This section explores the functions and interactions of these cytokines, highlighting how they contribute to the stages of OTM and emphasizing the regulatory mechanisms involved.

28.2.1 IL-1β and TNF-α: Key Pro-inflammatory Cytokines in OTM

IL-1β and TNF-α are among the earliest cytokines upregulated in response to orthodontic force application. These cytokines are critical in initiating the inflammatory cascade that promotes osteoclast activity and, consequently, bone resorption in the compression area of the PDL. Studies indicate that both IL-1β and TNF-α levels increase significantly within the first 24 hours after the application of force, facilitating rapid recruitment and activation of osteoclast precursor cells (Grant et al. 2013).

A study utilizing animal models demonstrated that inhibiting TNF-α and IL-1β via soluble receptors could reduce osteoclast activity and decelerate OTM, confirming the pivotal role of these cytokines in bone resorption (Qi et al. 2020). Moreover, it was found that the continuous expression of these cytokines, regulated through transcription factors such as NF-κB, sustains osteoclastogenesis and maintains bone resorption over the treatment (Nunes et al. 2017).

28.2.2 IL-6 and IL-8: Modulators of Inflammatory Response and Osteoclastogenesis

IL-6 is another cytokine that is significantly elevated during orthodontic treatment, promoting both local and systemic inflammatory responses. IL-6 has a dual role: it enhances osteoclast differentiation in the compression areas and modulates fibroblast and osteoblast activities in the tension areas, supporting new bone formation and PDL repair. Elevated IL-6 levels have been detected in gingival crevicular fluid (GCF) during the initial phases of OTM, suggesting that it also plays a role in regulating the inflammatory response across different tissues (Ren et al. 2007).

IL-8, a chemokine, is primarily involved in recruiting neutrophils to the site of inflammation. This recruitment helps sustain the inflammatory response necessary for tissue remodeling, although IL-8 levels tend to peak later than IL-1β and TNF-α, usually around 1 to 4 weeks after force application. The delayed peak of IL-8 suggests its role in maintaining the inflammatory environment required for sustained bone resorption (Yang et al. 2013).

28.2.3 RANKL and OPG: Key Regulators of Bone Remodeling

The RANKL and OPG signaling pathways are among the most crucial regulatory systems in bone remodeling during orthodontic treatment. RANKL, expressed by osteoblasts and stromal cells in the PDL, binds to RANK (receptor activator of nuclear factor kappa-B) on osteoclast precursors, promoting their differentiation into active osteoclasts. Conversely, OPG, a decoy receptor produced by osteoblasts, binds to RANKL, preventing it from interacting with RANK and thus inhibiting osteoclast formation.

The balance between RANKL and OPG levels determines the rate of bone resorption. During OTM, an increase in the RANKL/OPG ratio in the compression areas favors osteoclastogenesis, facilitating bone resorption. Studies measuring GCF biomarkers indicate that RANKL levels increase significantly at compression sites, while OPG levels remain stable, reinforcing the dominance of bone resorption signals in this area (Kapoor et al. 2014).

28.2.4 MMPs and TIMPs: Tissue Remodeling in OTM

Research demonstrates that during OTM, matrix metalloproteinases (MMPs), and tissue inhibitors of metalloproteinases (TIMPs) show increased expression in both the compression and tension sides of the PDL.

Specifically, MMP-9 plays a significant role in the degradation of the ECM in the PDL, which is necessary for remodeling the tissue during OTM. However, MMP activity is modulated by TIMPs to prevent excessive tissue degradation (Schröder et al. 2020). For instance, MMP-9 levels peak during the early phases of treatment and are then balanced by TIMP to maintain tissue integrity (Bildt et al. 2009). MMP-9 levels rise in the compression areas to assist in ECM degradation, while TIMP-1 expression in the tension areas helps maintain tissue integrity, supporting periodontal remodeling (Behm et al. 2021).

28.2.5 The Anti-inflammatory Role of IL-10 and TGF-β

To avoid excessive tissue damage, anti-inflammatory cytokines such as IL-10 and TGF-β are also upregulated during orthodontic treatment. IL-10 is known to inhibit the synthesis of pro-inflammatory cytokines like TNF-α and IL-1β, thus helping to regulate the inflammatory response and prevent excessive tissue destruction. TGF-β supports tissue repair by promoting collagen synthesis and fibroblast proliferation in the tension area of the PDL (Zhang et al. 2017).

28.2.6 Conclusion

The cytokine network in orthodontic treatment is a complex interplay of pro-inflammatory and anti-inflammatory signals, orchestrating the sequential stages of OTM. Pro-inflammatory cytokines like IL-1β, TNF-α, IL-6, and IL-8 initiate and sustain bone resorption in the compression areas, while anti-inflammatory cytokines such as IL-10 and TGF-β balance these effects, promoting tissue repair in the tension areas. The modulation of these cytokines, particularly through pathways involving RANKL and OPG, enables controlled bone remodeling that is essential for effective orthodontic outcomes. Understanding these molecular dynamics provides a foundation for developing therapeutic interventions that could enhance orthodontic treatment efficiency by fine-tuning the inflammatory response (Summarized in Fig. 28.1).

28.3 Immune Cells in Orthodontic Treatment

During OTM, immune cells play a pivotal role in initiating, sustaining, and resolving inflammation, leading to the remodeling of periodontal tissues. Key immune cells involved in OTM include T cells, macrophages, dendritic cells, and natural killer (NK) cells, each of which participates in different stages of tissue remodeling. Their actions are mediated by cytokines and chemokines that guide bone resorption in the compression areas and bone formation in the tension areas. This section delves into the roles of these immune cells in OTM and the molecular mechanisms underlying their functions, with insights from recent experimental studies.

28.3.1 T Cells and Their Cytokine Signaling in OTM

T cells, particularly CD4+ T-helper cells, are central to the immune response in OTM. They secrete pro-inflammatory cytokines, such as TNF-α and IFN-γ, which stimulate osteoclast formation and thus promote bone resorption in the compression areas. Experimental studies have shown that T-cell-deficient mice exhibit significantly reduced tooth movement, indicating that T cells are essential for optimal OTM. When T cells were reintroduced in these models, normal levels of osteoclast activity and tooth movement were restored, underscoring the role of T cells in orchestrating the inflammatory response necessary for bone remodeling (Yan et al. 2015).

Furthermore, recent studies have demonstrated that T cells not only initiate the inflammatory response but also contribute to its regulation. For instance, during the retention phase, T cells secrete regulatory cytokines that suppress excessive inflammation and promote tissue stabilization. This modulation is crucial for avoiding overt tissue destruction and ensuring that OTM results in stable and healthy periodontal structures (Liu et al. 2017).

28.3.2 Macrophages: Dual Role in Inflammation and Repair

Macrophages are among the first immune cells recruited to the site of mechanical stress in OTM. These cells exhibit plasticity, differentiating into two main types: pro-inflammatory M1 macrophages, which drive bone resorption, and anti-inflammatory M2 macrophages, which support tissue repair. In the initial stages of OTM, M1 macrophages dominate, releasing pro-inflammatory cytokines (e.g., IL-1β and TNF-α) that activate osteoclast precursors and accelerate bone resorption. Studies indicate that the ratio of M1 to M2 macrophages shifts as OTM progresses, with M2 macrophages becoming predominant during later stages to promote tissue resolution and homeostasis (Wang et al. 2023).

This dynamic regulation of macrophage phenotypes is critical for balancing tissue remodeling and repair. Experiments involving macrophage depletion have shown delayed

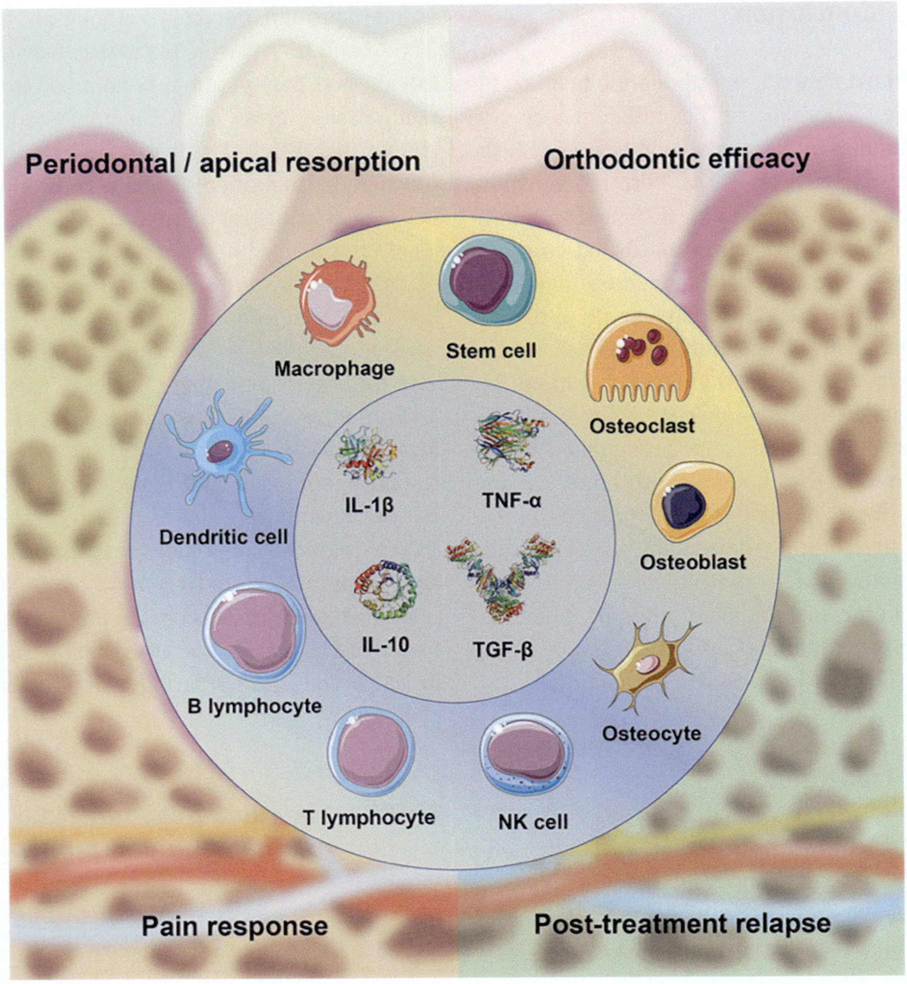

Fig. 28.1 This schematic illustrates the dual roles of immune signaling in orthodontic treatment. At the center, key cytokines coordinate bone remodeling. Surrounding icons represent the major cellular mediators: macrophages, dendritic cells, B lymphocytes, T lymphocytes, and NK cells (left semicircle), which drive immune activation and pain responses, and osteoclasts, osteoblasts, osteocytes, and mesenchymal stem cells (right semicircle), which execute bone resorption and formation for effective tooth movement. The four colored quadrants denote clinical outcomes. Together, the graphic highlights how fine-tuning immune pathways can enhance treatment efficiency, minimize discomfort, and promote periodontal health and long-term stability

OTM, suggesting that both types of macrophages are essential for effective tooth movement and periodontal stability (Wang et al. 2018).

28.3.3 Dendritic Cells and Antigen Presentation

Dendritic cells (DCs), as antigen-presenting cells, play a unique role in OTM by processing antigens and activating T cells. These cells are activated by mechanical stress and inflammation in the PDL and subsequently migrate to lymph nodes, where they prime T cells by presenting processed antigens (Song et al. 2018). Compressive forces in orthodontic treatment upregulate signaling pathways involving RANK and DC-STAMP in the PDL, showing that DCs contribute to osteoclast precursor differentiation and activation during OTM (Matsuike et al. 2018). This interaction fosters a local immune response that supports the bone resorption and formation necessary for OTM. A recent animal study highlighted that dendritic cells play a critical role in immune responses by presenting antigens and activating T cells. Inhibition of Dectin-1 on DCs prevented their maturation, which led to a reduction in T cell activation and subsequent immune response. This finding underscores the important role of DCs in immune modulation, a principle applicable in the context of OTM, where DC-T cell interactions are essential for bone remodeling (Ren et al. 2021).

28.3.4 NK Cells and Innate Immune Activation

NK cells, though less studied in the context of OTM, are recognized for their rapid response to inflammatory signals and their role in modulating the activity of other immune cells. NK cells respond to stress signals from periodontal tissues under mechanical force, releasing cytokines that can enhance macrophage and T cell functions. Recent research has shown that NK cells accumulate in the PDL during the early stages of OTM, where they assist in sustaining the inflammatory environment necessary for initiating bone remodeling (Gao et al. 2022).

Moreover, NK cells exhibit cytotoxic effects that can modulate tissue remodeling by removing senescent or damaged cells within the PDL. These mechanisms are integral to immune regulation within tissues and support tissue remodeling by ensuring cellular turnover, which aligns with NK cells' role in maintaining PDL health during OTM (Prager and Watzl 2019).

28.3.5 Regulatory Mechanisms and Immune Modulation in OTM

The immune response in OTM is not solely inflammatory but involves mechanisms that regulate and resolve inflammation to prevent excessive tissue damage. Treg cells, a subset of T cells, secrete anti-inflammatory cytokines such as IL-10 and TGF-β that inhibit the activities of pro-inflammatory T cells and macrophages. A study examining Treg cell function in an OTM model revealed that an increased presence of Treg cells was associated with reduced bone resorption and enhanced periodontal stability (Zhang et al. 2023).

The RANK/RANKL pathway is another significant regulatory mechanism in OTM. RANKL expression by osteoblasts and PDL cells promotes osteoclastogenesis, while its binding to OPG, a decoy receptor, prevents excessive bone resorption. Studies indicate that immune cells, particularly T cells and macrophages, influence the RANK/RANKL/OPG balance, thereby controlling the extent of bone remodeling during OTM (Wu et al. 2020).

28.3.6 Conclusion

Immune cells, including T cells, macrophages, dendritic cells, and NK cells, form a coordinated response that supports OTM through bone resorption and formation. Each cell type has a specific role, with T cells and macrophages primarily driving inflammation and bone remodeling, while dendritic cells and NK cells help sustain the immune response. Regulatory mechanisms, involving Treg cells and RANK/RANKL signaling, ensure that inflammation is controlled and resolved to prevent periodontal damage. The dynamic interplay of immune cells in OTM highlights the complexity of orthodontic-induced bone remodeling and provides insights into potential therapeutic targets to optimize treatment outcomes and minimize adverse effects.

28.4 Immune Response in Periodontal Tissues During Orthodontic Treatment

Orthodontic treatment, by applying mechanical forces to teeth, initiates a series of biological responses within the periodontal tissues that drive both inflammatory and immune reactions. These responses are critical for bone remodeling, as they coordinate bone resorption and formation on either side of the PDL (Zong et al. 2023b). The process involves various immune cells, cytokines, and molecular pathways that collectively maintain tissue homeostasis while ensuring effective tooth movement. This section explores the recent findings on the periodontal immune response to orthodontic forces, highlighting the involvement of both innate and adaptive immune systems.

28.4.1 Innate Immune Response: The Role of Neutrophils and Macrophages

Neutrophils and macrophages are among the first responders to mechanical force in the PDL, initiating the inflammatory response necessary for bone remodeling (Xu et al. 2022). Neutrophils, recruited rapidly to the site, release cytokines and reactive oxygen species that amplify the inflammatory environment, paving the way for tissue breakdown and osteoclast activity (Kim et al. 2020). These neutrophil responses have been shown to promote osteoclast differentiation in the compression area of the PDL, where bone resorption is required (Zeng et al. 2019).

Macrophages, which exhibit plasticity between pro-inflammatory (M1) and anti-inflammatory (M2) phenotypes, play a critical role in modulating inflammation (Veloso et al. 2020). During the initial stages of OTM, M1 macrophages are predominant and secrete pro-inflammatory cytokines such as TNF-α and IL-1β, which stimulate osteoclastogenesis. As treatment progresses, a shift occurs toward M2 macrophages, which secrete IL-10 and TGF-β, cytokines that suppress inflammation and facilitate tissue repair (Liu et al. 2022). This dynamic balance between M1 and M2 macrophages is essential for the regulation of bone remodeling and the resolution of inflammation (He et al. 2020).

28.4.2 Adaptive Immune Response: T Cells and B Cells in Periodontal Remodeling

T cells are instrumental in both amplifying and regulating the immune response in periodontal tissues during orthodontic treatment (Ogawa et al. 2019). CD4+ T helper cells (Th1) release cytokines such as IFN-γ and TNF-α, which promote osteoclast activity and thereby enhance bone resorption in the compression areas. The importance of T cells in bone remodeling during OTM has been demonstrated in animal models, where T cell depletion significantly reduces tooth movement due to diminished osteoclast activity (Wald et al. 2021).

B cells, while less studied in OTM, contribute to the immune response through their role in antibody production and cytokine secretion. Recent studies have indicated that B cells in periodontal tissues express receptor activators of nuclear factor kappa-B ligand (RANKL), which is critical for osteoclast differentiation and activation. By promoting RANKL expression, B cells further enhance bone resorption, a necessary process for tooth movement (Li et al. 2020a).

28.4.3 The RANK/RANKL/OPG Axis: Central Pathway in Periodontal Bone Remodeling

The RANK/RANKL/OPG pathway is a key signaling mechanism in the regulation of bone remodeling, particularly in response to orthodontic forces. RANKL, produced by osteoblasts, PDL cells, and immune cells, binds to its receptor RANK on osteoclast precursors, promoting their differentiation into active osteoclasts and facilitating bone resorption. OPG, a decoy receptor, binds RANKL to inhibit osteoclast activation, thus acting as a negative regulator in this pathway (Naoumova et al. 2024).

During OTM, the RANKL/OPG ratio increases in the compression areas, promoting osteoclast activity and bone resorption. In the tension areas, OPG levels remain relatively stable, which allows for a controlled bone formation process to occur. Experimental studies have shown that modulating this pathway can influence the rate and extent of tooth movement, suggesting that targeting RANKL/OPG signaling could provide therapeutic strategies to optimize orthodontic treatment outcomes (Wei et al. 2022).

28.4.4 Chemokine Signaling and Cellular Recruitment in the PDL

Chemokines, such as monocyte chemoattractant protein-1 (MCP-1) and regulated on activation, normal T cell expressed and secreted (RANTES), are essential for recruiting immune cells to the periodontal tissues. MCP-1 is upregulated in PDL cells under mechanical force, attracting monocytes that differentiate into macrophages upon arrival. RANTES plays a similar role, recruiting T cells to the PDL, where they participate in the immune and inflammatory response. Studies have demonstrated that chemokine inhibition reduces immune cell infiltration and delays OTM, underscoring the importance of chemokine signaling in coordinating the cellular response during bone remodeling (Mulholland et al. 2019; Rath-Deschner et al. 2020).

28.4.5 Toll-Like Receptors (TLRs) and Periodontal Immune Response

Toll-like receptors (TLRs), particularly TLR4, play a role in periodontal tissue response by recognizing bacterial components and other signals of tissue damage. Activation of TLR4 in the PDL leads to cytokine production and inflammation, which support bone remodeling (Koutouzis et al. 2013). TLR4 signaling pathways are activated not only by bacterial infection but also by endogenous ligands released from stressed cells under orthodontic forces. A recent study indicated that suppressing TLR4 can reduce inflammation and delay bone resorption in the PDL (Deng et al. 2020), highlighting the significance of TLRs in regulating periodontal immune responses during OTM (Roth et al. 2022).

28.4.6 Conclusion

The immune response in periodontal tissues during orthodontic treatment involves a finely balanced interplay between innate and adaptive immune cells, chemokines, and signaling molecules such as RANKL and TLRs. These components coordinate bone resorption in the compression areas and bone formation in the tension areas, supporting effective tooth movement. The recruitment of neutrophils, macrophages, T cells, and B cells, alongside the activation of pathways like RANK/RANKL and TLRs, ensures a controlled inflammatory response essential for OTM. Understanding these immune dynamics provides a foundation for developing therapeutic strategies to regulate inflammation, thereby optimizing treatment efficacy and minimizing adverse periodontal effects. Furthermore, it is also important to recognize that orthodontic force affects not only PDL but also the dental pulp, whereRNA-seq revealed immune activation and hypoxia at the early phase, followed by cell adhesion and repair pathways at the late phase (Zhao et al. 2024b)

28.5 Pain Response and Immune Modulation in Orthodontic Treatment

Pain is a frequent side effect of orthodontic treatment, often triggered by the mechanical forces exerted on periodontal structures, leading to an inflammatory response that activates various immune pathways. This pain results from an intricate interplay of immune cells and mediators, including cytokines, chemokines, and neural mechanisms. The following section explores recent research on the immunological and physiological basis of orthodontic pain, as well as immune modulation strategies aimed at managing pain during treatment.

28.5.1 Initial Inflammatory Response and Pain Mechanisms

The application of orthodontic force leads to compression of the PDL, resulting in vascular constriction and tissue hypoxia. This initiates an inflammatory cascade characterized by the recruitment of neutrophils, macrophages, and other immune cells, which release pro-inflammatory mediators such as IL-1β, TNF-α, and IL-6. These cytokines amplify the inflammatory response, sensitizing periodontal nociceptors and triggering pain perception. Research has shown that high levels of TNF-α correlate strongly with pain intensity during the early phases of orthodontic treatment (Long et al. 2016).

A clinical study examining pain levels in orthodontic patients found that pain peaked within the first 24 hours after force application and declined gradually as the inflammatory response resolved. This trend suggests that the release of cytokines following force application plays a crucial role in early pain generation (Santos et al. 2023).

28.5.2 Neurogenic Inflammation and Central Sensitization

Neurogenic inflammation, involving the release of neuropeptides such as substance P and calcitonin gene-related peptide (CGRP) from sensory neurons, further enhances pain during orthodontic treatment. These neuropeptides increase vascular permeability and attract immune cells to the site, sustaining the inflammatory environment. As these signals propagate along the trigeminal pathway, they contribute to central sensitization, heightening pain sensitivity not only in the treated area but also in adjacent tissues (Caviedes-Bucheli et al. 2021; Wang et al. 2019a).

28.5.3 Immune Cells and Cytokine Modulation of Pain

Macrophages are central to both the initiation and resolution of inflammation-induced pain. In the early stages, M1 macrophages produce elevated levels of pro-inflammatory cytokines IL-1β and TNF-α in response to inflammation, while blocking these cytokines reduces inflammatory effects. The findings highlight the roles of M1 and M2 polarization in modulating inflammatory responses in dental tissues, aligning with their roles in pain modulation during orthodontic treatment as described (Gopinath et al. 2023). However, as the treatment progresses, a shift toward M2 macrophages occurs, releasing IL-10 and TGF-β, which counteract inflammatory mediators and facilitate tissue repair (Arabpour et al. 2021). This aligns with the role of the M2 macrophages in transitioning to anti-inflammatory effects, which is important for pain reduction in tissue remodeling.

The RANK/RANKL signaling pathway also plays a significant role in pain modulation during orthodontic treatment. RANKL, which promotes osteoclastogenesis, indirectly contributes to pain by facilitating bone resorption in the pressure areas (El-sayed et al. 2022; Sakamoto et al. 2019).

28.5.4 Pharmacological and Non-pharmacological Pain Management

Pain management strategies during orthodontic treatment often involve nonsteroidal anti-inflammatory drugs (NSAIDs), which inhibit cyclooxygenase (COX) enzymes and reduce the synthesis of pro-inflammatory prostaglandins. However, NSAIDs can impede osteoclast activity and slow down tooth movement, presenting a dilemma for orthodontists. Recent studies on aspirin demonstrated that it can reduce both pain and inflammation in orthodontic patients, but careful dosing is essential to prevent interference with bone remodeling (Liu et al. 2017). Etoricoxib, the COX-2 selective NSAID, effectively reduces pain without significantly impairing PDL fibroblast-mediated periodontal inflammation, extracellular matrix remodeling, and osteoclastogenesis during simulated orthodontic compressive *strain* in vitro (Kirschneck et al. 2019) and OTM in vivo (Kirschneck et al. 2020b.

Low-level laser therapy (LLLT) is an emerging non-pharmacological approach for pain reduction. Studies have shown that LLLT can reduce cytokine expression, including IL-6 and TNF-α, and alleviate pain perception by modulating inflammatory pathways without affecting the pace of tooth movement (Prathapan Santhakumari et al. 2023). Recent clinical investigations reported that patients receiving LLLT experienced lower pain levels compared to those who did not, with significant effects observed up to 24–48 hours post-application (Brito et al. 2022; Doshi-Mehta and Bhad-Patil 2012).

28.5.5 Psychological Factors in Pain Perception and Immune Modulation

Pain perception in orthodontics is not solely influenced by biological factors but is also modulated by psychological elements such as anxiety and anticipation of pain. High levels of anxiety have been linked to increased pain perception, likely mediated by stress-induced immune responses (Reale et al. 2020). For instance, cortisol released in response to anxiety can modulate immune cell function, enhancing pro-inflammatory cytokine production, which may exacerbate pain (Munshi et al. 2020).

Studies have also shown that cognitive behavioral therapy (CBT) and other anxiety-reducing techniques may effectively lower perceived pain in orthodontic patients, suggesting that psychological interventions could complement biological strategies for comprehensive pain management.

28.5.6 Conclusion

The pain response during orthodontic treatment is a multifaceted phenomenon involving a cascade of immune responses and cytokine release, as well as neurogenic and psychological factors. Pro-inflammatory cytokines such as TNF-α and IL-1β, along with neuropeptides like substance P, contribute to pain perception, while regulatory mechanisms involving M2 macrophages and anti-inflammatory cytokines facilitate pain resolution over time. Strategies for pain modulation, including NSAIDs, LLLT, and psychological interventions, offer promising avenues for alleviating discomfort in orthodontic patients while preserving the efficacy of treatment.

28.6 Orthodontic Biomaterials and Immune Compatibility

The biomaterials used in orthodontics, including brackets, wires, adhesives, retainers, and aligners, play a significant role in treatment success but can also influence the immune response due to the material's composition and interactions with oral tissues. Understanding the biocompatibility of these materials, especially their potential to induce local immune reactions or hypersensitivity, is essential for optimizing treatment outcomes and patient comfort (Zong et al. 2023a).

28.6.1 Metal Brackets and Archwires

Metal brackets, typically made of stainless steel or nickel-titanium alloys, are widely used due to their durability and strength. However, nickel, a common component, is known to cause allergic reactions in sensitive individuals, leading to a type IV hypersensitivity response characterized by local inflammation, erythema, and, in some cases, systemic symptoms. A classic study found that approximately 28% of patients with nickel-containing brackets exhibited signs of hypersensitivity, such as mucosal inflammation (Janson et al. 1998), suggesting that pre-treatment testing for metal allergies may be beneficial (Zigante and Špalj 2022), especially for the potential for systemic hypersensitivity reactions (Stewart et al. 2023).

Archwires, commonly fabricated from nickel-titanium or stainless steel, also release trace metal ions into saliva, which can accumulate over time.

An in vitro study demonstrated that nickel and titanium ions from these wires significantly impact cellular morphology, with evidence of cellular stress in fibroblasts exposed to nickel-titanium surfaces. These findings emphasize the importance of monitoring nickel exposure in orthodontic patients, as chronic exposure may lead to localized immune reactions and inflammatory responses in the oral cavity (Sodor et al. 2015).

28.6.2 Ceramic Brackets

Ceramic brackets are a popular alternative for patients with metal sensitivities, offering aesthetic appeal and reduced allergenic potential. Despite these advantages, some ceramic brackets have been shown to increase bacterial adhesion, possibly due to the micro-roughness of ceramic surfaces. This can elevate the risk of gingival inflammation and biofilm formation, especially if oral hygiene is not maintained. Recent research compared ceramic brackets with different retention mechanisms and found that while they exhibited comparable bond strengths to metal brackets, the adhesive interface often retained bacteria, which could contribute to localized immune reactions (Mei et al. 2009).

28.6.3 Adhesives and Bonding Agents

Orthodontic adhesives, particularly those containing bisphenol A glycidyl methacrylate (Bis-GMA), have raised concerns regarding immune compatibility due to their potential estrogenic effects. Adhesives also promote bacterial adhesion at the bracket-enamel interface, a hotspot for bacterial colonization, leading to an increased risk of gingivitis and enamel demineralization. A study demonstrated that a novel adhesive incorporating niobium phosphate bioglass reduced Streptococcus mutans growth around brackets, thus lowering inflammation and demineralization risks associated with traditional adhesives (Degrazia et al. 2019).

28.6.4 Clear Aligners and Attachments

Clear aligners, primarily composed of thermoplastic polyurethane materials, are a preferred option for patients seeking aesthetic orthodontic solutions. These aligners are generally biocompatible; however, certain thermoplastics can induce mild local inflammation due to the material's degradation over time. Attachments, typically composite resin "buttons" bonded to teeth to facilitate aligner grip, can also increase bacterial adhesion, creating additional immune challenges for the surrounding gingival tissues. A recent clinical study indicated that aligners with attachments demonstrated higher levels of localized gingival inflammation compared to non-attached aligners, likely due to plaque accumulation around the attachment sites (Oikonomou et al. 2021).

28.6.5 Retainers and Fixed Appliances

Fixed retainers, usually constructed from stainless steel or nickel-titanium alloys, can remain in place for years, continually exposing the patient to trace metal ions. Research has shown that these retainers may release metal ions, potentially leading to allergic responses or cytotoxic effects in sensitive individuals. Nickel ions, in particular, have been associated with DNA damage in oral mucosa cells (Faccioni et al. 2003), highlighting the need for alternative materials for patients with known sensitivities (Zigante et al. 2020), particularly those exposed to fluoride (Chantarawaratit and Yanisarapan 2021).

28.6.6 Future Directions in Biocompatible Orthodontic Materials

Emerging materials, including nickel-free stainless steels, titanium alloys, and antibacterial coatings, are being developed to reduce immune compatibility issues associated with traditional orthodontic materials. Studies have demonstrated that nickel-free stainless steel brackets and archwires maintain similar strength and flexibility while significantly reducing the risk of allergic reactions. Additionally, 3D-printed ceramic brackets designed with patient-specific morphology are being explored for their potential to reduce bacterial adherence and improve overall biocompatibility (Yang et al. 2019). Recent advances in organoid technologies and controlled-drug-release hydrogels also offer a new paradigm for integrating biologically active scaffolds into orthodontic biomaterials, as exemplified by biomimetic-engineered hydrogels that prevent orthodontic root resorption. (Zong et al. 2025).

28.6.7 Conclusion

The biocompatibility of orthodontic biomaterials is crucial for minimizing adverse immune responses during treatment. Metal and ceramic brackets, archwires, adhesives, and retainers all interact with the oral environment and can influence immune reactions. Advances in material sciences, including the development of nickel-free alloys and antibacterial adhesives, offer promising avenues for enhancing immune compatibility, reducing hypersensitivity risks, and improving patient outcomes in orthodontic care.

28.7 Immune Response and Accelerated OTM

In OTM, immune responses play a pivotal role in regulating bone remodeling through the activation of osteoclasts and osteoblasts. By leveraging immune modulation, recent studies have shown potential methods to accelerate OTM, optimizing treatment duration without compromising periodontal health. This section explores the immune pathways involved in OTM and highlights emerging techniques for immunomodulation to enhance the rate of tooth movement.

28.7.1 Role of Cytokines and Chemokines in Bone Resorption and Formation

Cytokines and chemokines orchestrate the immune response in OTM, mediating bone resorption in the compression areas of the PDL and bone formation in the tension areas. Key cytokines involved in this process include IL-1β, TNF-α, and RANKL. These cytokines are upregulated during orthodontic force application, promoting osteoclast differentiation and activity, which is essential for bone remodeling (Yamaguchi and Fukasawa 2021).

In recent studies, IL-1β and TNF-α have been identified as crucial mediators in accelerating OTM by enhancing osteoclastogenesis. A study by Zheng and Yang (2021) demonstrated that LLLT, which increases IL-1β and RANKL levels in the gingival crevicular fluid, can significantly enhance the rate of tooth movement in patients undergoing orthodontic treatment. By adjusting the parameters of LLLT, the study achieved a safe modulation of cytokine levels, optimizing the inflammatory response required for faster bone resorption and OTM (Zheng and Yang 2021).

28.7.2 T Cells and Adaptive Immune Involvement in Accelerated OTM

Adaptive immune cells, particularly T cells, play a significant role in sustaining the inflammatory response during OTM. Research has shown that CD4+ T helper cells enhance osteoclast activity through the secretion of Th1-type cytokines like TNF-α and IFN-γ (Pradeep Yeola et al. 2020). A study involving T-cell-deficient mice demonstrated significantly reduced tooth movement, which was restored upon infusion of T cells. This highlights the necessity of T cell-driven immune responses in accelerating OTM through osteoclastic activation (Monasterio et al. 2019).

Additionally, anti-TNF-α therapies, commonly used for autoimmune conditions, have been explored as a way to modulate the inflammatory response. By transiently increasing TNF-α in periodontal tissues researchers were able to enhance osteoclastogenesis and thus accelerate tooth movement without compromising tissue integrity. This approach of controlled cytokine modulation offers a novel therapeutic avenue for accelerating orthodontic treatment duration while maintaining bone health.

28.7.3 Corticotomy and Micro-osteoperforation: Local Immune Modulation

Corticotomy and micro-osteoperforation (MOP) are minor surgical interventions designed to locally enhance the immune response, facilitating rapid OTM. These techniques create minor injuries in the alveolar bone, triggering a localized inflammatory response that increases osteoclast activity. Recent clinical trials have shown that MOP can increase cytokine levels, including IL-1β and RANKL, enhancing osteoclast recruitment and activity, which results in faster tooth movement (Alikhani et al. 2015).

A study on corticotomy by Liou et al. (2011) demonstrated that surgically induced trauma initiates a regional acceleratory phenomenon (RAP), where bone turnover rates increase significantly, reducing the time required for OTM by up to 50%. Corticotomy is particularly effective because it temporarily shifts the bone remodeling equilibrium toward resorption, allowing for rapid tooth movement. This approach, while invasive, provides a method for targeted immune activation that can effectively accelerate orthodontic treatment (Liou et al. 2011).

28.7.4 Low-Level Laser Therapy and Photobiomodulation

LLLT, or photobiomodulation, has gained attention as a noninvasive approach to modulating the immune response to accelerate OTM. LLLT stimulates mitochondrial activity in cells, promoting ATP production and increasing the release of inflammatory cytokines such as IL-1β and RANKL (Reis et al. 2022; Zheng and Yang 2021). In clinical studies, LLLT-treated sites demonstrated faster tooth movement due to elevated osteoclast activity, which was maintained at safe levels without adverse effects on the PDL (Sedky et al. 2019).

By modulating the inflammatory response, LLLT creates an optimized environment for bone remodeling, supporting rapid OTM. The efficacy of LLLT is dose-dependent, with higher frequencies and wavelengths promoting a stronger immune response, thereby enhancing OTM rates without compromising tissue health (Baser Keklikci et al. 2020; Fernandes et al. 2019).

28.7.5 Genetic and Molecular Modulation

With advances in gene therapy, molecular interventions to modulate immune response pathways during OTM have shown promise. Gene-based modulation of the RANK/RANKL pathway, which governs osteoclastogenesis, has been explored as a means to accelerate OTM. Gene transfer of RANKL or RANK in localized periodontal tissues could potentially enhance osteoclast activity, thus expediting the tooth movement process. Although these genetic approaches remain experimental, preliminary studies indicate their potential in achieving accelerated OTM with precise control over immune signaling (Tsolakis et al. 2023).

28.7.6 Conclusion

Immune modulation plays a pivotal role in accelerating OTM. Techniques such as LLLT, MOP, corticotomy, and gene therapy offer promising methods for harnessing immune responses to optimize OTM rates. By targeting cytokine pathways like RANKL, TNF-α, and IL-1β, and utilizing immune cells like T cells and macrophages, orthodontists can improve treatment efficiency and reduce overall duration. These immunomodulatory approaches, supported by recent clinical and experimental findings, represent a novel direction in orthodontics, with the potential to enhance patient outcomes while ensuring periodontal health and stability.

28.8 Immunological Factors in Orthodontic Retention and Relapse Prevention

Orthodontic relapse, defined as the tendency of teeth to return to their original positions after treatment, is a significant challenge in orthodontics. Integrating long-term radiographic evidence, such as post-treatment changes in third molar position, underlines the importance of extended follow-up beyond immediate retention (Butaye et al. 2023). This phenomenon is primarily influenced by immunological responses within periodontal tissues, which mediate bone remodeling through inflammatory and osteoclastic activities. Recent studies highlight the role of immune modulation in enhancing retention and preventing relapse, revealing potential therapeutic strategies to maintain treatment outcomes. However, it is important to recognize that many mechanistic studies have relied on split-mouth design in animal experiments. It has been demonstrated that moving teeth on one side induces adaptive movement on the contralateral side to maintain occlusal balance, so the untreated side cannot serve as a true control (Cadenas de Llano-Pérula et al. 2022). Consequently, all split-mouth design investigations, evaluating pharmacologic agents, underlying mechanisms, or orthodontic techniques, are likely to underestimate their effects.

28.8.1 Role of Osteoclastogenesis in Orthodontic Relapse

Osteoclasts, regulated by immune responses, play a critical role in the relapse process. Following orthodontic force removal, the PDL exhibits a rapid response characterized by the recruitment of osteoclasts, which mediate bone resorption and allow teeth to shift back toward their initial positions. A recent experimental study using an orthodontic mouse model demonstrated that retention duration is inversely proportional to relapse rates. Blocking the activity of macrophage colony-stimulating factor (M-CSF), a key factor for osteoclast formation, significantly reduced osteoclast numbers and minimized relapse, suggesting that osteoclastogenesis is a crucial target for relapse prevention (Qi et al. 2019).

Another approach to inhibiting osteoclast-mediated relapse involves targeting the receptor activator of RANKL, a molecule essential for osteoclast differentiation. Recent studies have shown that by modulating the RANKL/OPG pathway, osteoclast activity can be reduced, thus preventing relapse. OPG gene transfer into periodontal tissues has been demonstrated to decrease relapse rates in rat models, pointing to the potential of gene-based therapies for long-term retention stability (Zhao et al. 2012).

28.8.2 Immune Cells in Retention and Relapse Dynamics

The immune cell profile within periodontal tissues significantly influences orthodontic retention. T cells, particularly CD4+ helper T cells, play a key role in regulating inflammation and bone remodeling processes that can lead to relapse. A study involving animal models found that inhibiting CD4+ T cells with aspirin reduced relapse rates by modulating inflammatory cytokine production, specifically TNF-α and IFN-γ (Liu et al. 2017). Another animal study found that orchestrated T cells can regulate osteoclasts at the early stage of rapid palatal expansion and further facilitate bone formation during retention (Li et al. 2020b). These findings underscore the potential of targeting T-cell-mediated immune responses to achieve prolonged retention.

Moreover, Tregs have emerged as key players in maintaining periodontal homeostasis post-treatment. Tregs produce anti-inflammatory cytokines such as IL-10 and TGF-β, which help suppress osteoclastogenesis and promote tissue stability. Research indicates that increasing Treg activity can reduce the relapse tendency by maintaining a balanced immune environment conducive to bone preservation and reducing osteoclastic bone resorption (Levescot et al. 2021; Wang et al. 2019b).

28.8.3 Inflammatory Mediators and Cytokine Modulation

In recent studies, cytokines such as IL-1β, TNF-α, and IL-6 have been highlighted for their critical roles in mediating inflammatory responses that contribute to orthodontic relapse. Elevated levels of these pro-inflammatory cytokines in the PDL after the removal of orthodontic forces are known to stimulate osteoclast differentiation and activity, which can accelerate bone resorption. Interestingly, research by Kanzaki et al. (2015) demonstrated that activating the nuclear factor erythroid 2-related factor 2 (Nrf2) pathway can effectively reduce these cytokine-mediated effects. Their study found that Nrf2 activation attenuated both OTM and relapse by modulating the inflammatory response, specifically reducing osteoclastogenesis through suppression of cytokine production (Kanzaki et al. 2015). This suggests that targeting inflammatory pathways, such as through Nrf2 activation, may be a promising approach to control orthodontic relapse while preserving bone stability.

Emerging research is exploring targeted immunotherapy, including cytokine-specific inhibitors, as potential therapeutic options for retention (Kuo et al. 2021). These agents can locally modulate inflammatory pathways within periodontal tissues, providing a targeted approach to managing relapse while minimizing systemic side effects. This targeted immunotherapy offers a promising adjunct to traditional mechanical retention methods.

28.8.4 Advances in Immunomodulatory Therapeutics for Retention

Biological and pharmacological strategies that modulate the immune response have shown promise in enhancing orthodontic retention. For instance, the use of anti-c-Fms antibodies to inhibit M-CSF-driven osteoclastogenesis has demonstrated efficacy in reducing relapse rates in preclinical studies. This targeted approach not only addresses the underlying immune mechanisms responsible for bone resorption but also offers a less invasive method for maintaining orthodontic results (Qi et al. 2019).

Additionally, there is growing interest in gene therapy approaches, such as local OPG gene transfer, which specifically target osteoclastogenic pathways. By directly inhibiting RANKL-mediated osteoclast differentiation, these therapies have shown promising results in animal models for reducing orthodontic relapse without the need for prolonged mechanical retention (Li et al. 2019). Further research is needed to adapt these gene-based interventions for clinical use in humans, but they represent a significant step forward in orthodontic retention technology.

28.8.5 Conclusion

The immune system plays a pivotal role in the processes underlying orthodontic relapse and retention. By modulating immune responses, particularly those involving osteoclast activity and pro-inflammatory cytokines, it may be possible to improve retention outcomes and minimize relapse. Recent studies demonstrate the efficacy of approaches such as anti-c-Fms antibody treatments, aspirin therapy, and OPG gene transfer in reducing relapse rates, providing a foundation for future immune-based retention strategies. These immunomodulatory approaches, when combined with traditional retention devices, hold great potential for enhancing the stability of orthodontic outcomes and improving patient satisfaction in the long term.

References

Alikhani M, Alyami B, Lee IS, Almoammar S, Vongthongleur T, Alikhani M, Alansari S, Sangsuwon C, Chou MY, Khoo E, Boskey A, Teixeira

CC (2015) Saturation of the biological response to orthodontic forces and its effect on the rate of tooth movement. Orthod Craniofacial Res 18(S1):8–17. https://doi.org/10.1111/ocr.12090. https://pubmed.ncbi.nlm.nih.gov/25865529/

Alvarez C, Suliman S, Almarhoumi R, Vega ME, Rojas C, Monasterio G, Galindo M, Vernal R, Kantarci A (2020) Regulatory T cell phenotype and anti-osteoclastogenic function in experimental periodontitis. Sci Rep 10(1):19018. https://doi.org/10.1038/s41598-020-76038-w

Arabpour M, Saghazadeh A, Rezaei N (2021) Anti-inflammatory and M2 macrophage polarization-promoting effect of mesenchymal stem cell-derived exosomes. Int Immunopharmacol 97:107823. https://doi.org/10.1016/j.intimp.2021.107823

Babenko A, Kostiuk I, Oktysyuk Y, Avakov V, Shovkova N (2021) Cellular immunity indexes after therapeutic use of immunoactive and anti-inflammatory agents in patients with orthodontically induced gingival hyperplasia. Pharmacia 375. https://link.gale.com/apps/doc/A659180297/AONE?u=anon~fbed5e36&sid=googleScholar&xid=35dd6742

Bagheri Y, Babaha F, Falak R, Yazdani R, Azizi G, Sadri M, Abolhassani H, Shekarabi M, Aghamohammadi A (2019) IL-10 induces TGF-β secretion, TGF-β receptor II upregulation, and IgA secretion in B cells. Eur Cytokine Netw 30(3):107–113. https://doi.org/10.1684/ecn.2019.0434

Baser Keklikci H, Yagci A, Yay AH, Goktepe O (2020) Effects of 405-, 532-, 650-, and 940-nm wavelengths of low-level laser therapies on orthodontic tooth movement in rats. Prog Orthod 21(1):43. https://doi.org/10.1186/s40510-020-00343-3

Bawyan RF, Baajajah AJ, Alzahrani HA, Murad MA, Jamalellail HMH, Muthaffar SY, Alandijani AA, Khayat MA, Bahurmoz AA, Azab AT, Almalki MA (2021) Immunological and physiological responses related to orthodontic treatment. Int J Commun Med Public Health 9(1):394–398. https://doi.org/10.18203/2394-6040.ijcmph20214799

Behm C, Nemec M, Weissinger F, Rausch MA, Andrukhov O, Jonke E (2021) MMPs and TIMPs expression levels in the periodontal ligament during orthodontic tooth movement: a systematic review of in vitro and in vivo studies. Int J Mol Sci 22(13). https://doi.org/10.3390/ijms22136967

Bildt MM, Bloemen M, Kuijpers-Jagtman AM, Von den Hoff JW (2009) Matrix metalloproteinase inhibitors reduce collagen gel contraction and alpha-smooth muscle actin expression by periodontal ligament cells. J Periodontal Res 44(2):266–274. https://doi.org/10.1111/j.1600-0765.2008.01127.x

Bilgic F, Akinci Sozer O, Ozcan O, Gurpinar AB, Yilmaz H, Ay Y (2016) Evaluation of inflammation during fixed orthodontic treatment. Arch Oral Biol 71:54–58. https://doi.org/10.1016/j.archoralbio.2016.07.004

Brito MH, Nogueira CQ, Cotrin P, Fialho T, Oliveira RC, Oliveira RG, Salmeron S, Valarelli FP, Freitas KMS, Cançado RH (2022) Efficacy of low-level laser therapy in reducing pain in the initial stages of orthodontic treatment. Int J Dent 2022:3934900. https://doi.org/10.1155/2022/3934900

Butaye C, Miclotte A, Begnoni G, Zhao Z, Zong C, Willems G, Verdonck A, Jacobs R, Cadenas M (2023) Third Molar Position After Completion of Orthodontic Treatment: A Prospective Follow-Up. Dentomaxillofacial radiology, 20220432.https://doi.org/10.1259/dmfr.20220432

Cadenas de Llano-Pérula M, Zong C, Van Dessel J, et al (2022) 3D quantification of in vivo orthodontic tooth movement in rats by means of micro-computed tomography. Clinical Oral Investigations. https://doi.org/10.1007/s00784-021-04358-w

Caviedes-Bucheli J, Lopez-Moncayo LF, Muñoz-Alvear HD, Gomez-Sosa JF, Diaz-Barrera LE, Curtidor H, Munoz HR (2021) Expression of substance P, calcitonin gene-related peptide and vascular endothelial growth factor in human dental pulp under different clinical stimuli. BMC Oral Health 21(1):152. https://doi.org/10.1186/s12903-021-01519-x

Chantarawaratit PO, Yanisarapan T (2021) Exposure to the oral environment enhances the corrosion of metal orthodontic appliances caused by fluoride-containing products: cytotoxicity, metal ion release, and surface roughness. Am J Orthod Dentofacial Orthop 160(1):101–112. https://doi.org/10.1016/j.ajodo.2020.03.035

Chaushu S, Klein Y, Mandelboim O, Barenholz Y, Fleissig O (2022) Immune changes induced by orthodontic forces: a critical review. J Dent Res 101(1):11–20. https://doi.org/10.1177/00220345211016285

Degrazia FW, Altmann ASP, Ferreira CJ, Arthur RA, Leitune VCB, Samuel SMW, Collares FM (2019) Evaluation of an antibacterial orthodontic adhesive incorporated with niobium-based bioglass: an in situ study. Braz Oral Res 33:e010. https://doi.org/10.1590/1807-3107bor-2019.vol33.0010

Deng S, Hu Y, Zhou J, Wang Y, Wang Y, Li S, Huang G, Peng C, Hu A, Yu Q, Han X (2020) TLR4 mediates alveolar bone resorption in experimental peri-implantitis through regulation of CD45(+) cell infiltration, RANKL/OPG ratio, and inflammatory cytokine production. J Periodontol 91(5):671–682. https://doi.org/10.1002/jper.18-0748

Doshi-Mehta G, Bhad-Patil WA (2012) Efficacy of low-intensity laser therapy in reducing treatment time and orthodontic pain: a clinical investigation. Am J Orthod Dentofacial Orthop 141(3):289–297. https://doi.org/10.1016/j.ajodo.2011.09.009

El-sayed OH, Ibrahim SA, Attia MS, El Kabbany SM (2022) Assessment of pain sensation and release of RANKL in gingival crevicular fluid accompanying laser application in orthodontic treatment: a randomized controlled clinical trial. Al-Azhar Dent J Girls 9(2):355–364

Faccioni F, Franceschetti P, Cerpelloni M, Fracasso ME (2003) In vivo study on metal release from fixed orthodontic appliances and DNA damage in oral mucosa cells. Am J Orthod Dentofacial Orthop 124(6): 687–693.; discussion 693-684. https://doi.org/10.1016/j.ajodo.2003.09.010

Fernandes MRU, Suzuki SS, Suzuki H, Martinez EF, Garcez AS (2019) Photobiomodulation increases intrusion tooth movement and modulates IL-6, IL-8 and IL-1β expression during orthodontically bone remodeling. J Biophotonics 12(10):e201800311. https://doi.org/10.1002/jbio.201800311

Gao Y, Min Q, Li X, Liu L, Lv Y, Xu W, Liu X, Wang H (2022) Immune system acts on orthodontic tooth movement: cellular and molecular mechanisms. Biomed Res Int 2022:9668610. https://doi.org/10.1155/2022/9668610

Gopinath VK, Mohammad MG, Sheela S (2023) Immunomodulatory effect of IL-1RA in LPS-activated macrophage/dental pulp stem cells co-culture. Int Endod J 56(1):27–38. https://doi.org/10.1111/iej.13839

Grant M, Wilson J, Rock P, Chapple I (2013) Induction of cytokines, MMP9, TIMPs, RANKL and OPG during orthodontic tooth movement. Eur J Orthod 35(5): 644–651. https://doi.org/10.1093/ejo/cjs057

He D, Liu F, Cui S, Jiang N, Yu H, Zhou Y, Liu Y, Kou X (2020) Mechanical load-induced H(2)S production by periodontal ligament stem cells activates M1 macrophages to promote bone remodeling and tooth movement via STAT1. Stem Cell Res Ther 11(1):112. https://doi.org/10.1186/s13287-020-01607-9

Janson GR, Dainesi EA, Consolaro A, Woodside DG, de Freitas MR (1998) Nickel hypersensitivity reaction before, during, and after orthodontic therapy. Am J Orthod Dentofacial Orthop 113(6):655–660. https://doi.org/10.1016/s0889-5406(98)70226-4

Jeon HH, Yang CY, Shin MK, Wang J, Patel JH, Chung CH, Graves DT (2021) Osteoblast lineage cells and periodontal ligament fibroblasts regulate orthodontic tooth movement that is dependent on nuclear factor-kappa B (NF-kB) activation. Angle Orthod 91(5): 664–671. https://doi.org/10.2319/031520-182.1

Kanzaki H, Shinohara F, Itohiya-Kasuya K, Ishikawa M, Nakamura Y (2015) Nrf2 activation attenuates both orthodontic tooth movement and relapse. J Dent Res 94(6):787–794. https://doi.org/10.1177/0022034515577814

Kapoor P, Kharbanda OP, Monga N, Miglani R, Kapila S (2014) Effect of orthodontic forces on cytokine and receptor levels in gingival crevicular fluid: a systematic review. Prog Orthod 15(1):65. https://doi.org/10.1186/s40510-014-0065-6

Kikuta J, Tsukada M, Takagi K, Shimizu M, Hikida T, Nakayama E, Kasai K (2020) TGF-β1 stimulates bone resorption during orthodontic tooth movement. Int J Oral-Med Sci 19(3):193–199. https://doi.org/10.5466/ijoms.19.193

Kim AR, Kim JH, Choi YH, Jeon YE, Cha JH, Bak EJ, Yoo YJ (2020) The presence of neutrophils causes RANKL expression in periodontal tissue, giving rise to osteoclast formation. J Periodontal Res 55(6): 868–876. https://doi.org/10.1111/jre.12779

Kirschneck C, Küchler EC, Wolf M, Spanier G, Proff P, Schröder A (2019) Effects of the highly COX-2-selective analgesic NSAID etoricoxib on human periodontal ligament fibroblasts during compressive orthodontic mechanical strain. Mediat Inflamm 2019:2514956. https://doi.org/10.1155/2019/2514956

Kirschneck C, Thuy M, Leikam A, Memmert S, Deschner J, Damanaki A, Spanier G, Proff P, Jantsch J, Schröder A (2020a) Role and regulation of mechanotransductive HIF-1α stabilisation in periodontal ligament fibroblasts. Int J Mol Sci 21(24). https://doi.org/10.3390/ijms21249530

Kirschneck C, Wolf F, Cieplik F, Blanck-Lubarsch M, Proff P, Schröder A (2020b) Impact of NSAID etoricoxib on side effects of orthodontic tooth movement. Ann Anat 232:151585. https://doi.org/10.1016/j.aanat.2020.151585

Koutouzis T, Catania D, Neiva K, Wallet SM (2013) Innate immune receptor expression in peri-implant tissues of patients with different susceptibility to periodontal diseases. J Periodontol 84(2):221–229. https://doi.org/10.1902/jop.2012.120061

Kuo WC, Lee CC, Chang YW, Pang W, Chen HS, Hou SC, Lo SY, Yang AS, Wang AH (2021) Structure-based development of human interleukin-1β-specific antibody that simultaneously inhibits binding to both IL-1RI and IL-1RAcP. J Mol Biol 433(4):166766. https://doi.org/10.1016/j.jmb.2020.166766

Levescot A, Chang MH, Schnell J, Nelson-Maney N, Yan J, Martínez-Bonet M, Grieshaber-Bouyer R, Lee PY, Wei K, Blaustein RB, Morris A, Wactor A, Iwakura Y, Lederer JA, Rao DA, Charles JF, Nigrovic PA (2021) IL-1β-driven osteoclastogenic Tregs accelerate bone erosion in arthritis. J Clin Invest 131(18). https://doi.org/10.1172/jci141008

Li Q, Han G, Liu D, Zhou Y (2019) Force-induced decline of TEA domain family member 1 contributes to osteoclastogenesis via regulation of Osteoprotegerin. Arch Oral Biol 100:23–32. https://doi.org/10.1016/j.archoralbio.2019.01.020

Li S, Liu Q, Wu D, He T, Yuan J, Qiu H, Tickner J, Zheng SG, Li X, Xu J, Rong L (2020a) PKC-δ deficiency in B cells displays osteopenia accompanied with upregulation of RANKL expression and osteoclast-osteoblast uncoupling. Cell Death Dis 11(9):762. https://doi.org/10.1038/s41419-020-02947-3

Li J, Yu TT, Yan HC, Qiao YQ, Wang LC, Zhang T, Li Q, Zhou YH, Liu DW (2020b) T cells participate in bone remodeling during the rapid palatal expansion. FASEB J 34(11):15327–15337. https://doi.org/10.1096/fj.202001078R

Liou EJ, Chen PH, Wang YC, Yu CC, Huang CS, Chen YR (2011) Surgery-first accelerated orthognathic surgery: postoperative rapid orthodontic tooth movement. J Oral Maxillofac Surg 69(3):781–785. https://doi.org/10.1016/j.joms.2010.10.035

Liu Y, Zhang T, Zhang C, Jin SS, Yang RL, Wang XD, Jiang N, Gan YH, Kou XX, Zhou YH (2017) Aspirin blocks orthodontic relapse via inhibition of CD4(+) T lymphocytes. J Dent Res 96(5):586–594. https://doi.org/10.1177/0022034516685527

Liu J, Wang H, Zhang L, Li X, Ding X, Ding G, Wei F (2022) Periodontal ligament stem cells promote polarization of M2 macrophages. J Leukoc Biol 111(6): 1185–1197. https://doi.org/10.1002/jlb.1ma1220-853rr

Long H, Wang Y, Jian F, Liao LN, Yang X, Lai WL (2016) Current advances in orthodontic pain. Int J Oral Sci 8(2):67–75. https://doi.org/10.1038/ijos.2016.24

Matsuike R, Tanaka H, Nakai K, Kanda M, Nagasaki M, Murakami F, Shibata C, Mayahara K, Nakajima A, Tanabe N, Kawato T, Maeno M, Shimizu N (2018) Continuous application of compressive force induces fusion of osteoclast-like RAW264.7 cells via upregulation of RANK and downregulation of LGR4. Life Sci 201:30–36. https://doi.org/10.1016/j.lfs.2018.03.038

Mei L, Busscher HJ, van der Mei HC, Chen Y, de Vries J, Ren Y (2009) Oral bacterial adhesion forces to biomaterial surfaces constituting the bracket-adhesive-enamel junction in orthodontic treatment. Eur J Oral Sci 117(4):419–426. https://doi.org/10.1111/j.1600-0722.2009.00648.x

Monasterio G, Budini V, Fernández B, Castillo F, Rojas C, Alvarez C, Cafferata EA, Vicencio E, Cortés BI, Cortez C, Vernal R (2019) IL-22-expressing CD4(+) AhR(+) T lymphocytes are associated with RANKL-mediated alveolar bone resorption during experimental periodontitis. J Periodontal Res 54(5):513–524. https://doi.org/10.1111/jre.12654

Mulholland BS, Forwood MR, Morrison NA (2019) Monocyte chemoattractant protein-1 (MCP-1/CCL2) drives activation of bone remodelling and skeletal metastasis. Curr Osteoporos Rep 17(6):538–547. https://doi.org/10.1007/s11914-019-00545-7

Munshi S, Loh MK, Ferrara N, DeJoseph MR, Ritger A, Padival M, Record MJ, Urban JH, Rosenkranz JA (2020) Repeated stress induces a pro-inflammatory state, increases amygdala neuronal and microglial activation, and causes anxiety in adult male rats. Brain Behav Immun 84:180–199. https://doi.org/10.1016/j.bbi.2019.11.023

Naoumova J, Olofsson R, Almståhl A, Cevik-Aras H (2024) Salivary levels of Osteoprotegerin and receptor activator of nuclear factor-kappa ligand during orthodontic tooth movement-a prospective pilot study. Orthod Craniofac Res 27(1):78–83. https://doi.org/10.1111/ocr.12687

Nunes L, Quintanilha L, Perinetti G, Capelli JJ (2017) Effect of orthodontic force on expression levels of ten cytokines in gingival crevicular fluid. Arch Oral Biol 76:70–75. https://doi.org/10.1016/j.archoralbio.2017.01.016

Ogawa S, Kitaura H, Kishikawa A, Qi J, Shen WR, Ohori F, Noguchi T, Marahleh A, Nara Y, Ochi Y, Mizoguchi I (2019) TNF-α is responsible for the contribution of stromal cells to osteoclast and odontoclast formation during orthodontic tooth movement. PLoS One 14(10):e0223989. https://doi.org/10.1371/journal.pone.0223989

Oikonomou E, Foros P, Tagkli A, Rahiotis C, Eliades T, Koletsi D (2021) Impact of aligners and fixed appliances on oral health during orthodontic treatment: a systematic review and meta-analysis. Oral Health Prev Dent 19:659–672. https://doi.org/10.3290/j.ohpd.b2403661

Pradeep Yeola A, Akbar I, Baillargeon J, Mercy Ignatius Arokia Doss P, Paavilainen VO, Rangachari M (2020) Protein translocation and retro-translocation across the endoplasmic reticulum are crucial to inflammatory effector CD4(+) T cell function. Cytokine 129: 154944. https://doi.org/10.1016/j.cyto.2019.154944

Prager I, Watzl C (2019) Mechanisms of natural killer cell-mediated cellular cytotoxicity. J Leukoc Biol 105(6): 1319–1329. https://doi.org/10.1002/jlb.Mr0718-269r

Prathapan Santhakumari P, Varma Raja V, Joseph J, Devaraj A, John E, Oommen Thomas N (2023) Impact of low-level laser therapy on orthodontic tooth movement and various cytokines in gingival crevicular fluid: a split-mouth randomized study. Cureus 15(8):e42809. https://doi.org/10.7759/cureus.42809

Qi J, Kitaura H, Shen WR, Kishikawa A, Ogawa S, Ohori F, Noguchi T, Marahleh A, Nara Y, Mizoguchi I (2019) Establishment of an orthodontic retention mouse model and the effect of anti-c-Fms antibody on orthodontic relapse. PLoS One 14(6):e0214260. https://doi.org/10.1371/journal.pone.0214260

Qi J, Kitaura H, Shen W-R, Ogawa S, Ohori F, Noguchi T, Marahleh A, Nara Y, Adya P, Mizoguchi I (2020) Effect of a DPP-4 inhibitor on orthodontic tooth movement and associated root resorption. Biomed Res Int 2020:7189084. https://doi.org/10.1155/2020/7189084

Rath-Deschner B, Memmert S, Damanaki A, Nokhbehsaim M, Eick S, Cirelli JA, Götz W, Deschner J, Jäger A, Nogueira AVB (2020) CXCL1, CCL2, and CCL5 modulation by microbial and biomechanical signals in periodontal cells and tissues-in vitro and in vivo studies. Clin Oral Investig 24(10):3661–3670. https://doi.org/10.1007/s00784-020-03244-1

Reale M, Costantini E, D'Angelo C, Coppeta L, Mangifesta R, Jagarlapoodi S, Di Nicola M, Di Giampaolo L (2020) Network between cytokines, cortisol and occupational stress in gas and oilfield workers. Int J Mol Sci 21(3). https://doi.org/10.3390/ijms21031118

Reis CLB, de Souza Furtado TC, Mendes WD, Matsumoto MAN, Alves SYF, Stuani MBS, Borsatto MC, Corona SAM (2022) Photobiomodulation impacts the levels of inflammatory mediators during orthodontic tooth movement? A systematic review with meta-analysis.

Lasers Med Sci 37(2):771–787. https://doi.org/10.1007/s10103-021-03425-8

Ren Y, Hazemeijer H, de Haan B, Qu N, de Vos P (2007) Cytokine profiles in crevicular fluid during orthodontic tooth movement of short and long durations. J Periodontol 78(3):453–458. https://doi.org/10.1902/jop.2007.060261

Ren A, Li Z, Zhang X, Deng R, Ma Y (2021) Inhibition of dectin-1 on dendritic cells prevents maturation and prolongs murine islet allograft survival. J Inflamm Res 14:63–73. https://doi.org/10.2147/jir.S287453

Roth CE, Craveiro RB, Niederau C, Malyaran H, Neuss S, Jankowski J, Wolf M (2022) Mechanical compression by simulating orthodontic tooth movement in an in vitro model modulates phosphorylation of AKT and MAPKs via TLR4 in human periodontal ligament cells. Int J Mol Sci 23(15). https://doi.org/10.3390/ijms23158062

Sakamoto M, Fukunaga T, Sasaki K, Seiryu M, Yoshizawa M, Takeshita N, Takano-Yamamoto T (2019) Vibration enhances osteoclastogenesis by inducing RANKL expression via NF-κB signaling in osteocytes. Bone 123:56–66. https://doi.org/10.1016/j.bone.2019.03.024

Santos LLD, Conti A, Fernandes TMF, Garlet GP, Almeida MR, Oltramari PVP (2023) Influence of anxiety and catastrophizing on pain perception in orthodontic treatment and its association with inflammatory cytokines. Braz Oral Res 37:e010. https://doi.org/10.1590/1807-3107bor-2023.vol37.0010

Schröder A, Käppler P, Nazet U, Jantsch J, Proff P, Cieplik F, Deschner J, Kirschneck C (2020) Effects of compressive and tensile strain on macrophages during simulated orthodontic tooth movement. Mediat Inflamm 2020:2814015. https://doi.org/10.1155/2020/2814015

Sedky Y, Refaat W, Gutknecht N, ElKadi A (2019) Comparison between the effect of low-level laser therapy and corticotomy-facilitated orthodontics on RANKL release during orthodontic tooth movement: a randomized controlled trial. Lasers Dent Sci 3(2):99–109. https://doi.org/10.1007/s41547-019-00055-w

Sodor A, Ogodescu AS, Petreuş T, Şişu AM, Zetu IN (2015) Assessment of orthodontic biomaterials' cytotoxicity: an in vitro study on cell culture. Romanian J Morphol Embryol 56(3):1119–1125

Song L, Dong G, Guo L, Graves DT (2018) The function of dendritic cells in modulating the host response. Mol Oral Microbiol 33(1):13–21. https://doi.org/10.1111/omi.12195

Stewart CW, Hammond CM, Godat MS, Lew DB (2023) Delayed severe gingivitis after placement of orthodontic braces in an atopic teenager: a case report and literature review. Pediatr Allergy Immunol Pulmonol 36(2):46–49. https://doi.org/10.1089/ped.2022.0093

Tsolakis IA, Christopoulou I, Sitaras S, Lyros I, Rontogianni A, Dalampira M, Tsolakis AI (2023) Molecular and biological aspects of orthodontic tooth movement: possibilities for bioengineering intervention: a narrative review. Bioengineering (Basel) 10(11). https://doi.org/10.3390/bioengineering10111275

Veloso P, Fernández A, Terraza-Aguirre C, Álvarez C, Vernal R, Escobar A, Hernández M (2020) Macrophages skew towards M1 profile through reduced CD163 expression in symptomatic apical periodontitis. Clin Oral Investig 24(12):4571–4581. https://doi.org/10.1007/s00784-020-03324-2

Wald S, Leibowitz A, Aizenbud Y, Saba Y, Zubeidat K, Barel O, Koren N, Heyman O, Wilharm A, Sandrock I, Fleissig O, Tal Y, Prinz I, Aizenbud D, Chaushu S, Hovav AH (2021) γδT cells are essential for orthodontic tooth movement. J Dent Res 100(7):731–738. https://doi.org/10.1177/0022034520984774

Wang Y, Zhang H, Sun W, Wang S, Zhang S, Zhu L, Chen Y, Xie L, Sun Z, Yan B (2018) Macrophages mediate corticotomy-accelerated orthodontic tooth movement. Sci Rep 8(1):16788. https://doi.org/10.1038/s41598-018-34907-5

Wang XL, Cui LW, Liu Z, Gao YM, Wang S, Li H, Liu HX, Yu LJ (2019a) Effects of TRPA1 activation and inhibition on TRPA1 and CGRP expression in dorsal root ganglion neurons. Neural Regen Res 14(1):140–148. https://doi.org/10.4103/1673-5374.243719

Wang J, Jiang H, Qiu Y, Wang Y, Sun G, Zhao J (2019b) Effector memory regulatory T cells were most effective at suppressing RANKL but their frequency was downregulated in tibial fracture patients with delayed union. Immunol Lett 209:21–27. https://doi.org/10.1016/j.imlet.2019.03.018

Wang Y, Groeger S, Yong J, Ruf S (2023) Orthodontic compression enhances macrophage M2 polarization via histone H3 hyperacetylation. Int J Mol Sci 24(4). https://doi.org/10.3390/ijms24043117

Wei T, Shan Z, Wen X, Zhao N, Shen G (2022) Dynamic alternations of RANKL/OPG ratio expressed by cementocytes in response to orthodontic-induced external apical root resorption in a rat model. Mol Med Rep 26(1). https://doi.org/10.3892/mmr.2022.12744

Wu L, Su Y, Lin F, Zhu S, Wang J, Hou Y, Du J, Liu Y, Guo L (2020) MicroRNA-21 promotes orthodontic tooth movement by modulating the RANKL/OPG balance in T cells. Oral Dis 26(2):370–380. https://doi.org/10.1111/odi.13239

Xu H, Zhang S, Sathe AA, Jin Z, Guan J, Sun W, Xing C, Zhang H, Yan B (2022) CCR2(+) Macrophages promote orthodontic tooth movement and alveolar bone remodeling. Front Immunol 13:835986. https://doi.org/10.3389/fimmu.2022.835986

Yamaguchi M, Fukasawa S (2021) Is inflammation a friend or foe for orthodontic treatment?: Inflammation in orthodontically induced inflammatory root resorption and accelerating tooth movement. Int J Mol Sci 22(5). https://doi.org/10.3390/ijms22052388

Yan Y, Liu F, Kou X, Liu D, Yang R, Wang X, Song Y, He D, Gan Y, Zhou Y (2015) T cells are required for orthodontic tooth movement. J Dent Res 94(10):

1463–1470. https://doi.org/10.1177/0022034515595003

Yang JH, Li ZC, Kong WD, Zhang W, Jia YP, Zhang YL, Liu LB, Han XP (2013) Effect of orthodontic force on inflammatory periodontal tissue remodeling and expression of IL-6 and IL-8 in rats. Asian Pac J Trop Med 6(10):757–761. https://doi.org/10.1016/s1995-7645(13)60133-9

Yang L, Yin G, Liao X, Yin X, Ye N (2019) A novel customized ceramic bracket for esthetic orthodontics: in vitro study. Prog Orthod 20(1):39. https://doi.org/10.1186/s40510-019-0292-y

Zeng M, Kou X, Yang R, Liu D, Wang X, Song Y, Zhang J, Yan Y, Liu F, He D, Gan Y, Zhou Y (2015) Orthodontic force induces systemic inflammatory monocyte responses. J Dent Res 94(9):1295–1302. https://doi.org/10.1177/0022034515592868

Zeng MY, Miralda I, Armstrong CL, Uriarte SM, Bagaitkar J (2019) The roles of NADPH oxidase in modulating neutrophil effector responses. Mol Oral Microbiol 34(2):27–38. https://doi.org/10.1111/omi.12252

Zhang J, Zhang AM, Zhang ZM, Jia JL, Sui XX, Yu LR, Liu HT (2017) Efficacy of combined orthodontic-periodontic treatment for patients with periodontitis and its effect on inflammatory cytokines: a comparative study. Am J Orthod Dentofacial Orthop 152(4):494–500. https://doi.org/10.1016/j.ajodo.2017.01.028

Zhang KJ, Zhang HW, Yan B (2023) Research progress in immunomodulation in orthodontic tooth movement. Zhonghua Kou Qiang Yi Xue Za Zhi 58(4):380–385. https://doi.org/10.3760/cma.j.cn112144-20221031-00561

Zhao N, Lin J, Kanzaki H, Ni J, Chen Z, Liang W, Liu Y (2012) Local osteoprotegerin gene transfer inhibits relapse of orthodontic tooth movement. Am J Orthod Dentofacial Orthop 141(1):30–40. https://doi.org/10.1016/j.ajodo.2011.06.035

Zhao Z, Tarce M, Georgopoulou M, Zong C, Van Holm W, Attanasio C, Pedano MS, Cadenas M (2024a) Periodontal Ligament Reactions to Orthodontic Force: A Transcriptomic Study on Maxillary and Mandibular Human Premolars. J of periodontal research. https://doi.org/10.1111/jre.13352

Zhao Z, Attanasio C, Zong C, Pedano MS, Cadenas M (2024b) How Does Orthodontic Tooth Movement Influence the Dental Pulp? RNA-sequencing on Human Premolars. Int End J. https://doi.org/10.1111/iej.14131

Zheng J, Yang K (2021) Clinical research: low-level laser therapy in accelerating orthodontic tooth movement. BMC Oral Health 21(1):324. https://doi.org/10.1186/s12903-021-01684-z

Zigante M, Špalj S (2022) Clinical predictors of metal allergic sensitization in orthodontic patients. Cent. Eur J Pub Health 30(3):173–178. https://doi.org/10.21101/cejph.a7122

Zigante M, Rincic Mlinaric M, Kastelan M, Perkovic V, Trinajstic Zrinski M, Spalj S (2020) Symptoms of titanium and nickel allergic sensitization in orthodontic treatment. Prog Orthod 21(1):17. https://doi.org/10.1186/s40510-020-00318-4

Zong C, Van Dessel J, Vande Velde G, Willems G, Cadenas de Llano-Pérula M (2022) Dynamic changes in tooth displacement and bone morphometry induced by orthodontic force. Sci Rep 12(1):13672. https://doi.org/10.1038/s41598-022-17412-8

Zong C, Bronckaers A, Willems G, He H, Cadenas de Llano-Pérula M (2023a) Nanomaterials for periodontal tissue regeneration: progress, challenges and future perspectives. J Funct Biomater 14(6). https://doi.org/10.3390/jfb14060290

Zong C, Van Holm W, Bronckaers A, Zhao Z, Čokić S, Aktan MK, Castro AB, Van Meerbeek B, Braem A, Willems G, Cadenas de Llano-Pérula M (2023b) Biomimetic periodontal ligament transplantation activated by gold nanoparticles protects alveolar bone. Adv Healthc Mater 12(15):e2300328. https://doi.org/10.1002/adhm.202300328

Zong C, Tan X, Van Holm W, et al (2025) Biomimetic-Engineered Hydrogels Prevent Orthodontic Root Resorption. Advanced Functional Materials n/a (n/a):2502066. https://doi.org/10.1002/adfm.202502066

Evaluating the Impact of Clear Dental Aligners on Dental and Periodontal Conditions, Oral Microbial Patterns, and the Immune System

29

Saharnaz Esmaeili, Soheil Shahbazi, Heliya Ziaei, and Farnaz Younessian

Abstract

Clear aligner therapy (CAT) has gained widespread adoption for managing malocclusions, primarily due to its aesthetic advantages, improved access for hygiene, and enhanced patient comfort relative to fixed appliances (FAs). However, CAT introduces a distinct intraoral environment that can influence microbial composition, biofilm development, and host immune responses. This chapter explores the effects of clear aligners on dental and periodontal health, oral microbial communities, and immunological parameters. Evidence indicates that CAT may facilitate plaque and biofilm accumulation, particularly on composite attachments, thereby increasing the risk of white spot lesions (WSLs), gingivitis, and dental caries. Despite these risks, aligners are generally associated with lower levels of pathogenic bacteria compared to FAs. Microbiome analyses of subgingival plaque and saliva reveal shifts in bacterial genera and phyla, although these changes are frequently non-pathogenic. CAT has demonstrated minimal negative effects on clinical periodontal indices, including the plaque index and gingival bleeding index, particularly when effective oral hygiene is maintained. Furthermore, aligners may reduce sustained immune activation by limiting plaque retention and aseptic inflammation, potentially decreasing the expression of pro-inflammatory cytokines such as IL-1β, IL-6, and TNF-α. Although hypersensitivity reactions to aligner materials are infrequent, they warrant consideration. Overall, CAT appears to support periodontal health and promote a more stable oral microbial environment compared to FAs. Continued patient adherence to hygiene protocols and further long-term studies are necessary to fully evaluate systemic implications and optimize clinical outcomes.

Keywords

Clear aligner therapy (CAT) · Orthodontics · Oral microbiome · Periodontal health · Immune response · Dental biofilm

Saharnaz Esmaeili and Soheil Shahbazi contributed equally to this work.

S. Esmaeili · S. Shahbazi
School of Dentistry, University of California Los Angeles, Los Angeles, CA, USA

H. Ziaei
Universal Scientific Education and Research Network (USERN), Los Angeles, CA, USA

F. Younessian (✉)
Section of Orthodontics, School of Dentistry, University of California, Los Angeles, CA, USA
e-mail: FYounessian@dentistry.ucla.edu

N. Rezaei, H. Ziaei (eds.), *Oral Immunology*, Advances in Experimental Medicine and Biology 1492,
https://doi.org/10.1007/978-3-032-03176-1_29

29.1 Introduction

Malocclusion is defined as the misalignment of teeth due to abnormalities in bone structure, tooth loss, space limit actions, or traumatic injuries (Proffit et al. 2003). This misalignment can greatly influence oral health by impairing fundamental activities, including chewing, biting, and impaired self-cleaning function of saliva due to significant crowding (Proffit et al. 2003; Ruhl et al. 1994). Orthodontic treatments utilize fixed or removable appliances to address skeletal and dental malocclusions resulting from genetic or environmental factors (Contaldo et al. 2021). With advances in three-dimensional technology and the application of biomechanic principles, clear aligners (CAs) have become increasingly effective in correcting malocclusions (Alansari 2020; Song et al. 2023). The main advantages of clear aligner therapy (CAT) include easier hygiene maintenance, less diet restriction, and better aesthetics compared to traditional fixed appliances (FAs) (Weir 2017). The past decade has seen a marked rise in the adoption of this treatment modality, a trend that is expected to continue (Adobes-Martin et al. 2021; Llera-Romero et al. 2023).

The mouth is constantly exposed to plaque-forming microorganisms including bacteria and fungi. By protecting harmful microbes from treatment and bypassing the body's defense, plaque serves as the leading cause of dental caries, gingival inflammation, and periodontitis (Baybekov et al. 2023; Fernandes et al. 2018). Microbes inhabit different areas within the oral cavity, including teeth, tongue, lips, cheeks, and gingival tissues (Deo and Deshmukh 2019). Oral health relies on a delicate balance between these microbial communities and the host immune system. However, when this balance is disturbed, a condition known as dysbiosis, numerous oral diseases might develop (Kilian et al. 2016; Maier 2023) that may also contribute to systemic disorders (Pisano 2023).

Orthodontic appliances, whether fixed or removable, can influence the composition and dynamics of the oral plaque biofilm, thereby impacting overall oral health (Smolyar et al. 2022). Orthodontic treatments change the oral microbiota, elevating the risk of dental caries and gingivitis following food and bacterial accumulation (Ireland et al. 2014; Xu et al. 2015). These appliances increase the risk of plaque accumulation due to the additional surfaces on which plaque can adhere and the difficulty in maintaining optimal oral hygiene. Consequently, they increase the likelihood of periodontal disease due to plaque accumulation (Talic 2011). Moreover, plaque retention associated with orthodontic appliances has been shown to increase the risk of dental caries (Freitas et al. 2014; Giugliano et al. 2018). Studies indicate that oral health status deteriorates with elevated levels of plaque accumulation in individuals undergoing fixed orthodontic treatment; however, removable appliances, such as CAs, may attenuate these effects by facilitating improved oral hygiene (Cantekin et al. 2011; Jiang et al. 2018; Kaygisiz et al. 2014; Pandey et al. 2016). Despite their removability, CAs are recommended for continuous wear of 20–22 h per day, covering the entire surface of the teeth. This prolonged coverage creates a microenvironment that may promote increased plaque and biofilm accumulation, thereby contributing to oral health complications such as white spot lesions (WSLs), dental caries, and gingivitis (Miethke and Vogt 2005; Rossini et al. 2014). Additionally, the resin composite attachments utilized for CAs demonstrate a heightened susceptibility to biofilm accumulation, further increasing the potential for bacterial adherence and plaque formation (Jiang et al. 2018; Shpack et al. 2014).

Given the increasing prevalence of CAT, this chapter aims to critically evaluate the impact of clear dental aligners on oral health and microbiome, with a comparative analysis against fixed orthodontic appliances. Moreover, it will briefly discuss the influence of these appliances on immunological factors, particularly their role in orthodontic tooth movement, which elicits aseptic inflammation within periodontal tissues. Additionally, the potential cytotoxic effects of CA materials on oral tissues, hypersensitivity reactions associated with aligner polymers, and their influence on oral epithelial cell permeability and viability—determinants of the local immune response—will be explored. By analyzing the existing body of literature, this chapter will also delineate evidence-based strategies for effective cleaning and disinfection of CAs to optimize oral hygiene. This overview seeks to provide essential insights for dental professionals to enhance clinical decision-making in orthodontic treatment planning and aligner maintenance protocols.

29.1.1 Clear Aligners

The development and widespread adoption of CAT in orthodontic treatment have marked a significant advancement in the field of dentistry, providing a more discreet and aesthetically acceptable alternative to traditional FAs. The origins of this approach can be attributed to Dr. Harold Kesling, who first proposed the use of positioners for progressive tooth movement in 1946 (Kesling 1945, 1946). This foundational concept was further refined through subsequent innovations, including Ponitz's introduction of the "invisible retainer" in 1971, which laid the groundwork for the modern clear orthodontic appliances utilized today (Ponitz 1971).

The integration of computer-aided design (CAD) and computer-aided manufacturing (CAM) technologies has significantly transformed the field of orthodontics, facilitating the development of custom-fitted aligners designed with high precision to accommodate the unique dental anatomy of each patient. A pivotal milestone in this technological advancement was the introduction of the Invisalign® system by Align Technology (Santa Clara, CA) in 1998, marking the first digitally designed and manufactured orthodontic aligner system (Boyd et al. 2000). Invisalign®, along with other systems such as ClearCorrect® and ClearPath®, has played a crucial role in advancing orthodontic treatment by enabling comprehensive tooth movement and control with an unparalleled degree of customization and precision (Weir 2017).

The introduction of Invisalign® marked a significant turning point in orthodontics, demonstrating the potential of digital treatment planning and incorporating advanced features such as bonded resin attachments and precision cuts. These innovations have expanded the range of malocclusions that can be effectively addressed with aligners, setting new standards for comfort and aesthetic in orthodontic treatment. The transition to digital orthodontics has not only revolutionized treatment design and implementation but also significantly improved the patient experience by providing a more discreet, comfortable, and user-friendly alternative to conventional FAs (Meade and Weir 2023; Weir 2017). The preference for this approach over traditional braces is largely attributed to its superior aesthetic appeal and enhanced comfort (Macrì et al. 2022). Research has identified several advantages associated with CAT, including improved oral hygiene and periodontal health, reduced risk of root resorption following orthodontic treatment, and potential relief from pain and headaches linked to temporomandibular disorders (TMDs) (Festa et al. 2021; Li et al. 2020; Rossini et al. 2014; Rosvall et al. 2009). However, findings regarding periodontal outcomes remain inconclusive. While some studies suggest that CAs contribute to improved periodontal health compared to traditional FAs, others indicate a heightened risk of plaque accumulation and associated periodontal complications. Further research is required to clarify these effects and optimize treatment protocols accordingly (Lu et al. 2018; Raghavan et al. 2023; Rouzi et al. 2023).

Similar to traditional FA systems, CAT encompasses a diverse range of thermoformed plastic aligners, each distinguished by variations in material composition, fabrication techniques, and suitability for different types of malocclusions. While all systems employ transparent aligners designed to cover most or all of the dentition, there are substantial differences among these systems that significantly impact their clinical efficacy. These distinctions are critical in determining the capability of each specific system to address a broad spectrum of orthodontic conditions (Livas et al. 2023; Shrivastava et al. 2023; Weir 2017).

CA systems exhibit notable variations, particularly in material properties, the use of attachments, and the inclusion of proprietary mechanisms such as active surfaces designed to enhance precise tooth movement. The integration of orthodontic auxiliary aids, including elastics and mini-screws, has expanded the scope of aligner therapy, allowing the management of complex orthodontic cases that were previously unsuitable for CAT. The overall success of various aligner systems is influenced by multiple factors, including the mechanical properties of the thermoplastic material, the design of the gingival margin, the sequence of tooth movements, and the strategic placement and configuration of attachments, divots, or other orthodontic auxiliaries, all of which contribute to improved functionality and clinical effectiveness (Jindal et al. 2019; Kook et al. 2019; Liu and Hu 2018; Lombardo et al. 2018; Putrino et al. 2021; Rinchuse and Cozzani 2015). Despite advancements in CA technology, challenges remain, particularly in achieving predictable tooth movement. Recent studies indicate that the average predictability of planned movements with Cas remains below 50%, often necessitating frequent refinements or midcourse corrections to ensure successful treatment outcomes with CAs (Haouili et al. 2020).

29.1.2 Oral Microbiome

The study of the human microbiome, particularly within the oral ecosystem, has undergone a paradigm shift since Antonie van Leeuwenhoek's seminal observations in the eighteenth century (Baker et al. 2017; Gest 2004). Contemporary research has revealed the intricate interplay among over 700 microbial species inhabiting the oral cavity, including bacteria, fungi, viruses, archaea, and protozoans. Despite this remarkable diversity, a substantial proportion of these species remain uncharacterized and uncultivable, underscoring the vast, yet unexplored, complexity of the oral microbiome and its distinct microhabitats (Bäckhed et al. 2012; Caselli et al. 2020; Wade 2013). The individualized microbial fingerprint of each person is now recognized as a fundamental determinant of physiological, nutritional, and immunological processes, influencing both systemic and oral health outcomes (Franzosa et al. 2015).

The establishment of the oral microbiome begins prenatally, with emerging evidence challenging the long-held assumption of a sterile intrauterine environment (Sampaio-Maia and Monteiro-Silva 2014; Wade 2013). Postnatally, oral microbial colonization becomes increasingly complex, shaped by vertical transmission from the mother, dietary transitions, and social interactions. The eruption of primary teeth serves as a critical phase in microbial succession, reducing maternal microbial influence and fostering a more individualized microbiome. As individuals age, their oral microbiota undergoes dynamic shifts influenced by both genetic predisposition and environmental factors, including diet, psychological stress, oral hygiene practices, alcohol consumption, and smoking. These alterations are often associated with a decline in microbial diversity and compositional shifts, which have been correlated with an increased susceptibility to oral and systemic diseases (Baker et al. 2017; Costello et al. 2012; Pisano 2023; Wade 2013).

The oral microbiome is composed of both core and variable microbial communities, shaped by a combination of genetic factors and individual lifestyle choices. The oral cavity, with its diverse biological environments, supports distinct microbial communities in areas such as the gingival sulcus, tongue, buccal mucosa, and teeth. These site-specific communities contribute to the structural and functional complexity of the oral ecosystem, playing a crucial role in maintaining oral homeostasis (Deo and Deshmukh 2019; Zarco et al. 2012).

Within a healthy oral ecosystem, several key bacterial genera play essential roles in maintaining microbial homeostasis. Among the gram-positive cocci, notable representatives include *Abiotrophia*, *Peptostreptococcus, Streptococcus*, and *Streptococcus*. Additionally, gram-positive rods such as *Actinomyces*,

Bifidobacterium, *Corynebacterium*, *Eubacterium*, *Lactobacillus*, *Propionibacterium*, *Pseudoramibacter*, and *Rothia* are integral to the oral microbiota. The gram-negative spectrum is represented by cocci such as *Moraxella*, *Neisseria*, and *Veillonella*, alongside rods including *Campylobacter*, *Capnocytophaga*, *Desulfobacter*, *Desulfovibrio*, *Eikenella*, *Fusobacterium*, *Hemophilus*, *Leptotrichia*, *Prevotella*, *Selemonas*, *Simonsiella*, *Treponema*, and *Wolinella*. Beyond bacterial taxa, nonbacterial organisms also contribute significantly to the oral microbiome's diversity and ecological stability. *Protozoans* such as *Entamoeba gingivalis* and *Trichomonas tenax*, along with fungal species predominantly from the Candida genus, play crucial roles in oral microbial balance (Deo and Deshmukh 2019; Marsh 2000; Sharma et al. 2018).

The transition from traditional culture-based methodologies to advanced culture-independent approaches has dramatically enhanced our understanding of the oral microbiome. Modern "omics" technologies, including microbiomics and metagenomics, now enable the comprehensive characterization of microbial communities with ultrahigh resolution. State-of-the-art sequencing methods, such as 16S rRNA gene sequencing and shotgun metagenomic sequencing, have become indispensable tools for analyzing microbial composition, relative abundance, and functional interactions within the oral cavity. These cutting-edge techniques provide a deeper insight into the intricate microbial dynamics, facilitating the identification of previously unrecognized taxa and their potential implications in oral and systemic health (Caselli et al. 2020; Deo and Deshmukh 2019; Wade 2013; Zarco et al. 2012).

The role of the oral microbiome is crucial as it predominantly exists in the form of biofilms, playing an essential role in maintaining oral balance and safeguarding against diseases. The composition and interaction of the microbiota have significant implications for health outcomes and the progression of diseases (Jia et al. 2018). Through a dynamic balance, the oral microbiome contributes to the host's local and overall health. Any considerable change to the local environment may affect the host–microbe balance, increasing the risk of oral diseases (Deo and Deshmukh 2019; Kilian et al. 2016). This microbial shift is a major contributor to dental and periodontal disorders, such as WSLs, caries, and gingivitis. Therefore, maintaining the stability and ecological balance of the plaque biofilm is critical for ensuring effective treatment outcomes during orthodontic therapy (Gao et al. 2018; Rouzi et al. 2023).

Extensive evidence establishes a strong correlation between disruptions in the oral microbiome dysbiosis and the pathogenesis of various systemic diseases, including cardiovascular disorders, respiratory infections, diabetes mellitus, and gastrointestinal malignancies. Pathogenic microorganisms such as *Porphyromonas gingivalis (P. gingivalis)* and *Fusobacterium nucleatum (F. nucleatum)*, which are frequently implicated in periodontal disease, have been shown to translocate into the systemic circulation through inflamed periodontal tissues. Upon entering the bloodstream, these bacteria trigger systemic inflammatory responses that contribute to endothelial dysfunction and the pathophysiology of atherosclerosis, thereby increasing the risk of adverse cardiovascular events, such as myocardial infarction and stroke. Specifically, *P. gingivalis* induces the secretion of pro-inflammatory cytokines, including IL-1β, IL-6, and TNF-α, which exacerbate systemic inflammation and disrupt vascular homeostasis (Pathak et al. 2021; Sedghi et al. 2021). Beyond cardiovascular pathology, oral microbiome dysbiosis has also been implicated in pulmonary diseases. The aspiration of oral bacteria into the lower respiratory tract can contribute to the pathogenesis of bacterial pneumonia and exacerbate chronic respiratory conditions, such as chronic obstructive pulmonary disease (COPD). Notably, *F. nucleatum*, *Prevotella intermedia*, and *P. gingivalis* have been detected in the lungs of patients with respiratory infections, supporting the hypothesis that oral microbial imbalances significantly impact pulmonary health (Pathak et al. 2021; Peng et al. 2022).

The bidirectional relationship between periodontal disease and diabetes mellitus is well-documented. Periodontitis has been shown to exacerbate systemic inflammation and insulin resistance, thereby complicating glycemic control in diabetic patients. Conversely, hyperglycemia in diabetes promotes a pro-inflammatory oral environment, increasing susceptibility to periodontal infections. Notably, *P. gingivalis* stimulates the secretion of pro-inflammatory cytokines such as IL-6 and TNF-α, which interfere with insulin signaling and contribute to metabolic dysregulation, reinforcing the bidirectional link between periodontal disease and diabetes (Pathak et al. 2021; Peng et al. 2022).

Recent studies have also highlighted the role of *F. nucleatum* in the pathogenesis of gastrointestinal malignancies, particularly colorectal cancer. This microorganism has been shown to translocate from the oral cavity to the gastrointestinal tract, where it disrupts the gut microbiota, induces chronic inflammation, and facilitates neoplastic progression. *F. nucleatum* adheres to epithelial E-cadherin, promoting cancer cell invasion and metastasis, while also modulating immune responses by suppressing T-cell function, thereby facilitating immune evasion and cancer cell proliferation (Peng et al. 2022; Sedghi et al. 2021).

The implications of oral microbial dysbiosis extend beyond localized conditions such as dental caries and periodontitis, with profound consequences for systemic health. The maintenance of a balanced oral microbiome is critical not only for preserving oral homeostasis but also for reducing systemic inflammatory burdens and modulating host immune responses. Given the interplay between oral and systemic health, adherence to rigorous oral hygiene practices may serve as a modifiable factor in reducing the risk of systemic inflammatory diseases (Peng et al. 2022).

29.2 Effects of Clear Aligners on the Composition of Oral Microbial Communities

Multiple factors intricately influence the composition and dynamics of the oral microbiome in orthodontic patients, affecting both its qualitative and quantitative characteristics. These factors include plaque accumulation, metal corrosion, the host immune response, hormonal fluctuations, and the biomechanical forces associated with tooth movement (He et al. 2024; Peterson et al. 2024). Regardless of whether orthodontic appliances are fixed or removable, they facilitate the retention of food debris and serve as substrates for dental plaque formation, thereby complicating oral hygiene maintenance. Consequently, orthodontic treatment has been associated with an increased risk of oral pathologies, including gingivitis, periodontitis, WSLs, dental caries, and halitosis (Baybekov et al. 2023; Kilian et al. 2016; Rossini et al. 2014). Under physiological conditions, the oral immune system exhibits tolerance toward commensal microorganisms. However, the disruption of microbial homeostasis (dysbiosis), which is frequently exacerbated by orthodontic appliances, provokes inflammatory responses and sustained immune activation, particularly given the prolonged duration of orthodontic treatment (Peterson et al. 2024). Extensive investigations have characterized microbiological alterations in various oral niches, including supragingival plaque, subgingival plaque, aligner-associated biofilms, and salivary microbiota, providing valuable insights into the impact of CAs on oral microbial ecology.

Baybekov et al. (2023) conducted a comprehensive study to explore biofilm formation and microbial diversity associated with orthodontic aligner therapy. The study employed microbiological assessments of aligners collected from ten adolescent patients (aged 13–19 years) to identify microbial taxa and evaluate biofilm formation using densitometric analysis. Microbial strains were isolated from aligners that had not been cleaned for 24 h, followed by culturing on selective media to identify key microbial groups, including yeast-like fungi, *Bifidobacterium*, *Lactobacillus*, *Staphylococcus*, *Streptococcus*, *Enterobacteriaceae*, and *Clostridium* species. The growth conditions were optimized for each microorganism type, and species-level identification was conducted using matrix-assisted laser desorption/ionization-time of flight (MALDI-

TOF) mass spectrometry. Densitometric analysis using crystal violet staining was employed to quantify biofilm formation. The findings revealed the presence of 28 distinct microbial species, with a predominance of taxa such as *Bifidobacteria*, *Candida albicans*, *Escherichia coli*, *Peptostreptococcus anaerobius*, *P. gingivalis*, *Prevotella buccae*, *Staphylococcus aureus*, and *Streptococcus mitis*. Notably, several of these species exhibited substantial biofilm-forming capacity, underscoring their role in the microbial ecology of aligners. These findings emphasize the complexity and diversity of the microbial environment associated with orthodontic aligners and highlight the potential implications for oral health during CA therapy.

Caccianiga et al. (2022) conducted a comprehensive investigation to evaluate the impact of various oral hygiene protocols on the oral microbiota of patients undergoing orthodontic treatment with either multibracket appliances (MBs) or CAs. The study included 50 subjects, divided into two groups based on their orthodontic treatment modality, each following distinct oral hygiene regimens. Subgingival plaque samples were collected using a periodontal curette, specifically from interdental regions between the molars in both the maxilla and mandible, as these sites exhibit a high propensity for bacterial plaque accumulation. The collected samples were subsequently analyzed using phase-contrast microscopy, with specimens prepared in physiological solution and oil to enhance visualization. At baseline, both treatment groups exhibited a predominance of non-pathogenic bacterial flora. However, after 3 months, a significant microbiological shift was observed, particularly in the MB group, where a transition from non-pathogenic to pathogenic flora occurred. This pathogenic shift was characterized by an increased prevalence of motile bacteria, including spirochetes such as *Treponema denticola* and *Trichomonas tenax*, indicative of heightened periodontal disease risk. Notably, the implementation of a modified oral hygiene protocol, comprising the use of a sonic toothbrush with vertical motion (Broxo OraBrush, Santé Parodonte, Geneva, Switzerland), interdental brushes, and an oral irrigator (Broxo OraJets, Santé Parodonte, Geneva, Switzerland), applied at least twice daily, effectively restored the microbiota to a non-pathogenic state in all subjects by the subsequent assessment. These findings underscore the critical role of optimized oral hygiene protocols in mitigating dysbiosis and preserving periodontal health in orthodontic patients.

Similarly, Shokeen et al. (2022) conducted a study to elucidate the impact of FAs and CAs on supragingival plaque composition. This investigation included 24 participants, equally divided between FA and CA treatment groups. Plaque samples were systematically collected from both anterior and posterior dentition, with an additional set of samples obtained from the inner surfaces of recently worn aligners in the CA group. DNA extraction and 16S rRNA gene sequencing were employed to characterize microbial composition. The analysis identified ten dominant bacterial genera that constituted the core microbiota in both orthodontic treatment groups, including *Streptococcus*, *Actinomyces*, *Corynebacterium*, *Capnocytophaga*, *Leptotrichia*, *Neisseria*, *Rothia*, *Fusobacterium*, *Prevotella*, and *Veillonella*. Among these, *Leptotrichia* exhibited a statistically significant difference in abundance between the two modalities, with a higher prevalence in the FA group, particularly in posterior regions during follow-up assessments. Given that *Leptotrichia* has been implicated in dental caries and recognized as an emerging opportunistic pathogen (Peterson et al. 2013), its elevated presence in FA patients suggests an increased risk of dental caries and periodontal complications. In contrast, microbial communities associated with aligner trays exhibited distinct compositional features relative to tooth-associated plaque. Specifically, microbial diversity within the aligner tray biofilms was reduced, with notable shifts in the abundance of specific taxa: *Actinomyces*, *Corynebacterium*, and *Selenomonas* were significantly decreased, whereas *Streptococcus* and *Granulicatella* were markedly increased. These findings suggest that CAs establish a unique microenvironment that alters bacterial colonization patterns. Despite these compositional differences, alpha diversity—a measure of overall bacterial species richness—did not exhibit statistically significant variation between anterior and posterior plaque sites or between orthodontic treatment modalities. This suggests that while orthodontic appliances influence microbial composition, they do not induce substantial alterations in overall bacterial diversity.

Rouzi et al. (2023) conducted a longitudinal study to assess microbial compositional changes in CA trays and subgingival plaque over a 3-month period. The study cohort consisted of 15 individuals, from whom plaque samples were collected from both CA trays and subgingival sites. These samples were analyzed using 16S rRNA gene sequencing to evaluate shifts in microbial diversity and taxonomic composition. Prior to the start of the study, participants were instructed in the standardized application of the Bass brushing technique and proper aligner cleaning procedures. Additionally, they were instructed to refrain from brushing, aligner cleaning, eating, or switching aligners for at least two hours prior to each study appointment to minimize external variables that could affect microbiome assessment. The results revealed distinct microbial shifts within aligner trays throughout the treatment duration. Notably, there was a significant increase in the relative abundance of the phyla *Pseudomonadota* and *Bacillota*, while the abundance of *Actinomycetota* demonstrated a substantial decline by the end of the 3-month period. In contrast, subgingival plaque samples exhibited an increase in *Pseudomonadota* abundance but a decrease in *Bacillota*, with *Actinomycetota* displaying a fluctuating trend, suggesting divergent microbial dynamics between these two oral niches. At the genus level, aligner tray plaque demonstrated a significant increase in *Streptococcus* and *Haemophilus* populations, whereas *Actinomyces*, *Corynebacterium*, *Rothia*, and *Lautropia* exhibited marked reductions. In subgingival plaque, *Haemophilus* showed a steady increase throughout the treatment period, whereas *Lautropia* exhibited a transient elevation at the 1-month mark before returning to baseline levels by the 3-month assessment.

Zhao et al. (2020) investigated the impact of Invisalign® therapy on the oral microbiome of 25 adult patients over a 6-month period. In this prospective clinical study, salivary samples were collected at baseline and after 6 months of aligner wear, followed by bacterial composition analysis via Illumina MiSeq sequencing. The results indicated that overall microbial biodiversity remained stable between baseline and 6-month samples, suggesting that prolonged CA wear did not lead to substantial disruptions in the oral microbiome. The predominant bacterial phyla detected included *Firmicutes*, *Proteobacteria*, *Bacteroidetes*, *Fusobacteria*, *Actinobacteria*, and *Saccharibacteria*. While phylum-level composition remained largely consistent, significant genus-level variations were observed. Specifically, there was an increase in *Bacillus* abundance, while *Prevotella* populations declined after 6 months of treatment. Given that the presence of *Bacillus* species is often associated with a healthier oral environment, whereas *Prevotella* has been implicated in periodontal disease pathogenesis, these microbial shifts may reflect a positive modulation of oral health during treatment (Jebin and Suresh 2023; Kumar et al. 2003). At the species level, the study noted a significant reduction in the prevalence of *Prevotella pallens*, a bacterium associated with the oral microbiome, indicating potential shifts in microbial populations conducive to better oral health. The findings of Zhao et al. (2020) underscore the importance of rigorous oral hygiene in preserving microbial homeostasis during orthodontic treatment. Patients were instructed to adhere to evidence-based oral hygiene protocols, including the Bass brushing technique, regular flossing, and aligner cleaning with a soft-bristle toothbrush, water, and minimal toothpaste before reinsertion. Additionally, they were advised to brush and floss after every meal or snack to minimize bacterial accumulation. Collectively, these results suggest that, under optimal hygiene conditions, Invisalign® treatment does not exert detrimental effects on the oral microbiome within the first 6 months of therapy.

Guo et al. (2018) conducted a longitudinal study to assess alterations in the subgingival plaque biofilm microbiota among female patients undergoing CAT over a 3-month period. The study cohort consisted of ten female patients, from whom subgingival plaque samples were collected at three distinct time points: prior to orthodontic treatment, 1 month after treatment initiation, and 3 months post-initiation. DNA extraction and 16S rRNA gene sequencing were employed to characterize microbial community dynamics and taxonomic shifts. The findings revealed a slight decline in microbial diversity, as indicated by various alpha diversity indices; however, this trend did not reach statistical significance. Beta diversity analysis, based on weighted UniFrac distances, demonstrated significant differences in microbial community structures across the three time points, indicating compositional shifts in the subgingival microbiota over the course of treatment. The microbial community taxonomic analysis, performed at the phylum, genus, and species levels, revealed that while the overall microbial diversity decreased, significant shifts occurred in the microbial community structure. At the phylum level, the abundance of *Firmicutes* significantly decreased, whereas *Tenericutes* showed a similar trend from the beginning to the end of the study period. At the genus level, changes included significant reductions in the abundance of *Mycoplasma* and *Bergeyella* at 3 months compared to baseline. Despite these shifts, the abundance of known periodontal pathogens remained stable throughout the study. This study showed that CA treatment led to alterations in the subgingival microbial structure, with a decrease in microbial diversity; however, it did not increase the levels of periodontal pathogens. These findings suggest that CAs may induce nonpathogenic alterations in the subgingival microbiome during the initial 3 months of treatment. This supports the idea that CAT appears to be an effective option for maintaining periodontal health during orthodontic treatment.

Gujar et al. (2020) conducted a cross-sectional study to evaluate microbial profiles associated with different orthodontic appliances over a 1-month period using the checkerboard DNA-DNA hybridization technique. The study included 60 patients, stratified into three cohorts based on their orthodontic appliance type: CAs, metallic labial FA, and lingual FAs. The primary objective was to compare microbial colonization patterns, particularly focusing on the prevalence of bacterial species belonging to the orange and red complexes, which are known for their pathogenic potential in periodontal diseases. To ensure homogeneity across study groups, rigorous inclusion criteria were applied. All participants were required to be systemically healthy, have refrained from antibiotic use for at least 3 months prior to the study, and demonstrate clinically healthy periodontium, as evidenced by a gingival index (GI) score below 1 and an absence of periodontal pockets exceeding 3 mm in depth. Additionally, radiographic analysis was conducted to confirm the absence of alveolar bone loss, thereby ensuring comparable baseline periodontal conditions among participants. Microbial assessment was performed by collecting and analyzing plaque samples from orthodontic brackets and aligners after 30 days of wear. The results demonstrated notable differences in microbial colonization across the different orthodontic appliances. Metallic brackets, particularly those used in lingual appliances, exhibited the highest levels of microbial contamination. Specifically, lingual FAs harbored significantly elevated levels of *F. nucleatum*, *P. gingivalis*, and *Treponema denticola*. Labial FAs, in contrast, exhibited a higher prevalence of *Fusobacterium periodontium* and *Prevotella intermedia*. CAs demonstrated a significantly lower microbial burden compared to other orthodontic appliances, although slightly elevated levels of *Campylobacter rectus* and *Tannerella forsythia* were detected. The study concluded that microbial adhesion and biofilm formation are significantly influenced by the design and material composition of orthodontic appliances. Metallic brackets, particularly those used in lingual FAs, present a greater risk for microbial contamination, while CAs harbor fewer pathogenic bacteria, suggesting a potential advantage in maintaining oral hygiene during orthodontic treatment.

Lombardo et al. (2018) conducted a study to evaluate subgingival microbiological changes over the first 6 months of orthodontic treatment using CAs and FAs. The study tested the null hypothesis that there would be no significant microbiological differences between the two groups. The study included 27 subjects, stratified into CA and FA treatment groups based on the complexity of their malocclusion—assessed using the Index of Orthodontic Treatment Need (IOTN)—as well as patient preference for either treatment modality. Subgingival microbiological samples were collected at baseline 1, 3, and 6 months, with a focus on total bacterial load (TBL) and the presence of specific periodontal pathogens. Sampling was performed at defined anatomical sites: the distobuccal and distopalatal surfaces of the maxillary right central incisor and the mesiobuccal and mesiopalatal surfaces of the maxillary right first molar. The mandibular dentition was excluded from microbial analysis due to the higher incidence of accidental debonding in the lower arch, which could have compromised the reliability of subgingival sampling. The results indicated that while TBL significantly increased in the FA group over the 6-month period, it remained stable in the CA group. Among the targeted periodontal pathogens, *Aggregatibacter actinomycetemcomitans* was never detected, and species belonging to the "red complex" (*P. gingivalis*, *Treponema denticola*, and *Tannerella forsythia*) were rarely identified, with no significant variations over time in either treatment group. However, *F. nucleatum* and *Campylobacter rectus* exhibited statistically significant increases in the FA group compared to baseline, whereas in the CA group, microbial shifts were minimal, with the exception of a significant increase in *Campylobacter rectus* at the 6-month evaluation. The study concluded that FAs may contribute to a notable increase in TBL and specific periodontal pathogens, likely due to the challenges associated with maintaining adequate oral hygiene when using non-removable appliances. In contrast, the removability of CAs facilitated better hygiene management, thereby mitigating microbial accumulation and dysbiosis. The researchers recommended that future randomized controlled trials incorporating clinical periodontal assessments be conducted to determine whether different orthodontic treatment modalities necessitate tailored oral hygiene protocols.

Mummolo et al. (2020) presented detailed insights into the influence of orthodontic appliances on the salivary concentrations *of Streptococcus mutans* and *Lactobacilli*, two key microbial contributors to dental caries (Sounah and Madfa 2020). This observational cohort study included 90 orthodontic patients, categorized into three treatment groups: CAs, MBs, and removable positioners. The primary objective was to compare the impact of these appliances on the oral microbiome over a 6-month treatment period. Salivary samples were collected at baseline, 3 months, and 6 months, and bacterial concentrations were quantified using the CRT® bacteria test to assess *Streptococcus mutans* and *Lactobacilli* levels. The results demonstrated that after 6 months, 40% of patients in the MB group exhibited *Streptococcus mutans* concentrations indicative of a high caries risk, compared to 10% in the CA group and 13.3% in the removable positioner group. A similar trend was observed for *Lactobacilli*, with significant increases over time in the MB group. These findings suggest that the type of orthodontic appliance significantly influences the composition of the oral microbiome. FAs, particularly MBs, were associated with a substantial increase in cariogenic bacterial populations, thereby elevating the risk of caries development. In contrast, removable appliances such as CAs and removable positioners were linked to a more stable oral microbiome, suggesting a lower overall caries risk.

In conclusion, findings from multiple studies investigating the effects of CAs on oral microbial communities indicate that the type of orthodontic appliance plays a critical role in shaping the oral microbiome. Removable appliances, such as

CAs, appear to support a more stable microbial environment when compared to FAs like MBs. Extensive research has demonstrated that FAs create conditions favorable for the accumulation of cariogenic and periodontopathogenic bacteria, leading to increased plaque biofilm formation, gingivitis, and a higher risk of dental caries. Conversely, CAs are associated with reduced bacterial colonization and a lower likelihood of microbial dysbiosis. Additionally, the microbial dysbiosis associated with FAs has significant implications for the immune system. The persistent accumulation of plaque biofilms around FAs triggers chronic immune activation, resulting in sustained gingival and periodontal inflammation. This inflammatory response, exacerbated by the immune reaction to orthodontic tooth movement, stimulates the release of pro-inflammatory cytokines, including IL-1β, IL-6, and TNF-α, which contribute to gingivitis and may exacerbate bone resorption in susceptible individuals (Bilgic et al. 2016; Boas Nogueira et al. 2013). In contrast, CAs, by minimizing plaque accumulation, may mitigate excessive immune activation, thereby promoting a more balanced host–microbial interaction.

Furthermore, emerging research suggests that localized inflammation during orthodontic treatment may influence systemic inflammatory markers, potentially leading to broader systemic effects (Kloukos et al. 2023). However, additional well-designed studies are necessary to elucidate the long-term effects of both FAs and CAs on the oral microbiome, their immunological consequences, and potential systemic implications. These findings carry significant clinical relevance, underscoring the necessity for personalized preventive strategies tailored to the specific orthodontic appliance used by each patient. Orthodontists and dental professionals should consider appliance selection as a pivotal factor in managing the risk of caries and other oral health complications during treatment. This highlights the importance of regular monitoring and proactive oral hygiene interventions to maintain oral microbial homeostasis and optimize oral health outcomes in orthodontic patients.

29.3 Effects of Clear Aligners on Dental and Periodontal Health

WSLs are early manifestations of enamel demineralization, characterized by opaque, white patches that result from subsurface mineral loss. Their development is primarily influenced by plaque accumulation and oral hygiene practices (Rouzi et al. 2023; Sundararaj et al. 2015). These lesions are particularly prevalent among patients undergoing orthodontic treatment with FAs, which create additional surfaces for plaque retention and disrupt the oral microbial ecosystem, thereby increasing the risk of WSL development and other complications, such as periodontal disease. In contrast, CAs, due to their removability, provide a significant advantage by facilitating improved oral hygiene and potentially reducing the incidence of WSLs. CAs allow for more consistent mechanical plaque removal and promote a balance between enamel demineralization and remineralization, thereby minimizing plaque accumulation and its associated risks, including gingivitis (Ong and Wang 2002; Rossini et al. 2014). These findings suggest that CAT may be a preferable orthodontic treatment modality for minimizing the risk of WSLs. However, it is critical to emphasize the importance of adhering to dietary restrictions, particularly avoiding the consumption of acidic or sugary liquids while wearing aligners. Failure to follow these guidelines can result in prolonged exposure of the enamel to cariogenic substrates, effectively trapping sugars against the tooth surface and increasing the risk of enamel demineralization (Ehlen et al. 2008; Lanteri et al. 2020).

Periodontal health during orthodontic treatment is influenced by multiple factors, including individual host resistance, systemic conditions such as diabetes, deleterious habits such as smoking, anatomical characteristics such as buccal bone thickness, and the presence and composition of dental plaque. However, the most significant determinant remains the patient's adherence to optimal oral hygiene practices (Crego-Ruiz and Jorba-García 2023; Talic 2011). Several studies have proposed comprehensive oral hygiene protocols specifically tailored for orthodontic patients using CAs to mitigate plaque accumulation and bacterial proliferation, which can occasionally be exacerbated by aligner wear. These protocols integrate routine brushing with additional adjunctive methods to ensure optimal oral and appliance hygiene throughout treatment. A widely recommended strategy includes brushing both the teeth and aligners with a soft-bristle toothbrush after every meal to remove food debris and plaque accumulation beneath the aligners. To prevent surface abrasions, non-abrasive toothpaste or specialized aligner cleaning solutions are preferred over conventional toothpaste. Beyond standard brushing, the use of ultrasonic cleaning devices in combination with aligner cleaning crystals or rinsing with chlorhexidine mouthwash has been shown to significantly reduce bacterial colonization and biofilm formation, thereby maintaining both oral and appliance hygiene. Furthermore, it is imperative to remove aligners prior to consuming any food or beverages other than water, as the entrapment of sugary or acidic liquids beneath the aligners can contribute to enamel erosion and demineralization (Charavet et al. 2022; Lanteri et al. 2020; Shpack et al. 2014).

The role of various microbial pathogens, including viruses, fungi, and protozoa, in periodontal disease is well-documented. *Candida albicans* has been implicated in gingival erythema, particularly among immunocompromised pediatric patients, while *Entamoeba gingivalis* has been associated with periodontal tissue degradation. Additionally, *Candida* species contribute to both oral and systemic infections and have been shown to exacerbate dental caries in children due to their interactions with *Streptococcus mutans* within dental plaque, promoting biofilm formation and enhancing cariogenic potential (Contaldo et al. 2019, 2021; Dubar et al. 2020; Paoletti et al. 2013; Velegraki et al. 1999).

Orthodontic interventions must prioritize periodontal health to prevent the exacerbation of preexisting conditions such as periodontitis. Patients with active periodontal disease require comprehensive management before initiating orthodontic treatment to mitigate further deterioration. Standard clinical guidelines recommend postponing orthodontic therapy until a periodontist confirms that periodontal disease is stable and well-controlled. Once orthodontic treatment is deemed appropriate, CAs offer distinct advantages in maintaining periodontal health. Their removability facilitates improved plaque control, thereby reducing the risk of periodontitis progression. Additionally, CAs apply controlled and intermittent orthodontic forces, which support periodontal ligament recovery and promote even stress distribution, ultimately preserving gingival integrity (Rouzi et al. 2023; Wu and Hou 2019).

Song et al. (2023) investigated the effects of CAT on the salivary microbiome and metabolome in adolescents, specifically comparing individuals with and without WSLs. The primary objective was to elucidate microbial and metabolic alterations associated with WSL formation, which is often linked to suboptimal oral hygiene during CAT. The study cohort comprised adolescents aged 11–18 years who had been undergoing CAT for over a year. Participants with periodontal disease, generalized caries, systemic illnesses, or recent antibiotic use were excluded to minimize confounding variables. Saliva samples were analyzed using 16S rRNA gene sequencing to characterize microbial composition, while targeted metabolomic profiling was conducted to evaluate salivary metabolite levels. Key findings revealed that adolescents presenting with WSLs exhibited poorer oral hygiene practices and were more likely to consume carbonated beverages and eat while wearing CAs. Microbiome analysis demonstrated significant compositional differences between the WSL and control groups, with *Actinobacteria*, *Rothia*, and *Subdoligranulum* being more abundant in individuals with WSLs. Additionally, metabolomic profiling identified distinct metabolic signatures between the two groups, with significant variations in amino acid concentrations, suggesting a potential correlation between specific metabolites and WSL development. The findings highlight the importance of maintaining stringent oral hygiene, especially when using CAs, to prevent the development of WSLs and associated alterations in the oral microbiome and metabolome.

The prospective study conducted by Azeem and Ul Hamid (2017) evaluated the incidence of WSLs in patients undergoing orthodontic treatment with CAs. The study cohort consisted of 25 orthodontic patients (12 males and 13 females) with a mean age of 16.17 ± 1.76 years. Participants met specific inclusion criteria, including undergoing comprehensive CA treatment for both arches, being between 14 and 18 years of age at the initiation of treatment, and having no prior orthodontic therapy or systemic diseases that could confound the study outcomes. The primary variables assessed were the presence of sound enamel and the incidence of new WSLs, which were evaluated using pretreatment and posttreatment quantitative light-induced fluorescence imaging. The results demonstrated an overall incidence of new WSLs of 2.85% across all examined teeth, with 28% of patients developing at least one new lesion. No significant differences were observed in WSL prevalence across different age and gender groups. However, treatment duration was found to significantly influence WSL formation in maxillary incisors. The findings suggest that CAT is associated with a low overall incidence of new WSLs, making it a viable orthodontic option for individuals predisposed to enamel demineralization. The study further emphasized the critical importance of short treatment durations and maintaining rigorous oral hygiene protocols. The authors advocate for the integration of standard hygiene practices along with additional preventive strategies to minimize WSL risk during treatment.

Buschang et al. (2019) conducted a comparative analysis of WSL incidence in patients undergoing treatment with CAs versus those treated with traditional fixed appliances. No significant differences were found between the two groups concerning pretreatment age or gender distribution. However, the mean treatment duration was notably shorter for patients in the CA group than for those undergoing treatment with FAs. The results indicated that only 1.2% of aligner patients developed new WSLs, a significantly lower rate than the 25.7% observed in the FA group. Furthermore, aligner patients exhibited superior oral hygiene throughout the treatment period compared to those with traditional braces. The study concluded that CAs are associated with a substantially reduced risk of WSL development, attributed primarily to shorter treatment durations and enhanced oral hygiene facilitated by aligner removability. These findings highlight the advantages of CAs, particularly for individuals at an elevated risk of WSL formation. The ability to remove aligners enables more effective plaque control and enhances patient compliance with oral hygiene protocols during orthodontic therapy (Fig. 29.1).

Guo et al. (2018) examined the impact of CAs on periodontal health over a 3-month period, evaluating parameters such as the plaque index (PI) and the gingival bleeding index (GBI). While slight increases in PI and GBI were noted throughout treatment, these changes were not statistically significant, suggesting that CAs do not exert a detrimental effect on periodontal health in the short term.

The study by Levrini et al. (2015) aimed to compare the periodontal health status of patients treated with the Invisalign® system versus those using FAs over 3 months. This prospective clinical and microbiological evaluation included key periodontal health indicators, including PI, GI, probing pocket depth (PPD), and bleeding on probing (BoP), alongside total biofilm mass and the presence of periodontal pathogens, which were analyzed using real-time PCR. The results demonstrated that Invisalign® patients exhibited significantly better periodontal health outcomes, with lower PI, GI, PPD, and BoP scores compared to FA patients. Additionally, the Invisalign® group showed reduced total biofilm mass and a lower prevalence of periodontal pathogens, highlighting the potential benefits of CAT in maintaining periodontal health during orthodontic treatment.

The study by Sfondrini et al. (2021) evaluated the impact of CAT on oral microbiota and periodontal health over a 2-month period. The study included 40 participants divided into two groups: one undergoing CAT and a control group without

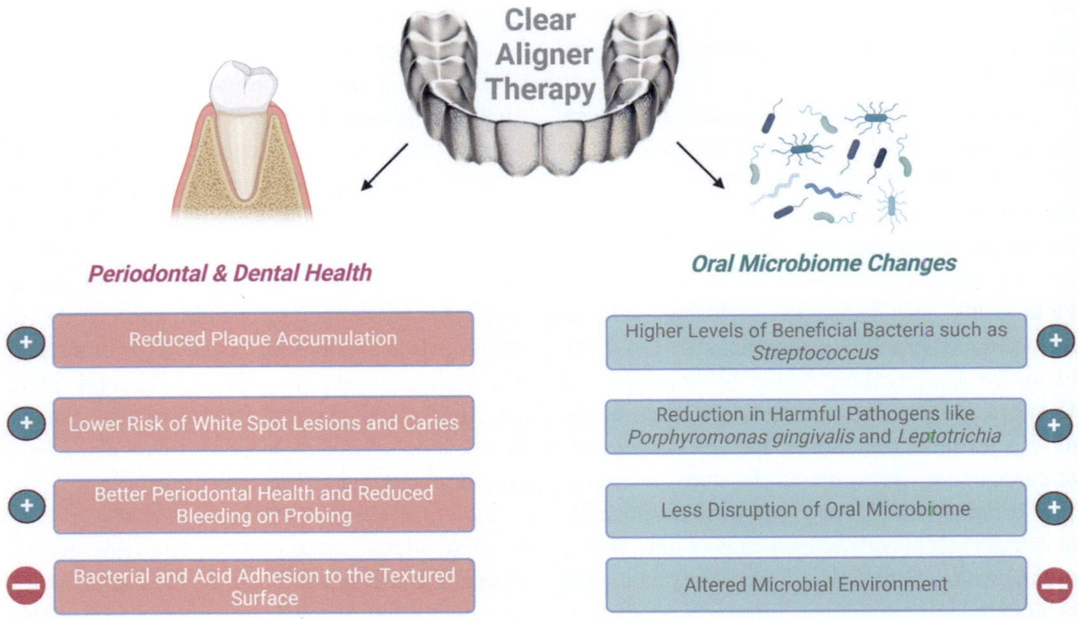

Fig. 29.1 An overview of the advantages and disadvantages of clear aligners compared to fixed orthodontic appliances

orthodontic treatment. Clinical parameters such as PPD, BoP, and PI were evaluated, and microbiological samples were analyzed for key periodontal pathogens, including *P. gingivalis* and *Tannerella forsythia*. The results indicated no significant differences in periodontal health status or pathogenic bacterial presence between the CA and control groups. Although a slight increase in TBL was observed in both groups, it was not significant to suggest an increased risk for periodontal disease. These findings suggest that CAT does not induce adverse microbial or periodontal changes in the early stages of treatment.

The study by Lanteri et al. (2020) examined the effects of CAs in the upper arch compared to FAs in the lower arch on periodontal health, focusing on PPD, PI, and BoP. The study followed a small patient cohort with assessments conducted at baseline, 1 month, 5 months, and treatment completion (9–12 months). The results demonstrated superior periodontal health outcomes in the CA-treated upper arch, where PPD remained stable or improved, while the FA-treated lower arch exhibited increased probing depths indicative of periodontal deterioration. Similarly, PI scores were significantly lower in the CA-treated arch, with reduced plaque accumulation, whereas the FA-treated arch showed higher plaque retention, particularly around brackets. BoP findings aligned with these results, with significantly lower bleeding rates in CA-treated sites compared to FA sites, where inflammation was more pronounced.

A recent study by Zhang et al. (2025) investigates the differences in gingivitis incidence between adolescents and adults undergoing CAT and evaluates prevention strategies. It included 120 patients divided into adolescent and adult groups, each further split into a study group (receiving personalized oral hygiene education) and a control group (maintaining their regular lifestyle). The results showed that adolescents were more likely to develop gingivitis compared to adults, and within the adolescent group, those who received oral hygiene education had a higher GI than those who did not. In contrast, the adult group's gingivitis levels did not differ significantly between the study and control groups. The study concluded that adolescents are more susceptible to gingivitis during CAT, and basic periodontal therapy is more effective than education alone in reducing its incidence.

In a systematic review by Lucchese et al. (2025), the impact of fixed orthodontic therapy and CAs on the oral microbiome was analyzed. Both treatments caused alterations in the oral microbiota, leading to dysbiosis. FAs, due to their plaque-retentive surfaces, resulted in increased bacterial loads and a higher prevalence of periodontopathogenic species, raising the risk for gingivitis, periodontitis, and caries. On the other hand, CAs, while also affecting the microbiome, allowed for better oral hygiene and exhibited a lesser impact on periodontal health. However, the reduced exposure of teeth to saliva during prolonged aligner wear may slightly increase the risk of WSLs and caries. The study concluded that CAs have a more favorable effect on oral health compared to FAs, although further research is needed to assess long-term changes.

Several other systematic reviews have also examined the periodontal health outcomes of patients treated with CAs versus traditional braces. These reviews generally support the notion that CAs contribute to improved periodontal health, including lower plaque levels and better gingival conditions. However, variations in study quality and methodological limitations have led some reviews to call for high-quality research to address potential biases. While current evidence suggests that CAs offer advantages in maintaining oral hygiene and reducing WSL risk, definitive conclusions remain hindered by heterogeneity among studies. Overall, CAs may be preferable for promoting periodontal health during orthodontic treatment, but further investigations are warranted to confirm these findings conclusively (Crego-Ruiz and Jorba-García 2023; Jiang et al. 2018; Llera-Romero et al. 2023; Rossini et al. 2014).

29.4 Hypersensitivity Reactions in Orthodontic Treatment

Hypersensitivity reactions in orthodontic patients, though rare, pose a clinical challenge due to the prolonged nature of treatment and potential systemic effects. FAs, due to their continuous intraoral presence, are associated with a greater likelihood of adverse inflammatory responses compared to CAs and other removable appliances (Di Spirito et al. 2024).

Hypersensitivity reactions in orthodontics can result from various allergens found in dental materials, with ion release from orthodontic appliances being a primary contributing factor. Nickel is the most implicated allergen, followed by cobalt and chromium, which are frequently used in FAs. In contrast, hypersensitivity reactions to acrylic and polymer-based materials used in CAs are less frequently reported than those associated with metal components in FAs (Di Spirito et al. 2024).

The predominant mechanism involved in orthodontic hypersensitivity is type IV hypersensitivity, a T-cell-mediated immune response that activates an inflammatory cascade upon exposure to metal ions. This delayed hypersensitivity reaction has been frequently reported in FA patients (Chakravarthi et al. 2012). However, whether these reactions are solely attributable to ion release remains a subject of ongoing investigation, with conflicting findings in the literature (Paschaei et al. 2024).

Another potential mechanism is type I hypersensitivity, an IgE-mediated reaction that can be triggered by metal ions, acrylic materials, or other dental allergens during the early phase of orthodontic treatment. Clinical manifestations of type I hypersensitivity may include bronchospasm, angioedema, and, in rare cases, anaphylaxis. Other reported symptoms include erythema, erosive–ulcerative lesions, mucosal desquamation, gingival hyperplasia, and localized discomfort, such as burning sensations, itching, and pain (Di Spirito et al. 2024). Poor oral hygiene has been shown to exacerbate the corrosion of orthodontic appliances, increasing ion release and heightening the risk of hypersensitivity reactions. Additional contributing factors include genetic predisposition to allergic responses and the duration of orthodontic treatment (Di Spirito et al. 2024).

29.5 Considerations for Patients and Practitioners

For both patients and practitioners, maintaining proper care of CAs throughout orthodontic treatment is crucial. Patients should adhere to prescribed aligner wear schedules, removing them only for eating and oral hygiene to prevent discoloration and maintain treatment efficacy. Practitioners should emphasize the importance of rigorous cleaning and disinfection protocols to reduce the risk of biofilm accumulation and associated oral health complications. Although CAs provide advantages over traditional braces by facilitating improved oral hygiene, they present challenges related to bacterial adhesion and plaque formation due to their textured surfaces. Patients must be informed of the potential consequences of inadequate aligner hygiene, including the risk of increased cariogenic bacterial colonization over time. Ongoing research is exploring antimicrobial coatings for CAs to mitigate bacterial growth and biofilm development; however, further clinical validation is required to assess the efficacy of these interventions (Worreth et al. 2022; Zhang et al. 2020).

In conclusion, while CAs offer notable benefits in orthodontic treatment, their effectiveness is contingent on diligent patient compliance and proper maintenance. By adhering to evidence-based oral hygiene practices and staying informed about emerging research, both patients and practitioners can optimize treatment outcomes and support long-term oral health.

References

Adobes-Martin M, Montoya-Morcillo ML, Zhou-Wu A, Garcovich D (2021) Invisalign treatment from the patient perspective: a Twitter content analyses. J Clin Exp Dent 13(4):e376–e382. https://doi.org/10.4317/jced.57835

Alansari RA (2020) Youth perception of different orthodontic appliances. Patient Prefer Adherence 14:1011–1019. https://doi.org/10.2147/ppa.S257814

Azeem M, Ul Hamid W (2017) Incidence of white spot lesions during orthodontic clear aligner therapy. J World Fed Orthod 6(3):127–130. https://doi.org/10.1016/j.ejwf.2017.07.001

Bäckhed F, Fraser CM, Ringel Y, Sanders ME, Sartor RB, Sherman PM, Versalovic J, Young V, Finlay BB (2012) Defining a healthy human gut microbiome: current concepts, future directions, and clinical applications. Cell Host Microbe 12(5):611–622. https://doi.org/10.1016/j.chom.2012.10.012

Baker JL, Bor B, Agnello M, Shi W, He X (2017) Ecology of the oral microbiome: beyond bacteria. Trends Microbiol 25(5):362–374. https://doi.org/10.1016/j.tim.2016.12.012

Baybekov O, Stanishevskiy Y, Sachivkina N, Bobunova A, Zhabo N, Avdonina M (2023) Isolation of clinical microbial isolates during orthodontic aligner therapy and their ability to form biofilm. Dent J (Basel) 11(1). https://doi.org/10.3390/dj11010013

Bilgic F, Akinci Sozer O, Ozcan O, Gurpinar AB, Yilmaz H, Ay Y (2016) Evaluation of inflammation during fixed orthodontic treatment. Arch Oral Biol 71:54–58. https://doi.org/10.1016/j.archoralbio.2016.07.004

Boas Nogueira AV, Chaves de Souza JA, Kim YJ, Damião de Sousa-Neto M, Chan Cirelli C, Cirelli JA (2013) Orthodontic force increases interleukin-1β and tumor necrosis factor-α expression and alveolar bone loss in periodontitis. J Periodontol 84(9):1319–1326. https://doi.org/10.1902/jop.2012.120510

Boyd RL, Miller RJ, Vlaskalic V (2000) The Invisalign system in adult orthodontics: mild crowding and space closure cases. J Clin Orthod 34(4):203–212

Buschang PH, Chastain D, Keylor CL, Crosby D, Julien KC (2019) Incidence of white spot lesions among patients treated with clear aligners and traditional braces. Angle Orthod 89(3):359–364. https://doi.org/10.2319/073118-553.1

Caccianiga P, Nota A, Tecco S, Ceraulo S, Caccianiga G (2022) Efficacy of home oral-hygiene protocols during orthodontic treatment with multibrackets and clear aligners: microbiological analysis with phase-contrast microscope. Healthcare (Basel) 10(11). https://doi.org/10.3390/healthcare10112255

Cantekin K, Celikoglu M, Karadas M, Yildirim H, Erdem A (2011) Effects of orthodontic treatment with fixed appliances on oral health status: a comprehensive study. J Dent Sci 6(4):235–238. https://doi.org/10.1016/j.jds.2011.09.010

Caselli E, Fabbri C, D'Accolti M, Soffritti I, Bassi C, Mazzacane S, Franchi M (2020) Defining the oral microbiome by whole-genome sequencing and resistome analysis: the complexity of the healthy picture. BMC Microbiol 20(1):120. https://doi.org/10.1186/s12866-020-01801-y

Chakravarthi S, Padmanabhan S, Chitharanjan AB (2012) Allergy and orthodontics. J Orthod Sci 1(4):83–87. https://doi.org/10.4103/2278-0203.105871

Charavet C, Gourdain Z, Graveline L, Lupi L (2022) Cleaning and disinfection protocols for clear orthodontic aligners: a systematic review. Healthcare (Basel) 10(2). https://doi.org/10.3390/healthcare10020340

Contaldo M, Romano A, Mascitti M, Fiori F, Della Vella F, Serpico R, Santarelli A (2019) Association between denture stomatitis, candida species and diabetic status. J Biol Regul Homeost Agents 33(3 Suppl. 1):35–41. Dental supplement

Contaldo M, Lucchese A, Lajolo C, Rupe C, Di Stasio D, Romano A, Petruzzi M, Serpico R (2021) The oral microbiota changes in orthodontic patients and effects on oral health: an overview. J Clin Med 10(4). https://doi.org/10.3390/jcm10040780

Costello EK, Stagaman K, Dethlefsen L, Bohannan BJM, Relman DA (2012) The application of ecological theory toward an understanding of the human microbiome. Science 336(6086):1255–1262. https://doi.org/10.1126/science.1224203

Crego-Ruiz M, Jorba-García A (2023) Assessment of the periodontal health status and gingival recession during orthodontic treatment with clear aligners and fixed appliances: a systematic review and meta-analysis. Med Oral Patol Oral Cir Bucal 28(4):e330–e340. https://doi.org/10.4317/medoral.25760

Deo PN, Deshmukh R (2019) Oral microbiome: unveiling the fundamentals. J Oral Maxillofac Pathol 23(1): 122–128. https://doi.org/10.4103/jomfp.JOMFP_304_18

Di Spirito F, Amato A, Di Palo MP, Ferraro R, Cannatà D, Galdi M, Sacco E, Amato M (2024) Oral and extra-oral manifestations of hypersensitivity reactions in orthodontics: a comprehensive review. J Funct Biomater 15(7). https://doi.org/10.3390/jfb15070175

Dubar M, Zaffino ML, Remen T, Thilly N, Cunat L, Machouart MC, Bisson C (2020) Protozoans in subgingival biofilm: clinical and bacterial associated factors and impact of scaling and root planing treatment. J Oral Microbiol 12(1):1693222. https://doi.org/10.1080/20002297.2019.1693222

Ehlen LA, Marshall TA, Qian F, Wefel JS, Warren JJ (2008) Acidic beverages increase the risk of in vitro tooth erosion. Nutr Res 28(5):299–303. https://doi.org/10.1016/j.nutres.2008.03.001

Fernandes GL, Delbem ACB, do Amaral JG, Gorup LF, Fernandes RA, de Souza Neto FN, Souza JAS, Monteiro DR, Hunt AMA, Camargo ER, Barbosa DB (2018) Nanosynthesis of silver-calcium glycerophosphate: promising association against oral pathogens. Antibiotics (Basel) 7(3). https://doi.org/10.3390/antibiotics7030052

Festa F, Rotelli C, Scarano A, Navarra R, Caulo M, Macrì M (2021) Functional magnetic resonance connectivity in patients with temporomadibular joint disorders [original research]. Front Neurol 12. https://doi.org/10.3389/fneur.2021.629211

Franzosa EA, Huang K, Meadow JF, Gevers D, Lemon KP, Bohannan BJ, Huttenhower C (2015) Identifying personal microbiomes using metagenomic codes. Proc Natl Acad Sci USA 112(22):E2930–E2938. https://doi.org/10.1073/pnas.1423854112

Freitas AO, Marquezan M, Nojima Mda C, Alviano DS, Maia LC (2014) The influence of orthodontic fixed appliances on the oral microbiota: a systematic review. Dental Press J Orthod 19(2):46–55. https://doi.org/10.1590/2176-9451.19.2.046-055.oar

Gao L, Xu T, Huang G, Jiang S, Gu Y, Chen F (2018) Oral microbiomes: more and more importance in oral cavity and whole body. Protein Cell 9(5):488–500. https://doi.org/10.1007/s13238-018-0548-1

Gest H (2004) The discovery of microorganisms by Robert Hooke and Antoni Van Leeuwenhoek, fellows of the Royal Society. Notes Rec R Soc Lond 58(2):187–201. https://doi.org/10.1098/rsnr.2004.0055

Giugliano D, d'Apuzzo F, Majorana A, Campus G, Nucci F, Flores-Mir C, Perillo L (2018) Influence of occlusal characteristics, food intake and oral hygiene habits on dental caries in adolescents: a cross-sectional study. Eur J Paediatr Dent 19(2):95–100. https://doi.org/10.23804/ejpd.2018.19.02.02

Gujar AN, Al-Hazmi A, Raj AT, Patil S (2020) Microbial profile in different orthodontic appliances by checkerboard DNA-DNA hybridization: an in-vivo study. Am J Orthod Dentofacial Orthop 157(1):49–58. https://doi.org/10.1016/j.ajodo.2019.01.026

Guo R, Zheng Y, Liu H, Li X, Jia L, Li W (2018) Profiling of subgingival plaque biofilm microbiota in female adult patients with clear aligners: a three-month prospective study. PeerJ 6:e4207. https://doi.org/10.7717/peerj.4207

Haouili N, Kravitz ND, Vaid NR, Ferguson DJ, Makki L (2020) Has Invisalign improved? A prospective follow-up study on the efficacy of tooth movement with Invisalign. Am J Orthod Dentofacial Orthop 158(3):420–425. https://doi.org/10.1016/j.ajodo.2019.12.015

He L, Zhang W, Liu J, Pan Y, Li S, Xie Y (2024) Applications of nanotechnology in orthodontics: a comprehensive review of tooth movement, antibacterial properties, friction reduction, and corrosion resistance. Biomed Eng Online 23(1):72. https://doi.org/10.1186/s12938-024-01261-9

Ireland AJ, Soro V, Sprague SV, Harradine NWT, Day C, Al-Anezi S, Jenkinson HF, Sherriff M, Dymock D, Sandy JR (2014) The effects of different orthodontic appliances upon microbial communities. Orthod Craniofacial Res 17(2):115–123. https://doi.org/10.1111/ocr.12037

Jebin AA, Suresh A (2023) Oral microbial shift induced by probiotic Bacillus coagualans along with its clinical perspectives. J Oral Biol Craniofac Res 13(3): 398–402. https://doi.org/10.1016/j.jobcr.2023.03.013

Jia G, Zhi A, Lai PFH, Wang G, Xia Y, Xiong Z, Zhang H, Che N, Ai L (2018) The oral microbiota – a mechanistic role for systemic diseases. Br Dent J 224(6): 447–455. https://doi.org/10.1038/sj.bdj.2018.217

Jiang Q, Li J, Mei L, Du J, Levrini L, Abbate GM, Li H (2018) Periodontal health during orthodontic treatment with clear aligners and fixed appliances: a meta-analysis. J Am Dent Assoc 149(8):712–720.e712. https://doi.org/10.1016/j.adaj.2018.04.010

Jindal P, Juneja M, Siena FL, Bajaj D, Breedon P (2019) Mechanical and geometric properties of thermoformed and 3D printed clear dental aligners. Am J Orthod Dentofacial Orthop 156(5):694–701. https://doi.org/10.1016/j.ajodo.2019.05.012

Kaygisiz E, Uzuner FD, Yuksel S, Taner L, Çulhaoğlu R, Sezgin Y, Ateş C (2014) Effects of self-ligating and conventional brackets on halitosis and periodontal conditions. Angle Orthod 85(3):468–473. https://doi.org/10.2319/041714-289.1

Kesling HD (1945) The philosophy of the tooth positioning appliance. Am J Orthod Oral Surg 31(6):297–304. https://doi.org/10.1016/0096-6347(45)90101-3

Kesling HD (1946) Coordinating the predetermined pattern and tooth positioner with conventional treatment. Am J Orthod Oral Surg 32:285–293. https://doi.org/10.1016/0096-6347(46)90053-1

Kilian M, Chapple IL, Hannig M, Marsh PD, Meuric V, Pedersen AM, Tonetti MS, Wade WG, Zaura E (2016) The oral microbiome – an update for oral healthcare professionals. Br Dent J 221(10):657–666. https://doi.org/10.1038/sj.bdj.2016.865

Kloukos D, Kalimeri E, Gkourtsogianni S, Kantarci A, Katsaros C, Stavropoulos A (2023) Impact of fixed orthodontic appliances on blood count and high-sensitivity C-reactive protein levels: a prospective cohort study. Am J Orthod Dentofacial Orthop 164(3):351–356. https://doi.org/10.1016/j.ajodo.2023.01.016

Kook M-S, Kim H-M, Oh H-K, Lee K-M (2019) Clear aligner use following surgery-first mandibular prognathism correction. J Craniofac Surg 30(6):e544–e547. https://doi.org/10.1097/scs.0000000000005491

Kumar PS, Griffen AL, Barton JA, Paster BJ, Moeschberger ML, Leys EJ (2003) New bacterial species associated with chronic periodontitis. J Dent Res 82(5):338–344. https://doi.org/10.1177/154405910308200503

Lanteri V, Poggi S, Blasi S, Angelis D, Gangale S, Farronato M, Maiorani C, Butera A (2020) Periodontal aspects of orthodontic treatment with invisalign® versus fixed appliances in the same patients – a pilot study ethical compliance, p 13

Levrini L, Mangano A, Montanari P, Margherini S, Caprioglio A, Abbate GM (2015) Periodontal health status in patients treated with the Invisalign(®) system and fixed orthodontic appliances: a 3 months clinical and microbiological evaluation. Eur J Dent 9(3):404–410. https://doi.org/10.4103/1305-7456.163218

Li Y, Deng S, Mei L, Li Z, Zhang X, Yang C, Li Y (2020) Prevalence and severity of apical root resorption during orthodontic treatment with clear aligners and fixed appliances: a cone beam computed tomography study. Prog Orthod 21(1):1. https://doi.org/10.1186/s40510-019-0301-1

Liu Y, Hu W (2018) Force changes associated with different intrusion strategies for deep-bite correction by clear aligners. Angle Orthod 88(6):771–778. https://doi.org/10.2319/121717-864.1

Livas C, Pazhman FS, Ilbeyli Z, Pandis N (2023) Perceived esthetics and value of clear aligner therapy systems: a survey among dental school instructors and undergraduate students. Dental Press J Orthod 28(3):e232225. https://doi.org/10.1590/2177-6709.28.3.e232225.oar

Llera-Romero AS, Adobes-Martín M, Iranzo-Cortés JE, Montiel-Company JM, Garcovich D (2023) Periodontal health status, oral microbiome, white-spot lesions and oral health related to quality of life-clear aligners versus fixed appliances: a systematic review, meta-analysis and meta-regression. Korean J Orthod 53(6):374–392. https://doi.org/10.4041/kjod22.272

Lombardo L, Colonna A, Carlucci A, Oliverio T, Siciliani G (2018) Class II subdivision correction with clear aligners using intermaxilary elastics. Prog Orthod 19(1):32. https://doi.org/10.1186/s40510-018-0221-5

Lu H, Tang H, Zhou T, Kang N (2018) Assessment of the periodontal health status in patients undergoing orthodontic treatment with fixed appliances and Invisalign system: a meta-analysis. Medicine 97(13):e0248. https://doi.org/10.1097/md.0000000000010248

Lucchese A, Marcolina M, Mancini N, Ferrarese R, Acconciaioco S, Gherlone E, Bonini C, Manuelli M, Polimeni A (2025) A comparison of the alterations of oral microbiome with fixed orthodontic therapy and clear aligners: a systematic review. J Oral Microbiol 17(1):2372751. https://doi.org/10.1080/20002297.2024.2372751

Macrì M, Murmura G, Varvara G, Traini T, Festa F (2022) Clinical performances and biological features of clear aligners materials in orthodontics [review]. Front Mater 9. https://doi.org/10.3389/fmats.2022.819121

Maier T (2023) Oral microbiome in health and disease: maintaining a healthy, balanced ecosystem and reversing dysbiosis. Microorganisms 11(6). https://doi.org/10.3390/microorganisms11061453

Marsh PD (2000) Role of the oral microflora in health. Microb Ecol Health Dis 12(3):130–137. https://doi.org/10.1080/089106000750051800

Meade MJ, Weir T (2023) Treatment planning protocols with the Invisalign appliance: an exploratory survey. Angle Orthod 93(5):501–506. https://doi.org/10.2319/111422-783.1

Miethke R-R, Vogt S (2005) A comparison of the periodontal health of patients during treatment with the invisalign® system and with fixed orthodontic appliances. J Orofac Orthop 66(3):219–229. https://doi.org/10.1007/s00056-005-0436-1

Mummolo S, Tieri M, Nota A, Caruso S, Darvizeh A, Albani F, Gatto R, Marzo G, Marchetti E, Quinzi V, Tecco S (2020) Salivary concentrations of Streptococcus mutans and Lactobacilli during an orthodontic treatment. An observational study comparing fixed and removable orthodontic appliances. Clin Exp Dent Res 6(2):181–187. https://doi.org/10.1002/cre2.261

Ong MM, Wang HL (2002) Periodontic and orthodontic treatment in adults. Am J Orthod Dentofacial Orthop

122(4):420–428. https://doi.org/10.1067/mod.2002.126597

Pandey V, Chandra S, Dilip Kumar HP, Gupta A, Bhandari PP, Rathod P (2016) Impact of dental neglect score on oral health among patients receiving fixed orthodontic treatment: a cross-sectional study. J Int Soc Prev Community Dent 6(2):120–124. https://doi.org/10.4103/2231-0762.178752

Paoletti I, Fusco A, Grimaldi E, Perillo L, Coretti L, Di Domenico M, Cozza V, Lucchese A, Contaldo M, Serpico R, Guida A, Donnarumma G (2013) Assessment of host defence mechanisms induced by candida species. Int J Immunopathol Pharmacol 26(3):663–672. https://doi.org/10.1177/039463201302600309

Paschaei N, Müller WD, Schmidt F, Hüsker K, von Baehr V, Pandis N, Jost-Brinkmann PG, Bartzela T (2024) Unveiling the role of metal ion concentration versus immune sensitization in orthodontic patients-a long-term prospective evaluation. J Clin Med 13(15). https://doi.org/10.3390/jcm13154545

Pathak JL, Yan Y, Zhang Q, Wang L, Ge L (2021) The role of oral microbiome in respiratory health and diseases. Respir Med 185:106475. https://doi.org/10.1016/j.rmed.2021.106475

Peng X, Cheng L, You Y, Tang C, Ren B, Li Y, Xu X, Zhou X (2022) Oral microbiota in human systematic diseases. Int J Oral Sci 14(1):14. https://doi.org/10.1038/s41368-022-00163-7

Peterson SN, Snesrud E, Liu J, Ong AC, Kilian M, Schork NJ, Bretz W (2013) The dental plaque microbiome in health and disease. PLoS One 8(3):e58487. https://doi.org/10.1371/journal.pone.0058487

Peterson BW, Tjakkes GH, Renkema AM, Manton DJ, Ren Y (2024) The oral microbiota and periodontal health in orthodontic patients. Periodontol 2000. https://doi.org/10.1111/prd.12594

Pisano M (2023) Oral dysbiosis and systemic diseases: a two-way relationship? Medicina (Kaunas) 59(11). https://doi.org/10.3390/medicina59111933

Ponitz RJ (1971) Invisible retainers. Am J Orthod 59(3):266–272. https://doi.org/10.1016/0002-9416(71)90099-6

Proffit WR, White RP, Sarver DM (2003) Contemporary treatment of dentofacial deformity. Mosby St, Louis

Putrino A, Barbato E, Galluccio G (2021) Clear aligners: between evolution and efficiency-a scoping review. Int J Environ Res Public Health 18(6). https://doi.org/10.3390/ijerph18062870

Raghavan S, Abu Alhaija ES, Duggal MS, Narasimhan S, Al-Maweri SA (2023) White spot lesions, plaque accumulation and salivary caries-associated bacteria in clear aligners compared to fixed orthodontic treatment. A systematic review and meta- analysis. BMC Oral Health 23(1):599. https://doi.org/10.1186/s12903-023-03257-8

Rinchuse DJ, Cozzani M (2015) Effectiveness and efficiency in clinical orthodontic practice. Int Orthod 13(4):507–524. https://doi.org/10.1016/j.ortho.2015.09.015

Rossini G, Parrini S, Castroflorio T, Deregibus A, Debernardi CL (2014) Periodontal health during clear aligners treatment: a systematic review. Eur J Orthod 37(5):539–543. https://doi.org/10.1093/ejo/cju083

Rosvall MD, Fields HW, Ziuchkovski J, Rosenstiel SF, Johnston WM (2009) Attractiveness, acceptability, and value of orthodontic appliances. Am J Orthod Dentofacial Orthop 135(3):276.e271–212. discussion 276–277. https://doi.org/10.1016/j.ajodo.2008.09.020

Rouzi M, Zhang X, Jiang Q, Long H, Lai W, Li X (2023) Impact of clear aligners on oral health and oral microbiome during orthodontic treatment. Int Dent J 73(5):603–611. https://doi.org/10.1016/j.identj.2023.03.012

Ruhl CM, Bellian KT, Van Meter BH, Hoard MA, Pham CD, Edlich RF (1994) Diagnosis, complications, and treatment of dentoskeletal malocclusion. Am J Emerg Med 12(1):98–104. https://doi.org/10.1016/0735-6757(94)90213-5

Sampaio-Maia B, Monteiro-Silva F (2014) Acquisition and maturation of oral microbiome throughout childhood: an update. Dent Res J (Isfahan) 11(3):291–301

Sedghi L, DiMassa V, Harrington A, Lynch SV, Kapila YL (2021) The oral microbiome: role of key organisms and complex networks in oral health and disease. Periodontol 87(1):107–131. https://doi.org/10.1111/prd.12393

Sfondrini MF, Butera A, Di Michele P, Luccisano C, Ottini B, Sangalli E, Gallo S, Pascadopoli M, Gandini P, Scribante A (2021) Microbiological changes during orthodontic aligner therapy: a prospective clinical trial. Appl Sci 11(15):6758. https://doi.org/10.3390/app11156758

Sharma N, Bhatia S, Sodhi AS, Batra N (2018) Oral microbiome and health. AIMS Microbiol 4(1):42–66. https://doi.org/10.3934/microbiol.2018.1.42

Shokeen B, Viloria E, Duong E, Rizvi M, Murillo G, Mullen J, Shi B, Dinis M, Li H, Tran NC, Lux R, Wu T (2022) The impact of fixed orthodontic appliances and clear aligners on the oral microbiome and the association with clinical parameters: a longitudinal comparative study. Am J Orthod Dentofacial Orthop 161(5):e475–e485. https://doi.org/10.1016/j.ajodo.2021.10.015

Shpack N, Greenstein RB, Gazit D, Sarig R, Vardimon AD (2014) Efficacy of three hygienic protocols in reducing biofilm adherence to removable thermoplastic appliance. Angle Orthod 84(1):161–170. https://doi.org/10.2319/012413-75.1

Shrivastava A, Mohanty P, Dash BP, Jena S, Sahoo N (2023) Proficiency of clear aligner therapy: a systematic review and meta-analysis. Cureus 15(9):e45072. https://doi.org/10.7759/cureus.45072

Smolyar N, Chukhray N, Lesitskiy M, Rybert Y, Musiy-Sementsiv K (2022) Assessment of oral hygiene maintenance in 12-18-year-old children and teenagers with fixed orthodontic appliances. Stomatologija 24(1): 21–25

Song Z, Fang S, Guo T, Wen Y, Liu Q, Jin Z (2023) Microbiome and metabolome associated with white spot lesions in patients treated with clear aligners. Front Cell Infect Microbiol 13:1119616. https://doi.org/10.3389/fcimb.2023.1119616

Sounah SA, Madfa AA (2020) Correlation between dental caries experience and the level of Streptococcus mutans and lactobacilli in saliva and carious teeth in a Yemeni adult population. BMC Res Notes 13(1):112. https://doi.org/10.1186/s13104-020-04960-3

Sundararaj D, Venkatachalapathy S, Tandon A, Pereira A (2015) Critical evaluation of incidence and prevalence of white spot lesions during fixed orthodontic appliance treatment: a meta-analysis. J Int Soc Prev Community Dent 5(6):433. https://doi.org/10.4103/2231-0762.167719

Talic NF (2011) Adverse effects of orthodontic treatment: a clinical perspective. Saudi Dent J 23(2):55–59. https://doi.org/10.1016/j.sdentj.2011.01.003

Velegraki A, Nicolatou O, Theodoridou M, Mostrou G, Legakis NJ (1999) Paediatric AIDS –related linear gingival erythema: a form of erythematous candidiasis? J Oral Pathol Med 28(4):178–182. https://doi.org/10.1111/j.1600-0714.1999.tb02020.x

Wade WG (2013) The oral microbiome in health and disease. Pharmacol Res 69(1):137–143. https://doi.org/10.1016/j.phrs.2012.11.006

Weir T (2017) Clear aligners in orthodontic treatment. Aust Dent J 62(S1):58–62. https://doi.org/10.1111/adj.12480

Worreth S, Bieger V, Rohr N, Astasov-Frauenhoffer M, Töpper T, Osmani B, Braissant O (2022) Cinnamaldehyde as antimicrobial in cellulose-based dental appliances. J Appl Microbiol 132(2): 1018–1024. https://doi.org/10.1111/jam.15283

Wu JL, Hou JX (2019) Influences on periodontal health and conditions of patients with periodontitis by clear aligner treatment. Zhonghua Kou Qiang Yi Xue Za Zhi 54(1):62–66. https://doi.org/10.3760/cma.j.issn.1002-0098.2019.01.012

Xu X, He J, Xue J, Wang Y, Li K, Zhang K, Guo Q, Liu X, Zhou Y, Cheng L, Li M, Li Y, Li Y, Shi W, Zhou X (2015) Oral cavity contains distinct niches with dynamic microbial communities. Environ Microbiol 17(3):699–710. https://doi.org/10.1111/1462-2920.12502

Zarco MF, Vess TJ, Ginsburg GS (2012) The oral microbiome in health and disease and the potential impact on personalized dental medicine. Oral Dis 18(2):109–120. https://doi.org/10.1111/j.1601-0825.2011.01851.x

Zhang M, Liu X, Xie Y, Zhang Q, Zhang W, Jiang X, Lin J (2020) Biological safe gold nanoparticle-modified dental aligner prevents the porphyromonas gingivalis biofilm formation. ACS Omega 5(30):18685–18692. https://doi.org/10.1021/acsomega.0c01532

Zhang H, Bi S, Zhang X (2025) Impact of clear aligners on gingivitis incidence and prevention strategies in adolescents and adults: a prospective observational study. BMC Oral Health 25(1):75. https://doi.org/10.1186/s12903-025-05439-y

Zhao R, Huang R, Long H, Li Y, Gao M, Lai W (2020) The dynamics of the oral microbiome and oral health among patients receiving clear aligner orthodontic treatment. Oral Dis 26(2):473–483. https://doi.org/10.1111/odi.13175

Part V

Therapeutic Strategies and Future Directions in Oral Immunology

Immunomodulation Strategies in Oral Diseases: From Microbiome Manipulation to Pro-resolution Therapies

30

Lina J. Suárez, Jessica Mera, Hernán Santiago Garzón, Roger M. Arce, João Gabriel S. Souza, and Jamil A. Shibli

Abstract

Host-response and inflammatory processes commonly triggered by dental biofilms play a crucial role in the pathogenesis pathways of prevalent oral diseases, such as periodontitis and implant-related infections, which have been widely explored by scientific evidence. The role of the oral microbiome and the interaction between microorganisms and immune components in the onset and development of such inflammatory conditions has been a primary focus for scientists and clinicians and has opened new possibilities for therapeutic targets to modulate disease progression. Such knowledge and new strategies with the aim of promoting microbial transition to a symbiotic state or modulating immune response have paved the way for innovative approaches to control, treat, and prevent oral diseases. This chapter presents the concepts of immunomodulation related to oral diseases, including the initial changes in resolving the inflammatory response (pro-resolution of inflammation) and how inflammatory processes can be modulated by immune modulatory agents in periodontal and peri-implant diseases.

Keywords

Periodontal diseases · Peri-implant diseases · Immune modulation · Resolvin E1 · Dental biofilm

L. J. Suárez
Dental Research Division, Department of Periodontology, Guarulhos University, Guarulhos, SP, Brazil

Facultad de Odontología, Departamento de Ciencias básicas y medicina oral, Universidad Nacional de Colombia, Bogotá, Colombia

J. Mera · H. S. Garzón
Facultad de Odontología, Departamento de Ciencias básicas y medicina oral, Universidad Nacional de Colombia, Bogotá, Colombia

R. M. Arce
Department of Periodontics and Dental Hygiene, School of Dentistry, University of Texas Health Science Center at Houston, Houston, TX, USA

J. G. S. Souza
Dental Research Division, Department of Periodontology, Guarulhos University, Guarulhos, SP, Brazil

Faculdade Israelita de Ciências da Saúde Albert Einstein, Hospital Israelita Albert Einstein, Sao Paulo, Brazil

J. A. Shibli (✉)
Dental Research Division, Department of Periodontology, Guarulhos University, Guarulhos, SP, Brazil

Department of Oral Medicine, Infection, and Immunity, Harvard School of Dental Medicine, Boston, US

Faculdade Israelita de Ciências da Saúde Albert Einstein, Hospital Israelita Albert Einstein, Sao Paulo, Brazil

Department of Oral Health Sciences, Periodontology, KU Leuven & Dentistry, University Hospitals Leuven, Leuven, Belgium
e-mail: jshibli@ung.br

N. Rezaei, H. Ziaei (eds.), *Oral Immunology*, Advances in Experimental Medicine and Biology 1492,
https://doi.org/10.1007/978-3-032-03176-1_30

30.1 Introduction

The significant increase in the prevalence of inflammatory diseases in recent decades has made them the central focus of numerous scientific investigations. According to the World Health Organization, inflammatory diseases such as cardiovascular disease account for the death of 17.9 million people annually, and these, together with cancer, respiratory diseases, and diabetes, account for more than 80% of early deaths caused by noncommunicable diseases. Notably, an estimated number of 0.1 billion cases of severe periodontitis occur worldwide, with higher prevalence rates observed in less developed countries (Chen et al. 2021).

Some inflammatory diseases, such as periodontitis, are caused by microbial dysbiosis, and failures in resolving and regulating the host response or controlling microbial accumulation often result in tissue damage. These failures in controlling inflammation may also mediate the link between localized infections and systemic involvement (Martínez-García and Hernández-Lemus 2021). Previous evidence suggests that impaired resolution mechanisms may compromise the prognosis of periodontal disease, while inflammatory processes may lead to the overgrowth of pathogenic microbial species (Sedghi et al. 2021). Therefore, understanding inflammatory diseases and the role of modulators of this process in controlling disease progression is crucial for developing new therapeutic strategies.

Different reasons that contribute to failures in controlling inflammatory processes have been identified, including those involving lipid mediators that are responsible for the resolution process and the restoration of tissue homeostasis (Panigrahy et al. 2021; Van Dyke and Sima 2020). In recent decades, evidence has been used to explore the use of pro-resolving substances as adjunct therapies for managing the inflammatory response, particularly in periodontal disease (Lee et al. 2016; Van Dyke 2017). Examining the systemic role of these substances derived from polyunsaturated fatty acids and their local effects is crucial for the treatment of periodontitis and related inflammatory diseases.

30.2 Host-Response, Inflammation, and Immune Modulation: Key Mediators of the Process

The inflammatory process is a natural response of the body to injury and is crucial for the physiological response to different types of damage. The body needs to manage the inflammatory process in its initial stages to prevent it from becoming chronic. Persistent infections caused by microbial biofilms can induce chronic inflammation, thereby impacting different biological systems and leading to systemic diseases such as diabetes, cardiovascular diseases, obesity, cancer, rheumatoid arthritis, and intestinal disorders. (Kotas and Medzhitov 2015).

Inflammation is a complex and highly regulated process characterized by interactions between soluble immunological factors and cells that can arise from traumatic responses, infections, toxic or autoimmune damage, or postischemic responses (Nathan 2002). During the acute inflammatory response phase, systemic immune cells migrate to the site of damage, guided by cytokines, chemokines, and acute-phase proteins (Germolec et al. 2018). The acute inflammatory response occurs with the appearance of polymorphonuclear cells, mainly neutrophils and proinflammatory lipid mediators (e.g., prostaglandins and leukotrienes). However, the ultimate stage of the inflammatory response is to restore homeostasis by resolving inflammation, thus preventing chronic inflammation and avoiding additional damage. These mechanisms of inflammation resolution are essential for maintaining tissue integrity and function.

Resolution can be defined histologically as the interval between the peak infiltration of neutrophils and their subsequent disappearance from the tissues (Serhan et al. 2011). Recently, natural resolution mechanisms have gained significant attention due to the recognition of inflammation as a central contributor in the pathogenesis of many diseases, including atherosclerosis, cancer, asthma, neurodegenerative disorders, and periodontitis (Kinney et al. 2018; Madaudo et al. 2024; Singh et al. 2019; Van Dyke 2017). The resolution of inflammatory infiltrates was initially considered to be a passive process, in which chemotactic stimuli and gradients gradually dissipate, allowing tissues to drain, repair, and return to normal function. However, it has been recognized that the resolution of inflammation in different microenvironments involves active biochemical and metabolic processes (Zhang et al. 2024). The number of infiltrating neutrophils decreases during the resolution of inflammation, triggering the secretion of a different set of anti-inflammatory and pro-resolving specialized lipid mediators, such as resolvins, lipoxins, protectins, and maresins. Thus, the resolution of inflammation occurs in very early stages through cellular pathways that actively biosynthesize specialized lipid mediators that function to promote the clearance of leukocytes and cellular debris from affected tissues and facilitate the restoration of homeostatic balance (Herrera et al. 2015; Panigrahy et al. 2021; Winkler et al. 2016).

Lipoxins were recognized as being the first mediators to have dual anti-inflammatory and pro-resolving activity. Other pro-resolving agents, such as resolvins, also exhibit strong dual-pathway potential, by limiting tissue damage through reducing vascular permeability and inhibiting the recruitment of polymorphonuclear cells (PMNs) (Serhan 2009). Resolvins also promote the recruitment of monocytes, activate the phagocytosis of apoptotic cells (efferocytosis), and inhibit the activation of transcription factor NF-kB without suppressing the overall immune function (Ariel and Serhan 2007). Studies have proposed that pro-resolving lipid mediators could stimulate lymphatic drainage of leukocytes. Interestingly, the signaling pathways that initially induce the production of prostaglandins (PGE2 and PGD2) to trigger inflammation can also actively switch to drive the production of lipid mediators, leading to a regulated transition from pro-inflammatory to pro-resolving (Serhan 2009).

These specialized pro-resolving mediators (SPM) that stimulate inflammation resolution can be derived from various fatty acids. These include arachidonic acid (AA) derivatives such as lipoxins, eicosapentaenoic acid (EPA) derivatives such as E-series resolvins, or docosahexaenoic acid (DHA) derivatives such as D-series resolvins, maresins, and protectins (Perez-Hernandez et al. 2021). In this way, physiological inflammation regulates its own resolution and promotes tissue homeostasis. SPMs act on different receptors distributed in various tissues. Some of these receptors mediate dual functions. Resolvin E1 (RvE1) acts through the BLT-1 receptor in neutrophils and ERV-1/ChemR23 receptor in monocytes and macrophages. RvE1 acts on these cells by decreasing the migration of PMNs and attenuating pro-inflammatory signaling. In macrophages, RvE1 stimulates the phagocytosis of apoptotic PMNs, while in dendritic cells, it acts by inhibiting their migration, the release of IL-12, and downregulating the expression of NF-kb (Fig. 30.1) (Perez-Hernandez et al. 2021).

Resolvin D1 (RvD1), a resolvin that plays a role in restoring tissue homeostasis, acts via the G protein-coupled receptor 32 (GPR32) and formyl peptide receptor type 2 (ALX/FPR2), both of which are expressed on phagocytes. As a result of these interactions, RvD1 promotes the inflammation resolution by the inhibition of leukocyte migration. It is also responsible for the regulation of microRNAs and their target genes, increasing spherocytosis (phagocytosis of apoptotic macrophages) and the polarization of M2 macrophages (Zarrough et al. 2023). RvD1 also has a direct action on fibroblasts, inhibiting the production of PGE2, and pro-inflammatory cytokines such as IL-1β and TNF-α, which leads to an increase in the release of fibroblast growth factors (Zarrough et al. 2023).

The protective role of RvD1 in cells such as odontoblasts has also been described. By inhibiting PGE2 production, as mentioned above, RvD1 acts directly on the RANKL/OPG40 receptor, which is related to alveolar bone resorption. Accordingly, RvD1 is considered a key pro-resolving lipid mediator in wound healing and tissue regeneration, including bone repair (Hasturk et al. 2007). Resolvin D2 (RvD2) is another type of specialized pro-resolving mediator of the D series, derived from docosahexaenoic acid. It is a potent regulator of bacterial sepsis through its G protein-coupled receptor for RvD2 (DRV2/GPR18); this

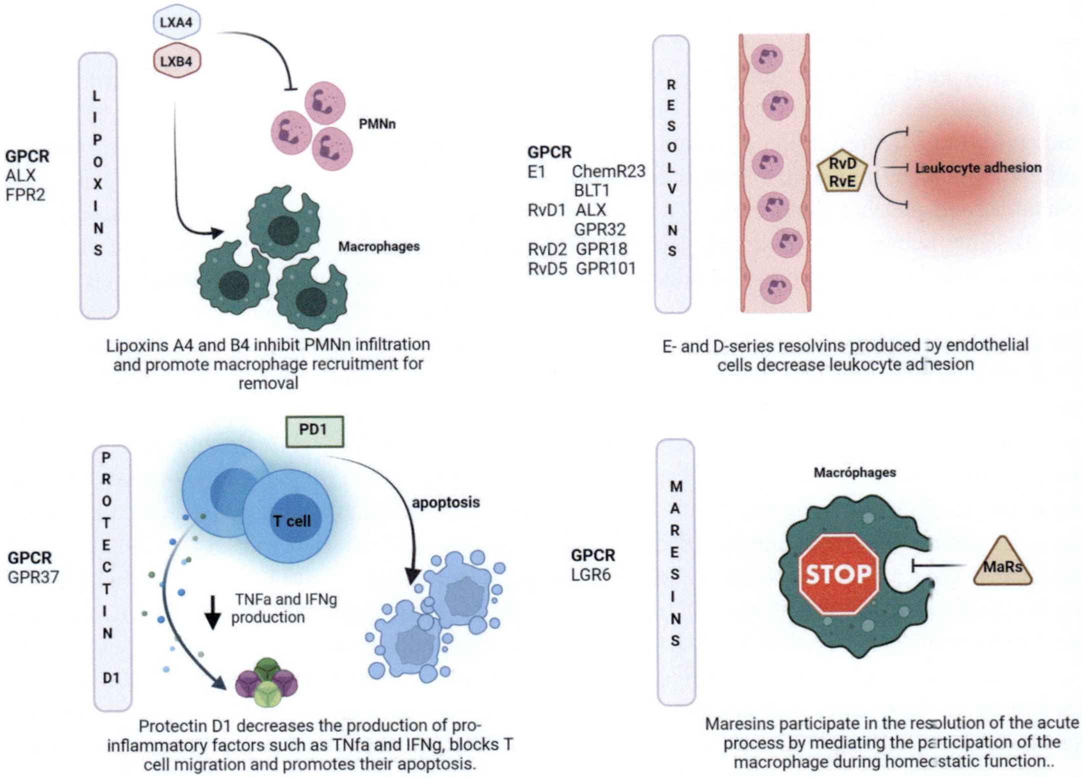

Fig. 30.1 Specialized pro-resolving mediator (SPM) functions and receptors. The ultimate goal of inflammation is to resolve the process, heal, and return homeostasis to the system. This happens through the production of specialized pro-resolving mediators (SPMs), including lipoxins, protectins, resolvins, and maresins that act on different cell groups through G protein-coupled receptors (GPCRs). Maresins participate in the resolution of the acute process by mediating the participation of macrophages during homeostatic function

binding allows the release of cAMP and the phosphorylation of specific kinases and transcription factors, essential for macrophage phagocytosis (Chiang and Serha 2020). Studies have also described that RvD2 suppresses tumor growth by improving the elimination of tumor cell debris. Similar to RvD1, RvD2 also promotes tissue protection, repair, and regeneration, in a cell- or organ-specific manner (for each entity involved).

Resolvin D3 (RvD3) and Resolvin D4 (RvD4) are also potent regulators of inflammation; they act by increasing the counter-regulation of NF-kB and stimulating efferocytosis. Studies have also described that their action decreases the release of neutrophil extracellular traps (Nets) by PMNs. Resolvin D5 (RvD5), derived from n-3 docosapentaenoic acid (DPA), was the first lipid mediator identified to regulate nociceptive signaling pathways, highlighting its potential as an innovative factor in pain management and pharmacotherapeutics (Chiang and Serha 2020). Maresins (MaR1), which are macrophage mediators in the resolution of inflammation, are one of the large families that are part of the specialized lipid mediators derived from DHA. Their action is mediated through the LGR6 receptor and, like other pro-resolving mediators, they promote efferocytosis and support wound healing (Bennett and Gilroy 2016). MaR1 has also been shown to be effective in the management of post-operative pain and promoting bone regeneration (Duvall and Levy 2016). Protectin D1 (PD1), also known as neuroprotectin D1, is also biosynthesized from DHA via a 15-LOX-initiated pathway in several human cell types, murine exudates, and brain tissues. It binds to the GPR37 receptor, exerting potent immunoregulatory effects including the inhibition of neutrophil migration and Toll-like receptor-mediated activation, suppression of Th2 inflammatory cytokines and pro-inflammatory lipid mediators, and upregulation of C-C chemokine receptor type 5 (CCR5) in PMNs. PD1 also blocks T cell migration in vivo and promotes apoptosis (Bennett and Gilroy 2016).

In the context of inflammation, multiple inflammatory diseases triggered by different stimuli are associated with dysregulated inflammatory processes mediated by impaired regulation of both innate and adaptive immune responses. The pathogenesis of diabetes, metabolic syndrome, arthritis, and cardiovascular diseases—all of which are linked to periodontitis via inflammatory processes—among other systemic conditions has been shown to involve such dysregulation (Hajishengallis and Chavakis 2021). As a result, these conditions might be responsive to therapeutic interventions through immunomodulation, which could help restore the dysregulated immune balance and modify the underlying inflammatory dysfunction.

Immunomodulation has been described as a process that involves the induction of changes in the immune system by either activating (immunostimulants) or suppressing (immunosuppressants) certain immune factors. Thus, immunomodulation promotes the "physiologically guided" resolution of pathological processes by influencing the relationship between microorganisms and cells of the immune system with the aim of reducing tissue damage. Thus, new therapeutic approaches have been developed and tested to treat such diseases by modulating the immune system and minimizing the reliance on systemic antibiotics.

The role of microorganisms in maintaining health is well known (Ataollahi Eshkoor and Fanijavadi 2024). However, different environmental factors may lead to the overgrowth of pathogens, disrupting the symbiotic state between the microbiome and host immune responses and triggering inflammatory conditions. In this context, both direct and indirect benefits of immunomodulatory therapies come into play, either due to their benefits in changing the inflammatory microenvironment or due to their effects on microorganisms. These effects are not limited to antibacterial actions, but they also have an impact on other groups of microorganisms, such as fungi and viruses, thereby making these therapies broad-spectrum agents with comprehensive biological activity.

In this context, different immunomodulators have been explored, including those with the aim of amplifying the innate immune response by expanding different effector cell populations in both arms of immunity. For the innate immune response, various approaches have been explored including the use of Toll-like receptor (TLR) agonists (Wang et al. 2018, 2020), growth factors (Ding et al. 2021), and therapies that increase the lifespan of certain cell populations by controlling apoptosis (Kayagaki et al. 2024). Other approaches in this category include ex vivo modification and reimplantation of host cells (Laskowski et al. 2022), commonly used in antitumor therapies, the use of synthetic "non-host" cells, and the regulation of activation and trafficking of early response cells, such as PMNs, which are critical for sepsis control (Lerman and Kim 2015). Moreover, other alternatives have been proposed for stimulating the resolution of inflammation (Djuricic and Calder 2021; Porcheray et al. 2005; Schmid et al. 2016). In the context of acquired immune responses, vaccination (Lynn et al. 2022) and antibody therapies (Saunders 2019) have been proposed. However, the effective implementation of these strategies requires a more detailed understanding of host–microbiota interactions, especially about regulatory mechanisms involved, to minimize the potential side effects of these types of modulators.

Furthermore, for nearly two decades, Workshops on Novel Antimicrobial Therapeutics have recommended the combination of antimicrobials and immunomodulators. In support of this concept, certain antibiotics have even been designed to be selectively activated through interaction with damage-associated signaling components of the immune system, thus serving a dual role in antimicrobial activity and immunomodulation.

30.3 Microbiome in the Context of Health and Disease

The advancements in metagenomic techniques have enabled a deeper understanding of microbial communities associated with both health and disease states. Although a diverse array of microorganisms inhabit the human body and perform essential functions in maintaining biological systems, these same organisms can also contribute to biofilm-associated diseases (Proctor 2016; Proctor et al. 2019). Like their host, microorganisms must adapt to constantly changing conditions while evolving and competing with each other. This strong host–microbiota correlation has motivated scientists to investigate how modulating the inflammatory response can benefit the host, with the main aim of influencing microbial communities without disrupting those that are beneficial to various organs of the human body. Such approaches hold promise for reducing susceptibility to multiple diseases and enhancing the efficacy of oncologic therapies (Fessler et al. 2019; Zhou et al. 2021), including conditions ranging from inflammatory bowel disease (Schirmer et al. 2019) to autoimmunity (Kim 2023).

From this perspective, it seems more practical to focus on controlling the microenvironment to support adaptation of microbial communities rather than developing antimicrobial agents that are not capable of distinguishing between commensals and "pathogens," and thereby affect non-pathogenic microorganisms. Therefore, a deeper understanding of intimate relationships between the microbiota and host should guide the development of therapeutic alternatives for treating diseases involving this type of interrelationship.

Although the role of host–microbiome interactions is well known in disease pathogenesis and targeting these interactions is a successful therapeutic alternative, direct microbiome modulation has also been a widely explored approach used as a strategy for managing inflammatory periodontal diseases of infectious origin. The composition of the oral microbiome is determined by factors such as the primary sources of nutrients (saliva and gingival fluid), immune response within the gingival sulcus microniche, and growth limitations due to daily mechanical cleaning, all of which contribute to the resilience of the microbial community (Wade 2021). Multiple factors have been associated with alterations in the microbiome, such as excessive consumption of fermentable carbohydrates reduced salivary flow, and the presence of gingival inflammation (Wade 2021). This opens up possibilities for new therapeutic alternatives (Iaconis et al. 2024), among which the most recognized approaches have been the use of pre- and pro-biotics (Teughels et al. 2013; Van Holm et al. 2023), nitrate-rich diets (Baab and Weinstein 1986; Qu et al. 2016), and strategies aimed at inhibiting bacterial adhesion and interfering with quorum sensing (Liu et al. 2018; Zhang et al. 2023).

30.3.1 Immune Modulation and the Microbiome

The oral cavity is constantly exposed to various stimuli, including diverse microorganisms. The expanded human oral microbiome database accounts for more than 770 species of prokaryotes, with bacteria being the most abundant (Verma et al. 2018). The high microbial diversity enhances the oral ecosystem's ability to resist and adapt to environmental changes and to be highly selective during colonization (Li et al. 2016). There is a dynamic bidirectional relationship between the microbiome and immune system, in which the host supports the symbiotic balance, while microorganisms "train" the immune system by modulating the induction of immune responses at the mucosal level through different mechanisms. These mechanisms include the secretion of IgA, antimicrobial peptides, and anti-inflammatory cytokines. Altogether, these processes minimize the contact between bacteria and epithelial cells, thereby maintaining the immune balance (Fomby and Cherlin 2011; Gagnière et al. 2016). Therefore, in a symbiotic state, the oral microbiome exists in equilibrium with the host and even performs beneficial functions. However, these microbial species can adhere to both biotic (e.g., oral tissues) and abiotic (e.g., dental materials and prosthetic devices) surfaces in the oral environment, where they accumulate to form biofilms. Moreover, various factors can disrupt the symbiotic state, leading to overgrowth of specific microbial species and increasing their potential to trigger biofilm-induced diseases.

Microbial communities comprising the human microbiome are formed through adaptive processes mediated by complex interactions among the microbes themselves, the environment, and the host immune system (van der Meulen et al. 2016). True commensals are microorganisms that have adapted to coexist in the human body by evolution and modulated their response without producing alterations in homeostasis, while others may be transient pathogens or microorganisms that acquire pathogenic characteristics (pathobionts) (Felix et al. 2019; Ivanov and Honda 2012).

In general, symbiosis in the microbiota of the oral cavity is determined by the predominance of facultative microorganisms, such as *Streptococci* and *Actinomyces*, whereas an increase in anaerobic microorganisms, including *phyla Firmicutes*, *Proteobacteria*, *Spirochaetes,* and *Bacteroidetes*, is indicative of dysbiotic communities. The changes mainly involve the structure of the microbial community rather than in its individual members, meaning that bacteria historically related to disease can also be present in healthy conditions, and vice versa, but in different proportions (Abusleme et al. 2013). These variations are mainly influenced by factors such as environment, genetics, age, and lifestyle (Hajishengallis and Lamont 2012; Hall et al. 2017). Dysbacteriosis, referring to changes in microbial communities due to environmental changes, is commonly known as dysbiosis and is often associated with the development of disease.

The organism must maintain an effective defense at the epithelial level of the oral cavity, including the oral, sulcular, and junctional epithelium, to ensure the preservation of balance essential for maintaining health. The expression of receptors in different structural cells plays an important role in this function, with particular interest in the expression of TLRs and Nucleotide oligomerization domain (NOD) receptors, as well as the regulation of their expression in different conditions and microenvironments (Groeger and Meyle 2019). For example, in the oral epithelium, the regulation of these receptors supports a state of tolerance, enabling controlled responses to stimuli recognized as non-harmful, or triggering an increase in β-defensin production in response to potentially harmful signals (Uehara and Takada 2008; Uehara et al. 2001, 2005, 2007).

Regarding the initiation of responses in the presence of bacteria in the subgingival microenvironment, the key role of innate immune cells, particularly polymorphonuclear neutrophils and macrophages in early-stage microbial control, has been well demonstrated. However, macrophages, the main effector cells of innate immunity, also play crucial roles in the initiation of adaptive responses by functioning as antigen-presenting cells, making them essential for maintaining immune system homeostasis. Thus, M1 macrophages are primarily responsible for "destroying pathogens" mainly through phagocytosis and the production of pro-inflammatory cytokines, such as IL-1β, IL-6, IL-12, IL-23, and TNF-α. In contrast, M2 macrophages play key immunoregulatory and anti-inflammatory roles, which mainly mediate processes of tissue repair and regeneration in response to injury. These functions are largely mediated by the production of cytokines such as IL-10 and TGF-β, which help modulate immune responses to that pathway (Shapouri-Moghaddam et al. 2018); however, the balance of these subpopulations is highly variable and can be significantly influenced by changes in the microenvironment, including those associated with dysbiosis.

The microbiota influences not only the host's overall immune response but also the innate immune response. The regulation of both local immune response, mediated by antigens present in tissues, and the systemic responses generated in the thymus, depends largely on regulatory T

(Treg) cells (Tregs, FOXP3+). The induction of these cells has been described as a potential mechanism that protects against unnecessary immune responses targeting commensal microorganisms (Pandiyan et al. 2019). Furthermore, both dendritic cells and other cell types release extracellular vesicles that are recognized as an early mechanism involved in the modulation of the immune response (Sáez et al. 2019). It has been shown that exosomes isolated from infected dendritic cells can reprogram immune cells responsible for experimental alveolar bone loss in vivo (Elashiry et al. 2020). Moreover, in the mucosa, the microbiota influences the phenotypes and functions of T and B cells. This interaction is best studied at the intestinal level, where the function of these cells is critical in maintaining immune homeostasis and supporting the epithelial barrier integrity. However, similar mechanisms are likely to occur in all epithelial barriers (Honda and Littman 2016).

Given the role of commensal microbiota in the development of immunity ((Fagundes et al. 2012; Sommer and Bäckhed 2013; Stappenbeck et al. 2002) and the regulation of the immune response against pathogens, the importance of immunomodulation must be recognized as a possible approach for treating dysbiotic conditions associated with inflammatory diseases.

30.4 Immune Modulation Through Resolution of Inflammation in the Treatment of Oral Diseases and Conditions

Chronic inflammation is characterized by the production of inflammatory cytokines, arachidonic acid (AA)-derived eicosanoids (prostaglandins, thromboxanes, and leukotrienes), reactive oxygen species, and changes in adhesion molecules. In periodontal disease, as in all inflammatory diseases, AA-derived eicosanoids play a key role in the initiation and pathogenesis of inflammatory lesions (Kohli and Levy 2009).

As previously described, in the presence of acute inflammation, many lipid mediators are responsible for mediating the inflammatory processes and actively leading to the resolution of the disease by generating a change in the pathogenesis of chronic inflammation. As in all other inflammatory diseases, AA plays a key role in the initiation and pathogenesis of inflammation in periodontitis, where lipid mediators contribute to both early pathophysiological events and the chronic nature of the disease (Engebretson et al. 2013). Periodontitis often develops from gingivitis, which is a reversible superficial inflammation. In this pathology, regeneration processes are activated, enabling complete replacement with new cells usually occurring without permanent tissue damage. However, in periodontitis, these repair mechanisms are impaired, resulting in irreversible scarring, fibrosis, and destruction of the cellular matrix, which usually requires lifelong treatment. Inflammation is central to the pathology of periodontal diseases, and a thorough understanding of its mechanism is key to developing new treatment strategies that can promote wound healing and restore tissue function (Ali et al. 2021).

Initial investigations into the role of pro-resolving mediators revealed a high production of lipoxin 4A (LXA4) by neutrophils in patients with periodontitis, compared to healthy individuals. Pro-inflammatory mediators such as LTB4 and PGE2 show a dramatic increase in periodontal tissues and gingival crevicular fluid as part of the early response to injury. The production of LXA4 leads to the downregulation of pro-inflammatory cytokines and affects neutrophil activity, often resulting in a state of hyperactivity. Despite this discovery, lipoxin levels were insufficient, and deficiencies in receptor binding were observed, preventing the resolution of periodontal disease. This finding was the starting point for further studies, investigating the protective role of pro-resolving mediators in periodontal disease (Van Dyke 2017).

The literature describes specific actions that SPM triggers, which vary depending on the cell type and the inflammatory process being targeted. Key mechanisms for deactivating inflammation include eliminating the inciting stimulus, inhibiting leukocyte trafficking, catabolism of pro-inflammatory mediators, phagocytosis of apoptotic cells, and eliminating the lesion. However, one or more of these pro-resolving pathways mentioned above may be insufficiently active, allowing the acute inflammatory response to persist and result in chronic pathologies (Van Dyke 2020).

In the case of lipoxins, their dysregulation has been associated with more severe forms of periodontal disease, such as aggressive or juvenile periodontitis. In previous research, it was found that patients with more severe stages of periodontitis showed decreased production of LXA4. This reduction was associated with a higher expression of superficial P-selectin in platelets and increased expression of CD18 in circulating neutrophils and monocytes, leading to the formation of platelet-neutrophil and platelet-monocyte aggregates. Defective phagocytosis of opsonized particles by macrophages was also observed, causing a failure in local resolution, which led to the progression of periodontal disease (Van Dyke 2017).

Another lipid mediator that has been widely studied in periodontal literature is RvE1. The establishment of periodontal disease is mediated by leukocyte recruitment and diapedesis of PMNs to inflamed periodontal tissues. Resolution of inflammation involves a class switch in the lipid mediators, transitioning from pro-inflammatory prostaglandins like PGE2 and PGD2 to pro-resolving mediators such as RvE1 in this case. RvE1 acts as an agonist depending on its receptor. As a result of this binding, PMN infiltration and the production of reactive oxygen species (ROS) are inhibited, improving apoptosis and spherocytosis, which ultimately leads to the resolution of inflammation. Despite this, studies have shown that failure in the resolution of inflammation is due to the persistent recruitment of PMNs to periodontal tissues, driven by the presence of anaerobic pathogens. Therefore, this persistent subgingival microbial challenge impairs RvE1's ability to effectively resolve the chronic inflammatory response (Van Dyke 2020).

Bone resorption is another major consequence of periodontal disease. At the cellular level, the failure to obtain resolution of inflammation results in the expansion of the inflammatory infiltrate into the alveolar bone. This process is characterized by a decrease in the synthesis of osteoprotegerin and an increase in the expression of RANK/RANKL, promoting bone loss. This is induced by the production of pro-inflammatory cytokines and chemokines, which stimulate the differentiation of pre-osteoclasts into mature osteoclasts, favoring bone loss promoted by local inflammation. In the case of RvE1, in vitro studies have demonstrated its action as a negative regulator in the differentiation of osteoclasts, induced by RANKL. This regulatory effect is mediated by binding to its receptor BLT1, which modulates inflammatory responses. Furthermore, by interacting with chemR23 receptor expressed in odontoblasts, RvE1 increases the expression of osteoprotegerin, a key protein inhibiting osteoclast activity, thereby limiting bone resorption (Balta et al. 2017).

By analyzing samples of gingival tissue and subgingival plaque in periodontally healthy subjects and those with periodontal disease, failures in the resolution of inflammation were associated with an imbalance between the levels of lipid mediators and the gene expression of their receptors, which led to an increase in the production of pro-inflammatory mediators and consequently an increase in periodontal inflammation. These changes in the oral microbiome include the presence of pro-inflammatory periodontal pathogens, such as those of the red complex: *Porphyromonas gingivalis, Tannerella forsythia and Prevotella intermedia.* This process triggers a deficiency in the expression of G protein-associated receptors, such as ChemR23 (RvE1), thereby preventing the transition from a pro-inflammatory immune profile to a pro-resolving type, contributing to the chronicity of the disease (Van Dyke 2017).

As previously mentioned, different types of SMP have been shown to control inflammatory diseases. However, in specific cases of periodontitis, the production and levels of these mediators present in the resolution process are associated with the inflammatory state of the disease, which is influenced by changes in the taxonomic composition of the subgingival microbiota. In other words, the profiles of these lipid mediators and their receptor genes are associated with the severity of inflammation (Lee et al. 2021).

It is noteworthy that in patients with both coronary artery disease and periodontitis, who also have a significant decrease in body mass index, improvements were observed in periodontal clinical parameters, leukocyte counts, LDL, protectin 1, and resolvin D1 levels at 6 months following medical and periodontal treatment. Moreover, there was a significant decrease in total cholesterol, HDL, resolvin E1, and marensin 1 at both 3- and 6-months post-treatment compared to baseline values ($p < 0.05$) (Güvenç et al. 2023). At the salivary level, the levels of protectins and marensins of patients with cardiovascular disease with gingivitis or periodontitis have been evaluated. The levels of protectins were elevated in the gingivitis group, while marensin levels were elevated in the group of patients with cardiovascular disease and periodontitis (Önal et al. 2021).

It is clear that in periodontitis, as in other chronic diseases, inflammation remains unresolved, and the activity of SMPs is reduced. To address this imbalance and compensate for the altered metabolism and the function of these mediators, several in vitro studies have been conducted, offering more promising approaches for the management of the disease.

30.4.1 Immunomodulation and Treatment of Periodontitis

As described throughout this text, the inability to restore the balanced immune environment at both local and systemic level is determined by sustained production of pro-inflammatory mediators, failures in different regulatory mechanisms, such as impaired apoptosis and defective phagocytosis, and the persistence of cells that perpetuate tissue damage. In local conditions like periodontitis, this pattern is no different. Therefore, an effective therapeutic approach may involve a combination of mechanical therapies and specialized lipid mediators, which together can lead to an improvement in outcomes and the resolution of inflammation both locally and systemically.

Over the course of more than two decades, the relationship between the host and the microbiota in the pathogenesis of periodontitis has evolved. It is now widely accepted that the presence of gram-negative microorganisms residing in subgingival plaque, under dysbiotic conditions and linked to an imbalance in the host immune response, triggers periodontal disease. A large population of these microorganisms residing in deep pockets can reach the peripheral circulation (bacteremia) and are capable of reaching remote sites, in addition to triggering low-grade systemic inflammation (Castillo et al. 2011). Given the inflammatory nature of periodontitis, SPMs have been studied for their potential in both the prevention and treatment of the disease. There is evidence that pro-resolving agents (series E and D) may help prevent the progression of periodontitis and promote bone regeneration (Osorio Parra et al. 2019). In vitro studies using stem cells have shown the potential to enhance the production of LxA4, which in turn causes an increase in RvD levels and a decrease in the secretion of PGE2, suggesting that LxA4 can possibly stimulate the production of lipid mediators by stem cells and their migration to the site of injury, thereby accelerating wound closure (Cianci et al. 2016).

One of the earliest and most important studies on the use of SPMs in the treatment of periodontitis was published almost two decades ago. In this experiment, periodontal disease was induced by ligatures and *P. gingivalis* infection in rabbits. These results demonstrated that the topical application of Resolvin E1 (RvE1) prevented the progression of the disease, in addition to promoting regeneration. The infection induced by *P. gingivalis* led to the overgrowth of gram-negative pathogens, which was effectively counteracted by RvE1. RvE1 reversed the biofilm-associated changes by eliminating *P. gingivalis* and promoted the resolution of dysbiosis. Moreover, RvE1 significantly reduced systemic inflammatory markers, such as C-reactive protein and IL-1B (Hasturk et al. 2007).

Later, studies in rabbits demonstrated that the overexpression of 15-lipoxygenase type I increases endogenous levels of lipoxin A4, protecting the host from the development of periodontal disease. Likewise, topical application of RvE1 decreased bone loss by 95% mainly by reducing neutrophil infiltration in periodontal tissues. Thus, it has been described that the combination of LxA4 and RvE1 contributed to a decrease in the production of pro-inflammatory mediators such as tumor necrosis factor-alpha and superoxide ions (Freire et al. 2017; Freire and Van Dyke 2013).

In murine models, it has been described that there are pro-resolving substances in the presence of periodontal disease. However, due to previously mentioned system failures, the granules secreted by neutrophils that activate the lipoxygenase pathways are either insufficient or unable to bind effectively to their receptors. Furthermore, mesenchymal stem cells treated with lipoxins and resolvins are resistant to apoptosis by the activation of the PI3K-Akt signaling pathway. Bone cells, like polymorphonuclear cells, are also affected, particularly osteoclasts, which express receptors for RvE1 such as ERV1 and BLT1. RvE1 inhibits osteoclastic differentiation by reducing the number of multinucleated cells and RANKL. This effect is associated with the translocation of the p50 subunit of the NF-kB factor without altering the apoptosis of surviving osteoclasts (Van Dyke 2017).

In the Hanford miniature pig, nano pro-resolving medications containing a stable analog of lipoxins such as benzo-lipoxin A4 (bLXA4) have been used. These treatments have been shown to reduce the inflammatory infiltrates in periodontitis lesions, while promoting bone and soft tissue regeneration. These effects were due to the stimulation of ALX/FPR2, the receptor for bLXA4. This treatment also had an effect on the regulation of systemic pro-resolving mediators, in addition to promoting the production of eicosanoids that promote bone formation (Van Dyke et al. 2015).

The role of resolvin A4 has also been evaluated as a possible target for the treatment of periodontitis. In a group of 21 patients with severe periodontitis, the levels of this resolvin in the gingival crevicular fluid were quantified before and after periodontal treatment. It was shown that these levels increased significantly ($p < 0.05$) after mechanical treatment, but they remained lower than those observed in samples from healthy individuals. Sites with higher levels of resolvin A4 after treatment showed greater reductions in probing depth, especially those that tested negative for *P. intermedia* or *Tannerella forsythia* (Ma et al. 2024).

In periodontitis, the influence of resolvin D2 (a product of docosahexaenoic acid) in the regulation of inflammation and the prevention of bone loss has also been demonstrated. In a mouse model of *P. gingivalis* infection, RvD2 was found to decrease the production of IFN-γ by CD4+ T cells and increase circulating neutrophil levels, while significantly reducing circulating neutrophil levels in tissues. It also decreases circulating monocytes and increases M2 macrophages in the gingival tissues. The mRNA levels of IFN-γ, IL-1β, and TNF-α were decreased in the gingival tissue after treatment with RvD2, while IL-10 expression was increased. RvD2 reduces the RANKL/OPG ratio in gingival tissues, thereby favoring bone preservation. The primary function of RvD2 is the regulation of excessive innate inflammatory responses by restricting adaptive Th1 responses at both systemic and gingival levels, which have been associated with the mediation of bone loss, and most interestingly, this modulation does not inhibit protective immunity (Mizraji et al. 2018).

In patients with periodontitis and systemic involvement, the use of SPMs or therapeutic strategies that increase their production is an emerging approach being explored. For instance, when using low-dose aspirin and omega-3 fatty acids as adjuvants to non-surgical periodontal therapy in patients with chronic periodontitis and diabetes, a statistically significant decrease in all periodontal parameters was seen, indicative of inflammation. Improvement was observed in both the control group (application of 75 mg aspirin 3 months after periodontal therapy) and the experimental group (500 mg of omega-3 fatty acids 3 months after periodontal therapy), with the response being significantly greater in the experimental group. This demonstrates that omega-3 derivatives, when used after non-surgical periodontal therapy, are effective adjuvant for managing and controlling periodontal disease (Rampally et al. 2019).

In a recent clinical trial, 30 patients with stage III or IV periodontitis underwent mechanical treatment. The test group ($n = 16$) additionally received a supplement of 2.6 g EPA and 1.8 g DHA daily for 3 months. There was a significant reduction in bleeding on probing and an improvement in clinical attachment loss in the experimental group compared to the control group. Levels of pro-inflammatory cytokines such as IL-8 and IL-17 were markedly lower, while the level of anti-inflammatory IL-10 was significantly higher in saliva samples from the omega-3 supplemented group (Stańdo et al. 2020).

Epoxyeicosatrienoic acids are short-lived lipid mediators involved in the resolution of inflammation. Their efficacy is limited by their rapid metabolism by soluble epoxide hydrolase (sEH). In a mouse periodontitis model study, sEH was pharmacologically inhibited to assess its effect on T cell activation and systemic SPM production. Mice received an sEH inhibitor starting before and continuing for 14 days after ligature placement. In parallel, gingival tissue samples from human periodontitis patients were analyzed for sEH levels. sEH inhibition prevented bone resorption and preserved bone architecture in mice. In addition, it increased the production of several SPMs, including lipoxin A4, lipoxin B4, resolvin E2, and resolvin D6. An increase in the ratio of regulatory T cells to Th17 cells was observed. In human samples, higher levels and increased expression of sEH were found in inflamed gingival tissues compared to healthy gingiva, with a positive correlation with the severity of periodontitis (Abdalla et al. 2024).

A similar effect was demonstrated in a clinical trial involving 40 smoking patients with periodontitis. The test group received 640 mg EPA and 480 mg DHA daily for 3 months, while the control group only received mechanical therapy. The results showed that the test group had significantly lower levels of clinical attachment loss, gingival index, and BOP ($p < 0.05$), as well as a greater reduction in the periodontal pocket depths of less than or equal to 7 mm (Savran and Sağlam 2024). However, other studies have shown that although there was a significant difference between the groups during the first 3 months, the 6-month follow-up showed a significant clinical improvement only in BOP for the group that consumed omega-3, but the number of periodontopathogenic bacteria at that time was significantly lower in this group (Stańdo-Retecka et al. 2023).

The publication of two systematic reviews with meta-analysis demonstrated that the use of omega-3 fatty acids as a complement to non-surgical periodontal therapy for at least 3 months, significantly contributed to improvement in clinical variables compared to conventional mechanical therapy. In a previous study by Castro dos Santos et al. (2022), a meta-analysis of eight randomized controlled clinical trials demonstrated a significant reduction in probing depth and a gain in clinical attachment levels associated with the use of omega-3 fatty acids. The pooled estimates revealed a significant overall reduction in probing depth of 0.42 mm (95% CI: 0.15; 0.68) and a gain in attachment levels of 0.58 mm (95% CI: 0.24; 0.92) (Castro dos Santos et al. 2022). The meta-analysis by Van Ravensteijn et al. (2022) demonstrated that the complementary use of fatty acids resulted in an additional reduction in probing depth of 0.39 mm (95% CI: −0.58; −0.21) and a gain in attachment levels of 0.41 mm (95% CI: −0.63; −0.19) compared with mechanical therapy without adjuvants (Van Ravensteijn et al. 2022).

Given the connection between inflammatory mechanisms in periodontitis and systemic inflammatory diseases, studies have explored the potential therapeutic benefits of SPMs in the management of both conditions. Omega-3 polyunsaturated fatty acid supplementation has been investigated in patients with diabetes mellitus and periodontitis. The effect of dietary supplementation with 4, 8, and 16 mL/kg of Sardinella longiceps fish oil on the expression of MMP-8 and tissue inhibitor of metalloproteinase 1 (TIMP-1) was evaluated in Wistar mice with induced diabetes mellitus. The treatment group that received a dose of 16 mL/kg showed a significant reduction in the expression of MMP-8 and a significant increase in the expression of TIMP-1 (Damaiyanti et al. 2019). Although systemic conditions, such as diabetes, can modulate immune responses, supplementation with immunomodulators has been shown to be an effective strategy to enhance the host response and prevent tissue damage under these conditions.

Recently, a controlled clinical trial evaluated the effects of omega-3 fatty acids and aspirins in diabetic patients with periodontitis. The supplements, omega-3 and aspirin, were administered for 2 months prior to scaling and root planing. Evaluation of clinical and immunological variables in the absence of periodontal treatment showed that this group had no additional benefit unless periodontal therapy was also performed. In the group of patients who received periodontal treatment along with a daily dose of 3 grams of fish oil plus 100 mg of aspirin for a period of 2 months, 36% of patients reached the treatment endpoint (≤4 sites with probing depths ≥5 mm), compared to only 16% in the group that received periodontal therapy alone. Moreover, this combined therapy also significantly reduced HbA1c levels by 0.51% which was a greater improvement compared with the other groups (8.14% at the first measurement and 7.63% at the final measurement) (Santos et al. 2020).

The effect of combinations of agents with a pro-resolving effect has also been investigated. Accordingly, the effect of the combination of periodontal therapy, omega-3, and cranberry juice in patients with diabetes mellitus and periodontitis showed that nonsurgical periodontal therapy plus cranberry juice enriched with omega-3 generated the most favorable results, by considerably decreasing HbA1c levels and probing depth values, while considerably increasing HDL levels (Zare Javid et al. 2018).

In other systemic repercussions of periodontitis such as atherosclerosis and atheromatous plaque formation, in rabbits with a high cholesterol diet and induced with periodontitis, it has been found that inflammatory markers remain persistently elevated at local and systemic levels. Topical application of resolvin E1 together with periodontal therapy was effective in the treatment of periodontitis, in addition to having a systemic influence on the reduction of inflammatory markers that prevented the appearance of cardiovascular events in susceptible individuals (Hamilton et al. 2017).

Therefore, since periodontal diseases are directly linked to local and systemic conditions, and given the well-established relationship between microbial challenges and host immune responses, disease management has evolved to address these conditions more effectively. The majority of therapies have focused on eliminating the microbial challenge. The inflammatory process can, however, promote the overgrowth of putative pathogens, leading to increased tissue damage. In this context, based on evidence, there has been increasing focus on developing and testing combinations of conventional therapies with the aim of eliminating polymicrobial biofilms while simultaneously controlling inflammatory processes, thereby enhancing host immune responses and promoting tissue healing. New agents, such as resolvins and omega-3 fatty acids, have shown promising results, even in clinical trials, suggesting significant advancements in disease management, particularly for patients with systemic conditions that compromise immune responses. The direct effects of these agents on inflammatory processes, molecular parameters, biofilm composition, and their underlying mechanisms need to be further investigated to guide their clinical application.

30.4.2 Immune Modulation and Peri-Implantitis

30.4.2.1 Pathogenesis of Peri-Implantitis

Peri-implantitis is an inflammatory disease induced by polymicrobial biofilms and characterized by progressive bone loss. It has been classified as one of the main reasons for dental implant failure. The prevalence of this disease has attracted the attention of scientists over the past few decades, since it is reported that approximately 25% of all patients with dental implants have experienced this condition. Thus, almost a quarter of patients who have dental implants face this pathology, for which the effective treatment continues to be a challenge (Derks and Tomasi 2015).

The accumulation of polymicrobial biofilms on the implant surface serves as a key stimulus for initiating the underlying inflammatory process. The composition of the peri-implant microbiota has also been a central focus, because it can explain the behavior and progression of the disease. In a healthy implant, there is a predominant colonization of bacterial species including *Streptococcus*, *Rothia, Neisseria*, and *Corynebacterium*, in addition to Gram-positive cocci, and facultative anaerobic bacilli, which maintain a microbial balance (Di Spirito et al. 2024). Microbiologically, it has been described that, in addition to the red complex microorganisms, other species are commonly found in peri-implantitis lesions, such as *Porphyromonas gingivalis*, *Fusobacterium nucleatum*, *Prevotella intermedia*, and *Tannerella forsythia*. *Treponema denticola*, *Prevotella nigrescens*, and *Campylobacter* species are also included. These microorganisms induce an exacerbated inflammatory response that results in causing loss of the surrounding peri-implant bone (Scarano et al. 2023).

The topographic characteristics of the implant and the perpendicular arrangement of the fibers favor the possibility of faster progression of the disease. Histological studies have revealed a significant accumulation of inflammatory infiltrates composed of pro-inflammatory macrophages (M1), with elevated levels of cytokines such as IL-1, IL-6, and TNF, factors associated with bone loss. However, the co-expression of iNOS and CD206 was also observed, which indicated a heterogeneous immune response that includes the presence of M2 macrophages in peri-implantitis lesions (Fig. 30.2) (Fretwurst et al. 2020).

Implants have also been shown to cause dysregulation of the immune response in the peri-implant mucosa. In general, studies have reported a higher expression of various immune cells, as well as increased expression of pro-inflammatory cytokines, an elevated IL17A/Foxp3 ratio, and higher RANKL expression, when compared with surrounding dental tissues. This altered immune profile in the peri-implant mucosa, called "dysregulated homeostasis," may explain the increased susceptibility to infection and the rapid bone loss observed in peri-implantitis (Salvi et al. 2017).

The levels of IL-1b, IL-4, G-CSF, TNF-α, IL-17A, and BAFF have also been shown to be significantly higher in the gingival crevicular fluid of tissues with peri-implantitis ($p < 0.05$). Although differences have been observed in the extent and progression of lesions in periodontitis and peri-implantitis, no significant differences have been seen between the composition or quantity of immune cells between these two conditions. Notably, both are characterized by an increased proportion of B cells (Malmqvist et al. 2024).

Therefore, as described for periodontal diseases, peri-implant diseases also show a direct link between microbial challenges and the host response, in which polymicrobial biofilms trigger inflammatory responses and tissue damage. Importantly, the implant topography presents a challenging surface to be treated with conventional mechanical therapies commonly used for teeth. Thus, there is no consensus relative to the optimal treatment protocol for these prevalent conditions. In this context, the combination of adjunctive therapies has been explored to enhance clinical treatment outcomes and promote the healing process. Immune modulators may, therefore, play a key role when combined with antimicrobial agents and mechanical therapies for the effective management of peri-implantitis.

30.4.2.2 Immune Modulation in Peri-Implantitis

Although bacterial biofilm removal and surface cleaning remain the current treatments of choice, the pathogenesis of peri-implantitis suggests that predictable results or expected clinical improvement are not always obtained (Heitz-Mayfield et al. 2018). Therefore, modulation of the immune response is currently one of the key targets in the treatment of local inflammatory pathologies, such as peri-implantitis. It has been hypothesized that this can be achieved through different mechanisms.

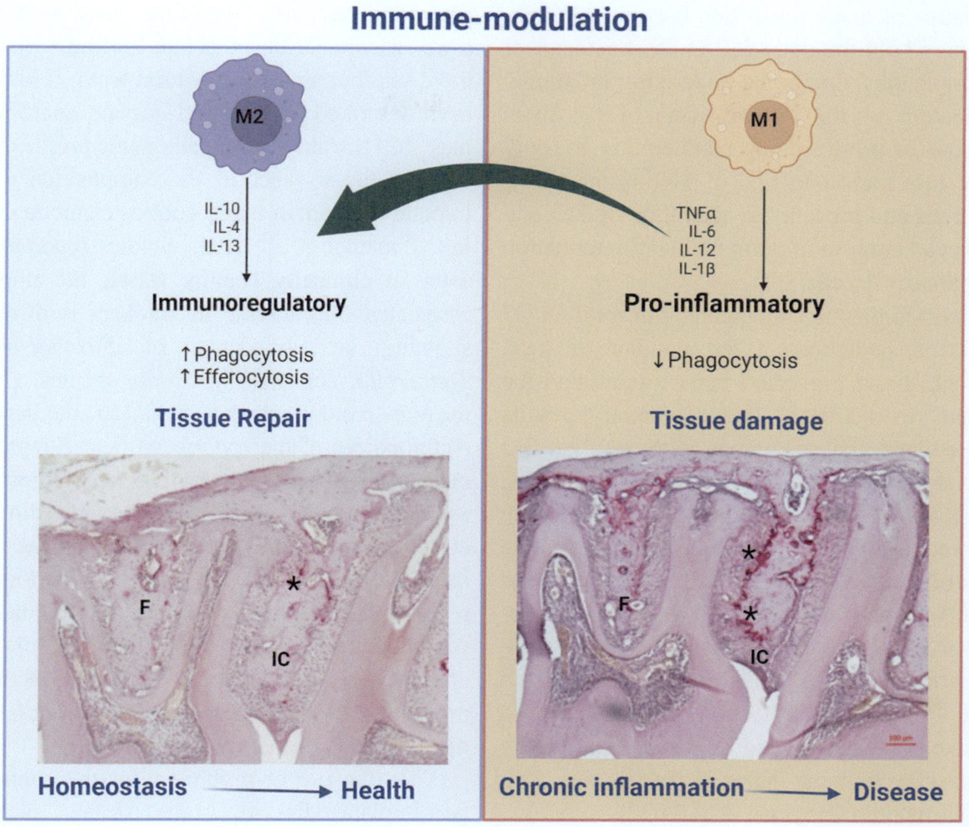

Fig. 30.2 Immune modulation. Immunomodulators seek to restore the immunological homeostasis characteristic of healthy states. Thus, by using immunostimulants or immunosuppressants, changes can be induced in immune response cells, such as macrophages, by shifting from a pro-inflammatory subpopulation (M1) to those that can promote pro-resolution, regulation, and healing of the damage (subpopulation M2). *: Trap + cells, *F* furcation region, *IC* interproximal bone crest

It has recently been suggested that CCR2 and its ligand CCL2 are not involved only in regulating osteoclastic activity but also in regulating the response of M1 macrophages. CCR2 is a key cytokine receptor for the recruitment of monocytes that subsequently differentiate into macrophages. In periodontitis, for example, the inhibition of CCR2 was marked by a very significant improvement in the monocyte/macrophage infiltrate and bone resorption (Jiang et al. 2022).

It has previously been determined that there are elevated levels of CCR2 and CCL2 in patients with peri-implantitis, suggesting their role in the increased monocyte/macrophage infiltration that would trigger bone resorption. This has led to investigating the application of inhibitors of these receptors, specifically CCR2/CCR5 (cenicriviroc), with potent anti-inflammatory and anti-fibrotic properties. Therefore, it has been suggested that cenicriviroc could potentially be helpful in the treatment of peri-implantitis (Jiang et al. 2022). In a murine model, the systemic application of cenicriviroc was shown to significantly decrease the pro-inflammatory monocyte/macrophage infiltration, decrease the amount of pro-inflammatory cytokines, and mitigate the bone loss around implants, demonstrating that CCR2 could be a therapeutic target for immunomodulatory purposes (Yuan et al. 2024).

The use of pro-resolving mediators, such as enhancers of the host immune response and facilitators of tissue repair and bacterial elimination, may hold the potential for the treatment of peri-implantitis, due to the underlying inflammatory nature of the disease. One of the main functions of pro-resolving mediators is not only to decrease the production of TNF-α and IFN-γ by T cells and prevent the activation of Th1 and Th17 cells but also to promote the differentiation of regulatory T cells.

In peri-implantitis, treatment with resolvin D2 showed a reduction in the number of local neutrophils and prevented bone loss around the implant. In addition, flow cytometry results demonstrated that treatment with resolvin D2 reduced the influx of leukocytes, T and B cells, and RANKL. No significant changes were observed in B cells expressing RANKL or in the influx of monocytes. To demonstrate the systemic effect, treatment was found to decrease the production of IFN-γ by splenocytes in the peri-implantitis group. Moreover, it was shown that a single *P. gingivalis* infection was sufficient to promote bone loss around implants, highlighting this fragile immune homeostasis in the peri-implant mucosa (Heyman et al. 2022).

Probiotics compete with other oral microbiota for nutrients and binding sites. Bifidobacterium has been described to inhibit peri-implant disease pathogens such as *P. gingivalis,* competitively. Furthermore, Bifidobacterium competes for certain binding sites, potentially disrupting pathogen coaggregation (PMID: 18312575). *Streptococcus salivarius* strain K12, a member of the oral microbiome, has been shown to exhibit the ability to inhibit pathogen biofilm formation on implant materials in vitro. This effect is due to the production of bacteriocins associated with *Streptococcus intermedius,* as one of the primary colonizers on dental implants. Therefore, this microorganism can be further studied as a possible preventive effect on peri-implant diseases (Vacca et al. 2020).

The microbiological composition of peri-implantitis has been characterized by the presence of aggressive and resistant species, showing a distinct profile compared with the microbiome related to periodontitis. This suggests that the peri-implantitis biofilm may be more heterogeneous and, therefore, may require a lower accumulation of dental biofilm to initiate the disease compared to natural teeth. It has also been described that the peri-implant microbiome may shift toward a more pathogenic profile due to external factors such as the composition of the implant titanium or habits such as cigarette smoking (Pimentel et al. 2018). Studies reported that even in clinically healthy states, the implant-associated microbiome in smokers is different, showing an abundance of *Fusobacterium, Tannerella,* and *Mogibacterium* species. Unlike the non-smoking group, they had an abundance of Actinomyces, Capnocytophaga, and Streptococcus, which are genera that are commonly associated with periodontal health (Pimentel et al. 2018). Thus, modification of the peri-implant microbiome as well as periodontal rebiosis are key approaches in the treatment of peri-implantitis. In this regard, multiple approaches have been tested, including probiotics, hyaluronic acid, antibiotics, and modifications of the implant surface.

With the use of probiotics for the control of peri-implant disease, some strains such as *Limosilactobacillus reuteri, Lactobacillus salivarius,* and *Weissella cibaria* have been shown to significantly inhibit the growth of pathogens associated with peri-implantitis such *as Fusobacterium nucleatum, Porphyromonas gingivalis, Tannerella forsythia, Prevotella intermedia*, and *Staphylococcus aureus* in in vitro models. This inhibition can be achieved because these probiotics can produce hydrogen peroxide, organic acids, and bacteriocins (for example reuterin, produced by Lactobacillus species that acts by oxidizing bacterial thiol groups) in order to directly inhibit pathogenic microorganisms. Some of the other probiotic microorganisms such as *Lactobacillaceae* and *Bifidobacterium*, including *Limosilactobacillus fermentum* and *Lactobacillus acidophilus*, also demonstrate this inhibitory action against pathogenic species (Xu et al. 2023).

Lactococcus lactis produces nisin (another bacteriocin that has been widely used for food preservation), which has also been tested for its ability to disrupt pathogenic peri-implant biofilm in vitro. Initially, it was determined that 105 CFU/ml of *L. lactis* considerably decreased the viability of the biofilm to 52.45 ± 3.41%. Remarkably, the action of nisin was to decrease the viability of pathogen-enriched biofilms in a dose-dependent manner, from 62.53 ± 3.69% to 54.26 ± 3.35%. It is important to highlight that these treatments may have a modulating effect, since they were able to change the relative abundance, diversity, and composition of microorganisms, directing them towards levels observed in healthy states, particularly in the *Proteobacteria* and *Firmicutes* phyla (Radaic et al. 2022).

Although some of the properties of probiotics, such as antibacterial action and modulation of the immune response, were tested in vitro and in animal models, and subsequently extrapolated to peri-implantitis models, clinical studies in this particular condition remain limited and inconclusive. Galofré et al. (2018) in a triple-blind randomized clinical trial evaluated the effect of L*actobacillus reuteri* in a non-surgical treatment of peri-implantitis in 22 patients for 30 days. The results showed improvements in bleeding on probing and probing depth indices. Microbiologically, however, no significant changes were observed in the peri-implant microbiota, because the only notable reduction was in the load of *P. gingivalis*–considering the trial was conducted in patients with peri-implant mucositis (Galofré et al. 2018).

Whereas, Gao et al. (2020), in a meta-analysis evaluating the effectiveness of *Lactobacillus* in peri-implant mucositis and peri-implantitis as an adjunct to non-surgical treatment, found that there was no significant difference in probing depths after 2 months of treatment with probiotics. A limited influence of the probiotic on the clinical variables, evaluated in both mucositis and peri-implantitis, was seen (Gao et al. 2020). Therefore, the immunomodulatory effect of probiotics in the treatment of peri-implantitis has not yet shown clinical outcomes superior to those of conventional treatments and should continue to be studied. Possible effects in the prevention of the disease have been described, but regarding the treatment, the evidence remains inconclusive (Xu et al. 2023).

Another approach to microbial modulation showed a statistically significant decrease in microorganisms of the genera *Streptococcus, Veillonella, Rothia, Granulicatella, Prevotella,* and *Campylobacter* in the control group ($p < 0.05$). Therefore, it is suggested that this change, induced by hyaluronic acid, may affect first colonizers. This type of modulation may be useful in the initial stages of the disease by targeting initial colonizers, and in advanced stages by decreasing microbiological alpha diversity. It may play a protective role against bacterial colonization (Soriano-Lerma et al. 2020).

Currently, implant surface modifications have emerged as another alternative for the management of peri-implantitis by generating cellular processes that modulate the response. Pulsed electromagnetic field therapy as an adjuvant method to the non-surgical treatment of peri-implantitis has been one of the approaches used. The principle of this therapy lies in its ability to modulate the immune response, a mechanism that has also been used in orthopedics to improve outcomes in osteoarthritis. Based on in vivo studies, it has been proven that through the adaptation of a new abutment that is exposed to radiofrequency fields of 1/500–1/5000, with intensities of 0.05–0.5 mT, and a frequency of 10–50 kHz for 30 days, patients in the experimental group had a statistically lower bone loss than the control group (2.48 mm vs. 3.73 mm, $p < 0.05$ and 2.39 vs. 3.37, $p < 0.01$). Furthermore, IL-1B levels decreased significantly in the experimental group after 2 weeks (72.86 pg/mL vs. 111.7, $p < 0.05$). However, it should be emphasized that in addition to these adjunct therapies, all measures for the control of dental biofilm are necessary (Mayer et al. 2023).

Another factor that can trigger immune system activation and negatively impact the condition of peri-implant tissues is the release of titanium particles into surrounding gingival tissues. This can occur due to corrosion of the implant surface, the process of implant placement, instrumentation during maintenance procedures, or even micromovements between the implant and its superstructure. In the oral environment, titanium faces constantly changing pH values due to the presence of bacterial biofilm, dietary products, and fluoride. This acidic environment causes titanium particles to dissolve and alter the TiO2 layer on the implant surface. These particles act as foreign bodies to immune cells, which are phagocytosed by macrophages that release pro-inflammatory cytokines such as IL-1β and TNF-α. These cytokines are associated with the activation of osteoclasts, leading to the development and progression of peri-implantitis. Imbalances in cytokine release levels may inhibit the resolution of inflammation and contribute to alveolar bone loss (Noronha Oliveira et al. 2018).

30.5 Conclusions

Inflammatory diseases are highly prevalent in the oral cavity. The majority of them are induced by polymicrobial biofilms. Although scientific evidence has focused on understanding biofilm formation, its composition, and how it leads to tissue damage, the interaction between microbial species and the host response plays a pivotal role in disease progression and inflammatory processes. The immune response is crucial in managing these conditions by recognizing potential threats, activating immune pathways, inducing inflammatory responses, and recruiting immune cells and mediators to manage disease progression.

Although the main focus of some therapeutic strategies is to target stress factors such as biofilms, immune modulators have shown strong potential to promote host resilience, control inflammation, and activate a cascade of immune events that support tissue healing and the restoration of a healthy state. Therefore, further studies should focus on developing these strategies, potentially in combination with antimicrobial agents, to identify new therapeutic agents, elucidate their mechanisms, and evaluate their potential effects on inflammatory diseases.

Source of Support JAS was supported by CNPq # 314479/2023-6.

References

Abdalla H, Puhl L, Rivas C, Wu Y, Rojas P, Trindade-da-Silva C, Hammock B, Maddipati K, Soares M, Clemente-Napimoga J, Kantarci A, Napimoga M, Van Dyke T (2024) Modulating the sEH/EETs axis restrains specialized Proresolving mediator impairment and regulates T cell imbalance in experimental periodontitis. J Immunol 212(3):433–445

Abusleme L, Dupuy AK, Dutzan N, Silva N, Burleson JA, Strausbaugh LD, Gamonal J, Diaz PI (2013) The subgingival microbiome in health and periodontitis and its relationship with community biomass and inflammation. ISME J 7(5):1016–1025. https://doi.org/10.1038/ismej.2012.174

Ali M, Yang F, Plachokova AS, Jansen JA, Walboomers XF (2021) Application of specialized pro-resolving mediators in periodontitis and peri-implantitis: a review. Eur J Oral Sci 129(1):1–16. https://doi.org/10.1111/eos.12759

Ariel A, Serhan CN (2007) Resolvins and protectins in the termination program of acute inflammation. Trends Immunol 28(4):176–183. https://doi.org/10.1016/j.it.2007.02.007

Ataollahi Eshkoor S, Fanijavadi S (2024) Dysbiosis-epigenetics-immune system interaction and ageing health problems. J Med Microbiol 73(11). https://doi.org/10.1099/jmm.0.001921

Baab D, Weinstein P (1986) Longitudinal evaluation of a self-inspection plaque index in periodontal recall patients. Clin Periodontol 13(4):313–318

Balta MG, Loos BG, Nicu EA (2017) Emerging concepts in the resolution of periodontal inflammation: a role for resolvin E1. Front Immunol 8(DEC):1–11. https://doi.org/10.3389/fimmu.2017.01682

Bennett M, Gilroy DW (2016) Lipid mediators in inflammation. Microbiol Spectr 4(6). https://doi.org/10.1128/microbiolspec.MCHD-0035-2016.Correspondence

Castillo DM, Sánchez-Beltrán MC, Castellanos JE, Sanz I, Mayorga-Fayad I, Sanz M, Lafaurie GI (2011) Detection of specific periodontal microorganisms from bacteraemia samples after periodontal therapy using molecular-based diagnostics. J Clin Periodontol 38(5):418–427. https://doi.org/10.1111/j.1600-051X.2011.01717.x

Castro dos Santos NC, Furukawa MV, Oliveira-Cardoso I, Cortelli JR, Feres M, Van Dyke T, Rovai ES (2022) Does the use of omega-3 fatty acids as an adjunct to non-surgical periodontal therapy provide additional benefits in the treatment of periodontitis? A systematic review and meta-analysis. J Periodontal Res 57(3):435–447. https://doi.org/10.1111/jre.12984

Chen MX, Zhong YJ, Dong QQ, Wong HM, Wen YF (2021) Global, regional, and national burden of severe periodontitis, 1990–2019: an analysis of the global burden of disease study 2019. J Clin Periodontol 48(9):1165–1188. https://doi.org/10.1111/jcpe.13506

Chiang N, Serha CN (2020) Specialized pro-resolving mediator network: an update on production and actions. Essays Biochem 64(3):443–462. https://doi.org/10.1042/EBC20200018

Cianci E, Recchiuti A, Trubiani O, Diomede F, Marchisio M, Miscia S, Colas RA, Dalli J, Serhan CN, Romano M (2016) Human periodontal stem cells release specialized Proresolving mediators and carry immunomodulatory and Prohealing properties regulated by Lipoxins. Stem Cells Transl Med 5(1): 20–32. https://doi.org/10.5966/sctm.2015-0163

Damaiyanti D, Mulawarmanti D, Parisihni K (2019) Protection against periodontal destruction in diabetic condition with Sardinella longiceps fish oil: expression of matrix-metalloproteinase 8 and tissue inhibitor of metalloproteinase 1. Dent J 52(1):51–56

Derks J, Tomasi C (2015) Peri-implant health and disease. A systematic review of current epidemiology. J Clin Periodontol 42(S16):S158–S171. https://doi.org/10.1111/jcpe.12334

Di Spirito F, Giordano F, Di Palo MP, D'Ambrosio F, Scognamiglio B, Sangiovanni G, Caggiano M, Gasparro R (2024) Microbiota of Peri-implant healthy tissues, Peri-implant Mucositis, and Peri-Implantitis: a comprehensive review. Microorganisms 12(6). https://doi.org/10.3390/microorganisms12061137

Ding T, Kang W, Li J, Yu L, Ge S (2021) An in situ tissue engineering scaffold with growth factors combining angiogenesis and osteoimmunomodulatory functions for advanced periodontal bone regeneration. J Nanobiotechnol 19(1):1–16. https://doi.org/10.1186/s12951-021-00992-4

Djuricic I, Calder PC (2021) Beneficial outcomes of omega-6 and omega-3 polyunsaturated fatty acids on human health: an update for 2021. Nutrients 13(7). https://doi.org/10.3390/nu13072421

Duvall MG, Levy BD (2016) DHA- and EPA-derived resolvins, protectins, and maresins in airway inflammation. Eur J Pharmacol 15(785):144–155. https://doi.org/10.1378/chest.114.4_supplement.290s

Elashiry M, Elashiry MM, Elsayed R, Rajendran M, Auersvald C, Zeitoun R, Rashid MH, Ara R, Meghil MM, Liu Y, Arbab AS, Arce RM, Hamrick M, Elsalanty M, Brendan M, Pacholczyk R, Cutler CW (2020) Dendritic cell derived exosomes loaded with immunoregulatory cargo reprogram local immune responses and inhibit degenerative bone disease in vivo. J Extracell Vesicles 9(1). https://doi.org/10.1080/20013078.2020.1795362

Engebretson S, Hyman L, Michalowicz B, Schoenfeld E, Gelato M, Hou W, Seaquist E, Reddy M, Lewis C, Oates T, Tripathy D, Katancik J, Orlander P, Paquette D, Hanson N, Tsai M (2013) The effect of nonsurgical periodontal therapy on hemoglobin A1c levels in persons with type 2 diabetes and chronic periodontitis: a randomized clinical trial. JAMA 310(23):2523–2532. https://doi.org/10.1001/jama.2013.282431.The

Fagundes CT, Amaral FA, Teixeira AL, Souza DG, Teixeira MM (2012) Adapting to environmental stresses: the role of the microbiota in controlling innate immunity and behavioral responses. Immunol Rev 245(1):250–264. https://doi.org/10.1111/j.1600-065X.2011.01077.x

Felix KM, Tahsin S, Wu HJ (2019) Host – microbiota interplay in mediating immune disorders. Ann N Y Acad Sci 1417(1):57–70. https://doi.org/10.1111/nyas.13508.Host

Fessler J, Matson V, Gajewski TF (2019) Exploring the emerging role of the microbiome in cancer immunotherapy. J Immunother Cancer 7(1):1–15. https://doi.org/10.1186/s40425-019-0574-4

Fomby P, Cherlin AJ (2011) Role of microbiota in immunity and inflammation. Natl Inst Health 72(2): 181–204. https://doi.org/10.1016/j.cell.2014.03.011.Role

Freire MO, Van Dyke TE (2013) Natural resolution of inflammation. Periodontol 63(1):149–164. https://doi.org/10.1111/prd.12034

Freire MO, Dalli J, Serhan CN, Van Dyke TE (2017) Neutrophil Resolvin E1 receptor expression and function in type 2 diabetes. J Immunol 198(2):718–728. https://doi.org/10.4049/jimmunol.1601543

Fretwurst T, Garaicoa-Pazmino C, Nelson K, Giannobile WV, Squarize CH, Larsson L, Castilho RM (2020) Characterization of macrophages infiltrating peri-implantitis lesions. Clin Oral Implants Res 31(3): 274–281. https://doi.org/10.1111/clr.13568

Gagnière J, Raisch J, Veziant J, Barnich N, Bonnet R, Buc E, Bringer MA, Pezet D, Bonnet M (2016) Gut microbiota imbalance and colorectal cancer. World J Gastroenterol 22(2):501–518. https://doi.org/10.3748/wjg.v22.i2.501

Galofré M, Palao D, Vicario M, Nart J, Violant D (2018) Clinical and microbiological evaluation of the effect of lactobacillus reuteri in the treatment of mucositis and peri-implantitis: a triple-blind randomized clinical trial. J Periodontal Res 53(3):378–390. https://doi.org/10.1111/jre.12523

Gao J, Yu S, Zhu X, Yan Y, Zhang Y, Pei D (2020) Does probiotic lactobacillus have an adjunctive effect in the nonsurgical treatment of Peri-implant diseases? A systematic review and meta-analysis. J Evid Based Dent Pract 20(1):101398. https://doi.org/10.1016/j.jebdp.2020.101398

Germolec DR, Shipkowski KA, Frawley RP, Evans E (2018) Markers of inflammation. Methods Mol Biol 1803:57–79. https://doi.org/10.1007/978-1-4939-8549-4_5

Groeger S, Meyle J (2019) Oral mucosal epithelial cells. Front Immunol 10(FEB):1–22. https://doi.org/10.3389/fimmu.2019.00208

Güvenç KK, Fentoğlu Ö, Calapoğlu M, Aksoy F, Orhan H (2023) Periodontal and cardiovascular therapies modify specialized pro-resolving lipid mediator (sPRLM) (LPXA4, PD1, RvE1, RvD1, and MaR1)-mediated pathway: the first pilot clinical study. Clin Oral

Investig 27(9):5549–5558. https://doi.org/10.1007/s00784-023-05174-0

Hajishengallis G, Chavakis T (2021) Local and systemic mechanisms linking periodontal disease and inflammatory comorbidities. Nat Rev Immunol 21(7):426–440. https://doi.org/10.1038/s41577-020-00488-6

Hajishengallis G, Lamont RJ (2012) Beyond the red complex and into more complexity: the polymicrobial synergy and dysbiosis (PSD) model of periodontal disease etiology. Mol Oral Microbiol 27(6):409–419. https://doi.org/10.1111/j.2041-1014.2012.00663.x

Hall MW, Singh N, Ng KF, Lam DK, Goldberg MB, Tenenbaum HC, Neufeld JD, Beiko RG, Senadheera DB (2017) Inter-personal diversity and temporal dynamics of dental, tongue, and salivary microbiota in the healthy oral cavity. NPJ Biofilms Microbiomes 3(1):1. https://doi.org/10.1038/s41522-016-0011-0

Hamilton JA, Hasturk H, Kantarci A, Serhan CN, Van Dyke T (2017) Atherosclerosis, periodontal disease, and treatment with Resolvins. Curr Atheroscler Rep 19(12). https://doi.org/10.1007/s11883-017-0696-4

Hasturk H, Kantarci A, Goguet-Surmenian E, Blackwood A, Andry C, Serhan CN, Van Dyke TE (2007) Resolvin E1 regulates inflammation at the cellular and tissue level and restores tissue homeostasis in vivo. J Immunol 179(10):7021–7029. https://doi.org/10.4049/jimmunol.179.10.7021

Heitz-Mayfield LJ, Aaboe M, Araujo M, Carrión JB, Cavalcanti R, Cionca N, Cochran D, Darby I, Funakoshi E, Gierthmuehlen PC, Hashim D, Jahangiri L, Kwon Y, Lambert F, Layton DM, Lorenzana ER, McKenna G, Mombelli A, Müller F et al (2018) Consensus report: risks and biologic complications associated with implant dentistry. Clin Oral Implants Res 29(May):351–358. https://doi.org/10.1111/clr.13307

Herrera BS, Hasturk H, Kantarci A, Freire MO, Nguyen O, Kansal S, van Dyke TE (2015) Impact of resolvin E1 on murine neutrophil phagocytosis in type 2 diabetes. Infect Immun 83(2):792–801. https://doi.org/10.1128/IAI.02444-14

Heyman O, Horev Y, Mizraji G, Haviv Y, Shapira L, Wilensky A (2022) Excessive inflammatory response to infection in experimental peri-implantitis: resolution by Resolvin D2. J Clin Periodontol 49(11):1217–1228. https://doi.org/10.1111/jcpe.13631

Honda K, Littman DR (2016) The microbiota in adaptive immune homeostasis and disease. Nature 535(7610): 75–84. https://doi.org/10.1038/nature18848

Iaconis A, De Plano LM, Caccamo A, Franco D, Conoci S (2024) Anti-biofilm strategies: a focused review on innovative approaches. Microorganisms 12(4):1–27. https://doi.org/10.3390/microorganisms12040639

Ivanov II, Honda K (2012) Intestinal commensal microbes as immune modulators. Cell Host Microbe 12(4): 496–508. https://doi.org/10.1016/j.chom.2012.09.009

Jiang W, Xu T, Yuan S, Wei Y, Song Z, Li Q, She S, Wang X, Wang C, Yang G, Cao J, Sun F, Shi M, Li S, Liu Z, Mo Y, Lv P, Zhang Y, Wang Y, Hu W (2022) Critical roles for CCR2 and the therapeutic potential of cenicriviroc in periodontitis: a pre-clinical study. J Clin Periodontol 49(11):1203–1216

Kayagaki N, Webster JD, Newton K (2024) Control of cell death in health and disease. Annu Rev Pathol 19:157–180. https://doi.org/10.1146/annurev-pathmechdis-051022-014433

Kim CH (2023) Complex regulatory effects of gut microbial short-chain fatty acids on immune tolerance and autoimmunity. Cell Mol Immunol 20(4):341–350. https://doi.org/10.1038/s41423-023-00987-1

Kinney JW, Bemiller SM, Murtishaw AS, Leisgang AM, Salazar AM, Lamb BT (2018) Inflammation as a central mechanism in Alzheimer's disease. Alzheimers Dement (N Y) 4:575–590. https://doi.org/10.1016/j.trci.2018.06.014

Kohli P, Levy BD (2009) Resolvins and protectins: mediating solutions to inflammation. Br J Pharmacol 158(4):960–971. https://doi.org/10.1111/j.1476-5381.2009.00290.x

Kotas M, Medzhitov R (2015) Homeostasis, inflammation, and disease susceptibility. Cell 160(5):816–827. https://doi.org/10.1016/j.cell.2015.02.010. Homeostasis

Laskowski TJ, Biederstädt A, Rezvani K (2022) Natural killer cells in antitumour adoptive cell immunotherapy. Nat Rev Cancer 22(10):557–575. https://doi.org/10.1038/s41568-022-00491-0

Lee C-T, Teles R, Kantarci A, Chen T, McCafferty J, Starr J, Neves Brito L, Paster B, Van Dyke T (2016) Resolvin E1 reverses experimental periodontitis and Dysbiosis. J Immunol 197(7):2796–2806. https://doi.org/10.1177/0022146515594631.Marriage

Lee CT, Li R, Zhu L, Tribble GD, Zheng WJ, Ferguson B, Maddipati KR, Angelov N, Van Dyke TE (2021) Subgingival microbiome and specialized pro-resolving lipid mediator pathway profiles are correlated in periodontal inflammation. Front Immunol 12(June):1–14. https://doi.org/10.3389/fimmu.2021.691216

Lerman Y, Kim M (2015) Neutrophil migration under normal and sepsis conditions. Cardiovasc Hematol Disord Drug Targets 15(1):19–28. https://doi.org/10.2174/1871529x15666150108113236

Li D, Wang P, Wang P, Hu X, Chen F (2016) The gut microbiota: a treasure for human health. Biotechnol Adv 34(7):1210–1224. https://doi.org/10.1016/j.biotechadv.2016.08.003

Liu X, Peng L, Meng J, Zhu Z, Han B, Wang S (2018) Protein-mediated anti-adhesion surface against oral bacteria. Nanoscale 10(6):2711–2714. https://doi.org/10.1039/c7nr08844b

Lynn DJ, Benson SC, Lynn MA, Pulendran B (2022) Modulation of immune responses to vaccination by the microbiota: implications and potential mechanisms. Nat Rev Immunol 22(1):33–46. https://doi.org/10.1038/s41577-021-00554-7

Ma R, Liu Y, Xu Y, Duan D (2024) Lipoxin A4 levels predict site-specific clinical improvements post scaling and root planing and correlate negatively with periodontal pathogens in severe periodontitis. BMC Oral Health 24(1):1–10. https://doi.org/10.1186/s12903-024-03948-w

Madaudo C, Coppola G, Parlati ALM, Corrado E (2024) Discovering inflammation in atherosclerosis: insights from pathogenic pathways to clinical practice. Int J Mol Sci 25(11). https://doi.org/10.3390/ijms25116016

Malmqvist S, Clark R, Johannsen G, Johannsen A, Boström EA, Lira-Junior R (2024) Immune cell composition and inflammatory profile of human peri-implantitis and periodontitis lesions. Clin Exp Immunol 217(2):173–182. https://doi.org/10.1093/cei/uxae033

Martínez-García M, Hernández-Lemus E (2021) Periodontal inflammation and systemic diseases: an overview. Front Physiol 12(October):1–26. https://doi.org/10.3389/fphys.2021.709438

Mayer Y, Khoury J, Horwitz J, Ginesin O, Canullo L, Gabay E, Giladi H (2023) A novel nonsurgical therapy for peri-implantitis using focused pulsed electromagnetic field: a pilot randomized double-blind controlled clinical trial. Bioelectromagnetics 44(7–8):144–155

Mizraji G, Heyman O, Van Dyke TE, Wilensky A (2018) Resolvin D2 restrains Th1 immunity and prevents alveolar bone loss in murine periodontitis. Front Immunol 9(APR):1–10. https://doi.org/10.3389/fimmu.2018.00785

Nathan C (2002) Points of control in inflammation. Nature 420(19/26):846–852

Noronha Oliveira M, Schunemann WVH, Mathew MT, Henriques B, Magini RS, Teughels W, Souza JCM (2018) Can degradation products released from dental implants affect peri-implant tissues? J Periodontal Res 53(1):1–11. https://doi.org/10.1111/jre.12479

Önal MA, Fentoğlu Ö, Aksoy F, Calapoğlu M, Varol E, Orhan H (2021) Salivary levels of last generation specific pro-resolving lipid mediators (SPMs) (protectin and maresin) in patients with cardiovascular and periodontal disease: a case-control study. J Periodontal Res 56(3):606–615. https://doi.org/10.1111/jre.12861

Osorio Parra MM, Elangovan S, Lee CT (2019) Specialized pro-resolving lipid mediators in experimental periodontitis: a systematic review. Oral Dis 25(5):1265–1276. https://doi.org/10.1111/odi.12979

Pandiyan P, Bhaskaran N, Zou M, Schneider E, Jayaraman S, Huehn J (2019) Microbiome dependent regulation of Tregs and Th17 cells in mucosa. Front Immunol 10(Mar):1–17. https://doi.org/10.3389/fimmu.2019.00426

Panigrahy D, Gilligan MM, Serhan CN, Kashfi K (2021) Resolution of inflammation: an organizing principle in biology and medicine. Pharmacol Ther 227. https://doi.org/10.1016/j.pharmthera.2021.107879

Perez-Hernandez J, Chiurchiù V, Perruche S, You S (2021) Regulation of T-cell immune responses by pro-resolving lipid mediators. Front Immunol 12 (November):1–13. https://doi.org/10.3389/fimmu.2021.768133

Pimentel SP, Fontes M, Ribeiro FV, Corrêa MG, Nishii D, Cirano FR, Casati MZ, Casarin RCV (2018) Smoking habit modulates peri-implant microbiome: a case-control study. J Periodontal Res 53(6):983–991. https://doi.org/10.1111/jre.12597

Porcheray F, Viaud S, Rimaniol AC, Léone C, Samah B, Dereuddre-Bosquet N, Dormont D, Gras G (2005) Macrophage activation switching: an asset for the resolution of inflammation. Clin Exp Immunol 142(3): 481–489. https://doi.org/10.1111/j.1365-2249.2005.02934.x

Proctor LM (2016) The National Institutes of Health human microbiome project. Semin Fetal Neonatal Med 21(6):368–372. https://doi.org/10.1016/j.siny.2016.05.002

Proctor LM, Creasy HH, Fettweis JM, Lloyd-Price J, Mahurkar A, Zhou W, Buck GA, Snyder MP, Strauss JF, Weinstock GM, White O, Huttenhower C (2019) The integrative human microbiome project. Nature 569(7758):641–648. https://doi.org/10.1038/s41586-019-1238-8

Qu XM, Wu ZF, Pang BX, Jin LY, Qin LZ, Wang SL (2016) From nitrate to nitric oxide: the role of salivary glands and oral bacteria. J Dent Res 95(13): 1452–1456. https://doi.org/10.1177/0022034516673019

Radaic A, Brody H, Contreras F, Hajfathalian M, Lucido L, Kamarajan P, Kapila YL (2022) Nisin and Nisin probiotic disrupt Oral pathogenic biofilms and restore their microbiome composition towards healthy control levels in a Peri-Implantitis setting. Microorganisms 10(7). https://doi.org/10.3390/microorganisms10071336

Rampally P, Koduganti R, Ganapathi S, Panthula V, Surya P (2019) Comparison of effectiveness of low-dose aspirin versus omega-3 fatty acids as adjuvants to nonsurgical periodontal therapy in type II diabetic patients with chronic periodontitis. J Indian Soc Periodontol 23(3):249–256

Sáez T, Toledo F, Sobrevia L (2019) Impaired signalling pathways mediated by extracellular vesicles in diabesity. Mol Asp Med 66(December 2018):13–20. https://doi.org/10.1016/j.mam.2018.12.001

Salvi GE, Cosgarea R, Sculean A (2017) Prevalence and mechanisms of Peri-implant diseases. J Dent Res 96(1):31–37. https://doi.org/10.1177/0022034516667484

Santos NC, Andere N, Araujo CF, Marco AC, Kantarci A, Dyke TE, Santamaria MP (2020) Omega-3 PUFA and aspirin as adjuncts to periodontal debridement in patients with periodontitis and diabetes. Physiol Behav 176(3):139–148. https://doi.org/10.1002/JPER.19-0613.Omega-3

Saunders KO (2019) Conceptual approaches to modulating antibody effector functions and circulation half-life. Front Immunol 10(Jun):1–20. https://doi.org/10.3389/fimmu.2019.01296

Savran L, Sağlam M (2024) Clinical effects of omega-3 fatty acids supplementation in the periodontal treatment of smokers and non-smokers with periodontitis: a retrospective study. Clin Oral Investig 28(8). https://doi.org/10.1007/s00784-024-05835-8

Scarano A, Khater AGA, Gehrke SA, Serra P, Francesco I, Di Carmine M, Tari SR, Leo L, Lorusso F (2023) Current status of Peri-implant diseases: a clinical review for evidence-based decision making. J Funct Biomater 14(4). https://doi.org/10.3390/jfb14040210

Schirmer M, Garner A, Vlamakis H, Xavier RJ (2019) Microbial genes and pathways in inflammatory bowel disease. Nat Rev Microbiol 17(8):497–511. https://doi.org/10.1038/s41579-019-0213-6

Schmid M, Gemperle C, Rimann N, Hersberger M (2016) Resolvin D1 polarizes primary human macrophages toward a Proresolution phenotype through GPR32. J Immunol 196(8):3429–3437. https://doi.org/10.4049/jimmunol.1501701

Sedghi LM, Bacino M, Kapila YL (2021) Periodontal disease: the good, the bad, and the unknown. Front Cell Infect Microbiol 11(December):1–26. https://doi.org/10.3389/fcimb.2021.766944

Serhan CN (2009) Systems approach to inflammation resolution: identification of novel anti-inflammatory and pro-resolving mediators. J Thromb Haemost 7(Suppl. 1):44–48. https://doi.org/10.1111/j.1538-7836.2009.03396.x

Serhan CN, Brain SD, Buckley CD, Gilroy DW, Haslett C, Neill LAJO, Perretti M, Rossi AG, John L (2011) Resolution of inflammation: state of the art , definitions and terms. FASEB J 21(2):325–332. https://doi.org/10.1096/fj.06-7227rev.Resolution

Shapouri-Moghaddam A, Mohammadian S, Vazini H, Taghadosi M, Esmaeili SA, Mardani F, Seifi B, Mohammadi A, Afshari JT, Sahebkar A (2018) Macrophage plasticity, polarization, and function in health and disease. J Cell Physiol 233(9). https://doi.org/10.1002/jcp.26429

Singh N, Baby D, Rajguru J, Patil P, Thakkannavar S, Pujari V (2019) Inflammation and cancer. Ann Afr Med 18(3):121–126

Sommer F, Bäckhed F (2013) The gut microbiota-masters of host development and physiology. Nat Rev Microbiol 11(4):227–238. https://doi.org/10.1038/nrmicro2974

Soriano-Lerma A, Magán-Fernández A, Gijón J, Sánchez-Fernández E, Soriano M, García-Salcedo JA, Mesa F (2020) Short-term effects of hyaluronic acid on the subgingival microbiome in peri-implantitis: a randomized controlled clinical trial. J Periodontol 91(6):734–745. https://doi.org/10.1002/JPER.19-0184

Stańdo M, Piatek P, Namiecinska M, Lewkowicz P, Lewkowicz N (2020) Omega-3 polyunsaturated fatty acids EPA and DHA as an adjunct to non-surgical treatment of periodontitis: a randomized clinical trial. Nutrients 12(9):1–14. https://doi.org/10.3390/nu12092614

Stańdo-Retecka M, Piatek P, Namiecinska M, Bonikowski R, Lewkowicz P, Lewkowicz N (2023) Clinical and microbiological outcomes of subgingival instrumentation supplemented with high-dose omega-3 polyunsaturated fatty acids in periodontal treatment – a randomized clinical trial. BMC Oral Health 23(1): 1–16. https://doi.org/10.1186/s12903-023-03018-7

Stappenbeck TS, Hooper LV, Gordon JI (2002) Developmental regulation of intestinal angiogenesis by indigenous microbes via Paneth cells. Proc Natl Acad Sci USA 99(24):15451–15455. https://doi.org/10.1073/pnas.202604299

Teughels W, Durukan A, Ozcelik O, Pauwels M, Quirynen M, Haytac MC (2013) Clinical and microbiological effects of lactobacillus reuteri probiotics in the treatment of chronic periodontitis: a randomized placebo-controlled study. J Clin Periodontol 40(11):1025–1035. https://doi.org/10.1111/jcpe.12155

Uehara A, Takada H (2008) Synergism between TLRs and NOD1/2 in oral epithelial cells. J Dent Res 87(7): 682–686. https://doi.org/10.1177/154405910808700709

Uehara A, Sugawara S, Tamai R, Takada H (2001) Contrasting responses of human gingival and colonic epithelial cells to lipopolysaccharides, lipoteichoic acids and peptidoglycans in the presence of soluble cd14. Med Microbiol Immunol 189(4):185–192. https://doi.org/10.1007/s004300100063

Uehara A, Sugawara Y, Kurata S, Fujimoto Y, Fukase K, Kusumoto S, Sasano T, Shibata K, Sugawara S, Takada H (2005) Chemically synthesized pathogen-associated molecular patterns increase the expression of peptidoglycan recognition proteins via toll-like receptors, NOD1 and NOD2 in human oral epithelial cells. Int Congr Ser 1284:163–168. https://doi.org/10.1016/j.ics.2005.06.032

Uehara A, Fujimoto Y, Fukase K, Takada H (2007) Various human epithelial cells express functional toll-like receptors, NOD1 and NOD2 to produce anti-microbial peptides, but not proinflammatory cytokines. Mol Immunol 44(12):3100–3111. https://doi.org/10.1016/j.molimm.2007.02.007

Vacca C, Contu MP, Rossi C, Ferrando ML, Blus C, Szmukler-Moncler S, Scano A, Orrù G (2020) In vitro interactions between streptococcus intermedius and streptococcus salivarius K12 on a titanium cylindrical surface. Pathogens 9(12):1–15. https://doi.org/10.3390/pathogens9121069

van der Meulen TA, Harmsen HJM, Bootsma H, Spijkervet FKL, Kroese FGM, Vissink A (2016) The microbiome–systemic diseases connection. Oral Dis 22(8):719–734. https://doi.org/10.1111/odi.12472

Van Dyke TE (2017) Pro-resolving mediators in the regulation of periodontal disease. Mol Asp Med 58:21–36. https://doi.org/10.1016/j.mam.2017.04.006

Van Dyke TE (2020) Shifting the paradigm from inhibitors of inflammation to resolvers of inflammation in periodontitis. J Periodontol 91(S1):S19–S25. https://doi.org/10.1002/JPER.20-0088

Van Dyke TE, Sima C (2020) Understanding resolution of inflammation in periodontal diseases: Is chronic inflammatory periodontitis a failure to resolve? Periodontology 2000 82(1):205–213. https://doi.org/10.1111/prd.12317

Van Dyke TE, Hasturk H, Kantarci A, Freire MO, Nguyen D, Dalli J, Serhan CN (2015) Proresolving nanomedicines activate bone regeneration in periodontitis. J Dent Res 94(1):148–156. https://doi.org/10.1177/0022034514557331

Van Holm W, Carvalho R, Delanghe L, Eilers T, Zayed N, Mermans F, Bernaerts K, Boon N, Claes I, Lebeer S, Teughels W (2023) Antimicrobial potential of known and novel probiotics on in vitro periodontitis biofilms. NPJ Biofilms Microbiomes 9(1):1–12. https://doi.org/10.1038/s41522-023-00370-y

Van Ravensteijn MM, Timmerman MF, Brouwer EAG, Slot DE (2022) The effect of omega-3 fatty acids on active periodontal therapy: a systematic review and meta-analysis. J Clin Periodontol 49(10):1024–1037. https://doi.org/10.1111/jcpe.13680

Verma D, Garg PK, Dubey AK (2018) Insights into the human oral microbiome. Arch Microbiol 200(4):525–540. https://doi.org/10.1007/s00203-018-1505-3

Wade WG (2021) Resilience of the oral microbiome. Periodontol 86(1):113–122. https://doi.org/10.1111/prd.12365

Wang D, Jiang W, Zhu F, Mao X, Agrawal S (2018) Modulation of the tumor microenvironment by intratumoral administration of IMO-2125, a novel TLR9 agonist, for cancer immunotherapy. Int J Oncol 53(3):1193–11203. https://doi.org/10.3892/ijo.2018.4456

Wang Y, Zhang S, Li H, Wang H, Zhang T, Hutchinson M, Yin H, Wang X (2020) Small-molecule modulators of toll-like receptors. Acc Chem Res 53(5):1046–1055

Winkler JW, Orr SK, Dalli J, Cheng CYC, Sanger JM, Chiang N, Petasis NA, Serhan CN (2016) Resolvin D4 stereoassignment and its novel actions in host protection and bacterial clearance. Sci Rep 6 (January):1–11. https://doi.org/10.1038/srep18972

Xu J, Chen C, Gan S, Liao Y, Fu R, Hou C, Yang S, Zheng Z, Chen W (2023) The potential value of probiotics after dental implant placement. Microorganisms 11(7):1–25. https://doi.org/10.3390/microorganisms11071845

Yuan S, Wei Y, Jiang W, Sun F, Li S, Li Q, Song Z, Liu Z, Mo Y, Wang X, Li N, Lv P, She S, Wang C, Zhang Y, Wang Y, Hu W (2024) CCR2 is a potential therapeutic target in peri-implantitis. J Clin Periodontol 51(3):354–364

Zare Javid A, Maghsoumi-Norouzabad L, Ashrafzadeh E, Yousefimanesh HA, Zakerkish M, Ahmadi Angali K, Ravanbakhsh M, Babaei H (2018) Impact of cranberry juice enriched with Omega-3 fatty acids adjunct with nonsurgical periodontal treatment on metabolic control and periodontal status in type 2 patients with diabetes with periodontal disease. J Am Coll Nutr 37(1):71–79. https://doi.org/10.1080/07315724.2017.1357509

Zarrough AE, Hasturk H, Stephens DN, Van Dyke TE, Kantarci A (2023) Resolvin D1 modulates periodontal ligament fibroblast function. J Periodontol 94(5):683–693. https://doi.org/10.1002/JPER.22-0462

Zhang H, Zhang Z, Li J, Qin G (2023) New strategies for biocontrol of bacterial toxins and virulence: focusing on quorum-sensing interference and biofilm inhibition. Toxins 15(9). https://doi.org/10.3390/toxins15090570

Zhang J, Liu S, Ding W, Wan J, Qin JJ, Wang M (2024) Resolution of inflammation, an active process to restore the immune microenvironment balance: a novel drug target for treating arterial hypertension. Ageing Res Rev 99(June):102352. https://doi.org/10.1016/j.arr.2024.102352

Zhou CB, Zhou YL, Fang JY (2021) Gut microbiota in cancer immune response and immunotherapy. Trends Cancer 7(7):647–660. https://doi.org/10.1016/j.trecan.2021.01.010

31 Modulating the Allergen-Specific Immune Response to Achieve Remission of Food Allergy Through Oral Immunotherapy

Sarah E. Ashley and Mimi L. K. Tang

Abstract

Food allergies represent a significant global health issue, affecting approximately 2% of children and up to 10% of 1-year-old infants. Oral immunotherapy has emerged as a promising treatment for food allergies. Oral immunotherapy (OIT) is effective at inducing desensitization, defined as a temporary increase in reaction threshold to food allergens, and a low-dose OIT approach (Palforzia) is now approved in several regions, including the United States, the United Kingdom, and the European Union. However, the ability for OIT to induce sustained unresponsiveness or remission, defined as the ability to tolerate the allergen without ongoing treatment or avoidance, is less certain and may depend on the OIT dosing regimen that is applied; for example, high-dose peanut OIT regimens have been shown to induce remission in ~30–74% of treated patients. Desensitization offers temporary protection against accidental exposure to allergens; however, drawbacks include the need for indefinite maintenance dosing and continuing allergen avoidance. Additionally, data show that desensitization does not improve health-related quality of life compared with a placebo. In contrast, remission allows participants to discontinue treatment, eat the allergen freely, and leads to substantial improvement in quality of life. This chapter examines the immune mechanisms driving food allergy pathogenesis and the mechanisms that support remission following OIT. Gene expression studies in patient samples from clinical OIT trials have provided insight into the immune pathways involved in redirecting the allergic response toward tolerance, with changes in allergen-reactive T cells, B cells, and baseline signatures. This chapter aims to provide an overview of the immune changes underlying food allergy and the immune mechanisms of OIT that lead to remission of allergy.

Keywords

Food allergy · Oral immunotherapy (OIT) · Desensitization and remission · Immune tolerance · Allergen-specific T cells

S. E. Ashley · M. L. K. Tang (✉)
Allergy Immunology Group, Murdoch Children's Research Institute, Parkville, VIC, Australia

The University of Melbourne, Parkville, VIC, Australia

The Royal Children's Hospital Melbourne, Parkville, VIC, Australia
e-mail: mimi.tang@rch.org.au

N. Rezaei, H. Ziaei (eds.), *Oral Immunology*, Advances in Experimental Medicine and Biology 1492,
https://doi.org/10.1007/978-3-032-03176-1_31

31.1 Immune Mechanisms Driving the Development of Food Allergy

31.1.1 Failure of Oral Tolerance

The gastrointestinal system interfaces directly with the external environment, encountering various dietary antigens and microbial stimuli. As the food necessary to sustain daily life is consumed, the gut immune system maintains immune homeostasis by establishing tolerance to innocuous food antigens while establishing protective responses against pathogens. The digestion process starts with chemical and mechanical degradation by enzymes in saliva (e.g., amylase), followed by further degradation by gastric acid and digestive enzymes in the stomach, which together diminish the allergenicity of food proteins through alterations in conformational epitopes. However, certain biochemical properties of food proteins may confer resistance to these digestive processes, including having a molecular weight of <70 kDa, glycosylation, solubility in water, and the presence of sequential epitopes (Vickery et al. 2011). Additionally, cooking processes may modify the allergenicity of food proteins; for example, roasting increases allergenicity, while boiling can reduce the allergenicity of peanut proteins (Beyer et al. 2001).

The process of appropriately recognizing food antigens as harmless and averting an unwanted pro-inflammatory immune response is called oral tolerance. Oral tolerance to food antigens is an active process initiated through the presentation of food antigens by antigen-presenting cells that deliver tolerogenic signals to naive CD4+ T cells. Appropriate antigen-presenting cell-naïve CD4+ T cell interactions under tolerogenic conditions lead to the differentiation of naïve T cells into antigen-specific regulatory T cells (Tregs) (Noval Rivas and Chatila 2016). The failure of immunological processes governing oral tolerance gives rise to aberrant immune reactions to otherwise harmless food molecules that manifest as food allergies. Oral tolerance is the default immune response to antigens encountered through the gut and processed via the gut-associated lymphoid tissues. Food allergy only occurs when this default process breaks down. Currently, the immunological pathways leading to defects in oral tolerance remain poorly understood (Pelaez-Prestel et al. 2021). Fortunately, failure of oral tolerance is not a common event, with food allergy occurring in only a small subset of individuals, and moreover to a limited number of food antigens in affected individuals.

31.1.2 Genetic and Environmental Factors Leading to Immune Dysregulation and Development of Food Allergy

There has been extensive research into the factors driving increased susceptibility to allergic disease. The pathogenesis of food allergy is multifaceted and includes both genetic and environmental factors. Higher rates of concordant peanut allergy among monozygotic twins compared to dizygotic twins (Kivisto et al. 2019; Sicherer et al. 2000) suggest a significant genetic component in food allergy. Having two or more family members with an allergy (e.g., asthma, eczema, food allergies, or allergic rhinitis) is also linked to higher rates of gold-standard challenge-proven food allergy (Koplin et al. 2013). A further example is that Australian-born children of East-Asian parents have higher rates of food allergy compared to Australian-born children of two Australian-born parents. The overriding influence of early life environmental exposures, beyond heredity, is evidenced by the finding that Australian-born children of East-Asian parents have a higher incidence of food allergy compared to their Asian-born parents (Koplin et al. 2014). Taken together, these findings provide compelling evidence that while both genetic and environmental factors contribute to the development of food allergy, environmental factors, particularly those encountered in early life, play a predominant role in the failure of oral tolerance that leads to food allergy. A major environmental factor contributing to the dysregulation of immune homeostasis is the intestinal microbiota (Savage et al. 2018). Further environmental factors identified as risk factors for food allergy include socioeconomic status, geographical location, vitamin D levels, and epigenetic modifications (Sicherer and Sampson 2018).

31.1.3 Antigen Presenting Cell-Naïve T Cell Interactions

Specialized epithelial cells of the gastrointestinal tract control the uptake, transmission, and presentation of luminal antigens via multiple pathways. The cells of this layer include goblet cells, microfold (M) cells, Paneth cells, tuft cells, enterocytes, enteroendocrine cells, and epithelial stem cells (Bevins and Salzman 2011; Gerbe and Jay 2016; Worthington 2015). As food antigens pass through the intestinal lumen of the gut, the particles cross into the intestinal mucosa or into specialized pockets of lymphatic tissue known as Peyer's patches. This process can be passive, where the antigen simply passes between enterocytes of the epithelial layer, or active, with directed transport of antigen across enterocytes, microfold (M) cells, or goblet cells. Food antigens can also be sampled by antigen-presenting cells (APCs) that extend dendrites between enterocytes into the lumen. APCs play a key role in determining the direction of an immune response to antigen, in terms of promoting the differentiation of T regulatory (Treg) cells that drive tolerance or T helper 2 (Th2) cells that drive allergy. The cognate and soluble signals presented by APC during initial interactions between APC and naïve CD4+ T cells can polarize differentiation of the naïve T cell into various T helper subsets, including Th1, Th2, Th17, and T follicular helper cells (Tfh), each having specialized effector functions and unique cytokine secretion profiles that drive specific immune responses. The cytokine milieu surrounding APC-naïve T cell interactions, along with costimulatory molecules expressed on the APC cell surface, can promote differentiation along selected T helper cell lineage pathways. For example, interleukin (IL)-12 and interferon (IFN)-γ production drive Th1 polarization; IL-4, IL-2, and IL-33 drive Th2 differentiation; IL-6, IL-1, and transforming growth factor-beta (TGF-β) induce Th17 differentiation; and IL-6, IL-21, and inducible costimulator ligand (ICOSL) drive differentiation to Tfh. These cytokines are not only important for the specialized effector function of these subsets but also provide additional positive feedback loops to further reinforce the commitment to these lineages by inhibiting polarization into other helper subsets. More recently, the nomenclature of the T helper cell polarization has been reframed according to the types of immune responses they elicit and the specific cells they target. These updated subsets include "type 1," targeting mononuclear phagocytes; "type 2," targeting primarily mast cells, eosinophils, and basophils; and "type 3," targeting non-immune cells, including epithelial cells. The cytokines driving the autocrine feed-forward loops that reinforce the polarization of these subsets include IFN-γ, IL-4, and IL-21, respectively (Tuzlak et al. 2021). The sources of these polarizing cytokines include dendritic cells and other innate immune cells. In the case of Th2/type 2 cell differentiation, stimulation of unpolarized T helper cells (Th0) with IL-4 leads to Signal transducer and activator of transcription 6 (STAT6) activation, which in turn leads to GATA3 expression. GATA3 is then upregulated within the cell, further promoting the production of the Th2 gene program and cytokine production. The breakdown of tolerogenic pathways supporting oral tolerance can lead to a Th2 skew and loss of Treg induction. This skewed Th2 environment creates the conditions for the failure of oral tolerance, causing allergic sensitization, characterized by the production of allergen-specific IgE (sIgE) (Fig. 31.1). sIgE antibodies can bind to high-affinity Fc receptor (FcεR1) on the surface of tissue-resident mast cells and circulating basophils. Upon re-exposure of a sensitized individual to the relevant food, food antigens bind to sIgE antibodies on the surface of mast cells and basophils. Crosslinking of sIgE

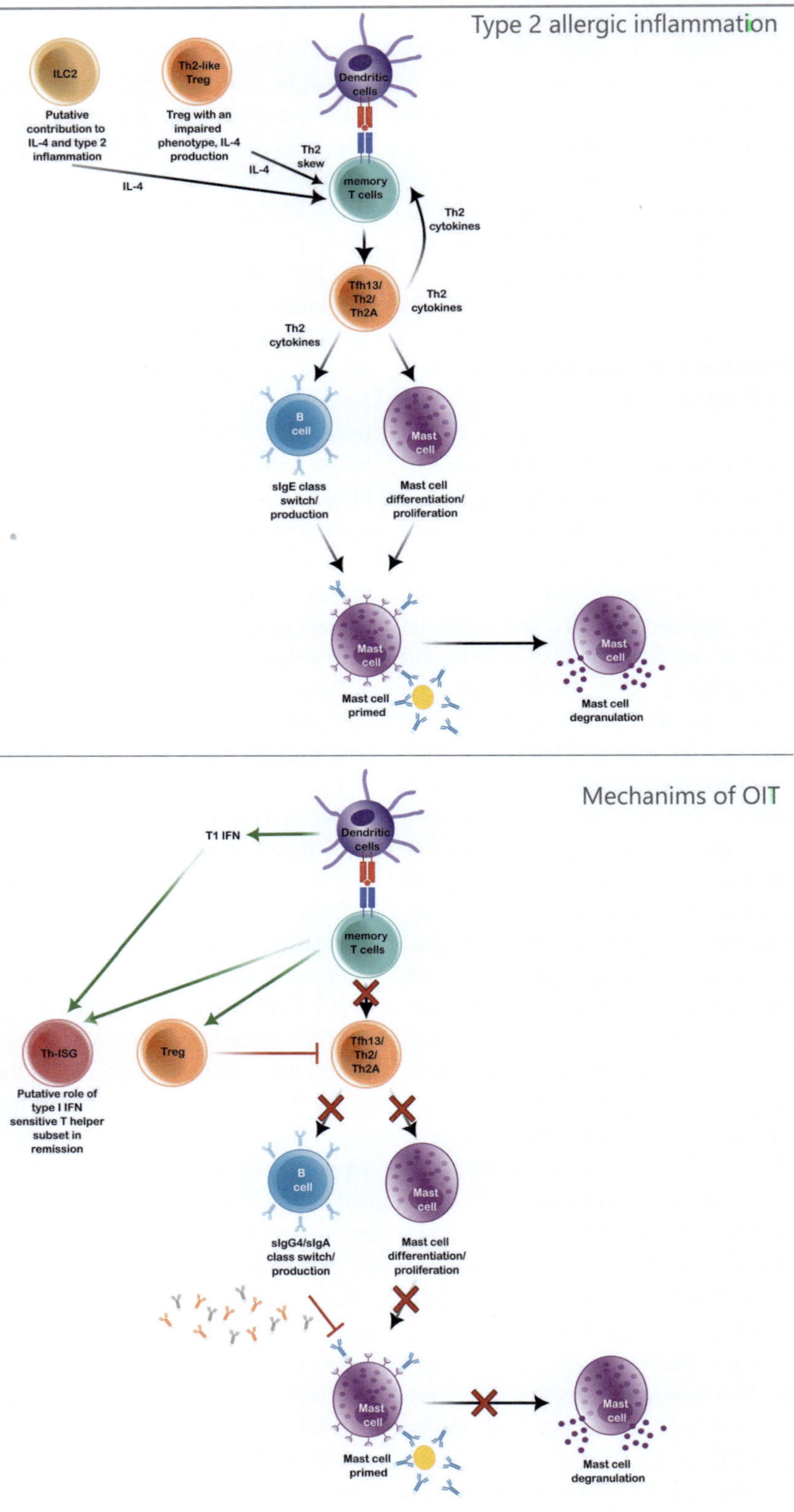

Fig. 31.1 Immune changes driving remission following oral immunotherapy for food allergy. Clonal anergy of T cells with a Th2 phenotype, decreased expression of Th2 genes in peanut reactive T cells and decreased production of antigen-specific IgE antibodies are all associated with clinical outcome following OIT. In parallel, increased production of antigen-specific sIgG4 during treatment functions to outcompete binding of sIgE, further

antibodies by antigens triggers signaling through high-affinity Fc receptor (FcεR1), which in turn leads to activation of these cell types (Fig. 31.1). This results in the exocytosis of preformed vasoactive and pro-inflammatory mediators, such as histamine, tryptase, and chymase (degranulation), as well as the production of newly formed mediators that lead to both immediate and delayed symptoms of an allergic reaction.

Consistent with the understanding that differentiation of T helper responses is critically determined by the activation status of APC and the cytokine milieu in which antigen presentation first occurs, the dual allergen exposure hypothesis has been proposed as a leading paradigm for allergic sensitization. This hypothesis suggests that skin barrier dysfunction leads to inflammation, and the presentation of food antigens in the inflamed skin environment results in the development of antigen-specific Th2 responses, ultimately resulting in a breakdown of tolerance. In this context, inflammatory signals arising due to epithelial damage, including but not limited to IL-25, IL-33, and thymic stromal lymphopoietin (TSLP), support allergic responses and a break in tolerance (Han et al. 2017).

31.1.4 Cell Populations-Driving Allergen-Specific T Helper 2 Allergic Responses

31.1.4.1 T Helper 2 Cells and the Allergic Response

Allergen-specific Th2 cells are characterized by the expression of the lineage-specifying transcription factor GATA3 and the production of IL-4, IL-5, IL-9, and IL-13. A subset of Th2 cells, known as peTh2/Th2A, which are phenotypically distinct from conventional Th2 cells, has been described in the context of allergy (Fig. 31.1) (Mitson-Salazar et al. 2016; Wambre et al. 2017). This subset is terminally differentiated ($CD45RB^{low}$ $CD27^{-}$) and co-expresses the chemoattractant receptor CRTH2, the natural killer cell marker CD161, and the homing receptor CD49d (Wambre et al. 2017). Within the allergen-specific T helper 2 compartments, these cells appear to uniquely contribute to Th2 immunity compared to conventional Th2 cells, likely through their ability to more prolifically produce cardinal Th2 cytokines (Wambre et al. 2017). In addition to the production of multiple cardinal Th2 cytokines simultaneously, Th2A cells also produce more IL-9 and IL-5 relative to conventional Th2. The transcriptomic profile of these cells shows higher expression of genes involved in allergic pathogenesis including IL-25 receptor (*IL17RB*), IL-33 receptor (*IL1RL1*), thymic stromal lymphopoietin-receptor (*CLRF2*), as well as genes linked to arachidonic acid signaling. Recently, a Th2-multipotent progenitor (Th2-MPP) high in TCF1, LEF1, CD27, and CD62L expression was described (Kratchmarov et al. 2024). These cells are likely to contribute to the sustained type 2 inflammation observed in those with allergic disease.

31.1.4.2 IgE+ B Cells and the Allergic Response

Production of IgE is a central downstream mediator in the allergic response. Nevertheless, IgE-producing IgE+ B memory cells represent only a small proportion of the overall memory B cell pool. Recent studies have identified subpopulations of cells within the IgG1+ B cell population that are implicated to play key roles in driving IgE-mediated allergic responses. Two independent research teams identified type 2-polarized B memory cells ($IL4R\alpha^{+}CD23^{+}$

Fig. 31.1 (continued) dampening pro-allergic, IgE-mediated responses. Demethylation of the *FOXP3* locus in T regulatory cells has been demonstrated, correlating with treatment outcome. Multiple studies of CD4+ T cell memory responses to allergen immunotherapy, using scRNAseq, have identified a novel subset of T helper cells that express type I interferon-stimulated genes (ISG) that emerge with treatment. Although the function of these cells is not fully elucidated, T1-IFN plays important roles in T cell differentiation and the regulation of Tfh and can suppress *GATA3*. It has also been shown to regulate Th2 and antibody production

Bmem) that are more abundant in children with high titers (>100 kU/L) of circulating sIgE than in children with low titers (<5 kU/L) and non-allergic children. Key features of this subset include the expression of germline *IGHE*, IL4 receptor gene (*IL4R*), low-affinity receptor *FCER2* (encoding CD23), and IgG (Koenig et al. 2024; Ota et al. 2024), and a highly diverse Ig gene repertoire. Allergen-specific type 2-polarized Bmem are poised to switch to IgE-producing plasma cells and thus serve as a reservoir of IgE. The relevance of this subset has been demonstrated in sublingual immunotherapy and murine models of allergic sensitization (Koenig et al. 2024).

As mentioned earlier, sensitized individuals who produce sIgE may not develop clinical allergies. A recent study suggests that IgE glycosylation may mediate progression to allergy (Shade et al. 2020). IgE sialic acid levels were elevated in children with peanut allergy compared to non-atopic individuals. Furthermore, in allergy models, removal of sialic acid ameliorated degranulation and subsequent anaphylaxis (Shade et al. 2020).

31.1.4.3 T Follicular Helper (Tfh) Cells

Tfh cells are a specialized subset of T helper cells that control the class switching, affinity, and longevity of B cells. They contribute to the production of high-affinity allergen-specific IgE by the production of Th2 cytokines, such as IL-4, and in some subsets, IL-13, which promote somatic class switching of B cells (Ballesteros-Tato et al. 2016). Although IL-4 is linked to IgE class switching, production of this cytokine alone is not sufficient. Tfh13 cells, a subset characterized by the production of IL-13, have been identified as the central player driving the allergic immune response (Gowthaman et al. 2019). Therefore, Tfh13 cells are thought to be a key reservoir of IL-13 contributing to allergic inflammation (Fig. 31.1).

31.1.4.4 Group 2 Innate Lymphoid cells

Group 2 innate lymphoid cells (ILC2) are a population of mucosal cells, which secrete Th2 cytokines and contribute to allergic pathology (Kim et al. 2013; Mjosberg et al. 2012). ILC2 express CD90 (Thy1), CD25 (IL-2Rα), IL-25R (IL-17RB), IL-33R (ST2), and CD127 (IL-7Rα) and are lineage-negative for markers of other lymphocytes (Walker et al. 2013). While they share characteristics of lymphocyte cells, they do not carry antigen-specific receptors. ILC2 was found to be enriched in allergic individuals and murine models of sensitization, contributing to allergic sensitization through IL-33R signaling, particularly through the production of IL-4 (Noval Rivas et al. 2016). Furthermore, IL-4 production by ILC2 was shown to impair antigen-specific Treg suppressive activity. ILC2 cells are, therefore, likely to have a twofold role in perpetuating the allergic response: first, by serving as an additional source of IL-4 production contributing to the polarization of Th2 cells, and second, by impairing Treg activity (Fig. 31.1).

31.1.4.5 Th2-Like T Regulatory Cells

A subset of impaired Treg cells, described as Th2-like Treg, has been identified in the blood of children with food allergies (Noval Rivas et al. 2015). These Treg cells were Foxp3+, yet produced IL-4 and were found to have limited capacity to suppress allergic inflammation. Furthermore, they were shown to play a role in the development of orally-induced allergic sensitization in a mouse model of food allergy (Noval Rivas et al. 2015). However, it is important to note that lineage tracing demonstrated that Th2-like Treg cells do not act as a reservoir for the production of Th2-conventional (Tcov Th2) cells and are unlikely to originate from the Tcov Th2 lineage (Noval Rivas et al. 2015). Th2-like Treg cells are considered to represent a unique phenotype of functionally impaired Treg in food-allergic individuals, contributing to the inflammatory milieu through their inability to appropriately suppress Th2-driven responses (Fig. 31.1).

31.2 Oral Immunotherapy and Clinical Outcomes That Can Be Achieved Following Treatment

Food oral immunotherapy (OIT) is now available for the treatment of food allergy. OIT involves the daily oral administration of incrementally increasing doses of the food, starting with quantities below the threshold that elicit a reaction. OIT comprises three phases: the day 1 rush induction, the build-up phase (over months), and the maintenance phase (over years). Rush induction involves a series of rapidly escalating doses administered over several hours with the aim of initiating desensitization mechanisms. The build-up phase continues the up-dosing process, with the daily dose of allergen increased gradually, typically every 2 weeks, over the course of months. Once the target maintenance dose is reached, patients continue the daily maintenance dose for months, years, or indefinitely, depending on the clinical outcome that is achieved.

There are several clinical outcomes that can be achieved following food OIT. Meta-analyses have shown that OIT induces desensitization in the majority of treated patients and leads to remission in a smaller subset (de Silva et al. 2022). Desensitization refers to an increased reaction threshold that is only maintained with continuing maintenance therapy. Drawbacks of desensitization are the need for indefinite maintenance dosing and continuing allergen avoidance. Furthermore, there is no convincing evidence that desensitization leads to improvement in the quality of life when compared to placebo treatment (standard care) (Galvin et al. 2022). Remission is the absence of clinical reactivity after treatment (allergen exposure) has been discontinued for a period of time, sufficient to ensure that desensitization effects have waned. Remission allows participants to discontinue treatment and eat the allergen freely in the diet, no longer having to adhere to allergen avoidance. Importantly, remission provides a substantial improvement in the quality of life compared to placebo (Loke et al. 2022, 2024).

Factors driving the attainment of remission remain poorly understood, but a lower pre-treatment allergen-specific sIgE level has been consistently reported to predict a greater likelihood of remission (Jones et al. 2016, 2022; Loke et al. 2022; Patil et al. 2019; Tsai et al. 2020; Vickery et al. 2014; Wright et al. 2016). Posttreatment levels of sIgE and the ratio of sIgE:sIgG4 are reported to correlate with clinical outcomes (Jones et al. 2022; Loke et al. 2022; Patil et al. 2019; Vickery et al. 2014); however, it is unknown whether sIgG4 is critically involved in supporting the attainment of remission or if it is merely indicative of allergen exposure during OIT (Hsiao et al. 2017). Younger age has been suggested to increase the likelihood of achieving remission (Jones et al. 2022), although findings are conflicting, with some studies reporting no age effects (Chinthrajah et al. 2019). The use of a higher maintenance dose and faster dose escalation during buildup (Patent application AU2020902339) or longer treatment duration (Jones et al. 2016) has also been suggested to enhance the likelihood of remission; however, evidence is limited. An open-label study of low-dose OIT (125 mg peanut protein) in 23 children, aged 3–13 years, reported remission of allergy in 13% (Blumchen et al. 2010). In contrast, two placebo-controlled randomized trials of high-dose OIT (2000 mg and 4000 mg) reported remission in 21% (20/96) of children, aged 1–3 years, and 35% (21/60) of patients, aged 7–55 years, compared with 2% (1/50) and 4% (1/25) of placebo participants, respectively (Chinthrajah et al. 2019; Jones et al. 2022). A high-dose (2000 mg) rapid escalation OIT regimen was shown to induce remission in 51% of children compared with 5% in the placebo group, suggesting a potential role for a faster buildup schedule in supporting remission (Loke et al. 2022). The importance of a higher maintenance dose for the attainment of remission may be less relevant in younger populations. One small study that compared low-dose (300 mg) compared to high-dose (3000 mg) peanut OIT in 37 children, aged 9–36 months, reported high rates of remission following both low-dose and high-dose OIT (85% vs. 71%) (Vickery et al. 2017), although some caution in interpreting findings is warranted, as there was no placebo control group. There are no randomized trials directly comparing low vs. high-dose peanut OIT in children >3 years of age.

31.3 Immune Changes Associated with Oral Immunotherapy

The immune mechanisms supporting outcomes following OIT are still under investigation. Most studies have reported immune changes associated with OIT without accounting for or distinguishing between the clinical outcomes of desensitization and remission, despite these distinct clinical states likely being supported by different immunological pathways. While OIT leads to desensitization in the majority of patients, only a subset of treated patients achieves remission. Immune changes following OIT, where the patient's clinical status has not been determined, are likely to reflect pathways that support desensitization (with or without remission) and cannot be assumed to reflect key changes driving remission. These findings are discussed in the following. Few studies have examined the immune changes specifically associated with remission; these are presented in Sect. 31.4.

Mast cells and basophils are the primary effector cells that mediate the symptoms and signs of an allergic reaction (see Sect. 31.1.3). The gradual increase in allergen dose during OIT leads to suppression of mast cell and basophil reactivity, and thereby the mitigation of IgE-mediated allergic symptoms that occur with desensitization. Basophil activation can be measured using an *ex vivo* antigen stimulation assay, the basophil activation test (BAT). BAT has high accuracy as a marker of clinical allergy status and effective desensitization during OIT (Kulis et al. 2018; Patil and Shreffler 2012). OIT is associated with strong suppression of mast cells and basophils, which can be demonstrated by reductions in the BAT. The increased threshold for basophil and mast cell degranulation (reduced BAT) occurs without associated modulation of circulating allergen-specific IgE levels; hence, changes in BAT can be seen in the absence of remission. As high-affinity IgE antibodies remain bound to the surface of mast cells, this suppressed activation state can recover rapidly once regular allergen exposure is discontinued (Ang et al. 2016), and mast cell reactivity has been shown to recover within 1.5–5 days after exposure to the desensitizing antigen is removed (Lewis et al. 2017).

Reductions in skin prick test wheal size are reported during OIT; however, these changes primarily reflect the strong suppression of mast cell activity rather than any effects on allergen-sIgE (Patil et al. 2019; Tsai et al. 2020). Changes in sIgE are best assessed by measurement of serum or plasma sIgE levels. A consistent observation with OIT is that circulating levels of allergen-sIgE increase during the early stages of desensitization before falling back to baseline levels (Jones et al. 2009; Patil et al. 2019; Varshney et al. 2011). It is currently unclear whether or how this initial increase in sIgE might be connected to the success or failure of achieving fundamental redirection of the underlying upstream allergic response that supports remission of allergy (Lopez-Sanz et al. 2022).

Antigen-specific IgG4 production is induced during OIT. sIgG4 has blocking properties, and some data suggest that it may inhibit effector cell activity via the inhibitory IgG receptor, FcγRIIb (Fig. 31.1). Importantly, sIgG4 levels do not differentiate desensitization and remission outcomes following OIT, suggesting that sIgG4 is not fundamental to or requisite for achieving remission or tolerance. Moreover, the functional properties of sIgG4, including its affinity, clonality, and specificity, influence the capacity for effector cell inhibition rather than the quantity of sIgG4 (Dodev et al. 2015; Holm et al. 2011). In other forms of allergen immunotherapy, the functional properties of sIgG4 were found to better correlate with clinical outcomes than levels of allergen-specific sIgG4 (Shamji et al. 2012).

31.3.1 Findings from Systems Biology Approaches

Systems biology is a powerful tool to decipher the processes governing complex immune disorders such as allergies. Systems biology enables the examination of the intricate interactions within biological systems, emphasizing the integration of data, such as multi-omics, from various levels of biological organization—from genes and proteins to cells and tissues. Systems biology pairs computational techniques with these whole systems approaches to understand the complex interplay of genes, proteins, and cellular interactions involved in biological states and processes. Whole transcriptome approaches involve the comprehensive sequencing of all RNA molecules expressed in a given cell or tissue at a specific time, providing a snapshot of gene expression profiles and revealing comprehensive cellular activity and regulatory networks. Transcriptomic data can be used to identify key genes and pathways that are activated or suppressed in response to specific treatments, enabling a deeper understanding of how therapies work at the molecular level. In the setting of OIT, transcriptomics can be used to study immune response mechanisms and identify biomarkers associated with successful treatment outcomes. The following sections examine the evidence from transcriptomic and systems biology approaches on mechanisms of OIT.

31.3.1.1 Differential Gene Expression

31.3.1.1.1 Peripheral Blood Mononuclear Cells (PBMCs)

Peripheral blood samples are often routinely collected in the clinical setting. Thus, gene expression profiling of PBMCs offers a valuable approach for identifying potential biomarkers associated with treatment outcomes in food allergies. However, variability in gene expression due to factors such as age, environmental influences, and individual genetic backgrounds can introduce noise and limit generalizability and usefulness. Additionally, the dynamic nature of immune responses means that gene expression profiles may change over time, potentially limiting the applicability of identified biomarkers. Nevertheless, some promising gene candidates have been identified in the setting of egg OIT. Gene expression profiling of PBMCs after 3 and 8 months of egg OIT identified specific signals linked to full and partial desensitization. Activation of NK-related gene activity was associated with full desensitization, defined as passing the end-of-treatment food challenge after 8 months of treatment, compared to those who reacted during the end-of-treatment food challenge (Karisola et al. 2021). After 8 months of treatment, egg OIT was associated with downregulation in IL-6, IL-17, and TREM1 pathways. Gene activity linked to IL-6, IL-17, and IL-10 signaling pathways was inversely correlated with antigen-specific IgG4 production. Partial desensitization was linked to a delayed regulation of the gene expression changes linked to full desensitization. As there was no additional challenge for remission following OIT, the full desensitization group is likely a mix of individuals with more lasting protection and desensitization without achieving remission; therefore, it cannot be determined whether the identified changes in inflammatory gene activity and NK activity are required for remission.

31.3.1.1.2 B Cells

B cells are a key mediator of the aberrant immunological response to allergens in those with food allergies. While studies on B cells during OIT are limited, a recent study in milk OIT has provided some novel insights. Allergen-specific B cells, profiled before and after milk OIT, exhibited partial shutdown of gene programs associated with desensitization without remission, in contrast to those with remission (Satitsuksanoa et al. 2024). The programs that were differentially expressed in the presence or absence of remission included ten pathways, spanning immunoglobulin-related immune responses, B cell activation, cytokine activity, receptor-mediated activity, and other core memory responses. Desensitization without remission was also associated with a unique profile at study entry; antigen-binding genes TAP1, ITGA4, and EP300 were upregulated in individuals who failed to achieve remission (Satitsuksanoa et al. 2024). Of note, in this study, modulation of gene expression profiles by milk OIT was only detected in the allergen-specific B cell compartment (Satitsuksanoa et al.

2024), with enrichment of IGHA and IGHG genes, encoding IgA heavy-chain and IgG heavy-chain subunits, respectively, consistent with reports of elevated serum levels of sIgA and sIgG following OIT (Fig. 31.1) (Jones et al. 2009, 2022; Kim et al. 2020; Loke et al. 2022; Patil et al. 2019; Tsai et al. 2020; Vickery et al. 2014).

31.3.1.2 Single-Cell Transcriptomics

Single-cell RNA sequencing (scRNA-seq) is rapidly transforming systems biology. Bulk RNA sequencing generates gene expression data averaged across composite cells arising from a sample, often containing a mix of cells in different unique states and/or of different cell types; in contrast, scRNA-seq provides information on the genes expressed within each cell of a profiled sample. This approach preserves the unique states and cellular identities of the sample, providing resolution of the unique gene expression response of each cell type and/or substrate. Applications of scRNA-seq data include, but are not limited to, cell clustering, differential expression, abundance analysis, and trajectory analysis. Clustering and abundance analysis can be used to identify unique sub-states or cell types associated with specific biological states such as those differentiating treatment responders. Differential expression analysis can highlight unique gene programs within given cell types that are enriched in a studied biological state or group. Trajectory analysis provides insights into the differentiation states of cells, helping to determine whether cells are directed toward a specific lineage at a specific time point or within a phenotype of interest. In the context of food oral immunotherapy, a handful of studies have been carried out on allergen-specific T helper and T regulatory cells.

31.3.1.2.1 Allergen-Specific T Cells

Although T cells are not the only source of type 2 immune cells, these populations, especially Tfh and Th2, but also potentially Th2-like Treg, are generally considered the most potent producers of Th2 cytokines and perpetuators of the allergic inflammation feed-forward loop. Thus, unsurprisingly, allergen-responsive T cells represent the best-studied population in the mechanisms of OIT. There are mixed reports on whether OIT modifies the frequency of allergen-specific Th2A. Data generated using scRNAseq of peanut-reactive T cells demonstrated that OIT reduced the expression of Th2 effector genes within peanut-specific Th2A subsets; however, this reduction was less pronounced in those who achieved desensitization without remission compared to those who achieved remission (Monian et al. 2022). The persistent reservoir of these cells in those who achieved desensitization without remission may contribute to the failure to achieve lasting protection.

Furthermore, a transient increase in TGF-beta expressing peanut-specific CD4 + T cells, measured using scRNAseq, has been observed 1 year into OIT, and this increase was associated with passing a food challenge 3 months after completing 2 years of treatment (Wang et al. 2020). However, the group was a mix of individuals who either followed a dietary exclusion of the eliciting allergen following the end of treatment or continued to receive daily peanut doses (300 mg) between the end-of-treatment food challenge and the 3-month post-treatment remission challenge. Therefore, while TGF-beta may support desensitization, it is not possible to conclude whether TGF-beta expression is a requirement for remission, since some of the groups continued on a daily desensitization therapy and may not have achieved remission.

Furthermore, the downregulation of gamma-delta T cells may play a role in OIT-related immune modulation as an early decrease, measured using scRNAseq, has been identified during treatment. However, it is unclear if this change is required to achieve remission, since the mechanistic data were not resolved by outcome following the treatment (Anvari et al. 2021).

31.3.1.2.2 Baseline Signatures Associated with Remission

Differences in immune status at baseline (pre-treatment) have been shown to predict the likelihood of achieving remission following OIT. In addition to a lower level of allergen-specific sIgE, differences in gene expression within certain cell subsets have been linked to success or failure in achieving remission. As already discussed above, individuals with lower allergen-specific IgE have greater odds of achieving remission (Chinthrajah et al. 2019; Jones et al. 2022). Among gene signatures, measured with scRNAseq, elevated inflammatory gene expression profiles in Th1 and Th17 cells in peanut-specific T cells at screening were associated with a failure to achieve remission (Monian et al. 2022). Furthermore, the inflammatory gene expression profiles in these individuals were not modulated by OIT.

31.4 Immune Changes Associated with Remission Following Oral Immunotherapy

Characterizing the immune mechanisms underlying remission will enable optimized OIT protocols that lead to higher remission rates. Delineating the pathways involved in the transition from allergy to tolerance may also lead to the identification of potential biomarkers for monitoring treatment success and may inform the development of targeted therapies that enhance the likelihood of achieving long-lasting remission.

31.4.1 Changes in Downstream Effector Cell Pathways

Due to the direct role of effector cells such as mast cells and basophils in establishing the inflammatory response to allergen, understanding how these cells behave during treatment may reveal important patterns that correlate with clinical outcomes, particularly the achievement of remission. Reduction in BAT is primarily an indicator of effective desensitization, and BAT does not distinguish between the presence or absence of remission at the time of testing. Furthermore, baseline BAT levels do not predict clinical outcomes following OIT. However, two studies suggest that the trajectory of BAT reduction during the early stages of OIT may correlate with the likelihood of achieving remission. Children who achieved remission following OIT had an early reduction in BAT, whereas those who only achieved desensitization demonstrated a more delayed reduction in BAT (Patil et al. 2019; Tsai et al. 2020). These patterns appear to correlate with an overall delayed immunomodulatory response experienced by individuals who are not on the pathway to remission.

31.4.2 Immune Changes Driving Remission Involve Redirection of Upstream Components of Allergic Response

Remission is underpinned by redirection of the allergic response toward tolerance. Applying fundamental immunological principles, this process is expected to depend upon abrogation of Th2 immune activity and subsequent emergence of a newly formed tolerogenic immune response. Successful transition from allergy to tolerance is expected to require initial silencing of the Th2 dominant response, because residual Th2 activity will continue to reinforce the established allergic response through potent positive feedback effects of Th2 cytokine production.

31.4.2.1 Allergen-Specific T Cells

While OIT appears to have suppressive effects on the activity of Th2 cells in the setting of both desensitization and remission, the durability of Th2 suppression with lasting abrogation of Th2 activity may be important for achieving remission. Targeted Th2 cytokine profiling in samples collected during OIT, regardless of clinical outcome, showed suppression of Th2 cytokine production, including IL-5 and IL-13, following OIT (Blumchen et al. 2010, 2019). Importantly, however, decreases in both Th2 clonal expansion (Monian et al. 2022; Ryan et al. 2016) and Th2 gene activity (Ashley et al. 2022; Blumchen et al. 2010, 2019; Monian et al. 2022) appear to be associated with remission and lasting protection following OIT (Fig. 31.1). Specifically, the data suggest that durable suppression of Th2, including clonal anergy, is associated with attainment of remission, while suppression of Th2 cytokines may occur with desensitization, regardless of whether remission is achieved.

The role of Treg cells in driving clinical outcomes following OIT remains unclear. A study tracked antigen-induced T cells from 23 participants undergoing OIT for 24 months by flow cytometry and identified a significant induction of peanut-specific memory Treg at the end of treatment compared to pretreatment. Remission was associated with demethylation of *FOXP3* gene ($n = 7$) compared to those who failed to achieve remission (Fig. 31.1) (Syed et al. 2014). Yet, numerous other studies using gene expression profiling have failed to detect an increase in the frequency of allergy-specific Treg with OIT and OIT-induced remission (Calise et al. 2023; Monian et al. 2022; Ryan et al. 2016).

31.4.2.2 Other Cell Types Driving Upstream Events in the Allergic Pathway

Dendritic cells are specialized APCs and have a central role in determining the type of immune responses to allergens. They direct polarization of T cells toward Th2-derived allergic responses or toward tolerogenic Treg. The inflammatory activity of DCs in infants has been linked with persistent food allergy (Neeland et al. 2018), whereas higher frequencies of plasmacytoid DC (pDC) were linked to the development of tolerance (Neeland et al. 2020). Thus, it is likely that DCs are key drivers of the tolerogenic T cell response leading to remission following OIT. Indeed, changes to the functional properties of pDCs during OIT have been detected. Increased production of IFN-α, concurrent with decreased production of IL-6 and subsequently IL-13 cytokines in pDC-T cell cocultures, was identified during treatment (Frischmeyer-Guerrerio et al. 2014). As pDC is the primary source of IFN-α (type I interferon) and type I interferon activity can both elicit Treg development and suppress Th2 activity, this OIT-induced modulation of functional pDC activity may represent a pathway through which Th2 activity is suppressed during OIT (Chen et al. 2008, 2020; Huber et al. 2014). Subjects with remission demonstrated early reductions in IL-6 production from pDC-T cell cocultures as early as 12 weeks into OIT, while reduced IL-6 production in cocultures was only detected at the end of treatment (60 weeks) in subjects who were desensitized without remission. Increased production of IFN-α to TLR7 occurred after 60 weeks of treatment in both groups and did not distinguish those who achieved remission ($n = 5$) from those who did not achieve remission ($n = 5$), although numbers in each group were small.

31.4.3 Findings from Systems Biology Approaches

31.4.3.1 Differential Gene Expression

31.4.3.1.1 B Cells

The kinetics of antibody production during OIT have been described in multiple studies (reviewed above), yet it is unknown if functional changes to B cells are required to support remission following OIT. For example, B cell studies in food-allergic individuals show that IgG can act as a reservoir for IgE, yet whether treatment can disrupt this process remains unknown. Gene expression studies have identified changes associated with treatment outcomes that are in line with the kinetics of antibody titers measured during OIT from patient plasma. They also reveal that memory B cell responsiveness to cytokine activity is inhibited by OIT, and the level of suppression correlates with treatment outcomes. Allergen-specific B cells profiled before and after milk OIT exhibited a decrease in cytokine receptor genes on B cells (IL1RAP, IL12RB2, IL18R1), as well as a high-level expression of immunoglobulin constant regions, and these changes were linked to remission and natural tolerance (Satitsuksanoa et al. 2024). Among these, *IGHD* encoding the IgD heavy chain was elevated in the serum of individuals with natural resolution; this finding is relevant because IgD can attenuate basophil degranulation by IgE (Shan et al. 2018). Downregulation of ten gene expression pathways, including those involved in B cell activation, cytokine/chemokine receptor signaling, responses mediated by immunoglobulins, and chemotaxis-related genes, was also reported in subjects who achieved remission, with complete abrogation (no expression) of these gene expressions in some cases (Satitsuksanoa et al. 2024).

31.4.3.2 Gene Co-expression Network Analysis

Gene co-expression network analysis is a powerful approach to studying genes within their full context. Genes work in coordinated pathways to achieve immune functions and network analysis is an approach capable of capturing this detail. Gene co-expression exhibits scale-free properties, with genes that are involved in a specific biological process being connected within modules. Within these gene co-expression modules, most genes only have a small number of connections, whereas a small number of so-called hub genes are highly interconnected and direct the biological response. As drivers of biological processes, hub genes represent prime targets for novel therapeutics. At the time of writing, only one gene co-expression study has been published in the context of food OIT (Ashley et al. 2022). Gene co-expression networks were examined in peanut-stimulated CD4 + T cells obtained before and after treatment from OIT-treated subjects, who achieved remission, and placebo-treated subjects who remained allergic. Findings showed that remission was linked to a complete shutdown of Th2 gene programs and uncoupling of Th2 and cell cycle gene co-expression. Following 18 months of peanut OIT, Th2 gene hubs that were dominant prior to treatment became integrated with a newly formed regulatory module driven by type I interferon (IFN) gene hubs (Fig. 31.1). This rewiring of the allergen-specific immune response remained stable up to 3 months post-treatment. In contrast, in placebo-treated individuals who remained allergic, the Th2-driven gene co-expression module in peanut-stimulated CD4 + T cells remained dominant at both ends of treatment and 3 months posttreatment. Strong Th2-cell cycle co-expression persisted at all three time points, while Th2 and type I IFN gene modules remained separated in the network topology and did not merge (Ashley et al. 2022).

31.4.3.3 Single-Cell Transcriptomics

ScRNA-seq approaches can provide greater resolution to reveal the gene programs and specific subsets that drive the redirection from allergy to remission induced by OIT. To date, peanut-specific T cell responses before, during, and after peanut OIT remain the most extensively studied. An early single-cell gene profiling study of allergen-specific T cells in five peanut-allergic individuals undergoing OIT profiled 22 allergy-related genes and tracked Ara h 2 Dextramer+ (Ara h 2-specific) CD4+ T clones over time (Ryan et al. 2016). OIT was associated with an expansion of Th2 (IL4 + IL13 + CD27^{low}) cells within the first 3 months of treatment, regardless of whether subjects became desensitized without remission or achieved remission (Ryan et al. 2016). Contemporaneously, a reduced number of Treg (CD25 + FOXP3 + IL-10+) cells has been reported during the first 3 months of OIT. This suggests that during the early phase of OIT, low-dose allergen exposure first exacerbates Th2-driven immunity, creating an inflammatory milieu high in IL-4 that inhibits Treg expansion and function. Thereafter, an expansion of anergic-type cells was observed, and this induction of anergic-type cells was associated with the successful attainment of remission (defined as clinical unresponsiveness to food challenge after 3 months of allergen avoidance) (Fig. 31.1) (Ryan et al. 2016).

Numerous studies have revealed that while the frequency of allergen-specific Th2 cells is not always depleted by OIT, their transcriptional program is often downregulated, particularly in those who achieved remission following OIT. Moreover, the downregulation of Th2 gene programs has been demonstrated to persist even after 12 weeks of allergen avoidance following OIT in those with remission (Monian et al. 2022). Interestingly, in the same study, T follicular helper-like cells were not modified by OIT (Monian et al. 2022). These findings provide compelling evidence that durable suppression of Th2 is a key feature of successful OIT-induced remission (Ashley et al. 2022; Monian et al. 2022; Ryan et al. 2016).

31.5 Knowledge Gaps

Recent studies of circulating cell populations have shed light on the immune mechanisms of OIT and specific changes linked to clinical outcomes following OIT. Nevertheless, the complex interplay between lymphocytes with Th2-supporting phenotypes and the kinetics of antibody production during OIT remain largely undescribed. Many of the pathogenic allergen-specific subsets that contribute to the development of food allergy reside in tissues and are difficult to study in humans. Mechanistic studies have largely focused on cell types that are easily accessible in the peripheral blood, including basophils, and circulating T and B cells, or relied on mouse models that may not fully reflect the pathology seen in humans. Because of this, the roles of tissue-resident cell types in the context of OIT remain poorly understood, leaving an incomplete picture of how OIT leads to remission. Systems biology approaches are beginning to elucidate some of the cellular changes leading to remission, but data from cell types aside from antigen-specific T cells are needed to complete the picture. OIT is on the path to global implementation, yet key questions remain regarding the mechanisms that support lasting protection through remission. More detailed understanding of the fundamental immune changes that support remission is needed to facilitate the refinement of existing treatments and the discovery of novel approaches that can achieve the ultimate goal of "tolerance."

References

Ang WX, Church AM, Kulis M, Choi HW, Burks AW, Abraham SN (2016) Mast cell desensitization inhibits calcium flux and aberrantly remodels actin. J Clin Invest 126(11):4103–4118. https://doi.org/10.1172/JCI87492

Anvari S, Watkin L, Rajapakshe K, Hassan O, Schuster K, Coarfa C, Davis CM (2021) Memory and naive gamma delta regulatory T-cell gene expression in the first 24-weeks of peanut oral immunotherapy. Clin Immunol 230:108820. https://doi.org/10.1016/j.clim.2021.108820

Ashley SE, Jones AC, Anderson D, Holt PG, Bosco A, Tang MLK (2022) Remission of peanut allergy is associated with rewiring of allergen-driven T helper 2-related gene networks. Allergy 77(10):3015–3027. https://doi.org/10.1111/all.15324

Ballesteros-Tato A, Randall TD, Lund FE, Spolski R, Leonard WJ, Leon B (2016) T follicular helper cell plasticity shapes pathogenic T helper 2 cell-mediated immunity to inhaled house dust mite. Immunity 44(2): 259–273. https://doi.org/10.1016/j.immuni.2015.11.017

Bevins CL, Salzman NH (2011) Paneth cells, antimicrobial peptides and maintenance of intestinal homeostasis. Nat Rev Microbiol 9(5):356–368. https://doi.org/10.1038/nrmicro2546

Beyer K, Morrow E, Li XM, Bardina L, Bannon GA, Burks AW, Sampson HA (2001) Effects of cooking methods on peanut allergenicity. J Allergy Clin Immunol 107(6):1077–1081. https://doi.org/10.1067/mai.2001.115480

Blumchen K, Ulbricht H, Staden U, Dobberstein K, Beschorner J, de Oliveira LC et al (2010) Oral peanut immunotherapy in children with peanut anaphylaxis. J Allergy Clin Immunol 126(1):83–91 e81. https://doi.org/10.1016/j.jaci.2010.04.030

Blumchen K, Trendelenburg V, Ahrens F, Gruebl A, Hamelmann E, Hansen G et al (2019) Efficacy, safety, and quality of life in a multicenter, randomized, placebo-controlled trial of low-dose peanut oral immunotherapy in children with peanut allergy. J Allergy Clin Immunol Pract 7(2):479–491 e410. https://doi.org/10.1016/j.jaip.2018.10.048

Calise J, DeBerg H, Garabatos N, Khosa S, Bajzik V, Calderon LB et al (2023) Distinct trajectories distinguish antigen-specific T cells in peanut-allergic individuals undergoing oral immunotherapy. J Allergy Clin Immunol 152(1):155–166.e159. https://doi.org/10.1016/j.jaci.2023.03.020

Chen W, Liang X, Peterson AJ, Munn DH, Blazar BR (2008) The indoleamine 2,3-dioxygenase pathway is essential for human plasmacytoid dendritic cell-induced adaptive T regulatory cell generation. J Immunol 181(8):5396–5404. https://doi.org/10.4049/jimmunol.181.8.5396

Chen F, Rowe RK, Gill MA, Farrar JD (2020) Type I interferon suppresses memory Th2 cell cytokine secretion from allergic subjects. Allergy 75(3):695–698. https://doi.org/10.1111/all.14056

Chinthrajah RS, Purington N, Andorf S, Long A, O'Laughlin KL, Lyu SC et al (2019) Sustained outcomes in oral immunotherapy for peanut allergy (POISED study): a large, randomised, double-blind, placebo-controlled, phase 2 study. Lancet

394(10207):1437–1449. https://doi.org/10.1016/S0140-6736(19)31793-3

de Silva D, Rodriguez Del Rio P, de Jong NW, Khaleva E, Singh C, Nowak-Wegrzyn A et al (2022) Allergen immunotherapy and/or biologicals for IgE-mediated food allergy: a systematic review and meta-analysis. Allergy 77(6):1852–1862. https://doi.org/10.1111/all.15211

Dodev TS, Bowen H, Shamji MH, Bax HJ, Beavil AJ, McDonnell JM et al (2015) Inhibition of allergen-dependent IgE activity by antibodies of the same specificity but different class. Allergy 70(6):720–724. https://doi.org/10.1111/all.12607

Frischmeyer-Guerrerio PA, Keet CA, Guerrerio AL, Chichester KL, Bieneman AP, Hamilton RG et al (2014) Modulation of dendritic cell innate and adaptive immune functions by oral and sublingual immunotherapy. Clin Immunol 155(1):47–59. https://doi.org/10.1016/j.clim.2014.08.006

Galvin AD, Vereda A, Rodríguez del Río P, Muraro A, Jones C, Ryan R, Norval D, Jobrack J, Anagnostou A, Wang J (2022) Children and caregiver proxy quality of life from peanut oral immunotherapy trials. Clin Transl Allergy 12(12):e12213. https://doi.org/10.1002/clt2.12213. PMID: 36573312; PMCID: PMC9762119

Gerbe F, Jay P (2016) Intestinal tuft cells: epithelial sentinels linking luminal cues to the immune system. Mucosal Immunol 9(6):1353–1359. https://doi.org/10.1038/mi.2016.68

Gowthaman U, Chen JS, Zhang B, Flynn WF, Lu Y, Song W et al (2019) Identification of a T follicular helper cell subset that drives anaphylactic IgE. Science 365(6456). https://doi.org/10.1126/science.aaw6433

Han H, Roan F, Ziegler SF (2017) The atopic march: current insights into skin barrier dysfunction and epithelial cell-derived cytokines. Immunol Rev 278(1):116–130. https://doi.org/10.1111/imr.12546

Holm J, Willumsen N, Wurtzen PA, Christensen LH, Lund K (2011) Facilitated antigen presentation and its inhibition by blocking IgG antibodies depends on IgE repertoire complexity. J Allergy Clin Immunol 127(4):1029–1037. https://doi.org/10.1016/j.jaci.2011.01.062

Hsiao KC, Ponsonby AL, Axelrad C, Pitkin S, Tang MLK, Team PS (2017) Long-term clinical and immunological effects of probiotic and peanut oral immunotherapy after treatment cessation: 4-year follow-up of a randomised, double-blind, placebo-controlled trial. Lancet Child Adolesc Health 1(2):97–105. https://doi.org/10.1016/S2352-4642(17)30041-X

Huber JP, Gonzales-van Horn SR, Roybal KT, Gill MA, Farrar JD (2014) IFN-alpha suppresses GATA3 transcription from a distal exon and promotes H3K27 trimethylation of the CNS-1 enhancer in human Th2 cells. J Immunol 192(12):5687–5694. https://doi.org/10.4049/jimmunol.1301908

Jones SM, Pons L, Roberts JL, Scurlock AM, Perry TT, Kulis M et al (2009) Clinical efficacy and immune regulation with peanut oral immunotherapy. J Allergy Clin Immunol 124(2):292-300, 300 e291-297. https://doi.org/10.1016/j.jaci.2009.05.022

Jones SM, Burks AW, Keet C, Vickery BP, Scurlock AM, Wood RA et al (2016) Long-term treatment with egg oral immunotherapy enhances sustained unresponsiveness that persists after cessation of therapy. J Allergy Clin Immunol 137(4):1117–1127 e1110. https://doi.org/10.1016/j.jaci.2015.12.1316

Jones SM, Kim EH, Nadeau KC, Nowak-Wegrzyn A, Wood RA, Sampson HA et al (2022) Efficacy and safety of oral immunotherapy in children aged 1-3 years with peanut allergy (the Immune Tolerance Network IMPACT trial): a randomised placebo-controlled study. Lancet 399(10322):359–371. https://doi.org/10.1016/S0140-6736(21)02390-4

Karisola P, Palosuo K, Hinkkanen V, Wisgrill L, Savinko T, Fyhrquist N et al (2021) Integrative transcriptomics reveals activation of innate immune responses and inhibition of inflammation during oral immunotherapy for egg allergy in children. Front Immunol 12:704633. https://doi.org/10.3389/fimmu.2021.704633

Kim BS, Siracusa MC, Saenz SA, Noti M, Monticelli LA, Sonnenberg GF et al (2013) TSLP elicits IL-33-independent innate lymphoid cell responses to promote skin inflammation. Sci Transl Med 5(170):170ra116. https://doi.org/10.1126/scitranslmed.3005374

Kim EH, Perry TT, Wood RA, Leung DYM, Berin MC, Burks AW et al (2020) Induction of sustained unresponsiveness after egg oral immunotherapy compared to baked egg therapy in children with egg allergy. J Allergy Clin Immunol 146(4):851–862 e810. https://doi.org/10.1016/j.jaci.2020.05.040

Kivisto JE, Clarke A, Dery A, De Schryver S, Shand G, Huhtala H et al (2019) Genetic and environmental susceptibility to food allergy in a registry of twins. J Allergy Clin Immunol Pract 7(8):2916–2918. https://doi.org/10.1016/j.jaip.2019.05.016

Koenig JFE, Knudsen NPH, Phelps A, Bruton K, Hoof I, Lund G et al (2024) Type 2-polarized memory B cells hold allergen-specific IgE memory. Sci Transl Med 16(733):eadi0944. https://doi.org/10.1126/scitranslmed.adi0944

Koplin JJ, Allen KJ, Gurrin LC, Peters RL, Lowe AJ, Tang ML et al (2013) The impact of family history of allergy on risk of food allergy: a population-based study of infants. Int J Environ Res Public Health 10(11):5364–5377. https://doi.org/10.3390/ijerph10115364

Koplin JJ, Peters RL, Ponsonby AL, Gurrin LC, Hill D, Tang ML et al (2014) Increased risk of peanut allergy in infants of Asian-born parents compared to those of Australian-born parents. Allergy 69(12):1639–1647. https://doi.org/10.1111/all.12487

Kratchmarov R, Djeddi S, Dunlap G, He W, Jia X, Burk CM et al (2024) TCF1-LEF1 co-expression identifies a multipotent progenitor cell (T(H)2-MPP) across human allergic diseases. Nat Immunol 25(5):902–915. https://doi.org/10.1038/s41590-024-01803-2

Kulis MD, Patil SU, Wambre E, Vickery BP (2018) Immune mechanisms of oral immunotherapy. J Allergy Clin Immunol 141(2):491–498. https://doi.org/10.1016/j.jaci.2017.12.979
Lewis A, MacGlashan DW Jr, Suvarna SK, Peachell PT (2017) Recovery from desensitization of IgE-dependent responses in human lung mast cells. Clin Exp Allergy 47(8):1022–1031. https://doi.org/10.1111/cea.12912
Loke P, Orsini F, Lozinsky AC, Gold M, O'Sullivan MD, Quinn P et al (2022) Probiotic peanut oral immunotherapy versus oral immunotherapy and placebo in children with peanut allergy in Australia (PPOIT-003): a multicentre, randomised, phase 2b trial. Lancet Child Adolesc Health 6(3):171–184. https://doi.org/10.1016/S2352-4642(22)00006-2
Loke P, Wang X, Lloyd M, Ashley SE, Lozinsky AC, Gold M et al (2024) Two-year post-treatment outcomes following peanut oral immunotherapy in the probiotic and peanut oral immunotherapy-003 long-term (PPOIT-003LT) study. Allergy 79(10):2759–2774. https://doi.org/10.1111/all.16262
Lopez-Sanz C, Jimenez-Saiz R, Esteban V, Delgado-Dolset MI, Perales-Chorda C, Villasenor A et al (2022) Mast cell desensitization in allergen immunotherapy. Front Allergy 3:898494. https://doi.org/10.3389/falgy.2022.898494
Mitson-Salazar A, Yin Y, Wansley DL, Young M, Bolan H, Arceo S et al (2016) Hematopoietic prostaglandin D synthase defines a proeosinophilic pathogenic effector human T(H)2 cell subpopulation with enhanced function. J Allergy Clin Immunol 137(3): 907–918 e909. https://doi.org/10.1016/j.jaci.2015.08.007
Mjosberg J, Bernink J, Golebski K, Karrich JJ, Peters CP, Blom B et al (2012) The transcription factor GATA3 is essential for the function of human type 2 innate lymphoid cells. Immunity 37(4):649–659. https://doi.org/10.1016/j.immuni.2012.08.015
Monian B, Tu AA, Ruiter B, Morgan DM, Petrossian PM, Smith NP et al (2022) Peanut oral immunotherapy differentially suppresses clonally distinct subsets of T helper cells. J Clin Invest 132(2). https://doi.org/10.1172/JCI150634
Neeland MR, Koplin JJ, Dang TD, Dharmage SC, Tang ML, Prescott SL et al (2018) Early life innate immune signatures of persistent food allergy. J Allergy Clin Immunol 142(3):857–864 e853. https://doi.org/10.1016/j.jaci.2017.10.024
Neeland MR, Andorf S, Manohar M, Dunham D, Lyu SC, Dang TD et al (2020) Mass cytometry reveals cellular fingerprint associated with IgE+ peanut tolerance and allergy in early life. Nat Commun 11(1):1091. https://doi.org/10.1038/s41467-020-14919-4
Noval Rivas M, Chatila TA (2016) Regulatory T cells in allergic diseases. J Allergy Clin Immunol 138(3): 639–652. https://doi.org/10.1016/j.jaci.2016.06.003
Noval Rivas M, Burton OT, Wise P, Charbonnier LM, Georgiev P, Oettgen HC et al (2015) Regulatory T cell reprogramming toward a Th2-cell-like lineage impairs oral tolerance and promotes food allergy. Immunity 42(3):512–523. https://doi.org/10.1016/j.immuni.2015.02.004
Noval Rivas M, Burton OT, Oettgen HC, Chatila T (2016) IL-4 production by group 2 innate lymphoid cells promotes food allergy by blocking regulatory T-cell function. J Allergy Clin Immunol 138(3):801–811 e809. https://doi.org/10.1016/j.jaci.2016.02.030
Ota M, Hoehn KB, Fernandes-Braga W, Ota T, Aranda CJ, Friedman S et al (2024) CD23(+)IgG1(+) memory B cells are poised to switch to pathogenic IgE production in food allergy. Sci Transl Med 16(733):eadi0673. https://doi.org/10.1126/scitranslmed.adi0673
Patil SU, Shreffler WG (2012) Immunology in the clinic review series; focus on allergies: basophils as biomarkers for assessing immune modulation. Clin Exp Immunol 167(1):59–66. https://doi.org/10.1111/j.1365-2249.2011.04503.x
Patil SU, Steinbrecher J, Calatroni A, Smith N, Ma A, Ruiter B et al (2019) Early decrease in basophil sensitivity to Ara h 2 precedes sustained unresponsiveness after peanut oral immunotherapy. J Allergy Clin Immunol 144(5):1310–1319 e1314. https://doi.org/10.1016/j.jaci.2019.07.028
Pelaez-Prestel HF, Sanchez-Trincado JL, Lafuente EM, Reche PA (2021) Immune tolerance in the oral mucosa. Int J Mol Sci 22(22). https://doi.org/10.3390/ijms222212149
Ryan JF, Hovde R, Glanville J, Lyu SC, Ji X, Gupta S et al (2016) Successful immunotherapy induces previously unidentified allergen-specific CD4+ T-cell subsets. Proc Natl Acad Sci USA 113(9):E1286–E1295. https://doi.org/10.1073/pnas.1520180113
Satitsuksanoa P, van de Veen W, Tan G, Lopez JF, Wirz O, Jansen K et al (2024) Allergen-specific B cell responses in oral immunotherapy-induced desensitization, remission, and natural outgrowth in cow's milk allergy. Allergy. https://doi.org/10.1111/all.16220
Savage JH, Lee-Sarwar KA, Sordillo J, Bunyavanich S, Zhou Y, O'Connor G et al (2018) A prospective microbiome-wide association study of food sensitization and food allergy in early childhood. Allergy 73(1): 145–152. https://doi.org/10.1111/all.13232
Shade KC, Conroy ME, Washburn N, Kitaoka M, Huynh DJ, Laprise E et al (2020) Sialylation of immunoglobulin E is a determinant of allergic pathogenicity. Nature 582(7811):265–270. https://doi.org/10.1038/s41586-020-2311-z
Shamji MH, Ljorring C, Francis JN, Calderon MA, Larche M, Kimber I et al (2012) Functional rather than immunoreactive levels of IgG4 correlate closely with clinical response to grass pollen immunotherapy. Allergy 67(2):217–226. https://doi.org/10.1111/j.1398-9995.2011.02745.x
Shan M, Carrillo J, Yeste A, Gutzeit C, Segura-Garzon D, Walland AC et al (2018) Secreted IgD amplifies humoral T helper 2 cell responses by binding basophils

via galectin-9 and CD44. Immunity 49(4):709–724 e708. https://doi.org/10.1016/j.immuni.2018.08.013

Sicherer SH, Sampson HA (2018) Food allergy: a review and update on epidemiology, pathogenesis, diagnosis, prevention, and management. J Allergy Clin Immunol 141(1):41–58. https://doi.org/10.1016/j.jaci.2017.11.003

Sicherer SH, Furlong TJ, Maes HH, Desnick RJ, Sampson HA, Gelb BD (2000) Genetics of peanut allergy: a twin study. J Allergy Clin Immunol 106(1 Pt 1):53–56. https://doi.org/10.1067/mai.2000.108105

Syed A, Garcia MA, Lyu SC, Bucayu R, Kohli A, Ishida S et al (2014) Peanut oral immunotherapy results in increased antigen-induced regulatory T-cell function and hypomethylation of forkhead box protein 3 (FOXP3). J Allergy Clin Immunol 133(2):500–510. https://doi.org/10.1016/j.jaci.2013.12.1037

Tsai M, Mukai K, Chinthrajah RS, Nadeau KC, Galli SJ (2020) Sustained successful peanut oral immunotherapy associated with low basophil activation and peanut-specific IgE. J Allergy Clin Immunol 145(3):885–896 e886. https://doi.org/10.1016/j.jaci.2019.10.038

Tuzlak S, Dejean AS, Iannacone M, Quintana FJ, Waisman A, Ginhoux F et al (2021) Repositioning T(H) cell polarization from single cytokines to complex help. Nat Immunol 22(10):1210–1217. https://doi.org/10.1038/s41590-021-01009-w

Varshney P, Jones SM, Scurlock AM, Perry TT, Kemper A, Steele P et al (2011) A randomized controlled study of peanut oral immunotherapy: clinical desensitization and modulation of the allergic response. J Allergy Clin Immunol 127(3):654–660. https://doi.org/10.1016/j.jaci.2010.12.1111

Vickery BP, Scurlock AM, Jones SM, Burks AW (2011) Mechanisms of immune tolerance relevant to food allergy. J Allergy Clin Immunol 127(3):576–584.; quiz 585-576. https://doi.org/10.1016/j.jaci.2010.12.1116

Vickery BP, Scurlock AM, Kulis M, Steele PH, Kamilaris J, Berglund JP et al (2014) Sustained unresponsiveness to peanut in subjects who have completed peanut oral immunotherapy. J Allergy Clin Immunol 133(2):468–475. https://doi.org/10.1016/j.jaci.2013.11.007

Vickery BP, Berglund JP, Burk CM, Fine JP, Kim EH, Kim JI et al (2017) Early oral immunotherapy in peanut-allergic preschool children is safe and highly effective. J Allergy Clin Immunol 139(1):173–181 e178. https://doi.org/10.1016/j.jaci.2016.05.027

Walker JA, Barlow JL, McKenzie AN (2013) Innate lymphoid cells–how did we miss them? Nat Rev Immunol 13(2):75–87. https://doi.org/10.1038/nri3349

Wambre E, Bajzik V, DeLong JH, O'Brien K, Nguyen QA, Speake C et al (2017) A phenotypically and functionally distinct human TH2 cell subpopulation is associated with allergic disorders. Sci Transl Med 9(401). https://doi.org/10.1126/scitranslmed.aam9171

Wang W, Lyu SC, Ji X, Gupta S, Manohar M, Dhondalay GKR et al (2020) Transcriptional changes in peanut-specific CD4+ T cells over the course of oral immunotherapy. Clin Immunol 219:108568. https://doi.org/10.1016/j.clim.2020.108568

Worthington JJ (2015) The intestinal immunoendocrine axis: novel cross-talk between enteroendocrine cells and the immune system during infection and inflammatory disease. Biochem Soc Trans 43(4):727–733. https://doi.org/10.1042/BST20150090

Wright BL, Kulis M, Orgel KA, Burks AW, Dawson P, Henning AK et al (2016) Component-resolved analysis of IgA, IgE, and IgG4 during egg OIT identifies markers associated with sustained unresponsiveness. Allergy 71(11):1552–1560. https://doi.org/10.1111/all.12895

Functions of Macrophages in Periodontal Disease Progression and Gingival Tissue Homeostasis

32

Tsuguno Yamaguchi

Abstract

Macrophages play pivotal roles in the immune response, acting as antigen-presenting cells that influence both inflammation and tissue homeostasis. This chapter explores the versatile functions of macrophages in periodontal disease, a chronic inflammatory condition characterized by the destruction of periodontal tissues. Macrophages are primarily derived from circulating monocytes and exhibit distinct polarization states, notably M1 and M2. In the context of periodontal disease, M1 macrophages are closely associated with pro-inflammatory responses and tissue destruction, while M2 macrophages facilitate tissue repair and resolution of inflammation. Evidence demonstrates that macrophage polarization is a dynamic continuum influenced by various environmental cues such as cytokines. Notably, an increase in M1 macrophages correlates with the active stages of periodontal disease, whereas M2 macrophages are more prevalent during the recovery phase. Furthermore, macrophages significantly contribute to bone metabolism by regulating osteoclastogenesis, with M1 macrophages promoting bone resorption through the secretion of inflammatory cytokines. This chapter also addresses the role of macrophages in the interaction between systemic diseases and periodontal disease and potential therapeutic strategies aimed at targeting macrophage polarization to mitigate inflammation and enhance periodontal regeneration. Overall, macrophages play various roles in the progression of periodontal disease and gingival homeostasis. Further understanding of the multifaceted roles of macrophages is essential for developing effective host modulation therapies that restore immune balance, improve treatment outcomes, and advance our comprehension of the complex mechanisms underlying oral health and disease. This knowledge holds promise for the development of innovative therapeutic approaches in the management of periodontal disease.

Keywords

Macrophage · Innate immunity · Periodontal disease · Host modulation · Inflammation

32.1 Introduction

Periodontal disease, especially periodontitis, is a chronic inflammatory condition that impacts the structures supporting the teeth, including the gingiva, periodontal ligament, and alveolar bone (Murakami et al. 2018). It is one of the most

T. Yamaguchi (✉)
Advanced Oral Health Science Research Laboratories, Lion Corporation, Tokyo, Japan
e-mail: tsuguno@lion.co.jp

N. Rezaei, H. Ziaei (eds.), *Oral Immunology*, Advances in Experimental Medicine and Biology 1492,
https://doi.org/10.1007/978-3-032-03176-1_32

prevalent oral diseases worldwide, with significant implications for both oral and systemic health. The etiology of periodontitis is multifactorial, with dysbiosis of the oral microbiome being a key factor that triggers aberrant immune responses. Among the various immune cells involved in periodontal disease, macrophages play a central role in the initiation, progression, and resolution of inflammation (Sima et al. 2019).

In this chapter, we will be reviewing the diverse role of macrophages in periodontal disease, examining their functions, polarization states, interactions with the oral microbiome, implications in systemic and periodontal diseases, and potential therapeutic approaches. Understanding the intricate relationship between macrophages and periodontal disease would help us gain insights into the underlying mechanisms and develop more effective treatment strategies.

32.2 Macrophages: Key Players in Innate Immune Response

32.2.1 Origin and Function

Macrophages are mainly derived from monocytes, circulating in the bloodstream, by differentiation induced by various stimuli and mechanisms, such as Colony Stimulating Factor-1 (CSF-1). Another source of macrophages is tissue-resident macrophages that reside in and play critical supportive roles within their resident tissues. Macrophages are one of the antigen-presenting cells, which serve as the first line of defense and play a crucial role in both innate and adaptive immune responses against bacteria and viruses, as well as dead or damaged cells (Hoebe et al. 2004; Martin and Garcia 2021). Their major roles include the following.

Phagocytosis Macrophages engulf and digest various harmful entities, including pathogens like bacteria and viruses, as well as cellular debris and dead cells. Once the macrophages internalize these particles, they utilize enzymes and reactive oxygen species to digest them.

Cytokine Production Macrophages produce a diverse array of cytokines and chemokines and orchestrate the proliferation, migration, and activity of other immune cells. By releasing specific cytokines, macrophages play a vital role in regulating both innate and adaptive immunity.

Antigen Presentation A key function of macrophages is their ability to present antigens to T cells, which is a fundamental aspect of the adaptive immune response. After ingesting pathogens, macrophages process the foreign material and display peptide fragments on their surface. This antigen presentation is crucial for the activation of T cells; thus, macrophages bridge the innate and adaptive immune systems, ensuring a coordinated response to infections.

Tissue Repair Beyond their roles in immunity, macrophages also function in tissue repair and remodeling. Following injury or infection, they contribute to the healing process by producing a variety of growth factors and extracellular matrix components. These substances not only promote cell proliferation and migration but also help rebuild damaged tissues and restore normal function.

32.2.2 Macrophage Polarization

Macrophages are heterogeneous immune cells with diverse tissue distribution (Gordon and Martinez 2010). This is partially due to the heterogeneity of undifferentiated circulating monocytes themselves, but the major reasons for their diversity lie in the trait of macrophages themselves, which is their polarization (Gordon and Taylor 2005).

Originally, macrophages have been classified into two groups by their phenotypes: M1 and M2 macrophages. These two types of macrophages differ in the expression of receptors, cytokines, and chemokines (Fig. 32.1). Lipopolysaccharide (LPS) and interferon-γ (IFN-γ) are the major factors that induce M1 polarization in macrophages, along with other factors such as

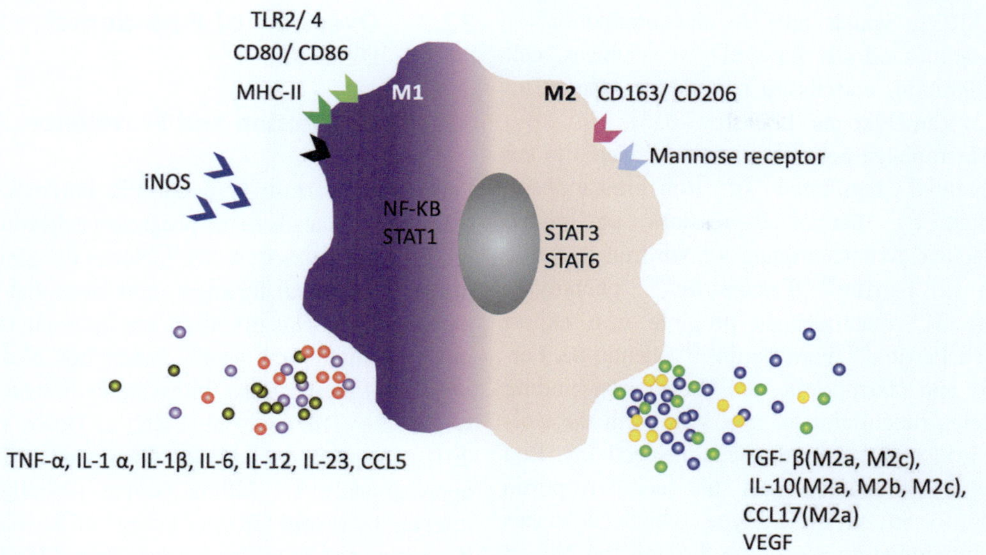

Fig. 32.1 Characteristics of polarized macrophages

Granulocyte-macrophage colony-stimulating factor (GM-CSF) and reactive oxygen species (ROS). They express CD86 and produce inflammatory cytokines, such as TNF-α, IL-1β, IL-6, IL-12, and IL-23, and chemokine (C-C motif) ligand 2 (CCL2) (Huang et al. 2019). They also express inducible nitric oxide synthase (iNOS), matrix metalloproteases-9-, and cyclooxygenase-2 (COX-2), which are utilized as M1 markers (Martinez et al. 2008). In terms of secretion, they produce high level of PGE2 from arachidonic acid, leading to induction of inflammation. M1 also shows a significant ability to phagocytose microorganisms and matrix debris, exhibiting a high antigen presentation capacity. On the contrary, M2 macrophages function as immunomodulators, producing anti-inflammatory cytokines and extracellular matrix playing a role in tissue regeneration (Chen et al. 2023). M2 macrophages have been further classified based on their function and expression of specific markers. These subsets include at least three types: M2a, M2b, and M2c, which are induced by IL-4 or IL-13, TLR agonists or IL-1 receptors, and IL-10 and glucocorticoid hormones, respectively (Mantovani et al. 2004).

Macrophage activation is accompanied by swift alterations in nutrient flow to support their immune activation and functional needs. Thus, M1 and M2 macrophages differ in their metabolic profiles as well (Zhu et al. 2015). M1 macrophages enhance glucose uptake and lactate production while exhibiting a reduced oxygen consumption rate. This shift toward anaerobic glycolysis enables them to fulfill their immediate energy needs and support the increased demand for biosynthetic precursors necessary for the production of pro-inflammatory proteins (Sun et al. 2012b). In contrast, M2 macrophages predominantly rely on oxidative glucose metabolism. Another metabolite closely related to the functions of polarized macrophage is L-arginine. In macrophages treated with LPS, the transport of L-arginine is upregulated, which triggers changes in the cationic amino acid transporter and favors the preferential uptake of L-arginine. The metabolism of L-arginine in polarized macrophages is tightly regulated by two enzymes: iNOS and ARG1. iNOS converts L-arginine into nitric oxide (NO) and L-citrulline, whereas ARG1 converts it into ornithine and urea (Messeri Dreissig et al. 2000). M1 macrophages are characterized by the expression of iNOS and the production of NO, which is essential for their microbicidal functions. On the other hand, IL-4-activated M2 macrophages show increased levels of ARG1, which aids in the production of polyamines critical for collagen synthesis, cell proliferation, and tissue remodeling (Hu et al. 2012; Van Dyke and Locksley 2013).

Macrophage polarization is also related to the differential regulation of iron metabolism, marked by distinct expressions of related molecules. M1 macrophages favor iron retention with a Ferritin^{high}/Ferroportin^{low} phenotype, while M2 macrophages promote iron export with a Ferritin^{low}/Ferroportin^{high} phenotype. Ferritin and ferroportin are both iron-binding proteins functioning in iron storage in the cells and iron export from the cells, respectively. Iron retention in macrophages that lack ferroportin leads to an M1 phenotype, which enhances pro-inflammatory responses through the NF-kB pathway and impairs tissue repair. In contrast, M2 macrophages maintain low intracellular iron due to elevated ferroportin and heme oxygenase 1 (HO-1) expression, which converts heme into beneficial byproducts like bilirubin and carbon monoxide. This reduced iron level also inhibits the expression of pro-inflammatory cytokines and iNOS. Additionally, the metabolism of fatty acids and vitamins is also linked to the polarization of macrophages (Zhu et al. 2015).

While much has been discovered on M1 and M2 macrophages, it is now known that macrophage polarization represents a continuum of functional states, and the classifications of M1 and M2 are merely an illustration of typical phenotype. Moreover, in vivo macrophages do not always exhibit the same transcriptional profiles as their in vitro counterparts, and surface markers identified in vitro may not accurately represent in vivo phenotypes. However, what the M1/M2 classification has proven is that they can exhibit different functional states depending on the signals they receive from their environment, which allows them to adapt to various situations in the body (Porcheray et al. 2005). Moreover, macrophage polarization has been known to be plastic and fully reversible, suggesting that one macrophage may first take part in inflammation and later function in its resolution, and this versatility makes them essential for the homeostasis of various tissues including periodontium.

32.3 Overview of Periodontal Disease

32.3.1 Definition and Prevalence

Periodontal disease is a chronic inflammatory condition that leads to the progressive destruction of periodontal tissue, which includes the gingival tissue, periodontal ligament, and bone that support the teeth. This disease is not only one of the most common oral health issues but also has significant implications for systemic diseases and well-being (Brown et al. 2002; O'Dowd et al. 2010). According to a study conducted in 2017, approximately 1.1 billion people globally are affected by severe forms of periodontitis, making it a major public health burden. Periodontal disease is the leading cause of tooth loss in adults, affecting their ability to chew, speak, and maintain proper nutrition, significantly impacting the quality of life, especially in the aged population. (Tonetti et al. 2017).

32.3.2 Etiology

Periodontitis is primarily caused by aberrant and chronic inflammatory responses and the resulting tissue destruction of the host, which are triggered by dysbiosis within the dental biofilm. Furthermore, periodontitis is a multifactorial disease influenced by a range of risk factors. Aging plays a significant role, as seen in the increased prevalence of this disease in the aged population. Smoking is another critical risk factor, along with systemic diseases, such as diabetes, cardiovascular diseases, autoimmune disorders, and immune dysfunctions, whether genetic or acquired. (Murakami et al. 2018; Van Dyke et al. 2020). The inflammatory responses in periodontal tissues are orchestrated by a variety of immune cells, such as neutrophils, macrophages, dendritic cells, and T and B cells. Macrophages, in particular, play a central role as a major source of inflammatory cytokines. (Clark et al. 2021; Sima et al. 2019).

32.3.3 Pathophysiology

The pathophysiology of periodontitis involves a complex interplay between the host immune response and pathogens. The initial accumulation of dental plaque triggers the activation of the immune system, leading to an influx of immune cells, including macrophages, into the periodontal tissues. This immune response is characterized by the release of pro-inflammatory cytokines, chemokines, and other mediators that promote inflammation to fight infection while also eliciting tissue destruction.

The chronic nature of periodontitis is due to the persistent presence of pathogens and the prolonged immune response employing both innate and adaptive immune systems. This chronic inflammation can lead to the degradation of the extracellular matrix, loss of periodontal attachment, and eventual tooth loss.

32.4 Macrophage Functions in Periodontal Disease Progression

32.4.1 Immune Response against Bacterial Infection and Dysbiosis

M1 macrophages are the main phenotype that responds to bacterial infection. They detect pathogen-associated molecular patterns (PAMPs) on pathogen surfaces via pattern-recognition receptors (PRRs). Upon interaction of PRRs and PAMPs, the macrophage membrane alters shape with the actin cytoskeleton, encasing the pathogens and forming phagosomes that subsequently fuse with lysosomes for digestion. Simultaneously, M1 macrophages aid in pathogen elimination by releasing pro-inflammatory cytokines, such as IL-1β and TNF-α, and initiating the adaptive immune response by upregulating chemokines and presenting antigens to T and B cells. Moreover, dysbiosis of the oral microbiome, which triggers the periodontal disease itself, may be one of the direct causes of macrophage polarization. Oral pathogens can activate different macrophage responses, depending on the species. For instance, a key pathogen for periodontal disease, *Porphyromonas gingivalis*, regulates the metabolism of macrophages and elicits polarization toward M1 macrophages and further inhibits the polarization toward M2. Another periodontal pathogen, *Aggregatibacter actinomycetemcomitans,* along with *Tannerella forsythia,* a member of the "red complex" pathogens, mainly elicit polarization toward M1 macrophages, while oral commensal bacteria primarily elicit M2-like macrophages (Ballini et al. 2020; Tamai et al. 2009).

While the inflammatory response of macrophages is to eliminate pathogens, occasionally it may work to favor the pathogens, as seen in several periodontal pathogens such as *Tannerella forsythia* (Schaffer and Andrukhov 2024). First, the inflammatory response of the periodontal tissue provides the oral microbiome with essential nutrients, inflammatory cytokines, and chemokines, which enable its survival, promote its further proliferation, and form a self-sustaining vicious cycle that drives periodontitis. Next, host tissue destructions, elicited by inflammatory reactions, allow pathogens to invade and colonize host tissue, in addition to the invasion mechanisms by proteases and other virulence factors of bacteria themselves. This shows the intricate interactions of macrophages with the oral microbiome as the first line of host defense.

32.4.2 Functions in the Initiation of Periodontal Inflammation

In gingival tissue, macrophages reside in the junctional epithelium and other periodontal tissues. In gingival tissue with periodontal disease, an increase in macrophage number is observed, and they play distinct roles during different stages of periodontal disease (Pejcic et al. 2024).

Upon the development of periodontal disease, the infiltration of M1 macrophages increases in periodontal tissue. The proportion of M1 and periodontal inflammation activity positively correlates. Contrarily, in the stationary phase of periodontal disease, the proportion of M2 macrophages is increased (Hussain et al. 2016; Yu et al. 2016). A detailed analysis of macrophage phenotype transition has been done in a ligature-induced periodontal disease mouse model. During the inflammatory phase, M1 macrophages expressing CD80, TNF-α, and TGF-β were prominent, while during the recovery phase, M2 macrophages expressing CD206 were observed (Viniegra et al. 2018). The importance of macrophages in periodontal inflammation is further demonstrated by the fact that the depletion of macrophages decreases the severity of inflammation upon periodontal disease induction in mice (Clark et al. 2021). The transition between M1 and M2 macrophages is an important factor regulating the transition between active and inactive periodontal disease.

Another aspect that highlights the significance of macrophages in the periodontium is aging. Macrophage functions and polarization are affected by aging, which may be associated with the increased prevalence of periodontal disease in older populations (Gibon et al. 2016). The possible traits of aging in macrophages include aggravated inflammatory reactions and the accumulation of damage/danger-associated molecular patterns (DAMPs) (Oishi and Manabe 2016). Macrophages derived from aged mice also had reduced capacity for anti-inflammatory functions, namely endotoxin tolerance and efferocytosis (Sendama 2020; Sun et al. 2012a), which will be discussed in the next section.

32.4.3 Functions in Tissue Destruction and Matrix Remodeling

The M1 macrophages, which release MMPs and other enzymes that degrade extracellular matrix components, characterize chronic inflammation associated with periodontitis. This tissue destruction is exacerbated by the sustained activation of macrophages, leading to the loss of periodontal attachment and bone resorption. In addition to MMPs, activated macrophages produce ROS and other inflammatory mediators that further contribute to tissue damage. The ongoing cycle of inflammation and tissue destruction ultimately results in the clinical manifestations of periodontal disease, including gingival recession, pocket formation, and tooth mobility (Graves 2008).

32.4.4 Functions in the Interaction Between Periodontal Disease and Systemic Diseases and the Role of Macrophages

Extensive research has been conducted on the connections between periodontal disease and various systemic diseases. While many gaps still exist, it is reported that periodontal disease interacts bidirectionally with systemic conditions such as diabetes, cardiovascular disease, and respiratory conditions (Hasturk et al. 2012). This dynamic response, which affects remote local tissues, relies on a complex network of both cellular and noncellular elements. Macrophages play a crucial role in this process due to their extensive functional interactions with other cells and their involvement in inflammatory processes, as outlined below.

In diabetes, one of the exacerbating factors is inflammation from Advanced Glycation Endproducts (AGE)-Receptor for AGEs (RAGE) axis. AGEs are compounds formed nonenzymatically by chemical reactions between carbonyl groups of reducing sugars and free amine groups, such as those found in proteins. They are recognized by several receptors, with the most extensively studied being RAGE, which triggers inflammation (Twarda-Clapa et al. 2022). Research shows that RAGE ligands can greatly enhance the hyper-responsive phenotype of monocytes and macrophages in diabetes. This effect may be reversible by blocking RAGE or by reducing RAGE ligands through better control of hyperglycemia (Ding et al. 2007).

Furthermore, increased oxidative stress, which is in part generated by the oxidative burst of macrophages against pathogens in periodontal tissue, promotes the production of AGEs, leading to a macrophage-mediated vicious cycle between the diseases.

Macrophages play important roles in the interaction between periodontal disease and cardiovascular disease as well. While macrophages are crucial in the development of atherosclerosis, especially in the early accumulation of cholesterol in the arterial walls, it has been proposed that infections and chronic inflammatory conditions, such as periodontitis, may affect the progression of atherosclerosis (Haraszthy et al. 2000). It has been suggested that bacteria or viruses can infect atherosclerotic lesions, contributing to the inflammatory process. Notably, *P. gingivalis* has been found in human atheromas, indicating a potential association between *P. gingivalis* infection and atherosclerosis. Infections in distant areas can intensify systemic inflammation by releasing toxins such as LPS or causing the leakage of chemical mediators into the bloodstream, yet the direct impact of periodontal disease or *P. gingivalis* infection on atherosclerosis remains unclear (Dorn et al. 2000). To investigate this, researchers studied a mouse model that developed both periodontitis and atherosclerosis. The ApoE-null mouse model showed that experimentally induced periodontal disease, through repeated inoculations of *P. gingivalis*, worsened early atherosclerotic lesions within four months. Additionally, serum IL-6 levels, aortic VCAM-1, and tissue factor antigen were increased in mice infected with *P. gingivalis*, indicating a potential causal relationship between the infection and inflammation in the development of atherosclerosis (Lalla et al. 2003).

Macrophages have been implicated in the interaction between periodontal disease and respiratory diseases as well. Macrophages in the lungs are essential for protecting against respiratory pathogens, and the presence of periodontal pathogens can impair this defense, making individuals more susceptible to infections.

32.5 Macrophage Functions in Gingival Homeostasis

While macrophages serve as mediators of inflammation to fight against external pathogens, they also reserve the function to terminate inflammation, initiate tissue repair, and suspend or inhibit excessive inflammation to protect periodontal tissue, thus playing important roles in gingival tissue homeostasis.

32.5.1 Functions Related to Resolution of Inflammation

A well-known macrophage function in the resolution of inflammation, i.e., termination of inflammation and initiation of tissue repair, is efferocytosis. Efferocytosis is the biological reaction by phagocytes, including macrophages, where apoptotic cells are cleared by phagocytosis. Since apoptotic cells or dying cells could cause inflammation by uncontrolled release of enzymes or overly activated inflammatory response, removal of these cells by efferocytosis is essential for the maintenance of tissue homeostasis. It also aids in the resolution of inflammation by prompting phenotypic changes in macrophages, which are initiated by molecules produced during the phagolysosomal degradation of apoptotic cells (Gerlach et al. 2021).

The importance of efferocytosis was originally recognized in the field of cardiovascular diseases (Cabrera and Makino 2022), and defects in efferocytosis have now been associated with various chronic inflammatory diseases (Doran et al. 2020; Martinez and Cook 2021). Recent studies have shown that it is also essential for the resolution of inflammation in periodontal disease. Not only can efferocytosis exert anti-inflammatory functions by eliminating apoptotic cells and inducing phenotypic changes in macrophages, but it also releases specialized pro-resolving mediators (SPMs), which play a crucial role in periodontal treatment.

The significance of efferocytosis is prominent in periodontal disease associated with congenital efferocytosis deficiency. In patients with leukocyte adhesion deficiency (LAD1), due to mutations in the CD18 gene, β2 integrin expression is diminished, which results in a defect in neutrophil recruitment and efferocytosis. This would lead to disrupted neutrophil homeostasis and further cause tissue destruction and inflammation (Kajikawa et al. 2020). The persistence of pathogens and periodontal disease is also observed in other congenital diseases with a lack or deficiency of neutrophils, such as Papillon-Lefèvre syndrome (Eick et al. 2014). On the contrary, overly active or excessive neutrophils also result in persistent inflammation and the damage of periodontal tissue (Aboodi et al. 2011; Ryder 2010). Prolonged neutrophil survival may also cause tissue destruction. During inflammation, apoptotic cells can receive signaling molecules from pathogens or their surroundings that interfere with their apoptotic process. This disruption can lead to a delay in neutrophil apoptosis and worsen various inflammatory conditions, including periodontal disease (Jia et al. 2004; Jonsson et al. 2005; Zaric et al. 2010).

The significance of efferocytosis is also observed in periodontal disease, not related to congenital neutrophil defects. In chronic periodontal disease, GM-CSF level is increased and inhibits neutrophil apoptosis (Gamonal et al. 2003). Reduction of neutrophil apoptosis is also induced by a periodontal pathogen, *Filifactor alocis* (*F. alocis*). *F. alocis* inhibits the apoptosis of neutrophils by upregulating the anti-apoptotic protein, resulting in prolonged ROS generation. Delayed efferocytosis also seems to be associated with the pathogenesis of periodontal disease. Gingipains (Kgp and Rgp) expressed by *P. gingivalis*, significantly reduce the expression of clearance receptor CD14 on apoptotic neutrophils. Thus, gingipains inhibit the directional recruiting and timely efferocytosis by macrophages, which may be a pathogenic mechanism of periodontal disease. Efferocytosis deficiency in periodontal disease may be evidenced by the accumulation of neutrophils in the saliva of patients with periodontal disease, compared to healthy subjects (Nicu et al. 2018).

32.5.2 Macrophage Functions to Inhibit Excessive Inflammation

Here, we will focus on two macrophage functions that inhibit excessive inflammation in periodontal tissue, endotoxin tolerance, and trained innate immunity.

Endotoxin tolerance is a function shared by myeloid cells, including macrophages, where upon long-term exposure to endotoxins, such as LPS, macrophages change their functions and refrain from further inflammation while promoting their tissue-repairing functions by alteration of signaling pathways (Collins and Carmody 2015). Endotoxin tolerance is a double-edged sword, making the host more susceptible to infections while preventing excessive tissue destruction due to inflammation (Biswas and Lopez-Collazo 2009; Larsson et al. 2022). However, increasing attention is being given to the impaired endotoxin tolerance of macrophages in chronic inflammatory conditions, such as periodontal disease (Matsushita et al. 2010; Steimle et al. 2019).

In patients with periodontal disease, the expression of endotoxin tolerance-inhibiting GM-CSF and TLR4 is increased, suggesting the significance of endotoxin tolerance in the inhibition of prolonged inflammation in periodontal disease (Nomura et al. 2000; Adib-Conquy and Cavaillon 2002; Sun et al. 2012a, b). Endotoxin tolerance may explain the increased prevalence of periodontal disease in older individuals. Endotoxin tolerance induced in peritoneal macrophages derived from aged mice resulted in higher inflammation and lower anti-inflammatory effects compared to those from young mice (Sun et al. 2012a). This proves that aging reduces the potential of endotoxin tolerance. As a possible mechanism for the inhibition of endotoxin tolerance due to aging, oxidative stress has been shown to inhibit endotoxin tolerance by altering the expression of signaling molecules, as demonstrated in in vitro and mouse model studies (Yamaguchi et al. 2023).

The capacity of endotoxin tolerance also depends on the endotoxin itself. Martin et al. reported that LPS from *P. gingivalis* had a limited capacity to induce endotoxin tolerance compared to *Escherichia coli* LPS, which may partially explain the sustained inflammation in periodontal disease (Martin et al. 2001).

Trained innate immunity refers to an enhanced state observed in innate immune cells, including macrophages, following previous encounters with immunological stimuli. Historically, the adaptive immune system was recognized as the only system with the memory of immune reactions. However, the advancements over the past decade have revealed that innate immune cells can develop and retain a "memory" of previous inflammatory events, such as those triggered by infectious agents like the fungal cell wall component β-glucan or by vaccines. This allows innate immune cells a heightened responsiveness upon re-exposure, meaning they can respond more quickly and vigorously to future threats. However, this memory-based response is not specific to the original stimulus, allowing it to be triggered by the same or even unrelated stimuli. This increased state of inflammatory activation, rooted in nonspecific innate immune memory, is referred to as "trained innate immunity" (Jia et al. 2004; Hajishengallis et al. 2022).

While innate immune cells circulating in blood have a relatively short lifespan of up to a few days, trained innate immunity lasts for months (Penkov et al. 2019; Yona et al. 2013). This is explained by findings showing that trained innate immunity can be initiated in the bone marrow, having a long-lasting influence on hematopoietic stem and progenitor cells, which are the precursors to differentiated innate immune cells (Mitroulis et al. 2018). The establishment of trained innate immunity also involves metabolic, epigenetic, and transcriptional reprogramming of the immune cells upon exposure to microbial and/or inflammatory stimuli. Trained innate immunity developed through the modulation of hematopoietic stem cells will lead to the formation of trained myeloid cell populations, such as macrophages, monocytes, or neutrophils, which can regenerate and enhance the innate immune system (Kalafati et al. 2020). While trained innate immunity is beneficial for the host upon re-infection or other future challenges, it may also cause aberrant immune responses that aggravate chronic inflammatory diseases, a phenomenon referred to as maladaptation (Chavakis et al. 2019; Leentjens et al. 2018).

Although the involvement of trained innate immunity in periodontal disease is still not fully understood, a study using a mouse model of periodontal disease shows that maladaptation may be involved in the onset and progression of the disease. Li et al. demonstrated the existence of trained innate memory in a murine model of ligature-induced periodontitis (LIP) originating from bone marrow hematopoietic progenitors. After 21 days of LIP training, mice showed increased myelopoiesis. To assess long-term memory, the ligatures were removed for 14 days, after which the mice were challenged with LPS. Mice previously exposed to LIP exhibited enhanced myelopoiesis and greater TNF-α and IL-6 production from neutrophils and monocytes in response to LPS. Analysis revealed epigenetic changes and increased chromatin accessibility of pro-inflammatory genes. The trained phenotype was transferable to naïve recipients via bone marrow transplantation, leading to increased inflammatory reactivity during arthritis. This suggests maladaptation may be involved not only in the progression of periodontal disease but also in the connections between periodontal disease and other diseases, such as arthritis (Li et al. 2022).

32.5.3 Functions to Promote Tissue Regenerations

Once the acute inflammatory phase resolves, the dominant macrophage population transitions from M1 to M2 macrophages (Porcheray et al. 2005). M2 macrophages are known for secreting growth factors such as VEGF and TGF-β, along with anti-inflammatory mediators, particularly IL-10. These factors support tissue repair by stabilizing angiogenesis, promoting the growth and proliferation of progenitor cells, and aiding in extracellular matrix (ECM) assembly and remodeling. While M2 macrophages have been categorized into several subtypes, among the defined M2 subtypes, M2a is most commonly linked to regenerative and fibrotic processes (Martin and Garcia 2021).

Effective wound healing relies on a carefully regulated transition of dominant macrophage phenotype from an M1 to an M2 macrophage response, eventually culminating in a return to homeostasis. The importance of macrophages in tissue regeneration is studied in various disease models apart from periodontal disease. While M2 macrophages are the key players in the tissue regeneration phase, the initial inflammatory response led by M1 macrophages is vital for effective tissue regeneration as well. If the immune response is suppressed too early, it can hinder tissue regeneration. Furthermore, the importance of initial M1 macrophages is also observed in the fact that macrophage depletion prior to or during the acute inflammatory phase can adversely affect tissue regeneration (Sinder et al. 2015). On the contrary, the existence of M2 macrophages during the acute inflammatory phase may also inhibit wound healing. Jetten et al. examined the effects of exogenously delivered macrophages, differentiated ex vivo into M2a and M2c phenotypes, on wound healing in a full-thickness cutaneous wound model. In wild-type mice, local injections of M2a macrophages or M2c macrophages during the acute inflammatory phase immediately following injury resulted in full wound healing, comparable to that observed in mice injected with saline or undifferentiated macrophages. However, in diabetic db/db mice, complete re-epithelialization occurred only in those injected with saline or undifferentiated macrophages, while wounds treated with M2a or M2c macrophages failed to achieve full closure (Jetten et al. 2014).

For effective wound healing, neo-angiogenesis, the process of forming new blood vessels at injury sites, is also vital. However, inadequate regulation of this process can result in abnormal vessel growth or insufficient vascularization, both of which are associated with various diseases (Hom and Maisel 1992). Macrophages play a critical role in the process of neoangiogenesis as well, facilitating different phases of vessel formation. Traditionally, M2 macrophages were believed to promote angiogenesis, while M1 macrophages were viewed as anti-angiogenic (Mantovani 1994). Recent studies, however, have shown that both macrophage types have important roles. These findings suggest that M1 macrophages are essential for the early stages of vascularization by the secretion of VEGF. Contrarily, M2 macrophages significantly contribute to later stages and maturation of vessels by the secretion of PDGF, a growth factor suggested to recruit pericytes and mesenchymal stem cells (Spiller et al. 2014).

Macrophages are essential for tissue regeneration and neo-angiogenesis, but they are also strongly associated with the promotion of tissue fibrosis. This is primarily due to their secretion of pro-fibrotic factors like TGF-β and galectin-3, which activate resident fibroblasts, stimulate their trans-differentiation into myofibroblasts, and increase collagen production and contractility in these myofibroblasts (Ruiz et al. 2003). Nevertheless, macrophages have also been shown to play key roles in resolving fibrosis, in part by releasing MMPs, which are enzymes that break down various extracellular matrix (ECM) proteins. During the injury phase, the MMPs released by macrophages are suppressed by the simultaneous secretion of tissue inhibitors of MMPs (TIMPs) by myofibroblasts, thereby fostering a pro-fibrotic environment. (Roderfeld 2018). Following the removal of injury stimuli, macrophages undergo a phenotypic transition to an anti-fibrotic, pro-resolution state, triggered by the uptake of cellular debris. These pro-resolution macrophages release various fibrolytic MMPs, such as MMP9, MMP12, and MMP13, which help break down the fibrotic extracellular matrix. Similar to tissue regeneration and neo-angiogenesis, the induction of fibrosis or maintenance of homeostasis is also regulated by the carefully timed transition between M1 and M2 macrophages (Ramachandran et al. 2012).

32.5.4 Functions on Bone Metabolism Regulations

The resorption of alveolar bone that leads to the loss of teeth is the most severe symptom of periodontal disease that has a detrimental effect on the patients' daily lives. Bone is an active tissue, regulated by bone formation and resorption by osteoblasts and osteoclasts, respectively. Physiologically, bone resorption and bone formation regulate each other, a process known as bone coupling, ensuring that the volume of newly formed bone matches the volume of the bone resorbed, leading to no net loss of bone (Graves et al. 2011). However, under bone lytic diseases such as periodontal disease, osteoclasts become both more prevalent in number and more active, leading to bone resorption. Osteoclasts are multinucleated cells that originate from hematopoietic stem cells. They adhere to the bone surface and secrete acid and lytic enzymes that degrade bone in a specialized, extracellular compartment (Boyle et al. 2003). The differentiation of these multinucleated osteoclasts relies on the Receptor activator of nuclear factor kappa-B ligand (RANKL), produced by periodontal ligament cells, osteoblasts, osteocytes, and immune cells. This RANK pathway, which is the main regulatory mechanism of osteoclast differentiation, i.e., osteoclastogenesis, is affected by various factors, including macrophages.

M1 macrophages facilitate osteoclastogenesis and their activation by releasing cytokines that enhance the Th1 response and stimulate osteoclast precursors. Additionally, M1 contributes to the production of key cytokines such as Prostaglandin E2(PGE2), TNF-α, Interleukin-6(IL-6), IL-1β, and IL-12, all of which play significant roles in the advancement of bone loss and periodontal disease. Notably, PGE2 is the most potent factor promoting periodontal bone loss, influencing several destructive processes in alveolar bone and cementum. This includes decreasing the mineralization and survival of osteoblasts and promoting the formation of osteoclasts, affecting bone metabolism from both sides (Oka et al. 2007; Ruiz-Heiland et al. 2021). TNF-α and IL-1β, which are also secreted by M1 macrophages upon LPS stimulation, induce RANKL production in T and B cells and enhance osteoclast activation and differentiation, respectively, leading to bone resorption (Becerra-Ruiz et al. 2022; Huynh et al. 2017). IL-6, another cytokine secreted by M1 macrophages, encourages osteoclasts to produce MMPs, which degrade the extracellular matrix and promote alveolar bone loss (Figueiredo et al. 2020). The significance of IL-6 in periodontal disease progression and bone resorption is evidenced by research using mouse models, where IL-6 receptor antibodies inhibit bone resorption and decrease periodontal inflammation (Balta et al. 2021; Matsuura et al. 2018).

These studies demonstrate that M1 macrophages play a crucial role in promoting bone absorption during the inflammatory response in periodontal disease. However, the role of M1 macrophages in bone metabolism is more complicated. M1 secretes CCL2 and vascular endothelial growth factor (VEGF), which recruit bone marrow mesenchymal stem cells, potentially supporting bone regeneration (Liang et al. 2021). M1-related cytokine IL-12 inhibits the differentiation of osteoclasts induced by RANKL and TNF-α, while IFN-γ also suppresses this differentiation and triggers osteoclast apoptosis (Amarasekara et al. 2018; Yamaguchi et al. 2016). Importantly, effective osteoclast function is necessary for bone regeneration, and the understanding of M1 macrophages' dual role, along with the timing and interactions of various factors, is crucial for further insights into the regulatory role of macrophages in bone metabolism.

32.6 Macrophages as the Target of Host Modulation Therapy in Periodontal Disease

Since periodontal disease is triggered by dysbiosis of the oral microbiome, emphasis on pathogenic treatment has been placed in its therapy. However, it is the excessive host response, i.e., uncontrolled and unresolved inflammation that causes tissue destruction in periodontal disease. Additionally, two primary findings were identified, which shed light on the importance of host modulation therapies. First is that the mechanism of action of antibiotic, tetracycline in the inhibition of periodontal tissue breakdown was mainly due to their previously unknown ability to block host-derived matrix metalloproteinases, especially collagenases, gelatinases, and macrophage metalloelastase, instead of their antimicrobial effects. Furthermore, nonsteroidal anti-inflammatory drugs (NSAIDs) alleviated the severity of periodontitis without affecting the microbiome. More recently, it has been reported that host inflammation exacerbates the dysbiosis of the oral microbiome, which leads to further inflammation (Lee et al. 2016). Hence, the importance of host modulation therapy has gained much attention in periodontal disease over the years. (Golub et al. 1998; Golub et al. 1994) Here, some of the major focuses for host modulation therapeutics when targeting macrophages will be discussed.

32.6.1 Macrophage Polarization Modulation

The main direction of current macrophage regulation strategies in host-modulation therapy of periodontal disease is to inhibit M1 macrophages while promoting the conversion to the M2 phenotype. This conversion, which plays an important role in reducing inflammation, promoting tissue repair, and providing overall anti-inflammatory benefits, can be achieved by several approaches (Sima et al. 2019). Research by Galarraga-Vinueza et al. demonstrated that cranberry concentrate, significantly reduced M1 polarization and increased M2 polarization, suggesting that cranberry-derived procyanidins have potential anti-inflammatory effects through their ability to modulate macrophage polarization (Galarraga-Vinueza et al. 2020). Specialized pro-resolving lipid mediators (SPMs) or peroxisome proliferator-activated receptor gamma (PPAR-γ) are also known to promote M2 macrophage polarization. Furthermore, biomaterials are increasingly used in periodontal therapy because of their customizable design and distinctive drug delivery system (Peng et al. 2024). Several studies have been carried out in the context of macrophage phenotype regulation. Shuai et al. have reported a nucleoside-based supramolecular hydrogel containing immunomodulatory agent 8-aminoguanosine (Shuai et al. 2025). Another research facilitates discoidal particles containing IL-4, that attach to the cell surface (Nakajima et al. 2025). However, much remains to be elucidated in this area, and further studies are needed to determine the effective treatment of periodontal disease.

32.6.2 Anti-inflammatory Therapies

Therapies aimed at modulating the inflammatory response of macrophages could help alleviate tissue destruction associated with periodontitis. NSAIDs and corticosteroids have been explored for their potential to reduce inflammation in periodontal disease.

Additionally, antibody-based inhibition of cytokines derived from M1 macrophages, such as TNFα, IL-1, and IL-6, has proven effective in other chronic inflammatory diseases such as rheumatoid arthritis and holds promise in periodontal disease too. Although these medications can effectively control inflammation and slow the progression of the disease, they may also lead to side effects like an increased risk of infection.

32.6.3 Regenerative Approaches

Regenerative approaches via macrophages that aim to restore lost periodontal tissues are also being explored. The use of growth factors, stem cells, and biomaterials to promote tissue regeneration may provide a comprehensive strategy for managing periodontitis.

By promoting M2 macrophage polarization and enhancing the resolution of inflammation, these regenerative approaches could facilitate healing and restore periodontal health.

32.7 Summary and Conclusions

Macrophages are immune cells known for their versatile functions, playing critical roles in both pro-inflammatory reactions and pro-resolution or tissue homeostasis. As discussed in this chapter, this versatility is also evident in the context of periodontal disease progression, influencing the inflammatory response and tissue destruction associated with the disease, and the maintenance of gingival tissue homeostasis. Their ability to polarize into M1 and M2 states is critical in determining the outcome of periodontal inflammation. In addition to their general immune functions, macrophages contribute significantly to the regulation of bone resorption during periodontal disease progression. This highlights their importance not only in inflammation but also in the structural integrity of periodontal tissues.

While substantial progress has been made in understanding the roles of macrophages in periodontal disease, there is still much to uncover. Over the past decade, advances in techniques such as single-cell sequencing (sc sequencing) and organoid cultures have opened new avenues for research, promising deeper insights into the behavior and functions of macrophages within gingival tissues and in the interaction with oral microbiome. These innovative methods allow for a more detailed exploration of the cellular dynamics at play, revealing potential new roles for macrophages in both health and disease.

As we continue to unravel the complex interactions involving macrophages in periodontal disease, it is crucial to integrate this knowledge into the development of more effective treatment and prevention strategies. A deeper understanding of macrophage functions could pave the way for novel host modulation therapies, which aim to restore balance within the immune system and improve patient outcomes in periodontal disease management. This research not only holds promise for advancing therapeutic approaches but also for enhancing our overall understanding of immune responses in oral health.

References

Aboodi GM, Goldberg MB, Glogauer M (2011) Refractory periodontitis population characterized by a hyperactive oral neutrophil phenotype. J Periodontol 82(5): 726–733. https://doi.org/10.1902/jop.2010.100508

Adib-Conquy M, Cavaillon JM (2002) Gamma interferon and granulocyte/monocyte colony-stimulating factor prevent endotoxin tolerance in human monocytes by promoting interleukin-1 receptor-associated kinase expression and its association to MyD88 and not by modulating TLR4 expression. J Biol Chem 277(31): 27927–27934. https://doi.org/10.1074/jbc.M200705200

Amarasekara DS, Yun H, Kim S, Lee N, Kim H, Rho J (2018) Regulation of osteoclast differentiation by cytokine networks. Immune Netw 18(1):e8. https://doi.org/10.4110/in.2018.18.e8

Ballini A, Dipalma G, Isacco CG, Boccellino M, Di Domenico M, Santacroce L, Nguyen KCD, Scacco S, Calvani M, Boddi A, Corcioli F, Quagliuolo L, Cantore S, Martelli FS, Inchingolo F (2020) Oral microbiota and immune system crosstalk: a translational research. Biology (Basel) 9(6). https://doi.org/10.3390/biology9060131

Balta MG, Papathanasiou E, Blix IJ, Van Dyke TE (2021) Host modulation and treatment of periodontal disease. J Dent Res 100(8):798–809. https://doi.org/10.1177/0022034521995157

Becerra-Ruiz JS, Guerrero-Velazquez C, Martinez-Esquivias F, Martinez-Perez LA, Guzman-Flores JM (2022) Innate and adaptive immunity of periodontal disease. From etiology to alveolar bone loss. Oral Dis 28(6):1441–1447. https://doi.org/10.1111/odi.13884

Biswas SK, Lopez-Collazo E (2009) Endotoxin tolerance: new mechanisms, molecules and clinical significance. Trends Immunol 30(10):475–487. https://doi.org/10.1016/j.it.2009.07.009

Boyle WJ, Simonet WS, Lacey DL (2003) Osteoclast differentiation and activation. Nature 423(6937): 337–342. https://doi.org/10.1038/nature01658

Brown LJ, Johns BA, Wall TP (2002) The economics of periodontal diseases. Periodontol 2000 29:223–234. https://doi.org/10.1034/j.1600-0757.2002.290111.x

Cabrera JTO, Makino A (2022) Efferocytosis of vascular cells in cardiovascular disease. Pharmacol Ther 229: 107919. https://doi.org/10.1016/j.pharmthera.2021.107919

Chavakis T, Mitroulis I, Hajishengallis G (2019) Hematopoietic progenitor cells as integrative hubs for adaptation to and fine-tuning of inflammation. Nat Immunol 20(7):802–811. https://doi.org/10.1038/s41590-019-0402-5

Chen S, Saeed A, Liu Q, Jiang Q, Xu H, Xiao GG, Rao L, Duo Y (2023) Macrophages in immunoregulation and therapeutics. Signal Transduct Target Ther 8(1):207. https://doi.org/10.1038/s41392-023-01452-1

Clark D, Halpern B, Miclau T, Nakamura M, Kapila Y, Marcucio R (2021) The contribution of macrophages in old mice to periodontal disease. J Dent Res 100(12): 1397–1404. https://doi.org/10.1177/00220345211009463

Collins PE, Carmody RJ (2015) The regulation of endotoxin tolerance and its impact on macrophage activation. Crit Rev Immunol 35(4):293–323. https://doi.org/10.1615/critrevimmunol.2015015495

Ding Y, Kantarci A, Hasturk H, Trackman PC, Malabanan A, Van Dyke TE (2007) Activation of RAGE induces elevated O2- generation by mononuclear phagocytes in diabetes. J Leukoc Biol 81(2):520–527. https://doi.org/10.1189/jlb.0406262

Doran AC, Yurdagul A Jr, Tabas I (2020) Efferocytosis in health and disease. Nat Rev Immunol 20(4):254–267. https://doi.org/10.1038/s41577-019-0240-6

Dorn BR, Burks JN, Seifert KN, Progulske-Fox A (2000) Invasion of endothelial and epithelial cells by strains of Porphyromonas gingivalis. FEMS Microbiol Lett 187(2):139–144. https://doi.org/10.1111/j.1574-6968.2000.tb09150.x

Eick S, Puklo M, Adamowicz K, Kantyka T, Hiemstra P, Stennicke H, Guentsch A, Schacher B, Eickholz P, Potempa J (2014) Lack of cathelicidin processing in Papillon-Lefevre syndrome patients reveals essential role of LL-37 in periodontal homeostasis. Orphanet J Rare Dis 9:148. https://doi.org/10.1186/s13023-014-0148-y

Figueiredo LC, Bueno-Silva B, Nogueira CFP, Valadares LC, Garcia KMM, Filho G, Milanello L, Esteves FM, Shibli JA, Miranda TS (2020) Levels of gene expression of immunological biomarkers in peri-implant and periodontal tissues. Int J Environ Res Public Health 17(23). https://doi.org/10.3390/ijerph17239100

Galarraga-Vinueza ME, Dohle E, Ramanauskaite A, Al-Maawi S, Obreja K, Magini R, Sader R, Ghanaati S, Schwarz F (2020) Anti-inflammatory and macrophage polarization effects of Cranberry Proanthocyanidins (PACs) for periodontal and peri-implant disease therapy. J Periodontal Res 55(6):821–829. https://doi.org/10.1111/jre.12773

Gamonal J, Sanz M, O'Connor A, Acevedo A, Suarez I, Sanz A, Martinez B, Silva A (2003) Delayed neutrophil apoptosis in chronic periodontitis patients. J Clin Periodontol 30(7):616–623. https://doi.org/10.1034/j.1600-051x.2003.00350.x

Gerlach BD, Ampomah PB, Yurdagul A Jr, Liu C, Lauring MC, Wang X, Kasikara C, Kong N, Shi J, Tao W, Tabas I (2021) Efferocytosis induces macrophage proliferation to help resolve tissue injury. Cell Metab 33(12):2445–2463 e2448. https://doi.org/10.1016/j.cmet.2021.10.015

Gibon E, Loi F, Cordova LA, Pajarinen J, Lin T, Lu L, Nabeshima A, Yao Z, Goodman SB (2016) Aging affects bone marrow macrophage polarization: relevance to bone healing. Regen Eng Transl Med 2(2):98–104. https://doi.org/10.1007/s40883-016-0016-5

Golub LM, Wolff M, Roberts S, Lee HM, Leung M, Payonk GS (1994) Treating periodontal diseases by blocking tissue-destructive enzymes. J Am Dent Assoc 125(2):163–169.; discussion 169–171. https://doi.org/10.14219/jada.archive.1994.0261

Golub LM, Lee HM, Ryan ME, Giannobile WV, Payne J, Sorsa T (1998) Tetracyclines inhibit connective tissue breakdown by multiple non-antimicrobial mechanisms. Adv Dent Res 12(2):12–26. https://doi.org/10.1177/08959374980120010501

Gordon S, Martinez FO (2010) Alternative activation of macrophages: mechanism and functions. Immunity 32(5):593–604. https://doi.org/10.1016/j.immuni.2010.05.007

Gordon S, Taylor PR (2005) Monocyte and macrophage heterogeneity. Nat Rev Immunol 5(12):953–964. https://doi.org/10.1038/nri1733

Graves D (2008) Cytokines that promote periodontal tissue destruction. J Periodontol 79(8 Suppl):1585–1591. https://doi.org/10.1902/jop.2008.080183

Graves DT, Li J, Cochran DL (2011) Inflammation and uncoupling as mechanisms of periodontal bone loss. J Dent Res 90(2):143–153. https://doi.org/10.1177/0022034510385236

Hajishengallis G, Li X, Divaris K, Chavakis T (2022) Maladaptive trained immunity and clonal hematopoiesis as potential mechanistic links between periodontitis and inflammatory comorbidities. Periodontol 2000 89(1):215–230. https://doi.org/10.1111/prd.12421

Haraszthy VI, Zambon JJ, Trevisan M, Zeid M, Genco RJ (2000) Identification of periodontal pathogens in atheromatous plaques. J Periodontol 71(10):1554–1560. https://doi.org/10.1902/jop.2000.71.10.1554

Hasturk H, Kantarci A, Van Dyke TE (2012) Oral inflammatory diseases and systemic inflammation: role of the macrophage. Front Immunol 3:118. https://doi.org/10.3389/fimmu.2012.00118

Hoebe K, Janssen E, Beutler B (2004) The interface between innate and adaptive immunity. Nat Immunol 5(10):971–974. https://doi.org/10.1038/ni1004-971

Hom DB, Maisel RH (1992) Angiogenic growth factors: their effects and potential in soft tissue wound healing. Ann Otol Rhinol Laryngol 101(4):349–354. https://doi.org/10.1177/000348949210100411

Hu X, Liu G, Hou Y, Shi J, Zhu L, Jin D, Peng J, Zhao Y (2012) Induction of M2-like macrophages in recipient NOD-scid mice by allogeneic donor CD4(+)CD25(+) regulatory T cells. Cell Mol Immunol 9(6):464–472. https://doi.org/10.1038/cmi.2012.47

Huang Y, Tian C, Li Q, Xu Q (2019) TET1 knockdown inhibits porphyromonas gingivalis LPS/IFN-gamma-induced M1 macrophage polarization through the NF-kappaB pathway in THP-1 cells. Int J Mol Sci 20(8). https://doi.org/10.3390/ijms20082023

Hussain QA, McKay IJ, Gonzales-Marin C, Allaker RP (2016) Detection of adrenomedullin and nitric oxide in different forms of periodontal disease. J Periodontal Res 51(1):16–25. https://doi.org/10.1111/jre.12273

Huynh NC, Everts V, Pavasant P, Ampornaramveth RS (2017) Interleukin-1beta induces human cementoblasts to support osteoclastogenesis. Int J Oral Sci 9(12):e5. https://doi.org/10.1038/ijos.2017.45

Jetten N, Roumans N, Gijbels MJ, Romano A, Post MJ, de Winther MP, van der Hulst RR, Xanthoulea S (2014) Wound administration of M2-polarized macrophages does not improve murine cutaneous healing responses.

PLoS One 9(7):e102994. https://doi.org/10.1371/journal.pone.0102994
Jia SH, Li Y, Parodo J, Kapus A, Fan L, Rotstein OD, Marshall JC (2004) Pre-B cell colony-enhancing factor inhibits neutrophil apoptosis in experimental inflammation and clinical sepsis. J Clin Invest 113(9): 1318–1327. https://doi.org/10.1172/JCI19930
Jonsson H, Allen P, Peng SL (2005) Inflammatory arthritis requires Foxo3a to prevent Fas ligand-induced neutrophil apoptosis. Nat Med 11(6):666–671. https://doi.org/10.1038/nm1248
Kajikawa T, Wang B, Li X, Wang H, Chavakis T, Moutsopoulos NM, Hajishengallis G (2020) Frontline science: activation of metabolic nuclear receptors restores periodontal tissue homeostasis in mice with leukocyte adhesion deficiency-1. J Leukoc Biol 108(5):1501–1514. https://doi.org/10.1002/JLB.5HI0420-648R
Kalafati L, Kourtzelis I, Schulte-Schrepping J, Li X, Hatzioannou A, Grinenko T, Hagag E, Sinha A, Has C, Dietz S, de Jesus Domingues AM, Nati M, Sormendi S, Neuwirth A, Chatzigeorgiou A, Ziogas A, Lesche M, Dahl A, Henry I et al (2020) Innate immune training of granulopoiesis promotes anti-tumor activity. Cell 183(3):771–785 e712. https://doi.org/10.1016/j.cell.2020.09.058
Lalla E, Lamster IB, Hofmann MA, Bucciarelli L, Jerud AP, Tucker S, Lu Y, Papapanou PN, Schmidt AM (2003) Oral infection with a periodontal pathogen accelerates early atherosclerosis in apolipoprotein E-null mice. Arterioscler Thromb Vasc Biol 23(8): 1405–1411. https://doi.org/10.1161/01.ATV.0000082462.26258.FE
Larsson L, Garaicoa-Pazmino C, Asa'ad F, Castilho RM (2022) Understanding the role of endotoxin tolerance in chronic inflammatory conditions and periodontal disease. J Clin Periodontol 49(3):270–279. https://doi.org/10.1111/jcpe.13591
Lee CT, Teles R, Kantarci A, Chen T, McCafferty J, Starr JR, Brito LC, Paster BJ, Van Dyke TE (2016) Resolvin E1 reverses experimental periodontitis and dysbiosis. J Immunol 197(7):2796–2806. https://doi.org/10.4049/jimmunol.1600859
Leentjens J, Bekkering S, Joosten LAB, Netea MG, Burgner DP, Riksen NP (2018) Trained innate immunity as a novel mechanism linking infection and the development of atherosclerosis. Circ Res 122(5): 664–669. https://doi.org/10.1161/CIRCRESAHA.117.312465
Li X, Wang H, Yu X, Saha G, Kalafati L, Ioannidis C, Mitroulis I, Netea MG, Chavakis T, Hajishengallis G (2022) Maladaptive innate immune training of myelopoiesis links inflammatory comorbidities. Cell 185(10):1709–1727 e1718. https://doi.org/10.1016/j.cell.2022.03.043
Liang B, Wang H, Wu D, Wang Z (2021) Macrophage M1/M2 polarization dynamically adapts to changes in microenvironment and modulates alveolar bone remodeling after dental implantation. J Leukoc Biol 110(3):433–447. https://doi.org/10.1002/JLB.1MA0121-001R
Mantovani A (1994) Tumor-associated macrophages in neoplastic progression: a paradigm for the in vivo function of chemokines. Lab Investig 71(1):5–16. https://www.ncbi.nlm.nih.gov/pubmed/7518882
Mantovani A, Sica A, Sozzani S, Allavena P, Vecchi A, Locati M (2004) The chemokine system in diverse forms of macrophage activation and polarization. Trends Immunol 25(12):677–686. https://doi.org/10.1016/j.it.2004.09.015
Martin KE, Garcia AJ (2021) Macrophage phenotypes in tissue repair and the foreign body response: Implications for biomaterial-based regenerative medicine strategies. Acta Biomater 133:4–16. https://doi.org/10.1016/j.actbio.2021.03.038
Martin M, Katz J, Vogel SN, Michalek SM (2001) Differential induction of endotoxin tolerance by lipopolysaccharides derived from Porphyromonas gingivalis and Escherichia coli. J Immunol 167(9): 5278–5285. https://doi.org/10.4049/jimmunol.167.9.5278
Martinez J, Cook DN (2021) What's the deal with efferocytosis and asthma? Trends Immunol 42(10): 904–919. https://doi.org/10.1016/j.it.2021.08.004
Martinez FO, Sica A, Mantovani A, Locati M (2008) Macrophage activation and polarization. Front Biosci 13:453–461. https://doi.org/10.2741/2692
Matsushita H, Ohta S, Shiraishi H, Suzuki S, Arima K, Toda S, Tanaka H, Nagai H, Kimoto M, Inokuchi A, Izuhara K (2010) Endotoxin tolerance attenuates airway allergic inflammation in model mice by suppression of the T-cell stimulatory effect of dendritic cells. Int Immunol 22(9):739–747. https://doi.org/10.1093/intimm/dxq062
Matsuura Y, Kikuta J, Kishi Y, Hasegawa T, Okuzaki D, Hirano T, Minoshima M, Kikuchi K, Kumanogoh A, Ishii M (2018) In vivo visualisation of different modes of action of biological DMARDs inhibiting osteoclastic bone resorption. Ann Rheum Dis 77(8):1219–1225. https://doi.org/10.1136/annrheumdis-2017-212880
Messeri Dreissig MD, Hammermann R, Mossner J, Gothert M, Racke K (2000) In rat alveolar macrophages lipopolysaccharides exert divergent effects on the transport of the cationic amino acids L-arginine and L-ornithine. Naunyn Schmiedeberg's Arch Pharmacol 361(6):621–628. https://doi.org/10.1007/s002100000241
Mitroulis I, Ruppova K, Wang B, Chen LS, Grzybek M, Grinenko T, Eugster A, Troullinaki M, Palladini A, Kourtzelis I, Chatzigeorgiou A, Schlitzer A, Beyer M, Joosten LAB, Isermann B, Lesche M, Petzold A, Simons K, Henry I et al (2018) Modulation of myelopoiesis progenitors is an integral component of trained immunity. Cell 172(1–2):147–161 e112. https://doi.org/10.1016/j.cell.2017.11.034
Murakami S, Mealey BL, Mariotti A, Chapple ILC (2018) Dental plaque-induced gingival conditions. J Clin

Periodontol 45(Suppl 20):S17–S27. https://doi.org/10.1111/jcpe.12937

Nakajima M, Kapate N, Clegg JR, Ikeda-Imafuku M, Park KS, Kumbhojkar N, Suja VC, Prakash S, Wang LL, Tabeta K, Mitragotri S (2025) Backpack-carrying macrophage immunotherapy for periodontitis. J Control Release 377:315–323. https://doi.org/10.1016/j.jconrel.2024.11.037

Nicu EA, Rijkschroeff P, Wartewig E, Nazmi K, Loos BG (2018) Characterization of oral polymorphonuclear neutrophils in periodontitis patients: a case-control study. BMC Oral Health 18(1):149. https://doi.org/10.1186/s12903-018-0615-2

Nomura F, Akashi S, Sakao Y, et al (2000) Cutting edge: endotoxin tolerance in mouse peritoneal macrophages correlates with down-regulation of surface toll-like receptor 4 expression. J Immunol 164(7):3476–3479. https://doi.org/10.4049/jimmunol.164.7.3476

O'Dowd LK, Durham J, McCracken GI, Preshaw PM (2010) Patients' experiences of the impact of periodontal disease. J Clin Periodontol 37(4):334–339. https://doi.org/10.1111/j.1600-051X.2010.01545.x

Oishi Y, Manabe I (2016) Macrophages in age-related chronic inflammatory diseases. NPJ Aging Mech Dis 2:16018. https://doi.org/10.1038/npjamd.2016.18

Oka H, Miyauchi M, Sakamoto K, Moriwaki S, Niida S, Noguchi K, Somerman MJ, Takata T (2007) PGE2 activates cementoclastogenesis by cementoblasts via EP4. J Dent Res 86(10):974–979. https://doi.org/10.1177/154405910708601011

Pejcic A, Andjelkovic Z, Marjanovic D, Minic I, Matvijenko V, Arsic Z, Jovanovic R, Subaric L (2024) Comparative analysis of antigen-presenting cells in gingival tissues in healthy and periodontitis patients. J Clin Pathol 77(10):702–708. https://doi.org/10.1136/jcp-2021-207975

Peng S, Fu H, Li R, Li H, Wang S, Li B, Sun J (2024) A new direction in periodontitis treatment: biomaterial-mediated macrophage immunotherapy. J Nanobiotechnol 22(1):359. https://doi.org/10.1186/s12951-024-02592-4

Penkov S, Mitroulis I, Hajishengallis G, Chavakis T (2019) Immunometabolic crosstalk: an ancestral principle of trained immunity? Trends Immunol 40(1):1–11. https://doi.org/10.1016/j.it.2018.11.002

Porcheray F, Viaud S, Rimaniol AC, et al (2005) Macrophage activation switching: an asset for the resolution of inflammation. Clin Exp Immunol 142(3):481–489. https://doi.org/10.1111/j.1365-2249.2005.02934.x

Ramachandran P, Pellicoro A, Vernon MA, Boulter L, Aucott RL, Ali A, Hartland SN, Snowdon VK, Cappon A, Gordon-Walker TT, Williams MJ, Dunbar DR, Manning JR, van Rooijen N, Fallowfield JA, Forbes SJ, Iredale JP (2012) Differential Ly-6C expression identifies the recruited macrophage phenotype, which orchestrates the regression of murine liver fibrosis. Proc Natl Acad Sci USA 109(46):E3186–E3195. https://doi.org/10.1073/pnas.1119964109

Roderfeld M (2018) Matrix metalloproteinase functions in hepatic injury and fibrosis. Matrix Biol 68–69:452–462. https://doi.org/10.1016/j.matbio.2017.11.011

Ruiz V, Ordonez RM, Berumen J, Ramirez R, Uhal B, Becerril C, Pardo A, Selman M (2003) Unbalanced collagenases/TIMP-1 expression and epithelial apoptosis in experimental lung fibrosis. Am J Physiol Lung Cell Mol Physiol 285(5):L1026–L1036. https://doi.org/10.1152/ajplung.00183.2003

Ruiz-Heiland G, Yong JW, von Bremen J, Ruf S (2021) Leptin reduces in vitro cementoblast mineralization and survival as well as induces PGE2 release by ERK1/2 commitment. Clin Oral Investig 25(4):1933–1944. https://doi.org/10.1007/s00784-020-03501-3

Ryder MI (2010) Comparison of neutrophil functions in aggressive and chronic periodontitis. Periodontol 2000 2000(53):124–137. https://doi.org/10.1111/j.1600-0757.2009.00327.x

Schaffer C, Andrukhov O (2024) The intriguing strategies of Tannerella Forsythia's host interaction. Front Oral Health 5:1434217. https://doi.org/10.3389/froh.2024.1434217

Sendama W (2020) The effect of ageing on the resolution of inflammation. Ageing Res Rev 57:101000. https://doi.org/10.1016/j.arr.2019.101000

Shuai F, Yin Y, Yao Y, Deng L, Wen Y, Zhao H, Han X (2025) A nucleoside-based supramolecular hydrogel integrating localized self-delivery and immunomodulation for periodontitis treatment. Biomaterials 316:123024. https://doi.org/10.1016/j.biomaterials.2024.123024

Sima C, Viniegra A, Glogauer M (2019) Macrophage immunomodulation in chronic osteolytic diseases-the case of periodontitis. J Leukoc Biol 105(3):473–487. https://doi.org/10.1002/JLB.1RU0818-310R

Sinder BP, Pettit AR, McCauley LK (2015) Macrophages: their emerging roles in bone. J Bone Miner Res 30(12):2140–2149. https://doi.org/10.1002/jbmr.2735

Spiller KL, Anfang RR, Spiller KJ, Ng J, Nakazawa KR, Daulton JW, Vunjak-Novakovic G (2014) The role of macrophage phenotype in vascularization of tissue engineering scaffolds. Biomaterials 35(15):4477–4488. https://doi.org/10.1016/j.biomaterials.2014.02.012

Steimle A, Michaelis L, Di Lorenzo F, Kliem T, Munzner T, Maerz JK, Schafer A, Lange A, Parusel R, Gronbach K, Fuchs K, Silipo A, Oz HH, Pichler BJ, Autenrieth IB, Molinaro A, Frick JS (2019) Weak agonistic LPS restores intestinal immune homeostasis. Mol Ther 27(11):1974–1991. https://doi.org/10.1016/j.ymthe.2019.07.007

Sun Y, Li H, Yang MF, Shu W, Sun MJ, Xu Y (2012a) Effects of aging on endotoxin tolerance induced by lipopolysaccharides derived from Porphyromonas gingivalis and Escherichia coli. PLoS One 7(6):e39224. https://doi.org/10.1371/journal.pone.0039224

Sun C, Sun L, Ma H, Peng J, Zhen Y, Duan K, Liu G, Ding W, Zhao Y (2012b) The phenotype and functional alterations of macrophages in mice with hyperglycemia for long term. J Cell Physiol 227(4): 1670–1679. https://doi.org/10.1002/jcp.22891

Tamai R, Deng X, Kiyoura Y (2009) Porphyromonas gingivalis with either Tannerella forsythia or Treponema denticola induces synergistic IL-6 production by murine macrophage-like J774.1 cells. Anaerobe 15(3): 87–90. https://doi.org/10.1016/j.anaerobe.2008.12.004

Tonetti MS, Jepsen S, Jin L, Otomo-Corgel J (2017) Impact of the global burden of periodontal diseases on health, nutrition and wellbeing of mankind: a call for global action. J Clin Periodontol 44(5):456–462. https://doi.org/10.1111/jcpe.12732

Twarda-Clapa A, Olczak A, Bialkowska AM, Koziolkiewicz M (2022) Advanced glycation end-products (AGEs): formation, chemistry, classification, receptors, and diseases related to AGEs. Cells 11(8). https://doi.org/10.3390/cells11081312

Van Dyke TE, Bartold PM, Reynolds EC (2020) The nexus between periodontal inflammation and dysbiosis. Front Immunol 11:511. https://doi.org/10.3389/fimmu.2020.00511

Van Dyken SJ, Locksley RM (2013) Interleukin-4- and interleukin-13-mediated alternatively activated macrophages: roles in homeostasis and disease. Annu Rev Immunol 31:317–343. https://doi.org/10.1146/annurev-immunol-032712-095906

Viniegra A, Goldberg H, Cil C, Fine N, Sheikh Z, Galli M, Freire M, Wang Y, Van Dyke TE, Glogauer M, Sima C (2018) Resolving macrophages counter osteolysis by anabolic actions on bone cells. J Dent Res 97(10): 1160–1169. https://doi.org/10.1177/0022034518777973

Yamaguchi T, Movila A, Kataoka S, Wisitrasameewong W, Ruiz Torruella M, Murakoshi M, Murakami S, Kawai T (2016) Proinflammatory M1 macrophages inhibit RANKL-induced osteoclastogenesis. Infect Immun 84(10): 2802–2812. https://doi.org/10.1128/IAI.00461-16

Yamaguchi T, Yamamoto Y, Egashira K, Sato A, Kondo Y, Saiki S, Kimura M, Chikazawa T, Yamamoto Y, Ishigami A, Murakami S (2023) Oxidative stress inhibits endotoxin tolerance and may affect periodontitis. J Dent Res 102(3):331–339. https://doi.org/10.1177/00220345221138523

Yona S, Kim KW, Wolf Y, Mildner A, Varol D, Breker M, Strauss-Ayali D, Viukov S, Guilliams M, Misharin A, Hume DA, Perlman H, Malissen B, Zelzer E, Jung S (2013) Fate mapping reveals origins and dynamics of monocytes and tissue macrophages under homeostasis. Immunity 38(1):79–91. https://doi.org/10.1016/j.immuni.2012.12.001

Yu T, Zhao L, Huang X, Ma C, Wang Y, Zhang J, Xuan D (2016) Enhanced activity of the macrophage M1/M2 phenotypes and phenotypic switch to M1 in periodontal infection. J Periodontol 87(9):1092–1102. https://doi.org/10.1902/jop.2016.160081

Zaric S, Shelburne C, Darveau R, Quinn DJ, Weldon S, Taggart CC, Coulter WA (2010) Impaired immune tolerance to Porphyromonas gingivalis lipopolysaccharide promotes neutrophil migration and decreased apoptosis. Infect Immun 78(10):4151–4156. https://doi.org/10.1128/IAI.00600-10

Zhu L, Zhao Q, Yang T, Ding W, Zhao Y (2015) Cellular metabolism and macrophage functional polarization. Int Rev Immunol 34(1):82–100. https://doi.org/10.3109/08830185.2014.969421

Immune and Stem Cell Interactions in Periodontal Disease: The Regulatory Role of Toll-Like Receptor 3

33

Mohamed Mekhemar and Tobias Hahn

Abstract

Toll-like receptor 3 (TLR3) is a vital component of the innate immune system known for its role in recognizing double-stranded RNA (dsRNA), which is commonly associated with viral infections. Within the periodontium, TLR3 is expressed by several cell types, including periodontal ligament cells, gingival fibroblasts, and resident mesenchymal stem/progenitor cells. When activated, TLR3 triggers a series of immunomodulatory responses, such as the upregulation of interferon-gamma (IFN-γ), indoleamine 23-dioxygenase (IDO), and human leukocyte antigen G (HLA-G). These responses collectively influence the local immune environment, promote regulatory T cell differentiation, and modulate the inflammatory process. Crucially, TLR3 activation also plays a significant role in shaping the microenvironment within the periodontium, affecting the regenerative potential of resident mesenchymal stem/progenitor cells. By altering the interactions between these cells and their surrounding environment, TLR3 can either facilitate or hinder tissue repair and regeneration, depending on the disease context. This chapter delves into the complex role of TLR3 in the periodontium, with a particular emphasis on how it influences microenvironmental dynamics in periodontal disease. Understanding these processes opens new avenues for developing targeted therapies that control inflammation, optimize the tissue environment, and enhance periodontal regeneration, ultimately improving the treatment of periodontal disease.

Keywords

Stem cells · Toll-like receptor 3 · Stemness · Inflammation · Immunomodulation · Differentiation

Abbreviations

AP-1	Activator Protein 1
AKT	Protein Kinase B
BM-MSCs	Bone Marrow-Derived Mesenchymal Stem Cells
c-Src	Proto-oncogene Tyrosine-Protein Kinase Src
CFU	Colony-Forming Unit
CXCL8 (IL-8)	C-X-C Motif Chemokine Ligand 8 (Interleukin-8)
DCs	Dendritic Cells
DP-MSCs	Dental Pulp Mesenchymal Stem Cells
dsRNA	Double-Stranded RNA

M. Mekhemar (✉) · T. Hahn
Clinic for Conservative Dentistry and Periodontology, Kiel University, Kiel, Germany
e-mail: mekhemar@konspar.uni-kiel.de

N. Rezaei, H. Ziaei (eds.), *Oral Immunology*, Advances in Experimental Medicine and Biology 1492,
https://doi.org/10.1007/978-3-032-03176-1_33

EBP	Cholestenol Delta-Isomerase
G-MSCs	Gingival Mesenchymal Stem Cells
GSK3β	Glycogen Synthase Kinase 3 Beta
HLA-G	Human Leukocyte Antigen G
IDO	Indoleamine 2,3-Dioxygenase
IFN-γ	Interferon-Gamma
IKK	IκB Kinase
IRF	Interferon Regulatory Factor
JNK	c-Jun N-Terminal Kinase
MAPK	Mitogen-Activated Protein Kinase
MCP-1	Monocyte Chemoattractant Protein-1
MMPs	Matrix Metalloproteinases
MSCs	Mesenchymal Stem Cells
MyD88	Myeloid Differentiation Primary Response 88
NF-κB	Nuclear Factor "Kappa-Light-Chain-Enhancer" of Activated B cells
PAMPs	Pathogen-Associated Molecular Patterns
PDLSCs	Periodontal Ligament Stem Cells
PI3K	Phosphoinositide 3-Kinase
PGE2	Prostaglandin E2
PRRs	Pattern Recognition Receptors
SCAP	Stem Cells from the Apical Papilla
SSEA-4	Stage-Specific Embryonic Antigen-4
TAB	TGF-Beta-Activated Kinase 1-Binding Protein
TAK1	Mitogen-Activated Protein Kinase Kinase Kinase 7
TGF-β	Transforming Growth Factor-Beta
TLR	Toll-Like Receptor
TLR3	Toll-Like Receptor 3
TRAIL	TNF-Related Apoptosis-Inducing Ligand
TRAF	TNF Receptor-Associated Factor
TRIF	TIR-Domain-Containing Adapter-Inducing Interferon-Beta
Wnt	Wingless/Integrated (Signaling Pathway)

33.1 Introduction

Periodontal disease is a multifaceted chronic inflammatory disorder characterized by the progressive destruction of the tooth-supporting structures, including the gingiva, periodontal ligament, and alveolar bone (Mekhemar et al. 2020a). Interactions between different cell types within the periodontium, particularly those involving stem cells and immune cells, are critical to the initiation, progression, and resolution of periodontal disease (Mekhemar et al. 2020b). Among these cell types, periodontal mesenchymal stem/progenitor cells (MSCs) are of particular interest because of their regenerative potential and their ability to modulate immune responses (Mekhemar et al. 2020a, b, 2021). These cells can differentiate into multiple lineages and contribute to tissue repair and regeneration (Mekhemar et al. 2020b). However, the inflammatory environment that characterizes periodontal disease can significantly influence the behavior of these stem cells, affecting their regenerative capacity and their interactions with the immune system (Mekhemar et al. 2020b).

The immune system, particularly the innate immune response, plays a central role in periodontal disease. Toll-like receptors (TLRs), which are key components of the innate immune system, are critical for recognizing microbial pathogens and initiating immune responses (Fawzy El-Sayed et al. 2016). Among the various TLRs, TLR3 has gained attention for its role in recognizing viral double-stranded RNA (dsRNA) and its involvement in modulating immune responses (Mekhemar et al. 2020b, 2022b). TLR3 is expressed by several cell types within the periodontium, including gingival fibroblasts, periodontal ligament cells, and resident mesenchymal stem/progenitor cells (Mekhemar et al. 2020b, 2022b).

Activation of TLR3 within the periodontium may influence both stem cell behavior and immune responses, thus playing a dual role in the context of periodontal disease (Mekhemar et al. 2020b, 2022a, b). TLR3 activation has been shown to modulate the immunomodulatory functions of mesenchymal stem cells (MSCs), enhancing their ability to regulate inflammatory responses and promote the secretion of cytokines and trophic factors that contribute to tissue repair (Mekhemar et al. 2020b, 2022a, b). In addition, TLR3-mediated signaling pathways, such as the NF-κB and PI3K-AKT pathways, can influence the differentiation and stemness properties of MSCs, further influencing regenerative outcomes in periodontal tissues (Mekhemar et al. 2020b, 2022a, b).

Understanding the role of TLR3 in the interactions between stem cells and immune cells in periodontal disease provides valuable insights into the underlying mechanisms of disease progression and healing. This chapter explores the complex interplay between stem cells, immune responses, and TLR3 activation in periodontal disease and highlights the potential of TLR3 as a therapeutic target for modulating inflammation and enhancing periodontal regeneration.

33.2 Periodontal Disease

33.2.1 Introduction to Periodontal Disease

Periodontal disease, particularly periodontitis, is an inflammatory disease that leads to the progressive degradation of the structures that support teeth, including the alveolar bone, cementum, and periodontal ligaments. This condition is driven by a complex interplay between pathogenic bacteria and the host's immune system, ultimately resulting in the loss of periodontal attachment and, if left untreated, can lead to tooth loss (Gasner and Schure 2024; Mekhemar et al. 2021). Periodontitis is a significant global health issue with a high prevalence and a substantial impact on both the quality of life and the economic burden associated with treatment and management (Mekhemar et al. 2021).

33.2.2 Pathophysiology of Periodontal Disease

The development of periodontitis involves the activation of host-derived molecular pathways triggered by bacterial pathogens. Key bacterial species linked to periodontitis include *Porphyromonas gingivalis*, *Treponema denticola*, *Tannerella forsythia*, *Prevotella intermedia*, *Fusobacterium nucleatum*, and *Aggregatibacter actinomycetemcomitans* (Mekhemar et al. 2020a). These pathogens colonize the gingival sulcus, forming biofilms on the tooth surface that initiate a local inflammatory response. This response involves the activation of immune cells, which subsequently induces the release of pro-inflammatory cytokines and other mediators that contribute to the destruction of periodontal tissues (Abdulkareem et al. 2023).

As the disease progresses, the junctional epithelium migrates apically, and the bacterial biofilm spreads along the root surface. This progression is often accompanied by a dysbiotic shift toward periodontopathogenic bacteria in the oral microbiome, which exacerbates the inflammatory response and further complicates the management of the disease (Abdulkareem et al. 2023). The chronic inflammation associated with periodontitis not only leads to the breakdown of periodontal tissues but also impacts systemic health, linking periodontitis to conditions such as cardiovascular disease, diabetes, and adverse pregnancy outcomes (Chatzopoulos et al. 2024) (Fig. 33.1).

33.2.3 Periodontitis as a Global Health Problem

Periodontitis ranks among the most prevalent chronic diseases globally, affecting millions of individuals worldwide (Kwon et al. 2021). Severe cases are estimated to affect approximately 11% of the global population, making periodontitis one of the leading causes of tooth loss in adults (Kwon et al. 2021). The economic implications of periodontitis are substantial, with both direct and indirect costs due to the need for extensive treatment and long-term management (Botelho et al. 2022). Furthermore, periodontitis has a profound impact on the quality of life, causing pain, discomfort, and difficulties with eating, speaking, and maintaining oral hygiene (Wong et al. 2021).

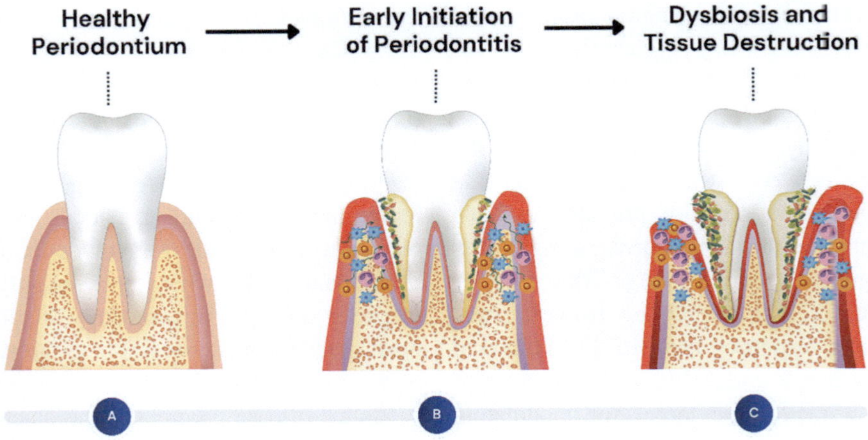

Fig. 33.1 Phases of periodontal disease development. The figure illustrates the three phases of periodontal disease progression: *Phase A—healthy periodontium*: A healthy periodontal state characterized by intact gingival tissues, firm attachment of the junctional epithelium to the tooth surface, and a balanced oral microbiome. Minimal inflammation is present, and no clinical signs of periodontal disease are observed. *Phase B—early initiation*: Bacterial pathogens begin to colonize the gingival sulcus, forming biofilms on the tooth surface. Key pathogens, including *Porphyromonas gingivalis, Treponema denticola, Tannerella forsythia, Prevotella intermedia, Fusobacterium nucleatum,* and *Aggregatibacter actinomycetemcomitans*, trigger a localized immune response. This phase is marked by increased production of pro-inflammatory cytokines and mediators, leading to initial signs of inflammation and mild tissue changes, such as gingival redness and swelling. *Phase C—dysbiosis and tissue destruction*: As periodontitis progresses, the junctional epithelium migrates apically, and the bacterial biofilm extends along the root surface. A dysbiotic shift in the oral microbiome exacerbates inflammation, resulting in the destruction of periodontal tissues, including connective tissue and alveolar bone. The chronic inflammatory state not only contributes to the breakdown of periodontal structures but also is linked to systemic health implications, such as cardiovascular disease, diabetes, and adverse pregnancy outcomes

33.2.4 Immune Responses and Inflammation in Periodontal Disease

The immune response plays a central role in the pathogenesis of periodontitis. While the host's immune system is intended to protect against bacterial invasion, it also contributes to the tissue destruction observed in periodontal disease (Han et al. 2023). The activation of immune cells results in the release of inflammatory mediators such as cytokines, chemokines, and matrix metalloproteinases (MMPs), which degrade the extracellular matrix and disrupt the structural integrity of periodontal tissues (Han et al. 2023).

In periodontitis, the delicate balance between pro-inflammatory and anti-inflammatory signals is disrupted, leading to chronic inflammation and progressive tissue damage (Bhuyan et al. 2022). Innate immune receptors, such as toll-like receptors (TLRs), are crucial in detecting microbial invaders and initiating immune responses. These receptors play a pivotal role in shaping the inflammatory milieu within the periodontium, influencing the progression and severity of the disease (Mekhemar et al. 2018, 2020b, 2022a, b; Song et al. 2017).

33.2.5 Therapeutic Strategies and Challenges in Periodontal Disease

The primary objectives of periodontal therapy are to eliminate microbial infections, regulate the inflammatory response, and restore the structure and function of the periodontium. Conventional treatments, including scaling and root planing, aim to mechanically remove plaque and calculus from tooth surfaces to reduce the bacterial load.

However, these treatments alone are often insufficient to fully resolve the inflammation and restore the damaged tissues (Mekhemar et al. 2020a, 2021).

Advanced therapeutic strategies focus on modulating the immune response and enhancing the regenerative capacity of resident stem/progenitor cells within the periodontium. By controlling inflammation, it is possible to create a more favorable environment for tissue regeneration. This approach involves not only reducing the destructive inflammatory signals but also promoting the recruitment, survival, and differentiation of stem/progenitor cells at the site of injury (Mekhemar et al. 2020a, b, 2021, 2022a, b).

The regeneration of periodontal tissues, including the bone-periodontal ligament-cementum complex, remains one of the most challenging aspects of periodontal therapy. The successful regeneration of these structures requires the precise coordination of multiple biological processes, including osteogenesis, cementogenesis, and the attachment of periodontal ligament fibers. Recent research has shown that enhancing the function of resident stem/progenitor cells through targeted therapeutic interventions holds significant potential for promoting effective periodontal regeneration (Mekhemar et al. 2020a, b, 2021, 2022a, b).

33.3 Mesenchymal Stem Cells (MSCs)

33.3.1 Introduction to Mesenchymal Stem Cells

Human mesenchymal stem cells (MSCs) were initially identified as non-hematopoietic cells with the ability to adhere to plastic surfaces, originally isolated from bone marrow. These cells are characterized by their ability to self-renew and differentiate into multiple lineages in vitro—a discovery initially made by Friedenstein et al. in 1970 (Bianco and Robey 2001; Chamberlain et al. 2007; Friedenstein et al. 1970; Pittenger et al. 1999). Subsequent research has expanded our understanding of MSCs, highlighting their presence in various tissues throughout the adult human body, including bone marrow, adipose tissue, placental tissue, umbilical cord blood, and Wharton's jelly (Mekhemar et al. 2022a, b). While bone marrow has long been recognized as a primary niche for MSCs, research has been focused on identifying alternative sources with fewer complications and lower donor site morbidity (Mekhemar et al. 2022a, b).

MSCs are not limited to a single tissue type and have been successfully isolated from several other tissues, including those found in the oral cavity (Roato et al. 2021). These include the alveolar bone (Liu et al. 2020), gingiva (Fawzy-El-Sayed et al. 2016; Mekhemar et al. 2018, 2020b, 2022a, b), periodontal ligament (Iwayama et al. 2022), dental follicle (Bi et al. 2021), and dental pulp (Yuanting Chen et al. 2022; Fawzy El-Sayed et al. 2023). Despite the phenotypic similarities between MSCs derived from different tissues, their functions and biological activities can vary significantly, emphasizing the uniqueness of MSCs depending on their tissue of origin (Mekhemar et al. 2020b).

33.3.2 Self-Renewal and Multipotency of MSCs

The capacity for self-renewal is a defining characteristic of MSCs, enabling them to proliferate and maintain their numbers over time, which is crucial for their role in tissue growth and repair (He et al. 2009). However, this ability is not indefinite; in vitro studies have shown that MSCs typically have a lifespan of up to 55 population doublings, equivalent to about 44 weeks (Mekhemar et al. 2022a, b). In addition to self-renewal, MSCs exhibit multipotency, meaning they can differentiate into various cell types, primarily those of the mesodermal lineage, such as chondrocytes, osteocytes, and adipocytes. Interestingly, MSCs also possess the potential to differentiate into cells of other embryonic lineages, expanding their versatility in regenerative medicine (Mekhemar et al. 2022a, b).

Identifying MSCs and verifying their "stemness" has been a focal point in stem cell research (Pittenger et al. 2019). In 2006 (Dominici et al. 2006), a consensus was reached on the minimum criteria for defining MSCs, which include their ability to adhere to plastic under standard culture conditions, their ability to undergo multilineage differentiation when exposed to specific stimuli, and the expression of a characteristic set of surface markers (Lv et al. 2014). Positive markers commonly associated with MSCs include CD29, CD44, CD71, CD73, CD90, CD105, CD106, CD120, CD124, CD146, CD166, CD271, Stro-1, and SSEA-4, while negative markers include CD86, CD11, CD80, CD14, CD18, CD31, CD34, CD40, CD45, and CD56. Another hallmark of MSCs is their ability to form colony-forming units (CFUs) when cultured, a property often used to assess their clonogenic potential (Mekhemar et al. 2022a, b).

33.3.3 Oral Stem Cells: A Unique Subset of MSCs

Oral stem cells represent a highly versatile and accessible subset of mesenchymal stem cells (MSCs) with significant potential for regenerative medicine (Poblano-Pérez et al. 2024). These cells can be derived from various tissues within the oral cavity, including the dental pulp (Sui et al. 2020), gingiva (Mekhemar et al. 2020b, 2022a, b), periodontal ligament (Queiroz et al. 2021), and apical papilla (Kang et al. 2019). Among these, the stem cells of the periodontium—specifically, gingival stem cells (G-MSCs) and periodontal ligament stem cells (PDLSCs)—are particularly noteworthy due to their critical roles in maintaining periodontal health and facilitating tissue repair (Mekhemar et al. 2020a, 2021).

G-MSCs, found within the gingival connective tissue, are known for their robust regenerative capacity and ability to promote scarless healing. These cells play a crucial role in the maintenance and repair of the gingival tissue, acting as a first line of defense against microbial invasion while also participating in wound healing processes (Mekhemar et al. 2020b). PDLSCs, located within the periodontal ligament, are essential for the homeostasis and repair of periodontal tissues. They contribute to the regeneration of the periodontal ligament, cementum, and alveolar bone, which are vital for the structural integrity of the tooth-supporting apparatus (Mekhemar et al. 2022a, b).

The periodontium-derived stem cells G-MSCs and PDLSCs not only express typical MSC markers but also demonstrate significant regenerative properties even under inflammatory conditions such as those present in periodontal disease (Mekhemar et al. 2018, 2020b). Their ability to differentiate into various cell types, coupled with their role in modulating immune responses (Mekhemar et al. 2018), underscores their potential in developing advanced therapies aimed at restoring the structure and function of periodontal tissues. These unique characteristics make stem cells of the periodontium a focal point in the research and development of regenerative treatments for periodontal and other oral diseases.

33.4 TLR3 and Its Activation Pathways

Pattern recognition receptors (PRRs) are expressed on cells known as "sensor cells," which play a crucial role in detecting pathogenic antigens (Gong et al. 2019). The presence of toll-like receptors (TLRs) on these cells enables them to recognize various pathogen-associated molecular patterns (PAMPs) (Henrick et al. 2019). TLRs are a family of single-pass transmembrane proteins, characterized by an α-helical domain transversing the membrane (Choe et al. 2005). Their function is often enhanced through dimerization, a process that strengthens antigen binding and triggers downstream signaling cascades (Wang et al. 2010). These dimers can either be homodimers composed of two same monomers (such as TLR3, TLR7, TLR8, and TLR9) or heterodimers consisting of different monomers (such as TLR1, TLR2, TLR4, TLR5, TLR6, and TLR10) (Janssens and Beyaert 2003).

TLRs can be classified based on their localization within the cell and their association with intracellular signaling pathways (Mekhemar et al. 2022a, b). Regarding localization, TLRs are primarily divided into those expressed on the cell surface (e.g., TLR1, TLR2, TLR4, TLR5, TLR6, and TLR10) and those predominantly found in intracellular endosomes (e.g., TLR3, TLR7, TLR8, and TLR9) (Mekhemar et al. 2022a, b). Interestingly, some TLRs, like TLR3 and TLR4, are capable of shifting their location between the cell surface and intracellular compartments—a process believed to facilitate rapid signal induction upon pathogen detection (Bugge et al. 2017). Extracellular TLRs are typically responsive to bacterial components such as lipopolysaccharides (TLR4), peptidoglycans (TLR1, TLR2, TLR6), and flagellin (TLR5) (Nagashima et al. 2015). While the ligand for TLR10 remains unidentified, its heterodimerization with TLR1 and TLR2 suggests a role in bacterial recognition (Nagashima et al. 2015). In contrast, endosomal TLRs are primarily involved in nucleic acid sensing, providing a defense mechanism against viral infections. These receptors are sensitive to double-stranded RNA (TLR3), single-stranded RNA (TLR7, TLR8), and viral DNA (TLR9) (Miyake et al. 2018).

The activation of toll-like receptors (TLRs) depends on intracellular adaptor molecules that facilitate receptor activation and signal transduction. Among these, myeloid differentiation primary response 88 (MyD88) and TIR-domain-containing adapter-inducing interferon-β (TRIF) are two main adaptor proteins. TLR1, TLR2, TLR6, TLR7, TLR8, and TLR10 predominantly trigger MyD88-dependent signaling pathways. However, TLR4, TLR5, and TLR9 are capable of signaling through both MyD88 and TRIF pathways (Mekhemar et al. 2022a, b). Notably, TLR3 is unique among TLRs as it exclusively signals through the TRIF pathway (Botos and Davies 2010).

Toll-like receptor 3 (TLR3) is the first identified antiviral TLR, recognizing viral double-stranded RNA (dsRNA) (Chen et al. 2021). Upon activation, TLR3 induces the transcription factor NF-κB, leading to the production of type I interferons (IFN-I) and pro-inflammatory cytokines. This response is a critical component of the antiviral defense and the broader activation of the innate immune response (Chen et al. 2021).

TLR3 is found in various immune and non-immune cells, including dendritic cells, monocytes, epithelial cells, fibroblasts, and neurons, which ensures the rapid detection of viral invasions (Chen et al. 2021). Additionally, TLR3 expression has been identified in mesenchymal stem cells (MSCs), including gingival mesenchymal stem cells (G-MSCs) (Mekhemar et al. 2020b), periodontal ligament stem cells (PDL-MSCs) (Chaikeawkaew et al. 2020), dental pulp stem cells (DP-MSCs) (Fawzy-El-Sayed et al. 2016), and stem cells from the apical papilla (SCAP) (Fehrmann et al. 2020). This expression indicates a potential association between the immune system's ability to detect pathogens and its role in facilitating tissue regeneration, as evidenced by studies showing a relationship between TLR3 ligation and the modulation of stem cell characteristics (Mekhemar et al. 2020b) (Fig. 33.2).

While TLR3 is predominantly located in intracellular endosomes, certain cells, such as dendritic cells and monocytes, have demonstrated surface expression of TLR3. These cells exhibit heightened sensitivity to extracellular pathogens, making them potential targets for pharmacological immune activation (Mekhemar et al. 2022a, b).

Structurally, TLR3 is composed of three main segments: an ectodomain located either extracellularly or within endosomes, a transmembrane domain, and an intracellular adaptor domain containing the toll-interleukin receptor (TIR) domain responsible for downstream signal transduction. The ectodomain, which is horseshoe-shaped and composed of 23 leucine-rich repeats, facilitates ligand binding. Upon binding to dsRNA, TLR3 forms a fragile complex with one monomer, which is stabilized by the formation of a TLR3 dimer complex (Behzadi et al. 2021).

The precise mechanisms by which TLR3 induces intracellular signaling remain to be fully elucidated. However, it is believed that changes

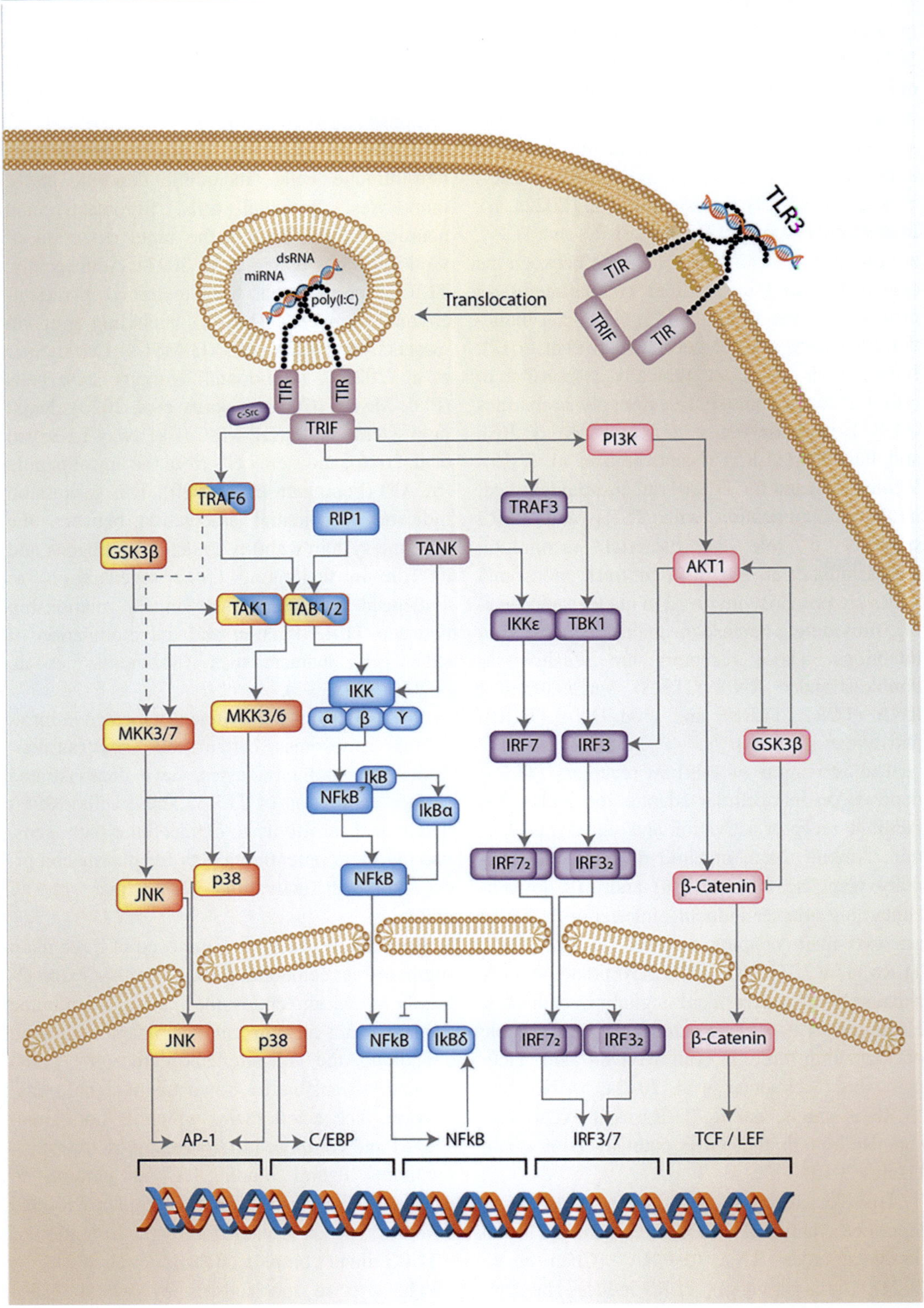

Fig. 33.2 TLR3 signaling pathways in MSCs. Illustration of the signaling cascades activated in mesenchymal stem cells (MSCs) following TLR3 priming by specific ligands. The pathways include *PI3K-AKT pathway*: mediates

in the ectodomain structure during antigen binding or the convergence of intracellular domains during dimerization play a role in signal initiation (Gosu et al. 2019; Liu et al. 2008). Unlike other TLRs, TLR3 exclusively signals through the TRIF adaptor protein.

In experiments where TRIF was inhibited, residual TLR3 activity was observed, which was attributed to the involvement of cellular sarcoma (c-Src) molecules. Short-term TLR3 activation leads to c-Src phosphorylation and induction, whereas prolonged stimulation results in c-Src sequestration and inactivation (Yamashita et al. 2012). Although the exact signaling pathway in MSCs has not been completely revealed, TLR3 stimulation appears to have multiple effects. The biphasic nature of TLR3 activation has been observed in studies on G-MSCs, indicating a complex regulatory role (Mekhemar et al. 2020b).

TRIF, as a key adaptor in TLR3 signaling, is subject to regulation, particularly through K27-linked polyubiquitination mediated by the RING E3 ubiquitin ligase complex Cullin-3-Rbx1-KCTD10 (X. Wu et al. 2019). Deficiencies in this complex can hinder TRIF-TLR binding and downstream signaling. TRIF functions as a central hub for four distinct signaling pathways, including the β-Catenin, IRF3/IRF7, NF-κB, and AP-1, that ultimately influence transcription factor activation and gene expression (Jiang et al. 2004).

The β-Catenin pathway, crucial for MSC pluripotency and cancer progression, is regulated by two major cascades: the PI3K/Akt pathway and the Wnt pathway (Hemmings and Restuccia 2012). PI3K activation, triggered by TRIF, converts PIP2 into its active form PIP3, enabling the phosphorylation and activation of the Akt complex (Hemmings and Restuccia 2012). This activation leads to the conversion of cytosolic β-Catenin to its active form, promoting cell proliferation (Gantner et al. 2012). The Wnt pathway further regulates β-Catenin accumulation and nuclear translocation, while its inactivation results in β-Catenin degradation (Fang et al. 2007; Gantner et al. 2012; Gerlach et al. 2014; Matsumoto et al. 2018).

TRIF also activates IRF3 and IRF7 via TRAF3 in a process that requires the formation of a TRIF-TBK1 complex. This complex inhibits K63-linked polyubiquitination of TRAF3 and TRIF, leading to the activation of kinases TBK1 and IKKε. This leads to the phosphorylation of IRF3 and IRF7, facilitating their dimerization and nuclear translocation. This process is critical for the activation of type I interferons (Mekhemar et al. 2022a, b).

Additionally, NF-κB and AP-1 are activated via TRAF6. The NF-κB pathway is influenced by the TRAF6-TAB1/2-TAK1-RIP1 complex, while AP-1 activation is mediated through MAPK pathways. Notably, AP-1 activation, unlike NF-κB, is dependent on GSK3β activity, which plays a role in the assembly of the TRAF6–TAK1–TAB complex, ultimately leading to the activation of transcription factors and gene expression (Mekhemar et al. 2022a, b).

In summary, the activation of TLR3 involves complex signaling pathways that influence various cellular processes, including immune responses, inflammation, and cell proliferation. Understanding these pathways is crucial for developing therapeutic strategies that target TLR3 and its downstream effects.

Fig. 33.2 (continued) cellular functions via β-catenin, contributing to survival, proliferation, and differentiation. *TRAF pathway*: influences MSC function through modulation by IRF, NF-κB, and AP-1, impacting immune response and inflammation. *TRIF-independent c-Src cascade*: regulates MSC proliferation through alternate signaling routes. *Abbreviations*: *TIR* toll/interleukin-1 receptor homology domain, *c-Src* proto-oncogene tyrosine-protein kinase Src, *TRIF* TIR-domain-containing adapter-inducing interferon-β, *PI3K* phosphoinositide 3-kinase, *AKT* protein kinase B, *GSK3β* glycogen synthase kinase 3 beta, *TRAF* TNF receptor-associated factor, *RIP* receptor-interacting protein, *TAB* TGF-beta-activated kinase 1 (MAP3K7)-binding protein, *TAK1* mitogen-activated protein kinase kinase kinase 7, *IKK* IκB kinase, *NF-κB* nuclear factor “kappa-light-chain-enhancer” of activated B cells, *IRF* interferon regulatory factor, *LEF* T cell factor/lymphoid enhancer-binding factor, *p38* p38 mitogen-activated protein kinase, *JNK* c-Jun N-terminal kinase, *AP-1* activator protein 1, *EBP* cholestenol delta-isomerase, *MKK* mitogen-activated protein kinase kinase

33.5 TLR3 Priming and MSC Immunomodulation in Periodontal Disease

33.5.1 The Role of TLR3 in Immune Cell Recruitment and MSC Interaction

Recent advancements in regenerative medicine have increasingly highlighted the therapeutic potential of mesenchymal stem cells (MSCs) in modulating immune responses, particularly in the context of chronic inflammatory conditions such as periodontal disease (Mekhemar et al. 2020a, b, 2021). The dynamic interaction between MSCs and the immune system plays a crucial role in the repair and protection of damaged tissues within inflamed environments. Among the various factors influencing MSC function, toll-like receptors (TLRs) are of significant interest, especially TLR3 due to their ability to modulate the immunomodulatory properties of MSCs (Mekhemar et al. 2020b, 2022a, b).

The recruitment of immune cells to sites of tissue damage is a fundamental aspect of the immune response, especially in inflammatory diseases like periodontal disease (Mekhemar et al. 2020a, 2021). The chronic nature of periodontal disease involves persistent inflammation driven by bacterial biofilms that result in the destruction of tooth-supporting structures such as the alveolar bone, periodontal ligament, and gingival tissues (Mekhemar et al. 2020a, 2021). In this context, MSCs can play a dual role by both modulating the immune response and contributing directly to tissue regeneration (Mekhemar et al. 2018, 2020b, 2022a, b).

TLR3 ligation on MSCs initiates a cascade of signaling pathways leading to the secretion of cytokines and chemokines, which are essential for immune cell migration. For instance, TLR3 activation in bone marrow-derived MSCs (BM-MSCs) has been shown to induce the secretion of key cytokines like IL-6, IL-8, and chemokines such as MCP-1, all of which play vital roles in recruiting immune cells to sites of inflammation (Mekhemar et al. 2022a, b). In periodontal disease, this recruitment is critical because it helps to orchestrate the immune response that combats bacterial invasion while simultaneously initiating tissue repair processes (Hajishengallis 2015).

In periodontal disease, the accumulation of bacterial biofilms on tooth surfaces leads to the activation of an inflammatory response. This response is mediated by the innate immune system, which recognizes pathogen-associated molecular patterns (PAMPs) through receptors such as TLRs, including TLR3 (Mekhemar et al. 2020b, 2022a, b). The activation of TLR3 by double-stranded RNA (dsRNA) or synthetic analogs like Poly(I:C) in MSCs triggers the recruitment of immune cells to the site of infection by increasing the secretion of chemokines such as CXCL8 (IL-8) and CCL2 (MCP-1) (Mekhemar et al. 2022a, b). These chemokines play a pivotal role in guiding neutrophils and monocytes to the site of infection in periodontal disease, where they contribute to the clearance of pathogens and the resolution of inflammation (Fig. 33.3).

33.5.2 MSCs as Modulators of Immune Responses in Periodontal Disease

Once immune cells have been recruited to the periodontal tissue, MSCs play a crucial role in modulating the immune response to either promote or suppress inflammation (Mekhemar et al. 2018). This immunomodulatory function of MSCs is vital in the context of periodontal disease, where chronic inflammation drives the progressive destruction of the periodontium. The ability of MSCs to regulate the local immune environment can significantly impact the outcome of periodontal treatment (Mekhemar et al. 2018, 2020b, 2022a, b).

TLR3 activation has been associated with enhanced immunosuppressive properties in MSCs, which can help control the excessive inflammatory response characteristic of periodontal disease. For example, TLR3-primed gingival

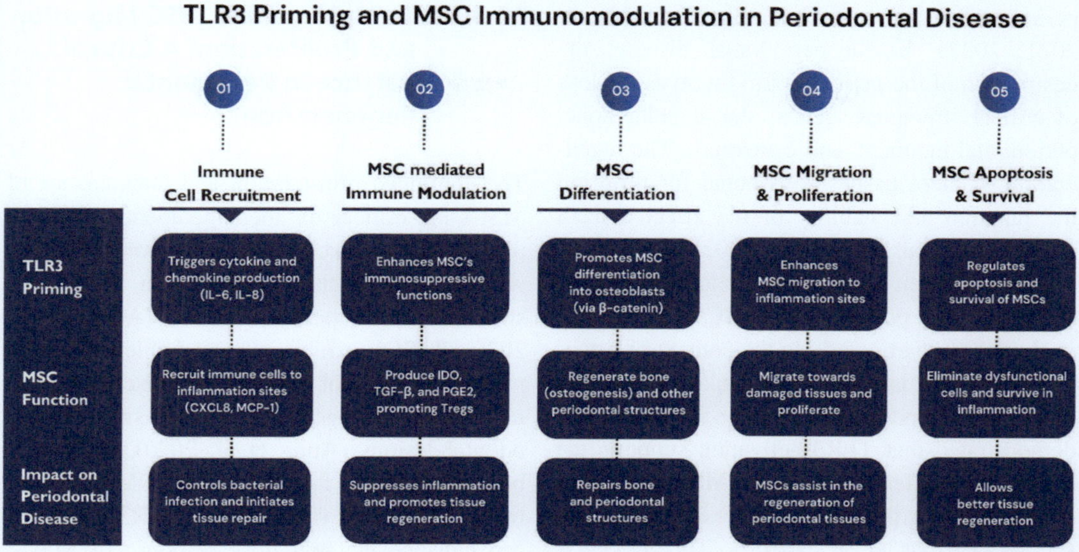

Fig. 33.3 Illustration of the different mechanisms by which TLR3 priming influences mesenchymal stem cell (MSC) functions, impacting the disease progression and regeneration in periodontal tissues. The five key mechanisms shown include 1. *immune cell recruitment*: TLR3 activation triggers cytokine and chemokine production (e.g., IL-6, IL-8, CXCL8, MCP-1), facilitating immune cell migration to inflammation sites, aiding in infection control, and initiating tissue repair. 2. *MSC-mediated immune modulation*: TLR3-primed MSCs enhance immunosuppressive functions, producing factors such as IDO, TGF-β, and PGE2, which help suppress inflammation and promote regulatory T cell expansion. 3. *MSC differentiation*: TLR3 signaling promotes osteogenic differentiation via pathways like β-catenin, aiding in the regeneration of bone and periodontal structures. 4. *MSC migration and proliferation*: Enhanced MSC migration toward damaged tissues supports the regeneration of periodontal tissues by delivering cells to sites of inflammation. 5. *MSC apoptosis and survival*: TLR3 regulates the balance between apoptosis and survival of MSCs, helping eliminate dysfunctional cells and maintain a regenerative environment

MSCs (G-MSCs) have been shown to produce higher levels of immunosuppressive mediators such as indoleamine-2,3-dioxygenase (IDO), transforming growth factor-beta (TGF-β), and prostaglandin E2 (PGE2) (Mekhemar et al. 2018). These mediators are critical in controlling the local immune environment by reducing the activity of pro-inflammatory T cells and promoting the expansion of regulatory T cells (Tregs). This shift toward an anti-inflammatory environment is essential for the regeneration of periodontal tissues as it limits ongoing tissue destruction and allows for the healing and repair of the periodontium.

In addition to their effects on T cells, TLR3-primed MSCs can modulate the activity of other immune cells involved in periodontal disease. For instance, dendritic cells (DCs) are key mediators in the initiation and modulation of immune responses by presenting antigens to T cells. TLR3 activation in MSCs has been shown to influence DC maturation and function, promoting the generation of tolerogenic DCs that support the induction of Tregs and the suppression of inflammatory T cell responses (Negi and Griffin 2020). This mechanism is particularly relevant in periodontal disease, where the control of T cell-mediated inflammation is crucial for preventing further tissue damage (Jiang et al. 2022).

33.5.3 TLR3's Influence on MSC Differentiation and Tissue Regeneration

The ability of MSCs to differentiate into various cell types is fundamental to their role in tissue regeneration, including the regeneration of periodontal tissues (Mekhemar et al. 2020a, b, 2021, 2022a, b). In periodontal disease, the destruction of the periodontium involves the loss of critical structures such as the alveolar bone, periodontal ligament, and cementum. The regeneration of these tissues is essential for restoring the function and stability of the affected teeth (Mekhemar et al. 2020a, 2021, 2022a, b).

TLR3 ligation has been shown to influence the differentiation potential of MSCs (Mekhemar et al. 2020b), particularly in the context of osteogenesis, which is vital for the repair of alveolar bone in periodontal disease. Research has demonstrated that TLR3 activation supports the osteogenic differentiation of MSCs, thereby enhancing their potential to regenerate bone tissue (Mekhemar et al. 2020b). This effect is facilitated by the activation of key signaling pathways, such as the β-catenin pathway, which plays a critical role in osteogenesis (Song et al. 2017).

In periodontal disease, the ability of TLR3-primed MSCs to differentiate into osteoblasts and to promote bone formation is particularly valuable. The destruction of alveolar bone is a hallmark of periodontitis, and the regeneration of this bone is crucial for restoring the structural integrity of the periodontium (Usui et al. 2021). The upregulation of β-catenin and the co-activation of NF-κB following TLR3 stimulation have been shown to promote the differentiation of MSCs into osteoblasts, leading to increased bone formation and mineralization (Liu et al. 2023).

Furthermore, TLR3 activation has been implicated in the regulation of other differentiation pathways, such as adipogenesis and chondrogenesis, which are also relevant to periodontal regeneration. The balance between osteogenic and adipogenic differentiation is particularly important in periodontal disease, as excessive adipogenesis can detract from the regeneration of hard tissues like bone (Wu et al. 2017). TLR3 activation has been shown to favor osteogenesis over adipogenesis in MSCs, which is beneficial for periodontal regeneration (Mekhemar et al. 2020b).

33.5.4 TLR3-Regulated MSC Migration and Proliferation: A Critical Balance in Periodontal Regeneration

The continued proliferation and maintenance of MSCs are critical for their therapeutic efficacy, particularly in the context of periodontal disease, where ongoing tissue destruction requires a sustained regenerative response (Aung et al. 2020). MSCs exhibit a remarkable ability to migrate to areas of inflammation, ischemia, and mechanical damage, including the sites of periodontal lesions (Aung et al. 2020). TLR3 has been identified as a key regulator of MSC migration, with studies showing that TLR3 activation can enhance the migratory response of MSCs, allowing them to efficiently reach and repair damaged tissues (Tolstova et al. 2023).

In periodontal disease, the ability of MSCs to migrate to the inflamed periodontium is crucial for their role in tissue regeneration (Chen et al. 2023). TLR3-primed MSCs exhibit enhanced migration toward inflammatory signals, which is essential for delivering the cells to the sites of tissue damage. Once at the site, these MSCs actively engage in the repair and regeneration of the periodontium, contributing to the restoration of lost structures and the stabilization of the affected teeth (Mekhemar et al. 2020b).

33.5.5 TLR3-Mediated Apoptosis and Survival: Protecting MSC Function in Periodontal Disease

TLR3 activation not only influences the differentiation and immunomodulatory functions of MSCs but also plays a significant role in regulating their survival and apoptosis (Waterman et al. 2010). In the context of periodontal disease, where the environment is often characterized by chronic inflammation and tissue stress, the ability of MSCs to survive and maintain their function is critical (Chew et al. 2019).

TLR3 ligands have been explored for their potential to induce apoptosis in various cell types, including MSCs (Sun et al. 2011). This apoptotic effect can be beneficial in eliminating damaged or dysfunctional cells from the periodontium, thereby facilitating the regeneration of healthy tissue (Cochran 2008). For example, TLR3-primed MSCs can induce apoptosis in inflammatory cells, aiding in the resolution of chronic inflammation, which is a key driver of periodontal tissue destruction. By eliminating these inflammatory cells, MSCs create a more favorable environment for tissue regeneration.

In periodontal disease, the removal of dysfunctional cells, including fibroblasts and inflammatory cells that contribute to tissue destruction, is essential for creating an environment conducive to tissue regeneration (Cochran 2008). TLR3-primed MSCs may enhance the clearance of these cells through the induction of apoptosis, thereby promoting the resolution of inflammation and the regeneration of periodontal tissues. This apoptotic effect may be mediated by the upregulation of proapoptotic factors such as Fas ligand (FasL) and TRAIL, which trigger the extrinsic apoptosis pathway in target cells (Mekhemar et al. 2022a, b).

However, TLR3 activation has also been shown to enhance the survival of MSCs under certain conditions, such as inflammation, by reducing apoptosis and promoting cell viability (Mekhemar et al. 2022a, b). This dual role of TLR3 in regulating MSC apoptosis and survival is particularly relevant in periodontal disease, where controlled cell turnover is necessary for effective tissue repair and regeneration.

33.5.6 MicroRNAs as Critical Modulators of TLR3-Mediated MSC Responses

MicroRNAs (miRNAs) are short, non-coding RNA molecules that modulate gene expression by interacting with messenger RNAs (mRNAs) and preventing their translation into functional proteins (Ratti et al. 2020). These molecules play a vital role in controlling various cellular functions, including the immune response and the differentiation of MSCs (Yang et al. 2021). In the context of TLR3-mediated signaling, miRNAs can significantly influence the behavior of MSCs, including their immunomodulatory and regenerative functions (Bayraktar et al. 2019).

In periodontal disease, miRNAs may modulate the response of MSCs to TLR3 activation, thereby influencing the outcome of the immune response and tissue regeneration. For example, miR-143 has been identified as a negative regulator of TLR3-mediated signaling in MSCs (Zhao et al. 2014). Downregulation of miR-143 following TLR3 activation has been associated with increased secretion of immunosuppressive cytokines and enhanced regenerative capacity of MSCs (Mekhemar et al. 2020b). This regulation of miRNAs could be particularly important in periodontal disease, where fine-tuning the immune response is essential for preventing further tissue destruction and promoting healing.

The interaction between miRNAs and TLR3 signaling in MSCs offers a potential avenue for therapeutic intervention in periodontal disease. By targeting specific miRNAs, it may be possible to enhance the beneficial effects of TLR3 activation on MSCs, thereby improving the outcomes of MSC-based therapies for periodontal regeneration (Santonocito et al. 2021).

33.6 Future Perspectives

The future of periodontal therapy is likely to focus on advancing stem cell-based approaches to improve regenerative outcomes and modulate immune responses (Nguyen-Thi et al. 2023). Among these, the exploration of TLR3's role in enhancing MSC therapies holds significant promise. TLR3 activation not only modulates immune responses but also enhances the regenerative potential of MSCs by influencing key cellular processes such as differentiation, proliferation, migration, and survival (Mekhemar et al. 2020b). Given the multifactorial nature of periodontitis, targeting TLR3 signaling pathways could provide a novel therapeutic approach that addresses both inflammation and tissue destruction. Future research should prioritize understanding the molecular mechanisms by which TLR3 regulates MSC functions in inflamed periodontal environments, focusing on signaling pathways such as NF-κB, PI3K/AKT, and Wnt (Mekhemar et al. 2020b). Optimizing these pathways could help balance anti-inflammatory benefits with regenerative outcomes.

Further investigation is needed to determine whether the timing and intensity of TLR3 activation produce different effects in acute versus chronic inflammatory conditions, allowing for precise therapeutic interventions tailored to the stage of the disease. This approach could potentially enable more effective management of inflammation and regeneration, depending on the specific needs of each patient. The integration of TLR3-targeted therapies with conventional periodontal treatments, such as scaling and root planing, may also enhance therapeutic efficacy by simultaneously addressing microbial dysbiosis and promoting tissue repair. This combination could offer a more comprehensive approach than traditional methods alone, potentially minimizing the need for invasive surgical procedures (Mekhemar et al. 2020a, 2021, 2022a, b). The use of TLR3 agonists, used as adjunctive agents in mechanical therapies, may help regulate local immune responses in a targeted manner, stimulating the repair process while mitigating excessive inflammation (Hajishengallis 2015; Mekhemar et al. 2018, 2020b, 2022a, b). Additionally, combining TLR3-based approaches with antimicrobial treatments could target both bacterial pathogens and host immune dysregulation, addressing the core aspects of periodontitis pathogenesis (Hajishengallis 2015).

An alternative strategy involves utilizing natural agents, such as thymoquinone, that have demonstrated the ability to upregulate TLR3 expression or modulate TLR functions (Chen and Yu 2016; Mekhemar et al. 2022a, b). These natural compounds could be used to enhance MSC responsiveness to TLR3 activation under inflammatory conditions, providing a complementary approach to synthetic TLR3 agonists. Thymoquinone, along with other bioactive plant-derived substances, could be incorporated into regenerative protocols to potentiate TLR3-mediated effects on MSCs, especially in chronic inflammation. The integration of natural agents into TLR3-modulating therapies aligns with the trend of using natural periodontal supportive treatments to boost the regenerative capacity of MSCs and improve clinical outcomes (Mekhemar et al. 2020a, 2021).

The development of advanced drug delivery systems presents exciting opportunities for optimizing TLR3-based therapies. Nanoparticles and hydrogels enable the targeted delivery of TLR3 agonists or natural compounds directly to the periodontal tissues, ensuring sustained and localized activation. These delivery platforms can be engineered to gradually release TLR3 activators, enabling prolonged therapeutic effects while minimizing systemic exposure. Furthermore, combining different bioactive agents within these delivery systems, such as TLR3 activators and growth factors, could enhance therapeutic outcomes by promoting cellular proliferation and matrix deposition while simultaneously controlling inflammation.

Another promising area for future research is the role of TLR3 in influencing other immune cells, such as macrophages, dendritic cells, and regulatory T cells. Investigating how TLR3-primed MSCs interact with these cells could lead to more comprehensive therapeutic strategies that not only regenerate tissues but also create a stable immune environment for long-term periodontal health. Furthermore, understanding the broader immunomodulatory effects of TLR3 activation may facilitate the development of holistic regenerative therapies that address the complexities of chronic periodontitis.

Overall, exploring TLR3-mediated effects on MSCs in periodontal disease represents a promising frontier in regenerative medicine. By integrating insights from molecular biology, immunology, and biomaterials science, future approaches could revolutionize the application of TLR3-primed MSCs in periodontal regeneration. Continued research into optimizing TLR3 signaling is essential for transforming treatment paradigms for periodontitis, providing effective solutions that target both the causes and consequences of tissue destruction. This multifaceted strategy could offer more comprehensive periodontal therapy that not only addresses tissue damage but also reduces the risk of disease recurrence, ultimately improving patient outcomes and quality of life (Al-Bitar et al. 2024).

33.7 Conclusion

The role of TLR3 in modulating the immunobiological and regenerative functions of MSCs is crucial for advancing periodontal disease treatment. By influencing key processes such as differentiation, proliferation, migration, and immune modulation, TLR3 activation holds promise for enhancing MSC-based regenerative therapies. Given the multifactorial nature of periodontitis, TLR3-targeted strategies could offer a comprehensive approach that addresses both inflammation and tissue destruction, leading to better therapeutic outcomes.

Further research is needed to elucidate the specific effects of TLR3 activation on MSCs within the periodontal environment and optimize treatment protocols. Understanding how TLR3 signaling interacts with other regulatory pathways and immune cells could provide insights into creating more tailored and effective interventions. Efforts should focus on translating knowledge from molecular biology, immunology, and tissue engineering into clinical applications, positioning TLR3-primed MSCs as a cornerstone of advanced periodontal therapies.

Combining TLR3 activation with conventional treatments and antimicrobial strategies could improve outcomes by providing targeted, multifaceted therapeutic effects. Novel delivery systems such as hydrogels and nanoparticles can facilitate localized and sustained activation, enhancing the therapeutic potential of TLR3-modulating agents. This could help create a more favorable environment for tissue regeneration while minimizing side effects.

The exploration of TLR3's role in periodontal regeneration holds significant promise for advancing MSC-based therapies. By leveraging its influence on immune regulation and tissue repair, TLR3-based strategies could offer durable and effective solutions for patients with chronic periodontal disease. The advancement of these strategies has the potential to transform periodontal treatment, leading to lasting improvements in oral health and quality of life.

References

Abdulkareem AA, Al-Taweel FB, Al-Sharqi AJB, Gul SS, Sha A, Chapple ILC (2023) Current concepts in the pathogenesis of periodontitis: from symbiosis to dysbiosis. J Oral Microbiol 15(1):2197779. https://doi.org/10.1080/20002297.2023.2197779

Al-Bitar KM, Garcia JM, Han S, Guentsch A (2024) Association between periodontal health status and quality of life: a cross-sectional study. Front Oral Health 5. https://doi.org/10.3389/froh.2024.1346814

Aung KT, Akiyama K, Kunitomo M, Mun AY, Tosa I, Nguyen HTT et al (2020) Aging-affected MSC functions and severity of periodontal tissue destruction in a ligature-induced mouse periodontitis model. Int J Mol Sci 21(21). https://doi.org/10.3390/ijms21218103

Bayraktar R, Bertilaccio MTS, Calin GA (2019) The interaction between two worlds: MicroRNAs and toll-like receptors. Front Immunol 10:1053. https://doi.org/10.3389/fimmu.2019.01053

Behzadi P, García-Perdomo HA, Karpiński TM (2021) Toll-like receptors: general molecular and structural biology. J Immunol Res 2021:9914854. https://doi.org/10.1155/2021/9914854

Bhuyan R, Bhuyan SK, Mohanty JN, Das S, Juliana N, Juliana IF (2022) Periodontitis and its inflammatory changes linked to various systemic diseases: a review of its underlying mechanisms. Biomedicine 10(10). https://doi.org/10.3390/biomedicines10102659

Bi R, Lyu P, Song Y, Li P, Song D, Cui C, Fan Y (2021) Function of dental follicle progenitor/stem cells and their potential in regenerative medicine: from mechanisms to applications. Biomolecules 11(7). https://doi.org/10.3390/biom11070997

Bianco P, Robey PG (2001) Stem cells in tissue engineering. Nature 414(6859):118–121. https://doi.org/10.1038/35102181

Botelho J, Machado V, Leira Y, Proença L, Chambrone L, Mendes JJ (2022) Economic burden of periodontitis in the United States and Europe: an updated estimation. J Periodontol 93(3):373–379. https://doi.org/10.1002/JPER.21-0111

Botos I, Davies DR (2010) Chapter 21 – toll-like receptors–structureand signaling. In: Bradshaw RA, Dennis EA (eds) Handbook of cell signaling, 2nd edn. Academic Press, San Diego, pp 139–143

Bugge M, Bergstrom B, Eide OK, Solli H, Kjonstad IF, Stenvik J et al (2017) Surface Toll-like receptor 3 expression in metastatic intestinal epithelial cells induces inflammatory cytokine production and promotes invasiveness. J Biol Chem 292(37): 15408–15425. https://doi.org/10.1074/jbc.M117.784090

Chaikeawkaew D, Everts V, Pavasant P (2020) TLR3 activation modulates immunomodulatory properties of human periodontal ligament cells. J Periodontol 91(9):1225–1236. https://doi.org/10.1002/jper.19-0551

Chamberlain G, Fox J, Ashton B, Middleton J (2007) Concise review: mesenchymal stem cells: their phenotype, differentiation capacity, immunological features, and potential for homing. Stem Cells 25(11): 2739–2749. https://doi.org/10.1634/stemcells.2007-0197

Chatzopoulos GS, Jiang Z, Marka N, Wolff LF (2024) Periodontal disease, tooth loss, and systemic conditions: an exploratory study. Int Dent J 74(2): 207–215. https://doi.org/10.1016/j.identj.2023.08.002

Chen L, Yu J (2016) Modulation of toll-like receptor signaling in innate immunity by natural products. Int Immunopharmacol 37:65–70. https://doi.org/10.1016/j.intimp.2016.02.005

Chen Y, Lin J, Zhao Y, Ma X, Yi H (2021) Toll-like receptor 3 (TLR3) regulation mechanisms and roles in antiviral innate immune responses. J Zhejiang Univ Sci B 22(8):609–632. https://doi.org/10.1631/jzus.B2000808

Chen Y, Huang H, Li G, Yu J, Fang F, Qiu W (2022) Dental-derived mesenchymal stem cell sheets: a prospective tissue engineering for regenerative medicine. Stem Cell Res Ther 13(1):38. https://doi.org/10.1186/s13287-022-02716-3

Chen L, Zhu S, Guo S, Tian W (2023) Mechanisms and clinical application potential of mesenchymal stem cells-derived extracellular vesicles in periodontal regeneration. Stem Cell Res Ther 14(1):26. https://doi.org/10.1186/s13287-023-03242-6

Chew JRJ, Chuah SJ, Teo KYW, Zhang S, Lai RC, Fu JH et al (2019) Mesenchymal stem cell exosomes enhance periodontal ligament cell functions and promote periodontal regeneration. Acta Biomater 89:252–264. https://doi.org/10.1016/j.actbio.2019.03.021

Choe J, Kelker MS, Wilson IA (2005) Crystal structure of human toll-like receptor 3 (TLR3) ectodomain. Science 309(5734):581–585. https://doi.org/10.1126/science.1115253

Cochran DL (2008) Inflammation and bone loss in periodontal disease. J Periodontol 79(8S):1569–1576. https://doi.org/10.1902/jop.2008.080233

Dominici M, Le Blanc K, Mueller I, Slaper-Cortenbach I, Marini F, Krause D et al (2006) Minimal criteria for defining multipotent mesenchymal stromal cells. The International Society for Cellular Therapy position statement. Cytotherapy 8(4):315–317. https://doi.org/10.1080/14653240600855905

Fang D, Hawke D, Zheng Y, Xia Y, Meisenhelder J, Nika H et al (2007) Phosphorylation of beta-catenin by AKT promotes beta-catenin transcriptional activity. J Biol Chem 282(15):11221–11229. https://doi.org/10.1074/jbc.M611871200

Fawzy El-Sayed KM, Klingebiel P, Dörfer CE (2016) Toll-like receptor expression profile of human dental pulp stem/progenitor cells. J Endodont 42(3):413–417. https://doi.org/10.1016/j.joen.2015.11.014

Fawzy El-Sayed KM, Rudert A, Geiken A, Tölle J, Mekhemar M, Dörfer CE (2023) Toll-like receptor expression profile of stem/progenitor cells from human exfoliated deciduous teeth. Int J Paediatr Dent 33(6):607–614. https://doi.org/10.1111/ipd.13080

Fawzy-El-Sayed K, Mekhemar M, Adam-Klages S, Kabelitz D, Dorfer C (2016) TLR expression profile of human gingival margin-derived stem progenitor cells. Med Oral Patol Oral Cir Bucal 21(1):e30–e38. https://doi.org/10.4317/medoral.20593

Fehrmann C, Dorfer CE, Fawzy El-Sayed KM (2020) Toll-like receptor expression profile of human stem/progenitor cells form the apical papilla. J Endodont 46(11):1623–1630. https://doi.org/10.1016/j.joen.2020.08.017

Friedenstein AJ, Chailakhjan RK, Lalykina KS (1970) The development of fibroblast colonies in monolayer cultures of guinea-pig bone marrow and spleen cells. Cell Tissue Kinet 3(4):393–403. https://doi.org/10.1111/j.1365-2184.1970.tb00347.x

Gantner BN, Jin H, Qian F, Hay N, He B, Ye RD (2012) The Akt1 isoform is required for optimal IFN-beta transcription through direct phosphorylation of beta-catenin. J Immunol 189(6):3104–3111. https://doi.org/10.4049/jimmunol.1201669

Gasner NS, Schure RS (2024) Periodontal disease. In: StatPearls. Ineligible companies. Disclosure: Ryan Schure declares no relevant financial relationships with ineligible companies. StatPearls

Gerlach JP, Emmink BL, Nojima H, Kranenburg O, Maurice MM (2014) Wnt signalling induces accumulation of phosphorylated beta-catenin in two distinct cytosolic complexes. Open Biol 4(11):140120. https://doi.org/10.1098/rsob.140120

Gong M, Wang F, Liu W, Chen R, Wu H, Zhang W et al (2019) Pattern recognition receptor-mediated innate immune responses in seminal vesicle epithelial cell and their impacts on cellular function. Biol Reprod. https://doi.org/10.1093/biolre/ioz136

Gosu V, Son S, Shin D, Song KD (2019) Insights into the dynamic nature of the dsRNA-bound TLR3 complex. Sci Rep 9(1):3652. https://doi.org/10.1038/s41598-019-39984-8

Hajishengallis G (2015) Periodontitis: from microbial immune subversion to systemic inflammation. Nat Rev Immunol 15(1):30–44. https://doi.org/10.1038/nri3785

Han N, Liu Y, Du J, Xu J, Guo L, Liu Y (2023) Regulation of the host immune microenvironment in periodontitis and periodontal bone remodeling. Int J Mol Sci 24(4). https://doi.org/10.3390/ijms24043158

He S, Nakada D, Morrison SJ (2009) Mechanisms of stem cell self-renewal. Annu Rev Cell Dev Biol 25:377–

406. https://doi.org/10.1146/annurev.cellbio.042308.113248

Hemmings BA, Restuccia DF (2012) PI3K-PKB/Akt pathway. Cold Spring Harb Perspect Biol 4(9): a011189. https://doi.org/10.1101/cshperspect.a011189

Henrick BM, Yao X-D, Zahoor MA, Abimiku AL, Osawe S, Rosenthal KL (2019) TLR10 senses HIV-1 proteins and significantly enhances HIV-1 infection. Front Immunol 10(482). https://doi.org/10.3389/fimmu.2019.00482

Iwayama T, Sakashita H, Takedachi M, Murakami S (2022) Periodontal tissue stem cells and mesenchymal stem cells in the periodontal ligament. Jpn Dent Sci Rev 58:172–178. https://doi.org/10.1016/j.jdsr.2022.04.001

Janssens S, Beyaert R (2003) Role of toll-like receptors in pathogen recognition. Clin Microbiol Rev 16(4): 637–646. https://doi.org/10.1128/cmr.16.4.637-646.2003

Jiang Z, Mak TW, Sen G, Li X (2004) Toll-like receptor 3-mediated activation of NF-kappaB and IRF3 diverges at Toll-IL-1 receptor domain-containing adapter inducing IFN-beta. Proc Natl Acad Sci USA 101(10):3533–3538. https://doi.org/10.1073/pnas.0308496101

Jiang Q, Huang X, Yu W, Huang R, Zhao X, Chen C (2022) mTOR signaling in the regulation of CD4+ T cell subsets in periodontal diseases. Front Immunol 13. https://doi.org/10.3389/fimmu.2022.827461

Kang J, Fan W, Deng Q, He H, Huang F (2019) Stem cells from the apical papilla: a promising source for stem cell-based therapy. Biomed Res Int 2019:6104738. https://doi.org/10.1155/2019/6104738

Kwon T, Lamster IB, Levin L (2021) Current concepts in the management of periodontitis. Int Dent J 71(6): 462–476. https://doi.org/10.1111/idj.12630

Liu L, Botos I, Wang Y, Leonard JN, Shiloach J, Segal DM, Davies DR (2008) Structural basis of toll-like receptor 3 signaling with double-stranded RNA. Science 320(5874):379–381. https://doi.org/10.1126/science.1155406

Liu Y, Wang H, Dou H, Tian B, Li L, Jin L et al (2020) Bone regeneration capacities of alveolar bone mesenchymal stem cells sheet in rabbit calvarial bone defect. J Tissue Eng 11:2041731420930379. https://doi.org/10.1177/2041731420930379

Liu X, Zhou Z, Zeng WN, Zeng Q, Zhang X (2023) The role of toll-like receptors in orchestrating osteogenic differentiation of mesenchymal stromal cells and osteoimmunology. Front Cell Dev Biol 11:1277686. https://doi.org/10.3389/fcell.2023.1277686

Lv FJ, Tuan RS, Cheung KM, Leung VY (2014) Concise review: the surface markers and identity of human mesenchymal stem cells. Stem Cells 32(6): 1408–1419. https://doi.org/10.1002/stem.1681

Matsumoto T, Yokoi A, Hashimura M, Oguri Y, Akiya M, Saegusa M (2018) TGF-beta-mediated LEFTY/Akt/GSK-3beta/Snail axis modulates epithelial-mesenchymal transition and cancer stem cell properties in ovarian clear cell carcinomas. Mol Carcinog 57(8): 957–967. https://doi.org/10.1002/mc.22816

Mekhemar MK, Adam-Klages S, Kabelitz D, Dorfer CE, Fawzy El-Sayed KM (2018) TLR-induced immunomodulatory cytokine expression by human gingival stem/progenitor cells. Cell Immunol 326:60–67. https://doi.org/10.1016/j.cellimm.2017.01.007

Mekhemar M, Hassan Y, Dorfer C (2020a) Nigella sativa and thymoquinone: a natural blessing for periodontal therapy. Antioxidants (Basel) 9(12). https://doi.org/10.3390/antiox9121260

Mekhemar M, Tolle J, Dorfer C, Fawzy El-Sayed K (2020b) TLR3 ligation affects differentiation and stemness properties of gingival mesenchymal stem/progenitor cells. J Clin Periodontol 47(8):991–1005. https://doi.org/10.1111/jcpe.13323

Mekhemar M, Geib M, Kumar M, Radha, Hassan Y, Dorfer C (2021) Salvadora persica: nature's gift for periodontal health. Antioxidants (Basel) 10(5). https://doi.org/10.3390/antiox10050712

Mekhemar M, Tölle J, Dörfer C, Fawzy El-Sayed KM (2022a) Toll-like receptor 3. In: Haider KH (ed) Handbook of stem cell therapy. Springer, Singapore, pp 1279–1302

Mekhemar M, Tolle J, Hassan Y, Dorfer C, El-Sayed KF (2022b) Thymoquinone-mediated modulation of toll-like receptors and pluripotency factors in gingival mesenchymal stem/progenitor cells. Cells 11(9). https://doi.org/10.3390/cells11091452

Miyake K, Shibata T, Ohto U, Shimizu T, Saitoh SI, Fukui R, Murakami Y (2018) Mechanisms controlling nucleic acid-sensing toll-like receptors. Int Immunol 30(2):43–51. https://doi.org/10.1093/intimm/dxy016

Nagashima H, Iwatani S, Cruz M, Jimenez Abreu JA, Uchida T, Mahachai V et al (2015) Toll-like receptor 10 in helicobacter pylori infection. J Infect Dis 212(10):1666–1676. https://doi.org/10.1093/infdis/jiv270

Negi N, Griffin MD (2020) Effects of mesenchymal stromal cells on regulatory T cells: current understanding and clinical relevance. Stem Cells 38(5):596–605. https://doi.org/10.1002/stem.3151

Nguyen-Thi TD, Nguyen-Huynh BH, Vo-Hoang TT, Nguyen-Thanh T (2023) Stem cell therapies for periodontal tissue regeneration: a meta-analysis of clinical trials. J Oral Biol Craniofac Res 13(5):589–597. https://doi.org/10.1016/j.jobcr.2023.07.001

Pittenger MF, Mackay AM, Beck SC, Jaiswal RK, Douglas R, Mosca JD et al (1999) Multilineage potential of adult human mesenchymal stem cells. Science 284(5411):143–147. https://doi.org/10.1126/science.284.5411.143

Pittenger MF, Discher DE, Péault BM, Phinney DG, Hare JM, Caplan AI (2019) Mesenchymal stem cell perspective: cell biology to clinical progress. npj. Regen Med 4(1):22. https://doi.org/10.1038/s41536-019-0083-6

Poblano-Pérez LI, Castro-Manrreza ME, González-Alva P, Fajardo-Orduña GR, Montesinos JJ (2024) Mesenchymal stromal cells derived from dental tissues: immunomodulatory properties and clinical potential. Int J Mol Sci 25(4). https://doi.org/10.3390/ijms25041986

Queiroz A, Albuquerque-Souza E, Gasparoni LM, de França BN, Pelissari C, Trierveiler M, Holzhausen M (2021) Therapeutic potential of periodontal ligament stem cells. World J Stem Cells 13(6):605–618. https://doi.org/10.4252/wjsc.v13.i6.605

Ratti M, Lampis A, Ghidini M, Salati M, Mirchev MB, Valeri N, Hahne JC (2020) MicroRNAs (miRNAs) and long non-coding RNAs (lncRNAs) as new tools for cancer therapy: first steps from bench to bedside. Target Oncol 15(3):261–278. https://doi.org/10.1007/s11523-020-00717-x

Roato I, Chinigò G, Genova T, Munaron L, Mussano F (2021) Oral cavity as a source of mesenchymal stem cells useful for regenerative medicine in dentistry. Biomedicine 9(9). https://doi.org/10.3390/biomedicines9091085

Santonocito S, Polizzi A, Palazzo G, Isola G (2021) The emerging role of microRNA in Periodontitis: pathophysiology, clinical potential and future molecular perspectives. Int J Mol Sci 22(11). https://doi.org/10.3390/ijms22115456

Song B, Zhang Y, Chen L, Zhou T, Huang W, Zhou X, Shao L (2017) The role of toll-like receptors in periodontitis. Oral Dis 23(2):168–180. https://doi.org/10.1111/odi.12468

Sui B, Wu D, Xiang L, Fu Y, Kou X, Shi S (2020) Dental pulp stem cells: from discovery to clinical application. J Endodont 46(9s):S46–s55. https://doi.org/10.1016/j.joen.2020.06.027

Sun R, Zhang Y, Lv Q, Liu B, Jin M, Zhang W et al (2011) Toll-like receptor 3 (TLR3) induces apoptosis via death receptors and mitochondria by up-regulating the transactivating p63 isoform alpha (TAP63alpha). J Biol Chem 286(18):15918–15928. https://doi.org/10.1074/jbc.M110.178798

Tolstova T, Dotsenko E, Kozhin P, Novikova S, Zgoda V, Rusanov A, Luzgina N (2023) The effect of TLR3 priming conditions on MSC immunosuppressive properties. Stem Cell Res Ther 14(1):344. https://doi.org/10.1186/s13287-023-03579-y

Usui M, Onizuka S, Sato T, Kokabu S, Ariyoshi W, Nakashima K (2021) Mechanism of alveolar bone destruction in periodontitis – periodontal bacteria and inflammation. Jpn Dent Sci Rev 57:201–208. https://doi.org/10.1016/j.jdsr.2021.09.005

Wang Y, Liu L, Davies DR, Segal DM (2010) Dimerization of Toll-like receptor 3 (TLR3) is required for ligand binding. J Biol Chem 285(47):36836–36841. https://doi.org/10.1074/jbc.M110.167973

Waterman RS, Tomchuck SL, Henkle SL, Betancourt AM (2010) A new mesenchymal stem cell (MSC) paradigm: polarization into a pro-inflammatory MSC1 or an immunosuppressive MSC2 phenotype. PLoS One 5(4):e10088. https://doi.org/10.1371/journal.pone.0010088

Wong LB, Yap AU, Allen PF (2021) Periodontal disease and quality of life: umbrella review of systematic reviews. J Periodontal Res 56(1):1–17. https://doi.org/10.1111/jre.12805

Wu Y, Wang Y, Ji Y, Ou Y, Xia H, Zhang B, Zhao Y (2017) C4orf7 modulates osteogenesis and adipogenesis of human periodontal ligament cells. Am J Transl Res 9(12):5708–5718

Wu X, Lei C, Xia T, Zhong X, Yang Q, Shu HB (2019) Regulation of TRIF-mediated innate immune response by K27-linked polyubiquitination and deubiquitination. Nat Commun 10(1):4115. https://doi.org/10.1038/s41467-019-12145-1

Yamashita M, Chattopadhyay S, Fensterl V, Zhang Y, Sen GC (2012) A TRIF-independent branch of TLR3 signaling. J Immunol 188(6):2825–2833. https://doi.org/10.4049/jimmunol.1103220

Yang C, Luo M, Chen Y, You M, Chen Q (2021) MicroRNAs as important regulators mediate the multiple differentiation of mesenchymal stromal cells. Front Cell Dev Biol 9:619842. https://doi.org/10.3389/fcell.2021.619842

Zhao X, Liu D, Gong W, Zhao G, Liu L, Yang L, Hou Y (2014) The toll-like receptor 3 ligand, poly(I:C), improves immunosuppressive function and therapeutic effect of mesenchymal stem cells on sepsis via inhibiting MiR-143. Stem Cells 32(2):521–533. https://doi.org/10.1002/stem.1543

Immunomodulatory Effects of Medicinal Plants 34

Wipawee Nittayananta

Abstract

Medicinal plants have long been recognized for their therapeutic benefits, particularly for their bioactive compounds that enhance quality of life. These plants and their derived compounds possess numerous pharmacological properties, including the modulation of immune system components. Immune responses are crucial for resolving oral infections caused by pathogens such as bacteria, fungi, and viruses, as well as for preventing or managing oral cancer. A robust immune system plays a central role in defending against these stimuli, making it essential to explore and promote the use of medicinal plants that can enhance immune responses or intervene in disease-related processes.

This chapter focuses on the immunomodulatory potential of specific plants and plant-derived compounds, specifying their bioactive chemical components responsible for their immune-boosting and anti-inflammatory properties. Many medicinal plants exhibit considerable immunomodulatory activities, primarily through the activation of signaling pathways like NF-κB and MAPK. These pathways are key regulators of inflammatory diseases and cancer, and their modulation by plant compounds holds promise for therapeutic applications. However, further studies are required to fully elucidate the underlying mechanisms of these effects and to explore additional pathways involved in plant-based immunomodulation.

Most studies on the immunomodulatory effects of medicinal plants have been conducted in vitro, so it is crucial to expand research into randomized clinical trials in diverse populations. Although many of the plants discussed in this chapter appear to be safe, their toxicological profiles in clinical settings remain underexplored and need further investigation. By advancing clinical research and assessing the safety and efficacy of medicinal plants, these natural products could play an increasingly important role in managing oral infections, cancer, and other immune-related conditions.

Keywords

Medicinal plants · Immunomodulation · Plant-derived bioactive compounds · Oral infections

34.1 Introduction

Plants and plant-derived compounds have been utilized for treating various diseases throughout human history. These medicinal plants are

W. Nittayananta (✉)
Faculty of Dentistry, Bangkokthonburi University, Bangkok, Thailand
e-mail: wipawee.nit@bkkthon.ac.th

N. Rezaei, H. Ziaei (eds.), *Oral Immunology*, Advances in Experimental Medicine and Biology 1492,
https://doi.org/10.1007/978-3-032-03176-1_34

renowned for their quality-of-life benefits and have long been employed for their healing properties. In recent decades, interest in plant-derived compounds, especially active substances of natural origin, has grown significantly due to their potential health benefits. Among the many therapeutic properties of medicinal plants, their immunomodulatory activity has attracted increasing attention. Numerous plant-derived compounds have been shown to possess nonspecific immunomodulatory properties, meaning they can influence immune functions in a broad manner, enhancing the body's ability to fight infections and diseases (Newman and Cragg 2016).

The immune system is a complex network, influenced by various factors, which in turn can impact other parts of the body, including the oral cavity. Medicinal plants' immunomodulatory properties may help in preventing disease and serve as novel, effective therapeutics. Chronic inflammation, particularly low-grade inflammation, is a known precursor to various diseases, and many medicinal plants possess potent anti-inflammatory properties.

There is a wide variety of plants with anti-inflammatory properties found around the world. These plants contain active compounds such as flavonoids, polysaccharides, alkaloids, lactones, diterpenoids, and glycosides, all of which demonstrate both immunomodulatory and anti-inflammatory effects. Some compounds inhibit pro-inflammatory stimuli by targeting pathways like cyclooxygenase (COX) activation (Tapsell et al. 2006). Specifically, the activation of NF-κB, a key regulator of COX-2, is linked to several inflammatory diseases, including cancer (Aggarwal and Shishodia 2006; Lim et al. 2001). Inhibition of COX activity leads to a reduction in NF-κB activation, which in turn helps mitigate inflammation and related symptoms (Shi et al. 2015).

While there are countless medicinal plants with immunomodulatory and anti-inflammatory properties, this chapter focuses on a selection of widely accessible and promising plants and plant-derived compounds. These include ellagic acid, turmeric (*Curcuma longa*), Andrographis (*Andrographis paniculata*), *Houttuynia cordata* extracts, ginger (*Zingiber officinale*), and various herbs and their essential oils, such as black cumin, lemon grass, clove, peppermint, lemon balm, marjoram, rosemary, thyme, and sage. These plants are known for their natural immune-boosting properties and are widely used in various formulations. They were selected based on their popularity, availability, and well-documented impact on the immune system (Kadiyska et al. 2023).

34.2 Plant-Derived Compounds with Immunomodulatory Activity

34.2.1 Ellagic Acid

Ellagic acid (EA) is a naturally occurring compound derived from gallic acid and exists as a dimer. It is found in many types of fruits, including pomegranates, blackberries, raspberries, strawberries, and walnuts (Cerda et al. 2005). These plants produce EA, which is then converted into ellagitannins, a form of tannin (Vattem and Shetty 2005). EA has a broad range of pharmacological activities, making it a promising compound for health promotion. Its properties include:

1. Anti-inflammatory: EA has been shown to inhibit the production of pro-inflammatory cytokines and reduce inflammation. It inhibits the activation of key signaling pathways, including NF-κB and MAPK, which play important roles in inflammation (Shi et al. 2015).
2. Antibacterial: EA has antibacterial activity, showing efficacy against a broad spectrum of pathogenic microorganisms. Its antibacterial activity can help prevent infections, particularly those related to oral health.
3. Antiviral: EA has been found to suppress the replication of specific viruses, providing a potential therapeutic option for viral infections, including those that affect the oral cavity.

4. Antitumor: EA has demonstrated antitumor activity by inhibiting the proliferation of cancer cells, making it a valuable compound in cancer prevention and therapy. Research indicates that ellagic acid may promote apoptosis in tumor cells and suppress tumor angiogenesis (Cerda et al. 2005).

Due to these diverse pharmacological effects, EA is considered a valuable compound in natural medicine, with potential applications in managing inflammatory diseases, infections, and even cancer. Further research on its mechanisms and clinical applications could expand its therapeutic use.

34.2.1.1 Effects of Ellagic Acid on Mucosal Innate Immunity

The epithelial cells lining the oral cavity are crucial for mucosal innate immunity, serving a key role in defending against infections caused by oral pathogens. (Dale 2002). Enhancing mucosal innate immunity can be an effective strategy to prevent such infections. EA, a plant-derived compound, has been demonstrated to modulate innate immune mediators, offering potential benefits for oral health.

A study by Promsong et al. (2015) revealed that EA enhances the expression of key components of the innate immune system, such as human beta-defensin 2 (hBD2) and secretory leukocyte protease inhibitor (SLPI), in oral epithelial cells. These proteins play a crucial role in the innate immune response, with hBD2 exhibiting antimicrobial activity, particularly against *Candida albicans*, an opportunistic pathogen often seen in immunocompromised individuals like cancer, diabetes, and HIV patients (Joly et al. 2004; Fidel Jr 2011). EA's ability to upregulate hBD2 and SLPI expression could potentially help in preventing oral infections, especially in immunocompromised patients (Promsong et al. 2015).

Moreover, hBD2 and SLPI have demonstrated anti-HIV-1 activities. hBD2 can suppress the replication of HIV-1, particularly X4 strains, while SLPI interferes with the viral fusion with host cells by binding to annexin II (Quiñones-Mateu et al. 2003; Ma et al. 2004). EA's effect in boosting the expression of these proteins could contribute to enhancing mucosal immunity against HIV-1. Furthermore, EA itself has been shown to inhibit the HIV-1 integrase enzyme (Panichayupakaranant et al. 2005). Recent studies have also suggested that microbicides containing EA could be effective in preventing HIV-1 transmission, especially given that more than 90% of global HIV-1 transmissions occur through mucosal surfaces (Kresina and Mathieson 1999; Nittayananta et al. 2022).

These findings emphasize the potential of EA not only in promoting oral immunity but also in providing therapeutic strategies to prevent infections, including HIV-1. The development of EA-based microbicides for topical use could be an innovative approach to reducing viral transmission, further supporting EA's potential as a therapeutic agent.

34.2.1.2 Immunomodulatory Property of Ellagic Acid

Normal oral epithelial cells play an important role in immune response by producing various cytokines and chemokines, including interferon-gamma (IFN-γ), interleukins (IL-2, IL-4), and tumor necrosis factor-alpha (TNF-α), which are essential for maintaining oral immunity (Fakhry et al. 2013). The effects of ellagic acid (EA) on cytokine secretion in oral epithelial cells have been investigated, revealing that EA modulates the production of pro-inflammatory and immunoregulatory molecules.

A study by Promsong et al. (2015) demonstrated that EA influences the secretion of cytokines like IL-2, IL-1β, and RANTES in oral epithelial cells. IL-1β plays an essential role in activating T lymphocytes by enhancing IL-2 and IL-2 receptor expression (Abbas et al. 2007). IL-2, in turn, is crucial for the activation and migration of immune cells, such as B cells, cytotoxic T cells, natural killer (NK) cells, and macrophages, to sites of infection (Borish and Steinke 2003). The study found that low concentrations of EA promoted a higher secretion of IL-2, while higher concentrations reduced IL-2 production. This suggests that EA at lower concentrations may stimulate oral epithelial cells directly to produce IL-2, or this effect may be indirectly mediated through IL-1β (Promsong et al. 2015).

Additionally, EA has been shown to improve mucosal immunity against HIV-1 by the upregulation of RANTES, a chemokine that acts as an entry inhibitor of the virus (Cocchi et al. 1995). The increased secretion of RANTES in response to the presence of EA may help prevent HIV-1 infection by interfering with the virus's ability to enter host cells.

EA also downregulates the expression of pro-inflammatory cytokines such as IL-6, IL-8, and TNF-α, which are typically regulated by the NF-κB signaling pathway. NF-κB is known to promote tumorigenesis by regulating the expression of inflammatory factors (Shen and Tergaonkar 2009). The downregulation of IL-6, IL-8, and TNF-α in response to EA suggests that EA may inhibit NF-κB activation, contributing to its anti-inflammatory and potential anti-cancer effects.

The ability of EA to induce the expression of antimicrobial peptides and other innate immune mediators highlights its potential therapeutic role in managing common immune-mediated oral diseases. Topical applications of EA-containing agents may be beneficial in treating conditions like aphthous ulcers, oral candidiasis, and oral lichen planus (Promsong et al. 2015). Therefore, EA's immunomodulatory properties make it a promising candidate for developing treatments aimed at enhancing oral immunity and managing oral infections.

34.2.2 *Houttuynia cordata*

Houttuynia cordata Thunb (*H. cordata*), also known as *heartleaf houttuynia*, is a plant native to several Asian countries and is often consumed as a fresh vegetable or side dish. It has gained recognition for its medicinal properties and is widely used in traditional medicine. H. cordata is known to have several pharmacological activities, including antimicrobial, anti-inflammatory, and antioxidant functions, making it an important plant in alternative medicine.

Research has shown that *H. cordata* exhibits antimicrobial activities, particularly against a variety of pathogens, including bacteria, viruses, and fungi (Fu et al. 2013). These antimicrobial properties are attributed to its bioactive compounds, which include flavonoids, terpenoids, and phenolic acids. The plant's ability to inhibit the growth of harmful microorganisms makes it useful for treating infections, including those affecting the oral cavity.

In addition to its antimicrobial effects, *H. cordata* is widely recognized for its anti-inflammatory properties, which contribute to the modulation of immune responses and the attenuation of inflammatory processes. This makes it a potential therapeutic agent for conditions like oral infections and inflammatory diseases.

Given its extensive use in traditional medicine and the growing interest in its therapeutic potential, *H. cordata* could play a significant role in developing natural remedies to enhance oral health and immune function.

34.2.2.1 Effects of *Houttuynia cordata* on Mucosal Innate Immunity

A study by Satthakarn et al. (2015) revealed that water extracts of *H. cordata* can significantly upregulate the expression of key components of the oral innate immune system, such as hBD2 and SLPI. The researchers found that the expression of hBD2 mRNA was induced in a dose-dependent manner without cytotoxic effects, and the secretion of hBD2 protein was slightly increased in the presence of the extract (Satthakarn et al. 2015).

hBD2 has well-documented antimicrobial properties, particularly against oral pathogens such as *Candida albicans* (Joly et al. 2004). Additionally, hBD2 is known for its ability to inhibit HIV-1 replication through various mechanisms (Quiñones-Mateu et al. 2003), and it also plays an important role in immune system modulation. It induces the production of cytokines and chemokines, as well as exhibiting chemotactic activity for immune cells like immature dendritic cells and memory T cells (Yang et al. 1999; Niyonsaba et al. 2007). Furthermore, hBD2 has been shown to support wound healing by promoting the proliferation and migration of keratinocytes (Niyonsaba et al. 2007).

Given these properties, the extract of *H. cordata* could be developed as an immunomodulatory agent to maintain oral health and as a potential new antimicrobial agent for the prevention or treatment of oral diseases. The enhancement of hBD2 expression and secretion suggests that *H. cordata* may offer significant therapeutic potential in addressing both infection and inflammation in the oral cavity.

34.2.2.2 Immunomodulatory Activity of *Houttuynia cordata*

A study by Satthakarn et al. (2015) validated that *H. cordata* water extract significantly upregulated the secretion of key cytokines and chemokines, including IL-2, IL-6, IL-8, CCL20, and IFN-γ, in gingival epithelial cells. These molecules play an essential role in both innate and adaptive immune responses. They help activate T cells, macrophages, and other immune cells, facilitating the killing of pathogens and the recognition of foreign molecules. Additionally, these cytokines and chemokines help modulate other immune mediators and promote antimicrobial activities, which are crucial for innate host defense (Borish and Steinke 2003; Yang et al. 2003; Schroder et al. 2004).

IL-8 functions as a potent chemoattractant and activator of neutrophils (Hedges et al. 1995). The upregulation of IL-8 expression induced by *H. cordata* may facilitate the neutrophil recruitment at mucosal surfaces, where they can target and eliminate pathogenic bacteria, thereby enhancing mucosal immunity. However, it is important to note that while the production of IL-8 could be beneficial in terms of immune cell recruitment, the increased expression of pro-inflammatory cytokines, including IL-6 and IL-8, may also contribute to undesirable inflammation (Hedges et al. 1995; Borish and Steinke 2003).

Thus, while *H. cordata* shows potential as an immunomodulator and antimicrobial agent; further research is needed to balance its pro-inflammatory effects and determine its clinical applicability for promoting oral health without triggering excessive inflammation.

34.2.2.3 Potential Role of *Houttuynia cordata* Against Inflammation

The major pro-inflammatory cytokines, such as TNF-α, various interleukins (ILs), and the reactive free radical nitric oxide (NO), play a critical role in the development of inflammatory diseases (Freeman and Natanson 2000). *H. cordata* has been reported to possess anti-inflammatory properties, particularly against bacterial and ionic toxins. For example, a study by Park et al. (2005) demonstrated that the aqueous extract of *H. cordata* inhibited the production of NO and TNF-α in LPS-induced mouse macrophages. Additionally, the ethanolic extract of *H. cordata* was shown to reduce the expression of IL-6 and NO in LPS-treated alveolar macrophages and lung epithelial cells (Lee et al. 2015). Furthermore, the ethanolic extract was found to suppress NF-κB signaling in human mast cells, leading to reduced levels of TNF-α, IL-6, and IL-8 (Lee et al. 2013).

Further studies have shown that *H. cordata* can inhibit histamine release and calcium ion uptake in a dose-dependent manner. This effect is likely due to an increase in intracellular cAMP levels and adenylate cyclase activity, which are involved in IgE-mediated allergic reactions that contribute to inflammation (Shim et al. 2009). Han et al. (2009) demonstrated that *H. cordata* water extract had an anti-inflammatory role against IgE-mediated allergic responses in rat mast cells.

The NF-κB and MAPK pathways are central to regulating inflammation. The ethyl acetate fraction of *H. cordata* was shown to inhibit the nuclear translocation of NF-κB p65 subunit and suppress MAPK activation in LPS-challenged RAW 264.7 cells. This resulted in a significant reduction in NO, prostaglandin E2 (PGE2), TNF-α, and IL-6 levels (Chun et al. 2014). Additionally, viral-induced inflammation was downregulated by the water extract of *H. cordata* through blocking NF-κB activation (Chen et al. 2011).

In addition to whole plant extracts, bioactive compounds such as volatile oils and polysaccharides derived from *H. cordata* have also demonstrated potent anti-inflammatory effects in vivo. *H. cordata* volatile oil was found to suppress LPS-induced TNF-α and NO secretion in mouse macrophages. This oil regulates iNOS and TNF-α at both the translational and transcriptional levels (Li et al. 2013). Moreover, *H. cordata* polysaccharides have been shown to prevent complement activation, macrophage migration, and the production of pro-inflammatory cytokines like TNF-α, IL-6, and IL-1β (Xu et al. 2015).

Notably, *H. cordata* exhibited anti-inflammatory effects similar to those of nonsteroidal anti-inflammatory drugs (NSAIDs) and the COX-2 inhibitor NS-398, particularly in preventing LPS-induced PGE2 production in mouse macrophages (Li et al. 2011). Furthermore, a study by Lee et al. (2015) demonstrated that the ethanolic extract of *H. cordata* was as effective as dexamethasone in alleviating LPS-induced airway inflammation in a mouse model. Oral administration of *H. cordata* extracts reduced the TNF-α/NO and cyclooxygenase-2/PGE2 pathways more effectively than dexamethasone. Additionally, *H. cordata* has been shown to reduce the expression of IL-6 and macrophage inflammatory protein-1 α (MIP-1α), and it can decrease inflammation via the PI3K/Akt/mTOR signaling pathway (Wan et al. 2016).

These findings highlight the potential of *H. cordata* as a natural anti-inflammatory agent, suggesting its use in managing inflammatory conditions, including oral diseases and systemic inflammation.

34.2.3 Curcumin

Curcumin, a yellow polyphenolic pigment extracted from the rhizome of *Curcuma longa* L. (turmeric), has been used historically for both culinary and medicinal purposes (Moghadamtousi et al. 2014). It has a long-standing use in traditional Asian medicine due to its anti-inflammatory, healing, and immunomodulatory effects.

Research has indicated that curcumin supplementation in the diet can significantly reduce the levels of pro-inflammatory biomarkers while elevating the production of anti-inflammatory mediators (Ferguson et al. 2021). This dual action of curcumin makes it a promising candidate for managing chronic inflammation and related diseases. Additionally, curcumin has been reported to attenuate the production of pro-inflammatory cytokines, such as IFN-γ, which suggests its potential role as an immunosuppressant in modulating immune responses (Skyvalidas et al. 2020).

Curcumin's ability to balance inflammatory responses by both suppressing pro-inflammatory signals and enhancing anti-inflammatory ones highlights its potential therapeutic value in treating inflammatory diseases, including those associated with immune dysregulation, such as autoimmune disorders and cancers. Its multifaceted action as an immunomodulator makes curcumin a compelling candidate for further research in immune-related therapies.

34.2.3.1 Immunomodulatory Activity of Curcumin

Curcumin's immunomodulatory and anti-inflammatory effects make it a potential therapeutic agent for a range of inflammatory and immune-related conditions. One of the key mechanisms through which curcumin exerts its anti-inflammatory effects is by inhibiting TNF-α-induced activation and nuclear translocation of NF-κB, a crucial transcription factor involved in inflammation (Olivera et al. 2012). Moreover, curcumin inhibits various pro-inflammatory cytokines, including IL-1β, IL-2, IL-5, IL-6, and IL-8, which are often implicated in chronic inflammation and immune responses (Fahey et al. 2007). Furthermore, curcumin can downregulate monocyte migration through MCP-1 and limit macrophage recruitment via macrophage inflammatory protein-1α (MIP-1α) (Zhou et al. 2011), further enhancing its anti-inflammatory effects.

Despite its promising effects, the therapeutic potential of curcumin is limited by its low bioavailability, which restricts its clinical application (Ireson et al. 2002). This issue has been addressed by combining curcumin with piperine, a compound found in black pepper, which has been shown to significantly increase the bioavailability of curcumin and thus improve its therapeutic efficacy (Shoba et al. 1998). This combination offers a potential solution to the bioavailability challenge and makes curcumin a more viable option for therapeutic use in treating inflammation-related diseases and immune dysregulation.

34.2.3.2 Effects of Curcumin on Mucosal Innate Immunity

A recent study by Nittayananta et al. (2024) underscored the diverse pharmacological properties of curcumin, including its antiviral, immunomodulatory, and anti-inflammatory properties. The study evaluated the impact of a film spray having curcumin on the expression of oral and nasal innate immune markers, revealing that curcumin modified mucosal innate immunity by increasing the production of antimicrobial peptides such as LL-37 and HD-5, in addition to pro-inflammatory cytokines, including IL-6 and TNF-α, by oral keratinocytes (Nittayananta et al. 2024). Both LL-37 and HD-5 have been shown to exhibit inhibitory effects against viruses such as SARS-CoV-2 and influenza A (Wang et al. 2021; Tripathi et al. 2015), suggesting that curcumin's antiviral properties may be enhanced through this mechanism.

The study also observed an increase in IL-6 and TNF-α production in response to the curcumin film spray, although no significant effects on other cytokines such as IL-1β, IL-10, and IFN-γ were detected. This suggests that curcumin may have a dual role, acting both as an anti-inflammatory and a pro-inflammatory agent, depending on the specific immune context (Chai et al. 2020). This complex balance could be leveraged for therapeutic purposes, particularly in enhancing mucosal immunity and combating infections, while also managing inflammatory responses.

34.2.4 Andrographis (*Andrographis paniculata*)

Andrographis, also known as the "King of Bitters," is an herbal plant recognized for its wide range of bioactivities, including antiviral, anti-inflammatory, and immunomodulatory properties (Li et al. 2022). The active compound andrographolide, extracted from the leaves of Andrographis, has been studied for its anti-inflammatory effects in vitro (Chandrasekaran et al. 2011; Choi and Kim 2022).

Research has shown that andrographolide can inhibit the release of various inflammatory mediators, such as nitric oxide (NO), prostaglandin E2, IL-1β, and IL-6, from macrophages in response to lipopolysaccharide (LPS) stimulation. Additionally, andrographolide has been found to affect the activity of cyclooxygenase enzymes COX-1 and COX-2, playing important roles in inflammation (Parichatikanond et al. 2010). Another study indicated that dehydroandrographolide, a derivative of andrographolide, modulated the secretion of pro-inflammatory cytokines like TNF-α, IL-6, IL-1β, and IL-10 in a dose-dependent manner, potentially through downregulation of genes involved in the inflammatory cascade.

Moreover, dehydroandrographolide has been reported to suppress NF-κB activation, a key regulator of inflammatory responses in animal models (Weng et al. 2018). Furthermore, studies in mice have reported that diterpenes like andrographolide and neoandrographolide significantly modulate both specific and nonspecific immune responses, highlighting their potential as immunomodulatory agents (Puri et al. 1993). These findings suggest that Andrographis and its active components could be valuable for managing inflammation and modulating immune function.

34.2.5 Ginger (*Zingiber officinale* Rosc)

34.2.5.1 Compounds in Ginger

Ginger, a widely used spice in Southeast Asia and known for its distinct flavor in Indian cuisine, belongs to the Zingiberaceae family. It has a variety of constituents, which can vary depending on the region and the form of its rhizomes. In general, ginger contains a rich array of compounds, including carbohydrates, minerals, phytochemicals, phenolic compounds, terpenes, polysaccharides, lipids, organic acids, and dietary fibers (Mao et al. 2019). It is also a good source of vitamins, minerals, and enzymes such as proteolytic enzymes and zingibain (Butt and Sultan 2011). Among its many compounds, ginger contains more than 40 with antioxidant activity, contributing to its potential in treating inflammatory conditions (Poles et al. 2021).

Ginger is commonly used in different forms to enhance the immune system, reduce nausea, and aid digestion, as demonstrated in multiple clinical trials. For instance, dietary supplementation with powdered ginger rhizome has been shown to increase testosterone levels (Nikkhah Bodagh et al. 2018; Banihani 2018). These properties make ginger a popular natural remedy for several health concerns, particularly those involving immune modulation and inflammation.

34.2.5.2 Immunomodulatory Effects of Ginger

Ginger has been shown to decrease TNF-α and IL-1β levels in patients with knee osteoarthritis and pain resulting from joint inflammation due to the excessive production of synovial cytokines, highlighting its ability to modulate inflammatory cytokine production in chronic diseases (Mozaffari-Khosravi et al. 2016). The active compounds gingerol and shogaol in ginger have been found to inhibit prostaglandin synthesis by preventing the biosynthesis of prostaglandin synthase. Additionally, ginger may exert effects on arachidonate 5-lipoxygenase, a key enzyme in leukotriene biosynthesis (Azeez and Lunghar 2021). Overall, ginger is believed to have anti-inflammatory effects by downregulating the expression of key pro-inflammatory cytokines, including IL-1, TNF-α, and IL-8, through mechanisms involving the modulation of inflammatory gene expression and related signaling cascades. (Tjendraputra et al. 2001; Verma et al. 2004).

Ginger has demonstrated significant immune-modulating properties, particularly in reducing inflammation. Studies have shown that it can decrease the levels of pro-inflammatory cytokines like TNF-α and IL-1β in patients with knee osteoarthritis, a condition characterized by joint inflammation and cytokine overproduction in the synovium (Mozaffari-Khosravi et al. 2016). This highlights ginger's potential to modulate immune function and suppress inflammation in chronic diseases.

The active compounds in ginger, particularly gingerol and shogaol potentially inhibit the production of prostaglandins by interfering with prostaglandin synthase biosynthesis. Additionally, these compounds may inhibit arachidonate 5-lipoxygenase, an enzyme involved in leukotriene biosynthesis, further supporting ginger's anti-inflammatory effects (Azeez and Lunghar 2021). Overall, ginger is believed to inhibit the production of several pro-inflammatory cytokines, including IL-1, TNF-α, and IL-8, through mechanisms that involve gene regulation (Tjendraputra et al. 2001; Verma et al. 2004). This makes ginger a promising candidate for managing inflammatory conditions and modulating immune responses.

34.2.6 Immunomodulatory Effects of Selected Herb Essential Oils

Essential oils, derived from a variety of medicinal herbs, are rich in bioactive compounds that possess numerous health-promoting properties, including immunomodulatory effects. Some of the most widely recognized medicinal herbs and their essential oils, known for their immunomodulatory activities, include:

1. *Black cumin* (*Nigella sativa*): Thymoquinone, the principal active component of black cumin

oil, exhibits anti-inflammatory, antioxidant, and immune-regulating properties that support the body's immunological responses.

2. *Lemongrass* (*Cymbopogon citratus*): Lemongrass oil is well known for its antimicrobial and anti-inflammatory effects, which may modulate the immune system function and defend the body against infections.
3. *Clove* (*Syzygium aromaticum*): Eugenol is a compound found in clove essential oil with potent anti-inflammatory, antimicrobial, and antioxidant activities that can help strengthen immune function.
4. *Peppermint* (*Mentha piperita*): Peppermint oil has immunomodulatory effects and can help boost immune responses while also exhibiting antimicrobial and anti-inflammatory properties.
5. *Lemon balm* (*Melissa officinalis*): Lemon balm oil is known for its calming, antiviral, and anti-inflammatory properties, supporting the immune system, especially during stress-related conditions.
6. *Marjoram* (*Origanum majorana*): Marjoram essential oil has anti-inflammatory and analgesic effects, which may contribute to regulating immune responses and reducing inflammation.
7. *Rosemary* (*Rosmarinus officinalis*): Rosemary oil is rich in antioxidants and has immunomodulatory and anti-inflammatory properties that can enhance the immune system's ability to fight infections.
8. *Thyme* (*Thymus vulgaris*): Thyme oil has potent antimicrobial and anti-inflammatory effects, which can help enhance immune function and combat infections.
9. *Sage* (*Salvia officinalis*): Sage oil is rich in antioxidant compounds and bioactives components that contribute to immune modulation, inflammation reduction, and respiratory system support.

These essential oils provide a natural source of immunomodulatory compounds that may support the defense system by regulating inflammation, enhancing antimicrobial activity, and promoting overall immune health.

34.2.6.1 Black Cumin

Black cumin (*Nigella sativa*) has been used for centuries in both culinary practices and traditional medicine because of its potent therapeutic properties. Its key bioactive compounds, particularly thymoquinone and α-hederin, have been broadly studied for their anti-inflammatory and immunomodulatory effects.

34.2.6.1.1 Thymoquinone

Thymoquinone, the principal bioactive compound in black cumin, has shown various anti-inflammatory actions in in vitro studies using LPS-stimulated macrophage-like cell models. These anti-inflammatory properties include:

1. Suppressing NO production and reducing the production of pro-inflammatory cytokines such as IL-6, IL-1β, and TNF-α.
2. Inducing the expression of nitric oxide synthase (iNOS) and COX-2, which are involved in inflammation.
3. Reducing the nuclear translocation of transcription factors and inhibiting the activation of key signaling pathways, including activator protein-1 (AP-1) and nuclear factor-kappa B (NF-κB), which are critical for the inflammatory response.
4. Suppressing IL-1 receptor-associated kinase 1-linked AP-1/NF-κB signaling pathways, further reducing inflammation (Hossen et al. 2017).

In in vivo studies, thymoquinone has shown strong anti-inflammatory effects, particularly in models of experimental gastritis and hepatitis. It has also been shown to inhibit iNOS and TGF-β1 and decrease serum immunoglobulin E (IgE) levels, reducing inflammation in experimental asthma models.

34.2.6.1.2 α-Hederin

α-Hederin, another important compound in black cumin, has demonstrated similar anti-inflammatory effects. In a study using sensitized rats, α-hederin affected the expression of miRNA-126 and interfered with the IL-13 secretory pathway, contributing to its anti-inflammatory properties (Fallahi et al. 2016).

34.2.6.1.3 Black Cumin Essential Oil

Black cumin essential oil has also been studied for its immunomodulatory and anti-inflammatory effects. Moreover, it has been shown to increase antioxidant activity, inhibit enzymes, and induce cytotoxicity in tumor cells (Silva et al. 2020). It also has protective effects against autoimmune encephalomyelitis in rat central nervous tissue, demonstrating neuroprotective effects. It modulated NO levels in the medulla spinalis while reducing ROS and NO concentrations in the brain (Ozugurlu et al. 2005).

A molecular docking study revealed the potential antiviral activity of black cumin seeds and their essential oil, indicating inhibitory activity against COVID-19 and SARS viruses. The results suggested that black cumin exhibited anti-inflammatory and antiviral activities comparable to, or exceeding, those of certain FDA-approved medications, highlighting its therapeutic potential for treating respiratory and viral infections (Bouchentouf and Missoum 2020).

In conclusion, black cumin, particularly its essential oil and key bioactive compounds like thymoquinone and α-hederin, demonstrates anti-inflammatory, immunomodulatory, antiviral, and antioxidant properties. These effects make it a promising natural therapeutic agent for the treatment of various inflammatory or immune-related diseases.

34.2.6.2 Lemongrass

Lemongrass (*Cymbopogon citratus*) has been widely used in traditional medicine for treating inflammatory diseases. Experimental evidence, derived from both in vitro and in vivo studies, supports its anti-inflammatory effects.

In Vitro Effects

Several studies have evaluated the inflammatory activity of lemongrass using LPS-stimulated macrophages:

- Pro-inflammatory mediators, including TNF-α and NO, produced by murine alveolar macrophages, were significantly reduced following exposure to lemongrass extracts (Tiwari et al. 2010).
- In mouse peritoneal macrophages, lemongrass caused a reduction in IL-1β and IL-6 production, as well as a decrease in the anti-inflammatory cytokine IL-10 secretion (Bachiega and Sforcin 2011).
- The NF-κB signaling pathway, a key mediator of inflammation, was inhibited by lemongrass extracts, reducing cytokine production in both human and mouse macrophages (Francisco et al. 2013).

In Vivo Effects

Lemongrass also demonstrated significant anti-inflammatory effects in animal models:

- In vivo studies on rats showed that lemongrass downregulated the expression of TNF-α and endothelin-1, while increasing the expression of the anti-inflammatory cytokine IL-10 and vascular endothelial growth factor (Said et al. 2019).
- In mouse models of paw and ear edema, both oral and topical administrations of lemongrass essential oil resulted in dose-dependent anti-inflammatory effects, significantly preventing edema (Boukhatem et al. 2014).
- A study on rat models of edema showed that lemongrass essential oil could significantly reduce edema over time in a dose-dependent manner (Gbenou et al. 2013).

Anti-inflammatory Mechanisms

The effects of lemongrass on pro-inflammatory cytokine production are partially attributed to its modulation of the NF-κB pathway, a key regulator of the immune response. In some studies, the application of lemongrass showed a reduction in the production of IL-1β, IL-6, and TNF-α in macrophages. It also showed a decrease in the secretion of IL-10, an anti-inflammatory cytokine, in certain macrophage models (Bachiega and Sforcin 2011; Sforcin et al. 2009).

Other Bioactivities of Lemongrass Essential Oil
In addition to its anti-inflammatory properties, lemongrass essential oil is also known for its antimicrobial, antioxidant, and antifungal activities. This makes it a versatile bioactive compound with potential therapeutic applications beyond inflammation, such as in combating infections and oxidative stress.

In conclusion, lemongrass, particularly its essential oil, exhibits significant anti-inflammatory properties, demonstrated through both in vitro and in vivo studies. It modulates key inflammatory cytokines like TNF-α, IL-1β, IL-6, and IL-10, primarily by downregulating the NF-κB pathway. Additionally, its antimicrobial and antioxidant activities further enhance its potential as a therapeutic agent for managing inflammatory diseases and other related conditions.

34.2.6.3 Clove

Clove (*Syzygium aromaticum*) is widely known for its culinary uses, but it also possesses notable immunomodulatory properties, mainly attributed to its active constituents, such as eugenol and isoeugenol. Several studies have explored the immunomodulatory effects of clove and its essential oil, demonstrating both pro-inflammatory and anti-inflammatory actions depending on the concentration and the biological context.

34.2.6.3.1 Immunomodulatory Effects of Clove and Eugenol

In Vitro Studies Numerous in vitro studies have examined the effects of clove extracts and eugenol on cytokine production. A study by Bachiega et al. (2012) showed that low doses of clove extract and eugenol increased the production of IL-1β, IL-6, and IL-10 in LPS-stimulated mouse macrophages, while higher doses reduced cytokine production. This suggests a dose-dependent immunomodulatory effect.

Pérez-Rosés et al. (2015) demonstrated that different concentrations of eugenol exerted immunomodulatory effects, with varying outcomes on cytokine production. Dibazar et al. (2015) reported that clove extract suppressed NO and TNF-α production in mouse peritoneal macrophages, but increased IL-6 production, and showed inconsistent effects on IL-2 production.

However, Mueller et al. (2010) reported that clove extract reduced IL-6 production, with no significant impact on the expression of IL-10 or TNF-α. This indicates variability in the effects of clove on cytokine expression, possibly due to differences in extract concentrations or cell types.

In Vivo Studies Studies on animal models also support the anti-inflammatory potential of clove. Rodrigues et al. (2009) found that clove extract inhibited the expression of IL-1β and IL-6 in LPS-activated mouse macrophages, supporting its anti-inflammatory potential.

34.2.6.3.2 Immunomodulatory Effects of Clove Essential Oil

In Vitro Studies Research on clove essential oil has also shown significant immunomodulatory effects. Rodrigues et al. (2009) demonstrated that clove essential oil strongly inhibited the expression of IL-1β and IL-6 in a dose-dependent manner in mouse peritoneal macrophages. This suggests clove essential oil's potential to modulate pro-inflammatory cytokine production.

Pérez-Rosés et al. (2015) assessed the effects of clove essential oil and eugenol on human neutrophils and the complement system. They discovered that eugenol inhibited the classical complement pathway, while clove terpenes showed no activity, suggesting that eugenol is the primary active compound in clove oil.

Han et al. (2017) reported that clove essential oil regulated signaling pathways that are correlated with the function of the immune system, cell cycle regulation, and cellular stress responses in human dermal fibroblasts. The oil also showed antiproliferative properties alongside a downregulation of inflammatory mediators, including vascular cell adhesion molecule-1 and interferon gamma-induced protein 10, emphasizing its role in immunomodulation and reducing inflammation.

In Vivo Studies The immune-stimulating effects of clove essential oil were further confirmed in animal models. According to Carrasco et al. (2009) study, clove essential oil activated cell-mediated immune mechanisms and effectively reestablished normal white blood cell counts in mice. It also enhanced humoral immunity, suggesting a potential effect on supporting the immune system.

34.2.6.3.3 Mechanisms of Action

Clove's immunomodulatory effects are partly due to its ability to influence key immune signaling pathways, such as the NF-κB pathway, COX-2, and iNOS expression.

Eugenol, the main active compound, is believed to play a central role in these effects, as evidenced by the inhibition of phagocytosis and complement system activation, along with modulation of pro-inflammatory cytokines like IL-1β, IL-6, and TNF-α.

In conclusion, clove and its essential oil containing eugenol exhibit significant immunomodulatory and anti-inflammatory properties. The effects vary depending on the concentration and biological context, but the overall evidence supports clove's role in both enhancing and regulating immune responses. Its actions include modulation of pro-inflammatory cytokines, suppression of COX-2, iNOS production, and modulation of immune cell functions. This makes clove a promising candidate for use in treating inflammatory conditions and supporting immune health.

34.2.6.4 Peppermint

Peppermint (*Mentha piperita*) is widely recognized for its aromatic and therapeutic properties, with several applications in food, cosmetic industries, and herbal medicine. Its immunomodulatory and anti-inflammatory effects have been extensively studied, primarily due to its main active compounds, such as menthol and menthone.

34.2.6.4.1 Immunomodulatory and Anti-inflammatory Effects of Peppermint

In Vitro Studies Several studies have shown the anti-inflammatory potential of peppermint extract. Li et al. (2017) demonstrated that ethanol extract of peppermint reduced the expression of NO, TNF-α, IL-6, and PGE-2 in murine macrophages in vitro. The reduction in LPS-induced NO production was observed in a concentration-dependent manner, highlighting the dose-dependent anti-inflammatory properties of peppermint.

In Vivo Studies Animal studies have also provided strong evidence for peppermint's immunomodulatory properties. Zaia et al. (2016) conducted an in vivo study using a commercial peppermint product and reported immunomodulatory effects, including a reduction in Th2-type immune responses (which are typically associated with allergic and inflammatory reactions) and an increase in IL-10 expression, alongside a reduction in IL-4 and IL-10 levels in the blood.

Bastaki et al. (2018) investigated the anti-inflammatory effects of menthol, the main compound in peppermint essential oil. The results demonstrated that menthol significantly reduced the level of pro-inflammatory cytokines such as IL-1, IL-23, and TNF-α, although IL-6 levels were unaffected, suggesting a targeted reduction in specific inflammatory pathways.

In another study, Osman et al. (2020) observed that an oral peppermint extract reduced the levels of IL-1, IL-6, TNF-α, C-reactive protein (CRP), and monocyte chemoattractant protein-1 (MCP-1). In contrast, the extract increased serum immunoglobulin levels (IgG, IgM, IgA), indicating a potential for immune system modulation through both anti-inflammatory and immune-enhancing effects.

34.2.6.4.2 Peppermint Essential Oil

The immunomodulatory properties of peppermint essential oil have also been studied in macrophage models. A study by Lang et al. (2019)

assessed the immunomodulatory effects of peppermint essential oil and found that pretreatment with peppermint essential oil significantly reduced IL-6 secretion in a murine macrophage model. Additionally, the study reported that peppermint oil inhibited phagocytosis but did not affect iNOS production, indicating its potential to modulate immune responses, particularly in the context of inflammation and immune cell activation.

34.2.6.4.3 Mechanisms of Action

The anti-inflammatory properties of peppermint, particularly its essential oil, may be attributed to its ability to influence key inflammatory biomarkers, such as NO, TNF-α, IL-6, and PGE-2, by inhibiting their production in response to LPS stimulation.

Menthol, a primary constituent, plays a critical role in these effects, potentially through its interaction with immune cells and cytokine production. Additionally, peppermint appears to promote a shift toward an anti-inflammatory immune response, enhancing IL-10 levels while reducing pro-inflammatory cytokines.

In conclusion, peppermint, especially its essential oil and menthol content, exhibits significant immunomodulatory and anti-inflammatory properties. These effects are demonstrated through a reduction in key pro-inflammatory cytokines, such as IL-1, IL-6, and TNF-α, as well as through the modulation of immune responses, including the activation of IL-10. Studies suggest that peppermint could be beneficial in managing conditions involving chronic inflammation and immune dysregulation, potentially offering therapeutic options for conditions like autoimmune disorders and inflammatory diseases.

34.2.6.5 Lemon Balm

Lemon balm (*Melissa officinalis*) has demonstrated a range of therapeutic properties, particularly regarding anti-inflammatory and immunomodulatory effects. These benefits have been explored in various studies. For example, Birdane et al. (2007) reported that lemon balm exhibits significant anti-inflammatory activities by reducing histamine- and carrageenan-induced paw edema in rats. This suggests that lemon balm can effectively block the formation of inflammatory mediators involved in such conditions.

In addition to its anti-inflammatory effects, lemon balm has demonstrated immunomodulatory properties. A study by Drozd and Anuszewska (2003) revealed that lemon balm extract modulated immune responses in mice immunized with sheep red blood cells. Furthermore, in vivo experiments by Hamza et al. (2016) showed that lemon balm extract considerably downregulated NF-κB expression, a key regulator of inflammation, and its downstream mediators such as TNF-α and COX-2.

These findings collectively highlight the potential of lemon balm as a natural anti-inflammatory agent. The essential oil of lemon balm has also been investigated for its anti-inflammatory effects, with results indicating strong activity in reducing edema, as shown in a study by Bounihi et al. (2013).

34.2.6.6 Majoram

Marjoram (Origanum majorana) has been shown to exhibit diverse biological activities, particularly with regard to its anti-inflammatory and immunomodulatory properties. Numerous studies have explored its potential in this area, with the major active component, rosmarinic acid, being a key contributor to its effects.

Experimental evidence has shown that marjoram can reduce inflammation, particularly through its effects on cytokines. For example, studies using a human THP-1 macrophage model demonstrated that marjoram significantly reduced the secretion of pro-inflammatory cytokines, including TNF-α, IL-1β, and IL-6 (Arranz et al. 2019; Villalva et al. 2018). These results indicate that marjoram can modulate immune responses and attenuate inflammation at the cellular level.

In in vivo studies, marjoram has also been shown to exert anti-inflammatory effects. Rababa'h et al. (2020) showed that marjoram treatment, both alone and in combination with metformin, led to significant reductions in IL-6 and TNF-α levels in the ovaries of rats with polycystic ovary syndrome. Similarly, Mansoury (2019) demonstrated that marjoram treatment resulted in decreased levels of TNF-α and IL-6 in a murine model.

The immunomodulatory properties of marjoram essential oil have also been investigated, with its bioactive constituents, including thymol, sabinene hydrate, terpinene-4-ol, and carvacrol, playing key roles. Research on human macrophages demonstrated that marjoram essential oil significantly reduced the production of IL-6 and IL-1β in LPS-stimulated cells (Arranz et al. 2015a, b). Furthermore, marjoram essential oil suppressed the expression of COX-2 and NF-κB genes, both of which play key roles in the process of inflammation. In studies involving human neutrophils, carvacrol, the main component of marjoram essential oil, was shown to inhibit phagocytosis and the classical complement pathway, although it did not affect the alternate complement pathway. These studies highlight marjoram as a promising natural agent for reducing inflammation and modulating immune responses.

34.2.6.7 Rosemary

Rosemary (*Rosmarinus officinalis*) is well known for its culinary uses, but it has also gained recognition in traditional medicine for its well-known anti-inflammatory and analgesic properties. The immunomodulatory effects of rosemary and its bioactive compounds, such as carnosol, carnosic acid, and rosmarinic acid, have been extensively studied.

34.2.6.7.1 Anti-inflammatory Effects of Rosemary

It has been demonstrated that rosemary extract can reduce levels of pro-inflammatory cytokines, including TNF-α, IL-1β, and IL-6, in human macrophages (Arranz et al. 2015a). This activity has been largely attributed to the compounds carnosol and carnosic acid found in rosemary. Another study by Mueller et al. (2010) showed that rosmarinic acid from rosemary extract reduced IL-6 secretion while promoting the release of anti-inflammatory IL-10.

Modulation of Immune Cell Activation Beyond cytokine regulation, rosemary extract has also been found to regulate immune cell proliferation and activation. A study by Arranz et al. (2017) indicated that rosemary extract inhibited the proliferation of activated murine splenocytes, which suggests its potential for improving immune regulation.

Mast Cell Activation Mast cells play a key role in allergic diseases by releasing inflammatory mediators. In a murine model, rosemary extract was shown to modulate bone marrow mast cell activation, inhibiting degranulation and reducing the secretion of pro-inflammatory mediators such as TNF-α, IL-6, and IL-13. This suggests that rosemary could be beneficial in treating allergic diseases associated with mast cell activation (Yousef et al. 2020).

Other murine models of inflammation in in vivo studies have further supported rosemary's anti-inflammatory effects. For example, a study using a mouse pleurisy model demonstrated that rosemary extract significantly reduced NO concentrations, leukocyte counts, myeloperoxidase activity, and the levels of IL-1β and TNF-α (Benincá et al. 2011).

Neuropathic Pain In another study examining the impact of rosemary extract on neuropathic pain resulting from sciatic nerve constriction, treatment with rosemary extract and its active compound rosmarinic acid was shown to reverse the inflammatory and oxidative stress markers in lumbar spinal cord tissue, post-injury (Rahbardar et al. 2017).

34.2.6.7.2 Rosemary Essential Oil

Phagocytosis Inhibition Rosemary essential oil has been shown to inhibit phagocytosis and the classical complement pathway in vitro (Pérez-Rosés et al. 2015). These findings suggest that rosemary oil may play a role in modulating immune cell activity.

In Vivo Anti-inflammatory Effects In animal models, rosemary essential oil demonstrated a range of anti-inflammatory actions. For instance, in a murine model of paw edema, rosemary essential oil initially increased the extent of edema but subsequently reduced it after 24 hours, along with a decrease in myeloperoxidase activity (Juhás et al. 2009). In another study, rosemary oil significantly reduced colonic inflammation and IL-6 levels in mice with colitis (Juhás et al. 2009).

Pain Relief and Inflammation Control Rosemary essential oil has been shown to inhibit granulomatous tissue formation and ear edema in murine models, demonstrating both anti-inflammatory and analgesic properties (Faria et al. 2011). Moreover, a study by Borges et al. (2018) found that rosemary essential oil, particularly in nanoemulsion form, was more efficacious than the essential oil alone in reducing rat paw edema.

These findings collectively support rosemary's potential as a therapeutic agent for managing inflammation and immune-related conditions, both as an extract and in essential oil form.

34.2.6.8 Thyme

Thyme (Thymus vulgaris), a Mediterranean herb, is widely used in culinary applications and traditional medicine. Its essential oil, primarily composed of thymol and carvacrol, has been studied for its anti-inflammatory and immunomodulatory effects.

34.2.6.8.1 Anti-inflammatory and Immunomodulatory Effects of Thyme

Several in vitro studies have demonstrated the anti-inflammatory effects of thyme and its extracts. In LPS- and IFN-γ-stimulated murine macrophages, thyme extract inhibited iNOS miRNA expression, thereby reducing inflammation (Vigo et al. 2004). In another study, thyme extract reduced the expression levels of pro-inflammatory cytokines including IL-6 and TNF-α, while also increasing the expression of anti-inflammatory IL-10 and strongly inhibiting iNOS expression in RAW 264.7 macrophages (Mueller et al. 2010). Additionally, thyme extracts exhibited dose-dependent inhibitory effects on mitogen-induced lymphocyte proliferation in human blood lymphocytes and the Jurkat cell line, suggesting a capability to regulate immune cell growth (Amirghofran et al. 2011).

Other in vitro studies have shown that thyme extract reduces pro-inflammatory cytokines IL-1β and TNF-α production in LPS-stimulated macrophages (De Oliveira et al. 2017). In a murine model of systemic inflammation, thyme extract increased serum immunoglobulin levels (IgA, IgG, IgM) while decreasing IL-1, IL-6, TNF-α, MCP-1, and CRP levels, indicating its immunomodulatory potential (Osman et al. 2020).

34.2.6.8.2 Thyme Essential Oil

Studies have shown that Thyme essential oil suppresses the expression of pro-inflammatory cytokines including IL-1β, IL-8, and TNF-α (Tsai et al. 2011). In a study using LPS-stimulated mouse mammary epithelial cells, thyme essential oil demonstrated a dose-dependent reduction in IL-6 and TNF-α, along with reducing the expression of iNOS and COX-2. These findings suggest that thyme essential oil may interfere with NF-κB and MAPK signaling pathways involved in inflammation (Liang et al. 2014).

In Vivo Anti-inflammatory Effects In rodent models, thyme essential oil and its components thymol and carvacrol were found to inhibit inflammatory edema. Specifically, thymol was particularly effective in reducing ear edema (Fachini-Queiroz et al. 2012). Other in vivo studies reported improved immune responses following supplementation with thymol and carvacrol (Hashemipour et al. 2013). Additionally, thyme essential oil has been shown to have therapeutic effects on hepatic coccidiosis in rabbits (Abu El Ezz et al. 2020).

34.2.6.8.3 Thyme Essential Oil and COVID-19

A recent randomized controlled trial on COVID-19 patients found that thyme essential oil could alleviate symptoms of SARS-CoV-2 infection. It helped reduce neutrophil counts and increase lymphocyte counts, suggesting potential benefits for immune modulation in viral infections (Sardari et al. 2021).

These findings indicate that thyme, particularly its essential oil and major components (thymol and carvacrol), has significant anti-inflammatory and immunomodulatory properties. These effects make thyme a promising candidate for therapeutic use in managing inflammatory and immune-related conditions.

34.2.6.9 Sage

Sage (*Salvia officinalis*) is well known for its therapeutic effects, including its immunomodulatory and anti-inflammatory effects. Several studies have investigated the bioactive compounds in sage and their potential to modulate inflammation, particularly through effects on macrophage activity and cytokine regulation.

34.2.6.9.1 Anti-inflammatory and Immunomodulatory Effects of Sage Extract: In Vitro Studies

Diterpenoid Compounds A study by Li et al. (2019) explored the anti-inflammatory effects of diterpenoid compounds isolated from sage in murine macrophages. Some of these fractions demonstrated a strong inhibitory effect on nitric oxide (NO) production and reduced the overexpression of inducible nitric oxide synthase (iNOS) and cyclooxygenase-2 (COX-2) in lipopolysaccharide (LPS)-stimulated macrophages. This was accomplished by inhibiting the phosphorylation of c-Jun N-terminal kinase (JNK) in the MAPK signaling pathway, which plays a critical role in inflammation.

Modulation of Cytokine Secretion in Macrophages In another study, sage extract was shown to reduce the secretion of pro-inflammatory cytokines including IL-6, IL-10, and TNF-α, while also inhibiting iNOS expression in LPS-stimulated RAW264.7 murine macrophages (Mueller et al. 2010).

NF-κB Inhibition A more recent study by Brindisi et al. (2021) reported that sage leaf and flower extracts decreased NO production via NF-κB inhibition. The study also found a reduction in pro-inflammatory cytokines like IL-1β, IL-6, and TNF-α, along with a decrease in intracellular reactive oxygen species (ROS), suggesting a comprehensive anti-inflammatory mechanism.

Sage Extract's Effects on Adipocytes and Neuroblastoma Cells The anti-inflammatory effects of sage extract appear to be cell-type dependent. In a study by Russo et al. (2021), sage extract was shown to decrease the production of IL-6, IL-8, TNF-α, and MCP-1 in human subcutaneous mature adipocytes. In contrast, the extract increased the expression of several cytokines and chemokines in IL-β-stimulated neuroblastoma cells, indicating a complex and cell-type-dependent response to sage treatment.

34.2.6.9.2 Anti-inflammatory Effects of Sage: In Vivo Studies

Systemic Immunomodulation in Murine Models A study by Ben Khedher et al. (2018) showed that the administration of a low-dose methanolic extract of sage was found to elevate plasma concentrations of anti-inflammatory cytokines IL-2, IL-4, and IL-10, while concurrently reducing levels of pro-inflammatory cytokines TNF-α and IL-12. These findings suggest that sage may help modulate immune responses in vivo.

Sage Essential Oil and Wound Healing The anti-inflammatory effects of sage essential oil, which is rich in monoterpenes, sesquiterpenes, and phenolic compounds, were investigated by Abu-Darwish et al. (2013) in LPS-stimulated murine macrophages. Results indicated a dose-dependent suppression of NO synthesis. Additionally, in a murine model of infected wounds, the topical application of sage essential oil reduced the expression of pro-inflammatory cytokines TNF-α, IL-1β, and IL-6. It also promoted cellular proliferation, collagen deposition, revascularization, and re-epithelization by increasing the expression of growth factors.

In conclusion, sage and its essential oil exhibit significant anti-inflammatory and immunomodulatory properties through multiple mechanisms, including the inhibition of pro-inflammatory cytokines, NO production, and the MAPK and NF-κB pathways. These effects are supported by both in vitro and in vivo studies, suggesting sage as a promising natural agent for inflammation-related disorders and immune modulation.

34.3 Conclusions and Future Perspectives

Essential oils derived from medicinal herbs have been widely utilized for health promotion and disease prevention for centuries, with increasing attention in recent decades on their potential bioactive compounds. Numerous studies have demonstrated that herbal extracts and their active components exhibit potent immunomodulatory and anti-inflammatory effects, offering promising therapeutic options for various systemic and local health conditions. However, most of the research has been conducted in in vitro and in vivo settings, with clinical studies still relatively limited.

34.3.1 Key Findings from Research

Regarding the immunomodulatory and anti-inflammatory effects, many medicinal herbs, including peppermint, lemon balm, marjoram, rosemary, thyme, and sage, have shown the potential to modulate immune responses and reduce inflammation. These effects are largely attributed to the bioactive components found in these herbs and their essential oils, such as menthol (peppermint), carvacrol (thyme), and rosmarinic acid (rosemary). The studies often show reductions in the level of pro-inflammatory cytokines, including IL-6, TNF-α, and IL-1β. They also exhibit inhibition of inflammatory mediators like NO and COX-2, and the modulation of pathways like NF-κB and MAPK.

Moreover, many studies in animal models (such as murine or rat models) have successfully demonstrated the anti-inflammatory and immunomodulatory properties of these herbs and their oils. These models provide important insights into how these compounds interact with the immune system, modulating immune responses and alleviating inflammation associated with various conditions.

34.3.2 Clinical Studies and Evidence Gap

While the experimental findings are promising, there is a significant gap in clinical studies on the safety and effectiveness of these herbal extracts in humans. To truly evaluate the advantages of these herbal extracts in human health, randomized clinical trials (RCTs) involving diverse populations are needed. These trials would help determine the optimal dosages, potential side effects, and long-term safety of these herbal remedies.

34.3.3 Toxicology and Drug Interactions

Despite the promising data, the toxicity of these herbs and their essential oils in humans remains underresearched. Most studies focus on the safety profiles of these herbs when used within the recommended dosage ranges, but toxicological data is sparse. Additionally, the drug interactions that might occur when using these herbs alongside conventional medications need further investigation. It is essential to assess how herbal extracts might influence drug metabolism through liver enzymes or other mechanisms, as this could impact the effectiveness or safety of prescribed medications.

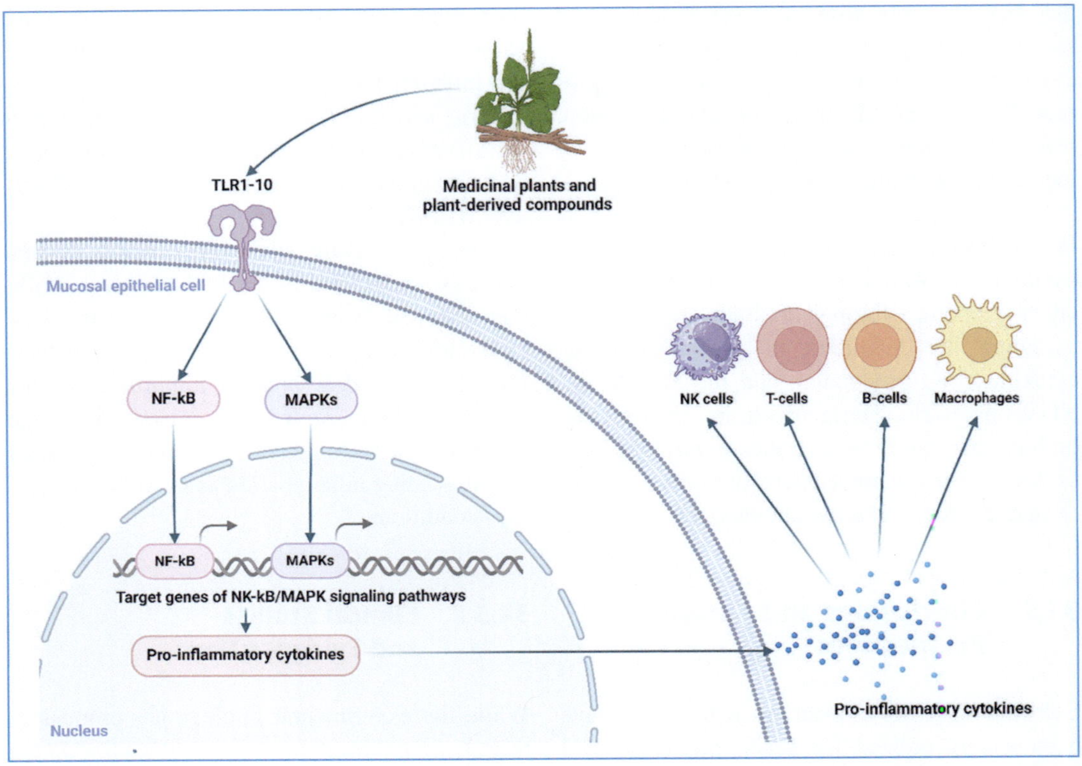

Fig. 34.1 General immunomodulatory effects of medicinal plants and plant-derived compounds on different immune cells. Mucosal epithelial cells are sensitized by medicinal plants and plant-derived compounds through toll-like receptors (TLRs) present on the cell surface. This interaction activates key signaling pathways, namely nuclear factor kappa-B (NF-κB) and mitogen-activated protein kinase (MAPK), leading to the production of various pro-inflammatory cytokines. These effects enhance the immune response, potentially benefiting the body's defense mechanisms against infections and diseases such as cancer. (Graphical figure created by Asst. Prof. Aornrutai Promsong, Ph.D., using BioRender (https://BioRender.com/r83y951))

In conclusion, the biological activities of medicinal herbs and their essential oils suggest significant therapeutic potential for immune system modulation and inflammation reduction (Fig. 34.1 *summarizes general immunomodulatory effects of medicinal plants on immune cells*). However, the limited availability of clinical trial data and toxicity assessments underscores the need for more comprehensive research to confirm their safety, effectiveness, and interactions with medications. Until then, while the existing evidence supports their use in traditional medicine, further research is essential to translate these findings into clinical practice for diverse populations.

References

Abbas A, Lichtman A, Pillai S (2007) Cellular and molecular immunology, 6th edn. Saunders, Elsevier, pp 267–301

Abu El Ezz NMT, Aboelsoued D, Hassan SE, Abdel Megeed KN, El-Metenawy TM (2020) Therapeutic effect of Moringa oleifera and Thymus vulgaris oils against hepatic coccidiosis in experimentally infected rabbits. Trop Biomed 37(4):1018–1028

Abu-Darwish MS, Cabral C, Ferreira IV, Gonçalves MJ, Cavaleiro C, Cruz MT, Al-bdour TH, Salgueiro L (2013) Essential oil of common sage (Salvia officinalis L.) from Jordan: assessment of safety in mammalian cells and its antifungal and anti-inflammatory potential. Biomed Res Int:538940

Aggarwal BB, Shishodia S (2006) Molecular targets of dietary agents for prevention and therapy of cancer. Biochem Pharmacol 71:1397–1421

Amirghofran Z, Hashemzadeh R, Javidnia K, Golmoghaddam H, Esmaeilbeig A (2011) In vitro immunomodulatory effects of extracts from three plants of the Labiatae family and isolation of the active compound(s). J Immunotoxicol 8(4):265–273

Arranz E, Jaime L, López de las Hazas MC, Reglero G, Santoyo S (2015a) Supercritical fluid extraction as an alternative process to obtain essential oils with anti-inflammatory properties from marjoram and sweet basil. Ind Crop Prod 67:121–129

Arranz E, Mes J, Wichers HJ, Jaime L, Mendiola JA, Reglero G, Santoyo S (2015b) Anti-inflammatory activity of the basolateral fraction of Caco-2 cells exposed to a rosemary supercritical extract. J Funct Foods 13:384–390

Arranz E, Guri A, Fornari T, Mendiola JA, Reglero G, Corredig M (2017) In vitro uptake and immune functionality of digested rosemary extract delivered through food grade vehicles. Food Res Int 97:71–77

Arranz E, Villalva M, Guri A, Ortego-Hernández E, Jaime L, Reglero G, Santoyo S, Corredig M (2019) Protein matrices ensure safe and functional delivery of rosmarinic acid from marjoram (Origanum majorana) extracts. J Sci Food Agric 99(5):2629–2635

Azeez TB, Lunghar J (2021) 6-Antiinflammatory effects of turmeric (Curcuma longa) and ginger (Zingiber officinale). In: Gopi S, Amalraj A, Kunnumakkara A, Thomas S (eds) Inflammation and natural products. Academic Press, pp 127–146

Bachiega TF, Sforcin JM (2011) Lemongrass and citral effect on cytokines production by murine macrophages. J Ethnopharmacol 137(1):909–913

Bachiega TF, de Sousa JP, Bastos JK, Sforcin JM (2012) Clove and eugenol in noncytotoxic concentrations exert immunomodulatory/anti-inflammatory action on cytokine production by murine macrophages. J Pharm Pharmacol 64(4):610–616

Banihani SA (2018) Ginger and testosterone. Biomolecules 8(4):119

Bastaki SM, Adeghate E, Amir N, Ojha S, Oz M (2018) Menthol inhibits oxidative stress and inflammation in acetic acid-induced colitis in rat colonic mucosa. Am J Transl Res 10(12):4210–4222

Ben Khedher MR, Hammami M, Arch JRS, Hislop DC, Eze D, Wargent ET, Kępczyńska MA, Zaibi MS (2018) Preventive effects of *Salvia officinalis* leaf extract on insulin resistance and inflammation in a model of high fat diet-induced obesity in mice that responds to rosiglitazone. PeerJ 6:e4166

Benincá JP, Dalmarco JB, Pizzolatti MG, Fröde TS (2011) Analysis of the anti-inflammatory properties of *Rosmarinus officinalis* L. in mice. Food Chem 124: 468–475

Birdane YO, Büyükokuroglu ME, Birdane FM, Cemek M, Yavuz H (2007) Anti-inflammatory and antinociceptive effects of *Melissa officinalis* L. in rodents. Revue de MédecineVétérinaire 158:75–81

Borges RS, Lima ES, Keita H, Ferreira IM, Fernandes CP, Cruz RAS, Duarte JL, Velázquez-Moyado J, Ortiz BLS, Castro AN, Ferreira JV, da Silva Hage-Melim LI, Carvalho JCT (2018) Anti-inflammatory and antialgic actions of a nanoemulsion of Rosmarinus officinalis L. essential oil and a molecular docking study of its major chemical constituents. Inflammopharmacology 26(1):183–195

Borish LC, Steinke JW (2003) 2. Cytokines and chemokines. J Allergy Clin Immunol 111:S460–S475

Bouchentouf S, Missoum N (2020) Identification of compounds from nigella sativa as new potential inhibitors of 2019 novel Coronasvirus (Covid-19): molecular docking study. ChemRxiv. https://doi.org/10.26434/chemrxiv.12055716.v1

Boukhatem MN, Ferhat MA, Kameli A, Saidi F, Kebir HT (2014) Lemon grass (Cymbopogon citratus) essential oil as a potent anti-inflammatory and antifungal drugs. Libyan J Med 9(1):25431

Bounihi A, Hajjaj G, Alnamer R, Cherrah Y, Zellou A (2013) In vivo potential anti-inflammatory activity of Melissa officinalis L. essential oil. Adv Pharmacol Sci 101759

Brindisi M, Bouzidi C, Frattaruolo L, Loizzo MR, Cappello MS, Dugay A, Deguin B, Lauria G, Cappello AR, Tundis R (2021) New insights into the antioxidant and anti-inflammatory effects of Italian *Salvia officinalis* leaf and flower extracts in lipopolysaccharide and tumor-mediated inflammation models. Antioxidants (Basel) 10(2):311

Butt MS, Sultan MT (2011) Ginger and its health claims: molecular aspects. Crit Rev Food Sci Nutr 51:383–393

Carrasco FR, Schmidt G, Romero AL, Sartoretto JL, Caparroz-Assef SM, Bersani-Amado CA, Cuman RK (2009) Immunomodulatory activity of Zingiber officinale Roscoe, Salvia officinalis L. and Syzygium aromaticum L. essential oils: evidence for humor- and cell-mediated responses. J Pharm Pharmacol 61(7): 961–967

Cerda B, Tomas-Barberan FA, Espin JC (2005) Metabolism of antioxidant and chemopreventive ellagitannins from strawberries, raspberries, walnuts, and oak-aged wine in humans: identification of biomarkers and individual variability. J Agric Food Chem 53:227–235

Chai YS, Chen YQ, Lin SH, Xie K, Wang CJ, Yang YZ, Xu F (2020) Curcumin regulates the differentiation of naïve CD4+T cells and activates IL-10 immune modulation against acute lung injury in mice. Biomed Pharmacother 125:109946

Chandrasekaran CV, Thiyagarajan P, Deepak HB, Agarwal A (2011) In vitro modulation of LPS/calcimycin induced inflammatory and allergic mediators by pure compounds of Andrographis paniculata (King of bitters) extract. Int Immunopharmacol 11:79–84

Chen X, Wang Z, Yang Z, Wang J, Xu Y, Tan RX, Li E (2011) *Houttuynia cordata* blocks HSV infection through inhibition of NF-κB activation. Antivir Res 92(2):341–345

Choi MJ, Kim YR (2022) Anti-allergic effect of fermented extracts of medicinal plants andrographis paniculate, salvia plebeia R. Br., canavalia gladiate, eleuthorococcus senticosus, ulmus davidiana var. japonica, and clerodendrum trichotomum thunb. ex murray. Microbiol Biotechnol Lett 50:512–521

Chun JM, Nho KJ, Kim HS, Lee AY, Moon BC, Kim HK (2014) An ethyl acetate fraction derived from *Houttuynia cordata* extract inhibits the production of inflammatory markers by suppressing NF-κB and MAPK activation in lipopolysaccharide-stimulated RAW 264.7 macrophages. BMC Complement Altern Med 10(14):234

Cocchi F, DeVico AL, Garzino-Demo A, Arya SK, Gallo RC, Lusso P (1995) Identification of RANTES, MIP-1 alpha, and MIP-1 beta as the major HIV-suppressive factors produced by CD8+ T cells. Science 270:1811–1815

Dale BA (2002) Periodontal epithelium: a newly recognized role in health and disease. Periodontol 30: 70–78

De Oliveira JR, de Jesus Viegas D, Martins APR, Carvalho CAT, Soares CP, Camargo SEA, Jorge AOC, de Oliveira LD (2017) Thymus vulgaris L. extract has antimicrobial and anti-inflammatory effects in the absence of cytotoxicity and genotoxicity. Arch Oral Biol 82:271–279

Dibazar SP, Fateh S, Daneshmandi S (2015) Immunomodulatory effects of clove (Syzygium aromaticum) constituents on macrophages: in vitro evaluations of aqueous and ethanolic components. J Immunotoxicol 12(2):124–131

Drozd J, Anuszewska E (2003) The effect of the Melissa officinalis extract on immune response in mice. Acta Pol Pharm 60(6):467–470

Fachini-Queiroz FC, Kummer R, Estevão-Silva CF, Carvalho MD, Cunha JM, Grespan R, Bersani-Amado CA, Cuman RK (2012) Effects of thymol and carvacrol, constituents of thymus vulgaris L. Essential oil, on the inflammatory response. Evid Based Complement Alternat Med:657026

Fahey AJ, Adrian Robins R, Constantinescu CS (2007) Curcumin modulation of IFN-beta and IL-12 signalling and cytokine induction in human T cells. J Cell Mol Med 11:1129–1137

Fakhry C, Marks MA, Gilman RH, Cabrerra L, Yori P, Kosek M, Gravitt PE (2013) Comparison of the immune microenvironment of the oral cavity and cervix in healthy women. Cytokine 64(2):597–604

Fallahi M, Keyhanmanesh R, Khamaneh AM, Ebrahimi Saadatlou MA, Saadat S, Ebrahimi H (2016) Effect of Alpha-Hederin, the active constituent of Nigella sativa, on miRNA-126, IL-13 mRNA levels and inflammation of lungs in ovalbumin-sensitized male rats. Avicenna J Phytomed 6(1):77–85

Faria LRD, Lima CS, Perazzo FF, Carvalho JCT (2011) Anti-inflammatory and antinociceptive activities of the essential oil from *Rosmarinus officinalis* L. (Lamiaceae). Int J Pharm Sci Rev Res 7:1–8

Ferguson JJA, Abbott KA, Garg ML (2021) Anti-inflammatory effects of oral supplementation with curcumin: a systematic review and meta-analysis of randomized controlled trials. Nutr Rev 79(9): 1043–1066

Fidel PL Jr (2011) Candida-host interactions in HIV disease: implications for oropharyngeal candidiasis. Adv Dent Res 23:45–49

Francisco V, Costa G, Figueirinha A, Marques C, Pereira P, Miguel Neves B, Celeste Lopes M, García-Rodríguez C, Teresa Cruz M, Teresa Batista M (2013) Anti-inflammatory activity of Cymbopogon citratus leaves infusion via proteasome and nuclear factor-κB pathway inhibition: contribution of chlorogenic acid. J Ethnopharmacol 148(1):126–134

Freeman BD, Natanson C (2000) Anti-inflammatory therapies in sepsis and septic shock. Expert Opin Investig Drugs 9(7):1651–1663

Fu J, Dai L, Lin Z, Lu H (2013) *Houttuynia cordata* Thunb: a review of phytochemistry and pharmacology and quality control. Chin Med 4:101–123

Gbenou JD, Ahounou JF, Akakpo HB, Laleye A, Yayi E, Gbaguidi F, Baba-Moussa L, Darboux R, Dansou P, Moudachirou M, Kotchoni SO (2013) Phytochemical composition of Cymbopogon citratus and Eucalyptus citriodora essential oils and their anti-inflammatory and analgesic properties on Wistar rats. Mol Biol Rep 40(2):1127–1134

Rahbardar MG, Amin B, Mehri S, Mirnajafi-Zadeh SJ, Hosseinzadeh H (2017) Anti-inflammatory effects of ethanolic extract of Rosmarinus officinalis L. and rosmarinic acid in a rat model of neuropathic pain. Biomed Pharmacother 86:441–449

Hamza AA, Ahmed MM, Elwey HM, Amin A (2016) Melissa officinalis protects against doxorubicin-induced cardiotoxicity in rats and potentiates its anticancer activity on MCF-7 cells. PLoS One 11(11): e0167049

Han EH, Park JH, Kim JY, Jeong HG (2009) *Houttuynia cordata* water extract suppresses anaphylactic reaction and IgE-mediated allergic response by inhibiting multiple steps of FcepsilonRI signaling in mast cells. Food Chem Toxicol 47(7):1659–1666

Han X, Parker TL, Dorsett J, Biro M (2017) An essential oil blend siginificantly modulates immune responses and the cell cycle in human cell cultures. Cogent Biol 3:1

Hashemipour H, Kermanshahi H, Golian A, Veldkamp T (2013) Effect of thymol and carvacrol feed supplementation on performance, antioxidant enzyme activities, fatty acid composition, digestive enzyme activities, and immune response in broiler chickens. Poult Sci 92(8):2059–2069

Hedges SR, Agace WW, Svanborg C (1995) Epithelial cytokine responses and mucosal cytokine networks. Trends Microbiol 3:266–270

Hossen MJ, Yang WS, Kim D, Aravinthan A, Kim JH, Cho JY (2017) Thymoquinone: an IRAK1 inhibitor with in vivo and in vitro anti-inflammatory activities. Sci Rep 7:42995

Ireson CR, Jones DJL, Orr S, Coughtrie MWH, Boocock DJ, Williams ML, Farmer PB, Steward WP, Gescher AJ (2002) Metabolism of the cancer chemopreventive agent curcumin in human and rat intestine. Cancer Epidemiol Biomarkers Prev 11:105–111

Joly S, Maze C, McCray PB Jr, Guthmiller JM (2004) Human beta-defensins 2 and 3 demonstrate strain-selective activity against oral microorganisms. J Clin Microbiol 42:1024–1029

Juhás Š, Bukovská A, Čikoš Š, Czikková S, Fabian D, Koppel J (2009) Anti-inflammatory effects of *Rosmarinus officinalis* essential oil in mice. Acta Vet Brno 78:121–127

Kadiyska T, Tourtourikov I, Dabchev K, Zlatarova A, Stoynev N, Hadjiolova R, Spandidos DA, Adamaki M, Zoumpourlis V (2023) Herbs and plants in immunomodulation (Review). Int J Funct Nutr 4:1

Kresina TF, Mathieson B (1999) Human immunodeficiency virus type 1 infection, mucosal immunity, and pathogenesis and extramural research programs at the National Institutes of Health. J Infect Dis 179(Suppl 3): S392–S396

Lang M, Ferron PJ, Bursztyka J, Montjarret A, Duteil E, Bazire A, Bedoux G (2019) Evaluation of immunomodulatory activities of essential oils by high content analysis. J Biotechnol 10(303):65–71

Lee HJ, Seo HS, Kim GJ, Jeon CY, Park JH, Jang BH, Park SJ, Shin YC, Ko SG (2013) *Houttuynia cordata* Thunb inhibits the production of pro-inflammatory cytokines through inhibition of the NFkappaB signaling pathway in HMC-1 human mast cells. Mol Med Rep 8(3):731–736

Lee JH, Ahn J, Kim JW, Lee SG, Kim HP (2015) Flavonoids from the aerial parts of *Houttuynia cordata* attenuate lung inflammation in mice. Arch Pharm Res 38(7):1304–1311

Li W, Zhou P, Zhang Y, He L (2011) *Houttuynia cordata*, a novel and selective COX-2 inhibitor with anti-inflammatory activity. J Ethnopharmacol 133(2): 922–927

Li W, Fan T, Zhang Y, Fan T, Zhou P, Niu X, He L (2013) *Houttuynia cordata* Thunb. volatile oil exhibited anti-inflammatory effects in vivo and inhibited nitric oxide and tumor necrosis factor-α production in LPS-stimulated mouse peritoneal macrophages in vitro. Phytother Res 27(11):1629–1639

Li Y, Liu Y, Ma A, Bao Y, Wang M, Sun (2017) In vitro antiviral, anti-inflammatory, and antioxidant activities of the ethanol extract of *Mentha piperita* L. Food Sci Biotechnol 26(6):1675–1683

Li L, Wei S, Zhu T, Xue G, Xu D, Wang W, Wang X, Luo J, Kong L (2019) Anti-inflammatory norabietane diterpenoids from the leaves of Salvia officinalis L. J Funct Foods 54:154–163

Li X, Yuan W, Wu J, Zhen J, Sun Q, Yu M (2022) Andrographolide, a natural anti-inflammatory agent: an update. Front Pharmacol 13:920435

Liang D, Li F, Fu Y, Cao Y, Song X, Wang T, Wang W, Guo M, Zhou E, Li D, Yang Z, Zhang N (2014) Thymol inhibits LPS-stimulated inflammatory response via down-regulation of NF-κB and MAPK signaling pathways in mouse mammary epithelial cells. Inflammation 37(1):214–222

Lim JW, Kim H, Kim KH (2001) Nuclear factor-kappaB regulates cyclooxygenase-2 expression and cell proliferation in human gastric cancer cells. Lab Investig 81: 349–360

Ma G, Greenwell-Wild T, Lei K, Jin W, Swisher J, Hardegen N, Wild CT, Wahl SM (2004) Secretory leukocyte protease inhibitor binds to annexin II, a cofactor for macrophage HIV-1 infection. J Exp Med 200(10):1337–1346

Mansoury MMS (2019) Marjoram (*Origanum majorana* L.) alleviate myocardial damage induced by doxorubicin in rats. J Biochem Tech 4:82–88

Mao QQ, Xu XY, Cao SY, Gan RY, Corke H, Beta T, Li HB (2019) Bioactive compounds and bioactivities of ginger (Zingiber officinale Roscoe). Foods 8(185)

Moghadamtousi SZ, Kadir HA, Hassandarvish P, Tajik H, Abubakar S, Zandi (2014) A review on antibacterial, antiviral, and antifungal activity of curcumin. Biomed Res Int:186864

Mozaffari-Khosravi H, Naderi Z, Dehghan A, Nadjarzadeh A, Fallah Huseini H (2016) Effect of ginger supplementation on proinflammatory cytokines in older patients with osteoarthritis: outcomes of a randomized controlled clinical trial. J Nutr Gerontol Geriatr 35:209–218

Mueller M, Hobiger S, Jungbauer A (2010) Anti-inflammatory activity of extracts from fruits, herbs and spices. Food Chem 122:987–996

Newman DJ, Cragg GM (2016) Natural products as sources of new drugs from 1981 to 2014. J Nat Prod 79(3):629–661

Nikkhah Bodagh M, Maleki I, Hekmatdoost A (2018) Ginger in gastrointestinal disorders: a systematic review of clinical trials. Food Sci Nutr 7:96–108

Nittayananta W, Promsong A, Levy C, Hladik F, Chaitaveep N, Ungphaiboon S, Tewtrakul S, Satthakarn S (2022) Microbicide containing ellagic acid can inhibit HIV-1 infection. Molecules 27(22): 7941

Nittayananta W, Lerdsamran H, Chutiwitoonchai N, Promsong A, Srichana T, Netsomboon K, Prasertsopon J, Kerdto J (2024) A novel film spray containing curcumin inhibits SARS-CoV-2 and influenza virus infection and enhances mucosal immunity. Virol J 21(1):26

Niyonsaba F, Ushio H, Nakano N, Ng W, Sayama K, Hashimoto K, Nagaoka I, Okumura K, Ogawa H (2007) Antimicrobial peptides human beta-defensins' stimulate epidermal keratinocyte migration,

proliferation and production of proinflammatory cytokines and chemokines. J Invest Dermatol 127(3): 594–604
Olivera A, Moore TW, Hu F, Brown AP, Sun A, Liotta DC, Snyder JP, Yoon Y, Shim H, Marcus AI, Miller AH, Pace TW (2012) Inhibition of the NF-κB signaling pathway by the curcumin analog, 3,5-Bis (2-pyridinylmethylidene)-4-piperidone (EF31): anti-inflammatory and anti-cancer properties. Int Immunopharmacol 12(2):368–377
Osman NN, Alsharari MA, Alsufiani HM (2020) Peppermint (Mentha piperita L.) and thyme (Thymus vulgaris) attenuate the immune and inflammatory disorders in rats consumed repeatedly heated palm oil. Int J Pharm Phytopharmacol Res 10(2):59–66
Ozugurlu F, Sahin S, Idiz N, Akyol O, Ilhan A, Yigitoglu R, Isik B (2005) The effect of Nigella sativa oil against experimental allergic encephalomyelitis via nitric oxide and other oxidative stress parameters. Cell Mol Biol (Noisy-le-Grand) 51(3):337–342
Panichayupakaranant P, Panyokaew T, Tewtrakul S, Yuenyongsawad S, Wang W (2005) Antioxidant and HIV-1 integrase inhibitory activities of ellagic acid isolated from Punica granatum fruit peel. In: The Fourth Indochina conference on pharmaceutical sciences, pp 261–263
Panichayupakaranant P, Tewtrakul S, Yuenyongsawad S (2010) Antibacterial, anti-inflammatory and anti-allergic activities of standardised pomegranate rind extract. Food Chem 123:400–403
Park E, Kum S, Wang C, Park SY, Kim BS, Schuller-Levis G (2005) Anti-inflammatory activity of herbal medicines: inhibition of nitric oxide production and tumor necrosis factor-alpha secretion in an activated macrophage-like cell line. Am J Chin Med 33(3): 415–424
Pérez-Rosés R, Risco E, Vila R, Peñalver P, Cañigueral S (2015) Effect of some essential oils on phagocytosis and complement system activity. J Agric Food Chem 63(5):1496–1504
Poles J, Karhu E, McGill M, McDaniel HR, Lewis JE (2021) The effects of twenty-four nutrients and phytonutrients on immune system function and inflammation: a narrative review. J Clin Transl Res 7:333–376
Promsong A, Chung WO, Satthakarn S, Nittayananta W (2015) Ellagic acid modulates the expression of oral innate immune mediators: Potential role in mucosal protection. J Oral Pathol Med 44:214–221
Puri A, Saxena R, Saxena RP, Saxena KC, Srivastava V, Tandon JS (1993) Immunostimulant agents from Andrographis paniculata. J Nat Prod 56:995–999
Quiñones-Mateu ME, Lederman MM, Feng Z, Chakraborty B, Weber J, Rangel HR, Marotta ML, Mirza M, Jiang B, Kiser P, Medvik K, Sieg SF, Weinberg A (2003) Human epithelial beta-defensins 2 and 3 inhibit HIV-1 replication. AIDS 17(16):F39–F48
Rababa'h AM, Matani BR, Ababneh MA (2020) The ameliorative effects of marjoram in dehydroepiandrosterone induced polycystic ovary syndrome in rats. Life Sci 15(261):118353
Rodrigues TG, Fernandes A Jr, Sousa JP, Bastos JK, Sforcin JM (2009) In vitro and in vivo effects of clove on pro-inflammatory cytokines production by macrophages. Nat Prod Res 23(4):319–326
Russo C, Edwards KD, Margetts G, Kleidonas S, Zaibi NS, Clapham JC, Zaibi MS (2021) Effects of Salvia officinalis L. and Chamaemelum nobile (L.) extracts on inflammatory responses in two models of human cells: Primary subcutaneous adipocytes and neuroblastoma cell line (SK-N-SH). J Ethnopharmacol 268:113614
Said AM, Atwa SAE, Khalifa OA (2019) Ameliorating effect of gum arabic and lemongrass on chronic kidney disease induced experimentally in rats. Bull Natl Res Cent 43:47
Sardari S, Mobaien A, Ghassemifard L, Kamal K, Khavasi N (2021) Therapeutic effect of thyme (*Thymus vulgaris*) essential oil on patients with COVID19: A randomized clinical trial. J Adv Med Biomed Res 29: 83–91
Satthakarn S, Chung WO, Promsong A, Nittayananta W (2015) *Houttuynia cordata* modulates oral innate immune mediators: potential role of herbal plant on oral health. Oral Dis 21:512–518
Schroder K, Hertzog PJ, Ravasi T, Hume DA (2004) Interferon-gamma: an overview of signals, mechanisms and functions. J Leukoc Biol 75:163–189
Sforcin JM, Amaral JT, Fernandes A Jr, Sousa JP, Bastos JK (2009) Lemongrass effects on IL-1beta and IL-6 production by macrophages. Nat Prod Res 23(12): 1151–1159
Shen HM, Tergaonkar V (2009) NFkappaB signaling in carcinogenesis and as a potential molecular target for cancer therapy. Apoptosis 14:348–363
Shi G, Li D, Fu J, Sun Y, Li Y, Qu R, Jin X, Li D (2015) Upregulation of cyclooxygenase-2 is associated with activation of the alternative nuclear factor kappa B signaling pathway in colonic adenocarcinoma. Am J Transl Res 7:1612–1620
Shim SY, Seo YK, Park JR (2009) Down-regulation of FcepsilonRI expression by *Houttuynia cordata* Thunb extract in human basophilic KU812F cells. J Med Food 12(2):383–388
Shoba G, Joy D, Joseph T, Majeed M, Rajendran R, Srinivas PS (1998) Influence of piperine on the pharmacokinetics of curcumin in animals and human volunteers. Planta Med 64:353–356
Silva AFC, Haris PI, Serralheir ML, Pacheco R (2020) Mechanism of action and the biological activities of Nigella sativa oil components. Food Biosci 38:100783
Skyvalidas DN, Mavropoulos A, Tsiogkas S, Dardiotis E, Liaskos C, Mamuris Z, Roussaki-Schulze A, Sakkas LI, Zafiriou E, Bogdanos DP (2020) Curcumin mediates attenuation of pro-inflammatory interferon γ and interleukin 17 cytokine responses in psoriatic disease, strengthening its role as a dietary

immunosuppressant. Nutr Res 75:95–108. https://doi.org/10.1016/j.nutres.2020.01.005

Tapsell LC, Hemphill I, Cobiac L, Patch CS, Sullivan DR, Fenech M, Roodenrys S, Keogh JB, Clifton PM, Williams PG, Fazio VA, Inge KE (2006) Health benefits of herbs and spices: the past, the present, the future. Med J Aust 185(S4):S1–S24

Tiwari M, Dwivedi UN, Kakkar P (2010) Suppression of oxidative stress and pro-inflammatory mediators by Cymbopogon citratus D. Stapf extract in lipopolysaccharide stimulated murine alveolar macrophages. Food Chem Toxicol 48(10):2913–2919

Tjendraputra E, Tran VH, Liu-Brennan D, Roufogalis BD, Duke CC (2001) Effect of ginger constituents and synthetic analogues on cyclooxygenase-2 enzyme in intact cells. Bioorg Chem 29:156–163

Tripathi S, Wang G, White M, Rynkiewicz M, Seaton B, Hartshorn K (2015) Identifying the critical domain of LL-37 involved in mediating neutrophil activation in the presence of influenza virus: functional and structural analysis. PLoS One 10(8):e0133454. https://doi.org/10.1371/journal.pone.0133454

Tsai ML, Lin CC, Lin WC, Yang CH (2011) Antimicrobial, antioxidant, and anti-inflammatory activities of essential oils from five selected herbs. Biosci Biotechnol Biochem 75(10):1977–1983

Vattem D, Shetty K (2005) Biological function of ellagic acid: a review. J Food Biochem 29:234–266

Verma SK, Singh M, Jain P, Bordia A (2004) Protective effect of ginger, Zingiber officinale Rosc on experimental atherosclerosis in rabbits. Indian J Exp Biol 42: 736–738

Vigo E, Cepeda A, Gualillo O, Perez-Fernandez R (2004) In-vitro anti-inflammatory effect of Eucalyptus globulus and Thymus vulgaris: nitric oxide inhibition in J774A.1 murine macrophages. J Pharm Pharmacol 56(2):257–263

Villalva M, Jaime L, Aguado E, Nieto JA, Reglero G, Santoyo S (2018) Anti-inflammatory and antioxidant activities from the basolateral fraction of Caco-2 cells exposed to a rosmarinic acid enriched extract. J Agric Food Chem 66(5):1167–1174

Wan CF, Zheng LL, Liu Y, Yu X (2016) *Houttuynia cordata* Thunb reverses oxaliplatin-induced neuropathic pain in rat by regulating Th17/Treg balance. Am J Transl Res 8(3):1609–1614

Wang C, Wang S, Li D, Chen P, Han S, Zhao G, Chen Y, Zhao J, Xiong J, Qiu J, Wei DQ, Zhao J, Wang J (2021) Human cathelicidin inhibits SARS-CoV-2 infection: killing two birds with one stone. ACS Infect Dis 7(6):1545–1554

Weng Z, Chi Y, Xie J, Liu X, Hu J, Yang F, Li L (2018) Anti-inflammatory activity of dehydroandrographolide by TLR4/NF-κB signaling pathway inhibition in bile duct-ligated mice. Cell Physiol Biochem 49:1124–1137

Xu YY, Zhang YY, Ou YY, Lu XX, Pan LY, Li H, Lu Y, Chen DF (2015) Houttuyniacordata Thunb. polysaccharides ameliorates lipopolysaccharide-induced acute lung injury in mice. J Ethnopharmacol 15(173):81–90

Yang D, Chertov O, Bykovskaia SN, Chen Q, Buffo MJ, Shogan J, Anderson M, Schröder JM, Wang JM, Howard OM, Oppenheim JJ (1999) Beta-defensins: linking innate and adaptive immunity through dendritic and T cell CCR6. Science 286(5439):525–528

Yang D, Chen Q, Hoover DM, Staley P, Tucker KD, Lubkowski J, Oppenheim JJ (2003) Many chemokines including CCL20/MIP-3alpha display antimicrobial activity. J Leukoc Biol 74(3):448–455

Yousef M, Crozier RWE, Hicks NJ, Watson CJF, Boyd T, Tsiani E, MacNeil AJ (2020) Attenuation of allergen-mediated mast cell activation by rosemary extract (Rosmarinus officinalis L.). J Leukoc Biol 107(5): 843–857

Zaia MG, Cagnazzo TD, Feitosa KA, Soares EG, Faccioli LH, Allegretti SM, Afonso A, Anibal FDEF (2016) Anti-inflammatory properties of menthol and menthone in schistosoma mansoni infection. Front Pharmacol 7:170

Zhou H, Beevers CS, Huang S (2011) Targets of curcumin. Curr Drug Targets 12:332–347

Immunomodulation Strategies to Enhance Oral Bone Regeneration

35

Miguel Cardoso, Farah Asa'ad, Teresa M. V. D. Pinho e Melo, Maria Filomena Botelho, Marta Pineiro, Carlos Miguel Marto, and Mafalda Laranjo

Abstract

Several volumetric and qualitative bone limitations may contraindicate the use of bone-supported oral rehabilitation, which remains the gold standard for edentulous ridges. However, oral bone regeneration is a multifaceted process influenced by immune responses, cellular dynamics, and systemic factors. Moreover, its predictability in clinical settings remains limited, leading to challenges in consistently achieving the desired regenerative outcomes. This chapter explores immunomodulation strategies as innovative approaches to enhance alveolar bone regeneration, addressing the challenges of bone healing and repair. Traditional methods such as grafting and biomaterials, while effective, face limitations, including donor site

Authors Carlos Miguel Marto and Mafalda Laranjo have equally contributed to this chapter.

M. Cardoso
University of Coimbra, Coimbra Institute for Clinical and Biomedical Research (iCBR), Area of Environment Genetics and Oncobiology (CIMAGO), Institute of Biophysics, Faculty of Medicine, Coimbra, Portugal

University of Coimbra, Coimbra Chemistry Centre-Institute of Molecular Sciences (CQC-IMS), Department of Chemistry, Coimbra, Portugal

F. Asa'ad
Department of Oral Biochemistry, Institute of Odontology, The Sahlgrenska Academy at University of Gothenburg, SEGöteborg, Sweden
e-mail: farah.asaad@gu.se

T. M. V. D. Pinho e Melo · M. Pineiro
University of Coimbra, Coimbra Chemistry Centre-Institute of Molecular Sciences (CQC-IMS), Department of Chemistry, Coimbra, Portugal
e-mail: tmelo@ci.uc.pt; mpineiro@qui.uc.pt

M. F. Botelho · M. Laranjo (✉)
University of Coimbra, Coimbra Institute for Clinical and Biomedical Research (iCBR), Area of Environment Genetics and Oncobiology (CIMAGO), Institute of Biophysics, Faculty of Medicine, Coimbra, Portugal

University of Coimbra, Center for Innovative Biomedicine and Biotechnology (CIBB), Coimbra, Portugal

Clinical Academic Centre of Coimbra (CACC), Coimbra, Portugal
e-mail: mfbotelho@fmed.uc.pt

C. M. Marto (✉)
University of Coimbra, Coimbra Institute for Clinical and Biomedical Research (iCBR), Area of Environment Genetics and Oncobiology (CIMAGO), Institute of Biophysics, Faculty of Medicine, Coimbra, Portugal

University of Coimbra, Center for Innovative Biomedicine and Biotechnology (CIBB), Coimbra, Portugal

Clinical Academic Centre of Coimbra (CACC), Coimbra, Portugal

University of Coimbra, Institute of Experimental Pathology, Institute of Integrated Clinical Practice and Laboratory for Evidence-Based Sciences and Precision Dentistry, Faculty of Medicine, Coimbra, Portugal

University of Coimbra, Centre for Mechanical Engineering, Materials and Processes (CEMMPRE), Advanced Production and Intelligent Systems (ARISE), Coimbra, Portugal
e-mail: cmiguel.marto@uc.pt

N. Rezaei, H. Ziaei (eds.), *Oral Immunology*, Advances in Experimental Medicine and Biology 1492,
https://doi.org/10.1007/978-3-032-03176-1_35

morbidity and immune rejection. Thus, immunotherapy approaches emphasizing cell-based therapies (macrophage polarization, cytokines, and mesenchymal stem cells), material-based strategies (bioactive scaffolds and nanoparticles), and other supporting approaches like anti-inflammatory drugs and lifestyle interventions will be outlined. Even though preclinical studies suggest that some of these techniques can be crucial in the maintenance or resolution of the inflammation and thereby directing bone healing, several uncertainties remain regarding their clinical validation. Additionally, insights on the role of emerging strategies, such as photodynamic therapy (PDT) and near-infrared light (NIR) therapy, will be explored, considering their impact on immunity, infection, and bone repair. All these strategies can be combined to synergistically and positively affect outcomes in bone regeneration.

Keywords

Alveolar bone regeneration · Immunomodulation · Bioactive scaffolds · Photodynamic therapy (PDT)

35.1 Introduction

Alveolar bone loss can occur due to multiple local (e.g., periodontal disease, trauma, and infection in the oral cavity) and systemic (e.g., osteoporosis and diabetes) acute or chronic conditions, leading to progressive bone defects. These defects compromise the periodontal supporting tissues, negatively affecting the longevity and stability of teeth and dental implants. Furthermore, such conditions have significant esthetic and functional implications since oral rehabilitation procedures, such as those involving bone-supported prostheses, are hindered by reduced bone volume. Bone-supported rehabilitation remains the gold standard for restoring oral function and esthetics in these cases (Giannobile et al. 2019).

Alveolar bone healing comprises a series of complex events, as represented in Fig. 35.1, starting with hemostasis and blood clot formation. During the inflammatory phase, various cellular and molecular events such as angiogenesis, recruitment of osteoprogenitor cells, and deposition of non-mineralized matrix take place, eventually leading to matrix mineralization and continuous tissue remodeling (Donos et al. 2019). In the absence of interventions aimed at bone preservation or regeneration, bone healing will likely lead to volumetric bone loss, both vertically and horizontally. To overcome these challenges, several clinical protocols have been proposed, mainly using biomaterials such as bone grafts, whether autologous, allogenous, or xenografts. Ideally, graft materials should be mechanically resilient, biocompatible, osteoconductive, and osteoinductive. Autologous grafts are the gold standard in bone regeneration. Moreover, growth factors and cells can be used, with or without complementary approaches, such as systemic factors administration and laser therapy. Unfortunately, these approaches have major limitations, such as the high morbidity of autologous grafts, limited biomaterial absorption, rejection issues, patients' refusal to receive xenografts, and the reduced efficacy of synthetic grafts in achieving the desired clinical outcomes (Donos et al. 2023; Elgali et al. 2017).

In recent years, increased attention has been given to leveraging the patient's systemic health and enhancing the patient's biological responses to improve clinical outcomes in bone regeneration. Immunomodulation strategies gained particular attention as they play an important role in regulating bone repair. Bone repair and regeneration are complex processes that involve a variety of cells, which are regulated by numerous factors such as immune cells, stem cells, cytokines, and growth factors, and are influenced by the individual's genetic profile and systemic health (Majidinia et al. 2018). Importantly, macrophages, T cells, and regulatory immune cells are now recognized for their pivotal role in orchestrating the transition from inflammation to tissue regeneration, with efforts increasingly focused on modulating these cellular responses to optimize the healing outcomes (Yu et al. 2022). However, tissue regeneration remains difficult to achieve due to the complex architecture

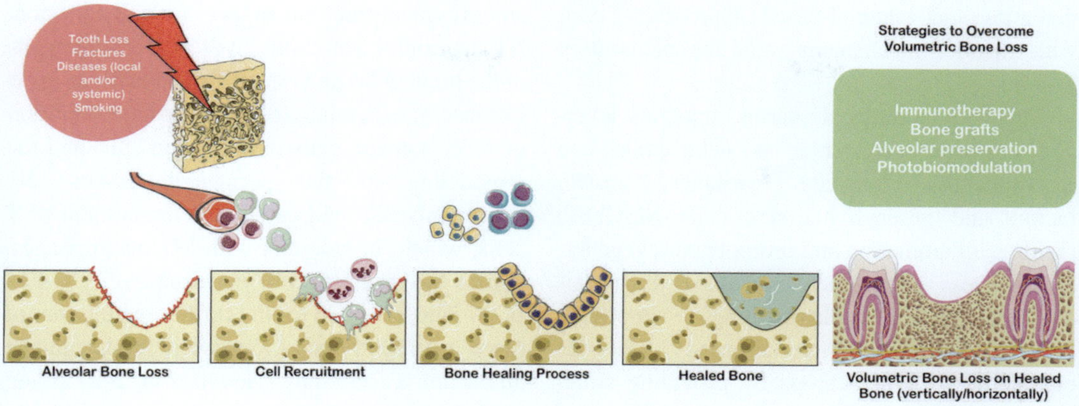

Fig. 35.1 Schematic representation of alveolar bone aggressions and the healing process. Alveolar bone loss is followed by the recruitment of monocytes and neutrophil progenitors from the vascular system, which stimulate an inflammatory response and further cell recruitment. In the affected surface, macrophages and neutrophils become active cells; subsequent cellular signaling further promotes osteoblast pre-osteoblasts' migration and differentiation and the formation of new bone. If no intervention is performed, the healing process often results in volumetric bone changes, which may compromise bone structure and function

of bone and its biomechanical requirements, which introduce great challenges for any material intended to be used in bone regeneration (Lee et al. 2019; Majidinia et al. 2018).

Emerging trends in the field of alveolar bone regeneration are focusing on novel approaches to enhance the immune response and promote bone healing. For example, the use of nanotechnology-based delivery systems has shown promise in improving the controlled release of growth factors, cytokines, and therapeutic agents at the defect site. Additionally, gene therapy is emerging as a strategy to directly modulate gene expression in situ, offering the potential for targeted manipulation of osteogenic and immunomodulatory pathways. Furthermore, the integration of biomaterial-based scaffolds that not only support bone tissue but also provide immunomodulatory cues is gaining attention as an advanced therapeutic strategy. These scaffolds are designed to deliver bioactive molecules, such as cytokines and growth factors, to influence the local immune microenvironment and, therefore, enhance tissue regeneration (Wen et al. 2023).

Additionally, photodynamic therapy (PDT), traditionally used in oncology, has potential as a noninvasive method to modulate immune responses and even enhance tissue regeneration in the oral cavity. PDT, combined with biomaterials, offers an emerging tool to promote a favorable environment for bone healing, particularly by reducing inflammation and enhancing the recruitment of regenerative cells (Xie et al. 2020). This innovative approach will be further discussed in subsequent sections.

In the following sections, we discuss some strategies for oral bone regeneration from an immunotherapy point of view, including cell-based approaches, biomaterials, and other therapies.

35.2 Immunotherapy in Oral Bone Regeneration

Immunotherapy has emerged as a promising approach to enhance oral bone regeneration by targeting the complex interplay between the immune system and the regenerative processes of bone tissue. This strategy leverages the immune system's ability to modulate the inflammatory response and promote tissue repair, facilitating the integration and effectiveness of regenerative treatments. Two major categories of strategies include cell- and biomolecule-based therapies and material-based approaches, each with distinct mechanisms and complementary benefits.

Cell and biomolecule-based strategies leverage immune cells, such as neutrophils and macrophages, alongside cytokines, growth factors, and mesenchymal stem cells (MSCs) to regulate inflammation and promote bone regeneration. These approaches enhance osteogenesis and angiogenesis while creating a supportive microenvironment. Material-based strategies complement these methods by providing structural scaffolds, such as bioceramics and composites, that mimic the bone extracellular matrix (ECM). Functionalized with nanoparticles or bioactive agents, these materials improve bioactivity, mechanical properties, and targeted delivery of regenerative factors.

All these strategies can be combined to achieve enhanced outcomes in bone regeneration. A popular and highly effective approach is the creation of scaffolds loaded with cells or cell components, as described by Lee et al. (2019). These scaffolds provide a biomimetic environment that supports cell adhesion, proliferation, and differentiation while delivering bioactive agents directly to the site of interest. Such combined strategies represent the forefront of immunotherapy in bone regeneration, offering a comprehensive solution to the challenges of restoring oral bone tissue.

35.2.1 Cell and Biomolecule-Based Strategies

35.2.1.1 Immune System Modulation

35.2.1.1.1 Neutrophils

Neutrophils are the first and most abundant cells to arrive at lesion sites. In the initial stage of acute inflammation, neutrophils show a pro-inflammatory phenotype (N1 neutrophils) and secrete cytokines, which result in the recruitment of other types of immune cells (Jiang et al. 2024). Simultaneously, fibronectin modulates degranulation and amplifies matrix metalloproteinase (MMP) release by neutrophils and provides structural support for cell adhesion. N1 neutrophil secretions hydrolyze components of the bone ECM and other MMP substrates. New fibronectin is synthesized to allow the infiltration of other somatic cells to the lesion sites and the remodeling of the peripheral tissues. N1 neutrophils also will enable the recruitment of T cells, which, in crosstalk with M1 macrophages, induce the production of interleukin 17 (IL-17). IL-17 is responsible for the further recruitment of more neutrophils and monocytes, linking innate and adaptive immunity (Herath et al. 2021; Jiang et al. 2024).

Along the healing process, the resolution of inflammation with the decrease of cytokines drives neutrophils polarization into N2 phenotype, which expresses a set of anti-inflammatory signals, thus leading other cells like macrophages to transition from a pro-inflammatory phenotype to an anti-inflammatory one, which later mediates MSC differentiation by secreting growth factors and resolving inflammation (Jiang et al. 2024). N2 neutrophils also promote the transformation of M1 macrophages into M2 macrophages through the neutrophil gelatinase-associated lipid carrier protein (NGAL) mechanism. Additionally, they exhibit strong signals that increase NGAL uptake by macrophages, thus establishing a synergistic environment, as shown by Shi et al. (2024). The advantages of macrophage polarization will be discussed in subsequent sections.

Interleukin 8 (IL-8) is the most effective cytokine for neutrophil recruitment and also influences the migration of bone marrow-derived MSCs (Matsushima et al. 2022), as shown by Cai et al. (2021); experimentally, a hematoma aged 2 or 4 days was treated with IL-8 and implanted in muscle. Neutrophils were recruited immediately after, but bone MSCs arrived at later stages. Also, when neutrophils and monocytes were depleted before the creation of the defect and remained absent till 7 days after injury, critical defects could not effectively heal, thus establishing the importance of these cells in the early stages of the healing process. Interestingly, in irradiated mice, transferring exogenous myeloid cells allowed the observation of neutrophil recruitment at the implantation site with slight ECM secretion.

However, when bone marrow-derived MSCs were transferred, although with increased cell recruitment, a high number of adipose vacuoles were identified, suggesting adipogenic differentiation in contrast to osteogenic differentiation. Transferring monocytes allowed the degradation of the implanted hematoma, indicating the stimulation of macrophage phagocytosis, and both monocytes and bone marrow-derived MSC transfers resulted in minimal ECM secretion. These findings further suggest that after the initial inflammation resolution, the decreased cytokine-level polarizes neutrophils into the N2 phenotype, which promotes the transformation of other cell types into anti-inflammatory phenotypes and, under specific IL-8 concentrations, promotes the recruitment of bone MSCs (Cai et al. 2021; Matsushima et al. 2022).

Importantly, the differentiation of myeloblasts into neutrophils and further polarization into N1 or N2 phenotypes is driven by the microenvironment. Some of the factors involved in favoring the transition of N1 into N2 neutrophils are interleukin 4 (IL-4), interleukin 10 (IL-10), and interleukin 13 (IL-13), and transforming growth factor beta (TGF-β). This transition can happen in a dependent manner, where any imbalance can shift the progression of the healing process and jeopardize the process, compromising both inflammation resolution and the osteogenic pathway of all the involved cell populations (Cai et al. 2021; Matsushima et al. 2022).

Therefore, neutrophils are indispensable at the early stage of bone healing and regeneration; these cells recruit and orchestrate the innate and adaptive immune cells and the MSC differentiation during the initial stage. Modulation of the neutrophil's activity can help control initial inflammation and favor the successive bone healing phases (Cai et al. 2021; Matsushima et al. 2022).

35.2.1.1.2 Macrophages

Macrophages are innate immunity cells found in most tissues and are among the first cell populations to arrive at bone lesion sites, playing a major role in normal tissue homeostasis. Infected areas present an increased number of macrophages through the recruitment of monocyte precursors to the affected regions, consequently leading to inflammation. In acute inflammation, macrophages will favor the restoration of tissue homeostasis by phagocytizing invading microorganisms and amplifying the inflammatory reaction, contributing to the recruitment of additional immune cells. Further in the healing process, macrophages contribute to tissue regeneration with the secretion of anti-inflammatory precursors, recruiting progenitor cells, such as MSCs, and producing growth factors that regulate and promote the differentiation of the recruited cells, contributing to angiogenesis. This functional plasticity of macrophages is known as macrophage polarization, where the pro-inflammatory phenotype is called M1 macrophages, and the anti-inflammatory phenotype is called M2 macrophages. Functional plasticity means that those two phenotypes can switch from one phenotype to the other, presenting both pro- and anti-inflammatory properties in response to the inflammatory microenvironment. This makes macrophages interesting targets for therapeutic actions (Schlundt et al. 2021).

Although most relations and cross talks between these immune cells and bone-forming cells are yet to be uncovered, some works have started to unveil the underlying mechanisms. It is assumed that the initial inflammatory reaction and pro-inflammatory macrophage activity result in the recruitment of MSCs, osteoprogenitor cells, and vascular cells to the lesion site. However, a critical balance exists, in which M1 macrophages contribute to the initial acute inflammation and the removal of debris, while M2 macrophages secrete growth factors supporting MSC-mediated bone formation necessary for bone healing.

M1 macrophages present a significant upregulation of pro-inflammatory cytokines, while M2 macrophages show an upregulation of growth factors, such as TGF-β, vascular endothelial growth factor (VEGF), and insulin-like growth factor 1 (IGF-1). M1 macrophage-mediated inflammatory reaction is essential at the initial stages of healing for optimal outcomes.

Still, its timing and duration are also critical, where their prolonged or delayed establishment is highly detrimental to the healing process since a chronic inflammatory condition results in increased osteoclast activity and the suppression of bone formation by osteoblasts, thus compromising bone regeneration (Kushioka et al. 2023).

Although macrophages seem to promote MSC differentiation, MSCs also have immunomodulatory functions in response to inflammatory microenvironments, where these cells become polarized into pro-inflammatory or anti-inflammatory phenotypes and regulate macrophage chemotaxis. MSC signaling pathways may favor bone regeneration by modulating macrophage function. An interesting hypothesis regarding macrophage-MSC crosstalk is that the initial MSCs, recruited to the lesion sites by pro-inflammatory macrophages, can modulate macrophage activity, polarizing them into M2 macrophages and thus favoring bone healing. Dysregulation in this balance may lead to failed bone regeneration by chronic maintenance of a pro-inflammatory microenvironment (Kushioka et al. 2023; Schlundt et al. 2021).

In experimental settings, implanted hydroxyapatite as bone mimetic particles was shown to interfere with macrophage polarization, and the size of these particles was critical for the provoked interaction. Larger particles promoted the production of pro-inflammatory factors like tumor necrosis factor-alpha (TNF-α) and IL-8, whereas nanoparticles were able to polarize macrophages into M2 phenotype and increase the production of IL-10. The implantation, ectopically to the site of origin, showed scarce regeneration in the larger hydroxyapatite particles, whereas nanoparticles could considerably repair the defects. IL-10 was also proven to be a dependent factor in the MSC osteogenesis pathway, being a key factor for calcification and mineralization within the bone morphogenetic protein 2 (BMP-2)-mediated osteogenic differentiation (Mahon et al. 2020).

The mechanism described in the previous section regarding neutrophils can also be applied to macrophages, where monocytes are provoked to differentiate into M1 or M2 macrophages in response to the microenvironment cues. This differentiation is influenced by factors like IL-4, IL-10, IL-13, TGF-β, and VEGF, which can happen in a concentration-dependent manner. Interestingly, N2 neutrophils and M2 macrophages cross talk lead to a synergic activation loop, promoting polarization toward inflammation resolution (Mihaila et al. 2024; Yang et al. 2024).

However, little research has been directly focusing on targeting macrophages to favor bone regeneration. Considering the recently uncovered interactions between macrophages and MSCs, strategies that target macrophages by adjusting their numbers at lesion sites and influencing their polarization may enhance MSC pathways and lead to improved outcomes in bone regeneration.

Neutrophil and macrophage differentiation pathways, as well as their polarization, are represented in Fig. 35.2.

35.2.1.1.3 T Cells

Treg cells are a specialized immune cell population with immunomodulatory functions, essential for the maintenance of immunological self-tolerance and the prevention of exaggerated inflammation in response to pathogens or tissue damage. This way, these cells play an important role in facing unfavorable pro-inflammatory conditions. One important feature of Tregs is their direct impact on ATP metabolism, resulting in the generation of strong immunomodulatory ATP derivatives like adenosine. Adenosine inhibits the switch of M2 macrophages to M1 macrophages, and the cytokine secretion by the proliferation of activated effector T (Teff) cells, thus promoting a less inflammatory environment that is favorable for bone regeneration (Li et al. 2018; Ribot et al. 2021; Schlundt et al. 2019).

Another T cell population is γδ T cells that were shown to have a positive impact on the osteoblastic differentiation of MSCs through the production of IL-17, a cytokine greatly present after tissue injury and characterized as an inflammatory cytokine that promotes the eradication of pathogenic microorganisms and starts autoimmune inflammation (Ono et al. 2016).

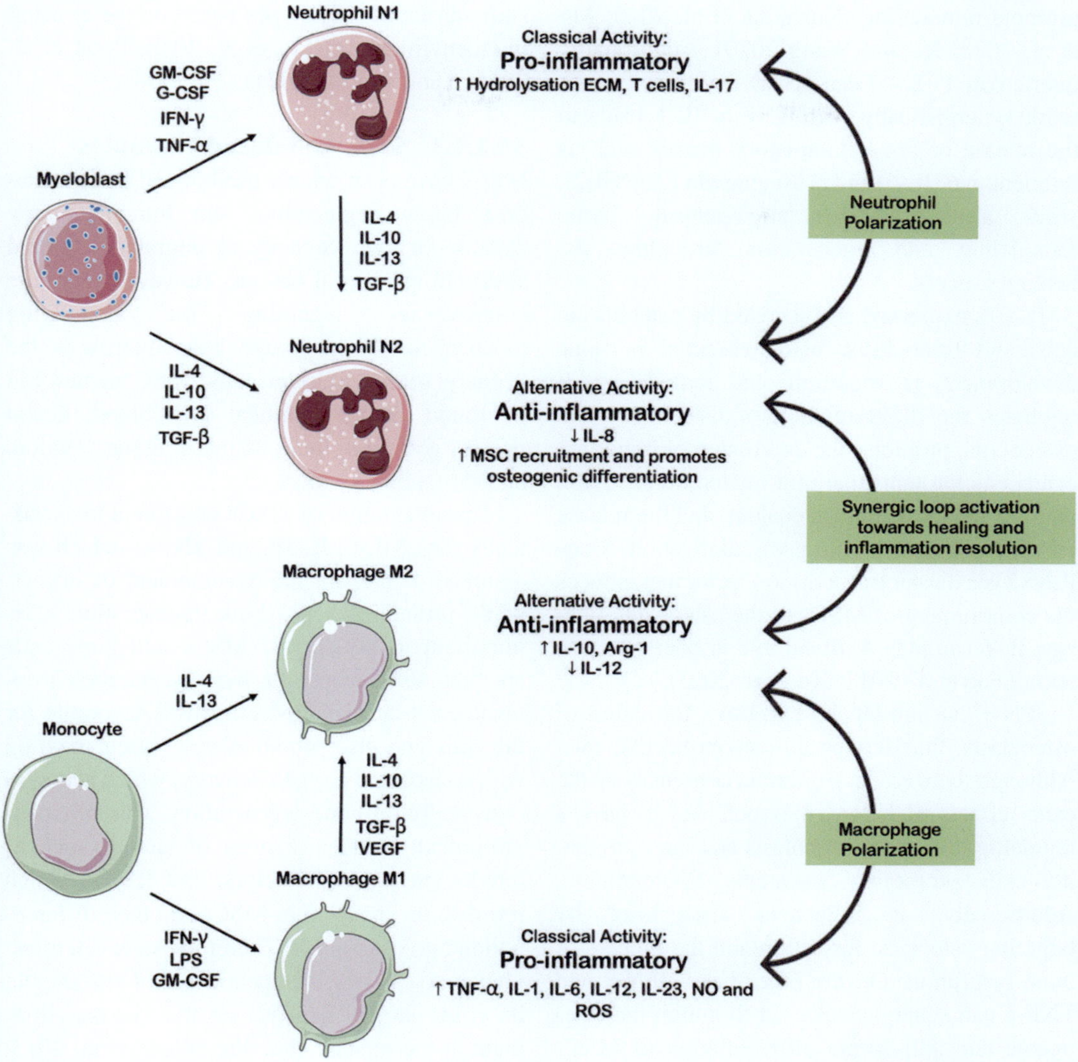

Fig. 35.2 Neutrophil and macrophage differentiation pathways, their polarization between N1 and N2 phenotypes, and the synergic activation loop of the anti-inflammatory phenotypes toward the resolution of inflammation and bone healing

35.2.1.2 Cytokine and Growth Factor Therapy

Cytokines are important effectors of immune cells and can influence bone regeneration, displaying different interactions according to their type, concentration, expression time, and the local microenvironment. Notable examples of them are TNF-α, BMPs, interleukins, and interferon-gamma (IFN-γ) (Moon et al. 2023).

TNF-α levels at bone lesion sites vary during the healing process, reflecting its dynamic role in modulating the local cellular environment. This cytokine not only influences the recruitment and activity of MSCs but also directly impacts osteocytes, the primary regulators of bone homeostasis. The effect of TNF-α on tissue healing is highly context-dependent, influenced by the differentiation state of osteocytes, their ability to orchestrate bone remodeling, and the duration of TNF-α exposure. Prolonged or excessive TNF-α levels can lead to osteocyte apoptosis, impairing bone regeneration, while controlled levels during the early stages of healing may support the recruitment of regenerative cells and promote remodeling (Kushioka et al. 2023; Mo et al. 2022; Ru and Wang 2020). Additionally, interleukin 1 (IL-1) can act as the basis for and work synergistically with TNF-α. IL-1 leads to the release of pro-inflammatory factors such as interleukin 6 (IL-6) and prostaglandin E2 (PGE2), while also promoting angiogenesis, tissue remodeling, and regeneration throughout the healing process.

IL-6 is expressed and secreted by osteoblastic cells and osteoclasts, being essential in bone development, regeneration, and remodeling. It regulates the differentiation of osteoblasts and osteoclasts, promotes the expression of important osteogenic proteins like runt-related transcription factor 2 (Runx2) and osteocalcin, and the release of VEGF, thus stimulating vascularization. Since Runx2 is a master transcription factor that induces the commitment of MSCs to the osteogenic lineage, IL-6 can play a critical role in bone regeneration (Hao et al. 2017; Xu et al. 2023).

IFN-γ can inhibit the excessive formation of osteoclasts, thus having an osteoprotective role. Although there is a lack of broad consensus on the exact effects of IFN-γ, it is believed to have a regulatory effect on osteoblasts and their progenitor cells, promoting osteogenic differentiation. Additionally, it influences osteoclasts by targeting osteoclast formation and the process of bone resorption. On the other hand, IFN-γ and TNF-α can synergistically act to inhibit the osteogenic and adipogenic differentiation of MSCs through the nuclear factor kappa B (NF-kB) pathway (Hao et al. 2017; Xu et al. 2023).

BMPs also represent an important signaling pathway for the regulation of osteogenesis. It is known to activate important proteins and upregulate the gene expression of *Runx2*, therefore promoting the differentiation of MSCs into osteoblasts. Of all, BMP-2 is one of the most studied BMPs and is reported to be osteoconductive, promoting cell chemotaxis, proliferation, and cell differentiation into an osteogenic lineage.

Even though the casualty nature of these interactions has been studied and identified, maintaining a balance is always required. Some interactions are concentration-dependent and may show different phenotypes based on the existing microenvironment (Liu et al. 2018; Poon et al. 2016; Yang and Liu 2021).

35.2.1.3 Stem Cell-Based Therapies

MSCs have been widely used in cell transplantation, tissue engineering, and immunotherapy because of their capacity to migrate to injured sites and their local actions. However, there are controversies regarding the underlying mechanisms. Even though bone marrow is the primary source of stem cells, MSC niches can be found in the umbilical cord blood, dental tissues, synovial fluid, adipose tissue, tendon, and skin, among others.

Transplantation of MSCs was found to specifically target IL-6, IL-1β, and TNF-α, which was assumed to prevent the development of fibrosis while promoting rapid bone regeneration. It is important to notice that MSCs and bone cells are both stimulated by a hypoxic microenvironment. Such a microenvironment is favorable for the cells and also promotes new vascularization with consequent oxygen delivery, which acts synergistically on bone regeneration. This microenvironment includes a series of factors, such as interleukins, growth factors, and BMPs, which promote the invasion of MSCs and their differentiation into osteoblasts. Therefore, stem cell transplantation can be a therapeutic tool of interest, but its effect largely depends on the microenvironment at the regeneration site (Wang et al. 2013; Kim et al. 2021; Torrecillas-Baena et al. 2023).

35.2.1.4 Gene Therapy

In theory, certain gene therapy technologies have potential applications in bone regeneration, as represented in Fig. 35.3. However, in practice, gene therapy for human applications is controversial due to the risk of unintended effects extending beyond the region of interest.

Nevertheless, among the available technologies, the application of gene-activated matrices found their way into clinical trials. This type of gene therapy consists of the use of matrices, which provide localized, slow, and controlled release of the gene of interest to the surrounding cells. They can also be complemented with

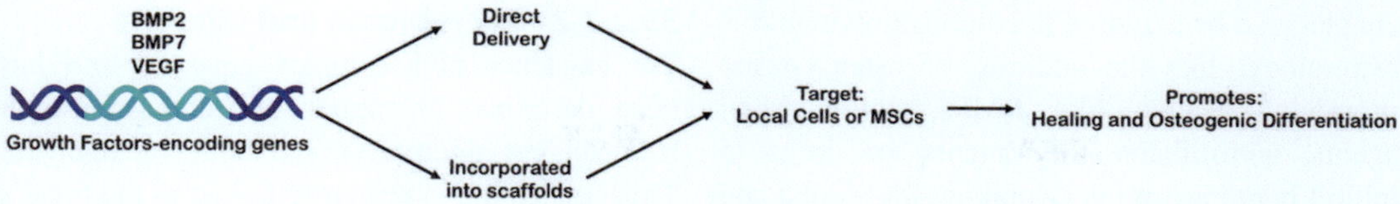

Fig. 35.3 Gene therapy strategies in bone regeneration

factors for microenvironment improvement and, in this way, supply the microenvironment with osteoconductive or osteoinductive features to promote bone regeneration. The primary interest in these technologies is more related to the local delivery of genes encoding growth factors like VEGF for vascularization enhancement or favorable proteins for bone regeneration like BMPs. On the other hand, this approach relies on scaffolds that are acellular, avascular, and aneural, which may be disadvantageous or impair the regeneration process (Shapiro et al. 2018).

35.2.2 Material-Based Strategies

35.2.2.1 Bioactive Materials

A biomimetic approach to enhance bone regeneration often involves the use of biodegradable scaffolds that promote osteoconductivity by replicating the extracellular matrix (ECM) of bone and facilitating the binding of bioactive factors. The ECM forms a complex and dynamic biological environment with specific mechanical and biochemical properties, regulating processes such as cell adhesion, proliferation, response to growth factors, and differentiation. Consequently, the condition of the ECM significantly influences the functional characteristics of mature bone.

Bioactive materials for bone regeneration include polymeric materials, bioceramics, and composite materials. Polymers are promising for this purpose, serving as active carriers for at least two components, a polymeric matrix and an inorganic filler. Despite their advantages, polymers typically exhibit low mechanical resistance. They can be derived from natural sources, such as polysaccharides (e.g., alginate, chitosan, hyaluronic acid) and protein-based compounds (e.g., collagen, gelatin), or synthesized artificially. Natural polymers generally induce minimal immunological responses in the host, enhancing their biocompatibility. However, synthetic polymers offer higher versatility for chemical modification and can be engineered to include specific features, such as controlled release of bioactive agents at the site of interest.

Bioceramics and composite materials share some characteristics but differ in their modification potential and bioactivity. Composites are more straightforward to customize in terms of mechanical properties, degradation rates, and bioactivity. In contrast, bioceramics are intrinsically more bioactive due to their strong bone-bonding capabilities and the release of ions that support bone regeneration. These materials actively modulate signaling pathways, provide mechanical support to the surrounding tissues, and help regulate the microenvironment to maximize healing and regeneration. A key aspect of their effectiveness lies in the incorporation of immunomodulatory agents into these scaffolds, which enhances their ability to support bone regeneration while favorably modulating the local immune response (Szwed-Georgiou et al. 2023; Tanvir et al. 2024).

35.2.2.2 Nanoparticles

Although bone regeneration materials are of great interest, currently available materials still often exhibit insufficient bioactivity, osteoinduction capacity, and osteointegration. Functionalization of these materials with nanoparticles may enhance their properties and overcome their disadvantages. This can be achieved using osteoinductive hydroxyapatite nanocrystals, resembling the inorganic portion of the bone, or using bioactive polymer nanoparticles, which will provide sustained delivery of pro-osteogenic agents directly into the site of interest. Other factors, discussed in the different sections of this chapter, can be explored in combination with this technology, like the addition of osteogenesis-promoting factors like BMPs, antimicrobial agents, anti-inflammatory agents, or drugs to inhibit bone resorption (Kupikowska-Stobba and Kasprzak 2021).

35.2.3 Other Strategies

35.2.3.1 Anti-inflammatory Drugs

Anti-inflammatory drugs (AIDs) are frequently used in clinical practice because of their safety and general pharmacodynamic properties. However, this class of medication comprises a variety of compounds with different mechanisms of action and efficacies on bone regeneration. Overall, most published works address the negative impact of AIDs on bone regeneration, but emerging research refers to aspects that may be positive toward bone regeneration.

The most common class of anti-inflammatory drugs studied is the cyclooxygenase 2 (COX-2) inhibitors. The study by Simon et al. (2002) showed that COX-2 is required for MSCs to differentiate into osteoblasts, whereas Yoon et al. (2010) demonstrated a decrease in the osteogenic potential associated with COX-2 inhibition. Additionally, Murnaghan et al. (2006) demonstrated that the administration of anti-inflammatory drugs inhibits angiogenesis; thereby, the use of COX-2 inhibitors would jeopardize bone regeneration.

Nonetheless, the use of ibuprofen was shown to improve bone quality via the reduction of osteoclast differentiation and inflammatory cytokines, which led to the reduction of bone catabolism and inflammation, having a protective role in bone hemostasis (Jain et al. 2014).

Alternatively, opioids could be considered for pain management, as they share this function with anti-inflammatory drugs. However, studies suggest that opioids may negatively affect bone density, making them a less suitable and less effective alternative to anti-inflammatory drugs for supporting bone regeneration (Marquez-Lara et al. 2016).

35.2.3.2 Microbiome and Lifestyle

The existence of a complex gut-bone axis has been described to support the communication between gut microbiota and bone metabolism. This connection is primarily linked to the bone's requirement of calcium absorption for mineralization, which occurs from diet through the intestine via Vitamin D and is essential for bone homeostasis. At the same time, the endocrine system is linked to bone metabolism and can secrete hormones or bioactive molecules that have an impact on bone health and growth. Ultimately, gut microbiota plays a crucial role in protecting against pathogenic microbes (Shavandi et al. 2020). This way, a synergistic system can be described, including gut microbiota, the endocrine system, and the immune system, which collectively regulate bone health and metabolism. Although little significance was reported about their direct impact on bone regeneration, nutrition, and lifestyle have an impact on the immune system, thus indirectly contributing to bone homeostasis (Najeeb et al. 2016).

35.3 Emerging Strategies and Future Perspectives

Emerging strategies in oral bone regeneration include innovative approaches that extend beyond established immunotherapy-based methods. Among these, techniques such as photodynamic therapy (PDT) and near-infrared (NIR) light therapy seem to have the potential to enhance bone healing through their ability to modulate biological responses, promote angiogenesis, and support cellular differentiation, as will be discussed below. Both PDT and NIR might offer not only direct regenerative benefits but also potential immunomodulatory effects. PDT, for instance, can generate reactive oxygen species that act as danger signals, recruiting and activating immune cells to create a pro-regenerative inflammatory environment. Similarly, NIR light has been shown to modulate immune responses by promoting the transition of macrophages to anti-inflammatory phenotypes, aiding in the resolution of inflammation and enhancing conditions favorable for bone repair.

Another possibility is hyperthermia. Hyperthermia, when applied within the mild range of 40–43 °C, has demonstrated effectiveness in promoting tissue regeneration. Compared to endogenous biological cues such as pH, oxidative stress, and enzymatic activity, hyperthermia offers a more controlled and desirable approach (Wang et al. 2022a). While extra-oral heat application is often limited by reduced tissue penetration, intraoral application has demonstrated higher efficiency in promoting the healing of bone defects. It stimulates the early differentiation of human mesenchymal stem cells (hMSCs) and supports the maturation of osteoblasts derived from MSCs (Tong et al. 2019).

35.3.1 Photodynamic Therapy

Photodynamic Therapy (PDT) is a minimally invasive treatment that relies on the interaction between a photosensitizer, light (typically in the blue or red regions of the visible spectrum), and oxygen (Cardoso et al. 2024). While its primary application is to induce cell death, particularly in cancer treatment, growing evidence highlights broader therapeutic applications, generating interest in its potential to address other conditions and diseases. In dentistry, in particular, PDT is a noninvasive treatment that has been used in oral and maxillofacial surgery, oral medicine, endodontics (root canal disinfection and inactivation of oral and intracanal biofilms), preventive dentistry, periodontology, and implantology (Stájer et al. 2020).

PDT has the ability to modulate both the innate and adaptive arms of the immune system, as demonstrated in the treatment of cancer (Chou et al. 2023; Donohoe et al. 2019). Additionally, there is evidence regarding the positive effect of PDT on bone regeneration. Therefore, PDT might be an interesting option to explore in the context of immunomodulatory strategies to enhance oral bone regeneration.

Specifically, clinical studies demonstrated the recruitment and activation of immune cells, including neutrophils, macrophages, and dendritic cells, to the treatment site following PDT (Dudzik et al. 2024). In addition, the reactive oxygen species generated during PDT trigger an inflammatory response. This includes the expression of heat-shock proteins, the release of danger-associated molecular signals, and the infiltration of immune cells. In some instances, these effects can also lead to the activation of long-term immune memory (Suresh et al. 2024; Warszyńska et al. 2023). These features open up new possibilities for using PDT combined with other therapeutic reagents designed to stimulate the immune response and enhance immunotherapy.

Nevertheless, despite years of research on PDT for cancer treatment, much remains to be understood about its interactions with the immune system. Key areas that require further exploration include how PDT affects immune cells, the potential benefits of combining tumor-targeted PDT with immunotherapy, the impact of integrating immune adjuvants with PDT, the role of PDT in modulating immune checkpoints, and the diverse effects of photosensitizers on the immune response (Dudzik et al. 2024).

As mentioned, PDT might be able to affect bone regeneration positively. A single PDT treatment applied to vertebrae in a rat model of spinal metastases demonstrated not only metastatic ablation but also overall vertebral mechanical stability, including improved vertebral bone strength, stiffness, and trabecular architecture (Won et al. 2010). PDT has also been investigated for the treatment of critical-sized bone defect fractures in animal models. Although variability in fracture formation was a limitation of the study, PDT showed promise for enhancing bone strength and healing, with increased local VEGF levels, suggesting a positive role in fracture repair. Still, the timing of PDT administration, particularly during the secondary stage of bone healing, appears to be critical for its effectiveness (Akens 2012).

A recently developed injectable and thermosensitive hydrogel was designed to address both antibacterial therapy and bone regeneration, which are crucial in managing periodontitis. The hydrogel acts as a carrier for the controlled release of an osteoinductive agent, bone morphogenetic protein-2 (BMP-2), and near-infrared (NIR-II) phototherapy agents, T8IC and H_2O_2. This system demonstrated bactericidal effects, osteogenic induction, and biosafety. Regarding bone regeneration, immunohistochemistry and micro-CT analyses confirmed its efficacy by alleviating inflammation and promoting bone healing (Wang et al. 2023).

In fact, it is now accepted in the literature that PDT is a useful approach to managing oral infections, including periodontal, peri-implant, and endodontic infections, as well as cariogenic, fungal, and viral pathogens (Gholami et al. 2022; Plotino et al. 2019). Still, variability in photosensitizers and irradiation parameters across studies prevents standardized clinical protocols (Gholami et al. 2022). Therefore, advances in photosensitizers hold potential, and further high-quality randomized trials are needed to establish consistent, evidence-based guidelines for their use in dentistry (Cardoso et al. 2024).

Regarding alveolar bone, several studies investigating the efficacy of PDT as an antimicrobial treatment have also provided evidence of ameliorated bone repair (Abdolrahmani et al. 2024; Derikvand et al. 2020; Park et al. 2019), reinforcing its potential use in bone regeneration. These examples include several pieces of evidence from preclinical and clinical cases of medication-related osteonecrosis of the jaw (Abdolrahmani et al. 2024), evidence from periodontitis preclinical models (Park et al. 2019), and a clinical case of a hopeless tooth where the loss of attachment was recovered with alveolar bone regeneration after several antimicrobial PDT sessions (Derikvand et al. 2020).

To fully realize the potential of PDT in bone regeneration, it is essential to develop photosensitizers designed explicitly for this purpose. These photosensitizers should possess optimal optical and photophysical properties to efficiently generate reactive oxygen species tailored to trigger desired biological responses, such as inflammation modulation, angiogenesis, and other regenerative processes, while addressing the unique characteristics of the oral cavity, including infection control. Additionally, systematic studies are needed to determine the optimal light parameters, such as wavelength and intensity, as well as the appropriate number of treatment sessions based on individual case presentations, to maximize the efficacy and applicability of PDT in bone healing and regeneration.

35.3.2 Near-Infrared Light

Infrared radiation is a form of electromagnetic energy with wavelengths ranging from 780 nm to 1000 μm. It is categorized into three main regions: Near-Infrared (NIR, 0.78–3.0 μm), Mid-Infrared (MIR, 3.0–50.0 μm), and Far-Infrared (FIR, 50.0–1000.0 μm). Research has shown that infrared radiation can promote the healing of skin wounds, aid in photoprotection, alleviate pain and symptoms associated with rheumatoid arthritis and ankylosing spondylitis, enhance the effects of photodynamic therapy, address ophthalmic, neurological, and psychiatric conditions, and stimulate the proliferation of mesenchymal and cardiac stem cells (Tsai and Hamblin 2017).

Near-infrared (NIR) light therapy is emerging as a noninvasive technique for promoting tissue regeneration, including bone healing, with notable implications for immune system modulation. Its ability to penetrate deeply into tissues allows it to influence underlying cellular processes, including mitochondrial activation and cytokine regulation. NIR light has been shown to reduce excessive inflammation by promoting the transition of macrophages from a pro-inflammatory (M1) to an anti-inflammatory (M2) phenotype, fostering an environment that supports tissue repair. This immunomodulatory capacity, combined with its ability to enhance cellular metabolism, makes NIR light a promising approach for addressing the complex interplay of immune responses and bone regeneration (Liao et al. 2021).

Currently, photobiomodulation with NIR light irradiation is being used clinically to reduce inflammation, edema, and pain, as well as to repair multiple damaged tissues. It is hypothesized that photobiomodulation might increase the activity of MSCs. Its primary site of action is the mitochondria, more specifically, cytochrome c oxidase. This process potentially involves the dissociation of inhibitory nitric oxide from cytochrome c oxidase, which may restore electron transport and increase the mitochondrial membrane potential. On the other hand, NIR light irradiation may trigger light or heat-gated ion channels and activate the change of transient receptor potential. These potential changes to mitochondria can increase the rate of respiration and ATP production, leading to a marked effect on stem cells via mitochondrial redox signaling (Tong et al. 2019; Wang et al. 2022b; Xue et al. 2024).

Regarding bone regeneration, different wavelengths have been explored, with improvements demonstrated from blue to NIR light. Irradiation with a high-frequency NIR diode laser increased cell division and migration of osteoblasts via the MAPK/ERK signaling pathway, thereby boosting bone regeneration (Kunimatsu et al. 2018). Equine MSCs irradiated with 1064 nm NIR showed enhanced cell viability and increased IL-10 and VEGF expression, leading to angiogenesis improvement and decreased inflammation (Peat et al. 2018). Recent research demonstrated that NIR light seems to accelerate bone regeneration by influencing the circadian clock protein cryptochrome 1 (CRY1). Specifically, NIR light enhances the ubiquitination of CRY1 within the nucleus of osteoblastic cells, a process that plays a key role in osteogenic differentiation. This process is linked to the activation of signaling pathways involving bone morphogenetic proteins (BMPs), which are essential for bone formation and repair (Peng et al. 2022). Overall, it seems that NIR light irradiation can have a positive effect on bone metabolism, and stem cells and bone progenitor cells appear to be particularly susceptible to photobiomodulation (PBM) effects. However, these multiple options have yet to be fully assessed, and the underlying mechanisms need to be understood.

Despite the encouraging results, PBM outcomes in bone regeneration remain inconsistent. While NIR light has shown potential benefits, other recent studies suggest that blue and green light may be more effective in promoting the osteogenic differentiation of MSCs, possibly through mechanisms involving calcium ion channel activation (Bölükbaşı Ateş et al. 2017; Wang et al. 2016, 2017). Additionally, studies have shown that 635 nm lasers are more effective than 808 nm and 405 nm lasers in promoting viability, proliferation, adhesion, and osteogenic differentiation of hMSCs. However, significant controversy remains regarding the optimal wavelengths for PBM strategies, as outcomes are influenced by multiple factors such as cell type, laser category, wavelength, energy density, and the duration of the PBM protocol (Tani et al. 2018).

Future perspectives linking PBM with PDT might take advantage of Platinum(II) Ring-Fused Chlorins, which function as NIR emitters at room temperature and are able to act as photosensitizers (Laranjo et al. 2020; Pereira et al. 2017). Specifically, a single molecule might be able to offer the possibility to combine these two modalities for enhanced therapeutic outcomes in achieving bone regeneration. These compounds could optimize the therapeutic benefits by simultaneously stimulating cellular metabolic activity and triggering reactive oxygen species (ROS) production, enhancing both regenerative processes and antimicrobial effects. Additionally, their NIR emission ensures deeper tissue penetration, making them suitable for applications in wound healing and bone regeneration. Future research may focus on investigating this hypothesis and tailoring protocols to harness their full potential.

35.4 Conclusion

Alveolar bone regeneration poses a complex challenge in dentistry due to the intricate structure and biomechanical requirements of bone tissue. Traditional approaches, such as grafting and biomaterials, while effective, face limitations like donor site morbidity, immune rejection, and reduced efficacy in some instances. Recent advancements in immunomodulation strategies

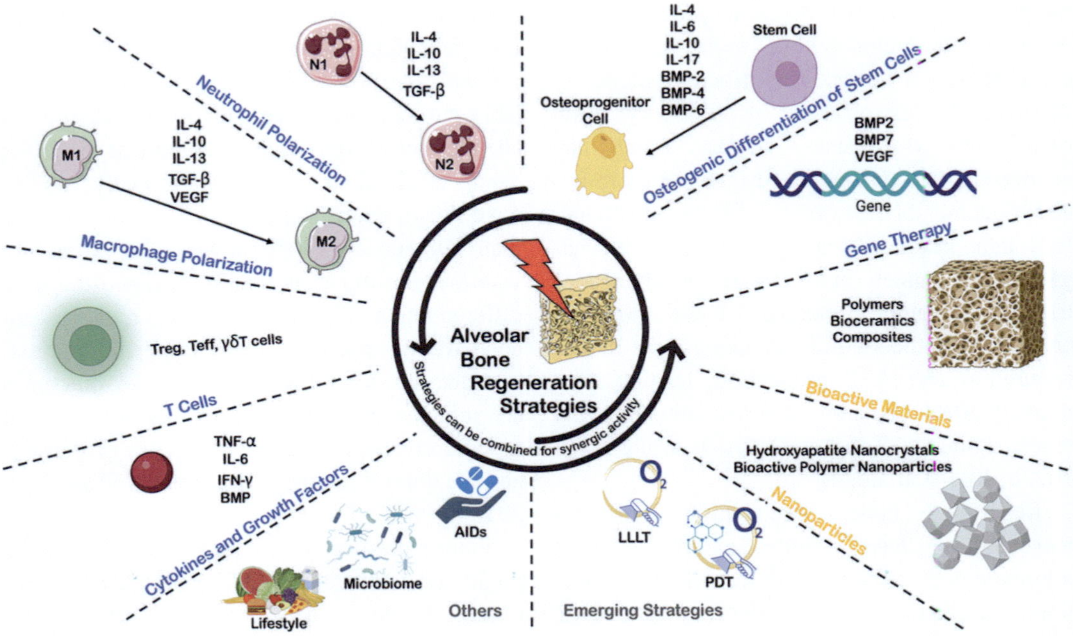

Fig. 35.4 Strategies for alveolar bone regeneration. Headings in blue represent immune system modulation strategies. Headings in yellow represent material-based strategies. Headings in grey represent emerging strategies as well as other additional techniques. The graphic organization suggests the possibility of a combination of techniques for improved results

are paving the way for innovative approaches that leverage the immune system's role in bone healing, such as macrophage polarization, cytokine therapies, and mesenchymal stem cell applications. Figure 35.4 provides an overview of current, innovative, and anticipated strategies for alveolar bone regeneration.

PDT and NIR might be emerging noninvasive solutions. PDT combines antimicrobial effects with immune system modulation, fostering an environment conducive to bone repair. Similarly, NIR therapy can enhance cellular metabolism and immune response, creating favorable conditions for tissue regeneration.

Combining traditional and cutting-edge techniques, supported by ongoing research, holds significant potential for improving alveolar bone regeneration. By addressing current limitations and leveraging innovative solutions, integrated strategies should offer more predictable and practical outcomes, ultimately improving oral rehabilitation and patient quality of life.

Conflicts of Interest The authors declare no conflicts of interest.

Funding Fundação para a Ciência e a Tecnologia (FCT) supports the Center for Innovative Biomedicine and Biotechnology through the UIDB/04539/2020 (10.54499/UIDB/04539/2020), UIDP/04539/2020 (10.54499/UIDP/04539/2020) and the Associated Laboratory funding LA/P/0058/2020 (10.54499/LA/P/0058/2020); the Coimbra Chemistry Centre (CQC) through the Strategic Projects UIDB/00313/2020 (10.54499/UIDB/00313/2020) and UIDP/00313/2020 (10.54499/UIDP/00313/2020) and the Institute of Molecular Sciences (IMS) through special complementary funds; and the Centre for Mechanical Engineering, Materials and Processes (CEMMPRE) through the projects UIDB/00285/2020 (10.54499/UIDB/00285/2020 and 10.54499/UIDP/00285/2020) and LA/P/0112/2020 (10.54499/LA/P/0112/2020). This work was also supported by the FCT-funded Project PTDC/QUI-QOR/0103/2021.

References

Abdolrahmani A, Epstein JB, Samim F (2024) Medication-related osteonecrosis of the jaw: evolving research for multimodality medical management. Support Care Cancer 32(4):212. https://doi.org/10.1007/s00520-024-08388-4

Akens MK (2012) Photodynamic therapy treatment to enhance fracture healing

Bölükbaşı Ateş G, Ak Can A, Gülsoy M (2017) Investigation of photobiomodulation potentiality by 635 and 809 nm lasers on human osteoblasts. Lasers Med Sci 32(3):591–599. https://doi.org/10.1007/s10103-017-2153-6

Cai B, Lin D, Li Y, Wang L, Xie J, Dai T, Liu F, Tang M, Tian L, Yuan Y, Kong L, Shen SGF (2021) N2-polarized neutrophils guide bone mesenchymal stem cell recruitment and initiate bone regeneration: a missing piece of the bone regeneration puzzle. Adv Sci (Weinh) 8(19):e2100584. https://doi.org/10.1002/advs.202100584

Cardoso M, Marto CM, Paula A, Coelho AS, Amaro I, Pineiro M, Pinho E Melo TMVD, Marques Ferreira M, Botelho MF, Carrilho E, Laranjo M (2024) Effectiveness of photodynamic therapy on treatment response and survival in patients with recurrent oral squamous cell carcinoma: a systematic review. Photodiagn Photodyn Ther 48:104242. https://doi.org/10.1016/j.pdpdt.2024.104242

Chou W, Sun T, Peng N, Wang Z, Chen D, Qiu H, Zhao H (2023) Photodynamic therapy-induced anti-tumor immunity: influence factors and synergistic enhancement strategies. Pharmaceutics 15(11). https://doi.org/10.3390/pharmaceutics15112617

Derikvand N, Hatami M, Chiniforush N, Ghasemi SS (2020) The use of antimicrobial photodynamic therapy to maintain a hopeless tooth with a periodonticendodontic lesion: a case report. J Lasers Med Sci 11(3):355–360. https://doi.org/10.34172/jlms.2020.57

Donohoe C, Senge MO, Arnaut LG, Gomes-da-Silva LC (2019) Cell death in photodynamic therapy: from oxidative stress to anti-tumor immunity. Biochim Biophys Acta 1872(2):188308. https://doi.org/10.1016/j.bbcan.2019.07.003

Donos N, Dereka X, Calciolari E (2019) The use of bioactive factors to enhance bone regeneration: a narrative review. J Clin Periodontol 46(S21):124–161. https://doi.org/10.1111/jcpe.13048

Donos N, Akcali A, Padhye N, Sculean A, Calciolari E (2023) Bone regeneration in implant dentistry: which are the factors affecting the clinical outcome? Periodontol 93(1):26–55. https://doi.org/10.1111/prd.12518

Dudzik T, Domański I, Makuch S (2024) The impact of photodynamic therapy on immune system in cancer – an update. Front Immunol 15:1335920. https://doi.org/10.3389/fimmu.2024.1335920

Elgali I, Omar O, Dahlin C, Thomsen P (2017) Guided bone regeneration: materials and biological mechanisms revisited. Eur J Oral Sci 125(5):315–337. https://doi.org/10.1111/eos.12364

Gholami L, Shahabi S, Jazaeri M, Hadilou M, Fekrazad R (2022) Clinical applications of antimicrobial photodynamic therapy in dentistry. Front Microbiol 13:1020995. https://doi.org/10.3389/fmicb.2022.1020995

Giannobile WV, Berglundh T, Al-Nawas B, Araujo M, Bartold PM, Bouchard P, Chapple I, Gruber R, Lundberg P, Sculean A, Lang NP, Lyngstadaas P, Kebschull M, Galindo-Moreno P, Schwartz Z, Shapira L, Stavropoulos A, Reseland J (2019) Biological factors involved in alveolar bone regeneration. J Clin Periodontol 46(S21):6–11. https://doi.org/10.1111/jcpe.13130

Hao Z, Ma Y, Wu J, Li X, Chen H, Shen J et al (2017) Osteocytes regulate osteoblast differentiation and osteoclast activity through Interleukin-6 under mechanical loading. RSC Adv 7(79):50200–50209. https://doi.org/10.1039/C7RA09308J

Herath TDK, Saigo L, Schaller B, Larbi A, Teoh SH, Kirkpatrick CJ, Goh BT (2021) In vivo efficacy of neutrophil-mediated bone regeneration using a rabbit calvarial defect model. Int J Mol Sci 22(23):13016. https://doi.org/10.3390/ijms222313016

Jain NX, Barr-Gillespie AE, Clark BD, Kietrys DM, Wade CK, Litvin J, Popoff SN, Barbe MF (2014) Bone loss from high repetitive high force loading is prevented by ibuprofen treatment. J Musculoskelet Neuronal Interact 14(1):78–94. http://www.ncbi.nlm.nih.gov/pubmed/24583543

Jiang T, Tang X-Y, Su H, Chen J-Y, Qin Y-Q, Qin Y-C, Ouyang N-J, Tang G-H (2024) Neutrophils are involved in early bone formation during midpalatal expansion. Oral Dis 30(7):4483–4494. https://doi.org/10.1111/odi.14849

Kim M, Shin DI, Choi BH, Min BH (2021) Exosomes from IL-1β-primed mesenchymal stem cells inhibited IL-1β- and TNF-α-mediated inflammatory responses in osteoarthritic SW982 cells. Tissue Eng Regen Med 18(4):525–536. https://doi.org/10.1007/s13770-020-00324-x

Kunimatsu R, Gunji H, Tsuka Y, Yoshimi Y, Awada T, Sumi K, Nakajima K, Kimura A, Hiraki T, Abe T, Naoto H, Yanoshita M, Tanimoto K (2018) Effects of high-frequency near-infrared diode laser irradiation on the proliferation and migration of mouse calvarial osteoblasts. Lasers Med Sci 33(5):959–966. https://doi.org/10.1007/s10103-017-2426-0

Kupikowska-Stobba B, Kasprzak M (2021) Fabrication of nanoparticles for bone regeneration: new insight into applications of nanoemulsion technology. J Mater Chem B 9(26):5221–5244. https://doi.org/10.1039/d1tb00559f

Kushioka J, Chow SK-H, Toya M, Tsubosaka M, Shen H, Gao Q, Li X, Zhang N, Goodman SB (2023) Bone regeneration in inflammation with aging and cell-based immunomodulatory therapy. Inflamm Regen 43(1):29. https://doi.org/10.1186/s41232-023-00279-1

Laranjo M, Aguiar M, Pereira N, Brites G, Nascimento B, Brito A, Casalta-Lopes J, Gonçalves AC, Sarmento-Ribeiro AB, Pineiro M, Botelho MF, Pinho e Melo T

(2020) Platinum(II) ring-fused chlorins as efficient theranostic agents: dyes for tumor-imaging and photodynamic therapy of cancer. Eur J Med Chem 200: 112468. https://doi.org/10.1016/j.ejmech.2020.112468
Lee J, Byun H, Madhurakkat Perikamana SK, Lee S, Shin H (2019) Current advances in immunomodulatory biomaterials for bone regeneration. Adv Healthc Mater 8(4):e1801106. https://doi.org/10.1002/adhm.201801106
Li J, Tan J, Martino MM, Lui KO (2018) Regulatory T-cells: potential regulator of tissue repair and regeneration. Front Immunol 9:585. https://doi.org/10.3389/fimmu.2018.00585
Liao W-T, Hung C-H, Liang S-S, Yu S, Lu J-H, Lee C-H, Chai C-Y, Yu H-S (2021) Anti-inflammatory effects induced by near-infrared light irradiation through M2 macrophage polarization. J Invest Dermatol 141(8): 2056–2066.e10. https://doi.org/10.1016/j.jid.2020.11.035
Liu H, Li D, Zhang Y, Li M (2018) Inflammation, mesenchymal stem cells and bone regeneration. Histochem Cell Biol 149(4):393–404. https://doi.org/10.1007/s00418-018-1643-3
Mahon OR, Browe DC, Gonzalez-Fernandez T, Pitacco P, Whelan IT, Von Euw S, Hobbs C, Nicolosi V, Cunningham KT, Mills KHG, Kelly DJ, Dunne A (2020) Nanoparticle mediated M2 macrophage polarization enhances bone formation and MSC osteogenesis in an IL-10 dependent manner. Biomaterials 239: 119833. https://doi.org/10.1016/j.biomaterials.2020.119833
Majidinia M, Sadeghpour A, Yousefi B (2018) The roles of signaling pathways in bone repair and regeneration. J Cell Physiol 233(4):2937–2948. https://doi.org/10.1002/jcp.26042
Marquez-Lara A, Hutchinson ID, Nuñez F, Smith TL, Miller AN (2016) Nonsteroidal anti-inflammatory drugs and bone-healing: a systematic review of research quality. JBJS Rev 4(3). https://doi.org/10.2106/JBJS.RVW.O.00055
Matsushima K, Yang D, Oppenheim JJ (2022) Interleukin-8: an evolving chemokine. Cytokine 153:155828. https://doi.org/10.1016/j.cyto.2022.155828
Mihaila AC, Ciortan L, Tucureanu MM, Simionescu M, Butoi E (2024) Anti-inflammatory neutrophils reprogram macrophages toward a pro-healing phenotype with increased efferocytosis capacity. Cells 13(3): 208. https://doi.org/10.3390/cells13030208
Mo Q, Zhang W, Zhu A, Backman LJ, Chen J (2022) Regulation of osteogenic differentiation by the pro-inflammatory cytokines IL-1β and TNF-α: current conclusions and controversies. Hum Cell 35(4): 957–971. https://doi.org/10.1007/s13577-022-00711-7
Moon S, Hong J, Go S, Kim B-S (2023) Immunomodulation for tissue repair and regeneration. Tissue Eng Regen Med 20(3):389–409. https://doi.org/10.1007/s13770-023-00525-0
Murnaghan M, Li G, Marsh DR (2006) Nonsteroidal anti-inflammatory drug-induced fracture nonunion: an inhibition of angiogenesis? J Bone Joint Surg Am 88(suppl 3):140–147. https://doi.org/10.2106/JBJS.F.00454
Najeeb S, Zafar MS, Khurshid Z, Zohaib S, Almas K (2016) The role of nutrition in periodontal health: an update. Nutrients 8(9):530. https://doi.org/10.3390/nu8090530
Ono T, Okamoto K, Nakashima T, Nitta T, Hori S, Iwakura Y, Takayanagi H (2016) IL-17-producing γδ T cells enhance bone regeneration. Nat Commun 7(1): 10928. https://doi.org/10.1038/ncomms10928
Park D, Choi EJ, Weon K-Y, Lee W, Lee SH, Choi J-S, Park GH, Lee B, Byun MR, Baek K, Choi JW (2019) Non-invasive photodynamic therapy against -periodontitis-causing bacteria. Sci Rep 9(1):8248. https://doi.org/10.1038/s41598-019-44498-4
Peat FJ, Colbath AC, Bentsen LM, Goodrich LR, King MR (2018) In vitro effects of high-intensity laser photobiomodulation on equine bone marrow-derived mesenchymal stem cell viability and cytokine expression. Photomed Laser Surg 36(2):83–91. https://doi.org/10.1089/pho.2017.4344
Peng J, Zhao J, Tang Q, Wang J, Song W, Lu X, Huang X, Chen G, Zheng W, Zhang L, Han Y, Yan C, Wan Q, Chen L (2022) Low intensity near-infrared light promotes bone regeneration via circadian clock protein cryptochrome 1. Int J Oral Sci 14(1):53. https://doi.org/10.1038/s41368-022-00207-y
Pereira N, Laranjo M, Casalta-Lopes J, Serra AC, Piñeiro M, Pina J, Seixas de Melo JS, Senge MO, Botelho MF, Martelo L, Burrows HD, Pinho e Melo TMVDVD (2017) Platinum(II) ring-fused Chlorins as near-infrared emitting oxygen sensors and photodynamic agents. ACS Med Chem Lett 8(3):310–315. https://doi.org/10.1021/acsmedchemlett.6b00476
Plotino G, Grande NM, Mercade M (2019) Photodynamic therapy in endodontics. Int Endod J 52(6):760–774. https://doi.org/10.1111/iej.13057
Poon B, Kha T, Tran S, Dass CR (2016) Bone morphogenetic protein-2 and bone therapy: successes and pitfalls. J Pharm Pharmacol 68(2):139–147. https://doi.org/10.1111/jphp.12506
Ribot JC, Lopes N, Silva-Santos B (2021) Γδ T cells in tissue physiology and surveillance. Nat Rev Immunol 21(4):221–232. https://doi.org/10.1038/s41577-020-00452-4
Ru J-Y, Wang Y-F (2020) Osteocyte apoptosis: the roles and key molecular mechanisms in resorption-related bone diseases. Cell Death Dis 11(10):846. https://doi.org/10.1038/s41419-020-03059-8
Schlundt C, Reinke S, Geissler S, Bucher CH, Giannini C, Märdian S, Dahne M, Kleber C, Samans B, Baron U, Duda GN, Volk H-D, Schmidt-Bleek K (2019) Individual effector/regulator T cell ratios impact bone regeneration. Front Immunol 10:1954. https://doi.org/10.3389/fimmu.2019.01954
Schlundt C, Fischer H, Bucher CH, Rendenbach C, Duda GN, Schmidt-Bleek K (2021) The multifaceted roles of

macrophages in bone regeneration: a story of polarization, activation and time. Acta Biomater 133:46–57. https://doi.org/10.1016/j.actbio.2021.04.052

Shapiro G, Lieber R, Gazit D, Pelled G (2018) Recent advances and future of gene therapy for bone regeneration. Curr Osteoporos Rep 16(4):504–511. https://doi.org/10.1007/s11914-018-0459-3

Shavandi A, Saeedi P, Gérard P, Jalalvandi E, Cannella D, Bekhit AE-D (2020) The role of microbiota in tissue repair and regeneration. J Tissue Eng Regen Med 14(3):539–555. https://doi.org/10.1002/term.3009

Shi W, Feng Y, Tang J, Xu Y, Wang W, Zhang L, Jiang X, Ding Z, Xi K, Chen L, Gu Y (2024) A genetically engineered "reinforced concrete" scaffold regulates the N2 neutrophil innate immune cascade to repair bone defects. Adv Healthc Mater 13(18):e2304585. https://doi.org/10.1002/adhm.202304585

Simon AM, Manigrasso MB, O'Connor JP (2002) Cyclo-oxygenase 2 function is essential for bone fracture healing. J Bone Miner Res 17(6):963–976. https://doi.org/10.1359/jbmr.2002.17.6.963

Stájer A, Kajári S, Gajdács M, Musah-Eroje A, Baráth Z (2020) Utility of photodynamic therapy in dentistry: current concepts. Dent J 8(2):43. https://doi.org/10.3390/dj8020043

Suresh N, Joseph B, Sathyan P, Sweety VK, Waltimo T, Anil S (2024) Photodynamic therapy: an emerging therapeutic modality in dentistry. Bioorg Med Chem 114:117962. https://doi.org/10.1016/j.bmc.2024.117962

Szwed-Georgiou A, Płociński P, Kupikowska-Stobba B, Urbaniak MM, Rusek-Wala P, Szustakiewicz K, Piszko P, Krupa A, Biernat M, Gazińska M, Kasprzak M, Nawrotek K, Mira NP, Rudnicka K (2023) Bioactive materials for bone regeneration: biomolecules and delivery systems. ACS Biomater Sci Eng 9(9):5222–5254. https://doi.org/10.1021/acsbiomaterials.3c00609

Tani A, Chellini F, Giannelli M, Nosi D, Zecchi-Orlandini S, Sassoli C (2018) Red (635 nm), near-infrared (808 nm) and violet-blue (405 nm) photobiomodulation potentiality on human osteoblasts and mesenchymal stromal cells: a morphological and molecular in vitro study. Int J Mol Sci 19(7). https://doi.org/10.3390/ijms19071946

Tanvir MAH, Khaleque MA, Kim G-H, Yoo W-Y, Kim Y-Y (2024) The role of bioceramics for bone regeneration: history, mechanisms, and future perspectives. Biomimetics (Basel) 9(4):230. https://doi.org/10.3390/biomimetics9040230

Tong L, Liao Q, Zhao Y, Huang H, Gao A, Zhang W, Gao X, Wei W, Guan M, Chu PK, Wang H (2019) Near-infrared light control of bone regeneration with biodegradable photothermal osteoimplant. Biomaterials 193:1–11. https://doi.org/10.1016/j.biomaterials.2018.12.008

Torrecillas-Baena B, Pulido-Escribano V, Dorado G, Gálvez-Moreno MÁ, Camacho-Cardenosa M, Casado-Díaz A (2023) Clinical potential of mesenchymal stem cell-derived exosomes in bone regeneration. J Clin Med 12(13):4385. https://doi.org/10.3390/jcm12134385

Tsai S-R, Hamblin MR (2017) Biological effects and medical applications of infrared radiation. J Photochem Photobiol B 170:197–207. https://doi.org/10.1016/j.jphotobiol.2017.04.014

Wang X, Wang Y, Gou W, Lu Q, Peng J, Lu S (2013) Role of mesenchymal stem cells in bone regeneration and fracture repair: a review. Int Orthop 37(12):2491–2498. https://doi.org/10.1007/s00264-013-2059-2

Wang Y, Huang Y-Y, Wang Y, Lyu P, Hamblin MR (2016) Photobiomodulation (blue and green light) encourages osteoblastic-differentiation of human adipose-derived stem cells: role of intracellular calcium and light-gated ion channels. Sci Rep 6:33719. https://doi.org/10.1038/srep33719

Wang Y, Huang Y-Y, Wang Y, Lyu P, Hamblin MR (2017) Red (660 nm) or near-infrared (810 nm) photobiomodulation stimulates, while blue (415 nm), green (540 nm) light inhibits proliferation in human adipose-derived stem cells. Sci Rep 7(1):7781. https://doi.org/10.1038/s41598-017-07525-w

Wang L, Hu P, Jiang H, Zhao J, Tang J, Jiang D, Wang J, Shi J, Jia W (2022a) Mild hyperthermia-mediated osteogenesis and angiogenesis play a critical role in magnetothermal composite-induced bone regeneration. Nano Today 43:101401. https://doi.org/10.1016/j.nantod.2022.101401

Wang S, Wang F, Zhao X, Yang F, Xu Y, Yan F, Xia D, Liu Y (2022b) The effect of near-infrared light-assisted photothermal therapy combined with polymer materials on promoting bone regeneration: a systematic review. Mater Des 217:110621. https://doi.org/10.1016/j.matdes.2022.110621

Wang W, Zhang G, Wang Y, Ran J, Chen L, Wei Z, Zou H, Cai Y, Han W (2023) An injectable and thermosensitive hydrogel with nano-aided NIR-II phototherapeutic and chemical effects for periodontal antibacteria and bone regeneration. J Nanobiotechnol 21(1):367. https://doi.org/10.1186/s12951-023-02124-6

Warszyńska M, Repetowski P, Dąbrowski JM (2023) Photodynamic therapy combined with immunotherapy: recent advances and future research directions. Coord Chem Rev 495:215350. https://doi.org/10.1016/j.ccr.2023.215350

Wen J, Cai D, Gao W, He R, Li Y, Zhou Y, Klein T, Xiao L, Xiao Y (2023) Osteoimmunomodulatory nanoparticles for bone regeneration. Nanomaterials (Basel) 13(4):692. https://doi.org/10.3390/nano13040692

Won E, Akens MK, Hardisty MR, Burch S, Bisland SK, Yee AJM, Wilson BC, Whyne CM (2010) Effects of photodynamic therapy on the structural integrity of vertebral bone. Spine 35(3):272–277. https://doi.org/10.1097/BRS.0b013e3181af8360

Xie L, Cai L, Wang F, Zhang L, Wang Q, Guo X (2020) Systematic review of prognostic gene signature in gastric cancer patients. Front Bioeng Biotechnol 8:805. https://doi.org/10.3389/fbioe.2020.00805

Xu J, Yu L, Liu F, Wan L, Deng Z (2023) The effect of cytokines on osteoblasts and osteoclasts in bone remodeling in osteoporosis: a review. Front Immunol:14. https://doi.org/10.3389/fimmu.2023.1222129

Xue P, Xi H, Tan X, Chen H, Peng C, Sun G, Ye Y, Jiang X, Liu X, Du B (2024) Near-infrared responsive properties of bone repair scaffolds facilitated by specific osteoinductive photothermal converters for highly efficient bone repair. ACS Appl Mater Interfaces 16(29):37581–37595. https://doi.org/10.1021/acsami.4c06346

Yang N, Liu Y (2021) The role of the immune microenvironment in bone regeneration. Int J Med Sci 18(16): 3697–3707. https://doi.org/10.7150/ijms.61080

Yang J, Xie Y, Xia Z, Ji S, Yang X, Yue D, Liu Y, Yang R, Fan Y (2024) HucMSC-exo induced N2 polarization of neutrophils: implications for angiogenesis and tissue restoration in wound healing. Int J Nanomedicine 19:3555–3575. https://doi.org/10.2147/IJN.S458295

Yoon DS, Yoo JH, Kim YH, Paik S, Han CD, Lee JW (2010) The effects of COX-2 inhibitor during osteogenic differentiation of bone marrow-derived human mesenchymal stem cells. Stem Cells Dev 19(10): 1523–1533. https://doi.org/10.1089/scd.2009.0393

Yu Y, Yue Z, Xu M, Zhang M, Shen X, Ma Z, Li J, Xie X (2022) Macrophages play a key role in tissue repair and regeneration. PeerJ 10:e14053. https://doi.org/10.7717/peerj.14053

Salivary Proteome Role in Infection and Immunity

36

Mailon Cury Carneiro, Natara Dias Gomes da Silva, Talita Mendes Oliveira Ventura, Paulo Sérgio da Silva Santos, Marília Afonso Rabelo Buzalaf, and Marcelo Freire

Abstract

This chapter is a comprehensive overview of the role of the salivary proteome in infection and immunity. It emphasizes saliva's multifaceted nature, highlighting its composition, functions, and origin from major and minor salivary glands. The dynamic nature of the salivary proteome is explored, emphasizing its complexity and the challenges in its analysis due to factors like high dynamic range and proteolytic activity. We underscore saliva's significance in oral and systemic health, detailing its role in maintaining oral homeostasis, its connection to systemic diseases, and its potential as a diagnostic tool. Advancements in salivary proteomic technologies, particularly mass spectrometry, are discussed, along with their applications in research and clinical settings for diagnostics, biomarker discovery, and therapeutic monitoring. The review also delves into the salivary proteome's role in the immune system, focusing on its innate and adaptive components, antimicrobial peptides, immunoglobulins, and immune response biomarkers. It further explores the connection between the salivary proteome and various oral and systemic diseases, highlighting its potential for noninvasive diagnostics and personalized medicine, including the novel knowledge database salivaryproteome.org. Finally, by summarizing key findings and emphasizing the promising future of salivary proteomics in healthcare dentistry and medicine, while also acknowledging challenges such as standardization and validation. Overall, our review thoroughly explores the salivary proteome's significance in understanding health and disease, its potential for clinical applications, and the ongoing advancements in this field.

M. C. Carneiro
Department of Surgery, Stomatology, Pathology, and Radiology Bauru School of Dentistry, University of São Paulo (FOB-USP), Bauru, São Paulo, Brazil

Department of Genomic Medicine & Infectious Diseases, J. Craig Venter Institute (JCVI), La Jolla, CA, USA

N. D. G. da Silva
Department of Orofacial Sciences, University of California San Francisco (UCSF), San Francisco, CA, USA

T. M. O. Ventura · M. A. R. Buzalaf
Department of Biological Sciences, Biochemistry Discipline School of Dentistry of Bauru, University of São Paulo (FOB-USP), Bauru, São Paulo, Brazil

P. S. da Silva Santos
Department of Surgery, Stomatology, Pathology, and Radiology Bauru School of Dentistry, University of São Paulo (FOB-USP), Bauru, São Paulo, Brazil

M. Freire (✉)
Department of Genomic Medicine & Infectious Diseases, J. Craig Venter Institute (JCVI), La Jolla, CA, USA

Department of Infectious Diseases and Global Health, School of Medicine, University of California San Diego (UCSD), La Jolla, CA, USA
e-mail: mfreire@jcvi.org

N. Rezaei, H. Ziaei (eds.), *Oral Immunology*, Advances in Experimental Medicine and Biology 1492,
https://doi.org/10.1007/978-3-032-03176-1_36

Keywords

Saliva · Proteins · Proteomics · Infection · Immunity

36.1 Introduction: Overview of the Salivary Proteome

36.1.1 Composition and Functions of Saliva

Saliva biofluid is a multifunctional and highly dynamic biological fluid, composed predominantly of water (approximately 99%) and a solute fraction (1%), including electrolytes (e.g., sodium, potassium, calcium, chloride, magnesium, bicarbonate, and phosphate), proteins (e.g., enzymes, immunoglobulins, glycoproteins), glucose, nitrogenous products (e.g., urea and ammonia), as well as cells, microorganisms, and gingival fluid. Primarily produced by the salivary glands, saliva also incorporates contributions from extrinsic sources such as blood, gingival crevicular fluid, oral mucosa, immune cells, nasopharyngeal secretions, and oral microbiota (Dodds et al. 2005; Llena-Puy 2006; Schulz et al. 2013; Magalhães et al. 2017; Lau et al. 2021).

This fluid performs essential functions such as lubrication, antimicrobial protection, pH regulation, dental remineralization, hydration of oral tissues, initial digestion, and taste perception. Specifically, saliva facilitates the mechanical and chemical processing of food through moistening, dissolution of taste components, and bolus formation, in addition to lubricating for swallowing (Lendenmann et al. 2000; Buzalaf et al. 2012; Ruhl 2012; Lau et al. 2021).

Additionally, saliva forms an acquired pellicle rich in glycoproteins over the tooth surface, protecting teeth against erosion, abrasion, and chemical agents (Vukosavljevic et al. 2014). It also modulates oral microbiome homeostasis, balancing interactions with resident microorganisms and defending the host against invasive pathogens (Cross and Ruhl 2018; Lau et al. 2021). These protective functions establish saliva as a crucial biological barrier for maintaining oral health.

Salivary composition is influenced by multiple factors, such as gustatory stimuli, mastication, circadian rhythms, hydration status, and systemic conditions (Dodds et al. 2005; Llena-Puy 2006; Magalhães et al. 2017). Moreover, saliva reflects various biomolecules present in blood and urine but at lower concentrations, making it a promising tool for clinical diagnostics (Lau et al. 2021; Torralba et al. 2021) due to its noninvasive and cost-effective collection (Christodoulides et al. 2005; Bouwman et al. 2011; Schulz et al. 2013; Lin et al. 2019; Buzalaf et al. 2020).

36.1.2 Salivary Glands and Their Secretions

The salivary glands, fundamental for saliva production, are classified into major glands, including parotid, submandibular, and sublingual glands, and minor glands, which are distributed throughout the oral cavity. Each type contributes specifically to the composition and functions of saliva:

- *Parotid*: The largest salivary gland, primarily secreting serous saliva, rich in amylase and ions, which play a crucial role in the initial digestion of carbohydrates
- *Submandibular*: Produces seromucous saliva, a combination of aqueous and viscous secretions, contributing to oral lubrication and protection
- *Sublingual*: Predominantly secretes mucous saliva with small amounts of serous fluid, aiding in hydration and protection of the oral mucosa
- *Minor glands*: These are distributed in regions such as the palate and labial mucosa. They mainly produce mucous saliva rich in glycoproteins, which protects and lubricates specific areas of the oral cavity (Magalhães et al. 2017)

These glands possess a specialized structure composed of acini, responsible for the initial

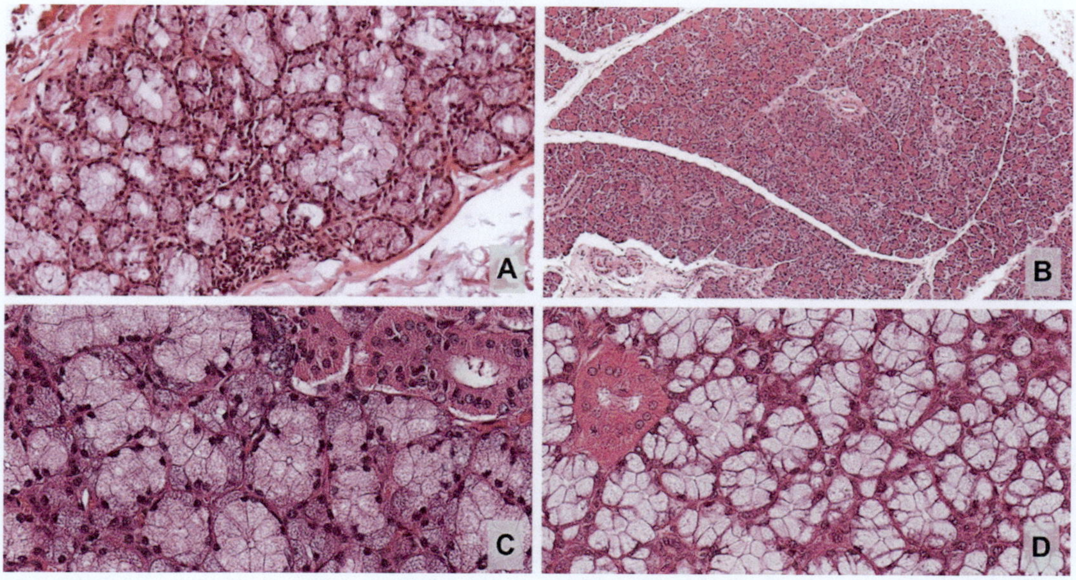

Fig. 36.1 Micrographs of human salivary glands (H&E staining). (**a**) *Minor salivary glands*, predominantly composed of mucous acini embedded in the connective tissue of the oral mucosa. (**b**) *Parotid gland*, mainly composed of serous acini with a well-defined lobular organization. (**c**) *Submandibular gland*, containing both serous and mucous acini. (**d**) *Sublingual gland*, characterized by the predominance of mucous acini. Courtesy of Prof. Dr. Denise Tostes Oliveira (Department of Pathology, Bauru School of Dentistry, USP) and Prof. Dr. Camila de Oliveira Rodini Pegoraro (Department of Histology, Bauru School of Dentistry, USP)

secretion of saliva, and a duct system that regulates its final composition (Fig. 36.1). Salivation is controlled by stimuli from the sympathetic and parasympathetic nervous systems, responding to chemical (e.g., taste) and mechanical (e.g., mastication) stimuli (Proctor and Carpenter 2007).

The molecular composition of saliva varies depending on the type of stimulus. Unstimulated saliva tends to be more viscous, whereas stimulated saliva, released during meals, has higher volume and lower viscosity, optimizing processes such as digestion and oral cleansing (Pedersen et al. 2005).

Once released into the oral cavity, whole saliva (passive drool) reflects a dynamic balance between the secretions of salivary glands and systemic components such as gingival fluid and bronchial secretions. This complex interaction sustains oral homeostasis and makes saliva a valuable tool for clinical diagnostics (Caporossi et al. 2010; Schulz et al. 2013).

36.1.3 The Dynamic Nature of the Salivary Proteome

The human salivary proteome is a complex collection of proteins derived from various sources, including glandular secretions, exfoliated epithelial cells, blood proteins from tissue injuries, dietary proteins from food, and products of microbial species present in the oral cavity. This collection, known as the metaproteome, is complemented by other biological components such as DNA, RNA, lipids, and metabolites delivered by exosomes from distal organs (Castagnola et al. 2012; Majem et al. 2015; Wei et al. 2020).

Saliva exhibits a wide dynamic range of molecules including proteins. A few dominant ones, such as amylase, albumin, and immunoglobulins, make it challenging to detect low-abundance proteins using mass spectrometry (Amado et al. 2013; Lin et al. 2019). Moreover, the high proteolytic activity in saliva can lead to rapid protein degradation at room temperature, posing a significant challenge in sample preparation for proteomic analysis (Esser et al. 2008; Thomadaki et al. 2011).

Salivary proteins undergo posttranslational modifications (PTMs), such as glycosylation, phosphorylation, and proteolytic cleavage, along with less common alterations like N-acetylhexosamine modification, which may be associated with cancer (Sondej et al. 2009; Vitorino et al. 2018). These complexities make the analysis of the salivary proteomic content a challenging and continuously evolving field.

Early studies using 2D gel electrophoresis and mass spectrometry identified fewer than 200 salivary proteins (Guo et al. 2006; Wei et al. 2020). More advanced methods, such as multidimensional fractionation and liquid chromatography coupled with mass spectrometry (LC-MS), have identified over 3000 proteins (Amado et al. 2013). These advancements are critical for improving proteomic coverage and identifying disease biomarkers at very low concentrations.

In addition, saliva contains immunoglobulins such as IgA, IgM, and IgG, which play crucial roles in oral immune defense. IgA, for instance, reduces bacterial adherence to gingival surfaces, while IgG, present in saliva via gingival fluid, inhibits bacterial colonization (Wei et al. 2020). Proteins like mucins, cystatin C, and mucin-7 also perform significant antimicrobial functions by binding to oral pathogens and inhibiting the growth of microorganisms such as *P. gingivalis* (Blankenvoorde et al. 1998; Lis et al. 2010).

Salivary enzymes, such as lactoferrin, lysozyme, peroxidase, and amylase, also protect against microbial invasions. Lysozyme, for example, was the first antibacterial protein identified in saliva, acting by degrading bacterial cell walls (Fábián et al. 2012).

With the evolution of proteomic technologies such as Orbitrap-based mass spectrometry and shotgun approaches, over 5000 salivary proteins have been identified in a single experiment, revealing the unexpected complexity of the salivary proteome (Grassl et al. 2016; Lau et al. 2021). Technical improvements, such as dual digestion, ultralong liquid chromatography separation, and optimized sample processing, have significantly enhanced the sensitivity and specificity of analyses (Thakur et al. 2011; Glatter et al. 2012; Kulak et al. 2014; Wei et al. 2020).

Beyond its immunological and antimicrobial functions, saliva reflects biomolecules present in blood plasma, making it a promising tool for noninvasive diagnostics of systemic diseases. However, challenges remain, including the high variability of salivary proteins and the influence of factors like salivary flow, gingival inflammation, and enzymatic degradation, which must still be addressed (Pfaffe et al. 2011; Helmerhorst et al. 2018; Lau et al. 2021).

A landmark in salivary proteome research was the 2008 multicenter study, which cataloged over 1000 proteins from glandular secretions and whole saliva. Since then, technological advances have enabled more robust datasets, providing detailed quantitative information on proteins essential for structural and antimicrobial functions (Denny et al. 2008; Siqueira et al. 2008; Bandhakavi et al. 2009; Jehmlich et al. 2013; Schulz et al. 2013).

Finally, saliva is highly adaptive, presenting compositional and functional variations influenced by salivary flow and received stimuli. Stimulated saliva, for instance, is rich in bicarbonate and sodium, while unstimulated saliva contains higher levels of mucins and ions such as calcium and phosphate. These differences impact buffering capacity, mineral balance, and dental biofilm formation (Dodds et al. 2005; Llena-Puy 2006; Magalhães et al. 2017).

36.2 The Role of Saliva in Oral and Systemic Health

36.2.1 Maintaining Oral Homeostasis

Saliva production levels and composition regulate oral homeostasis. By contributing to the chemical, physical, and biological balance of the oral cavity, saliva can impact the entire oral health.

One of its most critical functions is pH regulation, mediated by salivary buffering systems, particularly bicarbonate. These systems neutralize acids produced by bacterial metabolism, preventing enamel demineralization and reducing the risk of dental caries (Buzalaf et al. 2012; Magalhães et al. 2017).

In addition, the proteins in saliva are indispensable for mineral balance and enamel metabolism. They promote the remineralization of early enamel lesions by providing essential ions such as calcium, phosphate, and fluoride, which are crucial for the structural recovery of the tooth. Saliva also contains bicarbonate, which neutralizes acids, as well as proteins like statherin and trace elements such as magnesium that stabilize minerals and support remineralization. The acquired pellicle, formed on the tooth surface, acts as a protective barrier between the biofilm and the oral environment, preventing further damage (Buzalaf et al. 2012; Magalhães et al. 2017).

Other essential functions include the initial digestion of food, facilitated by the action of enzymes such as amylase and lipase, and immune defense. Immunoglobulins, particularly IgA, and antimicrobial proteins such as lysozyme, lactoferrin, and histatin play fundamental roles in protecting against pathogenic microorganisms (Dodds et al. 2005; Buzalaf et al. 2012; Magalhães et al. 2017).

36.2.2 The Oral-Systemic Connection

The interdependence between oral health and systemic health is widely recognized, and saliva plays a central role in this connection (Table 36.1). Examples include:

- *Systemic diseases*: Conditions such as Sjögren's syndrome, diabetes, cystic fibrosis, and Alzheimer's disease often cause xerostomia or hyposalivation, compromising saliva's protective capabilities. This reduction increases the risk of caries, oral infections, and taste alterations, highlighting the importance of saliva in systemic health (Magalhães et al. 2017).
- *Diagnostic monitoring*: Saliva offers a promising medium for noninvasive diagnostics, being utilized in pharmacological, endocrinological, and immunological studies. Salivary components, such as proteins and ions, can be analyzed to assess metabolic, hormonal, and infectious conditions, reinforcing their clinical potential (Spielmann and Wong; Hardy et al. 2012).
- *Protective effect against infections*: Saliva regulates the dental biofilm and inhibits the adhesion of pathogenic microorganisms to oral tissues. Proteins such as mucins and peroxidases play essential roles in neutralizing

Table 36.1 Connections between salivary function and systemic diseases

Disease	Connection	References
Sjögren's syndrome	Causes xerostomia or hyposalivation, compromising saliva's protective capabilities, increasing oral risks	Nadig et al. (2017), Turner (2016) and Magalhães et al. (2017)
Diabetes	Causes xerostomia or hyposalivation, leading to higher risks of caries, infections, and taste alterations	Turner (2016) and Magalhães et al. (2017)
Cystic fibrosis	Causes xerostomia or hyposalivation, impairing saliva's role in systemic and oral health	Magalhães et al. (2017)
Alzheimer's disease	Associated with xerostomia or hyposalivation, highlighting saliva's systemic health importance	Shi et al. (2011) and George et al. (2023)
Metabolic conditions	Assessed via salivary proteins and ions for diagnostic purposes	Spielmann and Wong (2011)
Hormonal conditions	Evaluated noninvasively using saliva as a diagnostic medium	Hardy et al. (2012)
Infectious conditions	Saliva is analyzed for components aiding in infection diagnostics	Spielmann and Wong (2011)
Pathogenic infections	Salivary proteins (e.g., mucins, peroxidases) inhibit the adhesion of pathogens and maintain balance	Contini et al. (2024)

infectious agents and maintaining microbiological balance (Magalhães et al. 2017).

Salivary proteins are dynamic and multifunctional, which in turn play essential roles for both oral and systemic health. Its unique composition enables a wide range of functions, from antimicrobial defense to the regulation of mineral balance and the oral microbiome. The in-depth study of the salivary proteome not only expands our understanding of its clinical importance but also highlights its diagnostic potential, emphasizing its relevance in preventive medicine and the treatment of systemic conditions related to oral health.

36.3 Salivary Proteomics: A Powerful Tool for Understanding Health and Disease

Saliva has emerged as a valuable tool in biomedical research due to its rich and dynamic composition, which reflects both oral and systemic health. Since the development of the open-source database at NIH (Lau et al. 2021), The Human Salivary Proteome database www.salivaryproteome.org was developed as an open-source independent database at the J. Craig Venter Institute through Freire's Laboratory and other collaborators at the ADA Forsyth Institute and University at Buffalo. This knowledge base aims to advance the field of salivary proteomics, dedicated to studying the proteins present in this fluid, which is rapidly growing, driven by technological advancements and the need for noninvasive diagnostic methods.

The analysis of the salivary proteome provides detailed insights into essential biological processes, enabling the identification of biomarkers associated with various health conditions and diseases. These advancements are particularly significant in personalized medicine, where saliva stands out as a practical and accessible alternative for disease monitoring, therapeutic response evaluation, and early screening of systemic and infectious conditions.

Recent progress in salivary proteome analysis highlights its clinical and research applications and its relevance in developing more effective diagnostic and preventive strategies. There are challenges to overcome, including setting parameters for analyses that are comparable across studies. Developing protocols that account for both pellet and fluid components and clear specifications for labeling and pretreatment processes, is essential. In addition to proteomics, many standardizations are also essential for bioinformatic pipelines to ensure consistency and reproducibility.

36.3.1 Advancements in Salivary Proteomic Technologies

The study of the salivary proteome has advanced significantly thanks to technological innovations, particularly in the field of Mass Spectrometry (MS). These technologies enable the identification and quantification of proteins at extremely low levels, ranging from nanograms to picograms per milliliter, overcoming the molecular concentration limitations that distinguish saliva from blood (Christodoulides et al. 2005; Schulz et al. 2013). MS is widely used to analyze proteins both in their intact and digested forms, revealing PTMs that are crucial for understanding their biological functions. Techniques such as liquid chromatography coupled with MS (LC-MS) and "top-down" and "bottom-up" approaches have transformed salivary proteome mapping (Jasim et al. 2016; Ventura et al. 2018).

36.3.1.1 Top-Down and Bottom-Up Approaches

Bottom-up proteomics (or shotgun proteomics) relies on the proteolytic digestion of proteins into peptides, which are then analyzed by mass spectrometry. This technique is widely used due to its ability to identify a large number of proteins, regardless of size. However, it may have limitations in fully covering protein sequences and detecting specific PTMs (Manconi et al. 2017; Ventura et al. 2018).

In contrast, the top-down approach analyzes intact proteins, avoiding alterations induced by digestion. This allows for a more precise characterization of PTMs, such as natural cleavages and alternative splicing. However, it is limited by technical challenges, including the need for prior sample separation and difficulties in analyzing proteins larger than 50 kDa (Manconi et al. 2017; Ventura et al. 2018).

36.3.2 Enrichment Techniques and Limitations

A major challenge in salivary proteomic analysis is the presence of highly abundant proteins, such as albumin and immunoglobulin G (IgG), which can mask low-abundance proteins. Strategies like affinity chromatography and depletion columns help mitigate this issue, enhancing the detection of low-abundance proteins (Krief et al. 2011, 2012; Ventura et al. 2018). Furthermore, the use of hexapeptide libraries and specific matrices for glycopeptides and phosphopeptides has revealed critical modifications associated with health and disease states (Zielinska et al. 2010; Bandhakavi et al. 2011; Schulz et al. 2013).

Despite these advancements, biological and methodological variability remains a challenge. Factors such as diet, circadian rhythms, and storage conditions can significantly impact results, necessitating rigorous protocols for sample collection and processing (Bardow et al. 2001; Messana et al. 2008).

36.3.3 Applications of Salivary Proteomics in Research and Clinical Settings

Recent studies have demonstrated the potential of salivary proteomics by identifying over 200 phosphorylation sites in salivary proteins, providing insights into their functional regulation and clinical relevance. These findings highlight saliva as a promising tool for detecting biomarkers associated with various health conditions (Stone et al. 2011; Schulz et al. 2013).

Salivary proteomics offers unique opportunities to understand biological processes and develop clinical applications, particularly in diagnostics, prognostics, and health condition monitoring.

36.3.3.1 Diagnostics and Biomarkers

Saliva contains biomolecules that reflect both systemic and local conditions, such as inflammation, infections, and autoimmune diseases. Examples include:

- *Oral and other cancers*: Biomarkers like c-erbB-2/Her2/neu have been investigated for the early detection of cancers, including breast and pancreatic cancers (Streckfus et al. 2000; Schulz et al. 2013). Furthermore, microorganisms such as *Fusobacterium nucleatum* and fibronectin-binding proteins from *Streptococcus gordonii* have been associated with oral squamous cell carcinoma, indicating their potential as markers for the early detection of this condition (Torralba et al. 2021).
- *Cardiovascular diseases*: C-reactive protein (CRP) in saliva has shown a strong correlation with its levels in the blood, making it a promising tool for tracking cardiac events (Punyadeera et al. 2011; Schulz et al. 2013).

36.3.3.2 Therapeutic Monitoring

The noninvasive collection of saliva enables frequent sampling, making it ideal for monitoring treatment responses in conditions such as diabetes and autoimmune diseases. Because we produce high amounts of saliva, it is feasible for longitudinal collection for clinical trials and others. Dynamic biomarkers provide real-time information about changes in health status, facilitating quicker and more effective therapeutic adjustments (Rao et al. 2009; Schulz et al. 2013).

36.3.3.3 Challenges and Opportunities

Despite progress, barriers such as the lack of standardized protocols, interindividual variations, and the need for validation in large clinical studies still limit the practical application of salivary proteomics. In addition, methods need to be optimized and standardized to make multistudy comparisons more reliable. However, technological advancements continue to expand their possibilities, offering greater precision in disease detection and personalized treatments (Manconi et al. 2017; Ventura et al. 2018).

36.3.3.4 Social and Economic Impact

Saliva represents an accessible and cost-effective diagnostic tool with the potential to greatly benefit low-income populations. Its noninvasive and straightforward collection process enables broader access to high-quality diagnostics, reducing reliance on specialized personnel and expensive equipment. This democratization of healthcare facilitates early detection of diseases and public health interventions, particularly in underserved regions. Additionally, the low risk of infectious disease transmission during collection and its cost-effectiveness makes saliva ideal for large-scale applications in resource-limited settings, addressing critical healthcare disparities (Schulz et al. 2013).

Figure 36.2 demonstrates the complete workflow of salivary proteomics, highlighting each step from sample collection to the identification of biomarkers and their application in disease diagnosis and monitoring.

36.4 Salivary Proteome and the Immune System

Saliva contains a variety of immune cells, including epithelial cells, neutrophils, monocytes, dendritic cells, and lymphocytes, which collectively contribute to robust innate and adaptive immune responses (Fábián et al. 2012). The innate immune system provides the first line of defense against pathogens, and it is a rapid, nonspecific response. The adaptive immune system is activated by the innate immune system and provides a slower, targeted response that develops immunological memory (Shang et al. 2023). The interplay between these two systems is essential for maintaining oral health.

36.4.1 Antimicrobial Peptides and Proteins in Saliva

Saliva houses a variety of antimicrobial peptides and proteins, such as lysozyme, lactoferrin, histatins, and defensins, which provide the first line of defense against invading pathogens. These substances exhibit various mechanisms of action, including disruption of bacterial cell walls, sequestration of essential nutrients, and modulation of immune responses (National Institute of Allergy and Infectious Diseases, n.d.).

Lysozyme is an enzyme that breaks down bacterial cell walls, leading to cell lysis and death. Lactoferrin is a protein that sequesters iron, an essential nutrient for bacterial growth, thereby inhibiting microbial proliferation. Histatins exhibit antifungal activity, preventing the overgrowth of *Candida albicans*, helping to maintain oral microbiome balance. Defensins are peptides that function by disrupting microbial cell membranes, leading to the elimination of pathogens, while also playing a role in immune response regulation.

36.4.2 Salivary Immunoglobulins and Their Functions

Secretory IgA (sIgA) is the predominant immunoglobulin in saliva, playing a crucial role in neutralizing toxins, agglutinating bacteria, and preventing their adherence to mucosal surfaces. SIgA functions primarily through immune exclusion, preventing pathogens from interacting with host tissues, but can also activate immune cells through specific FcαRI receptors, demonstrating its involvement in both innate and adaptive immunity (Alqedari et al. 2023).

36.4.3 Salivary Biomarkers of Immune Response

Saliva contains various biomarkers that reflect the state of the immune system, including cytokines, chemokines, and acute-phase proteins. These

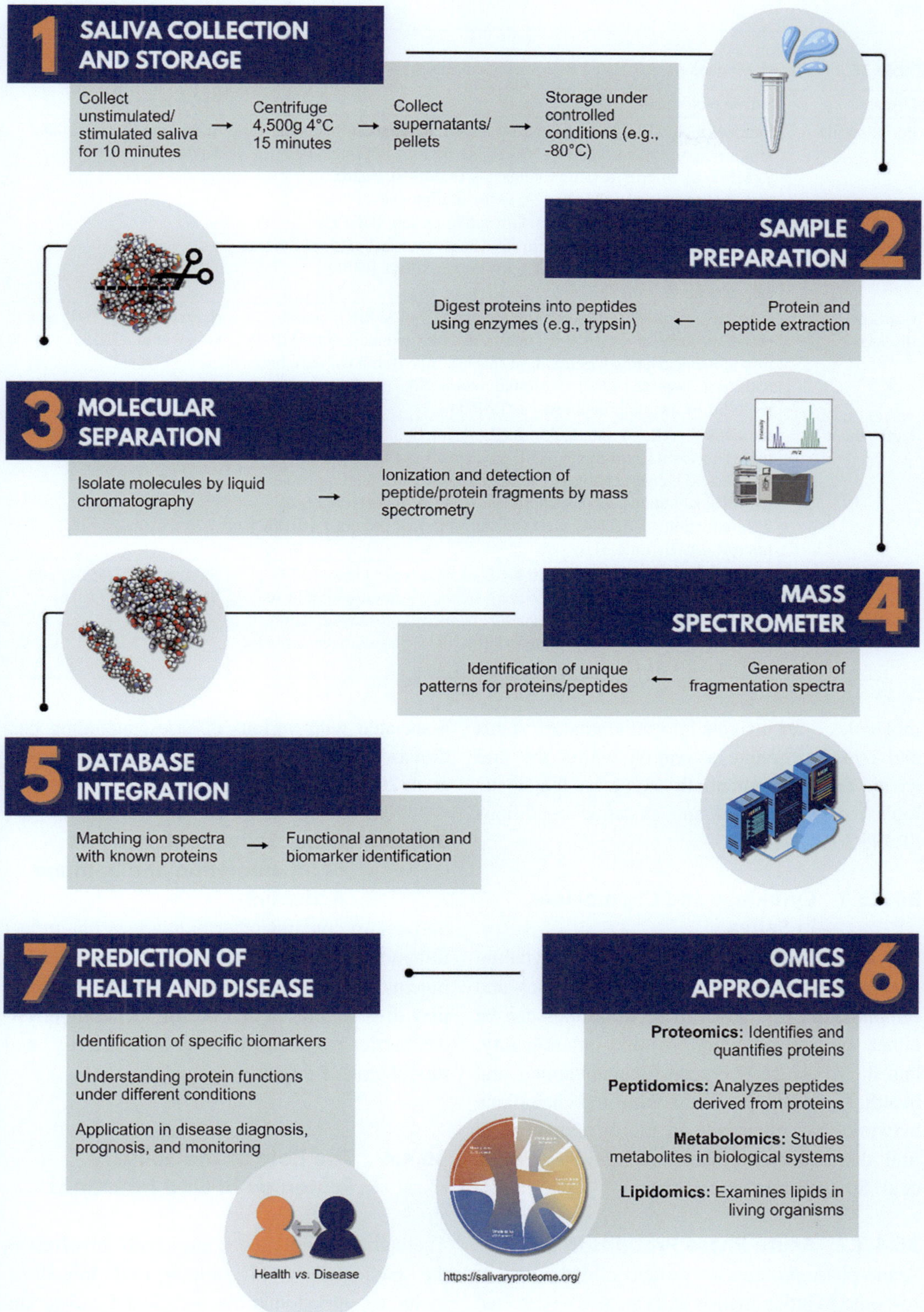

Fig. 36.2 The flowchart demonstrates the steps involved in the salivary proteomics analysis process. Collected saliva undergoes preparation procedures, such as centrifugation and enzymatic digestion, followed by molecular separation and analysis through mass spectrometry. The resulting data are integrated into databases for functional biomarker identification, enabling the application of omics approaches (proteomics, peptidomics, metabolomics, and lipidomics). These methods are used to identify and quantify specific biomarkers, supporting the diagnosis, prognosis, and monitoring of diseases

Table 36.2 Salivary biomarkers associated with various diseases

Disease	Salivary biomarkers	References
Periodontitis	Interleukin-1β (IL-1β), interleukin-6 (IL-6), interleukin-8 (IL-8), tumor necrosis factor-alpha (TNF-α), matrix metalloproteinase-8 (MMP-8), matrix metalloproteinase-9 (MMP-9), matrix metalloproteinase-13 (MMP-13), tissue inhibitor of metalloproteinases-1 (TIMP-1), C-reactive protein (CRP), β-glucuronidase, osteoprotegerin (OPG), receptor activator of nuclear factor-kappa B ligand (RANKL), myeloperoxidase (MPO), calprotectin (S100A8/S100A9)	Starmer et al. (1991) and Taylor (2014)
Cardiovascular diseases	C-reactive protein (CRP), creatine kinase-MB (CK-MB), matrix metalloproteinase-8 (MMP-8), matrix metalloproteinase-9 (MMP-9), troponin I (TnI), myeloperoxidase (MPO), myoglobin (MYO), brain natriuretic peptide (BNP), exosomal microRNAs (miRNAs), tissue inhibitor of metalloproteinase-1 (TIMP-1)	Starmer et al. (1991) and Gohel et al. (2018)
Oral cancer	Interleukin-6 (IL-6), interleukin-8 (IL-8), interleukin-1β (IL-1β), tumor necrosis factor-alpha (TNF-α), cytokeratin fragment 21-1 (Cyfra 21-1), cancer antigen 125 (CA 125), Mac-2 binding protein (M2BP), squamous cell carcinoma antigen (SCC-Ag), cluster of differentiation 44 (CD44), S100 calcium-binding protein P (S100P), microRNA-31 (miR-31)	Khurshid et al. (2018)
Lupus	Interleukin-1 (IL-1), interleukin-6 (IL-6), interferon-gamma (IFN-γ), tumor necrosis factor-alpha (TNF-α), salivary immunoglobulin A (salivary IgA), *Treponema denticola*, *Tannerella forsythia*, leptin, monocyte chemoattractant protein-1 (MCP-1), plasminogen activator inhibitor-1 (PAI-1)	Starmer et al. (1991) and Stanescu et al. (2018)

molecules offer insights into inflammatory status and overall immune activation within the oral cavity and can potentially serve as diagnostic tools for various oral and systemic conditions (Table 36.2).

36.4.3.1 Cytokines and Chemokines in Saliva

Cytokines and chemokines are signaling molecules that regulate immune responses and inflammation. In saliva, these molecules can be either pro-inflammatory or anti-inflammatory, and their balance is crucial for maintaining oral health. Dysregulation of cytokine and chemokine networks can contribute to the development of oral diseases and systemic conditions (Pessoa et al. 2019; Perera et al. 2020).

36.4.3.2 Acute-Phase Proteins in Saliva

Acute-phase proteins are produced by the liver in response to inflammation and can also be detected in saliva. These proteins, such as CRP and lactoferrin, can serve as noninvasive biomarkers of systemic inflammation and have been associated with various diseases, including cardiovascular diseases and periodontitis (Kleinstein et al. 2020).

36.4.3.3 Other Salivary Biomarkers of Inflammation and Immune Activation

Saliva also contains other biomarkers that reflect inflammation and immune activation, such as mucins, defensins, and histatins. These molecules play diverse roles in oral health, including physical barrier function, antimicrobial activity, and modulation of immune responses.

36.4.4 The Role of the Salivary Proteome in Oral Infections

The salivary proteome is intricately involved in the pathogenesis of various oral infections, including periodontal disease, dental caries, and oral candidiasis (Huang et al. 2021). Examples of biomarkers found in saliva have been outlined in Table 36.2.

36.4.4.1 Periodontal Disease

Periodontitis is a chronic inflammatory disease of the gingival tissue and supporting structures of the teeth that is strongly influenced by the oral microbiome and the host immune response (Kinane and Mark Bartold 2007). Saliva plays a critical role in modulating the oral microbiome and regulating inflammation in the periodontal tissues (Dawes and Wong 2019). Classically, the pathogenesis of periodontal disease is described in four stages after initiation of plaque accumulation: initial (2–4 days after plaque formation), early (4–7 days after plaque formation), established (12–21 days after plaque formation), and advanced lesion (Page and Schroeder 1976). The initial lesion is characterized by the recruitment and emigration of neutrophils through junctional epithelium (JE), combined with a localized inflammatory infiltrate immediately beneath the JE, occupying 5%–10% of connective tissue. This inflammatory infiltrate is mainly composed of monocytes/macrophages, lymphocytes, and neutrophils (Gemmell and Seymour 2004). This lesion is subclinical and can be observed only at the histopathological level. The only discrete clinical signs are the slight increase in gingival crevicular fluid (GCF) flow rate. *Streptococcus* species are the main bacterial components at this stage, and they are known to secrete several enzymes that increase sulcular epithelium (SE) and JE permeability. In addition, many pro-inflammatory substances, such as lipopolysaccharide (LPS), interleukin-8 (IL-8), tumor necrosis factor-alpha (TNF-α), and *Streptococcus* M protein, induce neutrophil extracellular trap (NET) formation (Brinkmann et al. 2004; Kolenbrander et al. 2006). Other cell types can also secrete extracellular traps, especially mast cells. These cells release mast cell extracellular traps, which limit the entrance of bacteria and their vesicles into the tissues, also contributing to localized tissue destruction. Polymorphonuclear neutrophils (PMNs) also secrete other cytokines, such as interleukin-1 (IL-1) and its antagonist receptor (IL-1RA) and interleukin-17(IL-17), which induces the secretion of IL-8 by stratified epithelial cells. IL-8 is a strong chemoattractant for PMNs and stimulates NET formation. IL-17 has a protective role in periodontal disease, maintaining the PMN barrier in the gingival sulcus.

The early lesion is characterized by increased neutrophil emigration and initial lateral proliferation of JE cells, occurring approximately 4–7 days after the initial plaque accumulation. A larger inflammatory infiltrate is observed at the gingival connective tissue, along with vascular proliferation, dilated vessels, and an increase in vessel permeability and collagen loss. Therefore, the GCF flow rate dramatically increases. Similarly, there is an increase in the permeability of SE and JE, allowing the entrance of bacterial products into gingival tissues. At this stage, the inflammatory infiltrate shifts from being predominantly composed of PMNs to lymphocytes and macrophages. Around 12–21 days after initial plaque accumulation, the lesion becomes clinically evident. By day 21, lymphocytes constitute around 70% of the inflammatory infiltrate, and a fourfold increase in the number of PMNs within JE occurs (Gemmell and Seymour 2004; Lang and Bartold 2018; Murakami et al. 2018).

In cases in which gingivitis progresses to periodontitis, the gingival inflammation can exert pressure for the development of a dysbiotic and inflammatory microbiota, including members that can subvert or evade the immune system. The severity of periodontal lesions may depend on host-related parameters that influence the host's immune, inflammatory, and regenerative response (Hajishengallis 2014). The advanced lesion corresponds to the advanced stage, in which the inflammatory infiltrate is mostly composed of plasma cells occupying more than 50% of the gingival connective tissue volume. An apical proliferation and migration of the JE along with detachment and ulceration of the pocket epithelium are observed (Nanci and Bosshardt 2006). In the presence of generalized, moderate, periodontal pockets, the ulcerated epithelium is estimated to correspond to the size of the palm (approximately 50–75 cm^2), constituting an important risk factor for the development of systemic diseases (Tonetti and Van Dyke 2013). The first line of defense of the gingival sulcus is the emigration of neutrophils through the JE, emerging via the sulcus into the oral cavity. The integrity of the inner environment is achieved by the cell attachment to enamel via hemidesmosomes and basal lamina, cell-cell junctions in the JE with desmosomes; and cell-connective tissue matrix attachment by basement membrane. The attachment of JE cells to the enamel surface and the underlying connective tissue is stronger than the cell-cell attachment, while the few desmosomes and wide intercellular spaces between JE cells allow the GCF and inflammatory and immune cells to transmigrate, protecting the body against microbial invasion.

During periodontitis, there is a conversion of JE to pocket epithelium. Detachment of JE from the tooth surface seems to occur after the increasing degree of inflammation and transmigration of neutrophils, eventually leading to focal disintegration of JE (Bosshardt 2018). Bacteria can also contribute to the detachment of JE and the formation of the pocket epithelium. *P. gingivalis* gingipains are capable of degrading components of cell-cell junctional complexes and cleaving intercellular adhesion molecule-1 (ICAM1, also known as CD54) on oral epithelial cells. This process disrupts the interaction of neutrophils and epithelial cells. Apically, the pocket epithelium is continuous with a reduced JE, which proliferates apically as the pocket deepens, maintaining the integrity of the gingival sulcus.

Dental plaque represents a continuous challenge to the innate immune system. In healthy sites, this challenge is beneficial, while in diseased sites there is an imbalance between host defenses and bacterial challenges. Due to the specific characteristics of JE, a highly orchestrated expression of mediators is released in the periodontal tissues, including e-selectin, ICAMs, and IL-8, which facilitates the transit of neutrophils from the gingival connective tissue to the gingival crevice (Tonetti et al. 1998; Darveau 2010). Other innate immune mediators expressed in periodontal tissues are ⍰-defensins (⍰D) 1, ⍰D2, ⍰D3, CD14, and lipopolysaccharide-binding protein (LBP). LBP, produced by gingival epithelial cells and the liver as an acute-phase response marker, and LBP mRNA are more highly expressed in healthy sites than diseased sites. As previously described, gingival tissues express TLRs, especially TLR2 and TLR4 in the development of periodontal diseases. Il-1⍰ was demonstrated in periodontal tissues of germ-free mice.

Many cytokines associated with inflammation are found at higher levels in the GCF of periodontitis patients compared to healthy individuals, including IL-1⍰, TNF, and PGE_2. In addition, soluble and membrane-bound CD14 and LBP are also present in higher concentrations in GCF from diseased than sites than healthy sites.

36.4.4.2 Dental Caries

Dental caries is one of the most prevalent chronic diseases worldwide, affecting individuals of all ages and different socio-economic backgrounds, with profound implications for oral health, quality of life, and overall well-being (Selwitz et al. 2007). It arises primarily from dental enamel demineralization due to acids generated during bacterial metabolism of dietary sugars, eventually leading to the breakdown of enamel (Selwitz et al. 2007; Bowen et al. 2018). Excessive sugar consumption fosters the proliferation of glycolytic microorganisms that thrive in anaerobic conditions, producing organic acids and extracellular polysaccharides. These polysaccharides enhance bacterial adhesion and create an acidic microenvironment (Buzalaf et al. 2011, 2015), disrupting biofilm homeostasis. Consequently, acidogenic and aciduric bacteria dominate, initiating demineralization of the hard dental tissues and increasing the risk of caries (Bowen 2002; Marsh et al. 2011; Takahashi 2015; Pitts et al. 2017; Klein et al. 2015). The acids lower the biofilm pH below the critical threshold for hydroxyapatite dissolution (pH 5.5) but not for fluorapatite (that remains stable above pH 4.5), leading to subsurface mineral loss that clinically manifests as white spot lesions. Early detection of these lesions enables noninvasive interventions to halt progression and avoid the need for restorative procedures (Takahashi and Nyvad 2011; Ten Cate and Buzalaf 2019).

The development of dental caries is a multifactorial process influenced by numerous factors, including diet, oral hygiene, microbial composition of the dental biofilm, and the salivary environment (Machiulskiene et al. 2020). Among these, saliva plays a pivotal role due to its protective functions (Hannig and Balz 2001; Lussi et al. 2004; Van Nieuw Amerongen et al. 2004; Hara et al. 2006). It facilitates the mechanical removal of food particles and bacterial load, buffers acids to neutralize pH, and supports enamel remineralization by supplying ions such as calcium, phosphate, and fluoride (Deng et al. 2015; Turner 2016). Additionally, saliva is rich in antimicrobial peptides, enzymes, and other proteins that contribute to oral health and form the acquired enamel pellicle (AEP) (Buzalaf et al. 2012; Vukosavljevic et al. 2014; Lin 2017). The AEP, a bacteria-free organic layer formed through the selective adsorption of salivary proteins onto the enamel, serves as a protective barrier against acids and regulates microbial colonization (Buzalaf et al. 2012; Lin 2017). Early colonizers, such as *Streptococcus salivarius, Streptococcus mitis, Streptococcus gordonii, Streptococcus sanguinis* and *Actinomyces* spp., initially attach to the pellicle through weak interactions, which are subsequently strengthened by adhesins-mediated binding. These early colonizers facilitate the adhesion of cariogenic bacteria like *Streptococcus mutans*, which are instrumental in the development of a cariogenic biofilm (Bowen et al. 2018). Disruptions in salivary flow or composition can impair these functions, altering the AEP's structure and composition. Such changes can influence microbial colonization and increase the susceptibility to caries, highlighting the critical interplay between saliva, AEP integrity, and oral health.

While dietary sugar control, fluoride application, and mechanical biofilm management remain the cornerstone of dental caries prevention (Buzalaf et al. 2018; Zhan 2018; Marshall 2019; Philip et al. 2019; Poorni et al. 2019; van Loveren 2019; Whelton et al. 2019), these measures have limitations, particularly in addressing individual variability in caries susceptibility. Recent research into the salivary proteome offers a promising avenue for overcoming these challenges by providing deeper insights into the molecular mechanisms underlying caries development (Nascimento et al. 2017; Zhang et al. 2024). Studies have observed differential expression of salivary proteins between caries-susceptible and caries-free individuals, suggesting their potential as biomarkers for assessing caries risk (Guedes et al. 2020; da Silveira et al. 2023). However, these findings are largely based on small populations, necessitating validation in larger, more diverse cohorts to establish their reliability. Currently, no specific salivary proteins have been conclusively identified as predictors of future caries development (Wang et al. 2019). Beyond advancing our understanding of caries biology, salivary proteomics holds significant promise for early detection and the creation of personalized preventive strategies. By identifying and validating robust biomarkers by protein-based assays, such as ELISA, immunoblot, and proteomics, clinicians could assess an individual's caries susceptibility with greater precision and tailor interventions accordingly. This approach has the potential to transform caries management, shifting it toward more predictive and personalized care. The role of microbiome and genetics was dissected in longitudinal twin studies (Freire et al. 2020), and microbes, which were discovered as influencing cavity risk, need to be further validated with proteomic analysis to advance their translational impact.

In this context, AEP engineering has emerged as a promising strategy to enhance the protective functions of the pellicle against dental carries (Araujo et al. 2021; da Silva et al. 2022; Araujo et al. 2024a, b). This innovative approach focuses on enriching the basal layer of the AEP with acid-resistant proteins, such as cystatins or statherin, to strengthen its resistance to demineralization and selectively modulate bacterial colonization. By promoting the growth of beneficial microbial species, AEP engineering may complement traditional caries prevention strategies, offering a biologically informed approach to oral health management (Marsh 2018). Furthermore, the integration of salivary proteomics with other "omics" technologies, including genomics and metabolomics, has the potential to revolutionize our understanding of caries dynamics (Moussa et al. 2022; Zhang et al. 2024). This multidisciplinary approach could uncover novel biomarkers and identify new molecular pathways for intervention, paving the way for more precise and effective caries prevention strategies. Together, these innovations could shift the paradigm of caries management, emphasizing prevention and personalization to improve long-term oral health outcomes.

36.4.4.3 Oral Candidiasis

The oral cavity is a dynamic and complex ecosystem that harbors a wide range of microorganisms, including bacteria, viruses, and fungi. Within this microenvironment, the microbiota plays a crucial role in maintaining oral health. However, external factors and systemic conditions can destabilize this balance, enabling the transition of commensal microorganisms to pathogenic states. A notable example is *Candida albicans*, a dimorphic fungus that typically exists as a commensal but can cause oral candidiasis when its growth becomes uncontrolled (Salvatori et al. 2016).

The transition of *C. albicans* from a commensal to a pathogenic state predominantly occurs in immunocompromised individuals, such as patients with AIDS, those undergoing chemotherapy, or individuals using immunosuppressive medications. Additionally, this shift can be triggered in situations of dysbiosis, salivary alterations, or the use of poorly fitted dental prostheses (Salvatori et al. 2016; Kawanishi et al. 2021). These factors create an environment conducive to the proliferation and pathogenicity of *C. albicans*, particularly due to their ability to switch between unicellular and filamentous forms—a characteristic that confers high adaptability and virulence (Muadcheingka and Tantivitayakul 2015).

Among the species of the *Candida* genus, in addition to *C. albicans*, others such as *C. glabrata*, *C. parapsilosis*, *C. tropicalis*, *C. krusei*, and *C. dubliniensis* are also associated with human infections. These species possess varying survival mechanisms, but *C. albicans* stand out due to their ability to form complex biofilms and efficiently invade tissues (Salvatori et al. 2016). Biofilms are structured microbial communities that exhibit increased resistance to antimicrobial agents, complicating treatment efforts.

Saliva plays a fundamental role in maintaining oral homeostasis. It contains antimicrobial enzymes, such as histatins, lactoferrin, and defensins, as well as antibodies like sIgA, which actively protect against pathogens. However, conditions such as hyposalivation compromise these natural defenses, facilitating colonization by *C. albicans* (Radfar et al. 2003; Salvatori et al. 2016; Nadig et al. 2017).

Patients with hyposalivation often exhibit alterations in salivary protein levels, such as mucins and sIgA, which aid in the mechanical removal of microorganisms (Radfar et al. 2003; Salvatori et al. 2016). This deficiency permits the unchecked growth of *C. albicans*, while reductions in key components such as histatins and calprotectin directly compromise antifungal defenses (Linden et al. 2008; Puri and Edgerton 2014). Systemic conditions like Sjögren's syndrome exacerbate this issue by significantly reducing salivary flow and promoting biofilm development and persistent infections.

Poorly fitted dental prostheses create environments conducive to biofilm formation, harboring *C. albicans* and other oral microorganisms, such as *Streptococcus oralis* and *Streptococcus gordonii*. These bacteria not only coexist within biofilms but also enhance the pathogenicity of *C. albicans* by stimulating hyphal formation, thereby facilitating tissue invasion (Bamford et al. 2009; Vílchez et al. 2010; Salvatori et al. 2016).

Conversely, beneficial bacteria such as *Lactobacillus acidophilus* exhibit antagonistic effects, inhibiting the growth of *C. albicans* and promoting a healthier oral environment (Elahi et al. 2005; Salvatori et al. 2016). Maintaining this balance between pathogenic and beneficial microorganisms is critical for preventing opportunistic infections. Saliva acts as a primary barrier against infections by promoting the mechanical removal of pathogens and directly combating them through its antimicrobial components. For instance, histatins induce osmotic stress in

C. albicans, leading to the disruption of its cell membrane (Puri and Edgerton 2014; Salvatori et al. 2016). However, biofilm formation reduces the effectiveness of these defenses, particularly on surfaces like dental prostheses (Kawanishi et al. 2021).

Oral candidiasis is a multifactorial disease requiring integrated approaches for its prevention and treatment. Effective measures include:

- *Maintaining salivary flow*: Achieved through hydration, masticatory stimulation, and, when necessary, medications to stimulate salivation
- *Proper adjustment of dental prostheses*: To reduce microenvironments that favor biofilm formation
- *Modulation of the oral microbiota*: Probiotic therapies, such as the use of *Lactobacillus* spp., show promise in reducing *C. albicans* colonization (Elahi et al. 2005; Salvatori et al. 2016; Kawanishi et al. 2021)
- *Strengthening host defenses*: Achieved through interventions that increase antimicrobial protein levels in saliva

In addition, treating established infections requires effective antifungal agents, such as nystatin or fluconazole, and addressing predisposing factors, such as immune dysregulation or the improper use of dental prostheses.

Oral candidiasis arises from a complex interplay of host factors, oral microbiota, and fungal properties. Maintaining salivary health, combined with proper dental practices and modern therapeutic interventions, is essential for preventing and managing this condition. Advances in probiotic-based therapies and antimicrobial biomaterials offer promising perspectives for effectively controlling *C. albicans* infections, with reduced risks of resistance and recurrence (Salvatori et al. 2016; Kawanishi et al. 2021).

36.4.4.4 Other Oral Infections

Saliva also plays a protective role against other oral infections, such as those caused by viruses and bacteria. The salivary proteome contributes to the overall immune surveillance and defense mechanisms within the oral cavity, preventing the establishment and progression of various infections.

In summary, saliva is an optimal biofluid that plays a critical role in maintaining both oral health and systemic health. Its diverse components, including immune cells, antimicrobial peptides, immunoglobulins, and various biomarkers, contribute to a dynamic and multifaceted barrier against pathogens. Understanding component interplay is essential for unraveling the complex mechanisms underlying oral infections and systemic diseases. Saliva offers a noninvasive diagnostic tool offering valuable insight into the immune system's status. Saliva holds immense potential for developing innovative diagnostic tools, personalized therapeutic interventions, and preventive strategies to improve overall health and well-being.

36.5 The Oral-Systemic Axis of the Salivary Proteins

Differentially expressed proteins play a role in systemic diseases such as cardiometabolic diseases, diabetes, insulin resistance, metabolic syndrome, altered cholesterol levels, steatohepatitis, glucose metabolism disorder, hepatic steatosis, and more.

In this topic, we will look at protein expression in saliva samples from people who have systemic disorders such as cardiovascular disease, diabetes, cancer, autoimmune disease, and neurodegenerative disease. The protein expressions of these disorders in saliva are being identified to better understand their pathogenesis and create therapeutic options based on their characteristics.

36.5.1 Cardiovascular Diseases

Cardiovascular diseases (CVDs) are a leading cause of death globally. Recent research has shown that saliva can potentially be used to diagnose and monitor CVDs (Zhang et al. 2016; Abdul Rehman et al. 2017). Salivary biomarkers, such as CRP, myeloperoxidase (MPO), and various cytokines, have been associated with CVDs

(Miller et al. 2010). For instance, elevated levels of CRP in saliva have been linked to an increased risk of heart attack and stroke (Punyadeera et al. 2011). Moreover, salivary MPO levels are elevated in patients with acute coronary syndrome (Mahmood et al. 2019). The relationship between obesity/overweight and cardiovascular disease may also be related to the differential expression of proteins in salivary plasma (Choi et al. 2015). For example, alpha-amylase activity was upregulated, including proteins such as QPCT, LCN2, PON2, DPP7, and CRKL, while other proteins, such as ARG1, CTSH, SERPINB6, OSMR, and ALDH3A, were downregulated (Khalifa et al. 2024).

Salivary biomarkers can also facilitate research on endothelins, which are found in blood vessels and play an important role in the modulation of vasomotor tone (Shah 2007). Systemic arterial hypertension and pulmonary arterial hypertension can be studied and potentially prevented through endothelin analysis (Montani et al. 2007). Studies on hypertension have demonstrated the relevance of several salivary indicators, including MMP-8 (Rathnayake et al. 2013) and lysozyme (Labat et al. 2013).

Salivary biomarkers, including cardiac troponin I (cTnI) (Mirzaii-Dizgah and Riahi 2013), CRP, MMP-9, IL-1beta, sICAM-1, adiponectin, MCP-1, Gro-alpha, E-selectin, IL-18, ENA-78, and sVCAM-1 (Miller et al. 2014), have been demonstrated to be upregulated within 12–48 h after the onset of acute myocardial infarction symptoms.

36.5.2 Diabetes

Type 2 diabetes mellitus (T2DM) is a chronic metabolic disease characterized by persistent high blood glucose levels. Saliva has emerged as a potential diagnostic fluid for diabetes due to its noninvasive nature and ease of collection. Studies have identified several salivary biomarkers that are altered in patients with diabetes, including alpha-amylase, glycosylated hemoglobin (HbA1c), and various inflammatory markers. For instance, salivary alpha-amylase activity has been found to be decreased in patients with T2DM (Aydin 2007). Moreover, salivary HbA1c levels have been shown to correlate with blood HbA1c levels, making it a potential noninvasive marker for monitoring glycemic control in diabetic patients (Panchbhai 2012; Fares et al. 2019). Furthermore, the proteomic discovery of salivary biomarkers will enable targeted therapies aimed at preventing or delaying the progression to frank diabetes and its complications (Rao et al. 2009).

Some studies have shown specific relationships between salivary biomarkers and manifestations and/or complications of diabetes that are relevant to the diagnosis, prevention, and possible future treatment of these clinical conditions. These include hyperglycemia, which is based on different levels of HbA1C. The proteomic changes were stronger in Type 1 diabetes mellitus (T1DM) than T2DM (Bencharit et al. 2013). Upon examining the possible salivary biomarkers that link the severity of T2DM and diabetic retinopathy, a number of biomarkers were discovered to have substantial expression when compared to a control group of diabetics (Chee et al. 2016). Given the severity and prognosis of diabetic retinopathy, lectin microarrays are the most commonly used salivary glycoparameters. The severity of diabetic retinopathy is correlated with the levels of the glycoparameters AAL, LEL, LCA, VVA, and NPA (Han et al. 2022).

36.5.3 Cancer

Carcinogenesis is a complex group of diseases characterized by uncontrolled cell growth and spread (Hanahan and Weinberg 2011). Salivary diagnostics offer a promising approach for the early detection and monitoring of various cancers (Lee and Wong 2009). Salivary biomarkers, such as proteins, DNA, RNA, and metabolites, have been investigated for their potential to distinguish between healthy individuals and cancer patients (Zhang et al. 2009). For example, elevated levels of certain proteins, such as CA125 and CA19-9, have been detected in the saliva of patients with ovarian and pancreatic cancer, respectively. Additionally, salivary microRNAs have shown promise as potential biomarkers for various cancers, including oral, breast, and lung cancers (Park et al. 2009; Zhang et al. 2010), as well as colorectal (Contran et al. 2024), pancreatic, and gastric cancers (Ghosh et al. 2024).

One of the most studied topics involving saliva, omics, and their various manifestations is oral cancer screening. When identifying biomarkers for oral cancer, genomics, transcriptomics, proteomics, metabolomics, and microbiomics are among the most frequently performed analyses (Nagi et al. 2016; Ishikawa et al. 2020). Liquid chromatography-mass spectrometry (LC-MS) for salivary biomarkers is one of the most widely used and effective technologies (Ishikawa et al. 2021). This approach can be crucial in identifying biomarkers for oral cancer, particularly when compared to samples from healthy individuals. Future research should focus on the six salivary proteins that were found to be significantly abundant in the study by Ishikawa et al. (2021), including α-2-macroglobulin-like protein 1, cornulin, hemoglobin subunit β, Ig κ chain V-II region Vκ167, kininogen-1, and transmembrane protease serine 11D (Ishikawa et al. 2021). In addition to these biomarkers, proteins like heat shock 70 kDa protein 1A/1B, pyruvate kinase PKM, and phosphoglycerate kinase 1 have been suggested to be differentially regulated and may serve as predictive markers for oral cancer (Remori et al. 2024).

36.5.4 Autoimmune Diseases

Autoimmunity occurs when the body's immune system mistakenly attacks its own tissues. Saliva can provide valuable insights into the immune system's activity and can potentially be used to diagnose and monitor autoimmune diseases (Chiappin et al. 2007; Javaid et al. 2016). Salivary biomarkers, such as autoantibodies, cytokines, and chemokines, have been associated with various autoimmune diseases, including Sjögren's syndrome, rheumatoid arthritis, and systemic lupus erythematosus. For instance, the presence of specific autoantibodies, such as anti-Ro and anti-La, in saliva can aid in diagnosing Sjögren's syndrome (Hu et al. 2007). Elevated levels of pro-inflammatory cytokines, such as TNF-α and IL-6, have also been identified in saliva, suggesting a correlation with systemic inflammatory responses (Kaczor-Urbanowicz et al. 2017).

Primary Sjögren's syndrome presents a diagnostic challenge, and the expression of biomarkers in whole saliva has become increasingly relevant. The most significant alterations are linked to glandular cells and inflammation in the oral cavity, which both contribute to immune response activation. Proteins, peptides, and mRNA are among the key elements reflecting these changes (Hu et al. 2007). Thymosin β (Tβ) can be expressed in saliva and minor salivary glands in patients with primary Sjögren's syndrome, with Tβ4 sulfoxide and Tβ10 being selectively present in patients with sicca symptoms (Bosello et al. 2016).

A key distinction between Sjögren's syndrome and non-Sjögren's Sicca syndrome lies in serum autoantibodies against the Ro60/SS-A and La/SS-B autoantigens, which are typically used for screening. An emerging approach for differentiating and tracking this illness involves studying salivary and lacrimal proteomes (George et al. 2023). Abrão et al. (2016) described the use of matrix-assisted laser desorption/ionization time-of-flight mass spectrometry (MALDI-TOF MS) technology to detect a set of proteins and peptides within a wide mass-to-charge (m/z) range, which has proven valuable for investigating salivary biomarkers in systemic lupus erythematosus. This technology offers a robust proteomic platform by enabling protein and peptide profiling in a single acquisition step, even without prior separation processes (Castagnola et al. 2012;). Proteins differently expressed in the C-chain region of immunoglobulin gamma-3 (IGHG3) have been used to characterize salivary proteins and identify promising biomarkers for systemic lupus erythematosus (Jung et al. 2021).

36.5.5 Neurodegenerative Diseases

Neurodegenerative diseases are characterized by the progressive loss of neurons in the brain and spinal cord (Dugger and Dickson 2017). Saliva has been identified as a promising biological fluid for the detection of biomarkers associated with neurodegenerative diseases, offering a noninvasive approach to aid in the early diagnosis, assessment of severity, and discrimination of conditions like Alzheimer's and Parkinson's diseases (Schepici et al. 2020; Contini et al. 2024). Salivary biomarkers, including tau protein and its phosphorylated form, have shown potential as accessible indicators for the early detection and monitoring of Alzheimer's disease, emphasizing saliva's role as a noninvasive diagnostic medium (Shi et al. 2011). For example, phosphorylated tau levels in saliva have been found to increase significantly in Alzheimer's disease patients, highlighting its utility as a biomarker for early diagnosis (Shi et al. 2011). Similarly, changes in salivary alpha-synuclein, including an increase in its oligomeric form, have been reported in patients with Parkinson's disease, indicating its relevance for disease monitoring (Shi et al. 2011; Schepici et al. 2020). Furthermore, salivary biomarkers may also aid in monitoring conditions such as Autism Spectrum Disorder and Huntington's disease (Farah et al. 2018).

Recent research on salivary biomarker expression linked Parkinson's disease to reduced levels of SN-cystatin and certain statherin proteoforms, while also showing increased levels of SA-cystatin and antileukoproteinase. In patients with Alzheimer's disease, a greater abundance of α-defensins and short-oxidized S100A9 was observed (Contini et al. 2024). Salivary proteins linked to inflammatory processes, exosome synthesis, and adipose tissue formation are less abundant in the proteomes of individuals with Parkinson's disease compared to healthy controls. In Parkinson's disease, S100-A16, ARP2/3, and VPS4B showed the lowest levels of expression. Future studies should further explore these differences between affected individuals and healthy populations (Figura et al. 2021).

36.6 Emerging Trends in Salivary Proteomics

36.6.1 Integration of Multiomics Data

The future of salivary diagnostics lies in the integration of proteomics with other "omics" sciences, such as genomics, transcriptomics, metabolomics, and others. This multiomics approach will provide a more comprehensive understanding of the molecular mechanisms underlying oral and systemic diseases. This will validate the origins and importance of saliva and other systemic parts of the human body.

For instance, combining salivary proteomic data with genomic information can help identify genetic predispositions to certain diseases, while integrating metabolomic data can reveal metabolic pathways involved in disease progression. This holistic approach will enable the development of more accurate diagnostic tools and personalized treatment strategies.

36.6.2 Development of Point-of-Care Salivary Diagnostic Devices

Another exciting trend is the development of portable, easy-to-use devices for point-of-care salivary diagnostics. These devices will enable rapid, on-site testing for a variety of diseases, eliminating the need for laboratory analysis and enabling faster diagnosis and treatment. Advances in microfluidics, nanotechnology, and biosensor technology are paving the way for the development of such devices, which have the potential to revolutionize healthcare delivery, especially in resource-limited settings.

36.6.3 Challenges and Limitations

Standardization of Salivary Collection and Processing One of the main challenges in salivary proteomics is the lack of standardized protocols for saliva collection and processing. Factors such as the type of collection device, the time of day, and the presence of food or contaminants can significantly affect the composition of the salivary proteome. Establishing standardized protocols will ensure the reproducibility and comparability of results across different studies and clinical settings.

Interindividual Variability in Salivary Proteome The salivary proteome exhibits significant interindividual variability, influenced by factors such as age, gender, genetics, lifestyle, and medication use. This variability can complicate the identification of disease-specific biomarkers and the development of universal diagnostic tools. Future research needs to focus on understanding the sources of this variability and developing strategies to account for it in data analysis and interpretation.

Validation of Salivary Biomarkers in Large-Scale Studies While numerous salivary biomarkers have been identified for various diseases, their clinical utility needs to be validated in large-scale clinical studies. These studies will help determine the sensitivity, specificity, and predictive value of these biomarkers, paving the way for their integration into clinical practice. Additionally, the development of robust and reliable assays for measuring these biomarkers in a clinical setting is essential for their widespread adoption.

In conclusion, while challenges remain, the field of salivary proteomics is poised for significant advancements in the coming years. The integration of multiomics data, the development of point-of-care diagnostic devices, and the ongoing efforts to standardize and validate salivary biomarkers hold immense promise for the future of healthcare. Longitudinal studies can validate markers during biomarker analysis or clinical in response to treatment in clinical trials. Salivary proteomics has the potential to transform the way we diagnose, monitor, and treat diseases, ultimately leading to improved patient outcomes and a more personalized approach to medicine.

36.7 Conclusions

36.7.1 Summary of Key Findings

This comprehensive review has highlighted the multifaceted role of the salivary proteome in oral and systemic health, emphasizing its intricate connection to infection and immunity. There is a need to dissect the composition of saliva in terms of cellular contribution of the proteins versus fluid composition. Saliva, a complex biofluid, serves as a dynamic reservoir of proteins, reflecting both local and systemic conditions. Advancements in proteomic technologies, particularly mass spectrometry, have enabled the identification and quantification of thousands of salivary proteins, revealing their diverse functions in maintaining oral homeostasis, regulating the immune response, and combating infections.

Key findings include the identification of specific salivary biomarkers associated with various oral and systemic diseases, such as periodontal disease, cancer, cardiovascular diseases, and autoimmune disorders. Also, there is a need for markers that are reliable and can be measured at home. These biomarkers offer promising potential for noninvasive diagnostics, disease monitoring, and personalized medicine approaches. Moreover, the salivary proteome's role in the immune system, through its innate and adaptive components, antimicrobial peptides, immunoglobulins, and inflammatory mediators, has been elucidated, providing insights into the complex interplay between oral and systemic health.

36.7.2 The Promising Future of Salivary Proteomics in Healthcare

The field of salivary proteomics holds immense promise for the future of healthcare, dental medicine, and biomedical fields. Its noninvasive nature, ease of collection, and potential for early disease detection and monitoring make it an attractive tool for clinical applications. Standardizations of methods, including protocols of collection, sample preparation, and analysis are needed to advance our understanding of populational proteomics. Ongoing research efforts are focused on developing point-of-care salivary diagnostic devices, integrating multiomics data for a deeper understanding of disease mechanisms, and validating salivary biomarkers in large-scale clinical studies.

While we highlighted the need to improve challenges such as standardization, interindividual variability, and the need for further validation, the continuous advancements in proteomic technologies and the growing understanding of the salivary proteome's complexity pave the way for a promising future. The advancement of open-source databases such as salivaryproteome.org will guide the future of clinicians and scientists to investigate biomarkers further. Salivary proteomics has the potential to revolutionize diagnostics, prognostics, and therapeutic monitoring, ultimately leading to improved patient outcomes and a more personalized approach to healthcare.

Acknowledgments We thank Prof. Dr. Denise Tostes Oliveira (Department of Pathology, Bauru School of Dentistry, USP) and Prof. Dr. Camila de Oliveira Rodini Pegoraro (Department of Histology, Bauru School of Dentistry, USP) for providing the microscopic images.

References

Abdul Rehman S, Khurshid Z, Hussain Niazi F, Naseem M, Al Waddani H, Sahibzada HA, Sannam Khan R (2017) Role of salivary biomarkers in detection of cardiovascular diseases (CVD). Proteomes 5(3): 21. https://doi.org/10.3390/proteomes5030021

Abrão AL, Falcão DP, de Amorim RF, Bezerra AC, Pombeiro GA, Guimarães LJ, Fregni F, Silva LP (2016) Salivary proteomics: a new adjuvant approach to the early diagnosis of familial juvenile systemic lupus erythematosus. Med Hypotheses 89:97–100. https://doi.org/10.1016/j.mehy.2016.02.010

Alqedari H, Altabtbaei K, Espinoza JL, Bin-Hasan S, Alghounaim M, Alawady A, Altabtabae A, AlJamaan S, Devarajan S, AlShammari T, Eid MB, Matsuoka M, Jang H, Dupont CL, Freire M (2023) Host-microbiome associations in saliva predict COVID-19 severity. In: bioRxiv: the preprint server for biology. https://doi.org/10.1101/2023.05.02.539155

Amado FM, Ferreira RP, Vitorino R (2013) One decade of salivary proteomics: current approaches and outstanding challenges. Clin Biochem 46(6). https://doi.org/10.1016/j.clinbiochem.2012.10.024

Araujo TT, Camiloti GD, Valle AD, Silva NDG, Souza BM, Carvalho TS, Câmara JVF, Shibao PYT, Henrique-Silva F, Cruvinel T, Magalhães AC, Buzalaf MAR (2021) A sugarcane cystatin (CaneCPI-5) alters microcosm biofilm formation and reduces dental caries. Biofouling 37(1). https://doi.org/10.1080/08927014.2021.1881065

Araujo TT, Dionizio A, Carvalho TS, Cmv BF, Vertuan M, Câmara JVF, Henrique-Silva F, Marchetto R, Chiaratti MR, Santos AC, Alves LO, Ferro M, Buzalaf MAR (2024a) Acquired enamel pellicle and biofilm engineering with a combination of acid-resistant proteins (CaneCPI-5, StN15, and Hemoglobin) for enhanced protection against dental caries – in vivo and in vitro investigations. Clin Oral Investig 28(5). https://doi.org/10.1007/s00784-024-05651-0

Araujo TT, Dionizio A, Carvalho TS, Debortolli ALB, Vertuan M, de Souza BM, Camara JVF, Henrique-Silva F, Chiaratti M, Santos A, Alves L, Ferro M, Magalhães AC, Buzalaf MAR (2024b) Acquired pellicle and biofilm engineering with CaneCPI-5: insights from proteomic and microbiomics analysis. Arch Oral Biol 166. https://doi.org/10.1016/j.archoralbio.2024.106025

Aydin S (2007) A comparison of ghrelin, glucose, alpha-amylase and protein levels in saliva from diabetics. J Biochem Mol Biol 40(1). https://doi.org/10.5483/bmbrep.2007.40.1.029

Bamford CV, d'Mello A, Nobbs AH, Dutton LC, Vickerman MM, Jenkinson HF (2009) Streptococcus gordonii modulates Candida albicans biofilm formation through intergeneric communication. Infect Immun 77(9). https://doi.org/10.1128/IAI.00438-09

Bandhakavi S, Stone O, G., Van Riper, S. K., & Griffin, T. J. (2009) A dynamic range compression and three-dimensional peptide fractionation analysis platform expands proteome coverage and the diagnostic potential of whole saliva. J Proteome Res 8(12). https://doi.org/10.1021/pr900675w

Bandhakavi S, Van Riper SK, Tawfik PN, Stone H, T., Rhodus, N. L., Carlis, J. V., & Griffin, T. J. (2011) Hexapeptide libraries for enhanced protein PTM identification and relative abundance profiling in whole human saliva. J Proteome Res 10(3). https://doi.org/10.1021/pr100857t

Bardow A, Nyvad B, Nauntofte B (2001) Relationships between medication intake, complaints of dry mouth, salivary flow rate and composition, and the rate of tooth demineralization in situ. Arch Oral Biol 46(5). https://doi.org/10.1016/s0003-9969(01)00003-6

Bencharit S, Baxter SS, Carlson J, Byrd WC, Mayo MV, Border MB et al (2013) Salivary proteins associated with hyperglycemia in diabetes: a proteomic analysis.

Mol BioSyst 9(11):2785–2797. https://doi.org/10.1039/c3mb70196d

Blankenvoorde MF, W v't H, Walgreen-Weterings E, van Steenbergen TJ, Brand HS, Veerman EC, Av NA (1998) Cystatin and cystatin-derived peptides have antibacterial activity against the pathogen Porphyromonas gingivalis. Biol Chem 379(11) https://pubmed.ncbi.nlm.nih.gov/9865612/

Bosello S, Peluso G, Iavarone F, Tolusso B, Messana I, Faa G, Castagnola M, Ferraccioli G (2016) Thymosin β4 and β10 in Sjögren's syndrome: saliva proteomics and minor salivary glands expression. Arthritis Res Ther 18(1):229. https://doi.org/10.1186/s13075-016-1134-7

Bosshardt DD (2018) The periodontal pocket: pathogenesis, histopathology and consequences. Periodontol 76(1):43–50. https://doi.org/10.1111/prd.12153

Bouwman FG, de Roos B, Rubio-Aliaga I, Katie Crosley L, Duthie SJ, Mayer C, Horgan G, Polley AC, Heim C, Coort SLM, Evelo CT, Mulholland F, Johnson IT, Elliott RM, Daniel H, Mariman ECM (2011) 2D-electrophoresis and multiplex immunoassay proteomic analysis of different body fluids and cellular components reveal known and novel markers for extended fasting. BMC Med Genet 4:24. https://doi.org/10.1186/1755-8794-4-24

Bowen WH (2002) Do we need to be concerned about dental caries in the coming millennium? Crit Rev Oral Biol Med Off Publ Am Assoc Oral Biol 13(2). https://doi.org/10.1177/154411130201300203

Bowen WH, Burne RA, Wu H, Koo H (2018) Oral biofilms: pathogens, matrix, and Polymicrobial interactions in microenvironments. Trends Microbiol 26(3). https://doi.org/10.1016/j.tim.2017.09.008

Brinkmann V, Reichard U, Goosmann C, Fauler B, Uhlemann Y, Weiss DS, Weinrauch Y, Zychlinsky A (2004) Neutrophil extracellular traps kill bacteria. Science. https://doi.org/10.1126/science.1092385

Buzalaf MAR, Pessan JP, Honório HM, Ten Cate JM (2011) Mechanisms of action of fluoride for caries control. Monogr Oral Sci 22. https://doi.org/10.1159/000325151

Buzalaf MA, Hannas AR, Kato MT (2012) Saliva and dental erosion. J Appl Oral Sci Revis FOB 20(5). https://doi.org/10.1590/s1678-77572012000500001

Buzalaf MA, Charone S, Tjäderhane L (2015) Role of host-derived proteinases in dentine caries and erosion. Caries Res 49(Suppl):1. https://doi.org/10.1159/000380885

Buzalaf MAR, Magalhães AC, Rios D (2018) Prevention of erosive tooth wear: targeting nutritional and patient-related risks factors. Br Dent J 224(5). https://doi.org/10.1038/sj.bdj.2018.173

Buzalaf MAR, Ortiz AC, Carvalho TS, Fideles SOM, Araújo TT, Moraes SM, Buzalaf NR, Reis FN (2020) Saliva as a diagnostic tool for dental caries, periodontal disease and cancer: is there a need for more biomarkers? Expert Rev Mol Diagn 20(5). https://doi.org/10.1080/14737159.2020.1743686

Caporossi L, Santoro A, Papaleo B (2010) Saliva as an analytical matrix: state of the art and application for biomonitoring. Biomark Biochem Ind Exp Resp Suscept Chem 15(6). https://doi.org/10.3109/1354750X.2010.481364

Castagnola M, Cabras T, Iavarone F, Fanali C, Nemolato S, Peluso G, Bosello SL, Faa G, Ferraccioli G, Messana I (2012) The human salivary proteome: a critical overview of the results obtained by different proteomic platforms. Expert Rev Proteomics 9(1). https://doi.org/10.1586/epr.11.77

Chee CS, Chang KM, Loke MF, Angela LVP, Subrayan V (2016) Association of potential salivary biomarkers with diabetic retinopathy and its severity in type-2-diabetes mellitus: a proteomic analysis by mass spectrometry. PeerJ 4:e2022. https://doi.org/10.7717/peerj.2022

Chiappin S, Antonelli G, Gatti R, De Palo EF (2007) Saliva specimen: a new laboratory tool for diagnostic and basic investigation. Clin Chim Acta Int J Clin Chem 383(1–2). https://doi.org/10.1016/j.cca.2007.04.011

Choi YJ, Nam YS, Yun JM, Park JH, Cho BL, Son HY et al (2015) Association between salivary amylase (AMY1) gene copy numbers and insulin resistance in asymptomatic Korean men. Diab Med 32(12): 1588–1595. https://doi.org/10.1111/dme.2015.32.issue-12

Christodoulides N, Mohanty S, Miller CS, Chris Langub M, Floriano PN, Dharshan P, Ali MF, Bernard B, Romanovicz D, Anslyn E, Fox PC, McDevitt JT (2005) Application of microchip assay system for the measurement of C-reactive protein in human saliva. Lab Chip 5(3):261–269. https://doi.org/10.1039/B414194F

Contini C, Fadda L, Lai G, Masala C, Olianas A, Castagnola M, Messana I, Iavarone F, Bizzarro A, Masullo C, Solla P, Defazio G, Manconi B, Diaz G, Cabras T (2024) A top-down proteomic approach reveals a salivary protein profile able to classify Parkinson's disease with respect to Alzheimer's disease patients and healthy controls. Proteomics 24(3–4): e2300202

Contran N, Arrigoni G, Battisti I, D'Incà R, Angriman I, Franchin C et al (2024) Colorectal cancer and inflammatory bowel diseases share common salivary proteomic pathways. Sci Rep 14(1):17711. https://doi.org/10.1038/s41598-024-68400-z

Cross BW, Ruhl S (2018) Glycan recognition at the saliva – oral microbiome interface. Cell Immunol 333. https://doi.org/10.1016/j.cellimm.2018.08.008

da Silva NDG, de Paiva PRB, Magalhães TVM, Braga AS, Psds S, Henrique-Silva F, Magalhães AC, Buzalaf MAR (2022) Effect of experimental and commercial artificial saliva formulations on the activity and viability of microcosm biofilm and on enamel demineralization for irradiated patients with head and neck cancer (HNC). Biofouling 38(7). https://doi.org/10.1080/08927014.2022.2111258

da Silveira EG, Prato LS, Pilati SFM, Arthur RA (2023) Comparison of oral cavity protein abundance among caries-free and caries-affected individuals-a systematic review and meta-analysis. Front Oral Health 4. https://doi.org/10.3389/froh.2023.1265817

Darveau RP (2010) Periodontitis: a polymicrobial disruption of host homeostasis. Nat Rev Microbiol 8(7): 481–490. https://doi.org/10.1038/nrmicro2337

Dawes C, Wong DTW (2019) Role of saliva and salivary diagnostics in the advancement of oral health. J Dent Res 98(2):133. https://doi.org/10.1177/0022034518816961

Deng J, Jackson L, Epstein JB, Migliorati CA, Murphy BA (2015) Dental demineralization and caries in patients with head and neck cancer. Oral Oncol 51(9). https://doi.org/10.1016/j.oraloncology.2015.06.009

Denny P, Hagen FK, Hardt M, Liao L, Yan W, Arellanno M, Bassilian S, Bedi GS, Boontheung P, Cociorva D, Delahunty CM, Denny T, Dunsmore J, Faull KF, Gilligan J, Gonzalez-Begne M, Halgand F, Hall SC, Han X, Fisher SJ (2008) The proteomes of human parotid and submandibular/sublingual gland salivas collected as the ductal secretions. J Proteome Res 7(5). https://doi.org/10.1021/pr700764j

Dodds MWJ, Johnson DA, Yeh C-K (2005) Health benefits of saliva: a review. J Dent 33(3):223–233. https://doi.org/10.1016/j.jdent.2004.10.009

Dugger BN, Dickson DW (2017) Pathology of neurodegenerative diseases. Cold Spring Harbor Perspect Biol 9(7). https://doi.org/10.1101/cshperspect.a028035

Elahi S, Pang G, Ashman R, Clancy R (2005) Enhanced clearance of Candida albicans from the oral cavities of mice following oral administration of Lactobacillus acidophilus. Clin Exp Immunol 141(1). https://doi.org/10.1111/j.1365-2249.2005.02811.x

Esser D, Alvarez-Llamas G, de Vries MP, Weening D, Vonk RJ, Roelofsen H (2008) Sample stability and protein composition of saliva: implications for its use as a diagnostic fluid. Biomark Insights 3. https://doi.org/10.4137/bmi.s607

Fábián TK, Hermann P, Beck A, Fejérdy P, Fábián G (2012) Salivary Defense proteins: their network and role in innate and acquired Oral immunity. Int J Mol Sci 13(4):4295. https://doi.org/10.3390/ijms13044295

Farah R, Haraty H, Salame Z, Fares Y, Ojcius DM, Said Sadier N (2018) Salivary biomarkers for the diagnosis and monitoring of neurological diseases. Biom J 41(2): 63–87. https://doi.org/10.1016/j.bj.2018.03.004

Fares S, Said MSM, Ibrahim W, Amin TT, Saad NES (2019) Accuracy of salivary glucose assessment in diagnosis of diabetes and prediabestes. Diab Metab Synd 13(2):1543–1547. https://doi.org/10.1016/j.dsx.2019.03.010

Figura M, Sitkiewicz E, Świderska B, Milanowski Ł, Szlufik S, Koziorowski D, Friedman A (2021) Proteomic profile of saliva in Parkinson's disease patients: A proof-of-concept study. Brain Sci 11(5): 661. https://doi.org/10.3390/brainsci11050661

Freire M, Moustafa A, Harkins DM et al (2020) Longitudinal study of oral microbiome variation in twins. Sci Rep 10:7954. https://doi.org/10.1038/s41598-020-64747-1

Gemmell E, Seymour GJ (2004) Immunoregulatory control of Th1/Th2 cytokine profiles in periodontal disease. Periodontology 2000, 35. https://doi.org/10.1111/j.0906-6713.2004.003557.x

George CT, Kurien BT, Scofield RH (2023) The potential utility of salivary and tear proteomics to discriminate Sjögren's disease from non-Sjögren's sicca. Int J Mol Sci 24(24):17497. https://doi.org/10.3390/ijms242417497

Ghosh AK, Nath A, Elangovan E, Banerjee A, Ramalingam K, Sethuraman S (2024) Exploring salivary biomarkers for tumor diagnosis: a narrative review. Cureus 16(7):e65725. https://doi.org/10.7759/cureus.65725

Glatter T, Ludwig C, Ahrné E, Aebersold R, Heck AJ, Schmidt A (2012) Large-scale quantitative assessment of different in-solution protein digestion protocols reveals superior cleavage efficiency of tandem Lys-C/trypsin proteolysis over trypsin digestion. J Proteome Res 11(11). https://doi.org/10.1021/pr300273g

Gohel V, Jones JA, Wehler CJ (2018) Salivary biomarkers and cardiovascular disease: a systematic review. Clin Chem Lab Med 56(9):1432–1442. https://doi.org/10.1515/cclm-2017-1018

Grassl N, Kulak NA, Pichler G, Geyer PE, Jung J, Schubert S, Sinitcyn P, Cox J, Mann M (2016) Ultra-deep and quantitative saliva proteome reveals dynamics of the oral microbiome. Genome Med 8(1):1–13. https://doi.org/10.1186/s13073-016-0293-0

Guedes SFF, Neves BG, Bezerra DS, Ghmf S, Lima-Neto ABM, Guedes MIF, Duarte S, Rodrigues LKA (2020) Saliva proteomics from children with caries at different severity stages. Oral Dis 26(6). https://doi.org/10.1111/odi.13352

Guo T, Rudnick PA, Wang W, Lee CS, Devoe DL, Balgley BM (2006) Characterization of the human salivary proteome by capillary isoelectric focusing/nanoreversed-phase liquid chromatography coupled with ESI-tandem MS. J Proteome Res 5(6). https://doi.org/10.1021/pr060065m

Hajishengallis G (2014) Immunomicrobial pathogenesis of periodontitis: keystones, pathobionts, and host response. Trends Immunol 35(1):3–11. https://doi.org/10.1016/j.it.2013.09.001

Han Q, Wang X, Ding X, Hao J, Li Q, Wang J, Yu H, Tang Z, Yang F, Cai G, Zhang D, Zhu H (2022) Salivary glycopatterns as potential non-invasive biomarkers for diagnosing and reflecting severity and prognosis of diabetic nephropathy. Front Endocrinol 13:790586. https://doi.org/10.3389/fendo.2022.790586

Hanahan D, Weinberg RA (2011) Hallmarks of cancer: the next generation. Cell 144(5). https://doi.org/10.1016/j.cell.2011.02.013

Hannig M, Balz M (2001) Protective properties of salivary pellicles from two different intraoral sites on enamel erosion. Caries Res 35(2). https://doi.org/10.1159/000047446

Hara AT, Ando M, González-Cabezas C, Cury JA, Serra MC, Zero DT (2006) Protective effect of the dental pellicle against erosive challenges in situ. J Dent Res 85(7). https://doi.org/10.1177/154405910608500706

Hardy J, Norris R, Anderson H, O'Shea A, Charles B (2012) Is saliva a valid substitute for plasma in pharmacokinetic studies of oxycodone and its metabolites in patients with cancer. Support Care Cancer Off J Multinat Assoc Support Care Cancer 20(4). https://doi.org/10.1007/s00520-011-1147-3

Helmerhorst EJ, Dawes C, Oppenheim FG (2018) The complexity of oral physiology and its impact on salivary diagnostics. Oral Dis 24(3). https://doi.org/10.1111/odi.12780

Hu S, Wang J, Meijer J, Ieong S, Xie Y, Yu T, Zhou H, Henry S, Vissink A, Pijpe J, Kallenberg C, Elashoff D, Loo JA, Wong DT (2007) Salivary proteomic and genomic biomarkers for primary Sjögren's syndrome. Arthritis Rheum 56(11). https://doi.org/10.1002/art.22954

Huang N, Pérez P, Kato T, Mikami Y, Okuda K, Gilmore RC, Conde CD, Gasmi B, Stein S, Beach M, Pelayo E, Maldonado JO, Lafont BA, Jang S-I, Nasir N, Padilla RJ, Murrah VA, Maile R, Lovell W, Byrd KM (2021) SARS-CoV-2 infection of the oral cavity and saliva. Nat Med 27(5):892–903. https://doi.org/10.1038/s41591-021-01296-8

Ishikawa S, Sugimoto M, Edamatsu K, Sugano A, Kitabatake K, Iino M (2020) Discrimination of oral squamous cell carcinoma from oral lichen planus by salivary metabolomics. Oral Dis 26(1):35–42. https://doi.org/10.1111/odi.13209

Ishikawa S, Ishizawa K, Tanaka A, Kimura H, Kitabatake K, Sugano A, Edamatsu K, Ueda S, Iino M (2021) Identification of salivary proteomic biomarkers for oral cancer screening. In Vivo 35(1):541–547. https://doi.org/10.21873/invivo.12289

Jasim H, Olausson P, Hedenberg-Magnusson B, Ernberg M, Ghafouri B (2016) The proteomic profile of whole and glandular saliva in healthy pain-free subjects. Sci Rep 6. https://doi.org/10.1038/srep39073

Javaid MA, Ahmed AS, Durand R, Tran SD (2016) Saliva as a diagnostic tool for oral and systemic diseases. J Oral Biol Craniof Res 6(1). https://doi.org/10.1016/j.jobcr.2015.08.006

Jehmlich N, Dinh KH, Gesell-Salazar M, Hammer E, Steil L, Dhople VM, Schurmann C, Holtfreter B, Kocher T, Völker U (2013) Quantitative analysis of the intra- and inter-subject variability of the whole salivary proteome. J Periodontal Res 48(3). https://doi.org/10.1111/jre.12025

Jung JY, Nam JY, Ryu KS, Son IO, Shin JH, Baek WY, Kim HA, Suh CH (2021) Salivary immunoglobulin gamma-3 chain C is a promising noninvasive biomarker for systemic lupus erythematosus. Int J Mol Sci 22(3):1374. https://doi.org/10.3390/ijms22031374

Kaczor-Urbanowicz KE, Martin C-PC, Aro K, Tu M, Garcia-Godoy F, Wong DT (2017) Saliva diagnostics – current views and directions. Exp Biol Med (Maywood, NJ) 242(5). https://doi.org/10.1177/1535370216681550

Kawanishi N, Hoshi N, Adachi T, Ichigaya N, Kimoto K (2021) Positive effects of saliva on Oral candidiasis: basic research on the analysis of salivary properties. J Clin Med 10(4):812. https://doi.org/10.3390/jcm10040812

Khalifa O, Al-Akl NS, Arredouani A (2024) Differential expression of cardiometabolic and inflammation markers and signaling pathways between overweight/obese Qatari adults with high and low plasma salivary α-amylase activity. Front Endocrinol 15:1421358. https://doi.org/10.3389/fendo.2024.1421358

Khurshid Z, Zafar MS, Khan RS, Najeeb S, Slowey PD, Rehman IU (2018) Role of salivary biomarkers in oral cancer detection. Adv Clin Chem 86:23–70. https://doi.org/10.1016/bs.acc.2018.05.002

Kinane DF, Mark Bartold P (2007) Clinical relevance of the host responses of periodontitis. Periodontol 43(1):278–293. https://doi.org/10.1111/j.1600-0757.2006.00169.x

Klein MI, Hwang G, Santos PH, Campanella OH, Koo H (2015) Streptococcus mutans-derived extracellular matrix in cariogenic oral biofilms. Front Cell Infect Microbiol 5. https://doi.org/10.3389/fcimb.2015.00010

Kleinstein SE, Nelson KE, Freire M (2020) Inflammatory networks linking Oral microbiome with systemic health and disease. J Dent Res. https://doi.org/10.1177/0022034520926126

Kolenbrander PE, Palmer RJ, Rickard AH, Jakubovics NS, Chalmers NI, Diaz PI (2006) Bacterial interactions and successions during plaque development. Periodontology 2000:42. https://doi.org/10.1111/j.1600-0757.2006.00187.x

Krief G, Deutsch O, Gariba S, Zaks B, Aframian DJ, Palmon A (2011) Improved visualization of low abundance oral fluid proteins after triple depletion of alpha amylase, albumin and IgG. Oral Dis 17(1). https://doi.org/10.1111/j.1601-0825.2010.01700.x

Krief G, Deutsch O, Zaks B, Wong DT, Aframian DJ, Palmon A (2012) Comparison of diverse affinity based high-abundance protein depletion strategies for improved bio-marker discovery in oral fluids. J Proteome 75(13). https://doi.org/10.1016/j.jprot.2012.05.012

Kulak NA, Pichler G, Paron I, Nagaraj N, Mann M (2014) Minimal, encapsulated proteomic-sample processing applied to copy-number estimation in eukaryotic cells. Nat Methods 11(3):319–324. https://doi.org/10.1038/nmeth.2834

Labat C, Temmar M, Nagy E, Bean K, Brink C, Benetos A, Bäck M (2013) Inflammatory mediators in saliva associated with arterial stiffness and subclinical atherosclerosis. J Hypertens 31:2251–2258

Lang NP, Bartold PM (2018) Periodontal health. J Periodontol 89(Suppl):1. https://doi.org/10.1002/JPER.16-0517

Lau WW, Hardt M, Zhang YH, Freire M, Ruhl S (2021) The human salivary proteome wiki: a community-driven research platform. J Dent Res 100(13). https://doi.org/10.1177/00220345211014432

Lee YH, Wong DT (2009) Saliva: an emerging biofluid for early detection of diseases. Am J Dent 22(4) https://pubmed.ncbi.nlm.nih.gov/19824562/

Lendenmann U, Grogan J, Oppenheim FG (2000) Saliva and dental pellicle–a review. Adv Den Res 14:22–28. https://doi.org/10.1177/08959374000140010301

Lin NJ (2017) Biofilm over teeth and restorations: what do we need to know? Dent Mat Off Publ Acad Dent Mater 33(6). https://doi.org/10.1016/j.dental.2017.03.003

Lin YH, Eguez RV, Torralba MG, Singh H, Golusinski P, Golusinski W, Masternak M, Nelson KE, Freire M, Yu Y (2019) Self-assembled STrap for global proteomics and salivary biomarker discovery. J Proteome Res 18(4). https://doi.org/10.1021/acs.jproteome.9b00037

Linden SK, Sutton P, Karlsson NG, Korolik V, McGuckin MA (2008) Mucins in the mucosal barrier to infection. Mucosal Immunol 1(3). https://doi.org/10.1038/mi.2008.5

Lis M, Liu TT, Barker KS, Rogers PD, Bobek LA (2010) Antimicrobial peptide MUC7 12-mer activates the calcium/calcineurin pathway in Candida albicans. FEMS Yeast Res 10(5):579–586. https://doi.org/10.1111/j.1567-1364.2010.00638.x

Llena-Puy C (2006) The rôle of saliva in maintaining oral health and as an aid to diagnosis. Med Oral Patol Oral Y Cirugia Bucal 11(5) https://pubmed.ncbi.nlm.nih.gov/16878065/

Lussi A, Jaeggi T, Zero D (2004) The role of diet in the aetiology of dental erosion. Caries Res 38(Suppl):1. https://doi.org/10.1159/000074360

Machiulskiene V, Campus G, Carvalho JC, Dige I, Ekstrand KR, Jablonski-Momeni A, Maltz M, Manton DJ, Martignon S, Martinez-Mier EA, Pitts NB, Schulte AG, Splieth CH, Tenuta LMA, Ferreira ZA, Nyvad B (2020) Terminology of dental caries and dental caries management: consensus report of a workshop organized by ORCA and cariology research group of IADR. Caries Res 54(1). https://doi.org/10.1159/000503309

Magalhães AC, de Oliveira RC, Buzalaf MAR (2017) Bioquímica básica e bucal. https://repositorio.usp.br/item/002897682

Mahmood Z, Enocsson H, Bäck M, Chung RWS, Lundberg AK, Jonasson L (2019) Salivary and plasma levels of matrix metalloproteinase-9 and myeloperoxidase at rest and after acute physical exercise in patients with coronary artery disease. PLoS One 14(2). https://doi.org/10.1371/journal.pone.0207166

Majem B, Rigau M, Reventós J, Wong DT (2015) Non-coding RNAs in saliva: emerging biomarkers for molecular diagnostics. Int J Mol Sci 16(4):8676. https://doi.org/10.3390/ijms16048676

Manconi B, Liori B, Cabras T, Iavarone F, Manni A, Messana I, Castagnola M, Olianas A (2017) Top-down HPLC-ESI-MS proteomic analysis of saliva of edentulous subjects evidenced high levels of cystatin A, cystatin B and SPRR3. Arch Oral Biol 77. https://doi.org/10.1016/j.archoralbio.2017.01.021

Marsh PD (2018) In sickness and in health-what does the oral microbiome mean to us? An ecological perspective. Adv Dent Res 29(1). https://doi.org/10.1177/0022034517735295

Marsh PD, Moter A, Devine DA (2011) Dental plaque biofilms: communities, conflict and control. Periodontology 2000, 55(1). https://doi.org/10.1111/j.1600-0757.2009.00339.x

Marshall TA (2019) Dietary implications for dental caries: a practical approach on dietary Counseling. Dent Clin N Am 63(4). https://doi.org/10.1016/j.cden.2019.06.005

Messana I, Inzitari R, Fanali C, Cabras T, Castagnola M (2008) Facts and artifacts in proteomics of body fluids. What proteomics of saliva is telling us? J Sep Sci 31(11). https://doi.org/10.1002/jssc.200800100

Miller CS, Foley JD, Bailey AL, Campell CL, Humphries RL, Christodoulides N, Floriano PN, Simmons G, Bhagwandin B, Jacobson JW, Redding SW, Ebersole JL, McDevitt JT (2010) Current developments in salivary diagnostics. Biomark Med 4(1). https://doi.org/10.2217/bmm.09.68

Miller CS, Foley JD, Floriano PN, Christodoulides N, Ebersole JL, Campbell CL et al (2014) Utility of salivary biomarkers for demonstrating acute myocardial infarction. J Dent Res 93:72S–79S. https://doi.org/10.1177/0022034514527796

Mirzaii-Dizgah I, Riahi E (2013) Salivary troponin I as an indicator of myocardial infarction. Indian J Med Res 138:861–865

Montani D, Souza R, Binkert C et al (2007) Endothelin-1/endothelin-3 ratio: a potential prognostic factor of pulmonary arterial hypertension. Chest 131(1):101–108

Moussa DG, Ahmad P, Mansour TA, Siqueira WL (2022) Current state and challenges of the global outcomes of dental caries research in the meta-omics era. Front Cell Infect Microbiol 12. https://doi.org/10.3389/fcimb.2022.887907

Muadcheingka T, Tantivitayakul P (2015) Distribution of Candida albicans and non-albicans Candida species in oral candidiasis patients: correlation between cell surface hydrophobicity and biofilm forming activities. Arch Oral Biol 60(6). https://doi.org/10.1016/j.archoralbio.2015.03.002

Murakami S, Mealey BL, Mariotti A, Chapple ILC (2018) Dental plaque-induced gingival conditions. J Clin Periodontol 45(Suppl):20. https://doi.org/10.1111/jcpe.12937

Nadig SD, Ashwathappa DT, Manjunath M, Krishna S, Annaji AG, Shivaprakash PK (2017) A relationship between salivary flow rates and Candida counts in patients with xerostomia. J Oral Maxillof Pathol 21(2). https://doi.org/10.4103/jomfp.JOMFP_231_16

Nagi R, Reddy-Kantharaj YB, Rakesh N, Janardhan-Reddy S, Sahu S (2016) Efficacy of light-based detection systems for early detection of oral cancer and oral potentially malignant disorders: systematic review.

Med Oral Patol Oral Cir Bucal 21(4):e447–e455. https://doi.org/10.4317/medoral.21104

Nanci A, Bosshardt DD (2006) Structure of periodontal tissues in health and disease. Periodontology 2000, 40. https://doi.org/10.1111/j.1600-0757.2005.00141.x

Nascimento MM, Zaura E, Mira A, Takahashi N, Ten Cate JM (2017) Second era of OMICS in caries research: moving past the phase of disillusionment. J Dent Res 96(7). https://doi.org/10.1177/0022034517701902

National Institute of Allergy and Infectious Diseases (n.d.) HCA dataset: 60ea42e1-af49-42f5-8164-d641fdb696bc. National Institute of Allergy and Infectious Diseases Data Portal. https://data.niaid.nih.gov/resources?id=hca_60ea42e1-af49-42f5-8164-d641fdb696bc

Page RC, Schroeder HE (1976) Pathogenesis of inflammatory periodontal disease. A summary of current work. Lab Invest J Tech Meth Pathol 34(3) https://pubmed.ncbi.nlm.nih.gov/765622/

Panchbhai AS (2012) Correlation of salivary glucose level with blood glucose level in diabetes mellitus. J Oral Maxillof Res 3(3):e3. https://doi.org/10.5037/jomr.2012.3303

Park NJ, Zhou H, Elashoff D, Henson BS, Kastratovic DA, Abemayor E, Wong DT (2009) Salivary microRNA: discovery, characterization, and clinical utility for oral cancer detection. Clin Cancer Res Off J Am Assoc Cancer Res 15(17). https://doi.org/10.1158/1078-0432.CCR-09-0736

Pedersen AM, Bardow A, Nauntofte B (2005) Salivary changes and dental caries as potential oral markers of autoimmune salivary gland dysfunction in primary Sjogren's syndrome. BMC Clin Pathol 5(1). https://doi.org/10.1186/1472-6890-5-4

Perera D, Kleinstein SE, Hanson B, Hasturk H, Eveloff R, Freire M, Ramsey M (2020) Impaired host response and the presence of Acinetobacter baumannii in the serum microbiome of type-II diabetic patients. iScience 24(1). https://doi.org/10.1016/j.isci.2020.101941

Pessoa L, Aleti G, Choudhury S, Nguyen D, Yaskell T, Zhang Y, Li W, Nelson KE, Neto LLS, Sant'Ana ACP, Freire M (2019) Host-microbial interactions in systemic lupus erythematosus and periodontitis. Front Immunol 10. https://doi.org/10.3389/fimmu.2019.02602

Pfaffe T, Cooper-White J, Beyerlein P, Kostner K, Punyadeera C (2011) Diagnostic potential of saliva: current state and future applications. Clin Chem 57(5). https://doi.org/10.1373/clinchem.2010.153767

Philip N, Leishman S, Walsh L (2019) Potential role for natural products in dental caries control. Oral Health Prev Dent 17(5). https://doi.org/10.3290/j.ohpd.a42741

Pitts NB, Zero DT, Marsh PD, Ekstrand K, Weintraub JA, Ramos-Gomez F, Tagami J, Twetman S, Tsakos G, Ismail A (2017) Dental caries. Nat Rev Dis Prim 3. https://doi.org/10.1038/nrdp.2017.30

Poorni S, Srinivasan MR, Nivedhitha MS (2019) Probiotic streptococcus strains in caries prevention: a systematic review. J Conserv Dent 22(2). https://doi.org/10.4103/JCD.JCD_505_18

Proctor GB, Carpenter GH (2007) Regulation of salivary gland function by autonomic nerves. Auton Neurosci Basic Clin 133(1). https://doi.org/10.1016/j.autneu.2006.10.006

Punyadeera C, Dimeski G, Kostner K, Beyerlein P, Cooper-White J (2011) One-step homogeneous C-reactive protein assay for saliva. J Immunol Methods 373(1–2). https://doi.org/10.1016/j.jim.2011.07.013

Puri S, Edgerton M (2014) How does it kill?: understanding the Candidacidal mechanism of salivary Histatin 5. Eukaryot Cell 13(8):958. https://doi.org/10.1128/EC.00095-14

Radfar L, Shea Y, Fischer SH, Sankar V, Leakan RA, Baum BJ, Pillemer SR (2003) Fungal load and candidiasis in Sjögren's syndrome. Oral Surg Oral Med Oral Pathol Oral Radiol Endodontics 96(3). https://doi.org/10.1016/s1079-2104(03)00224-5

Rao PV, Reddy AP, Lu X, Dasari S, Krishnaprasad A, Biggs E, Roberts CT, Nagalla SR (2009) Proteomic identification of salivary biomarkers of type-2 diabetes. J Proteome Res 8(1). https://doi.org/10.1021/pr8003776

Rathnayake N, Åkerman S, Klinge B, Lundegren N, Jansson H, Tryselius Y, Sorsa T, Gustafsson A (2013) Salivary biomarkers for detection of systemic diseases. PLoS One 8:e61356. https://doi.org/10.1371/journal.pone.0061356

Remori V, Airoldi M, Alberio T, Fasano M, Azzi L (2024) Prediction of oral cancer biomarkers by salivary proteomics data. Int J Mol Sci 25(20):11120. https://doi.org/10.3390/ijms252011120

Ruhl S (2012) The scientific exploration of saliva in the post-proteomic era: from database back to basic function. Expert Rev Proteomics 9(1). https://doi.org/10.1586/epr.11.80

Salvatori O, Puri S, Tati S, Edgerton M (2016) Innate immunity and saliva in Candida albicans-mediated oral diseases. J Dent Res 95(4). https://doi.org/10.1177/0022034515625222

Schepici G, Silvestro S, Trubiani O, Bramanti P, Mazzon E (2020) Salivary biomarkers: future approaches for early diagnosis of neurodegenerative diseases. Brain Sci 10(4):245. https://doi.org/10.3390/brainsci10040245

Schulz BL, Cooper-White J, Punyadeera CK (2013) Saliva proteome research: current status and future outlook. Crit Rev Biotechnol 33(3):246–259. https://doi.org/10.3109/07388551.2012.687361

Selwitz RH, Ismail A I, Pitts NB (2007) Dental caries. Lancet (Lond) 369(9555). https://doi.org/10.1016/S0140-6736(07)60031-2

Shah R (2007) Endothelins in health and disease. Eur J Intern Med 18(4):272–282

Shang YF, Shen YY, Zhang MC, Lv MC, Wang TY, Chen XQ, Lin J (2023) Progress in salivary glands:

endocrine glands with immune functions. Front Endocrinol 14:1061235. https://doi.org/10.3389/fendo.2023.1061235

Shi M, Sui Y-T, Peskind ER, Li G, Hwang H, Devic I, Ginghina C, Edgar JS, Pan C, Goodlett DR, Furay AR, Gonzalez-Cuyar LF, Zhang J (2011) Salivary tau species are potential biomarkers of Alzheimer disease. J Alzheimer's Dis 27(2):299. https://doi.org/10.3233/JAD-2011-110731

Siqueira WL, Salih E, Wan DL, Helmerhorst EJ, Oppenheim FG (2008) Proteome of human minor salivary gland secretion. J Dent Res 87(5):445. https://doi.org/10.1177/154405910808700508

Sondej M, Denny PA, Xie Y, Ramachandran P, Si Y, Takashima J, Shi W, Wong DT, Loo JA, Denny PC (2009) Glycoprofiling of the human salivary proteome. Clin Proteomics 5(1). https://doi.org/10.1007/s12014-008-9021-0

Spielmann N, Wong DT (2011) Saliva: diagnostics and therapeutic perspectives. Oral Dis 17(4). https://doi.org/10.1111/j.1601-0825.2010.01773.x

Stanescu II, Calenic B, Dima A, Gugoasa LA, Balanescu E, Stefan-van Staden RI, Baicus C, Badita DG, Greabu M (2018) Salivary biomarkers of inflammation in systemic lupus erythematosus. Ann Anat 219:89–93. https://doi.org/10.1016/j.aanat.2018.02.012

Starmer CF, Lastra AA, Nesterenko VV, Grant AO (1991) Proarrhythmic response to sodium channel blockade. Theoretical model and numerical experiments. Circulation 84(3). https://doi.org/10.1161/01.cir.84.3.1364

Stone C, Chen X, McGowan T, Bandhakavi S, Cheng B, Rhodus NL, Griffin TJ (2011) Large-scale phosphoproteomics analysis of whole saliva reveals a distinct phosphorylation pattern. J Proteome Res 10(4). https://doi.org/10.1021/pr1010247

Streckfus C, Bigler L, Dellinger T, Dai X, Kingman A, Thigpen JT (2000) The presence of soluble c-erbB-2 in saliva and serum among women with breast carcinoma: a preliminary study. Clin Cancer Res 6(6) https://pubmed.ncbi.nlm.nih.gov/10873088/

Takahashi N (2015) Oral microbiome metabolism: from "who are they?" to "what are they doing?". J Dent Res 94(12). https://doi.org/10.1177/0022034515606045

Takahashi N, Nyvad B (2011) The role of bacteria in the caries process: ecological perspectives. J Dent Res 90(3). https://doi.org/10.1177/0022034510379602

Taylor JJ (2014) Protein biomarkers of periodontitis in saliva. ISRN Inflamm 2014:593151. https://doi.org/10.1155/2014/593151

Ten Cate JM, Buzalaf MAR (2019) Fluoride mode of action: once there was an observant dentist. J Dent Res 98(7). https://doi.org/10.1177/0022034519831604

Thakur SS, Geiger T, Chatterjee B, Bandilla P, Fröhlich F, Cox J, Mann M (2011) Deep and highly sensitive proteome coverage by LC-MS/MS without prefractionation. Mol Cell Prot 10(8). https://doi.org/10.1074/mcp.M110.003699

Thomadaki K, Helmerhorst EJ, Tian N, Sun X, Siqueira WL, Walt DR, Oppenheim FG (2011) Whole-saliva proteolysis and its impact on salivary diagnostics. J Dent Res 90(11). https://doi.org/10.1177/0022034511420721

Tonetti MS, Van Dyke TE (2013) Periodontitis and atherosclerotic cardiovascular disease: consensus report of the joint EFP/AAP workshop on periodontitis and systemic diseases. J Periodontol 84(4 Suppl). https://doi.org/10.1902/jop.2013.1340019

Tonetti MS, Imboden MA, Lang NP (1998) Neutrophil migration into the gingival sulcus is associated with transepithelial gradients of interleukin-8 and ICAM-1. J Periodontol 69(10):1139–1147. https://doi.org/10.1902/jop.1998.69.10.1139

Torralba MG, Aleti G, Li W, Moncera KJ, Lin YH, Yu Y, Masternak MM, Golusinski W, Golusinski P, Lamperska K, Edlund A, Freire M, Nelson KE (2021) Oral microbial species and virulence factors associated with Oral squamous cell carcinoma. Microb Ecol 82(4). https://doi.org/10.1007/s00248-020-01596-5

Turner (2016) Hyposalivation and xerostomia: Etiology, complications, and medical management. Dent Clin N Am 60(2). https://doi.org/10.1016/j.cden.2015.11.003

van Loveren C (2019) Sugar restriction for caries prevention: amount and frequency. Which is more important? Caries Res 53(2). https://doi.org/10.1159/000489571

Van Nieuw Amerongen A, Bolscher JG, Veerman EC (2004) Salivary proteins: protective and diagnostic value in cariology? Caries Res 38(3). https://doi.org/10.1159/000077762

Ventura TMS, Ribeiro NR, Dionizio AS, Sabino IT, Buzalaf MAR (2018) Standardization of a protocol for shotgun proteomic analysis of saliva. J Appl Oral Sci 26:e20170561. https://doi.org/10.1590/1678-7757-2017-0561

Vílchez R, Lemme A, Ballhausen B, Thiel V, Schulz S, Jansen R, Sztajer H, Wagner-Döbler I (2010) Streptococcus mutans inhibits Candida albicans hyphal formation by the fatty acid signaling molecule trans-2-decenoic acid (SDSF). Chembiochem Eur J Chem Biol 11(11). https://doi.org/10.1002/cbic.201000086

Vitorino LM, Lucchetti G, Leão FC, Vallada H, Peres MFP (2018) The association between spirituality and religiousness and mental health. Sci Rep 8(1). https://doi.org/10.1038/s41598-018-35380-w

Vukosavljevic D, Custodio W, Buzalaf MAR, Hara AT, Siqueira WL (2014) Acquired pellicle as a modulator for dental erosion. Arch Oral Biol 59(6):631–638. https://doi.org/10.1016/j.archoralbio.2014.02.002

Wang K, Zhou X, Li W, Zhang L (2019) Human salivary proteins and their peptidomimetics: values of function, early diagnosis, and therapeutic potential in combating dental caries. Arch Oral Biol 99. https://doi.org/10.1016/j.archoralbio.2018.12.009

Wei CX, Yu Y, Aleti G, Torralba M, Edlund A, Nelson KE, Freire M (2020) Salivary bioscience and periodontal medicine. Salivary Biosci:419–447. https://doi.org/10.1007/978-3-030-35784-9_18

Whelton HP, Spencer AJ, Do LG, Rugg-Gunn AJ (2019) Fluoride revolution and dental caries: evolution of policies for global use. J Dent Res 98(8). https://doi.org/10.1177/0022034519843495

Zhan L (2018) Rebalancing the caries microbiome Dysbiosis: targeted treatment and sugar alcohols. Adv Dent Res 29(1). https://doi.org/10.1177/0022034517736498

Zhang L, Xiao H, Wong DT (2009) Salivary biomarkers for clinical applications. Mol Diagn Ther 13(4). https://doi.org/10.1007/BF03256330

Zhang L, Farrell JJ, Zhou H, Elashoff D, Akin D, Park NH, Chia D, Wong DT (2010) Salivary transcriptomic biomarkers for detection of resectable pancreatic cancer. Gastroenterology 138(3). https://doi.org/10.1053/j.gastro.2009.11.010

Zhang Y, Sun J, Lin CC, Abemayor E, Wang MB, Wong DT (2016) The emerging landscape of salivary diagnostics. Periodontology 2000, 70(1). https://doi.org/10.1111/prd.12099

Zhang JS, Huang S, Chen Z, Chu CH, Takahashi N, Yu OY (2024) Application of omics technologies in cariology research: a critical review with bibliometric analysis. J Dent 141. https://doi.org/10.1016/j.jdent.2023.104801

Zielinska DF, Gnad F, Wiśniewski JR, Mann M (2010) Precision mapping of an in vivo N-glycoproteome reveals rigid topological and sequence constraints. Cell 141(5). https://doi.org/10.1016/j.cell.2010.04.012

Dental Pulp-Derived Stem Cells in Mitigating Inflammatory Arthritis and Their Mechanisms in Conjunction with Oral Diseases

37

Prathyusha Naidu and Hiranmoy Das

Abstract

Rheumatoid arthritis (RA) is a chronic inflammatory condition influenced by genetic and environmental factors affecting joints, leading to deformity, loss of function, and disability. Similarly, chronic inflammation of the periodontium can lead to progressive periodontal tissue destruction and tooth loss. Emerging evidence suggested the bidirectional relationship between RA and periodontal disease (PD) as both share common inflammatory markers and immune dysregulation.

In this chapter, we discussed the connection between PD and RA, focusing on how the immune system responds to inflammatory stimuli. While inflammation is generally a protective mechanism, sustained inflammation leads to various pathogeneses and the development of diseases. We also discussed the cells involved in this process, the mechanisms by which they mediate inflammation, and the mechanisms of inflammation resolution. Additionally, we have highlighted the role of a novel transcription factor Krüppel-like factor 2 (KLF2), which plays an important role in anti-inflammatory processes and its regulatory mechanisms. Furthermore, the signaling pathways regulating inflammation and immune responses and their implications for future therapeutic developments of various medical conditions were discussed. We further explored the role of dental pulp stem cells (DPCSs), which have shown promise in reducing inflammation, promoting tissue regeneration, and reducing inflammatory mechanisms. We will discuss how DPSC can be used for future therapeutic developments targeting regenerative medicine.

Keywords

Periodontal disease · Rheumatoid arthritis · Inflammation · Kruppel-like factor 2 · dental pulp-derived stem cells

37.1 Factors Involved in Rheumatoid Arthritis

RA is a chronic and long-lasting inflammatory condition that affects synovial joints, impacting 0.5%–1% of adults worldwide (Aletaha et al. 2010). During the pathogenesis of RA, the immune system mistakenly attacks healthy

P. Naidu
Department of Pharmaceutical Sciences, Jerry H. Hodge School of Pharmacy, Texas Tech University Health Sciences Center, Amarillo, TX, USA

H. Das (✉)
Department of Pharmaceutical Sciences, Jerry H. Hodge School of Pharmacy, Texas Tech University Health Sciences Center, Amarillo, TX, USA

National Academy of Inventors, Tampa, FL, USA

Research Triangle Park, NC, USA
e-mail: hiranmoy.das@ttuhsc.edu

N. Rezaei, H. Ziaei (eds.), *Oral Immunology*, Advances in Experimental Medicine and Biology 1492,
https://doi.org/10.1007/978-3-032-03176-1_37

tissues, particularly the synovial lining of the joints (McInnes and Schett 2011). This inflammation can affect any diarthrodial joint lined with synovium, but it most commonly affects the small joints in the hands and feet (Imboden 2009). Moreover, larger joints like the knee and shoulder can also be affected, but the extent and severity vary among individuals (Bax et al. 2011). The prevalence of RA in the general population is less than 1%, but it can reach 2%–4% among siblings. Additionally, studies of twins have shown concordance rates for RA ranging from 12.3% to 15.4% for identical twins and 3.5% for nonidentical twins. With genetic factors estimated to contribute 50%–60% to the development of RA, twin and sibling studies highlight the significant role of genetics in the susceptibility to RA (Bax et al. 2011; MacGregor et al. 2000).

The cause of RA is not fully understood, but research suggests that it involves a combination of genetic and environmental factors. For instance, the human leukocyte antigens (HLA) DR Beta 1 (HLA-DRB1) genes are associated with a higher risk of developing RA. In individuals susceptible to the condition, environmental factors like smoking, infections, and hormonal changes may initiate autoimmune responses (Firestein and McInnes 2017). RA begins with a cascade of inflammatory events. T cells and B cells migrate to the synovium in response to pro-inflammatory cytokines such as tumor necrosis factor-alpha (TNF-α) and interleukin-6 (IL-6). This leads to synovial hyperplasia and angiogenesis, which results in the deterioration of cartilage and bone. At this stage, without early management and proper treatment, RA can progress to severe joint deformity, loss of function, and long-term disability (Conforti et al. 2021).

Different types of RA can be categorized based on the presence or absence of specific antibodies, age of onset, and treatment response. Patients with seropositive RA, characterized by rheumatoid factor (RF) and anti-citrullinated protein antibodies (ACPA), have a more severe progression and extra-articular manifestations (Aletaha et al. 2010). Conversely, those with seronegative RA lack these antibodies, including RF and ACPA, causing a slower progression yet still resulting in significant joint damage (Iqbal and Rattu 2019). Early RA refers to symptoms appearing within 6 months to 2 years of the onset, and early intervention has been shown to significantly improve long-term outcomes. Chronic RA lasts for more than 2 years and requires long-term management. Children under 16 are diagnosed with a variant of arthritis called juvenile idiopathic arthritis (JIA). This is with polyarticular JIA closely resembling adult RA, closely resembling adult RA (Ravelli and Martini 2007). On the other hand, palindromic rheumatism, characterized by repeated attacks of joint swelling, in some cases can progress into chronic RA. Apart from these, refractory RA is another form that fails to respond to standard treatments, often requiring advanced medical or surgical interventions. So, accurate classification of RA is essential for selecting appropriate treatments and predicting disease progression. However, given the individualized nature of RA, personalized treatment approaches are crucial for effective disease management (Firestein 2014).

37.1.1 Association of RA with Periodontal Diseases

Chronic inflammation of the periodontium, caused by dental plaque accumulation, is a highly prevalent oral disease, affecting 80%–90% of individuals (Saini et al. 2009). It damages both soft and hard periodontal tissues, resulting in tooth loss (Nascimento et al. 2017). The development and progression of periodontal damage rely on the virulence of the bacterial pathogens, as well as the host's resistance factors, including risk factors and immunological status (Detert et al. 2010a). Advances in molecular, cellular, and microbiological mechanisms have provided deeper insights into mechanisms that regulate the extent and balance of the humoral and cellular host response in periodontal tissues. Studies have revealed that the onset and progression of the disease are primarily due to a complex interplay between host immunity and the resident periodontal microbiome (Detert et al. 2010b).

Several epidemiological studies have demonstrated a clear and significant connection between these two conditions. This indicates that individuals with PD are more likely to have RA, and individuals with RA are more likely to have PD. For example, the prevalence of PD in RA patients is at least twice that of the general population (Pischon et al. 2008; Xiao et al. 2021). *Porphyromonas gingivalis* (*P. gingivalis*) is a bacterium known to cause chronic periodontitis, and it is thought to be a prominent factor linking RA and PD. A unique enzyme produced by *P. gingivalis*, called peptidyl-arginine deiminase (PPAD), modifies proteins by converting arginine to citrulline. This process is known as citrullination, and it can change the structure and function of proteins. Moreover, it could also lead to the development of autoantibodies against citrullinated proteins and cause a loss of immunological tolerance. These autoantibodies, which are specific to RA and can be detected even before the disease becomes clinically apparent, are also linked to the severity of the condition (Koziel et al. 2014).

Chronic periodontitis is triggered by *P. gingivalis* and its virulence factors, leading to local inflammation and promoting host protein citrullination. This, in turn, initiates an autoimmune process in susceptible individuals characterized by producing antibodies against citrullinated proteins, called ACPA. As indicated by a cross-sectional study of RA patients, those with severe periodontitis exhibit higher ACPA levels compared to those without periodontitis (De Pablo et al. 2008). Additionally, proteins such as fibrinogen, enolase, and vimentin are also targeted by autoantibodies in RA patients and can be citrullinated by the PPAD derived from *P. gingivalis*. In vivo studies provide further evidence that *P. gingivalis* contributes to the onset of RA. For example, mice with induced periodontitis are more likely to develop severe arthritis more quickly compared to normal animals. Moreover, *P. gingivalis* exacerbates collagen-induced arthritis (CIA) in mice, suggesting that the bacterium may play a potential role in modulating the body's immune response to RA (Fig. 37.1) (Fuggle et al. 2016; Potempa et al. 2017).

37.2 Mechanism of Inflammation

37.2.1 Types of Inflammation

The immune system of the body responds in various ways to harmful stimuli, such as chemicals, pathogenic agents, or damaged tissues, by triggering inflammation. This process serves as a protective mechanism aimed at eliminating the initial source of cell damage, removing injured tissues, and promoting tissue regeneration (Medzhitov 2008). Key components of the immune system, including white blood cells, blood vessels, and inflammatory mediators, play a role in the process of inflammatory response. The four cardinal signs of inflammation were initially described by Celsus in the first century as redness, swelling, heat, and pain (Hannoodee and Nasuruddin 2020). Later, the fifth sign, loss of function, or functio laesa, was introduced. Some of these symptoms arise due to increased blood flow to the inflamed region, fluid accumulation, and mediators of inflammation (Hannoodee and Nasuruddin 2020).

Inflammation is broadly classified into two main types of acute and chronic inflammation, depending on the severity and duration of the immune response. Acute inflammation is a brief, intense inflammatory response typically lasting from several hours to days. It is characterized by the rapid infiltration of granulocytes, mainly neutrophils, followed by the subsequent maturation of monocytes into inflammatory macrophages. This process helps eliminate the inflammatory stimulus and facilitates tissue recovery in the affected areas (Chen et al. 2018). On the other hand, chronic inflammation is a prolonged, persistent inflammatory reaction that can endure for several months to years. It is often associated with a variety of chronic diseases, including multiple malignancies, atherosclerosis, and rheumatoid arthritis. Tissue degradation, alterations in the predominant immune cell type in the inflamed area (mainly macrophages and lymphocytes), and ongoing tissue repair are characteristic features of chronic inflammation (Nathan and Ding 2010).

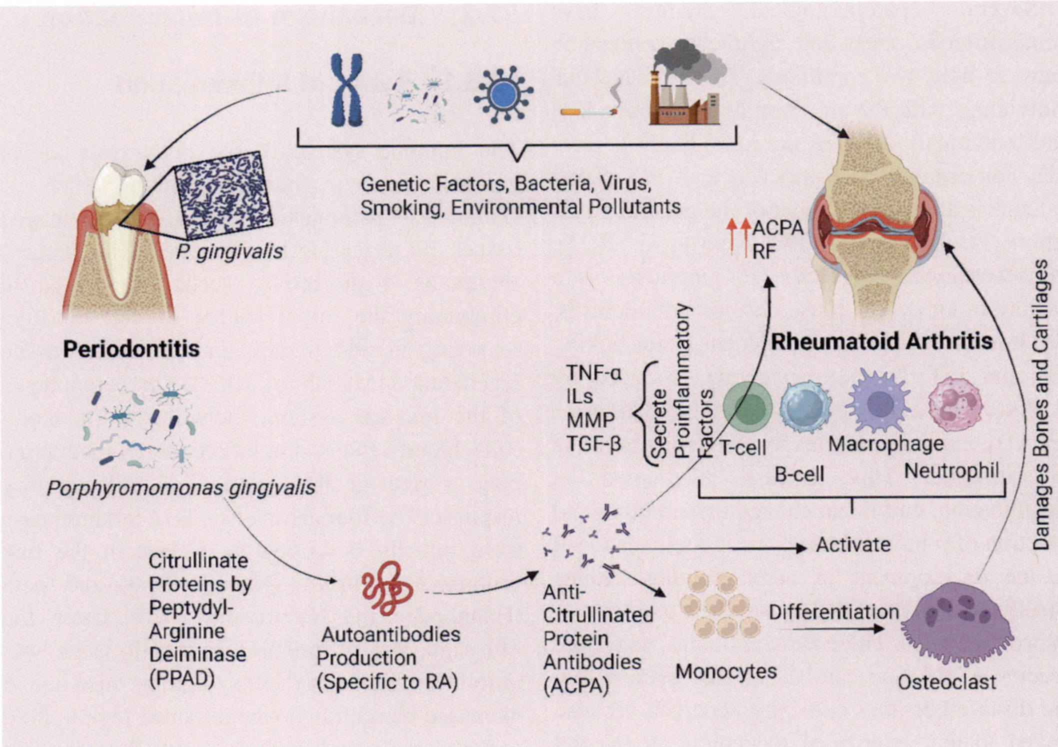

Fig. 37.1 *Association between periodontitis and rheumatoid arthritis*. Graphically shown is the pathway indicating various factors that mediate the pathogenesis of periodontitis, which could lead to the pathogenesis of rheumatoid arthritis (RA). This pathogenesis is mediated through the infection of bacteria called *porphyromonas gingivalis* that citrullinate proteins. Citrullinated protein produces autoantibodies called anti-citrullinated protein antibodies, which activate immune cells for the secretion of inflammatory factors, and differentiation of monocytes to osteoclasts that damage bones and cartilages in RA

37.2.2 Activation of Inflammation

Pattern-recognition receptors (PRRs) are the first to detect signals typically from pathogens. They are located on the surfaces of monocytes, macrophages, dendritic cells, and epithelial cells. PRRs recognize highly conserved microbial structures known as pathogen-associated molecular patterns (PAMPs) and molecules called damage-associated molecular patterns (DAMPs) that are released by stressed or injured cells (Rai et al. 2022). On the other hand, toll-like receptors (TLRs) play a role in identifying a wide variety of PAMPs, such as components of fungal cell walls, viral RNA and DNA, and bacterial lipopolysaccharide (LPS) (Nathan and Ding 2010). Additionally, C-type lectin receptors (CLRs), present on CLR cells, bind to carbohydrates on bacterial and fungal surfaces. The detection of viral RNA in the cytoplasm is carried out by five human RIG1-like receptors, including RIG-1 and MDA5. DAMPs and intracellular bacterial products are recognized by mNOD-like receptors (NLRs) (Kawai and Akira 2011).

Dysregulation of PRR signaling underlies PD and RA pathogenesis, both characterized by chronic inflammation and immune deregulation. Overexpression of TLR4 and TLR2 in macrophages and synovial fibroblasts in RA induces the production of cytokines TNF-α, IL-1β, and IL-6 (Kyburz et al. 2005). All these cytokines drive synovial inflammation and the degradation of joints. Likewise, in PD, in gingival tissue, overexpression of TLR4 and nucleotide-binding domain, leucine-rich repeat, and pyrin domain containing 3 (NLRP3) induces excessive immune response toward periodontal pathogens such as *Porphyromonas gingivalis* and *Aggregatibacter actinomycetemcomitans* (Song et al. 2017). Chronic stimulation of PRR pathways in PD induces excessive osteoclastogenesis and alveolar bone loss, similar to RA (Scher and Abramson 2013).

When a signaling molecule binds to PRRs, it triggers multiple signaling pathways that activate transcription factors. These transcription factors are essential for regulating the expression of genes associated with inflammation. Inhibitory IκB proteins bind to and sequester other proteins like nuclear factor-kappa B (NF-κB) in the cytoplasm and keep them in an inactive state. Upon stimulation, the inhibitor of NF-κB (IκB) is phosphorylated and degraded, allowing NF-κB dimers, such as p50/RelA, to move into the nucleus. After moving inside the nucleus, NF-κB binds to specific DNA sequences in the promoter region of target genes, including TNF-α, interleukin-1β (IL-1β), IL-6, and chemokines (Barnes and Karin 1997). These DNA sequences are known as κB sites. On the other hand, interferon regulatory factors (IRFs) promote the production of type I interferons (IFNs), such as IFN-β, in response to viral infections. This process involves the detection of viral genetic material in the cytoplasm, which leads to the phosphorylation and movement of IRF proteins, such as IRF3, into the nucleus (Ahmed 2011; Nathan 2002).

In RA and PD, the activation of NF-κB facilitates hyperinflammation. For instance, in RA, synovial fibroblasts exhibit constitutive NF-κB activity, resulting in persistent cytokine production and joint degradation (Lefevre et al. 2015). In PD, virulence factors of *P. gingivalis* modulate both NF-κB and IRF pathways, facilitating cytokine production and immune subversion (Abdulkareem et al. 2023). Furthermore, NLRP3 inflammasome, another PRR-related complex, is hyperactivated in both RA and PD, leading to the overproduction of IL-1β and IL-18, which drives tissue degradation and chronic inflammation (Marchesan et al. 2020).

Several mechanisms modulate epigenetically at a level of PRR-dependent inflammation, including histone methylation, acetylation, chromatin remodeling, and DNA methylation. All these modulate access for transcription factors, such as NF-kB, to promoter regions of inflammatory genes. Histone acetylation at genes is responsible for pro-inflammatory cytokines TNF-α and IL-6 in RA that induce more cytokine expression in maintaining synovial inflammation (Kouzarides 2007). Likewise, in PD, histone acetylation at the site of the IL-8 gene promoter is linked with increased infiltration of neutrophils and tissue injury (Shanmugam and Sethi 2012). Targeting epigenetic regulators, such as histone deacetylase inhibitors (HDACi) or DNA methyltransferase inhibitors (DNMTi), is a new therapeutic modality for controlling inflammation in RA and PD. In experimental studies, inhibition of HDAC activity suppresses the production of cytokines involved in inflammation and inhibits the degradation of joints in RA models (Grabiec et al. 2010). In PD, epigenetic therapy can normalize immune function via the inhibition of the overactivation of PRR and the expression of inflammatory genes (Zhang et al. 2022).

The coordinated action of numerous processes regulates the expression of proinflammatory genes. Transcription factors' stability and activation are controlled through phosphorylation, acetylation, methylation, and ubiquitination. The accessibility of transcription factor binding sites on DNA is determined by chromatin remodeling, histone modifications (such as acetylation and methylation), and DNA methylation. For example, histone acetylation can enhance the expression of genes regulated by NF-κB at the promoter regions of inflammatory genes (Abdulkhaleq et al. 2018).

37.2.3 Cells Mediate Inflammation

Inflammation in PD and RA involves a complex interrelationship between immune cells involved in both tissue degradation and tissue repair in a long-term manner. In both disease processes, mast cells, macrophages, lymphocytes, and neutrophils, with specific roles in disease pathogenesis, are the most important immune cells involved in causing inflammation.

Neutrophils are among the first immune cells to migrate into inflamed gingival tissue in PD and inflamed synovial tissue in RA. In RA, neutrophils migrate into the synovial tissue and induce tissue damage in joints through ROS, neutrophil extracellular traps (NETs), cartilage, and synovial lining-degrading proteolytic enzymes (Wright et al. 2010). Similarly, in PD, gingival crevicular fluid is rich in concentrations of neutrophils, and through them, releases MMPs and antimicrobial peptides and aids in destructive processes in the connective tissue (Abdulkareem et al. 2023). Overactivation of neutrophils in PD can even induce bystander tissue damage and exaggerate periodontal bone loss.

Macrophages play central roles in both RA and PD pathogeneses. In RA, chronic synovial macrophage production of pro-inflammatory cytokines, such as TNF-α, IL-1β, and IL-6, sustains ongoing joint inflammation and osteoclast differentiation and leads to bone loss (Udalova et al. 2016). Likewise, in PD, macrophages generate high amounts of pro-inflammatory mediators, such as IL-1β and prostaglandins, that orchestrate gingival inflammation and alveolar bone loss (Scher and Abramson 2013). Notably, macrophages play a dual role in the resolution of inflammation. Anti-inflammatory M2 macrophages facilitate tissue repair and wound healing by producing transforming growth factor-beta (TGF-β) and vascular endothelial growth factor (VEGF). Conversely, in chronic inflammatory diseases like RA and PD, an appropriate equilibrium between anti-inflammatory M2 macrophages and pro-inflammatory M1 macrophages is disrupted, and inflammation cannot be resolved (Liu et al. 2014).

Mast cells play a role in both RA and PD inflammatory cascades. In RA, histamine, tryptase, and chymase, mast cells' releasable contents when stimulated, induce hyperplasia of the synovium and white blood cell infiltration, contributing to joint enlargement and pain (Nigrovic and Lee 2004). In PD, TNF-α, and prostaglandins, mast cell-derived mediators promote blood vessel permeability and recruitment of immune cells, contributing to periodontal inflammation (Figueredo et al. 2019).

T and B lymphocytes have a critical role in both RA and PD immune dysregulation. T cells: CD4+ T helper (Th) cells, including Th1 and Th17 subsets, in RA cause inflammation through the production of IFN-γ and IL-17, with increased macrophage and neutrophil activation (McInnes and Schett 2017). Th17 cells promote osteoclastogenesis with increased bone loss. Similarly, in PD, an imbalance between Treg and Th17 cells underlies unregulated inflammation and alveolar bone loss (Dutzan et al. 2016). In RA, B cells contribute through the creation of autoantibodies such as rheumatoid factor (RF) and anti-citrullinated protein antibodies (ACPAs), which are responsible for joint degradation via immune complex deposition (Firestein 2003). In PD, B cells similarly produce cytokines and antimicrobial antibodies to pathogens in the mouth, but overactivation can cause persistent inflammation and tissue loss (Thorbert-Mros et al. 2015).

37.2.4 Cellular Factors Responsible for Inflammation

Inflammation is regulated through a variety of molecular mediators, modulating immune processes, causing tissue damage, and participating in disease pathology. Cytokines, chemokines, lipid mediators, complement proteins, ROS, and MMPs are the principal cellular mediators involved in inflammation. All these mediators have critical roles in RA, PD, and the inflammatory reaction of DPSC, impacting both tissue damage and healing.

Cytokines function as a kind of messaging between cells, regulating immune cell function. In RA and PD, excessive production of cytokines with a pro-inflammatory function, TNF-α, IL-1β, and IL-6, makes tissue susceptible to continuous inflammation and tissue loss (McInnes and Schett 2017). TNF-α and IL-1β, in RA, produced in large part by macrophages and synovial fibroblasts, drive osteoclast differentiation, with loss of bone. In PD, these cytokines drive alveolar bone loss via osteoclastogenesis and increased production of matrix metalloproteinases (MMPs), which are responsible for the degradation of the extracellular matrix (Di Benedetto et al. 2013). On the contrary, TGF-β and IL-10 are anti-regenerative and anti-inflammatory. DPSCs modulate inflammatory processes via IL-10 and TGF-β secretion, eliciting tissue regeneration and immune tolerance. DPSCs repress TNF-α and IL-6 expression in inflamed periodontal and synovial tissue and therefore have therapeutic value in both RA and PD (Wang et al. 2020).

Chemokines are a particular set of cytokines that guide immune cells to sites of inflammation. In RA and PD, C-C motif chemokine ligand 2 (CCL2, also known as MCP-1) recruits monocytes and macrophages, and C-X-C motif chemokine ligand 8 (CXCL8, also known as IL-8) facilitates the recruitment of neutrophils (Wright et al. 2010). Chemokines sustain inflammation, e.g., in RA synovium and periodontal lesions, by facilitating the infiltration of leukocytes and synthesis of inflammatory cytokines (Hajishengallis and Sahingur 2014). In hyperplasia, Prostaglandins (PGs) and Leukotriene Lipid-derived inflammatory mediators such as leuketerins (LTs), play a crucial role in regulating vascular permeability, immune cell activation, and pain sensitization. In RA and PD, prostaglandin E2 (PGE2) promotes synovial hyperplasia, periodontal tissue destruction, and osteoclast differentiation, exacerbating disease progression (Yang et al. 2013). Cyclooxygenase-2 (COX-2), the enzyme responsible for PGE2 synthesis, is upregulated in inflamed RA synovium and periodontium, making it a key therapeutic target for nonsteroidal anti-inflammatory drugs (NSAIDs) (Howell 1993).

DPSC, however, can suppress excessive production of PGE2 by regulating the expression of COX-2 and promoting the synthesis of anti-inflammatory lipid mediators, such as lipoxins and resolvins, which are involved in tissue repair and the resolution of inflammation (Lee et al. 2016). The complement system is involved in contributing to inflammation through enhanced phagocytosis, recruitment of immune cells, and direct pathogen lysis. Complement deregulation in RA synovial tissue and gingival crevicular fluid in PD has been observed, with augmented C3a and C5a levels potentiating neutrophil activation and tissue degradation (Lin et al. 2024). Additionally, reactive oxygen species (ROS) and reactive nitrogen species (RNS), such as hydrogen peroxide (H_2O_2), superoxide (O_2^-), and peroxynitrite ($ONOO^-$) are generated in RA and PD through stimulated macrophages and neutrophils. While ROS can kill pathogens directly, overproduced ROS induces oxidative damage, tissue degradation, and persistent inflammation. DPSCs have been shown to have antioxidant activity, preventing ROS-evoked damage and improving tissue regeneration in inflamed microenvironments (Hernández-Monjaraz et al. 2020).

Tissue Breakdown and Reconstruction MMPs are a family of enzymes that degrade ECM components, enhancing immune cell migration and tissue remodeling during inflammation. In RA, MMP-1, MMP-3, and MMP-9 in synovial fibroblasts have strong expression, enhancing cartilage degradation and joint destruction (Burrage et al. 2006). In PD, alveolar bone loss is caused by the degradation of collagen and periodontal ligament fibers through MMP-8 and MMP-9. DPSCs regulate the activity of MMP, reducing excessive ECM degradation and promoting wound healing by tissue inhibitors of metalloproteinases (TIMPs) secretions. This indicates their therapeutic potential in regenerative stem cell therapies for RA and PD (Nowwarote et al. 2022).

37.2.5 Role of Nuclear Factor-κB in Inflammation

NF-κB is a key modulator of inflammation at the level of gene expression, controlling a network of cascades of signals that coordinate immune function, tissue integrity, and resolution of inflammation. It is most often stimulated through two alternative pathways: the cytokine (TNF-α, IL-1β) and microbial (TLRs) stimuli, and the noncanonical, lymphotoxin-β and CD40L stimuli, participating in the stimulation of B cells and organogenesis of lymphoid tissue (Barnes and Karin 1997). Stimulation of NF-κB triggers the expression of cytokines (IL-6, TNF-α) and chemokines (CXCL8/IL-8, CCL2), and adhesion molecules, driving persistent inflammatory processes (McInnes and Schett 2017).

In RA, NF-κB is persistently activated in synovial fibroblasts, macrophages, and infiltrating immune cells, driving chronic synovial inflammation and joint destruction. NF-κB activation in synovial fibroblasts leads to the upregulation of TNF-α, IL-1β, IL-6, and matrix metalloproteinases (MMP-1, MMP-3, MMP-9), which degrade the extracellular matrix and contribute to cartilage erosion. In macrophages, NF-κB activation sustains inflammatory cytokine production, polarizing macrophages toward a pro-inflammatory (M1) phenotype, which further amplifies tissue damage (Roman-Blas and Jimenez 2006). Further, research has demonstrated that NF-κB inhibition reduces inflammation and joint destruction in RA models, and as such, it is a target for therapy (Roman-Blas and Jimenez 2006). Treatment with NF-κB inhibitors, such as IKK inhibitors, has been shown in clinical trials to block pro-inflammatory cytokine release and protect against synovial hyperplasia and bone erosion (Marcu et al. 2010).

NF-κB is central to the pathogenesis of periodontal disease, and its persistent activation leads to progressive periodontal tissue destruction. The bacterial pathogen *Porphyromonas gingivalis* exploits NF-κB signaling by manipulating TLR responses, inducing a dysregulated immune response and overproduction of inflammatory cytokines (Abdulkareem et al. 2023). Activated NF-κB promotes the secretion of pro-inflammatory cytokines (IL-1β, TNF-α, IL-6), chemokines (CCL2, CXCL8), and MMPs, which contribute to connective tissue destruction and alveolar bone loss. In addition, RANKL expression, an essential mediator of osteoclastogenesis, is increased through NF-κB, furthering periodontal bone resorption (Graves et al. 2019).

Blocking NF-κB signaling has been shown to inhibit osteoclast activation and expression of inflammatory cytokines in PD models, and thus, it can act as a therapeutic target for periodontal tissue degradation inhibition (Chen et al. 2024b). There is emerging translational research indicating that DPSCs have immunomodulatory actions, and in particular, target and downregulate NF-κB signaling pathways. DPSCs release anti-inflammatory cytokines (IL-10, TGF-β) and microRNAs (e.g., miR-146a, miR-155) that suppress IKK phosphorylation, resulting in decreased NF-κB activation (Fawzy El-Sayed et al. 2019). Studies have confirmed that DPSCs suppress TNF-α and IL-6 expression in RA synovial fibroblasts, reducing inflammatory signaling. They decrease osteoclast activity in PD models by inhibiting RANKL/NF-κB-dependent bone resorption. DPSCs also promote the anti-inflammatory (M2) macrophage phenotype, allowing for tissue healing. They enhance periodontal ligament (PDL) regeneration via inhibition of overexpression of MMP through NF-κB (Fawzy El-Sayed et al. 2019).

37.2.6 Role of KLF2 in Inflammation

KLF2, a member of the Kruppel-like family, is a transcription factor with potent anti-inflammatory properties in various immunological and endothelial cells. Its direct trans-repression of the promoter activity of several cytokines and chemokines involved in inflammation, such as IL-1, IL-8, TNF, MCP-1, and others, enables KLF2 to serve as an effective anti-inflammatory transcription factor in monocytes (Das et al. 2006, 2012, 2019; Ibrahim et al. 2021). It interacts with numerous coactivators, including PCAF and p300, crucial for NF-κB-dependent transcription, to fulfill this regulatory role. KLF2 aids in the displacement of these coactivators, which decreases the transcription of inflammatory genes by NF-κB. Furthermore, KLF2 also inhibits histone acetylation markers like H3K9 and H4K8, contributing to the suppression of proinflammatory signaling and participating in epigenetic regulation (SenBanerjee et al. 2004).

KLF2 plays a vital role in controlling inflammation and maintaining vascular homeostasis in endothelial cells. Normally, the phosphorylation and degradation of KLF2 at Ser 33 are tightly regulated, but pro-inflammatory cytokines like TNF-α can inhibit KLF2 by activating NF-κB. TNF-α triggers NF-κB, leading to reduced KLF2 expression through chromatin remodeling of the KLF2 promoter region mediated by histone deacetylases (HDACs) (Dabravolski et al. 2022). As a result, one important mechanism for regulating endothelial cell activation and subsequent inflammation involves the interplay between KLF2 and NF-κB. KLF2 deficiency significantly affects inflammation not only in endothelial and monocyte cells but also in various medical disorders (Jha and Das 2017). Studies using KLF2 hemizygous mice have demonstrated that the partial absence of this protein leads to altered immune responses in both acute and chronic inflammatory models. Decreased levels of KLF2 are associated with elevated inflammatory cytokines and increased severity of conditions in cases of sepsis or chronic inflammatory disorders such as atherosclerosis (Fig. 37.2) (Shi et al. 2018).

37.2.7 Role of Signaling Pathways in Inflammation

The mammalian target of rapamycin (mTOR), Janus kinase/signal transducers and activators of transcription (JAK/STAT), and mitogen-activated protein kinase (MAPK) pathways regulate inflammation and the immunological response. It was found that the mammalian target of the mTOR signaling pathway plays a critical role in several cellular functions, including inflammation, protein synthesis, and the breakdown of cellular components (Wei and Feng 2010). Therefore, any dysregulation of mTOR signaling may make an individual more susceptible to a range of illnesses, including cancer, autoimmune disorders, and neurodegenerative diseases. The mTOR pathway consists of two complexes: mTOR Complex 1 (mTORC1) and mTOR Complex 2 (mTORC2). mTORC1 comprises regulatory protein Raptor, mLST8, and inhibitory subunits like the 40 kDa proline-rich serine/threonine protein kinase (Akt) substrate and the mTOR-interacting protein with Disheveled, Egl-10, and Pleckstrin (DEP) homology domain (DEP) domain (Kipanyula et al. 2013). The mLST8 contributes to the stability of the kinase activation loop, while Raptor links mTORC1 to TOR signaling. Meanwhile, mTORC2, which is primarily associated with Akt phosphorylation and activation, is composed of DepTOR, Protor 1/2, mLST8, mTOR, and Rictor instead of Raptor (Nakagawa and Maeda 2012).

Concerning inflammation, mTORC1 regulates the development of T cells. Studies have demonstrated that rapamycin enhances T cell differentiation and immune regulation, particularly in $CD8^+$ memory T cells, while also inhibiting $CD8^+$ and $CD4^+$ effector T cells. As an immunosuppressant, rapamycin functions by preventing the proliferation of lymphocytes, which subsequently impacts T cells and prevents their activation (Zhao et al. 2021). Asthma is distinguished by increased levels of p13k, pAkt, p-mTOR, and p-p70s6K, as well as decreased regulatory T (Treg) cells and IFN-γ, and increased TH17 cells and interleukin-4 (IL-4). Additionally, mTOR inhibitors can ameliorate the imbalanced TH1/TH2 and TH17/Treg ratios in asthma. Thus, symptoms of osteoarthritis (OA) are alleviated when mTOR is genetically depleted. By reducing the expression of genes involved in producing pro-inflammatory cytokines and interleukins, mTOR inhibitors for IBD control boost the production of anti-inflammatory cytokines like IL-10 (Yeung et al. 2018).

The JAK/STAT pathway is an important signaling component linked to inflammation. It plays a role in cellular growth, differentiation, and immune response. Cytokines and growth hormones trigger this pathway. The MAPK pathway converts external signals to internal signals and regulates gene expression, cell division, growth, and differentiation, as well as cell survival and death. The important MAPKs include p38 MAPK, extracellular signal-regulated kinase (ERK), and JNK. Inflammatory cytokines and

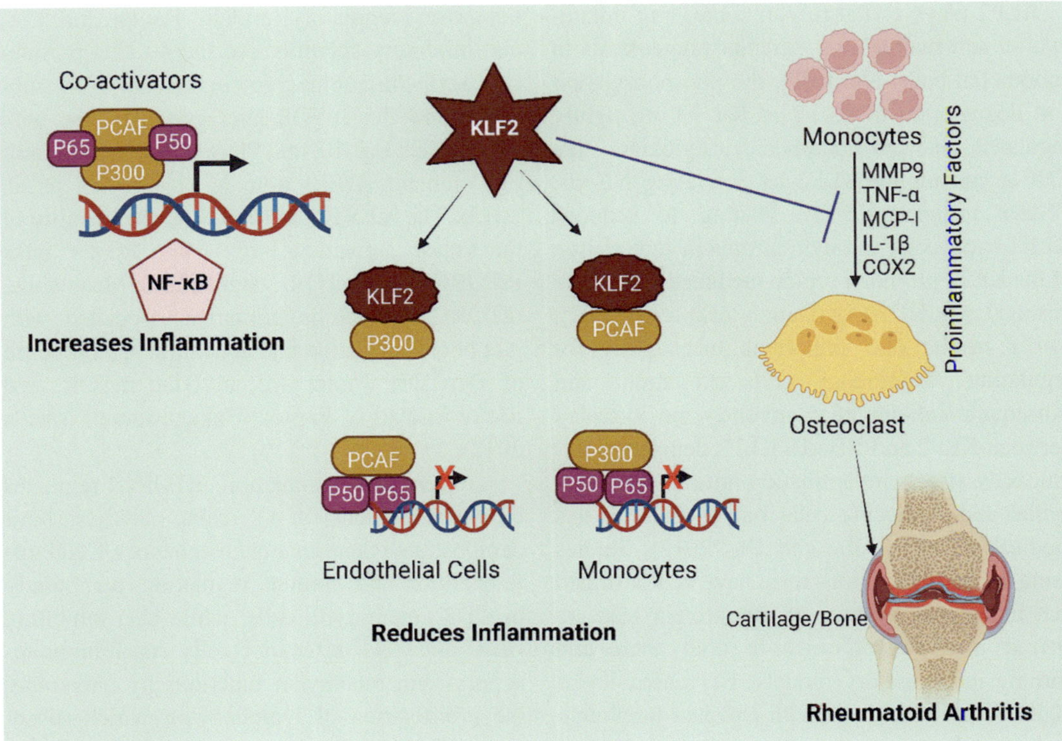

Fig. 37.2 *The interaction between the nuclear factor-κ B (NF-κB) and Kruppel-like factor 2 (KLF2).* Normally, NF-κB, the master regulator of inflammation, is activated by four cofactors, whereas KLF2 inactivates the function of NF-κB through binding with one of the cofactors in a cell-specific way to prevent the formation of a complex. KLF2 also inhibits proinflammatory secretions of monocytes and differentiation toward osteoclasts, that damages the bone and cartilage in arthritis

environmental stress specifically activate p38 MAPK to control the production of pro-inflammatory cytokines. Due to its role in triggering inflammatory responses, the MAPK pathway is closely associated with the development of various inflammatory diseases, such as arthritis and asthma (Babon et al. 2014; Purohit et al. 2023).

The dysregulation of the mTOR, JAK/STAT, and MAPK pathways has been most conspicuous in the pathophysiology of RA and PD and has accounted for tissue breakdown, immune dysregulation, and chronic inflammation. In RA, abnormal mTORC1 activation in synovial fibroblasts promotes excess growth, overproduction of cytokines (TNF-α, IL-6, IL-1β), and antiapoptotic activity, all perpetuating synovial hyperplasia and breakdown of joints (Pal et al. 2015). Similarly, in PD, mTOR signaling regulates osteoclast differentiation and breakdown of the bone, increasing alveolar bone loss (Zhu et al. 2022). Inhibitors of mTOR, such as rapamycin, have been shown to suppress cytokine production and tissue breakdown in models of RA and PD. Furthermore, the JAK/STAT pathway in RA is hyperactivated, wherein it fuels cytokine-dependent inflammatory signaling and arthritic inflammation. Blocking the JAK/STAT signaling (e.g., using JAK inhibitors like tofacitinib) has emerged as an effective therapeutic strategy in RA, while nascent research has suggested the inhibition of JAK could be used to reduce periodontal inflammation through blocking IL-6 and RANKL-dependent activation of osteoclasts (McInnes and Schett 2017).

Emerging evidence supports a role for DPSCs in regulating such inflammatory signaling cascades, with therapeutic potential for regenerative therapy in RA and PD. DPSCs secrete anti-inflammatory factors inhibiting mTORC1-dependent inflammation and thereby suppress excessive osteoclastogenesis and synovial fibroblast proliferation (Li et al. 2023). DPSCs have also been shown to activate JAK/STAT signaling through the secretion of anti-inflammatory cytokines (IL-10, TGF-β), and thereby antagonize inflammatory processes in both periodontal and synovial tissue. DPSCs suppress MAPK signaling, down-regulating p38 MAPK and JNK, central drivers of RA synovial inflammation and tissue destruction in PD (Chen et al. 2024a). This demonstrates that DPSCs, by suppressing such inflammatory cascades, may be a novel regenerative therapy for the resolution of chronic inflammation and tissue regeneration in RA and PD.

37.3 Role of Dental Pulp-Derived Stem Cells in Inflammation

A subset of adult mesenchymal stem cells located in the tooth pulp is known as dental pulp-derived stem cells (DPSC). These cells have sparked significant interest in regenerative medicine due to their mesenchymal characteristics and multipotential differentiation toward neural cells (Prateeksha et al. 2023b). DPSC can transform into various types of cells, including odontoblasts, osteoblasts, chondrocytes, adipocytes, and neurons. They are primarily derived from human permanent teeth, such as third molars or other healthy extracted teeth (Gronthos et al. 2000). These cells are considered a potential source for therapeutic and scientific use due to their ability to multiply and remain undifferentiated for a long period (Ogata et al. 2021).

One of the primary advantages of DPSCs is their relatively straightforward extraction process, often through tooth extraction. Additionally, these cells have a low risk of causing immune reactions, which suggests the potential for transplanting them between individuals with minimal risk of rejection (Pierdomenico et al. 2005). Because of the flexibility of DPSC, they have been explored in various regenerative medicine and tissue engineering applications. Dental tissue engineering has proven to be effective in regenerating dental pulps, dentin, and periodontal tissues (Anderson et al. 2022). Furthermore, DPSCs have exhibited promise in bone tissue engineering and in treating bone diseases, showing potential for use in orthopedic applications. The neurogenic potential of DPSC has also been studied, with research demonstrating that this capability could be beneficial in therapeutic applications of neurological conditions (Arthur et al. 2008). Moreover, these cells are recognized for their ability to stimulate the growth of new blood vessels, which is essential for tissue regeneration and repair. The anti-inflammatory and anticancer properties of DPSC have further enhanced their therapeutic potential, particularly in the treatment of autoimmune disorders. As ongoing research progresses, the characteristics of DPSC, including their multipotency, accessibility, and low risk of immune rejection, make them a viable source of stem cells for treating a variety of pathologies using a regenerative medicine approach (Gronthos et al. 2000).

DPSCs and human exfoliated deciduous tooth stem cells (SHED) have critical roles in regulating inflammation and immune function and have been positioned as future therapeutic candidates for treating RA and PD. Their anti-inflammatory cytokine-suppressing and immune tolerance-promoting actions have been documented in preclinical studies. DPSC and SHED modulate macrophage polarization, converting immune responses toward a switch between a pro-inflammatory (M1) and anti-inflammatory (M2) state, a critical function for tissue healing and resolution of inflammation (Li et al. 2023). This is facilitated, in part, through the production of immunoregulatory cytokines IL-10, TGF-β, and PGE2, which repress TNF-α, IL-6, and IL-1β production—leading drivers of inflammation in RA and PD (Yamaza et al. 2010). By modulating macrophage function, DPSC and

SHED inhibit RA joint synovitis and periodontal tissue loss in PD, showcasing their dual therapeutic potential for use in both inflammatory arthritis and oral disease.

One of the significant immunomodulatory actions of DPSC and SHED is their ability to activate the kynurenine pathway, an important metabolite in immune function. Indoleamine 2,3-dioxygenase (IDO), an important enzyme in the pathway, is augmented in DPSC following cytokine priming with cytokines involved in inflammation (TNF-α, IFN-γ, and IL-1β), with a concomitant increase in kynurenine, a molecule with antiproliferative actions in T cells and modulates function in dendritic cells (Chen et al. 2020). In RA, such a mechanism is significant, with the hyperactivation of T cells responsible for sustained synovial inflammation and bone loss. In a similar way, in PD, DPSCs downregulate NF-κB, an important mediator in the cascade of events following bacterial endotoxins and cytokine stimulation, and in consequence, inhibit expression of matrix metalloproteinases (MMP-9, MMP-13), responsible for tissue degradation and alveolar bone loss (Fawzy El-Sayed et al. 2019). Targeting such metabolite processes in immune function with DPSC-based therapies, therefore, could represent new therapeutic options for both systemic inflammatory arthritis and oral inflammatory disease.

Furthermore, DPSC and SHED modulate key inflammatory cascades, including Akt, ERK1/2, and MAPK, for tissue regeneration and immune homeostasis. In RA, hyperactivation of MAPK cascades in synovial fibroblasts induces persistent production of inflammatory cytokines and degradation of joints. DPSCs suppress p38 MAPK and JNK, and in the process, suppress the inflammatory response and induce cartilage regeneration (Chen et al. 2024b). Similarly, in PD, DPSCs suppress AKT/mTOR, an osteoclast differentiation cascade, and in the process, suppress periodontal bone loss (Peng et al. 2025). That they can maintain an anti-inflammatory microenvironment, induce angiogenesis, and induce tissue restoration makes them most effective in overcoming the burden of inflammation in RA and PD. Not only do these stem cells reverse pathology initiated by inflammation, but they create an environment conducive to long-term tolerance and restoration of tissue, and in the process, make them perfect candidates for therapeutic application in inflammatory arthritis and periodontal disease.

In animal studies, certain stem cells derived from dental pulp have shown promise in mitigating neurodegenerative conditions such as Parkinson's and Alzheimer's, which are primarily associated with persistent inflammation. Additionally, their ability to generate bone tissue in vivo clearly demonstrates their adaptability and potential for use in regenerative medicine. One of the key findings is the ability of DPSC to stimulate the transformation of monocytes and macrophages into a subtype known as M2, which is essential for tissue repair and the reduction of inflammation. The primary markers of the M2 subtype in these cells are YM-1 and Arg-1, and this process is significantly boosted when DPSCs are exposed to substances like LPS, TNF-α, or IFN-γ. These markers indicate that DPSC may have therapeutic potential for inflammatory conditions due to their tendency toward an anti-inflammatory nature (Tsuruta et al. 2018; Yamaguchi et al. 2015). Another key observation is the activation of the kynurenine pathway by DPSC, particularly when they are stimulated. This process results in the upregulation of the enzyme indoleamine 2,3-dioxygenase (IDO), which plays a role in converting tryptophan to kynurenine. Furthermore, the transcription factor peroxisome proliferator-activated receptor gamma (PPARγ), which is associated with kynurenine, facilitates the differentiation of monocytes and macrophages into the M2 subtype. This pathway illustrates how DPSC can modulate the immune response through metabolic processes, contributing to the shift toward a more anti-inflammatory immune response (Anderson et al. 2022).

The role of DPSCs in immunological polarization is complemented by their ability to regulate signaling pathways that govern cell migration and proliferation. They modulate signal transduction pathways such as Akt and ERK1/2, which play a key role in inflammation and wound repair. By modulating these signaling cascades, DPSCs contribute to general tissue healing by promoting tissue repair and reducing inflammatory responses. It is suitable to assume that DPSCs maintain their anti-inflammatory characteristics even after priming with TNF-α or IFN-γ, despite changes in mitochondrial activity. This adaptability suggests that DPSCs possess a highly adaptable metabolism, allowing them to fulfill their immunomodulatory role in diverse inflammatory environments.

Research into the potential use of DPSC for treating glioblastoma multiforme (GBM) yielded several significant findings. The results indicated that DPSC effectively reduced inflammation in GBM cells through multiple mechanisms. They notably decreased the levels of TNF-α, COX2, IL-1β, NF-κB p65, MMP9, and IL-6, all of which are crucial for sustaining inflammation and tumor development. Simultaneously, DPSC enhanced the expression of anti-inflammatory genes, including Arg1, IL10, and IL4R, which help in controlling inflammation. Furthermore, DPSC treatment led to a substantial reduction in intracellular and mitochondrial ROS levels, mitigating oxidative stress generated during cancer development and maintenance of chronic inflammation. Moreover, DPSC reduced ATP levels, mitochondrial respiration, and mitochondrial membrane potential in GBM cells, potentially impacting the metabolic processes associated with inflammation. Molecular target analysis revealed that DPSC inhibited the AKT and P38 MAPK signaling pathways, reinforcing their antiapoptotic and anti-inflammatory properties (Arthur et al. 2008; Prateeksha et al. 2023a).

37.4 DPSC-Mediated Signaling Pathways

To reduce the inflammation and facilitate tissue repair within the dental pulp, DPSCs employ various complex signaling pathways. One primary approach involves the utilization of the MEK/ERK and PI3K/AKT signaling pathways by DPSC to generate the necessary cellular response for tissue regeneration (Liang et al. 2021). By stimulating these pathways, the proliferation and differentiation of DPSCs are encouraged, which could help restore damaged pulp tissue and reduce inflammation. Therefore, DPSCs play an active role in promoting the healing process of the pulp environment and mitigating inflammation and other detrimental factors that affect tissue health and homeostasis by regulating PI3K/AKT and MEK/ERK signaling pathways (Bar et al. 2021).

Additionally, the regulation of inflammation by DPSC involves stromal-derived factor 1α (SDF-1α, also known as CXCL12), and its receptor CXC chemokine receptor 4 (CXCR4). As SDF-1α recruits DPSC to inflamed pulp tissue, it also contributes to their chemoattraction. DPSC must localize near the damaged site and actively participate in tissue repair for this recruitment mechanism to be effective. Through CXCR4 signaling, DPSCs migrate along SDF-1α gradients, settle in the inflamed region, and undergo differentiation into odontoblasts, neurons, and blood vessels. As a result, SDF-1α plays a dual role in regulating inflammation by promoting both pulp repair and inflammation resolution. Furthermore, DPSC activates wingless-related integration site (WNT) and bone morphogenetic protein (BMP) signaling pathways to enhance their anti-inflammatory function (Li et al. 2024). By activating these pathways, DPSC can directly aid in recruiting endogenous stem cells, which aid in regenerating the odontoblast-like and neurovascular stroma cells necessary for pulp tissue repair. Overall, DPSC contributes to maintaining optimal tissue health over time, decreasing inflammation, and restoring pulp structure and function through regulating WNT and BMP pathways (Boyle et al. 2014).

Exosomes derived from DPSC, known as DPSC-EXO, exhibit potent anti-inflammatory properties by suppressing inflammatory responses in periodontal ligament stem cells (PDLSC) that are exposed to LPS. Within PDLSC, these exosomes can significantly diminish the expression levels of important inflammatory markers such as IL-6, p-JAK2, and p-STAT3. Under inflammatory conditions, IL-6 stimulates the initiation of the JAK2/STAT3 signaling pathway, leading to STAT3 activation and the expression of genes linked to inflammation. DPSC-EXO mitigates inflammation in PDLSC by inhibiting the IL-6/p-JAK2/p-STAT3 signaling cascade. Additionally, by attenuating the acute inflammatory responses induced by LPS, DPSC-EXO and this molecular regulation present the potential for a promising therapeutic strategy for chronic inflammatory conditions such as periodontitis (Bar et al. 2021; Prateeksha et al. 2023a).

37.5 Current Therapies for Inflammation-Related Diseases

Therapies for chronic inflammatory diseases, other than autoimmune ones, involve various approaches aimed at modulating specific signaling pathways to reduce inflammation and prevent tissue damage. In the context of periodontitis and RA, these therapeutic strategies have the potential to limit immune overactivation and promote tissue regeneration.

Statins, commonly prescribed to reduce cholesterol levels and the risk of heart disease, also possess anti-inflammatory properties that could potentially prevent atherosclerosis. Despite their limited use in periodontitis, statins' anti-inflammatory properties indicate potential future strategies for controlling periodontal bone loss.

Similarly, Low-dose methotrexate (LD-MTX), classified as a disease-modifying antirheumatic drug (DMARD), has been tested in trials for psoriatic and rheumatoid arthritis. Based on the immune similarities between RA and PD, LD-MTX could provide insights into developing anti-inflammatory periodontal therapies. Furthermore, due to interactions between inflammasomes and IL-1β, and the role of IL-1β as the major drivers of chronic inflammation, drugs inhibiting IL-1β were also explored. These drugs include canakinumab, which is being studied for its potential to reduce inflammation related to atherosclerosis (Tabas and Glass 2013).

Although C-reactive protein (CRP) was initially identified as an inflammatory protein, it has diverse conformations and purposes. Human pentameric CRP (pCRP) is typically involved in defending against infections and targeting necrotic cells as well as other anti-inflammatory processes (Demaria et al. 2010). Nonetheless, pCRP can undergo conformational changes in its more physiologically active forms, such as its monomeric form (mCRP) and partially dissociated pentamer (pCRP*), in response to inflammation or tissue damage. These altered forms contribute to the enhancement of inflammatory responses through the production of cytokines, increased neutrophil and monocyte adherence to endothelial cells, and neutrophil extracellular trap (NET) formation (Hou et al. 2021). Targeting CRP as a therapeutic approach has been explored at different levels. For example, small molecule inhibitors like C10M bind to the phosphocholine binding site of pCRP, preventing it from changing into pro-inflammatory forms and reducing inflammatory activities. Additionally, some methods aim to lower serum CRP levels; these include CRP apheresis and antisense oligonucleotides (ASO), which may potentially reduce overall inflammation in different illnesses, including autoimmune disorders and cardiovascular diseases (Rizo-Téllez et al. 2023; Tabas and Glass 2013).

Endogenous lipid-derived molecules, including lipoxins, resolvins, protectins, and maresins, play a role in inflammation resolution. They are derived from omega-3 polyunsaturated fatty acids (resolvins, protectins, and maresins) or arachidonic acid (lipoxins). These molecules decrease the extravasation of white blood cells, reduce vascular permeability, and enable macrophage-mediated clearance of cellular debris and leukocytes (mainly through apoptosis). This process helps prevent long-term inflammation and aids in restoring tissue stability. In vitro, Resolvin E1 reduces inflammation in gingivitis. Additionally, human monocyte-derived macrophages respond to periodontal pathogens by producing Resolvin E1. This has the potential to influence the regulation of bone remodeling processes by promoting periodontal tissue regeneration and suppressing osteoclast activity (Passacquale et al. 2016; Preshaw 2018).

The levels of histone acetylation can be altered by histone deacetylase (HDAC) inhibitors, leading to an impact on the ability of transcription factors' binding to genes. Inhibition of HDAC activity can inhibit osteoclast differentiation and the production of inflammatory cytokines. Studies have demonstrated that specific HDAC inhibitors can reduce bone resorption caused by periodontopathogens like *P. gingivalis* in animal models of periodontitis; this outcome was a result of the decrease in specific transcription factors' binding, including c-Fos, which hinders the differentiation and activity of osteoclasts. While HDAC inhibitors have primarily been utilized in the treatment of cancer and other chronic inflammatory conditions, they may also have the potential to address periodontitis-related bone resorption. However, there are remaining concerns about their effectiveness and possible adverse effects in cases of periodontal disease, necessitating further investigation (Boyle et al. 2014; Tabas and Glass 2013).

Receptor activator of nuclear factor κ B ligand (RANKL) supports the formation and activation of osteoclasts and the subsequent absorption of bone matrix. One of the inhibitors is denosumab, a monoclonal antibody that hinders RANKL activity and helps prevent bone loss in conditions such as osteoporosis. Although there is limited clinical trial data on the use of osteoprotegerin (OPG), a mimic receptor of RANKL, in periodontitis, experimental models suggest that exogenous OPG administration effectively prevents the formation of alveolar bone. This suggests that the inhibition of RANKL signaling could be a viable treatment option for periodontal bone loss (Liu and Zhang 2015).

It is noted that azithromycin suppresses the synthesis of TNF-α and IL-1β, giving it anti-inflammatory properties. Due to its dual function, azithromycin can be utilized as an adjunctive therapy combined with traditional periodontal treatments. Moreover, the focus of first-generation medications targeting TLRs is on TLR-activating ligands and TLR-blocking antibodies. Modifying these TLRs may provide a broad-spectrum activity necessary for controlling inflammatory processes associated with periodontitis and other inflammatory diseases. Azithromycin, among other medications, can be used to address periodontal diseases due to its dual antibacterial and anti-inflammatory characteristics. These medications, by reducing inflammation and bacterial load, have the potential to improve the efficacy of standard periodontal therapy (Yazid et al. 2014; Zimmermann et al. 2018).

37.6 Proposed Future Therapeutic Interventions Using DPSC

Stem cells derived from dental pulp have the potential to be a valuable cell type for treating both inflammatory oral diseases and inflammatory arthritis at the same time. Due to the shared inflammatory processes of both conditions, this innovative approach aims to address inflammation by utilizing the unique properties of DPSCs, which may provide greater effectiveness compared to traditional treatments. DPSCs display impressive anti-inflammatory and immunomodulatory characteristics, and they can be easily isolated and collected from extracted teeth (Pisciotta et al. 2018). These multipotent stem cells can specifically target areas of inflammation in damaged oral tissues and joints. Later, these DPSCs employ multiple mechanisms to carry out their therapeutic effects (Liang et al. 2020). DPSCs have the ability to influence the immune response in conditions like RA by inhibiting pathogenic T cell function and promoting the production of Tregs. In a study, it was found that this type of immune system modulation can help manage synovitis and prevent the deterioration of bone and cartilage status. Additionally, DPSCs can differentiate into osteogenic and chondrogenic lineages, potentially assisting in the restoration of damaged joint tissue (Liu et al. 2011). Furthermore, DPSCs can decrease gingivitis and facilitate tissue healing in periodontitis by regulating the local immune response. They have been proven to effectively manage the inflammatory and destructive aspects of periodontal conditions by encouraging the regeneration of new alveolar bone, cementum, and periodontal ligament (Ferrarotti et al. 2018). The recommended treatment plan involves acquiring DPSCs from the patient or other compatible sources. Following this, the cells may undergo preconditioning and ex vivo expansion to enhance their potential for stem cell therapy. DPSCs can be administered in various ways, including topical application to affected oral tissues, intra-articular injection for joint treatment, and intravenous administration for systemic effects (Nakashima et al. 2017). One of the key benefits of this strategy is its potential for a dual-targeted therapy approach using cell-based therapy to simultaneously reduce both oral and arthritic inflammation. This is particularly crucial given the strong association between rheumatoid arthritis and periodontal disease, as both conditions are linked to common inflammatory processes and risk factors.

Moreover, preclinical and early clinical trials have indicated the safety profile of DPSC, with low risks of tumor formation and immune rejection. Unlike traditional immunosuppressive medications used in RA treatment, DPSC-based therapy has the ability to control inflammation without completely suppressing the immune system, which may offer distinctive advantages (Ducret et al. 2015). Although showing promising results, further investigation is required to improve the processing of DPSC, the methods of application, and the treatment protocol. Large-scale clinical trials are essential to demonstrate the effectiveness of DPSC-based therapy in treating inflammatory arthritis and periodontitis, as well as to confirm its long-term safety. While still in the early stages of regenerative medicine, this approach holds considerable potential for patients with these closely related inflammatory disorders, offering a promising outlook for future therapeutic advancements (Potempa et al. 2017).

37.7 Conclusion

Using stem cells derived from dental pulp to address inflammatory oral diseases and inflammatory arthritis may be seen as an innovative advancement in regenerative medicine. DPSCs possess the ability to transform and can be utilized to repair inflammatory-mediated oral tissue damage and modulate immune responses in cases of inflammatory arthritis. DPSCs demonstrate anti-inflammatory properties both locally and systemically through mechanisms such as cytokine-induced immunomodulation and tissue remodeling facilitated by various signaling pathways. They have shown the ability to lower levels of inflammatory markers and facilitate tissue regeneration, suggesting potential application in personalized and regenerative medicine for treating RA and periodontitis. Ongoing research and clinical trials will gradually reveal their effectiveness and any potential side effects, ultimately leading to the introduction of a broader spectrum of applications in clinical practice.

Funding This work was supported in part by National Institutes of Health grants, R01AR068279 (NIAMS), STTR R42EY031196 (NEI), and STTR 1R41AG057242 (NIA). The funders had no role in the preparation of the chapter.

References

Abdulkareem AA, Al-Taweel FB, Al-Sharqi AJ, Gul SS, Sha A, Chapple IL (2023) Current concepts in the pathogenesis of periodontitis: from symbiosis to dysbiosis. J Oral Microbiol 15(1):2197779

Abdulkhaleq L, Assi M, Abdullah R, Zamri-Saad M, Taufiq-Yap Y, Hezmee M (2018) The crucial roles of inflammatory mediators in inflammation: a review. Vet World 11(5):627

Ahmed AU (2011) An overview of inflammation: mechanism and consequences. Front Biol 6(4):274–281

Aletaha D, Neogi T, Silman AJ, Funovits J, Felson DT, Bingham CO III, Birnbaum NS, Burmester GR, Bykerk VP, Cohen MD (2010) 2010 rheumatoid arthritis classification criteria: an American College of Rheumatology/European league against rheumatism collaborative initiative. Arthritis Rheum 62(9): 2569–2581

Anderson S, Prateeksha P, Das H (2022) Dental pulp-derived stem cells reduce inflammation, accelerate wound healing and mediate M2 polarization of myeloid cells. Biomedicine 10(8):1999

Arthur A, Rychkov G, Shi S, Koblar SA, Gronthos S (2008) Adult human dental pulp stem cells differentiate toward functionally active neurons under appropriate environmental cues. Stem Cells 26(7):1787–1795

Babon JJ, Lucet IS, Murphy JM, Nicola NA, Varghese LN (2014) The molecular regulation of Janus kinase (JAK) activation. Biochem J 462(1):1–13

Bar JK, Lis-Nawara A, Grelewski PG (2021) Dental pulp stem cell-derived secretome and its regenerative potential. Int J Mol Sci 22(21):12018

Barnes PJ, Karin M (1997) Nuclear factor-κB—a pivotal transcription factor in chronic inflammatory diseases. N Engl J Med 336(15):1066–1071

Bax M, van Heemst J, Huizinga TW, Toes RE (2011) Genetics of rheumatoid arthritis: what have we learned? Immunogenetics 63:459–466

Boyle M, Chun C, Strojny C, Narayanan R, Bartholomew A, Sundivakkam P, Alapati S (2014) Chronic inflammation and angiogenic signaling axis impairs differentiation of dental-pulp stem cells. PLoS One 9(11):e113419

Burrage PS, Mix KS, Brinckerhoff CE (2006) Matrix metalloproteinases: role in arthritis. Front Biosci 11(1):529–543

Chen L, Deng H, Cui H, Fang J, Zuo Z, Deng J, Li Y, Wang X, Zhao L (2018) Inflammatory responses and inflammation-associated diseases in organs. Oncotarget 9(6):7204

Chen L, Liu G, Wu J, Zhou X, Zhao Y, Chen Z, Lin Z, Xiao Y (2020) Multi-faceted effects of mesenchymal stem cells (MSCs) determined by immune microenvironment and their implications on MSC/biomaterial-based inflammatory disease therapy. Appl Mater Today 18:100485

Chen W, Wang Q, Tao H, Lu L, Zhou J, Wang Q, Huang W, Yang X (2024a) Subchondral osteoclasts and osteoarthritis: new insights and potential therapeutic avenues: the role of OCs in OA. Acta Biochim Biophys Sin 56(4):499

Chen Z, Lang G, Xu X, Liang X, Han Y, Han Y (2024b) The role of NF-kappaB in the inflammatory processes related to dental caries, pulpitis, apical periodontitis, and periodontitis–a narrative review. PeerJ 12:e17953

Conforti A, Di Cola I, Pavlych V, Ruscitti P, Berardicurti O, Ursini F, Giacomelli R, Cipriani P (2021) Beyond the joints, the extra-articular manifestations in rheumatoid arthritis. Autoimmun Rev 20(2):102735

Dabravolski SA, Sukhorukov VN, Kalmykov VA, Grechko AV, Shakhpazyan NK, Orekhov AN (2022) The role of KLF2 in the regulation of atherosclerosis development and potential use of KLF2-targeted therapy. Biomedicine 10(2):254

Das H, Kumar A, Lin Z, Patino WD, Hwang PM, Feinberg MW, Majumder PK, Jain MK (2006) Kruppel-like factor 2 (KLF2) regulates proinflammatory activation of monocytes. Proc Natl Acad Sci USA 103(17): 6653–6658. https://doi.org/10.1073/pnas.0508235103

Das M, Lu J, Joseph M, Aggarwal R, Kanji S, McMichael BK, Lee BS, Agarwal S, Ray-Chaudhury A, Iwenofu OH, Kuppusamy P, Pompili VJ, Jain MK, Das H (2012) Kruppel-like factor 2 (KLF2) regulates monocyte differentiation and functions in mBSA and IL-1beta-induced arthritis. Curr Mol Med 12(2): 113–125. https://doi.org/10.2174/156652412798889090

Das M, Deb M, Laha D, Joseph M, Kanji S, Aggarwal R, Iwenofu OH, Pompili VJ, Jarjour W, Das H (2019) Myeloid Kruppel-like factor 2 critically regulates K/BxN serum-induced arthritis. Cells 8(8). https://doi.org/10.3390/cells8080908

De Pablo P, Dietrich T, McAlindon TE (2008) Association of periodontal disease and tooth loss with rheumatoid arthritis in the US population. J Rheumatol 35(1): 70–76

Demaria S, Pikarsky E, Karin M, Coussens LM, Chen Y-C, El-Omar EM, Trinchieri G, Dubinett SM, Mao JT, Szabo E (2010) Cancer and inflammation: promise for biologic therapy. J Immunother 33(4):335–351

Detert J, Pischon N, Burmester G, Buttgereit F (2010a) Pathogenese der Parodontitis bei rheumatischen Erkrankungen. Z Rheumatol 69:109–112

Detert J, Pischon N, Burmester GR, Buttgereit F (2010b) The association between rheumatoid arthritis and periodontal disease. Arthritis Res Ther 12:1–7

Di Benedetto A, Gigante I, Colucci S, Grano M (2013) Periodontal disease: linking the primary inflammation to bone loss. J Immunol Res 2013(1):503754

Ducret M, Fabre H, Degoul O, Atzeni G, McGuckin C, Forraz N, Alliot-Licht B, Mallein-Gerin F, Perrier-Groult E, Farges J-C (2015) Manufacturing of dental pulp cell-based products from human third molars: current strategies and future investigations. Front Physiol 6:213

Dutzan N, Konkel JE, Greenwell-Wild T, Moutsopoulos NM (2016) Characterization of the human immune cell network at the gingival barrier. Mucosal Immunol 9(5): 1163–1172

Fawzy El-Sayed KM, Elsalawy R, Ibrahim N, Gadalla M, Albargasy H, Zahra N, Mokhtar S, El Nahhas N, El Kaliouby Y, Doerfer CE (2019) The dental pulp stem/progenitor cells-mediated inflammatory-regenerative axis. Tissue Eng Part B Rev 25(5):445–460

Ferrarotti F, Romano F, Gamba MN, Quirico A, Giraudi M, Audagna M, Aimetti M (2018) Human intrabony defect regeneration with micrografts containing dental pulp stem cells: a randomized controlled clinical trial. J Clin Periodontol 45(7):841–850

Figueredo C, Lira-Junior R, Love R (2019) T and B cells in periodontal disease: new functions in a complex scenario. Int J Mol Sci 20(16):3949

Firestein GS (2003) Evolving concepts of rheumatoid arthritis. Nature 423(6937):356–361

Firestein GS (2014) The disease formerly known as rheumatoid arthritis. Arthritis Res Ther 16:1–3

Firestein GS, McInnes IB (2017) Immunopathogenesis of rheumatoid arthritis. Immunity 46(2):183–196

Fuggle NR, Smith TO, Kaul A, Sofat N (2016) Hand to mouth: a systematic review and meta-analysis of the association between rheumatoid arthritis and periodontitis. Front Immunol 7:80

Grabiec AM, Krausz S, De Jager W, Burakowski T, Groot D, Sanders ME, Prakken BJ, Maslinski W, Eldering E, Tak PP (2010) Histone deacetylase inhibitors suppress inflammatory activation of rheumatoid arthritis patient synovial macrophages and tissue. J Immunol 184(5):2718–2728

Graves DT, Corrêa JD, Silva TA (2019) The oral microbiota is modified by systemic diseases. J Dent Res 98(2):148–156

Gronthos S, Mankani M, Brahim J, Robey PG, Shi S (2000) Postnatal human dental pulp stem cells (DPSCs) in vitro and in vivo. Proc Natl Acad Sci 97(25):13625–13630

Hajishengallis G, Sahingur S (2014) Novel inflammatory pathways in periodontitis. Adv Dent Res 26(1):23–29

Hannoodee S, Nasuruddin DN (2020) Acute inflammatory response

Hernández-Monjaraz B, Santiago-Osorio E, Ledesma-Martínez E, Aguiñiga-Sánchez I, Sosa-Hernández NA, Mendoza-Núñez VM (2020) Dental pulp mesenchymal stem cells as a treatment for periodontal disease in older adults. Stem Cells Int 2020(1):8890873

Hou J, Karin M, Sun B (2021) Targeting cancer-promoting inflammation—have anti-inflammatory therapies come of age? Nat Rev Clin Oncol 18(5):261–279

Howell TH (1993) Blocking periodontal disease progression with anti-inflammatory agents. J Periodontol 64: 828–833

Ibrahim MA, Mohammed SA, Elhabak DM (2021) Role of KLF2: new insight in inflammatory acne pathogenesis. J Cosmet Dermatol 20(3):964–970

Imboden JB (2009) The immunopathogenesis of rheumatoid arthritis. Ann Rev Pathol Mech Dis 4(1):417–434

Iqbal S, Rattu MA (2019) Review of rheumatoid arthritis. US Pharm 44(1):8–11

Jha P, Das H (2017) KLF2 in regulation of NF-κB-mediated immune cell function and inflammation. Int J Mol Sci 18(11):2383

Kawai T, Akira S (2011) Toll-like receptors and their crosstalk with other innate receptors in infection and immunity. Immunity 34(5):637–650

Kipanyula MJ, Etet PFS, Vecchio L, Farahna M, Nukenine EN, Kamdje AHN (2013) Signaling pathways bridging microbial-triggered inflammation and cancer. Cell Signal 25(2):403–416

Kouzarides T (2007) Chromatin modifications and their function. Cell 128(4):693–705

Koziel J, Mydel P, Potempa J (2014) The link between periodontal disease and rheumatoid arthritis: an updated review. Curr Rheumatol Rep 16:1–7

Kyburz D, Brentano F, Schorr O, Michel B, Gay R, Gay S (2005) Toll-like receptor expression in synovial fibroblasts and normal skin fibroblasts and induction of matrix metallo-proteinase expression by various toll-like receptor ligands. Arthritis Res Ther 7:1–1

Lee S, Zhang Q, Karabucak B, Le A (2016) DPSCs from inflamed pulp modulate macrophage function via the TNF-α/IDO axis. J Dent Res 95(11):1274–1281

Lefevre S, Meier F, Neumann E, Muller-Ladner U (2015) Role of synovial fibroblasts in rheumatoid arthritis. Curr Pharm Des 21(2):130–141

Li F, Wang X, Shi J, Wu S, Xing W, He Y (2023) Anti-inflammatory effect of dental pulp stem cells. Front Immunol 14:1284868

Li W, Xiang Z, Yu W, Huang X, Jiang Q, Abumansour A, Yang Y, Chen C (2024) Natural compounds and mesenchymal stem cells: implications for inflammatory-impaired tissue regeneration. Stem Cell Res Ther 15(1):34

Liang Y, Luan X, Liu X (2020) Recent advances in periodontal regeneration: a biomaterial perspective. Bioactive Mat 5(2):297–308

Liang C, Liao L, Tian W (2021) Stem cell-based dental pulp regeneration: insights from signaling pathways. Stem Cell Rev Rep 17:1–13

Lin J, Hwang S, Luo H, Mohamud Y (2024) Double-edged sword: exploring the mitochondria–complement bidirectional connection in cellular response and disease. Biology 13(6):431

Liu W, Zhang X (2015) Receptor activator of nuclear factor-κB ligand (RANKL)/RANK/osteoprotegerin system in bone and other tissues. Mol Med Rep 11(5):3212–3218

Liu Y, Wang L, Kikuiri T, Akiyama K, Chen C, Xu X, Yang R, Chen W, Wang S, Shi S (2011) Mesenchymal stem cell–based tissue regeneration is governed by recipient T lymphocytes via IFN-γ and TNF-α. Nat Med 17(12):1594–1601

Liu Y-C, Zou X-B, Chai Y-F, Yao Y-M (2014) Macrophage polarization in inflammatory diseases. Int J Biol Sci 10(5):520

MacGregor AJ, Snieder H, Rigby AS, Koskenvuo M, Kaprio J, Aho K, Silman AJ (2000) Characterizing the quantitative genetic contribution to rheumatoid arthritis using data from twins. Arthrit Rheumat Off J Am Coll Rheumatol 43(1):30–37

Marchesan JT, Girnary MS, Moss K, Monaghan ET, Egnatz GJ, Jiao Y, Zhang S, Beck J, Swanson KV (2020) Role of inflammasomes in the pathogenesis of periodontal disease and therapeutics. Periodontol 82(1):93–114

Marcu KB, Otero M, Olivotto E, Maria Borzi R, Goldring MB (2010) NF-κB signaling: multiple angles to target OA. Curr Drug Targets 11(5):599–613

McInnes IB, Schett G (2011) The pathogenesis of rheumatoid arthritis. N Engl J Med 365(23):2205–2219

McInnes IB, Schett G (2017) Pathogenetic insights from the treatment of rheumatoid arthritis. Lancet 389(10086):2328–2337

Medzhitov R (2008) Origin and physiological roles of inflammation. Nature 454(7203):428–435

Nakagawa H, Maeda S (2012) Inflammation-and stress-related signaling pathways in hepatocarcinogenesis. World J Gastroenterol: WJG 18(31):4071

Nakashima M, Iohara K, Murakami M, Nakamura H, Sato Y, Ariji Y, Matsushita K (2017) Pulp regeneration by transplantation of dental pulp stem cells in pulpitis: a pilot clinical study. Stem Cell Res Ther 8:1–13

Nascimento GG, Leite FR, Scheutz F, López R (2017) Periodontitis: from infection to inflammation. Curr Oral Health Rep 4:301–308

Nathan C (2002) Points of control in inflammation. Nature 420(6917):846–852

Nathan C, Ding A (2010) Nonresolving inflammation. Cell 140(6):871–882

Nigrovic PA, Lee DM (2004) Mast cells in inflammatory arthritis. Arthritis Res Ther 7:1–11

Nowwarote N, Petit S, Ferre FC, Dingli F, Laigle V, Loew D, Osathanon T, Fournier BP (2022) Extracellular matrix derived from dental pulp stem cells promotes mineralization. Front Bioeng Biotechnol 9:740712

Ogata K, Matsumura-Kawashima M, Moriyama M, Kawado T, Nakamura S (2021) Dental pulp-derived stem cell-conditioned media attenuates secondary Sjögren's syndrome via suppression of inflammatory cytokines in the submandibular glands. Regen Ther 16: 73–80

Pal B, Endisha H, Zhang Y, Kapoor M (2015) mTOR: a potential therapeutic target in osteoarthritis? Drugs R&D 15:27–36

Passacquale G, Di Giosia P, Ferro A (2016) The role of inflammatory biomarkers in developing targeted cardiovascular therapies: lessons from the cardiovascular inflammation reduction trials. Cardiovasc Res 109(1): 9–23

Peng B, Li Z, Cheng Y, Jiang H, Ye Q, Han G (2025) Dental pulp stem cell-derived intracellular vesicles prevent orthodontic relapse by inhibiting PI3K/Akt/NF-κB-mediated osteoclast activity. Stem Cell Res Ther 16(1):22

Pierdomenico L, Bonsi L, Calvitti M, Rondelli D, Arpinati M, Chirumbolo G, Becchetti E, Marchionni C, Alviano F, Fossati V (2005) Multipotent mesenchymal stem cells with immunosuppressive activity can be easily isolated from dental pulp. Transplantation 80(6):836–842

Pischon N, Pischon T, Kröger J, Gülmez E, Kleber BM, Bernimoulin JP, Landau H, Brinkmann PG, Schlattmann P, Zernicke J (2008) Association among rheumatoid arthritis, oral hygiene, and periodontitis. J Periodontol 79(6):979–986

Pisciotta A, Bertoni L, Riccio M, Mapelli J, Bigiani A, La Noce M, Orciani M, De Pol A, Carnevale G (2018) Use of a 3D floating sphere culture system to maintain the neural crest-related properties of human dental pulp stem cells. Front Physiol 9:547

Potempa J, Mydel P, Koziel J (2017) The case for periodontitis in the pathogenesis of rheumatoid arthritis. Nat Rev Rheumatol 13(10):606–620

Prateeksha P, Howlader MSI, Hansda S, Naidu P, Das M, Abo-Aziza F, Das H (2023a) Secretome of dental pulp-derived stem cells reduces inflammation and proliferation of glioblastoma cells by deactivating Mapk-Akt pathway. Dis Res 3(2):74

Prateeksha P, Naidu P, Das M, Barthels D, Das H (2023b) KLF2 regulates neural differentiation of dental pulp-derived stem cells by modulating autophagy and Mitophagy. Stem Cell Rev Rep 19(8):2886–2900. https://doi.org/10.1007/s12015-023-10607-0

Preshaw PM (2018) Host modulation therapy with anti-inflammatory agents. Periodontol 76(1):131–149

Purohit M, Gupta G, Afzal O, Altamimi ASA, Alzarea SI, Kazmi I, Almalki WH, Gulati M, Kaur IP, Singh SK (2023) Janus kinase/signal transducers and activator of transcription (JAK/STAT) and its role in lung inflammatory disease. Chem Biol Interact 371:110334

Rai V, Mathews G, Agrawal DK (2022) Translational and clinical significance of DAMPs, PAMPs, and PRRs in trauma-induced inflammation. Arch Clin Biomed Res 6(5):673

Ravelli A, Martini A (2007) Juvenile idiopathic arthritis. Lancet 369(9563):767–778

Rizo-Téllez SA, Sekheri M, Filep JG (2023) C-reactive protein: a target for therapy to reduce inflammation. Front Immunol 14:1237729

Roman-Blas J, Jimenez S (2006) NF-κB as a potential therapeutic target in osteoarthritis and rheumatoid arthritis. Osteoarthr Cartil 14(9):839–848

Saini R, Marawar P, Shete S, Saini S (2009) Periodontitis, a true infection. J Global Infect Dis 1(2):149–150

Scher JU, & Abramson SB (2013) Periodontal disease, Porphyromonas gingivalis, and rheumatoid arthritis: what triggers autoimmunity and clinical disease? Arthritis Res & Ther 15(5):122

SenBanerjee S, Lin Z, Atkins GB, Greif DM, Rao RM, Kumar A, Feinberg MW, Chen Z, Simon DI, Luscinskas FW (2004) KLF2 is a novel transcriptional regulator of endothelial proinflammatory activation. J Exp Med 199(10):1305–1315

Shanmugam MK, Sethi G (2012) Role of epigenetics in inflammation-associated diseases. Dev Dis, Epigenetics, pp 627–657

Shi J, Zhou L-r, Wang X-s, Du J-f, Jiang M-m, Song Z, Han G-c, Mai Z-t (2018) KLF2 attenuates bleomycin-induced pulmonary fibrosis and inflammation with regulation of AP-1. Biochem Biophys Res Commun 495(1):20–26

Song B, Zhang Y, Chen L, Zhou T, Huang W, Zhou X, Shao L (2017) The role of Toll-like receptors in periodontitis. Oral Dis 23(2):168–180

Tabas I, Glass CK (2013) Anti-inflammatory therapy in chronic disease: challenges and opportunities. Science 339(6116):166–172

Thorbert-Mros S, Larsson L, Berglundh T (2015) Cellular composition of long-standing gingivitis and periodontitis lesions. J Periodontal Res 50(4):535–543

Tsuruta T, Sakai K, Watanabe J, Katagiri W, Hibi H (2018) Dental pulp-derived stem cell conditioned medium to regenerate peripheral nerves in a novel animal model of dysphagia. PLoS One 13(12): e0208938

Udalova IA, Mantovani A, Feldmann M (2016) Macrophage heterogeneity in the context of rheumatoid arthritis. Nat Rev Rheumatol 12(8):472–485
Wang M, Xie J, Wang C, Zhong D, Xie L, Fang H (2020) Immunomodulatory properties of stem cells in periodontitis: current status and future prospective. Stem Cells Int 2020(1):9836518
Wei J, Feng J (2010) Signaling pathways associated with inflammatory bowel disease. Recent Patents Inflamm Allergy Drug Discov 4(2):105–117
Wright HL, Moots RJ, Bucknall RC, Edwards SW (2010) Neutrophil function in inflammation and inflammatory diseases. Rheumatology 49(9):1618–1631
Xiao F, Li C, Lin Y, Peng Z, Xu X, Wen Y, Huang N, Zhang P (2021) Increased risk of periodontitis occurrence in patients with rheumatoid arthritis and its association with the levels of IL-1β and TNF-α in gingival crevicular fluid. Annal Palliat Med 10(8): 9078087–9079087
Yamaguchi S, Shibata R, Yamamoto N, Nishikawa M, Hibi H, Tanigawa T, Ueda M, Murohara T, Yamamoto A (2015) Dental pulp-derived stem cell conditioned medium reduces cardiac injury following ischemia-reperfusion. Sci Rep 5(1):16295
Yamaza T, Kentaro A, Chen C, Liu Y, Shi Y, Gronthos S, Wang S, Shi S (2010) Immunomodulatory properties of stem cells from human exfoliated deciduous teeth. Stem Cell Res Ther 1:1–11
Yang Y, Tang L-Q, Wei W (2013) Prostanoids receptors signaling in different diseases/cancers progression. J Recept Signal Transduct 33(1):14–27
Yazid FB, Gnanasegaran N, Kunasekaran W, Govindasamy V, Musa S (2014) Comparison of immunodulatory properties of dental pulp stem cells derived from healthy and inflamed teeth. Clin Oral Investig 18:2103–2112
Yeung YT, Aziz F, Guerrero-Castilla A, Arguelles S (2018) Signaling pathways in inflammation and anti-inflammatory therapies. Curr Pharm Des 24(14): 1449–1484
Zhang S, Meng Y, Zhou L, Qiu L, Wang H, Su D, Zhang B, Chan KM, Han J (2022) Targeting epigenetic regulators for inflammation: mechanisms and intervention therapy. MedComm 3(4):e173
Zhao H, Wu L, Yan G, Chen Y, Zhou M, Wu Y, Li Y (2021) Inflammation and tumor progression: signaling pathways and targeted intervention. Signal Transduct Target Ther 6(1):263
Zhu L, Zhou C, Chen S, Huang D, Jiang Y, Lan Y, Zou S, Li Y (2022) Osteoporosis and alveolar bone health in periodontitis niche: a predisposing factors-centered review. Cells 11(21):3380
Zimmermann P, Ziesenitz VC, Curtis N, Ritz N (2018) The immunomodulatory effects of macrolides—a systematic review of the underlying mechanisms. Front Immunol 9:302

38 Innovative Biomarkers for Oral Immunity: Advancements and Clinical Applications

Amin Gasmi, Asma Gasmi Benahmed, Amine Nehaoua, Sadaf Noor, and Alain Menzel

Abstract

Oral immunity is an essential barrier for protection against pathogens, supporting oral health, and overall health-related challenges. Nevertheless, dysregulation of oral immunity results in different pathological conditions that are often challenging to diagnose and manage. Biomarkers have revolutionized the management of oral health by offering information about molecular and cellular immune events, disease etiology, and treatment outcomes. This review focuses on the biomarkers for oral immunity in the context of the existing trends and includes genetic, proteomic, metabolomic, and microbiome biomarkers. Single-nucleotide polymorphism (SNP) and copy number variation (CNV) have improved the knowledge of the genetic risk factors for oral diseases and the treatment plan for patients. Cytokines, matrix metalloproteinases (MMPs), and antimicrobial peptides (AMPs) have enhanced the diagnostic accuracy of diseases, prognostication, and evaluation of therapeutic efficacy. Metabolomic biomarkers have provided valuable insights into the biochemical pathways related to oral health and diseases, while microbiome-derived biomarkers have revealed the intricate host-microbiota interactions. Recent developments in high-throughput screening, multi-omics, and computational biology have facilitated the identification of biomarkers of oral immunity. These molecular markers have been proven to have the potential for screening, prognosis, and targeted therapy of oral diseases such as periodontitis, oral cancer, and autoimmune diseases. However, some issues are still to be solved, such as standardization of biomarker usage, obtaining the approval of biomarkers by different authorities, and the economic issues regarding the application of biomarkers for clinical diagnostics.

A. Gasmi (✉)
Société Francophone de Nutrithérapie et de Nutrigénétique Appliquée, Villeurbanne, France

International Institute of Nutrition and Micronutrition Sciences, Saint-Étienne, France

A. G. Benahmed
Académie Internationale de Médecine Dentaire Intégrative, Paris, France

A. Nehaoua
Research Department, Precision Health International, Ain Oulmene, Algeria

Laboratory of Physical Activity Sciences and Public Health, Mohamed Lamine Debaghine Sétif 2 University, Setif, Algeria

S. Noor
Institute of Molecular Biology and Biotechnology, Bahauddin Zakariya University, Multan, Pakistan

A. Menzel
Laboratoires Réunis, Junglinster, Luxembourg

N. Rezaei, H. Ziaei (eds.), *Oral Immunology*, Advances in Experimental Medicine and Biology 1492,
https://doi.org/10.1007/978-3-032-03176-1_38

Keywords

Oral immunity · Biomarkers · Personalized medicine · Diagnostics · Oral diseases

38.1 Introduction

Oral immunity is a powerful defense against many pathogens and environmental factors that may affect oral as well as general health (Kozak and Pawlik 2023). The interactions between innate and adaptive immunity in the oral cavity are critical to prevent infections, and modulate tissue integrity and inflammation (Cheng et al. 2021). Nevertheless, dysregulation of oral immunity leads to several pathological states, such as periodontitis, oral lichen planus (OLP), and oral squamous cell carcinoma (OSCC), all of which present diagnostic, therapeutic, and management difficulties (Hajishengallis 2015; Nukaly et al. 2024). Previously, the diagnosis and treatment of oral immune-related diseases have been done clinically with the aid of simple investigations that are not very sensitive or specific (Al-Shehri 2023). It has also restricted our capacity to diagnose diseases in their early stages, estimate their progression rate, or prescribe therapies according to patients' characteristics. This emphasizes the need for the development of advanced techniques to boost our comprehension of oral immunity and enhance patient outcomes (Kornman 2018; Patil et al. 2023). Over the past few years, the idea of biomarkers has been proposed as a new approach to transforming oral health research (Korgaonkar et al. 2024). Biomarkers can be defined as biological indicators of normal physiological or pathogenic processes, as well as responses to therapeutic intervention (Ahmad et al. 2023; Strimbu and Tavel 2010). In terms of oral immunity, biomarkers can give valuable information regarding the molecular and cellular processes of immune response, disease pathogenesis, and therapeutic response (Cirillo 2024; Ebersole et al. 2013).

The prospect of biomarkers in the context of oral health is multifaceted and complex. First, they provide the potential for early diagnosis of a disease, even if the person does not experience any particular signs. These early interventions can enhance the prognosis and reduce the chance of exacerbating oral diseases to worse stages (Cirillo 2024). Second, biomarkers can help with the correct diagnosis because they offer measurable data in addition to the subjective evaluations of the clinician. This is especially useful where signs and symptoms are ambiguous or are seen in a variety of diseases or disorders (Ahmad et al. 2023; Kinney et al. 2011). In addition, biomarkers can revolutionize current treatment approaches by facilitating the development of individualized treatment plans. Biomarkers facilitate the determination of the risk of developing a certain disease, the evaluation of the possibility of relapse, as well as the identification of the optimal therapy for a given patient. Such specificity can help maximize the therapeutic values of a medication, while reducing its toxicity, thus increasing the patients' well-being (Giannobile et al. 2013; Laigle et al. 2023; Passaro et al. 2024). The study of oral immunity biomarkers is constantly developing, driven by the advancement in various fields such as genomics, proteomics, metabolomics, and microbiomics. Technological advancement has increased the opportunity to discover and validate new biomarkers, which have brought new hope for the management of immune-related oral diseases (Kumari et al. 2024). The key biomarker categories discussed in this review are summarized in (Fig. 38.1).

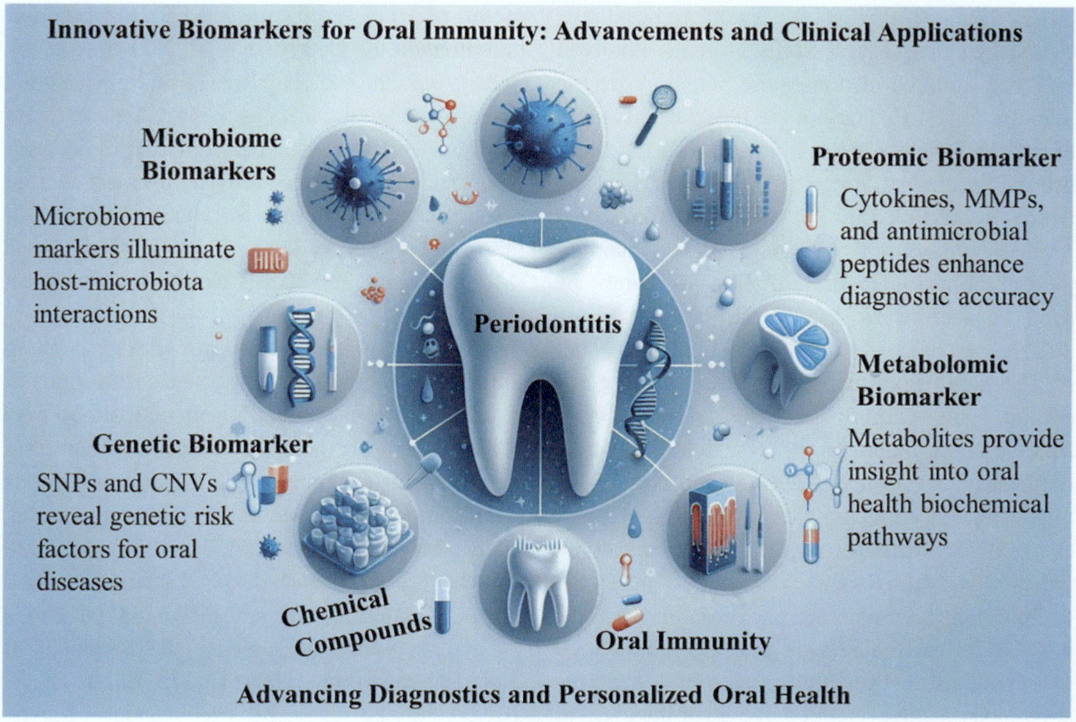

Fig. 38.1 This figure summarizes the key biomarker categories for advanced diagnosis and personalized oral health with the main focus on periodontitis as a key example. The surrounding icons represent various biomarker types including microbiome, proteomic, metabolomic, genetic, and chemical compounds. The purpose of this interdisciplinary biomarker approach is to facilitate the in-depth understanding of oral immunopathogenesis that will help devise highly sensitive as well as specific diagnostic tools and personalized oral healthcare strategies

38.2 Types of Biomarkers in Oral Immunity

38.2.1 Genetic Biomarkers

Genetic biomarkers have been established as crucial tools for explaining the relationship between an individual's genotype and immunological diseases of the oral cavity (Malcangi et al. 2023). Such biomarkers, including single SNPs, CNVs, and epigenetic changes, are useful in understanding the genetic vulnerability to several oral diseases and the possibility of individualized treatment strategies (Schafer et al. 2020; Ummer et al. 2024). Periodontitis is one of the most investigated subjects in the field of oral immunity genetics; it is a chronic inflammatory disease that affects the supporting structures of the teeth (Farhad et al. 2024). Many genetic polymorphisms have been implicated in the development and progression of periodontitis. For example, polymorphisms in genes coding for interleukin-1 (IL-1) have been reported to be associated with risk for aggressive periodontitis (D'Agostino et al. 2023). In a meta-analysis of genetic polymorphisms associated with chronic periodontitis, Nibali et al. (2016) identified the two polymorphisms *IL1A* −889 and *IL1B* +3954 as being significant among populations of Caucasian origin (Madruga González 2023). These findings have implications for the early screening for genetic markers of individuals who may be at a high risk of developing severe periodontal diseases, so that preventive measures can be taken. In the same way, genetic biomarkers have been used to explain the etiology of OLP, which is an immune-mediated disease affecting the mucocutaneous junctions (Danielsson et al.

2018; Liu et al. 2023). Two studies have established that several SNPs in genes involved in T cell immune responses are potential genetic markers for OLP. These were polymorphisms in the *FOXP3* gene which is important in the function of regulatory T cells and the *CTLA4* gene which is involved in T cell activation. Such genetic markers not only help in understanding the pathophysiology of OLP but also might be used as therapeutic targets (El-Howati et al. 2023).

The field of oral cancer has also not been left behind in terms of the revolutionary achievements made in the research of genetic biomarkers. *TP53* and *CDKN2A* are two of the most frequently mutated genes in OSCC (Hurník et al. 2023; Sriram et al. 2024). Besides, these genetic signatures are useful not only for deciphering the molecular background of OSCC but also for early diagnosis and risk stratification of patients. Breakthroughs in genomic technologies have enabled the identification as well as the validation of genetic biomarkers in oral immunity. Genome-wide association studies (GWAS) have become an important approach in the discovery of new genetic loci that are linked to oral immune-related diseases (Fujihara et al. 2023). For example, a recent GWAS on aggressive periodontitis by Fujihara et al. (2023) not only replicated previous findings but also revealed new risk loci in the neighborhood of *SIGLEC5* and *DEFA1A3* genes, which are associated with innate immunity. Not only do such studies contribute to the advancement of knowledge in the genetic pathology of oral diseases, but also they help in enhancing the potential of genetic risk profiling. Subsequent advancements in next-generation sequencing (NGS) technologies have only helped in identifying rare genetic variants and intricate genomic changes involved in the pathogenesis of oral immune disorders (Abdul et al. 2024). Among the targeted gene sequencing techniques, whole-exome (WES) and whole-genome sequencing (WGS) have been most effective in establishing the source of mutations that cause rare inherited disorders of oral immunity (Abdul et al. 2024). For example, Kapferer-Seebacher et al. (2017) employed WES to pinpoint mutations in the Cathepsin S (*CTSC*) gene as the underlying cause of Papillon-Lefèvre syndrome, a recessive genodermatosis that presents with severe periodontitis and palmoplantar hyperkeratosis. Such findings not only help in differential diagnosis but also point to potential targets for therapeutic interventions by gene therapy (McCarthy et al. 2023; Napiórkowska-Baran et al. 2024).

Genetic biomarkers are of great importance in the relatively new field of pharmacogenomics in oral health. An individual's genetic makeup may determine their response to drugs that are used in the treatment of immune-related oral conditions (Korczeniewska et al. 2024; Malcangi et al. 2023). For example, variations in the *CYP2C19* gene have been identified to influence the metabolism of nonsteroidal anti-inflammatory drugs (NSAIDs), which are used in the treatment of oral inflammation (Daly 2017; Zhou et al. 2023). In the case of oral cancer, genetic biomarkers are beginning to be applied in the management of the disease and its prognosis (Cirillo 2024). For instance, in OSCC, the genetic changes may include mutations in the *EGFR* gene, which, if present, may indicate the patient's potential response to targeted therapies such as tyrosine kinase inhibitors (Huang and O'Sullivan 2017; Veeraraghavan et al. 2024).

38.2.2 Proteomic Biomarkers

The discovery of proteomic biomarkers is seen as a significant factor in the immune responses in the oral cavity and a potential source of significant information on the molecular mechanisms of oral health and disease (Ahmad et al. 2024). Given that proteins are the molecular workhorses of the biological system, they are involved in various processes such as immune reactions, inflammation, and maintenance of homeostasis of the tissues in the oral cavity. Proteomic biomarkers have greatly enhanced our knowledge of oral immunity and have provided new ways of diagnosing, predicting, and managing several oral diseases (Fig. 38.2) (Ahmad et al. 2024;

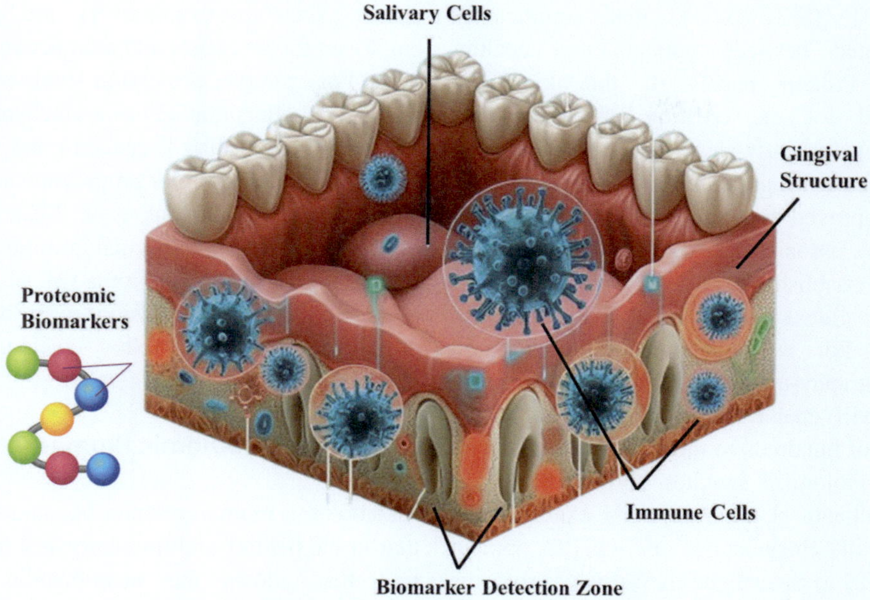

Fig. 38.2 Cross-sectional view of oral tissue illustrating key components of oral immunity and biomarker detection

Ebersole et al. 2015). Among the proteomic biomarkers of oral immunity, cytokines and chemokines have been the most extensively studied. These signaling molecules are involved in the coordination of immune responses and controlling inflammation in the oral environment (Einhaus et al. 2023). For example, IL-1β, tumor necrosis factor (TNF-α), and IL-6 have been identified to be implicated in the development of periodontal diseases, and are known to have a positive correlation with disease progression (Guarnieri et al. 2024; Hao et al. 2023). A systematic review and meta-analysis showed that the concentration of these pro-inflammatory cytokines was significantly higher in the gingival crevicular fluid (GCF) of periodontitis patients than in healthy individuals. These studies also emphasized cytokines as diagnostic markers and, more importantly, as targets for therapeutic interventions in periodontal inflammation. In addition to cytokines, other proteomic biomarkers that have been identified in oral immunity include MMPs, which have important functions in the degradation of the extracellular matrix and have been implicated in tissue pathology in several oral diseases (Pomella et al. 2024). Zalewska et al. (2024) have shown that salivary MMP-8 can be used to differentiate between various types of periodontitis and estimate outcomes of therapy. The authors suggested that MMP-8-based point-of-care tests could be useful for periodontal disease diagnosis and for the evaluation of disease activity in clinical practice.

Another functional proteomic biomarker in oral immunity is AMPs. These small and positively charged peptides are considered components of the innate immune system and have important functions in the regulation of oral microbiota (Morio et al. 2023). Human β-defensins (hBDs) and cathelicidins have been well-investigated in oral health (Atalay et al. 2024; Yousif and Qasim 2023). Ghosh et al. (2019) also reported that hBD-2 and cathelicidin (LL-37) were significantly downregulated in patients with recurrent aphthous stomatitis (RAS), suggesting that these two proteins could be potential diagnostic biomarkers and therapeutic targets for RAS, which is the most common oral mucosal disease. Csősz et al. (2017) employed mass spectrometry (MS)-based proteomics to establish a salivary protein signature for OSCC, consisting of complement factor H (CFH), complement component C (C3), complement component 4B (C4B), and serpin family A member 1 (SERPINA1), that significantly differentiated between patients and healthy controls. Protein panels of this kind are anticipated for the creation of noninvasive diagnostics for early-stage oral cancer. Recent developments in proteomic technologies have greatly improved the opportunity to identify new biomarkers associated with oral immunity. MS has been established as a high-throughput technique for global protein identification in oral biofluids. For example, methods like liquid chromatography-tandem mass spectrometry (LC-MS/MS) enable the identification and quantification of hundreds to thousands of proteins in complex biological samples (Grocholska et al. 2023; Kallianta et al. 2023). For example, the research of Bostanci et al. (2018) applied LC-MS/MS to investigate the salivary proteome in health and periodontitis and to discover more than 5000 proteins and potential biomarkers.

Microarrays of proteins are another significant technology in proteomic biomarker identification. These high-throughput platforms enable the identification and measurement of multiple proteins from small sample volumes (Poulos et al. 2023). Another study by Siting Chen et al. (2024b) used antibody microarrays to measure cytokine concentrations in the saliva of patients with OLP and revealed specific cytokine patterns characteristic of the disease. Beyond biomarker identification, these technologies facilitate the generation of multiplex assays to comprehensively evaluate the oral immune profile. For instance, in oral immunity, not only can the proteomic biomarkers be used to diagnose a condition, but they are also able to track the progression of the disease and its response to treatment (Zhou and Liu 2023). In periodontitis management, proteomic analysis in GCF or saliva in longitudinal studies would be useful in determining the effectiveness of the treatment plan and informing clinicians on the next course of action (Ebersole et al. 2024; George et al. 2024). Sharifi (2023) showed that the changes in GCF levels of IL-1β, MMP-8, and Osteopontin (OPN) during periodontal therapy were associated with clinical parameters; thus, they were used as surrogate markers for clinical trials and individualized treatment plans. In oral cancer, proteomic biomarkers are gradually employed for prognosis and therapeutic management. For example, expression levels of proteins associated with epithelial-to-mesenchymal transition (EMT), including E-cadherin and vimentin, have been linked to tumor progression and metastasis in OSCC (Ghantous et al. 2024; Lorenzo-Pouso et al. 2023). Molecular profiling of such markers can also be used in the risk stratification of patients and inform decisions regarding the use of adjuvant therapy and follow-up care.

38.2.3 Metabolomic Biomarkers

Metabolomic biomarkers have become more popular in the field of oral immunity as a new direction that allows the identification of key biochemical processes underlying oral health and disease (Zhou and Liu 2023). Metabolomics refers to the assessment of small molecule metabolites in biological systems; it is a functional genomics technique that gives a functional readout of cellular processes that are not available in the context of genomics and proteomics (Satrio et al. 2024). In the context of oral immunity, metabolomic biomarkers can provide information regarding immune status, microbial activity, and host-microbiome interactions, making them useful for diagnosis, prognosis, and monitoring treatment outcomes of oral diseases (Inchingolo et al. 2023; Wei et al. 2024a). The oral metabolome is an intricate and ever-changing system that is dependent on host genetics, immune system function, microbial processes, and environmental factors. Alterations in metabolite profiles are often the initial signs of dysbiosis and immune disorders before the onset of clinical symptoms (Wei et al. 2024a). For example, Rashid et al. (2024) looked at certain salivary metabolites that are linked to periodontal disease advancement. The authors discovered that the concentration of cadaverine, a polyamine synthesized by periodontal pathogens, was increased in patients with periodontitis and was related to clinical parameters of the disease. These metabolite signatures are not only biomarkers for diagnosis of the disease but also provide information about the microbial and host factors contributing to the disease progression (Qiu et al. 2023). Metabolomic analysis in oral cancer research has shown new biomarkers for early diagnosis and prognosis (Yao et al. 2023). Antonelli et al. (2024) conducted a large-scale metabolomic study to compare OSCC patients to healthy controls and identified salivary metabolites such as choline, betaine, and pipecolate as highly accurate biomarkers for accurately distinguishing between the two groups. These metabolic changes are indicative of the so-called Warburg effect, which is characteristic of cancer cells and can be used for the development of noninvasive diagnostic tests for oral cancer.

Molecular targets of metabolomics have also included inflammatory oral diseases like RAS and OLP. Rezazadeh et al. (2019) used nuclear magnetic resonance (NMR) spectroscopy to investigate the salivary metabolomics of patients with RAS. The researchers noted the changes in the levels of amino acids and carbohydrates, shedding light on metabolomic changes associated with the disease progression of this common oral mucosal disease and possible treatment modalities. Metabolomics has been used to investigate host-microbiome interactions within the oral ecosystem and has provided significant information on the contribution of the oral microbiome to health and disease.

38.2.4 Microbiome-Derived Biomarkers

The oral microbiome is the microbial community that inhabits the oral cavity and is involved in the regulation of oral health and the immune system (Kozak and Pawlik 2023). Innovations in molecular methods have revolutionized the knowledge of the oral microflora and its possible use as a source of biomarkers for oral immunity (Lamont et al. 2018). Oral microbiota and the immune system are in a mutually dependent and complex manner, which makes the understanding of their interactions challenging (Talapko et al. 2024). However, the host immune system is useful to maintain the microbial balance, while the microbiome also regulates the immune system in several ways. An imbalance of these microbial communities, known as dysbiosis, has been associated with different oral pathological diseases (OPDs) such as periodontitis, caries, and oral cancer (Zhang et al. 2024). Thus, the microbial molecules and their functional activities can serve as indicators of oral health conditions and the immune system function (Basic and Dahlén 2023). Among all the biomarkers, microbial metabolites are emerging as the most promising biomarkers of oral immunity. These small molecules are directly released from oral microorganisms and affect the host cells as well as immune system responses. For example, the short-chain fatty acids (SCFAs) derived from the commensal bacteria are known to regulate inflammation and maintain oral mucosal barrier integrity (Leonov et al. 2023). However, a study by Woo et al. (2021) revealed that the concentration of butyrate, one of the SCFAs, in the saliva of patients with periodontitis was significantly lower than that of healthy individuals. The authors suggested that such a decrease could lead to dysregulated immune responses and proposed salivary butyrate as a potential biomarker for periodontal health.

Another class of microbial metabolites that are recently being considered as biomarkers are quorum-sensing molecules (Markus et al. 2023). These signaling molecules, which are employed by bacteria for intercellular communication, can also modulate the host cell function and immune system. Autoinducer-2 (AI-2), a universal quorum-sensing signal, is involved in biofilm formation and the regulation of virulence in oral pathogens (Li et al. 2024). Wirth et al. (2022) demonstrated that AI-2 levels in saliva were correlated with the levels of periodontitis-associated bacteria and disease clinical parameters, highlighting its potential as a biomarker for periodontal disease activity. Another way of creating microbiome-derived biomarkers is the use of microbial DNA and RNA markers. Certain genes or transcripts can act as biomarkers of the existence and function of certain microbes or microbial communities in the oral ecosystem. For instance, Song Chen et al. (2024a) used

PCR-based identification of fungal internal transcribed spacer (ITS) regions in the oral mycobiome of HIV patients. The researchers established specific DNA profiles of fungi that were linked to oral candidiasis and proved that fungal DNA markers could be useful in the diagnosis of opportunistic infections in immunocompromised patients.

Compared to metagenomics, which captures all the genetic material in a sample, functional metagenomics provides insights into the metabolic and pathogenic potential of the oral microbiome. Ye et al. (2024) in their metagenomic analysis found that there were unique microbial genes and pathways associated with the subgingival microbiome of patients with periodontitis. The authors also observed that the genes in the lipopolysaccharide biosynthesis and flagellar assembly pathways were upregulated in disease states, suggesting their potential as functional biomarkers for periodontal inflammation.

38.3 Advancements in Biomarker Discovery and Validation

38.3.1 Technological Innovations

There is rapid progress in high-throughput screening technologies, bioinformatics tools, and the application of AI in biomarker research in the field of oral immunity (Chavda et al. 2023). These innovations have revolutionized the process of biomarker identification, qualification, and implementation into clinical practice, providing a unique and profound understanding of the molecular mechanisms of oral health and disease (Malcangi et al. 2023). Molecular biomarkers can be defined as molecules that are measurable indicators of pathophysiological processes, risk factors or responses to an intervention and high-throughput screening technologies have become particularly important as screening methods for biomarker discovery that allow the analysis of thousands of molecules in biological samples at the same time (Ahmad et al. 2023). These technologies have been applied in oral immunity research to identify various biomarkers, such as genetic variants, proteins, metabolites, and microbial profiles. For example, NGS platforms have revolutionized the possibilities of finding genetic biomarkers associated with oral immune disorders (Shreevatsa et al. 2024). In recent work, Xue et al. (2024) carried out whole-exome sequencing to discover new genetic risk factors for aggressive periodontitis in a group of patients. Not only did this approach validate previously identified risk loci, but it also discovered additional candidate genes involved in immune regulation and tissue homeostasis, thus providing new insights into the genetic predisposition to periodontal diseases. In the field of proteomic biomarker discovery, the application of mass spectrometers has been advanced in terms of sensitivity. Recent advances in data-independent acquisition (DIA) MS have improved the coverage and reproducibility of proteome analysis (Kitata et al. 2023).

Technological advancements have also facilitated metabolomic profiling through high-resolution MS and NMR spectroscopy for the precise identification and quantification of various metabolites in oral samples (Grootveld et al. 2023). Buss (2023) conducted an untargeted metabolomic study using ultra-high-performance liquid chromatography coupled with mass spectrometry (UHPLC-MS) to analyze saliva samples from patients suffering from OLP. This approach provided distinct metabolic profiles related to disease activity and treatment response, suggesting the feasibility of high-throughput metabolomics for biomarker identification in oral microbiome-related diseases. The employment of several high-throughput platforms in the multi-omics strategy has become well-accepted in biomarker identification. Through the integration of genomics, transcriptomics, proteomics, and metabolomics approaches, the investigators will be able to get an integrated picture of oral immunity (Peng et al. 2023). Wan et al. (2022) conducted a multi-omics analysis in OSCC using WGS, RNA-seq, and LC-MS/MS proteomics data to identify biomarkers for OSCC. The present integrated study provided new insights into the molecular changes of OSCC progression and potential therapeutic targets for individualized treatment.

Network analysis and systems biology approaches have also been used in identifying the relationships between the biomarkers and their functional roles in oral immunity. A recent systematic analysis of gingival tissues in health and periodontitis by Zhu et al. (2023) used a multi-omics approach of RNA-seq and quantitative proteomics. This study identified several key molecular networks and central genes associated with periodontal inflammation, which may help in understanding the disease mechanisms and finding new treatment targets.

38.3.2 Emerging Biomarkers in Oral Immunity

Oral immunity is a relatively young scientific area that has been rapidly developing leading to the discovery of new biomarkers for diagnosing oral health conditions and diseases in recent years (Cirillo 2024). These emerging biomarkers can be broadly classified into the following groups of molecular species: microRNAs (miRNAs), extracellular vesicles (EVs), and new protein posttranslational modifications. Their applications range from early diagnosis of diseases to developing suitable treatment plans, creating new opportunities for the enhancement of oral health care. miRNAs have recently gained significant attention as potential biomarkers for oral immune-related diseases. These small, non-coding RNA molecules are involved in the regulation of gene expression and are associated with several aspects of immune responses (Nag et al. 2023; Wu et al. 2023). Research on the subject of miRNA biomarkers in oral health and diseases was conducted by Mehdipour et al. (2023). The researchers also isolated miRNA profiles that were characteristic of periodontitis, oral cancer, and OLP. For example, miR-146a and miR-155 were identified to be upregulated in the periodontal tissues and GCF of patients with periodontitis and are involved in modulating inflammation. In the context of oral cancer, salivary miRNAs can be used as early diagnostic markers for OSCC, as revealed in a recent study (Sevagaperumal and Christy 2023). The authors identified a set of five miRNAs (miR-21, miR-125a, miR-31, miR-200a, and miR-375) capable of distinguishing OSCC patients from healthy controls with high sensitivity and specificity. This noninvasive method of cancer diagnosis could improve early diagnosis and treatment outcomes of oral cancer.

Exosomes and microvesicles, subclasses of EVs, have been identified as another promising source of biomarkers for oral immunity. These membrane-bound particles, secreted by various cell types, contain a protein, nucleic acid, and lipid complement that reflect the characteristics of the cells from which they originated (Minami et al. 2023). Among the early investigations, Byun et al. (2022) aimed at identifying and describing the salivary exosome proteome in individuals with periodontitis. The researchers also found specific proteins that correlated with disease severity and progression, indicating the possibility of using EV-based biomarkers to assess periodontal status. The application of EVs in oral cancer has also been a matter of significant interest (Kumar et al. 2024b). A recent study showed that salivary exosomes isolated from OSCC patients and healthy controls had distinct morphological and molecular profiles. The authors also found that exosomal proteins such as CD63 and IL-1β may be used to detect OSCC. To go further, these results bring new diagnostic prospects and contribute to the understanding of tumor-host interactions in the oral cavity. Protein modifications have also been identified as new biomarkers in oral immunity, and new modifications have been developed. Protein PTMs are the covalent modifications that occur after translation and can change the protein function and be used as markers of cellular processes and diseases. Yasuda et al. (2022) conducted a study to determine the implication of the PTM known as protein citrullination that is linked to autoimmune activity in periodontal disease. The researchers detected higher levels of citrullinated proteins in the gingival crevicular fluid of patients with periodontitis, and these levels were linked to disease severity. These observations raise the possibility that citrullinated proteins may be used both for the identification of periodontal diseases and for the assessment of the disease's progression.

Altered glycosylation patterns of salivary proteins have also been identified as potential biomarkers for immune-related diseases of the oral cavity. A recent glycoproteomic study by Goswami et al. (2024) showed that OLP patients have differential salivary glycosylation patterns. The study also defined certain specific glycoforms of salivary proteins, such as mucins and immunoglobulins, which are altered in various disease states. These glycoprotein biomarkers have the potential for diagnostic applications in oral mucosal diseases (OMDs) and for investigating the pathogenesis of these diseases. The application of multiple new biomarkers in multi-biomarker panels has been reported to enhance diagnostic performance as well as to offer a better understanding of the overall oral health status. Liu et al. (2021) conducted pioneering research on identifying a combined panel of salivary miRNA, protein, and metabolite biomarkers for the early diagnosis of oral cancer. This integrated strategy provided higher sensitivity and specificity than any single biomarker type, supporting the idea of multiple-biomarker identification in oral diagnosis. There is growing interest in the clinical applications of these novel biomarkers, as reflected in the growing number of case studies reported in the literature. For example, in a cross-sectional study by Sorsa et al. (2020), the authors assessed the feasibility of using salivary MMP-8, an already validated biomarker, together with newly discovered miRNA biomarkers for assessing periodontal treatment efficacy. In this study, the authors showed that this combined approach offered a better prediction of tissue healing and disease recurrence compared to traditional clinical parameters, which can be useful for individualized periodontal treatment.

In the oral cancer context, a case-control study by Kumar et al. (2024a) aimed to determine the diagnostic utility of a set of salivary exosomal miRNAs for the discrimination of early-stage OSCC from oral premalignant disorders (OPMDs). Their findings indicated that a panel of miR-21, miR-455-5p, and miR-1246 could be used to distinguish between these conditions and thus may serve as a promising diagnostic biomarker for early detection and management of patients with potentially malignant oral lesions.

38.4 Clinical Applications of Oral Immunity Biomarkers

38.4.1 Early Detection and Diagnosis

The use of oral immunity biomarkers in the diagnosis of oral diseases is one of the breakthroughs in dental and oral health care. These molecular indicators provide the possibility to reveal pathological processes before they turn into clinical signs and symptoms, thereby optimizing the treatment process and patients' prognosis (Cirillo 2024; Zhou and Liu 2023). The advancement of noninvasive diagnostic techniques based on these biomarkers has further contributed to their applicability in clinical practice, since these screening methods are more acceptable to patients (Kumar et al. 2024a). Periodontal disease is among one of the most promising conditions for the application of oral immunity biomarkers in early detection. The conventional clinical indices, including probing depth and bleeding on probing, diagnose periodontitis only after considerable tissue destruction has already occurred (Shi et al. 2023). However, molecular biomarkers can provide insights into the early onset of the disease, detecting it much earlier than any other diagnostic tool. The seminal work of Gürsoy et al. (2018) provided strong evidence of the use of multiple biomarkers for diagnosing periodontitis at an early stage. The researchers tested salivary MMP-8, IL-1β, and Pg in a large cohort of patients. They realized that clinical measures, when combined with host-derived and bacterial biomarkers, could point to a more accurate assessment of periodontitis risk than clinical measures alone. The study revealed a sensitivity of 87% and specificity of 91% in diagnosing early-stage periodontitis, emphasizing the role of this technique in screening the general population.

The advancement of point-of-care testing devices for oral immunity biomarkers has expanded the clinical utility of molecular markers. One of the best examples of chairside tests is the chairside test for active matrix metalloproteinase-8 (aMMP-8), which is considered to be very effective in periodontal diagnostics. Wei et al. (2024b) conducted a meta-analysis of studies to evaluate the diagnostic accuracy of aMMP-8 point-of-care tests. The authors identified that these rapid tests could separate healthy, gingivitis, and periodontitis regions in the oral cavity with high sensitivity and specificity. Furthermore, the aMMP-8 levels were significantly associated with the degree of periodontal inflammation and the efficacy of the treatment regime, thus proving the usefulness of this biomarker in the diagnosis and assessment of the state of periodontal disease (Lähteenmäki et al. 2023).

Therefore, noninvasive sampling techniques, particularly saliva sampling, have eased the use of biomarkers in diagnosing oral health conditions. Several benefits can be derived from saliva as a diagnostic fluid: easy to obtain, no pain or discomfort to the patient, and containing several biomarkers that can be indicative of the overall health of the individual. A systematic study by Liao et al. (2023) described the state of the art in salivary diagnostic tests in the context of oral health. The authors pointed out the possibility of the use of saliva-based diagnosis for the screening of numerous oral diseases, including periodontitis, oral cancer, and Sjögren's syndrome. They also discussed other diagnostic technologies for salivary biomarker analysis, including lab-on-a-chip and biosensor arrays, which hold the potential of making multiplexed biomarker testing as quick as routine dental visits. The combination of several biomarkers into panels has become one of the most effective strategies to enhance the sensitivity and specificity of diagnostic tests. According to Zhu et al. (2023), a new multi-omics panel was constructed for the early detection of OSCC. To develop the diagnostic signature, the researchers integrated salivary metabolomics, proteomics, and microbial profiling data. This integrated approach made it possible to obtain a sensitivity of 96% and specificity of 94% in distinguishing early-stage OSCC from healthy individuals and benign oral pathologies. The study not only proved the utility of multi-omics profiling in the diagnosis of oral lesions but also gave an understanding of the multiple molecular mechanisms involved in the development of oral cancer.

38.4.2 Prognostic Biomarkers

Diagnostic biomarkers are important in assessing the prognosis of the disease and predicting the patient outcomes in oral health. Such molecular markers offer useful data for clinical management, assisting in the decision-making process concerning the treatment of patients, risk categorization, and the implementation of more efficient and personalized patient management (Pekarek et al. 2023). The knowledge of biomarkers for assessing the prognosis of oral immunity has greatly expanded in recent years for various oral diseases. About periodontal disease, risk predictors are very useful for the early detection of patients who are at high risk of aggressive disease progression and unfavorable treatment outcomes. In a seminal study, Nascimento et al. (2019b) aimed to assess the accuracy of a set of salivary biomarkers in determining the progression of periodontal disease over 5 years of follow-up. In saliva samples from a large cohort of patients with chronic periodontitis, the researchers measured the levels of MMP-8, IL-1β, and osteoprotegerin (OPG). The authors discovered that MMP-8 and IL-1β levels were higher, while OPG levels were lower in the participants who experienced disease progression and tooth loss during follow-up sessions. Using this biomarker profile, they were able to predict the outcome in 87% of high-risk individuals, which was a better prediction than using clinical parameters alone. Nascimento et al. (2019a) also stressed the importance of prognostic biomarkers in treatment planning as well as the role of microbiota in disease progression and management. Patients with biomarker scores indicating a higher risk exhibited improved clinical outcomes when they received more invasive periodontal treatment, including systemic antibiotics, rather than when they were treated with mechanical debridement only. This finding also reflects the importance of prognostic biomarkers in the choice of the treatment strategy and in avoiding excessive intervention in patients with low risk of progression.

In the context of oral cancer, prognostic biomarkers have been developed to help estimate tumor behavior, ability to metastasize, and response to treatment. Liu et al. (2020) have also conducted detailed research on immune-related gene expression for OSCC and its prognostic value. The researchers conducted RNA sequencing of a large cohort of OSCC patients and found that an immune-related gene expression signature of 27 genes was associated with overall survival. This immune gene signature showed better prognostic value than clinicopathological factors in the assessment of patients' survival with a concordance index of 0.78 in the prediction of 5-year survival in patients. The techniques of liquid biopsy particularly involving circulating tumor DNA (ctDNA) and circulating tumor cells (CTCs) are very useful for prognostic evaluation in oral cancer without the need for invasive procedures. Research by Sarkar (2024) explored the feasibility of using ctDNA for evaluating treatment outcomes and for assessing disease relapse in OSCC patients. The researchers relied on NGS to monitor the evolution of tumor-specific mutations in plasma collected before, during, and after the therapy. They also discovered that the detection of ctDNA after the completion of therapy is highly correlated with the risk of relapse with a negative predictive value of 95% for disease-free survival at 2 years. This approach presents a promising strategy for monitoring treatment response and detecting disease relapse, enabling early follow-up and management.

The use of a combination of biomarkers in prognostic models has been demonstrated to have a great potential for enhancing predictive performance. Chen et al. (2022) identified genetic, epigenetic, and immune factors for the construction of a prognostic model for oral cancer. The researchers integrated somatic mutations, DNA methylation, and immune cell infiltration to develop a risk score to effectively predict high- and low-risk patients. This integrated approach produced a concordance index of 0.85 in the prediction of overall survival and was significantly better than the models constructed using single biomarker types or clinical factors. In the field of OPMDs, prognostic biomarkers are very useful for the detection of lesions that are likely to progress to malignancy. Speight et al. (2018) conducted a prospective study on a group of patients with oral leukoplakia to determine the accuracy of a set of molecular biomarkers in the prognosis of malignant transformation. The researchers looked at *TP53*, *MKI67*, and the loss of heterozygosity at chromosomes 3p and 9p in biopsy specimens taken from a group of patients with oral leukoplakia. The authors discovered that, when these markers were used in combination, they could predict high-risk lesions with a sensitivity of 85% and specificity of 80% over a median follow-up of 7 years. This molecular risk assessment concept has the advantage of providing a better opportunity to identify more specific surveillance and intervention plans for individuals with OPMDs.

38.4.3 Therapeutic Monitoring and Personalized Medicine

Therapeutic monitoring and personalized medicine based on oral immunity biomarkers is a shift in the management of oral diseases. These molecular indicators enable assessment of the effectiveness of the treatment, make corrections to the treatment in real time, and build an individual treatment approach for the patient. This approach has the potential to enhance therapeutic outcomes and optimize patient management in a variety of oral diseases and conditions. In periodontal disease treatment, biomarkers have been found to offer a good possibility in tracking the response to treatment and delivering individualized care. Haririan et al. (2021) in a large cross-sectional study analyzed the effectiveness of a set of salivary biomarkers for the assessment of periodontal treatment efficacy. MMP-8, IL-1β, and Pg are biomarkers of chronic periodontitis, and the researchers collected saliva samples before, during, and after nonsurgical periodontal therapy in patients with chronic periodontitis. They were able to determine that fluctuations in these biomarker concentrations throughout treatment were significantly associated with clinical outcomes in periodontal disease. However, as shown in the biomarker panel analysis, there was a higher sensitivity of the biomarkers in identifying early treatment response than the clinical parameters alone. In the treatment of oral cancer, biomarkers are particularly important for evaluating the effectiveness of the treatment and for tailoring individualized treatment plans. Zhang and Li (2023) conducted pioneering research focusing on the use of ctDNA analysis for the assessment of the treatment response in patients with OSCC. In this study, the researchers employed NGS to map the tumor-specific mutations in the plasma samples obtained during chemoradiotherapy. It was discovered that early alteration of ctDNA levels was a strong prognostic indicator of treatment responses, since patients who had a 90% decrease in ctDNA levels by the second week of the therapy had improved progression-free survival.

Yang et al. (2020) and Zhang and Li (2023) also showed the utility of ctDNA analysis in identifying patients who would benefit from adaptive treatment plans. Those patients who had a suboptimal decrease in ctDNA during the first-line treatment were selected as candidates for therapy escalation or change of strategy. The strategy of biomarker-guided adaptive therapy led to a 25% increase in overall response rates as compared to the standard of care. These findings highlighted the value of real-time molecular monitoring, which helps in personalizing cancer treatment to the patient's special needs, resulting in better outcomes. The use of several biomarkers in complex treatment monitoring panels has been demonstrated to be quite useful in capturing the complexity of treatment response. Ling et al. (2023) also designed a multimodal biomarker system in the assessment of immunotherapy response in recurrent/metastatic OSCC. CTCs, immune cells, and serum cytokine levels were used to develop a dynamic response index by the researchers. This combined biomarker panel was found to have 92% accuracy in terms of predicting clinical response to immune checkpoint inhibitor therapy, which indicates a much higher performance compared to the usage of single biomarkers or imaging-based techniques. Oral immunity biomarkers are not only used in cancer and periodontal disease but also in oral conditions. Biomarkers in the context of OLP, a chronic inflammatory mucosal disorder were found to have the ability to determine the appropriate treatment plan.

38.4.4 Biomarkers in Immunotherapy

Immunotherapy, as a new innovative approach, has been identified to be useful in the management of several oral diseases, specifically oral cancer and autoimmune diseases. Biomarkers are essential in this field, particularly in selecting patients who would benefit from immunotherapies, assessing the effectiveness of the treatment, and managing adverse effects. The incorporation of biomarker-guided approaches in the management of oral immunotherapy has greatly improved the therapeutic efficacy and has paved the way for the development of tailored medicine in oral health care (Qin and Wu 2023; Wang et al. 2023). Speaking of oral cancer, immune checkpoint inhibitors are considered to be quite effective in managing the advanced and recurrent forms of malignancies. Nevertheless, the response of the patients to these therapies is rather diverse. Cramer et al. (2019) conducted the first-ever research to determine the efficacy of using programmed death ligand-1 (PD-L1) expression as a biomarker for predicting response to pembrolizumab in patients with recurrent or metastatic head and neck squamous cell carcinoma (HNSCC). The studies identified that patients with tumors that had a high level of PD-L1 expression (CPS $\geq$ 20) had significantly better overall survival and progression-free survival than patients with lower expression. The findings of this study showed that PD-L1 could be a promising biomarker for the response to immunotherapy in patients with oral cancer, to make the treatments more personalized. On this basis, a more extensive study by Cristescu et al. (2022) further investigated the combination of several biomarkers for immunotherapy response prediction in HNSCC. The researchers used PD-L1 expression in addition to tumor mutation burden (TMB) and gene expression profiling to develop a biomarker score. This integrated approach was shown to provide enhanced prediction of the responsiveness to the anti-PD-1 treatment compared to the separate assessment of individual biomarkers with the reported positive predictive value of 85%. The study gave insight into how multi-biomarker panels could be used for guiding the immunotherapy approaches and patient selection.

Attention has also been given to biomarkers that could help to define the effectiveness of immunotherapy and the risk of side effects. Investigating circulating immune cell profiles as biomarkers for treatment response and toxicity in HNSCC patients receiving immune checkpoint inhibitors was performed by Hanna et al. (2024). The researchers noted that early alterations in circulating CD8+ T cell subsets, especially effector memory T cells, were positively correlated with the treatment response. Furthermore, they described certain biomarkers related to myeloid cell activation that could predict immune-related adverse events and potentially aid in early prevention and monitoring of treatment-related toxic effects. The combination of biomarker information with advanced imaging modalities has provided new methods of tracking the immunotherapy responses in oral cancers. Koukourakis et al. (2023) conducted a study to assess the integration of positron emission tomography (PET) imaging with ctDNA for early prediction of response to immune checkpoint inhibitors (ICIs) in HNSCC. The researchers reported that using the combination of the metabolic tumor volume on fluorodeoxyglucose-PET (FDG-PET) and ctDNA kinetics, it was possible to achieve 92% accuracy in the prediction of the treatment response within 6 weeks of therapy.

38.5 Challenges and Future Directions

38.5.1 Challenges in Biomarker Implementation

The application of biomarkers in the field of oral immunity and the clinical setting has several major issues that need to be solved to enhance the development of oral immunity. These challenges are across different domains such as standardization, regulation, and economics. Another important issue that should be addressed in the field of biomarker research is assay standardization to ensure that results obtained in one laboratory or clinic are comparable to those obtained in another. Because oral biological samples are rich in proteins and are contaminated easily, the procedures of sample collection, preparation, and analysis should be highly standardized. Schiegnitz et al. (2021) provided several examples of variability in biomarker measurement across the studies due to the differences in sample handling, storage conditions, and the analytical methods used. The authors emphasized the need to reach a consensus on the preanalytical and analytical protocols for oral biomarker measurements to facilitate the comparison between studies and ensure their applicability in clinical practice. To overcome this issue, international collaborations and consortia have been established to create guidelines for biomarker investigations in oral health. The Human Salivary Proteome Wiki started by Deutsch et al. (2016) is a good example of the attempt to establish the best practices in salivary proteomics and to provide access to the standardized data.

Another major barrier to the clinical application of oral immunity biomarkers is the regulatory issue. The path to the validation and approval of biomarker-based diagnostic tests is lengthy and complex, often lagging behind scientific progress.

Before biomarkers can be utilized in clinical practice, studies with a large number of patients and controls are needed to prove the analytical and clinical validity of the biomarker, and the clinical utility of the biomarker. Such investigations are expensive and may take a long time, a factor that poses a major challenge to the establishment of new biomarker-based tests. The National Academies of Sciences, Engineering, and Medicine (2020) in their report on biomarker validation also stressed the need for new approaches and partnerships for accelerating biomarker validation while also maintaining scientific rigor (Song et al. 2020; Yost et al. 2023).

38.6 Conclusion

Oral immunity biomarkers have been identified as an innovative tool to assess oral health and disease. The progress in the fields of genetics, proteomics, metabolomics, and microbiomics has revealed new diagnostic, prognostic, and therapeutic possibilities for oral diseases. This progress has been made possible by high-throughput screening technologies, multi-omics techniques, and sophisticated computational tools. Biomarkers are designed to identify pathophysiological changes at the molecular level before the onset of clinical symptoms, allowing for early intervention. Currently, these biomarkers are applied in risk assessment of oral cancer, periodontal disease, and other OMDs. The use of biomarkers in combination with AI and machine learning has opened new possibilities for oral care and better disease management and patients' quality of life. However, challenges remain in using oral immunity biomarkers in large-scale clinical practice. It is crucial to address issues related to standardization, approval, and cost in collaborative efforts. The next steps in this field may include single-cell analysis, spatial profiling, and the development of advanced computational tools to decipher the intricacies of the oral immune system.

References

Abdul NS, Shenoy M, Reddy NR, Sangappa SB, Shivakumar GC, Di Blasio M, Cicciù M, Minervini G (2024) Gene sequencing applications to combat oral-cavity related disorders: a systematic review with meta-analysis. BMC Oral Health 24(1):103

Ahmad A, Imran M, Ahsan H (2023) Biomarkers as biomedical bioindicators: approaches and techniques for the detection, analysis, and validation of novel biomarkers of diseases. Pharmaceutics 15(6):1630

Ahmad P, Hussain A, Siqueira WL (2024) Mass spectrometry-based proteomic approaches for salivary protein biomarkers discovery and dental caries diagnosis: a critical review. Mass Spectrom Rev 43(4): 826–856

Al-Shehri M (2023) The current state of clinical diagnostic algorithms for mucosal oral lesions: a scoping review. McGill University (Canada)

Antonelli R, Setti G, Treister NS, Pertinhez TA, Ferrari E, Gallo M, Bologna-Molina R, Vescovi P, Meleti M (2024) Salivary metabolomics in oral cancer: a systematic review. Oral Oncol Rep:100657

Atalay N, Balci N, Gürsoy M, Gürsoy UK (2024) Systemic factors effecting human beta-defensins in oral cavity. Pathogens 13(8):654

Basic A, Dahlén G (2023) Microbial metabolites in the pathogenesis of periodontal diseases: a narrative review. Front Oral Health 4:1210200

Bostanci N, Selevsek N, Wolski W, Grossmann J, Bao K, Wahlander A, Trachsel C, Schlapbach R, Öztürk VÖ, Afacan B (2018) Targeted proteomics guided by label-free quantitative proteome analysis in saliva reveal transition signatures from health to periodontal disease. Mol Cell Proteomics 17(7):1392–1409

Buss L (2023) Metabolic changes in the salivary gland following radiation treatment. The University of Arizona

Byun J-S, Lee HY, Tian J, Moon JS, Choi J, Lee S-H, Kim Y-G, Yi H-S (2022) Effect of salivary exosomal miR-25-3p on periodontitis with insulin resistance. Front Immunol 12:775046

Chavda VP, Anand K, Apostolopoulos V (2023) Bioinformatics tools for pharmaceutical drug product development. Wiley

Chen Q, Fan Y, Li Y, Wang J, Chen L, Lin J, Chen F, Wang J, Qiu Y, Shi B (2022) A novel nutritional risk score and prognosis of oral cancer patients: a prospective study. Oral Dis 28(1):108–115

Chen S, He F, Cheng X (2024a) Analysis of subgingival bacterial and fungal diversity in patients with peri-implantitis based on 16sRNA and internal transcribed spacer sequencing. Future Microbiol 19(5):397–411

Chen S, Wu X, Yang Y, Xu X, Xiong X, Meng W (2024b) Increased pathogenicity and pro-inflammatory capabilities of mucosal-associated invariant T cells involved in oral lichen planus. BMC Oral Health 24(1):829

Cheng X, Zhou X, Liu C, Xu X (2021) Oral osteomicrobiology: the role of oral microbiota in alveolar bone homeostasis. Front Cell Infect Microbiol 11: 751503

Cirillo N (2024) A roadmap for the rational use of biomarkers in oral disease screening. Biomolecules 14(7):787

Cramer JD, Burtness B, Ferris RL (2019) Immunotherapy for head and neck cancer: recent advances and future directions. Oral Oncol 99:104460

Cristescu R, Nebozhyn M, Zhang C, Albright A, Kobie J, Huang L, Zhao Q, Wang A, Ma H, Alexander Cao Z (2022) Transcriptomic determinants of response to pembrolizumab monotherapy across solid tumor types. Clin Cancer Res 28(8):1680–1689

Csősz É, Lábiscsák P, Kalló G, Markus B, Emri M, Szabó A, Tar I, Tőzsér J, Kiss C, Márton I (2017) Proteomics investigation of OSCC-specific salivary biomarkers in a Hungarian population highlights the importance of identification of population-tailored biomarkers. PLoS One 12(5):e0177282

D'Agostino S, Valentini G, Iarussi F, Dolci M (2023) Interleukins polymorphisms in type-1 diabetes and perio-dontitis: a review. https://doi.org/10.20944/preprints202306.1535.v1

Daly AK (2017) Pharmacogenetics: a general review on progress to date. Br Med Bull 124(1):65–79

Danielsson K, Nylander E, Sjöström M, Ebrahimi M (2018) Epstein-Barr virus is not detected in mucosal lichen planus. Med Oral Patol Oral Cir Bucal 23(5): e560

Deutsch EW, Csordas A, Sun Z, Jarnuczak A, Perez-Riverol Y, Ternent T, Campbell DS, Bernal-Llinares-M, Okuda S, Kawano S (2016) The ProteomeXchange consortium in 2017: supporting the cultural change in proteomics public data deposition. Nucleic Acids Res 45(Database issue):D1100–D1106

Ebersole JL, Dawson DR III, Morford LA, Peyyala R, Miller CS, Gonzaléz OA (2013) Periodontal disease immunology: 'double indemnity' in protecting the host. Periodontol 62(1):163–202

Ebersole JL, Nagarajan R, Akers D, Miller CS (2015) Targeted salivary biomarkers for discrimination of periodontal health and disease (s). Front Cell Infect Microbiol 5:62

Ebersole JL, Hasturk H, Huber M, Gellibolian R, Markaryan A, Zhang XD, Miller CS (2024) Realizing the clinical utility of saliva for monitoring oral diseases. Periodontology 2000 95(1):203–219

Einhaus J, Han X, Feyaerts D, Sunwoo J, Gaudilliere B, Ahmad SH, Aghaeepour N, Bruckman K, Ojcius D, Schürch CM (2023) Towards multiomic analysis of oral mucosal pathologies. Semin Immunopathol 45(1):111–123

El-Howati A, Thornhill MH, Colley HE, Murdoch C (2023) Immune mechanisms in oral lichen planus. Oral Dis 29(4):1400–1415

Farhad SZ, Karbalaeihasanesfahani A, Dadgar E, Nasiri K, Esfahaniani M, Nabi Afjadi M (2024) The role of periodontitis in cancer development, with a focus on oral cancers. Mol Biol Rep 51(1):814

Fujihara C, Hafiyyah OA, Murakami S (2023) Identification of disease-associate variants of aggressive periodontitis using genome-wide association studies. Jpn Dent Sci Rev 59:357–364

George AK, Malaiappan S, Joseph B, Anil S (2024) Biomarkers in gingival diseases: current insights and future perspectives

Ghantous Y, Mozalbat S, Nashef A, Abdol-Elraziq M, Sudri S, Araidy S, Tadmor H, Abu El-naaj I (2024) EMT dynamics in lymph node metastasis of oral squamous cell carcinoma. Cancer 16(6):1185

Ghosh SK, McCormick TS, Weinberg A (2019) Human beta defensins and cancer: contradictions and common ground. Front Oncol 9:341

Giannobile WV, Braun TM, Caplis AK, Doucette-Stamm-L, Duff GW, Kornman KS (2013) Patient stratification for preventive care in dentistry. J Dent Res 92(8): 694–701

Goswami GR, Patil R, Ghosh S, Kaur G, Abhijit G (2024) Oral microbiome modulates tumour micro-environment potentiating Oral cancer progression from benign stages-a scoping review. Available at SSRN 4853291

Grocholska P, Kowalska M, Bąchor R (2023) Qualitative and quantitative mass spectrometry in salivary metabolomics and proteomics. Meta 13(2):155

Grootveld M, Page G, Bhogadia M, Hunwin K, Edgar M (2023) Updates and original case studies focused on the NMR-linked metabolomics analysis of human oral fluids part III: implementations for the diagnosis of non-cancerous disorders, both oral and systemic. Metabolites 13(1):66

Guarnieri R, Reda R, Di Nardo D, Miccoli G, Pagnoni F, Zanza A, Testarelli L (2024) Expression of IL-1β, IL-6, TNF-α, and a-MMP-8 in sites with healthy conditions and with periodontal and peri-implant diseases: a case-control study. J Dent Res Dent Clin Dent Prospect 18(2):135

Gürsoy UK, Könönen E, Tervahartiala T, Gürsoy M, Pitkänen J, Torvi P, Suominen AL, Pussinen P, Sorsa T (2018) Molecular forms and fragments of salivary MMP-8 in relation to periodontitis. J Clin Periodontol 45(12):1421–1428

Hajishengallis G (2015) Periodontitis: from microbial immune subversion to systemic inflammation. Nat Rev Immunol 15(1):30–44

Hanna GJ, Dennis MJ, Scarfo N, Mullin MS, Sethi RK, Sehgal K, Annino DJ Jr, Goguen LA, Haddad RI, Tishler RB (2024) Personalized ctDNA for monitoring disease status in head and neck squamous cell carcinoma. Clin Cancer Res 30(15):3329–3336

Hao C-P, Cao N-J, Zhu Y-H, Wang W (2023) The impact of smoking on periodontitis patients' GCF/serum cytokine profile both before and after periodontal therapy: a meta-analysis. BMC Oral Health 23(1):60

Haririan H, Andrukhov O, Laky M, Rausch-Fan X (2021) Saliva as a source of biomarkers for periodontitis and periimplantitis. Front Dent Med 2:687638

Huang SH, O'Sullivan B (2017) Overview of the 8th edition TNM classification for head and neck cancer. Curr Treat Options in Oncol 18:1–13

Hurník P, Režnarová J, Chyra Z, Motyka O, Putnová BM, Čermáková Z, Blažek T, Fománek M, Gaykalova D, Buchtová M (2023) Enhancing oral squamous cell carcinoma prediction: the prognostic power of the worst pattern of invasion and the limited impact of molecular resection margins. Front Oncol 13:1287650

Inchingolo AM, Malcangi G, Piras F, Palmieri G, Settanni V, Riccaldo L, Morolla R, Buongiorno S, de Ruvo E, Inchingolo AD (2023) Precision medicine on the effects of microbiota on head–neck diseases and biomarkers diagnosis. J Personaliz Med 13(6):933

Kallianta M, Pappa E, Vastardis H, Rahiotis C (2023) Applications of mass spectrometry in dentistry. Biomedicine 11(2):286

Kapferer-Seebacher I, Lundberg P, Malfait F, Zschocke J (2017) Periodontal manifestations of Ehlers–Danlos syndromes: a systematic review. J Clin Periodontol 44(11):1088–1100

Kinney J, Morelli T, Braun T, Ramseier CA, Herr A, Sugai J, Shelburne C, Rayburn L, Singh A, Giannobile W (2011) Saliva/pathogen biomarker signatures and periodontal disease progression. J Dent Res 90(6): 752–758

Kitata RB, Yang JC, Chen YJ (2023) Advances in data-independent acquisition mass spectrometry towards comprehensive digital proteome landscape. Mass Spectrom Rev 42(6):2324–2348

Korczeniewska OA, Dakshinamoorthy J, Prabhakar V, Lingaiah U (2024) Genetics affecting the prognosis of dental treatments, vol 68. Dent Clin

Korgaonkar J, Tarman AY, Koydemir HC, Chukkapalli SS (2024) Periodontal disease and emerging point-of-care technologies for its diagnosis. Lab Chip 24(14): 3326–3346

Kornman KS (2018) Contemporary approaches for identifying individual risk for periodontitis. Periodontol 78(1):12–29

Koukourakis IM, Xanthopoulou E, Koukourakis MI (2023) Using liquid biopsy to predict relapse after radiotherapy in squamous cell head-neck and esophageal cancer. Cancer Diag Prog 3(4):403

Kozak M, Pawlik A (2023) The role of the oral microbiome in the development of diseases. Int J Mol Sci 24(6):5231

Kumar P, Gupta S, Das BC (2024a) Saliva as a potential non-invasive liquid biopsy for early and easy diagnosis/prognosis of head and neck cancer. Transl Oncol 40:101827

Kumar RMR, Joghee S, Puttaraju MK (2024b) Impact of exosomes in oral lichen planus: a review with insights into pathogenesis and biomarkers. J Dent Sci

Kumari S, Samara M, Ampadi Ramachandran R, Gosh S, George H, Wang R, Pesavento RP, Mathew MT (2024) A review on saliva-based health diagnostics: biomarker selection and future directions. Biomed Mater Devices 2(1):121–138

Lähteenmäki H, Pätilä T, Pärnänen P, Räisänen I, Tervahartiala T, Gupta S, Sorsa T (2023) aMMP-8 point-of-care-diagnostic methods and treatment modalities in periodontitis and peri-implantitis. Expert Opin Ther Targets 27(7):627–637

Laigle L, Chadli L, Moingeon P (2023) Biomarker-driven development of new therapies for autoimmune diseases: current status and future promises. Expert Rev Clin Immunol 19(3):305–314

Lamont RJ, Koo H, Hajishengallis G (2018) The oral microbiota: dynamic communities and host interactions. Nat Rev Microbiol 16(12):745–759

Leonov GE, Varaeva YR, Livantsova EN, Starodubova AV (2023) The complicated relationship of short-chain fatty acids and oral microbiome: a narrative review. Biomedicine 11(10):2749

Li C, Zhou H, Gou H, Fan Z, Zhang Y, Tang P, Huang J, Xu Y, Li L (2024) Autoinducer-2 produced by oral microbial flora and alveolar bone loss in periodontitis. J Periodontal Res 59(3):576–588

Liao C, Chen X, Fu Y (2023) Salivary analysis: an emerging paradigm for non-invasive healthcare diagnosis and monitoring. Interdiscip Med 1(3):e20230009

Ling X, Alexander GS, Molitoris J, Choi J, Schumaker L, Mehra R, Gaykalova DA, Ren L (2023) Identification of CT-based non-invasive radiomic biomarkers for overall survival prediction in oral cavity squamous cell carcinoma. Sci Rep 13(1):21774

Liu L, Wu Y, Li Q, Liang J, He Q, Zhao L, Chen J, Cheng M, Huang Z, Ren H (2020) METTL3 promotes tumorigenesis and metastasis through BMI1 m6A methylation in oral squamous cell carcinoma. Mol Ther 28(10):2177–2190

Liu D, Xin Z, Guo S, Li S, Cheng J, Jiang H (2021) Blood and salivary microRNAs for diagnosis of oral squamous cell carcinoma: a systematic review and meta-analysis. J Oral Maxillof Sur 79(5):1082. e1081–1082. e1013

Liu W, Li M, Zhang X, Zhou Z, Shen Z, Shen X (2023) Association of polymorphisms in Th1/Th2-related cytokines (IFN-γ, TGFβ1, IL-1β, IL-2, IL-4, IL-18) with oral lichen planus: a pooled analysis of case–control studies. J Den Sci 18(2):560–566

Lorenzo-Pouso AI, Silva FF-VE, Perez-Jardon A, Chamorro-Petronacci CM, Oliveira-Alves MG, Alvarez-Calderon-Iglesias O, Caponio VCA, Pinti M, Perrotti V, Perez-Sayans M (2023) Overexpression of e-cadherin is a favorable prognostic biomarker in oral squamous cell carcinoma: a systematic review and meta-analysis. Biology 12(2):239

Madruga González D (2023) Antimicrobial peptide LL-37 as a diagnostic biomarker in periodontal disease

Malcangi G, Patano A, Guglielmo M, Sardano R, Palmieri G, Di Pede C, de Ruvo E, Inchingolo AD, Mancini A, Inchingolo F (2023) Precision medicine in oral health and diseases: a systematic review. J Personal Med 13(5):725

Markus V, Paul AA, Teralı K, Özer N, Marks RS, Golberg K, Kushmaro A (2023) Conversations in the gut: the role of quorum sensing in normobiosis. Int J Mol Sci 24(4):3722

McCarthy RL, Gnanappiragasam D, Scorer M, Taylor M, O'Toole EA (2023) Cathepsin-C mutation in an individual with phenotypic features of Haim–Munk syndrome: a case report. Clin Exp Dermatol 48(11): 1298–1300

Mehdipour M, Shahidi M, Anbari F, Mirzaei H, Jafari S, Kholghi A, Lotfi E, Manifar S, Mashhadiabbas F (2023) Salivary level of microRNA-146a and microRNA-155 biomarkers in patients with oral lichen planus versus oral squamous cell carcinoma. BMC Oral Health 23(1):433

Minami S, Chikazu D, Ochiya T, Yoshioka Y (2023) Extracellular vesicle-based liquid biopsies in cancer: future biomarkers for oral cancer. Transl Oncol 38: 101786

Morio KA, Sternowski RH, Brogden KA (2023) Induction of endogenous antimicrobial peptides to prevent or treat oral infection and inflammation. Antibiotics 12(2):361

Nag S, Mitra O, Tripathi G, Samanta S, Bhattacharya B, Chandane P, Mohanto S, Sundararajan V, Malik S, Rustagi S (2023) Exploring the theranostic potentials of miRNA and epigenetic networks in autoimmune diseases: a comprehensive review. Immun Inflam Dis 11(12):e1121

Napiórkowska-Baran K, Darwish S, Kaczor J, Treichel P, Szymczak B, Szota M, Koperska K, Bartuzi Z (2024) Oral diseases as a manifestation of inborn errors of immunity. J Clin Med 13(17):5079

Nascimento PC, Castro MML, Magno MB, Almeida APCPSC, Fagundes NCF, Maia LC, Lima RR (2019a) Association between periodontitis and cognitive impairment in adults: a systematic review. Front Neurol 10:323

Nascimento GG, Leite FR, Peres KG, Demarco FF, Corrêa MB, Peres MA (2019b) Metabolic syndrome and periodontitis: a structural equation modeling approach. J Periodontol 90(6):655–662

Nibali L, Di Iorio A, Onabolu O, Lin GH (2016) Periodontal infectogenomics: systematic review of associations between host genetic variants and subgingival microbial detection. J Clin Periodontol 43(11):889–900

Nukaly HY, Halawani IR, Alghamdi SMS, Alruwaili AG, Binhezaim A, Algahamdi RAA, Alzahrani RAJ, Alharamlah FSS, Aldumkh SHS, Alasqah HMA (2024) Oral lichen planus: a narrative review navigating etiologies, clinical manifestations, diagnostics, and therapeutic approaches. J Clin Med 13(17):5280

Passaro A, Al Bakir M, Hamilton EG, Diehn M, André F, Roy-Chowdhuri S, Mountzios G, Wistuba II, Swanton C, Peters S (2024) Cancer biomarkers: emerging trends and clinical implications for personalized treatment. Cell 187(7):1617–1635

Patil S, Khan SS, Hosmani J, Khan ZA, Muruganandhan J, Mushtaq S, Yadalam PK, Bhandi S, Awan KH (2023) Identification of oral immune disorders–a review and a diagnostic algorithm. Dis Mon 69(1):101350

Pekarek L, Garrido-Gil MJ, Sánchez-Cendra A, Cassinello J, Pekarek T, Fraile-Martinez O, García-Montero C, Lopez-Gonzalez L, Rios-Parra A, Álvarez-Mon M (2023) Emerging histological and serological biomarkers in oral squamous cell carcinoma: applications in diagnosis, prognosis evaluation and personalized therapeutics. Oncol Rep 50(6):213

Peng Y, Yin D, Li X, Wang K, Li W, Huang Y, Liu X, Ren Z, Yang X, Zhang Z (2023) Integration of transcriptomics and metabolomics reveals a novel gene signature guided by FN1 associated with immune response in oral squamous cell carcinoma tumorigenesis. J Cancer Res Clin Oncol 149(9): 6097–6113

Pomella S, Melaiu O, Cifaldi L, Bei R, Gargari M, Campanella V, Barillari G (2024) Biomarkers identification in the microenvironment of oral squamous cell carcinoma: a systematic review of proteomic studies. Int J Mol Sci 25(16):8929

Poulos RC, Cai Z, Robinson PJ, Reddel RR, Zhong Q (2023) Opportunities for pharmacoproteomics in biomarker discovery. Proteomics 23(7–8):2200031

Qin L, Wu J (2023) Targeting anticancer immunity in oral cancer: drugs, products, and nanoparticles. Environ Res 116751:116751

Qiu S, Cai Y, Yao H, Lin C, Xie Y, Tang S, Zhang A (2023) Small molecule metabolites: discovery of biomarkers and therapeutic targets. Signal Transduct Target Ther 8(1):132

Rashid MH, Yellarthi SPK, Yellarthi PK, Didugu BGL, Mamillapalli A (2024) Combined assessment of lysine and N-acetyl cadaverine levels assist as a potential biomarker of the smoker periodontitis. Amino Acids 56(1):1–11

Rezazadeh F, Salehi S, Rezaee M (2019) Salivary level of trace element in oral lichen planus, a premalignant condition. Asian Pac J Cancer Prev 20(7):2009

Sarkar N (2024) Elucidating the role of miR-617 in oral squamous cell carcinoma pathogenesis

Satrio RD, Fendiyanto MH, Miftahudin M (2024) Tools and techniques used at global scale through genomics, Transcriptomics, proteomics, and metabolomics to investigate plant stress responses at the molecular level. In: Molecular dynamics of plant stress and its management. Springer, pp 555–607

Schafer MJ, Zhang X, Kumar A, Atkinson EJ, Zhu Y, Jachim S, Mazula DL, Brown AK, Berning M, Aversa Z (2020) The senescence-associated secretome as an indicator of age and medical risk. JCI Insight 5(12)

Schiegnitz E, Müller LK, Sagheb K, Theis L, Cagiran V, Kämmerer PW, Wegener J, Wagner W, Al-Nawas B (2021) Clinical long-term and patient-reported outcomes of dental implants in oral cancer patients. Int J Impl Den 7:1–8

Sevagaperumal A, Christy W (2023) microRNA—does it presage the malignant change of Oral potentially malignant disorders: a systematic review. J Scient Den 13(2):55–60

Sharifi A (2023) Targeted peri-implant crevicular fluid biomarkers in osteoporotic patients receiving a dental implant. Queen Mary University of London

Shi F, Liu W, Yao Y, Zhang Q, Chen Z, Xian Y, Sujanamulk B (2023) Predictive salivary biomarkers for early diagnosis of periodontal diseases–current and future developments. Turk J Biochem 48(4):335–344

Shreevatsa B, Abhigna N, Viveka S, Chakith MS, Dharmashekara C, Siddalingegowda SV, Mahesh K, Kollur SP, Shivamallu C (2024) Next genome sequencing in oral cancer. In: Bioinformatics for oral cancer. CRC Press, pp 83–91

Song X, Yang X, Narayanan R, Shankar V, Ethiraj S, Wang X, Duan N, Ni Y-H, Hu Q, Zare RN (2020) Oral squamous cell carcinoma diagnosed from saliva metabolic profiling. Proc Natl Acad Sci 117(28): 16167–16173

Sorsa T, Alassiri S, Grigoriadis A, Räisänen IT, Pärnänen P, Nwhator SO, Gieselmann D-R, Sakellari D (2020) Active MMP-8 (aMMP-8) as a grading and staging biomarker in the periodontitis classification. Diagnostics 10(2):61

Speight PM, Khurram SA, Kujan O (2018) Oral potentially malignant disorders: risk of progression to malignancy. Oral Surg Oral Med Oral Pathol Oral Radiol 125(6):612–627

Sriram RS, Anitha P, Balachander K, Chandra P, Paramasivam A, Priyadharsini JV (2024) Genetic alterations in CDKN2A interacting network and their putative association with head and neck squamous cell carcinoma. Hum Gene Ther 40:201276

Strimbu K, Tavel JA (2010) What are biomarkers? Curr Opin HIV AIDS 5(6):463–466

Talapko J, Erić S, Meštrović T, Stipetić MM, Juzbašić M, Katalinić D, Bekić S, Muršić D, Flam J, Belić D (2024) The impact of oral microbiome dysbiosis on the aetiology, pathogenesis, and development of oral cancer. Cancer 16(17):2997

Ummer H, Santosh ABR, Surboyo MDC, Chitipothu MD, Sivanandham S (2024) Single nucleotide polymorphisms: a paradigm in oral disease research, vol 57

Veeraraghavan VP, Daniel S, Dasari AK, Aileni KR, Patil SR (2024) Emerging targeted therapies in Oral oncology: focus on EGFR inhibition. Oral Oncol Rep 11: 100592

Wan Z, Xiong H, Tan X, Su T, Xia K, Wang D (2022) Integrative multi– omics analysis reveals candidate biomarkers for Oral squamous cell carcinoma. Front Oncol 11:794146

Wang T, Sun S, Zeng X, Li J (2023) ICI-based therapies: a new strategy for oral potentially malignant disorders. Oral Oncol 140:106388

Wei W, Li J, Tang B, Deng Y, Li Y, Chen Q (2024a) Metabolomics and metagenomics reveal the impact of γδ T inhibition on gut microbiota and metabolism in periodontitis-promoting OSCC. Msystems 9(2): e00777–e00723

Wei S, Lin T, Sáenz-Ravello G, Gao H, Zhang Y, Tonetti MS, Deng K (2024b) Diagnostic accuracy of salivary active matrix metalloproteinase (aMMP)-8 point-of-care test for detecting periodontitis in adults: a systematic review and meta-analysis. J Clin Periodontol 51: 1093

Wirth R, Maróti G, Lipták L, Mester M, Al Ayoubi A, Pap B, Madléna M, Minárovits J, Kovács KL (2022) Microbiomes in supragingival biofilms and saliva of adolescents with gingivitis and gingival health. Oral Dis 28(7):2000–2014

Woo JS, Min H-K, Choi J-W, Moon JH, Park M-J, Kwok S-K, Park S-H, Cho M-L (2021) Short-chain fatty acid butyrate induces IL-10-producing B cells by regulating circadian-clock-related genes to ameliorate Sjögren's syndrome. J Autoimmun 119:102611

Wu J, Liu G, Jia R, Guo J (2023) Salivary extracellular vesicles: biomarkers and beyond in human diseases. Int J Mol Sci 24(24):17328

Xue F, Zhao J, Gao X, Jiang X, Lan Z (2024) Potential susceptibility genes in patients with stage III and IV periodontitis: a whole-exome sequencing pilot study. Biomol Biomed 24(1):73

Yang W-Y, Feng L-F, Meng X, Chen R, Xu W-H, Hou J, Xu T, Zhang L (2020) Liquid biopsy in head and neck squamous cell carcinoma: circulating tumor cells, circulating tumor DNA, and exosomes. Expert Rev Mol Diagn 20(12):1213–1227

Yao Z, An W, Tuerdi M, Zhao J (2023) Identification of novel prognostic indicators for oral squamous cell carcinoma based on proteomics and metabolomics. Transl Oncol 33:101672

Yasuda T, Tahara K, Sawada T (2022) Detection of salivary citrullinated cytokeratin 13 in healthy individuals and patients with rheumatoid arthritis by proteomics analysis. PLoS One 17(3):e0265687

Ye H-L, Zhi M-F, Chen B-Y, Lin W-Z, Li Y-L, Huang S-J, Zhou L-J, Xu S, Zhang J, Zhang W-C (2024) Alterations of oral and gut viromes in hypertension and/or periodontitis. Msystems 9(1):e01169–e01123

Yost OC, Downey A, Ramos KS, National Academies of Sciences, Engeineering, and Medicine (2023) Future needs and opportunities for nonhuman primate models in biomedical research. In: Nonhuman primate models in biomedical research: state of the science and future needs. National Academies Press (US)

Yousif SH, Qasim AA (2023) Salivary Cathelicidins (LL-37) as a dental caries biomarker in relation to nutrition status among a group of adolescents. WORLD 14(4):353

Zalewska EA, Ławicka R, Grygorczuk P, Nowosielska M, Kicman A, Ławicki S (2024) Importance of metalloproteinase 8 (MMP-8) in the diagnosis of periodontitis. Int J Mol Sci 25(5):2721

Zhang X, Li B (2023) Updates of liquid biopsy in oral cancer and multiomics analysis. Oral Dis 29(1):51–61

Zhang Z, Wen S, Liu J, Ouyang Y, Su Z, Chen D, Liang Z, Wang Y, Luo T, Jiang Q (2024) Advances in the

relationship between periodontopathogens and respiratory diseases. Mol Med Rep 29(3):42
Zhou Y, Liu Z (2023) Saliva biomarkers in oral disease. Clin Chim Acta 548:117503
Zhou Y, Nevosadová L, Eliasson E, Lauschke VM (2023) Global distribution of functionally important CYP2C9 alleles and their inferred metabolic consequences. Hum Genomics 17(1):15
Zhu J, Zhang S, Shi J, Ning N, Wei Y, Zhang Y (2023) Periodontitis is associated with the increased levels of visfatin: a meta-analysis. BMC Oral Health 23(1):799

MIX
Papier aus verantwortungsvollen Quellen
Paper from responsible sources
FSC® C105338

If you have any concerns about our products,
you can contact us on
ProductSafety@springernature.com

In case Publisher is established outside the EU,
the EU authorized representative is:
Springer Nature Customer Service Center GmbH
Europaplatz 3, 69115 Heidelberg, Germany

Printed by Libri Plureos GmbH
in Hamburg, Germany